PRYOR TRE
PLANT

THE HANDBOOK OF CHLORINATION

THE HANDBOOK OF CHLORINATION
Second Edition

Geo. Clifford White
Consulting Engineer

VNR VAN NOSTRAND REINHOLD COMPANY
New York

Copyright © 1986 by Van Nostrand Reinhold Company Inc.

Library of Congress Catalog Card Numer: 85-5377
ISBN: 0-442-29285-6

All rights reserved. Certain portions of this work copyright © 1972 by Van Nostrand Reinhold Company Inc. No part of this work covered by the copyright hereon may be reproduced or used in any form or by any means—graphic, electronic, or mechanical, including photocopying, recording, taping, or information storage and retrieval systems—without permission of the publisher.

Manufactured in the United States of America

Published by Van Nostrand Reinhold Company Inc.
135 West 50th Street
New York, New York 10020

Van Nostrand Reinhold Company Limited
Molly Millars Lane
Wokingham, Berkshire RG11 2PY, England

Van Nostrand Reinhold
480 Latrobe Street
Melbourne, Victoria 3000, Australia

Macmillan of Canada
Division of Gage Publishing Limited
164 Commander Boulevard
Agincourt, Ontario MIS 3C7, Canada

15 14 13 12 11 10 9 8 7 6 5 4 3 2 1

Library of Congress Cataloging in Publication Data

White, George Clifford.
　The handbook of chlorination.

　Includes bibliographies and index.
　1. Water—Purification—Chlorination. I. Title.
TD462.W47　　1985　　　628.1'662　　85-5377
ISBN 0-442-29285-6

For my wife Charlotte

Preface to the Second Edition

More than a decade has passed since the first edition was published in 1972. In those years we have witnessed more interest, more development, more controversy, more manufacturers and more published documents on the subject of chlorination than in the 50 years preceding the publication of the first edition. Realizing this at the outset, I purposely eliminated three chapters (8, 9, and 10) from the first edition which amounted to 150 pages. This was done to keep the size of the book manageable. The chapters eliminated were: Chlorination of Swimming Pools, Chlorination of Cooling Water, and Other Applications of Chlorine.

In view of all that has occurred in the intervening years, I felt it was incumbent upon me to bring the first edition up to date. It was more difficult than I had imagined. As in the preparation of the first edition I read and digested all of the pertinent chlorination–dechlorination and disinfection literature collected for my library, which has grown tenfold since the first edition. During this time I also maintained an active correspondence with knowledgeable people in the industry.

However, the most rewarding of my activities has been my consulting assignments which took me around the world. These tasks allowed me to broaden my expertise as I had to delve into areas not available to me before I launched myself as a consultant in 1972.

In the early 1970s there was a great surge of activity to safeguard the "environment." This was developed largely by dedicated environmentalists and public health agencies. In 1976 the U.S. Congress passed the Clean Water Act giving regulatory powers to the Environmental Protection Agency. The emphasis was on wastewater treatment and disinfection. This inspired me to produce a second book—*Disinfection of Wastewater and Water for Reuse* (1978). All the material in that book has been updated and is included in the present volume.

As in the first edition, it has been my intention that this book be a reference book for the designer, a manual for the operator, a textbook for the student, and a guide for the regulatory agencies.

I have had help from my colleagues, from all over North America, Western Europe, Israel, and the British Isles. I am most grateful to all of you. Your contributions have improved the quality and broadened the scope of this text. I will be eternally grateful and I hope this work will please you.

San Francisco   G.C.W.

Preface to the First Edition
(1972)

In preparing this book, I have attempted the monumental task of bringing together under one cover all the pertinent information necessary to provide a practical "real world" approach to the entire subject of chlorination as it is applied to potable water, wastewater, cooling water, industrial water, and swimming pools.

Several factors inspired me to write this book. First and foremost was the fact that to my knowledge such a book had never been written. The quest for information on this subject has grown rapidly, particularly by the water and wastewater industries. Whenever I gave a lecture or held a seminar on the subject I was usually asked if a reference text were available. This, coupled with the rising interest and demands for water pollution abatement seemed reason enough for such a text to be made available. I hoped that my thirty-five years of active participation in the field of chlorination could be turned into a significant contribution to the environmental sciences.

The first step in this project was to make a systematic review of the literature. This took well over a year, and fortunately I had at my disposal a fine array of scientific libraries in the San Francisco Bay area. I read and studied some thousand or more references scattered through upward of fifty different scientific periodicals. Whenever I was unable to resolve the contents of a particular reference, and this did occur several times, I corresponded with the author for amplification or explanation. This proved most valuable, as all the authors were most co-operative.

In addition to the above reference material, I have collected for my own use over the years a vast amount of information of great practical value derived from personal experience in all the various applications of chlorine in the United States, Canada, Great Britain, and the European continent. Some is from private communications with colleagues here and abroad. A great deal of the information contained in this book has never been published before to my knowledge.

It is intended that this book be a handbook for the designer, a manual for the operator, a textbook for the student, and a guide for the regulatory agencies. I have tried to present this information so that anyone with a background of high school chemistry and mathematics will have no difficulty in understanding the text.

The reader will note that in some instances, particularly in Chapter 10, the references are listed without page numbers. This means that I worked from my collection of reprinted articles which do not show the page numbers as they appear in the periodicals.

Producing this book has been a formidable but enjoyable task, and I am most appreciative of the help I have had along the way from my colleagues. I hope this work will please them.

San Francisco G.C.W.

Contents

Preface to Second Edition/vii
Preface to First Edition/ix

1. Chlorine: History, Manufacture, Properties, Hazards and Uses/1
2. Hypochlorination/62
3. On-Site Generation of Chlorine/120
4. Chemistry of Chlorination/150
5. Determination of Chlorine Residuals in Water and Wastewater Treatment/214
6. Chlorination of Potable Water/256
7. Chlorination of Wastewater/394
8. Disinfection of Wastewater/476
9. Chlorination Facility Design/585
10. Dechlorination/755
11. Operation and Maintenance of Chlorination and Dechlorination Equipment/789
12. Chlorine Dioxide/833
13. Ozone/893
14. Bromine, Bromine Chloride, Iodine and UV Radiation/962

Appendix/963
Index/1053

THE HANDBOOK OF CHLORINATION

1
Chlorine: History, Manufacture, Properties, Hazards, and Uses

HISTORICAL BACKGROUND

Elemental Chlorine

Chlorine is an element of the halogen family, but it is never found uncombined in nature. It is estimated to account for 0.15 percent of the earth's crust in the form of soluble chlorides, such as common salt (NaCl), carnallite (KMgCl$_3 \cdot$ 6H$_2$O), and sylvite (KCl). In nature, therefore, it exists only as the negative chloride ion with a valence of -1.

Chlorine is a most unusual and versatile chemical, since its properties differ so widely in the gaseous, liquid, and aqueous states. For this reason, each phase will be treated separately.

Chlorine Gas

Chlorine was discovered in its gaseous state in 1774 by Karl W. Scheele, a Swedish chemist, when he heated a black oxide of manganese with hydrochloric acid:[1]

$$MnO_2 + 4HCl \xrightarrow{heat} MnCl_2 + Cl_2 + 2H_2O \tag{1-1}$$

The chlorine thus liberated is a strong-smelling, greenish-yellow gas with a pungent odor; it is extremely irritating to mucuous membranes.

This element must certainly have been known to the medieval Arab chemist Geber (ca. 720–810), as he was the discoverer of aqua regia (3 parts HCl, 1 part HNO$_3$) used to dissolve the noble metals. When this mixture is heated, it gives off fumes similar to "Scheele's gas," about which there was a great controversy when it was studied by the great chemists Berthollet, Lavoisier, Gay-Lussac, Berzelius, Therrard, and Davy.[1]

Scheele called the gas he discovered dephlogisticated muriatic acid on the theory that manganese had displaced phlogiston (as hydrogen was then called) from the muriatic acid (HCl). Scheele also observed that the gas was soluble in water, that it had a permanent bleaching effect on paper, vegetables, and flowers, and that it acted on metals and oxides of metals.

During the decade following Scheele's discovery, Lavoisier successfully attacked and, after a memorable struggle completely upset the phlogiston theory of Scheele. Lavoisier was of the opinion that all acids contained oxygen. Berthollet found that a solution of Scheele's gas in water, when exposed to sunlight, gives off oxygen and leaves muriatic acid behind. Considering this residue proof of Lavoisier's theory, Berthollet called it oxygenated muriatic acid.[2] However, Humphry Davy (1778–1829) was unable to decompose Scheele's gas. On July 12, 1810, before the Royal Society of London, he declared the gas to be an element, which in muriatic acid is combined with hydrogen. Therefore, Lavoisier's theory that all acids contain oxygen had to be discarded. Davy proposed the name "chlorine" from the Greek "chloros," variously translated "green," "greenish yellow," or "yellowish green," in allusion to the color of the gas.[1]

Pelletier in 1785 and Karsten in 1786 succeeded in forming yellow crystals of chlorine hydrate by cooling Scheele's gas in the presence of moisture. From this they inferred that it was not an element. In 1810, Davy proved that these crystals could not be formed by cooling the gas even to $-40°F$ in the absence of moisture. It is now known that these crystals are in fact chlorine hydrate ($Cl_2 \cdot 8H_2O$) and will form under standard conditions with chlorine gas in the presence of moisture beginning at $49.3°F$.

Chlorine Liquid

In 1805, Thomas Northmore liquified Scheele's gas by compression. He noted that it became a yellowish amber liquid under pressure and upon release of the pressure, volatilized rapidly and violently into a green gas. He further noted its pungent odor and severe damage to machinery.

Michael Faraday (1791–1867) also observed liquid chlorine. On March 5, 1823, he was visited in his laboratory (he worked as an assistant to Davy) by J. A. Paris while he was working with chlorine hydrate in a sealed tube. Paris noted a yellowish, oily-appearing substance in the tube and chided Faraday for working with dirty apparatus. When Faraday tried to open the tube, it shattered and the oily substance vanished. Aftery studying the accident, Faraday wrote Paris: "Dear Sir: The oil you saw in the sealed tube yesterday turned out to be liquid chlorine. Yours faithfully, Michael Faraday."[1]

MANUFACTURE OF CHLORINE

History

From 1805 to 1888, Scheele's gas remained a laboratory curiosity—and a dangerous one—until it was produced on a commercial scale by cooling and compression in a suitable apparatus. This was made possible by Kneitsch's discovery in 1888 that dry chlorine did not attack iron or steel, thus making it possible to package chlorine as a liquid under pressure.

Until this time, chlorine was used as a bleaching agent in the form of a solution. In 1785, Berthollet prepared this solution by dissolving Scheele's gas in water and adding it to a solution of caustic potash. This was done at a chemical plant in Javel, then a small French town, now a part of Paris. Hence, the solution is known as Javelle water. James Watt, the inventor, obtained from Berthollet a license for the manufacture of Javelle water and brought it to Scotland for Charles Tennant, founder of the English chemical company. In 1789, Tennant, through extensive experimentation, produced another liquid bleaching compound, a chlorinated milk of lime. A year later, he improved it greatly by making it a dry compound, which has been known ever since as bleaching powder. The chlorine for the manufacture of bleaching powder was obtained by Berthollet's method of heating sodium chloride, manganese, and sulfuric acid in lead stills. During this time, chlorine was also being made on a limited scale by the Weldon process, which employed the basic reaction of Scheele's discovery, hydrochloric acid and manganese dioxide. This method was given considerable impetus when in 1836 Gossage invented his coke towers for the absorption of waste hydrochloric acid, resulting from the LeBlanc soda process. This invention resulted in cheap hydrochloric acid for the Weldon process.[1] However, in 1868, Scheele's method by the Weldon process became obsolete.* Henry Deacon and Ferdinand Hurter patented a process for producing chlorine by decomposing hydrochloric acid with atmospheric oxygen in the presence of a catalyst.[2] A mixture of hydrochloric acid and air is heated. About 70 percent of the hydrogen chloride can be converted to chlorine as it mixes with the air and steam. This mixture is condensed, and the steam absorbs the hydrogen chloride, forming a very strong muriatic acid mixed with hydrogen chloride gas. This mixture is passed through a superheater, raised to about 430°C, and then passed through a decomposer consisting of a brick- or pumice-lined chamber impregnated with cupric chloride, the catalyst. After this step, the mixture is washed first with water and then with sulfuric acid. The remaining mixture consists of nitrogen and oxygen containing 10 percent chlorine gas which can be utilized without any difficulty in the manufacture of bleach liquids and powders. The original HCl is recycled so that the only additional materials are heat and air. Considering the amount of chlorine produced, the plant is extremely bulky. The reaction employed is:

$$4HCl + O_2 \xrightarrow{heat} + \xrightarrow{catalyst} 2Cl_2 + H_2O \qquad (1\text{-}2)$$

This reaction is reversible and incomplete. The rate of reaction is made satisfactory by the addition of heat and the catalyst, cupric copper.

The hydrogen chloride in this process is largely a byproduct of the LeBlanc soda process. With the advent of the Solvay sodium ammonia process in 1870,

---

* The Weldon process or Scheele's method is still used to advantage for making hypochlorites in remote areas (See Chapter 2).

4 HANDBOOK OF CHLORINATION

the LeBlanc process fell into a sharp decline, causing the abandonment of the Deacon process in favor of the electrolytic methods which were then emerging from the experimental state.

Electrolytic Processes

History. In 1883, after years of research, Faraday postulated the laws that govern the action of passing an electric current through an aqueous salt solution. He coined the word "electrolysis" to describe the resulting phenomenon. These fundamental laws are among the most exact in chemistry and are as follows:

The weight of a given element liberated at an electrode is directly proportional to the quantity of electricity passed through the solution. (The unit of electrical quantity is the coulomb.)

The weights of different elements liberated by the same quantity of electricity are proportional to the equivalent weights of the elements.

Charles Watt obtained an English patent for the electrolytic manufacture of chlorine in 1851. However, at that time electric current generators of sufficient size were not available, and so the patent was only of academic interest. When this equipment became available, interest in electrochemistry was greatly stimulated. In 1890 the first commercial production of chlorine by the electrolytic method was introduced by the Elektron Company (now Fabwerkie-Hoechst A.G.) of Griesheim, West Germany.[1] The first electrolytic plant to go into production in America was at Rumford Falls, Maine, in 1892 for the Oxford Paper Company. In 1894, Mathieson Chemical Company acquired the rights to the Castner mercury cell and began the first commercial production of bleaching powder at a demonstration plant in Saltville, Virginia. In 1897, this operation was moved to Niagara Falls, New York. These original Castner rocking cells operated successfully until they were shut down in 1960.[3,4]

At first, the electrolytic process for the commercial production of chlorine was primarily for the manufacture of caustic. Chlorine was a byproduct. At the Niagara Falls plant, a small amount was used for bleach and for the making of hydrochloric acid, the rest being discharged into the Niagara River. Not until 1909 was liquid chlorine manufactured on a commercial basis. It was first packaged in 100-lb capacity steel cylinders made in Germany. The business grew slowly but steadily, most consumers using it for bleaching textiles, pulp, and paper. The first American tank car with a capacity of 15 tons, was manufactured in 1909. The next year, 150-lb cylinders came into use, and in 1917 ton containers were made for the chemical warfare service.[1]

Current Practice. Most of the chlorine today (1980's) is manufactured by three types of electrolytic cells: diaphragm, mercury, and membrane. There are other methods of production which are designed to fit the raw material containing the chloride ion. These methods include the electrolysis of hydrochloric acid, the salt process, and the HCl oxidation process. Chlorine is frequently produced as a byproduct of heavy metal recovery such as the tungsten sponge process or the extraction of magnesium from magnesium chloride ore. While the methods of electrolytic production of chlorine have not changed the equipment use has changed dramatically.

ELECTROLYTIC CELL DEVELOPMENT

The Ideal Electrochemical Cell

This would be a cell composed of an anode, a cathode, and a separator which isolates the liquids contained in the anode chamber and the cathode chamber. The function of the separator is to isolate the two chambers while allowing the migration of selected ions from the anode chamber to the cathode chamber. Brine composed of sodium chloride and water is introduced into the anode chamber, where oxidation of the sodium chloride takes place. Chlorine gas is released at the anode. The sodium ions are attracted to the negatively charged cathode and transported through the separator. If the separator is doing a perfect job, which in an ideal cell it would, all of the chloride would be contained on the anode side of the cell.

Water is reduced at the cathode and hydrogen gas is evolved. The remaining hydroxide ion combines with the sodium ion to form sodium hydroxide solution (caustic), which exits the cathode chamber. The ideal separator (membrane) would keep all the OH-ions on the cathode side of the cell. See Fig. 1-1.

Recent Developments

In practice electrolytic cells are plagued with a variety of problems such as corrosion, erosion of electrodes and plugging of the separator. All of these factors generate high maintenance costs. There is also great concern over the cost and availability of energy required to drive the oxidation reaction of the brine solution.

In recent years the industry has made important advances on all of these fronts. Energy consumption has been reduced significantly by improved cell design and improved electrical equipment required to convert AC to DC. Motor generator sets, mercury arc, and mechanical rectifiers, have been largely replaced by highly efficient semiconductor rectifiers which operate at efficiencies on the order of 97 percent. This compares with efficiencies of 90 percent experienced with mercury arc rectifiers at lower cell circuit voltages 30 years ago.

Another example is the case of centrifugal chlorine compressors. They are now available with unit capacities in excess of 600 tons/day chlorine and the ability to operate at 10 atm. discharge pressure. These units have superseded the 30–50 ton/day compressors with discharge pressure capability of only 3–4 atm.

The major advances in the industry have been in the electrolytic cell design.

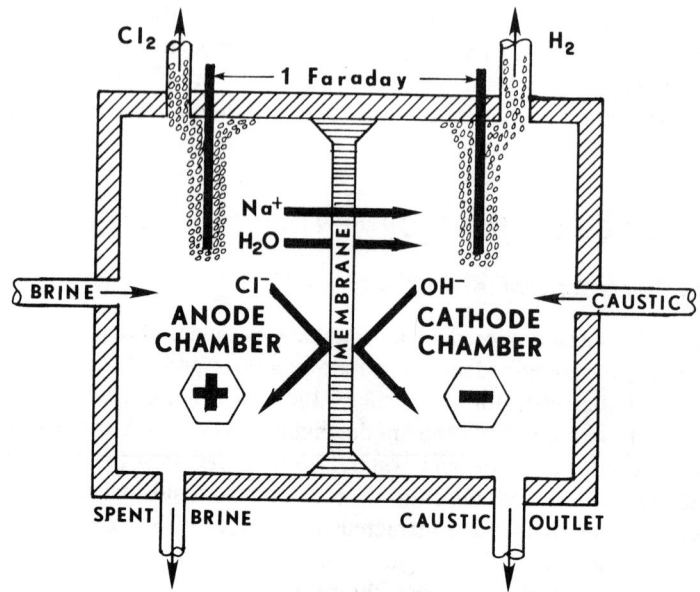

Fig. 1-1. The ideal membrane cell.

New materials of construction have revolutionized cell design. Such materials as valve metal (titanium or tantalum), various noble metal oxides, plasticized diaphragms, and special cladding processes that eliminate or retard the corrosion of cell components. Two major achievements have been the development of dimensionally stable anodes and the membrane cell.

Dimensionally Stable Anodes

The chlor-alkali industry has sought for years a stable metal anode for chlorine production. Aerospace industry developments in the 1960s provided the material at a reasonable cost. In 1968 Diamond Shamrock Corporation, a world leader in the production of chlorine and caustic announced the development and commercialization of new types of metallic anodes. These they have named DSA® dimensionally stable anodes.[5,6,50,51]

These anodes have reduced capital and operating costs. As their name implies, they retain their size and shape during use and have a life longer than 10 years in diaphragm cells, compared to about 180 days for the graphite anode. Moreover, replacement of graphite anodes removes the major source of hydrocarbon contamination of chlorine, which poses a health problem in potable water chlorination (see Impurities in Chlorine this chapter).

These anodes can be reactivated by redeposition of the metal coating. In addition to the saving in downtime and labor for anode replacement, the saving in graphite consumption alone is said to run as high as $4.00 per ton of chlorine. The DSA®s have played an important role in the future of the chlor-alkali industry. They are an important factor in the improvement and expansion of existing facilities.

Approximately 70 plants in North America are using DSA®s producing 7.3 million tons of chlorine per year. The conversion to DSA® from other anode systems have resulted in a power saving of 16 billion Kwh per year.[6,7]

Anodes used before the introduction of the DSA®s were made of graphite or some form of impregnated carbon. They belong to the soluble type of anodes. Their life was short, about 5–6 months. They eroded in an irregular pattern which resulted in variable and increased voltage requirements. This phenomenon reduced significantly the efficiency of the cell.

Dimensionally stable anodes (DSA®) are based on inventions using mixed metal oxide coatings. The metal to which these oxide coatings are applied is called valve metal and is either titanium or tantalum. Tantalum is typically not used because of higher cost and toxicity. The type of coating to be used is generally determined by the parameters of the process in which it is to be used. For chlorine electrolysis the coatings are usually ruthenium–titanium oxides.[5] However, several types of coatings and a wide variety of substrate geometries, thicknesses, and shapes are available. Therefore DSA® can be custom made to fit the application whether it be cell modification or process requirement.

Mixed oxide coatings of the DSA® type exhibit stable operation for long periods of time at low operating voltages. In the production of chlorine this results in a significant reduction in labor required to maintain cell operation, a reduction in downtime for cell cleaning and adjusting, plus an energy savings of 20 percent over the formerly used soluble graphite anodes. Although operating experience of DSA® anodes in membrane cells is relatively short, the coating life could approach the coating life in diaphragm cell service.

Anode lifeime depends upon the current density at which the cell operates. At relatively low current densities of about 1.5–2.6 kA/m^2 which is typical for diaphragm cells, the DSA®s have operated well in excess of ten years. At 10 kA/m^2, which is common for mercury cells, these anodes operate for about two years before the structure must be removed from the cell and a new coating applied.

The mixed oxide coatings do not dissolve in mercury, and the coating surface is not easily wetted by mercury. These unique features permit operation of mercury cells with extremely small gaps between the anode and the cathode, commonly as low as 3 mm. (See Fig. 1-8.) Anode–cathode gaps of 12–15 mm are common in diaphragm cells. With the use of newly developed polymer-modified asbestos diaphragms this gap can be reduced to 3–5 mm. In the development of the membrane cell, research is directed to the design of a zero gap cell. The smaller the gap the greater the power savings.

Membrane Cell

History. Hooker Chemicals and Plastics, and Diamond Shamrock Corp. both began membrane cell development programs in the 1950s.[8,9] The largest producer of chlorine in the U.K., the ICI corporation, began membrane cell development twenty years later.[10] Catalytic Inc. of Philadelphia is licensed in the USA to sell ICI membrane cells. The membrane separator concept was made feasible by the availability of selective ion exchange membranes. These membranes became available as a result of Du Pont's research on fuel cells. The first of this family of membranes was Nafion which consisted of 2–10-mil-thick film of perfluorosulfonic acid resin.[11] This is a copolymer of tetrafluoroethylene and another monomer to which negative sulfuric acid groups are attached.[8,11]

The world's first commercial membrane cell chlor-alkali plant rated to produce 40,000 metric tons/year caustic soda began successful operation April 1975 for Asahi Chemical Co. at Nobeoka, Japan on Kyushu Island.[21] This plant produced caustic soda with concentrations varying from 18 to 21 percent at a current density of 5 kA/m^2, 4.2 V, and a current efficiency of 80.5 percent. The power consumption was 3496 kWh/metric ton NaOH and 0.89 metric ton chlorine.

It was discovered that the Nafion 315 membrane could not attain high current efficiency when producing high concentration caustic soda because the hydrophilic sulfonic acid groups lead to counter migration of hydroxide ions from the cathode to the anode compartments. In view of this, research began on other membranes. Asahi Chemical Co. developed a perfluorocarboxylic acid membrane that demonstrated a 90+ percent current efficiency at high concentrations of caustic soda in the electrolyzer. This was announced in 1976.[9] DuPont followed with a series of improved membranes. Their third generation membrane and presumably their best one so far for the chlor-alkali industry is the Nafion series which has high current efficiency for a 35 percent caustic concentration in the electrolyzer.

In the meantime the Asahi Glass Co. developed its third generation membrane, Flemion 753 in 1981. This is the membrane used by Asahi Chemical Industry Co Ltd., ICI, and Uhde in their chlor-alkali electrolyzers.[9,12,13] Both DuPont and Asahi use a combination of both sulfonic and carboxylic groups to minimize the water hydration which allows the back-migration of the hydroxyl ions. Dow Chemical Co., one of the largest chlor-alkali producers, is currently developing its own membranes which they believe will result in a membrane structure that will eliminate the back-hydration problem and furthermore will allow the construction of a "zero gap" electrolyzer.[14]

Theory of Operation. Referring to Fig. 1-2, the membrane cell is a simulation of the ideal cell. The operation of any electrolytic cell is limited by the ability of the components to perform as required. The necessity for research and development is to find the proper materials for the electrodes and separator (membrane) to provide the greatest yield and purity of product with the least amount of energy and maintenance.

Salt, water, and electric current are the raw materials. Solid salt is dissolved in a saturator with fresh water and depleted brine. The saturated brine is chemically treated to precipitate impurities which are removed through clarification and filtra-

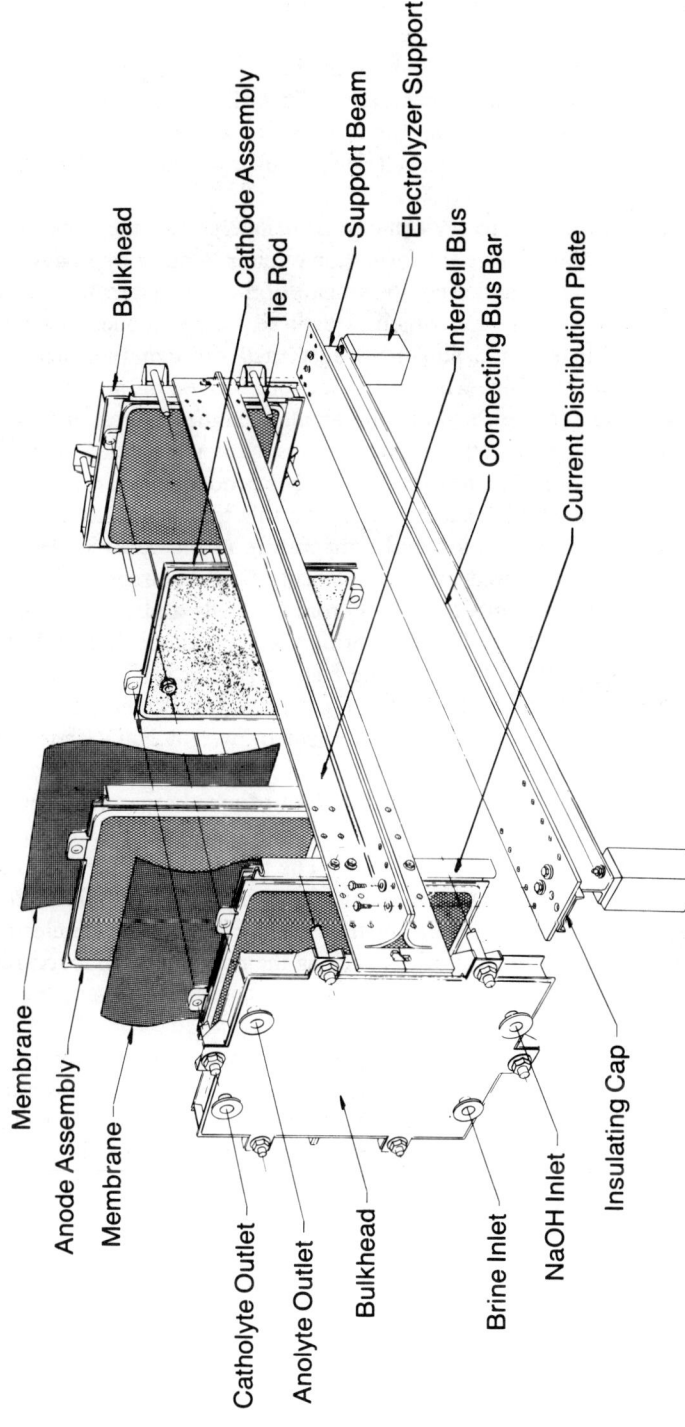

Fig. 1-2. The membrane cell (courtesy ELTECH Systems Corporation).

tion. The brine is then further treated by an ion exchange process specific for brine. This additional purification is necessary in order to limit the calcium ion content to *less than* 0.05 mg/l. The membranes are greatly affected by calcium deposits.[45] If this limit is not achieved cell performance will suffer and maintenance costs will escalate.

Hydrochloric acid is added to the brine to neutrilize part of the back-migrating hydroxide ions. These ions reduce the formation of objectionable byproducts before the purified brine is fed to the anode compartments of the cells. The salt is electrically decomposed to form chlorine gas which is evolved at the anode. The resulting sodium ions remain in solution and are transported through the membrane to the cathode where they combine with hydroxide ions formed in the cathode chambers of the cells. The depleted brine in the anode chamber is treated to separate out any remaining chlorine. After this reprocessing it is recycled to the brine saturator. Caustic solution that has formed in the cathode compartment flows to a caustic surge tank. Most of the caustic entering this tank is cooled and recycled to control the caustic concentration and temperature in the cells. Softened water is added to maintain the desired product concentration. Hydrogen which is also produced in the cathode chambers, passes from the cells and may be vented or recovered for use in the plant. The feed brine contains about 320g/l NaCl and 30g/l max. of chlorate ion. Depleted brine has a pH range of 2–5 and contains about 170 g/l NaCl. Chlorine is 97–99.5 percent pure. It contains 0.5–3.0 percent oxygen. The caustic product which is 30 percent by weight NaOH contains 40–50 ppm NaCl and 5–15 ppm ClO_3^-. Hydrogen is 99.9 percent pure.

Tables 1-1 and 1-1a give typical operating data of membrane cells now in use.

Monopolar vs. Bipolar Configuration. The term "electrolyzer" has an important meaning in the electrolytic process of making chlorine.[16] It refers to a cluster of individual cells. The arrangement of these cells to form an electrolyzer varies from one manufacturer to another, such as the potential production requirements.

Monopolar. When the cells of an electrolyzer are arranged electrically in parallel it is a monopolar design. All of the anodes are electrically joined together and so are all the cathodes. The electrical energy flows into all of the anodes simultaneously. It then passes through the membranes to the cathodes, and then exits the electrolyzer. The amperage supplied to the electrolyzer is not limited by the electrode area of one cell but by the electrode area of the entire electrolyzers.

This configuration yields a high amperage electrolyzer that exhibits low voltage. Also all the cells within the electrolyzer will operate at the same voltage regardless of variable resistances within the individual cells. Therefore it is not necessary to electrically isolate each cell from the remaining cells in the electrolyzer since they are all at the same potential.

The monopolar arrangement allows simplification of feed and discharge piping.

Table 1-1 Operating Data—Membrane Cells*

	Manufacturer & Model No. of Cell				
	Asahi	Catalytic FM-21	Olin OM-PAC	PPG BIZEC	Uhde HUMM
Cell current (kA)	2030	220	300	15.3	70
Current density (kA/m^2)		3.93	5.0	4.0	3
Current efficiency (%)	93–95	94–96	95–97	95	93–95
Voltage/cell	3.11	3.3	3.32	3.25	3.23
Chlorine production (ton/day)	69.4	6.05	9.9	23.0	15
Caustic production (ton/day)	75.13	6.82	11.1	26.0	2.5
Power consumption (kWh/ton Cl$_2$)	2100	2070	2080	2034	2570
Caustic concentration (%)	20–40	25–35	32	36	33–35
H$_2$ production (lb/day/cell)	40.26	1.64	62	29	940
Catholyte temp (°C)	90	90–95	95	90	85–90
Anode life (yrs.)	>5	>5	>4	>5	2–5
Membrane life (yrs.)	>2	>2	>2	>2	2
Membrane brand name	perfluoro-carboxylic	Flemion 723	varies	permionic	Nafion
No. of cells	100	24		50	100

Typical analyses, ASAHI electrolyzer:
  Chlorine 99.4%, 0.5% O$_2$ (vol.)
  Caustic soda 50% (wt), 50 ppm (wt) NaCl
  Hydrogen 99.9% (vol.)
  Brine feed 305–315 g/l NaCl 60°C
* See Table 1-1a for Eltech MGC membrane electrolyzer

Table 1-1a

Eltech MGC Electrolyzer Design Range

Cells per electrolyzer	2–30
Current density (kA/m^2)	2–5
Circuit current (kA):	6–202
Production per electrolyzer:	
Chlorine (MTPD)	0.2–6.3
NaOH (MTPD)	0.2–7.0
Caustic concentration (wt. %)	30–35

Expected MGC Electrolyzer Operating Characteristics

Basis: 3.1 kA/m^2
    32 wt. % NaOH from Cells

Electrolyzer voltage	3.1 volts
Power consumption	2165 DCkWh/MT NaOH
Catholyte temperature	90°C
Anode coating life	>5 years
Membrane life	>2 years
Membranes available	DuPont and Asahi Glass

Electrical short circuits which cause corrosion is greatly reduced. It also provides a safer environment for the operators. Electrical arcing between the anode and the cathode does not readily occur so the membrane remains intact. This prevents the formation of an explosive hydrogen–chlorine mixture.

A more important advantage of the monopolar electrolyzer is that it minimizes the danger caused by a loss of brine flow in the cell. When this occurs the resistance to the flow of electricity increases, but the current flow redistributes itself throughout the electrolyzer. This maintains equal but higher voltages in the cell; but not enough to cause the arcing described above. The operator, recognizing an abnormal voltage on the electrolyzer, has the opportunity to seek out and correct the brine flow problem before a hazardous condition results.

Bipolar. When the cells of an electrolyzer are arranged electrically in series, it is a bipolar design. The electric current enters one end of the electrolyzer and passes through each membrane cell in series until it exits the electrolyzer. The current enters into an anode, passes through the membrane to a cathode, through electrical connections into an anode, through another membrane into a cathode, and so on throughout the entire electrolyzer. Because the electrical resistances within the cells are additive, this configuration yields a high voltage–low amperage electrolyzer. The total amperage is limited by the electrode area of a single cell. For this design the voltage drop between cells is low since the cathode of one cell is connected directly to the anode of the next cell.

The rectifier for the bipolar electrical system is slightly less expensive than for the monopolar design.

The companies who favor the monopolar design cite the following disadvantages of the bipolar configuration:

1. Each cell must be electrically isolated.
2. Design of feed and discharge piping is complex.
3. Extra care must be exercised to prevent the failure of seals or welds and to prevent hydrogen penetration of bimetallic bonds.
4. If the brine feed to one cell is interrupted and not corrected quickly the electrical resistance in the system becomes so high that the electrical energy arcs from the anode to the cathode, passing through both the membrane and the newly formed vapor space in the cell. This arcing will burn holes in the membrane permitting hydrogen gas to mix with chlorine gas in the anode chamber, forming an explosive mixture.

This is one of the most dangerous situations that can occur in a chlorine plant. For additional information on bipolar design see Ref. 22.

Choice of Design. It is fair to say that both monopolar and bipolar electrolyzers have been thoroughly field tested by the various manufacturers. In spite of the dismal picture painted by the proponents of the monopolar design itemized above, one of the most successful membrane cell is bipolar. However each manufacturer has a good reason for a particular choice. In some cases local conditions might favor one over the other. As of 1983, the favorite choice seems to be the monopolar design.

Advantages of the Membrane Cell. The most impoirtant feature of the membrane cell is the savings in energy. Fig. 1-3 illustrates this by comparing it to the mercury cell and the diaphragm cell.

The membrane cell is easy to build, easy to operate and easy to maintain.

The synthetic membrane separator and the DSA® in combination with coated metal cathodes provide an electrolytic cell with characteristics that approaches those of the ideal cell. About the only feature that is lacking is its ability to produce a 50 percent caustic solution as can the amalgam (merc.) cell. To date the most concentrated caustic solution produced by a membrane cell is 30–40 percent. Note that in Fig. 1-3 the power consumption has been adjusted to the amount required to concentrate the caustic in the membrane and diaphragm cells up to 50 percent.

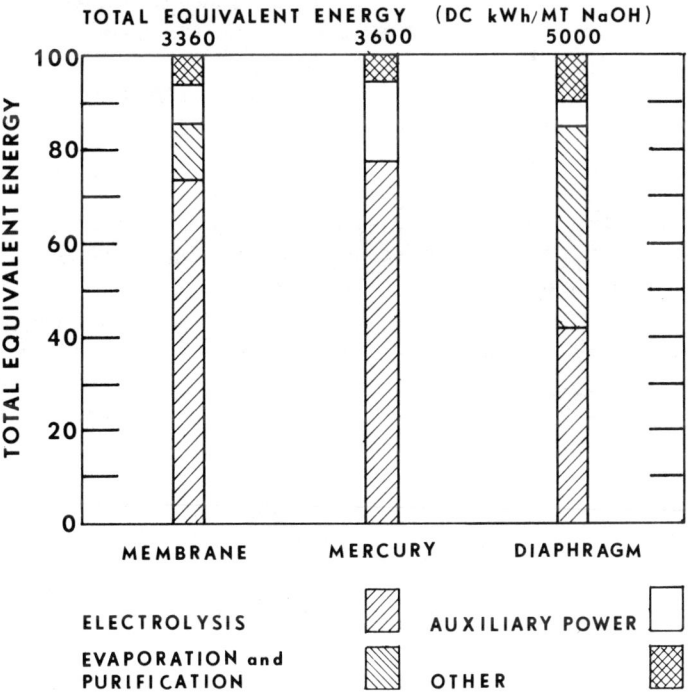

Fig. 1-3. Energy consumption comparison between membrane, mercury and diaphragm cell processes.

Current Developments

Membrane Gap Cell. Most membrane cells in operation in 1983 are defined as finite gap cells. This means that there is an established distance or space between the membrane and each of the electrodes. When the electrodes are separated from each other by only the thickness of the membrane, as is achieved, for example, in Eltech's MGC electrolyzer, significant savings in power results. It is estimated that cell voltage could probably be reduced by 0.6 V at 3.1 kA/m^2.[15] Eventually cell operating potentials of about 2.9 V may be achieved by elimination of interelectrode gaps and membrane improvement.

Air Cathodes. Research has indicated that cell performance would be greatly improved by substituting the air cathode for the hydrogen cathode.[15] It is further predicted that electrolyzers equipped with the air cathode should be able to operate in the 2.1 V region at 3.1 kA/m^2. To accomplish this goal a totally new electrode in a new cell configuration will be required.

In a hydrogen cathode system the water in the catholyte need only to reach the electrode surface to be reduced to hydrogen gas and the hydroxide ion. Air cathode, by comparison, must receive the reactants from two different phases. Oxygen from air supplied to the cathode must penetrate the porous electrode. Likewise water from the catholyte must also penetrate the cathode from the electrolyte side. Oxygen is reduced to the hydroxide ion at the electrode in the presence of suitable catalysts. The air in the cathode is now depleted in oxygen and must diffuse back out of the electrode into the air chamber. The hydroxide ion must then move back into the catholyte.

The electrode used to accomplish all of this has a three-layer structure. The porous carbon layer is the region in which the reaction occurs. The second layer which is also porous and located between the active layer and the air chamber prevents the penetration of electrolyte through the electrode. This prevents flooding of the air chamber by the catholyte. A third layer is the conductive grid located on the electrolyte side. It is used to distribute current to the active layer immediately adjacent to it.

Figure 1-4 illustrates schematically the operation of an air cathode system in a chlor-alkali electrolytic cell.[15]

There are three major differences in the air cathode cell compared to a conventional cell. The first is the inclusion of an air chamber behind the electrode to provide the air for the cathodic reaction. Second, since hydrogen is not produced by the reduction of oxygen, hydrogen handling equioment is not required. Third, the cell operates at approximately 1.0 V lower potential than a conventional hydrogen evolving cell.

Diaphragm Cells

History. The bulk of the chlorine produced in the USA and North America is by the diaphragm cell. Approximately 70 percent of the total U.S. production

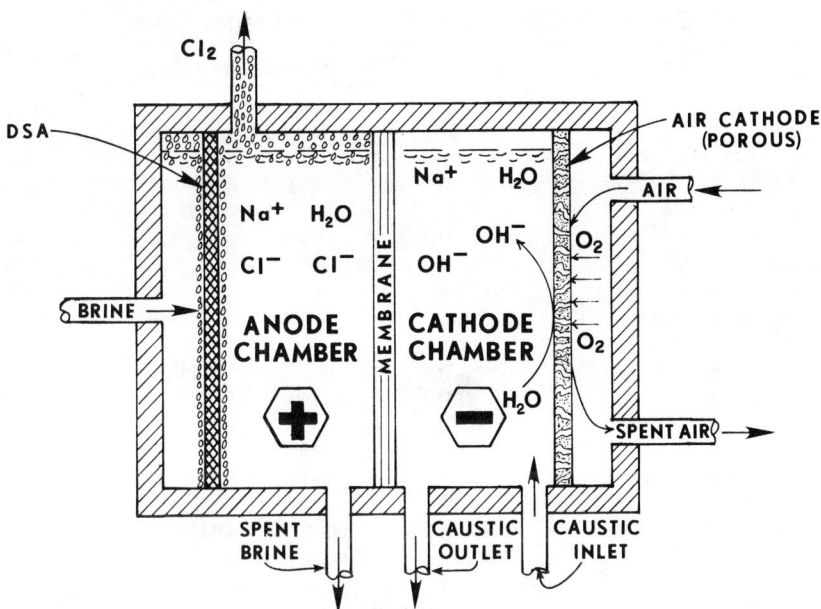

Fig. 1-4. Air cathode cell (courtesy ELTECH Systems Corporation).

is by these cells.[16] Although several hundred types of diaphragm cells have been designed, those most widely used are listed in Table 1-2, which tabulates typical operating data of the modified cells. These cells have undergone a transformation related to the introduction of DSA®s and polymer-based improvements of the diaphragms. These innovations occurred in the 1970s. The replacement of graphite anodes with DSA®s was a most significant improvement in the art of chlor-alkali production. Next in importance has been the improvement in the original asbestos slurry diaphragms. The new modified asbestos diaphragms incorporate polymer application technology which results in lower power consumption, a more rugged diaphragm, dimensional stability of the diaphragm which produces a longer diaphragm life and more reproducible diaphragm operating characteristics.

Table 1-3 illustrates the improvement in diaphragm cell operating characteristics from 1920 to 1982. The improvements are due largely to coated metal anodes and better diaphragm material.

Chemistry of Electrolysis. A typical diaphragm cell is illustrated in Fig. 1-5. Fig. 1-6 illustrates the chemistry involved, which is described below.

The overall chemical reaction is:

$$2NaCl + H_2O + \text{electric current} \longrightarrow NaOH + Cl_2 + H_2 \qquad (1\text{-}3)$$

The principal anode reaction is:

$$2Cl^- \longrightarrow Cl_2 + 2e^- \qquad (1\text{-}4)$$

Table 1-2 Operating Data; Diaphragm Cells

Characteristics	Manufacturer and Cell			
	Eltech MDC-55	Hooker H-4	PPG V-1144	Uhde HU-60
Current (kA)	135	150	72	80–120
Current efficiency (%)	95.8	96.6	95–96	
Voltage/cell	3.46	3.47	3.5	3.41
Power (kWh/ton Cl_2)	2475	2492	2500	2700
Cl_2 production (lb/day/cell)	9053	10024	4900	4780–7280
NaOH production (lb/day/cell)	10323	11309	5418	5380–8200
S/C ratio (lb NaCl/lb NaOH)	1.3	1.25		
NaOH in cell liquor (%)	11	11.5	11–12	
NaOH in cell liquor (g/l)	135	140	135–145	
Catholyte temp. (°C)	95	97	95	
H_2 production (lb/day/cell)	257.4	287	142.6	149.6–226.6
H_2 cu.ft/day/cell (760 mm 0°C)	45866	51118	25403	
Anode life (mos.)	>120	132	84–132	
Diaphragm life (days)	300–600	300–1000	356–730	

Typical analyses—Eltech (MDC-55)

Chlorine Gas, percent by vol:		Caustic Cell Liquor, grams/liter		Hydrogen, percent by vol.	
Cl_2	96.5–98.0	NaOH	130–150	H_2	99.8
H_2	0.1– 0.3	NaCl	175–210	N_2	0.07
CO_2	0.1– 0.3	$NaClO_3$	0.05–0.30	O_2	0.10
O_2	1.0– 3.0	Fe	<0.00005	CO_2	0.04
		Na_2SO_4	0.00–7.00		

Chlorine formed at the anode saturates the anolyte, and an equilibrium is established, as follows:

$$Cl_2 + OH \longrightarrow Cl^- + HOCl \qquad (1\text{-}5)$$

$$HOCl \longrightarrow H^+ + OCl^- \qquad (1\text{-}6)$$

Table 1-3 Examples of Changes in Diaphragm Cell Characteristics

Characteristic	Producer and Cell				
	Allen-Moore	Hooker S	Hooker S-38	Hooker S4	Eltech MDC-55
Year	1920	1938	1952	1965	1981
Anode material	G	G	G	G	DSA®
Current (kA)	1.3	6	27	55	135
Current density (kA/m²)	0.4	0.6	1.3	1.3	2.5
Voltage/cell	3.6	3.3	3.9	4.0	3.46
Ton Cl_2/cell/day	0.04	0.20	0.90	1.83	4.53

G = Graphite

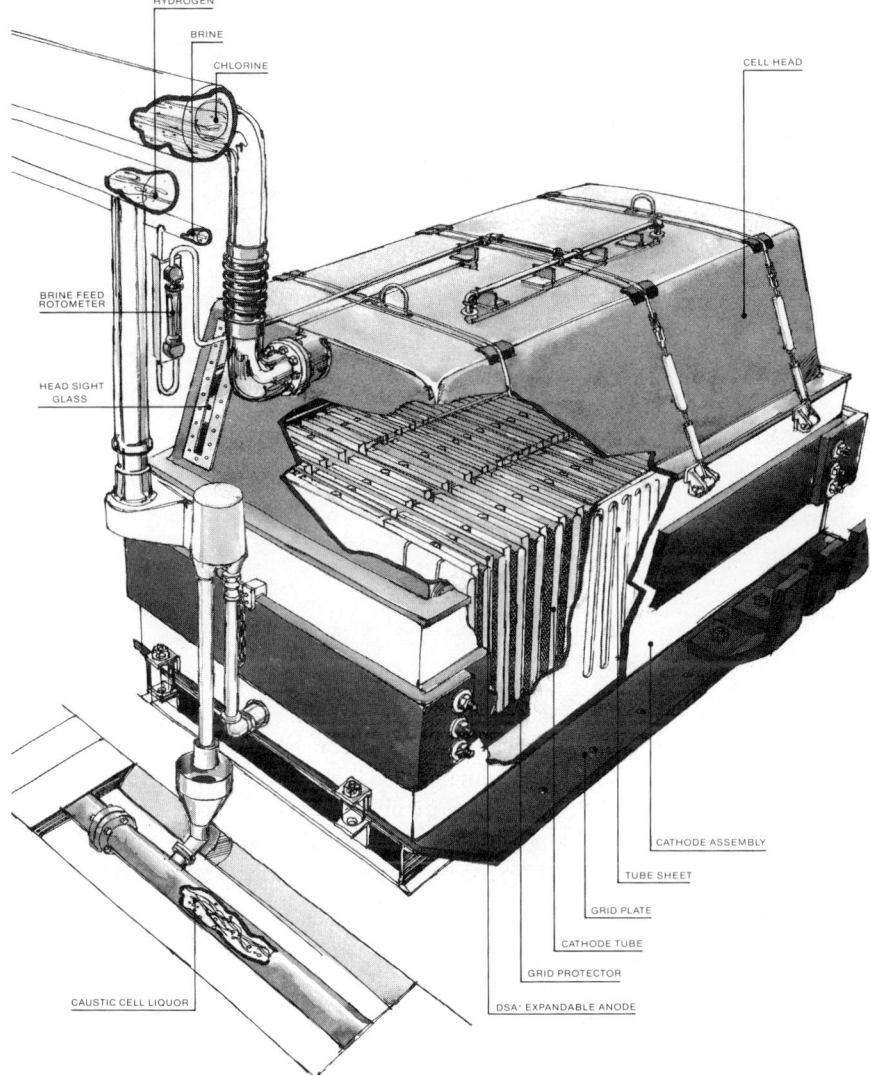

Fig. 1-5. ELTECH MDC-55 diaphragm cell (courtesy ELTECH Systems Corporation).

The principal cathode reaction is:

$$2H^+ + 2OH^- + 2e^- \longrightarrow H_2 + 2OH^- \qquad (1\text{-}7)$$

Therefore the H^+ ion present as H_2O in the catholyte evolves at the cathode as hydrogen gas, leaving the hydroxyl ion (OH^-) behind in the catholyte. Since

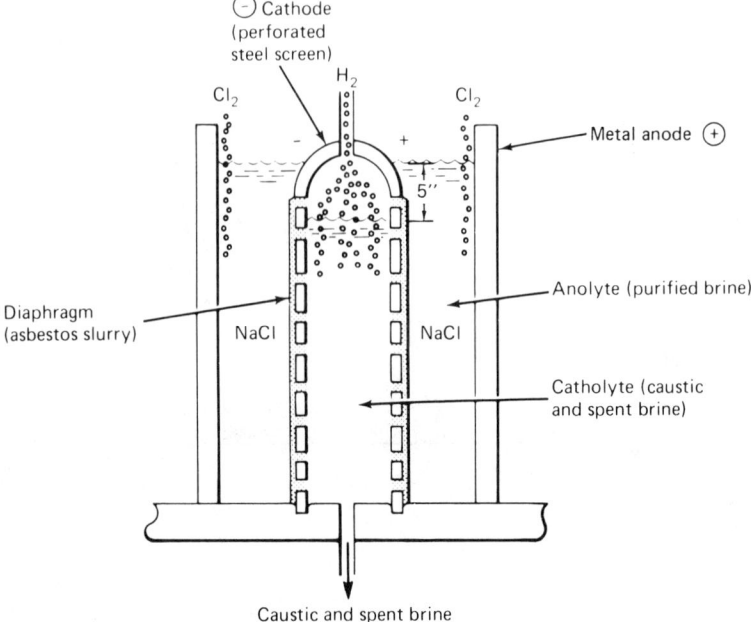

Fig. 1-6. Schematic diaphragm cell.

chlorine has evolved at the anode, the sodium ion is free to join the hydroxyl ion as it migrates from the anolyte chamber to the catholyte chamber by means of differential hydraulic head. The porous diaphragm is used to inhibit the migration of the OH⁻ ions from the cathode to the anode; otherwise the mixing of these solutions would result in the formation of hypochlorite instead of the evolution of chlorine.

By maintaining the pH of the anolyte solution between 3 and 4 and a differential head of five inches between the anolyte and catholyte solution levels, the operation of the cell can be kept in chemical equilibrium to produce a 97 percent chlorine gas at the anode, almost 100 percent hydrogen gas at the cathode and an 11–12 percent NaOH content in the catholyte effluent, which is replaced by purified brine at 60–70°C at the inlet.[17] The diaphragm is usually made by introducing an asbestos slurry into the brine and drawing a vacuum on the inside of the cathode compartment. This deposits the asbestos slurry on the perforated steel screens that serve as the cathode. In time this diaphragm material becomes clogged with impurities and must be removed and replaced.

Fig. 1-7 illustrates the flow sheet of a typical modern chlor-alkali plant producing chlorine by electrolysis in a diaphragm cell.

The ingredients required are brine, water, and electric power. Brine is made available in one of two ways: (1) Rock salt is delivered to the plant and dissolved

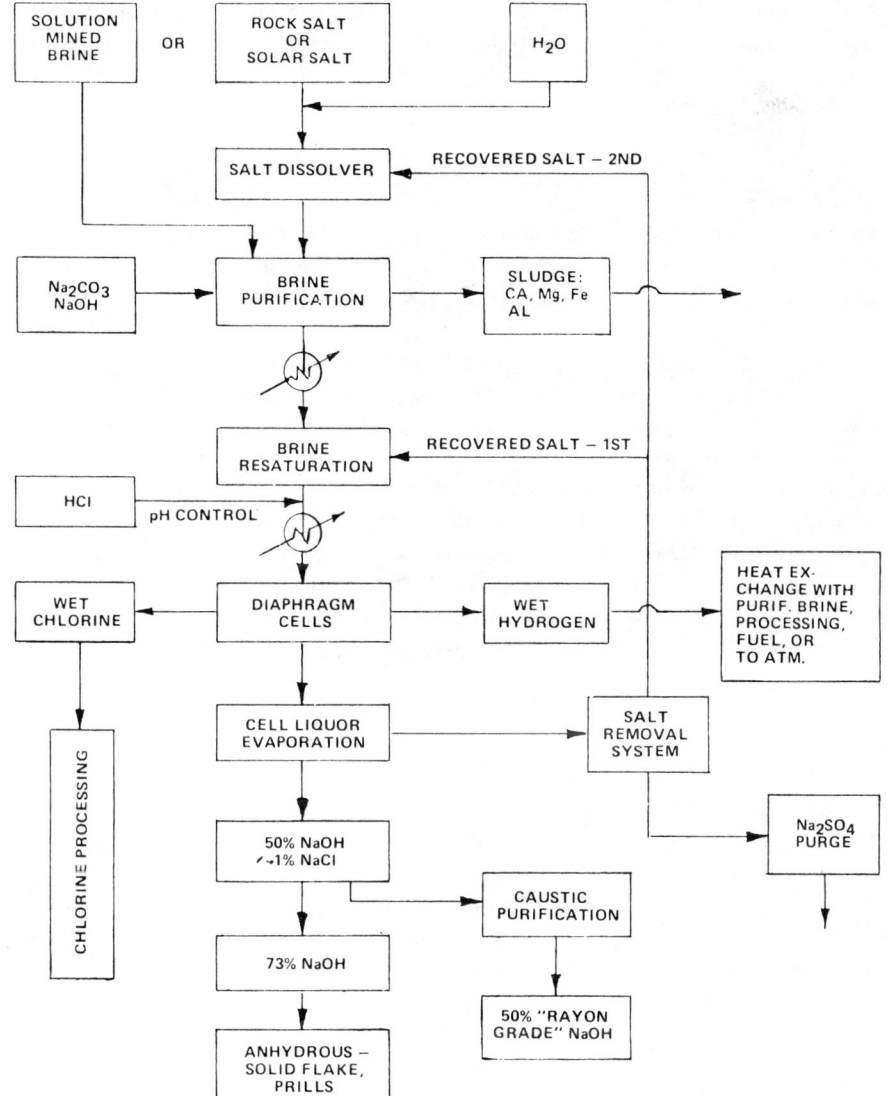

Fig. 1-7. Chlorine-caustic soda plant by the Hooker diaphragm cell process (courtesy Occidental Chemical Corporation).

in water, or (2) the plant may be located adjacent to underground deposits of salt from which the brine is produced by water injection into a well. Impurities such as the calcium and magnesium ions of the sulfates and chlorides must be removed from the brine solution. This is accomplished by the addition of sodium carbonate and sodium hydroxide in a carefully controlled treatment process followed by sedimentation and filtration. The salt content of this brine must be increased to the optimum level of approximately 26.6 percent (322 g/l) of NaCl at 70°C. This is accomplished by heating the brine and saturating it with purified salt. The latter is obtained from the evaporation step in the processing of the cell liquor.

The water supply used for dissolving the salt should be completely softened for the removal of Ca, Mg, and Fe ions to minimize blockage in the cell diaphragm. This water should also be free of any ammonia ion to minimize the nitrogen content. This reduces the hazard of forming the potentially explosive nitrogen trichloride in the production of chlorine.

The power requirement is approximately 3000 kWh per short ton of chlorine. This is usually a high-voltage AC source, which must be put through various pieces of switch gear to step down the voltage and rectify it to DC. Power to the cells is low voltage (4–5 V DC) and, depending upon the size and particular design of the cell, the current consumption ranges between 10,000 and 50,000 amperes for modern installations.[17]

The products of this electrolysis process are chlorine gas, hydrogen gas (both of which are saturated with water vapor), and the spent cell liquor (caustic).

The gas leaving the cathode is about 99.8 percent hydrogen. It is scrubbed with water both to cool and to remove any traces of salt or caustic, and it is then compressed for supplying various processes or is used as fuel.

The gas leaving the anode is about 97.5 percent chlorine. The remainder usually consists of a mixture of water vapor, oxygen, nitrogen, and carbon dioxide. At this point the gas is hot (210°F), moist, and extremely corrosive. It must be cooled and dried. Oddly enough, it is usually cooled by direct contact with water in a packed tower and then dried by scrubbing with sulfuric acid. The dried chlorine is then compressed to about 60 psi. After compression, it is fed into a special fractionating tower to remove impurities such as chloroform and chlorinated hydrocarbons. This innovation was introduced about 1930 in an effort to clean the chlorine so that it could be more easily handled by conventional chlorination equipment used for potable water and wastewater treatment. The impurities are removed at the bottom of the tower as a solution containing little or no chlorine. Usually these impurities are less than 0.2 percent by weight in untreated liquid chlorine.

After leaving the fractionating tower, the chlorine gas is then liquified by refrigeration and pumped to storage tanks, from which it is then pumped into tank cars and ton containers. The trend today is for the manufacturer to ship liquid chlorine in tank cars to packagers, who use the cars as storage from which to fill 150-lb and ton cylinders, which are then shipped directly to the consumer. Each distributor has a liquid bleach manufacturing operation in order to utilize the "snift gas" that would otherwise be wasted in the cylinder-filling operation.

Manufacturing plants also have the problem of recovering the snift or blow gas as well as the chlorine lost in the water used for cooling the gas. Recovery

CHLORINE: HISTORY, MANUFACTURE, PROPERTIES, HAZARDS AND USES 21

of chlorine from these two waste sources is usually accomplished by some patented process. The Hooker process uses water to absorb the chlorine in the snift gas. This water is then used in the cooler. Upon leaving the cooler, it is heated with steam and then acidified, thus stripping the water of chlorine, which is then put back into the packaging cycle. The Diamond–Alkali process uses carbon tetrachloride to absorb chlorine; the carbon tetrachloride is then heated and stripped of chlorine.[17]

The spent liquor from the cells usually contains about 11.5 percent NaOH and 16 percent NaCl. Before this can be marketed, the salt must be removed and the concentration of caustic must be raised to the optimum 50 percent. To raise to this concentration, the liquid is first passed through a double- or triple-effect evaporator. As it proceeds through the evaporation cycle, the NaOH concentration increases and the salt crystallizes and is separated from the caustic liquor by decantation and filtration. The salt is then washed free of caustic, dissolved, and recycled through the brine system. More salt is removed from the 50 percent caustic by cooling and settling.

According to Faraday's laws, one coulomb of electricity deposits exactly 0.00111801 g of silver; or, said another way, a current of one amp deposits 0.00111801 g of silver in one second. According to the second law, the quantity of electricity which liberates one gram equivalent weight of an element is the same for all elements. Since the equivalent weight of silver is 107.880, this quantity must be:

$$\frac{107.880}{0.00111801} = 96{,}493 \text{ coulomb} \tag{1-8}$$

Therefore, from Faraday's law we know that 96,493 coulomb (1 faraday) will liberate 1.0080 g of hydrogen and 35.457 g of chlorine in the electrolysis of salt. Converting this to amperes required per pound of chlorine per day, we get:

$$\frac{\text{Amp} \times 86{,}400 \text{ sec/day}}{96{,}493 \text{ coulomb}} \times \frac{35.457 \text{ g Cl}_2}{454 \text{ g/lb}} \tag{1-9}$$

$$\text{Amp} \times 0.07 = \text{lb/day Cl}_2$$

Referring now to Table 1-2, we see that 135,000 amperes (Eltech) with a current efficiency of 95.8 percent will produce 9053 lb chlorine per day.
Check: $135{,}000 \times 0.07 \times 95.8 = 9053$

Mercury Cells

History. The method of producing chlorine and caustic utilizing a mercury cathode was discovered simultaneously by two men on different continents. Each discoverer was unaware of the other's efforts. One was an American, Hamilton Y. Castner; the other an Austrian, Karl Kellner. Both applied for patents in 1892.

The first Castner cell installation (a demonstration plant for Mathieson Chemical Company, at Saltville, Virginia, in 1897, designed for 550 ampere operation), was later moved to Niagara Falls, New York, where it was operated successfully until 1960. At that time it was replaced by the E.11, the 1961 version of the Olin-Mathieson mercury cells, with a capacity of 100,000 amperes.[17,20]

In the 1970s mercury cell operations in North America were found to be discharging effluents containing mercury in excess of safe limits established by environmental agencies. Some installations were shut down, but those that did not improved the process to a point where the mercury loss was well below the minimum contamination levels.

Although 70 percent of the chlorine produced today in the United States is by the diaphragm cell, the mercury cell is still a formidable competitor because of its ability to produce a 50 percent caustic solution without any further concentration procedures. This is one of the major advantages of the mercury cell.

The mercury cell has been improved considerably. Replacement of the "soluble" graphite anodes with DSA®s has increased the power efficiency of the cell as well as other overall operating features such as longer anode life, a smaller gap between the anode and the amalgam cathode, and less maintenance. Cleaning up the mercury loss has also contributed to the increased efficiency and lower operating cost of the cell.

It is interesting to note that Uhde of West Germany, one of the leading manufacturers of mercury cells, lists 91 installations world wide producing 10886 tons chlorine per day. Moreover 8 of these installations were commissioned in 1983 with a production of 1075 tons of chlorine/day.

The following is a list of mercury cell manufacturers currently listed by the Chlorine Institute:

de Nora	Milan, Italy
ICI	Cheshire, England
Krebskosmo	Berlin, W. Germany
Krebs	Zurich, Switzerland
Kureha	Tokyo, Japan
Olin	Charleston, Tenn. USA
Solvay	Brussels, Belgium
Toso	Tokyo, Japan
Uhde	Dortmund, W. Germany

Chemistry of Electrolysis. The mercury cell has two essential parts: (1) the electrolyzer, and (2) the amalgam decomposer. In the electrolyzer, a salt solution is electrolyzed, making use of a DSA® anode and a flowing mercury cathode. Chlorine gas is liberated at the anode, and sodium is deposited at the surface of

CHLORINE: HISTORY, MANUFACTURE, PROPERTIES, HAZARDS AND USES

the flowing mercury cathode, in which it dissolves to form a liquid amalgam. This amalgam flows into the decomposer, where it is decomposed with water to form sodium hydroxide and hydrogen gas. (See Fig. 1-8)

The principal reactions are as follows:

1. *Electrolyzer*
   a. At the anode:

   $$Cl^- = \tfrac{1}{2}Cl_2 + e^- \qquad (1\text{-}10)$$

   b. At the cathode:

   $$Na^+ + (Hg)^+ + e^- = Na(Hg) \qquad (1\text{-}11)$$

Overall reaction:

$$NaCl + (Hg) \xrightarrow{1\ \text{faraday}} Na(Hg) + \tfrac{1}{2}Cl_2 \qquad (1\text{-}12)$$

2. *Decomposer*
   a. At the anode:

   $$Na(Hg) = Na^+ + (Hg) + e^- \qquad (1\text{-}13)$$

   b. At the cathode:

   $$H_2O + e^- = OH^- + \tfrac{1}{2}H_2 \qquad (1\text{-}14)$$

Overall reaction:

$$Na(Hg) + H_2O \xrightarrow{1\ \text{faraday}} NaOH + \tfrac{1}{2}H_2 + (Hg) \qquad (1\text{-}15)$$

The fundamental differences between this and the diaphragm cell are as follows: The spent brine or anolyte is withdrawn separately from the mercury amalgam. The caustic is produced in the decomposer as a byproduct resulting from the preparation of the amalgam to be returned as the mercury cathode. The amalgam is the catholyte and does not mix with the anolyte. Both processes uses about 1.7 tons of salt per ton of chlorine produced.

The net result is essentially the same as with diaphragm cells. The ingredients are the same, except for the inventory of mercury required.

Electrolyzer. The purified, saturated (305 g/l) alkaline brine solution is fed to the electrolyzer portion of the cells when the pH is adjusted to a range of 2.5–5 with HCl. This is somewhat dependent upon the amount of $CaSO_4$ that can be tolerated in the brine. The pressure on the anode side is kept at atmospheric ±15 mm Hg. The direct current in the specified amount of the cell rating (Olin-

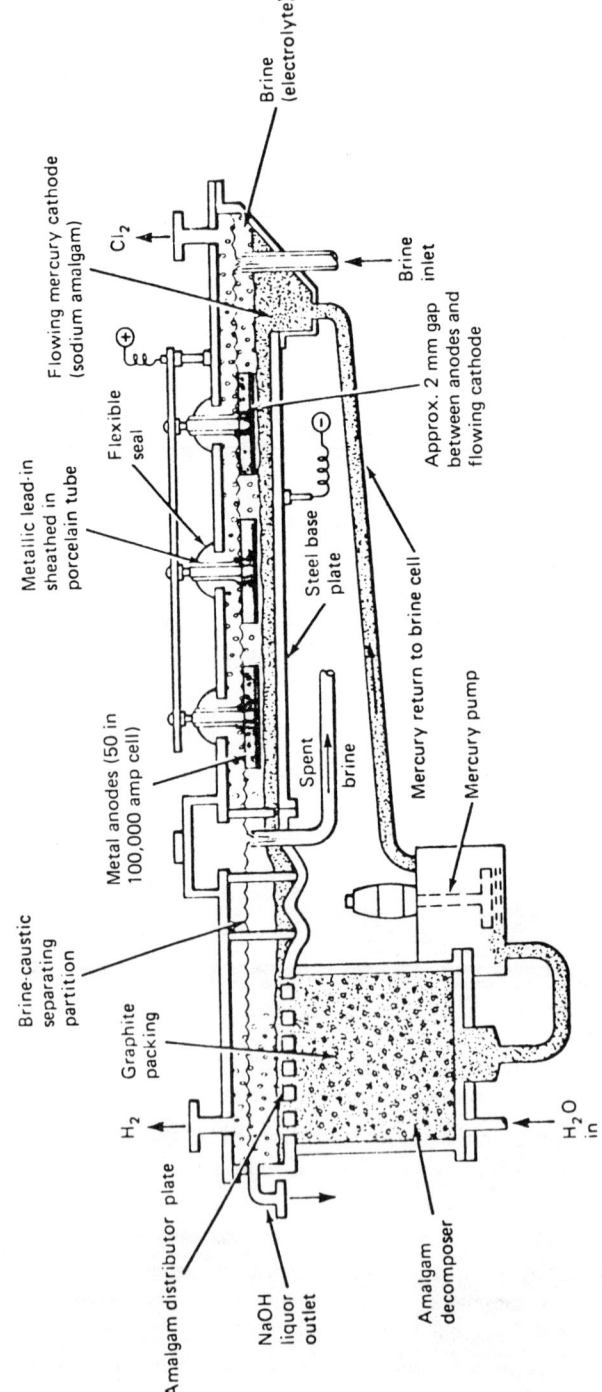

Fig. 1-8. Olin E-510 mercury cell with metal anodes (courtesy Olin Corporation).

Mathieson E-11 is 100,000 amperes) is applied at a voltage of 4–4.5 V between metal anode and mercury cathode, with the chlorine being liberated at the anode. The spent brine is monitored so that it contains 260–280 g/l of NaCl as it leaves the cell and so that the temperature does not exceed 85°C. This is accomplished by regulation of brine feed to the cells. The spent brine is dechlorinated by stripping, concentrated by contact with solid salt, treated with NaOH to pH 10, settled, filtered, and recycled to the cells.

Decomposer. The decomposer is a vertical tower packed with lumps of broken graphite. A distributor plate spreads the amalgam over the top of the packing. Purified and softened water enters the bottom of this vertical tower below the packing and overflows as 50 percent sodium hydroxide above the packing. Sodium amalgam is decomposed in contact with the graphite packing and water, to form sodium hydroxide and hydrogen. Hydrogen is collected from the top of the decomposer. It is wet and contains some mercury vapor and entrained caustic spray. These contaminants are removed mainly by cooling in a scrubber or condenser.

The caustic solution produced by this method is 50 percent (diaphragm cells produce 11–12 percent caustic and membrane cells 30–40 percent). It is customary to filter the 50 percent caustic solution for the removal of graphite particles picked up in the decomposer. These particles also contain some mercury which is recovered from the sludge of the filtering operation.

The chlorine produced from this cell is treated in the same way as that produced from the diaphragm cells.

Operating Characteristics. Table 1-4 is a tabulation of mercury cell operating data. These data were furnished by the electrolyzer manufacturers shown in this table.

OTHER PROCESSES

Salt Process

The salt process for the production of chlorine is based upon the reaction between sodium chloride and nitric acid. It has been known for a long time that chlorides are changed to nitrates. Aqua regia was made by the alchemists in the 8th Century from a mixture of niter, salt, and sulfuric acid. Evolution of chlorine in this reaction was simply a byproduct.

Variations in fertilizer industry requirements and other lesser factors prompted investigation into the production of nitrate and chlorine from salt. This process was put on a commerical basis by Allied Chemical Company at Hopewell, Virginia, in 1936. The overall reaction of this process is:

$$3NaCl + 4HNO_3 \longrightarrow 3NaNO_3 + Cl_2 + NOCl + H_2O \qquad (1\text{-}16)$$

Table 1-4 Operating Data: Mercury Cells

	Manufacturer and Model			
	Krebs ZTE-100-10M	Uhde	deNora 30M2	Olin E-812
Operating Current (kA)	100	300–350	330	300
Cell to cell voltage (metal anodes)	4.15	4–4.25	3.9	4.24
Cathode current density (kA/m^2)	10	10–12	10	10
Power (kWh/ton Cl$_2$)	3000	3100–3300	3070	3310
Current efficiency (%NaCl)	96	96–98	96	97
Anode material	activated titanium	coated titanium	DSA®	metal
Anode Life (yrs.)	3–5	5	2	2
Mercury inventory (lb/cell)	2900	9130	9500	7304
Caustic strength (percent)	50	50	50	50
Mercury loss (lb/ton Cl$_2$)	0.01 0.02	<.003	<.001	0.23
Chlorine production (ton/day)	3.3		11.0	9.9
Caustic production (ton/day)	3.75		12.39	11.11

Typical analyses—Uhde cell with metal anodes

Chlorine Gas (by vol.)		Caustic Soda Solution		Hydrogen	
Cl$_2$	99.5%	NaOH	50%	H$_2$	99/9%
CO$_2$	0.2%	NaCl	30 ppm	Hg	1–10 μg/m^3
H$_2$	0.1%	Na$_2$CO$_3$	200 ppm		
Air	0.2%	Fe	2 ppm		
		Hg	0.05–3 mg/l		

Refinements of this process include oxidation of the nitrosyl chloride for further recovery of chlorine.

$$2NOCl + O_2 \longrightarrow N_2O_4 + Cl_2 \qquad (1\text{-}17)$$

The nitrogen tetroxide can be taken as a product or recycled for acid for sodium nitrate manufacture.

In this process, dilute nitric acid (55 percent or less) is first concentrated by evaporation to 63–66 percent, mixed with sodium chloride, and heated with steam where the reaction takes place. Nitrosyl chloride, chlorine, and sodium nitrate are produced in equal molar quantities. The solution is stripped of nitrosyl chloride and chlorine, which are then scrubbed, dried, and liquefied with refrigerated brine. The nitrosyl chloride–chlorine mixture is then put through a separating column; the chlorine leaves as a gas at the top, and the nitrosyl chloride leaves at the bottom as a liquid. The latter is sent to a recovery operation, and the chlorine is reliquefied and sent to storage.[17]

HCl Oxidation Processes

In recent years the market for chlorine has increased at a much greater rate than that for caustic, while the market for hydrochloric acid has declined. This situation has created a real demand for the production of chlorine from byproduct hydrochloric acid. This demand has revived the Deacon process, which is attractive because of its simplicity. It has a mildly exothermic reaction, with low electrical power and thermal needs. The reaction—HCl oxidation—takes place in the vapor phase over a copper base catalyst as follows:[17]

$$4HCl + O_2 \xrightarrow{450-650°C} 2Cl_2 + 2H_2O \qquad (1\text{-}18)$$

There are no side reactions or competing reactions: the principal problem is to develop operating conditions which best balance the increased rate of reaction at higher temperatures against higher yields at lower temperatures. The Air Reduction Company has developed an improved Deacon process that makes it practical to produce chlorine at the one ton per day level from byproduct HCl at about 27 percent chlorine by volume with air and about 90 percent chlorine with 95 percent oxygen. For economical commerical production, the size of the plant is limited to about 15 tons per day.[3]

The utilization of metal chlorides formed by hydrochloric acid with the metal oxide is followed by decomposition of the metallic chlorides by oxygen and heat.

The Grosvenor Miller process, with a fixed bed system, utilizes the following reactions:[17]

$$Fe_2O + 6HCl \xrightarrow{250-300°C} 2FeCl_3 + 3H_2O \qquad (1\text{-}19)$$

$$2FeCl_3 + 1\tfrac{1}{2}O_2 \xrightarrow{475-500°C} 2Fe_2O_3 + 3Cl_2 \qquad (1\text{-}20)$$

A product gas composition approaching a maximum of 70 percent chlorine can be obtained in this process using a three- to five-bed continuous system. The procedure is briefly described as follows: Reactor I is laden with dry ferric chloride at 250–300°C and is fed oxygen.[3] The ferric chloride is converted to chlorine and ferric oxide. The gas contains approximately 30 percent chlorine, 70 percent unreacted oxygen, and some air. This mixture passes into Reactor II, which is maintained as a mixed bed of ferric oxide and ferric chloride at 500°C to serve as both chlorinator and oxidizer. Both hydrochloric acid vapor and oxygen (with excess HCl) are passed into Reactor II so that the HCl reacts with Fe_2O_3. The chlorine gas does not react, and so it and the excess HCl pass through Reactor II. The effluent gas from this reactor, containing chlorine gas, HCl vapor, steam, and excess oxygen, is passed into Reactor III (maintained at 250–400°C). This reactor was previously Reactor I, and thus is oxygen-laden. Reactor III strips the HCl vapor from the gas. Chlorine does not react, and so it passes through, carrying with it steam and excess oxygen until this reactor becomes saturated with chlorides. At this point, the functions of Reactor III are transposed to those of Reactor I, and gas flow is III to II to I. There are other variations of this process, such as the Dow Moving Bed process and others using molten metallic chlorides as catalysts.

The Kel-Chlor process[23] is a modification of the Deacon process which has always been plagued with chemical equilibrium problems which resulted in low chlorine yields.

The Kel-Chlor process solves the equilibrium problems by combining a very active catalyst (nitrogenous compounds) with a powerful dehydrating agent (sulfuric acid) which reduces the activity of the steam to a negligible value and thus allows the reaction to proceed to completion. The result is a better HCl to chlorine conversion ratio. Plant experience has also demonstrated that corrosion problems have been solved as well. The Kellog Co. claim that chlorine from waste hydrochloric acid can be produced for $20/ton. Chlorine made by this process can be refined to a higher purity than ordinary commercial chlorine prepared by electrolysis.[23]

Electrolysis of Hydrochloric Acid Solutions

The Hoechst–Uhde Process.[24] Considerable quantities of aqueous hydrochloric acid or hydrogen chloride gas result each year as byproducts from a variety of chemical manufacturing activities. These byproducts are a form of industrial waste and are difficult to dispose of; therefore a process utilizing them is of special interest. The I. G. Farben Industries began to develop a process in 1938 at the Bitterfield, Germany plant using bipolar diaphragm cells. The early cells had a limited production capacity, usually less than 50 tons chlorine per day. A system producing one ton of chlorine per day will yield 10,000 cu ft of hydrogen; it requires 2060 lb of 100 percent muriatic acid, 1900 kWh AC, and 10,000 gal. of cooling water at 15°C for Cl_2 and H_2. This process has had many fabricating and operating problems. A joint experimental effort by Farbwerke Hoechst AG and Friedrich Uhde GmbH resulted in a successful design which was first constructed in 1963.

The first electrolysis unit in the U.S.A. was built in 1971–72 at the Mobay Chemical Corp. at Baytown, Texas.[24] This electrolysis plant was built by the Hoechst–Uhde Corp. and has a capacity of 180 metric tons per day chlorine (198 short tons). The Mobay operation uses waste hydrogen chloride gas from the manufacture of isocyanate. In addition to the manufacture of isocyanates, hydrochloric acid is a waste product in the manufacture of chlorine bearing solvents, production of raw materials for detergents, production of chlorination products, production of frigens and silicones.

Today's Hoechst–Uhde electrolyzers consist of 30–36 single elements working at current densities from 4–5 kA/m^2. The chlorine capacity varies from 90–180 metric tons chlorine per day. A typical 30 element unit occupies a space of only 12 × 16 ft. Several of these plants are operating in Germany, France, Japan, and the U.S.A.

IMPURITIES IN CHLORINE

Historical Background

In view of the fact that only a small fraction (5 percent) of all chlorine production is used for potable water and wastewater disenfection, the chlorine manufacturers have done a commendable job in providing a clean and moisture-free product. Most of the chlorine produced does not match the purity required for chlorine metered through conventional chlorinators. Chlorine users in industry think in terms of hundreds or thousands of pounds per hour whereas in potable water treatment it is more likely to be hundreds of pounds per day.

In the early days of chlorination (1920s), metering equipment was continually plagued by fouling from what was commonly called "gunk" or "taffy." It was difficult to prove the origin of this gunk until Wallace and Tiernan developed the bell-jar chlorinator (1922). This chlorinator employed a self-cleaning type of chlorine pressure-reducing valve (visible through a glass dome) which operated under a vacuum. When the gunk plugged this valve, stopping the flow of chlorine, the valve would automatically shift from a throttled position to a wide-open position by raising the water level, which in turn raised a float-operated mechanism. In most cases, the valve in this wide-open position would purge itself of the gunk, thus allowing the inrush of gas to spew it all over the bell jar. This would force the water level down, and the chlorinator would automatically resume operation. In the meantime, the customer was able to back up his complaint with a sample of gunk. After many of these complaints, the chlorine manufacturers introduced into their manufacturing process a fractionating tower[18] in an attempt to clean up their product and eliminate the gunk (ca. 1935). At no time were the impurities in chlorine used for chlorination suspected of having any health significance.

Nothing much was heard about the purity of chlorine until it became possible to identify more than 500 compounds in water using sophisticated techniques and instrumentation such as gas chromatography and mass spectrometry. This development occurred in the 1970s. After the Harris report* on the levels of pollution in the Mississippi River at New Orleans in 1974 the first accusation related to chlorine manufacture was the concern over mercury pollution of the environment from chlorine produced by mercury cells. This was followed by the discovery that chlorination of these and other waters using the free residual process could and did form chloroform in the finished water. Two compounds that caused immediate concern were chloroform ($CHCl_3$) and carbon tetrachloride (CCl_4).

---

* "The Implications of Cancer Causing Substances in Mississippi River Water." Environmental Defense Fund, 1525 18th St. N.W., Wash. D.C. 20036.

In July 1977 occasional concentrations of CCl_4 above the detectable level (1 mg/L) were found in the finished water at Belmont and Queen Lane,[52] two Philadelphia plants receiving waters from the Schuylkill River. A subsequent investigation traced the source of these high concentrations to the chlorine supplied to these plants. When the chlorine supply was changed to another manufacturer the problem disappeared. It was revealed that the manufacturer of the chlorine in question was using a carbon tetrachloride scrubbing system in the chlorine–hydrogen air separation system. The problem was solved by segregating the chlorine to be used for potable water sales from the chlorine extracted from the separation system. This episode was thoroughly examined on a national level by the Chlorine Institute and representative chlorine manufacturers in order to establish an interim maximum level for CCl_4 in chlorine used in potable water treatment. This interim level was established at 100 mg/l by EPA and agreed to by the manufacturers' association pending an assessment of what the chlorine industry is currently producing and what it is capable of producing with available technology.[52]

The current AWWA chlorine purity standard is contained in the ANSI/AWWA report of June 7, 1981.[53] This report limits the maximum concentration level of carbon tetrachloride to not more than 150 mg/l (0.015 percent). This report states that the chlorine supplied under this standard shall be "99.5 percent pure" by volume as obtained by vaporizing the chlorine as determined by ASTM Standard E 412–70 Assay of Liquid Chlorine (Zinc Amalgam Method). Other limitations listed are: moisture not to exceed 150 ppm (0.015 percent) by weight; heavy metals not to exceed 30 ppm (0.003 percent) expressed as Pb; mercury not to exceed 1 ppm (0.0001 percent); and nonvolatile total residue not to exceed 50 ppm (0.005 percent) by weight in chlorine tank cars and tank trucks and 150 ppm (0.015 percent) by weight as packaged in ton containers or cylinders. It is significant to note that no mention is made of nitrogen trichloride.

Sources of Impurities

The major sources of impurities found in chlorine are:

1. Moisture entrapment during packaging.
2. Ammonia nitrogen in brine fed to electrolytic cells. The source of nitrogen can be either the make-up water for the brine or in the salt supply or both.
3. Organic impurities in the salt.
4. Graphite or carbon anodes in the electrolytic cell.
5. Hydrocarbon sources introduced from valve lubricant, pump seals, and various packings used throughout the manufacturing process.
6. Recovery systems used to separate chlorine produced from entrained hydrogen and air.
7. Recycling system where chlorine is recovered from the tail-gas of a chlorinated hydrocarbon process.

These impurities are classified as follows:[25]

Gases	Volatile Liquids	Volatile Solids	Nonvolatile Solids
CO_2	bromine	hexachlorbenzene	$FeCl_3 \cdot 6H_2O$
H_2	carbon tetrachloride	hexachlorethane	
O_2	carbonyl chloride (phosgene)		$Fe(SO_4)_3 \cdot 9H_2O$
N_2	chloroform		nitrosyl chloride
NCl_3	HCl		nitrogen tetroxide
	methylene chloride		H_2SO_4
	moisture		

Consequences of Impurities

Public Health. Chlorine reacts with hydrocarbons to form both carbon tetrachloride and chloroform. In the manufacture of chlorine the use of graphite or carbon anodes can be a source of the hydrocarbons. These two compounds are known to be animal carcinogens and therefore should be rigorously controlled and monitored in the production of chlorine. Another source of chlorinated hydrocarbon production associated with the manufacture of chlorine are those instances where tail-gas recovery systems are used in the manufacture of chlorinated hydrocarbon compounds. This source can and should be eliminated where chlorine production is used in part for potable water treatment.

Chlorine Control Equipment. Of all the impurities listed above moisture in the chlorine is the worst offender because it makes the chlorine highly corrosive, which leads to the formation of ferric chloride. This causes gross fouling in the metering equipment (chlorinators).

The next most offensive impurities are hexachloroethane and hexachlorobenzene. These compounds form a taffy-like substance commonly called "gunk." They also cause serious equipment fouling. Their origin can be from deteriorating graphite or carbon anodes, or from valve lubricating compounds, packings, gaskets, etc. used in the various manufacturing components.

Most impurities in chlorine are soluble in liquid chlorine. Ferric chloride seems to plate out in thin layers on the contact surface between the liquid chlorine and metal piping, etc. Deposition is most notable in areas of chronic flashing, usually where there is a restriction or a turbulent flow regime. Precautions can be taken to minimize the difficulty from ferric chloride. Ferric chloride appears to have the ability to spread from the liquid phase to the vapor phase so that the contamination carries through the chlorination equipment. The largest deposits occur at points of pressure reduction and areas of reliquefaction.

Both hexachlorethane and hexachlorbenzene tend to sublime at room temperature, and are usually found deposited at points of pressure reduction in the metering and control equipment.

For the benefit of consumers using chlorine control equipment in the sanitary field, chlorine producers have long made it a practice to limit the total impurities in liquid chlorine to about 200 mg/l and the nonvolatile impurities to 20–30 mg/liter.[25]

Nitrogen Trichloride in Liquid Chlorine

Occurrence, Formation and Significance. Nitrogen trichloride was first obtained in 1811 from the action of chlorine on a solution of ammonium chloride by Dulong, who lost an eye and three fingers as the result of an explosion. When generated in the laboratory this compound appears as a yellow oil with a pungent odor resembling somewhat the odor of chlorine. It is practically insoluble in water, but easily soluble in most organic solvents, such as benzene, carbon tetrachloride, ether, etc. It has been reported that a drop of the oil explodes violently when touched with a feather dipped in turpentine.[54,59]

Nitrogen trichloride is not listed as an impurity in chlorine produced in the U.S.A. or Canada, where producers have long been aware of the dangers of NCl_3 in liquid chlorine. Its occurrence in the manufacture of liquid chlorine in the U.S.A. and Canada has been virtually nil since about 1930.

Formation of nitrogen trichloride in the production of chlorine occurs if ammonia nitrogen is present in the brine fed to the electrolytic cells. It is soluble to the extent of 7.3 mg/l in liquid chlorine, whereas it is not soluble to any extent in either water or concentrated sulfuric acid. The latter is used in the production of chlorine to remove the moisture content. Therefore, once NCl_3 forms in the electrolytic cells, it will pass with the chlorine gas through the coolers, scrubbers (H_2SO_4), and acid sealed pumps, and be condensed with the liquid chlorine. The greatest danger of explosion occurs at the point when the liquid chlorine vessel (ton container, evaporator etc.) becomes exhausted of liquid chlorine and only the chlorine vapor remains. This is the result of the solubility of NCl_3 in liquid chlorine. It concentrates itself in the layer of liquid chlorine next to the vapor phase. As the liquid is gradually used up the concentration of NCl_3 keeps increasing in the small area of the vapor–liquid interface.

Explosions Caused by Nitrogen Trichloride. The last reported chlorine explosions from NCl_3 in liquid chlorine in the U.S.A. and Canada were the two ton container explosions at chlorination stations operated by the New York City Water Department Circa 1929. These explosions occurred when the cylinders had been emptied to zero gage pressure. The cause of these explosions was traced to the presence of nitrogen trichloride in the liquid chlorine. The ammonia nitrogen concentration in the prepared brine solution for the electrolytic cells was found to be about 15 mg/l, and the NCl_3 in the liquid chlorine was 500 mg/l (0.5 g/l).[54] Theoretically each mg/l of ammonia nitrogen in the brine will generate 50–60 mg/l of NCl_3 in the liquid chlorine. Owing to the fact that it is difficult to predict the "safe" allowable concentration of NCl_3 in chlorine, the manufacturers' consensus is a 5 mg/l limit.[54]

There are many chlorine producers outside of U.S.A. and Canada who are unaware of the dangers of NCl_3, its occurrence, and its prevention. An evaporator explosion was reported in India in 1965. In 1981 White investigated several evaporator explosions in South America. These were all the result of NCl_3 in concentrations between 50 and 300 mg/l. The ammonia nitrogen in the brine make-up water varied on a weekly basis from about 1.5 to 6 mg/l. The sudden appearance of ammonia nitrogen into the plant water supply was the result of a new sewage collection system which discharged untreated domestic wastewater into the river supply upstream from the plant.

Prevention of Nitrogen Trichloride Formation in Liquid Chlorine. The most positive method of dealing with the NCl_3 problem is to remove all of the ammonia nitrogen from the brine solution before it reaches the cells. This takes care of those situations where the brine solution is prepared from rock salt that contains ammonia nitrogen.

Two methods can be used to solve the NCl_3 problem. One is to remove the ammonia nitrogen from the brine water by breakpoint chlorination; the other is to subject the chlorine gas exiting the cells to ultraviolet light within the spectrum of 3600–4400 angstrom wavelength. The UV application has to be done before the chlorine enters the scrubbers. As it is difficult to monitor the effectiveness of the NCl_3 removal by UV, it is more practical and reliable to remove the ammonia nitrogen from the brine water by breakpoint chlorination, followed by aeration to remove as much as possible the combined chlorine residual resulting from the breakpoint procedure. This method has proved very effective and reliable in North America. The UV method has been used successfully where small quantities of NCl_3 are involved. One such plant is the PPG Industries plant at Natrium, West Virginia, U.S.A. The decomposition of NCl_3 by UV is only about 50 percent complete.[54]

Silica Contamination

This is a rare situation but worth mentioning since its occurrence caused considerable aggravation to one consumer.

The end result of chlorine contaminated with silica is the formation of white crystals (SiO_2) at the entrance of the chlorine gas into the injector assembly. The injector inlet port plugs up rapidly and renders the chlorinator completely inoperable.

The source of silica is thought to be from silica-contaminated brine water, or silicone grease used in valves by the chlorine packager. Silica-contaminated brine water is most likely to occur when the production of chlorine is a byproduct of a metal refining process such as the extraction of magnesium from a magnesium chloride ore.

If the brine is contaminated with silica, electrolysis will convert the silica present to silicon tetrachloride. When the liquid chlorine is vaporized the chlorine vapor carries the $SiCl_4$ through the chlorinator and into the injector. When chlorine comes in contact with the water in the injector the $SiCl_4$ is immediately transformed into SiO_2 crystals, which eventually plug the injector inlet.

PHYSICAL AND CHEMICAL PROPERTIES

General

Chlorine has an atomic number of 17 (number of excess positive charges on the atomic nucleus) and an atomic weight of 35.457. Molecular chlorine, Cl_2 has a weight of 70.914. Two isotopes of chlorine, Cl^{35} and Cl^{37}, occur naturally, and at least five other isotopes have been artificially produced.[1] Ordinary atomic chlorine consists of a mixture of about 75.4 percent Cl^{35} and 24.6 percent Cl^{37}. Chlorine usually forms univalent compounds, but it can also combine with a valence of 3, 4, 5, and 7. (See Chapter 4.)

In its elemental form, chlorine is a greenish yellow gas which can be readily compressed into a clear, amber-colored liquid which solidifies at atmospheric pressure at about 150°F. Chlorine gas forms into a soft ice upon contact with moisture at 49.3°F and at atmospheric pressure. (This is chlorine hydrate, $Cl_2 \cdot 8H_2O$.)

In commerce, chlorine is always packaged as a liquefied gas under pressure in steel containers. The liquid is about one and one-half times as heavy as water (denser) and the gas is about two and one-half times as heavy as air. The liquid vaporizes readily at normal atmospheric temperature and pressure. It has an unmistakable irritating, penetrating, and pungent odor. The properties of chlorine gas and liquid are listed in Tables 1-5 and 1-6 respectively, which are supplemented by Figs. A-1–A-7 in the Appendix.

Some of these properties merit comment.

Critical Properties

The *critical temperature* is that above which chlorine exists only as a gas (291.2°F), despite the pressure. The *critical pressure* is the vapor pressure of liquid chlorine at this critical temperature. The *critical density* is the mass of a unit volume of chlorine at the critical pressure and temperature.

Compressibility Coefficient

The compressibility coefficient of liquid chlorine is greater than that of any other liquid element. It represents the percent decrease in volume corresponding to a unit increase in pressure when the liquid is held at constant temperature. This

CHLORINE: HISTORY, MANUFACTURE, PROPERTIES, HAZARDS AND USES 35

Table 1-5 Properties of Chlorine Gas

	Ref.
Symbol: Cl_2	
Atomic wt.: 35.457	
Atomic number: 17	
Isotopes: 33, 34, 35, 36, 37, 38, 39	60
Density (see appendix) at 34°F and 1 atm: 0.2006 lb/ft³	61
Specific gravity at 32°F and 1 atm: 2.482 (air = 1)	62
Liquefying point at 1 atm: −30.1°F (−34.5°C)	67
Viscosity (see Appendix) at 68°F: 0.01325 centipoise (approximately the same as saturated steam between 1 and 10 atm)	63, 74, 76
Specific heat at constant pressure of 1 atm and 59°F: 0.115 Btu/lb/°F	64
Specific heat at constant volume at 1 atm pressure and 59°F: 0.085 Btu/lb/°F	64
Thermal conductivity at 32°F: 0.0042 Btu/hr/ft²/ft	65
Heat of reaction with NaOH: 626 Btu/lb Cl_2 gas	
Solubility in water at 68°F and 1 atm: 7.29 g/L.	66

Combining Quantities

1 lb chlorine gas combines with
  1.10 lb commercial hydrated lime (95% $Ca(OH)_2$
  $$2Ca(OH)_2 + 2\ Cl_2 = Ca(OCl)_2 + Cl\ Cl_2 + 2H_2O$$
  0.83 lb commercial quicklime (95% CaO)
  $$2Ca + H_2O + Cl_2 = Ca(OCl)_2 + CaCl_2 + 2H_2O$$
  1.13 lb caustic soda (100% NaOH)
  $$2NaOH + Cl_2 = NaOCl + NaCl + H_2O$$
  2.99 lb soda ash
  $$2\ Na_2CO_3 + Cl_2 + H_2O = NaOCL + NaCl + 2NaHCO_3$$

Table 1-6 Properties of Liquid Chlorine

		Ref.
Critical temperature	144°C; 291.2°F	67
Pretical pressure	1118.36 psia	67
Critical density	573 g/1; 35.77 lb/ft³	67
Compressibility	0.0118% per unit vol per atm increase at 68°F	68
Density (see Appendix) at 32°F	91.67 lbs/ft³	62
Specific gravity at 68°F	1.41 (water = 1)	69
Boiling point (liquefaction point) 1 atm	−34.5°C; −30.1°F	67
Freezing point	−100.98°C; −149.76°F	70
Viscosity (see Appendix) at 68°F	.345 centipoise (approx. 0.35 × water at 68°F)	71, 75
1 volume liquid at 32°F and 1 atm	457.6 vol gas	62
1 lb liquid at 32°F and 1 atm	4.98 ft³ gas	72
Specific heat	0.226 Btu/lb/°F	73
Latent heat of vaporization	123.8 Btu/lb at −29.3°F	70
Heat of fusion	41.2 Btu/lb at −150.7°F	73

physical characteristic is the reason why the volume–temperature relationship of chlorine is so important. This is described below.

Volume–Temperature Relationship

The volume of liquid chlorine increases rapidly as its temperature increases. Because of this characteristic, coupled with noncompressibility, extreme care must be taken to prevent the possibility of hydrostatic rupture of containers or pipe lines by expanding liquid chlorine due to a rise in temperature. All containers are filled to their prescribed weight of chlorine at 60°F so that 15 percent of the container volume is vapor space. Actually the Chlorine Institute Manual (4th ed., 1969) says on p. 5, paragraph 2.1.5b, "The maximum permitted filling density is defined by the D.O.T. [173.300(g)] as '. . . the percent ratio of the weight of gas (sic) in a container to the weight of water that the container will hold at 60°F. (One pound of water equals 27.737 cubic inches at 60°F)' " The practical approach to an overly complicated D.O.T. definition is to know the density of water at 60°F (62.366 lb./ft^3) and the water volume of the container.

Example: A 55 ton car holds 10,564 gal of water. At 7.48 gal/ft^3 the car volume is 1412.30 ft^3 at 60°F. This weight of water amounts to 88,080 lb. Multiplying water weight by the D.O.T. factor of 1.25 gives 110,100 lb. This equals 55.05 tons of liquid chlorine. (Not gas, as described in the D.O.T. proclamation.)

The vapor space provided by the above requirement is designed to accommodate a temperature rise sufficient to melt the fusible plug in the 150 lb cylinders and ton containers. The fusible plug metal is designed to melt between 158°F and 165°F, thus relieving pressure and preventing rupture of the container in case of fire or other exposure to high temperature. An inspection of Fig. A-7 in the Appendix illustrates the volume–temperature relationship in a container loaded to its authorized limit. From this curve the container will be "skin-full" when the liquid chlorine temperature reaches 153.64°F. At this temperature (Fig. A-6 in the Appendix) the vapor pressure is 290 psi, and at the upper melting temperature of the fusible metal (165°F) the vapor pressure would be about 330 psi. This appears to demonstrate that the D.O.T. definition for filling a container does not allow for enough space to match the fusible metal melting temperature of 165°F. However, in chlorine cylinders and ton containers there exists an elastic volumetric expansion of the metal. When the containers are hydrostatically tested at 500 psig it is not uncommon for a 3 percent expansion to be observed.[56] This would easily permit a temperature higher than 160°F without fear of rupturing. Ton containers have an added expansion factor in the dished heads. They can reverse from the concave posture to the convex position before rupturing. This has been observed in several cases of overpressuring ton containers.[57]

The case of tank cars and stationary tanks is quite different than vessels with fusible plugs. Tank cars and stationary storage tanks used for potable water treatment or wastewater treatment are usually fitted with a spring-loaded Chlorine Institute safety valve combined with a breaking pin assembly that breaks at 225 psig. On cars used in pulp and paper chlorination, the safety valve relieves at 375 psig because it is air padded at a higher pressure.

Density of Chlorine Vapor

The density of chlorine vapor varies widely over changes in pressure and modestly over changes in temperature. This is a most important variable when calculating gas flow pressure drop in both vacuum and pressure systems. The relationship of vapor density for various pressures and temperatures is shown in Figs. A-1 and A-2 in the Appendix.

Density of Chlorine Liquid

The density of liquid chlorine varies only slightly with temperature. At 40°F it is 90.85 lb/ft^3 and at 140°F it is 79.65 lb/ft^3. See Fig. A-4 in the Appendix.

Viscosity of Chlorine

This is the measure of internal molecular friction when a substance is in motion. It is necessary to know this property for both liquid and gaseous chlorine, as it is a variable in calculating the Reynolds number for the determination of friction losses in pipelines. The temperature–viscosity relationship for both liquid and gas is shown in Fig. A-3 in the Appendix.

Latent Heat of Vaporization

This is the heat required to change one mass of liquid to vapor without a change of temperature. If the liquid chlorine is at 70°F it requires about 100 BTU to vaporize one pound of liquid chlorine. (See Fig. A-5 in the Appendix).

Vapor Pressure

This is the pressure of chlorine gas above liquid chlorine when the vapor and the liquid are in equilibrium. This pressure varies widely with temperature. It is necessary to know this relationship particularly when the consumer is transferring chlorine from tank cars to vaporizing equipment. (See Fig. A-6 in the Appendix).

Specific Heat

This is the amount of heat required to raise the temperature of a unit weight of chlorine vapor one degree F. At atmospheric pressure and 59°F it requires 0.085 BTU/lb.

Solubility of Chlorine Gas in Water

Chlorine has a limited solubility in water. At atmospheric pressure and 68°F its solubility in water is 7.29 g/l. This does not represent the conditions that surround the application of chlorine in this text. Operation of chlorination equipment which produces chlorine solution is always at partial pressures (vacuum). At the vacuum levels currently being used the maximum solubility is about 5000 mg/l. The upper limit of solubility recommended by all chlorinator manufacturers is 3500 mg/l.

This arbitrary figure has been successful in preventing solution discharge systems from being adversely affected by gas pockets in the solution piping and off-gassing at the point of application.

Chemical Reactions

Liquid chlorine in the absence of moisture will not attack ferrous metals, hence the use of steel containers. Since there is no such thing as absolutely "dry" liquid chlorine, extra wall thickness is provided to offset corrosion.

Liquid chlorine will attack and very quickly destroy PVC materials and rubber, hard or soft.

Dry chlorine gas will not attack ferrous metals, copper, or ferrous alloys. Dry chlorine gas will support combustion of carbon steel at 483°F.[27] Chlorine exists only as a gas above 291.2°F regardless of pressure (critical temperature).

Moist chlorine gas will destroy all ferrous metals including stainless steel, copper, and ferrous alloys. Gold, platinum, and tantalum are the only metals that are totally inert to moist chlorine gas. Silver is widely used with moist chlorine gas, because the silver chloride formed upon contact with the moist gas is inert.

Aqueous solutions of chlorine are extremely corrosive. For this reason PVC, Fiberglass, Kynar, polyethylene, certain types of rubber, Saran, Kel-F, Viton, and Teflon are commonly used where both moist gas and chlorine solutions are encountered.

Chlorine reacts with ethyl alcohol and ether in trace amounts to form solid, waxy hexachloroethane. It also reacts with grease and oils to form a voluminous frothy substance. Solid complex hydrocarbons are formed by the reaction of chlorine and the various petroleum distillates.[28] At normal temperatures there are no reactions between chlorine gas and the methane derivatives, chloroform, wood alcohol, and carbon tetrachloride.

The chemical reactions of chlorine gas and chlorine solutions in the various phases of potable water and wastewater treatment are discussed in considerable detail in other chapters.

HAZARDS

Toxic Effects

Liquid chlorine is a skin irritant and can cause severe damage, resembling a burn, to body tissues. Since the liquid vaporizes rapidly to gas at atmospheric temperature and pressure, it is difficult to attribute this damage to the liquid or the gas phase. The gas in low concentrations is an irritant to mucous membranes and the respiratory system. The amount of gas inhaled determines the severity of the impairment of the respiratory system. Although it has not been recognized in the literature, there are two types of gassings by chlorine. The one usually referred to is from the fumes of chlorine gas in the dry state. The other type of gassing, and probably the more dangerous of the two, is by chlorine fumes from an aqueous solution. This occurs in the case of a *chlorine solution* leak or exposure in a confined area if the concentration of the solution is in excess of approximately 750 ppm titrable chlorine. Because the fumes are laden with moisture, they seem more tolerable to the respiratory tract, and the victim will unwittingly inhale excessive amounts of molecular chlorine. This results in a slow but significant production of pulmonary edema that could cause death by "drowning" while the victim is asleep. In this type of gassing, the victim should place his head downward as far as possible to drain the edema before retiring.

Gassing from dry gas is more common and much more disagreeable. As soon as the gas enters the throat area, the victim will sense a sudden stricture in this area—nature's way of signaling to prevent passage of the gas to the lungs. At this point, the victim must attempt to do two things: to get out of the area of the leak, proceeding upwind, and to take only very short breaths through the mouth. Normal breathing will cause coughing, which must be prevented if possible.

First Aid. In severe cases of inhalation, the first thing to do is to call the local fire department to administer a mixture of CO_2 and oxygen under an exhalation pressure not to exceed 4 cm water. While awaiting their arrival, the patient should be kept in the open, because in cases of severe exposure the victim's clothing will have absorbed a considerable amount of chlorine, which would be further inhaled if he is confined to a room. If blankets are available, the clothing should be removed and the patient kept warm and quiet. Never let a rescue squad use a Pulmotor on a gassed victim.

In an extremely severe case, the victim may stop breathing and may also turn blue (cyanosis). Obviously a physician should be summoned and artificial respiration started immediately. The Chlorine Institute recommends the Nielson armlift-back pressure method.

A physician can take steps to reduce the formation of pulmonary edema, arrest falling blood pressure, and supervise administration of oxygen. Other than that, there is little to be done in severe cases. The Chemical Warfare Service of the United States Army[28] has done considerable research on phosgene gas ($COCl_2$) poisoning, and has found that adrenalin, Pituitrin, and morphine are injurious and should never be used as they lower blood pressure, which is already falling fast in cases of severe exposure.

After the severely gassed victim has recovered from anoxemia and acute pulmonary edema, he will need careful watching and nursing to prevent development of pneumonia.

For additional details on the medical care of severely gassed patients, see the reference for the article by Joyner and Durel at the end of this chapter.[34]

Treatment of mild cases occurring from time to time around chlorination equipment is considerably different. The first step is of course to get clear of the fumes, breathe lightly, move slowly without exertion, remain quiet, keep warm, and resist the impulse to cough.

The victim will be at first seized with fear and may become panicky because of the sensation of the closing of the throat and a feeling of suffocation. In a mild case, immediate relief can be achieved by sipping a teaspoon of either Anti-Chlor[30] or whisky.

Anti-Chlor (2 liters)	
Water	1485 cc
Sugar	45 g
Tincture lavender, Co.	37 cc
Sp. ammonia aromatic	56 cc
Ethyl alcohol	333 cc
Oil of peppermint	27 cc
Sp. chloroform	55 cc

As equivalent of one tablespoon every fifteen minutes until relief is obtained, or for one hour, should be administered. In preparing the mixture, add oil of peppermint to the alcohol, then the spirits of chloroform, spirits of ammonia, and lavender, in the order given. Stir after each addition. Next, add this mixture to the water, to which the sugar has been added. To make spirits of chloroform, add 60 cc of chloroform to 940 cc of grain alcohol.

Since alcohol is the principal ingredient in the formula the same results can be obtained by sipping a similar amount of straight whisky. The alcohol counteracts the shock and panic, relaxes the muscles in the throat, eliminates the desire to cough, and restores normal breathing.

So far as is now known, there are no residual symptoms attributable to severe gassing by chlorine for most patients.[31] There are, however, some anxiety reactions which linger for many months. Victims with a history of asthma have had lingering effects of distress; mild cases have been known to cure bronchial ailments in young persons. Further clinical research is necessary to determine the long-range effects of gassing by chlorine.

Physiological Response. The United States Bureau of Mines gives the following physiological responses to various concentrations of chlorine gas:

Effect	Parts of Chlorine per Million Parts of Air by Volume
Least amount required to produce slight symptoms after several hours of exposure	1
Least detectable odor	3.5
Maximum amount that can be inhaled for one hour without serious disturbances	4
Noxiousness, difficulty in breathing, several minutes	5
Least amount required to cause irritation of the throat	15.1
Least amount required to cause coughing	30.2
Amount dangerous in 30 minutes to one hour	40–60
Kill most animals in a very short time	1000

Characteristics of a Major Chlorine Release

It is important to be aware of what is liable to take place in the event of a major chlorine release. A major release is one where a tank car is ruptured or where a guillotine break in a liquid piping system might occur. In postulating such a leak, those responsible for coping with such a situation assume that if a tank car is involved, the liquid excess flow valves will fail and that the car outlet valves will not close. Under these conditions 20 percent of the liquid in the car will flash off as vapor and during this period another 5% will vaporize.[32]

For the case of a 55 ton car leaking liquid chlorine through a break in a ¾ inch line, about 27,500 lb of chlorine will flash off in about 25 minutes, assuming the liquid temperature is about 70°F. During this flash-off period the liquid is cooling rapidly, thereby reducing the car pressure and retarding the rate of liquid flow from the car. This initial flash-off will cause the liquid to cool to the atmospheric boiling point (−30°F). From this point forward about 1–10 lb/sec will be emitted continuously as the liquid evaporates from the heat picked up from the surroundings. If the liquid is spilled on the ground it might vaporize more quickly than it would if it were contained in the car. This depends largely upon the ambient temperature. Vaporization from a pool of liquid chlorine can be inhibited significantly if a source of cold water is available. Pouring cold water (about 40–45°F) on the pool will form a thin sheet of insulating chlorine ice on the surface of the pool. This ice is chlorine hydrate which forms when the chlorine and water are at a temperature of 49°F or less.

A great many researchers have investigated the phenomenon of major releases of hazardous chemicals. The equation used by most investigators is the Gaussian Plume Model equation. This predicts the length and shape of the cloud formed from the initial release provided weather conditions are known. The cloud that emerges from this model is shaped like a cone sliced in half with the flat part at ground level and the apex at the source of the release. The value of the mathematical model is to assist authorities to set reasonable boundaries for evacuation after a release has occurred. The Gaussian equation takes into account release rate, the standard deviation of the crosswind plume, width and height of plume, height of initial source, and downwind, crosswind and vertical distances, and chlorine concentration as follows:

$$C_{xyz} = \left[\frac{Q}{\pi \cdot U \cdot \Sigma y \cdot \Sigma z} \right] e^{-\left(\frac{h^2}{2\Sigma z^2} + \frac{y^2}{2\Sigma y^2} \right)} \quad (1\text{-}21)$$

when

C = concentration units/m³
Q = release rate, units/sec
Σy, Σz = the standard deviation of the crosswind plume (width and height in meters)

U = mean wind speed velocity (m/sec) at h.
h = release source height (meters)
x,y,z = downwind, crosswind, and vertical distances (meters).

There are three factors not accounted for in the above plume model. These are: ambient temperature, relative humidity, and local terrain. These factors contributed significantly to the fatalities in the Youngstown accident. A release in a fog-shrouded area is probably the worst case. Air movement in a low-lying fog-shrouded area is usually nil. The relative humidity is at the saturation point, which allows the moisture-seeking chlorine gas to saturate the fog shroud. Clothing on the people in the release area will absorb the chlorine-laden moist air, thus multiplying the inhalation of chlorine. In such cases it is not sufficient to merely evacuate the area quickly. Exposed persons must remove their clothing as soon as possible. This adds another dimension of risk because a fog usually occurs in an area where the ambient temperature is quite low.

While fire is to be avoided at all costs where chlorine containers are stored, a brisk fire adjacent to a chlorine release can be a big help. This was demonstrated in a recent derailment when a tank car was ruptured by a following propane car, which exploded and caught on fire. The heat from the burning propane produced a chimney effect and the entire contents of the 90 ton tank car escaped without anyone suffering from chlorine inhalation.

CHLORINE ACCIDENTS

History

A chlorine accident occurs when fumes of the gas or liquid are inadvertently released into the atmosphere. However, a great many so-called chlorine accidents that receive wide coverage by the news media are those where no chlorine emission occurs—the chlorine tank cars involved in freight train derailments or the sinking of chlorine barges in rivers. It would seem that every string of derailed freight cars contains at least one chlorine tank car. This of course is not the case. Approximately 30 chlorine tank cars are involved in wrecks each year. Compared with over 9000 tank cars in service (1982), each making from 12 to 16 trips per year, the number of wrecks is small. In 1981 there were 24 tank cars involved in rail incidents in the U.S., mostly from derailments. Of these incidents, six were reported to have resulted in chlorine emissions.[33]

A major cause of large releases of chlorine in the early years was the cracking of the tank at the anchor, which is the attachment of the tank body to the railroad car frame. These early designs used forged anchors. The problem was corrected by using welded anchors. There has been no report of an anchor failure since 1963.

Usually the most serious damage that is inflicted on a tank car is to the protective dome assembly that houses the outlet valves and safety relief valve. On a few occasions in the past 20 years, one or more valves on the dome containing two liquid chlorine and two gaseous chlorine outlets have been sheared off. This has resulted in only minor spills. Such a situation automatically triggers into operation the excess flow valves inside the tank. These valves act as automatic checks, thereby containing the liquid in the car.

The most serious accidents occur when the tank car is punctured in a derailment by the coupler of an adjoining car. To avoid puncture damage, all new 90 ton cars are built with thicker steel shells and head ends, and are provided with *shelf couplers*. These couplers prevent disengagement in the vertical plane, which usually occurs in a derailment. Existing 55 ton and 90 ton cars are being provided with shelf couplers. How well this retrofitting has progressed is not known now (1983).

The first incident of tank car tupture occurred in 1961 and caused the death of an 11-month-old infant.[29] This was the result of a serious spill from a 30 ton car in a rural area of Louisiana. The chlorine cloud resulting from this spill covered an area of about six square miles. It was about 80–90 feet high, 2½ miles wide, and extended 3½ miles downwind. Measurements of chlorine concentration were made that revealed levels exceeding 400 ppm seven hours after the accident. There were 451 animal fatalities, including cows, horses, mules, hogs, and other domestic animals. The infant child died after about 15 min exposure 150 ft from the tank car. The mother who had eleven other children said that during the panic to get all her family away from the wreck she became confused and lost count. About 100 people were treated for vapor burns and respiratory irritation. All of the 15 hospitalized patients were discharged on the sixteenth day.

Fear of chlorine gas in large cities has led to bizarre situations worthy of comment. For example, the city of Chicago will not allow a chlorine tank car to pass through the Loop in order to get to the water treatment plants on the shores of Lake Michigan, but will allow tank cars carrying 15 ton containers to pass. Each container has six fusible plugs and two valves, each a source of leakage. The hazard is obviously greater from a load of such containers than from a tank car. New York City has restrictions against liquid chlorine which produce absurd situations for designers. Most have been a result of a single accident that should have gone unnoticed but actually developed into a serious situation nearly resulting in tragedy. In June, 1944, a delivery truck carrying chlorine cylinders was traveling through the streets of Brooklyn.[31] When a leak in one of the 150-lb cylinders was discovered, the driver pulled up at the side of the street only a few inches from a series of gratings covering ventilating shafts leading to an adjacent subway station. The truck was parked in this position for approximately twenty minutes. Some 400 people were overexposed to chlorine. All but two were in the subway station. Two hundred and eight persons of the 400 examined were admitted to eight different hospitals. Thirty-three persons showed evidence of moderate or severe poisoning and required hospital care for one or two weeks. All the rest were released after two or three days. If the driver had kept his truck moving, probably little or no overexposure would have resulted. This statement is substantiated by known cases of chlorine leaks occurring in tank cars in transit. One occurred in Indiana in 1935 when a tank car developed a leak en route and lost 60,000 pounds of chlorine without property damage or personal injury.[34] The point here is that the tank car kept moving.

Transport Accidents

Mexico. Some of the worst chlorine spills have occurred during the movement of chlorine in tank cars and tank trucks. One of the worst spills ever occurred August 1, 1981 at Estación Montana, San Luis Potosí, Mexico, population 400. A 38-car freight train hauled by two diesels and with a caboose left the tracks about 300 yards from the station. This train was hauling 32 fifty-five ton chlorine tank cars. The official report listed 17 people dead from this accident. Four of the train crew died as a result of the derailment and 13 people died as a result of chlorine vapor inhalation. The accident was caused by excessive train speed in a terrain with a descending slope of 4 percent plus the failure of the braking system. Seven tank cars lost all of their contents. Two of these lost their contents almost immediately. One car had a 20-inch hole in one of the heads and lost all of its contents. Four of the cars lost their valves and their protective housing. The cars leaked for several days and eventually lost all of their contents. Vegetation and some corn crops were damaged by the chlorine. Trees and shrubs turned yellow; however, several weeks after the accident it was noticed that the vegetation was growing again. The Mexican government no longer allows this many chlorine cars in a freight train.[35]

Youngstown, Florida (2/26/78). At 1:55 A.M., February 26, 1978 a freight train traveling at 45 mph with a string of 140 cars hauled by 5 diesel locomotives derailed near Youngstown, Florida. The train was 1½ miles long and had no radio communication. All of the locomotives and the seven lead cars derailed, as did the 9th through the 44th car. The latter group contained 10 tank cars of hazardous materials, including two of chlorine. One of the chlorine cars was punctured by a car originally located seven cars behind. The puncture was about 0.75 sq ft in size, with several long radial cracks extending as much as 2.5 ft from the puncture. The other derailed chlorine car was badly damaged but did not leak. Both were 90 ton cars. The chlorine that leaked out of the ruptured car was the direct cause of 8 fatalities and 138 injuries. It is estimated that about 50 tons of chlorine were lost in the first few minutes in both the liquid and gas phase. About twelve hours after the wreck visual observation indicated thirty to forty tons of liquid chlorine remained in the tank car. This gassed off slowly.

Seven hours after the derailment, the chlorine cloud was 3 miles wide, 4 miles long, and 1000 ft high. Wind velocity varied from 2 to 3 knots. The temperature was between 50 and 55°F. A systematic evacuation of residents within a 10-mile radius was made. The cloud was so dense it obscured the highway 300 ft away.

Before it could be blocked off, a number of motorists entered the area of the gas cloud. Their automobiles kept stalling due to the heavy concentration of chlorine in the air. Seven motorists, mostly teenagers, abandoned their cars and fled into the adjacent swamp, but died in vain attempts to flee the gas. It was later believed that it was this group who were responsible for the derailment, which was most likely due to sabotage. An eighth motorist drove successfully through the cloud, only to succumb to the effects of the gas a short distance down the road. Others suffered extreme nausea and respiratory distress. One hundred twelve persons were treated and released from five hospitals. Twenty-two were so severely injured that they were hospitalized. The injured included motorists, train crew members, and law enforcement personnel. Witnesses reported that at night the chlorine cloud appeared as fog and conveyed no indication of danger. The only warning conveyed by the cloud was its pungent odor. By then it was too late. The punctured car was not equipped with shelf couplers or head shields.[36]

Crestview, Florida (4/8/79). At 8:00 A.M., April 8, 1979, 29 cars, including 26 placarded tank cars of hazardous materials, derailed while traveling at 40 mph on an S curve near Crestview, Florida. This was a 116 car Louisville and Nashville freight train containing cars loaded with chlorine, ammonia, acetone, sulfur, methanol, carbolic acid, and carbon tetrachloride. Two tank cars of anyhydrous ammonia ruptured and rocketed. Twelve other cars containing five kinds of hazardous material ruptured, including a car of chlorine which had an 18 inch puncture hole in the side of the tank. Chlorine and ammonia vapor dispersed, forming a cloud that grew until it threatened a 300 square miles downwind. More than 4,500 persons were evacuated. The cloud posed a threat to the health of the population, and to wreck-clearing and emergency personnel, for nine days. There were no fatalities and only 14 persons were slightly injured by exposure to the variety of hazardous chemicals. Three were hospitalized. At the time of the accident it was daylight, with 7 mile visibility, temperature 57°F, overcast, misting rain, and winds at 5 mph, gusting to 20 mph. Throughout the nine days of the emergency, the gases reacted with the moisture of mucous membranes of the victim's lungs, eyes, and noses, producing various acids which became trapped in the affected organs. Ten wreck-clearing workers were overcome despite the use of self contained breathing apparatus and short work shifts. It is believed that failure of the breathing apparatus might have been due to beards on some of the work crew.

The cloud movement was observed by an Air Force aircraft and entry into the air space was prohibited. The vapor cloud reached a 5000 ft altitude and extended 28 miles in four hours. The next day, the cloud height was 1000 feet and chlorine was estimated to be leaking at a rate of 18,000 lb per hour. Emergency response and post-emergency investigation of this accident involved local, regional, state, and private industry forces and at least six federal agencies. The emergency activities reportedly were so uncoordinated that the first team of chlorine emergency specialists was actually turned back en route to the scene.[37]

As is usually the case, poor road-bed conditions were blamed for the accident. There were other factors that could have been more significant than the road-bed conditions. The train was probably going too fast (40 mph) to negotiate the S curve safely. The train was probably too long for the curve, and the power thrust of the engines might have been improperly applied. It is worthy to note that none of the derailed cars were equipped with shelf couplers or head shields.

Mississauga, Ontario, Canada (11/12/82). This town is a suburb of Toronto. Late Saturday night, November 10, 1979, a 106-car freight train carrying tank cars of chlorine, caustic soda, and flammables derailed in Mississauga, Ontario, 20 miles west of Toronto, Canada in 20°F weather. A 90-ton car of chlorine was upended adjacent to eight burning cars of propane. Heat from the fire and the presence of chlorine vapor in the top of the car caused the steel in the chlorine tank car to burn spontaneously. This allowed the release of about 70 tons of chlorine. In this case the chimney effect caused by the burning propane drove the escaping chlorine upward into the atmosphere, where it was carried away by natural air movement. Evacuation boundaries were extended through Sunday, until at 8:30 A.M. (20 hours after the accident) police announced that virtually the entire 50 square mile area encompassing Mississauga was to be evacuated. That order affected 250,000 residents and was made due to a wind shift. It was reportedly the biggest evacuation in North America. The wire services reported that anyone breathing 3 ppm of chlorine for 15 minutes would receive medical treatment.

Taking advantage of the low ambient temperature (20°F) water was poured into the remaining 20 tons of liquid chlorine in the ruptured car. This formed a thin sheet of chlorine hydrate ice on top of the liquid chlorine. This inhibited vaporization and thus expedited and simplified the transfer of this chlorine to other containers. Six days later all of the liquid was removed from the car and the evacuation was terminated. There were no reports of injuries due to chlorine exposure or other hazardous materials involved in the wreck. The low ambient temperature was an important factor that helped control the chlorine emissions.[38]

Tanker Trucks

The only chlorine tank truck accident that has been reported to date is the one that occured in Alexandria, Egypt in December 1965. Five people died, including the truck driver and two would-be rescuers.[34]

A chlorine tank truck, loaded with seven tons of liquid chlorine, swerved to avoid hitting a passenger car, overturned, and sheared off a gas valve. Civil defense procedure was inadequate, the truck construction had neither protective dome over the valves nor excess flow valves, and knowledge of emergency procedure was probably nonexistent. The most seriously exposed were those who tried to rescue the unconscious truck driver injured by the crash. Approximately two tons of chlorine leaked out before the gas valve had been sealed off. By this time some 500 persons had been exposed to the fumes.

Notable Consumer Accidents

General. There have been a variety of lost time accidents from chlorine leaks at water and wastewater treatment plants and industrial plants. In California alone there are about 200 reported cases of lost time accidents annually due to chlorine and its compounds and about forty due to sulfur dioxide, according to the California Department of Industrial Relations. The bulk of these accidents occur in industry. This is not surprising, since the chlorine used at water and wastewater treatment plants and at power stations accounts for a mere 5 percent of the annual production of chlorine.

Only one employee fatality has been reported for a water or wastewater treatment plant operation. A utility man employed by a Municipal Sanitation District in California died from inhalation of chlorine. His vocal chords were burned and he died two weeks later from pulmonary edema. Details of this accident have never been made available.

Some of the most serious of consumer accidents are detailed below.

A Fatal One-Ton Cylinder Liquid Leak. This accident resulted in the most serious consequences of any chlorine leak in North America associated with the handling of chlorine containers at the consumer level. It caused the untimely and quite unnecessary deaths of two people close to but not associated with the handling of the chlorine. This particular accident occurred in May 1959. At the time, the leak was quickly blamed on a "faulty gasket." A team of newspaper reporters made a comprehensive investigation of this accident and arrived at the following conclusions:

1. The accident probably would not have occurred if there had not been a power failure during a brief rain squall.
2. The accident probably would not have occurred if the city had not removed the auxiliary municipal light plant service to the filtration plant one month before the accident.
3. If working gas masks were conveniently available the leak could have been halted immediately and no injuries would have resulted.
4. Residents adjacent to the plant would not have been affected if the chlorine container room had been farther than 65 ft away from their homes.

An analysis of this accident and the newspaper reporters' conclusions leads to the following observations.

The power failure plunged the chlorine container room into practically total darkness (a rain squall at 4 P.M. shut out most of the light). This situation caused a great delay in the operator's response to all of his duties required during the power failure. It is easy to imagine the panic which struck his heart at smelling a severe chlorine leak in the dark! He could not find the gas masks quickly so he left the area immediately (as he should have) to go for help, as he was the sole operator on duty.

With the power out, the operator was unable to relieve the chlorine piping system of pressure, even though the container valve at the leaking connection might have been closed. During the time of a power failure, all of the liquid chlorine in the piping system between the containers and the metering equipment can exit through any leaking joint unless there are appropriate isolating valves. When power is available the metering equipment is able to withdraw the liquid chlorine from the supply piping system very quickly, thereby averting a massive chlorine leak.

Therefore, auxiliary lighting to show the operator the way and auxiliary power to operate the chlorine withdrawal system are essential for safety in chlorine handling.

Easy access to chlorine gas masks is of top priority. Just as soon as the fire department arrived with the necessary assistance to provide gas masks, and emergency lights, the operator was able to locate the leaking cylinder connection. By this time power had been restored and the operator was able to relieve the system of pressure and repair the leak.

In these situations where an operator faces overwhelming odds he needs all the isolating valves available to him to reduce the length of piping under pressure adjacent to the leak. This is why every container should be connected with an auxiliary cylinder valve attached to the cylinder outlet valve. The outlet of the auxiliary valve is then connected to the inlet of the flexible connection and the outlet to the stationary header valve. Some operators prefer an auxiliary header valve at the outlet of the flexible connection so that for liquid withdrawal the flexible connection can be shut off at each end and removed during each cylinder change without the fear of discharging liquid chlorine. The best way to operate is to use duplicate header systems so that one header can be completely emptied of liquid chlorine during a cylinder change.

The deaths which occurred in this accident could have been avoided had the residents been told to get out of their houses. The children of the deceased did run away from the house, which undoubtedly saved their lives. By staying in a house through which the chlorine gas has passed, any occupant is subject to continuous chlorine exposure because the gas becomes absorbed immediately by upholstered furniture, drapes, carpets, and clothing. Body heat and perspiration enhances the chlorine absorption by the clothing. This exposes the victim to continuous chlorine inhalation from both the clothing and house furnishings.[39]

A Four-Ton Liquid Leak. This leak allowed escape of the liquid chlorine contents of four one-ton chlorine cylinders which were "on line" at the time of the leak. Fortunately, there was no loss of life, but many residents nearby were "treated" for various degrees of chlorine inhalation.

This massive leak was the result of the following factors, in their order of importance:

1. The operator attempted to stop a leak at the stem of a chlorine header valve while the system was under full pressure from the four cylinders.
2. This leak was caused by the structural failure of a bushing (¾" × 1") in the 1-inch chlorine header into which the leaking header valve was threaded.
3. The failure of the bushing was the result of pernicious corrosion over a long period of time.

The lessons to be learned from this accident are as follows:

1. Never use bushings in the chlorine supply system.
2. Operators should be warned never to attempt repair of a chlorine leak while the system is under supply pressure.
3. The first step in repairing a leak is to relieve the system of the chlorine cylinder pressure. Necessary gauges must be included in the design to provide the operator with this vital information.
4. *Duplicate header systems should be provided* so that these piping systems can be replaced in their entirety on a regular basis (i.e., every 5 years for systems passing 2 tons or more per day and every 10 years for all others passing less than 2 tons per day).

A 1600-lb Gas Leak. This case was a freak accident compounded by mistakes resulting from lack of experience in coping with a massive chlorine leak.

A workman for a natural gas utility company was busy with a cutting torch working at several isolated locations in an industrial area. He was cutting into sections of empty, unused natural gas pipelines preparatory to sealing them off. By mistake he cut into a 6-in. underground chlorine pipeline about 7000 ft long. This pipeline was used to transport chlorine gas from a chemical plant to a nearby plastics plant. The pipe was under tank pressure of about 85 psi at the time of the accident.

This calculates to about 1600 lb of chlorine gas which escaped into the atmosphere. Fortunately, the line was equipped with automatic shut-off valves at both ends. When the pressure dropped as a result of the hole made by the cutter's torch, these valves automatically closed.

Coincidentally with the action of the pipe cutter, which occurred at about 7:45 A.M., some 300–400 people were waiting to go to work at a nearby refinery. These people were all gathered in a small cluster waiting for the gates to open at 8:00 A.M. They were a short distance downwind from the leak. The morning air was cool and the humidity was high—about 70 percent. The general location is only a few miles from the ocean surf and the wind blowing in their direction was 10–15 mph. Even though this is considered a massive leak only 30–40 of these people were taken to the hospital. It is probable that the high humidity coupled with a strong wind was responsible for rapid dilution and dispersal of the chlorine into the atmosphere. However, the humidity had an adverse effect. The combination of body moisture and high humidity conspired to absorb chlorine into the workers clothing. The people taken to the hospital were herded into a small room and were immediately exposed to more chlorine coming from their clothing. A hospital attendant realized what was happening and they were ushered outside and then taken in one by one and had their clothing removed. They remained in the hospital long enough for a thorough observation and time to get a change of clothes. None stayed more than one night in the hospital.[39]

The lessons to be learned from this accident are as follows:

1. Pipelines carrying chlorine liquid or gas should never be buried in the ground. The preferred method is a grate-covered concrete channel at grade level or in overhead support systems.
2. Whenever a person is exposed to chlorine, that person's clothing should be removed as soon as possible and the body showered thoroughly with warm water if available to avoid further shock.
3. Never allow a person subjected to chlorine exposure to remain in a confined area, particularly a room with rugs, carpets, drapes, or upholstered furniture. Get them into the open air as soon as possible.

Help in Chlorine Emergencies

This includes all hazardous chemicals:

PHONE—TOLL FREE—DAY OR NIGHT

(800) 424-9300

CHEMTREC/MCA

Effect of Chlorine Accidents on the Community

That locally stored chlorine is a potential hazard is undeniable. However, the risk of chlorine storage for use at potable water plants and wastewater plants must be put in perspective. There have been many more serious accidental emissions from ton containers than from tank cars while stored at these plants. A tank car is not nearly as vulnerable as a ton container. In spite of this, there is much more chlorine to worry about in a tank car or stationary storage tank than in a battery of ton containers.

The City of San Francisco has three wastewater treatment plants within the city limits. Two of these used tank cars for more than 25 years. The third plant used a battery of 12–15 ton containers from 1938 to 1981. Moreover there were 50-ton stationary chlorine storage tanks located at the plants using tank cars. The safety record for chlorine handling at these plants was exemplary. There were a few minor accidental emissions, mostly at the plant using ton containers where there was one known lost time accident.

In 1978 a newspaper reporter wrote a story about the hazards of transporting chlorine tank cars across San Francisco Bay and storing these cars at the wastewater plants. The evening paper ran the story under the front page headline "S.F. SITS ON A GAS DISASTER" (*San Francisco Examiner*, Sept. 5, 1978). Although the reporter focused largely on the way the cars were transported, the San Francisco County Board of Supervisors voted to stop using chlorine gas as soon as possible and switch to sodium hypochlorite. This was accomplished in March 1980. Sulfur dioxide was also discontinued in favor of sodium bisulfite at the same time. The city had to pay out of its own funds about 3 million dollars for this project. Federal grant money was denied the city on the argument that their existing chlorination–dechlorination facilities were safe and reliable.

Knowing that the hypochlorite was going to be significantly more expensive than chlorine gas, the City of San Francisco attempted to secure other large municipal users of chlorine to enter into a joint venture of making hypochlorite from liquid chlorine in tank cars and caustic solution. They all declined. Several of these users who were using tank cars made separate investigations of the safety aspects and the hazard potential of their use of tank cars. All of these investigations resulted in the decision to stay with the tank cars. The two overriding factors turned out to be the high cost of hypochlorite (40¢/lb vs. chlorine gas at 6–7¢/lb) and the fact that of the total amount of chlorine transported in and around these treatment plants only 10 percent was being used at these treatment plants.

On November 16, 1979 a corroded bolt on the flange of the tank car *flexible connection failed,* spewing liquid chlorine at a sodium hypochlorite manufacturing plant across the tracks from one of the above wastewater plants which used tank cars. This hypochlorite manufacturing plant is located adjacent to the freeway which is the interchange between the San Francisco–Oakland Bay Bridge and the East Bay cities. This main thoroughfare had to be closed during the morning rush hour until the leak was secured. Although the leak was of fairly short duration, 10 people were admitted to the hospital and released after a short time. The incident did not provoke any negative community reaction.

Frequency and Magnitude of Chlorine Leaks

A list of the chlorine accidents reported to the Chlorine Institute[34] over the past 50 years indicates by frequency and magnitude of chlorine emission the following category of causes in order of potential hazard:

1. Fire.
2. Flexible connection failure.
3. Fusible plug failure.
4. Freak accidents caused by carelessness and ignorance.
5. Valve packing failure.
6. Gasket failure.

7. Piping failure.
8. Equipment failure.
9. Collision accidents causing physical damage to containers.
10. Container failure.

1. Fire in the area of chlorine containers is by far the most serious hazard, although not the most frequent. Fire will raise the temperature so that the fusible plug on 150 lb cylinders and ton containers will melt at 158°F and the entire contents of the containers will be discharged. In some cases during a fire, chlorine containers have been ruptured by the heat while the fusible plugs have remained intact. The exact cause of such ruptures has not been explained.

2. Probably the most frequent cause of chlorine emissions resulting in overexposure to personnel is the failure of connecting lines between the container and the metering and control equipment. These connecting lines are traditionally made of annealed copper, 2000 psi strength, and cadmium plated. Copper is used because it is flexible and has the proper structural strength; however, at each container change, the chlorine remaining inside the tubing will absorb moisture from the atmosphere, and a cycle of corrosion will begin. Therefore, any flexible connection has a life dependent upon how many times it is disconnected and left open to the atmosphere. To check the worthiness of a flexible connection, bend it carefully; if it screeches slightly, it is due for replacement. The noise produced by the bending is indicative of the magnitude of corrosion products on the inside of the tube.

3. Fusible plug failure without any evidence of elevated temperature caused by fire or direct sunlight is next in order of hazard magnitude. A fusible plug which is supposed to melt at 158°F may leak from corrosion or a poor bond between the lead alloy and the plug retainer. There is only one fusible plug on a 150-lb container, at the base of the outlet valve. There are three in each of the dished heads of a ton container. Gas emissions through faulty fusible plugs have been numerous enough to question their safety value. The AWWA Committee report[28] *recommends elimination of fusible plugs.*

4. Carelessness is high on the list of causes of chlorine accidents, sometimes involving unusual situations. One case involved a 6-inch buried chlorine gas line[40] originally entirely within the property of two chemical plants. Adjoining property became subdivided 20 years later, and some of the underground piping, including that for chlorine gas, became part of the pavement area of new streets. This chlorine gas line was first cut into by mistake by a welder who was supposedly inactivating obsolete natural gas lines. The heat from the torch burned a small hole in the pipe, so that the chlorine could support combustion of the carbon steel pipe. Within seconds, the hole was approximately eight inches in diameter, resulting in an almost immediate discharge of the contents of 7,000 feet of six-inch pipe. A few months later, on successive days, this same line was broken twice again by a backhoe excavating for additional underground lines. After these experiences, the decision was made to lower the pipe from its former depth of approximately 2 feet to approximately 5 feet.

5. Valve packing failures have never caused any serious problems. If the leak is minor, it can often be corrected by tightening the packing nut. Serious leaks are taken care of by the application of a proper container safety kit.

6. Gasket failures are serious only when they occur on the seal between the dome and tank of a chlorine tank car. Other gaskets are located so that the supply can be secured, the system emptied of gas, and the gasket replaced in a routine fashion. Gasket failures on tank cars have occurred, but these are rare and the emission of chlorine is minor.

7. Piping failures have been rare, and are sometimes the results of using improper materials. Lines carrying liquid chlorine can be a potential hazard. One of the few fatalities attributed to chlorine in recent years was caused by the failure of a liquid chlorine line. The worker's death was the result of his constricted breathing caused by the amount of chlorine fumes, which made it impossible for him to climb out of the confined location of the leak.

8. Equipment failure usually refers to vaporizers used between the containers and the chlorine metering and control equipment. These failures are due primarily to corrosion by the chlorine, and are a function of the amount of wall thickness versus the amount of chlorine passing through the vaporizer. The frequency of this type of accident is low.

9. Chlorine emission accidents caused by collisions involving containers are rare.

10. Container failures, except those caused by fire, are rare. The chlorine packagers throughout the United States and Canada are keenly aware of the potential hazards of handling chlorine containers. This has resulted in a strict monitoring of container condition. Perusal of the Chlorine Institute accident reports indicates two container failures over a period of fifteen years. Considering that shipments in 150-lb and ton containers amount to nearly 100,000 per year, this sort of accident is a rarity and is usually a result of using an over-age cylinder. One of these accidents occurred at a military base. A ton container was being loaded at a dock. It slipped out of the sling and fell on its end. The tank split at the "chime," which is the joint between the cylinder and the dished head. Investigation of this accident revealed that this cylinder had been in service by the military for 40 years.

Chlorine tank car container failures are practically nonexistent for a very good reason: these containers are readily accessible for interior inspection on a programmed basis.

Safety Precautions

Handling chlorine need not be a serious hazard if the personnel working with it are properly educated and trained in its handling.

The following are some guidelines for assuring the safe handling of chlorine:

1. Install leak detector sensors at appropriate locations.

2. Provide proper instruction and supervision to workers charged with responsibility of chlorination equipment.

3. Provide proper and approved self-contained breathing apparatus for persons working where there is a possibility of exposure to chlorine gas fumes. Locate the breathing apparatus close to the potential leak area but far enough away so that they are accessible in case of a major leak.

4. Survey the areas of most likely chlorine emissions with an attempt to predict the downwind travel in case of accidental release. Use a *wind sock* to determine air movement so as to establish upwind areas.

A convenient aid in formulating this prediction is the use of "Downwind Vapor Hazard Nomographs." These are based on equations developed by Sutton of the United Kingdom and modified by Calder and Milly of the U.S. Army Chemical Corps.[41] Additional information on these nomographs may be obtained from the office of the Chief Chemical Officer, Department of the Army, Washington 25, DC.

Another paper of more recent vintage often used by engineers to predict the spread of toxic fumes is the one published by the Chlorine Institute titled "Estimating Area Affected by a Chlorine Release" by A. E. Howerton[32] March 1969. This publication contains other pertinent references.

5. Never store combustible or inflammable materials in or near chlorine containers.

6. Never apply direct heat to a chlorine container. Never attempt any welding operation on an "empty" chlorine gas line without having purged it with air.

7. Always keep available close to the chlorine containers a water supply which can be used to keep containers cool in case of fire or by personnel if they accidentally come into bodily contact with chlorine gas or liquid.

8. In case of a leak, determine if the rupture is in a faulty container or in the control apparatus or connecting pipe.

9. If the leak is in a container, an appropriate emergency kit should be brought into action.

10. If the leak is in the control apparatus or connecting pipe, at least two persons should don breathing apparatus, find the leak by means of ammonia fumes, and secure the valves at the containers. The operation of the control equipment will drain the system of chlorine pressure. When the system is down to atmospheric pressure (zero gage), steps can be made to make the necessary repairs.

11. After a worker has been exposed to a chlorine leak of sufficient magnitude while working with self-contained breathing apparatus, his clothes should be removed and his body showered. The clothes should be aired adequately. The danger here is that the normal perspiration absorbed by clothing retains a tremendous amount of chlorine gas, which will be released continuously after the exposure. When the worker has left the leak area, he may think he is still being exposed to a leak because of the chlorine given off by his clothing. Therefore, always remove breathing apparatus in an open area—never in a room or confined location.

12. Spraying water on leaking containers may make the leak worse as a result

of corrosion. Water in sufficient quantity and velocity can be used to confine or limit the spreading of moderate leaks.[42]

13. Never try to disperse chlorine gas directly from a container to an open body of water. Chlorine gas in only very slightly soluble in water at atmospheric pressure. The author has demonstrated[43] that it is far more beneficial to attempt the dispersal of liquid chlorine in a body of water if the conditions are favorable. Liquid chlorine goes into solution much more rapidly than gaseous chlorine does at atmospheric pressure.

14. Chlorine leaks must be given prompt attention.

15. When entering a chlorine equipment area, always be on the alert. Take shallow breaths when entering and until it is ascertained that no leaks are in progress.

16. Keep upwind of chlorine leaks. Although chlorine gas is two and a half times as heavy as air, it will always follow air currents, as it has an affinity for moisture in the air. It is therefore a fallacy that chlorine always settle to the ground or to the floor.

17. Be aware of proper first aid procedures. Never give anything by mouth to an unconscious person.

18. If a container develops a leak in transit keep the vehicle moving. Conversely if a stationary container develops a leak try to transport it quickly to a predetermined disposal site until the emergency response team arrives.

19. It is advisable to rely upon chlorine control and metering equipment for direct disposal into a natural stream or treatment plant facilities.

20. Do not attempt to rely upon direct disposal methods of the contents of chlorine containers, evaporators or liquid chlorine piping unless these systems are connected to a predesigned chlorine absorption tank. (See Chapter 9.)

PRODUCTION AND USES OF CHLORINE

Annual Production

The amount of chlorine produced in the U.S.A. is reported by the Chlorine Institute in two categories: gaseous and liquid. Gaseous production generally refers to the captive generation by industries that use the gas in the manufacture of their products. This has a tendency to distort somewhat the chlorine usage figures. The most productive year to date (1983) was 1979. In that year 12.3 million tons of gas was produced as compared to 7.3 million tons of liquid chlorine. The latter was packaged and delivered to consumers.[16] This compares with the 1950 production of 2 million tons of gas and one million tons of liquid chlorine.[16]

Energy Consumption in the Production of Chlorine

Table 1-7 shows the energy input to produce chlorine. The figures shown are for both English and Metric units.[58]

Table 1-7 Energy Consumption in Chlorine Production

Energy Input	English Units	Metric (SI) Units
Steam amount	10,500 lb/ton	5,250 kg/kkg[a]
Steam equivalent fuel[b]	14.1 × 10⁶ Btu/ton	16.4 GJ/kkg[c]
Electricity amount	3.043 kWh/ton[d]	12.1 GJ/kkg
Electricity equivalent fuel	26.2 × 10⁶ Btu/ton[e]	30.5 GJ/kkg
Total	40.3 × 10⁶ Btu/ton	46.9 GJ/kkg

[a] 1 kkg = 1000 kg = 2,204.6 lb
[b] Based upon 1340 Btu/lb steam
[c] 1 GJ = Giga Joule = 10⁶ Joules; 1 Btu = 1,054.8 J
[d] 1 kWh = 3.6 × 10⁶ Joules
[e] Based upon 50 percent self-generated electricity; overall: 8,610 Btu/kWh

End Uses of Chlorine

General. The best estimate of end uses that are significant is based upon a 1981 first quarter study by the Diamond-Shamrock Corporation[44] shown in Table 1-8. A 1981 tabulation of chlorine end use by the Pennwalt Corporation[45] is shown in Table 1-9.

Table 1-8 End Uses of Chlorine

Market	Chlorine Use (1000 S.T.)	Percent Distribution
PVC–VCM	2067	18.5
Pulp and paper	1160	10.5
Organic chemicals	1758	16.0
Chlorinated ethanes	1327	12.0
Propylene oxide	814	7.5
Inorganic chemicals	664	6.0
Water and wastewater treatment	600	5.5
Fluorocarbons	828	7.5
Chlorinated methanes	442	4.0
Miscellaneous	1418	12.5
Total	11078	100.0

Water and Wastewater Treatment. It should be of interest to the environmentalists who are forever complaining about the potential hazards to the environment caused by the chlorination of drinking water and wastewater that this use accounts for only 5–6 percent of the market, whereas the chemicals industry accounts for about 50 percent of the market. The chlorinated hydrocarbons used in pesticides which have direct access to the environment in rainfall runoff are estimated at about 15 percent of the market.

Table 1-9 End Uses of Chlorine

Market	Percent
PVC production	20
Solvent manufacture	24
Organic chemicals	20
Inorganic chemicals	12
Pulp and paper	12
Water and wastewater (Municipal 2% industrial 3%)	5
Miscellaneous	7
Total	100

The Power Industry. The electric power industry uses a substantial amount of chlorine to prevent biofouling in heat exchangers. This biofouling seriously affects the heat exchange efficiency of the condensers. A 1978 survey determined that this industry used approximately 10,800 lb of chlorine and hypochlorite.[46] This amounts to 1.8 percent of the figure shown in Tables 1-8 and 1-9 for water treatment. This makes the industrial percentage in Tables 1-8 look credible, because there are many industrial users of chlorine for biofouling control in condensers other than the power industry. There are also many industrial users of chlorine in process water treatment as well as in wastewater treatment.

The Chemical Industry. Chlorine is a most unusual and versatile substance with more diverse uses than any other chemical known—from rocket fuels to the manufacture of food products. One of its most unusual products is monochloroacetic acid, used to make carboxylmethyl cellulose, which, in turn, is used as a detergent builder, as a filler in ice cream,[17] and in the manufacture of thioglycollic acid (used in home permanents). Since 1950, the largest single use for chlorine has been in the manufacture of ethylene oxide and glycol, used to make antifreeze fluids and synthetic fibers.

Before the tremendous increase in the chemical industry, most of the chlorine was used in the textile industry for bleaching purposes. In 1965, 30,000 tons were used in textile bleaching.[47] This was less than 1 percent of the total United States production in 1965.

Most chlorine manufacturing plants are situated where inexpensive sources of either power or salt, or both, are available. It is also necessary to plan for geographical demand, since chlorine must be sold on an FOB freight equalized basis. While location is an important consideration, it is estimated that between 70 and 75 percent of chlorine production is captive. Since chlorine and its co-products must be carefully balanced, it sometimes becomes advantageous to utilize methods other than the electrolysis of brine for production. For example, the HCl oxidation processes are advantageous when there is an overabundance of hydrochloric acid.

As the chlorine industry continues to expand toward a possible production of fifteen million tons per year by 1984, all these factors play an important part in location and manufacturing methods.

Since the chemical industry accounts for the largest share of the total United States production, it is of interest to list the end use of these chlorine compounds. There are ten inorganic and 25 organic compounds derived from chlorine which find more than fifty end uses.[48] The following is a list of these compounds and some of their end uses:

Aluminum chloride	aircraft, automotive cutting oils, paints, metal processing, pigments, plastics
Allyl chloride	adhesives, aircraft, anesthetics, drugs, electronics, explosives, food additives, insecticides, paints, soft drinks, solvents, water repellants
Chlorinated Benzenes	adhesives, aircraft, bactericides, detergents, drugs, fumigants, fungicides, household chemicals, protective coatings, solvents
Chlorinated paraffins	automotive, cutting oils, metal processing, plastic sanitizing
Chloral	aerosols, anesthetics, dry cleaning, insecticides, mothproofing
Chlorine Trifluoride	rocket fuel, missiles
Chloroform	aerosols, anesthetics, drugs, dyestuffs, fumigants, perfumes, plastics, refrigeration
Dichlorobenzene	insecticides, organic solvents
Epichlorhydrin	detergents, dyestuffs, electronics, explosives, missiles, nuclear reactors, pigments, synthetic rubber
Ethyl chloride	anesthetics, automotive, drugs, oil processing, plastics, refrigeration
Ethylene chlorhydrin	antifoaming agents, bactericides, cutting oils, dry cleaning, explosives, food additives, germicides, paints, pulp and paper, polyfoams, rocket fuels, soft drinks, solvents, synethetic fibers, textiles, tobacco
Ethylene dichloride	adhesives, aircraft, automotive, electronics, plastics
Ferric chloride	water treatment, wastewater treatment
Fluorocarbon	aerosols, aircraft, anesthetics, drugs, fire extinguishers, food additives, germicides, paints, plastics, polyfoams, synthetic fibers
Glycerine	synthetic resins, tobacco, explosives, electronics, food processing
Hexachlorocyclopentadiene	fire retardants, household chemicals, insecticides, plastics
Hexamethylene diamine	aircraft, automotive, household chemicals, missiles, textiles
Hydrazine	aircraft, drugs, electronics, metal processing, refrigeration
Hydrochloric acid	aircraft, electronics, food processing, drugs, gasoline, household chemicals, missiles, pulp and paper, pigments, plastics, solvents, water treatment
Methyl chloride	adhesive, aerosols, aircraft, anti-foaming agents, bactericides, drugs, food additives, oil processing, plastics, refrigeration, silicones

Methylene chloride	aerosols, drugs, plastics, paint remover, solvents
Monochloroacetic acid	herbicides, detergents, pharmaceuticals, food additives, textiles
Perchloroethylene	drycleaning, metal processing, missiles, nuclear reactors, solvents
Phosgene	adhesives, aircraft, automotive, electronics, paints, protective coatings, synthetic fibers
Phosphorous trichloride	aircraft, automotive, disinfectants, drugs, electronics, fire retardants, fungicides, gasoline, household chemicals, hydraulic fluids, missiles, paints, perfumes, pigments, plastics, protective coatings
Propylene glycol	polyester resins, cellophane
Propylene oxide	aircraft, automotive, disinfectants, drugs, electronic, food additives, fungicides, paints, plastics, protective coatings, solvents
Titanium tetrachloride	aircraft, automotive, dyestuffs, electronics, fire retardants, gasoline, missiles, pigments, plastics, titanium
Trichlorocyanurate	bactericides, bleaches, detergents, drugs, household chemicals, swimming pools
Trichloroethylene	dry cleaning, food processing, electronics, missiles, oil processing, solvents
Trichlorethane	aerosols, aircraft, dry cleaning, food processing, solvents
Vinyl chloride	aircraft, electronics, household chemicals, paints, pulp and paper, plastics, synthetic fibers, textiles
Zirconium tetrachloride	aircraft, disinfectants, drugs, dyestuffs, fire retardants, gasoline, pigments, plastics, synthetic fibers

REFERENCES

1. Baldwin, R. T., "History of the Chlorine Industry," *J. Chem. Ed.,* 313 (1927).
2. Mond, L., "History of the Manufacture of Chlorine," *J. Soc. Chem. Ind.* (London), 713 (1896).
3. Sommers, H. A., "The Chlor-Alkali Industry," *Chem. Engr. Prog.* **61**, 94 (1965).
4. Gardiner, W. C., "Castner, A Pioneer Inventor in Alkali–Chlorine," *Proc. Chlorine Bicentennial Symposium,* Electrochem. Soc., Princeton, NJ, 1974.
5. Holden, H. S., and Kolb, J. M., "Metal Anodes," *Kirk-Othmer Encyclopedia of Chemical Technology,"* 3rd. ed., Vol. 15, p. 172, John Wiley & Sons, Inc., New York, 1981.
6. Anon., "Dimensionally Stable Anodes," ES-EC-1 Eltech (Diamond Shamrock Corp.) Chardon, Ohio 1979.
7. Thomas, V. H., and Rudd, E. J., "Energy Saving Advances in the Chlor-Alkali Industry," paper presented at the Chlorine Conference, Soc. of Chem. Ind., London, UK, June, 1982.
8. Dahl, S. A. "Chlor-alkali Cell Features New Ion-Exchange Membrane," *Chem. Engineering,* 60 (Aug. 18, 1975).
9. Anon. "Membrane Cell Technology," Eltech, Chardon, Ohio, Feb. 1983.
10. O'Brien, T. F., and Gilliatt, B. S., "The FM 21, A Novel Approach to Membrane Cell Design," paper presented to the Chlorine Plant Operations Seminar, Atlanta, Georgia, Feb. 17, 1982.
11. Hora, C. J., and Maloney, D. E., "Nafion Membranes Structured for High Efficiency Chlor-Alkali Cells," paper presented at 152nd National Mtg. Electrochemical Soc., Atlanta, Georgia, Oct 10–14, 1977.

12. Isfort, H., "Uhde's Membrane Cell Technology," paper presented at the Chlorine Conference, Soc. of Chem., Ind., London, UK, June 1982.
13. Udagawa, H., private communication, Asahi Chemical Co., Feb., 1983.
14. Anon. "A Revolution in Chlor-Alkali Membranes," *Chem. Week,* 35 (Nov. 17, 1982).
15. Gestaut, L. J., Thomas, V. H., and Moomaw, J. A., "Air Cathodes for the Chlor/Alkali Industry," Eltech Systems Corp., Chardon, Ohio, 1982.
16. Anon. "North American Chlor-Alkali Industry Plants and Production Data Book," Chlorine Institute Pamphlet 10, New York, NY, Jan. 1982.
17. Sconce, J. S., *Chlorine: Its Manufacture, Properties and Uses,* Reinhold, New York, 1962.
18. Penfield, W., and Cushing, R. E., "Bathing the Green Goddess," *Ind. Eng. Chem.* **31,** 377 (1939).
19. Sisler, H. H., *College Chemistry,* p. 201, Macmillan, New York, 1935.
20. Schultze, H. W., "The Chlorine Industry; Past, Present and Future," *Proc. Chlorine Bicentennial Symposium,* Electrochem. Soc., Princeton, NJ, 1974.
21. Iammartino, N. R. "New Ion-Exchange Membrane Stars in Chlor-Alkali Plant," *Chem. Engineering,* 86 (June 21, 1976).
22. Kienholz, P. J. "Bipolar Chlorine Cell Development," *Chlorine Bicentennial Symposium,* San Francisco, CA, p. 198, 1974.
23. Anon., Kel-Chlor, M. W. Kellog Co., Houston, Texas, company bulletin, 1980.
24. Anon. "Recovering Chlorine from Waste HCL," *Env. Sci. Technol,* **9,** 16 (Jan. 9, 1975).
25. Laubusch, E. J., "Standards of Purity for Liquid Chlorine," *J. AWWA,* **51,** 742 (1959).
26. Adams, F. W., and Edmonds, R. G., "Absorption of Chlorine by Water in a Packed Tower," *Ind. Eng. Chem.,* **29,** 447 (1937).
27. Heinemann, G., Garrison, F. G., and Haber, P. A., "Corrosion of Steel by Gaseous Chlorine," *Ind. Eng. Chem.,* **38** 497 (1946).
28. Committee Report, "Chemical Hazards in Waterworks Plants," *J. AWWA,* **27,** 1225 (1935).
29. Joyner, R. E., and Durel, E. G., "Accidental Liquid Chlorine Spill in a Rural Community," *J. Occup. Med.,* **4,** 3 (Mar. 1962).
30. Hedgepeth, L. L., "Handling Chlorine to Avoid Trouble," *J. AWWA,* **26,** 1602 (1934).
31. Chasis, H., Zapp, J. A., Bannon, L. H., Whittesberger, J. L., Helm, J., Doheny, J. J., and McLeod, C. M. "Chlorine Accident in Brooklyn," *Occup. Med.,* **4,** 152 (Aug. 1947).
32. Howerton, A. E. "Estimated Area Affected by a Chlorine Release," Chlorine Institute Report 71, March 1969.
33. Anon., unpublished accident reports, 1981, Chlorine Institute, New York, 1983.
34. Anon., unpublished accident reports, 1930–1982, Chlorine Institute, New York, 1983.
35. Perez, F. J., Plant Mgr., Celulosa y Derivados, S. A. Monterrey, Mexico, Letter to Chlorine Institute, Oct. 22, 1981.
36. Anon., National Transportation Safety Board Accident Report No. NTSB-RAR-78-7, Wash., DC.
37. Anon., National Transportation Safety Board Accident Report No. NTSB-RAR-79-11, Wash., DC.
38. Anon., unpublished accident reports, 1982, Chlorine Institute New York, 1982.
39. Unpublished chlorine accident correspondence, G. C. White, San Francisco, CA, 1950–1983.
40. *Los Angeles Times,* Sept. 13, 1966.
41. Anon., "The Handling and Storage of Liquid Propellants," Office of the Director of Defense Research and Engineering, Washington, DC, 1961.
42. Hopkins, E. W., and Faber, H. A., "Chlorine Dispersion with Fog Nozzles," *Water Sew Wks,* **98,** 350 (Aug. 1951).
43. White, G. C., unpublished data on discharging liquid chlorine from 150-lb cylinder.
44. Krupp, R. C., Diamond Shamrock Corp., private communication, Dec. 11, 1981.
45. Trojak, G. F., Pennwalt Corp., private communication, Nov. 25, 1981.
46. Chow, Winston, Electric Power Research Inst., private communication, Jan. 8, 1982.
47. Sconce, J. S., private communication, 1968.

48. Anon. "Chlorine Facts," Chlorine Institute, New York, NY, 1968.
49. Molnar, C. J., and Dorio, M. M., "Effects of Brine Purity on Chlor-Alkali Membrane Cell Performance," paper presented at the 152nd National Mtg. Electrochemical Soc., Atlanta Georgia, Oct. 10–14, 1977.
50. U.S. Patent No. 3,632,498, "Electrode and Coating Therefor," H. B. Beer, filed Feb. 2, 1968.
51. Horacek, J., and Puschaver, S., "A Comparison of Economic Factors Involved in the Choice of Chlorine–Caustic Cell Anode System," paper presented at Chlorine Institute annual conf., 1972.
52. Cairo, P. R., Lee R. G., Aptowicz, B. S., and Blankenship, W. M., "Is Your Chlorine Safe to Drink?" *J. AWWA,* **71,** 450 (Aug. 1979).
53. Anon., Amer. Nat. Std. for Liquid Chlorine, ANSI/AWWA B301–81, June 7, 1981.
54. Unpublished Chlorine Institute file on nitrogen trichloride (ca. 1935).
55. White, G. C., unpublished investigation of chlorinator injector fouling, Orange Co. Water District, Fountain Valley, CA, June 16, 1982.
56. Doyle J. H., Private communication, Chlorine Inst., New York, NY, Feb. 14, 1983.
57. White, G. C., Unpublished report on over pressure in ton containers causing gross deformation without rupture. Bogotá, Colombia, 1981.
58. Meyers, J. G., "Energy Consumption in Manufacturing," The Conference Board, Energy Information Center; The Ford Foundation, Energy Policy Project, NSF-RANN 1973.
59. DuLong, P. L., *Schweigger's J. Chem. Pharm.,* **8,** 32 (1812).
60. Sconce J. S., *Chlorine, Its Manufacture, Properties, and Uses,* Amer. Chemical Society Monograph Series No. 154, Reinhold, New York, 1962.
61. National Research Council, *International Critical Tables,* Vol. 1, McGraw-Hill, New York, 1926.
62. Kapoor, R., and Martin, J., "Thermodynamic Properties of Chlorine," Engr. Res. Inst. Univ. of Mich., 1957.
63. Wobser, R. and Muller, F., *Kolloid-Beihefte,* **52,** 162 (1941).
64. Lange, N. A., *Handbook of Chemistry,* Handbook of Publishers Inc., Sandusky, Ohio, 10th. ed., 1961.
65. Eucken, A., and Hoffman, *Z. Physik. Chem.* **B5,** 422 (1929).
66. Adams, F. W., and Edmonds, R. G., "Absorption of Chlorine by Water in a Packed Tower," *Ind. Eng. Chem.,* **29,** 447 (1937).
67. Pellaton, M., "Constantes Physiques du Chlorine," *Jour. de Chemie Physique,* **13,** 426 (1915).
68. National Research Council, *International Critical Tables,* Vol. 3, McGraw-Hill, New York, 1928.
69. Lacey, J. I. (Hooker Electrochemical Co.), *Anal. Chem.,* **20,** 379 (1949).
70. Giaque, W. F., and Powell, T. M., "Chlorine, The Heat Capacity, Vapor Pressure, Heats of Fusion and Vaporization, and Entropy," *J. Am. Chem. Soc.,* **61,** 1970 (1948).
71. National Research Council, *International Critical Tables,* Vol. 7, McGraw-Hill, New York, 1930.
72. Lange, N. A., "Euber einige Eigenschaften des Verflussigten Chlor," *Z. Angewandte Chemie.,* **13,** 683 (1900).
73. National Research Council, *International Critical Tables,* Vol. 1, McGraw-Hill, New York, 1926.
74. National Research Council, *International Critical Tables,* Vol. 5, McGraw-Hill, New York 1929.
75. Steacie, E. W. R., and Johnson, F. M. G., "The Viscosities of the Liquid Halogens," *J. Am. Chem. Soc.,* **47,** 756 (1926).
76. Trautz, V. M., and Ruf. F., "Die Reibung, Warmeleitung und Diffusion in Gasmischungen," *Annalen Physik,* **20,** 127 (1934).

2
Hypochlorination

HISTORICAL BACKGROUND

One of the first known uses of chlorine for disinfection was in the form of hypochlorite, known as chloride of lime. Snow used it in 1850 in an attempt to disinfect the Broad Street Pump water supply in London after an outbreak of cholera caused by sewage contamination. Sims Woodhead used "bleach solution" as a temporary measure to sterilize potable water distribution mains at Maidstone, Kent (England) following a typhoid outbreak in 1897.[1]

Subsequent to Scheele's discovery of the element chlorine in 1774, the compounds made from this element were used expressly for bleaching (particularly textiles) and not for sanitation. Prior to the discovery of chlorine as a bleaching agent, textiles were bleached by the cumbersome method of using the sun's rays known as crofting. The textiles to be bleached were spread on large grass fields.

In 1785, Berthollet prepared a bleaching agent by dissolving "Scheele's gas" in water. In 1789, he improved this bleaching liquid by mixing it with a solution of caustic potash (KOH). This work was carried out at a French chemical plant in Javel, a former town in France, now a part of Paris. This solution was called Javelle water and still goes by that name today.[2]

A short while later, Labarraque replaced the expensive potassium hydroxide with caustic soda obtained from soda ash. This development resulted in what was probably the first use of sodium hypochlorite as a bleach. Because of common usage, it too became known as Javelle water. It completely replaced Berthollet's expensive potassium hypochlorite, and soon crofting went out of existence. About 1810, the introduction of the Leblanc process for the manufacture of soda ash made the sodium hypochlorite process even more secure by making available more soda ash at a greatly reduced price. Other methods for producing sodium hypochlorite were developed, but the chlorination of caustic soda remains the most popular method.

In 1798, Tennant[3] patented a successful liquid bleach for the Union Alkali Company by passing chlorine through a milk of lime solution. This is known as "bleach liquor." Tennant continued his experimentation and developed a bleaching powder by passing chlorine gas over slaked lime. This process was patented in 1799 as Tennant's bleach, and had a tremendous impact upon society. Here was a solid form of chlorine bleach which could be easily transported and which needed only to be dissolved in water to be available for use. It was a boon to the textile industry, and became a cornerstone of the early chemical industry. Much of our chemical engineering heritage dates to early efforts to make soda ash, caustic soda, chlorine, and bleaching powder for use in bleaching cloth.

Freshly prepared bleaching powder contains about 36 percent available chlorine.[1] Its storage life is short, especially in warm climates. This degradation at higher temperatures demonstrated a need for a more stable compound. The addition of quicklime to bleaching powder produced what is known as "tropical bleach," which is fairly stable at tropical temperatures and contains 25 to 30 percent available chlorine. This product, developed about 1920, is still used in significant amounts in the underdeveloped areas of the world.

The first United States–produced dry calcium hypochlorite appeared on the market in 1928. This bleaching compound contains about 70 percent available chlorine, and is sold under the trade names of HTH, Perchloron, Pittchlor, and others. This product has largely replaced Tennant's bleaching powder in the United States.

Liquid bleach (sodium hypochlorite) came into widespread use about 1930 for laundry, household, and general disinfecting uses. Today it is the most widely used of all the chlorinated bleaches. More than 150 tons per day are now used in the United States alone. Its preparation is a modification of Labarraque's method, using lower residual alkali, which simplifies purification and sedimentation while maintaining a pH of about 11 for stability.

CHEMISTRY OF HYPOCHLORITES

Potable Water

The application of hypochlorite in potable water and waste treatment achieves the same result as does that of chlorine gas:

sodium hypochlorite:

$$NaOCl + H_2O \longrightarrow HOCl + Na^+(OH)^- \qquad (2\text{-}1)$$

or calcium hypochlorite:

$$Ca(OCl)_2 + 2H_2O \longrightarrow 2HOCl + Ca^{++}(OH^-)_2 \qquad (2\text{-}2)$$

The active ingredient is the hypochlorite ion (OCl^-), which hydrolyzes to form hypochlorous acid. The only difference between the reactions of the hypochlorites and chlorine gas is the side reaction of the end products. The reaction with the hypochlorites increases the hydroxyl ions by the formation of sodium hydroxide (Eq. 2-1) or calcium hydroxide (Eq. 2-2); the reaction with chlorine gas and water increases the H^+ ion concentration (see Chapter 4) by the formation of hydrochloric acid. There is reason to speculate that a chlorine gas solution at pH 2 to 3 will always be somewhat more effective than a solution of hypochlorite at pH 11 to 12 at the immediate area of the point of application, simply because there is more of the active ingredient HOCl and possibly some extremely active molecular chlorine on account of the low pH of the chlorine gas solution. It is a well-known fact that at pH 11 or 12 the HOCl is almost completely dissociated to the ineffective hypochlorite ion as follows:

$$HOCl \rightleftharpoons H^+ + OCl^- \qquad (2\text{-}3)$$

This high pH condition will exist only momentarily at the interfaces of the hypochlorite solution and the water or waste to be treated.

Wastewater

There is reason to speculate that in wastewater chlorination, where much higher dosages are used than in potable water, better germicidal efficiency would result from chlorine gas, particularly in poorly buffered wastewater. In these cases chlorine gas solution tends to lower the pH of the treated wastewater, which increases the effectiveness of the HOCl.

However Sawyer[4] reported in 1957 greater efficiency from hypochlorite solutions in the disinfection of wastewater than with chorine gas solutions. He theorizes there may be undesirable side reactions occurring in the lower pH environment of the chlorine gas solutions, robbing it of some of its disinfecting powers. These side reactions may be in the nature of the formation of organic chloramines, with little or no germicidal efficiency, and yet will appear as part of the total chlorine residual. Sawyer's laboratory work indicated that 1 lb available chlorine in the form of hypochlorite was equivalent in disinfecting power to about 1.5 lb of aqueous chlorine gas solution. This fact was not substantiated in treatment plant application. However, the most impressive findings of Sawyer's work showed significantly higher amounts of residual chlorine—as measured by amperometric titration—after 15 min contact time in the case of hypochlorite treated samples for all dosages.

EXAMPLE: At a chlorine dosage of 10 mg/liter (Providence sewage) the amperometric residual for aqueous chlorine gas solution at the end of 15 min contact was 0.80 mg/liter versus 1.40 mg/liter for sodium hypochlorite solution. Similarly, sewage from Worcester, Massachusetts, showed, for the same dosage and contact time as above, 1.75 mg/liter for aqueous chlorine gas solution versus 2.75 for sodium hypochlorite solution.

Unfortunately these observations when applied to field conditions were not of the same magnitude or consistency, but they indicate that any given wastewater is likely to demonstrate a higher chlorine demand when aqueous chlorine gas solution is used than when sodium hypochlorite is used. This becomes an economic consideration for wastewaters of poor quality and high chlorine demand. These situations call for highly efficient mixing to eliminate the possibility of undesirable side reactions. Such reactions are more prone to occur in waters of high chlorine demand in the low pH area surrounding the point of application of aqueous chlorine gas solutions. This theory has yet to be proved but should be investigated.

The next consideration is the total effect on the pH of the effluent due to chlorination. This depends upon three factors: 1) the buffering capacity of the wastewater; 2) the chlorine dosage; and 3) whether chlorine is in the form of hypochlorite or aqueous gas solution. Sawyer concluded that the variation in pH of the final sewage mixture is not a significant factor. However, as the pH drops below 7.5 the ratio of monochloramine to dichloramine shifts toward the formation of more dichloramine; therefore, the disinfection efficiency should improve because dichloramine has at least twice the germicidal efficiency of monochloramine.[5] The pH of the effluent in the 36 plants investigated by White[6] was usually between 6.9 and 7.3. Only a few plants noted any appreciable downward shift because of chlorination. One plant using a very high dose of chlorine (25–30 mg/liter), which lowers the pH to about 6.3 from 7.0, reported considerable improvement in disinfection efficiency. Such increased efficiency has been reported by others.[7] In another case (primary effluent), experiments were carried out to determine the difference between the efficiency of hypochlorite and aqueous gas solution. A hypochlorite dose of 15 mg/liter raised the pH from 7.1 to 7.4 but was lowered to 6.85 with the same dose of an aqueous chlorine gas solution. Likewise, another primary effluent in the same city, but with a high chlorine demand and a pH of 7.4 prior to chlorination, reached a pH of approximately 8.0 after a hypochlorite dose of 40 mg/liter. After a similar dose of aqueous chlorine gas solution it reached a final pH of approximately 6.5. In both of these instances the disinfection efficiency was significantly increased as the pH was lowered due to the use of the aqueous gas solution.[8]

However, of the 36 plants White observed in 1972 and an additional 20 from 1973 to 1976 the final pH was not in itself a major factor contributing to disinfection efficiency with the two exceptions mentioned above.[6]

An additional consideration must be made comparing hypochlorite versus aqueous chlorine gas solution, and that is the concentration of available chlorine in the solution. By design the chlorine concentration in the injector discharge of a conventional chlorinator is limited to 3500 mg/liter as this is the upper limit of containment of the molecular chlorine (off-gassing) inherent in these solutions. However, a hypochlorite solution of 15 percent available chlorine has an available chlorine concentration of 150,000 mg/liter or about 43 times the concentration of a conventional chlorinator discharge. This may be of great significance in the practical application of chlorine because it relates to the segregation phenomenon.[9] This phenomenon briefly states that to mix two liquids, one a chemical to be intimately mixed with the process flow, the mixing is most efficient when the chemical added is in the smallest possible amount as compared to the process flow. In other words, a chemical solution of the highest possible concentration is the optimum procedure for better mixing. To date this phenomenon has not been investigated for disinfection efficiency relative to the concentration of the chlorine solution applied.

Sodium Hypochlorite

Characteristics. This is often referred to as liquid bleach or soda bleach liquor, and is the most widely used of the hypochlorites for potable water and wastetreatment purposes. While it requires much more storage space than does the high-test calcium hypochlorite and is more costly to transport long distances, it is more easily handled and gives the least maintenance problems with pumping and metering equipment.

The preparation of sodium hypochlorite is a relatively simple procedure, involving the reaction of chlorine with caustic soda. This reaction proceeds as follows:

$$\underset{\substack{\text{caustic}\\\text{soda}}}{2\,NaOH} + \underset{\text{chlorine}}{Cl_2} \longrightarrow \underset{\substack{\text{sodium}\\\text{hypochlorite}}}{NaOCl} + \underset{\substack{\text{sodium}\\\text{chloride}}}{NaCl} + H_2O + \text{Heat} \qquad (2\text{-}4)$$

On the basis of molecular weight, 1 lb chlorine reacts with 1.128 lb caustic soda to produce 1.05 lb sodium hypochlorite and 0.83 lb sodium chloride. In practice, an excess of caustic soda is used as a stabilizer. Table 2-1 shows the quantities of chlorine and caustic soda required to make 1000 gallons of sodium hypochlorite of various strengths.[10,11]

The strength of a soda bleach solution is commonly expressed in terms of its available chlorine content as "trade percent" or "percent by volume." A more accurate expression is the actual weight percent of the available chlorine or sodium hypochlorite. The relationship between these values is:

$$\text{Trade percent (percent by volume)} = \frac{G/l \text{ available } Cl_2}{10} \qquad (2\text{-}5)$$

$$\text{Weight \% available } Cl_2 = \frac{\text{Trade percent}}{\text{Specific gravity of solution}} \qquad (2\text{-}6)$$

$$\text{Weight percent sodium hypochlorite} = \frac{G/l \text{ sodium hypochlorite}}{\text{Specific Gravity of solution} \times 10} \qquad (2\text{-}7)$$

$$\text{Weight percent sodium hypochlorite} = \frac{Gm/l \text{ available } Cl_2 \times 1.05}{\text{specific gravity of solution} \times 10} \qquad (2\text{-}8)$$

The weight in terms of the lb/gal for a particular strength is not a constant. It varies, depending on the amount of excess sodium hydroxide the manufacturer uses to promote stability.

Table 2-1 Quantities of Chlorine and Caustic Required to Make 1000 Gallons of Sodium Hypochlorite of Various Strengths[10]

Available chlorine		As NaOCl			Excess NaOH Gm/l	C/2 Req lbs.	Caustic soda required			Water required for diluting caustic (gal.)	
Gm/l	Trade %	Wt. %	Gm/l	Wt %	S.G. 70°F			lbs solid	Gals 50% liq	Solid	50% liq
10	1.0	.98	10.5	1.0	1.014	1.0	83	104	16	991	979
25	2.5	2.4	26.2	2.5	1.035	2.5	209	261	40	979	949
50	5.0	4.7	52.5	4.9	1.070	5.0	417	523	80	958	897
100	10.0	8.8	105.0	9.2	1.140	10.0	834	1045	160	915	794
120	12.0	10.3	126.0	10.8	1.168	12.0	1001	1254	192	898	753
150	15.0	12.4	157.5	13.0	1.210	15.0	1251	1568	240	873	691

Table 2-2 gives approximate values of weights suitable for calculating dosages of sodium hypochlorite. In Table 2-2, to find the available chlorine in lb per gallon, multiply the trade strength by 0.08345.

Stability of Solutions. Sodium hypochlorite solutions are vulnerable to a significant loss of available chlorine in a few days. This is a major problem with this type of chlorination system. The user must dedicate laboratory time to monitoring the decay rate in available chlorine. This serves two purposes: (1) it establishes an understanding with the supplier to arrive at the optimum cost for a given trade strength of solution and (2) it will establish the most cost-effective quantity per delivery and frequency of delivery to minimize loss of chlorine in the stored hypochlorite solution. Of several users that were questioned about this surveillance, none were monitoring the available chlorine decay rate.

The stability of hypochlorite solutions is greatly affected by heat, light, pH, and the presence of heavy metal cations. These solutions will deteriorate at various rates depending upon the following factors:

1. The higher the concentration the more rapid the deterioration.
2. The higher the temperature the faster the rate of deterioration.
3. The presence of iron, copper, nickel, and cobalt catalyzes the rate of deterioration of hypochlorite.

Iron is the worst offender. In minute quantities it causes rapid deterioration of these solutions.[12] The source of iron is usually the caustic used in the making of these solutions. Iron in quantities as low as 0.5 mg/l will cause rapid deterioration of a 15 percent solution in a few days.

Copper should be kept as low as possible; not in excess of 1 mg/l in the finished solution. It is generally present because of the copper flexible connections and brass body chlorine line valves used in the chlorine supply system. Great care must be taken *by the producer* to prevent, insofar as is possible, active corrosion of these parts. This can be done by keeping them internally free of moisture. This is a difficult task.

Table 2-2

Trade % Avail. Cl_2 $\frac{gm/l}{10}$	Approx wt. of one gallon (lb)	Available Cl_2 (lb/gal.)	Gpd req'd to dose 1 mgd to 1.0 ppm	Gpm req'd to dose 1 mgd to 1.0 ppm
1.00	8.45	0.0083	100	0.0694
5.00	8.92	0.42	20	0.0138
5.25	8.95	0.44	19	0.0132
10.0	9.50	0.83	10	0.0069
15.0	10.1	1.25	6.6	0.0046

The most stable solutions are those of low hypochlorite concentration (10%), with a pH of 11 and iron, copper, and nickel content less than 0.5 mg/l, stored in the dark at a temperature about 70°F.

Fig. 2-1 illustrates the chlorine strength decay rate over a period of 60 days for three different sources of hypochlorite under the controlled conditions as indicated. These data reflect a "best case" situation.

In 1942 the Transportation corps of the U.S. Army made a thorough investigation of sodium hypochlorite solution in an effort to determine the most effective strength over a period of 30 days. This hypochlorite was to be used with the hypochlorination systems on all troop transports used in the Pacific Theatre of World War II. The Army investigation determined that a 10 percent available chlorine solution was the strongest solution with the least decay for a 30 day period for the temperatures encountered.

Caustic (NaOH) is used in the making of these solutions purely as a stabilizing factor. A large excess of alkalinity, however, does not stabilize a hypochlorite solution any more than the slight excess given in Table 2-1. However, *if the pH drops below 11* decomposition becomes more rapid.

Another way of defining the decay of various concentrations of sodium hypochlo-

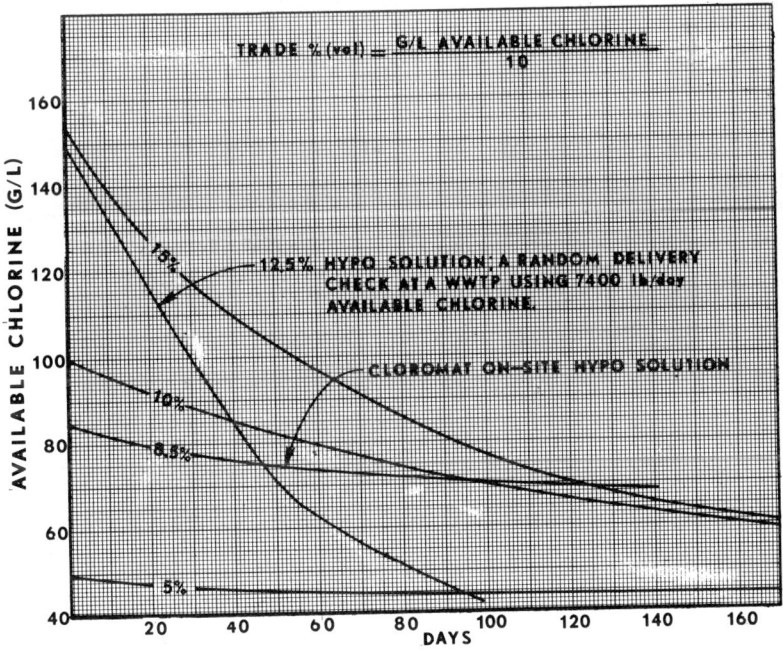

Fig. 2-1. Decay rate sodium hypochlorite solutions.

rite solutions at different temperatures is by the half-life method[13] shown in Table 2-3. When the rate of decay is nonlinear, as in hypochlorite solutions, this method is the best way of expressing "shelf life."

Hypochlorite does not lose its strength at a constant rate per day, but at a decreasing rate as it loses its strength. It is estimated that the rate of decomposition of 10 and 15 percent solutions nearly doubles with every 10°F temperature rise.

It is important to note that a 167 g/liter solution (16.7 trade percent) stored at 80°F will decay in strength 10 percent in 10 days, 20 percent in 25 days, and 30 percent in 43 days.[14]

The influence of light on a solution of sodium hypochlorite is easy to demonstrate by putting one portion in a clear container and the other in an amber container and exposing both to sunlight. The half-life of a 10–15 percent available chlorine solution will be reduced about three or four times by sunlight. For stronger solutions up to 20 percent, the result is a reduction of half-life of about six times.[13]

These solutions will freeze, but at temperatures considerably lower than the freezing point of water. (See Fig. 2-2.)

Sodium hypochlorite is commercially available in strengths of approximately 5–15 percent. If stronger solutions are attempted, solid sodium chloride is formed, presenting an additional problem of purifying; so most makers limit the strength to 15 percent to avoid this problem.

These solutions are packed in various ways: in 5-gallon carboys, in rubber-lined 55-gallon steel drums, and in some localities in tank trucks. However, the most popular method is a corrugated-type carton with four one-gallon plastic jugs (usually colored red), all of which are returnable for refill.

There is no fire hazard connected with the storage of this chemical, as there is with some of the powdered hypochlorite. It should, however, be kept away from any equipment which could be damaged by corrosion caused by spillage in the normal handling process. Corrosion from these solutions is far more devastating than that caused by an aqueous solution of chlorine made from chlorine gas. Handling of sodium hypochlorite solutions is at best a messy situation. The handling and storage area should have a concrete deck that can be flushed with copious amounts of water. Otherwise, the installation becomes impossible to maintain properly.

Table 2-3

Percent Available Chlorine		Half-life, Days		
	212°F	140°F	77°F	59°F
10.0	0.079	3.5	220	800
5.0	0.25	13.0	790	5000
2.5	0.63	28.	1800	
0.5	2.5	100.	6000	

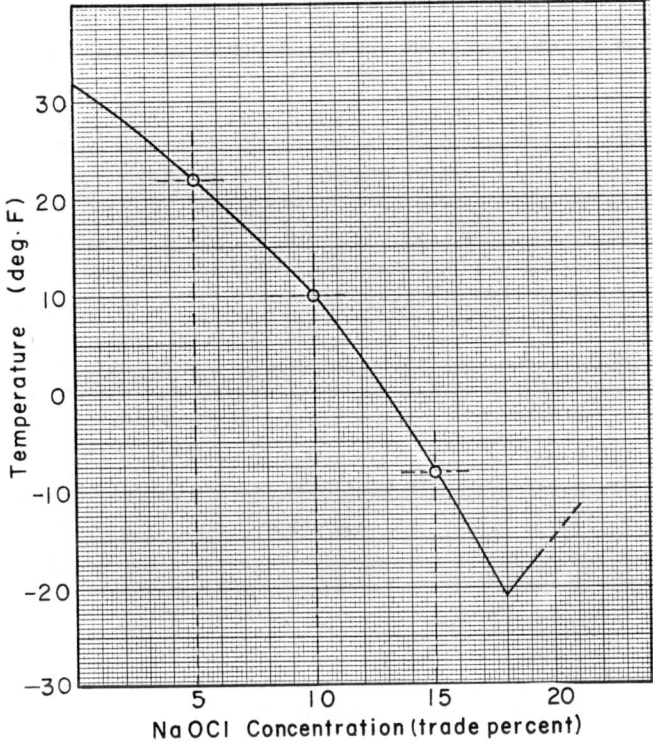

Fig. 2-2. Freezing temperatures of hypochlorite solutions (courtesy Dow Chemical U.S.A.).

Suggested Specifications. All bulk sodium hypochlorite shipments for large installations should be purchased on the basis of specifications delineating the available chlorine, iron and copper content, and excess caustic. Copper should be limited to a maximum of 0.5 mg/liter and iron to 1.0 mg/liter. Excess caustic should be limited by a pH not to exceed 11.2. If these limitations cannot be met by the suppliers, then compromises with price and chlorine strength will have to be made. Furthermore, it should be specified that the material be free of sediment and suspended solids. All shipments of bulk sodium hypochlorite should be analyzed upon receipt for available chlorine, iron, copper, and excess caustic.

Granular Calcium Hypochlorite

Manufacturing Techniques. This, the prevailing form of dry bleach in North America, is marketed at 70* percent available chlorine content. The first calcium

---

*Since Feb. 1979 there has been an industry reduction to 65 percent to reduce fire hazards.

hypochlorite marketed was German Perchloron, which consisted of hemibasic calcium hypochlorite and impurities. The available chlorine content was about 65 percent.

Mathieson Chemical Company developed a more stable calcium hypochlorite by mixing equivalent amounts of sodium hypochlorite and calcium chloride. In this process, a slurry of lime and caustic soda is chlorinated and then cooled to $-10°F$. At this temperature, the crystals which form are centrifuged to remove the mother liquor and the insoluble impurities. These crystals are then added to a slurry of chlorinated lime which contains calcium chloride in an amount equivalent to the sodium hypochlorite content of the crystals. When this mixture is warmed, a calcium hypochlorite dehydrate precipitate is formed, and the sodium chloride remains in solution. The precipitate is then filtered, and the resulting cake is granulated, sized, and dried. The result is a product containing over 70 percent available chlorine and less than 3 percent lime.[1]

The Pennwalt Corp (formerly the Pennsylvania Salt Co) developed the American version of the Perchloron process, an improvement over the German process by the same name. This process forms calcium hypochlorite dihydrate crystals and also retains a sufficient amount of the large hemibasic crystals to permit filtration of the material. This product has an available chlorine content in excess of 70 percent. This process reduces the lime content of the German method from 12–16 percent to 3–5 percent.

PPG Industries (formerly Columbia Southern) utilizes an entirely different approach. First, hypochlorous acid is made by the addition of chlorine monoxide to water. This acid solution is then neutralized with a lime slurry to produce a solution of calcium hypochlorite. The clear liquor is first spray-dried and then granulated and vacuum-dried to yield a product containing 70 percent available chlorine and 4 to 6 percent lime. All of these products contain some calcium carbonate and other insolubles which cause precipitates when dissolved in tap water. The amount of precipitates formed depends on the dilution factor and total hardness of the diluting water.

Storage of Granular Calcium Hypochlorite. The storage of granular calcium hypochlorite, whether the 65 percent or the 35 percent available chlorine, is a major safety consideration. It should never be stored where it is subject to heating or allowed to contact any organic material of an easily oxidized nature. The decomposition of calcium hypochlorite is exothermic, and will proceed rapidly if any part of the material is heated to 350°F. Many fires of spontaneous origin have been caused by improperly stored calcium hypochlorite. The decomposition releases oxygen and chlorine monoxide.

Under normal conditions, it will lose 3 to 5 percent available chlorine per year. Contact with water or moisture from the atmosphere induces the decomposition. The oxygen released in contact with a corrugated cardboard carton can result in spontaneous combustion of the carton, depending on its proximity to the oxygen being released by the decomposing hypochlorite.

Lithium Hypochlorite

Lithium, an excellent granular hypochlorite, is prepared by mixing a strong solution of lithium chloride with a strong solution of sodium hypochlorite. A large portion of sodium chloride precipitates, and the 30 to 35 percent solution of lithium hypochlorite is evaporated and drum-dried. The finished product is white, free-flowing, granular, and dust-free.

Lithium Corporation of America, manufacturers of this product, lists the following ingredients:

	Percent by Wt.
Available chlorine	35
LiOCl	30
NaCl	34
Na_2SO_4 & K_2SO_4	20
LiCl	3
$LiClO_3$	3
LiOH	1
Li_2CO_3	2
H_2O	7

Lithium hypochlorite is readily soluble in water, and has the distinct advantage over the calcium hypochlorite solutions of being clear when prepared, thereby eliminating the need for the settling and discarding of sludge. When stored properly in closed containers at 75°F and 75 percent relative humidity for four months, it retains more than nine-tenths of its available chlorine.[15] It does not affect the alkalinity or pH as much as do the other hypochlorites. The pH of a 100 ppm solution of available chlorine at 25°C is only 10.0.

This product was introduced in 1964 as a laundry bleach and dishwashing compound because of its excellent solubility characteristics. Then it was introduced to the swimming pool industry in competition with all the other disinfectants.[16] While the cost of this material is somewhat greater than that of the other hypochlorites, on the basis of available chlorine, it is gaining in popularity because of the ease with which it can be handled. It has the same disinfecting characteristics as any other chlorine compound which produces HOCl as an hydrolysis product.

The same safety precautions must be followed for the storage of granular lithium hypochlorite as those for granular calcium hypochlorite.

Liquid Calcium Hypochlorite

Calcium hypochlorite liquor can be made directly from the reaction of chlorine and hydrated lime in solution. However, the maximum strength of these liquors is limited to about 35 percent available chlorine. These solutions are made by passing chlorine into a milk of lime suspension according to the equation:

$$2Ca(OH)_2 + 2Cl_2 \longrightarrow Ca(OCl)_2 + CaCl_2 + 2H_2O \quad (2\text{-}9)$$
$$148.192 \quad 148.828 \quad\quad 142.994 \quad 110.994 \quad 36.032$$

Theoretically, 1.043 lb actual Ca(OH)$_2$ or 0.791 lb actual CaO will react with 1 lb chlorine to produce 1.008 lb calcium hypochlorite and 0.782 lb calcium chloride. In actual practice, it is usually assumed that the hydrated lime contains 95 percent Ca(OH)$_2$ and that the quicklime contains 95 percent CaO. On this basis, the equivalents of the chemical reaction, which do not include the excess lime required for stability are: 1.0 lb Cl$_2$ + 1.10 lb commercial hydrated lime or 1.0 lb Cl$_2$ + 0.833 lb commercial quicklime will produce 1.008 lb calcium hypochlorite and 0.78 lb calcium chloride.

In practice, an excess of from 10 to 20 lb Ca(OH)$_2$ (hydrated lime) or 7 to 14 lb CaO (quicklime) should be used per 1000 gallons of calcium hypochlorite to insure proper stability by keeping the pH at 11.2 or above.

The following formulae apply to making 1000 gallons of calcium hypochlorite of various strengths:[11]

$$\text{lb Cl}_2 = \text{g/l available chlorine} \times 8.34 \tag{2-10}$$

$$\text{lb Ca(OH)}_2 = \text{lb Cl}_2 \times \frac{1.043 \times 100}{\% \text{ Ca(OH)}_2 \text{ in lime}} + 10 \text{ to } 20 \text{ lb} \tag{2-11}$$

$$\text{lb CaO} = \text{lb Cl}_2 \times \frac{0.791 \times 100}{\% \text{ CaO in quicklime}} + 7 \text{ to } 14 \text{ lb} \tag{2-12}$$

Table 2-4 shows the theoretical minimum and maximum amounts of lime required to make liquid calcium hypochlorite:

Table 2-4 Pounds Required per 1000 gallons Ca(OCl)$_2$

Avail. Cl$_2$ gm/l	lb Cl$_2$	Hydrated Lime Ca(OH)$_2$ 95%			Quicklime CaO 95%		
		Theor.	Min.	Max.	Theor.	Min.	Max.
10	83.4	91.7	101.0	110.3	69.4	76.4	83.5
20	166.8	183.5	192.7	202.0	138.8	145.9	152.9
30	250.2	275.2	284.5	193.7	208.2	215.3	222.3
40	333.6	367.0	376.2	385.5	277.7	284.7	291.7
50	417.0	458.7	468.0	477.2	347.1	354.1	361.1
60	500.4	550.4	559.7	568.9	416.5	423.5	430.5
70	583.8	642.2	651.4	660.7	485.9	492.9	499.9

The preparation of liquid calcium hypochlorite is not an easy process. First, the quality of lime available must be considered. Lime with a low magnesium or aluminum oxide content is desirable since these impurities cause excessive sludge and poor settling properties. The problem of getting the hydrated lime to settle out of solution is the most difficult to overcome, and depends upon both the lime and the arrangement of equipment.

CHOICE OF EQUIPMENT

Introduction

There is no engineering limitation to the quantity of hypochlorite that can be metered in potable water and wastewater treatment. The hypochlorites are universally applied as aqueous solutions, although there is no reason why the granular material cannot be metered through gravimetric or volumetric dry chemical feeders. The latter method does pose monumental problems of corrosion if the material is not kept completely free of moisture. This also raises the question of possible spontaneous combustion of granular hypochlorites. At the moment the recommendations are exclusively for metering and controlling the *solutions* of sodium, calcium, or lithium hypochlorite, rather than the granular forms of calcium or lithium hypochlorites.

Feeding equipment for the hypochlorites is in direct competition with chlorine gas feeding equipment. The latter is always the chemical of choice because it is compact, flexible, easy to maintain and operate, and adapts well to automation. Furthermore, chlorine gas is so much less expensive than hypochlorite that the use of the latter has been limited to the situations below the range of control possible with chlorine gas equipment. Recently, however, considerable attention has been given to the use of hypochlorite on large installations strictly on the basis of safety. Proponents claim that handling chlorine in tank car quantities is much too hazardous in populated areas. But even when contracts have been made for the furnishing of large quantities of sodium hypochlorite, the cost to the user is as much as double or triple the cost of using chlorine gas.

Therefore the use of hypochlorite for potable water and wastewater treatment is usually confined to the very small water supply or the very large installation in which the latter user is primarily concerned with the hazard of a chlorine gas accident.

Small Water Supplies

The first question to be decided in selecting chlorination equipment for small water supplies is whether to use chlorine gas in cylinders or hypochlorite. Gas chlorinators do have practical limitations as to minimum feed rates. Chlorinators that utilize rotameters for indicating chlorine feed rate can be obtained with ranges as low as 0.1–1.2 lb/24 hr. However, the smallest metering orifice consistent with good accuracy and a minimum amount of maintenance is from 0.5 to 10.0 lb/24 hr. The one exception to this is a chlorinator that uses a bubbling type meter still manufactured by Wallace & Tiernan, Division of Pennwalt Corporation, which has a practical operating range as low as 0.1 lb/24 hr.

A good rule of thumb in selecting hypochlorite or gas feeding equipment is based upon the daily usage of the chemical. This is related to the economics of chemical cost, storage space, and maintenance of equipment. A hypochlorinator installation usually requires more attention, owing to the necessity of refilling solution containers with chemical. Generally the dividing point is close to 3 lb/day.

If the usage is below this figure, hypochlorite equipment should be seriously considered. At current prices, one pound of chlorine gas in 150-lb cylinders costs approximately fifteen cents; the equivalent in hypochlorite purchased in 4-gallon lots is between fifty and seventy-five cents per pound of chlorine.

If three pounds of available chlorine consumption is accepted as the upper limit for hypochlorite feeding equipment, let us relate this to water and wastewater to be treated.

If a dose of one ppm for water treatment is considered adequate, then 3 lb/day will treat approximately 360,000 gpd (1 ppm = 8.33 lb/mg; $\frac{3.0}{8.33} = 0.36$ mg).

However, there is a growing trend to size chlorination equipment for application bordering on superchlorination, primarily to be certain that possible virus contamination may be dealt with. Therefore a design criterion of 2 ppm might be more realistic. This would decrease the amount of water treated by 3 lb/day chlorine from 360,000 gal to only 180,000 gpd or 125 gpm. On this basis, it is conceivable to use hypochlorite feeding equipment on supplies utilizing an intermittent pumping arrangement.

Pumped Supplies. For example, supposing such a system utilized a 250 gpm pump but operated only twelve hours per day. This would be a candidate for a hypochlorinator, but it would also be in competition with a small gas chlorinator. If it were a deep well, and the chlorine could not be applied down the well (for construction reasons), then the hypochlorinator would have the advantage, because the gas chlorinator would require a booster pump (to inject chlorine solution against the pressure in the main) as an additional piece of equipment to maintain and operate. In general then, it can be considered that up to about 200 gpm for a pumped supply is the practical limit for hypochlorinators.

Fig. 2-3 shows schematically the installation of a hypochlorinator on a pumped surface supply. Situations like this raise a question about the point of application. If the pump suction is accessible, always apply the chlorine in the open body of water adjacent to the foot valve on the pump suction. This is preferable to the discharge side of the pump. The latter point of application, because of the static head involved, results in more wear and tear on the reciprocating parts of the hypochlorinator.

Never tap the pump suction line for the application of chemical. This becomes a source for an air leak, which will impair the efficiency of the water pump.

If the hypochlorite solution discharge line is much longer than fifteen feet to the point of application at the pump suction, it is well to consider the advisability of installing a back pressure valve at the point of application. This keeps the solution line from emptying each time the pump shuts down.

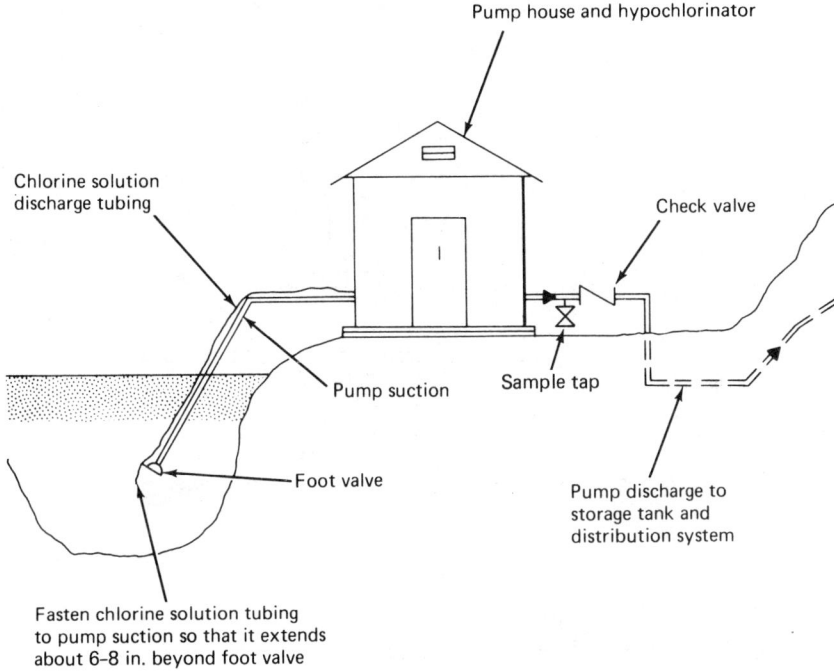

Fig. 2-3. Hypochlorinator installation for a pumped supply (Hypochlorinator is an electric-drive diaphragm pump with manual dosage control arranged to start and stop automatically both water pumps.

Fig. 2-4 shows the installation of a hypochlorinator with a well pump. For simplicity, the point of application is made on the discharge side of the pump. In some cases where submersible pumps are used, this may not be possible, and so it becomes necessary to run the solution line down the well. When this alternative point of application is used, always determine as accurately as possible the location of the pump bowls and then terminate the solution line ten feet below this point.

Electrically driven hypochlorinators can be arranged to provide continuous automatic proportional control, as shown in Fig. 2-5. In this case, there are three pumps, each of a different capacity, but discharging into a common line. A propeller type flow meter can be utilized to transmit a milliamp signal in proportion to the flow. This signal is then used to operate an SCR unit, which drives a DC motor at variable speeds in accordance with the milliamp signal. While this is presented here to illustrate the possibilities, such a situation probably would not occur in actual practice. There are very few small water systems that find it necessary to resort to multiple pumping systems.

Gravity Systems. These systems usually present the designer with the most difficult problems in selecting a proper chlorination system. There is usually no

78 HANDBOOK OF CHLORINATION

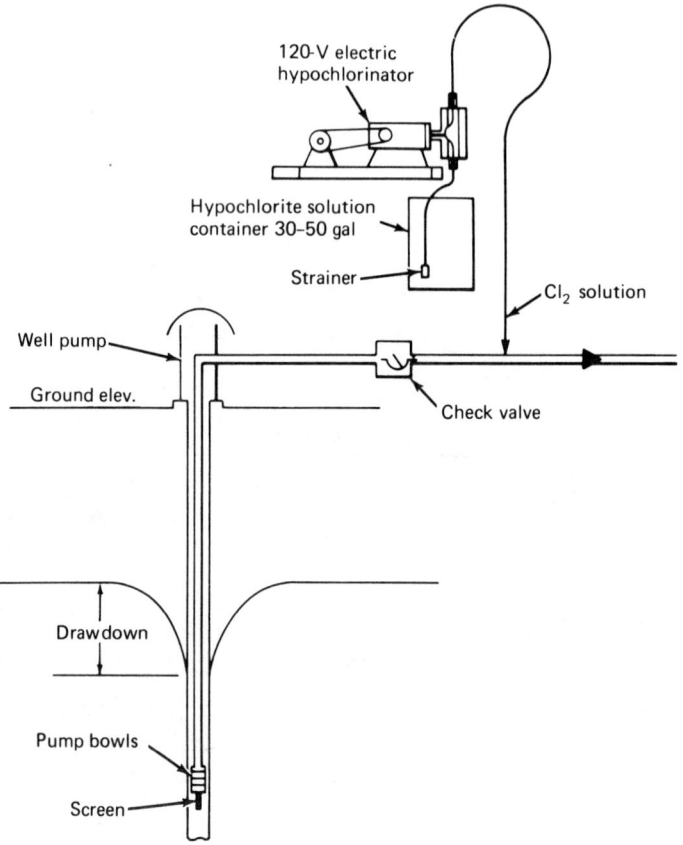

Fig. 2-4. Hypochlorinator installation with a well pump (Hypochlorinator is arranged to start and stop with well pump).

electric power available. The size of the distribution and transmission piping is usually designed for fire flows that are far in excess of the normal domestic consumption. For these gravity systems, particularly those without electric power, the equipment of choice is a water-operated (hydraulic ram) hypochlorinator that can be paced by any of several types of water meters.

The meters to be selected from are as follows:

1. The disk type meter is a positive displacement type, used especially where chlorination of low minimum flows is required. As a group, these meters have the highest head loss. (See Fig. 2-6.) For sizes under two inches, this type meter is used exclusively.
2. The crest or current type meter is for heavy flows where low minimum flows

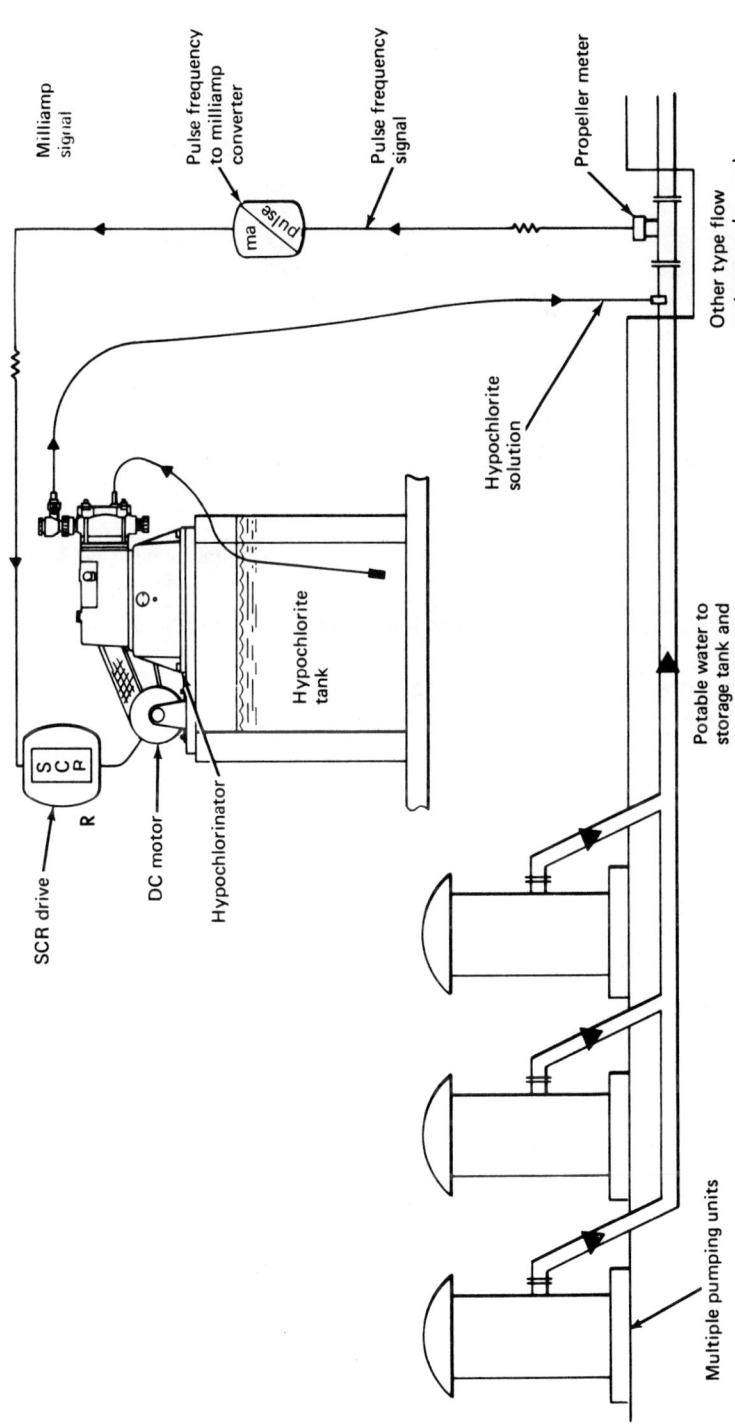

Fig. 2-5. Electric-drive hypochlorinator arranged for automatic flow-proportional control (The DC motor provides hypochlorinator with variable speed drive from SCR unit.)

80 HANDBOOK OF CHLORINATION

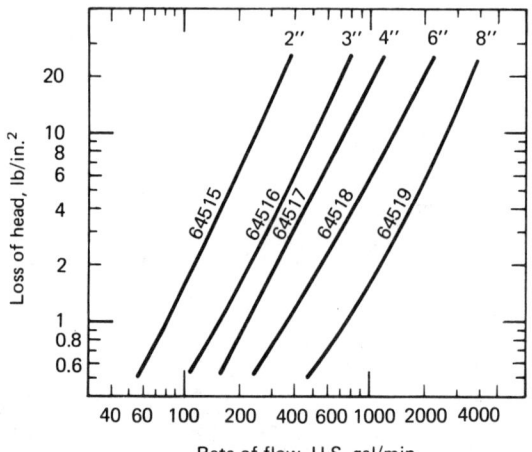

Fig. 2-6. Loss of head through crest meters (courtesy Neptune Meter Company).

are of no concern. These have much lower head losses than do the disk type meters. (See Fig. 2-7.)

3. The propeller type meter is universally used in main line installations where head loss is of concern and where low minimum flows are of no concern. These meters exhibit the lowest head loss characteristics. When low minimum flows are of concern, the compound type unit must be used.

4. The compound propeller meter consists of a full-size main line meter followed by a smaller meter capable of handling the low flows plus a compounding

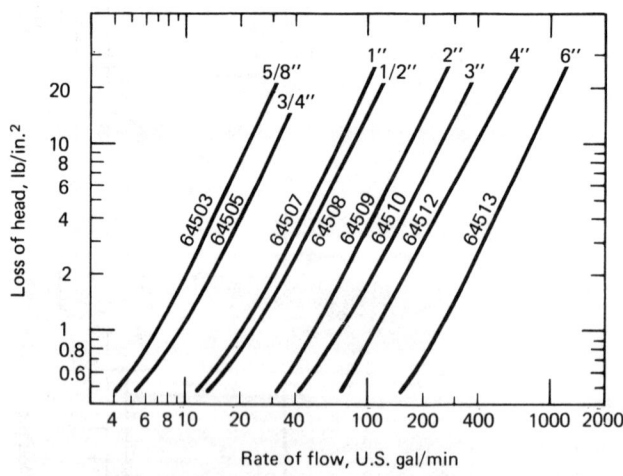

Fig. 2-7. Loss of head through disk meters (courtesy Neptune Meter Company).

valve that routes the flow through the appropriate meter depending on the flow. Both meters drive through ratchet clutches to a common totalizer shaft which is thus operated in step with the faster turning meter. The flow through either one or the other is registered, but the two are never duplicated. Only one hypochlorinator is required with a Sparling compound meter. Other compound meters consist of a disk type meter for low flows, a compounding valve which remains closed until a certain differential is reached and a crest type meter for large flows. Each meter operates independently and therefore requires two hypochlorinator units. One is controlled from the disk section, the other from the crest section.

There are three types of chlorination systems using water meters for direct proportional pacing. In the systems to be described, the water meter rotates a cam, operating a flapper valve, which, in turn, actuates the hydraulic ram portion of the hypochlorinator. Activation of this flapper by the meter requires a negligible amount of torque, so that no extra load is placed on the water meter which might affect its accuracy or reliability. The operation of this type of hypochlorinator is based on the balanced diaphragm principle, utilizing water power to actuate the forward (or pumping) stroke. The suction or return stroke is by a spring mechanism that has been compressed during the forward or pumping stroke. By utilizing the balanced diaphragm principle, these units can operate on water pressures as low as 10 psi.

It is better to rely on a minimum water pressure of 20 psi. The water motor uses about 1 to 1.25 gpm continuously. Of the total elapsed time for each stroking cycle, about 80 percent is on the pumping stroke. The maximum number of strokes is usually limited to sixteen per minute.

The three ways that the meter-paced hypochlorinator can be arranged are as follows (electric power not required):

1. Main-Line Meter. This is the preferred type of installation, where all the water to be treated passes through the meter. Fig. 2-8 shows a typical installation. The big decision for the designer in this case is the proper selection of the meter as to size and type.

Condition A:
When maximum flow is not more than 200 gpm and a head loss of 10 psi through the meter can be tolerated, always choose disk type meters.

Condition B:
Where maximum flows are in excess of 200 gpm, and 10 psi head loss can be tolerated, choose crest type meters.

Condition C:
Where maximum flows are 100 gpm or greater, and only 2 or 3 psi head loss can be tolerated, and the measuring of minimum flows is not critical, choose a propeller type meter.

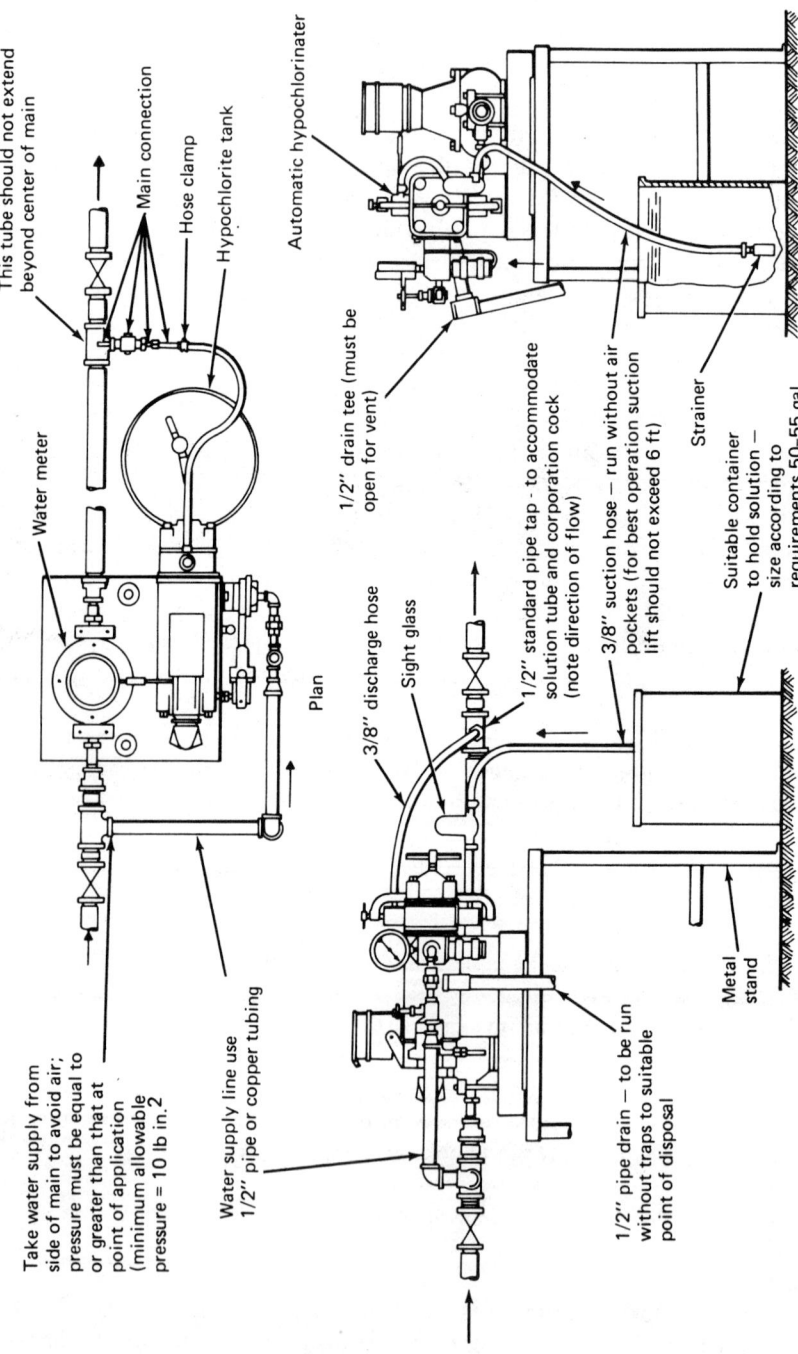

Fig. 2-8. Installation of an automatic hypochlorinator paced by a main-line meter (courtesy Wallace & Tiernan Div. Pennwalt Corporation).

Condition D:
The same as Condition C except that, if low flow measurement is critical, choose a compound type propeller meter.

2. Bypass Installation. Fig. 2-9 shows such an installation, utilizing a fixed orifice plate in the main line. This orifice creates an artificial pressure drop, forcing a certain proportion of water through the water meter in the bypass. The idea here is that the ratio of the proportion of water going through the main line to that going through the meter remains constant, thereby giving accurate chlorination to the total flow. This ratio, however, does not remain fixed throughout the range of flow through the meter, but the deviation results in a negligible error in the overall chlorination. The range of chlorination is based on the pressure drop through the orifice for a given flow. The meter size is always limited to either ⅝", ¾",

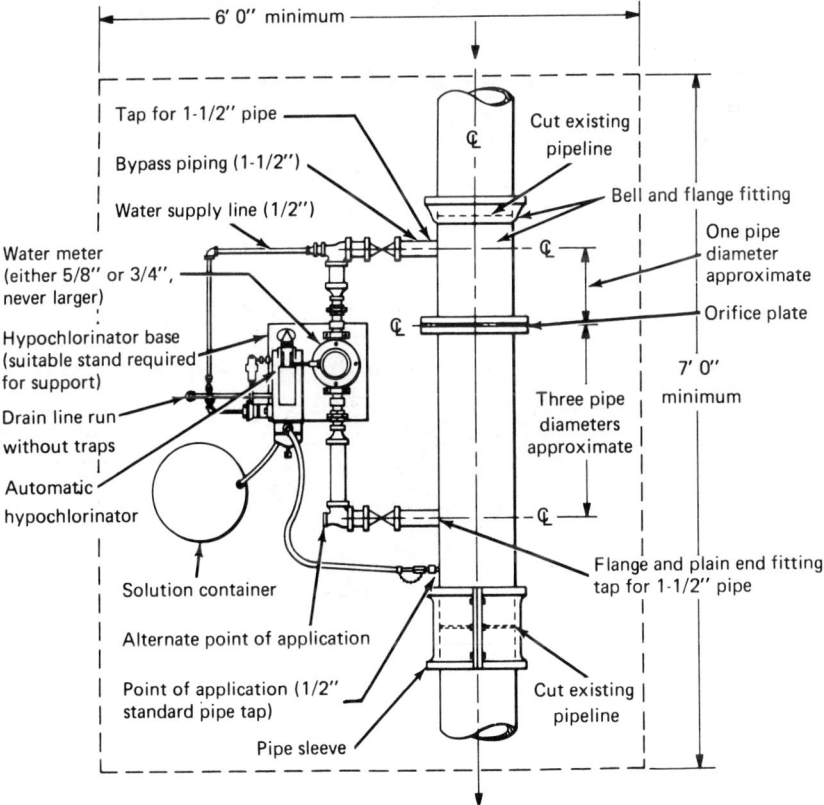

Fig. 2-9. Installation of an automatic hypochlorinator around orifice plate (courtesy Wallace and Tiernan Div. Pennwalt Corporation).

or 1". The bypass piping should always be twice the size of the meter, to keep pressure losses in this piping as low as possible.

In order to get the widest range of chlorination, it is best to consider only a ⅝" disk meter.

EXAMPLE 1. EFFECT OF METER SIZE. If the maximum main line flow is to be 300 gpm and the orifice plate is to be designed to give a pressure drop of 3 psi at this flow, and assuming a 1 psi drop in the bypass piping, this would allow a 2 psi drop through the water meter. In Fig. 2-7 a 2 psi drop would allow 30 gpm to pass a 1" meter, 14 gpm to pass a ¾" meter, and 10 gpm through a ⅝" meter. *The range of chlorination is going to be only as effective as the range of the flows through the water meters.* These are as follows:

$$\begin{array}{ll} ⅝" & 1\text{--}24 \text{ gpm} \\ ¾" & 2\text{--}42 \text{ gpm} \\ 1" & 3\text{--}63 \text{ gpm} \end{array}$$

In this particular case, because of the pressure drop chosen, the range of chlorination will be limited as follows:

1" meter	30 gpm down to 3 or 10–1
¾" meter	14 gpm down to 2 or 7–1
⅝" meter	10 gpm down to 1 or 10–1

Therefore, if a 1" meter is chosen, the ratio between the main line at maximum flow of 300 gpm will be 300 : 30 gpm, for a total of 330 gpm. If this ratio (10 : 1) can be assumed to stay fixed at the low flows, then the lowest flow that can be properly chlorinated will be 33 gpm.

If a ⅝" meter is chosen, the range will be the same as for the 1" meter, because the ⅝" meter shows a 1–10 gpm range. However, at minimum flow, the meter is passing only 1 gpm for a total of $\frac{300}{10} = 30 + 1$, or 31 gpm as the lowest flow that could be properly chlorinated. While there is little difference between the minimum flow of 33 gpm with the 1" meter versus 31 gpm for a ⅝" meter, the point to be made is that it is never necessary to use larger than a ⅝" meter in a fixed pressure drop bypass arrangement.

EXAMPLE 2. EFFECT OF PRESSURE DROP. Assume that the pressure drop across the orifice was arbitrarily chosen to be 10 psi. If we assume 2.5 psi drop through the bypass piping, this leaves 7.5 psi drop across the water meter. In Fig. 2-7 a 1" meter will pass 61 gpm, and a ⅝" meter 19 gpm at this pressure drop. Therefore the range of chlorination will be 3–61 gpm or approximately 20 : 1 with a 1" meter, 1–19 gpm or 19 : 1 with a ⅝" meter. Since the ⅝" meter can control down to 1 gpm, whereas the 1" meter can only control down to 3 gpm, it is always wisest to choose a 5/8" meter for a bypass installation.

The next most important decision is the selection of an allowable pressure drop. A 10 psi drop will give a 20–1 operating range for a 5/8" meter with ease, therefore it is never necessary to use a larger meter. Nothing less than a 3 to 5 psi drop across the orifice plate is practical.

3. Weighted Check Valve. When it becomes necessary to use any type of bypass installation, this is the preferred method. Fig. 2-10 illustrates such an installation. This arrangement is predicated on the assumption that, for all normal flows for domestic and other purposes, the check valve is closed and all the water treated passes through the meter. When an abnormal or fire flow occurs, the demand becomes so great that the check valve opens. In this instance, the hypochlorinator is operating at its maximum rate, so that underchlorination will occur during these periods. The weighted check valve is set by means of the difference in pressure of the two gauges when a fire flow in the system is simulated by opening the desired number of hydrants. The weighted check valve should be limited to a 5 to 7 psi differential across the valve in sizes 6" or smaller. In larger size lines, this differential should be kept down to about 3 psi.

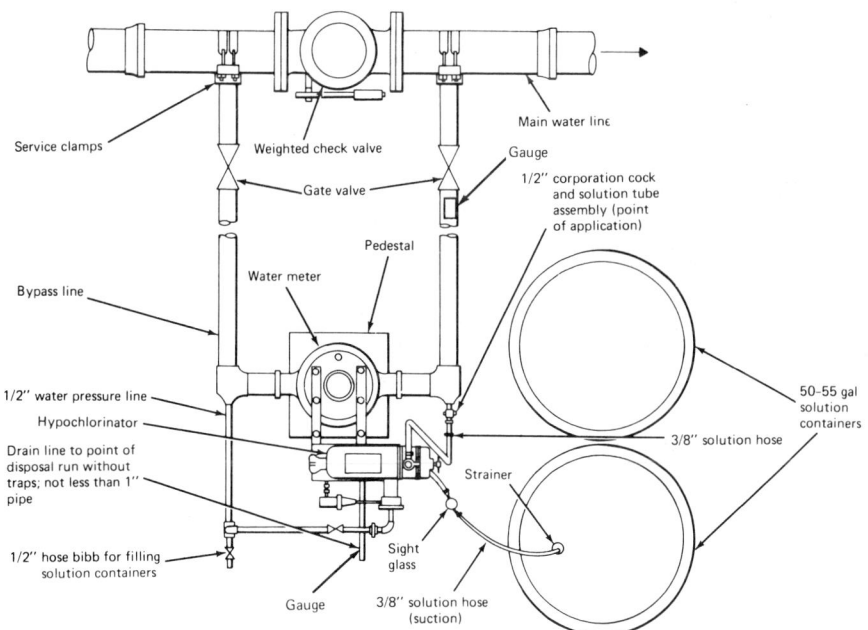

Fig. 2-10. Installation of an automatic hypochlorinator in by-pass around weighted check-valve (courtesy Wallace and Tiernan Div. Pennwalt Corporation).

EXAMPLE. The water supply system has a 6" transmission line to a distribution system. The design fire flow is three streams at 250 gpm, for a total of 750 gpm. The normal domestic flow is expected to range from 10 to 150 gpm. A 2" disk meter in the bypass would be suitable for this range, but would require a 7 psi drop across the check valve to pass 150 gpm. Use 4" piping for the bypass and simulate a 150 gpm flow through the meter. Set the weight on the check valve to keep it closed at this flow. Note the differential pressure between the two gauges. Now increase the flow by opening up additional outlets in the system, and adjust the weight so that the check valve begins to open when more than 150 gpm is going through the meter.

Designers' Check List. The following must be accounted for in order to provide an adequate chlorination facility:

1. Water pressure at point of installation
2. Point of application (keep solution discharge line as short as possible)
3. Proper contact time between point of chlorination and first consumer (minimum 10 min)
4. Chlorine residual test tap from 10 to 50 diameters (main line) downstream from point of application
5. Suitable drain facilities
6. Hose bib for filling solution containers
7. Range of normal flows
8. Allowance for fire flow
9. On basis of 7, selection of meter size
10. On basis of 8, selection between bypass and main line types of installation
11. If it is to be a bypass type, keep the bypass piping as short as possible. In other words, keep the chlorination unit as close to the main transmission line as is feasible.

Gravity Systems with Power Available. Whenever power is available, it is often possible to use a system that treats a constant flow either steadily or intermittently. The difference here is that an electric motor-driven hypochlorinator may be utilized, eliminating the need of a pacing meter. This simplifies the installation.

Fig. 2-11 illustrates a typical gravity supply that originates from a stream or spring and discharges to a holding tank. The flow through this tank is constant on a daily or weekly basis, varying only seasonally. The entire flow should be treated, as the chemical cost to treat the overflow is negligible—about two dollars or three dollars per million gallons. The hypochlorinator is simply plugged into 120 V single-phase power and allowed to run continuously. The chlorine residual should be measured at the tap shown and not at the overflow. The water in the

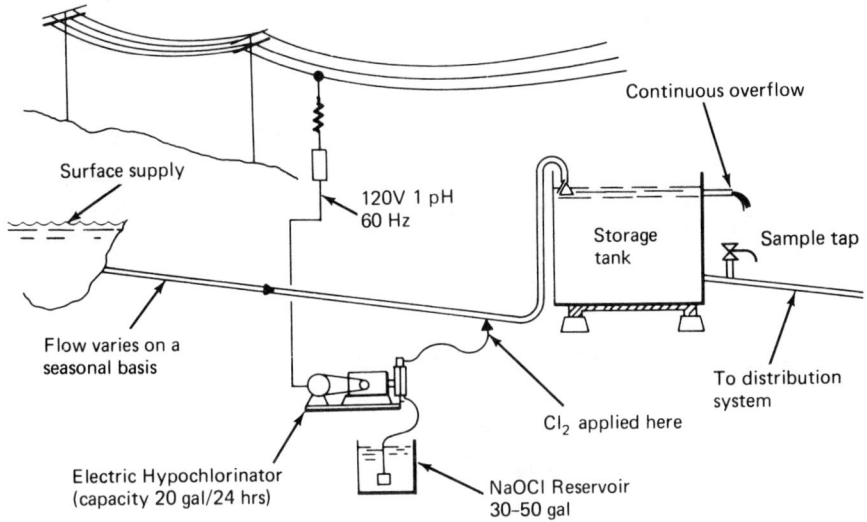

Fig. 2-11. Electric hypochlorinator in gravity supply. (Entire flow, including tank overflow is chlorinated).

overflow line does not have the benefit of any contact time, since it is short-circuited directly across the tank from the inlet.

If the client does not want to treat any overflow, an arrangement such as that shown in Fig. 2-12 can be used. Here the water entering the tank is controlled by an inlet valve that operates from a fully closed position to a fully open position without any throttling action in between. This valve (Cla-Val Company) can be arranged with an electrical contact so that when it opens it will energize a relay that starts the electric drive on the hypochlorinator.

Assuming the tanks are open, the residual at the sampling point should be ample to allow for any aftercontamination in the open tank. If the chlorine residual loss due to sunlight on the open tank is appreciable, then the tank should be covered. Most health authorities recommend closed tanks to avoid contamination by birds, rodents, and other wildlife.

In the general area of water treatment, other chemicals are often required, such as hexametaphosphate for sequestering iron or manganese deposition, alum for filters, and soda ash for pH and alkalinity correction. These can also be metered by the same type of equipment used to meter hypochlorite.

Gravity Filter Systems. In those cases where the filtration system rides on the line with no clearwell facilities, the chlorination should precede the filtration to keep the filter free from slime and organic deposits. The hypochlorinator in this case would have to be the water-operated type paced by a meter, or, if power

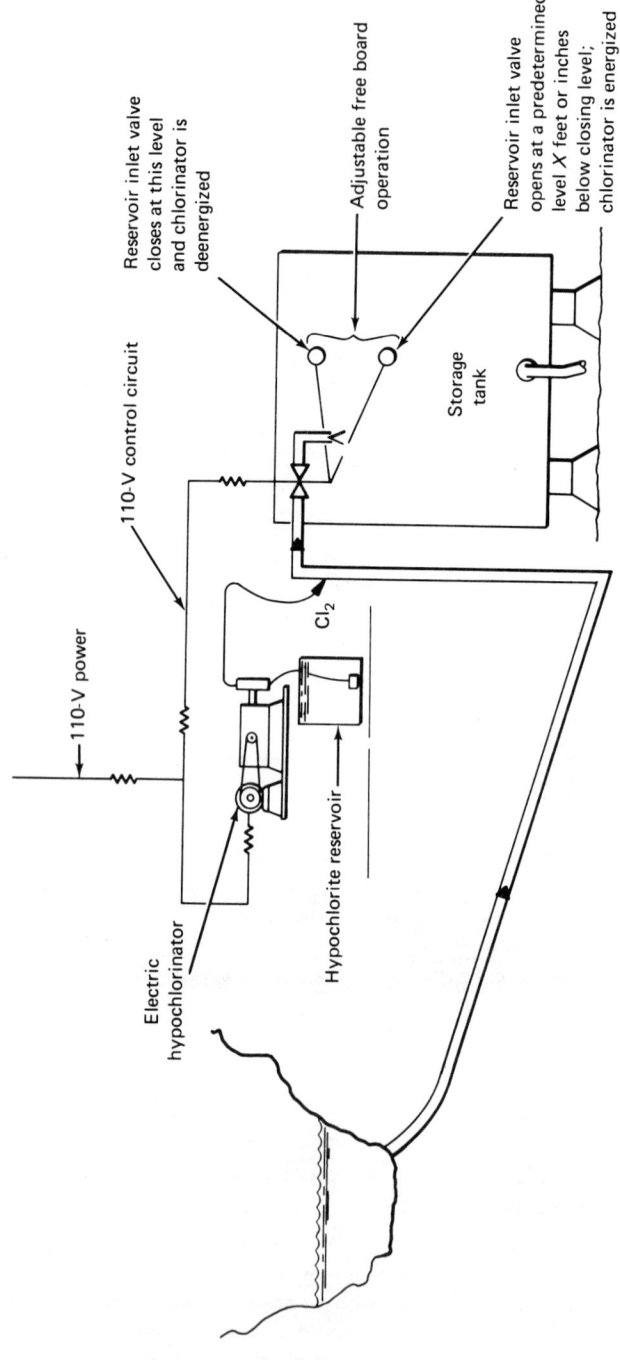

Fig. 2-12. Electric hypochlorinator in gravity supply. This system is unique in that the chlorinator unit is in operation only when the reservoir inlet valve is open. This valve opening is always full open, so the chlorinator always treats the same flow at each opening of the reservoir valve.

is available, the electric type paced electrically from the meter through a variable speed drive SCR unit which requires a 10 to 50 mA control signal from the water meter. Pacing an electric drive hypochlorinator from a water meter on a pulse frequency start-stop basis is not recommended, as it does not provide continuous chlorination proportional to flow. The SCR drive unit does provide proper chlorination for these conditions.

If the filtration system utilizes clear well storage facilities, then it is possible to arrange the clearwell to fill and draw on an intermittent basis, so that by means of an altitude type valve with an electrical contact, the electric hypochlorinator may be started and stopped automatically.

This type of system creates a pressure break by reason of the operation of the clearwell, and so it is limited to those cases where a loss of head due to the pressure break is of no significance.

SAMPLE CALCULATIONS. Any hypochlorinator installation necessitates calculating the amount of hypochlorite to be added to the solution container. Usually the capacity of these containers is thirty or fifty-five gallons. The amount of hypochlorite to be added is only approximate. Final adjustment of dosage is done on the hypochlorinator.

Case I. Let us assume that a start-and-stop hypochlorinator operates with a 50 gpm well pump. Also assume that a 1 ppm dosage is adequate, that the capacity of the solution container is thirty gallons, and that the hypochlorinator can pump twenty gallons solution per day.

Step 1. convert 50 gpm into mgd = .072 mgd

1 ppm = 8.33 lb chlorine per mg = 8.33 × .072 = 0.599 lb/day *rate*

Let us further assume that the hypochlorinator will operate at 50 percent capacity, or 10 gpd, if running continuously. Then there must be approximately 0.6 lb chlorine in every ten gallons of solution. So to each full 30-gallon container, 0.6 × 3 = 1.8 lb available chlorine must be added. If 10 percent sodium hypochlorite is used, this will yield approximately 0.83 lb chlorine per gallon. (See Table 2-2.) So adding 2 gallons (1.66 lb available chlorine) would be close enough. The hypochlorinator could then be adjusted to give the desired residual.

Case II. Let us assume that an automatic hypochlorinator is paced by a 1½" disk type meter in a main-line installation. The hypochlorinator is geared to give a maximum of twelve strokes at 85 gpm, and a 1.0 ppm dosage is desired at this rate. The hypochlorinator instruction book shows that at twelve strokes per minute the maximum pumping rate is 60 gpd hypochlorite solution.

85 gpm × 1440 min/day = 122400 gpd = 0.1224 mgd

1 ppm = 8.33 × 0.1224 = 1.019 lb chlorine per day

This means that there must be an equivalent of 1.019 lb available chlorine in every sixty gallons of solution. Allowing some freeboard for a fifty-five-gallon container—say fifty gallons of available solution—there must be added to the container $1.019 \times 50 = 0.85$ lb Cl_2. If 10 percent sodium hypochlorite is used, the addition of one gallon to every container full (50 gal) will give approximately 0.85 lb chlorine. The final adjustment can be made on the stroke adjustment of the hypochlorinator.

Shipboard Installations

Historical Background. During World War II, it was the consensus that the quality of the water in the Southwest Pacific was questionable at all times. Therefore it soon became mandatory that all troop ships be equipped with chlorination facilities for any water used by the crew and troops. Since amoebic dysentery was a potential hazard in this area, it was decided that all installations be capable of dosing 10 ppm chlorine followed by forty-five minutes detention, after which it was dechlorinated by means of an activated carbon filter.[17] Part of the chlorinated water was bypassed, so that water going to the ship's system carried about 0.3–0.5 O-T residual chlorine.

In the light of S. L. Chang's work at Harvard in 1941 and 1944,[18,19] it develops that these requirements were adequate for the destruction of cysts, providing of course that the original chlorine demand of the raw water was not excessive.

This method of treatment provided not only a safe water but one that was extremely palatable. Passage through the activated carbon filter enhanced the taste of the treated water considerably.

Fig. 2-13 illustrates a typical shipboard chlorinator installation. Note that there are dual hypochlorite solution tanks with a capacity of approximately fifty-five gallons. These are rubber-lined, with a compression fit inspection hatch with handhold and an integral mounting stand for the hypochlorinator. The tanks are fitted with special supports that are secured to the deck. The cover is bolted to the tank so that there is no possibility of any spillage. The hypochlorinator is usually located between the retention tank and carbon filter. The retention tank is constructed with special end-around baffles to prevent any short-circuiting. The hypochlorinators were paced by either 1½" or 2" meters, depending upon the size of the troop ship (Class C1 or C2). These installations were always main-line systems in which all the water treated passed through the meter. The maximum pressure drop through the system was by design 10 psi, which limited the maximum amount of water which could be chlorinated by a 1½" meter unit to 85 gpm and that of a 2" meter to 175 gpm.

In the early part of the Pacific war, shipboard chlorinator installations were confined to troop and supply ships. It was assumed that the fighting ships (battle cruisers, aircraft carriers, destroyers) had no water supply problems because they were able to fulfill all their water requirements by means of the shipboard evapora-

Fig. 2-13. Installation of automatic hypochlorinator aboard ship.

tors. This turned out to be an erroneous assumption. During the invasion of Tarawa, most of the Marine assault forces were suffering from dysentery when they hit the beach. Investigation into the cause of this epidemic indicated that it was probably contaminated drinking water. Just prior to the invasion (which witnessed the largest amassing of ships in one place up to that time—approximately six hundred vessels), it was concluded that the seawater in the vicinity of the vessels became grossly polluted from their own waste discharges. Even though the shipboard evaporators were functioning properly, it was discovered that at the end of each evaporation cycle a small amount of seawater siphoned over into the distilled portion, just enough to cause gross contamination of the entire potable water supply.

Following this discovery, drastic changes were made in supplying these ships with water. Several fuel tankers were immediately converted into water supply ships, each with a 300 lb/day gas chlorinator installation. These water tankers were able to chlorinate all the water as it came aboard or as it was being transferred to another vessel. Provision was made to recirculate and rechlorinate en route if necessary. The filling and loading rates were so high (400 to 1000 gpm) that gas chlorinators were mandatory. This system prevailed until the end of the war.

Shortly after World War II, the U.S. Navy Bureau of Ships began to equip all fighting ships with shipboard chlorination systems to treat the entire water system. Details of these requirements are covered in the *U.S. Navy Manual of Preventive Medicine*.[10] All water, evaporated or otherwise, must show a 0.2 ppm free chlorine residual at the end of thirty minues contact time. The equipment used is functionally identical with that illustrated in Fig. 2-13.

These units are specially made to meet the U.S. Navy shock-test requirements, and must be practically nonmagnetic. These installations do not utilize the special detention tank or carbon purifier used on the troop ships, since it is necessary to chlorinate the evaporator effluent and the water in the holding tanks only as it is being transferred from one tank to another. This water is low in organic matter and other pollutants, and so elaborate treatment is not necessary.

In recent years the U.S. Navy and the Coast Guard have stepped up their potable water disinfection program so that all of their vessels from the largest aircraft carrier down to the smallest patrol boat are equipped with disinfection systems. A great many of these ships are equipped with brominators that utilize a polybromide impregnated cartridge. See Chapter 14 for further details.

In 1946 the U.S. Public Health Service decreed that all present-day ocean going vessels carrying passengers and utilizing skin tanks for potable water storage must chlorinate this water as it is pumped to the ships' water systems. A skin tank is defined as one that utilizes the ship's hull as a part of the tank. The theory here is that an opening in the hull or the "skin" of the ship would cause instant contamination. In installations on the American President Lines, an electrically operated, constant feed rate hypochlorinator can be used, as it is a usual practice to chlorinate only while the water stored in the skin tanks is being transferred to the double-bottom or other tanks. The proper contact time is achieved by the amount of storage available in these transfer tanks.

Shipboard installations must be carefully analyzed to determine whether or not a meter-paced unit or a constant-feed unit is required. A much better choice is always the water-powered meter-paced unit, which eliminates the dependence on the ship's power system. During times of electrical aberrations, the chlorination system is certain to be neglected, and there is no easy way to compensate for water not chlorinated during the electrical outages.

The one disadvantage of the water-powered unit is the necessity for providing a suitable drain for the power water, which must go to waste at atmospheric pressure of about 1 to 1.5 gpm, equivalent to about seventy tons of water per day. Obviously the hypochlorinator must be located at a point in the ship where this wastewater can flow back into the ship's storage system, usually to the double-bottom tanks.

Other special considerations for a shipboard installation are rleated to the possible high ambient temperatures in the region of the installation and the elevated temperature of the water passing through the system. Ambient temperatures of 120°F have been reported aboard ship, while water temperatures have been known to go as high as 90°F. This requires that the water meter be equipped with a special hot-water disk. The suction and discharge valves of the hypochlorinator pumping head must be made of material that will not only resist corrosion but will also not expand owing to this temperature rise. Certain ceramics are suitable for the ball and seats of these valves.

Suction and discharge tubing of the high pressure type, with natural or reclaimed rubber lining, is the most suitable. Neoprene is not at all suitable for this service. Certain types of plastics that "flow" or otherwise deteriorate at these temperatures must not be used in any of the hypochlorinator components.

For the operator's assistance, a liquid level staff that measures the amount of solution remaining in the hypochlorite tanks should be provided, together with a chart that shows the quantity of hypochlorite per inch of tank height required to provide the prescribed solution strength and a 10- to 16-ounce measuring cylinder, so that the proper amount of sodium hypochlorite can be added for any given amount of solution required to fill the tank.

Recent Waterborne Outbreaks on Cruise Ships. One of the more popular vacation adventures today is a trip aboard a passenger cruise vessel. In 1966 about 300,000 passengers sailed on these vessels. By 1973 the number had reached about 750,000, 75 percent of whom sailed from New York City and two Florida ports—Miami and Port Everglades.[46] This number is still growing at a rapid rate.

The first large outbreak reported in the period 1970-75 aboard a passenger cruise vessel occurred in January 1970, where 10 passengers and 36 crew members, aboard the British ocean liner *Oronsay* contacted typhoid fever; one passenger died.[44] An investigation revealed that the most likely source of the outbreak was the ship's water. Later that year the Center for Disease Control* initiated a Vessel Sanitation Program in an effort to prevent waterborne and foodborne outbreaks aboard cruise ships.

This inspection program is under the direction of the U.S. Public Health Service. The standards used resemble those adopted by the World Health Organization for seagoing vessels. The vessel inspection consists of a 100 point checklist that covers water, refrigeration, food preparation, and the personal cleanliness of the food handlers. John Yashuk, chief sanitation officer for the Public Health Service Miami based quarantine division advises that the inspection program is a voluntary one between the U.S. Public Health Service and the cruise industry. The Health Service has no regulatory powers, they can only recommend. However, when a ship fails the inspection and the Health Service advises against the vessel sailing, the vessel usually complies.

This of course is not the case where U.S. Flag ships are concerned. These ships must comply with all Health Service requirements.

A 1972 survey by the Center for Disease Control found that food handling

---

* Atlanta, GA.

practices and water systems aboard selected ships demonstrated a significant potential for the transmission of foodborne and waterborne disease.[45] Outbreaks of illness on passenger cruise vessels during 1970–1975 were caused by *Shigella flexneri, Salmonella* and *Vibrio parahaemolyticus*. Vehicles for the etiological agents were water, multiple foods, seafood cocktail, and shrimp and lobster. *Staphlycoccus aureus* caused two outbreaks of foodborne illness on aircraft during the same period. The vehicles were custard and ham.

The most notorious outbreak was on the TV "Love Boat," otherwise known to unsuspecting passengers as the *M/S Sun Viking*.[47] This ship was commissioned in 1972 and is one of three 18,500 ton Finnish built ships of Norwegian registry that comprise the fleet of the Miami based Royal Caribbean Cruise Line. The ship has a passenger capacity of 800 and a total staff of 324. The ship's potable water was found to contain an unacceptable concentration of coliform organisms. An extensive study of the drinking water supply revealed that this was the source of the gastrointestinal illness. The CDC investigated four outbreaks of gastrointestinal illness aboard the *M/S Sun Viking* in 26 months.[47]

The water purification system was of course suspect. Here is a synopsis of the system: Water was bunkered through the fresh water filling line into three fresh water skin tanks and the laundry tank. The water delivered to the three fresh water tanks was batch chlorinated to a level of 2 ppm. (The fresh water supply is from the salt water evaporators). After the so-called batch chlorination system the water is pumped without filtration through two UV disinfecting units and then stored in three potable water tanks.

The illness suffered by the passengers resembled acute gastroenteritis. The water system was suspect as the common source of the four outbreaks because at the time of the last outbreak the ship's UV disinfection system was not operating properly and the ship's drinking water was found to be contaminated with coliform bacteria. Moreover the UV system had no verifiable backup system that could be related to a chlorine residual measurement system.

An illustration of how elusive the causes of a diarrheal illness outbreak might be is the case of the *T.S.S. Fairsea*.[48] This is a 25,000 ton vessel of Liberian registry staffed by an Italian crew and based in Los Angeles, California. The ship carries 950 passengers and a crew of 426. On five consecutive cruises in the period April 23 to June 18, 1977, outbreaks of diarrheal illness occurred. On the five different cruises the illness attack rate among the passengers varied from 20 to 60 percent, which is a shocking situation. Unfortunately all of the investigations were unable to uncover the etiology of these outbreaks. To this day the *Fairsea* problem remains a mystery.

Three days after she failed the U.S. Public Health inspection for the sixth time in 1977, the luxury Dutch cruise ship *Statendam* sailed out of Miami on December 2 for a 10 day Caribbean cruise.[50] Aboard were 671 passengers and a crew of 372. By the time the vessel docked again in Miami, at least 82 passengers and 13 crew members had been stricken with gastrointestinal afflictions. Investigators found that there was a serious malfunction in the chlorination system of the potable water supply. This was directly related to the chlorine residual analysis system used to monitor the chlorine residual. Apparently the low residual alarm system was inoperative.

Many other violations of the Public Health Service code for cruise ships occur on a regular basis. The question of "how safe is it to sail on a cruise liner that fails to meet the sanitary standards of the U.S. Public Health Service" continues to be asked because of the following examples.

A January 30, 1979 inspection of the luxury liner *Queen Elizabeth 2* made by the Los Angeles Quarantine Station cited the ship for poor food handling and preparation practices.[51] At the same time the Public Health Service in Miami advised that the Swedish American Line's *Linblad Explorer* flunked the test because of problems with its potable water distribution system.[51]

In February 1979, more than 120 vacationers on a week-long cruise through the Caribbean aboard the Miami-based cruise ship *Festivale* came down with stomach cramps, nausea, and diarrhea.[53] There were so many things wrong with the vessel's method of food preparation and storage that the investigators never did pinpoint the exact origin of the salmonella poisoning, which caused the Carnival Cruise Lines to cancel the next cruise at a loss of $1 million. However, it is encouraging to note that the current chief of all these ship inspections, John Yashuk, believes that now 80 percent of all cruise ships that call at U.S. ports regularly meet the standards, while in July 1975 when the new codes went into effect, not a single ship could meet the standard. In spite of these encouraging statistics, the *Festivale* failed all seven of the service's sanitary inspections between July 1975 and July 1979.

Choice of Chemical

Three factors govern the choice of hypochlorite to be used: (1) cost; (2) type of metering equipment to be used; and (3) availability. Other factors being equal, the chemical of choice would always be sodium hypochlorite. There are fewer maintenance problems with metering equipment, and there is no fire hazard connected with its storage. It is costly to transport over long distances and takes greater storage space. Per pound of available chlorine, it is more expensive than the calcium hypochlorites purchased in moderate quantities.

In remote areas where transportation cost is significant, the granular hypochlorites [$Ca(OCl)_2$] are the chemicals of choice. However, the potential insolubles of these products are sufficient to cause maintenance problems with metering equipment. If the hardness of the water available for dissolving this granular material is in excess of 75 to 100 ppm, the water should be softened before attempting to make a hypochlorite solution. This can be done by adding equal amounts of soda ash with the calcium hypochlorite, allowing it to settle for twenty-four hours, and then decanting the clear liquor; or the makeup water can be softened by a conventional ion exchange unit. Another possibility is to sequester the potential insolubles (calcium ions) by the addition of 2 to 5 ppm of a sodium hexametaphosphate (sodium polyphosphate) to the makeup water before adding the calcium hypochlorite. This will yield a sodium hypochlorite solution with very little insoluble material to cause equipment fouling. There are commercial laundry preparations of granular calcium hypochlorite and sufficient sodium polyphosphate to yield a nearly complete conversion to sodium hypochlorite when dissolved for use. But all these are expensive. It is therefore worthwhile to investigate the cost of lithium hypochlorite, which does not have these potential insolubles that plate out and plug up the metering equipment.

If for one reason or another (usually transportation and storage expense) a granular hypochlorite is the chemical of choice, lithium hypochlorite (LiOCl) may be the most economical, considering the time and expense required to remove or sequester the insolubles of $Ca(OCl)_2$. No matter what precautions are taken with the preparation of calcium hypochlorite solutions, the operator of the metering equipment should be prepared to recirculate muriatic acid from the point of entry to the point of discharge of such equipment to remove the calcium deposit that will eventually deposit to some extent throughout the travel of the hypochlorite solution. The intervals of purging with acid can be more widely spaced if more precautions are taken to remove the insolubles first.

Maintenance

Maintenance of these and similar systems using hypochlorite require the periodic flushing of the hypochlorinator with muriatic acid to remove the white, scaly deposit resulting from the interaction of hypochlorite with the hardness in the makeup water as well as the deposition of the inherent impurities in the hypochlorite. These deposits interfere with the efficiency of the internal valve action, causing errors in pumping rates and clouding the suction sight glass. When the sight glass becomes so coated with a deposit that its operation is obscured, then it is time to flush the apparatus with muriatic acid.

LARGE HYPOCHLORITE SYSTEMS

Current Practices

In recent years (1970–1980) the stress on safety and the fear of a chlorine accident has caused large metropolitan areas to consider the use of hypochlorite rather than chlorine gas systems where large amounts of the liquid-gas chlorine is stored in either stationary tanks, rail cars, or ton cylinders. This has occurred in spite of the good safety record of such installations. Since 1908, when chlorine gas was first used in the U.S.A., there has been only one fatality from a chlorine accident at a water or wastewater installation in the U.S.A. There have been 9 transportation related fatalities resulting from massive derailment of tank cars. Between 1908 and 1961 there was only one fatality that was transport related.

Despite this record and the considerable additional cost of hypochlorite over chlorine gas (two to four times) and its inherent unwieldy and cumbersome handling problems, the city of New York changed from gas to hypochlorite at certain of their wastewater treatment plants in 1969.

The first to experiment with the use of large amounts of hypochlorite was the power industry. Electric generator stations use considerable quantities of chlorine for slime control of the condenser cooling water. In one such case, a hurricane along the northeast coast of the United States ripped loose a ton cylinder of chlorine at one of these power stations and carried it a considerable distance. No damage was done, but the management decided it was time to eliminate any possible hazard from chlorine gas.

Sometime in 1955, the Narragansett Electric Company of Providence, Rhode Island, effected the switchover to hypochlorite.[22] Other stations in Chicago and New York followed suit.

A notable experiment in the application of hypochlorite for wastewater disinfection has been done by the Metropolitan Sanitary District of Greater Chicago (MSDGC). Installation of disinfection facilities using sodium hypochlorite at 15 percent "trade" strength were begun in 1969 at the 330 mgd North Side Plant, the 900 mgd West-Southwest Plant and the 220 mgd Calumet Plant.[21,30,39,40,41] Since July 1972, the District has been continuously chlorinating all the effluents from these plants.[42] The newest facility of MSDGC is the John E. Egan plant at Schaumburg, Illinois, which was first used in 1976.

Currently the city of Houston, Texas uses sodium hypochlorite at its largest plants and liquid-gaseous chlorine at the smallest plants.[41]

In 1980, the city of San Francisco switched from chlorine tankcars to sodium hypochlorite. This move prompted three of the largest wastewater treatment plants in the San Francisco Bay Area to evaluate the situation of chlorine gas versus hypochlorite. After separate and independent investigations these plants decided not to change to hypochlorite for the following reasons: (1) the reliability and safety procedures of the chlorine storage system was satisfactory; (2) the amount of chlorine delivered to these plants was less than 10 percent of the total amount of chlorine moving into the area; and (3) the cost of hypochlorite was too great compared to chlorine.

In the last decade (1970–1980) some of the large users have switched back to chlorine gas. After several years of trial some power plants have given up hypochlorite because of the inherent difficulties in handling it in large amounts. Others have done so to save money. In 1975 Cleveland decided to make the change from hypochlorite to gas at the Easterly Wastewater Treatment Plant.

The interest shown in the use of hypochlorites generated a great deal of interest in the use of on-site chlorine generating apparatus. This equipment is described in Chapter 3.

Hypochlorite Quantities Required

To get an idea of quantities involved, let us examine the chlorine requirement for disinfection of a secondary treated effluent discharging into a receiving water. Proper disinfection to maintain the receiving waters safe for water contact sports is usually about 100–125 lb chlorine per million gallons of treated effluent.

Using a high-strength sodium hypochlorite of 10 percent by weight chlorine would require the following amount of sodium hypochlorite:

$$\% \text{ available } Cl_2 \text{ by wt.} = \frac{10\%}{\text{s.g.} = 1.14} = 8.8\%$$

Each gal NaOCl contains 9.5 × 8.8% = 0.84 lb Cl_2. If the dosage is 100 lb/mg, then 100/0.84 = 119 gal 10 percent NaOCl/mg. Therefore a 200 mgd plant would require 200 × 119 = 23,809 gal 10 percent NaOCl/24 hr. Assuming peak rate of 2½ times average = 500 mgd × 119 = 59,500 gpd or 59,500/1,440 = 41 gpm. So the metering equipment should be sized to handle a maximum of 50 gpm of 10 percent hypochlorite solution.

Comparing the half-lives of various strengths of hypochlorite, it appears that 10 percent strength is the most economical. Large installations are probably suited for a maximum storage period of one week. There would be very little deterioration in the strength of a 10 percent solution in this length of time. Manufacturers of sodium hypochlorite are able to provide strengths as high as 15 percent.

The choice of one over the other is primarily a matter of economics. The 10 percent solution has a greater stability than the higher strengths, and so, other things being equal, it should be favored. However, storage facilities are such a large cost factor in the overall installation that the economy of the 15 percent solution must be considered as well as the deterioration due to age.

To calculate dosages and storage facilities, the reader is referred to Table 2-2.

Hypochlorite Facility Design

Pumped Systems. Feeding and control systems should be categorized by the quantity of hypochlorite solution to be applied. Up to rates of approximately 200 gal/hr the best choice would be a positive displacement diaphragm pump. An appropriate example would be the Wallace and Tiernan 44 Series metering pumps, with single or dual heads.[23] These pumps can be automatically controlled by either a variable speed drive (SCR control) or by a stroke length positioner (electric or pneumatic) or both. Therefore, a compound loop control system is practical. In practice the flow signal is sent to the SCR controller, which provides a 20 : 1 metering range and the chlorine residual signal is sent to the stroke positioner, which has a 10 : 1 range. This is more than ample for any wastewater disinfection application.

BIF of Providence, Rhode Island, makes a line of diaphragm metering pumps capable of handling hypochlorite solutions. One group utilizes a mechanical diaphragm, and is limited to 48 gal/hr in the dual head version. Automatic control is limited to variable speed drive. Their hydraulic diaphragm series metering pump has a maximum capacity of 830 gal/hr in the dual head version. These pumps can be arranged for compound loop control (i.e., flow signal to variable speed drive and dosage signal to automatic stroke positioner).[24]

Pulsafeeder of Rochester, New York, also makes a suitable diaphragm pump with an automatic feed rate control for hypochlorite solutions. The Pulsafeeder control approach is to resolve the entire variation of solution discharge as a function of stroke length, maintaining the stroke speed constant at all times. This means that a system using the Pulsafeeder would have to have a multiplier to combine the flow and residual signals into a single signal. Therefore, the concept of compound loop control is not yet available with a Pulsafeeder system.[25]

Diaphragm metering pumps suitable for this service can tolerate back pressures as high as 150 psi. This is far in excess of any need that could occur in a wastewater system.

Systems requiring pumping rates in excess of the capacities available in diaphragm pumps can use a centrifugal pump. Feed rate control in these situations is by the use of a modulating rate valve downstream from the pump discharge.

Two classes of pumps are available for hypochlorite service. One is the precious metal alloy and the other is the fiberglass type. The Duriron Co. of Dayton, Ohio, makes a titanium pump which is excellent for this service. This pump has a long and satisfactory record of pumping chlorinator injector discharge solutions into transmission lines where pressures are too high for the usual injector booster pump upstream from the injector. This pump can also be used as a transfer pump or a low-lift pump for any hypochlorite installation. The "Durco Titanium" line of pumps comes in a variety of sizes for either low- or high-head purposes.

The Fybroc Division of Metpro Corporation, Hatfield, Pennsylvania makes a line of fiberglass pumps that are also suitable for corrosive solutions. They are ideal for low-head service such as at a wastewater treatment plant.

Pumps other than those mentioned above should be carefully investigated for their suitability for hypochlorite service. Recently a chlorine manufacturer put out a bulletin on the advantages of hypochlorite which recommended the use of graphite pumps as suitable for use with high strength (15 percent) hypochlorite solutions. Upon investigation the manufacturer of such a pump (Union Carbide Corp.) advised that their Karbate line of graphite pumps is *not recommended* for this purpose.[26]

Plunger- or gear-type pumps are not recommended owing to their high maintenance costs and poor reliability. There is insufficient operating experience available on these types of pumps.

Gravity Systems. The preferred system is to meter the hypochlorite through a modulated rate valve and allow the solution to flow by gravity to the point of application. If the hydraulic gradient from the storage tanks to the point of application cannot provide gravity flow, the hypochlorite could be pumped to an intermediate head tank. This can be done with a sonic-type level control switch allowing automatic intermittent operation of the supply pumps which would always pump at a constant rate. The modulating rate valve would be somewhere on the downstream side from the intermediate head tank.

The Metropolitan Sanitary District of Greater Chicago, has used all three systems of hypochlorite delivery. They have found the gravity system to be the most reliable and the one that requires the least amount of maintenance.

Eductor System. Eductors can be used effectively up to about 10–15 psi total back pressure. These are normally powered by pumps using the plant effluent. The principle of operation is similar to that of a chlorinator injector system. Of the three different systems described, this one is the least reliable for controlling flows. Another disadvantage is the loss of available chlorine caused by the presence of ammonia nitrogen in the plant effluent. The breakpoint reaction when using a mixture of hypochlorite solution of about pH 10 takes place almost instantaneously. Therefore, in the eductor system there will be a loss of chlorine to complete this reaction by a factor of 10 parts (by wt) chlorine for each part of ammonia nitrogen. As an example, the eductor system at the Calumet treatment plant of the MSDGC requires about 65 GPM of water to provide the necessary minor flow of sodium hypochlorite (about 2–4 GPM). If the treated effluent is used for powering the eductor system and if the effluent is not nitrified, the ammonia nitrogen content may be as high as 20 mg/liter. Therefore, the chlorine consumption in the eductor water would be as follows.

$$65 \text{ GPM} \times 1440 = 0.0936 \text{ mgd}$$
$$20 \text{ mg/liter} \times 8.34 = 167 \text{ lb/mg}$$
$$167 \text{ lb/mg} \times 0.094 \text{ mgd} \times 10 \text{ mg/liter Cl}_2 = 157 \text{ lb Cl}_2/\text{day}$$

So regardless of the chlorine dosage 157 lb of available chlorine will be used in the reaction with the effluent eductor water. Assuming a 15 percent trade strength hypochlorite solution this would amount to about 126 gal. of hypochlorite.*

Control System. The elements of an adequate control system are the utilization of both flow pacing and chlorine residual information plus the chlorine residual monitoring of the final effluent. The flow and residual signals can be cascaded or they can be a compound loop. Figure 2-14 illustrates an electronically controlled system utilizing the cascade principle where the flow signal is combined with the dosage signal from the chlorine residual analyzer. An alternate system would be to substitute diaphragm pumps for the eductors and change the control system

---

* One gal. of 15 percent hypochlorite contains 1.25 lb available chlorine.

HYPOCHLORINATION 101

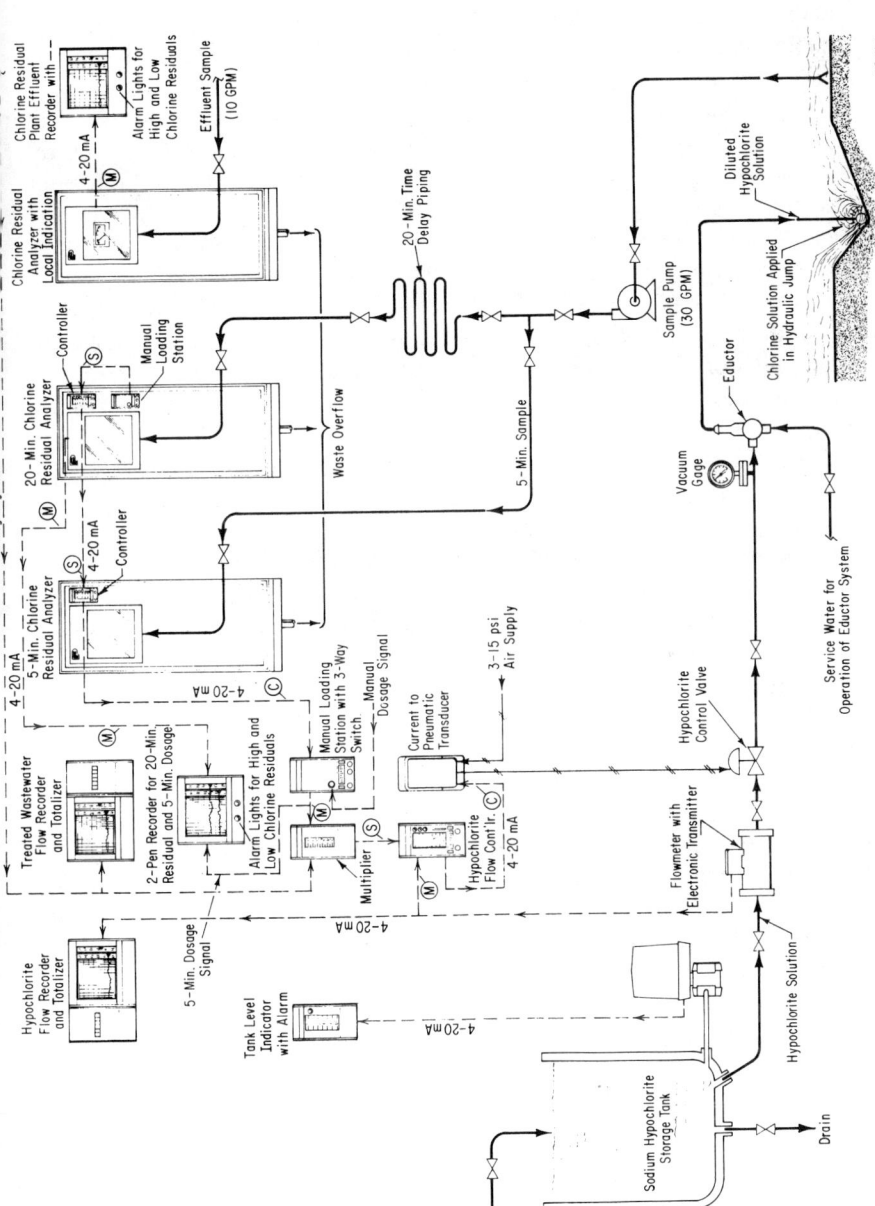

Fig. 2-14. Layout of hypochlorite closing system using electronic controls. Key: S = set point; M = measured output; C = control signal; [twenty minute chlorine residual analyzer output signal (measured) is used to control the output of the five-minute analyzer. This output becomes the chlorine dosage signal]. (Courtesy Fischer & Porter Co., Warminster, Pa.)

to a compound loop. If either centrifugal-type pumps or a gravity system were used the control principle as shown in Fig. 2-14 would be used.

The system illustrated has the added control feature of a second residual analyzer operating on a 20-min. sample detention time to "set" the control point on the 5 min. sample time analyzer. The 5-min. analyzer provides the dosage control signal to the system. A third analyzer is used to monitor the plant effluent and control the dechlorination system (sulfonator) if required. To provide the operator with all the elements necessary for proper operation, the system should include the following functions: continuous recording of hypochlorite flow; tank-level indicator; continuous recording of the chlorine control residual and the 20- or 30-min. residuals; and continuous recording of residual in the final effluent. Alarms should be provided for low-hypochlorite supply level, low- and high-chlorine residual on all analyzers, loss of hypochlorite flow, and loss of water supply to eductor.

It is of interest to note that in 1977 Fischer and Porter announced the marketing of their "Feedrator II Liquid Chemical Feed Dispenser"[27] which is a packaged system for the continuous feeding (automatic or manual) of chemicals normally used in potable water and wastewater. This includes high-strength hypochlorite. Referring to Fig. 2-14 the Fischer and Porter package mounts in a single cabinet the magnetic flowmeter with transmitter, the hypochlorite control valve, the eductor, the hypochlorite flow controller, the current to pneumatic transducer, and the signal multiplier. One of the main features of this system is the incorporation of their new Chloromatic-type valve constructed of PVC and teflon. This unit is used up to 5 GPM of hypochlorite. Above 5 GPM a Saunders-type diaphragm valve is used. In this system the Chloromatic operator along with the magmeter becomes the flow controller. The plant flow meter and residual signal are fed into this operator which multiplies these two signals to provide the control signal to the chloromatic valve. This eliminates the need for the external controller as shown in Fig. 2-14. This system is also equipped with a potentiometer to manually set the hypochlorite flow rate. In this mode of operation (without chlorine residual signal) the control action of the chloromatic valve is adequate (when the magmeter combination is still operable). The hypochlorite flow will remain proportional to the flow signal regardless of changing vacuum conditions in the eductor. This results in a subsequent change of pressure across the rate valve.

These systems can be either electronic or pneumatic. The advantages of each are:

Electronic System
- Has a rapid response.
- Can operate over long distances without a time lag.
- Is not subject to cold weather difficulties.
- Is compatible with telemetering and data processing systems.
- The medium is always clean.

Pneumatic System
- Low first cost of equipment.
- Easy to service and understand.
- No explosion or fire hazard.
- Valve positioners operate directly from the control medium without the need for conversion.

Monitoring the System. All hypochlorite systems should be equipped to indicate and totalize the flow of hypochlorite to each point of application. Frequent checks of the solution strength should be made so that the operator can easily verify the chlorine dosage. It is imperative that the operator be able to verify the chlorine dosage from a visual indicator. This is one of the most important tools an operator can have. Therefore the selection of a hypochlorite flow measuring device should be made with great care and consideration. Two types of flow meters are available. One is the mag-meter and the other is the rotameter. When a rotameter is used it should be equipped with both an indicator and a transmitter.

When diaphragm pumps are used, pulsation dampeners should be installed on both the suction and discharge lines adjacent to the pumps. These devices are necessary to minimize the pulsation effect which interferes with the operation of the flow meter. Wallace and Tiernan Div. of Pennwalt Corporation added a new tubular-type diaphragm pump in 1982 which is available in either single, double, or triple heads. This design when using the duplex or triplex models minimizes the pulsation effect.

Materials of Construction

Storage Tanks. The first large storage tanks built for the Chicago projects were of the filament-wound fiberglass type. These tanks proved unsatisfactory having a life of only about four years.[38] About 1975 these tanks were converted to hand lay-up fabricated fiberglass tanks utilizing a vinyl resin binder. This change was made based upon positive results obtained from another operating installation. The performance of these new tanks appears to be acceptable.[28]

The experience of the Metropolitan Sanitary District of Greater Chicago with underground concrete tanks has indicated that the fiberglass lining of concrete is undesirable. Laminate failures have caused clogging of valves, pumps, and diffusers. The use of these tanks has been discontinued. For the newest MSDGC at the John E. Egan wastewater plant, which was put into operation in 1976, the hypochlorite storage tanks were put underground because of their size (12 ft in diameter and 35 ft long, approximately 30,000 gal). These are plastic lined, continuous weld, full weight carbon steel tanks. The lining, which was applied on site, consisted of two coats of fiberglass reinforced polyester material applied at a nominal thickness of 35 mils.

Rubber lined steel tanks would be equally as satisfactory as any type of PVC lining. This method has been used in various aspects of chlorination for many years.

Storage tanks should be equipped with a level gauge and transmitter for continuous readout of its contents. They should also have vents and some method of manually sampling both the incoming delivery and the contents of the tank.

The fill piping system should discharge through the top of the tank. A pressure relief and overflow pipe should also be provided. The fill pipe connection to the delivery vehicle should be a Hastelloy C or a titanium nipple securely braced to the tank. The fill pipe itself can be PVC, R.L. steel, or Saran lined steel, or Resistoflex (Kynar®). A high capacity drain or sewer with a proper drainage gradient adjacent to the storage tanks should be provided in the event of a storage tank rupture. Underground tanks should be equipped with equivalent facilities consisting of a completely corrosion resistant sump pump.

Piping. Above ground piping can be Sch 80 PVC, Kynar, hard rubber lined steel, or Saran lined steel. The steel lined pipe presents some installation difficulties. The preferred material is Kynar for both pipe and fittings. This pipe is also known as Fluoroflex-K. The latter is a trademark of Resistoflex Co.,[29] and Kynar is a Pennwalt Corp. trademark.

Underground piping should be some type of lined steel pipe. Chicago favors either a PVC or polypropylene lining. Hard rubber or Saran lining is also acceptable and is preferred by others.

Valves. Plug valves are preferable to ball valves. Ball valves are subject to stem breakage. This limits a ball valve operating life to about six months.[28] Plug valves made of steel and lined with PVC or polypropylene are preferable.

Diffusers. The distribution of the hypochlorite solution at the point of application needs special attention because of the very high concentration of the solution (15 percent = 150,000 mg/liter). This compares to the chlorine solution concentration of 3,500 mg/liter (maximum) in a conventional chlorinator discharge. Although the segregation phenomenon favors better mixing with a higher concentration of the process chemical, efficient dispersion is the goal for a hypochlorite diffuser. The diffusers should be designed with perforations for across-the-channel installations using about a 25–30 ft/sec velocity at the perforations.

The diffusers should be made of either PVC, Kynar, or rubber lined and covered steel pipe.

Hypochlorite Flow Meters. These can be either teflon lined magnetic flow meters manufactured by Fischer and Porter or PVC Straight-Through Vareameters with transmitter (electric or pneumatic) manufactured by Wallace and Tiernan Div. of Pennwalt Corp.

Rate Control Valves. The most common control valve in use is the Saunders-type using a rubber diaphragm with teflon facing. The valve body can be all PVC or a steel body lined with either PVC, Kynar, hard rubber, or Saran. These valves can be equipped with electric or pneumatic actuators. Plug-type valves are also available for this service. PVC ball-type construction is not recommended for this service. Maintenance cost is too high and reliability is low.

For flows up to 5 GPM Fischer and Porter uses their Chloromatic PVC and teflon valve; above 5 GPM a Saunders-type diaphragm valve is used.

Eductors. These must be all PVC construction because of the corrosivity of the hypochlorite solution. The eductor used at the MSDGC Calumet wastewater-treatment plant is a Penberthy Model No. 168P (PVC). This unit operates at about 35 psi and uses approximately 70 GPM wastewater for a maximum back pressure of 10 ft.[30] It can deliver about 3 GPM of 15 percent hypochlorite.

Hazards of Hypochlorite

The use of hypochlorite as an alternative to liquid (gaseous) chlorine in reasonably large quantities is primarily for safety reasons. However the hazards due to the presence of hypochlorite must not be overlooked. These hazards derive from storage accidents.

One such accident occurred in Knoxville, Tenn. April 8, 1983.[53] A lethal cloud of chlorine gas escaped from sodium hypochlorite tanks used in the disinfection system for the wastewater treatment plant. They also use ferric chloride as a coagulant which is normally shipped in rail cars. When rail cars are not available, ferric chloride is shipped by tank trucks. The hypochlorite is always delivered by tank trucks similar to those used for shipping ferric chloride.

In this instance the ferric chloride was shipped by truck and since the truck connections were compatible, the driver, who had never made a delivery to the plant before, made the connection, pressurized the truck and unloaded approximately 600 gal of ferric chloride, which mixed with approximately 3000 gal of 10–12 percent sodium hypochlorite.

Owing to the low pH of the $FeCl_3$ and the concentration of the reactants, molecular chlorine was released instantaneously from the hypochlorite. A cloud of Cl_2 began rolling out of the hypochlorite tank vent. Fortunately an emergency response plan that had been worked out by the city was implemented as soon as the cloud was sighted. This included local evacuation and rerouting of all mobile traffic near the area.

Precautions must be taken to make certain that only hypochlorite can be put into a hypochlorite storage tank. Any acidic chemical will generate the release of molecular chlorine from a sodium hypochlorite solution.

When the City and County of San Francisco switched to hypochlorite at all three of their wastewater treatment plants they also switched from SO_2 to bisulfite solution. Precautions were taken in the design of these systems to prevent the possibility of unloading bisulfite into a hypochlorite tank or vice versa. This mixture produces a heat of reaction sufficient to cause disintegration of a fiberglass tank. The heat generated is so great that an explosive force would surely be produced.

Another hazardous situation could occur at installations where both chlorine and ammonia are used for treating potable water. Adding ammonium hydroxide to hypochlorite will generate lethal quantities of nitrogen trichloride. Moreover this mixture might produce a violent explosion if the right concentrations were reached.

Still another possibility is the generation of chlorine dioxide on site. The ingredients for generating chlorine dioxide are commonly chlorine, acid (HCl) and sodium chlorite solution ($NaClO_2$). If an acid delivery gets into the sodium chlorite, molecular chlorine will be released instantaneously. If hypochlorite is used as the chlorine source, the acid-hypochlorite mixture will be an additional hazard. Mixing the sodium chlorite (pH 12) with the hypochlorite (pH 12) may not be a lethal combination but precautions should prevent such an accidental mixture.

Operating Costs

The operating cost of any imported hypochlorite system will depend entirely upon the amount of chemical to be delivered at one time and the total amount to be consumed over a contract period. The following are examples of hypochlorite and chlorine liquid-gas prices in various metropolitan areas.

CHICAGO, ILLINOIS. The Metropolitan Sanitary District of Greater Chicago (MSDGC) purchases 15 percent (trade) sodium hypochlorite under contract agreements with two suppliers: K. A. Steele and Barton Chemical. The unit price paid for sodium hypochlorite is adjusted each month based upon quantity requirements of a facility, published prices of certain raw materials, transportation, fixed costs, and the Consumer Price Index. Utilizing this rather involved calculation, one of the potential suppliers in the Chicago area supplied 15 percent hypochlorite on the basis of one tank truck (4000 gal) minimum delivery at a cost of 38¢/gal for March 1983. The other supplier delivered hypochlorite to the smaller plants at a cost of 60¢/gal. The 38¢/gal figure calculates to 31¢/lb available chlorine. The 60¢/gal figure calculates to 48¢/lb available chlorine.[31]

Chlorine gas in the Chicago area is grossly dependent upon the size of the containers, because of a peculiar safety ruling of long standing. The various water-treatment facilities for the city of Chicago are compelled to receive their chlorine supply by the truck-load (14-ton containers). By law, chlorine tank cars are not allowed to have access to the Lake Michigan water treatment plants. Therefore, the cost per ton of chlorine gas at these plants is $140/ton (7¢/lb) in 1982, down from $204/ton (10¢/lb) in 1976. If 90 ton tank cars were allowed access to these plants the chlorine would cost $120/ton (6¢/lb) in 1982 down from $140/ton in 1976.[32]

HOUSTON, TEXAS. Sodium hypochlorite is available from local suppliers at 48¢/lb available chlorine in tank truck quantities.

Local suppliers are quoting liquid chlorine in ton containers at $240/ton (12¢/lb) and 90 ton cars at about $150 ton (8¢/lb). These are December 1982 prices.[33]

SAN FRANCISCO BAY AREA. Sodium hypochlorite at 14.8 percent available chlorine is being used at the three San Francisco wastewater treatment plants. It is delivered in 5000 gallon tanker trucks. The cost of chlorine delivered at each plant calculates to 33¢/lb (1982).[43] Jones Chemical will deliver 12.5 percent hypochlorite anywhere in the bay area for 48¢/gal (46¢/lb available chlorine).

Continental Chemical Co. will deliver ton containers for $176/ton at 5 tons per week. PPG and Hooker Chemical are delivering 90 ton tankcars at $65 ton in a glut market (1982). This is only 8¢/lb.

As can be seen from the above discussion, the cost of chlorine gas and hypochlorite varies considerably depending upon the locality, demand, and availability.

The price spread between hypochlorite and chlorine gas increases significantly as the distance from the source of chlorine gas manufacture and the user increases.

The most optimistic estimate is that imported hypochlorite will cost at least three, or more likely, four times that of liquid-gas chlorine.

Cost comparisons of the chlorination facility between liquid chlorine and hypochlorite should include storage and supply facilities, metering equipment, instrumentation, and monitoring equipment.

Generally speaking, the metering and feeding equipment for chlorine gas is more expensive than that for hypochlorite, but the expense of storage facilities for hypochlorite are far greater and more than offset the equipment difference. Maintenance of a hypochlorite system requires more man hours than the gas system.

Reliability

For maximum reliability the hypochlorite flow control system should consist of two separate and independent systems which accurately control the flow of hypochlorite to provide a predetermined chlorine residual in the plant effluent.

If pumps are required to deliver the hypochlorite or operate eductors, both standby equipment and standby power should be available to prevent interruption of hypochlorite flow to the point of application.

ON-SITE MANUFACTURE OF HYPOCHLORITE

Introduction

The on-site manufacture of hypochlorite is normally based upon the interaction of chlorine gas and either sodium hydroxide or lime. This usually depends upon the quantity of hypochlorite to be produced. Two systems are described below.

108 HANDBOOK OF CHLORINATION

The first example is a small plant application using less than 4000 lb/day chlorine. The other system involves massive amounts of chlorine; 50,000 lb/day or more.

Potable Water System (4000 lb/day)

Sometimes in certain extraordinary cases in the application of chlorine for water treatment, such as in bleaching organic color, the amount of chlorine required depresses the pH, thereby interfering with the bleaching action of chlorine. In one such instance[35] calcium hypochlorite was made in situ. Fig. 2-15 shows a schematic diagram of how this was accomplished. Here it is interesting to see how a conventional installation of a lime feeder and chlorinator was arranged to produce, in an efficient manner, a calcium hypochlorite solution so that the chlorination process of bleaching could be carried out without depressing the pH of the treated water. Here too, the most difficult problem to solve was to properly settle out of solution the hydrated lime and its associated impurities. This flash mixing process of making calcium hypochlorite consists of three basic elements: (1) a premix tank, (2) a calcium hypochlorite sedimentation tank, and (3) a pH detection system.

In Fig. 2-15, chlorine solution discharging at approximately 15 gpm from a

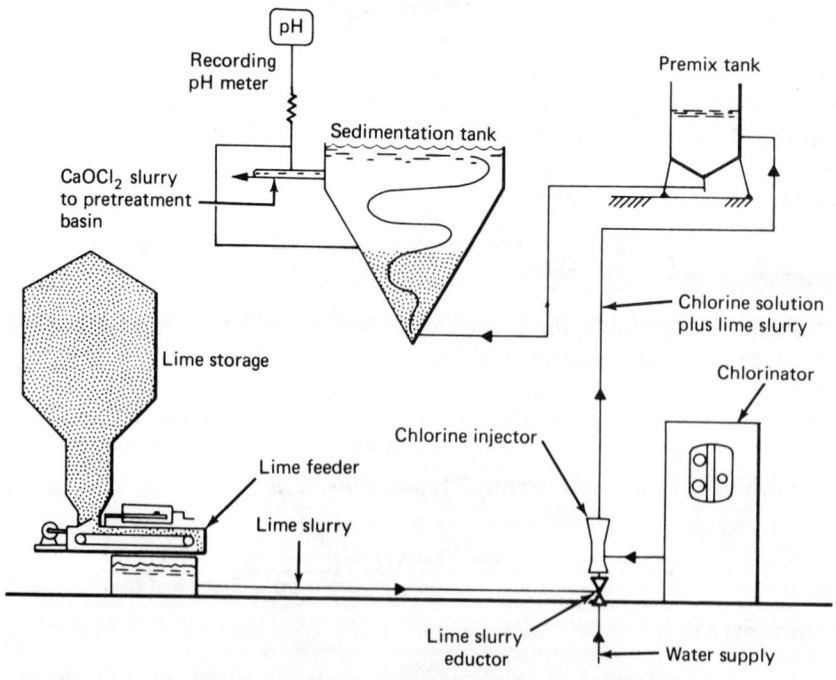

Fig. 2-15. On-site manufacture of hypochlorite.

chlorinator is made to flow through the premix tank to the sedimentation tank and thence to raw water in one continuous operation. At a point located well upstream from the premix tank, lime slurry is fed into the chlorine solution at approximately 15 gpm, the concentration of which is equivalent to the quantity of free chlorine present. The purpose of the premix tank is to remove the turbulence caused by the lime injector and the chlorine injector prior to the entrance into the sedimentation tank. The gradually sloping sides of the sedimentation tank cause a decrease in velocity of the solution. This gradual slowing down of the upward movement of the solution toward the overflow weir allows the settleable material to form a sort of mat, allowing clear liquor above to enter the raw water. This lime suspension hydraulically grades itself with the large particles composing the lower section of the suspension and the fine particles forming the top or upper layer. The effluent from the sedimentation tank contains less than 0.1 percent total solids, owing to the small velocity at the crest of the tank. The solution flows by gravity to the point of application. After the desired feed rate is set on the chlorinator, the lime feeder is set in accordance with the figures shown in Table 2-4. The final determining factor in setting the lime feeder is the ratio of the pH of the samples taken from within the sedimentation tank and the effluent of the sedimentation tank. When the two values of pH are in equilibrium, the lime feeder is set correctly, indicating that the quantity of lime collecting in the "lime layer" is not in excess of that required by the reaction. Underfeeding of lime is noted by both a drop in pH and a free chlorine odor coming off the surface of the two tanks. The average dosage by this system amounts to approximately 13 ppm calcium hypochlorite to 26 mgd water treated, or approximately 2800 lb chlorine and 2900 lb lime per day.

Wastewater System (50,000 lb Cl_2/day)

Introduction. This concept of a chlorination facility is entirely different in method and scope from that of on-site generation.

About 1957, the Sanitation Districts of Los Angeles county, California, pioneered a unique concept of chlorination for wastewater disinfection. This chlorination system was designed for the Joint Water Pollution Control Plant adjacent to South Figueroa St. in Carson, California. This treatment plant currently discharges about 400 mgd primary effluent through a sophisticated outfall system consisting of a special ocean outfall and dispersion piping system. The outfall tunnels under the San Pedro Hills and then discharges into the Pacific Ocean about two miles offshore from the Palos Verdes peninsula.

This hypochlorite generating and control system is unique in that it produces a stable hypochlorite solution (approximately 8000 mg/liter available chlorine) directly from liquid chlorine in tank cars without the use of intermediate vaporizing equipment. Conventional chlorination systems always require vaporization for the accurate control of chlorine feed rates.

This particular installation compelled the designers to consider a constant strength solution because the point of application of chlorine is at too great a distance from the point of generation to consider remote injectors.

This system's desirability is related in part to the incompressible characteristic of water. Therefore, regardless of chlorine solution flow variation, the impact of this change occurs instantaneously at the point of application several thousand feet away. This is accomplished by the production of a constant strength chlorine solution at the chlorine station.

System Description. The on-site manufacturing process is a simple one. It merely involves the simultaneous injection of liquid chlorine and lime slurry or caustic solution into a transmission line carrying the makeup water. The hypochlorite solution produced is about 8000 mg/liter in strength. It is normally poised at a pH of 7.5 to 8.5. Although this is an alkaline solution it is preferable to transport this solution in corrosion resistant piping such as PVC, KYNAR, or rubber lined, or saran-lined steel pipe. The secret of a successful installation of this type is the ability to achieve complete and immediate mixing of liquid chlorine to prevent evolution of any chlorine gas.

The complete system includes storage facilities for chlorine tank cars, a 175 psi air-pad system with air driers, liquid chlorine flow control system, liquid chlorine injectors, makeup water supply system, pH control system, hypochlorite solution lines, and lime or caustic facilities.

Chlorine Facilities

Chlorine Supply System. The layout of railroad siding and the design of chlorine headers, loading platforms, air padding, air drying, expansion tanks etc., is the same as for liquid-gas systems using conventional chlorination equipment described in Chapter 9 with one exception: the air pad on the tank car must be continuous at 175 psi for reasons described below.

Reserve Tank. The same applies here to the use of the reserve tank concept described in Chapter 9. This insures a continuous uninterrupted supply of liquid chlorine. Moreover, it eliminates the need for weighing devices.

It is also practical to use a resettable reverse totalizer on the readout signal provided by the liquid chlorine rotameter signal. With an accurate signal from the liquid chlorine rotameter to the flow meter totalizer, the reverse counter on the totalizer can be set to alarm at 500–700 lb left in the car. Then the operators should check with the totalizers on the flow meter to verify the amount left in the car. The alarm on the reserve tank serves as the final verification that the car is empty of liquid chlorine.

Chlorine Control System. It has been stated elsewhere in this text that it is impractical to measure liquid chlorine with conventional metering devices. The pressure drop, although very small, across the meter causes the liquid to flash to gas which destroys the accuracy of the meter. This is true at the pressures usually encountered in conventional practice. The flashing phenomenon can be eliminated if the liquid supply is pressurized to 175 psi. At this pressure, Los Angeles County operators have confirmed that the liquid chlorine rotameter reading is as accurate as a similar meter would be if measuring Cl_2 gas or water flow. (See Fig. 2-16.)

The chlorine control system can be designed to respond to either a plant-flow signal or a combination of flow and chlorine residual. This is the same as a conventional system. The major difference is that the hypochlorite makeup water flow rate, rather than the chlorine feed rate is the primary response to changes in plant flow and/or chlorine residual. The chlorine feed rate follows as a secondary response.

The flow rates of liquid chlorine and lime slurry (or caustic) are varied automatically to comply with changes in the makeup water flow, so as to maintain a constant concentration of hypochlorite solution going to the point of application.

Fig. 2-16. On-site hypochlorite production system showing metering and control system (courtesy Sanitation Districts of Los Angeles County California).

Chlor-Alkali Mixing. This is the key to a successful operating system. In order to achieve a stable hypochlorite solution without the evolution of molecular chlorine, a constant back pressure of 35–40 psi must be maintained at the point of liquid chlorine injection. The alkali solution (CaO slurry or NaOH) is injected diametrically opposite to the chlorine injection. Immediately downstream from this injection point is a turbine mixer.

The high head loss from the artificially padded liquid chlorine line upon discharge from the special chlorine injection valve permits the chlorine to vaporize immediately inside the makeup waterpipe. If the pressure drop across the injection valve and the back pressure in the hypochlorite solution line are not maintained, a complete and stable solution will not be produced. The necessary back pressure can be sustained by the installation of an automatic pressure regulating valve in the solution line downstream from the mixer. If the hypochlorite solution is discharged through more than one diffuser, each diffuser header must be equipped with a pressure regulator. A single regulation in the mainline is not sufficient to keep the proper back pressure and provide uniform flow when more than one diffuser is involved.

Solution Line and Diffusers. These appurtenances should be constructed of the same materials and designed for the same hydraulic conditions as for a conventional chlorination system. (See Chapter 8.)

Although the alkaline hypochlorite solution is mildly corrosive, it will be necessary at prescribed intervals to purge the calcium scale deposit in the solution lines caused by the use of lime slurry. Removal of this scale deposit is easily accomplished by lowering the pH of the hypochlorite solution (temporarily) to about 4 or 5. Obviously these procedures require corrosion resistance piping, valves, and fittings.

Makeup Water System. The water to make the hypochlorite solution should be chlorinated plant effluent, even if the ammonia nitrogen content is high. A small portion of chlorine injected into this water will be consumed immediately by the ammonia nitrogen in the breakpoint reaction. This reaction between Cl_2 and NH_3-N is complete in a few seconds at the pH and chlorine concentrations used in this method. For example, 350 GPM of makeup water containing 20 mg/liter NH_3-N will consume 834 lb/day chlorine in the breakpoint reaction. For an 8000 mg/liter hypochlorite solution this represents 3 percent chlorine loss with 20 mg/liter NH_3-N present in the makeup water. (See Fig. 2-17.)

A variable speed booster pump is essential for the makeup water system. Instantaneous response of chlorine dosage rates is made possible by holding a constant hypochlorite concentration in the solution line. Therefore, the rate of flow in the solution line is varied according to plant flow and residual changes. Regardless of solution line length, chlorine dosage changes at the chlorination station are reflected immediately at the point of application.

Fig. 2-17. On-site hypochlorite production system; makeup water system and chemical injection points (courtesy Sanitation Districts of Los Angeles County).

If the system is designed to produce a hypochlorite solution with a chlorine concentration of 7000–8500 mg/liter, the makeup water flow would have to be about 240–290 GPM/1000 lb/hr chlorine feed rate.

Essential instrumentation for this system requires that a flowmeter be installed in the common discharge of the makeup water line. The flow signal from this meter is combined through a ratio station with the chlorine flow signal to provide a signal to the hypochlorite concentration flow recorder. (See Fig. 2-16.)

Alkali System. The chemical of choice is pebble lime; however, some systems designed for nitrogen removal where peaks cannot be handled by biological processes would be better suited to a caustic installation, particularly if the intermittent operation is 70–75 days or less per year. The choice is mainly one of economics, but from a chemical standpoint lime may be preferable because the calcium ion present provides a protective coating on the hypochlorite solution line.

Burned pebble lime costs about $40/ton as compared to caustic soda at about $170/ton. Even though the storage and handling system for lime is more expensive than for caustic, the cost-effectiveness is heavily in favor of lime. In the quantities required for this type of operation, quicklime is preferable to hydrated lime; however, source of supply and chemical cost would still govern the choice. Less mechanical equipment is required for the hydrated lime system, and the least mechanical equipment is required for the caustic soda system. However, the overall costs invariably favor the quicklime system.

Quicklime is readily available by truck or in railroad carload lots. It can be conveyed pneumatically to storage silos with ease and without dust or other air pollution problems. Proper quicklime (pebble) handling equipment includes the following:[36]

1. Air compressors for vacuum and pressure requirements.
2. Transfer piping.
3. Pressure-vacuum separation device (lime is removed from delivery vehicle by vacuum and delivered to storage by pressure).
4. Bulk storage facilities.
5. Filters on air intakes and discharges.
6. Monitoring, alarm, and control devices.

The pneumatic conveyor system for unloading quicklime should have interlocking safety devices to shut down the system if excessive vacuum or pressure develops. It should be further interlocked with level measuring devices in the lime storage tank to cause the conveyor to shut down when a full tank is reached. Low-level warning devices should also be installed in the storage tanks.

Feeding of bulk lime to the slaker is normally by gravity. The slaker package consists of a slaker, lime feeder, control panel, and accessory control valves. The lime feeder should be equipped with a variable speed drive motor so that it can be manually adjusted to run at any of the numerous lime feed rates. Automatic operation of the slaker-feeder combination can be achieved, by a level probe system in the lime slurry storage tank, to activate or deactivate the combination as low- or high-levels are attained. Manual adjustment of the lime feeder to the slaker is sufficient to supply slurry of the proper concentration. The slaker equipment normally includes grit removal components and the overall system must include belt or other conveying equipment to move this material to storage facilities for disposal. The bulk storage and slaking facilities can be eliminated by purchasing liquid slurry, but at greatly increased costs.

Lime slurry from the slakers flows by gravity to lime slurry storage tanks where the lime is kept in suspension by turbine mixers. Lime slurry from the storage tanks is pumped by a variable speed progressive cavity positive displacement pump into the hypochlorite pipeline for combination with chlorine. A piston or diaphragm positive displacement pump is not satisfactory for this application. The lime slurry pump, pumps lime as demanded by a pH control system, which samples the contents of the hypochlorite pipeline and signals the lime pump to pump as required to maintain a pre-set pH level required for satisfactory hypochlorite production. To maintain a slightly alkaline pH, approximately 0.80 ton of CaO are required per ton of Cl_2.

Reliability. The low initial cost of all the mechanical equipment and piping lends itself to complete duplication of the entire system. This provides a system reliability comparable to or better than most conventional systems. Power failures effect this system with equal results as a conventional system, whether it be on-site generation or imported hypochlorite.

Cost Comparison. The system described above was designed for the Sacramento County Regional Plant (1974) as a nitrogen removal process of a secondary effluent to supplement the biological nitrification–denitrification system. This was considered necessary only during the cannery season when biological treatment could not handle the cannery load. The chlorination facility was designed for 110 mgd. The nitrogen removal process by breakpoint chlorination was estimated to require 70 days operation per year.

The comparison plant was designed for 60 mgd to produce a commonly nitrified secondary effluent using on-site generated chlorine for nitrogen removal.

Scaling the above two plants to a common size, the following is an economic evaluation of the on-site hypochlorite manufacturing (Los Angeles System) versus the on-site generated system. This comparison is of considerable interest owing to the enormous dollar savings in capital cost of chlorination equipment, particularly where chlorine usage approaches tank car quantities.

The comparative costs are as follows:[37]

Amortized capital cost (25 yr. at 7 percent) plus operating* and maintenance costs based on 1974 prices—$/mg (annual average):

On-site hypochlorite manufacturing $37.00.
On-site hypochlorite generation $107.00.

A similar plant which would reduce the ammonia nitrogen (biologically) continuously by nitrification followed by denitrification, the following are the relative costs based on the Sacramento analyses:

	Cap. Cost $	Annual Oper. Cost $
Nitrification-denitrification	5,640,000	1,250,000
Breakpoint Chlorination	32,000	1,295,000

---

* Costs are based on chlorine @ $144/ton and caustic soda @ $168/ton. A system using lime for caustic buffer would be even less costly.

Based on these figures, if nitrogen removal by chlorine is a requirement for a given effluent, or if the chlorine dosage requirements for disinfection approaches a maximum of 10,000–12,000 lb/day, the *on-site hypochlorite manufacturing* system is of considerable interest. All of the cost figures for the Sacramento design (which was for only 70 days operation per year) were based upon the most expensive alkali system, i.e., caustic soda instead of lime. It should further be noted that this system of hypochlorite manufacture must be as described because it is not feasible without the addition of alkali.

Practical Considerations. Since only one operating system falls into this descriptive category, one may hesitate regarding its feasibility. The answer to this is straightforward and unequivocal. The first such unit was put into operation about 1958. This was a much smaller system than the prototype, however, all of the instrumental anomalies were sorted out in the first system so that the current system enjoys either continuous or intermittent operation, whichever is required, with the same flexibility that is enjoyed by conventional chlorination systems.

Manufacture in Underdeveloped Areas

Many underdeveloped areas of the world suffer great losses in human life from such waterborne diseases as cholera, typhoid, dysentery, and virus infections. In these areas, the cost of transporting a disinfectant over long distances makes the use of chlorine a luxury. Therefore an alternative method for producing hypochlorite with materials available locally is presented here to show how it can be done without electric power and without chlorine gas as a raw material. The following is based upon an operating installation described by Stone[38] in 1950. The materials available were common salt and manganese dioxide, mined nearby. Fig. 2-18 illustrates this installation.

About eighty-five miles from where the hypochlorite was to be made was a sulfuric acid plant. There was also available locally a supply of low-grade slaked lime. With these materials it is possible to make a bleaching powder with about 35 percent available chlorine. Here is how it is done:

The manganese dioxide and common salt are mixed and placed in a reaction tank which is suspended above an open type water boiler. Above the chemical reaction tank is a sulfuric acid tank with a control valve which allows regulation of the flow of acid to the reaction tank (Fig. 2-18). The rate of chlorine gas generation is regulated by the flow of sulfuric acid and the temperature of the water. The heated water, in turn, heats the chemicals in the reaction tank, accelerating the following reaction:

$$MnO_2 + 2NaCl + 2H_2SO_4 \longrightarrow Cl_2 \uparrow + MnSO_4 + Na_2SO_4 + 2H_2O \quad (2\text{-}13)$$

The proportion of these chemicals required is directly related to their molecular weights.

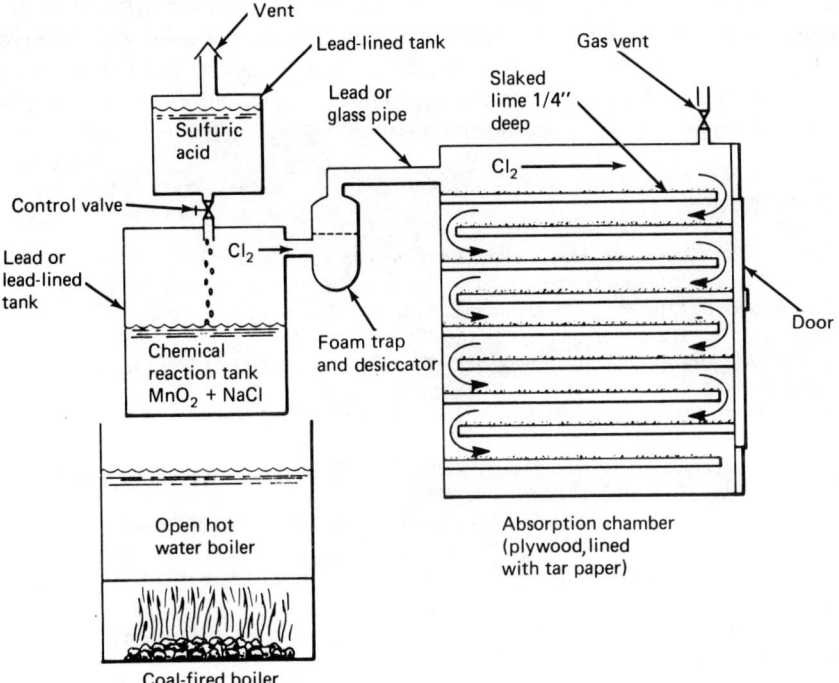

Fig. 2-18. Manufacture of hypochlorite in underdeveloped areas.[38]

The chlorine gas generated is passed through a foam trap and desiccator, where it is dried before going to the absorption chamber. The absorption chamber consists of several trays containing a quarter-inch layer of carefully slaked lime. The chlorine enters at the top of the chamber and, being heavier than air, passes downward over the trays, reacting to form bleaching powder:

$$Cl_2 + Ca(OH)_2 \longrightarrow CaOCl_2 + H_2O \qquad (2\text{-}14)$$

The resulting product contains about 70 percent calcium oxychloride; therefore when this powder is dissolved in water, it hydrolyzes to form calcium hypochlorite and calcium chloride:

$$2\ \underset{\text{calcium oxychloride}}{CaOCl_2} \longrightarrow \underset{\text{calcium hypochlorite}}{Ca(OCl)_2} + \underset{\text{calcium chloride}}{CaCl_2} \qquad (2\text{-}15)$$

An outlet or vent valve in the top of the absorption chamber is left open until the chlorine gas has displaced the air. It is closed when a chlorine gas odor can be detected at this outlet. The generation of gas is allowed to proceed until the chemical reactions are completed. It was found that this method, with all its primitiveness, would produce about thirty pounds of bleaching powder with 35 percent available chlorine in twelve hours operation. This particular installation, limited only by its physical dimensions, could produce enough chlorine to treat a little more than 1 mgd at 1.0 ppm chlorine dose.

REFERENCES

1. Sconce, J. S., *Chlorine: Its Manufacture, Properties and Uses,* Reinhold, New York, 1962.
2. Mond, Ludwig, "History of Manufacture of Chlorine," *J. Soc. Chem. Ind.,* 713 (1896).
3. Baldwin, R. T., "History of the Chlorine Industry," *J. Chem. Ed.,* 313 (1927).
4. Sawyer, C. N. "Hypochlorination of Sewage," *Sewage Ind. Waste,* **29,** 978 (1957).
5. Fair, G. M., Morris, J. C., and Chang, S. L. "The Dynamics of Water Chlorination." *J. NEWWA,* **61,** 285 (1947).
6. White, G. C. "Disinfection Practices in the San Francisco Bay Area." *J. WPCF,* **46,** 89 (Jan. 1974).
7. Krusé, C. W., Olivieri, V. P., and Kawata, K. "The Enhancement of Viral Inactivation by Halogens," *Water and Sewage Works,* **118,** 187 (June 1971).
8. White, G. C., private communication, K. Fraschina and A. Bagot, City and County of San Francisco, 1971.
9. Rietema, K. "Segregation in Liquid-Liquid Dispersions and its Effect on Chemical Reactions," *Chem. Eng. Sci.,* **8,** 103 (1958).
10. Anon., "Chlorine Bleach Solutions," Solvay Div. Allied Chem. Co. Bull. 14, New York, Second Ed., 1960.
11. Anon., "Chlorine," Columbia Southern Chemical Corp., Pittsburgh (now PPG Ind.), 1952.
12. Baker, R. J., "Characteristics of Chlorine Compounds," *JWPCF,* **41,** 482 (1969).
13. "Chlorine Bleach Solutions," Solvay Div. Chem. Co. Bull. 14, New York, 2nd Ed., 1960.
14. "Chlorination of Sewage With Hypochlorites," Dow Chem. Co. Form #125-1086-68, 1968.
15. Anon., Lithium Corp. of America, Product. Bull., New York, 1966.
16. Kirk, B. A., and Lindeke, W. A., "A New Pool Sanitizer Discussed by Maker," *Sw. Pool Age,* 36 (Jan. 1968).
17. White, G. C., unpublished data, shipboard chlorination systems, 1942.
18. Chang, S. L., and Fair, G. M., "Destruction of *Entamoeba histolytica,*" *J. AWWA,* **33,** 1705 (1941).
19. Chang S. L., "Destruction of Microorganisms," *J. AWWA,* **36,** 1192 (1944).
20. Anon., "Manual of Naval Preventive Medicine," Dept. of Navy, Bur. of Med. and Surgery, Chap. 6, part IV, Washington DC, Dec. 1962.
21. Bacon, V. W., "Chicago MSD Progress Report on Chlorination," *Water and Sew. Wks.,* **114,** 350 (Sept. 1967).
22. Springs, J. D., "Hypochlorination for Slime Control," *Power,* 102 (June 1957).
23. "Wallace and Tiernan 44 Series Solution Metering Pumps," Cat. File 440, 100 Rev. 5–76.
24. Peterson, R. BIF, private communication, Walnut Creek, CA, 1977.
25. Pulsafeeder Catalog No. 7120 "Pulsa 7120," Rochester, NY, 1976.
26. Straight, W. S., Union Carbide Co., personal communication, Oct. 12, 1976.
27. "Feedrator II Liquid Chemical Feed Dispenser," Fischer and Porter Co. Warminster, Pa., Cat. 70 FR 1000, 1977.
28. Barbolini, R. R., Metropolitan Sanitary District of Greater Chicago, personal communication, Nov., 4, 1976.
29. Resistoflex Corp., "Flexible and Rigid Piping Accessories," Bull. Sk-5 Roseland, NJ, 1975.
30. Barbolini, R. R., private communication, Aug. 4, 1970.

31. K. A. Steel, Chemicals Inc., Melrose Park, Illinois, private communication, Feb. 15, 1983.
32. Halter, C., Deputy commissioner of Water Operations, City of Chicago, private communication, Dec. 23, 1982.
33. Knox, D., Wallace and Tiernan Div. Pennwalt Corp., private communication, Houston, TX, Dec. 1982.
34. Jones Chemical Co, Milpitas CA, private communication, 1983.
35. Murray, B. W., "A Calcium Hypochlorite Manufacturing Process for Water Treatment Plant Use," Water and Sew. Wks., **100**, 318 (Sept. 1963).
36. Nagel, Carl, private communication, Sanitation Districts of Los Angeles County, 1972.
37. In-house analysis by Brown & Caldwell Engineers and G. C. White acting for Sacramento Area Consultants, County of Sacramento Regional Plant, 1975.
38. Stone, R., "The Small Scale Manufacture of Bleaching Powder in Backward Areas," *J. AWWA*, **42**, 283 (1950).
39. Bacon, V. W. "How Chicago Saved $2.5 Million," *American City*, 16, (Oct. 1967).
40. Dorolek, R. J., "Wastewater Plant Effluent Chlorination Made Easy and Inexpensive," *Water and Wastes Eng.*, 48 (Oct. 1968).
41. Tech. Practice Committee, Water Pollution Control Federation MOP No. 4, "Chlorination of Wastewater," Washington, DC, 1976.
42. Lue-Hing, C., Lynam, B. T., and Zenz, D. R., "Wastewater Disinfection: A Case Against Chlorination," Paper Presented at Forum on Disinfection With Ozone, Chicago, IL, June 2–4, 1976.
43. Chinn, R. Supt. North Point Treatment Plant, San Francisco, private communication, Nov. 1982.
44. Davies, J. W., Cox, K. G., and Symon, W. R., "Typhoid at Sea: Epidemic aboard an Ocean Liner," *Can. Med. Assoc. J.*, **106**, 877 (1972).
45. Merson, M. H. MD, Hughes, J. M. MD, Wood, B. T., Yashuk, J. C., and Wells, J. G. "Gastrointestinal Illness on Passenger Cruise Ships," *J. Am. Med. Assoc.*, **231**, 723 (Feb. 17, 1975).
46. Merson, M. H., Hughes, J. M., Lawrence, D. N., Wells, J. G., D'Agnese, and Yashuk, J. C., "Food and Waterborne Disease Outbreaks on Passenger Cruise Vessels and Aircraft," *J. Milk Food Technology,* **39**, 285 (April, 1976).
47. Center for Disease Control "Diarrheal Illness Among Passengers on the Cruise Ship *M/S Sun Viking,*" Public Health Service, CDC, Atlanta, GA, March 8, 1977.
48. Center for Disease Control, "Gastrointestinal Illness, *TSS Fairsea,*" Public Health Service, CDC, Atlanta, November 7, 1977.
49. Davis, W. A., "Many Luxury Liners Fail U.S. Sanitation Test," *Chicago Tribune,* April 16, 1978.
50. Blumenthal, R., "Many Cruise Ships Fail Health Tests," *New York Times,* Dec. 19, 1977.
51. Borcover, A., "Ship Violations Listed," *Chicago Tribune,* Jan. 1979.
52. Clary, M., "U.S. Inspections Help to Improve Ship Sanitation," *Chicago Tribune,* Sept. 30, 1979.
53. Brower, G. R., "A Chlorine Gas Cloud From Sodium Hypochlorite?", *AWWA Op Flow,* **10**, 3 (March 1984).

3
On-site Generation of Chlorine

HISTORICAL BACKGROUND

The Beginning

From time to time interest is revived in the possibility of producing chlorine electrolytically at the point of use, thereby eliminating the potential hazard of chlorine stored on-site in containers. One of the first installations of this kind was at Brewster, New York in 1893. The installation was known as the Woolf process. This process is particularly efficient where saline water is present. Therefore The Central Electricity Generating Board of London experimented with these systems as an alternative method of chlorinating steam power plant condenser cooling water at electric generating stations. They found that it was inefficient and beset by a great many operating problems, and the experimentation was terminated.

On-site generation utilizing the electrolytic process dates back to the 1930s, when Wallace and Tiernan made electrolytic chlorinators for YMCA swimming pools. Success in this field has always been marginal because the cost of the electrolytic equipment is always much greater than conventional equipment using bottled chlorine gas.

Electrodes were always a major source of trouble in the electrolytic systems. The most common of these have been platinum-coated; others have been made of carbon and iron, graphite with a lead shield inside a stainless steel sheath, and titanium electrodes coated with rare metals. The process is inefficient at best but it does have interesting possibilities, in particular the safety factor and the fact that raw materials might all be at the point of application, thereby eliminating storage requirements. However one of the current difficulties is the ever increasing cost of the electric power required for electrolytic production.

Early Experience in the USA

On-site generation of hypochlorite in the United States was largely inspired by the use of hypochlorite solutions during World War I (1914–1918). This solution became known as the Carrell-Dakin solution.[1] The success of this antiseptic treatment of open wounds led to the on-site generation of this solution in hospitals. One of the first electrolytic cells for this purpose was developed by Van Peursem et al.[2] This cell was designed to produce the equivalent of the Carrel-Dakin solution.[3]

Wallace and Tiernan first made electrolytic chlorinators to provide a safe means of swimming pool chlorination for those installations where pools were located in buildings where people slept. As early as 1939 Wallace and Tiernan established a policy that chlorine gas equipment should not be installed in such buildings. For this purpose they developed an electrolytic chlorinator.[4]

This unit aroused the interest of Pan American Airways who, at the time, (1936) were developing refueling sites on their San Francisco to Sydney and Orient flights. The electrolytic chlorinator for water supplies at these way stations was ideal. World War II changed all of these plans.

Current Interest

The enthusiasm for on-site generation disappeared until the hazard potential of chlorine gas stored in containers was evaluated owing to the proliferation of chlorine gas installations at wastewater and potable water treatment plants. At about the same time small electrolytic generators began appearing on the market (1950s) for use in backyard swimming pools. The cost of these units and the manufacturer's inability to provide satisfactory service discouraged the use of this equipment.

In the 1970s the popularity of on-site generation began to rise once again, largely because of the potential hazards of liquid–gas systems using chlorine stored in containers, and the availability of Federal funds for the necessary research and development of reliable equipment.

IMPORTANCE OF RAW MATERIAL

Seawater Systems

All of the on-site generators described in this text are designed primarily to use seawater, but can also use brine. There is one exception. This is the Ionics Inc. Cloromat unit, which uses a conventional membrane cell that produces both chlorine and caustic. This results in a conventional sodium hypochlorite solution. The seawater units can use a brine solution instead of seawater. The brine solution concentration is limited to 30,000 mg/l, and whatever goes into these units is discharged into the process stream that is to be treated. Therefore whether it is seawater or brine the TDS concentration in the generator discharge is going to be about 30,000 mg/l. This limits the application of these "seawater" units to disinfection of wastewater, cooling water chlorination, off-shore well drilling installations using chlorine to control marine growth and other such marine installations. Only under special circumstances can they be used for potable water treatment and water reuse applications (see the Summary of this chapter). They are not applicable to hydrogen sulfide control in sewage collection systems because seawater exacerbates the formation of H_2S in sewage.

These units are designed to be able to use saline waters without any pretreatment except screening or microstraining. However, in some instances filtration has proved to be necessary.

Seawater varies in composition from ocean to ocean. The TDS ranges from 30,000 to 36,000 mg/l, and 19,000 mg/l chlorides is about average. Table 3-1 shows the composition of seawater according to the Hydrographic Laboratory of Copenhagen.

Brine Systems

The quality of the raw material is of great concern in the operation of any electrolytic process. Chlorine manufacturers have long realized that successful cell operation is dependent upon the use of a pure brine. The best product, which needs very little pretreatment except for the brine makeup water, is food-grade salt which is mostly refined solar salt extracted from seawater by evaporation. Solar salt which is not refined must be treated by the on-site system. This is known as "stack" salt.

Then there is mined salt, or brines naturally occurring in the earth's crust. This material, unless it is of exceptional quality, also has to be treated at the site. Any underground brines should be checked for ammonia nitrogen content, as this is most undesirable in the electrolytic production of chlorine.

Salt or brine impurities seriously affect the operation and maintenance of any type of membrane cell. The allowable calcium ion content in the brine must be less than 0.05 mg/l according to Eltech. See Chapter 1.

EQUIPMENT DEVELOPMENT

Membrane Cell

Circa 1968–1970, the U.S. Environmental Protection Agency became actively concerned over the environmental impact of storm water overflows (from large combined sewer systems) discharging into confined receiving waters such as the Charles River, San Francisco Bay, Chesapeake Bay, etc. These massive overflows have long been considered hazardous to health (water contact sports and shellfish growing

Table 3-1 Composition of Standard Seawater

Cations	mg/l	Anions	mg/l
Na^+	11,035	Cl^-	19,841
Mg^{++}	1,330	$SO_4^=$	2,769
Ca^{++}	418	HCO_3^-	146
K^+	397	Br^-	68
SR^+	14	F^-	1.4

Total salinity, 36047 mg/l
Total alkalinity, 119.8 mg/l

areas) in the absence of disinfection. Since the application of chlorine would have to be at the point of storm water overflow, the chlorine containers would surely have to be transported through congested areas, thus creating the potential hazard of a chlorine spill.

The EPA, considering these factors, funded a study for on-site generation of chlorine carried out by Ionics, Inc., of Watertown, Massachusetts.[5,6] This project resulted in the development of an extremely efficient electrolytic cell similar to, but more advanced than, some of the older designs currently used by manufacturers of chlorine gas. Details covering the development and features of this cell are well documented.[5]

The two-compartment membrane cell with expanded electrodes as developed by Ionics, Inc. is illustrated in Fig. 3-1. The most important feature of this cell is the membrane which separates the anode compartment from the cathode compartment. This membrane separator concept is not new. Hooker Chemicals and Plastics Corp. began a program in 1950 which recently culminated in the introduction of the MX chlor-alkali cell for commercial production of liquid-gas chlorine.[7] The concept reached economic feasibility with the recent availability from DuPont of Nafion membrane material. The Nafion-family membranes consist of a 2–10 mil-thick film of perfluorosulfonic acid resin; a copolymer of tetra fluoroethylene; and another monomer to which negative sulfonic groups are attached. (For more detailed information on the development of membrane cells, see Chapter 1.)

The world's first *commercial membrane cell* chlor-alkali plant has been operating successfully since April 1975 for Asahi Chemical Co. at Nobeoka, Japan using Nafion 315 membrane cells.[8] Recently, however, a perfluorocarboxylic acid membrane developed by Asahi Chemical is reported to give a higher current efficiency and is being used to replace the Nafion membranes.[8]

In a membrane cell (Fig. 3-1) the anode and cathode compartments are separated by the cation-exchange membrane. This membrane inhibits negative ions (anions) from moving through the membrane but allows the positive ions (Na^+, K^+, H_3O^+ cations) to move freely. This effect is known as the Donnan exclusion.

There is no direct hydraulic flow from the anode compartment to the cathode compartment. The only water that passes through the membrane is endosmotic water, which is associated with the ions being transferred. The anolyte liquor (spent brine) can be sent to waste or partially recycled. Chlorine from the anode chamber is sent to a water cooled reactor where it is mixed with the caustic solution.

Brine is fed to the anode compartment. Water is fed to the cathode compartment to sweep out the sodium hydroxide (caustic) that is produced. The cathode compartment is cooled by water flowing through a heat exchanger.

The caustic and hydrogen produced in the cathode compartment discharges from a common port. The hydrogen is vented to the atmosphere.

The cathode compartment cooling water cannot be scale-forming. The electrodes are described as "expanded." This refers to their shape. They are rectangular

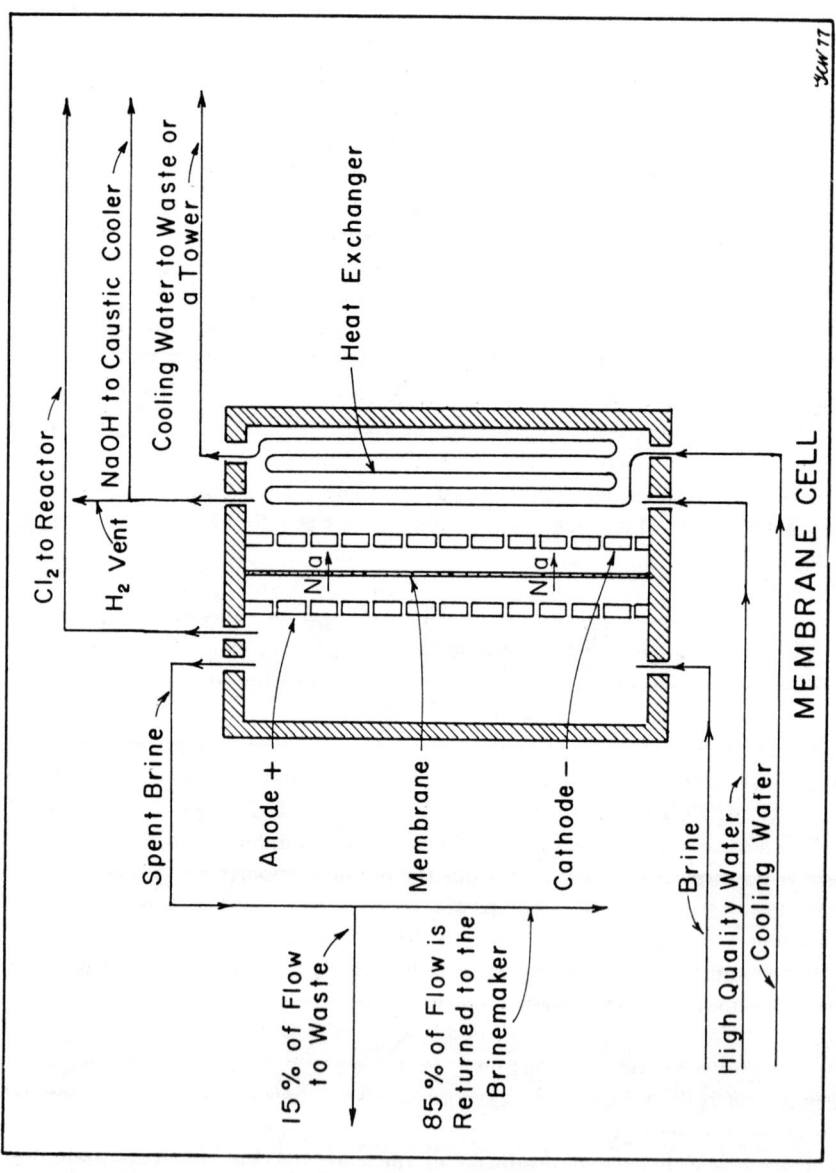

Fig. 3-1. Cloromat membrane cell with expanded electrodes.

pieces of metal perforated so that they look like a grating. The anode is dimensionally stable (DSA®), i.e., it is nonsacrificial. It is made of pure titanium with a platinized coating. The cathode has the same configuration as the anode except that it is sacrificial, as it is made of iron.

One of the major advantages of a membrane cell is that the anode and cathode can be placed close to the membrane. This increases the current efficiency and reduces the space required in stacking the cells side by side.

The Cloromat System (Ionics, Inc.)

Introduction. Ionics, Inc. use the membrane cell described above. The cells are assembled in modular form as in a filter press. The cells within a module are connected in series electrically, but connected in parallel hydraulically. Manifolds molded into the cell frames provide the connections for parallel fluid flow to and from the cells. The module and individual cells are designed for easy dismantling and reassembly required for annual maintenance. To increase the production of a unit, cells are added to the cell module in increments of about 165 lb/day chlorine per cell. A 2500 lb unit consists of 15 cells. The system is designed to produce an 8 percent hypochlorite solution. For this strength solution a 2500 lb/day unit requires about 2 GPM of high-quality water to sweep out the caustic from the catholyte compartment of the 15 cells involved. The flow diagram of a 2500 lb/day unit is shown in Fig. 3-2. The brine maker requires about 1 GPM so that the total amount of water to be produced by the water treatment unit is on the order of 3–4 GPM.

A water supply of 25 GPM is required for cooling the cells, the caustic cooler, and the chlorine-caustic reactor (see Fig. 3-2). This must be a water supply completely free of any scale forming substance.

The brine solution pumped from the brinemaker to the cells is about 1 GPM containing 3 lb/gal. NaCl (4400 lb salt/day). This calculates to 1.75 lb salt/lb chlorine.

The spent brine should be discharged to waste unless recycling is planned as part of the system. Recycled spent brine must be subjected to special treatment before it is allowed to return back into the system. The spent brine flow is about 0.6 GPM for a 2500 lb/day unit (see Fig. 3-2).

All of the mass balance figures shown in Fig. 3-2 are based upon food-grade salt. This represents 1.75 lb NaCl/lb chlorine, at 2.0 kWh/lb of chlorine. In the San Francisco area electrical energy costs about 8.5¢/kWh.

When considering this system the designer must be careful to specify the grade of salt to be supplied. According to Ionics, Inc.[9] food-grade salt, which is refined, is so free from the usual impurities normally associated with sodium chloride that the hypochlorite solution thus produced is characterized as "Rayon Grade Bleach." This means that this hypochlorite does not contain the usual impurities

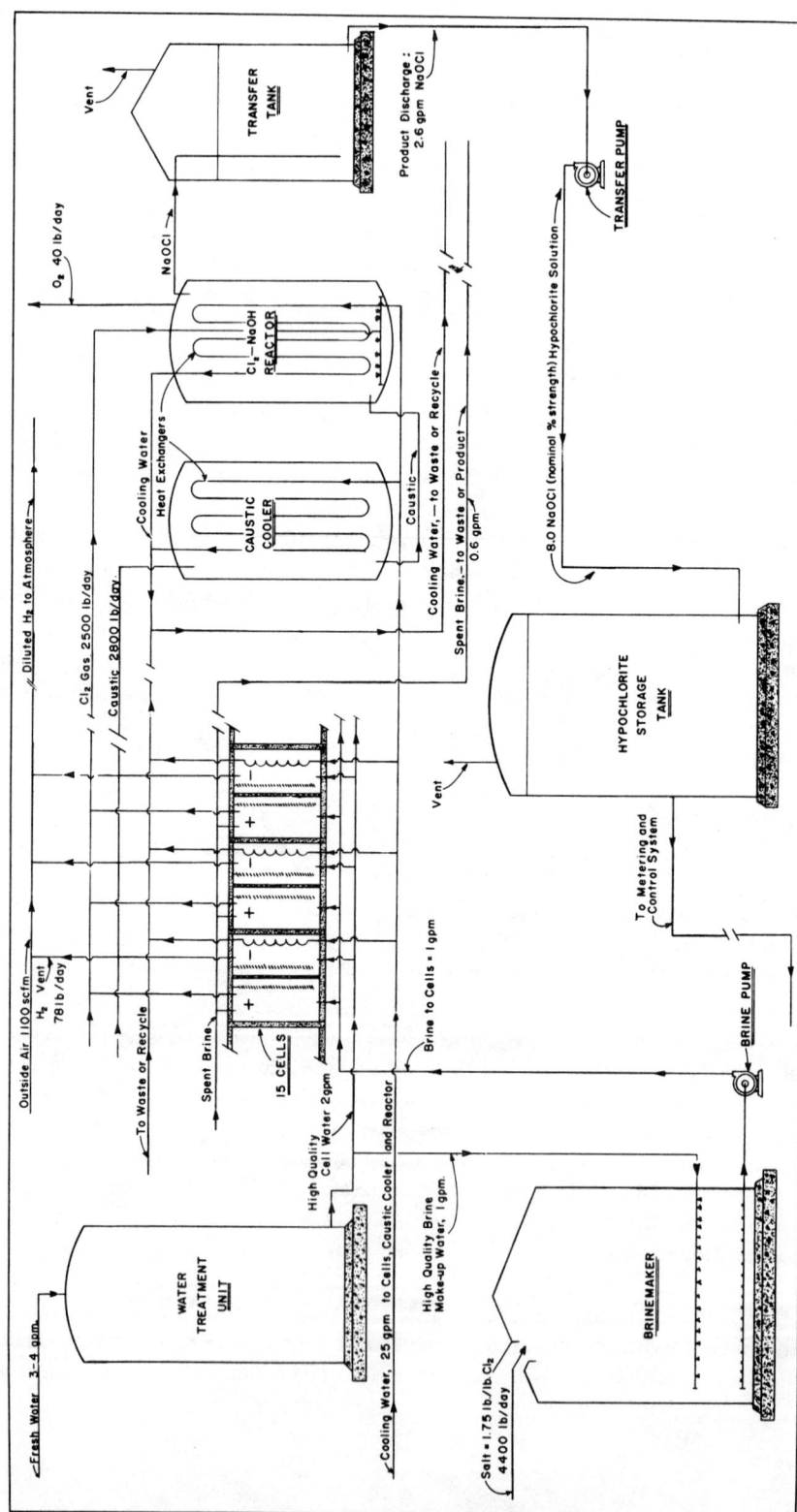

Fig. 3-2. Ionics, Inc. Cloromat hypochlorite generating system 2500 lb/day chlorine capacity.

(heavy metal ions) which contribute to the instability of the hypochlorite solution strength during average storage conditions. For example, the usual 10 percent commercial bleach (NaOCl) deteriorates from 10 to 8.5 percent in 40 days as compared to the Cloromat 8.43 percent solution which deteriorates to 7.5 percent in the same length of time. However, it becomes a question of economics whether or not food-grade or stack-grade (unrefined) solar salt should be used.

Stack-grade salt requires a pretreatment system which, in terms of power, is estimated at 0.32 kWh/lb chlorine.

Food-grade salt in the San Francisco Bay Area at the harvester costs (1977) 2.0¢/lb while stack grade is 1.0¢/lb.

The Cloromat system is of course a proprietary manufacturing process. When it is compared to other on-site processes there will always be claims and counter-claims for the various processes.* The designer is therefore warned to consider in detail all of the suggested pretreatment requirements of raw materials presented and recommended by the various manufacturers. These will include, but will not be limited to: the treatment of cell water, cooling water (if any), brine, and recycled brine. Other factors are the variations required in the materials of construction. These can be dependent upon the degree of pretreatment and the use of reclaimed raw material.

For example, there will always be a small amount of molecular chlorine in recycled brine, so the brinemaker tank should be made of filament wound fiberglass if brine recycling is involved.

Space Requirements. Any comparison between alternative methods of chlorination should include the space required to house both the equipment and chemical storage facilities. For example, a 5000 lb/day conventional chlorinator facility using ton containers requires about 1000 sq ft for all the equipment (chlorinators, evaporators, analyzers, etc.) plus space for 16 containers.

Based upon two Cloromat installations (ca. 1980), a facility capable of producing 5000 lb/day including standby equipment requires about 5100 sq ft. This includes space for the electrolyzer cells, metering pumps, analyzers, chemical storage and a brine purification system. This does not include much needed laboratory space to carry out monitoring of the chlorine generation process.

Operating Costs. The following costs are based upon a user analysis report covering a 12 month span of continuous operation.[19] All costs analyzed cover only those pertaining to the production of chlorine. These costs are separated into seven separate categories described below.

---

* One of the most important considerations of any on-site generation system is the strength of hypochlorite solution produced. This involves size of storage tanks, pumps, and piping systems.

Raw Materials

Salt consumption	1.75 lb/lb Cl	@ $.02/lb	= $.035/lb Cl
Feed water to cells	9 gal/lb Cl	@ $.0007/gal	= $.006/lb Cl
Sodium hydroxide	0.01 lb/lb Cl	@ $.16/lb	= $.002/lb Cl
Soda Ash	0.02 lb/lb Cl	@ $.14/lb	= $.003/lb Cl
Water softeners salt beads	0.02 lb/lb Cl	@ $.05/lb	= $.001/lb Cl
		Total	$.047/lb Cl

Electrical Energy

Electrolytic cells:	2 kWh/lb Cl	@ $.085/kWh	= $.170/lb Cl
Brine purification system	.32 kWh/lb Cl	@ $.085/kWh	= $.027/lb Cl
		Total	$.197/lb Cl

Anode and Membrane Replacement
Calculated cost from 12 months experience $ .01/lb Cl

Mechanical and Electrical Labor Cost
Routine maintenance $.005/lb Cl
Periodic maintenance $.009/lb Cl

Operating Labor Cost (6 hrs per shift)
Routine daily inspection and calibration plus electrolyzer and brinemaker maintenance: $.054/lb Cl

Summary of Chlorine Production Cost

Raw materials	$.047
Electrical energy	.197
Anode and membrane replacement	.010
Routine maintenance	.005
Periodic maintenance	.009
Operating labor	.054
Parts and materials	.006
Total chlorine cost per lb	$.328

Comparison with Purchased Hypochlorite. At the same time that the above report was made two local suppliers were quoting 14 percent trade strength sodium hypochlorite at $.47 per gallon delivered. To calculate the amount of "available chlorine" in one gallon, multiply trade strength by 0.08345. Therefore local hypochlorite at 1.17 lb chlorine per gal = $.40/lb Cl. However, in actuality the cost of chlorine will be more than this because of the high decay rate of these solutions. Purchases of hypochlorite in quantity should be negotiated to arrive at a fair price on the basis of a two week decay rate (see Chapter 2).

Engelhard Industries

Introduction. Early in the 1970's, the United States Department of Interior Office of Saline Water awarded a contract to construct, install, and operate a seawater hypochlorite generator at their desalting test facility, Wrightsville Beach, North Carolina. The contract for this project was awarded to Engelhard Industries following an 18 month test of a prototype unit at the OSW facility.[10]

The unit developed by Engelhard for desalting plants is also applicable to other seawater situations, the most important being the control of marine fouling organisms in seawater-piping systems. Originally the Engelhard Chloropac system was designed specifically for shipboard use to combat slime and marine growths of all kinds which normally thrive in the seachests and the ship's seawater systems. Protection from the proliferation of these growths is needed wherever seawater is used (i.e., condenser cooling, general engine room services, circulating water in the ship's air conditioning system, fire system, and other seawater piping throughout the vessel). Moreover, chlorination is now mandatory for the potable-water supply aboard ship even though the water is produced by the ship's distillation process.

One other important application which gives validity to this system and thereby enhances its reliability is the widespread use of the Engelhard Chloropac on seawater drilling platforms and seawater supertanker storage platforms. These storage systems require reliable prevention of marine growth in the pump passages and storage-system piping.

Projecting this concept of making hypochlorite from saline water will, therefore, surely include its use for wastewater disinfection whenever a reasonable supply of seawater is available.

Raw Material. All of the Engelhard systems are designed to handle an electrolyte equivalent in strength to a normal seawater (i.e., 19,000 mg/liter chlorides and 30,000 mg/liter TDS). This holds true for all of their systems whether it be recycled seawater or brine. The brine is always diluted to the optimum seawater chloride content before it is sent to the cells.

This presents a significant parameter, namely, the amount of raw product required to provide the chlorine generating capacities. This amounts to a minimum of 20 gpm of electrolyte (seawater) for a module containing a series grouping of 2 to 10 cells. Each ten-cell module can produce up to 240 lb/day chlorine as hypochlorite. The modules are connected hydraulically in parallel so a 480 lb/day unit consisting of two 10-cell modules requires 40 gpm seawater and so on. The maximum concentration of chlorine in the hypochlorite solution generated in the once through seawater system is about 1200 mg/liter and the TDS will be on the order of 30,000 mg/liter. If refined brine is used instead of seawater, the TDS will be about 20,000 mg/liter.

Cell Configuration. The Engelhard cell is a flow-through or "open-type" cell where the saline water is subjected to electrolytic decomposition on an incremental basis as the salt water flows through a series of these cells. The cell is designed so that the salt water flows through an annular opening at a velocity of 5–7 ft/sec. This velocity is supposed to continuously flush out precipitates of insoluble anions normally found in seawater or other brackish waters. The patented Chloropac® cell is illustrated in Fig. 3-3. It is constructed of three titanium cylinders. Two cylinders are placed axially in line and connected by their flanges with an insulating cylindrical spacer to form a smooth-bore pipe. The third cylinder is small in diameter and longer than the first two pipes. This third cylinder is sealed at each end and placed inside the pipe formed by the first two (see Fig. 3-3.). The salt water electrolyte flows in the annular space between the inside of the outer cylinders and the outside of the inner cylinder.

The inside surface of the outer cylinder on the left is coated with a proprietary platinum alloy which allows it to respond as though the entire cylinder were a solid platinum alloy. This cylinder connected to the positive terminal generates molecular chlorine on its inner surface. The outside of the inner cylinder adjacent (to the right) receives the electric current as a cathode and releases the cathode products (i.e., sodium hydroxide and hydrogen). The right hand half of the cell operates in the same fashion except the roles of the anode and cathode surfaces are reversed. Here the outside of the inner titanium surface is coated with platinum alloy and the current passes from the outside of the inner cylinder to the inside of the second outer cylinder. The flowing stream of electrolyte mixes the products produced at the anodes and cathodes which produces a weak sodium hypochlorite solution and minute bubbles of hydrogen gas. This chemical reaction is summarized by the equation shown on Fig. 3-3.

It should be noted that both the inner and outer pipes of this cell are made of titanium, whether platinized or not. Titanium is resistant to salt water corrosion. Moreover, titanium possesses the unique chemical characteristic of being able to form a protective oxide coating so that it will receive but not emit a current in the 8–12 V dc range. The platinized anodes are consumed at the rate of 6 mg/A year. This calculates to an expected anode life of about 6 years.

As long as the system stays within the power design specified by the manufacturers, the titanium cathode will not be consumed in the electrolytic process. It is therefore labeled as an "infinite life" electrode by the manufacturer.

Chloropac Generator. The Engelhard system is usually arranged for 10–12 cells in series. This is identified as a module, and the arrangement of these modules is further identified with a model number as to equivalent chlorine capacity in lb/day. Figure 3-4 illustrates a 960 lb/day generator. This unit consists of 40

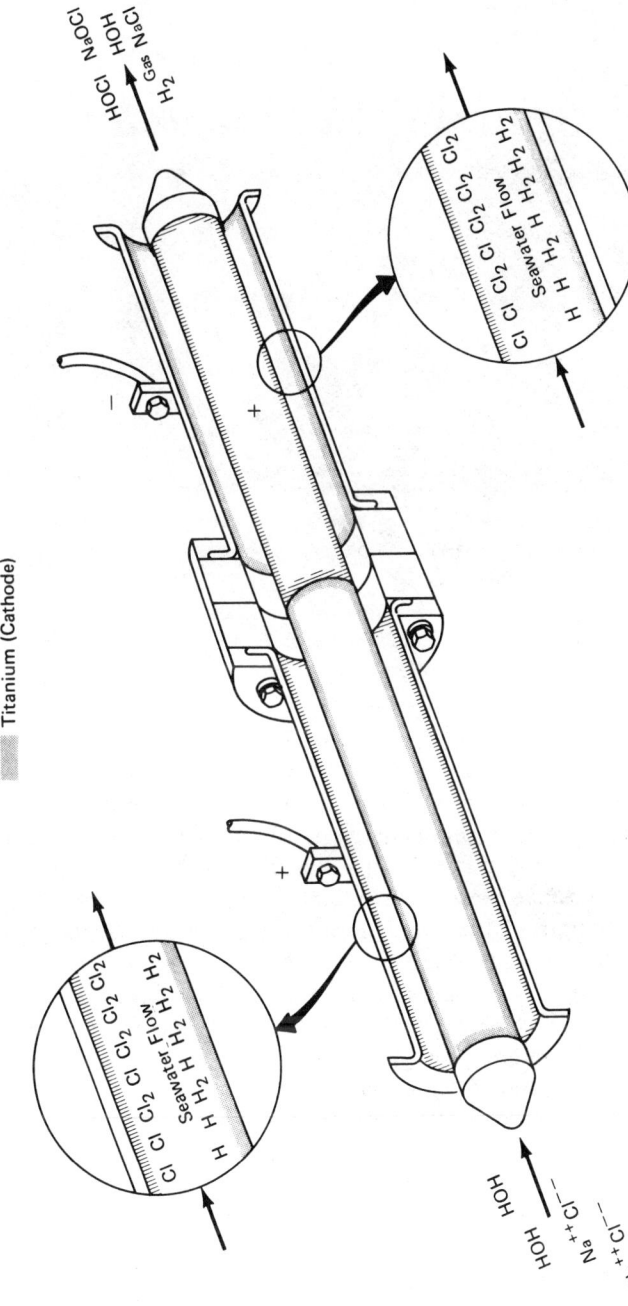

Fig. 3-3. Englehard seawater cell anode reaction $2Cl \rightarrow Cl_2 + 2e$, cathode reaction $2H + 2OH \rightarrow H_2\uparrow + 2OH + 2e$, overall reaction $NaCl + H_2O + D.C. \rightarrow NaOCl + H_2$ gas (courtesy Engelhard Industries, Systems Dept.)

Fig. 3-4. Engelhard seawater cell system of 40 cells in series; capacity 960 lb/day available chlorine (courtesy Engelhard Industries, Systems Dept.).

cells arranged so that there are two groups of 20 cells in series. However, each of these two groups is connected in parallel hydraulically.

The individual cells are arranged in pairs so that the first and last cells of any group are properly grounded to eliminate the possibility of stray current corrosion. Each cell is then electrically connected in series to the center pair. These anodes are connected to the positive power source. Since the cells are connected in series hydraulically, the strength of the hypochlorite solution produced increases from one cell to another.

Assuming a seawater with a total salt content of 32,000 mg/liter will contain about 19,000 mg/liter of chloride ions, each cell will generate about 100 mg/liter of hypochlorite. Therefore, each group of 10–12 cells produces a 1000–1200 mg/liter hypochlorite solution. At this concentration there is no need for product cooling or hydrogen gas venting (H_2 venting is required if a hypochlorite storage tank is used).

The electrolyte flow through any given cell is limited to 20 GPM. Therefore, assuming a seawater content of 19,000 mg/liter chloride ions, any saline water of this concentration will produce 1.0 lb Cl_2/hr/cell at the 20 GPM flow of electrolyte. Therefore, the Chloropac generator system is arranged so that the modules of hypochlorite production are hydraulically connected in parallel consisting of series connected units of 10 lb/hr capacity each. A 20 lb/hr system would consist of two groups of 10 cells in series arranged in parallel.

Chloropac Systems

Once-Through Seawater System. This is the simplest system used for an unlimited supply of saline water containing from 1.5 to 4.0 percent salt and where a weak sodium hypochlorite solution is acceptable. Such a salt supply will result in hypochlorite solution strengths varying from 100 to 1000 mg/liter available chlorine. Typical users of this salt water source include: seaboard utilities; industrial condenser cooling water systems directly dependent upon a salt water source; offshore oil production facilities; desalting facilities; and sea-going vessels.

Once-Through Brine System. This system is attractive if there is an abundant supply of either salt or brine, and the cost is lower than for the more efficient recycling system. Brine solution is prepared in a salt dissolver, then mixed with feedwater until it reaches the approximate salinity of sea water, 3–3.5 percent salt, which is the optimum strength for electrolytic decomposition. This solution is fed to the electrolytic generator and converted to a weak NaOCl solution as in the seawater system above. The typical user might be an inland industrial plant or waterflood facility located in an area of natural salt beds or strong brackish ground water.

Seawater Recycle System. In this system, the initial weak hypochlorite solution is discharged from the generator to a holding tank and then recycled to join the incoming brine flow. Recycling on a continuous basis gradually raises the strength and temperature of the solution. The maximum strength of solution is about 4000 mg/liter available chlorine. The rise in temperature of this process tends to speed up the decay of the hypochlorite solution, so a minimum storage time of the final product should be considered.

Seawater or Brine Recycle with Cooling of Product. Whenever recycling is involved, maximum extraction of chlorine is dependent upon the optimum temperature of the product. This requires a cooling system.

The system is designed to provide maximum economy of chlorine extraction from the brine. The brine is first diluted to nominal seawater salinity (about 19,000 mg/liter chlorides), and then it is recycled through the Chloropac generator until the hypochlorite generated reaches about 1.0 percent concentration. The amount shown in the Engelhard literature is 7520 mg/liter.[11] Figure 3-5 illustrates the operation of the brine recycle system. The entire process of brine dilution, product recycling, and product cooling is automatically controlled.

Automatic Dosage Control. The Engelhard systems are responsive to automatic control. The concentration of the hypochlorite solution can be controlled by a saturable reactor which responds to a 0–5 mV signal. This controller, which is a transformer within a transformer, can utilize a cascade system whereby a flow signal is combined through an electronic multiplier with a chlorine residual

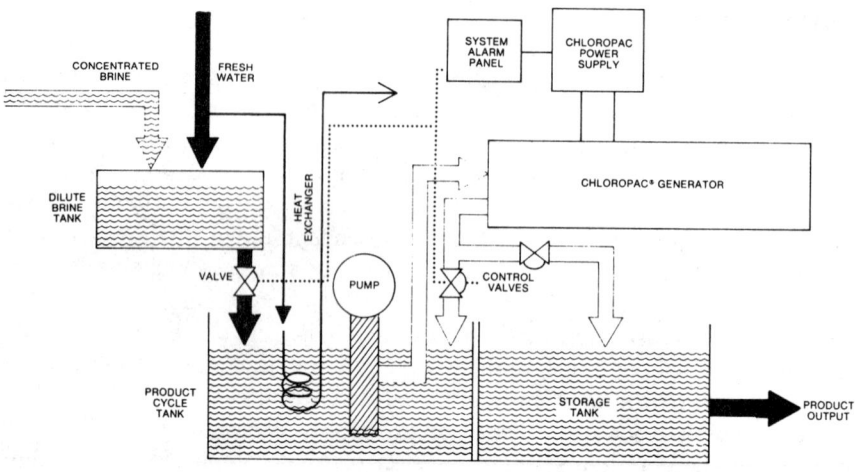

Fig. 3-5. Engelhard Brine System (courtesy Engelhard Industries, Systems Dept.).

analyzer signal to produce a final control signal. This signal changes the power input which in turn changes the number of Faraday units applied to the constant flow of brine through the cells. This changes the amount of chlorine generated in accordance with the control signal.

Cost Comparisons. Engelhard claims 2.3 kWh/lb of chlorine at 70°F and 2.6 kWh/lb at 50°F, assuming a nominal seawater concentration of 19,000 mg/liter of chloride ion. The estimated salt requirement for the Engelhard brine system is 3 lb salt/lb of chlorine.

The Houston analysis by Matson and Coneway[12] compared on-site generation of a system producing an 0.8 percent hypochlorite solution and one producing an 8.0 percent solution. Both of these systems were evaluated on the basis of a concentrated brine system. Cost comparisons for individual systems are burdened with many unknown pitfalls; however, it is pertinent to note that the 8 percent hypochlorite system was by far the best buy for the dollars invested. This system demonstrated a 30 percent advantage over the 0.8 percent system for a 5-ton per day facility. This is not meant to be an indictment against the seawater systems but a warning to the designers of the problems involving system selection.

Sanilec[13] claims that using 80 percent strength seawater at 25°C will require 2 kWh to produce 1 lb of chlorine at full load production.

Much more operating experience is needed with both the seawater and brine systems to determine the chlorine production costs of these systems.

Sanilec Systems

Introduction. On site hypochlorite generating systems by this trade name have an interesting origin. The parent company of Sanilec is Diamond Shamrock, one of the leading producers of chlorine over the past 20 yr. Previously, it was known as the Diamond Alkali Co. The parent company must be credited with one of the most significant improvements in the chlor-alkali industry for many years: "the Dimensionally Stable Anode" DSA®. After acquiring the basic patents, trademarks, technologies, and laboratories of Electronor Corporation, Diamond Shamrock formed the Electrode Corp. of America and developed this new type of metallic electrode in cooperation with the DeNora interests in Milan, Italy.* Through Electrode Corp., Diamond Shamrock has licensed over 55 percent of the North American chlorine capacity to use the energy saving DSA® technology. The continuing development of chlorine production efficiency led to the development of cells for the production of chlorine from seawater. Diamond Shamrock is also responsible for the successful development of an ORP (Oxidation-Reduction Potential) control system used in the production of hypochlorite solutions.

With this background and demonstrated expertise, Diamond Shamrock launched the Sanilec systems for on-site generation of hypochlorite to capture whatever market was available on the basis of either expediency or safety.

Sanilec offers two types of systems for on-site generation. One is for the use of saline waters similar to seawater situations or various dilutions thereof, depending on local conditions. The other is the use of a concentrated brine as raw material. These systems have been described in detail by Bennett and Cinke.[14] These authors emphasize that many important facts distinguish electrolytic cells which use seawater from those which use prepared brine solutions. Therefore, Sanilec set about to design two distinctly different types of cells; one for seawater and one for pure brine.

Seawater System. These systems suffer from impurities inherent in seawater; therefore, the cell configuration must be designed accordingly. These impurities cause bulky deposits which interfere with the electrolyte flow. These deposits are a result of natural seawater hardness, which is about 1800 mg/liter, caused by calcium and magnesium in seawater containing 19,000 mg/liter chloride. The precipitates of these ions not only interfere with the hydraulics of the cell system but also act to insulate the cathode. This results in a reduction in the process efficiency. By using a turbulent flow regime through the cell Sanilec claims 3–6 months continuous operation before cell cleaning is required. Cleaning consists of flooding the cells with a 10 percent muriatic acid solution for 1–2 hr. This therefore requires the availability of standby cells to accomplish this routine maintenance. The Sanilec seawater cell is shown in Fig. 3-6. The electrodes of the seawater cell are classed as dimensionally stable. The anode is the expanded metal type while the cathode is a thin solid sheet of metal which is a proprietary alloy.

The seawater system is shown in Fig. 3-7. This system requires filtration of the seawater before it is sent to the cells. Each cell module receives a constant

---

* Now the chlor-alkali technology is operated by Eltech Corp., a subsidiary of Diamond Shamrock Co.

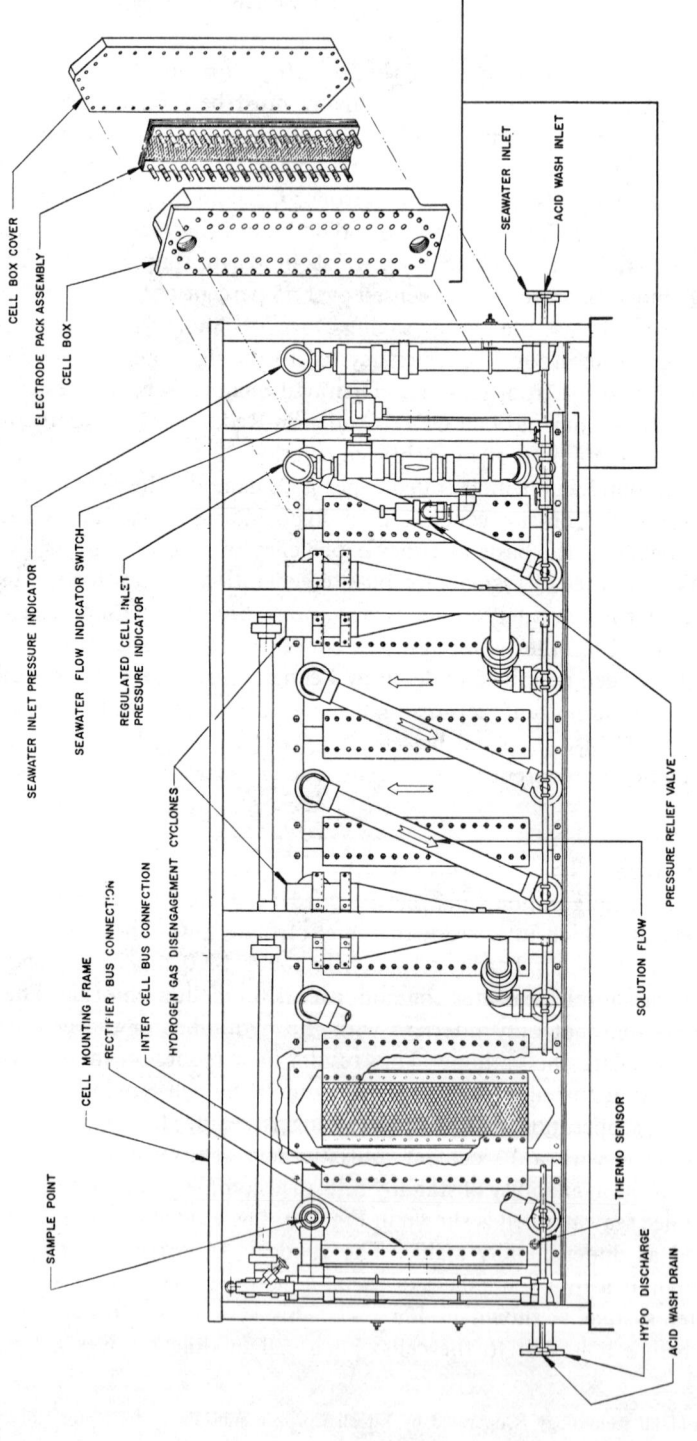

Fig. 3-6. Sanilec seawater cell (courtesy Diamond Shamrock, Electrode Corp.)

ON-SITE GENERATION OF CHLORINE 137

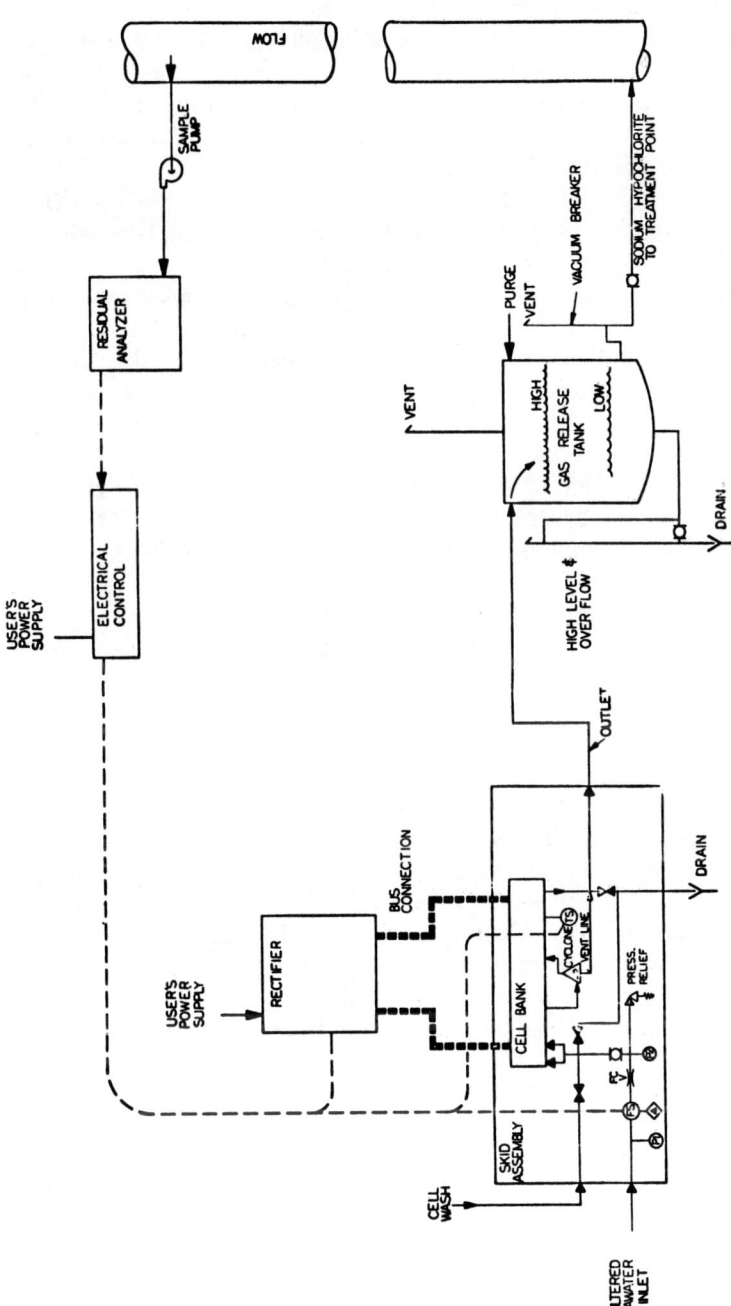

Fig. 3-7. Sanilec seawater system (courtesy Diamond Shamrock, Electrode Corp.)

flow of seawater. Hypochlorite production is controlled by current variation from the rectifier. Production control may be accomplished either manually or paced automatically by plant flow and trimmed by residual chlorine concentration analysis as shown in Fig. 3-7. The concentration of the hypochlorite solution produced from seawater is limited to a range of 200–1250 mg/liter hypochlorite as available chlorine (0.02–0.125 percent).

Hydrogen scavenging of the cathode is provided for each pair of cells. Otherwise the hydrogen formed at the cathodes would form an insulating layer which would result in an abnormal voltage rise across the electrodes.

The hypochlorite solution leaving the electrolytic cells passes to a gas release tank where the hydrogen produced in the electrolytic process is vented to the atmosphere by air dilution or is used as a by-product for its heat value.

These systems can provide hypochlorite solutions at up to 15 psi back pressure without the necessity of pumping the hypochlorite solution. The proper amount of seawater must be delivered to cells at a minimum of 40 psi pressure. Water flow varies with the size of the system. The Sanilec 1800 lb/day unit which is their Model S-1800 produces a 950 mg/liter available chlorine solution using six 300 lb/day cells and requires a seawater flow rate of 160 GPM. A 150 lb/day system requires two 75 lb/day cells and 20 GPM seawater flow.

Recycling is not used in the Sanilec seawater system. Although recycling can be used to lower cell voltage, this advantage cannot balance the accompanying inherent loss in current efficiency.

Sanilec points out one important factor in the production of hypochlorite from seawater: these solutions generated from seawater are inherently unstable. They deteriorate significantly within 48 hr because the seawater brine contains a vast array of ions which catalytically decompose the hypochlorite solution. Therefore, recycling seawater systems is not recommended by Sanilec.

The power consumption of a standard Sanilec system is a function of three variables; seawater salinity, water temperature, and percent of maximum production. A seawater salinity of 100 percent is generally defined as 18,900 mg/liter of chloride ion. In practice, seawater strength can be and is significantly diluted, especially in harbors or waterways with large inflows of surface water run-off. The Sanilec system is not designed to operate on seawater with less than about 9500 mg/liter of chloride ion. The example given by Sanilec for a seawater containing 80 percent salinity ($0.8 \times 18,900$) at a water temperature of 25°C will produce available chlorine at 2 kWh/lb chlorine at full load.

The use of seawater as cell feed contributes to a certain amount of chlorides and total dissolved solids to the plant effluent which may sometimes be a significant factor. Additionally, electrolysis of seawater will produce a certain amount of suspended solids in the form of magnesium and calcium hydroxides, and carbonates. For example, a 10 mg/liter dose of chlorine will add to the treated effluent; 189 mg/liter chlorides, 355 mg/liter TDS and 2 mg/liter SS.

Present operating experience indicates that one of the most crucial parts of any seawater system, whether it be Sanilec, Engelhard, De Nora or others, is the seawater supply system. *First,* the seawater must be filtered; *second,* the pumping system and piping should be in duplicate because the intake piping is subject to marine life biofouling and has to be taken out of service for periodic cleaning; *third,* the piping system from the intake suction to the electrolyic cells must be designed to provide a flushing velocity of 4–5 ft/sec. Moreover, all of this piping should be PVC, Kynar, fiberglass, or saran or rubber lined steel pipe.

Brine System. As in all electrolytic systems of chlorine and/or hypochlorite production, prepared brine solutions rather than seawater or other brackish waters are much easier to deal with. This leads to a contrasting cell design. Electrolysis with selected grades of salts makes it practical to remove all hardness involved with cation resin exchangers. This allows lower electrolyte flow rates which results in higher concentration of hypochlorite solutions.

The Sanilec brine electrolysis system requires the use of purified salt (food grade) to be dissolved in softened water as do competitive systems. The resulting salt solution is optimum at 28 g/l NaCl. The hypochlorite solution produced by the electrolysis of this salt solution will contain about 8000 mg/liter (0.8 percent) available chlorine. The Sanilec cell used in the brine system is shown in Fig. 3-8. Owing to the design of the brine system cell and the low concentration of hypochlorite, no system cooling is required. This system produces hypochlorite directly in the cell. No molecular chlorine is evolved. Both of the electrodes are the expanded metal type. The anode is coated with a precious metal oxide. This cell is considerably different from the seawater cell.

The Sanilec purified brine system compares with other similar systems as follows: 3.5 lb of salt, 15 gal. of water, and 2.5 kWh of electrical power will produce 1 lb available chlorine.[14]

Wallace and Tiernan (OSEC) On-Site Electrolytic Chlorination Systems

Introduction. Wallace and Tiernan, one of the leaders in chlorination technology for 70 years, has introduced two innovative systems for the electrolytic generation of hypochlorite.[15] Standard systems are available in capacities from 25 to 2500 lb/day. See Fig. 3-9. Special systems are available for higher capacities. Each system is custom-designed and engineered for each installation. The operating characteristics of each of these systems is based upon experience with both brine and seawater. Of all the suppliers of on-site chlorine generating technology Wallace and Tiernan has the longest record.[4] White inspected and operated one of the first of these units to be used at Pan American Airways refueling stations on the San Francisco to Sydney and Orient run, circa 1939.

The Electrolyzer. The electrolyzer (Fig. 3-10) is a once-through flow unit designed for generating efficiency and simplicity of field maintenance. The anodes,

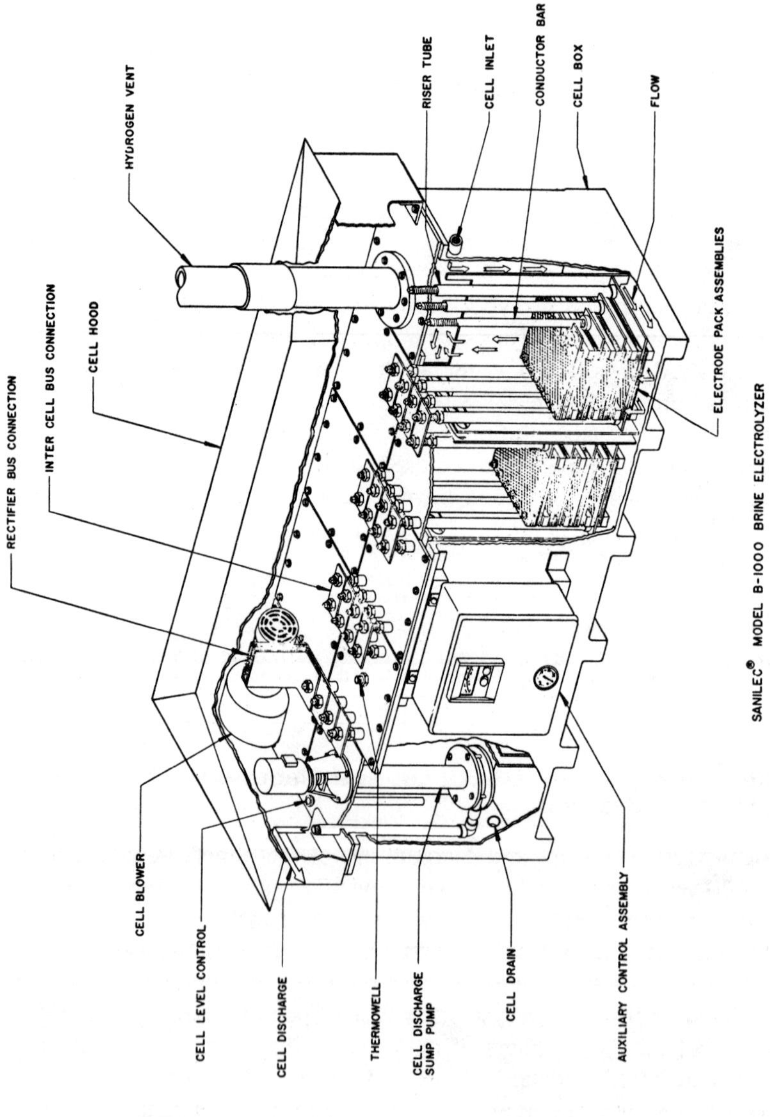

140 HANDBOOK OF CHLORINATION

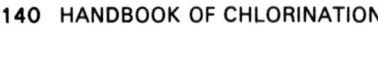

Fig. 3-8. Sanilec brine cell (courtesy Diamond Shamrock, Electrode Corp.)

Fig. 3-9. W&T series 85.510 OSEC seawater system capacity 2500 lb/day available chlorine. (Courtesy Wallace and Tiernan Div. Pennwalt Corp.).

which are the only parts susceptible to passivation (wear through electron transfer), are made of oxide-coated titanium (valve metal substrate). A minimum service period of two years can be expected of these anodes. The cathodes are made of Hastelloy "C." The electrode chassis is light weight and is arranged so that it can be easily removed from the electrolyzer casing for servicing without disturbing the plumbing.

The electrolyzer consists of 1, 2, 4, or 8 casings depending upon the chlorine production capacity. Each casing is divided into four cells, each with a set of electrodes. The cells are arranged in a bipolar configuration. Brine enters the casings and floods the cells. A DC current impressed upon the electrolyzer converts the sodium chloride to molecular chlorine and sodium hypochlorite. There also remains in solution unreacted brine and hydrogen gas, which is a product of electrolysis. This is shown in the following diagram:

142   HANDBOOK OF CHLORINATION

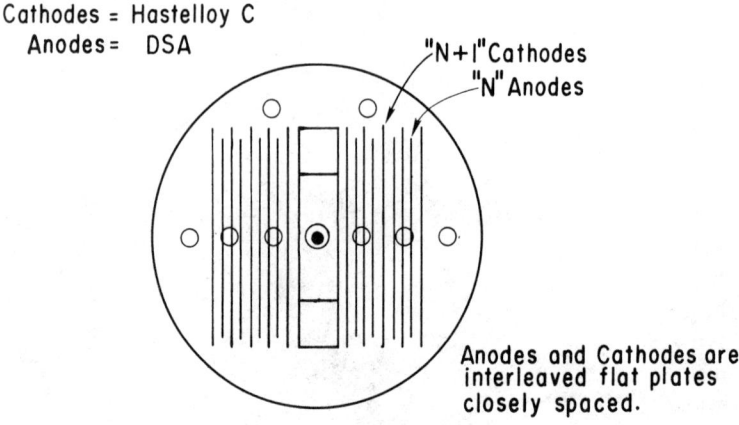

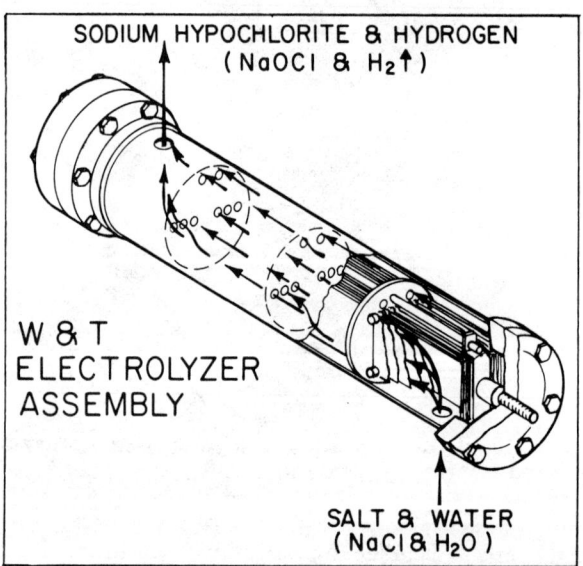

Fig. 3-10.   OSEC electrolyzer (courtesy Wallace and Tiernan Div. Pennwalt Corp.).

$$2NaCl + 2H_2O \xrightarrow[\text{Cathode }(-)]{\overset{\text{Anode }(+)}{\underset{2NaOH + H_2}{\overset{Cl_2}{\rightleftarrows}}}} NaOCl + H_2 + NaCl + H_2O$$

seawater
or   + energy + sodium hypochlorite
prepared brine

The design configuration of the electrolyzer accelerates the removal of hydrogen gas from the generation zone by thermal convection. The separated hydrogen gas passes through gas ports in the compartment partitions, travelling laterally to a discharge connection. The brine, electrolyte, and hypochlorite pass from one cell to the next through ports located below the solution level. In multiple casing arrangements, the electrolyte and hypochlorite pass through an outlet connection in the first cell of the next casing. In the final casing the electrolyte and hypochlorite are discharged together with the hydrogen gas to a storage tank which is designed to scavenge the hydrogen from the final product.

As brine progresses through the successive electrolyzer cells, its degree of conversion to hypochlorite increases. Concentration of the final product is about 8000 mg/l available chlorine when generated from prepared brine; and about 1800 mg/l from seawater.

Most salts, and practically all waters contain some degree of hardness in the form of calcium and/or magnesium. These impurities are prone to form insulating precipitates between the electrodes which inhibits the electrolytic process. This reduces the conversion efficiency and increases maintenance costs. Wallace and Tiernan claim their design minimizes this ubiquitous problem if properly maintained.

Figure 3-11 illustrates all of the components of a typical on-site electrolytic chlorination system, Wallace and Tiernan version. This illustration shows both or either the brine and seawater applications.

Differences Between Seawater Model and Brine Model. This is an important consideration in the choice of these models, regardless of manufacturer.

The Wallace and Tiernan brine system produces about 1 lb of chlorine for every 3.5 lb of salt; i.e., about 8,000 mg/l available chlorine solution using a 28,000 mg/liter TDS brine. With a brine system the NaCl is efficiently used at the expense of a little less efficient power utilization.

A seawater system provides an almost "free" source of salt; however power efficiency is paramount. The Wallace and Tiernan seawater system will generate an 1800 mg/liter available chlorine solution using seawater at 34,000 mg/liter TDS or about 1 lb chlorine for every 19 lb salt. In this model the water velocity is kept high and the electrode spacing is increased in order to minimize problems with scale formation that is inherent in seawater use. The scale formation is a formidable problem owing to the high concentration of calcium and magnesium ions.

Manufacturers Recommendations

Seawater Systems. Seawater temperature, min. 40°F; Salinity, min. 17 g/l. Seawater must be filtered to remove all particles larger than 0.045 in. These systems must be acid flushed with 2 percent hydrochloric acid. This removes the inherent

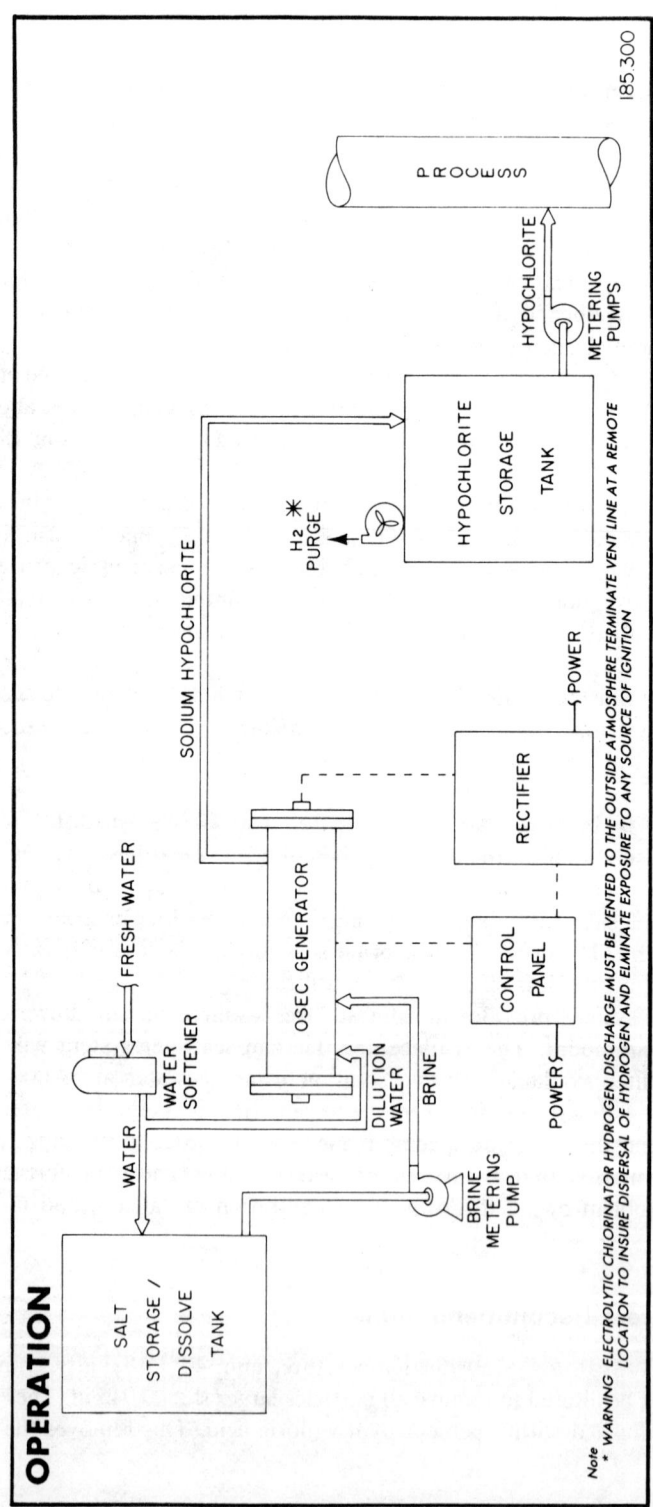

Fig. 3-11. Wallace & Tiernan OSEC brine system (courtesy Wallace & Tiernan Div. Pennwalt Corporation).

build-up of calcium and magnesium precipitates on the electrode surfaces. Length and frequency of flushing is empirical.

Capacities are available in two arrangements: (1) low capacity systems are 25, 50, 75, and 100 lb/day units; (2) high capacity systems are 575, 1200, and 2400 lb/day units. From the lowest to the highest capacity unit the seawater requirement is from 7.5 to 120 gpm. Power consumption will vary depending upon temperature and salinity of the seawater. This variation is probably on the order of 1.7–3.0 kWh/lb chlorine.

Brine Systems. To minimize the formation of calcium and magnesium precipitates within the electrolyzer, the salt should be as free from hardness as possible. Evaporated salt, Southern rock salt, stack and/or soda salts, and foodgrade salt are all satisfactory materials for making the cell brine. The supply water for brine preparation should not exceed 25 mg/l as $CaCO_3$ hardness. The temperature of this water should lie between 35 and 80°F.

Power consumption will be about 2.5 kWh per lb chlorine and salt consumption about 3.5 lb per lb chlorine.

Operating Experience. Several important features have been either confirmed or corrected by the use of pilot plant test installations in an attempt to fulfill the design expectations. One was the determination of the flushing period duration and frequency. The optimum time, based upon seawater on opposite coasts of the U.S.A., appeared to be a 2 hour recycled flushing of the electrolyzer cells with 2 percent hydrochloric acid every two weeks. This calls for back-up equipment or storage of hypochlorite to span the flushing period.

During these tests the manufacturers claim that using 3 percent saline seawater at 50°F temperature the power requirement of 2.5 kWh/lb chlorine was verified.

Also during these tests design changes were made that resulted in successful operation in seawater at 40°F on the shores of the U.K.

Observation of product strength decay at an ambient temperature of 65°F and at initial product strength of 700 mg/l, 9 percent decay occurred after 4 hours storage, 24 percent after 8 hours, and 61 percent after 48 hours.[20]

De Nora Seaclor® Systems

Introduction. The name of Oronzio De Nora, Milan, Italy, is one of the oldest names in the technology of electrolytic production of chlorine. They pioneered the development and use of the mercury cell which still accounts for a major proportion of world-wide chlorine production. They have recently developed a seawater electrolyzer that is being used extensively in the Mediterranean area and Middle East countries. These systems have been supplied in capacities up to 5000 lb/day available chlorine. The electrolyzers are arranged in a vertical position

(See Fig. 3-12) as contrasted to the Wallace and Tiernan and Engelhard units which operate in a horizontal position.

Electrolyzer. The electrolyzers are of modular construction formed by electrolytic cells arranged electrically and hydraulically in series and firmly bound together to constitute an electrode assembly which is placed in a closed container of corrosion resistant materials. The cells are arranged in the bipolar configuration. The anodes are dimensionally stable DSA® coated metal which are widely used in all electrolytic chlorine production processes. The cathodes are made of valve metal (titanium).

Standard electrolyzers are available in unit capacities from 40 to 5800 lb per day of equivalent chlorine. This allows the possibility of standard systems with capacities up to 3400 lb/day of equivalent chlorine without the multiplicity of small cells.[16,17]

Seawater Systems. Seawater must be screened to prevent particles in excess of 0.04 in. diameter (1 mm) from entering the electrolyzer. The seawater should contain a minimum of 10 g/l NaCl concentration; 25 g/l is preferred. The temperature of the seawater should be not less than 40°F nor more than 86°F.

The maximum hypochlorite concentration that can be generated from seawater containing 19,000 mg/l chloride ion and 36,000 mg/l total salinity is 2,500 mg/l available chlorine. This is without a recycle.

Fig. 3-12. DeNora Seaclor® electrolyzers (courtesy Oronzio de Nora Impianti Elettrochimici, Milan, Italy).

De Nora claims a power consumption of 1.7–2.0 kWh per lb equivalent chlorine from a standard seawater at 68°F.

Hydrogen gas is scavenged at the hypochlorite exit of each electrolyzer.

Prepared Brine Systems. The De Nora electrolyzers are equally adaptable to prepared brine solutions as they are to seawater. When using a prepared brine solution the hypochlorite solution produced can achieve a concentration of 8000 mg/l of available chlorine. This is similar to other seawater electrolyzers. To minimize maintenance problems, the brine dilution water hardness should not exceed 25 mg/l as $CaCO_3$, and be at a temperature not less than 40°F. The raw salt should similarly be as free of calcium and magnesium hardness as is possible. Any significant concentration of these salts will result in precipitation of the salts in the electrolytic components. The precipitates formed inhibit the electrolytic process and thereby significantly reduce the efficiency of the system. Additionally, the maintenance time required to overcome process efficiency reduction due to precipitates is directly proportional to the concentration of calcium and magnesium salts.

Power consumption in these systems will be about 2.25 kWh per lb available chlorine and salt consumption about 3.5 lb per lb chlorine. This is consistent with other manufacturers of similar equipment for this purpose.

SUMMARY

On-Site Capabilities

All of the leading manufacturers of this equipment claim nearly equivalent chlorine production per kWh depending upon seawater salinity and temperature—about 2.5 kWh per lb of available chlorine when the seawater salinity is 3 percent and the temperature about 68°F.

The hypochlorite solutions produced from seawater are usually limited to about 1800 mg/l available chlorine and those produced from brine about 8000 mg/l. These strengths vary from the once-through systems to the recycled systems and from manufacturer to manufacturer.

Choice of Systems

General Considerations. The primary consideration for an on-site generating system might be the availability of raw material such as salt and/or electric power. Or it could be the superior safety inherent in the use of on-site generated hypochlorite. Evaluating these factors will require a choice between seawater and prepared brine systems, presuming both raw materials are available.

Hypochlorite generated from a prepared brine solution is generally quite stable and can be stored at 8000 mg/l concentration for significantly longer periods than high concentration commercial hypochlorite.[15]

Chlorine solutions generated from seawater are highly unstable because of the heavy metal ions present in seawater. Normal decay of solution strength is about 2–3 percent per hour. Therefore storage of solution is not practical.

Health Considerations. The use of either brine or seawater for chlorine generation will add sodium to the process stream. Health officials are in general against any process that increases the sodium content of potable water. The seawater system will add 13 parts sodium for each part chlorine. The brine system adds only 2.3 parts sodium per part chlorine.

In the Middle East where fresh water is scarce, the only reliable source of potable water is desalted seawater. Therefore the use of seawater chlorine generating systems is widespread in this area.[18] They are preferred over brine systems because they eliminate the necessity of obtaining solar salt by evaporation. Moreover, chlorine is not readily available in these countries.

In areas where desalted seawater is the only potable water source disinfection is imperative in order to protect the quality of the finished water from possible aberrations that usually occur in a desalting process. Seawater is a potential health hazard particularly when the source is adjacent to a community.

Wastewater Treatment. The broadest spectrum of use in North America is probably for the disinfection of wastewater where the potential hazard of storing chlorine gas is the overriding factor.

The next consideration in making a decision between a brine system and a seawater system is the resultant effect of TDS in the treated wastewater. The dilution factor when using a seawater system will be about 4–5 to 1, meaning that there would be 4 to 5 times more TDS using seawater instead of brine. This would lead to the conclusion that a seawater system should not be used for any water reuse situation.

It is also questionable whether or not a seawater system should be considered for chlorine production when used for odor control in sewage collection systems. The large amount of seawater required, because of the low chlorine concentration in the hypochlorite, would undoubtedly cause an additional load on the sewage flow which in turn would promote the generation of hydrogen sulfide thereby causing more and more odor. Force mains can generate hydrogen sulfide in concentrations sufficient to require 20–30 mg/liter of chlorine for proper control. Seawater hypochlorite generators would be largely self-defeating in these cases.

Typical Applications. In addition to the applications described above, seawater systems have found particular favor for the following uses:

- Offshore oil platforms
- Ships water systems

- Air scrubber systems
- Electric utility sites
- Slime control in seawater cooling systems
- Remote unattended locations
- Food processing

REFERENCES

1. Sweeney, O. R., and Baker, J. E. "An Electrolytic Apparatus for the Production of Antiseptic Sodium Hypochlorite Solution," Iowa State College Bulletin 111, Ames, Iowa, Jan. 1933.
2. Van Peursem, R. M., Pospishu, B. K., and Harris, W. D. "Antiseptic Hypochlorite by Electrolysis," *Iowa State College J. Sci.,* **4,** 37 (1929).
3. Griffith, I. "The Dakin or Carrel–Dakin Solution," *Am. J. Pharm.,* **89,** 497 (1917).
4. "W & T Electrolytic Chlorinator Type EVC-M," Wallace and Tiernan Company, Inc., Tech. Pub. No. 201 Newark, NJ, 1941.
5. Michalek, S. A., and Leitz, F. B. "On-Site Generation of Hypochlorite," *J. WPCF,* **44,** 1697 (Sept. 1972).
6. Leitz, F. B. "On-Site Hypochlorite Generator for Treatment of Combined Sewer Overflows," Report No. 11023 DAA 03/72, EPA, Washington, DC, 1972.
7. Dahl, S. A., "Chlor-Alkali Cell Features New Ion-Exchange Membrane," *Chem. Eng.,* 60 (Aug. 18, 1975).
8. Iammartino, N. R., "New Ion-Exchange Membrane Stars in Chlor-Alkali Plant," *Chem. Eng.,* 86 (June 21, 1976).
9. D'Elia, R. A., Mfg. Rep., Ionics, Inc., private communication, San Mateo, Calif., 1977.
10. Baur, F. "Contract Granted for Sodium Hypochlorite Generator," *Water and Sewage Works,* **119,** 76 (June 1972).
11. "Engelhard Brine Chloropac®," Catalog 75.001, Union, NJ, 1977.
12. Matson, J. V., and Coneway, C. R. "Economics of Disinfection," paper presented at the IOI Forum on Ozone Disinfection, Chicago, IL, June 2–4, 1976.
13. Seawater Data Systems Tech. Information Bulletin E-SC-21, Diamond Shamrock-Sanilec Systems, Electrode Corp., Chardon, Ohio, 1976.
14. Bennett, J. E., and Cinke, J. E. "On-Site Hypochlorite Generation for Water and Wastewater Disinfection," Electrode Corp., Chardon, Ohio, 1975.
15. Anon., "Wallace and Tiernan On-Site Electrolytic Chlorination Systems (OSEC)," Cat. File No. 85.500, Jan. 1980.
16. Anon., "Seaclor® Systems for On-Site Generation of Hypochlorite Solution from Seawater," Oronzio De Nora Impianti Elettrochimici Spa, Milan, Italy, 1983.
17. Spinelli, F., private communication, March 11, 1983.
18. Kott, Y., private communication, Technion University, Haifa, Israel, Sept. 1982.
19. Staff Report, "Hypochlorite Generator System Comparative Cost Analysis," Union Sanitary District In-House Report, Fremont, CA, 1982.
20. Bryant, J. L., private communication, Wallace and Tiernan, Seattle, WA, July 1983.

4
Chemistry of Chlorination

FUNDAMENTALS OF CHLORINE CHEMISTRY

Introduction

This chapter will describe the chemistry of chlorine gas molecules and their reactions when dissolved in aqueous solutions. The purpose of this presentation will be to show all of the fundamental reactions so that the practical application of chlorine to potable water, industrial process water, and wastewater can be better understood and analyzed.

Our knowledge of the fundamental chemistry of chlorination has been enlarged considerably in the past fifty years. This has contributed to significant advancement in the field. However, the more we learn, the more we realize how fortuitous it is that chlorine, applied in its simplest form (Cl_2), can be such a potent disinfectant. This phenomenon of chemical simplicity must surely be an important contributing factor to its germicidal efficiency. As a disinfectant, it is without equal, despite its shortcomings.

It is well known that the amount and complexity of pollutants reaching our potable water supplies are increasing at an alarming rate. This has a direct effect on the chemical reactions of chlorine in aqueous solutions. In general it can be said that the following compounds are of significance in their reactions with chlorine, insofar as water and waste treatment are concerned,

1. Ammonia
2. Amino acids
3. Proteins
4. Total organic carbon (TOC)
5. Nitrites
6. Iron
7. Manganese
8. Hydrogen sulfide
9. Cyanides
10. Organic nitrogen

Before discussing the reactions of these compounds with chlorine, it is desirable to become acquainted with how the chlorine molecule is handled as a disinfectant. Chlorine gas is dissolved either directly in water (by a chlorinator) to form hypochlo-

CHEMISTRY OF CHLORINATION   151

rous acid, or by a specially controlled process in a solution containing caustic to yield a hypochlorite bleach solution. When the latter solution is used as a disinfectant it is diluted with water to form hypochlorous acid, as in the first method. The first part of this discussion will concern the fundamental reaction of chlorine and water and the formation of the oxidizing agent (hypochlorous acid).

Hydrolysis of Chlorine Gas

When chlorine gas is dissolved in water, it hydrolyzes rapidly according to the following equation:

$$Cl_2 + H_2O \longrightarrow HOCl + H^+ + Cl^- \qquad (4-1)$$

The rapidity of this reaction has been studied by many investigators. Complete hydrolysis occurs in a few tenths of a second at 18°C; at 0°C only a few seconds are needed.[1] This unusually rapid rate of reaction is best explained if the mechanism is a reaction of the chlorine molecule with the hydroxyl ion rather than with the water molecule. This can be represented as follows:

$$Cl_2 + OH^- \rightleftharpoons HOCl + Cl^- \qquad (4-2)$$

The rate constant for this reaction is about 5×10^{14} indicating that the reaction occurs at almost every collision of ions.[2] This reaction is of great practical importance because it relates to the chemistry of aqueous chlorine solutions discharging from conventional chlorination equipment. The resulting solution in a chlorinator discharge is limited by design to 3500 mg/liter. At this concentration the most highly buffered injector water would result in a pH of no higher than 3. At this pH the amount of molecular chlorine in equilibrium with HOCl is substantial. Concentrations higher than 3500 mg/liter cause excessive chlorine gas release at the point of application which is extremely undesirable. Likewise if negative pressures exist in the chlorine solution piping, this contributes to the release of molecular chlorine at the point of application. In addition to the degassing effect, the release of gas in the solution piping has been known to adversely affect the hydraulic gradient between the injector and the point of application. Injector systems are usually designed to maintain at least 2 psig at the injector discharge. At this pressure and a temperature of 20°C, the solubility of chlorine in water is only about 7.5 g/liter.[3]

To demonstrate the relationship of the molecular chlorine-hypochlorous acid equilibrium for both buffered and unbuffered water, the following tables have been compiled from a computer printout provided by the Bioengineering Research and Development Lab, U.S. Army, Fort Detrick, Maryland.[4] The results are based upon the Cl_2-HOCl equilibrium; the Cl_3^- ion formation from Cl_2 and the chloride ion; a mass balance for all chlorine species; and an ion balance on Cl^-. Thus the

152 HANDBOOK OF CHLORINATION

Table 4-1 Percent Molecular Chlorine and Hypochlorous Acid in a Water Solution Buffered from pH 1–6 at 15°C at Atmospheric Pressure

	Solution Concentration (mg/liter)									
	500		1000		1500		2000		3500	
pH	Cl_2	HOCl	Cl_2	HOCl	Cl_2	HOCl	Cl_2	HOCl	Cl_2	HOCl
1	54.30	45.65	64.67	35.25	69.94	29.95	73.29	26.57	78.91	20.89
2	17.66	82.31	27.41	72.52	33.95	65.93	38.78	61.05	49.70	49.97
3	2.48	97.51	4.73	95.25	6.79	93.17	8.68	91.26	13.57	86.28
4	0.26	99.72	0.52	99.46	0.77	99.20	1.02	98.45	1.76	98.19
5	0.026	99.74	0.05	99.71	0.078	99.68	0.104	99.66	0.181	99.58
6	0.000	97.68	0.005	97.67	0.008	97.67	0.010	99.67	0.018	97.66

mole percent for HOCl in the following tables is based upon a lengthy and complex cubic equation* which is best described as follows:

$$\text{Percent HOCl} = \frac{100 \times (\text{HOCl})}{[(\text{HOCl}) + (\text{Cl}_2) + (\text{OCL}^-) + (\text{Cl}_3^-)]} \quad (4\text{-}3)$$

Table 4-1 illustrates what happens in the chlorine solution discharge from a chlorinator ranging in feed rates to produce concentrations varying from 500–3500 mg/liter. It also demonstrates the necessity for maintaining a constant high concentration of chlorine (e.g., 1500–2000 mg/liter at a low pH in the generation of chlorine dioxide). The molecular chlorine present in the solution coming in contact with the sodium chlorite provides the impetus for a fast and complete reaction.

Table 4-2 demonstrates the stability of a chlorine water solution buffered with either sodium hydroxide or calcium hydroxide. These figures are of interest for on-site manufacture as well as on-site generation of hypochlorite.

In any hypochlorite solution the active ingredient is always hypochlorous acid.

$$\text{NaOCl} + \text{H}_2\text{O} \longrightarrow \text{HOCl} + \text{Na}^+ + \text{OH}^- \quad (4\text{-}4)$$

$$\text{Ca(OCl)}_2 + 2\text{H}_2\text{O} \longrightarrow 2\text{HOCl} + \text{Ca}^{++} + (\text{OH})^= \quad (4\text{-}5)$$

When a chlorine solution, such as the solution discharge of a conventional chlorinator (unbuffered), is subjected to negative pressure conditions, the solubility is reduced which usually results in the release of molecular chlorine at the point of

---

* This equation is shown in the Appendix.

Table 4-2 Percent Molecular Chlorine, Hypochlorous Acid and OCl⁻ Ion in a Water Solution Buffered from pH 6–9 at 20°C

	Solution Concentration (mg/liter)								
	5000			7000			10000		
pH	Cl_2	HOCl	OCl⁻	Cl_2	HOCl	OCl⁻	Cl_2	HOCl	OCl⁻
6.5	.0063	92.28	7.71	.0088	92.28	7.71	.0126	92.28	7.71
7.0	.0017	79.10	20.89	.0024	79.10	20.89	.0034	79.10	20.89
7.5	.0004	54.84	45.51	.0005	54.49	49.51	.0007	54.49	45.51
8.0	.0001	27.46	72.54	.0001	27.46	72.54	.0001	27.46	72.54
8.5	.0000	10.69	89.31	.0000	10.69	89.30	.0000	10.69	89.30
9.0	.0000	3.65	96.35	.0000	3.65	96.35	.0000	3.65	96.35

application provided the diffuser is in an open body of water such as an open channel.

For example, at atmospheric pressure and 20°C, the maximum solubility of chlorine is about 7395 mg/liter. However, if the solution is subjected to a negative head of 9 in. Hg the solubility is reduced to about 5560 mg/liter.[4] Therefore, all systems that are not closed should be designed to avoid negative pressure conditions in the chlorinator solution lines.

Chemistry of Hypochlorous Acid

Effect of pH. The next most important reaction in the chlorination of an aqueous solution is the formation of hypochlorous acid. This species of chlorine is the most germicidal of all chlorine compounds with the possible exception of chlorine dioxide.

Hypochlorous acid is a "weak" acid which means that it tends to undergo partial dissociation as follows:

$$HOCl \rightleftharpoons H^+ + OCl^- \tag{4-6}$$

to produce a hydrogen ion and a hypochlorite ion. In waters of pH between 6.5 and 8.5 the reaction is incomplete and both species are present to some degree. The extent of this reaction can be calculated from the equation

$$K_i = \frac{(H^+)(OCl^-)}{(HOCl)} \tag{4-7}$$

K_i, the ionization constant, varies in magnitude with temperature. The values of this constant shown in Table 4-3 have been computed from the acid dissociation constant, pK_a, based on J. C. Morris,[5] best fit formula developed in 1966 as follows:

$$pK_a = \frac{3000.00}{T} - 10.0686 + 0.0253T \qquad (4\text{-}8)$$

where $T = 273 +$ degrees centigrade.

Table 4-4 shows the percent undissociated HOCl species for the various temperatures and pH values from 4–11.7. The percent OCl⁻ ion is the difference between these numbers and 100.

The percent distribution of the OCl⁻ ion (hypochlorite ion) and undissociated hypochlorous acid can be calculated for various pH values as follows:

$$\frac{(HOCl)}{(HOCl)+(OCl^-)} = \frac{1}{1+\frac{(OCl^-)}{(HOCl)}} = \frac{1}{1+\frac{K_i}{(H^+)}} \qquad (4\text{-}9)$$

Example: At 20°C and pH 8, the percent distribution of HOCl is:

$$100 \times \left[1 + \frac{2.61 \times 10^{-8}}{10^{-8}}\right]^{-1} = \frac{100}{3.61} = 27.65\%$$

Effect of Ionic Strength (*I*)

The concentration of positive and negative ions in solution affects the ability of molecules to dissociate into their respective ions. In the case of HOCl, it is the H⁺ ion and the OCl⁻ ion. There is a powerful mutual attraction of oppositely charged ions, therefore the more ions (the more TDS) the stronger the forces are for the dissociation of the molecules. The more ions, the more TDS, the greater the ionic strength ±. Figure 13 in the Appendix illustrates the HOCl dissociation curve for three ionic strengths with water at 25°C. Examination of this curve shows 50 percent HOCl at pH 7.5 for $I = 0.001$. The TDS at this ionic strength is 40 mg/liter. At the same pH but in a water with an ionic strength of $I = 0.01$ (TDS = 400 mg/l) the HOCl concentration drops to 45 percent. Waters with a TDS concentration of 40 mg/l would be those obtained from melting snow packs. Those with very high TDS concentrations are to be found in many groundwa-

TABLE 4-3   HOCl Ionization Constant

Temperature (°C)	0	5	10	15	20	25	30
$K_i \times 10^{-8}$ (moles /liter)	1.488	1.753	2.032	2.320	2.621	2.898	3.175

CHEMISTRY OF CHLORINATION 155

Table 4-4[a]

pH	Percent HOCl						
	0°C	5°C	10°C	15°C	20°C	25°C	30°C
5.0	99.85	99.82	99.80	99.79	99.74	99.71	99.68
5.5	99.53	99.45	99.36	99.27	99.18	99.09	99.00
6.0	98.53	98.28	98.00	97.73	97.45	97.18	96.92
6.1	98.16	97.84	97.50	97.16	96.82	96.48	96.15
6.2	97.69	97.29	96.88	96.45	96.02	95.60	95.20
6.3	97.11	96.62	96.10	95.57	95.05	94.53	94.04
6.4	96.39	95.78	95.14	94.49	93.84	93.21	92.61
6.5	95.50	94.75	93.96	93.16	92.37	91.60	90.87
6.6	94.40	93.47	92.51	91.54	90.58	89.65	88.78
6.7	93.05	91.92	90.75	89.58	88.43	87.32	86.27
6.8	91.41	90.03	88.63	87.23	85.85	84.54	83.31
6.9	89.42	87.77	86.10	84.43	82.82	81.29	79.86
7.0	87.04	85.08	83.10	81.16	79.29	77.53	75.90
7.1	84.22	81.92	79.63	77.39	75.26	73.27	71.44
7.2	80.91	78.25	75.64	73.11	70.73	68.52	66.52
7.3	77.10	74.08	71.15	68.35	65.75	63.36	61.22
7.4	72.78	69.42	66.20	63.18	60.39	57.87	55.63
7.5	67.99	64.33	60.88	57.68	54.77	52.18	49.90
7.6	62.79	58.89	55.27	51.98	49.03	46.43	44.17
7.7	57.27	53.23	49.54	46.23	43.32	40.77	38.59
7.8	51.57	47.48	43.81	40.58	37.77	35.35	33.30
7.9	45.82	41.79	38.25	35.17	32.53	30.28	28.39
8.0	40.18	36.32	32.98	30.12	27.69	25.65	23.95
8.1	34.79	31.18	28.10	25.50	23.32	21.51	20.01
8.2	29.77	26.46	23.69	21.38	19.46	17.88	16.58
8.3	25.19	22.23	19.78	17.76	16.10	14.74	13.63
8.4	21.10	18.50	16.38	14.64	13.23	12.07	11.14
8.5	17.52	15.28	13.46	11.99	10.80	9.84	9.06
8.6	14.44	12.53	11.00	9.77	8.77	7.97	7.33
8.7	11.82	10.22	8.94	7.92	7.10	6.44	5.91
8.8	9.62	8.29	7.23	6.39	5.72	5.18	4.75
8.9	7.80	6.70	5.83	5.15	4.60	4.16	3.81
9.0	6.29	5.39	4.69	4.13	3.69	3.33	3.05
9.5	2.08	1.77	1.53	1.34	1.19	1.08	0.98
10.0	0.67	0.57	0.49	0.43	0.38	0.34	0.31
10.5	0.21	0.18	0.15	0.14	0.12	0.11	0.10
11.0	0.07	0.06	0.05	0.04	0.04	0.03	0.03
11.5	0.02	0.02	0.015	0.013	0.012	0.01	0.01
11.7	0.01	0.01	0.01	0.01	0.007	0.007	0.006

[a]Computer printout *courtesy* D. S. Cherry, N.C. State Univ., Raleigh, N.C.

ters and some surface waters. Those with the highest TDS concentration; i.e., greater than 400 mg/l, would probably be highly polished reclaimed wastewater effluents.

The ionic strength of any solution can be calculated from a known TDS concentration by use of the following equation:

$$K_i = \frac{[(f_{H^+})\,(H^+)]\,[(f_{OCl^-})\,(OCl^-)]}{(HOCl)} \qquad (4\text{-}9a)$$

where f is the activity coefficient of the ionic species, H^+, and OCl^-, and K_i is the ionization constant. For very dilute solutions (TDS 10 mg/l) $f = 1.0$. So as I approaches zero, f approaches unity.

To determine the ionic strength of a particular water, the activity coefficients have to be determined. These can be approximated by the Debye–Hückel equation. This equation and the appropriate constant needed to solve this equation are to be found in the Appendix.

Hypochlorite Solutions

The exact same chemical reaction occurs when hypochlorite solutions are used instead of aqueous chlorine solutions. If, for example, common bleach (sodium hypochlorite) is used it disperses in water to form hypochlorous acid:

$$NaOCl + H_2O \longrightarrow HOCl + NaOH \qquad (4\text{-}10)$$

The hypochlorous acid formed by this reaction proceeds to dissociate as in Eq. (4-6).

Familiarity with these factors is essential to an understanding of the behavior of dilute chlorine solutions, because $HOCl$ and OCl^- ion have very different germicidal efficiencies, as will be discussed later in this chapter.

To properly understand the chemistry of the basic reaction of chlorine in an aqueous solution, Eq. (4-1), it is helpful to understand the structure of the chlorine atom.

Chlorine has the periodic number 17, which indicates that there are seventeen positively charged protons in the nucleus of the chlorine atom. These protons in the nucleus are neutralized by seventeen negatively charged electrons in a series of outer shells. Two of the outer shells contain an irrevocable number of electrons, two in the first and eight in the second. These electrons are not available for reaction. This, then, leaves seven electrons of a possible eight in the last outer shell. Since this shell lacks one electron, it is not closed. This means that there are seven electrons available for reaction. If the chlorine atom were to lose these seven electrons in a chemical reaction, the atom would then have an excess of seven positive protons, giving the chlorine atoms a valence of +7. Furthermore,

CHEMISTRY OF CHLORINATION 157

the nature of the chlorine atom is such that this outer open shell can take on a maximum of only eight electrons in its outer shell to give one negative electron in excess over the protons; this results in a valence of −1 on the chlorine atom. When this situation of −1 valence occurs, the outer shell of the chlorine atom is filled to its maximum of eight electrons, thus giving a stable chlorine radical Cl^{-1}, which is recognized as the chloride radical.

Valence of Chlorine

Example of the various valences of the chlorine atom in different compounds are as follows:

Sodium perchlorate	$Na^{+1} Cl^{+7} O_4^{-8}$
Sodium chlorate	$Na^{+1} Cl^{+5} O_2^{-6}$
Chlorine dioxide	$Cl^{+4} O_2^{-4}$
Sodium hypochlorite	$Na^{+1} O_2^{-2} Cl^{+1}$
Hypochlorous acid	$H^{+1} O^{-2} Cl^{+1}$
Hydrochloric acid	$H^{+1} Cl^{-1}$
Monochloramine	$N^{-3} H_2^{+2} Cl_2^{+1}$
Dichloramine	$N^{-3} H^{+1} Cl_2^{+2}$
Nitrogen trichloride	$N^{-3} Cl_3^{+3}$

Therefore the chlorine atom can have a valence ranging from +7 to −1.

Next, we see that all molecules and compounds must have a sum valence of zero, which means that all of the electrons are balanced by the protons. The chlorine molecule can therefore be considered as two chlorine atoms, $Cl^{+1}Cl^{-1}$, each having a positive and a negative valence balancing to zero. When the chlorine molecule becomes completely oxidized to two Cl^{-1} ions, each with a valence of −1, only the positive chlorine atom of the molecule has undergone a change. It changes from valence +1 to valence −1, gaining two electrons for the chlorine molecule. Thus, chlorine is by definition an oxidizing agent. The rule is as follows: Any radical that loses electrons is being oxidized and is therefore a reducing agent. Conversely, any radical that gains electrons is being reduced and is an oxidizing agent.

When chlorine reacts with water, a special type of oxidation-reduction reaction takes place. The chlorine molecule with the sum valence of zero enters into what is known as a disproportionation reaction with water to form HOCl with a Cl^{+1} radical and HCl with a Cl^{-1} radical.[6] The sum valence of the reaction

$$Cl_2 + H_2O \longrightarrow H^{+1}O^{-2}Cl^{+1} + H^{+1}Cl^{-1} \qquad (4\text{-}11)$$

is still zero, indicating that no electrons have been gained or lost, thus signifying that no oxidation or reduction has occurred and that no available chlorine has been lost in forming HOCl. Now, when oxidation of a substance by HOCl takes place, the Cl^{+1} radical in the hypochlorous acid will steal two electrons from the substance being oxidized and will then become a chloride radical with a valence of -1. This gain of two electrons by definition shows that the oxidizing capacity of the HOCl is equal to two equivalents of chlorine or one mol of Cl_2.

$$H^+ + Cl^- + HOCl + 2e^- \longrightarrow H_2O + 2Cl^- \qquad (4\text{-}12)$$

Available Chlorine

The term "available chlorine" has no place in the field of water and waste treatment. This term was established as the basis for comparing the potential bleaching or disinfecting power of chlorine compounds. To compare one bleaching compound with another, it was necessary to be able to establish the oxidizing power of any such compound. This had to be accomplished by a quantitative analysis of the chlorine that was available for oxidizing, or, as seen above, how much chlorine with a valence of greater than -1 is present in the compound. At that time, the only quantitative method in use was the starch-iodide test, now known as the iodometric method.[7] When potassium iodide is added to a solution containing chlorine available for an oxidizing reaction, it will liberate iodine quantitatively from the potassium iodide, as follows:

Hypochlorous Acid:

$$HOCl^{+1} + 2KI^{-2} + HAc^* = I_2^0 + KCl^{-1} + KAc + H_2O \qquad (4\text{-}13)$$

Sodium Hypochlorite:

$$NaOCl^{+1} + 2KI^{-2} + 2HAc = I_2^0 + NaCl^{-1} + 2KAc + 2H_2O \qquad (4\text{-}14)$$

Monochloramine:

$$NH_2Cl^{+1} + 2KI^{-2} + 2Hac = I_2^0 + KCl^{-1} + KAc + NH_4Ac \qquad (4\text{-}15)$$

Dichloramine:

$$NHCl_2^{+2} + 4KI^{-4} + 3HAc = 2I_2^0 + 2KCl^{-1} + 2KAc + NH_4Ac \qquad (4\text{-}16)$$

From the foregoing basic chemistry, two things are evident: (1) Each chlorine radical with a valence of $+1$ will liberate elemental iodine I_2 on a quantitative basis; and (2) since all of the chlorine compounds (those that contain at least

---

* Ac = acetic acid.

one chlorine radical with ±1 valence) are made from elemental chlorine Cl_2, it really takes one molecule of Cl_2 to liberate one molecule of I_2.

To show how the available chlorine percentage is determined, let us examine the case of sodium hypochlorite (NaOCl):

$$\begin{array}{ccc} Na & O & Cl \\ 23 & 16 & 35.5 \end{array} \quad \begin{array}{l} \text{mol wt} = 74.5 \\ = 74.5 \end{array}$$

$$\text{available chlorine (by wt)} = \frac{35.5}{74.5} = 47.7\%$$

In Table 4-5, multiply the percent by weight of chlorine actually present by 2, to give 95.4 percent available chlorine. This shows that the term "available chlorine" is a misnomer, as it is the calculated weight of elemental chlorine (Cl_2) that is required to liberate the same amount of elemental iodine. Since only half of the elemental chlorine molecule is of positive valence, when in solution, the available chlorine content of any chlorine compound which has a Cl^+ radical will always be twice the amount of this radical present.

Further proof of this lies in the manufacture of hypochlorite. For example, 10 percent by wt of active chlorine in a sodium hypochlorite solution contains 0.834 lb of active chlorine per gallon. This means that the manufacturer had to use 0.834 lb of chlorine gas (elemental chlorine, Cl_2) to make every gallon of the solution. Furthermore, if ten gallons of this solution were required to give a 0.5 ppm chlorine residual to one million gallons of water treated, it would simply

Table 4-5 Available Chlorine

Compound	Mol wt.	Mol of Equivalent* Chlorine	Percent of Chlorine by Weight	
			Actually Present	Available
Cl_2	71	1	100	100
HOCl	52.5	1	67.7	135.4
NaOCl	74.5	1	47.7	95.4
$Ca(OCl)_2$	143	2	49.6	99.2
NH_2Cl	51.5	1	69.0	138.0
$NHCl_2$	86	2	82.5	165.0
NCl_3	120.5	3	88.5	265.5

* Number of mol of chlorine having an oxidizing capacity equivalent to one mol of the compound.

require 8.34 lb of chlorine gas dissolved in one million gallons of water to produce the same residual.

In summary, the term "available chlorine" is a misnomer. Actually, it refers to the oxidizing power of the compound tested, and is always twice the value of the active chlorine by weight present in that compound. The term "available chlorine" is analogous to alkalinity as "calcium carbonate." There could even be a case of a compound being tested which does not contain any chlorine at all. For example, it is proper to use the term "available chlorine" when comparing the oxidizing power of ozone. The reader is cautioned that this term is not be confused with the term "free available chlorine," which is the concentration of hypochlorous acid and hypochlorite ions existing in chlorinated water.

Chlorine and Nitrogenous Compounds

The most important and undoubtedly the most complex chemistry of water and wastewater chlorination is its reaction with various forms of nitrogen naturally occurring in water.

If the water to be treated did not contain nitrogenous compounds, the chlorination of water would be extremely simple. The total residual would always be free available chlorine. There would be no problem with quantitative determination of residuals. The disinfecting efficiency of chlorine could be predicted and controlled within a negligible margin of error. Problems of taste and odor from chlorination would probably be nonexistent.

However, this is not the case. Nitrogen appears in most natural waters and in varying amounts as either organic or inorganic nitrogen. These compounds of nitrogen and their relationship to chlorination will be considered in the general grouping as follows:

Inorganic Nitrogen *Organic Nitrogen*
Ammonia Amino Acids
Nitrites Proteins
Nitrates

The chemical stae of any nitrogen compound found in nature is a function of time in the overall life processes of all plants and animals. The amounts of these various forms of nitrogen relate directly to the sanitary quality of the water to be treated. These compounds fit very definitely in time on the nitrogen cycle of nature's own processes of purification. A look at the nitrogen cycle (Fig. 4-1) will show the relationship which exists between the various nitrogen compounds and the changes that are likely to occur in nature.[8] Consider the nitrogen cycle as a clock. Organic nitrogen from plant protein occurs at, say, 6 A.M., while organic nitrogen from animal protein occurs at 8 A.M. The first appearance of inorganic

CHEMISTRY OF CHLORINATION 161

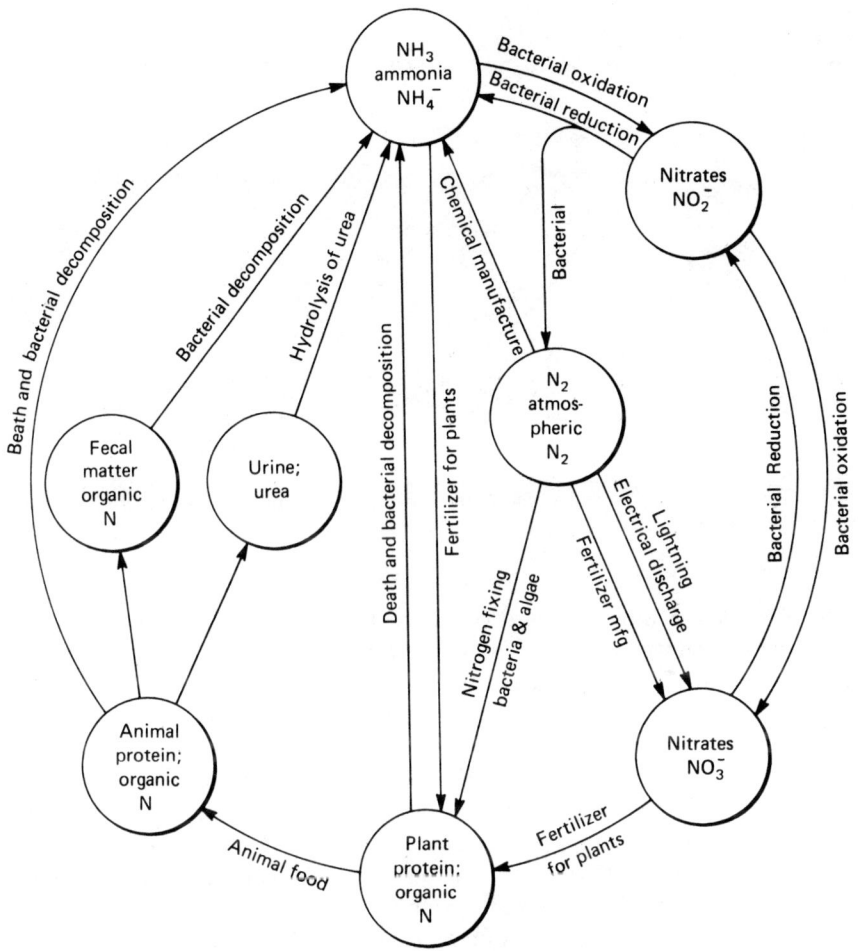

Fig. 4-1. The Nitrogen cycle. After Sawyer.[8]

nitrogen in the form of ammonia occurs at Noon. The completion of the stabilization of the nitrogen compounds occurs at about 4 P.M. in the form of nitrates. These are then available to start the process all over again. This is obviously an oversimplification of the process. It is the relationship in time that is important. Further, it should be remembered that nitrogen and its compounds are in a continuous state of flux throughout this cycle. The various reactions are continuously competing with one another. Nature has conveniently set up a stored supply of nitrogen in the atmosphere, which can reach the earth's surface in two ways. One is by electrical discharge during a storm. In this case, the nitrogen is converted to N_2O_5, which hydrolyzes on contact with water to HNO_3, and falls to earth in solution with

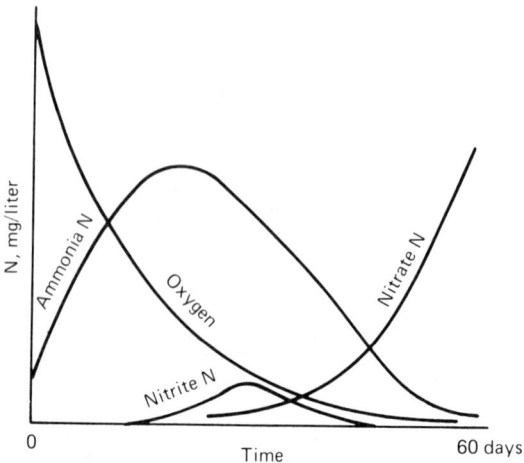

Fig. 4-2. Relationship of nitrogen compounds occurring in polluted water under aerobic conditions. After Sawyer.[8]

the rain. The second way is by nitrogen-fixing bacteria and algae which have the ability to extract nitrogen directly from the atmosphere.

Certain conclusions can be drawn about the degree of pollution of water if the chemical composition of the nitrogen compound is known. For example, it is safe to say: that water containing only nitrates is rather remote in time from any pollution; that water containing nitrites will be highly suspicious; and that waters containing mostly organic nitrogen and ammonia will have been subjected to recent pollution. However, the nitrogen method of pollution evaluation is not sensitive enough to predict whether a polluted body of water constitutes a public health hazard. The general relationship of these compounds occurring in polluted water undergoing purification by aerobic processes is best illustrated in Fig. 4-2.

The reaction of chlorine with any compound containing the nitrogen atom with one or more hydrogen atoms attached will form a compound broadly classified as an N-chloro compound, or, more commonly, as chloramine. There are two distinct classes of chloramines—organic, and inorganic. The inorganic chloramines are formed by the reaction of chlorine in an aqueous solution with free ammonia naturally occurring in the potable water or wastewater being treated. These chloramines are relatively simple compounds.

THE BREAKPOINT PHENOMENON

Introduction

The chemistry of this phenomenon is based upon the inorganic reaction of chlorine with ammonia nitrogen. In dilute aqueous solutions (1–50 mg/l) the reaction between ammonia nitrogen and chlorine forms three types of chloramines in the following competing reactions:

$$HOCl + NH_3 \longrightarrow NH_2Cl \text{ (monochloramine)} + H_2O \quad (4\text{-}17)$$

$$NH_2Cl + HOCl \longrightarrow NHCl_2 \text{ (dichloramine)} + H_2O \quad (4\text{-}18)$$

$$NH_2Cl + HOCl \longrightarrow NCl_3 \text{ (trichloramine*)} + H_2O \quad (4\text{-}19)$$

These reactions are in general by steps, so that they all compete with each other. A series of complex reactions with all of these substances involves the chlorine substitution of each of the hydrogen atoms in the ammonia molecule. These competing reactions are grossly dependent upon pH, temperature, contact time, initial chlorine to ammonia ratio, and most of all upon the initial concentrations of chlorine and ammonia nitrogen. Note that in all three equations the chlorine atom is positively charged.

The reaction of Eq. 4-17 will convert all of the free chlorine to monochloramine at pH 7 to 8 when the ratio of chlorine to ammonia is equimolar (5:1 by wt) or less—that is, 4:1, 3:1, and so on.

The rate of this reaction Eq. (4-17) is extremely important, since it is pH-sensitive. According to reaction rates established by Morris,[9,10] the fastest conversion of HOCl to NH_2Cl occurs at pH 8.3. The following are calculated reaction rates for 99 percent conversion of free chlorine to monochloramine at 25°C with a molar ratio of 0.2×10^{-3} mol/l HOCl and 1.0×10^{-3} mol/l NH_3:

pH	seconds
2	421
4	147
7	0.2
8.3	0.069
12	33.2

The reaction slows appreciably as the temperature drops. At 0°C, it requires nearly five minutes for 90 percent conversion at pH 7.

The pH dependence of this reaction is described accurately on the basis of the $HOCl - OCl^-$ equilibrium and the $NH_3 - NH_4^+$ equilibrium.

The reaction of Eq. (4-18) will form dichloramine between pH 7 and 8 if the ratio of chlorine to ammonia is 2 mol chlorine to one mol ammonia nitrogen (10:1 by wt). The rate of this reaction is much slower than that of Eq. 4-17. It may take as long as one hour for 90 percent conversion[9] and up to five hours at pH 8.5 and above when ammonia nitrogen concentrations are very low. As the pH approaches 5 the reaction speeds up appreciably. This reaction is dependent on pH, initial ammonia nitrogen, and temperature. The reaction time of Eq. (4-18) is known to be minutes when the initial nitrogen concentration is in excess of 1 mg/l and the pH is favorable.

---

* More commonly called nitrogen trichloride.

The reaction of Eq. (4-19) will form some nitrogen trichloride when the pH is between 7 and 8 if the chlorine to ammonia nitrogen ratio is 3 mol of chlorine to 1 mol of ammonia nitrogen (15:1 by wt). At present very little is known about the kinetics of this reaction, particularly in concentrations of less than 10 ppm (10^{-4} M). Nitrogen trichloride does form, even at equimolar ratios of chlorine to ammonia nitrogen, if the pH is depressed to 5 or less. At one time it was thought that it would not form above pH 5. It is known to exist in water treatment plants when the pH is as high as 9.[11] This occurs at very high chlorine to ammonia nitrogen ratios (25:1 by wt).

In waterworks practice, if the chemistry of Eq. (4-17) is practiced, it is known as either the chlorine-ammonia process, the chloramine process or chloramination. Eqs. (4-18) and (4-19) are related to the "breakpoint" phenomenon. It was Griffin's work which led to the discovery of this phenomenon in 1939.[12,13] Griffin was attempting to explain the sudden loss of chlorine residuals and the simultaneous disappearance of ammonia nitrogen at treatment plants which were experimenting with higher than usual chlorine residuals (2–15 mg/l).

These high residuals were used in an attempt to destroy obnoxious taste and odors. Griffin was startled to find that increasing the chlorine dose in certain waters not only did not increase the residual but reduced it significantly. He called this point of maximum reduction of residual the breakpoint. Fig. 4-3 illustrates this phenomenon, with point A designated as the breakpoint.

The importance of the breakpoint phenomenon is its relationship to the control of taste and odors either naturally present in the water or caused by the addition of chlorine. Griffin found that tastes and odors occurring in the region on the left of A disappeared on the right of A in Fig. 4-3. Other significant benefits were experienced in plant practice, the most important being the increased germicidal efficiency. The killing power of chlorine on the right of A is 25-fold or more than on the left of A.

Attempts to explain Griffin's findings[13,14] led to serious scientific investigations between 1946 and 1950. The most extensive laboratory investigation was accomplished at Harvard University. The work of Fair, Morris, Chang, Weil, Burden, Culver, and Granstrom is largely responsible for our present knowledge of the physical chemistry of chlorination.[9,15-22] Palin's work in England during the same period is also noteworthy.[23,24] Williams' work on a plant scale substantiates the findings of both Palin and the Harvard group.[25-27] Williams' work is of significant importance because it documents a period of 20 years and because it demonstrates the practical value of this knowledge. It is of historical interest to note that all of these investigations substantiated that of Holwerda done almost 20 years earlier.[28] The work at Harvard was an effort directed primarily toward the explanation of the mechanism that triggers the breakpoint. These investigators used the most sophisticated and accurate method available for the identification of the various chlorine residual fractions—namely, the light absorbance principle of the spectrophotometer.

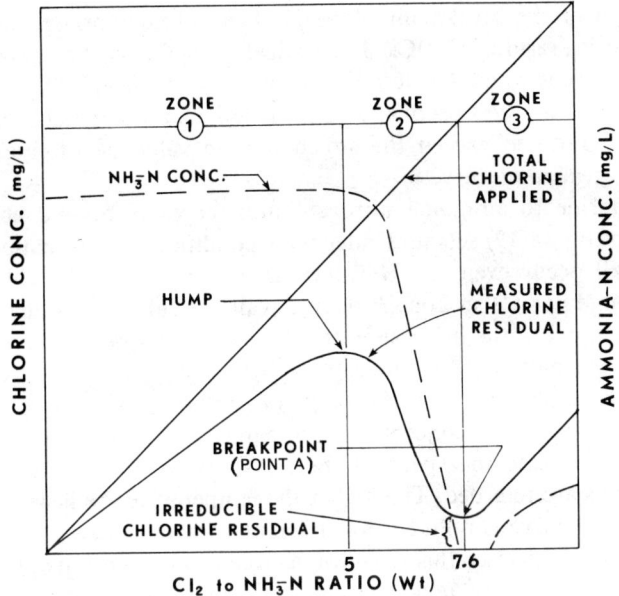

Fig. 4-3. Theoretical breakpoint curve.

The Breakpoint Curve

The breakpoint curve is a graphic representation of chemical relationships which exist as varying amounts of chlorine are added to waters containing small amounts of ammonia nitrogen. The theoretical breakpoint curve is shown in Fig. 4-3. It was originally developed as a result of Griffin's work in 1941–44.[13,14] This curve has several characteristic features. The principal reaction in Zone 1 is the reaction between chlorine and the ammonium ion indicated in Eq. (4-17). This results in a chlorine residual containing only monochloramine all the way to the hump in the curve. The hump occurs, theoretically, at a chlorine to ammonia nitrogen weight ratio of 5:1 (molar ratio of 1:1). This ratio indicates the point where the reacting chlorine and ammonia nitrogen molecules are present in solution in equal numbers. As the molar ratio begins to exceed 1:1 some of the monochloramine starts a disproportionation reaction* to form dichloramine in accordance with Eq. (4-18).[22]

---

* A disproportionation reaction is one that transforms a substance into two dissimilar compounds by a process involving simultaneous oxidation and reduction. Therefore the chemical equilibrium in Zone 2 favors the formation of dichloramine and the oxidation of ammonium ion according to Eqs. (4-17), (4-18), and (4-19). This results in a marked reduction of ammonia nitrogen caused by the oxidation of chlorine. The above reactions proceed in competition with each other to produce a breakpoint at a theoretical Cl_2 to NH_4^+ weight ratio of 7.6 to 1. At the breakpoint (Point A) ammonia nitrogen is either at a minimum or disappears entirely.

To the right of the breakpoint, Zone 3, chemical equilibria require the build-up of free chlorine residual (HOCl). In practical applications of breakpoint chlorination, reactions occur which result in the formation of nitrogen gas, nitrate, nitrogen trichloride, and other end products. These reactions consume chlorine and cause the $Cl_2:NH_4^+$ ratio to exceed the stoichiometric value of 7.6:1 and affect the shape of the breakpoint curve.†

As the chlorine to ammonia nitrogen ratio increases beyond about 12–15:1 the reaction of Eq. (4-19) sets in. Under these conditions the formation of nitrogen trichloride will occur even at pH values as high as 9. As the chlorine dose is increased beyond point A in Zone 3 the free available chlorine residual will increase in an amount equal to the increase in the dosage. Therefore, the breakpoint curve in Zone 3 should plot at a 45° angle.

It should be emphasized that the shape of the breakpoint curve is affected by contact time, temperature, concentration of chlorine and ammonia, and pH. High concentrations increase the speed of the reactions. As the pH decreases below 8.3 the reactions are retarded. The higher the temperature the faster the reactions. The shape of the curve is different for different contact times.

In potable water practice this is known as free residual chlorination rather than the breakpoint process. In potable water treatment the practical significance of the curve is briefly as follows:

Zone 1. The residuals in this zone up to the hump are all monochloramine. The residuals in this zone do not form trihalomethanes nor do they usually contribute to tastes and odors.

Zone 2. As the hump is passed the monochloramine plus the addition of more free chlorine begins to form dichloramine, which is about twice as germicidal as monochloramine. However, this may not be the best part of the curve for the production of a palatable water. A pure dichloramine residual has a noticeable disagreeable taste and odor, while monochloramine does not. It is generally considered best to avoid this part of the curve in order to avoid taste and odor problems.

Zone 3. At the dip of the curve (point A) and beyond, free chlorine residual will appear. The total residual will be made up of the nuisance residuals plus free chlorine. If nitrogen trichloride is formed it will appear in this zone. In practice it has been found that a ratio of free chlorine to total residual of 85 percent or greater will result in the most palatable water.

---

† The following values have been observed at the breakpoint: Griffin and Chamberlin observed 10–12.5:1 at pH values between 6 and 9;[13] Palin observed 9.5:1 at pH 6, 8.25:1 at pH 7, and 8.4:1 at pH 9;[24] Moore, Megregian, and Ruchhoft observed an average value of 9.0:1 at pH 6–9;[24] Metropolitan Water District of Southern California has observed values of 9:1 at pH 7.9 and 11:1 at pH 7.5 in their two different supplies.[35] This is also the point at which there exists an irreducible minimum of chlorine residual commonly described as "nuisance residuals." These residuals titrate as predominantly dichloramine with a trace of free chlorine and monochloramine. Griffin also discovered that in Zone 3 there will be a return of ammonia nitrogen.

The Breakpoint Reaction

The mechanism by which the breakpoint phenomenon occurs was first proposed by Rossum in 1943.[30] He calculated the distribution of mono- and dichloramine by means of mass action equations. Although he recognized the possible occurrence of nitrogen trichloride, he considered the breakpoint reaction as one of the relationships of mono- and dichloramine. He confirmed Griffin's finding that ammonia does reappear to the right of the dip in the curve, and was one of the first to observe the destruction of free chlorine (HOCl) by sunlight.

The equilibrium condition between mono- and dichloramine in equimolar concentrations was established by the Harvard group, using spectrophotometric methods.[31] This work demonstrated that the equilibrium condition between mono- and dichloramine can be expressed as follows:

$$\frac{(NH_4^+)(NHCl_2)}{(H^+)(NH_2Cl)} = 6.7 \times 10^5 = K_{eq} \qquad (4\text{-}20)$$

Curve A in Fig. 4-4 shows this equilibrium relationship for mono- and dichloramine at different pH levels based on Eq. (4-20).

The work of other investigators is also shown. It is interesting to note that Chapin's work[32] using a variation of the starch–iodide technique for the residual fractions compares more closely to the expression in Eq. (4-20) than do the others. However, Palin's[24] and Baker's[33] work tend to agree with each other. Palin separated the residual fractions by neutral O-T FAS titration. Baker used the amperometric titration method, as did Kelly and Sanderson.[34] For unknown reasons, the work of Kelly and Sanderson does not compare favorably with the rest. Williams' work,[36,37] which is on a plant operation, more nearly substantiates the Harvard investigation.

From all of this information, Morris[9] suggested that the key to the final breakpoint reaction was the formation and decomposition of dichloramine. This work, in turn, inspired the investigation of the disproportionation reaction of monochloramine in the breakpoint phenomenon.[21] This investigation has led to the following information about the breakpoint phenomenon:

As the chlorine to ammonia nitrogen ratio proceeds from 5:1 to 10:1 and greater, two important reactions take place which tend to shift the relative concentrations of mono- and dichloramine:

1. The spontaneous conversion of monochloramine forms dichloramine and ammonia. This is known as the disproportionation of monochloramine.
2. The decomposition of dichloramine decreases its concentration.
  These two reactions are thought to occur only in the presence of excess free chlorine.

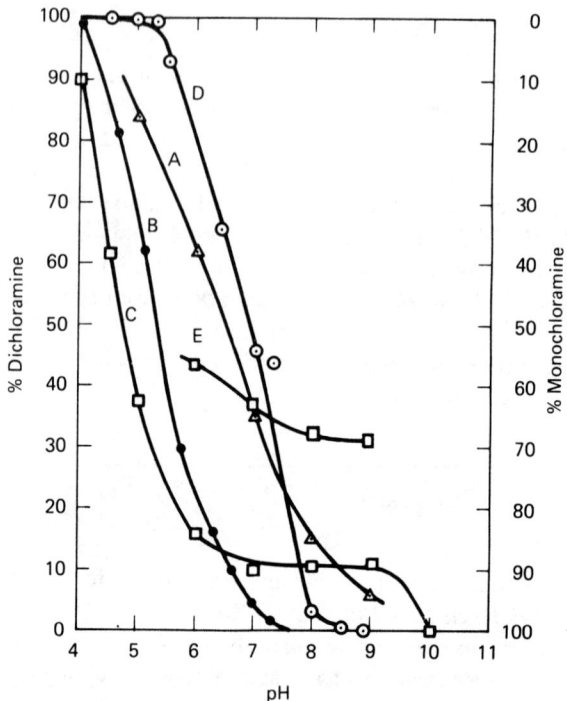

Fig. 4-4. Distribution of monochloramine and dichloramine with relation to pH. Curve A = Fair et al[31] Cl–NH$_3$ wt. ratio 5:1; curve B = Palin[24] Cl–NH$_3$ 5:1, 2 hr contact; curve C = Baker[33] Cl–NH$_3$ 4:1 2-hr contact; curve D = Chapin[32] (excess NH$_3$); curve E = Kelly and Sanderson,[34] Cl–NH$_3$ 2:1 (Curve A computed from EQ20, K$_{eQ}$ = 6.7 × 10^5).

The conversion of mono- to dichloramine takes place by means of hydrolysis of a monochloramine molecule to form HOCl, which then reacts with another monochloramine molecule to form dichloramine as follows:

$$NH_2Cl + H_2O \longrightarrow HOCl + NH_3 \quad (4\text{-}21)$$
$$\downarrow$$
$$NH_2Cl \longrightarrow NHCl_2 + H_2O$$

$$HOCl + NH_2Cl \longrightarrow NHCl_2 + H_2O \quad (4\text{-}22)$$

This is considered to be a first-order reaction, which does not seem to be affected by the pH or the buffering of the solution.

There is also a second-order reaction, which proceeds in parallel with the previous one, but is pH- and buffer-dependent. It is thought to proceed by means of acid catalysis. This reaction probably occurs between a monochloramine acid complex catalyst, which then reacts with another monochloramine molecule to form dichloramine, as follows:

$$NH_2Cl + Acid \rightleftharpoons [NH_2Cl \cdot Acid] \quad (4\text{-}23)$$
$$[NH_2Cl \cdot Acid] + NH_2Cl \rightleftharpoons NHCl_2 + NH_3 + Acid$$

(In any catalytic reaction, the catalyst is never lost.)

These two reactions then make up the spontaneous conversion of mono- to dichloramine, known as the disproportionation of monochloramine.

While the rate of formation of monochloramine is dependent on the concentration of ammonia and free chlorine, it is dependent also on pH (because of the equilibria $H^+ + OCl^- \rightleftharpoons HOCl$ and $NH_4 \rightleftharpoons H^+ + NH_3^-$) but is not affected by the buffering of the solution, whereas the rate of formation of dichloramine is acid-catalyzed—in other words, it is dependent upon the buffering of the solution.

An excess of ammonia suppresses the disproportionation reaction of monochloramine. There must be an excess of HOCl for Eqs. (4-22) and (4-23) to proceed.

The next step in the breakpoint reaction is the decomposition of the dichloramine, which is dependent on the hydroxyl ion activity. Morris[9] has proposed the following mechanism for the decomposition of dichloramine:

First, the dichloramine ionizes as a weak acid:

$$NHCl_2 \longrightarrow H^+ + NCl_2^- \quad (4\text{-}24)$$

This proceeds to react with the hydroxyl ion (OH^-) in either of two possible ways:

$$NCl_2^- \xrightarrow{(OH)} N-Cl + Cl^- \quad \text{(slow)} \quad (4\text{-}25)$$
$$N-Cl + OH^- \longrightarrow NOH + Cl^- \quad \text{(fast)} \quad (4\text{-}26)$$

or:

$$(NCl_2)^- + (OH)^- \longrightarrow NCl(OH)^- + Cl^- + H^+ \quad (4\text{-}27)$$
$$NCl(OH)^- \longrightarrow NOH + Cl^- \quad (4\text{-}28)$$

Eq. (4-28) shows the formation of an intermediate reaction product, the nitroxyl radical NOH. This can proceed to decompose into the end products of the reaction in three ways. Each way is valid since the end products depend on many variables, such as molar ratio of chlorine to ammonia, initial ammonia nitrogen concentration, pH, temperature, and unknown side reactions of chlorine and organic matter.

The first way assumes that the predominant end point of the breakpoint reaction might be N_2O as first postulated by Chapin.[32] This would require the formation of hyponitrous acid from two nitroxyl radicals through dimerization:

$$2 \text{ NOH} \longrightarrow H_2N_2O_2 \tag{4-29}$$

The hyponitrous acid slowly decomposes to give nitrous oxide:

$$H_2N_2O_2 \longrightarrow N_2O + H_2O \tag{4-30}$$

This mechanism calls for two moles of chlorine for every atom of nitrogen that is oxidized. Morris[39] points out that Palin found that it required less than two moles of chlorine to oxidize each atom of nitrogen, which suggests the possibility that nitrogen gas (N_2) is the predominant end product. In view of this statement it is interesting to note that the recent work (1970) on nitrogen removal from wastewater by Pressley, Bishop, and Roan[40] concludes that it requires less than two moles of chlorine to oxidize one atom of nitrogen and that the end products of the reaction were found to be N_2, NO_3^-, and NCl_3.

The second way the breakpoint reaction might proceed would be to assume that the main end product is nitrogen gas (N_2). The decomposition of dichloramine to produce the intermediate reactive product NOH would be as follows:[39]

$$NHCl_2 + H_2O \longrightarrow NOH + 2H^+ + 2Cl^- \tag{4-31}$$

This would be followed by three competing reactions involving the intermediate, NOH, as follows:

$$NOH + NH_2Cl \longrightarrow N_2 + H_2O + H^+ + Cl^- \tag{4-32}$$

$$NOH + NHCl_2 \longrightarrow N_2 + HOCl + H^+ + Cl^- \tag{4-33}$$

$$NOH + 2 \text{ HOCl} \longrightarrow NO_3^- + 3H^+ + Cl^- \tag{4-34}$$

These hypothetical reactions are compatible with the end products of the breakpoint reaction.

Eq. (4-19) is a part of this reaction as it accounts for the formation of nitrogen trichloride (NCl_3) when an excess of chlorine is present. The reverse of this reaction:

$$NCl_3 + H_2O \longrightarrow NHCl_2 + HOCl \tag{4-35}$$

is required to demonstrate the well known fact that NCl_3 is quite stable under conditions of water chlorination only when an excess of chlorine is present, and to account for the slow decomposition of NCL_3. It is, however, easily aerated because of its low solubility in water.

The third breakpoint reaction could occur in the presence of a large excess of free chlorine.

$$H_2N_2O_2 + HOCl \longrightarrow 2NO + H_2O + H^+ + Cl^- \qquad (4\text{-}36)$$

This reaction would require 2.5 mol of chlorine for every nitrogen atom oxidized. This is what Griffin[13] found for large excesses of free chlorine in the pH range 6–9. (To convert ratios in moles to weight ratios multiply by 5, i.e., $Cl:N = 71:14 = 5$.)

It can be concluded then that the breakpoint phenomenon will occur when the ratio of chlorine to ammonia nitrogen by weight is from 9 to 1 or greater providing the pH of the environment is favorable. Reaction times are a function of initial ammonia nitrogen concentration but will range from minutes to hours for a given pH and temperature conditions. The optimum pH for the fastest reaction is between 7 and 8. Above and below this range the reaction slows appreciably. Lowering the temperature retards the reaction.

Other conclusions of academic interest may also be drawn. Free available chlorine is converted from its positive valence state to its negative valence, or chloride, state as a result of oxidizing the nitrogen atom. The end products are most probably released as gases of nitrogen compounds. If there is any carbonaceous material, chlorine will react with these compounds, with carbon exerting a chlorine demand and being released as carbon dioxide. This release of gases as a result of the breakpoint reaction is readily observed in rotameters installed in the chlorine solution lines. These rotameters are downstream from the injector and will carry a solution of HOCl at a concentration in the range of 500 to 3500 ppm. The pH is usually about 2. Copious quantities of gas bubbles are usually observed. Most of these bubbles will consist of CO_2 and be related to the alkalinity of the injector operating water. The nitrogen gases will be related to the amount of ammonia nitrogen present in the water. In treated wastewater, the organic nitrogen probably will not be released at this point, since the reaction time is not sufficient. A scientific investigation of this reaction might enhance our knowledge of chlorination chemistry. It might also serve to establish some design criteria, because the gas released in this phenomenon is sufficient to cause abnormal friction losses in solution lines and because in some waters the bubbles are of such magnitude that rotameters will not function.

Nitrites have been suggested as a possible product of decomposition. This is unlikely, since nitrites readily react with free chlorine and are oxidized rapidly to nitrates:[38]

$$NO_2^- + HOCl \longrightarrow NO_3^- + H^+ + Cl^- \qquad (4\text{-}37)$$

Nitrites will not react with chloramines in the pH range of 6 to 9. However, at pH 4 there is some indication of a reaction with dichloramine.

Selleck–Saunier Breakpoint Chemistry

In an attempt to fully understand the chemistry of the breakpoint phenomenon, Prof. Selleck and graduate student Saunier decided to review this reaction using both potable water and wastewater. At the time there were disagreements in the literature concerning the chemical pathways to the end products. This work by Saunier[41,42] under the guidance of Prof. Robert E. Selleck, University of California, Berkeley, CA, resulted in the creation of a computer model based upon more than 80 experimental runs, within a pH range of 6–9; ammonia nitrogen from 1 to 20 mg/liter; and chlorine to nitrogen molar dose ratios of 1.6:3.5; and temperatures between 12°C and 21°C. The model as developed by this research can be used to compute the production of the various species of chlorine compounds formed during the breakpoint reaction; the pH of the water immediately after chlorine addition; the amount of chemical required for buffering the reaction to the optimum pH for the greatest speed of the reaction, as well as the decrease in pH as the reaction proceeds. Based on Saunier's[41] research the following set of reactions appears to be the most reasonable:[43]

$$NH_4^+ + HOCl \Rightarrow NH_2Cl + H_2O + H^+ \qquad (4\text{-}38)$$

$$NH_2Cl + HOCl \Rightarrow NHCl_2 + H_2O \qquad (4\text{-}39)$$

$$0.5\ NHCl_2 + 0.5\ H_2O \Rightarrow 0.5\ NOH + H^+ + Cl^- \qquad (4\text{-}40)$$

$$0.5\ NHCl_2 + 0.5 NOH \Rightarrow 0.5 N_2 + 0.5\ HOCl + 0.5 H^+ + 0.5\ Cl^- \qquad (4\text{-}41)$$

and finally:

$$NH_4^+ + 1.5\ HOCl \longrightarrow 0.5 N_2 + 1.5\ H_2O + 2.5\ H^+ + 1.5\ Cl^- \qquad (4\text{-}41a)$$

The above reactions are based on the formation of NOH, apparently a catalytic intermediary compound which Saunier and Selleck believe is a result of the formation of hydroxylamine (NH_2OH) as an intermediate reaction.[41,42] These reactions are as follows:

$$NHCl_2 + 2H_2O \longrightarrow NH_2OH + HCl + HOCl \qquad (4\text{-}42)$$

$$NH_2OH + HOCl \longrightarrow NOH + HCl + H_2O \qquad (4\text{-}43)$$

$$NOH + NHCl_2 \longrightarrow N_2 + HOCl + HCl \qquad (4\text{-}44)$$

The formation of the hydroxylamine and the resulting formation of the catalytic compound NOH greatly increases the speed of the breakpoint reaction as the ammonia nitrogen concentration increases with the greater NOH production.

CHEMISTRY OF CHLORINATION

The breakpoint occurs through the sequential formation of monochloramine and dichloramine with the subsequent catalytic decomposition of dichloramine to produce an end product of nitrogen gas, with a partial return of free chlorine residual (HOCl) to the solution. These reactions confirm that 1.5 moles (gram molecular weight) of chlorine are required to oxidize 1.0 mole of ammonia to nitorgen gas.

Stoichiometrically, the breakpoint reaction requires a weight ratio of chlorine to ammonia nitrogen ($Cl_2:NH_4^+-N$) at the breakpoint of 7.6:1 as shown below:

$$\text{Molecular wt HOCl} = 70.9 \text{ (as } Cl_2)$$

$$\text{Moles HOCl required} = 1.5$$

$$\text{Molecular wt } NH_4^+ = 14 \text{ (as N)}$$

$$\text{Moles } NH_4^+ \text{ required} = 1.0$$

Therefore, $Cl_2:NH_4^+-N = (1.5)(70.9):(1.0)(14.0) = 7.6:1$; so for each mg per liter of NH_4^+-N, 7.6 mg per liter of chlorine is required to reach the breakpoint. In actual wastewater-treatment practice, as was demonstrated by the Rancho Cordova project, it required 10 mg per liter of chlorine for each 1.0 mg per liter ammonia nitrogen present in the process influent.[41] Approximately 70 percent of this breakpoint dosage was consumed to produce nitrogen gas (N_2) from the ammonium ion (NH_4^+) at pH set points between pH 7 and 8. The oxidation of NH_4^+ to NO_3^- consumed 8 to 19 percent of the total chlorine dosed to the system. Overall, about 96 percent of the total chlorine dosage was accounted for in reactions between chlorine and nitrogeneous species in specific chemical pathways and free chlorine residual remaining in solution following breakpoint.

Side Reactions of Breakpoint Chlorination

The following are some of the chemical reactions other than the direct oxidation of ammonia to nitrogen gas. The reaction products and chlorine consumption for such reactions are governed by factors such as the type and degree of pretreatment, initial $Cl_2:NH_4^+-N$ ratio, pH, and alkalinity. These reactions are as follows:

Description	Reaction Stoichromerity	
Breakpoint reaction	$NH_4^+ + 1.5 HOCl \longrightarrow 0.5N_2 + 1.5H_2O + 2.5H^+ + 1.5Cl^-$	(4-45)
NCl_3 formation	$NH_4^+ + 3HOCl \longrightarrow NCl_3 + 3H_2O + H^+$	(4-46)
Nitrate formation		
1-From ammonia	$NH_4^+ + 4HOCl \longrightarrow NO_3^- + H_2O + 6H^+ + 4Cl^-$	(4-47)
2-From nitrite	$NO_2 + HOCl \longrightarrow NO_3^- + H^+ + Cl^-$	(4-48)

pH and Alkalinity Considerations

The nature and concentration of the breakpoint chlorination end products, the chlorine dosage required to reach breakpoint, and the rate of the breakpoint reaction are all affected by the initial pH (following chemical addition) and the pH change which occurs as the breakpoint reaction proceeds. The initial pH in the reaction zone and pH change through breakpoint depends upon the pH and alkalinity of the process influent stream, the ammonia concentration, the chlorine dosage, and the amount of alkalinity supplementation.

Acidity is generated in breakpoint chlorination applications (nitrogen removal) from both the hydrolysis and dissociation of chlorine gas (when Cl_2 gas solutions are used), and the oxidation of ammonia nitrogen.

When the acidity generated is from the hydrolysis and dissociation of chlorine gas the following reaction occurs:

$$1.5\ Cl_2 + 1.5\ H_2O \longrightarrow 1.50\ OCl^- + 3H^+ + 1.5\ Cl^- \tag{4-49}$$

If acidity is from the oxidation of ammonia the following two reactions prevail:

$$NH_4^+ + 1.5\ OCl^- \longrightarrow 0.5\ N_2 + 1.5\ H_2O + H^+ + 1.5\ Cl^- \tag{4-50}$$

and

$$NH_4^+ + 1.5Cl_2 \longrightarrow 0.5\ N_2 + 4\ H^+ + 3Cl^- \tag{4-51}$$

To counteract the acidity generated in the above reactions either lime or caustic can be used as follows:

$$\text{Lime:}\quad 2\ CaO + H_2O \longrightarrow 2Ca^{++} + 4OH^- \tag{4-52}$$

$$\text{Caustic:}\quad 4\ NaOH \longrightarrow 4Na^+ + 4OH^- \tag{4-53}$$

Stoichiometrically, three moles of hydrogen ions are liberated in the hydrolysis and dissociation of chlorine gas to provide sufficient chlorine for the oxidation of one mole of ammonia nitrogen; assuming the initial pH in the reaction zone is alkaline. One mole of hydrogen ion is liberated in the oxidation of ammonia to nitrogen gas.

Saunier's Research

Saunier studied the kinetics of the breakpoint reaction in both tap water and tertiary effluent.[41] Only the facets of this research will be discussed here as they may relate to tertiary effluents.*

One of the most important findings by Saunier was the confirmation that there

---

* This includes water for reuse.

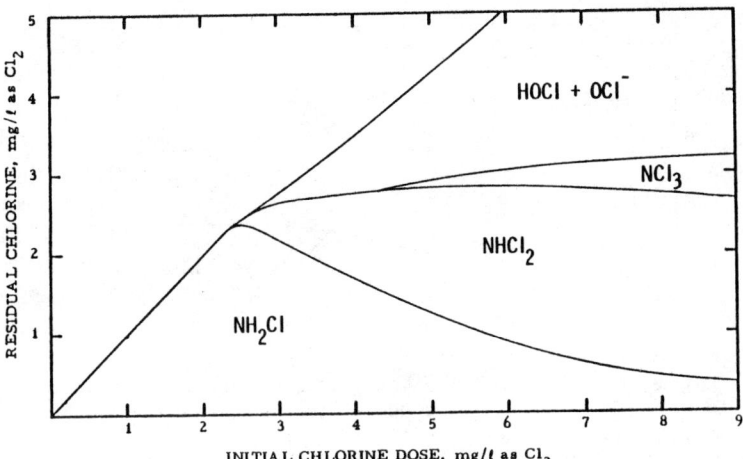

Fig. 4-5. Chlorine dose residual curves predicted by the model after 2.5 min. contact time (pH = 7.4, NH_3–N = 0.5 mg/liter, Temp = 15°C).

is a considerable time difference required to complete the breakpoint reaction as related to the ammonia nitrogen concentration, other factors being equal.

Fig. 4-5 illustrates the distribution of the chlorine residual species (as predicted by Saunier's model) when the NH_3–N is 0.5 mg per liter and the contact time is 2.5 min.; Fig. 4-6 is the same situation except that the contact time is 20 min. Fig. 4-7 is the predicted breakpoint kinetics but with an ammonia nitrogen concentration of 2.5 mg per liter.

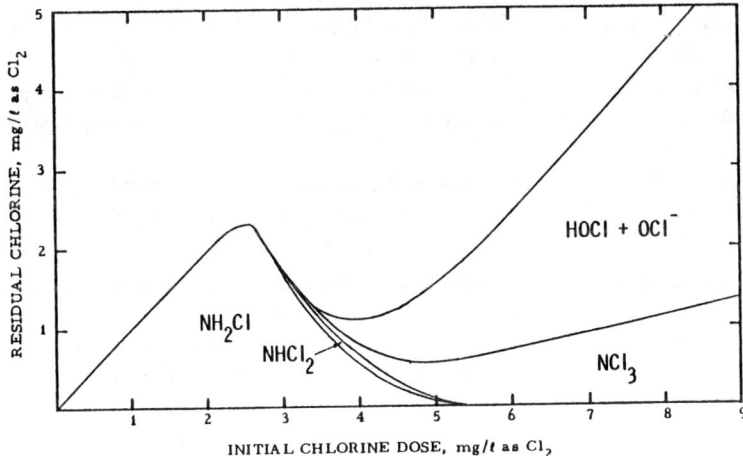

Fig. 4-6. Chlorine dose residual curves predicted by the model after 20 min. contact time (pH = 7.4 NH_3–N = 0.5 mg/liter, Temp = 15°C).

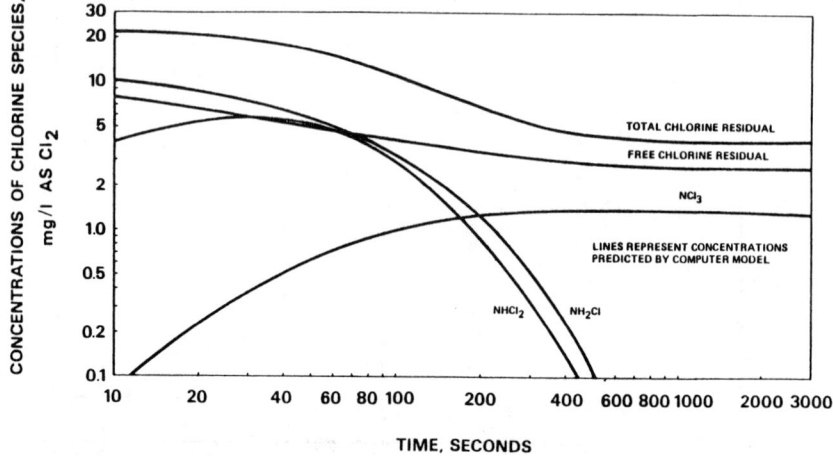

Fig. 4-7. Predicted breakpoint kinetics in a plug flow reactor (pH = 7.5, NH_3–N = 2.5 mg/liter, Temp = 15°C and Cl_2/N = 9.0 by wt). The lines represent concentrations predicted by the model.

Compare these curves with Fig. 4-8 which is a tertiary effluent with an NH_3 –N content of 18 mg/liter and molar ratio of Cl:N of 1.77. The breakpoint reaction in Fig. 4-8 occurs in less than 3 min. (at the disappearance of NH_2Cl) while it takes about 20 min. for water containing less than 1.0 mg per liter NH_3 –N and about 8–10 min. when the NH_3 –N concentration is 2.5 mg per liter, other factors such as pH and temperature being equal. The model prediction of Fig. 4-8 is in close agreement with the Rancho Cordova project findings[43] where molar ratios of Cl_2 –NH_3 –N of 2:1 produced the breakpoint reaction in 60–90 sec. at NH_3 –N initial concentrations of 15 to 20 mg per liter. So for tertiary effluents nitrified to NH_3 –N concentrations of less than 2 mg per liter, it will require 15–20 min. to complete the breakpoint reaction and produce a controllable, stable, free chlorine residual.

Further findings by Saunier showed that after the breakpoint is reached, the resulting *dichloramine fraction* of the *residual is nearly as germicidal as the free HOCl residual,* and furthermore the HOCl residual is 30 times more germicidal than the OCl^- ion. These findings are indeed different from previous findings.

Nitrogen gas and nitrate are the final end products of the breakpoint reaction. Nitrogen trichloride* formed during the reaction decreases slowly with time. While NCl_3 does not appear to be an end product of the reaction, for all practical purposes it is an end product; because in the time frame of the process in practice, the remaining NCl_3 (at the end of the contact time) will revert to ammonia nitrogen after dechlorination.

Saunier also confirmed the speed of the breakpoint reaction to occur most rapidly

---

* Nitrogen trichloride is a most effective germicide.

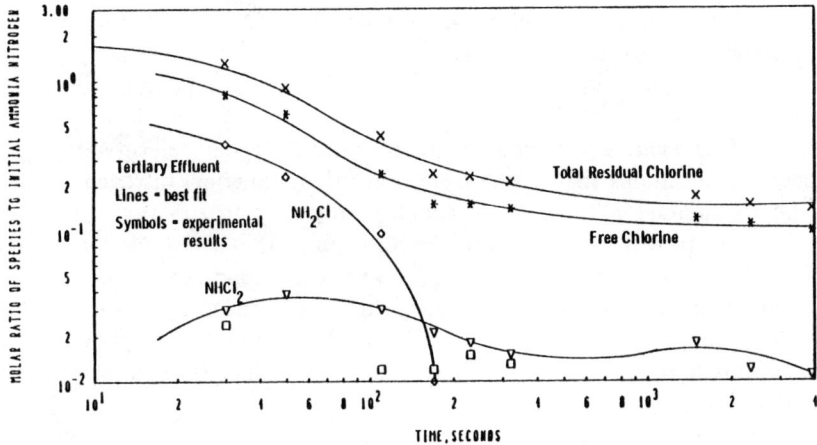

Fig. 4-8. Breakpoint chlorination kinetics of a tertiary effluent in a plug flow reactor (pilot plant experiment). pH = 7.50 after 150 sec., Temp. = 18.8°C, N = 18.03 mg/liter and Cl_2/N = 1.77 molar ratio. Lines = best fit; symbols = experimental results.

at pH 7.5. He also confirmed that the NCl_3 concentration increased with increasing Cl to N ratios at all levels of pH. Also that the organochloramines interfered mainly with the dichloramine and indirectly with the NCl_3 DPD-FAS analysis, primarily at pH values above 7.5. Traces of organic nitrogen (0.02 mg per liter) formed organochloramines in sufficient amount to give false DPD-FAS dichloramine readings at pH values above pH 7.5. The oxidation of the organic nitrogen proceeded faster at pH values of less than 7.5.

In summarizing Saunier's findings, particularly as they relate to tertiary effluents, it can be concluded that if it is desired to achieve a free chlorine residual for bacteria and virus destruction, when the ammonia nitrogen concentration is less than 2.0 mg per liter, the process should be carried out at a control pH of 7.0–7.2 with a minimum contact period of 30 min. Moreover, Saunier's work shows some instability of the free chlorine residual in the usual control sampling time of 1 to 3 min. after the point of application. Therefore, residual control in situations where the initial ammonia nitrogen concentration is less than 2.0 mg per liter should be based on the total chlorine residual measurement.

At this level of chlorination with most wastewater effluents, pH adjustment due to the addition of chlorine would probably not be required. Most effluents have an alkalinity of 150 to 200 mg per liter which is easily capable of maintaining the pH at status quo with chlorine dosages as high as 25 mg per liter.

Significance of Organic Nitrogen

Chlorination of waters containing organic nitrogen presents a variety of problems not encountered in waters containing ammonia nitrogen. Organic nitrogen looms as a formidable stumbling block in the production of an acceptable water, because of the complexity of the compounds that contain organic nitrogen.

A little background on the source and nature of these compounds will be helpful in understanding the problem.

Standard Methods[7] gives analytical procedures for the determination of two nitrogen classifications that derive from other than ammonia nitrogen:

1. One is albuminoid nitrogen, which gives only an approximation of the proteinaceous matter present in the water. The nitrogen in this determination is thought to be related to the nitrogen content of the various amino acids.

2. The other is total organic nitrogen, which is contributed in various degrees by proteins, polypeptides, and amino acids. In other words, the total organic nitrogen determination measures the nitrogen associated with proteins and all the hydrolytic products thereof.

The proteins are the most difficult to deal with by chlorination. Proteins are a constituent of plant and animal life normal to the aquatic life and are found in both human and animal waste. All proteins are made up of nitrogen, carbon, oxygen, and hydrogen, and some contain sulfur. Of these elements nitrogen, carbon, and sulfur exert a chlorine demand. They are complex high molecular weight organic compounds found in various stages of hydrolysis. The hydrolysis is accomplished by hydrolytical enzymes, and the products of degradation are as follows:

$$\text{protein} \longrightarrow \text{proteoses} \longrightarrow \text{peptones} \longrightarrow \text{polypeptides}$$
$$\downarrow$$
$$\alpha \text{ amino acids} \longleftarrow \text{dipeptides} \longleftarrow$$

The α amino acids are then deaminized by enzymic action, and free fatty acids and other acids result; the free acids serve as food for the microorganisms, which, in turn, are converted to carbon dioxide and water.[8]

From the products of hydrolysis, we see that the proteins represent one end of the spectrum and the amino acids the opposite end. Therefore the chlorination of organic nitrogen compounds can be classified into these two categories. Likewise, the determination of loss of total organic nitrogen will represent the measure of oxidation of proteins by chlorine, while the loss of albuminoid nitrogen will measure the oxidation or decomposition of the amino acids involved.

Taras[44] found that the total organic nitrogen consumed after chlorination follows a well-defined general pattern. The ammonia nitrogen is lost completely within one hour of contact time; the simple and unsubstituted amino nitrogen of many common amino acids is consumed more slowly over an extended period of time (many hours); protein nitrogen shows only a negligible loss, even after many days. Taras also demonstrated that this total nitrogen consumption is an exponential function of time.

By way of contrast and clarification, it should be noted that the albuminoid nitrogen content of many simple amino compounds is reduced by as much as 75 percent within the first hour of contact with chlorine, but from proteinaceous matter the reduction is only very slight.

Chlorination of this group of compounds is further complicated by the many varieties encountered. The particular protein involved will have a variable chlorine reaction, depending on the number of amino groups available for reaction with chlorine, which is a function of the peptide linkage in the protein. Therefore the chemistry of chlorination of organic nitrogen compounds is at best extremely complex because of the various hydrolysis products which may be encountered. Fortunately, most of the proteins are colloids that are partially removed in normal water treatment processes.

The significant difference between the reaction of chlorine with organic nitrogen compounds and that of ammonia is one of time, the comparative loss of nitrogen, and the formation of complex organic chloramines.

Both Griffin[13,14] and Williams[11] found that waters containing a mixture of both ammonia and organic nitrogen did not display the classic dip of the breakpoint reaction but, rather, a plateau effect. Fig. 4-9 illustrates what may be expected with a water containing 0.3 ppm ammonia nitrogen and 0.3 ppm organic nitrogen, with about 50 percent of the latter attributable to the simple amino acids and the rest proteinaceous matter. The results shown here are based largely on observa-

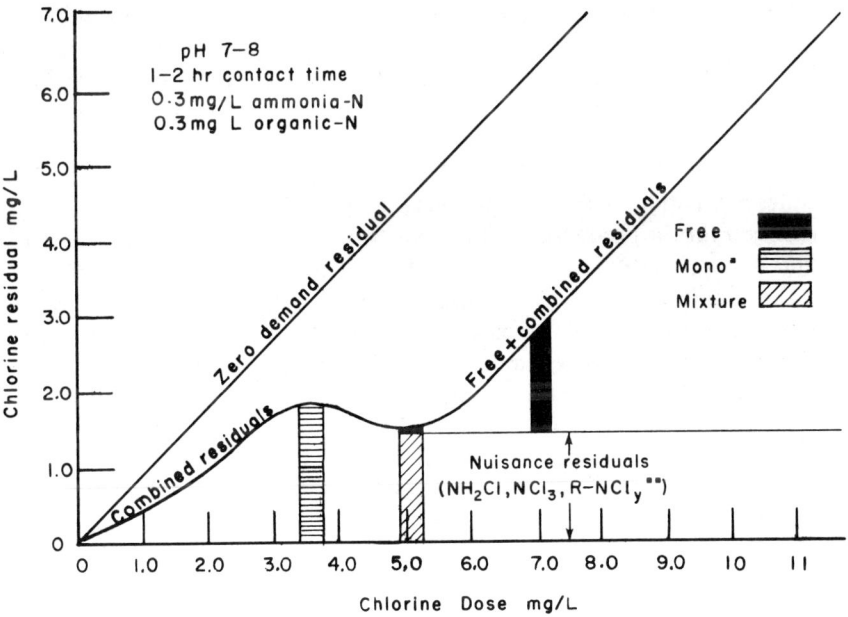

*Predominantly monochloramine; R-NCl$_y$ titrates as dichloramine
**R-NCl$_y$ = organochloramines

Fig. 4-9. Relationship between ammonia nitrogen and organic nitrogen and chlorine.

tions by me and by Williams[11,27] and Palin.[24] Additional scientific evidence is needed to better predict the free available residual curve with waters characterized by significant amounts of organic nitrogen. Fig. 4-9 indicates very little drop in chlorine residual beyond the hump. This signifies two things: (1) continuing and competing reactions between mono- and dichloramine, and (2) very little loss of nitrogen—hence the plateau effect. Beyond the plateau, free available residual begins to appear; but note that the irreducible minimum residual is considerably greater than in the reaction with ammonia nitrogen (Fig. 4-5). Further, it is to be noted that the combined available residual can be made up of equal amounts of mono- and dichloramine, and as the ratio of chlorine to nitrogen increases, significant quantities of nitrogen trichloride can begin to form. This usually occurs when the $Cl:NH_3-N$ ratio exceeds 14:1. Therefore these waters have the ability to produce: proportionately greater nuisance residuals; dichloramine, which contributes to taste problems; and nitrogen trichloride, which has an obnoxious odor. In some cases, the ratio of di- to monochloramine has been known to be 2 or 3:1 in the presence of excess free chlorine. The residuals in these systems are unstable, inasmuch as the free chlorine continues to react to decompose the mono- and dichloramine that are forming with the continued reaction with the organic nitrogen.

Another source of organic nitrogen which might play a significant part in the chlorination chemistry is the urea present in the raw water resulting from sewage discharge. Urea will hydrolyze, with the nitrogen breaking down to ammonia as one of the end products. This hydrolysis usually takes place in 3 to 4 hours under normal conditions; but if there is a lack of the urease enzyme, which breaks down urea, the formation of ammonia nitrogen is greatly inhibited. Urea does not form urea-chloramine at low concentrations, owing to its high dissociation constant,[45] but if a significant quantity of urea-N is present and hydrolysis proceeds at a slow rate, a situation of unstable residuals could easily result. The urea-N would be a reservoir for the production of ammonia to keep reacting with the free available chlorine. The role of urea in water chemistry is still rather obscure.

Another significant factor is the chlorine demand exerted by the carbon content of some of the simple amino acids, such as cysteine and glycine.[9] For glycine, Palin[24] noted that there were similarities to the chlorine-ammonia breakpoint phenomenon. Morris[9] work on the kinetics of chlorine and glycine reveal that up to a 1:1 molar ratio there is only a small loss of oxidizing chlorine. Beyond this ratio the loss of chlorine increases rapidly until it is complete at a molar ratio of 1.5 chlorine to 1 glycine. This is a lower ratio than the 2:1 for the chlorine-ammonia breakpoint system (10:1 by weight). At higher chlorine concentrations, there is a distinct departure from the chlorine-ammonia pattern. Significant amounts of oxidizing chlorine are not found in the solutions again until a molar ratio of 4:1 is reached. Upon analyzing the solutions, Morris[9] found it was the carbon, and not the nitrogen, that was being oxidized, when the molar ratios were 1.5:1 and 2:1. The original nitrogen was still present either as ammonia or as organic nitrogen. The oxidized carbon is released as carbon dioxide.

The reaction between chlorine and organic nitrogen, if given sufficient time, is certain to form organic chloramines of one kind or another. These chloramines display characteristics of dichloramine because they always appear in the dichloramine fraction when subjected to present amperometric and DPD-Ferrous titrimetric methods of analysis. Recent experience with completely nitrified and highly polished reclaimed wastewater reveals that a small fraction of these organic chloramines will intrude into the monochloramine reading.

Another reaction involving organic nitrogen and chlorine is the relationship between the organic chloramines and the monochloramine formed when ammonia-N is present. White et al.[70] found that there is a definite shift of the monochloramine fraction to the dichloramine (organic chloramine) fraction with time. This is probably caused by the disproportionation reaction of monochloramine, whereby HOCl is produced by the monochloramine, which then reacts with more organic N to form more organic chloramines. This causes a simultaneous reduction of monochloramines.

Some organic chloramines are known to hydrolyze to produce HOCl as one end product. The net result of organic chloramines is an overall reduction in the effectiveness of the total chlorine residual. In the free residual process they become a major portion of the nuisance residuals and are known to cause taste and odor problems.

Chlorine Demand Research by Feben and Taras

General Discussion. To help recognize these systems as well as the general overall relationship between chlorine and nitrogenous compounds, Feben and Taras[46,47] in 1950 made a notable contribution in detailed study of the chlorine demand of Detroit water.

Taras pursued this work[44,48] during the next few years, and revealed a definite and significant relationship between chlorine demand and the character of the nitrogen compounds present. These findings are of such practical importance that they are presented here to demonstrate not only how the chemist can identify these troublesome compounds but also how the problem of organic nitrogen can be controlled.

Taras' findings are as follows:

1. There is a definite mathematical relationship between the amount of chlorine consumed in one contact period and that amount consumed in any other period.

2. A study of the constants in this mathematical relationship of chlorine consumed provides clues as to the nature of the nitrogen compounds reacting with the chlorine.

3. This mathematical relationship makes it possible to maintain specific chlorine residuals in the finished water.

4. The source of the chlorine-consuming nitrogenous compounds can be categorized as ammonia, amino acids, and proteins.

182 HANDBOOK OF CHLORINATION

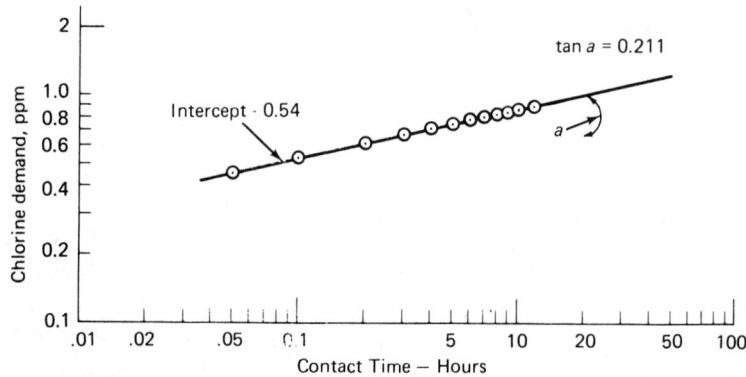

Fig. 4-10. Relationship between chlorine consumed and contact time.

5. The total nitrogen consumed follows a well-defined general pattern.

To find the mathematical relationship between chlorine consumed as a function of time (chlorine demand), a series of chlorine determinations are made on a sample of water with varying contact times from thirty minutes to thirty-six hours. These results are plotted on logarithmic paper, as seen in Fig. 4-10. In order to achieve such a plot, it is necessary for the chlorine dose to be sufficient always to provide a measurable residual at the end of the contact time.

Mathematical Relationship. It is preferable to use the amperometric method for determining chlorine residual. Two curves can be plotted: one for free chlorine residual; the other, for total chlorine residual. The results should plot in a straight line, which is of the general form:

$$D = kt^n \qquad (4\text{-}54)$$

where

$D =$ chlorine demand ppm (chlorine dose $-$ chlorine residual)
$t =$ chlorine contact time in hours
$k =$ chlorine demand after one hour ppm
$n =$ slope of curve (tan a)

For the example illustrated in Fig. 4-8, the equation then becomes:

$$D = 0.54 t^{0.21} \qquad (4\text{-}55)$$

This equation can be put into a more practical form if the 30-minute and 1-hour chlorine demands are known. Eq. (4-54) can be rewritten to read:

CHEMISTRY OF CHLORINATION

$$Dt = D_1 t^n \tag{4-56}$$

where D_1 is the chlorine demand after one hour contact, and Dt is the chlorine demand after t hours of contact. Then, when $t = 0.5$ hours, Eq. (4-56) can be written:

$$D_{0.5} = D_1 \times 0.5^n \tag{4-57}$$

and expressed logarithmically as:

$$\log D_{0.5} = \log D_1 + n \log 0.5 \tag{4-58}$$

then:

$$n = \frac{(\log D_1 - \log D_{0.5})}{-\log 0.5} \tag{4-59}$$

or:

$$n = \frac{(\log D_1 - \log D_{0.5})}{0.30103} \tag{4-60}$$

or:

$$n = 3.322 \, (\log D_1 - \log D_{0.5}) \tag{4-61}$$

Substituting Eq. (4-61) in Eq. (4-56) it becomes:

$$D_t = D_1 t^{3.322(\log D_1 - \log D_{0.5})} \tag{4-62}$$

Eq. (4-62) may then be expressed in the logarithmic form:

$$\log D_t = \log D_1 + 3.322 \log t \, (\log D_1 - \log D_{0.5}) \tag{4-63}$$

or:

$$D_t = D_1 \left(\frac{D_1}{D_{0.5}}\right)^{3.322 \log t} \tag{4-64}$$

where

D_t = chlorine consumed in t hours (chlorine dose minus chlorine residual)
D_1 = chlorine consumed in one hour
$D_{0.5}$ = chlorine consumed in thirty minutes

With this equation, the chlorine demand at the end of any contact time t can be determined, providing the chlorine demands at one hour and at thirty minutes are known.

Practical Significance. The usefulness of the basic equation derived from measuring chlorine residuals as plotted in Fig. 4-10 is the variation in the exponent value n. This reveals the speed of the reaction and gives a clue to the nature of the organic material involved in the reaction with chlorine. The inorganic ions (NH_3^- Fe^{++}, $S^=$ etc.) all react almost instantaneously, causing rapid initial chlorine demand. This decreases the slope of the curve, so that the exponent n approaches zero.

The higher the value of the exponent, the more complicated the organic material. For example, well waters generally show lower exponents (0.01 to 0.05) than do treated surface waters (0.10 to 0.20). This is because the nitrogen content of well water is mainly in the form of the inorganic ammonium ion, which reacts rapidly with chlorine, whereas surface waters generally suffer from sewage pollution and therefore contain the more complex compounds of amino acids and proteins.

In amino acids, the fifteen-minute chlorine demand is usually directly related to the amount of albuminoid nitrogen present. The exponential reaction constant as a function of time is dependent on the individual structure of the amino acid. An increase in the structural complexity results in higher values of the reaction constant n, Eq. (4-56), and will therefore exhibit prolonged chlorine demand.

Since the proteins are among the most complex compounds found in nature, they are difficult to decompose by chlorination. They have a high molecular weight and consist generally of 16 percent nitrogen, 50 percent carbon, 22 percent oxygen, and 7 percent hydrogen. Some contain as much as 2 percent sulfur; others, phosphorous and small amounts of iron. The presence of significant amounts of proteinaceous material in chlorinated water is indicated by a relatively high value of the reaction constant n. Some proteinaceous substances have yielded values of n as high as 0.90.[47]

Therefore, by plotting on logarithmic paper the chlorine demand of various dosages versus contact time, the determination of the value of n will serve as a clue to the nature of the nitrogen content present. A significant rise in the value of the exponent would indicate a rise in the organic nitrogen present and, further, a deterioration in the raw water quality.

The practical significance of organic nitrogen increase and its relation to chlorine demand, becomes a major parameter of raw water quality as follows:

1. Waters containing inorganic nitrogen (usually NH_3) are the easiest to handle. Stable chlorine residuals can be achieved after one hour or less contact time, depending on the nitrogen concentration.

2. Waters containing organic nitrogen will not produce stable residuals except after many hours of contact. This fact presents difficulties, because it means that the residuals leaving the confines of the plant are most likely unstable and that

special treatment procedures are required. The stability of these residuals depends upon the complexity of the proteinaceous compounds as well as upon their concentration. The amino acid group is identified by the albuminoid nitrogen, which is available for fairly rapid destruction by chlorination. This can usually be entirely destroyed by free residual chlorination in one to two hours; ammonia nitrogen, in thirty minutes to one hour. The proteins are identified by the total organic nitrogen, and are the most resistant to decomposition by chlorine. The residuals produced are the least stable of the three, inasmuch as oxidation reaction of chlorine continues for days.

These reactions are characterized by the loss of nitrogen when subjected to the appropriate analyses. For example, when the total nitrogen is in the form of the ammonium ion, the reaction with chlorine will produce a loss of almost 90 percent of the nitrogen. When the total nitrogen is in the form of both ammonia and albuminoid nitrogen, the loss may be only 75 percent or less, depending on the relative concentration of each. In sharp contrast to this is the almost negligible loss of nitrogen when it is present in the form of proteins.[44,49]

CHLORINE AND OTHER INORGANICS

Alkalinity

Since chlorine solutions are highly corrosive the application of chlorine to a process stream will often raise the question of chlorine corrosion. Corrosion by chlorine is related directly to pH, which is dependent upon alkalinity. Therefore it is appropriate to know the effect of chlorine on the alkalinity of a water.

If a water is dosed with chlorine to the extent of the chlorine demand of the water, all of the chlorine applied will end up as chloride ion (Cl^-) as follows:

$$Cl_2 + H_2O \longrightarrow HOCl + HCl \qquad (4\text{-}65)$$

$$HOCl + Cl \text{ demand} + HCl \longrightarrow 2HCl \qquad (4\text{-}66)$$

This calculates as 1.4 parts alkalinity for each part chlorine:

$$2HCl + H_2O + CaCO_3 \longrightarrow CaCl_2 + CO_2 + 2H_2O \qquad (4\text{-}67)$$

$$\text{Alkalinity} + CaCo_3 = \frac{100}{Cl_2} = \frac{100}{71} = 1.4$$

The reaction in Eq. (4-67) occurs at pH 4.3, which is the endpoint of the alkalinity titration, and where CO_2 exists.

When HOCl is not reduced, only one Cl goes to HCl and the reaction consumes just half of the alkalinity, or 0.7. As can be seen from the above, the subject of alkalinity destruction by chlorine is not a simple one.

Take the case of a water having sufficient alkalinity to maintain a 7.0 pH and none of the chlorine applied is consumed by chlorine demand. In this case 50 percent of the chlorine applied will go to HCl. 80 percent of the remaining 50 percent will be undissociated HOCl and unreactive in the alkalinity reactions (see Table 4-4). However, the other 20 percent of the remaining 50 percent at pH 7 is H$^+$ and OCl$^-$.

The alkalinity reduction is calculated as follows:

$$50\% = (0.20 \times 50\%) = 60\% = 0.60$$

$0.60 \times 1.4 = 0.84$ parts alkalinity reduction by chlorine at pH 7.0.

The rule of thumb for alkalinity correction by the use of caustic to maintain pH is a pound of caustic to a pound of chlorine.[82] One part of caustic produces 1.25 parts of alkalinity. If a water at pH 7 is dosed with chlorine, and the demand is 6 mg/l, then the alkalinity reduction is 1.3 parts of alkalinity for each part of chlorine. This rule is reliable.

Carbon

Chlorine reacts with inorganic carbon in much the same manner as it reacts with organic carbon. The reaction is rapid, and can be expressed:

$$C + 2Cl_2 + 2H_2O \longrightarrow 4HCl + CO_2 \qquad (4\text{-}68)$$

This is the reaction that takes place in the dechlorination process using granular carbon filter beds. The production of HCl consumes approximately 2.1 parts of alkalinity as $CaCO_3$ for each part of chlorine removed. The weight ratio of chlorine to carbon is 1.0:0.0845.

Cyanide

In an alkaline solution of pH 8.5 or higher, chlorine reacts with cyanide to form cyanate. This is a primary step in treating plating wastes for the destruction of cyanides, and may be described as follows:

$$2\ Cl_2 + \underset{\text{(caustic)}}{4\ NaOH} + \underset{\text{(cyanide)}}{2\ NaCN} \longrightarrow \\ \underset{\text{(cyanate)}}{2\ NaCNO} + 4\ NaCl + 2\ H_2O \qquad (4\text{-}69)$$

This reaction requires 2.73 parts of chlorine and 3.07 parts of caustic for each part of cyanide to convert to cyanate.

The complete destruction of cyanide by chlorine is usually carried out at a pH of 8.5 to 9.5, and follows the following equation:

$$5 \text{ Cl}_2 + 10 \text{ NaOH} + 2 \underset{\text{(cyanide)}}{\text{NaCN}} \longrightarrow \qquad (4\text{-}70)$$

$$2 \text{ NaHCO}_3 + 10 \text{ NaCl} + \text{N}_2 \uparrow + 4 \text{ H}_2\text{O}$$

The cyanide decomposes, liberating nitrogen as a gas; the carbon atom joins to form sodium bicarbonate. The other decomposition product is water. The weight ratio is 6.82 parts of chlorine and 7.69 parts of caustic to decompose one part of cyanide (CN).

Hydrogen Sulfide

Hydrogen sulfide gas is frequently found dissolved in underground waters (rarely in surface supplies). It is also a constituent of septic sewage and is characterized by its obnoxious odor, best described as that of rotten eggs. It reacts instantaneously with chlorine either to precipitate free sulfur or to form dilute sulfuric acid. The determining factor in the formation of these end products is the $p\text{H}$.

Table 4-6 by Nordell,[50] shows how chlorine and hydrogen sulfide react at different $p\text{H}$ levels.

The complete oxidation of hydrogen sulfide to the sulfate form is as follows:

$$\text{H}_2\text{S} + 4 \text{ Cl}_2 + 4 \text{ H}_2\text{O} \longrightarrow \text{H}_2\text{SO}_4 + 8 \text{ HCl} \qquad (4\text{-}71)$$

Therefore 8.32 ppm of chlorine are required to oxidize 1 ppm of hydrogen sulfide to the sulfate form. However, the chlorine required to oxidize hydrogen sulfide to sulfur and water is only 2.08 ppm chlorine to 1 ppm hydrogen sulfide:

$$\text{H}_2\text{S} + \text{Cl}_2 \longrightarrow \text{S} \downarrow + \text{H}_2\text{O} \qquad (4\text{-}72)$$

Table 4-6 Oxidation of Sulfides by Chlorine at Different pH Levels. (Contact time 10 minutes)

Final pH	Sulfides as ppm H_2S	Cl_2 Added mg/l	Cl_2 Residual mg/l	Cl_2 Consumed mg/l
3.2	5	50	6	44
5.0	5	50	7	43
6.2	5	50	7	43
6.4	5	50	9	41
6.8	5	50	11	39
7.1	5	50	18	32
7.6	5	50	18	32
9.0	5	50	25	25
10.1	5	50	25	25

(Amount of chlorine required to oxidize 5 ppm sulfides expressed as H_2S is 8.32 × 5 = 41.6 mg/l.)

In this instance the sulfur precipitates as finely divided white particles which are sometimes colloidal in nature.

In Table 4-6 it can be seen that the sulfides were all oxidized to sulfates at pH values below 6.4. At pH around 7.0, about 70 percent was oxidized to sulfate and 30 percent to water and sulfur;* at pH values of 9 and 10, approximately 50 percent was oxidized to sulfate and 50 percent to sulfur and water.

It should be noted that for complete oxidation both sulfuric acid and hydrochloric acid are end products which will consume alkalinity. Each part of H_2S oxidized in this reaction will consume 10 ppm alkalinity as $CaCO_3$. For the reaction which precipitates sulfur, only 2.6 mg/l alkalinity as $CaCO_3$ are consumed.

In actual practice, the destruction of hydrogen sulfide in potable water is not as simple as portrayed by the stoichiometric reactions of Eqs. (4-71) and (4-72)—there are side reactions which interfere with the production of the end products. (See Chapter 6.)

Iron

The reaction of chlorine with iron is a very convenient one, since it can be used for two different purposes: to remove iron from water; to produce a coagulant for both water and sewage treatment.

In water supplies, iron is usually associated with underground waters and is generally in the form of ferrous bicarbonate, which is slightly soluble in water. Chlorine reacts with the ferrous ion and converts it to the ferric form. Depending upon the hydroxyl ion activity, the ferric chloride formed will quickly hydrolyze to ferric hydroxide. The latter precipitates as a reddish fluffy mass, depending on the concentration of ferric ion. Omitting the intermediate reaction of the formation of ferric chloride, the reaction is as follows:

$$2Fe(HCO_3)_2 + Cl_2 + Ca(HCO_3)_2 \longrightarrow 2Fe(OH)_3 + CaCl_2 + 6CO_2 \quad (4-73)$$

This reaction produces a rapid release of carbon dioxide, which causes a significant rise in pH. This reaction can proceed over a wide pH range (4 to 10), but the optimum is above 7. Each part of iron to be removed requires 0.64 parts of chlorine. This reaction consumes 0.9 parts of alkalinity as $CaCO_3$ for each part of ferrous ion oxidized to ferric ion. Either free or combined chlorine will produce this reaction, which is instantaneous for inorganic iron in solution. If the iron is in the form of an organic compound, the reaction is considerably inhibited.

It is often desirable to use the ferric ion as a coagulant in the water treatment process. This is most conveniently accomplished by converting the easy-to-handle ferrous ion to ferric by oxidation with chlorine. For example, pickle liquor, a

---

* When chlorine is used to control H_2S formation the treated water will likely appear milky due to the formation of free sulfur. This is known as the Tyndall effect.

by-product of steel mills, is the liquid form, and copperas the granular form, of ferrous sulfate commonly used with chlorine to make a coagulant, as follows:

$$6\ FeSO_4 \cdot 7\ H_2O + 3\ Cl_2 \longrightarrow 2\ FeCl_3 + 2\ Fe_2(SO_4)_3 + 42\ H_2O \quad (4\text{-}74)$$

Each of the ferric salts will hydrolyze further to ferric hydroxide ($Fe(OH)_3$), utilizing the hydroxyl ion of the water and thus forming a brown gelatinous floc. To complete this reaction requires 7.8 parts of ferrous sulfate to each part of chlorine. The resulting solution will contain about 29.9 percent ferric chloride and 70.1 percent ferric sulfate. Each part of chlorinated ferrous sulfate consumes 0.54 parts of alkalinity as $CaCO_3$. This reaction proceeds over a wide pH range. With pickle liquor, adequate floc can be produced at as low as pH 4; however, the optimum should be kept at 7 or higher, since the resulting ferric hydroxide hydrolyzes more rapidly in this range.

Manganese

Free available chlorine reacts to oxidize soluble manganese compounds, but chloramines or combined chlorine residuals have little effect. The oxidized manganese drops out as a precipitate in the form of manganese dioxide, as follows:

$$MnSO_4 + Cl_2 + 4\ NaOH \longrightarrow MnO_2 + 2\ NaCl + Na_2SO_4 + 2\ H_2O \quad (4\text{-}75)$$

This reaction requires 1.3 parts of free chlorine for each part of manganese oxidized, and 3.4 mg/l of alkalinity as $CaCO_3$ will be consumed. The reaction will proceed in a pH range of 7 to 10, with the optimum close to 10. It usually takes from two to four hours to reach completion. If the manganese is in the form of an organic complex, the reaction time will be even longer and unpredictable.

Methane

Methane gas is sufficiently soluble in water to be both explosive and a fire hazard. Waters in this category are known to exist, and one of their characteristics is a high chlorine demand (10 to 25 mg/l). It was thought that chlorine reacted directly with methane to form carbon tetrachloride, as follows:

$$CH_4 + 4\ Cl_2 \longrightarrow CCl_4 + 4\ HCl \quad (4\text{-}76)$$

This reaction with elemental chlorine gas and methane gas is possible at temperatures of 300°C and higher. However, there is no reaction between methane and hypochlorous acid. The high chlorine demand of these waters is attributed to the presence of high amounts of oxidizable organics.

Nitrites

Chemistry of Formation. The effect of nitrites on the chlorination process is often overlooked and misunderstood. Nitrites appear as transitory compounds when for a variety of reasons ammonia nitrogen is being oxidized to nitrates. The following equations illustrate the chemical pathways of this reaction.

$$2NH_4^+ + 3O_2 \longrightarrow 2NO_2^- + 4H^+ + H_2O \qquad (4\text{-}77)$$

Generally the *Nitrosomas* genera are involved in conversion of ammonium to nitrite under aerobic conditions. This is the reaction that often occurs within the biological slime on filter media.

The nitrites can be oxidized to nitrate generally by *Nitrobacter* according to the following reaction

$$2NO_2^- + 2O_2 \xrightarrow{\text{bacteria}} NO_3 + 2N^+ + H_2O \qquad (4\text{-}78)$$

To oxidize 1 mg/l of ammonia nitrogen requires about 4.6 mg/l of oxygen (excluding synthesis of nitrifiers).

Reaction with Chlorine. Nitrites exert a significant chlorine demand in the presence of free chlorine as follows:

$$HOCl + NO_2^- \longrightarrow NO_3^- + HCl \qquad (4\text{-}79)$$

Since it takes two atoms of chlorine to make one molecule of HOCl it takes five parts of chlorine to oxidize one part of nitrite (as nitrogen). Therefore each mg/l of nitrite represents 5 mg/l chlorine demand in the presence of free chlorine. *However, a combined (chloramine) chlorine residual will not oxidize or react with nitrites.*

Nitrites are a deceptive factor in the chlorination process because they interfere with chlorine residual measurements. Apparently nitrites can oxidize KI to I_2. This is a reaction similar to that of free chlorine and chloramines. Therefore since there is no reaction between nitrites and chloramines, any residual measurement using the acid–iodide ($pH < 4$) procedure will show higher residuals when nitrites are present with chloramines.

When the phenylarsine oxide–iodine procedure is used (back titration) for measuring chloramine residuals the nitrites will be oxidized by the iodine solution used for back-titrating the PAO. Therefore the apparent chlorine residual will be higher than the true value (see Chapter 5).

Seawater Chemistry

Effect of Bromides. The chemistry of seawater is unique when chlorination is involved owing to the presence of bromides in the seawater. These bromide compounds result in a bromide ion content of 50–70 mg/l which will always be well in excess of any chlorine applied for chlorination purposes. A good example is the chlorination of seawater used for cooling water in an electric power generating station. Chlorination is used to control the marine growths in the steam condenser tubes because these growths interfere with the heat transfer efficiency of the condensers.

When the chlorine solution from a conventional chlorinator mixes with the seawater (pH 8.3) the chlorine reacts immediately with the bromide ions to form hypobromous acid[60,83]

$$HOCl + Br^- \longrightarrow HOBr + Cl^- \qquad (4\text{-}80)$$

The rate constant (K_o) for this reaction is 2.95×10^3 liter per mol per sec. The hypobromous acid will dissociate, as does hypochlorous acid depending upon the pH, temperature, and ionic strength as follows:

$$K = \frac{(H^+)(OBr^-)}{(HOBr)} \qquad (4\text{-}81)$$

where

$$K = 2 \times 10^{-9} \text{ at } 20°C$$

The reaction rate between chlorine and the bromide ion, Eq. (4-80), is rapid. Using documented rate constants, Selleck[84] calculated the time required to convert 99 percent of the chlorine to bromine at a temperature of 25°C and a pH of 8.3* for various percentages of seawater. Fig. 4-11, taken from Selleck's calculations, shows that the reaction of Eq. (4-80) proceeds to 99 percent completion within 10 seconds in highly saline waters; i.e., 19,000 mg/l chlorides.

Effect of pH. The importance of pH in the reaction of chlorine species and the bromides in seawater must be emphasized. Free bromine (Br_2) is extracted from seawater and from the Dead Sea by chlorination:

$$Cl_2 + 2Br^- + 2H^+ \longrightarrow Br_2 + 2HCl \qquad (4\text{-}82)$$

However this reaction will not proceed to completion until the seawater is acidified to pH 4. (See Chapter 14.)

Chloramines will not oxidize bromides in either the neutral pH zone of 7–9 or in the acidic zone of 4–6. This is why the chlorine–ammonia process does not increase the TTHM potential when bromides are present in potable water.

At pH 4, chloramines will not oxidize the bromide ion to free bromine but

---

* Seawater is always at pH 8.3 in the salinity range of 12,000–19,000 mg/l chlorides.

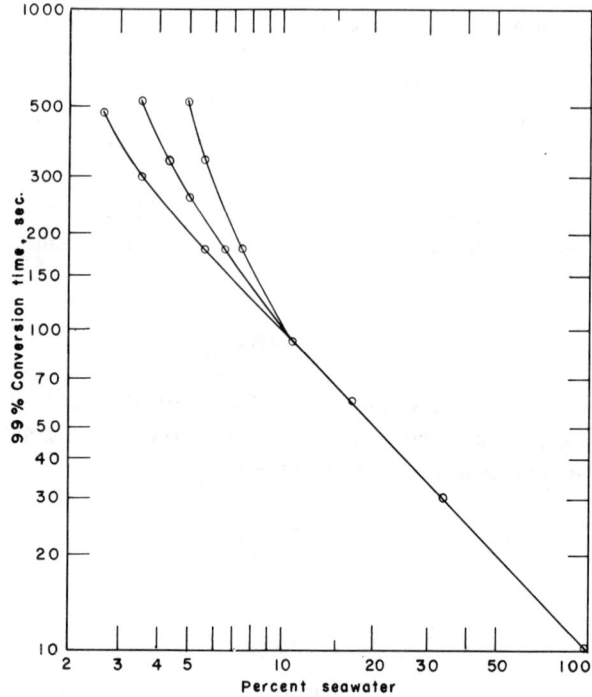

Fig. 4-11. Time required for 99 percent conversion of free chlorine to HOBr at 25°C and pH 8.3[86] (courtesy *Journal WPCF*).

HOCl will. This bit of fundamental chemistry is the basis for the invention of a free chlorine residual analyzer that is not affected by the presence of chloramines. (See Chapter 9.)

Dissociation of Hypobromous Acid. This is a salient factor in the chlorination of seawater because HOBr is an effective biocide. This must be assumed because all seawater chlorination studies have reported residuals as chlorine, when in reality the residuals are most likely to be a mixture of bromine species. Seawater chlorination has been practiced by the electric power industry for six decades, and the success of this practice of reporting residuals as chlorine for the control of marine growths is sufficient evidence to conclude that the resulting bromine species are potent biocides. The dissociation of HOBr plays an important role. Some comparative experiments have been made on coliform destruction using chlorine generated from seawater versus chlorine solution with freshwater and conventional chlorination equipment. These experiments tended to show a better coliform kill with on-site generated hypochlorite from seawater.

The explanation might be found in the dissociation of HOCl vs. HOBr. From

Eq. (4-81) it can be calculated that at pH 8, the undissociated HOBr is 83 percent vs. HOCl at only 28 percent.

Effect of Ionic Strength. Since seawater chlorination produces species of bromine residuals owing to excess bromide ions present in seawater, the dissociation values for HOCl based upon ionic strength of seawater cannot be used. From the dissociation value for HOBr described above the ionic strength of seawater ($I = 1.0$) will have little effect on the potency of the undissociated HOBr.

The Role of Ammonia Nitrogen. The presence of ammonia nitrogen in seawater complicates the chemical reactions between the chlorine species and the bromides. When this situation occurs there will be two competing reactions operating simultaneously. These reactions are: (1) chlorine reacts with the ammonia N practically instantaneously at pH 8.3 to form chloramines, and (2) the chlorine will react at about the same speed with the bromide ion. However, since the bromide ion concentration is on the order of 50–70 mg/l and the ammonia-N not more than 2–3 mg/l, the chlorine–bromide reaction might be the faster. The result is probably the immediate formation of both monochloramine and hypobromous acid. These two compounds start another round of reactions as follows:

$$NH_2Cl + Br^- + 2H_2O \longrightarrow HOBr + NH_4OH + Cl^- \qquad (4\text{-}84)$$

$$HOBr + NH_4OH \longrightarrow NH_2Br + 2H_2O \qquad (4\text{-}85)$$

$$NH_2BR + HOBr \longrightarrow NHBr_2 + H_2O \qquad (4\text{-}86)$$

$$NHBr_2 + H_2O \longrightarrow NOH + 2H^+ + 2Br^- \qquad (4\text{-}87)$$

The intermediary NOH formed in Eq. (4-87) is the catalyst which allows the dibromamine to trigger the breakpoint as follows:

$$NHBr_2 + NOH \longrightarrow N_2 + HOBr + H^+ + Br^- \qquad (4\text{-}88)$$

The reactions described by Eqs. (4-84) through (4-88) are sequential and rapid. Very little monobromamine is formed. Moreover there is no evidence of tribromamine formation. The most significant aspect of the above reaction is the rapid formation of dibromamine, which triggers the breakpoint. Selleck points out that due to the presence of excess bromide ion the removal of ammonia nitrogen from seawater by these reactions is independent of the molar ratio of HOBr to NH_3 —N.[85] Therefore the presence of ammonia-N in seawater does not present any problems for the chlorination process.

The above reactions take place only in the range of seawater pH, 8.3.

GERMICIDAL SIGNIFICANCE OF CHLORINE RESIDUALS

Introduction

Studies have been made on the nature of the killing mechanism of chlorine on bacteria,[20,51-54] cysts,[20,55,56] and spores.[20,57] The inactivation mechanism of viruses by chlorine and other oxidants has never been resolved. An infective virus particle or virion consists of a core composed of or containing one kind of nucleic acid which is encased in a protein shell or capsid. The capsid is built from a number of morphological subunits identified as capsomeres. It is thought that the oxidant penetrates the capsid by chemical transformation and then attacks the nucleic acid.

Exactly how chlorine kills bacteria, cysts, and spores is still an academic puzzle. During the period of its early use (circa 1910) as a disinfectant, it was believed that the germicidal power of chlorine was due solely to the liberation of nascent oxygen from the hypochlorous acid formed by the reaction of chlorine and water (HOCl → HCl + O). This theory led to considerable confusion on the subject and thwarted many serious investigations. It was not until 1944 that Chang[56] dispelled this belief by showing that hydrogen peroxide and potassium permanganate liberated considerable quantities of nascent oxygen but showed little germicidal efficiency. He further proved that there was no liberation of oxygen involved in the chlorine reaction, and that the disinfecting agent is hypochlorous acid. He favored the formation of the toxic substance theory.

In 1946, Green and Stumpf[51] concluded that chlorine reacted irreversibly with the enzymatic system of bacteria, thereby killing the bacteria. They were able to show that a bacterial suspension became sterile when bacteria lost the power to oxidize glucose. They also found that when the enzymes which contain sulfhydril groups were oxidized by chlorine, the oxidation was irreversible, thus abolishing enzyme action and resulting in the destruction of the bacteria, and that the most significant of this group is the enzyme triosephosphate dehydrogenase. In 1953, Ingols et al.[58] concurred with this theory, but, using monochloramine, they found that even though they restored the sulfhydril groups the bacteria were not restored. This led them to believe that while the sulfhydril group may be the most vulnerable to a strong oxidant like hypochlorous acid or chlorine dioxide, there were changes in other groups caused by monochloramine and dichloramine which may be important in bringing about death to bacteria.

More recently (1962), Wyss[54] has pointed out that probably the effective mechanism of death to microorganisms is the phenomenon of unbalanced growth. That is, destruction of part of the enzyme system will so throw the cell out of balance that by progress of its own metabolism the cell dies before the necessary repairs are made.

Whatever the chemical action, it is generally agreed that the relative efficiency of various disinfecting compounds is a function of the rate of diffusion of the active agent through the cell wall. Fig. 4-12 shows the essential elements of a bacterial cell. It is assumed that after penetration of the cell wall is accomplished, the disinfecting compound has the ability to attack the enzyme group whose destruction results in death to the organism. Factors which affect the efficiency of destruction are:

1. Nature of disinfectant (kind of chlorine residual fraction)
2. Concentration of disinfectant
3. Length of contact time with disinfectant
4. Temperature
5. Type and concentration of organisms
6. pH

Hypochlorous Acid

HOCl is the most effective of all the chlorine residual fractions. This fraction is known officially in the industry as free available chlorine residual.[59,60] Hypochlorous acid is similar in structure to water; hence, the formula HOCl is preferred to HClO. The germicidal efficiency of HOCl is due to the relative ease with which it can penetrate cell walls. This penetration is comparable to that of water, and can be attributed to both its modest size (low molecular weight) and to its electrical neutrality (absence of an electrical charge).

Other things being equal, the germicidal efficiency of a free available chlorine residual is a function of the pH, which establishes the amount of dissociation of HOCl to H^+ and OCl^- ions. Table 4-4 shows the percentage of undissociated HOCl in a chlorine solution for various pH values and temperatures. Lowering

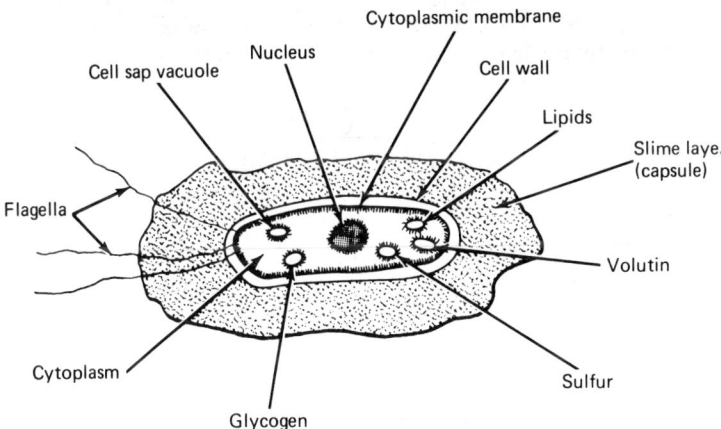

Fig. 4-12. Schematic diagram of a bacterium.

the temperature of the reacting solution suppresses the dissociation; conversely, raising the temperature increases the amount of dissociation.

The rate of dissociation of HOCl is so rapid that equilibrium between HOCl and the OCl$^-$ ion is maintained, even though the HOCl is being continuously used. For example, if water containing 1 mg/liter of titrable free available chlorine residual has been dosed with a reducing agent that consumes 50 percent of the hypochlorous acid, the remaining residual will redistribute itself between HOCl and OCl$^-$ ion according to the values shown in Table 4-4. This is commonly referred to as the "reservoir" effect.

Hypochlorite Ion

The OCl$^-$ ion, which is a result of the dissociation phenomenon, is a relatively poor disinfectant, because of its inability to diffuse through the cell wall of microorganisms. The obstacle to this passage is the negative electrical charge, as substantiated to some extent by the fact that the activation energy for disinfection by HOCl is in the range of those for diffusion ($E = 7,000$ calories), whereas that of the OCl$^-$ ion is more characteristic of a chemical reaction ($E = 15,000$ calories). The relative efficiencies of HOCl and the OCl$^-$ ion have been well documented by Professor Fair and his colleagues at Harvard.[20,37] These findings took into consideration the work done by Butterfield et al.[61] in 1943 and by Butterfield and Wattie[62] in 1944. This was a concerted effort to relate disinfection to chemistry.

It is well known that the disinfecting efficiency of free available chlorine residual decreases significantly as the pH rises. At a pH above 9 there is little disinfecting power. At this pH level and at 20°C, 96 percent of the titrable free available chlorine will consist of the OCl$^-$ ion. This is an indication of the low germicidal efficiency of the OCl$^-$ ion. The Harvard group developed an equation devised to calculate the total free available chlorine residual required to kill a given percentage of a specified organism. This equation requires the use of a constant, which is the ratio of germicidal efficiency of the OCl$^-$ ion to HOCl:

$$R = A \frac{1 + [K_i/(H^+)]}{1 + B[K_i/(H^+)]} \qquad (4\text{-}89)$$

where

$R =$ the required total titrable available chlorine residual.
$A =$ the amount of HOCl alone (expressed as titrable chlorine) required to destroy the bacteria
$K_i =$ the dissociation constant of HOCl for a given temperature
$B =$ the ratio of the efficiency of OCl$^-$ to HOCl.

Fair et al.[59] observed concentrations of titrable free available chlorine required to kill 99 percent *Escherichia coli* in 30 minutes at 2–5°C at different pH values:

$$R = .005 \frac{1 + [2.2 \times 10^{-8}/(H^+)]}{1 + 0.012[2.2 \times 10^{-8}/(H^+)]} \quad (4\text{-}90)$$

The value of B used (0.012) indicates that the OCl^- ion is about 1/80 as efficient as HOCl for the conditions stated. This is not in very close agreement with investigations by Selleck, University of California, Berkeley. His work demonstrates that the OCl^- ion is only 20 times less effective than HOCl rather than 80 times.[63]

A similar investigation was made on the destruction of cysts, and further demonstrated the greater efficiency of HOCl over the OCl^- ion. Experimental cysticidal residuals were obtained for a 100 percent destruction of thirty cysts per ml of *Endamoeba histolytica* over a temperature range of 3–23°C and with a contact time of 30 minutes. The relative efficiencies of the OCl^- ion to HOCl for inactivation of cysts are summarized as follows:

Temperature, °C	OCl^- to HOCl Relative Effective Ratio
3	1/150
10	1/200
18	1/250
23	1/300

Similarly, Eq. (4-89) can be used to calculate the amount of R required to destroy this organism under specific conditions of temperature contact time and pH. Since cysts are the most difficult miroorganisms to destroy with a chlorine residual, it is reasonable to expect that other organisms, such as spores and viruses, would show a relative effective ratio of the OCl^- ion to HOCl somewhere between the extremes of *Esch. coli* at 1/20 and *E. histolytica* at 1/300. It is evident that because of this effective ratio the low germicidal efficiency of OCl^- must be taken into account and that the effectiveness of free available chlorine residual is seriously impaired when the pH exceeds 9.0. The reason for this is obvious. At pH 9 and a temperature of 15°C the percentage of HOCl in a given solution is 4.13; at pH 7.5 it is 57.68. (See Table 4-4.) This prevails in spite of the HOCl "reservoir" effect.

Chloramines

Background Discussion. That chloramines are slower to kill microorganisms than free available chlorine has been common knowledge for some time. Probably the first person to become aware of this was Alexander Houston[64] in 1925. The first scientific laboratory work on the germicidal efficiency of chloramine was done by K. Holwerda[28] at the Laboratory for the Purification of Water at Manggarai in the Dutch East Indies. His work investigated chloramines at pH 4.5, 6.8, and 8.5. At pH 4.5, the chlorine compound would most likely be 100 percent dichloramine (see Fig. 4-4); at pH 8.5, it would probably be 100 percent monochloramine.

Holwerda's findings indicated a time factor as high as 80:1 required for monochloramines over free available chlorine, and as low as 20:1 for the same germicidal efficiency. This compares with Butterfield and Watties' findings[62,65,66] that chloramine required approximately a hundredfold increase in contact time over free available chlorine at pH 9.5. Kabler[67] also concluded that, to obtain the same kill with same amounts of free available and combined available chlorine residual, the latter required 100 times more exposure time.

The above historical data have categorized chloramines as a poor disinfectant compared to free chlorine. However, recent studies indicate that if chloramines are allowed contact times of 45–60 minutes they are able to match the efficiency of free chlorine based upon the destruction of coliform organisms. (See Chapter 8.)

Monochloramine. From the work by Butterfield and Wattie mentioned above, and later confirmed by Kabler, it might be concluded that for the same conditions of contact time, temperature, and a pH range of about 6–8, it will take at least 25 times more combined available chlorine than free available chlorine to produce the same germicidal efficiency. Furthermore, it can be assumed that if the chlorine to ammonia nitrogen ratio is less than 5:1 and if the pH is 7.5 and higher, the combined residual will probably be 100 percent monochloramine. This difference in potency of monochloramine and HOCl might be explained by the difference in their oxidation potentials, assuming that the action of chloramine is of an electrochemical nature rather than one of diffusion, as seems to be the case with HOCl. (See Fig. 4-16.) The work of Ingols et al.[58] might also explain the difference between free available and combined available chlorine residuals. They indicated that by using monochloramine only, its reaction with vulnerable enzyme groups (in bacteria) may be a reversible reaction. Another explanation of the low potency of monochloramine is that it may be a function of hydrolysis. The hydrolysis constants of chloramine in general can be expressed as follows:

$$K_n = \frac{(HOCl)(RR'NH)}{(RR'NCl)} \quad (4\text{-}91)$$

From this equation it can be seen that one of the products of hydrolysis is free available chlorine (HOCl), which tends to be low in strong solutions, but which increases in weak solutions.[16] The hydrolysis constant of monochloramine, as determined by Corbett, Metcalf, and Soper, was found to be $K_n = 2.8 \times 10^{-10}$ at 15°C in accordance with:

$$NH_2Cl + H_2O \rightleftharpoons NH_3 + HOCl \quad (4\text{-}92)$$

This is in fair agreement with more recent work by Morris[69] who found that $K_n = 1.0 \times 10^{-11}$ (approximately). These two investigations were carried out under quite different conditions: Corbett et al. performed their study at pH 14 and $10^{-2}M$; Morris, at pH 5 and $10^{-4}M$. Morris calculates that a pure solution of NH_2Cl at a concentration of 2.0 mg/l as chlorine ($2.82 \times 10^{-5}M$) will be 0.58 percent hydrolyzed at pH 7 and 25°C.

However, the work done at the San Jose/Santa Clara Water Pollution Control Plant[70] over a two-year period (1980-1982) indicates that monochloramine has a much greater germicidal efficiency than described above. This investigation demonstrated clearly that in a completely nitrified effluent monochloramine residuals based upon a chlorine to ammonia-N ratio of 6:1 can achieve coliform kills equivalent to free chlorine residuals of the same concentration and the same contact time, i.e., 50 minutes.*

Other experiences quoted elsewhere indicates that monochloramine can be equivalent to free chlorine provided the contact time is at least 45-60 min. and the Cl:N ratio is 6:1. The potency of monochloramine may be a function of its slow rate of diffusion through the bacteria cell wall, the hydrolysis product HOCl, or its redox potential.

It is generally concluded that ammonia chloramines formed when NH_3-N is present in the water (or wastewater) will slowly hydrolyze with time to produce HOCl, which then reacts with organic N present to form organic chloramines. Since the organic chloramines are considered to be relatively ineffective germicides this would explain the phenomenon of decreasing (with time) germicidal efficiency of a combined chlorine residual in a wastewater effluent.

Dichloramine. Although Holwerda was probably not aware of it, his work definitely demonstrated that dichloramine is a more potent germicide than monochloramine. At pH 4.5 and with 0.5 mg/l chloramine (probably 100 percent dichloramine) disinfection was achieved in twenty minutes, whereas at pH 8.6 (probably 100 percent monochloramine) it required sixty minutes. This was based upon the observance of gas formers. On another series using plate counts, it was ten minutes at pH 4.5 and sixty minutes at pH 8.6 to achieve a one plate count with the control showing 388.

Fair et al.[20] found that at pH 4.5, which they observed to be 100 percent dichloramine (see Fig. 4-4), a residual of 2.0 mg/l gives a 100 percent kill of cysts (*E. histolytica*); whereas at pH 10.0, when only monochloramine is present, the required cysticidal residual is approximately 7.5 mg/l. They used a concentration of 30 cysts/ml at a temperature of 23°C and a contact time of thirty minutes. Their overall results with cysts show that dichloramine is about 60 percent and monochloramine about 22 percent as efficient as HOCl. Because of the formation of the relatively ineffective OCl^- ion in HOCl solutions at high pH values, the chloramines are more efficient cysticidal agents above a pH of about 9. On the limited number of experiments performed with spores of *B. anthracis*, dichloramine showed an

---

* NH_4OH added to nitrified effluent.

efficiency of 15 percent that of HOCl for a thirty-minute contact period, while the efficiency of monochloramine was too low to be measured.[20]

Kelly and Sanderson[34] attributed the inactivation of poliomyelitis and coxsackie viruses to dichloramine rather than to monochloramine. They were able to achieve 99.7 percent inactivation with a combined available chlorine residual in three hours at pH 6, while at pH 10 it required six to eight hours for polio virus. They determined that dichloramine was 43 percent of the total at pH 6 and, at pH 10, was 32 percent. (See Fig. 4-4.) Although these fractions were determined by the amperometric method, the results vary widely from the curve by Fair et al., which was determined by a more sophisticated method—light absorbence as measured by a spectrophotometer.

It can probably be concluded from all these data that dichloramine is twice as potent as monochloramine as a germicidal agent. However, additional scientific data are necessary to establish this as a fact. This would be a welcome contribution to the field of wastewater chlorination. However, in water treatment it is of only academic interest, since dichloramine is to be avoided because it contributes to taste and odor problems. Moreover it is doubtful that there ever would be a condition whereby a pure dichloramine residual could be isolated and measured as such. Those residuals that exhibit themselves as dichloramine via all of the well known analytical procedures are probably a mixture of organochloramines and not a specific specimen of dichloramine. This is the weakness and failure of the analytical procedures. A pure dichloramine made in the laboratory is unstable in the presence of HOCl. It is quite conceivable that in waters lacking organic nitrogen but containing ammonia-N, pure dichloramine will form in the free residual process. However, it will disappear when the competing reactions pass beyond the breakpoint zone.

To date there are no analytical methods that can identify pure dichloramine as differentiated from the organochloramines that appear in the dichloramine procedure.

Nitrogen Trichloride. This chlorine compound is known to be an effective oxidizing agent. It has been used for over thirty years in the bleaching of flour, as a fungicide, and for control of insect pests on fruit in storage rooms and cars during shipment. Its oxidation potential and its bactericidal efficiency has never been evaluated, but this compound may possibly contribute substantially to the oxidation of organic matter and destruction of microorganisms in water and wastewater chlorination in those instances when it does appear. Nitrogen trichloride imparts an objectionable taste and a foul odor when present in potable water. Fortunately it is unstable in sunlight and highly insoluble in water, and aerates easily. It is characterized by its pungent chlorine-like odor. It causes severe eye tearing in low concentrations. It is difficult to capture by any analytical procedure—it is too volatile. Nitrogen trichloride cannot exist without the presence of HOCl. It has been known to form in the distribution system long after leaving the treatment plant. This situation can be corrected by converting the free chlorine residual to chloramines by post-ammoniation. However, the preferred method is complete dechlorination of the finished water followed by rechlorination.

It is never necessary to determine the concentration of NCl_3 present. The operator needs only to know if it is there. This is easily determined by human responses, either by the eyes tearing or by its odor. It is the only chlorine species that causes tearing of the eyes.

Organic Chloramines

General Discussion. This discussion *excludes* the manufactured organic chloramines used for sanitizing purposes and swimming pool treatment; i.e., the chloroisocyanurics, chlorinated hydantoins, etc. These compounds have hydrolysis constants as high as 10^{-4}. They will yield HOCl concentrations equivalent to hypochlorites of the same available chlorine content.[16] This is achieved by hydrolysis at a slow but steady rate.

The organic chloramines of interest here are those that are formed during the chlorination of potable water or wastewater. These organochloramines are directly related to the organic nitrogen content in potable waters, which varies from a trace up to a maximum of 3 mg/l. Any concentration greater than 0.25 is certain to cause taste and odor problems.

Well oxidized secondary effluents will contain from 10 to 15 mg/l organic N. Well polished filtered and nitrified effluents will contain from 1 to 3 mg/l organic N.

These chloramines can be measured by forward titration using either the amperometric or DPD-Ferrous titrimetric method. They will appear in the dichloramine fraction. These chloramines have relatively little or no germicidal efficiency, hence the term: nuisance residuals.

Recent Studies of Germicidal Efficiency. As early as 1966 Feng[87] discovered a great disparity in the germicidal efficiency between ammonia chloramines and those found in an environment of pure organic nitrogen compounds. He has reported that methionine, which is one of the indispensable amino acids for biological growth, and is expected to be present in wastewater, forms a measurable chlorine residual with no germicidal power. Feng also investigated the lethal activities of the glycine, taurine, and gelatin chloramines. His work shows that taurine chloramines are as lethally active as ammonium chloramines at pH 9.5, but that their germicidal efficiency falls off as the pH decreases. The glycine chloramines are as germicidally active as monochloramine at pH 4 but are totally inert at pH 7, and the gelatin chloramines are active at pH 9.5 but are inert at pH 7 and 4. There are certain to be other such organic nitrogenous compounds which contribute to the total chlorine residual which have little or no germicidal effect.

Sung[88] made a controlled laboratory study of fifteen organic compounds representing seven groups to evaluate their individual and combined effect upon the chlorination process. Nine of the fifteen compounds were found to interfere with the germicidal efficiency of the chlorination process. Of these nine compounds five were organic nitrogen compounds. Cystine and uric acid were the severest inhibitors of the nitrogen group. When five of the interfering compounds were mixed together, their combined effect was found to be more pronounced than any of their individual effects, but did not equal the sum of their individual effects. Sung compared the germicidal efficiency of a simulated wastewater with and without the interfering organic compounds. He found that the germicidal efficiency of wastewater containing the interfering compounds by themselves and the resulting chlorine residuals had little or no germicidal effect. The greatest interference was observed to be caused by cystine, tannic acid, humic acids, uric acids and arginine.

Cystine is an amino acid connected by two sulfur groups that is known to react with chlorine. Tannic, humic and uric acids are capable of exerting a significant chlorine demand when present in water or wastewater. *Arginine* is a basic amino acid. The reaction between chlorine and arginine is almost instantaneous.

The organic compounds that had little or no interfering effects on the chlorination process were: acetic acid, cellubiose, dextrose, glutamic acid, uracil, and lauric acid.

The significance of the above findings by Sung[88] confirms the theory of interference in the chlorination process by the presence of organic nitrogen. It further points to the fact that present analytical techniques do not provide for separating the chlorine residual fractions into those of equal germicidal efficiency.

It is also interesting to note that Esvelt, Kaufman and Selleck[89] found that the toxicity of combined chlorine residuals diminished with time. This finding together with those of Sung[88] demonstrates quite convincingly that there are a significant number of organic compounds in wastewater which will react with chlorine to form organic chloramines of little or no germicidal potential, and moreover, these compounds appear to increase in concentration with time. This apparent increase of this chlorine residual fraction with time would also explain the loss of germicidal efficiency of combined chlorine residuals in wastewater with time. The supposition is that the predominantly monochloramine residual present hydrolyzes to form free chlorine (HOCl), and both the monochloramine and HOCl react with the organic nitrogen compounds to form the ineffective organic chloramines.

Nuisance Residuals. These are the residuals, other than free chlorine (HOCl), which appear beyond the breakpoint. This is Zone 3 in Fig. 4-4. They usually consist of chloramines that titrate in the dichloramine fraction. As described before they are probably made up of mostly organic chloramines.

At the breakpoint there are at least three competing reactions which tend to produce free chlorine, monochloramine, dichloramine, and nitrogen trichloride. Pure dichloramine is unstable in the presence of free chlorine; nitrogen trichloride cannot form without free chlorine; and monochloramine is almost always present of the germicidal efficiency of the different fractions of available chlorine residual.

when nitrogen trichloride is formed. Therefore a reasonably good guess as to the composition of nuisance residuals in actual practice might be: (1) When NCl_3 is formed, 85–90 percent organic chloramines, 8–12 percent monochloramine, and 2–3 percent NCl_3; (2) when NCl_3 is absent, 90 percent organic chloramines and 10 percent monochloramine.*

Investigators Consensus of Germicidal Efficiency

In 1964 Clarke, Berg, Kabler, and Chang[71] prepared graphics (see Fig. 4-13) showing the relative germicidal efficiency of HOCl, OCl⁻, and monochloramine. These curves represent a composite of available data adjusted to yield a common base for comparison, and are presented here to summarize the preceding discussion

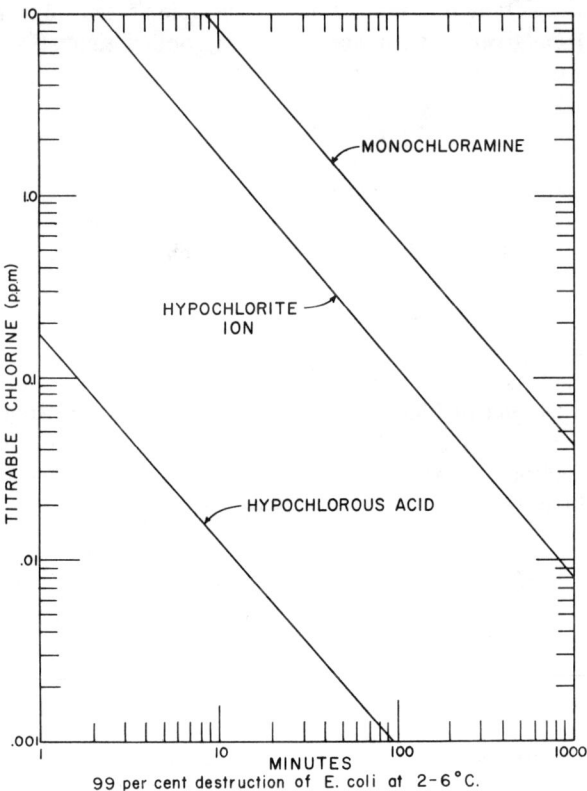

Fig. 4-13. Comparison of germicidal efficiency of hypochlorous acid, hypochlorite ion, and monochloramine.

---

* Organic chloramines that appear in the dichloramine fraction of the forward amperometric titration procedure are known to intrude into the monochloramine fraction. This distorts the monochloramine reading.

of the germicidal efficiency of the different fractions of available chlorine residual. Scientific data are lacking to properly locate on this curve the germicidal efficiency of the dichloramine and nitrogen trichloride fractions.

It should be noted that these curves represent chlorine dosages applied to chlorine demand-free solutions sufficient to achieve only a 2 log reduction at very low temperatures and unreported contact times. The lower end of the curves may not be completely reliable. Full scale plant operation on water reuse situations have demonstrated that for contact times of 45–60 minutes and 4–5 log reductions of coliforms, monochloramine residuals are almost equal to free chlorine residuals and in some cases superior to free chlorine. This anomaly is described in Chapter 8.

Morris' Lethality Coefficient

In 1967 Morris[72] presented a tabulation of germicidal concentrations giving 99 percent (2 logs) inactivation within ten minutes contact time. From this Morris derived a lethality coefficient:

$$\lambda = 0.46/C_{99:10}$$

where C is the concentration of the chlorine compound in mg/liter. The values of λ computed from this 1967 tabulation are shown in Table 4-7.

This tabulation illustrates the superiority of free chlorine over both monochloramine and the hypochlorite ion. This is an interesting and useful comparison, but here again the organism destruction is based upon a short contact time (10 min) and only 2 logs removal. From plant operation experience White believes that longer contact times and greater total coliform removals (4–5 logs) closes the gap between the germicidal efficiency of free chlorine and monochloramine. This has been demonstrated by Prof. R. E. Selleck, University of California,[73] Sanitation Districts of Los Angeles County,[74] and the San Jose/Santa Clara, CA Water Pollution Control Plant.[70] (See Chapter 8.)

ELECTROCHEMICAL PROPERTIES OF CHLORINE RESIDUALS

Introduction

General dissatisfaction with the orthotolidine method of measuring chlorine residuals led a number of investigators to try the oxidation–reduction potential concept for evaluating the germicidal efficiency of the various species of chlorine residuals. This concept is commonly referred to as ORP or redox systems. In its simplest terms it is an electrode potential reading.

The originators of this method as it related to the chlorination of water were Rideal and Evans,[75] who suggested it as early as 1913. They believed that the germicidally active chlorine could best be measured by the potential created across

Table 4-7 Values of λ at 5°C*

Species	Enteric Bacteria	Amoebic Cysts	Viruses	Spores
HOCl as Cl_2	20	0.05	1.0 up	0.05
OCl^- as Cl_2	0.2	0.0005	<0.02	<0.0005
NH_2Cl as Cl_2	0.1	0.02	0.005	0.001

* λ in (mg/liter)$^{-1}$ (min)$^{-1}$

an electrode system in the solution. Most investigators studying the disinfecting power of chlorine found that its efficiency varies widely under different conditions. The chief factor influencing the efficiency of free chlorine (HOCl) is the pH of the solution.

In 1933 Schmelkes[76] demonstrated that the level of potential of a chlorine residual does in fact affect its germicidal efficiency and that the potential is related to pH, chlorine concentration, and the ratio of chlorine to ammonia.

Fig. 4-14 shows that free available chlorine has a much higher potential than do chloramines and it varies with pH.

Fig. 4-15 shows the relationship between the chlorine–ammonia ratio and potential level, indicating again the lower potential of the chloramines. During these

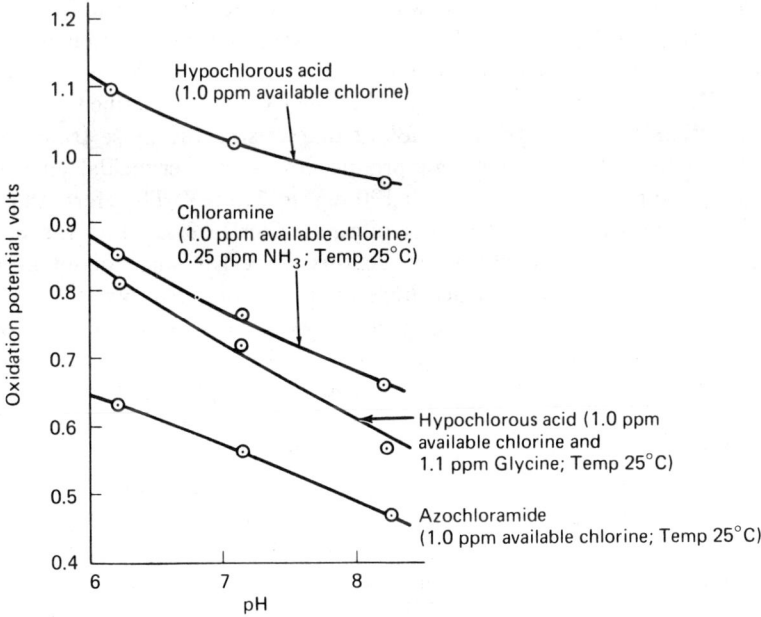

Fig. 4-14. Oxidation potentials of chlorine compounds.

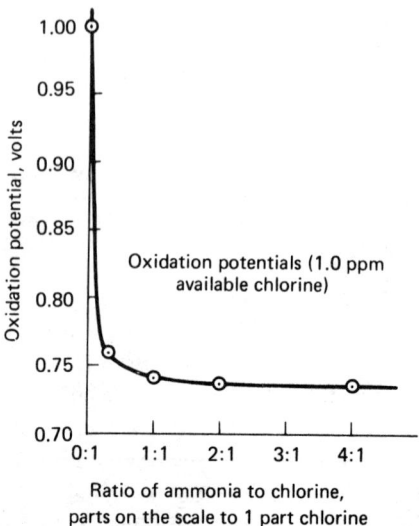

Fig. 4-15. Relationship of chlorine to ammonia ratio and oxidation potential.

early investigations it was generally believed that the germicidal efficiency of chloramines was less than free chlorine. Studies had suggested that the electrochemical properties of chlorine might be directly related to its germicidal efficiency.

In 1939 Schmelkes, Horning, and Campbell,[77] pursued this concept by investigating the "differential" potential of various types of free and combined chlorine residuals. Their work which confirmed the previous investigations is shown in Fig. 4-16. However they considered an additional variable described as the "base potential" of different waters. They found that raising the poise from 170 mV base potential to 190 mV might not produce the same germicidal efficiency as raising the poise in another water from 190 mV to 210 mV. Therefore, the degree to which a water was poised had a significant effect in evaluating the potential as related to germicidal efficiency. While this work left some doubt as to the reliability of the concept, it provided hope for future investigators.

In the next few years, S. L. Chang of Harvard University did some notable work on the destruction of microorganisms by chlorine.[55,56,78] Part of this work was an investigation of the oxidation potential concept of germicidally active chlorine.[79] Chang concluded that because of interfering substances, the uncertain amount of dissolved oxygen in raw waters, and the uncertainty of the reversibility of these systems, it was impossible to evaluate the potential measurements of residual chlorine. He did find that free chlorine, inorganic chloramines, and organic chloramines all had their own characteristic oxidation potentials but that the potential for one compound at cysticidal concentrations gives no information on the cysticidal efficiency of any other compound. This in itself renders this concept useless for evaluating the germicidal efficiency of different chlorine compounds.

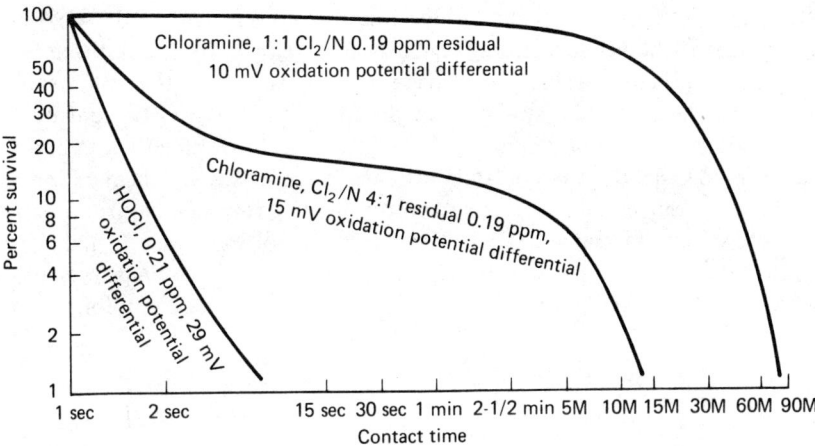

Fig. 4-16. Relationship of germicidal efficiency and oxidation potential differential.

Practical Application of the ORP Concept

In spite of the inability to relate the ORP concept of chlorination to germicidal efficiency it has been tried in a variety of applications.

The practical value of ORP control of chlorination is in qualitative and not quantitative control. The foremost success of this method is in the destruction of cyanide wastes by chlorination. Here the operation depends upon the qualitative appearance of a chlorine residual that is clearly defined by plateaus in the ORP potential curve. The cyanide destruction process is not controlled by the amount of chlorine residual. (See Chapter 7.)

This method was thought to have great promise in the control of hydrogen sulfide odors in wastewater treatment processes. It has been tried over and over again without success. At present there is little or no interest in this application.

The ORP concept has been used successfully in the manufacture of hypochlorite solutions. This is a special case but must be mentioned here to avoid confusion on the subject.

The Diamond Shamrock Company developed a special electrode system that is made up of a single plastic rod containing a platinum measuring electrode and a silver electrode which becomes a silver–silver chloride half cell in the presence of a brine solution in contact with the silver electrode. This silver electrode develops a silver chloride coating due to the presence of high concentration of chloride ions. This half cell works well under these conditions as it is always in the same environment, i.e., a high concentration of same type of dissolved solids. The manufacture of hypochlorite by this electrode system, which is essentially an ORP system, is successfully accomplished because it is controlling to an alkalinity end-point by the addition of chlorine, not to an ORP value due to chlorine concentration.

ORP control of swimming pool chlorination has been used in the USA for at least ten years. The first successful system was developed by Frank Strand under the name of Stranco.[80,81] This system has proved so successful that other suppliers have entered the field. The system is made up of two parts: ORP control of the chlorine addition and ORP control of the pH. A control panel starts and stops the chlorinator and the pH control starts and stops a $NaCO_3$ chemical metering pump. The system maintains precise control of the pH which assures a constant percentage of undissociated HOCl. The ORP control maintains a high level of free chlorine (1.5–2.5 mg/l) which helps account for its success. A survey of three of these operating installations verifies their success and practicality for swimming

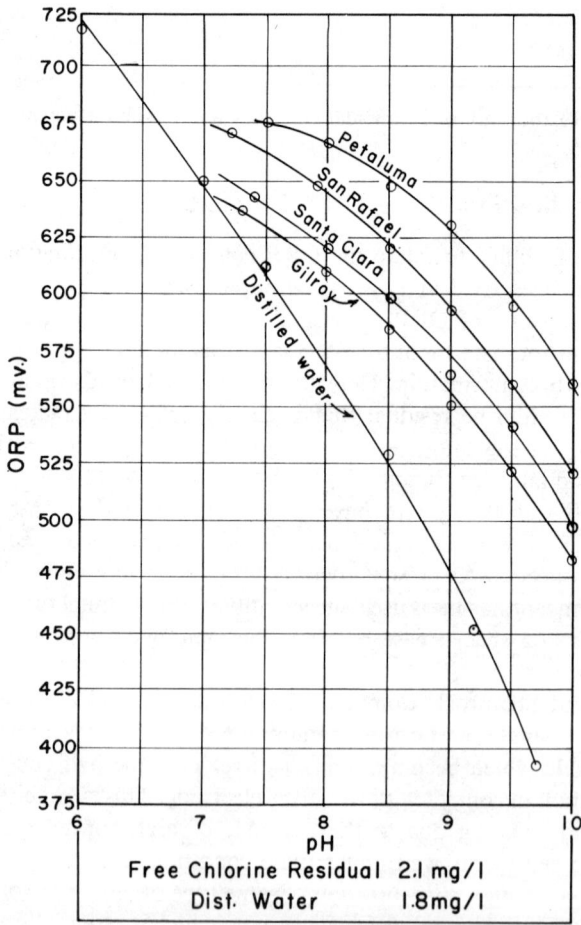

Fig. 4-17. Relationship between oxidation potential and pH at a fixed chlorine residual.

pool application (see below). In every case the Strantrol system, if with faults at all, errs on the side of safety. The free chlorine residual was always in the 1.5–2.5 mg/l bracket and the pool water appeared brilliant and sparkling. At pH 7.5 this is an ideal situation.

In 1975 White investigated four swimming pools. Three of these pools were using the Strantrol system and one was using pH control and conventional chlorine residual control. The following relationships were investigated:

1. At a constant free chlorine residual of 2.1 mg/l, ORP measurements were made at pH levels from 6 to 10 (see Fig. 4-17). It is significant to note that the difference between the Gilroy and Petaluma waters is on the order of 60 mV for

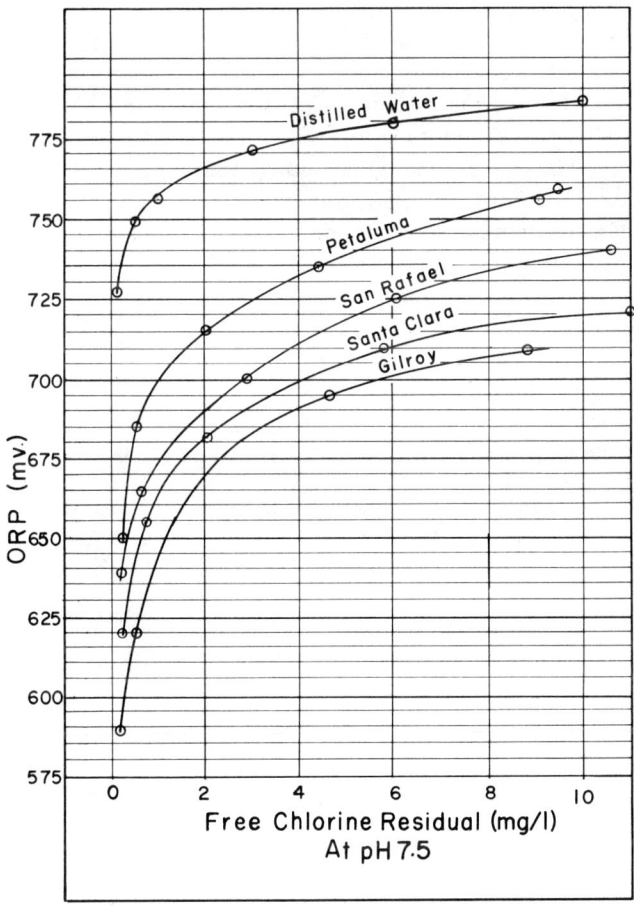

Fig. 4-18. Relationship between oxidation potential and free chlorine residual at a constant pH in five different waters.

identical pH levels. These curves also show that any given water with a constant chlorine residual will show an ORP shift of about 60 mV when there is a one point shift in pH.

2. With the pH constant at 7.5, ORP measurements were made with free residuals ranging from 0.1 to 10 mg/l (see Fig. 4-18). The curves shown demonstrate the logarithmic nature of the chlorine residual versus ORP values and the low sensitivity of this method of chlorine residual measurement. There is only 37 mV difference in the ORP values between 1 and 4 mg/l chlorine residual. It is obvious that such a system will be inherently difficult to control. However there is another factor to be considered. A free chlorine residual tends to "poison" the ORP cell. This reduces the sensitivity of the cell which retards the cell response to changes in chlorine residual and pH. In swimming pools where lag times may be long this can be a benefit.

Since the above investigation illustrates that every water has a different ORP "poise," each installation has to be calibrated to that poise. If the water source or the character of the water is changed a new calibration must be made. This further demonstrates the fact that the ORP control method is qualitative and not quantitative. The success of this system is the close control of the pH and the maintenance of high free chlorine residuals. The close control of the pH insures the presence of a constant amount of undissociated HOCl which is the most active chlorine residual species.

REFERENCES

1. Shilov, E. A., and Soludushenkov, S. N., "Hydrolysis of Chlorine," *Comptes Rendus Acad. Sci. l'URSS,* **3,** 17, no. 1 (1936), and *Acia Physiochim URSS,* **20,** 667, No. 5 (1945).
2. Morris, J. C. "The Mechanism of the Hydrolysis of Chlorine," *J. Am. Chem. Soc,* **68,** 1692 (1946).
3. Adams, F. W. and Edmonds, R. G. "Absorption of Chlorine by Water in a Packed Tower," *Ind. Eng. Chem,* **29,** 447 (1937).
4. Rosenblatt, D. H. and Small, M. J. A Private Communication. U.S. Army Medical Bioengineering Research and Development Lab., Fort Detrick, MD, 1976.
5. Morris, J. C. "The Acid Ionization Constant of HOCl from 5 to 35°C," *J. Phys. Chem.,* **70,** 3798 (Dec. 1966).
6. Chamberlin, N. S., and Snyder, H. B., "Technology of Treating Plating Wastes," Tenth Annual Waste Conf., Purdue Univ., 1955.
7. Anon., *Standard Methods for the Examination of Water and Wastewater,* 12th ed., pp. 91, 185, 186, 208, 209, and 210, American Public Health Assoc., New York, 1965.
8. Sawyer, C. N., *Chemistry for Sanitary Engineers,* Chapter 25, McGraw-Hill, New York, 1960.
9. Morris, J. C., Weil, Ira, and Culver, R. H., "Kinetic Studies on the Break-Point with Ammonia and Glycine," unpublished copy from senior author, Harvard Univ., 1952.
10. Morris, J. C., "Kinetic Reactions between Aqueous Chlorine and Nitrogen Compounds," Fourth Rudolphs Res. Conf., Rutgers Univ., June 15–18, 1965.
11. William, D. B., private communication, Brantford, Ontario, Canada, 1967.
12. Griffin, A. E., "Effect of Applying Chlorine Before Ammonia and Vice Versa," private communication, 1938.
13. Griffin, A. E., and Chamberlin, N. S., "Some Chemical Aspects of Break-Point Chlorination," *J. NEWWA,* **55,** 371 (1941).

14. Griffin, A. E., "Chlorine for Ammonia Removal," Fifth Annual Water Conf. Proc. Engrs. Soc. Western Penn., p. 27, 1944.
15. Morris, J. C., "The Chemistry of the pH Factor in Pools and Its Relation to Reactions with Nitrogenous Substances," paper presented at Water Chemistry Seminar N.S.P.I. Convention Chicago, Illinois, Jan., 1964.
16. Robson, H. L., *Encyclopedia of Chemical Technology,* Vol. 4, pp. 908–928, John Wiley & Sons Inc., New York, 1964.
17. Weil, Ira, and Morris, J. C., "Kinetic Studies on Chloramines," *J. Am. Chem. Soc.,* **71,** 1664 (1949).
18. Weil, Ira, and Morris, J. C., "Equilibrium Studies on N-Chloro Compounds," *J. Am. Chem. Soc.,* **71^3,** 3123 (1949).
19. Morris, J. C., Salazar, J. A., and Wineman, Margaret, "Equilibrium Studies on the N-Chloro Compounds," *J. Am. Chem. Soc.,* **70,** 2036 (1948).
20. Fair, G. M., Morris, J. C., and Chang, S. L., "The Dynamics of Water Chlorination," *J. NEWWA,* **61,** 285 (1947).
21. Granstrom, M. L., "The Disproportionation of Monochloramine," Ph.D. dissertation in Sanitary Engineering, Harvard Univ., 1954.
22. Morris, J. C., Weil, Ira, and Burden, R. P., "The Formation of Monochloramine and Dichloramine in Water Chlorination," paper 117th meeting, Am. Chem. Soc. Detroit, Mich., April 16–20, 1950.
23. Palin, A. T., "Chemical Aspects of Chlorine," *Inst. of Water Engrs. (England),* p. 565 (1950).
24. Palin, A. T., "A Study on the Chloro-Derivatives of Ammonia and Related Compounds with Special Reference to Their Formation in the Chlorination of Natural and Polluted Waters," *Water and Water Engineering (England),* p. 151 (Oct. 1950), p. 189 (Nov. 1950), p. 248 (Dec. 1950).
25. Williams, D. B., "How to Solve Odor Problems in Water Chlorination Practice," *Water and Sew. Wks.,* **99,** 358 (1952).
26. Williams, D. B., "Control of Free Residual Chlorine by Ammoniation," *J. AWWA,* **55,** 1195 (1963).
27. Williams, D. B., "Elimination of Nitrogen Trichloride in Dechlorination Practice," *J. AWWA,* **58,** 248 (1966).
28. Holwerda, K., "On the Control and Degree of Reliability of the Chlorinating Process of Purifying Drinking Water, Especially in the Relation to the Use of Chloramines for This Purpose," *Meddeelingen van den Dienst der Volksgezondheid in Nederlandsch-Indie,* **17,** 251 (1928) and **19,** 325 (1930).
29. Moore, W. A., Megregian, S., and Ruchhoft, C. C., "Some Chemical Aspects of the Ammonia-Chlorine Treatment of Water," *J. AWWA,* **35,** 1329 (1943).
30. Rossum, J. R., "A Proposed Mechanism for Break-Point Chlorination," *J. AWWA,* **35,** 1446 (1943).
31. Fair, G. M., Morris, J. C., Chang, S. L., Weil, Ira, and Burden, R. P., "The Behavior of Chlorine as a Water Disinfectant," *J. AWWA,* **40,** 1051 (1948).
32. Chapin, R. M., "Dichloramine," *J. Am. Chem. Soc.,* **51,** 2112 (1929).
33. Baker, R. J., "Types and Significance of Chlorine Residuals," *J. AWWA,* **51,** 1185 (1959).
34. Kelly, S. M., and Sanderson, W. W., "The Effect of Chlorine in Water on Enteric Visuses II., The Effect of Combined Chlorine on Poliomyelitis and Coxsackie Viruses," *Amer. J. Publ. Health,* **59,** 14 (1960).
35. Barrett, S., private communication, Metropolitan Water District of Southern Calif., June 1983.
36. Williams, D. B., "Monochloramine and Dichloramine Determinations in Water," *Water and Sew. Wks,* **98,** 429 (1951).
37. Williams, D. B., "Free Chlorine, Monochloramine and Dichloramine in Water," *Water and Sew. Wks.,* **98,** 475 (1951).
38. Hulburt, R., "Chlorine and Orthotolidine Test in the Presence of Nitrites," *J. AWWA,* **26,** 1638 (1934).
39. Morris, J. C., and Wei, Irving, "Chlorine-Ammonia Breakpoint Reactions: Model Mechanisms

and Computer Simulation," paper Annual Meet. Am. Chem. Soc., Minneapolis, MN, April 15, 1969.
40. Pressley, T. A., Bishop, D. F., and Roan, S. G., "Nitrogen Removal by Breakpoint Chlorination," Paper, Am. Chem. Soc. Chicago, Sept. 1970.
41. Saunier, B. M. "Kinetics of Breakpoint Chlorination and Disinfection." PhD. Thesis, Univ. of Calif., Berkeley, CA (1976).
42. Saunier, B. M., and Selleck, R. E. "The Kinetics of Breakpoint Chlorination in Continuous Flow Systems," paper presented at the AWWA Ann. Conf., New Orleans, LA, June 22, 1976.
43. Stone, R. W. "Rancho Cordova Breakpoint Chlorination Demonstrations," report by Sacramento Area Consultants, Sept. 1976.
44. Taras, M. J., "Effect of Free Residual Chlorine on Nitrogen Compounds in Water," *J. AWWA,* **45,** 47 (1953).
45. Chang, S. L., private communication, 1968.
46. Feben, D., and Taras, M. J., "Chlorine Demand of Detroit Water," *J. AWWA,* **42,** 453 (1950).
47. Feben, D., and Taras, M. J., "Chlorine Demand Constants," *J. AWWA,* **43,** 922 (1951).
48. Taras, M. J., "Chlorine Demand Studies," *J. AWWA,* **42,** 462 (1950).
49. Williams, D. B., "The Organic Nitrogen Problem," *J. AWWA,* **43,** 837 (1951).
50. Nordell, E., *Water Treatment for Industrial and Other Uses,* 2nd ed., p. 213, Reinhold, New York, 1961.
51. Green, D. E., and Stumpf, P. K., "The Mode of Action of Chlorine," *J. AWWA,* **38,** 1301 (1946).
52. Knox, W. E., Stumpf, P. K., Green, D. E., and Auerbach, V. H., "The Inhibition of Sulfhydril Enzymes as the Basis of the Bactericidal Action of Chlorine," *J. Bact.,* **55,** 451 (1948).
53. Marks, H. C., and Strandkov, R. B., "Halogens and Their Mode of Action," *Annals of NY Academy of Sci.,* 163 (1950).
54. Wyss, O., "Disinfection by Chlorine, Theoretical Aspects," *Water and Sew. Wks.,* **109,** R155 (1962).
55. Chang, S. L., and Fair, G. M., "Viability and Destruction of the Cysts of *Endamoeba histolytica,*" *J. AWWA,* **33,** 1705 (1941).
56. Chang, S. L., "Destruction of Microorganisms," *J. AWWA,* **36,** 1192 (1944).
57. Marks, H. C., Wyss, O., and Strandkov, F. B., "Studies on the Mode of Action of Compounds Containing Available Chlorine," *J. Bact.,* **49,** 299 (1945).
58. Ingols, R. S., Wyckoff, H. A., Kethley, T. W., Hogden, H. W., Fincher, E. L., Hildebrand, J. C., and Mandel, J. E., "Bacterial Studies of Chlorine," *Ind. and Eng. Chem.,* **45,** 996 (1953).
59. Anon., *Water Quality and Treatment,* 2d ed., p. 206, American Water Works Assoc., New York, 1951.
60. Farkas, L., Lewin, M., and Bloch, R., "The Reaction Between Hypochlorites and Bromides," *J. Amer. Chem. Soc.,* **71,** 1988 (1949).
61. Butterfield, C. T., Wattie, E., Megregian, S., and Chambers, C. W., "Influence of pH and Temperature on the Survival of Coliform and Enteric Pathogens When Exposed to Free Chlorine," *Pub. Health Rpts.,* **58,** 1837 (1943).
62. Butterfield, C. T., and Wattie, E., "Relative Resistance of *E. coli* and *E. typhosa* to Chlorine and Chloramines," *Pub. Health Rpts.,* **59,** 1661 (1944).
63. Selleck, R. E., private communication, Univ. of Calif. Berkeley, CA, July 1981.
64. Houston, Sir A. C., "19th and 20th Annual Reports of the Metropolitan Water Board, London, England," 1925 and 1926.
65. Butterfield, C. T., and Wattie, E., "Influence of pH and Temperature on the Survival of Coliforms and Enteric Pathogens When Exposed to Chloramine," *Pub. Health Rpts.,* **61,** 157 (1946).
66. Butterfield, C. T., "Comparing the Relative Bactericidal Efficiencies of Free and Combined Available Chlorine," *J. AWWA,* **40,** 1305 (1948).
67. Kabler, P. W., "Relative Resistance of Coliform Organisms and Enteric Pathogens in the Disinfection of Water with Chlorine," *J. AWWA,* **43,** 553 (1953).

68. Corbett, R. E., Metcalf, W. S., and Soper, F. G., "The Reaction Between Ammonia and Chlorine in Aqueous Solutions," *J. Chem. Soc. (London)*, 1927, Part II (1953).
69. Faust, S. D., and Hunter, J. V., *Principles and Application of Water Chemistry*, p. 31, John Wiley & Sons, New York, 1969.
70. White, G. C., Beebe, R. D., Alford, V. F., and Sanders, H. A., "Problems of Disinfecting Nitrified Effluents," paper presented at Second National Symposium on Municipal Wastewater Disinfection, Orlando, Florida, Jan. 26–28, 1982.
71. Clarke, N. A., Berg, G., Kabler, P. W., and Chang, S. L., "Human Enteric Viruses in Water: Source, Survival and Removability," International Conf. Water Pollution Research, Pergamon Press, London, Sept. 1964.
72. Morris, J. C. "Aspects of the Quantitative Assessment of Germicidal Efficiency," *Disinfection: Water and Wastewater*, J. D. Johnson (Editor), Ann Arbor Science, Ann Arbor, MI, p. 1, 1975.
73. Selleck, R. E., Saunier, B. M., and Collins, H. F., "Kinetics of Bacterial Deactivation with Chlorine," *J. Env. Eng. Div. ASCE*, p. 1197 (Dec. 1978).
74. Selna, M. W., Miele, R. P., and Baird, R. B. "Disinfection for Water Reuse," paper presented at the Disinfection Seminar at the AWWA Annual Conf. Anaheim, CA, May 8, 1977.
75. Rideal, S., and Evans, U. R., *J. Soc. Pub. Analysts and Other Analytical Chemists*, London (Aug. 1913).
76. Schmelkes, F. C., "The Oxidation-Reduction Potential Concept of Chlorination," *J. AWWA*, **25**, 695 (1933).
77. Schmelkes, F. C., Horning, E. W., and Campbell, G. A., "Electro-Chemical Properties of Chlorinated Water," *J. AWWA*, **31**, 1524 (1939).
78. Chang, S. L., "Studies on *Endamoeba histolytica*. III, Destruction of Cysts of *Endamoeba histolytica* by a Hypochlorite Solution, Chloramines in Tap Water and Gaseous Chlorine in Tap Water of Varying Degrees of Pollution," *War Medicine*, **5**, 46 (1944).
79. Chang, S. L., "Applicability of the Oxidation Potential Measurements in Determining the Concentration of Germicidally Active Chlorine in Water," *J. NEWWA*, **59**, 79 (1945).
80. Hubbard, L. S., "Electronic Robot Controls Water Quality at Eisenhower Pool," *Park Maintenance* (March 1970).
81. Abbott, R., "Automation Solves Problems at Michigan Univ. Pool." *Swimming Pool Age* (May 3, 1971).
82. Baker, R. J., private communication, March 1, 1983.
83. U.S. Patent No. 2,443,429, "Procedure for Disinfecting Aqueous Liquids," Marks and Strandkov, and Wallace and Tiernan Co., Inc. Belleville, N.J., June 15, 1948.
84. Selleck, R. E., et al., "Optimization of Chlorine Application Procedures and Evaluation of Chlorine Monitoring Techniques," Univ. of California Publication No. UCB-ENG-4180, Berkeley, CA, 1976.
85. Selleck, R. E., private communication, Univ. of Calif., Berkeley, CA, Sept. 1983.
86. Hergott, S. J., Jenkins, David, and Thomas, J. F., "Power Plant Cooling Water Chlorination in Northern California," *J. WPCF*, **50**, 2590 (Nov. 1978).
87. Feng, T. H. "Behavior of Organic Chloramine in Disinfection." *J. WPCF*, **38**, 614 (1966).
88. Sung, R. D. "Effects of Organic Constituents in Wastewater on the Chlorination Process," Ph.D. Thesis, Univ. of Calif., Davis, CA, 1974.
89. Esvelt, L. A., Kaufman, W. J., and Selleck, R. E., "Toxicity Assessment of Treated Municipal Wastewater," paper presented at the 44th Ann. Conf. of WPCF, San Francisco, CA, Oct. 4–8, 1971.

5
Determination of Chlorine Residuals in Water and Wastewater Treatment

DEVELOPMENT OF ANALYTICAL METHODS

When chlorine compounds were first introduced, about 1902, as a means of water disinfection, the only method available for testing residual chlorine was the starch-iodide one. Between 1902 and 1908, the proponents and investigators of the process were merely feeling their way and were not ready to accept such a concept as residual chlorine. As a matter of fact, considerable doubt existed as to the bactericidal efficiency of chlorine, so that results were evaluated entirely on the total bacteria and coliform kill. The latter was determined by the coliform presumptive test.

After the momentous Jersey City litigation in 1910, wherein the court ruled unequivocally in favor of chlorine as a successful disinfecting process for potable water, the workers in this field turned their attention to the proper control of chlorination. In 1909 E. B. Phelps[1] first proposed orthotolidine as a qualitative indicator of residual chlorine. Then in 1913 Ellms and Hauser[2] developed the quantitative test for residual chlorine using orthotolidine. In addition, they developed colorimetric standards for its use. This was a great contribution, since it paved the way for a scientific approach to the study of chlorination. Between 1917 and 1920 Wolman and Enslow[3] made studies on chlorine absorption in water and demonstrated the suitability and reliability of the orthotolidine test. Their work was the first scientific approach to the use of chlorine in water treatment. Until about 1930 the use of chlorine in water treatment was increasing at a tremendous rate. This inspired more and more studies of the various phenomena of controlling the application of chlorine. Refinements in the use of orthotolidine for measuring residual chlorine continued. The color standards formulated by Ellms and Hauser were modified, first by Meur and Hale[4] in 1925, then by Scott[5] in 1935 and again in 1939.[6] A final refinement of these standards was made in 1943 by Chamberlin and Glass.[7]

In 1940 the bombing of Britain focused attention on the need for an adequate field test for high chlorine residuals to be used in emergency sterilization of water mains. Contributions on the measurement of high residuals by Chamberlin[8] and Griffin[9] are the most notable. As a result of this work, it appeared necessary to overhaul the entire procedure of chlorine residual measurement by orthotolidine.

This led in 1943 to the *AWWA Joint Committee Report*,[10] which was based on the research work of Chamberlin and Glass.[7,11] This work is also the basis for both the procedure and color standards of the orthotolidine test appearing in the twelfth edition of *Standard Methods*.[12]

In 1939 the discovery of the breakpoint phenomenon revealed that there was more than one kind of chlorine residual. These were identified as free available and combined chlorine residuals. Further work indicated that the breakpoint phenomenon could not be properly evaluated or controlled without an adequate method of differentiation between these two types of residual. This led to more important contributions to the technique of chlorination. First came the orthotolidine flash test by Laux[13,14] in 1940, followed by the OTA (orthotolidine arsenite) test developed by Hallinan[15,16] in 1944. In the meantime, an even more important development in the method for measuring chlorine residuals was being explored. This was the amperometric method, first introduced for measuring chlorine residuals by Marks[17] in 1942.

Subsequent refinements resulted in changes in both the application of chlorine and its measurement. The orthotolidine test gave considerable ground to the OTA test, while the amperometric test was refined[18] to give the various fractions of free available and combined chlorine residuals. Marks[19,20] in 1951 presented the amperometric method capable of further differentiation of the total chlorine residual into free chlorine, monochloramine, and dichloramine by a series of titrations.

While Marks in the United States was investigating the amperometric method, Palin in England was exploring different colorimetric procedures to differentiate the various fractions of free available and combined available chlorine residuals. Palin[21] first reported on the use of p-aminodimethylaniline in 1945 as a method superior to orthotolidine. While this indicator had been reported on by Moore[22] in 1943 in the United States, Palin's work was considerably more extensive. Further work by Palin convinced him that he should discard this indicator in favor of a new approach. In 1949 Palin[23] announced a new and improved method, using ferrous ammonium sulfate as the titrating agent and neutral orthotolidine as the indicating reagent. This approach showed considerable promise in that it was shown to be valid for measuring free available chlorine, monochloramine, dichloramine, and even nitrogen trichloride. Further refinements were reported by Palin[24] in 1954, but owing to limitations of water temperature and errors due to interferences Palin explored other colorimetric indicating reagents. In 1957[25] he reported on his investigation of diethyl-*p*-phenylene diamine, soon to be known as DPD. This method using the various reagents in tablet form is the most widely accepted method in England.

In 1966 the Chester Beatty Research Institute at the Royal Cancer Hospital, London, England, issued a report that led to the abandonment in 1969 of orthotolidine as a residual chlorine test in the British Isles. The report, "Precautions for Laboratory Workers Who Handle Carcinogenic Aromatic Amines," gave a list of chemicals—including orthotolidine—regarded as potential causes of tumors in the urinary tract. Consequently, the only colorimetric procedure now used in the British Isles is Palin's DPD method.*

Other noteworthy contributions toward finding better methods for determining chlorine residuals have been made. The orthotolidine titration technique was introduced by Connell[26] in 1947 as an attempt to differentiate free available from combined available residual chlorine. This method was based on the development of a cherry red color when insufficient O-T is present, followed by the conversion of the red chlorine substituted holoquinone to the yellow holoquinone by rapid addition of sufficient O-T. This method, of questionable practical value, has never been adequately evaluated. This method assumes no reaction with chloramines present with the O-T reagent in the time it takes to perform the titration. Since O-T is the reagent, the usual limitations apply: the temperature of the sample is a limiting factor affecting the accuracy of the free chlorine residual. Furthermore, other investigators do not agree that it is proper procedure to convert the red color of the chlorine substituted holoquinone to the yellow holoquinone.[8,9,11]

Chamberlin[11] lists as one of the conditions for the accuracy of the O-T test that the sample should be added to the reagent and not vice versa, as in the above procedure.

Next came the investigation of Methyl Orange as a substitute for orthotolidine. Chang[36] used this method in 1944 for the measurement of free chlorine in his investigation of the relative cysticidal power of free and combined chlorine.

Taras[27,28] investigated this method and described both the volumetric and colorimetric procedures in 1946 and 1947. As recently as 1965 the methods proposed by Taras were more fully evaluated by Sollo and Larson[29] in their quest for a better and simpler field method of differentiating free and combined chlorine residual.

This method never gained popularity nor widespread credibility. It is mentioned here as an historical item and also because the EPA "Analytical Reference Service Report No. 40," 1971, showed that the Methyl Orange procedure demonstrated the best accuracy for measuring both free and combined chlorine (separately). The other methods tested were: Leuco Crystal Violet SNORT, DPD-Tritrimetric, DPD-Colorimetric, Amperometric and OTA.**

The Methyl Orange method is based upon the almost instant decolorization of methyl orange by chlorine on a quantitative basis, whereas combined chlorine (chloramines) is much slower in its bleaching effect on methyl orange. Free chlorine bleaches methyl orange quantitatively on the basis of two molecules of chlorine to one of methyl orange. Therefore the weight ratio of one molecule of M.O. to two molecules of chlorine is 2.34:1. One milliliter of 0.005 percent M.O. is instantly decolorized by 21.9 mg of chlorine and 1 ml of 0.005 percent M.O. solution contains

---

* Owing to its carcinogenic properties, *Standard Methods* eliminated all orthotolidine procedures in the 15th edition. However O-T methods are still used in the U.S.A.
** The EPA Report 40 cited above is listed as a reference on page 300, *Standard Methods*, 15th edition.

50.0 mg of M.O. This yields an M.O. to chlorine weight ratio of 2.28 : 1 as compared to the molecular weight ratio of 2.34:1. This is well within the margin of experimental error. Therefore the arithmetical relationship of this M.O. reaction with chlorine is:

$$\text{mg/l Cl}_2 = 0.217 \times \text{ml } 0.005\% \text{ M.O. solution} + 0.04 \qquad (5\text{-}1)$$

(1.0 ml M.O. is equivalent to 0.23 mg/l free chlorine.)

The analyst has the choice of the volumetric method,[26,27] where a titration is carried to an end point using an exact amount of M.O. titrant, or the colorimetric method of Sollo and Larsen,[29] using an excess of M.O. and comparing to permanent color standards, or by determining the light absorbence of the color with a spectrophotometer properly calibrated for the test.

The reagent, methyl orange, is a primary standard, is stable, maintains its titer indefinitely, and reacts with chlorine over a wide range of temperatures. However, the useful range of this method is limited to 2 mg/l free chlorine. The EPA Report No. 40 was based upon free chlorine concentrations of 0.44 and 0.98 mg/l and combined chlorine of 0.66 mg/l. This method has not been used except in a few laboratory projects.

In 1967, Black and Whittle[30,52] reported on a new indicator solution identified as leuco crystal violet. It possesses excellent properties for the quantitative measurement of free chlorine. As a matter of fact it was rated second to M.O. in the EPA Report No. 40. This scheme of measuring chlorine residuals was developed as a result of investigating the efficiency of free iodine versus free chlorine in both potable water and swimming pools. This method has not been widely acclaimed in spite of the fact that it is superior to the O-T method for either free or combined residuals. It is described on page 293, *Standard Methods,* 15th edition.

In 1963, the U.S. Army Medical Research and Development Command began supporting research to develop an improved field method for measuring free available chlorine (FAC). In 1967 the Army adopted the modified orthotolidine–arsenite method (MOTA) for measuring FAC in water.[37] This method is an improvement over Hallinan's OTA method.[15] The MOTA method was an attempt to eliminate or minimize interfering substances; however, the performance was not successful enough for the army. Combined chlorine, nitrites, and iron and manganese compounds continued to cause significant errors in the determination of FAC.

In 1965 the School of Public Health at the University of North Carolina began an investigation sponsored by the U.S. Army to find a method that could accurately measure FAC in the presence of combined chlorine under all conditions of military operations.[31-33] Moreover the method had to be a simple field operation, like the O-T method.

The method chosen to be developed involved the use of a stabilized neutral O-T solution. It is known as the SNORT method. It is based upon the development of the blue meriquinone colors of O-T in the neutral range of sample–reagent mixture. This method was first used by Harrington[34] at Montreal and then by Caldwell[35] at Springfield without benefit of a stabilizer. The SNORT procedure developed for the Army, which included a test kit with disk like the Wallace and Tiernan and Hellige O-T kits, was presented at the 1976 fall meeting of the Armed Forces Epidemiological Board. After reviewing the test results the board recommended against adoption of this method because of the false positive FAC reading in the presence of combined chlorine. This factor erased any advantage of SNORT over the MOTA method.

In 1972 Bauer et al.[38] reported on the use of a treated strip of paper and a color chart which would offer the user a rapid means of measuring free chlorine. This test was based upon the reaction of free chlorine with the compound syringaldazine. The developer, Miles Laboratory, claimed no interference from combined chlorine. The U.S. Army Medical Environmental Research Unit was suitably impressed with the possibilities of this method. They engaged Guter and Cooper to evaluate this method.[39,40] This investigation concluded that a syringaldazine solution was the most specific for FAC of the five hand-held portable field test kits that were evaluated. However, as might be expected, organically polluted water resulted in appreciable false positive FAC readings for the majority of the kits.[40]

In the meantime the U.S. Army Medical Bioengineering Research and Development Laboratory, Aberdeen Proving Ground, Md., continued its own investigation of tests for FAC. This they did in the presence of significant concentrations of combined chlorine compounds. These tests concluded that the syringaldazine (FACTS) method was the most specific for FAC.[41,42] The Army pursued their search for an FAC test accurate under field conditions. They had long ago decided from World War II and Viet Nam experiences that only a free chlorine residual of a given magnitude would provide the drinking water safety needed. In 1975 the Army investigative unit suggested that the DPD method might be modified to decrease the combined chlorine breakthrough into the FAC reading.[43] Palin followed through on this suggestion by proposing the addition of glycine immediately after mixing the DPD tablet with the sample. Glycine is supposed to suppress the combined chlorine reaction with DPD reagent thereby inhibiting the combined chlorine breakthrough into the FAC reading. This modification was examined by Meier et al.[50] in 1978 for the Army. The Army was not satisfied with this modification as compared with the syringaldazine method. Palin made a further modification which he called Steadifacs.[44,45] This modified DPD procedure utilizes the addition of thioacetamide solution. When added in the correct amount it will immediately dechlorinate completely a sample containing combined chlorine without affecting a previously developed DPD color representing the FAC in the same portion of the same sample.

Both the U.S. Army and the Drinking Water Research Center at the Florida International University, Miami, FL carried out more testing and observations of various methods of FAC detection and measurement.[47,48] The investigators concluded that the measurement of FAC by the FACTS method was acceptable in the presence of significant concentration of chloramines, and was superior to the DPD method, which suffers appreciably from chloramine intrusion. (FACTS is an acronym which stands for: Free Available Chlorine Tests by Syringaldazine.)

In 1982 EPA approved the FACTS method for the determination of free chlorine residuals.[49] In 1983 Gibbs et al.[53] and Cooper et al.[55] summarized their work exploring for the Army the best method for measuring FAC in the presence of chloramine under military field conditions. This work concluded that the FACTS method was superior.

Finally in 1983 Cooper et al reported the FAC residuals measured by FACTS and amperometric titrations equivalent.[54]

CURRENT STATUS OF ANALYTICAL METHODS

Colorimetric

General Considerations. The most important factor to consider when using colorimetric methods is that the user will probably perform field measurements with a test kit that utilizes a rotating disk or a slide with permanent glass color standards. Therefore the operator of these kits must depend upon visual acuity to determine the residual. However, the investigators who probed the accuracy of these methods used sophisticated laboratory equipment (spectrophotometer) that can identify residuals to the second decimal place. There are currently available field test kits that use spectrophotometers, but they are not yet widely used. One of the most popular and probably the most sophisticated of these test kits is described below.

Another factor is the shelf life of chemicals used to perform the tests. This must be determined in order to have confidence in the test results. Chemical solutions or tablets should always be stored in a cool, dark, and dry place.

Test equipment should be selected based upon the expected accuracy of the measurements and the range of residual concentrations.

It should be noted that *Standard Methods* calls for the use of either a spectrophotometer or a photometer for all colorimetric methods. Permanent color standards are an alternative and are acceptable if range and readibility fit the user's requirements.

Orthotolidine

General Discussion. This method is obsolete and its use is diminishing year by year. It may disappear in a few years because it is a health hazard to the people making the powder and the solutions. It is a potential carcinogen in humans—it affects the bladder and urinary tract.

Owing to its use over so many years it is discussed here because much of the investigative work in the past relied upon O-T as a method for measuring residuals.

It is a simple and reliable method for measuring total residual chlorine in potable water. Only one reagent is involved. It covers a range of 0–10 mg/l, and in the 0–5 mg/l range residuals can be estimated with reliability to the nearest 0.25 mg/l.

It is not an acceptable method for measuring chlorine residuals in wastewater because the sample reagent mixture is in such a low pH environment (<2). At this pH the chlorine residual species becomes highly active and is partially consumed before the O-T color develops. This results in residual measurements lower than the true residual. Investigations have shown that the discrepancy is 2–2.5 times lower for O-T than amperometric measurement in primary effluents.

The free chlorine species can be distinguished by the O-T method provided the sample reagent mixture is chilled to 1°C. This takes time and therefore the additional contact can produce a slight error due to total chlorine residual decay.

One other factor to be aware of while examining investigations of germicidal efficiency, particularly those comparing chlorine and ozone, is to determine if the medium used contains ammonia nitrogen. Some bacteriological researchers inadvertently used bacterial cultures that contained ammonia nitrogen which converted the free chlorine to chloramines, thus distorting the results.

Measuring Residuals Greater than 10 mg/l. The upper limit of residual measurement with O-T is 10 mg/l. However there always exists the necessity to measure residuals in the 50 mg/l range. These residuals are required for water main sterilization, reservoirs, subdivisions, ships' tanks, hospital water systems, new office buildings, etc. This is accomplished by the following method.

The Drop Dilution Method. This method was developed about 1941 for determining residuals from 10 to 100 mg/l.[9] This method consists of the addition of one or more drops of the chlorinated water sample to a cell or vessel of known volume containing orthotolidine and distilled water. This method is simple and rapid for approximating residuals between 10 and 100 mg/l. Using a standard chlorine residual comparator, calibrate the dropper furnished with the comparator. Use this dropper exclusively for measuring the chlorinated water sample. One drop from such a dropper usually equals 0.05 ml. The standard comparator utilizes cells that have a 15 ml capacity. With this information, the procedure is as follows:

1. Collect sample in a small glass container.
2. Add 0.5 ml O-T to center tube and fill to the scribed mark with distilled water. (The standard comparator cell holds 15 ml sample to the scribed mark.)
3. Fill the other tube with distilled water. (This tube compensates for the natural color of the water sample.)
4. Add one drop of chlorinated sample to the center tube: mix and read immediately.

If no color appears, add additional drops of chlorinated sample, one at a time, until a reading within the range of the color disk is obtained.

Computation of Residual. First, it shall be assumed that the dropper capacity is 0.05 ml per drop and that the cell capacity is 15 ml. Then we have

$$\text{mg/l chlorine present} = \frac{\text{capacity of cell}}{\text{ml of chlorinated sample added}} \times \text{comparator reading} \quad (5\text{-}2)$$

EXAMPLE: If two drops of sample produced a comparator reading of 0.25 mg/l Cl_2, then

$$Cl_2 \text{ mg/l} = \frac{15}{(0.05 \times 2)} \times 0.25 = 37.5 \text{ mg/l}$$

Interfering Substances. The 1943 *Joint Committee Report on the Control of Chlorination*[10] declared that the color developed by the reaction of chlorine and O-T could be considered residual chlorine if the water sample contained no more than the following amounts of interfering substances: 0.3 mg/l iron, 0.01 mg/l manganese (manganic), and 0.10 mg/l nitrite. These limits were never changed. These substances interfere with other colorimetric methods.

The substance that is most likely to contribute significant interference is the nitrite ion NO_2^-.[60] If the presence of nitrites is suspected, the test should be carried out so that the color development proceeds in total darkness. Then the test should be repeated in artificial light. In this way the color interference by nitrites will be nearly eliminated. Colors due to nitrites will develop even in the dark if nitrites are present in excess of 1.0 mg/l. Nitrites are the intermediary between the conversion of ammonia nitrogen to nitrate by biological action. The difficulty arises when there is sufficient nitrogen present to maintain all the chlorine residual as chloramine. The nitrites present causes false residual readings which deceive the operator, and the chloramine residual is unable to oxidize the nitrites. The result is an algal bloom in a swimming pool, and in a water treatment plant it is a proliferation of bacteria. The reaction between nitrite and orthotolidine is not well understood, but it is certain the reaction is one of oxidation by nitrites. This confirms previous reports that nitrites are an intermediary state in the conversion of ammonia to nitrates. This would indicate that nitrites can act as either an oxidizing or reducing agent.

Leuco Crystal Violet Method

Chemistry of the Method. The details of this method appear on page 293, *Standard Methods,* 15th edition. This method measures separately the FAC and the TRC (total residual chlorine). The LCV reagent reacts instantaneously with free chlorine to form a bluish color. Interference from combined chlorine is avoided by completing the test for free chlorine (taking a color depth reading) within 5 min. after adding the reagent.

The total residual chlorine reading involves the reaction of free and combined chlorine with iodide ion to produce hypoiodus acid, which in turn reacts instantaneously with LCV to form the dye, crystal violet. The color of this dye is stable for days. This reaction requires the addition of pH 4 buffer and potassium iodide solutions to perform the release of I_2 which forms the hypoiodous acid.

Two separate sets of color standards are required for this method: one for free chlorine (bluish colors) and one for total residual chlorine (violet color). Commercially prepared standards are available in test kits for free chlorine determinations but not for TRC. Color values expressed as chlorine residual can be measured by a calibration curve established for either a filter photometer or a spectrophotometer.

The minimum detectable concentration is 10 μg FAC and 5 μg TRC. The practical range for this method is 0–10 mg/l for both FRC and TRC.

Interference with FRC occurs when the combined chlorine concentration is 5.0 mg/l or greater. The addition of 5.0 ml of sodium arsenite solution ($NaAsO_2$ at 5 g/l) will minimize this interference.

The major interference in the FRC measurement is from the *manganic ion* which increases the apparent chlorine residual reading. When the manganic ion is known to be present the photometric procedure is used to determine the absorbance due to the manganic ion separately. This measurement is subtracted from the total absorbance to yield that produced by FAC alone.

In the presence of monochloramine, nitrites will cause serious interference in the determination of free chlorine. Addition of sodium arsenite tends to minimize this interference.

Procedure. See pages 293–300, *Standard Methods,* 15th edition.

FACTS Method (Syringaldazine). See page 298, *Standard Methods,* 15th edition.

General Discussion. A saturated solution of syringaldazine is used as the indicating reagent. It is stable when stored as a solid or as a solution. It is oxidized by free available chlorine on a 1:1 molar basis, yielding a colored product with an absorption maximum at 530 nm. The color product is only slightly soluble in water; therefore at chlorine concentrations greater than 1 mg/l, the final reaction mixture must contain 2-propanol to prevent product precipitation and color fading. The pH of reagent and sample mixture is critical. It must be held between 6.5 and 6.8. Otherwise color develops too slowly or too fast, which results in fading. A buffer reagent which produces a pH of 6.7 must be used along with the indicator. Care must be taken when chlorinated samples derive from predominantly acid or alkaline water so that the amount of buffer added is in the amount necessary to produce a reagent–sample mixture pH within the 6.5–6.8 limitation.

Interferences common to other methods for determining FAC do not affect the FACTS procedure. Monochloramine up to 18 mg/l, dichloramine up to 10 mg/l, and oxidized forms of manganese up to 1 mg/l do not interfere. Concentrations of ferric iron up to 10 mg/l do not interfere. Nitrite concentrations less than about 250 mg/l do not interfere. Strong oxidizing agents such as iodine, bromine, and ozone will produce a color.

The minimum detectable FAC is 0.1 mg/l or less. The range of FAC measurement is 0.1–10.0 mg/l.

Temperature has a minimal effect on color reaction. The maximum error observed over a range of 5–35°C is ± 10 percent.

It is necessary to use either a filter photometer or a spectrophotometer to convert depth of color (light to dark pink) to mg/l FAC. *This method measures only FAC.*

Some commercial color kits are available with 4 permanent glass color standards of 0.5, 1.0, 1.5, and 2.0 mg/l FAC.

Procedure. See page 299, *Standard Methods,* 15th edition.

DPD Colorimetric Method I. See page 292, *Standard Methods,* 15th edition.

General Discussion. This is the only colorimetric method with the capability of differentiating the various species of chlorine residuals. The indicator which produces gradations of red color in the presence of chlorine residuals is N,N-diethyl-p-phenylenediamine (DPD). The most significant interference is oxidized manganese. To compensate for this a blank must be used. The minimum detectable concentration is about 10 μg/l as chlorine. The range of the test is 0–4 mg/l without sample dilution.

Palin Tablet Method. All of the reagents required are incorporated into four tablets which have to be completely dissolved before executing the residual measurements. These are as follows:

Tablet	Contents
DPD No. 1	DPD indicator with EDTA and buffer
DPD No. 2	Stabilized KI for monochloramine activation
DPD No. 3	Stabilized KI for dichloramine activation
DPD No. 4	All reagents in a single tablet

Depending upon the tests desired the appropriate procedure is selected thus:

FAC	Use Tablet No. 1
FAC + combined chlorine	Use Tablets No. 1 and 3
Monochloramine	Use Tablet No. 2
Dichloramine	Use Tablet No. 3
Total available chlorine	Use Tablet No. 4

The colorimetric procedures using the Palin tablet method require the use of a calibrated spectrophotometer to measure the color development of the indicator, or the use of permanent glass color standards available in a variety of commercial kits. These kits should be carefully chosen for range and residual interval spread.

Monochloramine Breakthrough. This phenomenon occurs when monochloramine is present in significant concentrations. Palin has suggested the use of thioacetamide solution to inhibit or suppress the monochloramine breakthrough into the FAC reading. When this reagent is used Palin calls it the DPD Steadifac method.[44] Details of this alternative are given in the section "DPD Ferrous Titrimetric Method" below.

DPD Colorimetric Method II

The Residometer Method. This is the basic DPD method described above, except that it measures only free and total available chlorine residuals. It utilizes a calibrated spectrophotometer and solution reagents instead of tablets. The measuring instrument is the Wallace and Tiernan Residometer with a digital readout to the second decimal place (see Fig. 5–1). It was developed by Wallace and Tiernan in Great Britain where it is widely used. It is equipped with batteries for portable use and 110 V power for laboratory use. It is mounted in a carrying case the size of a conventional attaché case. One model is equipped with a pH electrode for pH measurements from 0 to 14. Three reagents are supplied for chlorine residual measurements and two for pH. This instrument has been tested extensively in the USA.[56] The reagents have been found to have a shelf life of at least 18 months under ordinary conditions. The dropper bottles which are used to measure the reagent dosage have proved to be well within the accuracy of the DPD method. The solution reagents allow the user to perform the analytical tasks with greater speed and ease than when tablets are used. Speed of color development in the DPD measurements enhances the accuracy of the method. Testing in the USA indicates that this method using the Residometer has an accuracy equivalent to the amperometric titration method.[56]

This method is affected by the same interferences as those described in Method I; namely oxidized manganese.

Procedure. The Residometer comes with complete instructions on how to use the solutions for both free and total chlorine residuals.

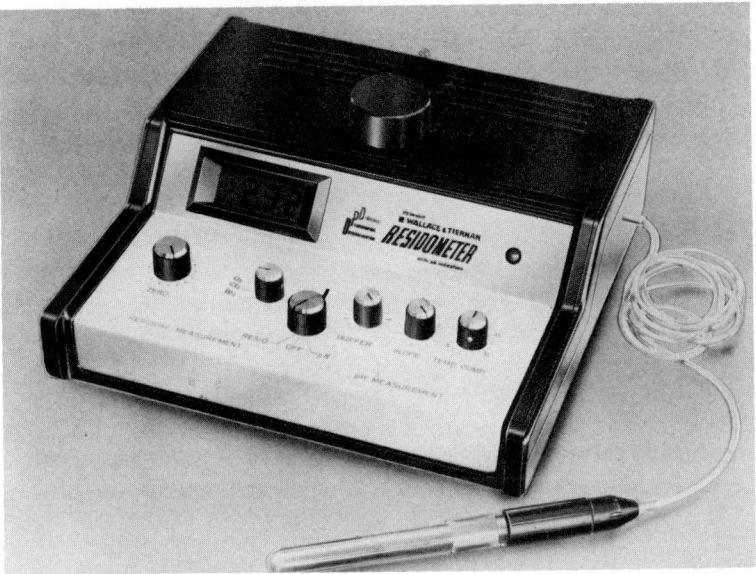

Fig. 5-1. Residometer DPD (Courtesy Wallace and Tiernan Div. Pennwalt-England).

Titrimetric Methods:

Iodometric Method I. See page 280, *Standard Methods*, 15th edition.

General Discussion. The principle of this method is based upon the phenomenon of the release of elemental iodine from an iodide solution, such as potassium iodide, in the presence of both free and combined chlorine residual if the pH is 8 or less. Furthermore, the release of elemental iodine is quantitatively proportional to the chlorine residual present. The liberated iodine is titrated with a standard solution of sodium thiosulfate using starch as the indicator.[61,62] However, there are certain conditions that limit and control this phenomenon of iodine release.

The procedure for the test is briefly as follows:

1. The pH of the sample to be tested is adjusted to the desired range by a buffer solution.
2. The addition of potassium iodide is immediately followed by the release of elemental iodine, which will cast a brown color to the sample.
3. A reducing agent is added only until the brown cast disappears.
4. Then a starch solution is added, which turns the sample blue in the presence of elemental iodine.
5. The titration with the reducing agent is then continued only until the blue color disappears.

The following equations show the chemistry of the reaction:
Free chlorine residual:

$$HOCl + KI \longrightarrow KCl + I_2 + H_2O \qquad (5\text{-}3)$$

Chloramine or combined residual:

$$NH_2Cl + KI \longrightarrow KCl + I_2 + NH_4OH \qquad (5\text{-}4)$$

(NOTE: The above equations are not balanced, because the other products in the reaction, HCl and H_2O are purposely omitted.) Titration by a reducing agent (thiosulfate) is as follows:

$$I_2 + 2Na_2S_2O_3 \underset{\text{(thiosulfate)}}{\longrightarrow} 2Na_2S_4O_6 + 2NaI \qquad (5\text{-}5)$$

or

$$I_2 + 2S_2O_3^= \longrightarrow S_4O_6^= + 2I^- \qquad (5\text{-}6)$$

It is preferable to carry out the above reactions at pH 3 or 4 because at neutral pH the reaction is not stoichiometric owing to the oxidation of some thiosulfate to sulfate. This method can be carried out with phenylarsene oxide or thiosulfate. PAO is preferable because it reacts much faster than thiosulfate, which reacts in a stepwise fashion.

This starch–iodide iodometric titration procedure is the oldest method used to determine chlorine residuals. This method was used at one time to distinguish only those residuals (monochloramine) that released iodine from KI at neutral pH. This is no longer practiced; now the pH of the system must be reduced to 4 to measure all of the chlorine residual species. Acetic acid was found to be more desirable than either sulfuric or hydrochloric acid because it gave the best accuracy of pH control at pH 4. If ferric or manganic compounds are present these will give false residual readings. To compensate for these the procedure should include a blank titration. Nitrites are a serious interference but can be overcome by the sulfamic acid procedure described below.

This method has been popular for establishing temporary chlorine standards in spite of the fact that the minimum practical reading is 1.0 mg/l TRC. However, its popularity is limited by several drawbacks: the large sample required, low sensitivity, the relative instability of the thiosulfate and starch solution reagents, together with the tricky endpoint which is sensitive to temperature.

In spite of these difficulties it is still widely used for standardizing chlorine water used in laboratory studies of chlorine demand, tastes and odor control, and coliform destruction. This is the most convenient method for measuring chlorine concentrations in the range of 50–2500 mg/l.

For the convenience of laboratory technicians, the procedure is outlined below:

Standardization of Chlorine Water

Reagents:

1. Glacial acetic acid (concentrated).
2. KI Crystals, USP.
3. Starch solution, 1 percent starch (water soluble). Boil ten minutes and decant after standing overnight.
4. Standard sodium thiosulfate $0.1N$. Dissolve 24.82 g $Na_2S_2O_3 \cdot 5H_2O$ in 1 liter freshly boiled distilled water.

Procedure:

1. Place a few crystals of KI in a wide-mouthed Erlenmeyer flask.
2. Add approximately 10 ml distilled water.
3. Add about 2 ml glacial acetic acid.
4. Add 10–50 ml chlorine water by means of a volumetric pipette. The tip of pipette should almost touch surface of water in order to avoid surface loss.
5. Titrate with standard $0.1N$ sodium thiosulfate, using starch as an indicator near the end of the titration.

NOTE: In step 4, the chlorine will liberate iodine (I_2) from the KI present, giving a brown cast to the sample. The brown cast will be discharged upon addition of the reducing agent, sodium thiosulfate, and at this point 1 ml of starch indicator is added. The starch will produce a blue color in the presence of free iodine, and the titration is continued until this color disappears. Read the amount of thiosulfate used at first disappearance of blue color, and disregard reappearance of color upon standing.

Calculation: ml $0.1N$ thiosulfate used × 3.545, divided by ml of sample tested × 1000 = mg/l chlorine concentration of chlorine water, or

$$\frac{\text{ml } 0.1N \text{ thiosulfate} \times 3.545 \times 1000}{\text{ml sample}} = \text{mg/l Cl} \qquad (5\text{-}7)$$

This procedure will be applicable to waters generally in the range of 500 to 2000 mg/l chlorine

Measuring Residuals in Wastewater (Iodometric Method II)

General Discussion. When determination of residuals in wastewater became an important consideration in the 1960s, the majority of the effluents examined had only primary treatment. In the usual procedure, such as Iodometric I, the iodine release is performed first and the titrant is added to reduce the iodine to iodide, which becomes the measure of the chlorine residual. It was found that during the time required to neutralize the iodine by the titrant, some of the iodine was being consumed by the organic matter in the sewage. To overcome this occurrence the back-titration method was devised. It is applicable to both iodometric and amperometric procedures, as follows.

Back Titration Method. See Iodometric Method II, page 283, *Standard Methods,* 15th edition. This is the preferred method for primary effluents and highly polluted or poorly treated effluents. See Chapter 8 for other procedures for well oxidized effluents.

Procedure. The amount of sample to be taken for titration is governed by the concentration of chlorine in the sample. For residual chlorine of 10 mg/l or less, 200 ml sample should be titrated. For greater concentrations, proportionately less of the sample should be used.

Titration. Place 5.0 ml 0.00564N phenylarseneoxide (PAO) solution in a flask or white casserole. Add excess KI (approximately 1 g) and 4 ml pH 4 acetate buffer solution, or sufficient to reduce the pH to between 3.5 and 4.2. Pour in the sample and mix with a stirring rod. Just prior to titration with 0.0282N iodine, add 1 ml starch solution for each 200 ml sample. Titrate to the first appearance of blue color which persists after mixing. Since 1 ml 0.00564N PAO consumed by a 200 ml sample represents 1 mg/l available chlorine, 5 ml PAO solution is sufficient for residual chlorine concentrations up to 5 mg/l. For residual chlorine concentrations of 5–10 mg/l, 10 ml PAO solution is required.

Calculation.

$$\text{mg/l chlorine} = \frac{(A - 5B) \times 200}{C}$$

A = ml 0.00564N PAO solution
B = ml 0.0282N iodine
C = ml of sample

This method is least desirable when there may be interference from color or turbidity in the wastewater. Otherwise it gives results comparable to the amperometric method.

Control of Nitrite Interference. Serious interference from nitrites was reported by the California State Dept. of Health Bureau of Sanitary Engineering in 1972.[57] They found residuals as high as 16 mg/l with *no chlorine added*. This interference occurred with the iodometric method. Based upon the reaction in Eq. (5-4) it would appear that nitrites (NO_2^-) must act to release I_2 from KI.

Control of this interference is relatively simple. it depends upon the use of sulfamic acid, which will convert the nitrites to nitrates.

To a 200 ml sample of wastewater add 1 ml of 5 percent solution of sulfamic acid (NH_2SO_3H). Mix and let stand for ten minutes. Then add excess standard phenylarsene oxide (0.00564N) and excess KI (1 g) and titrate with standard iodine (0.00282N) using starch as the indicator.

This is the same as the procedure outlined in Method I except that sulfamic acid is used instead of phosphoric or acetic acids to depress the pH of the sample mixture. Sulfamic acid will depress the pH to the level necessary to perform the quantitative release of I_2 from KI in proportion to the total chlorine residual. The final pH will be about 4.3.

When using this procedure it is desirable to check the pH of the sample–sulfamic acid mixture; if the pH is above 4 add some pH 4 buffer (acetic acid).

Standardizing 0.0282N Iodine Solution with PAO. This solution will remain stable for months if it is stored in the dark, in a brown bottle, and in moderate temperature conditions. It is only necessary to check the normality on a weekly or biweekly frequency. This can be done with PAO solution as described below in the amperometric titration section.

DPD Ferrous Titrimetric Method. See page 289, *Standard Methods*, 15th edition.

General Discussion. This method, developed by A. T. Palin,[25,63] is the only one of the three he developed that has survived. (The other two were the p-aminodimethylanaline[21] and the neutral O-T–ferrous ammonium sulfate titration procedure.[24] The method was first introduced in 1956. It utilizes an indicator solution of diethyl-p-phenylene diamine. When used as a colorimetric test it is much preferred over either the acid or neutral O-T methods or the OTA method for its sharper free chlorine–chloramine differentiation. This method requires five reagents and has a range up to 4 mg/l without dilution. The minimum detectable concentration is about 18 μg/l as chlorine.

Principle. The titrimetric procedure requires the use of a standardized ferrous ammonium sulfate as the titrant and N,N-diethyl-p-phenylenediamine as the indicator (DPD). In the absence of iodide ion, FAC (free available chlorine) reacts instantly with DPD indicator to produce a red color. Subsequent addition of a small amount of iodide ion acts catalytically to cause monochloramine to produce color. Further addition of potassium iodide to excess evokes a rapid response from dichloramine.* The color at each stage is titrated to a colorless endpoint.

---

* In wastewater practice this fraction is made up mostly of organochloramines that have "dichloramine" characteristics. These chloramines are also known to intrude into the monochloramine fraction.

Procedure:
Reagents.

1. Standard ferrous ammonium sulphate (FAS) solution 1 ml = 0.100 mg available chlorine
2. DPD No. 1 powder
3. Potassium iodide crystals

The above DPD No. 1 powder is a combined buffer–indicator reagent. If liquid reagents are used buffer solution and DPD solution must be kept separate.

Procedure for Free Available Chlorine: To 100 ml sample add approx. 0.5 g DPD No. 1 powder. Mix rapidly to dissolve and titrate immediately with FAS solution (Reading A).

Procedure for Combined Available Chlorine Compounds. Add to the above one very small crystal of potassium iodide. Mix and continue titration immediately (Reading B). Add several crystals (approx. 0.5 g) of potassium iodide. Mix to dissolve and after standing about two minutes continue titration (Reading C). If this fraction is fairly high, use double the quantity of potassium iodide (i.e., approx. 1.0 g).

Procedure for Nitrogen Trichloride. To 100 ml sample add one small crystal of potassium iodide. Mix, then add approx. 0.5 g DPD No. 1 powder. Titrate immediately with FAS solution (Reading D).

Calculations. For 100 ml sample 1 ml FAS solution = 1 mg/l available chlorine:

A = free available chlorine
B = free + mono
C = free + mono + di + ½NCl_3
D = free + mono + ½NCl_3*
E = total available chlorine

Free chlorine	= A
Monochloramine	= $B - A$
Dichloramine	= $D - C$
Nitrogen trichloride	= $2(D - B)$
Total available chlorine	= E

Total Available Chlorine. This may be obtained in one step by adding the DPD No. 1 powder and the full quantity of potassium iodide to the sample at the start and standing for about 2 minutes. A combined DPD and potassium iodide

---

* Monochloramine is always present whenever NCl_3 is formed in a potable water or wastewater chlorination reaction.

reagent known as DPD No. 4 powder is available thus providing a single reagent for total available chlorine. For 100 ml sample use about 0.5 g of this powder or, if high concentrations of available chlorine are present, use about 1.0 g. After standing the required 2 minutes or so, titrate with the FAS solution to obtain total available chlorine (Reading E).

Monochloramine Breakthrough. In the event of a substantial amount of monochloramine present with free chlorine, an alternative procedure is recommended to prevent monochloramine breakthrough into the free residual reading. This breakthrough can be as high as 2 percent per minute of the monochloramine present. To overcome this, add 0.5 ml 0.25 percent solution of thioacetamide to the 100 ml sample immediately after mixing the DPD reagent. This stops further reaction with the combined chlorine in the free measurement. Continue immediately with FAS titration to obtain free chlorine. Then obtain total available chlorine from the above procedure (Reading E) without thioacetamide. When thioacetamide is used with the DPD procedure (either colorimetric or titrimetric) it is called the Steadifac method by Palin.[44]

The reaction of the Steadifac solution occurs in a molar ratio of 1:1, from which it may be calculated that one drop (0.05 ml) of 0.25 percent solution of this acetamide added to a 10 ml sample can eliminate 11.8 mg/l of chloramine.

Interferences. The most significant interfering substance likely to be encountered in water is oxidized manganese. This can be corrected by the addition of 0.5 ml of sodium arsenite solution with 5 ml of buffer solution into the titration flask. Add 5 ml DPD indicator solution, mix, and titrate with standard FAS titrant until red color is discharged. Subtract this reading from Reading A (free chlorine) obtained by the normal forward titration procedure (described above) or from the total available chlorine, Reading E, also described above.

If the combined reagent in powder form is used, first add KI and arsenite to the sample and mix, then add combined buffer–indicator reagent.

Amperometric Titration Method. See page 286, *Standard Methods,* 15th edition.

General Discussion. This is one of the most widely used methods of measuring chlorine residuals in North America. The titrator is easy to use, accurate, and reliable so long as reasonable care is exercised in maintaining electrode sensitivity. This is accomplished simply by soaking the electrodes in a dilute chlorine solution for a few minutes before operating the titrator. If the titrator is to be used in the back-titration mode, then soak the electrodes in a weak (light brown) iodine solution.

Principle of Titrator Operation. The amperometric method is a special adaption of the polarographic principle. This is a class of electrometric titration whereby the current which passes through the titration cell between an indicator electrode and an appropriate depolarized reference electrode (at suitable applied EMF) is measured as a function of the volume of a suitable titrating solution.[64] In general, the endpoint of an amperometric titration can appear either as the cessation of current at the equivalence point (complete reduction of titrant), or as the sudden increase in current (when titrant is the oxidant and overcomes the reducing agent in the electrolyte), or as an abrupt change in current at the equivalence point.

In any cell consisting of two electrodes contacting an electrolyte, the impressed voltage will be opposed by a countervoltage, due to the accumulation (polarization*) of the electrolysis products, or the exhaustion (depolarization) of the material being electrolyzed at the electrode surfaces.

The current that can flow in the electrolyte between the electrodes is said to be decreased by concentration polarization, which is sometimes described as concentration overpotential or activation potential. The flow of current in a cell will cause the accumulation of reducing agents at the cathode (negative electrode) and oxidizing agents at the anode (positive electrode). The addition of an oxidizing agent to the cathode solution or a reducing agent to the anode solution will decrease the accumulation at the respective electrodes. This phenomenon, known as depolarization of the electrodes, reduces the counter EMF, thereby allowing more current to flow.

Historical Development. The principle of amperometric titrations was proposed as early as 1897 by Salomon.[65] It was used in 1905 by Nernst and Merriam,[66] who showed that the diffusion current is proportional to the concentration of the electroreducible substance. Amperometric titrations are classified into two categories: one-indicator electrode with reference electrode, and dual-indicator electrodes.

Marks first made this method available for measuring chlorine residual in 1942.[17] In his investigation and development of this method, he used the single-indicator electrode (gold) with a silver chloride reference electrode. The titrating agent used was sodium arsenite. He later found a more discriminating reducing agent (phenylarsene oxide), which allowed modification of the procedure to give better differentiation of the various chlorine residual fractions.[18-20] In 1949 the first amperometric titrator for residual chlorine developed by Marks was made available. A current version of this unit, utilizing a single-indicator electrode, is illustrated in Fig. 5-2 and is shown schematically in Fig. 5-3.

The dual-indicator electrode system was introduced by Foulk and Bawden in 1926 for the titration of iodine with thiosulfate.[67] Not until 1966 was the dual-

---

* Polarization is defined as an effect produced on the electrodes of a cell by the deposition on them of the gases liberated by the current. It is caused chiefly by hydrogen, which increases the resistance and sets up a counter EMF, and can be visualized as numerous tiny bubbles of gas covering the surface of the electrode. These tiny bubbles insulate the electrode, thereby increasing the resistance to current flow through the electrolyte.

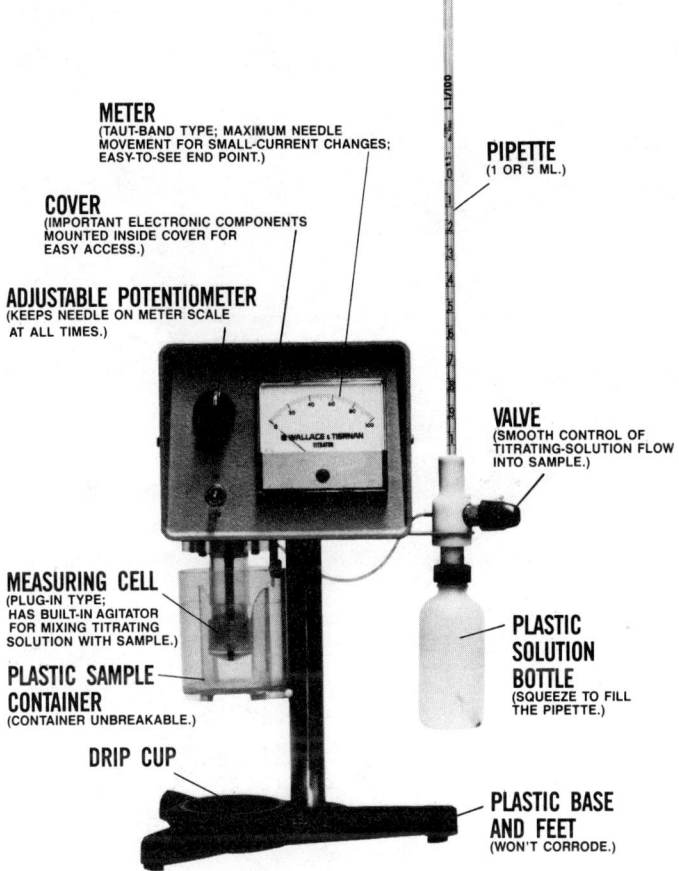

Fig. 5-2. Amperometric titrator (Courtesy Wallace & Tiernan Div. Pennwalt Corporation).

electrode system proposed for residual chlorine determination.[68] This system is illustrated in Fig. 5-4.

The Single-Indicator Electrode. Fig. 5-3 illustrates schematically the operation of this unit.[20] This titrator consists of a platinum (cathode) indicating electrode (which is polarized to some extent) and a silver electrode immersed in a saturated salt solution. This is the reference electrode and is the nonpolarizable anode. The potential of this silver electrode is such that it provides an internal voltage and makes it unnecessary to impress an external voltage. Because of the low concentrations of chlorine to be measured, the surface area of the indicating electrode is made large to increase the sensitivity of the cell. In this case the output of the cell is such that a 0.01 ppm change in chlorine concentration will produce a current

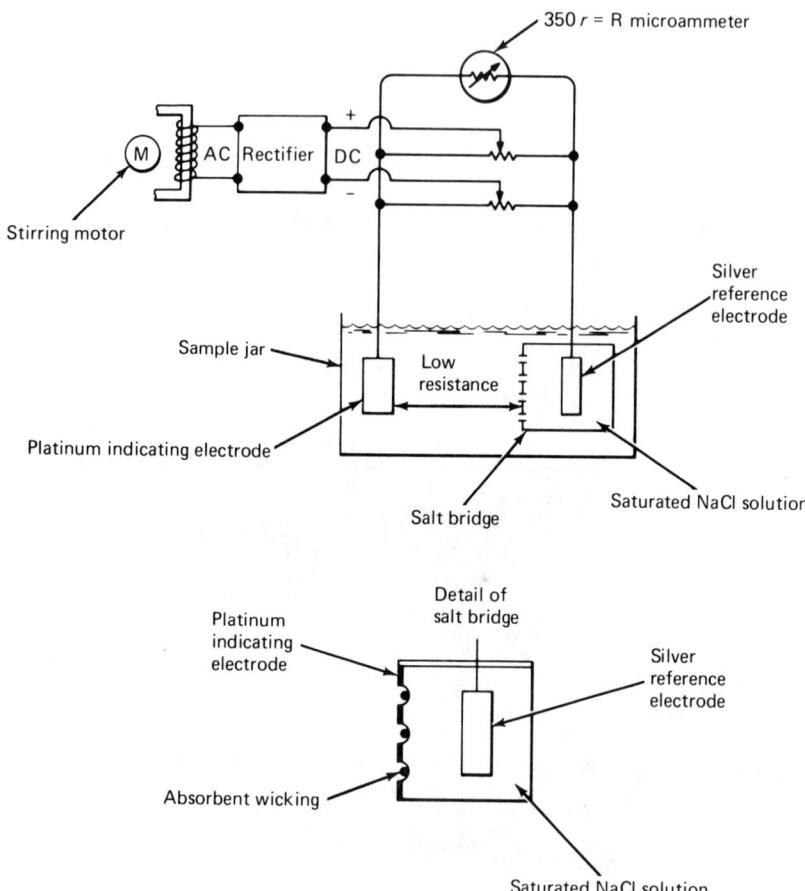

Fig. 5-3. Schematic amperometric titrator with single-indicating electrode (Courtesy Wallace and Tiernan Division Pennwalt Corporation).

of about 2.5 microamps, which will move the pointer about 7 divisions. Efficient agitation at the indicating electrode surface, which is necessary to obtain a higher and more uniform diffusion current, is achieved by a high-speed rotating Lucite sleeve which fits over the cylinder containing the platinum electrode with minimum clearance. The electrical path through the water sample (electrolyte) is made short and of low electrical resistance by putting a salt bridge in the intervals in the narrow space between the turns of the platinum spiral electrode. To perform a titration, the sample which contains the oxidizing agent—either chlorine (HOCl) or iodine (I_2)—is placed in position and the agitator is started. The microammeter will deflect to the right, upscale, depending upon the concentration of the oxidizing agent. The reducing agent (phenylarseneoxide) is then added, which decreases

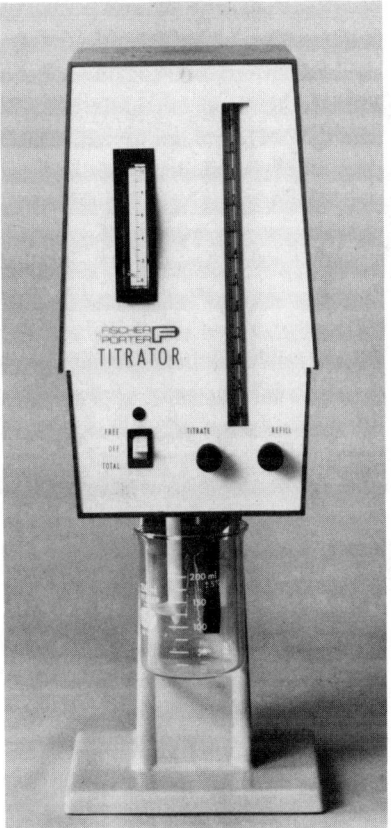

Fig. 5-4. Amperometric titrator with dual indicating electrodes (Courtesy Fisher and Porter Co.).

the concentration of the oxidizing agent and accordingly decreases the current through the cell. When the oxidizing agent has been completely reduced by the titrating agent (PAO), the endpoint is indicated by no change in the current upon further addition of reagent.

The reverse of this procedure, known as back titration, is performed by using an oxidizing agent (iodine) as the titrating solution to overcome an excess of reducing agent previously added to the sample. In this case the endpoint is observed as the first appearance of a current caused by the exhaustion of the oxidizing agent by the reducing agent.

The Dual-Indicator Electrode. The Fischer and Porter portable amperometric titrator (17T2000) utilizes dual-indicator electrodes. (see Fig. 5-4). Fig. 5-5 illustrates the essential components of the titrator. It uses a platinum and copper electrode

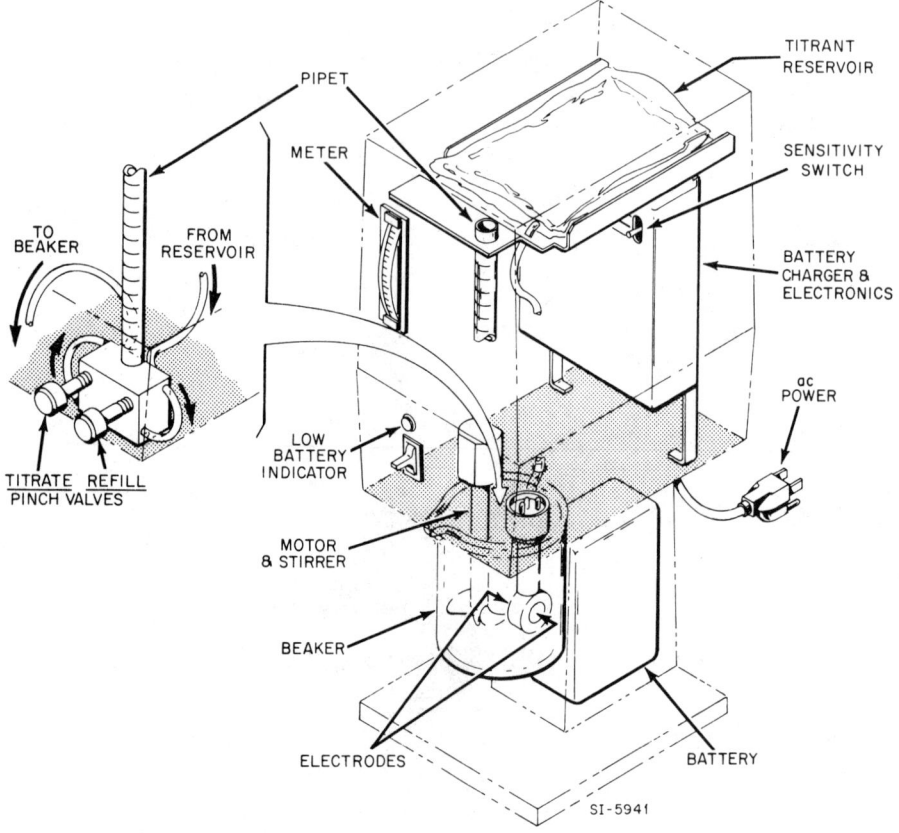

Fig. 5-5. Diagram showing essential components of F&P titrator (Courtesy Fischer and Porter Co.).

pair rather than the previously used dual platinum electrodes. This new arrangement incorporating the Pt–Cu pair provides a good current–voltage curve, yielding a usable plateau, wherein small voltage variations do not produce significant current changes. See Fig. 5-6. At the selected electrode potential (200 mV, TRC; 100 mV, FAC) the cell generates a stable current level, as a function of the chlorine concentration, which can be applied to several stages of amplification. This amplification provides the unit with a sensitivity of 0.005 mg/l. A Pt–Pt pair produces a current vs. voltage curve which is essentially a straight line for free chlorine (FAC). Therefore a small shift in electrode potential generates a significant change in current making amplifier input–output unpredictable and of limited use.

The final display output (meter reading) is made more readable in the critical endpoint region by applying the cell output to a log-linear amplifier circuit. The upper curve in Fig. 5-6 compares the log function to the conventional display.

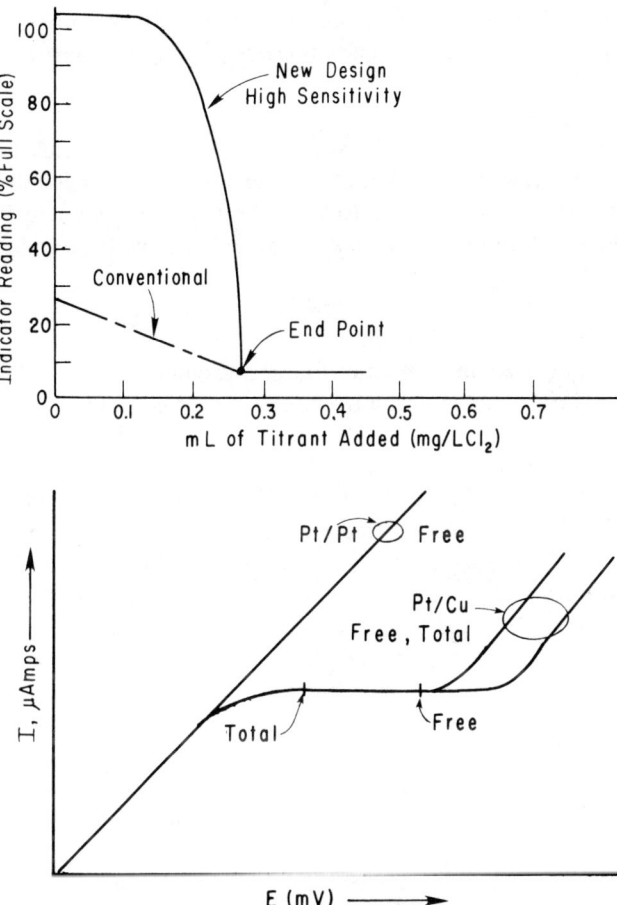

Fig. 5-6. Relationship between current and applied voltage across the Pt/Cu electrodes and log function readability versus conventional display (Courtesy Fischer & Porter Co.).

For any given chlorine concentration the meter response is minimal at the beginning of the titration, increasing drastically as the endpoint is approached.

This system is known as biamperometric titration. The procedure is the same as that used in the single-electrode system. When the oxidizing agent is being titrated by a reducing agent, both electrodes remain depolarized and a current flows in proportion to oxidizing agent and is indicated on the output meter.

When the last traces of the oxidizing agent have been destroyed by the reducing agent, the cathode is without a depolarizer and the flow of current is immediately arrested, signifying the endpoint. This is indicated by the milliammeter. Conversely if the titrant is an oxidizing agent, such as iodine, which is added to overcome a reducing agent, as in the back-titration procedure, the endpoint occurs at the first appearance of a current. This occurs as soon as there is an excess of iodine, which depolarizes the cathode. The anode is already depolarized by the iodide ion.

Precision and Sensitivity. The precision of the single-indicator electrode described above is on the order of ±0.05 mg/l, while the sensitivity is 0.01 mg/l. These units can be made supersensitive to measure residuals down to 0.001 mg/l by the following modification, which was performed on a Wallace and Tiernan Model A-790 Titrator:[58]

1. Dilute the PAO reagent 4 times.
2. Encase sample jar in aluminum foil and ground it.
3. Remove ammeter and insert a Rochester converter that will put out a milliamp signal.
4. Connect milliamp output to an electrician's type of voltmeter.
5. This will measure the electrical potential across the electrodes.
6. This will allow the user to read residuals down to 0.001 mg/l.

Operating Characteristics. The following statements appear in *Standard Methods*, 15th edition: "The method is not as simple as the colorimetric methods and requires greater operator skill to obtain the best reliability. Loss of chlorine can occur because of rapid stirring in some commercial equipment. Electrode cleanliness and conditioning are necessary for sharp end points," p. 286. "Amperometric titration requires a higher degree of skill and care than colorimetric methods. Chlorine residuals over 2 mg/l are measured best by means of smaller samples or by dilution with water that neither has residual chlorine nor a chlorine demand," p. 278.

None of these statements can be supported by the present author's personal experience nor by operators with whom he has worked over the past twenty years on both potable water and wastewater. Measuring residuals up to 20 mg/l without dilution is a routine task with an amperometric titrator. As for skill required, the only skill needed is how to handle a pipette and a burette and how to interpret an ammeter.

The one technique that is often overlooked is care of the electrodes. Each manufacturer has specific instructions for electrode care; however, the most important task to perform is to sensitize the electrodes. For forward titrations, sensitize the electrodes with a dilute free chlorine residual (5 mg/l) by letting them soak in this solution in the sample jar. When back titrations are made in wastewater situations let the electrodes soak in a dilute iodine solution (light brown color in sample jar). When routine residual measurements are being made for *both* forward and back titrations, separate titrators should be used because it reduces the rinsing time required to remove the iodine when performing the free chlorine measurement.

The electrodes should be sensitized as described above for about 10 minutes before using if the titrator is not being used on a daily basis. Otherwise, they should be sensitized at weekly intervals. Frequent use of the titrator is very rewarding because its repeatability builds the operator's confidence. Colorimetric methods always leave a doubt in the operator's mind because color depth acuity varies from individual to individual. Therefore errors will often occur in colorimetric methods that do not occur in the amperometric method. Moreover, all colorimetric methods that do not use a spectrophotometer readout require that the operator estimate visually the color depth between the color standards, such as the difference between 1.0 and 1.5 mg/l, etc. Quite often operators have a difficult time distinguishing between 1.0 and 1.5 mg/l.

Chemistry of the Amperometric Method. The success of this method is largely due to the characteristics of the reducing agent phenylarsene oxide (C_6H_5AsO). Other reducing agents were tried and discarded for a variety of reasons. For example, thiosulfate, which is used in the starch–iodide method, does not sufficiently discriminate between free and combined available chlorine. Further, it acts in a reducing process by steps, causing a time lag that interferes with the endpoint. Sodium arsenite, which was used in some of the early investigations of this method, had possibilities because it would react with HOCl only in the absence of potassium iodide and with any iodine released by chloramines in the presence of iodide. However, it would not react quantitatively below pH 6, which is essential in order to pick up dichloramine, which liberates iodine only at a pH much below 6.

Phenylarsene oxide reacts only with free chlorine at pH 7 in the absence of potassium iodide (KI). PAO reacts with monochloramine in the presence of 50 mg/l KI at pH 7 and with dichloramine in the presence of 250 mg/l KI at pH 4. Actually PAO does not directly titrate either mono- or dichloramine. Iodine is liberated quantitatively by these fractions in the presence of the amounts of potassium iodide shown and at the pH levels indicated. The PAO then titrates the iodine liberated.

These two general reactions are described by the following equations:

$$C_6H_5AsO + HOCl + H_2O \longrightarrow C_6H_5AsO(OH)_2 + HCl \quad (5\text{-}8)$$

$$C_6H_5AsO + I_2 + 2H_2O \longrightarrow C_6H_5AsO(OH)_2 + HI \quad (5\text{-}9)$$

The strength of PAO was selected at $0.00564N$ so that in using a 200 ml sample 1 ml of PAO is equivalent to 1 ppm of chlorine. Therefore when the endpoint of the titration is reached, the volume of PAO used represents the chlorine concentration in milligrams per liter. One mole PAO reacts with two equivalents of halogen.

In determining free chlorine, the pH must not be greater than 7.5, because of sluggish reactions at higher pH values, or less than 6.5 because at lower pH values some combined chlorine may react even in the absence of iodide. When determining combined available chlorine, the pH must not be less than 3.5, because substances such as oxidized manganese interfere at pH values lower than 3.5, or greater than 4.5, because the reaction of combined residual chlorine is not quantitative at higher pH values.

The tendency of monochloramine to react more readily with iodide than does dichloramine provides a means of further differentiation. The addition of a small amount of potassium iodide in the neutral pH range enables the estimation of monochloramine content. Lowering the pH into the acid range and increasing the KI concentration allows the separate determination of dichloramine.

Preparation for Titration

Apparatus. An amperometric titrator, the endpoint detection instrument, consists of a two-electrode cell connected to a microammeter with an adjustable potentiometer. Accessories should include a 200 ml sample jar with displacement cup for accurately measuring 200 ml, a 1.0 ml buret with 0.01 ml graduations, and a finger pump for charging the buret with the reducing reagent.

To prepare the titrator for operation, the operator should check the following:

1. If the unit utilizes a silver-silver chloride electrode (Fig. 5-4), make certain that there are sufficient salt tablets in this cell (2/3 full) and that enough distilled water has been added to cover the tablets.
2. Make certain that the electrical contacts between cell and microammeter are clean and making proper contacts.
3. Be sure that the platinum electrode surface is free of any deposits. If it is dirty clean by lightly rubbing the platinum surface with scouring powder, using only the fingers. Be sure to avoid disturbing the porous wicking which lies between the turns of the platinum ribbon.
4. If the titrator has been used routinely for free chlorine measurements only and it is desired to measure combined or total chlorine residual, the platinum electrode must be sensitized to iodine. This will be noticed during a titration when the potassium iodide is added. When KI is added and the needle deflects downscale momentarily and remains there, the cell is said to have lost its sensitivity to iodine. The sensitivity is easily restored by adding enough free iodine to the water in a sample jar to create a yellowish color, agitating sample for two or three minutes, and then allowing it to stand in this solution for ten or fifteen minutes. After this treatment, the cell unit should be rinsed thoroughly to remove all traces of free iodine.
5. All glassware used in this procedure should be chlorine demand-free. The sample jar should be treated with water containing at least 10 mg/l of chlorine for three hours or more before using and then rinsed with chlorine demand-free chlorine water.
6. To prevent errors in titration from contaminants, it is good laboratory practice

to fill the pipette to the top and to run the contents to waste before beginning a series of titrations.

Reagents

1. Standard phenylarsene oxide titrant (a $0.00564 N$ solution of PAO, C_6H_5AsO)
2. pH 4 acetate buffer solution
3. pH 7 phosphate buffer solution
4. Potassium iodide solution (5 parts CP grade potassium iodide in 95 parts distilled water)
5. Electrolyte tablets (made from USP sodium chloride)

Procedure

I. Free Available Chlorine

1. Connect the titrator to a source of 115 V single-phase power.
2. Fill the pipette with phenylarsene oxide solution. Remove all air from the pipette and tubing and then discharge the PAO. Refill the pipette to the top calibration mark.
3. Measure a 200 ml sample of the water to be tested in the sample container and place in position on the titrator.
4. Add 1 ml of phosphate buffer solution pH 7 to the water sample. If it is known that the natural pH of the sample lies between 6.0 and 7.5, it is not necessary to use the pH 7 buffer solution. NOTE: Droppers are usually furnished for these reagents and are scribed to the 1 ml mark; therefore a dropperful of solution should be used whenever 1 ml of solution is called for.
5. Start the agitator by turning the switch to "on."
6. Adjust the potentiometer so that the microammeter needle reads near maximum on the scale. If the needle is above maximum when the adjusting knob is rotated completely counterclockwise, then the titration must be started with the knob in this position.
7. Begin adding the phenylarsene oxide solution just under the surface of the sample. If there is free chlorine present as the titrant is being added, the cell current will decrease, resulting in a downscale movement of the microammeter needle. If the needle is above maximum at the beginning of the titration, the needle will remain above maximum until enough phenylarsene oxide solution has been added to reduce the free chlorine residual (and hence the current) to a point where the needle will read less than maximum. As the titrant is added, it will be necessary to adjust the potentiometer from time to time to bring the needle back on scale. As the endpoint is approached, the response of the microammeter to each increment becomes more sluggish, and smaller increments of titrant should be added. The endpoint is just passed when a

single drop of PAO no longer causes a downscale (decrease of current) movement of the needle.* The buret is then read, and the last increment (one drop) of titrant is subtracted from the reading. The result is the free available chlorine concentration in mg/L.

II. Monochloramine

This follows immediately and in the same sample as in I.

1. Note the buret reading at the endpoint in I.
2. Add 0.2 ml of potassium iodide to the original sample. The furnished dropper usually delivers 20 drops per ml. Therefore 4 drops correspond to 0.2 ml. If monochloramine is present, the needle will deflect to the right (probably off scale) immediately upon addition of the potassium iodide.
3. Proceed with the addition of PAO to the endpoint as in I. Note buret reading. The difference between the buret readings at the endpoints of I and II is the amount of monochloramine present.

Note that this titration is performed at pH 7.

III. Dichloramine

This determination follows the procedure in II and with the same sample.

1. Decrease the pH of the sample to 4 (3.5 to 4.5) by adding 1 ml buffer solution pH 4.
2. Add 1 ml potassium iodide solution. If dichloramine is present, the needle will deflect to the right at this point.
3. Proceed with addition of PAO until the endpoint is reached as before.

Note the reading on the buret. The difference between this reading and the previous reading at the end of II is the amount of dichloramine present in mg/l.

IV. Total Available Chlorine Residual

This procedure bypasses all the separate fractions described previously and lumps all these into one total reading.

Starting with a fresh 200 ml sample, proceed as follows:

1. Add 1 ml buffer solution pH 4.
2. Add 1 ml potassium iodide solution. At this point the needle will deflect

---

* This endpoint is true for nearly all waters. There have been some cases when the needle will deflect downscale after the endpoint has been reached. In such cases the true endpoint occurs when for equal amounts of PAO added the amount of deflection changes from large to small. At this point the last increment of titrant is subtracted from the total reading as before.

first to the left and then go upscale (probably off scale). Any chlorine residual, free or combined, will release iodine from the potassium iodide quantitatively.
3. Proceed with the addition of PAO to the endpoint as before.

Note the reading of the buret. This represents the total of all fractions of available chlorine residual present in the sample.

The indicating electrodes must be sensitized to iodine as explained previously.

V. Free and Combined Available Chlorine Residual

It is often desirable to make separate determinations of free and combined available chlorine residual. This can be done with only one sample.

1. Measure free chlorine as in I.
2. Add 1 ml buffer solution pH 4.
3. Add 1 ml potassium iodide.

At this point all mono- and dichloramine liberates iodine from potassium iodide quantitatively. If combined chlorine residual is present, the needle will deflect first to the left then sharply to the right upon the addition of potassium iodide.

4. Add PAO until the endpoint is reached as before.

Note the reading on the buret. This reading represents the total available chlorine, and the difference between this reading and the reading in I is the combined available chlorine residual—all in mg/l.

NOTE: If free available chlorine residuals are made following procedures using potassium iodide, the cell unit must be rinsed thoroughly to remove traces of potassium iodide and pH 4 buffer solution.

VI. Nitrogen Trichloride

It is a well-established fact that nitrogen trichloride can exist simultaneously with free chlorine at a pH as high as 8.5 or even 9[25,72,73] if free ammonia is available to react with the chlorine and if chlorine is added in sufficient quantity to produce the breakpoint phenomenon. In the amperometric method it is believed that any nitrogen trichloride titrated will appear partly in the first fraction as free chlorine and partly in the third fraction as dichloramine.[20] Dowell and Bray[69] found that the starch-iodide titration includes only 80 percent of the total nitrogen trichloride present, while the acid-orthotolidine method includes only about 60 percent of the total. Palin's methods have been shown tentatively to account for all the nitrogen trichloride present. However, sufficient reliable data are lacking to establish the ability of the amperometric method to differentiate the total nitrogen trichloride from the other fractions.

Williams[71] uses the following method:

1. A carefully collected sample, subjected to a minimum amount of agitation to prevent aeration of nitrogen trichloride, is placed in the titrator. Before the agitator is started, an excess of pH 4 buffer and an excess of potassium iodide are added, to "fix" all the chlorine fractions, including NCl_3.

2. The above is followed by the procedure for determining total available chlorine residual. This is reading A.

3. Next, the titrator cell, agitator, and sample jar are thoroughly rinsed. Another sample is placed in titrator jar; with the sample jar held in the hand, to achieve the maximum amount of turbulence, the agitator to turned on, to aerate out all the nitrogen trichloride.

4. The sample jar is quickly returned to the normal position. The rest of the procedure is the same as that for $HOCl$, NH_2Cl, and $NHCl_2$.

The sum of these fractions will be reading B. the difference $(A - B)$ will be the nitrogen trichloride present, assuming that all the NCl_3 will appear in the total available chlorine determination (step 1).

If so desired, the nitrogen trichloride can be extracted by the use of carbon tetrachloride rather than by aeration.* This method, however, is more cumbersome.

The amperometric method can also be used to determine total available iodine and bromine. The procedure for measuring chlorine dioxide is fully described in Chapter 12.

FAC Residuals at Short Contact Times. In some special instances it is desirable and necessary to be able to determine the free available chlorine residual fraction when the contact time is only a matter of seconds.** This method is based on the ability of phenylarsene oxide to react with free chlorine at pH 7 but not with combined chlorine.

By adding an excess of PAO solution to a properly buffered sample, the amount of free chlorine residual in the sample at the time the PAO is added can be determined by titrating the remainder of the excess PAO with a standardized chlorine solution.

The apparatus and reagents are the same except for the standardized chlorine solution (0.02 percent $HOCl$). This solution should be standardized with PAO solution so that 1 ml chlorine solution is equivalent to 1 ml of PAO solution.

Procedure

1. To the empty but clean titrator jar add 1 ml pH 7 buffer solution and 5 ml PAO.
2. Add precisely 200 ml of sample to the jar and mix as thoroughly and rapidly as possible, probably not in the vicinity of the titrator.

---

* Power stations and other cooling water systems.
** NCl_3 is highly soluble in carbon tetrachloride.

3. Place sample jar in the titrator and proceed with titration using the standardized chlorine solution (0.02 percent HOCl) as the titrant.
4. At each addition of the titrant the microammeter needle will make a small momentary deflection to the right, returning each time. The endpoint is reached when the needle deflects discernibly to the right, returning each time. The endpoint is reached when the needle deflects discernibly to the right, indicating that the excess PAO has all been oxidized by the HOCl solution and that the last addition of titrant has produced a current. This last addition should be subtracted from the buret reading. This is reading A.

Calculation. The free chlorine residual mg/l = ml PAO − A.

NOTE: If reading A is greater than 4.0, repeat the procedure using less PAO solution; if reading A is less than 0.2, repeat procedure using more PAO solution.

The purpose of this procedure is to determine the free chlorine residual after a contact time of only seconds. The contact time between the chlorine and sample is measured from the time the chlorine is applied to the water being treated until the time the sample reacts with the PAO in the sample jar. The time it takes to transport the sample jar to the titrator and to run the titration is not a part of the contact time.

Determination of Residual Chlorine in Wastewater. In 1948 when it became commonplace to chlorinate primary effluents to enhance their acceptability, Wallace and Tiernan Company suggested a new approach to amperometric titration of chlorine residuals.[59] Primary effluents presented a problem because they exerted an immediate iodine demand by reducing agents that were unaffected by the chloramine residual present. To avoid this loss of iodine the titration procedure was reversed as follows: An excess of titrant (PAO) is added to the sample before adding the iodide. Then when the sample pH is reduced to 4 with buffer and the iodide added, the iodine released is immediately consumed by the PAO. The remaining PAO is back titrated with a standardized iodine solution.

Back-Titration Procedure.

1. Sensitize electrodes with a dilute solution of iodine—a light brown color is satisfactory. This will produce sharp end points.
2. Allow electrodes to soak in this solution in the sample jar for 15–20 min.
3. Thoroughly rinse out the sample jar after sensitizing is completed.
4. Place a 200 ml sample of wastewater in the titrator.
5. Start the agitator.
6. Add 5 ml phenylarsene oxide (PAO) solution to sample and mix. If total residuals are known to be greater than 5 mg/l add 10 ml PAO.
7. Add 4 ml pH 4.0 buffer solution (or sufficient to insure a sample pH between 3.5 and 4.2) to sample and mix.
8. Finally add 1 ml KI solution and mix.

9. Adjust microammeter pointer so that it reads about 20 on the scale.
10. Add $0.0282N$ iodine solution in small increments from a 1 ml pipette or a 1 ml buret.
11. As iodine is added to the sample, the pointer remains practically stationary until the endpoint is approached. Just before the true endpoint, each increment of iodine solution causes a temporary deflection of the ammeter to the right, but the pointer drops back to about its original position. The true endpoint is reached when a small addition of iodine solution gives a definite and permanent pointer deflection to the right (upscale).
12. Note the volume of iodine solution used to reach the endpoint, and calculate the total residual chlorine as follows:

$$\text{Total chlorine residual mg/l} = \text{ml phenylarseneoxide} - 5 \times \text{ml iodine solution} \qquad (5\text{-}14)$$

This calculation assumes a 200 ml sample and a PAO solution that is $0.00564N$ and an iodine solution that is $0.0282N$.

Monitoring the Normality of Iodine Solution. In the past the $0.0282N$ iodine solution was reported to be so unstable that it should be made up fresh daily. Experience spanning many years discloses that this was a gross misrepresentation. If stored in the dark (in a brown bottle) at moderate temperatures it will remain stable for months.

However, it is prudent to check the normality on a weekly or biweekly basis. The following is a simple field procedure that can be accomplished quickly with an amperometric titrator:

1. Sensitize the titrator electrodes to iodine by placing the sample jar in place containing a weak iodine solution—one that gives a light brown cast. Leave the electrodes immersed for 15 minutes.
2. Rinse electrodes thoroughly with tap water.
3. Add 5 ml PAO to 200 ml distilled water and place in titrator.
4. Titrate with $0.0282N$ iodine solution.
5. The endpoint is reached when a small addition of iodine produces an ammeter needle deflection to the right (upscale) which holds for 15–20 seconds. At this point all of the PAO has been oxidized.
6. Now read the amount of iodine used from the buret (or pipette). If 1.00 ml iodine solution neutralizes the 5 ml PAO solution, the iodine solution is $0.0282N$. If the iodine solution has deteriorated, the volume of iodine solution reaching the end point will be somewhat greater than 1.00.

Sample Calculation. Upon standardizing the iodine solution, it is found that 1.2 ml $0.0282N$ I_2 is required to neutralize 5.0 ml PAO in a 200 ml sample and that the PAO is $0.00564N$.

The chlorine residual is calculated from the following equation:

$$\text{chlorine residual mg/l} = \text{ml PAO} - (5 \times \text{ml } 0.0282N \text{ I}_2) \quad (5\text{-}10)$$

Since it requires 1.2 ml I_2 to neutralize 5 ml PAO, then 1.0 ml of PAO will only neutralize

$$5.0/1.2 = 4.17 \text{ ml PAO}$$

Therefore the calculation for residual, using Eq. (5-10), becomes:

$$Cl_2 \text{ residual (mg/l)} = \text{ml PAO} - (4.17 \times \text{ml I}_2)$$

Nitrogen Trichloride. Nitrogen trichloride is produced in two ways outside of the laboratory in potable water and wastewater treatment systems: (1) in the manufacture of chlorine when the electrolytic cell water contains ammonia nitrogen (see Chapter 1); (2) in water systems containing ammonia nitrogen when the chlorine to ammonia-N ratio is about 12:1 or greater at *p*H up to 9 (see Chapter 4). Its occurrence is rare in potable water treatment. At the Rancho Cordova 3 mgd nitrogen removal study using chlorine the nitrogen trichloride generation was not a problem because of its rapid decay.[51] Palin[23,73] and Cooper et al.[48] have spent an enormous amount of time investigating the detection and measurement of NCl_3 in potable water. It is a tantalizing exercise but a seemingly futile one because NCl_3 solutions are made up in the laboratory under controlled conditions. NCl_3 does not occur in plant operation in the same way that the solutions are made in the laboratory, so there is little relevance between detection under laboratory conditions compared to field conditions.

Two factors relating to nitrogen trichloride formation that should be kept in mind are: (1) HOCl must always be present for NCl_3 to exist; and (2) monochloramine is always found in the presence of NCl_3. This is the result of the simultaneous competing breakpoint reactions described in Chapter 4.

Nitrogen trichloride is a gas. It is *practically* insoluble in water, but it is soluble in liquid chlorine and in most solvents like carbon tetrachloride, etc. The strongest solution of NCl_3 achieved by Cooper et al.[47] was 13.8 mg/l. The solutions were stored at 0°C or used immediately. These solutions and those by Palin were made under carefully controlled conditions and reacting concentrations substantially different from any field conditions.

From a practical viewpoint the operator is more interested in whether or not NCl_3 formation is occurring in the treatment process, rather than how much is formed. Moreover the operator can detect the presence of NCl_3 quicker than a laboratory technician can. The olfactory nerve can detect concentrations as low as 0.02 ppm,[71,72] but the eyes can detect it before the nose can. At the slightest concentration the eyes will smart and prolonged exposure will cause severe tearing. As the concentration increases the effect is the same as tear gas. Unlike with chlorine, the respiratory system is only slightly affected even at high concentrations.[74]

The only documentation covering the measurement of NCl_3 generated under field conditions at a treatment plant was by Williams.[71,72] His method of NCl_3 measurement by amperometric titration is described in this chapter.

Chemistry of Nitrite Interference. In the early 1970s the Bureau of Sanitary Engineering, California State Department of Health, carried out numerous investigations related to disinfection of wastewater effluents by chlorination. At that time the starch–iodide forward titration was a popular method. The procedure called for the sample to be acidified to a pH between 1 and 2 to perform the I_2 release by the combined chlorine residual. In several instances false residuals were reported, i.e., residuals were being measured when the chlorinators had been shut down.[57] An investigation discovered that the false residuals were due to the presence of nitrites. It has since been confirmed that at a pH below 3.0 nitrite will oxidize KI to I_2. The rate depends upon acidity and KI concentration.[78]

For a very long time the starch–iodide titration was carried out at a pH of 1 or 2 to eliminate the iodine–organic matter reactions.[78]

Nitrite interference in both the starch–iodide and amperometric endpoint procedures is eliminated by the use of pH 4 buffer (acetic acid).

The question of nitrite interference in the back-titration procedure using PAO has been raised. In this method excess PAO is added to the sample containing chlorine residual. The pH is depressed to 4 after the addition of KI. The iodine released by the KI–chlorine residual reaction is immediately consumed by the PAO. I_2 will react with nitrite to form nitrate but this reaction is so slow at pH 4 that nitrite interference is nonexistent.

The short-term solution to the nitrite interference described above by the California State Dept. of Health was to destroy the nitrites present in the samples with sulfamic acid. However, sulfamic acid cannot be used for a long-term solution because it hydrolyzes to form HOCl. This adds another dimension to the interference.

Baker's Alternative Procedure.[75] About 1974 chlorine residuals in the receiving waters were discovered to be toxic to aquatic life. Since that time the measurement of these residuals has come under close scrutiny. Wastewater and cooling water discharges became the prime targets of this scrutiny. The basis of Baker's procedure is the assumption that chlorine species capable of any significant performance in the chlorination of potable water, wastewater, cooling water, and swimming pools have an ORP high enough to oxidize iodide to iodine at pH 7.[76] Driving this reaction to completion at pH 4 does not represent reality. One of the implications of this hypothesis concerns dechlorination. Current practice may require dechlorination of nontoxic residuals appearing in the "dichloramine" fraction (see Chapter 6).

This procedure demonstrates the hypothesis described above most dramatically with nitrified wastewater effluents where only a trace of ammonia-N is present and the organic N content is 3.0 mg/l or greater.

Procedure

1. Tritrate for free chlorine with PAO.
2. Titrate for mono at pH 7 in the usual way with PAO.
3. Take another sample, add excess KI (1 ml) and PAO, back-titrate at pH 7 with standardized chlorine solution. One cannot back-titrate with I_2 because it will be consumed by organic matter at pH 7—but not at pH 4.
4. Then take sample 3 and drop the pH to 4. Whatever titrates in this procedure will be the "dichloramine" species.

In wastewaters with organic N concentrations greater than 1.0 mg/l, the "dichloramine" species is most likely to be impotent organochloramines (see Chapter 8).

SUMMARY AND RECOMMENDATIONS

General Considerations

The selection of analytical method involves several factors based upon local conditions and requirements. This involves a variety of situations, such as potable water or wastewater; field or laboratory use; field monitoring or analyzer calibration; private, semi-private, or public water supplies. There are methods to cover all of these situations. They are described below with situation recommendations based upon many years of experience.

Potable Water Treatment

Small Supplies: Private and Semi-Public. These supplies serve summer camps, resorts, public campgrounds, motels, highway restaurants, private subdivisions, etc.

Monitoring chlorine residuals for this type of water system is similar to U.S. Army "field conditions." Therefore it would follow that the method selected by the Army, after intensive investigation, would be the method of choice. This is the FACTS method, using a testing kit with permanent glass color standards (0.2, 0.4, 0.7, 1.0, 2.0, 5.0 and 10.0). Made by the Ames Division of Miles Laboratory. Two solutions are required: buffer and syringaldazine. This measures only free chlorine. This is important for the small nonregulated water supply because a free chlorine residual is imperative for the short contact times inherent in these water systems.

Municipal Supplies

Chlorination Stations. There are many water utilities that do not have a central treatment plant but have chlorination stations scattered throughout the system. A good example would be a city that has a multiple well field with several chlorinator stations. The field test choice in a case like this could be either a portable amperometric titrator (Fig. 5-2 or 5-4) or a portable DPD colorimetric test kit with a spectrophotomer like the Residometer (Fig. 5-1). Operating requirements for these supplies would probably rely on either free or total chlorine residual monitoring.

Treatment Plants. Practically all treatment plants have laboratory facilities. Therefore operations have an additional choice. This is the DPD–ferrous titrimetric method. The amperometric titrator and the Residometer would remain as options. All of these measure either free or total chlorine residual.

In the case where the plant needs to monitor all the chlorine species, the choice is either the DPD titrimetric or the amperometric methods.

Continuous Chlorine Residual Analyzers. These instruments provide intelligence only on a secondary level. Their information output is not absolute. The primary source of intelligence is derived from the information provided by the analytical method used for calibrating the analyzer. Therefore the calibration method must be quantitatively precise and qualitatively definitive. Operating personnel have three choices: amperometric titration, DPD-FAS titration or DPD colorimetric based upon spectro-photometric analysis. Other methods lack the precision and quality requirements for accurate determinations of residual levels.

Chloramines. All chloramine installations should have testing equipment capable of monitoring free, monochloramine and dichloramine residuals. This is for process control of THMs and taste and odor control. Ability to measure these species provide helpful information relating to the Cl to N ratio location on the breakpoint curve. Appearance of dichloramine will affect the taste and odor of the treated water. Therefore to measure all chlorine residual species use either an amperometric titrator or the DPD–ferrous titrimetric method.

Wastewater Treatment

Primary Treatment Plants. The quality of these effluents is such that the alternative procedure identified as the back-titration method should always be used. Two methods are available, the amperometric titrator or the iodometric (starch–iodide–iodate) method. The latter is acceptable when total residuals are 1 mg/l or greater. Colorimetric methods have never proved useful in measuring chlorine residuals for primary effluents.

Secondary Treatment or Well Oxidized Effluents. The back-titration method, using the amperometric endpoint, is probably the most popular today. Next in popularity is the long-standing iodometric (starch–iodide) II method using the back-titration procedure. Many users of the starch–iodide (iodometric II) route have found the DPD–ferrous titrimetric method preferable.

For treatment plants with moderate to severe disinfection requirements (2.2 to 240/100 ml MPN), the forward titration procedure, either amperometric or DPD–ferrous titrimetric methods should be used for the following reasons: White et al.[77] found that greater efficiency of mixing, at the point of application of chlorine, resulted in a greater proportion of mono- to dichloramine in the final residual, resulting in greater disinfection efficiency. Therefore the forward titration procedure has an important place in monitoring the disinfection process. It is further recommended that the forward titration procedure to determine mono- and dichloramine be supplemented by running total chlorine residuals by the back-titration procedure. The mono- to dichloramine relationship and its effect upon disinfection efficiency are described in Chapter 8.

Tertiary (Filtered) Effluent

Nitrified. These effluents will show both free and combined chlorine residuals. To properly monitor the disinfection process it is of significant benefit to run forward titrations supplemented by total chlorine residuals if dechlorination is required. If the effluent is completely nitrified, monochloramine will be absent, so the second step can be omitted. However the third step is important because the ratio of free to dichloramine residual becomes important to disinfection efficiency. The two methods of choice are amperometric and DPD–ferrous titrimetric.

Non-Nitrified. Residual measurements for these effluents should be by the forward titration procedures to quantify mono- and dichloramine fractions, followed by a separate total chlorine residual either by back titration or forward titration. The latter serves as a check on the forward titration for mono- and dichloramine. The methods of choice are amperometric and DPD–ferrous titrimetric.

Organic Chloramine Interference

Of significant interest concerning the various methods of FAC determination is the critique by Wajon and Morris.[46] They pointed out that all of the methods currently used to measure FAC are subject to interference from organic chloramines. Therefore the measurement of FAC in potable water, reclaimed water, or wastewater containing organic nitrogen can not always be defined as FAC unless the types of organic nitrogen compounds present are known. See Chapter 8, "Free Residual Chlorination of Nitrified Effluents."

REFERENCES

1. Phelps, E. B., Testimony in re: In Chancery of New Jersey: Jersey City vs Jersey City Water Supply Co., Vol. 12, pp. 6921 et seq; decision rendered May, 1910.
2. Ellms, J. W., and Hauser, S. J., "Orthotolidine as a Reagent for the Colorimetric Estimation of Small Quantities of Free Chlorine," *J. Ind. and Eng. Chem.*, **5,** 915 and 1030 (1913).
3. Wolman, A., and Enslow, L. H., "Chlorine Absorption and Chlorination of Water," *J. Ind. and Eng. Chem.*, **11,** 209 (1919).
4. Meur, H. F., and Hale, F. E., "Present Orthotolidine Standards for Chlorine," *J. AWWA,* **13,** 50 (1925).
5. Scott, R. D., "Improved Standards for the Residual Chlorine Tests," *Water Wks. and Sew.*, **82,** 399 (1935).
6. Scott, R. D., "Refinement in the Preparation of Permanent Chlorine Standards," *Ohio Conf. Water Purif.*, **18,** 39 (1939).
7. Chamberlin, N. S., and Glass, J. R., "Colorimetric Determination of Chlorine Residuals Up to 10 ppm, Part II, Modified Scott Permanent Chlorine Standards," *J. AWWA,* **35,** 1205 (1943).
8. Chamberlin, N. S., "The Determination of High Chlorine Residuals," *Water Wks. and Sew.*, **89,** 496 (Nov., 1942).
9. Griffin, A. E., and Chamberlin, N. S., "Estimation of High Chlorine Residuals," *J. AWWA,* **35,** 571 (1943).
10. Anon. Committee Report, "Control of Chlorination," *J. AWWA,* **35,** 1315 (1943).
11. Chamberlin, N. S., and Glass, J. R., "Colorimetric Determination of Chlorine Residuals Up to 10 ppm, Part I, Production of Orthotolidine-Chlorine Colors," *J. AWWA,* **35,** 1065 (1943).
12. Anon., "Standard Methods for the Examination of Water and Wastewater," 12th ed., American Public Health Assoc., New York (1965).
13. Laux, P. C., "Break-Point Chlorination at Anderson, Indiana," *J. AWWA,* **32,** 1027 (1940).
14. Laux, P. C., and Nickel, J. B., "A Modification of the Flash Color Test to Control Break-Point Chlorination," *J. AWWA,* **34,** 1785 (1942).
15. Hallinan, F. J., "The OTA Test of Residual Chlorine," *J. AWWA,* **36,** 296 (1944).
16. Hallinan, F. J., and Gilcreas, F. W., "Use of OTA Test for Residual Chlorine," *J. AWWA,* **36,** 1343 (1944).
17. Marks, H. C., and Glass, J. R., "A New Method for Determining Residual Chlorine," *J. AWWA,* **34,** 1227 (1942).
18. Marks, H. C., Bannister, G. L., Glass, J. R., and Herrigel, E., "Amperometric Methods in the Control of Water Chlorination," *Ind. and Eng. Chem. Anal. Ed.*, **19,** 200 (1947).
19. Marks, H. C., Williams, D. B., and Glasgow, G. U., "Determination of Residual Chlorine Compounds," *J. AWWA,* **43,** 201 (1951).
20. Marks, H. C., "Residual Chlorine by Amperometric Titration," *J. NEWWA,* **66,** 1 (Mar. 1952).
21. Palin, A. T., "The Determination of Free Chlorine and of Chloramine in Water Using *p*-Aminodimethylaniline," *Analyst* (England), **70,** 203 (1945).
22. Moore, W. A., "The Use of *p*-Aminodimethylaniline for the Determination of Chlorine Residuals," *J. AWWA,* **35,** 427 (1943).
23. Palin, A. T., "The Estimation of Free Chlorine and Chloramine in Water," *Inst. of Water Engrs.* (England), **3,** 100 (1949).
24. Palin, A. T., "Determination of Chlorine Residuals in Water by the Neutral Orthotolidine Method," *Water Wks. and Sew.*, **101,** 74 (Feb. 1954).
25. Palin, A. T., "Determination of Free Chlorine and Combined Chlorine in Water by the Use of Diethyl-*p*-phenylene Diamine," *J. AWWA,* **49,** 873 (1953).

26. Connell, C. H., "Orthotolidine Titration Procedure for Measuring Chlorine Residuals," *J. AWWA*, **39**, 209 (1947).
27. Taras, M. J., "The Micro-Titration of Free Chlorine with Methyl Orange," *J. AWWA*, **38**, 1146 (1946).
28. Taras, M. J., "Colorimetric Determination of Free Chlorine with Methyl Orange," *Ind. and Engr. Chem. Anal. Ed.*, **19**, 342 (1947).
29. Sollo, F. W., Jr., and Larsen, T. E., "Determination of Free Chlorine by Methyl Orange," *J. AWWA*, **57**, 1575 (1965).
30. Black, A. P., and Whittle, G. P., "New Methods for the Colorimetric Determination of Halogen Residuals," *J. AWWA*, **59**, 471 (1967).
31. Johnson, J. D., Overby, R., and Okun, D. A., "Analysis of Chlorine, Monochloramine, and Dichloramine with Stabilized Neutral Orthotolidine," paper presented at 85th Ann. Conf. Amer. Water Works Assoc., Portland, Oregon, June 28, 1965.
32. Johnson, J. D., and Overby, R., "Stabilized Neutral Orthotolidine, (SNORT) Colorimetric Method for Chlorine," paper presented Ann. Conf. Am. Chem. Soc., Detroit, Michigan, April 1965. Publication #163, School of Public Health, Univ. North Carolina, Chapel Hill, NC.
33. Johnson, J. D., and Overby, R., "Stabilized Neutral Orthotolidine Method for Free Available Chlorine," private communication, 1967.
34. Harrington, J. H., "Photo-Cell Control of Chlorination," *J. AWWA*, **32**, 859 (1940).
35. Caldwell, D. H., "Automatic Residual Chlorine Indicator and Recorder," *J. AWWA*, **36**, 771 (1944).
36. Chang, S. L., "Destruction of Microorganisms," *J. AWWA*, **36**, 1192 (1944).
37. Anon., Evaluation of the Stabilized Neutral Orthotolidine Test (SNORT) for Determining Free Chlorine Residuals in Aqueous Solutions, Sanitary Engineering Special Study No. 24-3-68/70 USA EHA, Edgewood Arsenal, MD, 1970.
38. Bauer, R., Phillips, B. F., and Rupe, C. O. "A Simple Test for Estimating Free Chlorine," *J. AWWA*, **64**, 787 (Nov. 1972).
39. Guter, K. J., and Cooper, W. J. "The Evaluation of Existing Field Test Kits for Determining Free Chlorine Residuals in Aqueous Solutions," U.S. Army Medical Environmental Engineering Research Unit, Edgewood Arsenal, MD, Report No. 73-03, Oct. 1972.
40. Guter, K. J., Cooper, W. J., and Sorber, C. A., "Evaluation of Existing Field Test Kits for Determining Free Chlorine Residuals in Aqueous Solutions," *J. AWWA*, **66**, 38 (Jan. 1974).
41. Cooper, W. J., Meier, E. P., Highfill, J. W., and Sorber, C. A., "The Evaluation of Existing Field Kits for Determining Free Chlorine Residuals in Aqueous Solutions, Final Report," U.S. Army Medical Bioengineering Research and Development Laboratory, Technical Report 7402, Aberdeen, MD, April 1974.
42. Meier, E. P., Cooper, W. J., and Sorber, C. A., "Development of a Rapid Specific Free Available Chlorine Test with Syringaldazine (FACTS)," U.S. Army Medical Bioengineering Research and Development Laboratory, Aberdeen Proving Ground, MD 21010, May 1974.
43. Cooper, W. J., Sorber, C. A., and Meier, E. P., "A Rapid Specific Free Available Chlorine Test with Syringaldazine (FACTS)," *J. AWWA*, **67**, 34 (Jan. 1975).
44. Palin, A. T., "A New DPD-Steadifac Method for the Specific Determination of Free Available Chlorine," *J. Inst. Water Engrs. and Scientists*, **32**, 327 (1978).
45. Palin, A. T. "A New DPD-Steadifac Method for Specific Determination of Free Available Chlorine in the Presence of High Monochloramine." *J. AWWA*, **72**, 12 (1980).
46. Wajon, J. E., and Morris, J. C., "The Analysis of Free Chlorine in the Presence of Nitrogenous Organic Compounds," Paper Presented at the Division of Environmental Chemistry, ACS, Anaheim, CA, March 13-17, 1978.
47. Cooper, W. J., Roscher, N. M., and Slifker, R. A., "Determining Free Available Chlorine by DPD Colorimetric, DPD-Steadifac (Colorimetric) and FACTS Procedures," *J. AWWA*, **74**, 362 (July 1982).
48. Cooper, W. J., and Gibbs, R. P., "Equivalency Testing of the Free Available Chlorine Test with

Syringaldazine (FACTS)," U.S. Army Medical Bioengineering Research and Development Laboratory, Fort Detrick, Frederick, MD 21701, July 1, 1982.
49. Kimm, V. J., Director, Office of Drinking Water (WH-550), Directive Sept. 7, 1982.
50. Meier, E. P., Cooper, W. J., and Highfill, J. W., "Evaluation of the Specificity of the DPD–Glycine and FACTS Test Procedures for Determining Free Available Chlorine (FAC)," U.S. Army Medical Bioengineering Research and Development Laboratory, Fort Detrick, Frederick, MD, 1978.
51. Stone, R. W., "Rancho Cordova Breakpoint Chlorination Demonstration," report prepared by the Sacramento Area Consultants, Sept. 1976.
52. Whittle, G. P. and Lapteff, A. Jr., "New Analytical Techniques for the Study of Water Disinfection," paper presented at National ACS meeting, Dallas, TX, April 1973.
53. Gibbs, P. H., Cooper, W. J., and Ott, E. M., "A Generalized Statistical Experimental Design for Comparison Testing of Analytical Procedures," report in process to *J. AWWA*, April 1983.
54. Cooper, W. J., Gibbs, P. H., Ott, E. M., and Patel, P., "Equivalency Testing of Test Procedures for Free Available Chlorine: Amperometric Titration, DPD and Facts," *J. AWWA*, in press.
55. Cooper, W. J., Mehran, M. F., Slifker, R. A., Smith, D. A., and Villate, J. T., "Comparison of Several Methods for the Determination of Chlorine Residuals in Drinking Water," U.S. Army Medical Bioengineering Research and Development Laboratory, Fort Detrick, Frederick, MD 21701, April 1982.
56. Carns, K. E., private communication, East Bay Municipal Utility District, Mgr. Water Quality, Jan. 13, 1983.
57. Jopling, Wm., "Measurement of Chlorine Residuals," *Calif. Water Poll. Control Bulletin*, p. 25 (Jan. 1972).
58. Brooks, A. S., private communication, Univ. Wisconsin, Milwaukee, WI, Oak Ridge Conference, Oct. 23, 1975.
59. Marks, H. C., Joiner, R. R., and Strandskov, F. B., "Amperometric Titration of Residual Chlorine in Sewage," *Water and Sew. Wks.*, **95**, 175 (May 1948).
60. Hulbert, R., "Chlorine and the Orthotolidine Test in the Presence of Nitrite," *J. AWWA*, **26**, 1638 (1934).
61. Hallinan, F. J., "Thiosulfate Titration for Determining Chlorine Residuals," *J. Am. Chem. Soc.*, **61**, 265 (1939).
62. Willson, V. A., "Determination of Available Chlorine in Hypochlorite Solution by Direct Titration with Sodium Thiosulphate," *Ind. and Engr. Chem. Anal. Ed.*, **7**, 44 (1935).
63. Palin, A. T., "A Study on the Chloro-Derivatives of Ammonia and Related Compounds With Special Reference to Their Formation in the Chlorination of Natural and Polluted Waters," *Water and Water Engineering* (England), p. 151 (Oct. 1950); p. 189 (Nov. 1950); p. 248 (Dec. 1950).
64. Kolthoff, I. M., and Lingane, J. J., *Polarography*, Vols. 1 and 2, 2nd ed., chapter 47, Interscience, New York, 1952.
65. Salomon, E., "On a Galvanometric Titration Method," *Z. Physik. Chem.*, **24**, 55 (1897); **25**, 366 (1898); *Z. Elektrochem.*, **4**, 71 (1897).
66. Nernst, W., and Merriam, E. W., "Amperometric Titrations," *Z. Physik. Chem.*, **53**, 235 (1905).
67. Foulk, C. W., and Bawden, A. T., "A New Type of End Point in Electrometric Titration and Its Application to Iodimetry," *J. Am. Chem. Soc.*, **48**, 2045 (Aug. 1926).
68. Morrow, J. J., "Residual Chlorine Determination with Dual Polarizable Electrodes," *J. AWWA*, **58**, 363 (Mar., 1966).
69. Dowell, C. T., and Bray, W. C., "Experiments with Nitrogen Trichloride," *J. Amer. Chem. Soc.*, **39**, 896 (1917).
70. Stock, J. T., *Amperometric Titrations*, Interscience, New York, 1965.
71. Williams, D. B., private communication, 1968.
72. Williams, D. B., "Elimination of Nitrogen Trichloride in Dechlorination Practice," *J. AWWA*, **58**, 248 (1966).

73. Palin, A. T., "Determination of Nitrogen Trichloride in Water," *Proc. Soc. for Water Treatment and Examination,* **16,** 127 (1967).
74. White, G. C., unpublished reports on operation and maintenance of Wallace and Tiernan Agene Gas (NCl$_3$) systems used in flour mills for flour bleaching, 1937–1941.
75. Baker, R. J. "Measurements of Chlorine Compounds," paper presented at Electric Power Research Institute and Maryland Power Plant Siting Program, Johns Hopkins Univ., Baltimore, MD, Jan. 16–17, 1975.
76. Baker, R. J., private communication, Wallace and Tiernan Div., Pennwalt Corp., April 1975.
77. White, G. C., Beebe, R. D., Alford, V. F., and Sanders, H. A. "Problems of Disinfecting Nitrified Effluents," *Municipal Wastewater Disinfection,* Proc. of Second Ann. Nat. Symposium, Orlando, Fla., EPA Report 600/9-83-009, July 1983.
78. Baker, R. J., private communication, Wallace and Tiernan Div., Pennwalt Corp., Jan. 13, 1984.

6
Chlorination of Potable Water

PURPOSE

Just as water is close to being a universal solvent, so chlorine is nearly a universal water treatment chemical.

The primary objective of water supply chlorination is disinfection. Because of chlorine's oxidizing powers, it has been found to serve other useful purposes in water treatment, such as taste and odor control, prevention of algal growths, maintaining clean filter media, removal of iron and manganese, destruction of hydrogen sulfide, color removal by bleaching of certain organic colors, maintenance of distribution system water quality by controlling slime growths, restoration and preservation of pipeline capacity, restoration of well capacity, main sterilization, and improved coagulation by activated silica.

None of the so-called alternatives to chlorine can compete with its versatility.

HISTORICAL BACKGROUND

Use as a Germicide

Chlorine was first introduced to water treatment as a disinfectant about the turn of the century. Since that time it has become by far the predominant method used for this purpose. This popularity is deserved because of its potency and range of effectiveness as a germicide. It is easy to apply, measure, and control; it is free from toxic or physiological effects; it persists reasonably well; and it is relatively inexpensive. Other agents may equal or even excel aqueous chlorine in any one of these characteristics, but there is none that combines them in such an advantageous way. Ozone, bromine, iodine, chlorine dioxide, silver ions, ultraviolet, and ultrasonics have been investigated. Some of these have found important uses in special situations, but none of them so far has been a serious competitor of chlorine.

As a result, of all the municipal water supplies that are being chemically disinfected, at least 99 percent use chlorine.[1] Very few processes enjoy this kind of monopoly. Even in countries where ozone is preferred, theoretically for physiological reasons, chlorination is almost universally employed in practice as an adjunct to ozone.

Probably the first known use of chlorine as a germicide was by Semmelweis, when in 1846 he introduced the scrubbing and cleansing of hands in chlorine water between contacts with each mother after he found that child bed fever (puerperal) was being transmitted from mother to mother in the maternity wards of the Vienna General Hospital. Then, in 1881 the German bacteriologist Koch demonstrated under controlled laboratory conditions that pure cultures of bacteria could be destroyed by hypochlorites. The earliest record of a suggestion to chlorinate water, even before water was known to be a carrier of disease germs, is a statement by Dr. Robley Dunlingsen in his *Human Health,* published in Philadelphia in 1835: "To make the water of the marshes potable, it has been proposed to add a small quantity of chlorine or one of the chlorides in small but sufficient amounts to destroy the foulness of the fluid."[2] Who originally made this proposal is not known. That water is a mode of transmission of disease has been known for only a little more than a hundred years. Dr. John Snow first theorized in 1849 that water was the mode of communication of cholera.[3] He was later able to demonstrate his theory in the Broad Street pump episode in London in 1854 by removing the pump handle, thereby staying an epidemic of cholera that had already claimed some five hundred lives. Dr. Snow also found the source of the infection: a soldier recently returned from active service in India, residing in a rooming house adjacent to the broken sewer which contaminated the water drawn from the Broad Street pump, was a carrier of cholera.[4]

Early Uses in Water Treatment

The first use of chlorine as a continuous process in water treatment was probably in the small town of Middelkerke, Belgium, in 1902. This was known as the ferrochlor process. Ferric chloride for coagulation was mixed with calcium hypochlorite, which hydrolyzes to form a ferric hydroxide floc and hypochlorous acid as the disinfectant.[5] At Ostende, Belgium, in 1903, chlorine gas was generated by mixing potassium chlorate and oxalic acid. These installations, designed by Maurice Duyk, a chemist for the Belgian ministry of public works, are the earliest known continuous applications of chlorine for disinfection of potable water.

Chlorine in the form of hypochlorites (bleaching powder and chloride of lime) was first used as a temporary expedient to stay typhoid epidemics. The first recorded use was in 1896 when a typhoid epidemic occurred at the Austria-Hungary naval base of Pola on the Adriatic Sea.[2] In 1897 Sims Woodhead used a solution of bleaching powder for disinfecting Maidstone, England, water supply during a typhoid epidemic.[6] The first known continuous use of chlorine in England was at Lincoln in 1905.[7,8] A serious typhoid fever epidemic had occurred at Lincoln late in 1904, and was traced to the water supply. The advice of A. C. Houston, bacteriologist, and G. McGowan, chemist to the Royal Commission on Sewage Disposal, was sought. The source of the water supply was the River Witham and two small tributaries. Houston and McGowan recommended that the water be treated with chloros (alkaline solution of sodium hypochlorite of about 10 to 15 percent available chlorine), the most easily available chlorine compound at the time.[8] Treatment began on February 11, 1905, under the direction of D. B. Byles, starting with a dose of 10 ppm and finally settling on one of 1 ppm. The amount of water treated was 1 to 1½ mgd. This treatment continued until 1911, when a new supply of water came into use.

First Uses in United States

The introduction of chlorination in the United States occurred on May 22, 1888, when patents on electrolytic treatment of water were issued to Albert R. Leeds, Professor of Chemistry at Stevens Institute of Technology at Hoboken, N.J.[2] Webster's British patent of January 27, 1887, was the forerunner and possibly the model of later patents issued in Great Britain and the United States for the use of electrode-generated current to disinfect potable water and wastewater. Leeds' theory of purifying water was to remove organic impurities found in natural waters or wastewaters which cannot be removed by mechanical filtration or chemical precipitation by treating the water with the gases obtained by the decomposition of water containing an acid or a salt in solution, the decomposition being effected by use of an electric current. The acid employed was to be either hydrochloric, nitric, phosphoric, chromic, or sulfuric, or the salts of these acids, or a mixture of these acids or salts. Leeds' thought that the best results would come from the hydrochloric acid, or its salt sodium chloride. So far as is known, Leeds did not follow up his patent with any operating installations.

Also in 1888 a patent was granted to Omar H. Jewell and his son William M. Jewell, both of Chicago, for a combination of electrodes with a mechanical filter. The patent was assigned to the Jewell Pure Water Company. The apparatus consisted of electrodes mounted vertically in the dome of a pressure sand filter. Passage of electric current through these electrodes and a solution of carbonic acid was supposed to produce an insoluble blanket of sodium bicarbonate on top of the filter sand to remove material that the sand could not. Nothing came of this device, which could not be demonstrated to be of any value.

Some years later William Jewell designed another electrolytic process which was used by George W. Fuller in January, 1896 for experimental work at Louisville, Kentucky. Twelve of Jewell's electrolytic generators were installed. These units required ten pounds of salt to produce one pound of chlorine. This installation was capable of providing a chlorine dosage of 0.25 mg/1 into the filtered effluent. This same device was used again in 1899 on an experimental basis at Adrian, Michigan. Jewell experimented briefly with chlorine gas by this process but abandoned it in favor of chloride of lime.

In 1893 Albert E. Woolf used an electrolytic process, similar to the one used by Leeds, to generate chlorine in the sewage effluent at Brewster, New York. Disinfection was required here because the discharge was into the east branch of the Croton River, a part of the New York City water supply. This chlorination system operated successfully for 18 years, until it was destroyed by fire in 1911.

The first notable and successful chlorination installation in the United States is credited to George A. Johnson at the Bubbly Creek Filter Plant in Chicago in 1908, to serve the Chicago stockyards.[8-10] The raw water here contained a large amount of sewage and treatment consisted of filtration in conjunction with copper sulfate application. The large stock suppliers complained that animals drinking this water gained less weight than when city water was supplied to them.

Under pressure of a law suit brought by the city of Chicago against the Union Stockyard Company, the contractors who built the filter plant were enabled to fulfill their guarantee when Johnson substituted chloride of lime for the copper sulfate treatment. The rate of application was 45 pounds of chloride of lime (approximately 30 percent available chlorine) per mg seven and one-half hours before filtration. The treated water showed better bacteriological results than the city water: Bubbly Creek water showed 0.34 percent cases of $E.$ $coli$ versus 12.8 percent for city water.

The Jersey City Case

Again in the same year, Johnson, who was a consulting engineer for the firm headed by George Fuller (Herring and Fuller), was hired by Dr. Leal to chlorinate the Jersey City water supply.[6,11] This installation is significant because it followed litigation over the quality of the new water supply. Jersey City had entered into a contract with the builders, who later identified themselves as the Jersey City Water Supply Company, to provide a new water supply. This work consisted mainly of constructing a large reservoir to impound water from the Rockaway River, a tributary of the Passaic, and a pipeline to carry 40 mgd of water from the reservoir at Boonton to Jersey City, twenty-three miles away. The contractors fell into financial difficulties, and the project had to be completed by the bonding company five years after the contract was awarded in 1899. Prior to its completion in 1904, Jersey City started a suit against the Jersey City Water Company, claiming that the work was not being performed according to the specifications, which read in part:

> The water to be furnished must be pure and wholesome for drinking and domestic purposes; and the works shall be so constructed and maintained by the contractor that the water delivered therefore shall be pure and wholesome and free from pollution deleterious for drinking and domestic purposes.

The city's claim was based on the pollution in the reservoir, which occurred two or three times a year during high water caused by storm runoff. The city claimed that the pollution could be corrected by a filtration plant. The defendants countered that if proper sewage disposal facilities were constructed by the city, this pollution would not occur.

The court rendered an opinion that the construction of a filter plant was not the responsibility of the defendants, and suggested, not as its own opinion but from evidence given by the city, that the water could be made to comply with the contract by the construction of sewers and sewage disposal works for various towns in the watershed.[6]

Dr. Leal was convinced that this would not entirely eliminate the pollution, and strongly advised that the company be allowed to suggest to the court its own method for the complete fulfillment of the contract requirements. He pointed out that the greater percentage of bacteria and *B. coli* found at the point of delivery in Jersey City was due primarily to the washing of soils, roads, streets, and manured fields rather than from any sewage contamination.

Leal's request was granted in a court decree filed June 4, 1908. Dr. Leal had in mind the electrolytic process of making chlorine, and on June 16, 1908, hired the firm of Herring and Fuller to build and operate a hypochlorite plant under the direction of G. A. Johnson. The plant itself was started on September 26, 1908. It was an enormous and cumbersome affair, as might be expected, consisting of three 10,500 gallon concrete chloride of lime solution storage tanks as well as a large building at the Boonton reservoir outlet.[11] After several months of operation, the dosage was reduced from approximately 1.0 to 0.2 ppm, requiring a high of 0.35 ppm during times of high water. Bacterial reduction in the raw material was from 200 to 20,000 per cc to from 20 to 30 per cc in the delivered water.

After several months of practical operation, extended testimony was taken in the court of chancery by ex-Chancellor Magie from a score of the foremost experts of that time. In his opinion, handed down in May, 1910,[11] Magie said:

> From the proofs taken before me, of the constant observations of the effect of this device, I am of the opinion and find that it is an effective process which destroys in the water the germs, the presence of which is deemed to indicate danger, including the pathogenic germs, so that the water after this treatment attains a purity much beyond that attained in water supplies of other municipalities. The reduction and practical elimination of such germs from the water was shown to be substantially continuous.

In the search for witnesses in this case, a toxicologist was sought to declare the process of chlorination a poisonous one. No one could be found who had this opinion.

It is interesting to note that the expert testimony developed the following theory of chlorination chemistry: Hypochlorous acid formed in the reaction of chloride of lime and water liberates oxygen in a very active state and leaves hydrochloric acid. The latter drives off the weaker carbonic acid and unites with calcium to form calcium chloride. Hypochlorous acid will form whether or not the process is electrolytic production of molecular chlorine or hydrolysis of chloride of lime.

However, the nascent oxygen theory of the killing action of chlorine presented in testimony in this case prevailed as the accepted theory until 1944, when it was finally disproved by S. L. Chang. (See Chapter 4.) The nascent theory was also responsible for the popularity of ozone.[5] Another point of interest developed was the fear of producing "free chlorine." This term referred to molecular chlorine and not hypochlorous acid, which is present-day terminology for free chlorine. This fear apparently stemmed from the knowledge of the toxicity of molecular chlorine, particularly in its contact with the respiratory tract of man.

The ruling by Magie in the Jersey City case gave chlorination a much needed boost but did not settle the argument as to the efficiency or desirability of chlorination as a water treatment process. There was a great deal of public sentiment as well as that of engineers against the idea of chlorination of potable water.[5,12] Many prominent persons in the water works industry proposed outright abandonment of a water supply source if it had to be disinfected, regardless of the method. These same persons were also against any kind of pollution, for which they must be commended. In the meantime, they gained considerable support from the public, who took great pride in a pure and wholesome water supply. This attitude was largely responsible for the revival of the Jersey City case some ten years after chlorination had been in successful operation.

In March, 1919, Jersey City, still irked by the two decisions against them in 1908 and 1910, appealed to the highest state court, the Court of Errors and Appeals, contending that chlorination does not afford the remedy claimed for it, asserting that in cold weather, "bacteria are stunned but not killed." Expert testimony and scientific evidence presented by the defendant clearly demonstrated to the court that chlorination was a reliable disinfectant that did in fact effectively destroy pathogenic organisms under all weather conditions.

Prior to the use of chlorine for disinfection and after the discovery of bacteria in water—thus making water as a carrier of disease—filtration was investigated as a practical means of mechanically removing these bacteria. This was accomplished at first by allowing the water to slowly settle through formations of specially graded sand, which trapped some of the bacteria but not all. Filtration became greatly improved by chemical precipitation and sedimentation prior to sand filtration. The first known use of a municipal filtration system was the one built in London in 1829.[11] The apparent success of the filtration system in Hamburg, Germany, during the 1892 cholera epidemic convinced sanitationists that filtration should be seriously considered as a treatment process for polluted water. Sand filter plants were built as early as 1874 at Poughkeepsie, New York. George Johnson, one of the outstanding proponents of chlorination for disinfection, stressed the inadequacy of filtration as the sole means of treatment for polluted water, and was able to amply demonstrate his opinions by scientific data collected at the Bubbly Creek filter plant in Chicago.[10,11]

Subsequent statistical evidence made public health experts aware of the necessity for establishing disinfection as a process either by itself or in conjunction with filtration or to abandon polluted supplies altogether. The statistics included data not only on death rates from typhoid fever but on the general community health.

Typhoid Fever and Waterborne Outbreaks

The death rate from typhoid fever became a yardstick in the United States for public health authorities. Just before the turn of the century it was firmly established that this disease could be waterborne.

Military campaigns often produced virulent outbreaks. In the American Civil War, typhoid casualties were estimated at more than 75,000 cases, with more than 27,000 deaths. In the Franco-Prussian War, typhoid accounted for more than 73,000 cases, with more than 8,000 deaths; this latter death toll was higher than that from combat. The British suffered severe outbreaks in the Boer War, totaling more than 57,000 cases, with some 8,000 deaths.[13] A fateful individual case was that which took the life of Albert, Prince Consort to Queen Victoria in 1861.

The French and German armies adopted vigorous sanitation procedures as a result of their typhoid experience, so that on the eve of World War I the typhoid morbidity was down to eight cases per 100,000. It was in the Russo-Japanese War of 1904-5 that for the first time troops went to the front with units for boiling drinking water as a method of disinfection. By this time it had been established that boiling for a given length of time would kill the causative organism. In 1910 Whipple estimated that causes of typhoid fever could be grouped as: waterborne 40 percent; food 25 percent; ordinary contagion, including fly transmission 30 percent; shellfish and all others 5 percent.[8]

In 1900 the average death rate from typhoid in the United States was 36 per 100,000 population. Thus, more than 25,000 persons died of typhoid fever that year. The death rate dropped to twenty per 100,000 in 1910[13] and to three per 100,000 in 1935.[18] By 1960, less than twenty persons died from typhoid fever in the entire United States.[14]

Between 1920 and 1936, 470 waterborne outbreaks of typhoid fever in the United States and Canada caused nearly 1,200 deaths and illness to approximately 125,000 persons. During this same period, there were reported more than 16,000 cases of waterborne typhoid fever, more than 100,000 of gastroenteritis, more than 200 of bacillary dysentary, 1400 of amoebic dysentery, and 28 of jaundice. More than 30 percent of the United States outbreaks originated in well water. The cases of amoebic dysentery and jaundice were unique in water supply and public health history.[15]

The principal causes of these outbreaks, in order of magnitude, were: surface pollution of shallow wells; cross-connection with a polluted supply; contamination of spring or infiltration gallery by pollution of a watershed; contamination of a brook or stream by pollution on a watershed; use of polluted water from a river or irrigation ditch without treatment; inadequate chlorination as the only treatment; inadequate control over the filtration and chlorination process; seepage of surface water or sewage into a gravity conduit.

One of the most notable examples of chlorination of water supply for the safeguarding of public health is reflected in the efforts made by the Kansas State Health Department.[16] Between 1908 and 1917, at least thirty Kansas towns started using hypochlorite solutions for disinfection. In 1917, chlorine gas equipment was installed at Coffeyville and Valley Falls; by 1919, Atchison, Leavenworth, Herington, Independence, Iola, Kansas City, Ottawa, Sabetha, Topeka, and Wichita followed suit, by 1942, all sixty of the surface water supplies and forty of the ground water supplies in the state were so treated. Then in 1942, a disastrous waterborne bacillary dysentery epidemic of 3000 cases occurred in Newton, Kansas. Eventually, in 1956, the state health department ordered the chlorination of all public water supplies. By 1960, 462 of the 463 public supplies were being chlorinated.

Another factor of great importance in favor of safeguarding a water supply by chlorination was the tremendous cost to the cities in damages resulting from these waterborne epidemics. One example was the 1929 typhoid fever epidemic at Olean, New York, which cost the city $350,000.[15] Chlorination proved, among other things, to be cheap insurance.

The Mills–Reincke Phenomenon and the Hazen Theorem

As early as 1893, two public health researchers, Mills and Reincke, discovered after studying a great many communities that when a polluted water supply was replaced by a pure supply, the general overall health of the community was greatly improved—to a greater degree than could be accounted for by the reduced prevalence of typhoid fever and other recognized typical waterborne diseases. This discovery became known as the Mills-Reincke phenomenon. In 1903, Allen Hazen, a pioneer in the water works industry, discovered that when a community water supply was changed from bad to excellent by adequate treatment, for every person saved from death by typhoid three other persons would be saved from death by other causes, many of which were probably never thought to have any connection with, or to be especially affected or influenced by, the quality of the public water supply. This change in the death rate was known as the Hazen theorem. Therefore, disinfection of public water supplies goes farther than just the control of waterborne diseases.

WATERBORNE DISEASES

The sole purpose of disinfection of potable water is to destroy pathogenic organisms and thereby eliminate and prevent waterborne diseases. The following diseases

Typhoid Fever

This disease was once a scourge that destroyed armies more effectively than weaponry and a pestilence that haunted towns because of its mystery. Typhoid masked itself in symptoms common to other ills and was often mistaken for typhus or was classified merely as one of many fevers. In North America it has been responsible for more sickness and death than any other waterborne disease. One frightening aspect is that the causative organism may be carried for a lifetime by an individual who has been infected but has recovered.* Not everyone who has contracted the disease and recovered becomes a carrier, but a great many do.

The causative organism of typhoid fever, *Salmonella typhi,* was discovered by Karl J. Eberth in 1880 (in some textbooks, it is referred to as *Eberthella typhosa*). This durable organism resides and proliferates solely in the intestines of man (not animals). It is readily destroyed by heat or such disinfectants as chlorine, iodine, bromine, and ozone. Under natural conditions in soil, water or feces, it can remain alive for weeks or months. Beard[19] confirmed that typhoid can survive up to five months in frozen debris and ice. When the ice was allowed to melt, the water became infected again.[19] Typhoid bacilli enter the body through the mouth, invade the mucosa of the small intestine, and traverse the intestinal lymphatic and mesenteric nodes to reach the blood. Subsequent proliferation occurs in the liver, gall bladder, biliary tract, and duodenum, with particular concentration in the lymphoid tissue of the small intestine mucosa. The bacillus is discharged in both the urine and feces, with stools becoming bacteriologically positive about two weeks after onset.

Since the organisms can retain their virulence and remain viable for long periods of time, any water supply contaminated by sewage is potentially a carrier of the disease if a carrier of the causative organism contributes to the sewage. Gradually over the years as the spread of this organism has been controlled by chlorination and other means of disinfection (boiling, iodine, ozone), the number of carriers

---

* America's best-known typhoid carrier was Mary Mallon, otherwise known as Typhoid Mary, who came to symbolize a fearsome drama in the public mind. She worked as a cook for several upper-class New York families before being tracked down in 1907 as a carrier responsible for several outbreaks of the disease. When her feces were examined bacteriologically, it was found that she had practically a pure culture of the typhoid bacterium, *Salmonella typhi.* She remained a carrier for many years, probably because her gall bladder was infected and organisms were continuously being excreted from there into her intestine. She was imprisoned for 3 years when she refused to have her gall bladder removed. She was released from prison when health authorities accepted her pledge that she would not cook or handle food for others and that she would report to the health department every three months. She disappeared, changed her name, and cooked in hotels, restaurants, and sanitariums, leaving behind a wake of typhoid fever victims. After 5 years she was captured as a result of the investigation of an epidemic at a New York hospital. She was promptly arrested and imprisoned. She remained in prison for 23 years where she died in 1938, 32 years after she was discovered to be a typhoid carrier.[11,13]

has been diminished. The incidence of typhoid in public water supplies is practically nil. In 1981 the CDC in Atlanta, Georgia did not report a single case. When a case is reported it is usually a foreign traveler. Between 1946 and 1976, 610 cases were reported; 507 of these cases were traced to private water supply systems.[20] Continued diligence is necessary because the populace has not developed any natural resistance to *Salmonella typhi*.

Typhoid fever has put a strange blotch across the pages of history. It was a scourge that destroyed armies more effectively than weapons. This pestilence haunted towns because of its mystery. It masked itself in symptoms common to other illnesses. In the 19th Century mortality rates rose as high as 300 per 100,000. Military campaigns produced virulent outbreaks. In the American Civil War there were more than 27,000 deaths from typhoid, and in the Franco-Prussian War there were more than 8,000 deaths. The death toll in the latter war was more by typhoid than by gunfire.

Cholera

Cholera was originally known as Cholera Morbus in its death-dealing form. It has been known in India since about 400 BC. The name was known the world over because of the magnitude of its attack rate which resulted in a high mortality rate. It ravaged entire cities. The organism stayed in India until 1817, when it escaped for the first time.[21] It spread at the leisurely pace of human travel in those times to China, Russia, Europe, and Britain. It took some 20 years to cross the Atlantic into North America. It came to Britain in 1831 and killed 21,000 people. There have been several epidemics in London caused by returning British soldiers who became carriers after contracting the disease while on military duty in India.[21]

In the 20th Century it has spread to the Philippines, Indonesia (Sulawesi province), the Middle Eastern countries, Africa, and Italy.

At one time cholera was common in Europe and North America, but the disease has been virtually eliminated in those areas by effective water treatment practices.

The causative organism was originally called *Vibrio comma* because of its shape. It is now labeled *Vibrio cholerae*. There is a new vicious strain of cholera organism, known as the El Tor *Vibrio*. It is slightly different from other strains. It is named after the Arab refugee camp in the Sinai Peninsula, which was hit by the disease in the 1960s. The El Tor cholera pandemic that reached Naples, Italy in 1973, began on the island of Sulawesi (Celebes) in 1960 and 1961. Then it turned up in the Philippines, and two years later in Korea, where it spread to China.

In 1981 there were 913 culture-confirmed cases of cholera El Tor, serotype Ogawa, which occurred in Bahrain. The overall attack rate was 27 per 10,000.[21] This outbreak was caused by drinking water. The Naples outbreak was caused by shellfish contaminated by sewage containing the El Tor organism.

Houston found in the 1930s that the virulence of *V. cholerae* is reduced 99 percent by one week's storage in water, and that the organism is easily killed by chlorine.[23]

The countries affected by cholera could easily wipe out this haunting scourge simply by practicing better hygiene, improved sanitation, and more effective water treatment practices.

Amoebic Dysentery

Often referred to as amebiasis, this disease is a potential threat throughout the world. It is terrifying because it is almost impossible for a patient to experience complete recovery unless correct diagnosis and proper medical attention are immediately available. This disease was one of the most feared by our armed forces serving in the Pacific during World War II. The causative organism (*Entamoeba hystolytica*) is a cyst, and can therefore lie dormant in the intestinal tract and go undetected by the carrier for long periods of time, only to initiate a recurring illness without warning. The best protection against this disease is removal of the organism from water supplies by filtration. *E. histolytica* is large enough to be trapped by sand filters; high doses of chlorine (10 ppm) and a long contact time (one hour or more) are required to effect disinfection. One of the most notable of the several outbreaks of this disease in the United States was at the Congress Hotel in Chicago during the 1933 World's Fair, when the water system became contaminated with sewage through defective flushing valves on the toilet bowls. The most recent outbreak in the United States was in 1955.[25]

Bacterial Gastroenteritis

There are sporadic outbreaks of this intestinal disorder which produces a wide variety of symptoms, such as nausea, vomiting, diarrhea, abdominal cramps, fever, and headache. The organisms usually responsible for such waterborne outbreaks belong mainly to four genera of organisms: *Shigella, Salmonella, Campylobacter,* and *Yersina.*

Shigellosis. The genus *Shigella* include *S. dysenteriae, S. paratyphi* (both responsible for paratyphoid in man), *S. shiga, S. sonnei,* and *S. flexneri.* The mode of transmission of these bacteria is by the fecal–oral route, which means they can be waterborne. In the period 1946–1975 there were 49 outbreaks of shigellosis in the U.S.A. which affected 10,869 people. The most common serotype isolated in these outbreaks was *S. sonnei.* The species *S. flexneri* has been isolated from patients following an outbreak on a cruise ship. This too was traced to the ship's water supply.[26] This genus is also responsible for most foodborne outbreaks of gastroenteritis. In general these bacteria are not long lived in the environment, so they are rarely isolated from bathing beaches or large bodies of water.

Salmonellosis. Salmonellosis is the generic name of the diseases caused by organisms of the genus *Salmonella*. These organisms are most usually pathogenic to man and warm-blooded animals. The genus *Salmonella* contains over 1,000 distinct types having different antigenic specificities in their O antigens.

Until the Riverside, California outbreak in 1965, it was thought that *Salmonella typhi* (typhoid fever, described above), *Salmonella paratyphi*, and *S. schottmuelleri* (paratyphoid) were the only significant organisms of the genus *Salmonella* capable of carrying waterborne diseases. This outbreak was an epidemic of acute gastroenteritis that affected 18,000 people. There were three deaths. After a lengthy investigation it was concluded that the causative organism was *Salmonella typhimurium*, which was transported through the city water system.[27-29] The water supply was underground well water. Until this outbreak occurred it was generally believed that *Salmonella* species were only associated with food poisoning (excepting *S. typhi*).

The dominant organism in food poisoning is *S. enteriditis*. There are so very many strains of *Salmonella* that it is difficult to identify the particular strain in a foodborne outbreak that might be associated with a contaminated water supply. The CDC listed 18 foodborne *Salmonella* outbreaks affecting 1573 people in 1975 and a similar number of outbreaks in 1974 affecting 5499 people. The common factors in these outbreaks indicate that food prepared away from home has a potential for causing gastroenteritis.

Toxigenic *E. coli*. This bacterial pathogen has been found to cause travelers' diarrhea in Mexico, commonly called "Montezuma's Revenge." Median duration of the illness is about 5 days and occurs about 6 days after the travelers arrive in Mexico. There are, of course, other contributing pathogens such as *Shigella* and *Salmonella*. Enterotoxigenic *E. coli* has also been found to be responsible for diarrhea in travelers to Brazil and Kenya and in persons living in Brazil, Asia, Japan, and the U.S.[30] The World Health Organization believes that 80 percent of all intestinal illnesses in Third World countries is due to lack of sanitation and proper hygiene. The highest disease rate and death rate is from diarrhea, which is commonly waterborne.[31] The causative agent is most likely enterotoxigenic *E. coli*. In 1960 the diarrhea death rate in New York state was 2.4 per 100,000; in India it was 312 per 100,000.[31] Death from diarrhea is thought to be the result of dehydration. Other symptoms produced by this coliform pathogen include abdominal cramps, nausea, headache, vomiting, and fever.

In all cases investigated, the primary cause is either poor or no sanitation, absence of disinfection, malfunction of equipment, or improper design.

***Campylobacter* Enteritis.** Outbreaks of this disease have been reported as waterborne.[22,32] One in Sweden during October 1980 that affected about 2000 people was found to be caused by the species *C. jejuni*. This organism was isolated in the stools of patients. Neither food nor drink was found to be a common factor.

Epidemiologic evidence indicated that the infection was spread via the water mains, but the total coliform count gave no indication of fecal contamination.

The first outbreak of this disease occurred in 1978 in Bennington, Vermont. There were 3000 cases of diarrheal illness representing 19 percent of the population. Typical symptoms were abdominal cramps, diarrhea, and headache. The illnesses were associated with drinking water from the town water supply, which obtained unfiltered, but chlorinated, water from a nearby brook. *C. fetus* and *C. jejuni* were cultured from the feces of 42 percent of the cases, but not from 23 people who served as controls. Oddly enough, coliforms were not found in the water at the time of the outbreak. However, after chlorination dosage was increased and a boil-water order was issued, no further cases were reported. Another outbreak occurred in Britain at a boarding school, affecting 234 students and 23 staff members. The causative agent was again *C. jejuni,* which was traced to an unchlorinated water supply.

Yersina Enteritis. An outbreak of gastroenteritis which affected 87 persons was reported from the state of Washington in January 1982.[32] An investigation identified *Y. enterocolitica* as the causative agent. The water supply used for the packing of soybean curd (tofu) was found to contain the causative organism. A water purification system which included disinfection was required to eliminate this health problem.

Pseudomonas. *Pseudomonas aeruginosa* is considered to be an opportunistic pathogen. It is commonly found in water systems. It is primarily a soil organism, and frequently grows in filter systems. It has been known to cause gastroenteritis in newborn babies.[33] This "nonpathogenic" bacterium is a particular problem in hospitals, where infections are acquired from catheterizations, trachostomies, and intravenous intrusions. An epidemic of diarrhea in a hospital premature-baby nursery was traced to *Pseudomonas* bacteria in an aerating cap on a spigot used by ward personnel.[34,35] This raises an important point about the role of disinfection in water supplies: The advent of antibiotics has broadened or increased the range of conditions in which gram negative bacteria, such as *Pseudomonas,* may become pathogenic. Therefore the presence of these organisms in water supplies assumes greater importance than it has in the past because of the increased virulence of these organisms.

Schistosomiasis

This serious parasitic disease is produced by a blood fluke, such as *Schistosoma mansoni,* which lives in human abdominal veins and expels eggs through the urine or feces. Within a few hours after discharge into fresh or brackish water, these ova mature into a free-swimming embryo known as *Miracidia* which escape the ova and then seek and penetrate a snail host. Within the tissues of the snail they grow and reproduce large numbers of larvae known as *Cercariae.* Literally thousands of these larvae emerge from a single snail each morning when the body of water in which the snail is living is warmed by the sun. The *Cercariae* are considerably larger than the *Miracidia*, can swim vigorously but not very far, but can be carried a long way by water currents. They tend to remain near the water surfaces, where they can easily make contact with swimmers or waders. Upon contact with human skin the *Cercariae* attach themselves by means of suckers and then start a process of penetration. Within hours they are in the blood stream and are carried to the liver, where they grow to maturity in a few weeks. They mate and move to small blood vessels in the walls of the intestine or bladder.

It is estimated that this disease may affect some 200 million persons who live in tropical climates throughout the world.[36] It is thought that the disease can be contracted by drinking water contaminated with the *Cercariae*. This is a serious threat to public health because of the possibility of its being widespread by immigration of persons from areas where the disease is endemic. Once infected with the blood flukes, a victim never gets rid of the disease but finally succumbs after his vital organs are incapacitated. This cycle may take many years of gradual debilitation of the patient, who can function adequately most of the time and who is always a carrier of the potential infection to others. Chlorination appears to provide the most effective treatment of drinking and bathing water, but its application to streams or other uncontrolled waters is not practical.[38] Control may be best accomplished by breaking the life cycle of the *Schistosome* through environmental control, the principles of which include: preventing discharge of *Schistosome* ova into water courses; applying molluscicides to rid streams and ditches of snail hosts; killing the *Cercariae;* and protecting populations from contact with infected waters.[36,37]

Giardia

The flagellated protozoan *Giardia lamblia* is responsible for the disease giardiasis in man. These parasitic flagellates are distributed worldwide and are the most commonly reported human intestinal parasites in the United States and Great Britain.[39,40] This is a debilitating febrile disease. The fever is accompanied by severe cramps, headache, and diarrhea. Children are more vulnerable than adults.[41] This organism has eight flagella and a ventral sucker which it attaches to the intestinal mucosa. This sucker causes diarrhea in the human species. Giardia cysts are carried by dogs, cats, and wildlife in addition to man. The disease has been referred to as "beaver fever" because it is thought to be spread by beavers who make their homes in the watershed. The primary source of *Giardia* infestation, however, is human feces. Giardiasis is found all over the world, primarily in places that have primitive hygienic practices and poor sanitary conditions. It is believed that leaking septic tanks and poorly treated sewage together with the proliferation of backpackers and campers has been responsible for the widespread outbreaks of giardiasis from coast to coast in the United States. People have passed it along by practices as diverse as having poor hygiene, participating in oral and anal sex or drinking polluted water.[23]

In the years 1970–1980 outbreaks of giardiasis have tended to occur in small mountain communities in the Northeast, Pacific Northwest, and Rocky Mountains. However, in the 1980s larger communities have suffered outbreaks—Wilkes-Barre, Pennsylvania and Reno, Nevada.[23] More recently *Giardia* cysts have been found in the San Francisco water supply at the source. This is one of the highest quality waters in the U.S.A. It derives from melted snow high in the Sierra Nevada mountains adjacent to Yosemite National Park 185 miles east of San Francisco. This discovery is a stern warning that the natural barriers against transmission of waterborne diseases have finally been broken. This means that the pristine mountain water supply is a thing of the past and that disinfection of these supplies by current chlorination practices without coagulation, sedimentation, and filtration is no longer sufficient.

Inactivation by ultraviolet irradiation is not effective. Therefore UV is not an appropriate method of disinfection for Giardia cysts.[49]

Giardia cysts are extremely resistant to chlorine. Jarroll et al.[43] found that it required a chlorine concentration of 4 mg/l and a 60 min. contact time to kill all cysts in a chlorine-demand-free solution at pH levels of 6, 7, and 8.* This in effect means that in a system with chlorine demand the chlorine dosage would have to be sufficient to produce a 4 mg/l free residual at the end of 60 min. These findings compare favorably with those of Meyer and Jaroll.[40]

Giardia cysts are even more resistant to chloramines. This is to be expected, as they generally follow the disinfection resistance pattern of *Entamoeba* cysts. Bingham and Meyer[318] found that complete destruction of *Giardia* cysts required 4.5 hours contact time with 2.6 mg/l chloramine residual at a water temperature 50°F. Sample water pH was not an important factor. Cyst death was slightly more rapid at pH 7.0 than at pH 8.5.

Inactivation of *Giardia* is best achieved by the multiple barrier approach. This would include removal of cysts by conventional water treatment processes such as flocculation, coagulation, sedimentation, and filtration. Any remaining cysts can then be inactivated by terminal disinfection. It is strongly recommended that prechlorination be applied as far upstream in the treatment train as is possible so as to achieve the optimum contact time of 60 min. This is similar to the recommendation for inactivation of *Entamoeba histolytica* cysts.

A chlorination system for a high quality surface water that does not require filtration can be designed to inactivate *Gardia* cysts, as has been done in some instances for amoeba cyst inactivation. In either case partial dechlorination after terminal disinfection must be considered.

Legionnaire's Disease

Since 1977 the Center for Disease Control in Atlanta, Georgia has investigated 30 outbreaks of this disease.[44] A clear mode of transmission was not found. However, potable water as the basic source of the causative organism is strongly implicated in these and other investigations here and abroad.

---

* At 5°C.

The name of the disease derives from a mysterious explosive outbreak of pneumonia following an American Legion Convention in Philadelphia, Pennsylvania in 1976.[45,49] Pneumonia occurred primarily in persons attending the convention. Twenty-nine of the 182 cases were fatal. The causative pathogen was previously unknown in the annals of microbiology. The spread of the bacterium appeared to be airborne, but the source was not found. The epidemiologic analysis suggested that exposure may have occurred in the lobby of the headquarters hotel or in the area immediately surrounding the hotel.

Subsequent investigations of Legionnaire's disease at institutions have shown that multiple sites have yielded the causative agent, *Legionella pneumophila*. These include cooling towers, evaporative condensers, water heaters, faucets, and showerheads. Therefore it is highly probable that the source in the American Legion outbreak was the aerosols from the air conditioning system.

The earliest documented outbreak of Legionnaire's Disease occurred at St. Elizabeths Hospital (Washington, DC) in July and August 1965.[46] It was thought that this outbreak was in some way related to the excavation of the hospital grounds for the installation of a lawn sprinkler system. It was found that patients whose beds were near windows in buildings close to the excavation and patients who had access to the excavation were the ones who became ill. No attempt was made to recover the disease agent from the soil, which would have been most difficult at that time. However, in 1977 stored serum-pairs were tested and *L. pneumophila* were documented in 85 percent of the pairs. Since 1977, *L. pneumophila* has been discovered in mulch and soil.[46]

This disease is now known as legionellosis, which is the generic term for a group of diseases caused by the bacteria in the family *Legionellaceae*, genus *Legionella*. This family has one genus which contains at least six species: *L. pnuemophila, L. dumoffii, L. bozemanii, L. gormanii, L. micdadei,* and *L. longbeachii*. Although the species are genetically and antigentically distinct, they are morphologically and biochemically similar. *L. pneumophila* is the most commonly isolated species and has been found to have six serogroups.[46]

Legionellosis is currently recognized in two distinct forms: Legionnaire's disease, characterized by pneumonia with significant mortality rates; and Pontiac fever, which is a non-pneumonic, self-limiting febrile illness with a very high attack rate.[47] The incubation period for Legionnaire's disease is 2–20 days; for Pontiac fever it is only 5–66 hours.[48] Though the disease is respiratory in nature, person to person transmission has not been conclusively established. It is still not known why *L. pneumophila* causes two such different illnesses, or why the bacterium affects only 1 percent of the people exposed to it in the case of legionellosis, and as many as 95 percent in the case of Pontiac fever.[46]

It is paradoxical that an organism so fastidious in its nutritional requirements and so difficult to grow in the laboratory apparently persists and succeeds so well in an inanimate environment. The precise ecological niche and nutritional supply have not been defined.

L. pneumophila is now considered to be a common environmental contaminant.

It may be ubiquitous. It has been isolated from many aquatic sites, including ponds, lakes, potable water, and equipment using potable water such as cooling water and evaporative condensers. *L. gormanii* has been isolated from a creek bank and *L. micdadei* has been isolated from nebulizer water.

The identification of legionella in 23 lakes in Georgia and 67 lakes (793 samples) in North Carolina, Florida, and Alabama[48] is cause for concern, although legionellosis does not occur by ingestion of water. It apparently attacks via the airborne route, as a result of aerosolization, and most likely utilizes algae or amoebae as transport mechanisms. Investigations have also demonstrated that *L. pneumophila* is able to colonize and grow in certain parts of water distribution systems. Warm water seems to enhance this ability. Legionellae have been found growing in certain species of free-living freshwater amoebae and soil amoebae. These amoebic species have been isolated from humidifiers, and if they are inhabitants of slime in showerheads, this could explain the presence of large numbers of *Legionellae* in samples from such fixtures. It has been estimated that a person inhaling a single infected amoeba could receive a dose of up to 1000 *Legionella* cells.[48]

It can be reasonably concluded that potable water probably is no more than the transport vehicle of this pathogen. However this can have serious implications because it tends to colonize and grow to high densities in components usually found in hotels and institutions. These components are storage tanks, cooling towers, evaporative condensers, showerheads, and water faucets.

Legionellae are susceptible to chlorination. Chlorination has been found to be effective at the 1–2 mg/l dosage level. Precise data for contact time and free residuals are not available. Institutions, hotels, etc., should make it a practice to sterilize their water systems on a regular basis similar to the recommended practice of main sterilization. This could be as frequently as once each six months.

Water suppliers whose systems have large holding reservoirs should be aware that *Legionellae* can work in such environments.

Worms

Since the disease Schistosomiasis described above is carried by an organism sometimes referred to as a worm, it is pertinent to mention other worms that occur in water supplies. Free-living nematodes (NAIS), visible to the naked eye and resembling a tiny white worm, have been reported to occur in water supplies. In 1961 a survey of twenty-two cities indicate that these nematodes are fairly common in areas using surface waters. Although these nematodes are nonpathogenic, it is thought they might be able to ingest pathogenic organisms that could later transmit a waterborne disease.[50] Large numbers of one species, *Diplogaster medicapitus* of the Rhabitidae family, are present in the effluent of trickling filters and activated sludge process of sewage treatment and have been isolated in treated water supplies. They have been found to be capable of ingesting *Salmonella* and *Shigella* bacteria and Coxsackie virus. Of the ingested pathogens, 6 to 16 percent survived one day and 0.1 to 1 percent survived two days. These nematodes resisted 2.5 to 3.0 ppm chlorine residuals in contacts of up to two hours. It is unlikely, however, that the nematodes could ingest pathogens at a sewage plant and be carried by the polluted water through the water system in one day. Even though the chance is remote, the threat exists.

While the problem of nematodes is primarily aesthetic, they can impart a strong earthy or musty odor to a water supply if allowed to propagate within the treatment plant. When the finished water shows more than ten worms per gallon, it should be investigated.

The most practical approach to preventing nematode infestation is to prechlorinate the raw water to a free residual of 0.5 ppm and six hours contact time. Although many of the nematodes will not be killed, they will be sufficiently affected so that they cannot survive and will therefore be settled out in the flocculation process.[50]

Tularemia

This disease of rodents, man, and some domestic animals derives its name from Tulare County, California, where it was discovered. In man it produces an irregular fever which continues for several weeks and results in moderate to severe debilitation.

The usual means of transmission is by direct contact, although the causative organism *Pasteurella tularensis* has been isolated in certain streams in Montana that are sources of water supplies.[51] These organisms are easily destroyed by chlorination.

Acute Gastrointestinal Illnesses (AGI)

All reported outbreaks of unknown etiology are classified by the CDC and EPA as AGI. In the period 1971 to 1977 a total of 192 outbreaks of waterborne disease affecting 36,757 persons was reported in the U.S. There was no etiology in 57 percent of the outbreaks and 58 percent of the illnesses.[68] These illnesses are therefore listed as AGI. Before the cooperative effort between the CDC and EPA to document, investigate, and report waterborne outbreaks in 1971, AGI outbreaks were not always documented. For example, in the 10 year period 1961–1970, of 130 waterborne outbreaks affecting 46,374 people, 30 percent were classified as a gastrointestinal outbreak of unknown origin.[20,69] In 1981, 44 percent of the outbreaks, affecting 43 percent of the cases, were classified as AGI in the CDC Annual Summary.

Recently Described Conditions with Conspicuous GI Symptoms

At a 1983 gastroenterology seminar held in London,[297] several important new microbial agents and infective conditions were discussed as to the clinical features,

Table 6-1 Recently Described Conditions with Prominent Gastrointestinal Symptoms

Syndrome	Causative Organisms	Site of Infection	Main Gut Symptoms
Toxic shock syndrome	*Staphylococcus aureus*—enterotoxin F-producing strains	Vaginal and other sites	Diarrhea
Legionnaire's disease	*Legionella pneumophila*	Lower respiratory tract	Diarrhea and vomiting: abdominal distension
Traveller's diarrhea	*E. coli*—enterotoxigenic strains; *Campylobacter* spp.	Gut	Diarrhea
Gay-bowel syndrome	*Giardia lamblia Entamoeba histolytica*	Gut	Diarrhea
Infant botulism	*Clostridium botulinum*	Gut (?)	Constipation

pathogenesis, epidemiology, occurrence and infection site. These recent syndromes are listed in Table 6-1.

The gay-bowel syndrome is of interest in this text because the causative organisms *Giardia lambia* and *Entamoeba histolytica* are waterborne diseases. This syndrome refers to sexually active homosexual males. Whether or not the persons suffering from this malaise are potential carriers is not known. It is well known, however, that the causative organisms are virulent and difficult to destroy. Since this situation involves the fecal-oral route of transmission, foodborne infections are possible.

A 1982 survey of a large sector of the San Francisco gay community was made by a medical task force from the Kaiser-Permanente Medical Center.[298,299] This survey studied the intestinal parasitic infections in homosexual males and found, among other things, that 68 percent had intestinal parasites (*G. lamblia* and *E. histolytica*). This finding compares with a national background level of about 3 percent. This should be of concern to those involved in water production and food handling.

Viral Diseases

General Background. All viral diseases are not waterborne. The human enteric viruses (poliomyelitis, Coxsackie, and echo) do not produce their associated diseases via the waterborne route.[52] It is also possible to have these viruses in the blood stream without having the disease. These facts appear to contradict the nature of the diseases involved, which complicates the understanding of viral diseases, especially in potable water supply.

Adenovirus, which is not an enteric virus, has been reported to have caused a disease outbreak in a water supply.[53]

The viruses of known concern to potable water producers are: Hepatitis A, Norwalk, and Rotavirus. It has been said that the latter two are responsible for about 77 percent of acute waterborne gastroenteritis.[54] These three viruses can and do travel the waterborne route. It is claimed by some investigators that Rotavirus can survive modern wastewater treatment processes. These viruses travel the fecal–oral route.

Coping with Viruses in Potable Water. It is the consensus of virologists that the raw water of all surface water supplies in the U.S.A. and Canada contains naturally occurring human enteroviruses. These investigators and public health agencies believe that it is necessary to adopt the multiple barrier approach in order to protect the public from sickness attributable to viruses.

A comprehensive study by O'Connor et al., published in 1982,[55] concluded that virus removal and inactivation are essentially complete when modern treatment practices are followed. This means the use of prechlorination, flocculation, coagulation, sedimentation, and filtration followed by terminal disinfection. Prechlorination will enhance the reliability of the treatment plant to remove and inactivate the viruses because of the extended contact time through the treatment processes. If the coagulation step is to be effective the water must have a reasonable amount of turbidity.

In those instances where riverbank filtration is used to simulate slow-sand filtration, flooding the banks will nullify the purpose of this treatment step and cause a rapid increase in the virus concentration in the treatment plant.

Another situation which causes virus population increase is sludge blanket breakthrough in the sedimentation step.

The effect of chlorination system failures is well known and need not be repeated here.

Another promising method of virus removal is by storage. For this method to be effective, the water to be stored must be biologically active. Virus removal by storage can easily achieve levels of one plaque-forming-unit.[53] Storage is reliable if there is no short circuiting. The use of lagoons has been studied in Israel for many years. Three log reductions of viruses in wastewater after 28 days storage have been observed.[56]

If every effort is made to remove or destroy all viruses in potable water it seems unrealistic to set zero virus count as a goal.

Consensus Organism Indicator. It is not possible at this time to isolate an organism that would serve as a virus indicator. Moreover, the problem is compounded by the fact that there is no observed relation between the microbial indicators, such as total coliforms, fecal coliforms, or standard plate counts, and the presence of viruses. Also there is no observed relation between the detection of *Salmonellae* and viruses and vice-versa. Nor can the enteroviruses or coliphages be used as virus indicators with our present knowledge of the organisms.

For example, a potable water could be contaminated by a small source which introduces viruses but no fecal indicators like total or fecal coliforms. Likewise, if there is a fecal contamination from a populace which is not "shedding" the virus there would be fecal indicators but no viruses.

Adenoviruses. These viruses, originally isolated from tissue culture of human adenoids, are responsible for febrile pharyngitis, swimming pool conjunctivitis, and acute respiratory catarrh. They are not enteric viruses but apparently can be waterborne, as they have been detected in surface waters. There are some 28 serologic types of this group of viruses. Types 4 and 7 are found in most cases of acute respiratory disease. Types 1, 2, and 5 are frequently associated with febrile respiratory infections in young children. Types 6 and 10 cause conjunctivitis, which can be transmitted in swimming pools and possibly other surface waters. It is common in recruit camps where attack rates have been reported as high as 70 percent. It is also seen in children's institutions due to the crowding together of susceptible young hosts.[57] While these viruses are resistant to antibiotics[41] they are the most susceptible to chlorination of all the viruses normally detected in surface waters.

Enteroviruses. These viruses derive their name from the fact that the alimentary tract forms a reservoir where they may be found. They enter the body through the mouth. Infection is spread mainly by the fecal–oral route caused by poor hygiene, poor sanitation, or a poor-quality water supply. These viruses cause the greatest concern in water supply practice because of the severity of the diseases they carry as well as their resistance to disinfection. The enteric viruses of least concern to the water producer are the poliomyelitis, coxsackie, and echoviruses, for reasons described above. These are not considered to be waterborne, in spite of the fact that they are excreted in the feces of infected patients.

Poliovirus. This virus is the cause of what used to be a common infectious disease of the central nervous system known as infantile paralysis, or poliomyelitis. The virus was thought at one time to have been transmitted by water supplies because it is readily detected in the feces of infected patients. However, a great many studies have failed to confirm this theory. The introduction of mass vaccination in the United States about 1960 has reduced the occurrence of poliomyelitis to something less than 5 cases per 100,000 per year. It appears that permanent immunity may be possible provided booster doses are taken at specified intervals. Therefore it is practically impossible to prove whether or not the disease is waterborne without the occurrence of an epidemic.

Coxsackie. This virus is so called because it was first isolated at Coxsackie, New York. There are many types of this virus which cause Bornholm disease, named after a Danish island in the Baltic Sea, where there was an epidemic of pleurodynia (an inflammation of fibrous tissue causing acute pain in the side), herpangina (a contagious disease of children characterized by fever, headache, and vesicular eruption in the throat), and epidemic pharyngitis. Some types of Coxsackie viruses are the most resistant of all viruses to chlorination. Here again this is of no concern, because the diseases caused by this virus are not known to be waterborne.

Echoviruses (Enteric, Cytopathogenic, Human Orphan). These are so called because on first isolation they appeared to be associated with no known human infection. They have since been shown to cause acute respiratory symptoms, enteritis, and meningitis.[58] There are more than twenty different types.

Infectious Hepatitis (Hepatitis A). The etiological agent is a virus whose characteristics are not yet completely known. Another type, serum hepatitis, is clinically indistinguishable from infectious hepatitis, although the virus is different. Infectious hepatitis is an acute viral inflammation of the liver characterized by fever, malaise, gastrointestinal symptoms (nausea and vomiting), and jaundice. The fact that infectious hepatitis can be waterborne is well documented. Its incidence has been rising to such an alarming extent since about 1950 that it is recognized as a disease of considerable environmental significance and one which is of great concern in water supply practice. The most explosive and devastating waterborne outbreak of infectious hepatitis occurred in New Delhi, India, in 1955–56, resulting in more than 50,000 cases.[59] The raw water was known to have been contaminated by domestic sewage, but the water was treated by conventional methods. The failure of the treatment process was traced to an inadequate chlorination facility.

Viral Gastroenteritis

AGI. This is the label for acute gastrointestinal illness of unknown etiology. CDC reported 4,430 cases of waterborne illness from 32 outbreaks for the calendar year 1981. AGI accounted for 1893 cases from 14 outbreaks, while Rotavirus accounted for 1761 cases from only one outbreak. It is speculated that the majority of AGI cases are the result of Norwalk-like viruses.

Norwalk Virus. This virus has been associated with epidemics of acute gastrointestinal illness occurring in families, schools, and communities.[60–62] It is regarded as a common pathogen of older children and adults in the United States. This is an interesting observation, because the original outbreak caused by this virus occurred in an elementary school in Norwalk, Ohio, October 30–31, 1968. Its role in pediatric diarrhea has not been thoroughly investigated. The incubation period is 18–48 hours, and symptoms last from 24 to 48 hours. Usual symptoms are abdominal cramps, headache, malaise, low grade fever, nausea, and either vomiting or diarrhea or both. This sickness has often been characterized as 24-hour flu.[60,62,63]

An investigation of 25 separate outbreaks of nonbacterial gastrointestinal illnesses were studied serologically for evidence of infection with the Norwalk virus and the rotaviruses. Eight of the 25 appeared to be related to the Norwalk virus. In one of the 25 there was evidence of Rotavirus infection. These observations suggest that the Norwalk virus or serologically related agents play an important role in epidemic nonbacterial gastroenteritis in adults and older children.[62,63]

The outbreaks studied occurred in communities, cruise ships, schools, summer camps, and colleges. The 1968 Norwalk outbreak was studied by means of stool swabs. These swabs failed to detect *Salmonella, Shigella,* enteropathogenic *E. coli, Staphylococcus aureus, Aeromonas,* noncholera *Vibrio,* or enterococci. The attack rate was 50 percent (116 of 232) of students and teachers. There were several schools in the town, and food for lunches was distributed to all schools. There was no illness at any of the other schools. It was concluded that the source was the water well at the affected school, which was the only school that had its own well.[64]

In August 1980 there was a potable water–related outbreak of acute gastrointestinal illness in the community of Lindale, Georgia.[65] The attack rate was determined in 10 neighborhoods. This rate increased significantly near a textile plant where there was known to be a cross-connection between an industrial water supply system (which contained fecal coliforms) and the community potable water system. The infectious agent was thought to be waterborne. There is ever-increasing evidence that acute gastrointestinal illness is indeed waterborne and one of the more important infectious agents is Norwalk virus.[66]

Rotavirus. This virus is known to cause both sporadic and epidemic cases of gastroenteritis in infant and young children throughout the world. Approximately one-half the infants and young children hospitalized with diarrhea are infected with rotavirus. In temperate climates the peak prevalence is in the winter months. Infection is uncommon in the summer. Transmission is thought to be by the fecal–oral route. Diarrhea of 5–8 days' duration is the main feature of rotaviral disease. It is frequently accompanied by vomiting and fever. Death due to this virus is rare but has been reported. Rotaviruses have been observed in diarrheal feces from the young of numerous mammalian species.[60] Waterborne rotavirus infection has been reported in Europe.[57,67] This virus is not believed to be as important as the Norwalk virus as the cause of acute gastrointestinal illness (AGI).

Summary

Examination of the statistics tends to indicate that waterborne outbreaks are on the upswing. This may be due to better reporting or it may be that some of the barriers are being overstressed. Public health agencies are deeply committed to the theory of multiple barriers or multiple points of control between sewage discharges and water supply intake. These barriers or points of control include wastewater treatment, land confinement, dilution, time, distance, and potable water treatment. Any type of treatment is fallable, so reliance on natural barriers should be maintained as long as possible. However, it is obvious that disinfection of potable water is the last line of defense.

The Safe Drinking Water Act of December 1974 resulted in improved surveillance of drinking water supplies. The provisions of this act apply to all water systems serving 15 connections or at least 25 individuals for a minimum duration of 60 days. This act was followed by the National Interim Primary Drinking Water Regulations created by the EPA in 1975 and implemented in 1977. These regulations established maximum contaminant levels for specified microbiological and chemical contamination, as well as turbidity limits and monitoring frequency.

A perusal of the annual reports by Gunther F. Craun in the *J. WPCF* June issue, Literature Review, under the heading "Disease Outbreaks Caused by Drinking Water," 1971–1980, reveals some interesting data about the supplies involved in these outbreaks.

The supplies that suffered the most outbreaks were the noncommunity supplies such as cruise ships, resorts, summer camps, boarding schools, and other small private supplies. In one year these supplies accounted for as much as 75 percent of the outbreaks. However it is the community supplies that account for the greatest number of cases, about 85 percent of cases each year. The traveling public suffers greatly from waterborne illnesses, particularly in Mexico.

The causes of these outbreaks fall into the following categories: contaminated ground water—65 percent of outbreaks and 63 percent of cases; inadequate or interrupted chlorination, 31 percent of outbreaks and 44 percent of cases. In some years contamination of the municipal distribution system causes as much as 70 percent of cases.

It remains to be seen whether or not waterborne diseases are on the upswing. It may be that the reporting is more comprehensive.

EARLY DEVELOPMENT OF CHLORINATION EQUIPMENT

Historical Background

The next important development in the progress of chlorination was the invention of apparatus that could use chlorine gas directly from a cylinder. Until about 1910 the only successful practical method was using hypochlorite solution. The one successful electrolytic installation at Brewster, New York, which operated continuously for eighteen years on sewage effluent, was replaced by a hypochlorite installation after the original one was destroyed by fire in 1911. At best the hypochlorite method was cumbersome, extremely awkward, and messy to handle and apply. In June, 1910, C. R. Darnall, a major in the Medical Corps, U.S. Army, began to experiment with liquid molecular chlorine compressed in steel cylinders.[17] The cylinders, which were not then available in the United States, had to be shipped here from Germany. Major Darnall not only made the first practical chlorinator using chlorine gas,[18] but also published the first work on the bactericidal efficiency of elemental chlorine dissolved in water.[17,70] Darnall's apparatus (U.S. Patent 1,007,647), illustrated in Fig. 6-1 consisted of a pressure-reducing mechanism, a metering device, and an absorption chamber. Only one of these units was ever installed—at Youngstown, Ohio.

Although Darnall never entered into the equipment field, he intended to set up license agreements with communities to install his equipment for waste disinfection. Under such an arrangement, the community would pay him a royalty based on per capita per year or total amount of water treated. He approached the Electro Bleach Gas Company of Niagara Falls, New York, which in 1909 was one of the first commercial producers of liquid chlorine in the United States, and proposed that all the liquid chlorine used in his installations be supplied by the company and that, in return, the company should agree to sell liquid chlorine for this purpose to no one else. E. D. Kingsley, president and founder of the company agreed, provided that Darnall furnish the chlorine feeding device in working form within six months. When Darnall failed to do so, Kingsley assigned the job of developing such a device to George Ornstein, then a consulting chemist to the company, employed to develop applications for liquid chlorine in bleaching textiles and paper pulp.[38]

Meanwhile, in December, 1912, John Kienle, chief engineer of the Wilmington, Delaware, water department, began experimenting with the application of chlorine gas from cylinders.[70] He fed the chlorine gas to an absorption tower, through which a counter flow of water absorbed the chlorine, which was then applied as an aqueous solution. A balanced, weighted pressure-reducing valve was connected to the cylinder for chlorine feed-rate control but became inoperative and so was discarded; only the valve on the cylinder was available for control. Kienle recommended to his water board that installation of such chlorination equipment be made, but only by a manufacturer that agreed to demonstrate efficiency of control for thirty days and guarantee the apparatus against deterioration for one year.

Following Kienle's experiments at Wilmington, Ornstein developed a chlorine control apparatus (U.S. Patent 1,142,361) which consisted of a high- and a low-pressure gage, the latter being calibrated to indicate the flow of gas through a fixed orifice. The metered gas was then directed into a stoneware absorption tower, where it became dissolved in the water flowing by gravity out of the bottom of the tower as an aqueous chlorine solution. This unit was first tested on a plant scale in November, 1912, at the Western New York Water Company plant at Niagara Falls. Under the direction of Harry F. Huy, the bacteriological results using this apparatus were documented and compared with the chloride of lime method.[38] The results exceeded expectations. This experimental apparatus was greatly improved, and on February 1, 1913, the Ornstein chlorinator, as manufactured by the Electro Bleach Gas Company, was put into operation at Kienle's Wilmington, Delaware, plant. This is considered the first commercial solution feed chlorinator installation in the United States.[18,38] John Kienle shortly thereafter

CHLORINATION OF POTABLE WATER 281

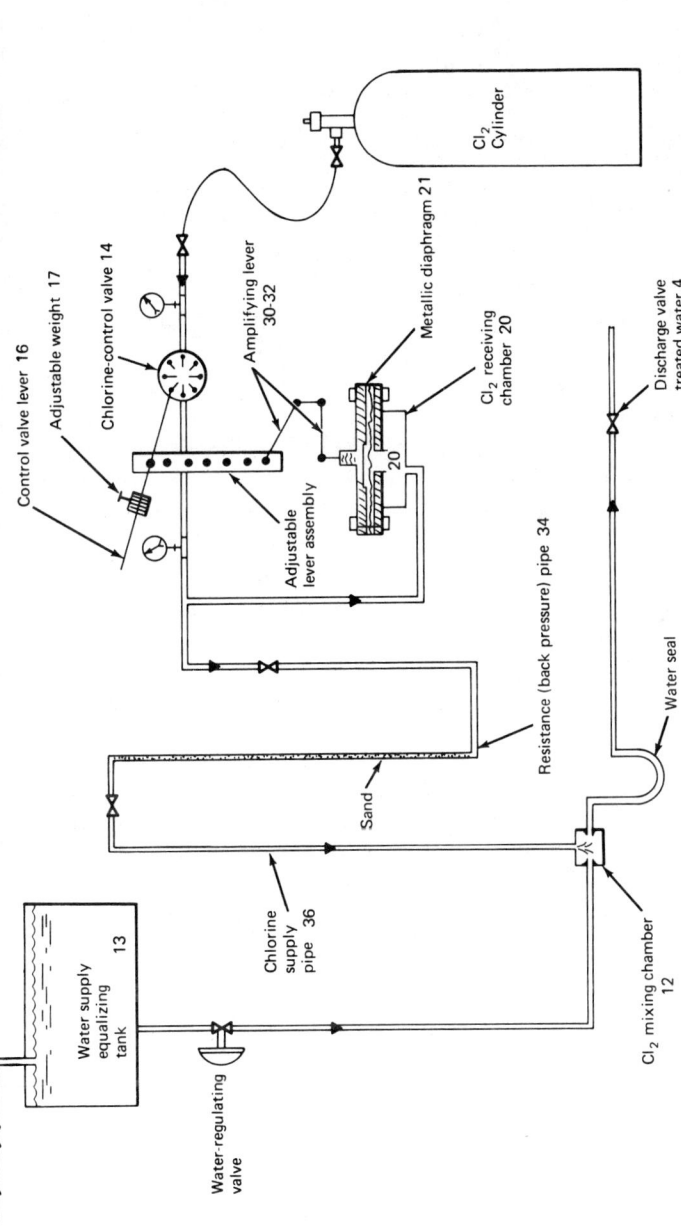

Fig 6-1 Major Darnall's chlorinator. When valve on chlorine cylinder is opened the chlorine gas passes through piping and control valve to the receiver 20 in the diaphragm-type chlorine pressure-reducing valve and simultaneously through the back-pressure assembly 34. Because of the pressure buildup in 34, the pressure in 20 rises and lifts the metallic diaphragm 21, which operates the lever assembly 30–32 so as to close the chlorine control valve 14. As the gas passes through chamber 34, the pressure in chamber 20 falls and the lever assembly 30–32 operates to open valve 14, thus permitting more gas to pass.

The quantity of chlorine may be increased or decreased by shifting weight 17 on lever 16. Once adjusted, the chlorine flow is presumed uniform. The water flow is maintained constant by means of the equalizing tank 13.

Supposedly this system was meant to be automatic start-and-stop based on opening and closing valve 4 as follows:

If valve 4 is closed, the flow of water out of the tank will stop and presumably raise the pressure in pipe 34 via pipe 36, thereby reacting on the chlorine diaphragm assembly 21, causing the flow of chlorine through valve 14 to shut off.

became associated with the Electro Beach Gas Company to manage the sales of its chlorine feeding equipment. Under his guidance, installations of this equipment were made in both potable water and wastewater plants, as far west as Salt Lake City, Utah, Monterey, California, and Seattle, Washington.

About this time, Wallace and Tiernan entered the field. The firm set about making a gas chlorinator for an installation at Dover, Delaware, on the Rockaway River, a tributary which fed the Boonton, New Jersey, reservoir. This, their first direct-feed gas chlorinator, was installed February 22, 1913.[18] This installation included a gas pressure gage, a temperature-pressure compensating device, and an ingenious inverted glass siphon for measuring low rates of chlorine gas flow which is still in use today in modern equipment. The chlorine gas was applied directly to the water being treated, where it was dissolved by means of porous ceramic diffusers. This new chlorinator was simple and dependable in operation, and so it soon gained popular acceptance. Late in 1913, Wallace & Tiernan began developing solution-feed chlorinators, which proved more reliable than the direct-feed units. The direct-feed gas diffusers were plagued by cold weather. Within a year, Wallace and Tiernan began to dominate the field. By September, 1914, they had twenty-three chlorinators installed in eighteen different cities; in 1916 they installed nine direct-feed units in Croton, New York, aqueducts. In 1917, Wallace and Tiernan became sole licensee of the Ornstein patent and the Electro Bleach Gas Company discontinued marketing its liquid chlorine feeding equipment. By 1918, Wallace & Tiernan equipment was installed as far west as California; by 1920, they were represented by sales and service facilities in major cities throughout the United States; and in 1925 the British company was established. After World War II the company was represented around the world; its impact on the water works industry was extraordinary. In the late 1920s, when it was confirmed that a combination of chlorine and ammonia could eliminate many tastes and odors due to chlorination, Wallace and Tiernan promptly supplied the appropriate metering and control equipment to handle anhydrous ammonia.

In the years that followed, the name Wallace and Tiernan became synonymous with chlorination. They were responsible for all the significant developments in chlorination apparatus. Among these achievements were: the unique vacuum solution-feed chlorinator utilizing the aspirator-type injector; the development of the visible vacuum principle of equipment (ca 1920), which resulted in an industry-wide change in the manufacturing of chlorine gas to produce a cleaner, less troublesome product. Fig. 6-2 is an illustration of the bell-jar chlorinator, invented by C. F. Wallace. This particular model (A-255) was the first automatic flow-paced solution-feed chlorinator (ca 1925).

This was followed by the development of the amperometric titrator in 1942 by Henry Marks of Wallace and Tiernan. This invention provided the analytical techniques required for the control of the free residual process in accordance with the breakpoint phenomenon. Shortly thereafter the continuous amperometric chlo-

Fig 6-2   Flow paced automatic bell-jar chlorinator Wallace and Tiernan Series A-255 (courtesy Wallace and Tiernan Div. Pennwalt, England).

rine residual analyzer was made available (ca 1950), followed in 1953 by apparatus capable of automatic residual control. Other significant developments were: the first chlorine leak detector (ca 1955) and the compound-loop principle of automatic chlorine residual control (U.S. Patent 2,929,393) introduced in 1960.

In the early days of chlorination (1920s) Wallace and Tiernan devoted a considerable amount of time to residual control research. Systems were developed which could continuously control to a specific O-T residual. The first of these was installed at Little Falls, NJ in 1929; the second at Rahway, NJ, in 1930.[71] A modification of these units was installed in 1932 and 1934 on the Los Angeles System.[72] The Lower San Fernando installation is shown in Fig. 6-3.

Fig 6-3 Colorimetric chlorine residual controller lower San Fernando. Reservoir Los Angeles Department of Water and Power (courtesy Wallace and Tiernan Div. Pennwalt).

Advances in Equipment

During these early years the most successful competitors of Wallace and Tiernan were Pardee and Everson Manufacturing. Wallace and Tiernan controlled about 90 percent of the chlorinator business until Fischer and Porter entered the field in the early 1950s. At that time plastic pipe and fittings were beginning to replace rubber hose and rubber-lined pipe for chlorine solutions. The use of injection-molded plastic components transformed the manufacturing concepts of chlorinators overnight. Fischer and Porter capitalized on these developments. They displayed a reasonably comprehensive line of cabinet-style molded plastic chlorinators in the 1000–2000 lb/day capacity range. Wallace and Tiernan bell jar chlorinators were dedicated to using expensive molded hard rubber parts along with other expensive parts made with silver and tantalum. Their only recourse was to abandon the bell jar line and follow the path of Fischer and Porter, which proved prudent and successful. Some years later one of the key chlorinator design engineers left Fischer and Porter to start his own chlorinator company, which is now known as Capital Controls. It is estimated that these three companies share 97 percent of the chlorinator business in 1983.

The most significant advances in equipment have resulted in wider range of chlorine flow control, a greater variety and sophistication of automatic controls and the development of reliable continuous chlorine residual measuring equipment. In 1970, Fischer and Porter developed an analyzer that can measure free chlorine alone in the presence of significant amounts of combined chlorine.[300] Analyzers that measure total chlorine residual have been available since 1950. These have been greatly improved for better reliability.

Advances in safety include remote vacuum chlorine supply systems. This allows the gas from the container to travel to the chlorination equipment under a vacuum, thereby minimizing opportunities for leaks. Development of the remote injector concept has greatly simplified hydraulic problems and piping configurations in chlorinator installations.

Installations are designed to be compatible with strict safety procedures, which include use of container safety kits, leak detectors, and air and nitrogen purge systems of chlorine headers. Safety drills with breathing apparatus and emergency kits are routine exercises at many water departments.

DEVELOPMENT OF CHLORINATION PRACTICE

Historical Background

It was found that in addition to destroying disease-producing organisms chlorine can also destroy the nuisance organisms that abound in water supply: those that cause taste and odor, foul the filter media, and degrade the quality of water in the distribution system. Although one of the primary objections to chlorine was that it was enforced medication, a more obvious objection soon rallied many more enemies of chlorination—the chlorinous taste, a byproduct of early disinfection practices. Initially chlorination for disinfection was on the basis of total dosage. Little or no account was taken of water quality or the reactions from the added chlorine. A determination of the effectiveness of the chlorine added was not possible for at least 24 hours, when bacteriologic test results were available.

In 1919, Wolman and Enslow[73] made a most significant contribution to chlorination practice by demonstrating that chlorine absorption could vary widely from water to water. Thus began the first concept of chlorine demand.*

In the same year Alexander Houston reported to the Metropolitan Water Board of London that variable dosages of chlorine will destroy tastes and odors and, moreover, that, the more the chlorine dose is increased, the more certain will be the absence of tastes following dechlorination. These discoveries encouraged an examination of the effects of different chlorine dosages. While there were those who either had no problems resulting from the use of chlorine as a disinfectant or chose to ignore them, there were others who found that they could produce a more palatable water with much higher dosages of chlorine.

---

* Chlorine demand is defined as the difference between the amount of the chlorine added to the water and the amount of chlorine (free available or combined available) remaining at the end of a specified contact period.

In 1926 Howard and Thompson[75] reported on an exhaustive study began in 1922 on taste and odor control of the Toronto water supply by means of what they termed superchlorination followed by dechlorination. Experiments at Cleveland[75] also tried heavy doses of chlorine to eliminate tastes resulting from industrial waste discharges into the receiving waters of the Cleveland water supply. The most objectionable of these tastes derived from phenols, the same compounds that have plagued most of the water supplies of Europe, particularly those from the Seine and Marne in France and the Rhine and Ruhr in Germany.

Cox[76] reported in 1926 that heavy doses of chlorine could control algae, tastes, and odors in the New York City water supply system. Cox called his method double chlorination, consisting of splitting the chlorine solution discharge from one chlorinator to two points of application instead of one. The heavier dose of chlorine for taste and odor control was ahead of all other treatment. The second point of application was postchlorination for disinfection.

At about the same time Cox and Braidech were experimenting with heavy doses of chlorine, McAmis introduced a new approach to chlorinous taste and odor control: the use of ammonia along with chlorine.[77] The success of the ammonia-chlorine process was so extensive that it practically replaced plain chlorination. This process enjoyed its greatest popularity from about 1930 until 1942. The reason for its decline was twofold: the discovery by Griffin in 1939 of the breakpoint phenomenon, which provided a means for more reliable chlorinous taste and odor control; and the shortage of ammonia nitrogen due to U.S. participation in World War II.

The discovery of the breakpoint phenomenon marked the beginning of the free residual chlorination process. It also sparked a great deal of interest in the kinetics of chlorination and the discovery of the various chlorine residual fractions and analytical methods for their differentiation. It also revived interest in the concept of superchlorination (to a free residual) followed by dechlorination. However, conventional water treatment plants continued to be designed with a variety of locations for the addition of chlorine. In addition to provisions for chlorinating the raw water and the finished water, there was usually a point of application ahead of the filters and another at the entrance to the sedimentation basins. Multiple chlorine application points usually contribute to an overcomplicated chlorine solution piping system, which requires elaborate and expensive flow splitting devices.

The Chlorination–Dechlorination Concept

System Description. When reliable chlorine residual controllers became available in the 1960s, the superchlorination–dechlorination approach was an attractive method for controlling tastes, odors, and sedimentation tank and filter biofouling, as well as providing reliable high-quality water in the distribution system. The term "superchlorination," which was used in the early days, has been replaced by "free residual chlorination."

Using the chlor–dechlor method can achieve the modern approach to the various objectives of chlorination described in the preceding text. The system must be able to apply enough chlorine at the influent of the treatment process to maintain an adequate free chlorine residual into the treated water storage. Then at this point the residual should be controlled automatically by dechlorination.[293] This is in lieu of manually adjusting the prechlorination dosage so that it does not interfere with the postchlorination dosage. Experience has shown that the chlorination process has suffered unnecessarily in a great many instances because the operators were unable, by reason of design, to achieve proper flexibility of application. First, if the plant experiences seasonal problems with tastes and odors, the operator is reluctant to increase the dosage to achieve control of these tastes and odors for fear of having too high a residual entering the distribution system. Second, the operator knows that when he has achieved a substantial free chlorine residual there will be considerable loss from the action of sunlight on the open basins. This causes a wide variation of residual chlorine leaving the plant, which cannot be tolerated. This phenomenon is complicated further by the weather. Changes in the cloud cover introduce another variable, which contributes to additional variation in the residual. Under these conditions it is impossible to calculate hours ahead what the prechlorination dosage must be to achieve the desired residual in the treated water effluent. The chlor–dechlor method eliminates all of these problems.

The Chlor–Dechlor Facility. The usual arrangement of this type of system is to use two flow-paced automatic chlorinators for prechlorination. One unit is for standby. One sulfonator is used to apply sulfur dioxide to the finished water. This unit can be operated as a compound-loop system or by direct residual control. The success of the system depends upon the proper maintenance of the chlorine residual analyzer which is used to control the sulfonator. The analyzer can be either the amperometric type or the membrane type. Colorimetric analyzers are not recommended because their response time is much too long.

The arrangement of this equipment to control the residual leaving the plant is simple. Instead of having to bother with postchlorination, the postchlorinator becomes a sulfonator. This unit dechlorinates the treated water to a specified fixed residual. This residual is controlled by the chlorine residual analyzer equipped with a controller. The sulfonator is usually fitted with a flow-paced chlorine metering orifice which receives a flow signal from the plant effluent flow meter. Therefore the sulfonator is arranged to receive two separate signals (flow and residual). This defines the compound loop system. If a flow meter in the treated water line is not available it will be necessary to rely entirely on direct residual control. This method of control has been successful because it has good accuracy over a 7 to 1 range. Never attempt to utilize the plant influent flow meter signal to provide flow pacing for the sulfonator.

The degree of success of this method will depend largely upon the ability to provide the chlorine residual analyzer controller with a well mixed sample. Mixing is described in Chapters 8 and 9.

If there is too much loss of free residual in the sedimentation tanks and filters due to sunlight, and if it is not practical or cost-effective to cover these open structures with fiberglass, etc., it may be necessary to superimpose an automatic prechlorination dosage change between day and night operations.

If the primary control is the residual signal, and if the flow of water through the plant does not follow the usual smooth diurnal flow change pattern, as in the case of step rate flow due to pump sequencing, then the chlorine metering orifice should be controlled by an electric step rate controller. This step rate flow signal is imperative even if residual control is not involved.

An Operating Installation. The following description illustrates the salient features of a chlor–dechlor prototype installation that has been in operation for many years. Fig. 6-4 shows the plant layout and the points of application of chlorine and sulfur dioxide.

The importance of the continuous record of the chlorine residual is illustrated graphically in Figs. 6-5 and 6-6, which show two charts: one from the residual analyzer, and one from a corresponding SO_2 flow record of the sulfonator. It can be seen that when the SO_2 feed rate has varied from a low of 20 lb/24 hours to a high of 85 lb/24 hours the chlorine residual is practically constant at 0.9 mg/l. Note that at 8:00 A.M. there is an upset in the chlorine residual, due to the backwashing of some filters, with water having zero residual going by the analyzer. At this particular installation the SO_2 is mixed in approximately 25 feet of 24-inch line. The steady line on the analyzer chart reveals that it is a good mix. The plant flow rate averages 5 mgd. The prechlorination dosage is changed automatically by a time clock from 6–7 mg/l in the daytime to 3.5–4.5 mg/l at night. This change in prechlorination dosage is to effect an approximate compensation for the free chlorine residual destruction by sunlight. On a bright sunny day (Lat. 35°) the free residual chlorine decay due to sunlight was measured to be as high as 0.5 mg/l per hour.* The residence time in the open basins is long, about 4 hours at average flow. Cloudy days can be distinguished from sunny ones by mere inspection of these charts. The prechlorinator is paced by a flow signal from a metering unit at the influent to the plant. Because there is no metering equipment at the effluent of the plant, the sulfonator is controlled solely by the chlorine residual analyzer.

Plants using this system in the western part of the United States use from 2 to 7 mg/l prechlorination dosage to solve seasonal taste and odor problems, and are easily able to have enough free residual at the effluent to be trimmed by SO_2 to a final residual of 0.5–0.75 mg/l.

Controlling THM Formation. Owing to the success of the chlor–dechlor method it is necessary to consider the problem of THM formation that may result

---

* A comprehensive examination of this phenomenon in 1984 at 35° N lat. and 5000 ft elev. revealed that free chlorine destruction by sunlight on a bright sunny day can be as high as 2.5 mg/l per hour.

CHLORINATION OF POTABLE WATER 289

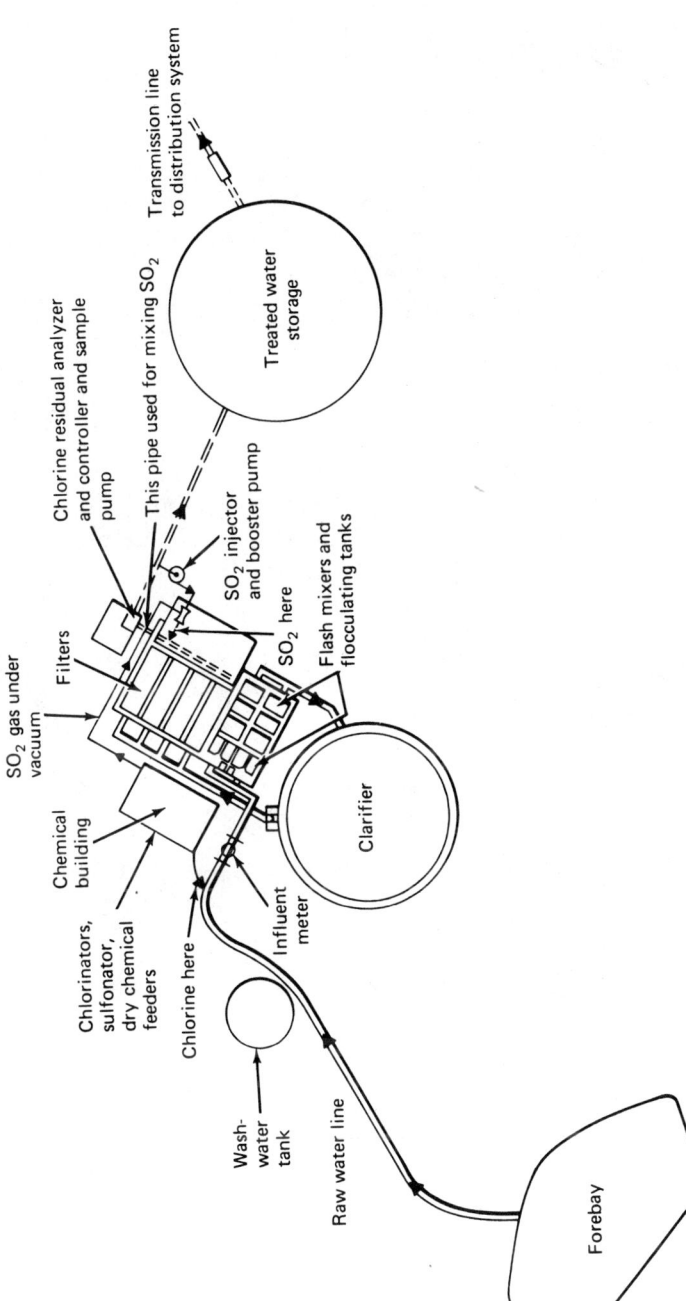

Fig 6-4 The modern approach: Chlorination and dechlorination using residual control. Chlorinators are controlled by electric signal from influent meter plus automatic dosage control. A time clock is used to select two basic prechlorination dosages—one for daytime, and one for nighttime. The sulfonator trims the outgoing residual either up or down in accordance with the set point on the residual analyzer.

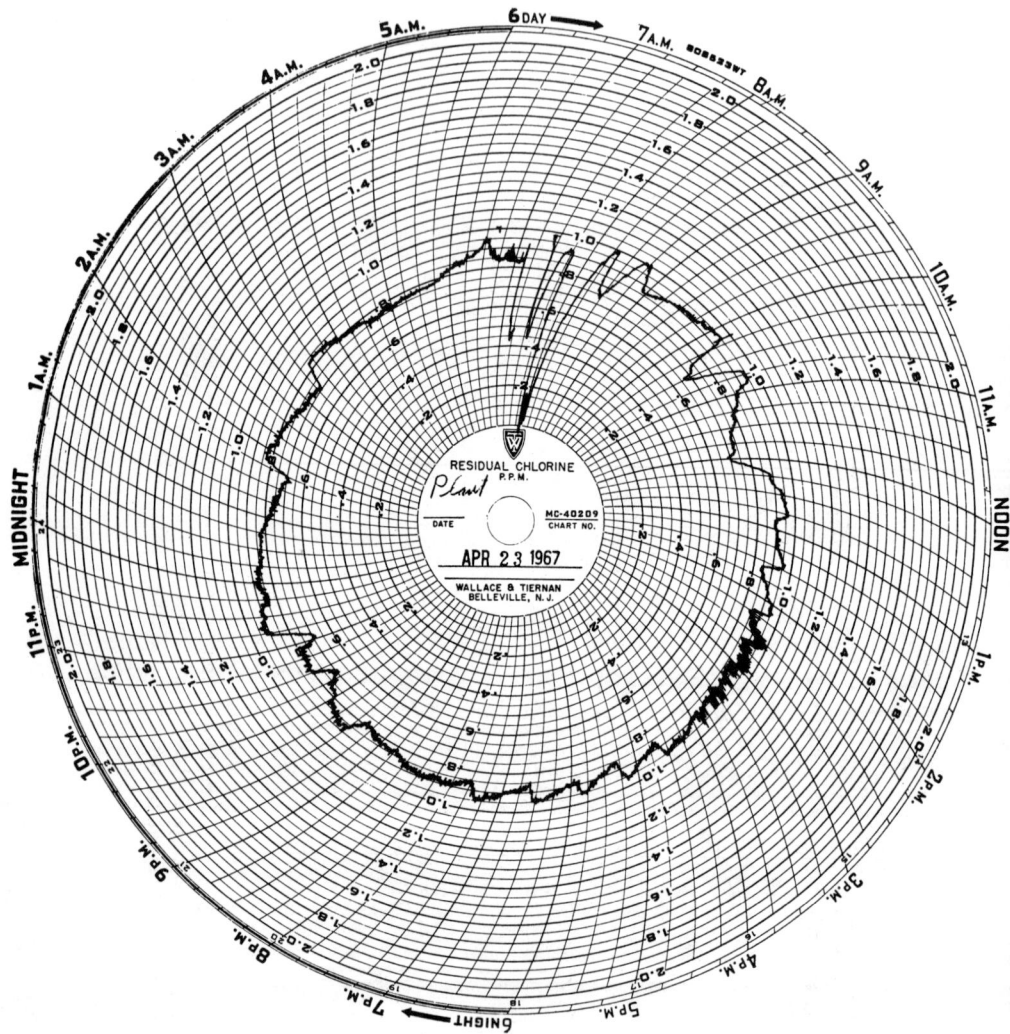

Fig 6-5 A chart from the analyzer showing free chlorine residual.

from a free chlorine residual. To avoid this problem it is only necessary to substitute chlorine dioxide for chlorine in the chlor–dechlor facility, but it remains to be seen whether or not the chlorite and chlorate ions remaining after dechlorination will pose a health hazard. See Chapter 12. The most practical way to control THM (trihalomethane) formation is to use the ammonia–chlorine process. This obviates the use of the free residual process.

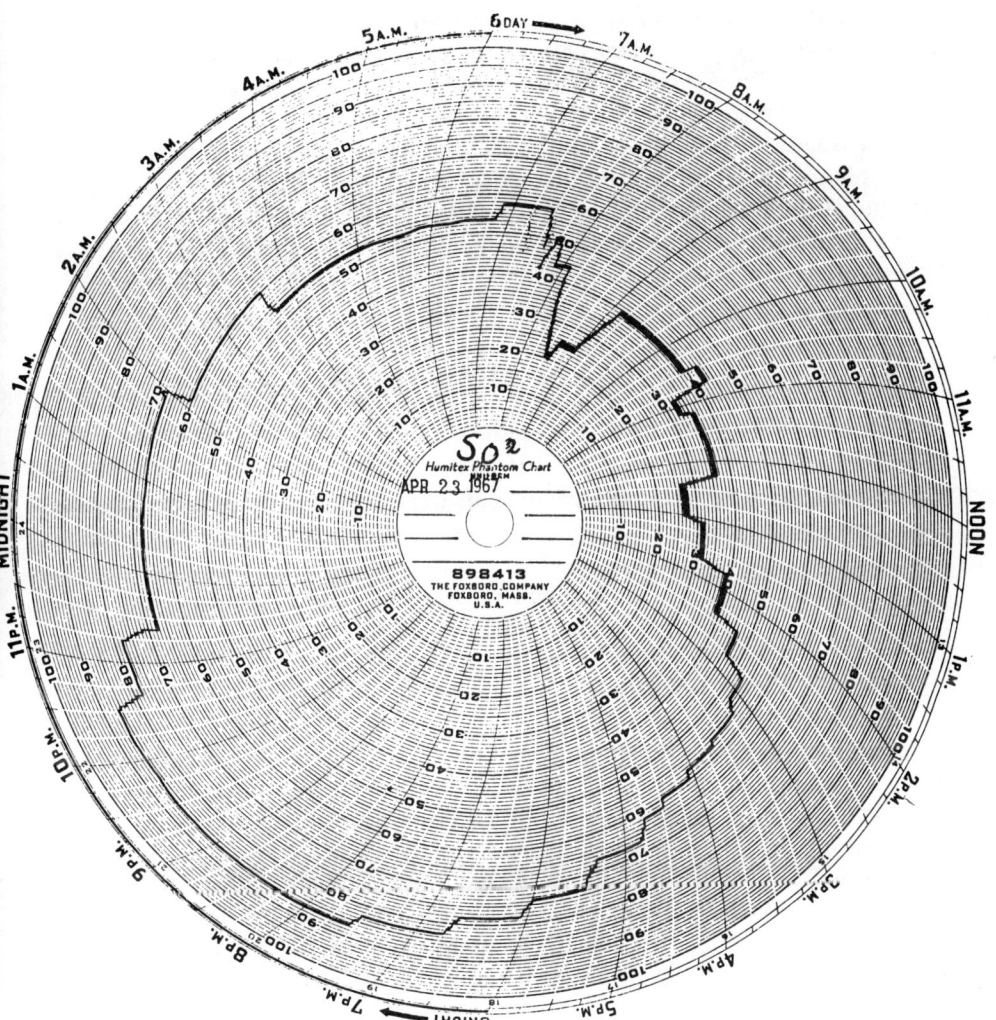

Fig 6-6 Chart from the sulfonator showing record of SO_2 feed rate corresponding in time to analyzer chart. Reprinted from *Journal American Water Works Association* **60** (5) May 1968, by permission of the Association.

Complete Dechlorination Followed by Rechlorination. There are many waters that could benefit from this technique, in particular those that develop taste and odor problems caused by objectionable concentrations of the nuisance residuals beyond the breakpoint. In this procedure dechlorination does not remove the ammonia-N bound up in these residuals. It remains to form true monochloramine in the rechlorination step. The objectionable nuisance residuals are removed, which is the great advantage of this procedure. Since rechlorination is designed to be in the range of 3–5 to 1 chlorine to the remaining ammonia nitrogen there is no danger of any nuisance residuals forming. The end result is in fact the ammonia–chlorine process preceded by the free residual process.

This is an extension of common practice in many plants which follow the rule: free chlorine through the plant and chloramine in the distribution system. There are of course exceptions, particularly in view of the necessity to avoid excessive THM formation. There are some waters that cannot be cleaned up with free residual chlorination.

The Ammonia–Chlorine Process

Historical Background. The use of ammonia with chlorine in water treatment has an unusual history. Both its discovery and use were largely by accident. Fritz Raschig[80] in 1907, while working with aniline and hypochlorite, noticed that no color developed when these two compounds were reacted in the presence of ammonia. This aroused his wish to experiment further with these compounds. By reacting two parts by weight of chlorine and one part of ammonia, he was able to produce a compound resembling a faint yellow oil. He termed this compound chloramine in accordance with the following equation:

$$NaOCl + NH_3 \longrightarrow NH_2Cl + NaOH \qquad (6\text{-}1)$$

Raschig's experimental work provided the theoretical basis for the later application of chlorine and ammonia in the water treatment field. Race[81,83] was probably the first to use the ammonia-chlorine process. He was in search of a method that would reduce the cost of chlorination, and was influenced by some observations by Rideal,[294] who noted that during the chlorination of sewage bactericidal action continued even after all the hypochlorite had "disappeared." Rideal attributed this continuing action to the influence of ammonia in the sewage. After performing successful laboratory experiments with ammonium hypochlorite, Race adopted this method in 1916 at the principal treatment plant at Ottawa, Ontario. Purely by an accidental error of some sort, his findings indicated that ammonium hypochlorite had three times the germicidal action that bleaching powder (calcium hypochlorite) did. Plant scale operations convinced Race that it was a more economical method than chlorine alone and, furthermore, that the finished water was free from objectionable tastes and odors, which he attributed primarily to the decrease in dosage.

In view of present-day knowledge, both of his conclusions were incorrect. Race also noted the elimination of aftergrowths following the prolonged use of ammonium hypochlorite. The first installation in the United States was made at Denver in 1917.[295] Alexander Houston[24] paved the way, in the development of preammoniation, to prevent taste formation caused by the reactions of chlorine and organic matter.

Although the first successful application of the ammonia-chlorine process for taste and odor control in the United States was by McAmis[77] in 1926, it was not until the work by Harold[86] and Adams[87] in England, and by Lawrence[88] and Braidech[89,90] at Cleveland, Ohio, that the possibilities of the process were fully realized. After the intensive investigation (five months and nine thousand tests) of Lawrence and Braidech, this process for taste and odor control achieved nationwide prominence. This was due in part to a previous investigation by Ellms and Lawrence[91] in 1921 on the same water, in which they submitted that waters polluted by phenols did not respond to superchlorination followed by dechlorination with activated carbon.

The Cleveland investigation concluded that preammoniation prevents tastes and odors with phenol concentrations as high as 1.0 mg/l, and that combined residuals up to at least 0.6 mg/l can be maintained without producing chlorinous taste and odors. This was considered a high chlorine residual because at this time an acid O-T residual of 0.2 to 0.3 mg/l was considered adequate.

The installations at Cleveland began in November, 1929, at the Division plant, followed two months later at the Baldwin Street plant. Other successful installations were reported at Springfield, Illinois, by Spaulding in 1929,[92] at Lansing, Michigan, in 1929 by Harrison,[93] and at Lancaster, Pennsylvania, by Ruth in 1931.[94] These successes created a great demand for the process, which resulted in some miserable failures because of ignorance of the limitations of the process.

This process enjoyed its greatest popularity between 1929 and 1939. Two years after the Cleveland installations, there were 190 other installations. In 1938, based on replies to a questionnaire from 2541 supplies in thirty-six states, 407 of these supplies used ammonia with chlorine.[98] In 1958 Griffin and committee sent out a questionnaire to 114 municipalities. Among the eighty-five answers, only seven reported preammonia; twenty-six reported postammonia; two, both pre- and postammonia. The surprising number of users of postammoniation was probably for treatment of high pH waters. The process fell into a rapid decline shortly after the discovery of the breakpoint phenomenon in 1939. This, coupled with the inability to purchase ammonia during the war years (1941–45), relegated its use to cases of special treatment.

General Discussion. This process involves the addition of ammonia and chlorine compounds separately to a water processing system. These compounds are usually anhydrous ammonia and hypochlorous acid. The combination forms chloramines. This procedure can also be called chloramination or the chloramine process. In general, when the ammonia is applied first it tends to prevent the formation of compounds that would otherwise produce chlorinous taste and odors. However, there are many instances where ammonia is added to a water containing a combination of free and combined residual for the sole purpose of converting all the free residual to combined residual. This has been well documented by Williams.[78,79] This technique of free residual conversion to chloramines has been called dechlorination by ammonia. This is a misnomer, because it does not remove any chlorine compounds. One disadvantage of this procedure is the inability of the ammonia addition to remove any of the nuisance residuals that may have formed in the free residual process. However, it is common practice to add ammonia to a water containing free residual to purposely provide chloramine residual in the distribution system.

Conversion of free residual chlorine to chloramines is standard practice in high pH waters, including waters that are softened by the excess lime process.

The preferred ratio of chlorine to ammonia-N has been documented as 3:1. This ratio appears to produce the best tasting water. The ratio for maximum disinfection efficiency is 6:1; therefore, each application should be ratio tested in the laboratory to find acceptable taste close to maximum disinfection efficiency.

There has been a great surge in the interest in the ammonia–chlorine process since 1979. This is the result of the discovery that chloramines do not form THMs.

Before the discovery of the breakpoint phenomenon this process was used primarily for the control or prevention of chlorinous tastes and odors in the finished waters. Its widest use has been in the maintenance of chlorine residuals in a distribution system. It is almost universally successful in preventing the degradation of water quality that might otherwise result in tastes and odors and discolored water from bacterial action. Its success is due to the fact that a chloramine residual can persist many times longer than a free chlorine residual, and therefore is able to penetrate into stagnant areas of a system. It is also very useful for the same purpose on long transmission lines, where it is necessary to preserve carrying capacity.

The application of ammonia with chlorine should be evaluated on the merits of each individual case. In some instances it has been known to successfully prevent the formation of chlorinous tastes by being used to *induce a breakpoint.* * In another instance ammonia was applied to three separate raw water sources from the same watershed at the rate of 1 lb/mg, followed ten minutes later by 19 to 20 lb/mg chlorine. This produced a water free of taste and odor with a free chlorine residual between 1 and 2 ppm. When chlorine was used alone on this water, a chlorophenol-like taste developed that did not respond to super doses of chlorine, but persisted for months. The source of the phenol-like compounds contributing to these chlorinous tastes was unknown.† Therefore each water that develops obnoxious tastes and odors resulting from chlorination may benefit by the use of the ammonia–chlorine process for the prevention of such tastes. This procedure began 35 years ago and is still being used today (1983).

---

* This procedure, used by O'Connell and Harvill,[95,97] proved successful in the Houston water supply problems in the 1940s.
† The source of these compounds was found to be from the Quaking Aspen tree (Poplar trimulus) on the watershed.

After the discovery of the breakpoint phenomenon in 1939, the ammonia–chlorine process fell into a sharp decline because the free chlorine residual produced by the breakpoint procedure proved to be equally effective in either controlling or eliminating taste and odors. Other additional benefits that were not necessarily common to chloramine residuals were also discovered. The demise of the ammonia–chlorine process was nearly complete during World War II. It was almost impossible to get any ammonia nitrogen compounds during the war (1941–1945). But its use did continue on a limited basis during these times because there was still a sizeable segment of the practitioners in water quality control who believed in chloramine residuals to maintain water quality in the distribution system. This and other practices using chloramines continue today (1983).

Germicidal Efficiency. The fate of this process was in some doubt when the EPA labeled chloramines a secondary disinfectant in 1978.[99] The EPA outlined certain conditions that would allow the process to be used. The most acceptable situation was to apply the chlorine first, then apply the ammonia after 10 minutes contact time with chlorine alone. As of 1983 the ammonia–chlorine process is now accepted by the EPA as a primary disinfectant provided contact time is adequate and that there is proof of disinfection. This is largely due to the fact that the addition of ammonia before chlorine prevents the formation of trihalomethanes. There are some waters that cannot comply with EPA MCL limitations of THMs when the chlorine is applied before the ammonia; therefore these waters must rely on the ammonia–chlorine process to meet EPA standards.

The EPA originally labeled chloramines as secondary disinfectants because of extensive evidence that free chlorine residuals are 100 times more germicidal than chloramines. This conclusion was based upon studies involving short contact times, 5 minutes or less, and some of the methodology and residual detection methods used are suspect. But in 1967 Morris derived a lethality coefficient which showed that HOCl is 200 times more lethal to enteric bacteria at 5°C than is monochloramine after 10 minutes contact time.[100] This is why the EPA allowed the use of chloramines as a primary disinfectant provided there was a full chlorine residual for ten minutes prior to the addition of ammonia.

Treatment Plant Research. Recent research in the use of chloramines for water reuse situations reveals an entirely different concept of the germicidal efficiency of chloramines. The plant scale investigations by the Sanitation Districts of Los Angeles County[101] developed criteria for disinfection of wastewater to be used for bathing and reuse. Their work revealed that chloramines were almost as good as free chlorine both as a bactericide and as a virucide. This efficiency is based upon good mixing and contact times of 50–60 minutes with the same chlorine dose.

Of equivalent significance is a study by Selleck et al.[102] at the University of California Sanitary Engineering Research Laboratory, Richmond, California. This study revealed that the germicidal efficiency of chloramines is practically equal to that of free chlorine provided the contact time is long enough and provided the ratio of chlorine to ammonia is such that the residual falls on the slope of the curve between the hump and the breakpoint.

The Metropolitan Water District of Southern California has made a lengthy pilot plant study to evaluate THM formation and disinfection efficiency of chloramine using a chlorine to ammonia ratio of 3 to 1 by weight.[103] Taste and odor formation and other chlorine to ammonia ratios were not investigated. They studied three schemes of ammonia–chlorine application in addition to free chlorine. These schemes were:

1. Sequential chloramination: chlorine upstream from the rapid mixer, then ammonia 17 minutes later downstream from flocculation and upstream from sedimentation
2. Concurrent chloramination: chlorine upstream from rapid mixer, then ammonia 50 seconds later in the rapid mix basin
3. Preammoniation: ammonia added in the rapid mix basin followed by chlorine in the flocculation basin

The lowest levels of THM formation were observed for the concurrent and preammoniation schemes.

The point of application of chlorine and ammonia did not affect the occurrence of coliform bacteria in the pilot plant effluent. They were uniformly absent. This finding agrees with others: given long contact times free chlorine and chloramine residuals are equivalent disinfectants (based upon coliform kill).[101,102,170-172] The contact time from influent to effluent in the pilot plant was 3 hours 21 minutes. However, coliforms were observed on 13 occasions upstream from the sedimentation basins (17 minutes contact time) using the preammoniation scheme.

The use of the Millipore standard plate count procedure (m-SPC) bacteria indicated that preammoniation produced higher pilot plant effluent total plate counts than concurrent or sequential addition of ammonia or free chlorine. This disparity between the total plate counts of the different chloramination schemes was thought to be the result of preammoniation interference with the physical removal of particulate material (including bacteria). However, no statistical difference in effluent turbidity was found for the ammoniation schemes. Therefore the higher plate counts appear to be due to the reduced bactericidal efficiency and not compromised filtration efficiency.

Factors Affecting the Efficiency of the Process. The preferred sequence of chemical addition is usually ammonia first and then the chlorine. The ammonia must be well dispersed before the chlorine is added. This is a critical factor. When the chlorine is added it must be mixed as quickly as possible (2 sec). This is necessary because the reaction between ammonia and chlorine solution at the pH

range of 7–8.5 is practically instantaneous. If mixing is slow or poor, side reactions between the chlorine and organic matter can interfere with the formation of true chloramines. An example would be organic matter subject or prone to bleaching by chlorine solution.

White et al.[172] found that in a high-quality tertiary effluent, mixing at the point of application greatly affected the bactericidal efficiency of the chloramine process. In this particular case the ammonia was added to a completely nitrified and filtered effluent which passed through a pumping cycle before the chlorine was added. The chlorine solution was mixed with turbine-type mechanical mixers. The total coliform count with the mixers was usually about 13–20/100 ml MPN. With no mixing the count was 1300/100 ml MPN or greater.

Water temperature affects the reaction time between ammonia and chlorine. The reaction is significantly retarded at temperatures below 50°F. Treatment plant operators in cold areas should be cognizant of this because it can effect the location of the point of application of each chemical.

Pretreatment Problems. When using the ammonia–chlorine process as a pretreatment process in which rapid sand filters are involved, certain precautions must be taken. An addition of ammonia in excess of chlorine can promote the growth of nitrifying bacteria in the filter beds. Since chloramines will not react to destroy nitrites, and since the excess of ammonia acts as a nutrient, the nitrifying bacteria will proliferate. Then it becomes necessary to destroy the bacteria and the nitrites by free chlorine. If a proper ratio of chlorine to ammonia is maintained, including the natural ammonia content of the water, and if an adequate chloramine residual (0.5–1.0 ppm) is maintained at the discharge of the filters, such difficulties with nitrifying bacteria will not occur. To eliminate all possibility, the operator should not use the acid O-T method for residual determination when using a combination of chlorine and ammonia, because if nitrites do develop, they cause a significant false O-T reading. Use instead the amperometric titration procedure or the DPD–ferrous titrimetric method.

Disadvantages of the Ammonia–Chlorine Process

Effect upon Water Quality in Distribution Systems. A significant disadvantage of the chloramine process is that the addition of ammonia to a water does provide nutrients sufficient to cause algal blooms in reservoirs and an increase in distribution system bacteria population, which might not occur otherwise. This situation can be avoided by the continual presence of a chlorine residual regardless of contact time.

Fortunately, contact time has no effect upon THM formation by chloramines. Therefore the nutrient factor can be overcome by maintaining a residual regardless of the length of contact time. If the system is too spread out to maintain a residual, then rechlorination may be necessary. In this instance only chlorine is required.

When the residual disappears the chloramines revert to ammonia in the form of NH_4OH.

Effect upon Kidney Dialysis Patients. Kjellstrand et al.[300] reported that chloramine residuals in tap water pass through reverse osmosis membranes in dialysis machines quite easily. Furthermore, they directly induce oxidant damage to red blood cells, resulting in methemoglobin formation and damage to the HMPS with which red cells defend themselves against oxidant damage, which induces hemolysis and short red cell survival time. Moreover, they sensitize the patients to oxidant drugs like primaquine, sulfonamides, etc.

The effect upon the patient is apparently directly from the chloramines and not from any nitrogen compound associated with the chloramines. When Kjellstrand experimented with ascorbic acid as a dechlorinating agent there was no detectable methemoglobin formation. Ascorbic acid reduces chloramines* to hydrochloric acid and ammonia.[300]

Activated carbon has been thoroughly investigated as an acceptable method for the dechlorination of both free chlorine (HOCl) and chloramines ($NHCl_2$, NH_2Cl, and NCl_3).[302-304] The most difficult species to dechlorinate on a predictable basis are the chloramines. Meyer and Klein[301] investigated the use of granular activated carbon (GAC) as a practical method to remove chloramines from dialysis water. They found that the dechlorination capacity of three different carbons differed as much as one order of magnitude. They further estimated that a 1.5 mg/l chloramine residual could be reliably removed during 156 5-hour dialyses with a GAC column.

Ascorbic acid is a more positive method and is appealing because of its simplicity. There is no danger of any chloramine breakthrough into the dechlorinated water provided the ascorbic acid dosage is adequate and the application is reliable. The GAC method requires careful monitoring to avoid chloramine breakthrough as the carbon approaches exhaustion.

Experience has shown that the use of GAC for dechlorinating chloramines is site-specific. Therefore any water containing chloramines to be dechlorinated should be bench tested to determine precisely the equipment required for a particular carbon source to be used.

Effect upon the Aquarium. Chloramines represent a two-pronged toxicity dilemma to aquatic life. Both chloramine residuals and unionized ammonia are toxic to fish in very low concentrations. The mechanism of chloramine toxicity is as follows: Chloramines pass readily through the permeable gill epithelium with an insignificant amount of cell damage. Once the chloramines have entered the bloodstream they chemically bind to iron in hemoglobin in red blood cells. This cripples the ability of the cells to bind oxygen. This condition is known as methemoglobinemia. It is similar to the oxidation of hemoglobin caused by nitrite toxicity.[305,306]

---

* The types of chloramines were not identified.

CHLORINATION OF POTABLE WATER

When monochloramine is dechlorinated by a reducing compound which produces sulfurous acid (H_2SO_3), ammonium chloride is formed as follows:

$$NH_2Cl + H_2SO_3 \longrightarrow NH_4Cl + H_2SO_4 \qquad (6\text{-}2a)$$

Ammonium chloride ionizes to form ionized NH_4^+ and unionized NH_3 nitrogen (ammonia). Therefore all of the ammonia utilized to produce monochloramine is returned to the water as shown in Eq. (6-2a), $NH_4^+ + Cl^-$. This represents total ammonia. To avoid fish toxicity unionized ammonia must be limited to 0.025 mg/l. Unionized ammonia concentration can be calculated from total ammonia. See Chapter 7. The maximum allowable total ammonia is about 0.4 mg/l at pH 8.3 and 70°F in a marine aquarium.

Therefore ammonia removal is also a must for the aquarium when chloramines are dechlorinated chemically. In aquatic systems such as an aquarium, ammonia accumulates from nitrogenous wastes released directly from the fish and from deamination of protein in food and wastes by heterotrophic bacteria. However, nature provides a mechanism that serves to control ammonia buildup. It is called nitrification. It is the primary means of ammonia removal in fish culture systems.[307] By this process, ammonia is oxidized to nitrate in two separate steps, each step dependent upon and controlled by a specific group of bacteria:

$$NH_4^+ + \frac{3}{2}O_2 + \textit{Nitrosomonas}\ \text{Group} \longrightarrow NO_2^- + 2H^+ + H_2O \qquad (6\text{-}2b)$$

$$NO_2^- + \frac{1}{2}O_2 + \textit{Nitrobacter}\ \text{Group} \longrightarrow NO_3^- \qquad (6\text{-}2c)$$

Simply placing fish into an aquarium with a recirculation filter will initiate the growth of these two groups of bacteria because they exist naturally in the environment. Unfortunately, these bacteria grow slowly, so they need a conditioning period to attain the necessary population to limit the production of the ammonia produced by the aquatic culture.

Removal of the ammonium ion (NH_4^+) is best accomplished by naturally occurring zeolites known also as clinoptilolites. Zeolites have a high selectivity for NH_4^+ removal, which decreases the toxic unionized ammonia concentration. They are relatively inexpensive and commercially available. They have been used for many years in industry as a freshwater demineralizer.[305,307]

Finally it should be pointed out that the toxicity of ammonia decreases with pH. If the pH of the system is below 7 the ammonia produced will not be a problem in the aquarium. For example: in water at pH 6.9 and 75°F, 99.58 percent of the ammonia will be in the nontoxic ionic form. However, at the same temperature, but at a pH of 8, ammonia will be 9.49 percent ionic.[305]

Activated carbon has been used with moderate success to dechlorinate chloramines in potable water treatment processes, but its principle role is to adsorb organics in order to eliminate taste and odors. It is a popular filter material used in the aquarium industry for the filtration of fresh water and marine aquaria. It is a porous material that removes molecules from solution by binding them to the carbon's surfaces by a process known as *adsorption*. When carbon removes chloramines, ammonia is formed in the early stages of the adsorption process, followed by a decrease in ammonia as it is converted to nitrogen gas. This reaction occurs with predictable results if the carbon has been seasoned by exposure to residual chlorine. When chloramines come in contact with virgin carbon the chloramines are destroyed, but some of them revert to ammonia as the surface of the carbon oxidizes and chloride ions are released into solution.[305] If the carbon has been previously exposed to chlorine the carbon surface is coated with oxides which prevent the formation of ammonia by chloramine reversion. Therefore the surface reaction of the chloramines and carbon results in a combination of adsorption and catalytic oxidation, releasing nitrogen gas and chloride ions and lowering the water pH.

General Summary. For dialysis systems ascorbic acid for dechlorination seems to have all the practical and reliability assets. Activated carbon and some green sand zeolite are also capable of achieving complete dechlorination.

Aquaria require much more treatment because both chlorine residuals and ammonia are toxic to aquatic life. Dechlorination plus ammonia removal can be accomplished in one step using activated carbon provided the system is carefully planned and properly designed. Dechlorination can be easily accomplished by chemicals such as the sulfite ion species or thiosulfate. The latter is not recommended, however, because it is too expensive and has a longer stepwise reaction. Following chemical dechlorination the ammonia can be removed by the use of zeolites. The zeolite removal process is similar to a water softening system.

American vs. European Practice: A Paradox. The paradox is the current European limitation on the allowable ammonia nitrogen concentration in the finished water (0.05 mg/l) vs. the allowable addition of ammonia nitrogen to U.S. drinking water to form chloramines which prevents the formation of THMs.

European Practice. The ammonia–chlorine process was never practiced as such in European treatment schemes because most of the surface waters contained an abundance of ammonia nitrogen. The source of ammonia nitrogen derived mainly from untreated sewage and industrial wastes. Therefore the European treatment schemes used prechlorination for nitrogen removal. This in turn allowed the practice of free residual chlorination. In some instances as much as 20 mg/l chlorine were required to achieve the breakpoint.

The detection of trihalomethanes where free residual chlorination of polluted waters was practiced and the general undesirability of ammonia nitrogen in potable water resulted in some significant changes in European water treatment practices. There is no question that ammonia interferes considerably with the disinfection process, particularly when that process is free residual chlorination. This circumstance, coupled with the observations that trihalomethane concentrations increase as free chlorine residuals increase, was considered by the European community of water producers as grounds for elimination of ammonia nitrogen in potable water supplies. Although it is never mentioned as such, one of the dominating factors against ammonia nitrogen, particularly in U.S. supplies, is the nutrient factor, which eventually contributes to water quality degradation in reservoir systems, long transmission lines, and widespread distribution systems unless adequate chlorine residuals can be maintained.

European water purveyors were thus made subject to an agreement by the ministers of the European Economic Community on December 19, 1978, limiting the allowable ammonia-nitrogen concentration in water delivered to the consumer to 0.05 mg/l. Moreover, the ministers decreed that the method of nitrogen removal to comply with the limitation must be biological nitrification. Realizing that such a drastic change in the treatment procedures would take time, the use of chlorine for nitrogen removal was allowed until installation of the recommended system was found practicable.

As of 1979 the agreed upon treatment train for nitrogen removal where applicable and possible included: raw water ozonation at a fixed dosage; groundwater storage followed by coagulation, sedimentation, and filtration followed by ozonation; and then biological nitrification followed by chlorine dioxide for effluent disinfection.

Actually, effluent disinfection by either chlorine or chlorine dioxide was required, for ozone was not generally accepted as the final disinfectant. Intermediate points of chlorination or chlorine dioxide are acceptable provided they do not interfere with biological nitrification. This is a two-step process of ammonia oxidation which involves the formation of nitrites. The conversion of nitrites to nitrates is the final step in ammonia removal. This is accomplished by the *Nitrobacter* species of bacteria. Oxidation of nitrites and/or destruction of the *Nitrobacter* bacteria by chlorine or chlorine dioxide may either interrupt or destroy the nitrification process. Therefore the addition of either chlorine or chlorine dioxide at intermediate points must be carefully selected.

Rittman and Snoeyink[309] (1984) presented a comprehensive review of European practice together with the microbiological theory demonstrating how biological processes within a water treatment plant can remove the organic and inorganic substrates that promote biological instability, the root cause of biofouling in a distribution system. A biologically stable water is one which does not support the growth of microorganisms to a significant extent, whereas an unstable water results in high numbers of microbes in the distribution system unless persisting chlorine residuals are maintained.

Rittman and Snoeyink[309] investigated both French and British practices of nitrification using biological filters, fluidized bed filters (also called biological sedimentation filters), rapid sand filters, and granular activated carbon beds. All of these processes are biofilm reactors that achieve nearly 100 percent ammonia removal even at low temperatures.

While the primary objective of these processes is to remove ammonia nitrogen they simultaneously remove a significant proportion of the natural chlorine demand of the raw water. This allows the Europeans to disinfect at lower chlorine dosages, which eliminates, in their view, the THM hazard. If chlorine dioxide is used its demand is likewise lowered, thereby eliminating fears of high concentrations of chlorate and chlorite ions.

United States Practice. In 1980 the EPA changed its stance on the use of chloramines as a disinfectant for potable water. It was allowed to become a primary instead of a secondary disinfectant. This provided an easy solution to those water producers who were plagued with unacceptable THM levels caused by free chlorine residuals. Understandably this revived the ammonia–chlorine process in the U.S.A.—so much so that in the long run it may do more harm than good. A 1984 report of a 50-state survey by Hack[310] revealed that 53 percent of chloramine users claimed distribution system residuals in the 3–4 mg/l range. Such residuals would be appalling to Europeans. This practice of high residuals to maintain water quality will eventually lead to greater problems.

It is unfortunate that chloramines do not contribute to the formation of THMs. Otherwise water producers in the U.S.A. would have to clean up the raw water, thereby reducing the chlorine demand. This would allow significantly lower doses of chlorine and/or chlorine dioxide and would provide biologically stable water to enter the distribution system.

What the water industry needs now is to put a ceiling on chlorine dosages. White has been advocating for years that an arbitrary limit of about 2.0 mg/l be placed on the 15-min chlorine demand for any raw water supply. When compared to European practice this may be too high. To put chlorine demand in perspective one only has to look at the 15-min chlorine demand of this planet's natural water source—seawater. It rarely exceeds 1.5 mg/l. A limitation on *chlorine demand* should be a surrogate parameter for raw water quality. If the limit is exceeded then the water would have to undergo special pretreatment to reduce the 15-min chlorine demand to the acceptable level.

Blending Chloraminated Water with Chlorinated Water. The Metropolitan Water District of Southern California changed its primary disinfectant from free chlorine to chloramine in 1984 to help its member agencies control THM formation. Problems of taste and odor complaints were anticipated so they designed a laboratory study to determine the effects of blending chloramine residuals with chlorine residuals by developing blend-residual curves to simulate the systems involved.[311] The potential problems of blending are related to a mixture that proceeds through the breakpoint curve which allows the formation of dichloramine and possibly nitrogen trichloride. These two compounds are notorious for causing consumer complaints. MWD decided on a 3:1 Cl–N ratio for chloramination. Using this ratio and accounting for the almost negligible concentration of naturally occurring ammonia-N they developed an induced breakpoint curve for each of their two supplies: (a) The Northern California water supply known as state project water (SPW) and (b) the Colorado River water labeled CRW. The induced breakpoint of the SPW at pH 7.5 occurred at a Cl–N ratio of 11:1 and for the CRW at pH 7.9 the ratio was 9.2:1. These breakpoints occurred after 4 hours in the dark at 25°C. They followed this with a series of breakpoint curves due to blending. Two of these are shown in Figs. 6-6a and 6-6b, which illustrates the same two waters at contact times of 5 minutes and 4 hours. These figures demonstrate that at short contact times all of the chlorine species may exist together in the breakpoint region.

From this experimental work the MWD staff developed a computational model that allows the prediction of acceptable blends of chloraminated and chlorinated waters without experimentally determining each case. They anticipate the need of further analyses as a check on the distribution system to see if there may be occurrences of Cl–N ratios greater than 5:1. If such a situation exists it becomes bacteriologically significant at ratios of about 6.5:1. From this point forward the chlorine residual decreases rapidly to nearly zero at the breakpoint where disinfection ceases entirely. This is a more serious situation than a consumer taste and odor complaint.

Advances in Chlorination

Attempts to explain Griffin's findings[104,105] led to serious scientific investigations during 1946–1950. The work of Fair, Morris, Chang, Weil, Burden, and Culver[106-108] is largely responsible for our present knowledge of the kinetics of chlorination.

In 1950 Palin[109,110] in England did some masterful work on the kinetics of the breakpoint reaction and the distribution of various chlorine residual fractions. In 1954 Granstrom[111] explained the breakpoint phenomenon. Then Williams of Canada, working with a most difficult water, proceeded to substantiate the laboratory findings of both Palin and the Harvard group.[112-114]

In the pursuit of his research work on chlorination, Palin directed his attention toward the analytical measurement of chlorine residuals. He wanted to be able to quantify each chlorine residual fraction such as free chlorine, monochloramine, dichloramine, and nitrogen trichloride. His many years of work resulted in the DPD method of residual differentiation.[115,116] In addition to the chlorine fractions, Palin developed procedures for chlorine dioxide, chlorite ion, bromine and ozone.[117] The titrimetric procedure using ferrous ammonium sulfate as the titrant is a favorite with researchers.

However, the amperometric titrator remains the most popular method among

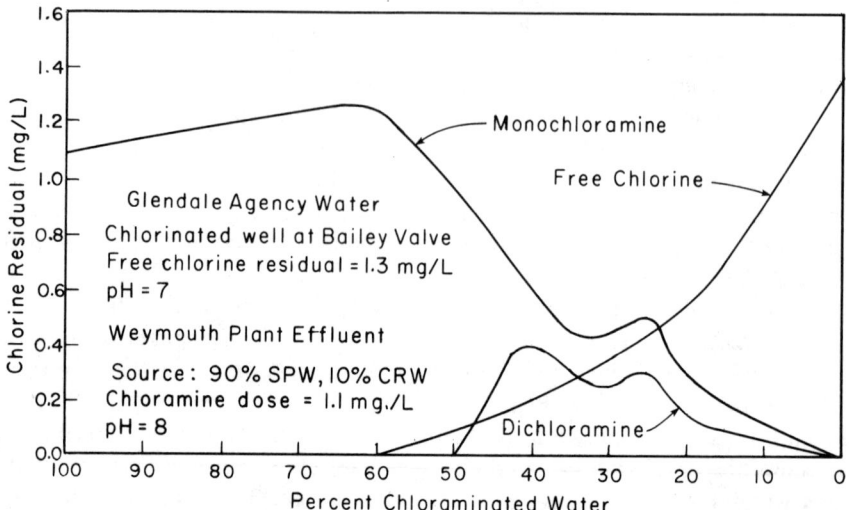

Fig 6-6a Chlorine residual curves for Glendale California chlorinated well water blended with Weymouth treatment plant chloraminated effluent after 5-min contact time.[311] (courtesy Metropolitan Water District of Southern California).

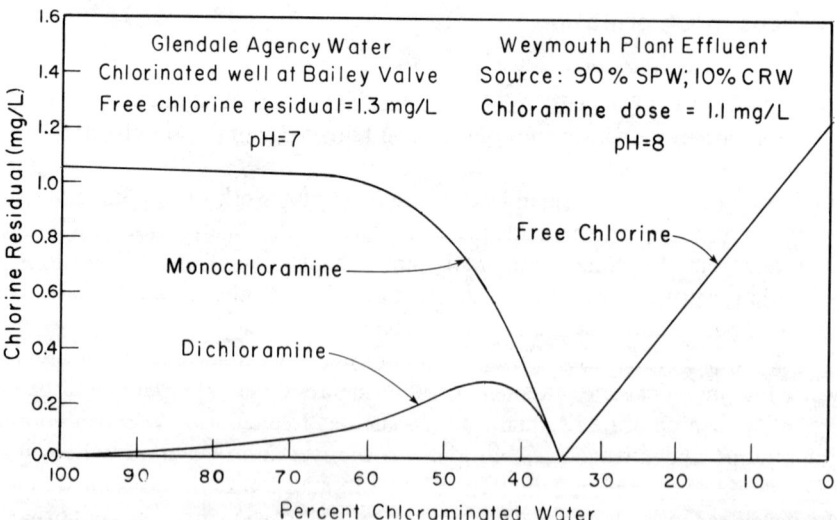

Fig 6-6b Chlorine residual curves for Glendale California chlorinated well water blended with Weymouth treatment plant chloraminated effluent after 4-hour contact time.[311] (courtesy Metropolitan Water District of Southern California).

operators who need to differentiate between free and combined chlorine. As the titrator is portable it is well suited to field work for on-site determination of chlorine residuals. It is also the most popular method in wastewater and water reuse situations. Orthotolodine is still being used but is rapidly being replaced by the colorimetric DPD tablet method and the FACTS method. Orthotolidine lost its popularity when it was discovered that the orthotolidine powder was a carcinogen that affected the urinary tract of workers who prepared the indicator solution.

The most recent laboratory research on the dynamics of chlorination and the breakpoint phenomenon was by Wei and Morris[118] in 1972 at Harvard and by Saunier and Selleck[119] at the University of California, Berkeley in 1976. This work explained in much greater detail the kinetics of chlorination and provided a more precise explanation of what actually does trigger the breakpoint. Moreover computer models were developed which provide the practitioners with a tool for predicting what will occur in a given situation.

Although the laboratory research work taught the industry about the theoretical aspects of chlorine and ammonia reactions, it was the work by Williams[120] on a plant scale basis which showed the practical effects of chlorinating a water containing both ammonia nitrogen and organic nitrogen. His work demonstrated that chlorination in the presence of organic nitrogen formed an unusually stable chloramine compound which differentiates as the dichloramine fraction in either the amperometric or DPD–FAS procedures. This has since been confirmed by many others.

Some of the most important advances in chlorination have been the improvement in the metering and control equipment. Automatic residual control was not available in North America until about 1960. Before that time Wallace and Tiernan had made some custom-designed residual controllers (using orthotolidine color as the control) for special situations. About 1950 Wallace and Tiernan of Great Britain began offering automatic residual control. Even though Wallace and Tiernan U.S.A. had already developed the amperometric titrator and the amperometric residual recorder by 1950, it was another decade before they believed that their new line of chlorination equipment was capable of combining flow pacing with residual control. Since 1960 a wide variety of control modes have been available from both Wallace and Tiernan and Fischer and Porter. Since about 1976 Capital Controls has been offering flow pacing combined with residual control. These developments have increased the reliability of the chlorination process.

The Free Residual Process

Role of Ammonia Nitrogen. This process is the implementation of the breakpoint phenomenon. The chemistry of this phenomenon is described in Chapter 4.

Free residual chlorination as a process was a direct result of the search for a better way to produce a palatable water than by the ammonia–chlorine process.

This search was accomplished in the field and laboratory almost simultaneously on opposite ends of the U.S.A by Griffin[121,122] and O'Connell[123] in 1939. After literally thousands of tests on different waters, with different dosages and contact times it was found that the optimum residual for the most palatable water should contain 85 percent free chlorine.

This process is based upon the fact that a free residual (<5 mg/l HOCl) does not impart any off-flavor to the water. Moreover this species of chlorine residual fraction is by coincidence the most germicidal.

In the free residual process, sufficient chlorine is added to destroy the ammonia nitrogen. This occurs when the ratio of chlorine to ammonia nitrogen (as N) is about 10:1 by weight. However, the stoichiometric ratio (theoretical) is 7.6:1. In practice it is known to vary from 8.5:1 to 11:1.

If ammonia nitrogen is present in the water, chlorine reacts rapidly with it to form mono- and dichloramine. These compounds will appear in the combined residual. Depending upon the factors of concentration and ratio of chlorine to ammonia nitrogen, NCl_3 (nitrogen trichloride) may form, which causes an offensive odor at the tap. Chlorine reacts slowly with most organic nitrogen compounds to form organic chloramines which are stable even in the presence of high concentrations of free residual. Unfortunately these N-chloro compounds have little if any germicidal power. Moreover, most of the N-chloro compounds impart objectionable off-flavors to the treated water, depending upon their concentration. These compounds make up part of the combined residual fraction, the other fraction is the free chlorine.

Fig. 6-7 illustrates the chlorine and ammonia reaction that produces the breakpoint phenomenon. At the breakpoint (bottom of the curve) there exists an irreducible minimum of total chlorine residual. These are called the "nuisance residuals." They consist of free chlorine, mono- and dichloramine, and organochloramines. To the right of the dip there will be found increasing amounts of free residual together with the nuisance residuals and possibly some nitrogen trichloride. Nitrogen trichloride is not soluble in water, so it is extremely volatile. Therefore, water that generates nitrogen trichloride in the distribution system will produce an offensive odor at the instant of withdrawal from a tap. Reaction time between the chlorine and ammonia nitrogen is all-important. The only satisfactory way to provide proper control is to perform a bench scale study using different Cl:N ratios, different dosages, and different contact times. If the water tends to form NCl_3, try to find a contact time that occurs before the water leaves the plant, then the finished water can be subjected to postaeration, which readily removes any NCl_3. It is also readily destroyed by sunlight.

The success of the free residual process depends upon maintaining a free residual that is always at least 85 percent of the total residual, because free residual is taste free and is the most germicidal species.

The speed of reaction between chlorine and ammonia nitrogen to produce a free residual varies from water to water because this reaction is not only temperature

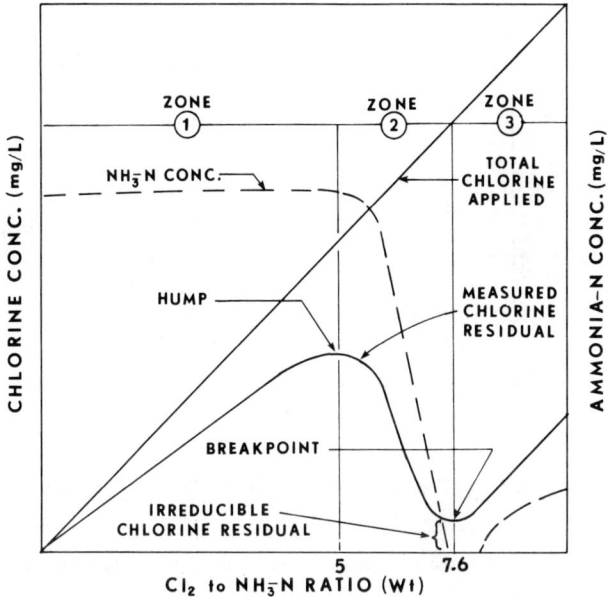

Fig 6-7 Effect of the chlorine and ammonia reaction illustrating the breakpoint phenomenon.

and pH dependent, it is grossly affected by the concentrations of both chlorine and ammonia-N. If organic nitrogen is present in the untreated water it compounds the problems associated with this process. This occurrence is described below.

Effect of pH. The free residual process is grossly affected by the pH of the water. This is due to the chemistry of hypochlorous acid (see Chapter 4). It dissociates in water as follows:

$$HOCl \longrightarrow H^+ + OCl^- \tag{6-3}$$

All analytical techniques for measuring free residual include the sum of the undissociated hypochlorous acid (HOCl) and the hypochlorite ion (OCl$^-$). However the efficiency of the free residual process relates directly to the concentration of the undissociated HOCl. The reason for this lies in the fact that the germicidal efficiency of HOCl is believed to be about 100 times more than the OCl$^-$ ion.[100] The ratio of formation of each of these species is a function of pH (see Table 4-4, Chapter 4). For example, at pH 5 an aqueous solution of free chlorine contains 99.74 percent undissociated HOCl at 20°C. At pH 8 the undissociated HOCl is only 27.69 percent, the remainder is the OCl$^-$ ion. As can be seen from Table 4-4, dissociation is also affected by the temperature of the water but to a lesser extent.

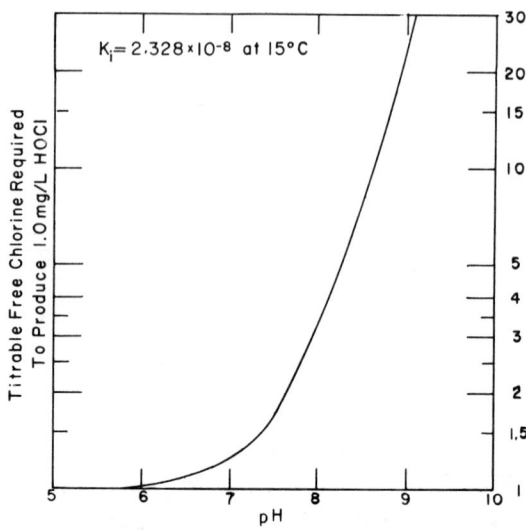

Fig 6-8 Effect of pH on the formation of hypochlorous acid.

Fig. 6-8 illustrates the amount of titrable free chlorine residual required to produce 1.0 mg/l HOCl. At pH 8 it requires 5 mg/l titrable free residual to produce 1 mg/l undissociated HOCl. If the pH were 7.5, only 2 mg/l of titrable free chlorine residual would be required to yield 1 mg/l HOCl, etc. The effect of pH on the free residual process affects the selection of chlorine dose as can be seen from the above example. Consider the example of virus destruction. It is generally conceded that a free chlorine residual is the most reliable for virus destruction. Clark et al.[124] believe that it takes about 0.3 mg/l of undissociated HOCl residual to inactivate Coxsackie A2 virus in about 20 minutes at 5°C.

If the pH of the water to be treated is 8.5 then 10 mg/l of titrable chlorine is required to produce 1.0 mg/l undissociated HOCl (see Fig. 6-8). The free chlorine residual required at this pH level will be 10 × 0.3 or 3 mg/l. This then is the minimum dose that will be required to provide 0.3 mg/l of 100 percent undissociated HOCl at the end of 20 minutes.[124,125]

Another consideration of the free residual process is the effect of sunlight on HOCl. While it is a relatively stable compound in that it resists loss due to aeration, losses due to bright sunlight have been reported as high as 2 mg/l HOCl in 4 hours. Chloramines act in reverse fashion; they do not suffer appreciable loss due to sunlight but are subject to losses as high as 15 percent of the residual by aeration. If nitrogen trichloride is present, all of it is quickly lost due to either sunlight or aeration or both.

Reservoir Effect of OCl− Ion. The notion is valid that at high pH values the distribution of the two species of titrable chlorine—HOCl and OCl−—is such

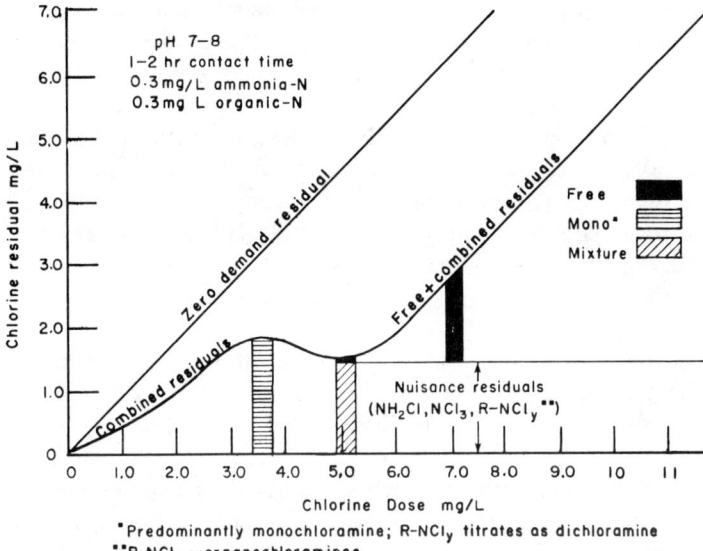

Fig 6-9 Effect of chlorine-ammonia reaction in a water containing organic nitrogen.

that dissociation of the OCl⁻ ion and the hydrogen ion provides a reservoir for the formation of HOCl. In accordance with Le Chatelier's principle,* as soon as the HOCl in Eq. (6-3) is used up, i.e., reduced to HCl, more HOCl is immediately formed from the OCl⁻ ion and H⁺ ion to maintain the chemical equilibrium in Eq. (6-3).

Role of Organic Nitrogen. Waters containing organic nitrogen in addition to ammonia nitrogen are quite another matter. Organic nitrogen in concentrations as low as 0.3 mg/l will seriously interfere with the chlorination process, and looms as a formidable obstacle in producing an acceptable water. Two areas that have been troubled with the organic nitrogen problem are Little Falls, New Jersey[126] and Brantford, Ontario. The latter is well documented.[114,120] Griffin[104] found that waters high in albuminoid ammonia (organic nitrogen) displayed a plateau rather than the typical sharp dip associated with breakpoint curves. Fig. 6-9 illustrates this characteristic. Palin[110] and others have confirmed these findings. In Chapter 4 Taras was able to clearly demonstrate how continuing chlorine demand tests could provide clues that can alert an operator to anticipate organic nitrogen problems derived from increasing pollution in the raw water supply.

The significant difference between ammonia nitrogen and organic nitrogen is

---

* Le Chatelier's principle is defined as follows: If some stress is brought to bear upon a system in equilibrium (such as the concentration of one or more of the components), a reaction occurs which displaces the equilibrium in the direction which tends to undo the effect of the stress.

the enormous varieties of organic nitrogen compounds that may occur in a water supply. When any of these are present in excess of 0.3 mg/l this will complicate the control of the free residual process. Each of these organic N compounds may have widely varying reaction times with chlorine to form organochloramines. The simple ammonia nitrogen compound (NH_3) reaction with chlorine rarely takes longer than 30 min for the complete destruction of the ammonia and a 75–80 percent loss of nitrogen. Depending upon the temperature this would be for an ammonia nitrogen concentration of 0.3–0.5 mg/l. However the reaction between chlorine and organic nitrogen may go on for days before going to completion with the chlorine.[120] Depending upon the particular nitrogen compound the reaction may never go to completion.[127]

Organic nitrogen in potable water is a direct result of wastewater contamination. However there is thought to be a significant amount entering surface supplies due to natural runoff that leaches and dissolves organic compounds from industrial waste dump sites. The source of most organic nitrogen compounds is in proteinaceous matter and urine, both of which are present in copious quantities in all domestic wastewater discharges. Urine contains substances that react with chlorine to form extremely stable N-chloro compounds that exhibit characteristics of dichloramine when differentiated by either the amperometric method or the DPD method of residual analysis. One of these constituents of urine is creatinine, which forms a chlor-creatinine compound and shows up on the curve of Fig. 6-9 as a nuisance residual. There are other N-chloro compounds (organic chloramines) in this same category that have little or no bactericidal value since they appear to be unreactive.[127] Their significance is not clearly understood, but the consensus is they are undesirable. In Fig. 6-9 it is to be noted that there is very little drop in the chlorine residual beyond the hump, signifying continuing and competing reactions between mono- and dichloramine, no perceptible loss of nitrogen in the reactions, and that the irreducible minimum residual is considerably larger than in water containing only ammonia nitrogen. Here again, it is the ratio of the free chlorine beyond the plateau to the total chlorine residual that is of importance. The combined fraction will contain N-chloro compounds responding to analysis as mostly dichloramine, and always some monochloramine. As the ratio of chlorine to nitrogen approaches 20:1, obnoxious quantities of nitrogen trichloride are certain to develop. Williams[113,114] has discovered that best results occur when enough chlorine is applied to produce a total of approximately 6.0 mg/l combined chlorine residual, resulting in about 5.0 mg/l free residual (HOCl), 0.4 mg/l monochloramine, and 0.6 mg/l dichloramine, as differentiated by the amperometric forward titration method. To prevent formation of nitrogen trichloride in the distribution system, this water is subsequently treated with an excess of ammonia to convert the outgoing residual to monochloramine 80 percent and dichloramine 20 percent. The water can be dechlorinated and rechlorinated, whichever makes the most palatable water at the least cost.

Another important factor in the kinetics of chlorine and nitrogenous compounds is the relationship of monochloramine to organochloramines. In water reuse systems where contact times can be monitored and are designed for maximum effectiveness, it has been discovered that, after a certain point, as the contact time is lengthened the germicidal efficiency starts to decrease. This phenomenon seems to occur when the contact time exceeds one hour—at least in some systems. In these particular systems it was discovered that the combined residual showed a decrease in the monochloramine fraction and a simultaneous increase in the dichloramine fraction which is suspected of being ineffective organochloramines. Therefore it appears that given time, monochloramine will be gradually converted to organochloramines.

Whenever organic nitrogen is present in concentrations higher than 0.3–0.5 mg/l the operator can expect difficulties with the free residual chlorination process. This much organic nitrogen indicates a highly undesirable low-quality water with recent heavy sewage contamination. Special pretreatment processes should be considered for these waters. White has suggested that a water quality parameter for any potential drinking water supply be limited to 0.3 mg/l organic N.[128]

Formation of Potentially Hazardous Compounds. The 1970s witnessed the first serious indictment of the free chlorine residual process. This discovery was the consequence of the U.S. Public Health Service concern over the widespread use of herbicides, pesticides and other petrochemical products. They feared the intrusion of these organic chemicals into the water supplies. A 1970 survey discovered that the levels of dissolved organics in many water supplies exceeded the Public Health Service's recommended limit for carbon chloroform extractable organics, which at that time was 200 micrograms per liter.[129] In 1972, the EPA reported that 46 organic chemicals were present in trace amounts in both the raw and finished water supplies at three locations along the Lower Mississippi,[130] and a 1974 EPA study identified 66 organic compounds in the New Orleans drinking water.[131] On the same day that the New Orleans report was issued (Friday, Nov. 8, 1974), Russell E. Train, Administrator of the EPA, announced that he was ordering an immediate nationwide survey to determine the concentration and potential health effects of certain organic chemicals in drinking water.

On December 16, 1974 President Ford signed into law, Public Law 93–523, The Safe Drinking Water Act. Section 1442(a)(9) of this Act states:

> The Administrator shall conduct a comprehensive study of public water supplies and drinking water sources to determine the nature, extent, sources of, and control of contamination by chemicals or other substances suspected of being carcinogenic. Not later than 6 months after the enactment of this title, he shall transmit to Congress the initial results of such study together with such recommendations for further review and corrective action as he deems appropriate.

Then on Dec. 18, 1974, Administrator Train named the 80 cities to be included in the National Organics Reconnaissance Survey.

The issue of chlorinated organics formed in the treatment of drinking water became a priority topic to the U.S. EPA beginning in January, 1975. In 1974 the work of Rook[132] in the Netherlands and Bellar, Lichtenberg, and Kroner,[133] in the U.S.A. demonstrated that chlorine used for disinfection did *in some waters* react with organic precursors to produce some trihalomethanes (mostly chloroform) in the finished water which were not found in the respective raw water at the locations of study. This was indeed an unsettling observation. To assess the general situation across the United States, the National Organics Reconnaissance Survey was conducted in 1975 by Symons et al.[134]

This survey revealed that four trihalomethanes (THMs): chloroform (trichloromethane, $CHCl_3$), bromodichloromethane ($CHCl_2Br$), dibromochloromethane ($CHClBr_2$), and bromoform (tribromomethane, $CHBr_3$) are widespread in chlorinated drinking water in the U.S.A. Table 6-2 shows the THM concentration in water supplies of heavily populated areas picked at random from the NORS report described above.[134] There are only three metropolitan areas that exceed the EPA interim maximum contaminated level.

A review of the current knowledge of trihalomethane formation by Trussell and Umphries is an informative document.[135] It discusses in detail the chemistry of the formation of chloroform by chlorine and the chemical path of bromoform formation by the reaction of chlorine with the bromide ion.[136]

Chloroform, which seems to be present in the greatest concentration of the four THMs detected, is a known carcinogen to animals and is therefore of immediate concern. However it is not considered an acute hazard to man, particularly at the low concentrations detected. Nevertheless it became clear that chlorination practices should be reviewed in order to minimize THM formation.

Morris et al.[137] reported in 1980 that a large number of nitrogenous organic compounds occur naturally in water supplies and react readily with aqueous chlorine. This results in the formation of N-chloro compounds that exert a significant chlorine demand and some can produce chloroform, particularly in the pH range of 8.5–10.5. The compounds identified by Morris et al, are adenine, 5-chlorouracil, cytosine, guanine, purine, thymine, and uracil.

This same investigation also revealed that summer blooms of blue-green algae can result in a significant increase in the organic N content of a potable water supply. This raises concern about the potability of a finished water during occurrences of summer blue-green algae. Aside from the taste and odors associated with such occurrences, high levels of organic N material released by these algae, could result in increased THM formation as well as combined N-chloro species that demonstrate little or no germicidal value. Moreover some of these N-chloro compounds yield false positive free chlorine residuals.*

---

* The precise level of interference is not known. The amperometric titration method seems to demonstrate the least interference from the N-chloro compounds in the measurement of free chlorine. Morris et al.[137] indicated the interference for some compounds was less than 0.5 mg/l residual and some were slightly more.

Table 6-2 Total THM Concentration in Treated Water from a Selected List of Metropolitan Areas from NORS Report[134]

Water Supply and Raw Water Source	TTHM, µg/l	Water Supply and Raw Water Source	TTHM, µg/l
Boston, Mass Norumbego Treatment Sta.	4.8	Concord, CA Contra Costa Co. Water Dist. Calif. Water Proj. & San Joaquin Riv. Bollman Plant	55.8
New York City Croton Reservoir	29.9	Atlanta GA Chattahoochee River Plant	48
Little Falls, NJ Passaic Valley Water Co. Passaic River	71	Chattanooga, TN Chattanooga River	40
Philadelphia Water Dept. Torresdale Plant Delaware River	106	Nashville TN Lawrence Plant Cumberland River	21.3
Washington DC Dalecarlia Plant Potomac River	51.2	Cincinnati, OH Ohio River	62.3
Baltimore, MD Loch Haven Reservoir Montebello Plant No. 1	45	Chicago, IL So. District Filter Plant	29.3
Fairfax Co. Water Authority Anandale, Virginia New Lorton Plant Occoquan River Impoundment	73.5	Indianapolis, IN White Riv. Plant & Wells White River	42
Miami, FL Preston Plant Groundwater	427	Detroit, MI Waterworks Park Plant Detroit River at Belle Isle	24.4
Ottumwa, IA Des Moines River	0.8	Milwaukee, WI Howard Ave. Plant Lake Michigan	19.1
St. Louis, MO Central Plant Missouri River	72.2	Dallas, TX Bachman Plant Trinity River, Elm Fork	23
Denver CO Marston Plant Marston Lake	27	Los Angeles, CA Owens River Aqueduct	51
Cape Girardeau, MO Mississippi River	141.3	San Diego, CA Miramar Plant Colorado River Aqueduct	104
Salt Lake City, UT Mountain Dell Res.	42	San Francisco, CA O'Shaugnessy Lake Yosemite Calif. San Andreas Filter Plant	60.6
Phoenix, AZ Verde Plant Salt & Verde Rivers	44	Seattle WA Cedar River System	15.9
Calif. Water Project California Aqueduct Sacramento & San Joaquin Riv.	50	Cleveland, OH Division St. Filter Plant	31

The EPA generated many investigations at several treatment plants in the hopes of modifying the chlorination process towards the reduction of THMs. It was found among other things, that it was the presence of free chlorine (HOCl) which enhanced the formation of chloroform. It was also found that the THM formation was as much a result of the organic content of the water as it was the result of the chlorine being used. It was also confirmed that neither ozone, chlorine dioxide nor chloramines contributed to the formation of THMs.

In 1978, the EPA set maximum concentration limits on various compounds in potable water. Chloroform formation was given a maximum concentration level (MCL) of $100/\mu g/l$. A dosage limit for chlorine dioxide of 1.0 mg/l was suggested and it was stipulated that chloramines could only be used as a secondary disinfectant.[99] The EPA stance on chloramines changed in the 1980s; see the section on the "Ammonia–Chlorine Process" in this chapter.

In 1977, the EPA published a state of the art report on the use of ozone, chlorine dioxide and chloramines as alternatives to chlorine for disinfection of potable water.[138] These studies are discussed for chlorine dioxide in Chapter 12 and for ozone in Chapter 13.

Control and Removal of Trihalomethanes

Current Control Strategies. In spite of the fact that free chlorine can in some instances produce unacceptable concentrations of THMs there are a great many waters that remain either unaffected or are amenable to simple procedures such as moving the point of application or reducing the chlorine dosage.

The generation of THMs by chlorine is related to the precursors such as humic and fulvic acids, and a variety of organics. The quality and quantity of precursors is in turn related to the chlorine demand of the water which reflects upon the quality of the untreated water. This is why a continuous record of the chlorine demand of the untreated water is so valuable in evaluating any change in water quality.

Since THM formation is time dependent, some water systems follow free residual chlorination with ammonia application after some specified contact time with the free chlorine.[139] The time interval between the chlorine–ammonia sequence is usually on the order of 10–20 min. This utilizes the germicidal efficiency of HOCl, while the ammonia inhibits the formation of THMs. Moreover postammoniation has been practiced for many years for quality control in the distribution system.

Barnett and Trussell[140] reported in 1978 their investigation of THM formation in the Casitas, California, water system. They found a wide seasonal fluctuation in the total organic carbon (TOC) and potential THM content in the raw water supply (Lake Casitas). They found that THM levels can be maintained within acceptable limits by either adding ammonia downstream from the first point of chlorine application or by free chlorine followed by activated carbon filtration and that TOC content and THM precursors can be reduced significantly by adding 2 mg/l ozone prior to chlorination. Rook[141] reported in 1976 on the removal and prevention of haloform formation that the combination of ozone (8 mg/l) followed in two minutes by chlorine (7 mg/l) is an effective way to remove the fulvic acid precursor. He found that anion exchange resins following coagulation, sedimentation, and filtration is effective but relatively expensive for the removal of organic precursors. Rook also reported that the removal of haloforms generated during chlorination was accomplished by activated carbon or air stripping. The carbon method was disappointing because the haloforms broke through after only two to three weeks operation.

Aeration is effective in reducing chloroform since it is a volatile compound. Some investigators have reported removals better than 95 percent[142] when using an ammonia stripping tower at pH 11, while others can do no better than 50–55 percent using a countercurrent flow aerator.[143]

Absorption by either powdered activated carbon or granular activated carbon (GAC) filters will remove THMs. The powdered form of carbon is not an attractive choice because huge doses (50–60 mg/l) are usually required, which in addition to being expensive, create a sludge problem.[150] In 1981 Anderson et al.[144] examined the feasibility of trace organics control with a specially formulated powdered activated carbon, at dosages considerably less than previously reported. THM formation levels were reduced more than 50 percent. A new ruling by the EPA in March 1982 will allow the use of PAC as an accepted technology provided its use is intermittent or seasonal at dosages not to exceed 10 mg/l on an annual average basis.

Granular activated carbon (GAC) and biologically activated carbon (BAC) have been investigated for both precursor removal and TTHM (Total THM) removal. Carbon is effective, but has a low reliability factor. In TTHM removal, chloroform invariably breaks through in two to three weeks. BAC is discussed in Chapter 13. GAC has been the subject of many investigations.[145-149]

The most overpowering objection to GAC is the cost. In 1980–81 the Metropolitan Water District of Southern California hired James M. Montgomery Engineers to make a comprehensive investigation that scrutinized all of the possible alternatives to control the formation of THMs and/or to remove their precursors. This project was a joint effort between the client and the Engineers.[151] During this investigation pilot plant studies demonstrated that GAC was 100 times more expensive than chloramines. For various valid reasons all of the many alternative methods studied were discarded in favor of chloramines.

Sometime in 1981 the American Water Works Association filed a lawsuit against the U.S. EPA which challenged the EPA standard of 100 micrograms per liter for total trihalomethanes (TTHM). This suit was settled in March 1982. The EPA made two important changes which represented a victory for AWWA. First, chloramines were raised to alternate primary disinfectant status and second, the GAC filtration and BAC (biologically activated carbon) filtration technologies as requirements were eliminated in order to obtain a variance under the proposed new rules by the EPA.[152]

A report similar to the one developed for MWD of Southern California was completed in 1982 by Kennedy Jenks Engineers for the Alameda (California) County Water District. This investigation concluded that treatment with chloramines was the best alternative.[153] Others who have already switched from free chlorine to chloramines have reported successful control of THMs and equally successful water quality control.[154,155]

Max THM Potential. The potential for the formation of THMs is directly related to the chlorine demand. Trihalomethanes are formed by the reactions between natural organic compounds (humic and fulvic acids) in the environment, naturally occurring bromides, and chlorine used in the water treatment processes. One of the techniques to determine the potential THM formation is a procedure long known as the "chlorine demand" of a water.

In the case of THM formation the recommended procedure is to add sufficient chlorine to the sample so that a measurable free chlorine residual will persist for at least seven days. The samples should be stored in screw-top or crimp-top bottles filled to exclude air space and placed in a 25°C water bath in the dark. Depending upon local circumstances the detention time may be more or less than seven days.[156]

The total trihalomethane (TTHM) concentration is measured at daily intervals. These concentrations are then plotted to generate a TTHM curve as shown on Fig. 6-10. When the curve reaches a plateau that is the maximum THM potential (MTP).

This extended chlorine demand test can serve as a useful tool for the water producer. It is a significant water quality parameter that should be evaluated annually. To make use of Taras' chlorine demand concept described in Chapter 4, chlorine residuals should be measured after 30 min and 60 min contact time and then at hourly intervals for the first 24 hours.

Each water supply will react differently to this test. Chlorine dosages to provide a measurable free chlorine residual to produce a plateau will vary from 2 to 20 mg/l and the reaction time may take from 5 to 15 days. These variations will be a function of the water quality and the nature and concentration of the precursors.

Chlorine Demand. There is sufficient data from more than fifty years of operating experience to declare that the 15 minute chlorine demand of a raw water be used as a surrogate parameter for water quality. It would be proposed then that if the chlorine consumed in this 15-min period exceeded the arbitrary allowable level, the water in question would have to undergo some pretreatment process to reduce the natural chlorine demand to the acceptable level.

To put the chlorine demand level in perspective it is important to mention that the source of all the water on this planet—seawater—has a 15-min chlorine

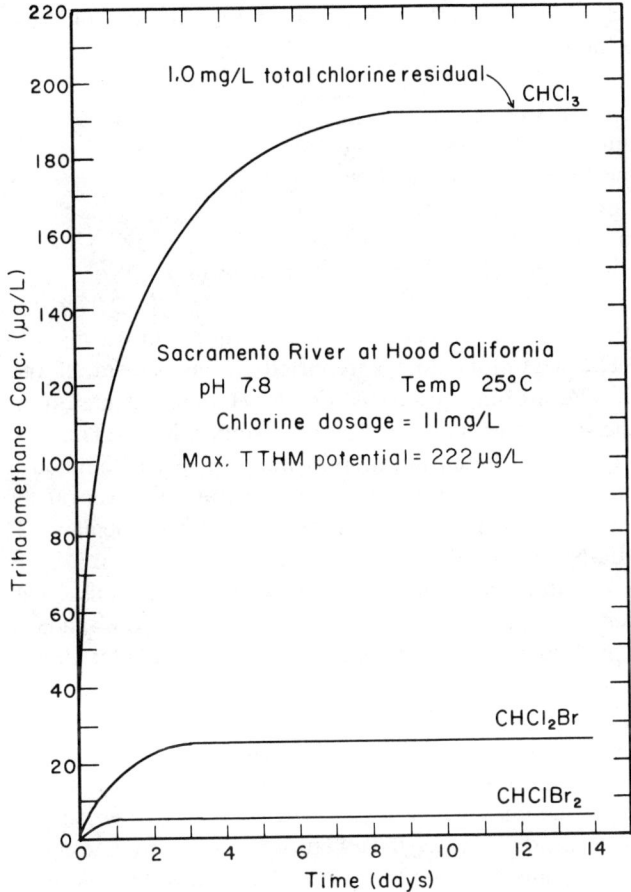

Fig 6-10 Total trihalomethane formation potential due to the free residual process.[156]

demand that rarely exceeds 1.5 mg/l. White suggests the maximum allowable raw water chlorine demand should not exceed 2 mg/l.

CURRENT PRACTICES IN THE U.S.A.

Review of Modern Practices

Chlorine Dosages. In March 1978 a questionnaire developed by the AWWA Disinfection Committee was mailed to numerous water utilities in an attempt to document current practices.* There were some 350 respondents to this questionnaire.

---
* This questionnaire was under the leadership of Geo. Clifford White, chairman of the committee.

The respondents indicated that the points of application could be a combination of any of the following:

- Pre–raw water storage
- Precoagulation/post–raw water storage
- Presedimentation/postcoagulation
- Postsedimentation/prefiltration
- Postfiltration (disinfection)
- Distribution system (disinfection/water quality)

The chlorine dosage range reported was a minimum of 0.2 mg/l to a maximum of 40 mg/l. The maximum dosage reported by 96 percent of the respondents was 15 mg/l. The all-time record is 120 mg/l reported at Ottumwa, Iowa in 1950.[157,158] The raw water was from the Des Moines River. This range of dosages demonstrates dramatically the variation in water quality throughout North America. When practicing the free residual process, winter ice cover on surface supplies will almost automatically double the summertime chlorine demand. This is a result in the ammonia-N increase due to the ice cover.

To combat pollution and taste and odor problems various combinations and dosages of chemicals are in use today in both the U.S. and Canada. These include combinations of chlorine and chlorine dioxide, ammonia, potassium permanganate, and activated carbon, sometimes followed by dechlorination with either sulfur dioxide or carbon.

Prechlorination of low quality water is most often the operator's salvation. It is one of the most important tools for maintaining the efficiency of a water treatment plant. In these situations, which are numerous, it would be virtually impossible to turn out an acceptable water if it were not for the unique ability of chlorine to maintain a persisting residual throughout the process. In these cases disinfection is just a side effect. However a lot of these water suppliers will have to reevaluate the chlorination process in order to meet mandatory TTHM regulations.

Two examples of low quality water that require chlorine dosages up to 16.0 mg/l are the Grand River at Brantford, Ontario, Canada[159] and the Passaic River at Little Falls, New Jersey.[160]

In North America, poor quality water has been treated for decades by a prechlorination dose sufficient to provide a substantial free residual in the flocculation or sedimentation basins. Many of these plants follow this by an intermediate dose of chlorine or chlorine dioxide sufficient to carry a residual through the filters. This practice will certainly have to be modified in some instances. Switching from free residual chlorine to chloramines or chlorine dioxide might be the answer in some cases. Other cases might need to change the entire pretreatment processes, which could include ozone. (See Chapter 13.)

In those cases where free residuals through the treatment plant do not generate objectionable levels of THMs in the finished water, postammoniation to convert the free chlorine to chloramines can be beneficial. This has been practiced at Brantford Ontario and other plants for decades.

Dechlorination by sulfur dioxide to trim the chlorine residual as it enters the distribution system is being practiced in many places.

The case of clean waters where chlorine is used solely for disinfection is remarkably different from low quality waters. Two typical examples are the supplies for the City of San Francisco and the East Bay Communities in Alameda and Contra Costa counties in California. These waters are derived directly from melted snow in separate runoff areas high in the Sierra Nevada mountains. The San Francisco supply is transported by a 165-mile aqueduct and the East Bay supply travels about 90 miles to various local storage reservoirs. The chlorine dosage required for disinfecting these waters is 0.8–1.2 mg/l. The San Francisco supply was plagued from the start by assorted difficulties. The most humiliating one occurred a few months after the system was put in operation. Part of the 165 mile aqueduct consists of a concrete lined tunnel 25 miles long through the Coast Range mountains. Although the concrete tunnel lining is up to 12 inches thick, many cracks developed due to heavy ground caused by groundwater. The groundwater is laden with filamentous bacteria (*Crenothrix*) which infiltrates the tunnel and gives rise to luxuriant growths on the tunnel lining. This created a debacle in the distribution system. Industries were forced to shut down because of this biofouling and the sloughing of the filamentous debris. To combat the problem, the city of San Francisco, after having investigated several treatment methods, concluded that a persistent chlorine residual was the only answer. Therefore a chloramine station was built (1937) to inject chlorine and anhydrous ammonia at the entrance to the tunnel. A 1.5 mg/l dose produced a combined residual of about 1.0 mg/l at the end of the tunnel 25 miles away. When ammonia became difficult to get during World War II the system was converted to the free residual process which is still in use (1983). San Francisco water does not generate THMs above the EPA minimum contamination limits.

There are many areas in the U.S.A. which are not as fortunate as the San Francisco Bay Area. Two of these are St. Louis and Kansas City, Missouri: St. Louis, using Mississippi River water, prechlorinates at an average 5 mg/l. This is supplemented by intermediate doses of chlorine to preserve the efficiency of the filter system. Kansas City, using Missouri River water, prechlorinates at about 12 mg/l. These chlorine applications are not for disinfection. This is to keep the water treatment processes from deteriorating and for taste and odor control.[161]

The City of Chicago, using a cleaned up Lake Michigan water, prechlorinates at only 1–1.2 mg/l. This is comparable to the melted snow waters of California. This prechlorination dose is designed to provide an absolute minimum of 0.25 mg/l free chlorine through the filters. It is supplemented by a postchlorination dose of 0.25–0.5 mg/l to give a 0.75 mg/l free chlorine residual entering the distribution system. To exemplify the effect of pollution on chlorination, whenever there is an upset with the Chicago Ship Canal, amounting to a reversal of flow which dumps canal water into Lake Michigan, the chlorine demand escalates to 10 mg/l, and even at this dosage there is so much NH_3-N in the raw water that a free chlorine residual is not attainable.[162]

Surface waters in some areas of Pennsylvania require prechlorination doses of 7–8 mg/l and the City of Philadelphia provides sufficient chlorination equipment capacity to dose as high as 30 mg/l at the raw water basin outlet (ice cover situation) and up to 4 mg/l chlorine dioxide for pretreatment.[161]

It is significant to note the quality of the water which is provided by the billion dollar California Water Plan. This water is a mixture of several rivers, primarily the Sacramento and San Joaquin. This water requires only 2–4 mg/l prechlorination dose to produce a 0.7 mg/l free residual in the filter plant finished water without any intermediate chlorination. This is a tribute to the diligence and surveillance of both the California State Department of Health and the California Water Resources Quality Control Board for their programs designed to preserve the water quality of the receiving waters.

Chlorine Residuals. To this question there were 226 respondents to the AWWA Disinfection Committee Questionnaire.[163] The highest reported was 5 mg/l and the mean was 1.4 mg/l (total Cl). The figures for free residuals were about the same: maximum 4.75 mg/l and mean 1.0 mg/l.

Out of 332 respondents, 235 were using DPD, 165 amperometric titrator, 127 acid O-T, 11 neutral O-T, 1 FACTS, and 1 starch–iodide.

Contact Time. One of the questions asked in this questionnaire was designed to determine the contact time between the point of application of chlorine and the first consumer. Strange as it seems, 90 percent reported 10 hours, while 6 percent reported 2 minutes. The median was 60 min, while the mean was 237 min.

Chlorine Demand. Chlorine demand tests reportedly take place in about 40 percent of the respondents facilities. Twenty of the utilities reported maximum demands between 5 and 10 mg/l chlorine, and another 11 reported demands between 10 and 22 mg/l. Of 123 respondents the median demand was 1.8 mg/l. However the maximum reported was 65 mg/l at a maximum contact time of 12 hours.

Deficiencies in Current Practices

The Community Water System Survey. The most significant information available on the production of potable water in the U.S. is the Community Water System Survey of 1970.[129] This survey points directly to the deficiencies in current practices. As is always the case, whether it be potable water or wastewater treatment, it is the small communities and the small producers who have the greatest deficiencies. The large producers are endowed with more sophisticated treatment techniques, adequate funding, and more qualified operating personnel. This report reveals that in communities of less than 500 population, only 50 percent met the Drinking Water Standards of 1962 (USPHS). At the same time in communities of 100,000 or larger, 73 percent met these standards. Of equal or greater significance the survey revealed that 19 percent of surface and mixed-source water not practicing disinfection were not properly protected at the source, and of those facilities practicing disinfection, 16 percent were inadequate and 7 percent had inadequate control of the disinfection process. Overall, this survey indicates that too many Americans are not being provided with potable water that meets contemporary standards of good practice.

Too many unnecessary risks are being taken in the production and distribution of potable water, particularly those small but numerous water systems, e.g., ski and other resorts, roadside restaurants, bus stops, motorway rest areas, trailer camps, farms, suburban homes, isolated institutions, organized summer camps, and other similar small water systems. Many of these small supplies are pumped into a pressure tank and thence to a close-coupled distribution system with practically zero contact time. Obviously the only line of defense for these systems is the disinfection facility. Therefore, this facility must be designed with a proper factor of safety to allow for variation in water-quality parameters affecting disinfection. The same rationale would also hold true for large systems under similar circumstances.[165]

Viruses. One of the major causes for concern is that the frequency of viral diseases and acute gastrointestinal illnesses (AGI) has been on the rise in the past four decades, while typhoid has almost disappeared in the U.S.A. Until 1971 gastroenteritis as such was not a reportable disease. Since 1971 the EPA and CDC have entered into a cooperative reporting scheme classifying such illnesses along with other outbreaks of known etiology. This has gone a long way in solving the origin of these outbreaks. During the period 1961–1970 a total of 26,546 cases of gastroenteritis were definitely attributed to water.[166] Of 52 waterborne outbreaks in the U.S.A. in 1971–72 there were 22 outbreaks of AGI amounting to 5615 from a total of 6817 cases of all waterborne illnesses.[167] The concern here is that these cases may all be of viral origin. See the section on "Viral Diseases" in this chapter. There are over 100 viruses excreted in the feces of man which have been reported to be in contaminated water. Any of these could cause a waterborne disease and some have done so.

In order to cope with viruses, the disinfection process will have to address itself to longer minimum contact times and higher residuals. Surface waters that rely solely on disinfection will have to be reevaluated. In these cases it may be necessary to adopt filtration as an additional barrier.

Proof of Disinfection. To date we have been relying upon the coliform group of organisms as proof of disinfection. So far this indicator has served us well for bacterial infection. However it cannot be relied upon as an indicator for viral infection. Microbiologists are in total agreement that there is no observed relation between total coliforms, fecal coliforms, or standard plate counts and the presence of viruses. Also there is no observed relation between the detection of *Salmonella* and viruses and vice versa. With our present knowledge of viruses, the coliphages or enteroviruses cannot be used as indicators. This means that proof of disinfection must derive from a properly treated water; this would involve filtration and the assurance of substantial chlorine residual in the distribution system. It has been demonstrated that water treatment processes such as those used to treat Missouri River water downstream from Kansas City, where fecal coliform densities are as high as 16,000 per 100 ml, are a reliable barrier for viruses.[55] This investigation was unable to isolate any viruses in the treated water.

Sanitary biologists have expressed dissatisfaction with coliforms as indicator organisms for more than a decade. They insist that this group of organisms is not resistant enough to chlorine or other disinfectants to allow any factor of safety. There have been many cases of enteric organisms such as the genera *Klebsiella* and *Enterobacter* (*cloacae*) found responsible for positive samples occurring in the routine coliform sampling procedure. The discovery of such potentially hazardous organisms in a distribution system is unsettling, particularly when the water has been coagulated, settled, filtered, and chlorinated sufficiently to carry a residual (0.5 mg/l) throughout the distribution system. There have been cases where the residual was kept much higher in an attempt to destroy these organisms, but without complete success. It is thought that these encapsulated organisms, once they are in the distribution system, may be harbored in protective slime, scale, or sediment. The *Klebsiella* germs are encapsulated organisms and can cause severe enteritis in children. In adults they can cause pneumonia and upper respiratory tract infections, septicemia meningitis, peritonitis, and urinary tract infections. *E. cloacae* present in a water system implies that there could be contamination of fecal origin. *E. cloacae* is not known to have caused any waterborne disease. However, *Klebsiella* is a hazardous organism.[169]

In view of the above, the work of Engelbrecht et al. is of considerable importance.[168] Their investigation evaluates two promising groups of organisms. These include two acid-fast cultures, *Mycobacterium fortuitum* and *M. phlei*, and a yeast, *Candida parapsilosis*. The resistance of these two groups of organisms to chlorine is believed to be in the range necessary to inactivate both bacillary pathogens and waterborne viruses. Further research for a better indicator organism is recommended.

The precise reason why *Klebsiella* and *E. cloacae* can be isolated in distribution systems in the presence of a significant free chlorine residual after the water has been coagulated, settled, filtered, and disinfected remains a mystery. In this chapter, under "Distribution System," some probable causes are discussed.

In spite of the above anomalies it is still the consensus that the best method of assuring the microbiological safety of drinking water is to maintain good clarity, provide adequate disinfection (which includes maintenance of a disinfectant residual in the distribution system), and make frequent measurement of the total coliform density and general bacterial population (SPC) in the distributed water.[174] The emphasis is always on a persisting disinfectant residual in the distribution system.[173,174] The Community Water Supply Study[164] undertaken in 1969 to investigate the status of the surveillance program in each of 969 water supply systems found that only 10 percent met the sampling criteria, whereas 90 percent either did not collect sufficient samples or collected samples that showed poor bacterial quality or both. In general the survey showed that the probability of finding coliform bacteria in a distribution system decreases as the residual chlorine increases. Overall the Community Survey specifically showed that in chlorinated water supplies a chlorine residual must be maintained throughout the distribution system in order to have confidence that disinfection has been accomplished.

Robeck[174] has been able to demonstrate with data that when properly used, the chlorine residual test satisfies the problems occurring with bacteriological sampling. The sampling problems in many utilities occur when there is too high a proportion of the sampling in the free flowing, high turnover area of a system, and too small a percentage in the troublesome areas: reservoirs, dead ends, and low flow areas.

It could be concluded then that the safest drinking water system is one that maintains an adequate chlorine residual throughout the distribution system which is verified by frequent sampling or continuous residual monitoring. This notion presupposes that this water meets the turbidity standards as well.

OBJECTIVES OF CHLORINATION

Disinfection Guidelines

Free Residual Process. Disinfection practices should be governed by the chlorine residual–contact time envelope, based upon the destruction of a consensus organism.[128] This concept was first proposed by Baumann and Ludwig in 1962.[175] The major factors affecting the germicidal efficiency of the free residual process are: chlorine residual concentration, contact time, pH, and water temperature. Increasing the chlorine residual, the contact time, or the water temperature increases the germicidal efficiency. Increasing the pH above 7.5 drastically decreases the germicidal efficiency of free chlorine. (See Chapter 4.)

A review of the literature shows the resistance of pathogens to free residual chlorination varies over a wide spectrum of organisms. Most of the viral pathogens are considerably more resistant than the bacterial pathogens. Cysts in most instances are the most resistant to free chlorine. Varma and Baumann compiled the chlorine residuals and contact times needed to kill vegetative bacteria, viruses and amoebic cysts from a comprehensive review of the literature.[176] From this review they plotted the available data shown graphically in Fig. 6-11 for the chlorine concentra-

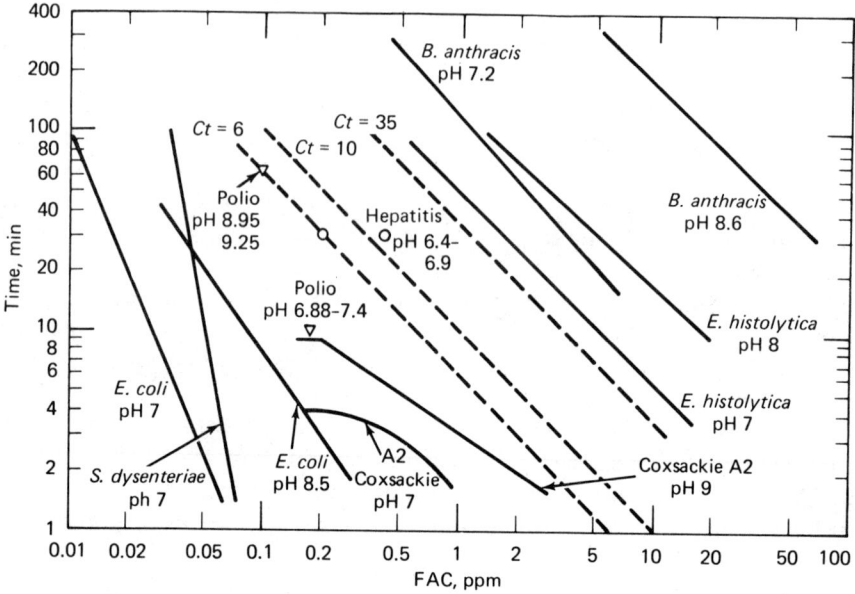

Fig 6-11 Disinfection versus free available chlorine residuals (Time scale is for 99.6 to 100 percent kill. Temperature was in the range of 20 to 29°C, with pH as indicated.) Reprinted from *Journal American Water Works Association* **54**, 1379, Nov. 1962, by permission of the Association.

tion–contact time relationship to achieve the destruction of these four classes of organisms. Table 6-3 shows the source of the data for the plots in Fig. 6-11 which were considered to be the most up-to-date and reliable. Fig. 6-12 represents the plot of disinfection versus free chlorine residuals at 0 to 5°C as compared to Fig. 6-11 which is at a temperature range of 20 to 29°C. Since there are many locales where the minimum water temperature for both well water supplies and surface waters may never fall as low as 0°C, Baumann and Ludwig[175] constructed Fig. 6-13 which represents a transposition of the plots on Fig. 6-11 at 20 to 29°C and Fig. 6-12 at 0 to 5°C to a plot representing a water temperature of 10°C. This transposition was accomplished by means of the Van't Hoff–Arrhenius reaction rate equation:

$$\log_e \frac{t_1}{t_2} = \frac{E(T_2 - T_1)}{R\,T_1\,T_2} \tag{6-4}$$

in which T_1, T_2 = upper and lower temperatures between which reaction rates are compared. t_1, t_2 = times in minutes required for equal percentage of kill to be effected at temperatures T_1 and T_2 at a fixed concentration of disinfectant. E = activation energy (calories). R = gas constant, 1.99 cal/°K, 10°C = 283°K.

CHLORINATION OF POTABLE WATER

Table 6-3 Sources of Data for Disinfection Time— Chlorine Concentration

Organism	0°C Temp. Range	pH	Source
E. coli	2–5	7, 8.5	Butterfield[177]
E. coli	20–25	7, 8.5	Butterfield[177]
Salm. typhosa	20–25	9.8	Butterfield[177]
S. dysenteriae	20–25	7.0	Butterfield[177]
Poliomyelitis	20–30	6.85–9.25	Lensen[178]
Poliomyelitis	0	7, 8.5	Weidenkopf[179]
Coxsackie A2	3–6	7–9	Clark and Kabler[180]
Coxsackie A2	27–29	7–9	Clark and Kabler[180]
E. histolytica	3	7, 8	Fair[181]
E. histolytica	20–25	7, 8	Snow[182]
B. anthracis	4	7.2, 8.6	Brazis[183]
B. anthracis	22	7.2, 8.6	Brazis[183]
P. tularensis	15.5–18.5	7.3	Foote[51]
Hepatitis	room	6.4–6.9	Neefe[184]

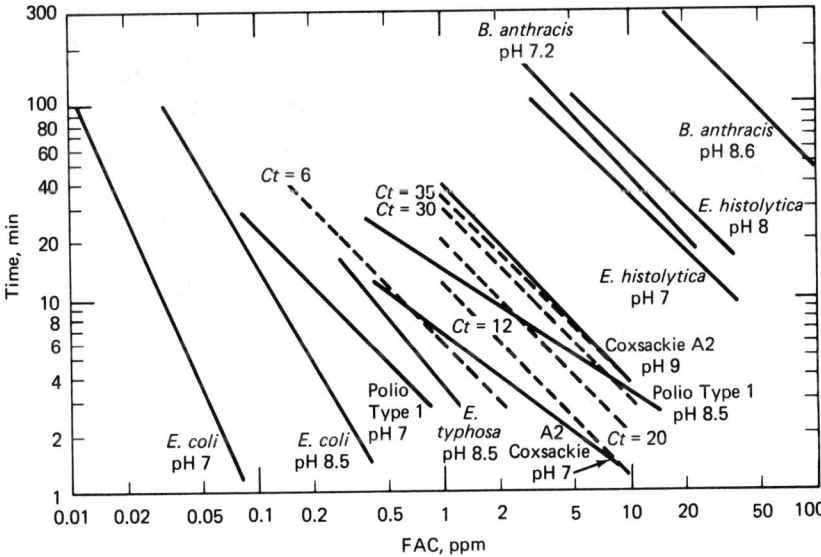

Fig 6-12 Disinfection versus free available chlorine residuals. (Time scale is for 99.6 to 100 percent kill. Temperature was in the range of 0 to 5°C, with pH as indicated.) Reprinted from *Journal American Water Works Association* **54**, 1379, Nov. 1962, by permission of the Association.

The result of the application of this mathematical transformation procedure is Fig. 6-13.

The curves illustrated in Figs. 6-11, 6-12 and 6-13 are plotted on log—log paper to give straight lines. These lines follow the general equation $y = ax^b$, where b is the slope. Since b is negative, the curve is hyperbolic. The curves shown in solid lines have slopes that approximate -1. If y is plotted as time t, and x as free chlorine concentration C, the equation may be written:

$$t = aC^b \qquad (6\text{-}5)$$

b is a positive number expressing the relationship between C and t, a is a constant for a given organism, water pH and water temperature. The slope is expressed as $-b$; then if the slope approximates -1, the equation may be written:

$$t = aC^{-1} \qquad (6\text{-}6)$$

or

$$a = Ct \qquad (6\text{-}7)$$

The dotted lines in Figs. 6-11, 6-12 and 6-13 are constructed from the above equation to represent envelopes of disinfection time-chlorine concentration of various pH values. For example, in Fig. 6-12 the envelope $Ct = 35$ represents the envelope for the organism Coxsackie A2 at pH 9, and the envelope $Ct = 12$, for pH 7. Other envelopes for intermediate pH values are constructed by interpolation.

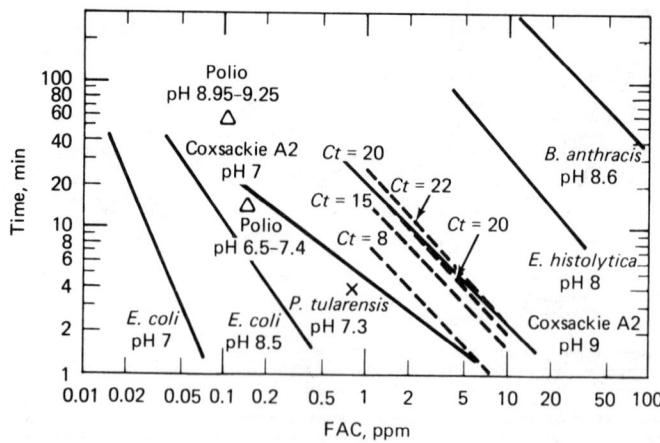

Fig 6-13 Disinfection versus free available chlorine residuals. (Time scale is for 99.6 to 100 percent kill. Temperature was 10°C, with pH as indicated.) Reprinted from *Journal American Water Works Association* **54**, 1379, Nov. 1962, by permission of the Association.

This is based upon the HOCl concentration available from the total titrable chlorine at a given pH.

From all of the data shown in Figs. 6-11, 6-12, and 6-13 the question now becomes, how much chlorine should be applied and for how long should the residence (contact) time be. Baumann has suggested that the minimum dose be sufficient to inactivate Coxsackie A2 virus in the treatment of small water supplies. If this reasoning is applicable to small water supplies, it should be extended to all water supplies simply to encourage better disinfection practices and provide a wider margin of safety.

Baumann[175,183] selects Coxsackie A2 virus as the indicator organism: if it is destroyed by chlorination, then all other pathogenic bacteria and viruses would also be killed. This would not include *Entamoeba histolytica, B. anthracis, Giardia lamblia*, or possibly Hepatitis A. Table 6-4 gives the required values for the values of a in $Ct = a$. These are shown as the dotted lines in Figs. 6-11 and 6-12.

Application of this information is most useful for small supplies that suffer from the problem of short contact time.

EXAMPLE: A well supply with a pH of 8.2 and a possible minimum temperature of 10°C shows $Ct = 22$. The well pump discharges to a pressure tank and thence to a small distribution system, so that the estimated contact time when the pump is on is six minutes

Therefore $C = 22/6 = 3.7$ mg/l free available chlorine residual (not dosage). This would call for dechlorination. For a small supply, such as bus stops, trailer courts, camp sites, motels, the activated carbon pressure filter would be the preferred choice as a method of dechlorination. However, there is an alternative: A holding tank could be placed just downstream of the pump and pressure tank to provide a minimum of twenty minutes contact time. This then would require a free chlorine residual of only $22/20 = 1.1$ mg/l which would not require dechlorination.

Larger supplies, such as municipally operated systems, usually have considerably greater contact time; however, the envelopes are just as valid for large municipal

Table 6-4 Disinfection Time–Chlorine Concentration Envelopes for 0–5°C and 10°C

Water pH Range	Value of a in Ct = a	
	0–5°C	10°C
7.0–7.5	12	8
7.5–8.0	20	15
8.0–8.5	30	20
8.5–9.0	35	22

supplies as they are for small supplies. Part of the reasoning behind higher doses of chlorine for small supplies is that proper supervision and laboratory control are not usually available; the systems are usually designed without any alternative to long enough contact time, and are not under strict public health supervision. Most treatment plants have from one and one-half to three hours contact time from influent to treated water storage at average flows, so that the contact time for disinfection is usually easy to achieve.

Whether or not Coxsackie virus A2 could be considered the consensus organism is of course debatable. However, the idea is appealing, since it embodies all the elements of a positive and thorough approach to disinfection. In the free residual process it would be most helpful in cases of THM control to know precisely the chlorine concentration–contact time envelope to achieve the desired disinfection before the addition of ammonia and how much if any dechlorination might be required. However, so long as the health agencies and the EPA are satisfied with the total coliform concentration as proof of disinfection there will be no need to develop a consensus organism.

Chlorine–Ammonia Process. This disinfection process has not received much research attention since the 1940s. Most of the kinetic investigations of chloramines have been with naturally ammoniated wastewater effluents.

The kinetics of this disinfection process need some serious study now that it appears to be one of the most desirable methods of controlling THMs. Any kinetic study should include the following variables: effect of pH, effect of water temperature, stability in sunlight and the germicidal efficiency over a reasonably broad spectrum of conditions. The germicidal efficiency should be measured for a wide range of organisms including, but not limited to, enteric viruses, amoebic cysts, *Giardia,* hepatitis, and enteric bacteria. The objective of the germicidal efficiency investigation would be to provide a chlorine residual concentration–contact time envelope for several different organisms similar to the information provided by Baumann described above.

There are three other areas that need careful study: (1) evaluate germicidal efficiency for a variety of chlorine to ammonia-N ratios; (2) compare tastes and odors for these different ratios; and (3) compare areas 1 and 2, applying ammonia first vs. chlorine first allowing 10 min contact with chlorine before ammonia addition.

Taste and Odor Control

Introduction. The term "taste and odor" should more logically be replaced by the term "flavor" as applied to the palatability of water or for that matter anything that is ingested by humans whether food or drink. Taste itself is a sensation caused by buds on the tongue and limited to acid, bitter, salt, and sweet. All other sensations of "taste" are either combinations or are modified by smell. An additional sensation which the tongue can detect is feel or touch—slick or oily as well as metallic, dry, or astringent. "Flavor" could be used as an all-inclusive term.

The origins of taste and odors in water supply are of two categories: natural and man-made. Natural sources include: aquatic growth, such as algae and diatoms and other organic compounds from decaying vegetative matter; and inorganic compounds, such as hydrogen sulfide, sulfates, and other sulfurous compounds.

Man-made sources of taste and odor in water supplies are industrial and domestic waste discharges. The worst are the discharges from manufacturing plants that contain phenols, cresols, certain amines, and mercaptans. Pulp and paper mills that discharge sulfites also cause serious taste and odor problems. With the possible exception of hydrogen sulfide, the worst offenders by far are the organic substances.

Substances causing odor are volatile and usually liquids. Most odorous substances are soluble in both water and organic liquids. Only a very few inorganic substances have odor, but they do provide the nutrients to promote the growth of odor-producing algae.

The first step in attacking a taste and odor problem is to review the literature. Several hundred articles have been published in various journals dealing with tastes and odors in the past thirty years. To save time, the serious student, researcher or operator should first read a 1977 review by S. D. Lin[186] and another by E. J. Middlebrooks in 1965.[187] These two reviews are replete with references. Another valuable reference is the AWWA handbook of specific experiences of taste and odor control in the U.S. and Canada.[188]

The next step is to learn about the Threshold Odor Test detailed in *Standard Methods*. This reference and papers by Gerstein[189] and Sigworth[190] describe in detail some of the ways to construct continuous and batch type odor monitoring devices.

There are three ways to deal with odorous contaminants: keep them out of the water, remove them from the water, or destroy them in the water. The tastes and odors produced by algae have been generally described as aromatic, fishy, grassy, earthy, musty, septic. Threshold odors caused by algae may be in some areas as low as 1–14, but in other areas go up to 30–40 and occasionally as high as 90 or more. Algal odor is generally objectionable, even when the threshold number is low. For satisfactory results the threshold odor should be reduced to 5 or less.[191]

Identification and quantitative analysis of the offending organism are important facets of taste and odor control, since the number of standard areal units per ml varies with the particular species of alga. For example, for *Asterionella* it may be 3000; for *Synura*, 200;[191] and for Dinobryon, 30.[192]

Principal Odor-Producing Algae. *Synura* is a very potent odor producer. A comparatively few colonies per ml will cause a perceptible odor resembling ripe cucumber or muskmelon. It can also produce a bitter taste leaving a persistent dry, metallic sensation on the tongue. When present in large numbers, it may cause a fishy odor.

Dinobryon imparts a prominent fishy odor when the standard areal count reaches 30 per ml. This organism develops in the southern end of Lake Michigan in June and July of almost every year.

Asterionella (500 units or more) imparts an aromatic geranium-like odor that changes to fishy when present in large numbers.

Tabellaria has a similar effect.

Synedra produces an earthy to musty odor, and also inhibits proper floc formation.

Stephanodiscus contributes a vegetable oily taste but produces very little odor.

Ceratium is an armored flagellate that is abundant in California and is responsible for odors varying from fishy to septic. It is likely to proliferate rapidly during any season.

Anabena, Anacystis (formerly known as *Microcystis, Polycystis,* and *Clathrocystis*), and *Aphanizomenon* are blue-green algae well known for developing very foul "pigpen" odors in water. In small concentrations these three algae impart a grassy to moldy odor to the water. They may in large enough masses cause luxuriant blooms. The foul odor undoubtedly develops from products of decomposition as the algae begin to die off in large numbers.

Green algae are less often associated with tastes and odors in water. Their growth may help to somewhat inhibit or keep in check the blue-green algae and the diatoms, and thereby be helpful in the control of water quality.

Dictyosphaerium, one of the worst offenders among the green algae, will produce a grassy to nasturtium odor as well as a fishy odor in larger concentrations. Some of the swimming green algae, such as *Volvox,* may also produce fishy odors.

Jenkins et al.[193] have reported several odorous organic sulfur-containing compounds produced in decaying blue-green algae cultures and reservoir waters containing blue-green algal blooms. These compounds included methyl mercaptan, dimethyl sulfide, isobutyl mercaptan, and N-butyl mercaptan.

Actinomycetes is an order of filamentous, branching bacteria. It was formerly identified with the general grouping of blue-green algae now called cyanobacteria, and encompasses several families of bacteria.[26] This group of organisms has long been suspected as the source of earthy odors in water supplies, and therefore has been subjected to intensive investigation.[194-197] These earthy-musty T/O compounds are considered to be the major source of consumer complaints.

Appreciable progress has been made in evaluating the relationship between volatile products of actinomycetes and the musty-earthy odor problems affecting water supplies across the nation. Modern research techniques have led to isolating two major earthy-musty smelling compounds: geosmin and 2-methylisoborrieol (MIB). They are metabolites of actinomycetes and blue-green algae (cyanobacteria) and show very low threshold odor concentrations.[198-200] Work done by Dougherty and Morris[201] in 1961 on the Cedar River identified the causative agent, which they called mucidone. They classified this compound as a metabolite of actinomycetes, so it was probably geosmin. Their work using chlorine and activated carbon indicated what others have found, that oxidation by chlorine is relatively ineffective but that activated carbon adsorption (granulated), to be effective (lower the threshold number to an acceptable level), requires a 25 mg/l dose. They also concluded that to be effective an oxidant would have to be able to break the carbon–carbon double bond of the T/O compound produced by actinomycetes.

Others who have attempted to control or remove the taste and odor from actinomycetes have had moderate success with combinations of activated carbon and potassium permanganate.[188] Silvey et al.[194] reported in 1950 that activated carbon and chlorine dioxide were somewhat effective but that ozone, bromine, and oxygen were not. The Metropolitan Water District of Southern California had no success in removing geosmin or MIB by air stripping.[197] The realization that both oxidation and adsorption are only marginally effective in controlling the T/O compounds of actinomycetes began to focus attention on preventing proliferation of actinomycetes. This, however, is a formidable task, since these organisms are ubiquitous in the environment. They are widely distributed in terrestrial, freshwater, and marine habitats. Moreover, the terrestrial organisms have easy access to surface waters during runoff conditions. The control of actinomycetes taste-and-odor-producing compounds remains in doubt. The Metropolitan Water District of Southern California has had some success controlling the blooms of these organisms that occur in the bottom of the reservoirs. They have designed and built a mobile chlorination system using a long snakelike chlorine solution line and diffuser that can reach into all areas of these reservoirs.

Man-made Sources of Taste and Odor. In the following text, certain values will be given for various organic compounds as threshold taste and odor concentrations. This means that the values given are the minimum amounts that will produce a response to the taste and odor detection system of the observer—the consumer complaint level of response.

Table 6-5 gives the threshold concentration of various chemicals that produce objectionable tastes and odors in water supplies.[202] The tastes are variously described as phenolic, iodoform, medicinal, and so on.

The potentially worst offenders for taste and odor production are the discharges from the manufacture of chemicals, dyes, medicinal products, coke (quench water), ammonia recovery, wood oil, phenols, cresols, petroleum products, textiles, and paper products. Of all the various chemicals that are contributory to off flavor in potable water, the ones that have received the most attention are the phenols. This was a result of the intensification of the natural tastes of such chemicals by marginal chlorination in the early years of its use. Table 6-5 illustrates how micro quantities of certain chemicals can produce objectionable tastes and odors in water supplies. Almost from the very beginning of chlorination practice, the supplies for Chicago, Cleveland, and Toronto, among others, have been plagued with off

Table 6-5 Chemicals that Produce Tastes and Odors

Chemical	Threshold Odor Level (ppm) Av.	Range
Acetic acid	24.4	5.07–81.2
Acetophenone	0.17	0.0039–2.02
N-amyl acetate	0.08	0.0017–0.86
Aniline	70.1	2.0–128
Benzene	31.3	0.84–53.6
N-Butanol	2.5	0.012–25.3
P-Chlorophenol	1.24	0.02–20.4
o-Cresol	0.65	0.016–4.1
m-Cresol	0.68	0.016–4.0
Dichloroiso propylether	0.32	0.017–1.1
2,4-Dichlorophenol	0.21	0.02–1.35
Ethylacrylate	0.0067	0.0018–0.0141
2-Mercaptoethanol	0.64	0.07–1.1
Methylamine	3.33	0.65–5.23
Methyl ethyl pyridine	0.05	0.0017–0.225
β-Naphthol	1.29	0.01–11.4
Phenol	5.9	0.016–16.7
Pyridine	0.82	0.007–7.7
Quinoline	0.71	0.016–16.7
Trimethylamine	1.7	0.04–5.17
N-Butyl mercaptan	0.006	0.001–0.06

flavor from these chemicals. These problems were studied, evaluated, and eventually corrected by the diligence and wisdom of men like Baylis and Vaughn at Chicago, Ellms and Braidech at Cleveland, Howard and Thompson at Toronto, and Vaughn and Besozzi at Whiting and Hammond, Indiana.

It is beyond the scope of this text to include all the organic compounds that find their way into the nation's drinking water and are potential taste and odor producers, or those that are toxic when ingested. As of 1983, more than 65,000 organic compounds have been manufactured in the U.S.A. since World War II. This number increases by 3000 per year. There is great concern that the proliferation of these chemicals could devastate the nation's water supply. The U.S. EPA is concentrating its efforts to find ways that will effectively treat the volatile organics in drinking water.[235]

Role of Chlorination in Taste and Odor Control

Algal Tastes and Odors. Early chlorination practice consisted of the application of the absolute minimum amount to insure the maximum bacteria kill consistent with good public health practice. This in most cases consisted of dosages as low as 0.25 ppm with fifteen minutes O-T residuals recorded as 0.05 ppm. Compare this with current practice of dosages from 1.0 to 5.0 ppm. Earlier it was also thought that the lower the dose, the lower the threshold taste from chlorination. During this time public opinion was very much against chlorination as being forced medication, and therefore the most direct and obvious mode of attack by the public was the complaint of the terrible taste it caused. As a result widespread public opinion made bad-tasting water synonymous with chlorination. This attitude inspired many able men in the waterworks profession to explore more vigorously the mechanics of chlorination. The first assault was of course directed at improving or eliminating tastes and odors. Howard of Toronto[203-205] led the way, in 1925, in exploring doses of chlorine beyond the marginal bactericidal doses. He described this approach as superchlorination. At about the same time Bushnell[206] reported in 1925 that by raising marginal doses to 3 to 5 ppm, described as overchlorination, eliminated foul odors thought to be caused by gnats breeding inside a covered reservoir. Griffin[121] pursued the exploration of superdoses of chlorine (up to 25 to 30 ppm) in a wide variety of conditions and practically revolutionized chlorination practice. Not only did his work lead to the discovery of the breakpoint phenomenon, but it proved that the free chlorine residual fraction was not necessarily in itself a producer of off flavor. In those cases where it was demonstrated that the residual in the finished water was too great or seemed offensive, the situation was usually corrected by dechlorination. The success of heavy doses of chlorine in the raw water for the correction and control of tastes and odors that presumably originated in algae growths and/or seasonal reservoir situations is well documented.[207-211] It is evident from this information that offensive tastes from algae blooms caused by *Synura, Synedra, Dinobryon, Asterionella, Anabena, Ceratium,* and *Anacystis* can be controlled by proper prechlorination that will produce a sizable free chlorine residual—1 to 5 ppm. Taste and odor problems are not necessarily caused entirely by either odor-producing algae or trade wastes but usually by a combination that manifests irself in a continuing degradation of raw water quality. While it is desirable to control the plankton count in storage reservoirs before it reaches the treatment plant, Riddick[211] has pointed out that there are so many contributing factors of geology and terrain that prevent such control that proper facilities at the treatment plant are the best assurance of taste and odor control. This would include a chlorination facility capable of carrying a free residual through the entire treatment process.

Industrial Wastes. Heavy doses of chlorine were first used by Howard[203] not to correct only off flavors caused by algae but primarily those caused by trade wastes. It was apparent that the worst offenders were the phenolic compounds that had their source in the wastes from coke and natural gas manufacturing plants. The addition of low dosages of chlorine to a water bearing phenols will produce a chlorophenol compound that imparts an extremely objectionable medicinal taste to the water. The reaction between chlorine and phenolic compounds has been thoroughly investigated.[212-215]

Ettinger and Ruchhoft[213] demonstrated that the result of chlorinating phenols was a reaction by steps whereby the taste-producing intensity was enormously increased by partial chlorination until a maximum intensity was reached, after which the further addition of chlorine resulted in a progressive decrease of the taste intensity until the chlorophenol tastes disappeared completely. These phenolic materials include cresol, naphthol, and dichlorophenol. In a later investigation, using more sophisticated research tools, such as gas chromatography, Burttschell et al.[214] discovered that the phenolic compound that was always present when there was a chlorophenolic taste was identified as 2,6-dichlorophenol.

It was further discovered in these investigations that, in order to eliminate tastes from these industrial wastes, the system would have to achieve a truly stable free chlorine residual and be allowed sufficient time for the reaction to go to completion. This means that time and pH environment are influencing factors. Ammonia in the raw water definitely slows down the reaction, but this can be overcome by increasing the chlorine dosage. As in all chemical reactions, water temperature is also a factor. The system behaves much differently in the winter months than it does in the summer months. The colder the water, the longer the contact time and the more chlorine required.

Others have demonstrated also that there is a definite ratio between chlorine required and phenolic-like substances present. However, each different water to be treated must be considered unique, and the problem has to be evaluated on an individual basis for dosage, time of contact, and so on. It has to be concluded, however, that the application of chlorine can definitely be used to destroy off flavors caused by phenolic compounds because of the many successes reported in the literature.[205,216-218]* However, the subject of taste and odor control has become so complex, owing to the general deterioration of potable water supplies throughout the United States and Canada, that there is no single solution to the problem. Heavy doses of chlorine with plenty of contact time is not a cure-all. Supplies that have a history of taste complaints are generally equipped to combine chlorine with some form of activated carbon treatment. For example, the city of Chicago, at its South Water filtration plant, is able to apply either chlorine or activated carbon to the raw water tunnel about 1100 feet ahead of coagulation. When trade wastes produce a threshold odor value of 10 or more, carbon is added at this point; however, when the *Dinobryon* count reaches 3000 or more per ml, chlorine is usually applied here. The reason for this flexibility is that sometimes it is more desirable to apply carbon first and sometimes chlorine. Depending on the trade waste, particularly if it is predominantly phenols, it may be more economical to let the carbon do the work of absorbing the phenols to prevent chlorophenol tastes than it will be to destroy the phenols with heavy doses of chlorine and

---

* Also see Chapter 12 the use of Chlorine dioxide as a specific treatment for the destruction of phenols.

use the carbon for the absorption of the off flavors resulting from chlorination. It must be remembered that if phenols are only partially chlorinated to produce chlorophenols, very high doses of activated carbon are required to absorb the resulting chlorophenols.

Several added benefits have been discovered in the application of heavy doses of chlorine to raw water. This type of treatment controls algae growth that tend to clog filter media and inhibit flocculation. Residual chlorine in the sedimentation basins prevents septicity in the sludge blanket. Therefore the maintenance of a free chlorine residual in these areas improved sedimentation and filtration. Another effect was color reduction in some waters caused by chlorine bleaching the organic matter, giving the treated water a polished look.

There is no definite upper limit on the chlorine dosages required. Usually the range is from 5 to 10 mg/l with a few instances requiring as high as 25 to 30 mg/l the record is 120 mg/l at Ottumwa, Iowa.[219,220] It is interesting to note that the successful users of heavy prechlorination doses for taste and odor control had to rely on dechlorination facilities, and those who failed either did not have enough chlorination capacity to achieve high enough doses or were not equipped to dechlorinate and therefore chose not to release water to the consumers with an abnormally high chlorine residual.

Use of Carbon with Chlorine. Another lesson learned in those cases where heavy prechlorination doses were used was that the consumption of powdered activated carbon was reduced significantly. Also, the application of carbon following prechlorination did not interfere with either process. Hence off flavors caused by chloro products were eliminated by the carbon while the consumption of chlorine by the carbon amounted to only about one pound of chlorine for twenty pounds of carbon applied. When phenolic wastes are known to be in the water, it takes heavier doses of carbon to absorb the chlorophenols than it does to absorb the raw phenol wastes. The use of carbon or chlorine first is a question of economics: Should the phenols be destroyed by chlorine, or should they be absorbed first by the carbon?

Seeking the Solution to a Taste and Odor Problem. The way to success in dealing with taste and odor problems is anticipation, by considering local conditions. Some waters at certain times of the year cannot tolerate any combination of chlorine dosage and contact time without producing objectionable tastes and odors. Some of these are discussed in the chlorine-ammonia process. Knowledge of raw water quality—such as threshold odor, organic nitrogen, ammonia nitrogen, degree of pollution, concentrated chloroform extract, pH, and temperature—is of great importance. The threshold odor data are the most important, there being no substitute for this information.

The mechanism provided by a chlorination facility to control taste and odors is to: remove taste- and odor-causing substances by oxidizing them to odorless and tasteless compounds; and to control or prevent the growth of odor-producing algae and microorganisms during the treatment process and in the distribution system.

The following course of action is recommended when the T/O problem occurs in the treated water before it enters the distribution system.

Experiment first in the laboratory, because if it won't work in the lab it won't work in the plant.

Chlorine Dosage. Run a series of chlorine demand curves for contact times of 5, 15, 30, 45, and 60 min with a dose of chlorine normally used. If the curves plotted from the residuals at this time indicates a possible breakpoint, run some intermediate contact times to locate the breakpoint. Check the T/O for the original dose and compare to other dosages, higher and lower than the normal dose.

*Ammonia.** If the problem still persists, try adding ammonia (before chlorine) in dosages to correspond to Cl:N ratios of 3, 4, 5, 6, 7, 10, and 12 to 1. Pick two contact times: 30 and 60 min. Check the residuals at these times and compare the T/O with a control sample.

Activated Carbon and Potassium Permanganate. If the combination of free chlorine and chloramines fails to achieve the desired results, try some combinations of chlorine and activated carbon (before and after chlorination). Also try some combinations with potassium permanganate. In addition to these combinations it will be necessary to try a variety of dosages. The combination might seem awesome, so it is always prudent to get some advice from the chemical suppliers for carbon and potassium permanganate.

Chlorination System Requirements for Taste and Odor Control. In order to provide the maximum choice and flexibility of control by the operator, a chlorination facility must include the following: (1) chlorination equipment adequate to produce a sufficient free chlorine residual throughout the treatment process; (2) separate automated equipment for each point of application of prechlorination, to avoid the so-called split-feed of the chlorine solution from one chlorinator; (3) automated dechlorination facilities based on residual control; (4) provision for possible preammoniation application. (5) automated postchlorination and postammoniation if distribution system residuals are required; (6) alternate points of application for activated carbon and chlorine—carbon first followed by a detention period of twenty to forty-five minutes and then chlorine for those special cases where hydrocarbon wastes are a predominate factor; (7) for taste and odor control, a minimum contact time of at least one or two hours if any ammonia is present—longer if the organic nitrogen concentration is significant—0.3 to 0.5 mg/l; (8)

---

* If ammonia is used to induce a breakpoint use a Cl:N ratio of 12:1 or greater.

provisions for adequate chlorine dosage and sufficient contact time prior to the raising of the pH for waters that undergo softening at high pH values (10 to 11); (9) provision for application of chlorine dioxide when indicated; (10) provision for the application of both activated carbon and potassium permanganate to the raw water if filtration is part of the process (potassium permanganate should not be used unless it is followed by filtration); (11) a continuous taste- and odor-monitoring system in a special "odor-free" room. Since taste and odor problems do not necessarily end at the treatment plant, just as much consideration must be given to the off flavors that may develop in the distribution system. It is practically axiomatic that, in systems with a tendency to develop off flavors, these unpalatable flavors will develop as soon as the available chlorine residual disappears in the distribution system. Tastes and odors from the application of chlorine are not likely to occur from the chlorine compounds themselves up to the limits listed below:

Free chlorine (HOCl)	20.0 mg/l
Monochloramine (NH_2Cl)	5.0 mg/l
Dichloramine ($NHCl_2$)	0.8 mg/l
Nitrogen trichloride (NCl_3)	0.02 mg/l

Nitrogen trichloride can cause a severe odor problem because its solubility in water is negligible. It is difficult to capture and measure in a sample as it will aerate with the slightest bit of agitation. However there is no mistaking the presence of NCl_3. In concentrations too low to get a response from the olfactory system it will cause the eyes to tear quite profusely. The odor of NCl_3 is entirely different from those of free chlorine and the other chloramines. It does not require much energy to aerate large quantities of NCl_3. It can, however, become an air pollution problem.

Williams[114,221,222] has demonstrated over a period of 20 years that if the ratio of monochloramine to dichloramine is kept at or greater than 2:1, objectionable tastes due to dichloramine will be at a minimum. This ratio of chloramines to achieve palatability is site-specific. It is not a hard and fast rule.

Studies by Ryckman and Grigoropolous[215] and Erdei[195] indicate the necessity for the removal of ammonia, organic nitrogen, phenolic compounds, and other organic extracts, all of which interfere with the chlorination process, with the result that it is becoming virtually impossible to provide palatable water in cases where these compounds are present in significant concentrations. These compounds are abundant in man-made wastes and should be prevented from reaching the potable water supplies. This is an impossible task; however, every effort should be made to minimize their entry into these supplies. The proliferation of toxic wastes threatening our nation's water supplies is awesome. Since World War II, 64,000 new organic compounds have been manufactured, and this number increases by about 3,000 per year. It is inevitable that these compounds will reach our water supplies, either into the ground or into the rivers and streams. The consequences can be devastating.

Distribution Systems and Transmission Lines

General Discussion. The problem of delivering a palatable and safe water to the consumer does not end as the water leaves the treatment plant, pumping station, or well discharge. The difficulties of maintaining water quality in transmission conduits and distribution systems are legion. The problems of water quality control in distribution systems are of two kinds. Probably the most prevalent is that of taste, odor, and dirty water complaints. The other is deterioration of bacteriological quality.

The question of the proper solution to these problems is one of the most controversial subjects of modern water treatment practice.

There seem to be more proponents of free residual chlorine as the proper treatment than those favoring chloramine residuals. Then there are those who favor either ammonia-induced breakpoint or ammonia-controlled free residual chlorination as well as those who favor chlorine dioxide. Now that THM formation is an important consideration there will be increasing numbers of proponents favoring chloramines. It should be emphasized, however, that it is imperative in a chloramine application to maintain a residual throughout the system and to enforce a flushing program, because when the chlorine residual disappears the ammonia nitrogen returns, which is a nutrient that promotes bacterial proliferation.

Nature of the Problem. The deterioration of water quality in a distribution system is usually attributed to three phenomena: (1) biofouling, due to the proliferation of microorganisms, which cause tastes, odors, and dirty water, with loss of carrying capacity of the pipes and sometimes severe corrosion; (2) chemical and electrolytic corrosion, resulting in undesirable end products, such as metallic and brackish tastes as well as failure of hot water heaters and residence water piping; (3) appearance of coliform organisms in the distribution system, indicating recontamination of an otherwise safe water.

Each of these phenomena contributes to consumer complaints. The most difficult problem to handle is the proliferation of microorganisms.

Photographic evidence that a large and diverse microbe community can adhere to and colonize the interior walls of underground water distribution mains and the surfaces of suspended particulate matter in drinking water systems has been provided by Ridgway and Olson[236] and Ridgway, Means, and Olson[237] using a scanning electron microscope. Such microbial colonization can apparently take place in summer months in spite of intermittent low-level free chlorine residuals (0.1–0.2 mg/l). There is ample evidence that microorganisms attached to pipe walls or present in partially anaerobic sediments in the invert of the pipe in low-flow and dead-end portions of the distribution system are responsible for taste, odor, and color problems of potable water systems.

Proliferation of bacterial colonies on pipe surfaces and suspended particulate matter leads to a type of biofouling that can harbor and encapsulate organisms such as *Klebsiella* p. and *Enterobacter cloacae*. These organisms have been isolated in distribution systems in the presence of free chlorine residuals. This phenomenon is not yet understood but it is believed to occur in older systems where slime layers of bacterial colonies are layered with particulate matter followed by an inorganic scale of silica and/or calcium, as shown in Fig. 6-14. This sequence keeps repeating itself until a sloughing process occurs which tends to keep the layering in equilibrium. It is thought that the reappearance of coliform organisms in these systems is caused by sloughing adjacent to the inlet of the sample line.

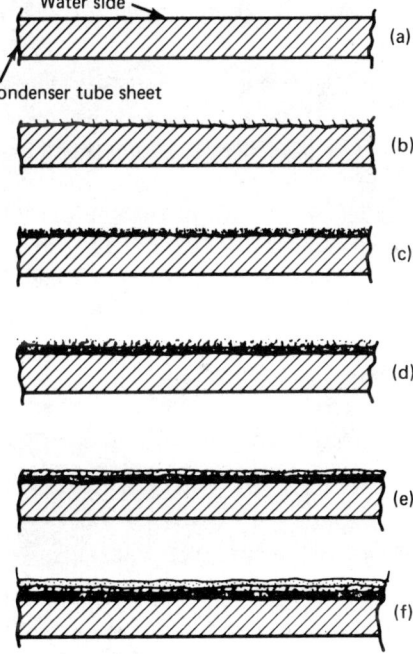

Fig 6-14 Mechanics of biofowling: (a) the pipe is new, i.e.; smooth and clean so that nothing can attach or accumulate. (b) The pipe surface becomes abraded or rough, followed by attachment of gelatinous slime forming organisms. (c) The gelatinous film serves as a microstrainer and entraps sediment. Some bacteria can promote deposition of inorganic salts as well, which sometimes is mistaken for a scale deposit. (d) Successive layers of slime and particulate matter form. (e) As time passes and if bacteria are allowed to proliferate, the mass becomes more dense. (f) It is believed that microbial colonies establish themselves in these protective layers and in time the layers spell-off discharging organisms that are well protected by this debris.

This sloughing involves the release of a piece of the top layer of the scale, thereby exposing the slime layer which harbors the coliforms that are drawn into the sample line.

This notion if valid would suggest that all sampling in any distribution system should provide a representative sample of the flowing stream. To determine the possibility of the type of contamination that reflects the reappearance of indicator organisms, sampling should be done at the consumer's tap. Sample connections to the distribution main should extend into the pipe about one-quarter of a pipe diameter. If this is not convenient, then use a fire hydrant.

The photographs taken with the electron microscope by Ridgway and Olson illustrates graphically many of the organisms discussed below. They also show clearly the scale-type layer, inorganic salts, bacteria attachment material, filaments coated with debris, actinomycetes adhering to the mineral layer (scale), *Gallionella* attached to pipe surface, extracellular slime, and capsular material.

Bacteriology of Water Systems. Bacteria are always present in water, even in the effluent from the most efficient and modern treatment plant. The kinds and numbers of these bacteria are dependent upon their environments. Many bacteria that may not be able to flourish in a particular environment are able to survive in a dormant state for indefinite periods of time and proliferate rapidly when the environment changes to a more favorable one. The most important factor for a favorable environment is the available food supply. Waters vary within wide limits as to the kind and quantities of food substances they carry, and likewise the type of bacteria which flourish in the pipes of these systems is equally variable.

Impounded surface waters that are allowed to produce abundant growths of algae and that are not filtered will contain great quantities of bacterial food in the bodies of algae that die in the pipes of the distribution system. Filtration effectively removes solid foods but not the soluble ones, such as proteinaceous matter, nitrates, phosphates, sulfur, ferrous and manganous compounds, and many others.

Underground supplies seem to be the worst offenders. They are low in dissolved oxygen, which contributes to anaerobic conditions, and are more likely to contain significant quantities of soluble iron and sulfur compounds, which are energy for the two most offensive species of bacteria that infest water distribution systems. These are the so-called iron bacteria (*Crenothrix, Leptothrix, Clonothrix, Cladothrix,* and *Gallionella*) and sulfur bacteria (*Beggiatoa, Thiobacillus thioxidans,* and *Sphaerotilus*).

Iron Bacteria. The iron bacteria obtain energy by oxidizing soluble ferrous compounds to insoluble ferric compounds. The amount of deposition of these insoluble compounds is large in comparison to the enclosed cells. These compounds may be deposited in a sheath that surrounds the organism or is secreted so as to form stalks or ribbons attached to a cell. The iron may be obtained from the pipe itself or from the water in the system.

There are also bacteria which do not oxidize ferrous iron but may indirectly cause it to be dissolved or deposited. In their growth they either liberate iron by utilizing organic radicals to which the iron is attached or they alter environmental conditions to permit solution or deposition of iron. Less ferric hydrate may be produced, but taste, odor, and fouling may be promoted by these bacteria.

Iron bacteria are normal inhabitants of soils, and so they readily find their way into water supplies. These are never found as a single species in a pure culture. Analysis of slimes and tubercles consist of many bacteria intermingled with the crenoform filamentous iron bacteria.[223] Most of these iron bacteria can equally utilize the soluble manganese compounds, which cause black deposits as compared to the reddish brown deposits associated with iron deposition.[224,225]

The filamentous types most commonly found are *Crenothrix, Cladothrix, Clonothrix, Leptothrix,* and *Sphaerotilus*. These "crenoform" organisms cause the most serious type of biofouling if they are allowed to proliferate. These long filamentous organisms have been known to grow "massive curtains" up to two and three feet in length in concrete tunnels.

Since iron is the predominant metallic deposit they are classified as iron bacteria. These bacteria are considered a special group because their appearance closely resembles certain species of algae and fungi. There are two general types: filamentous and stalked.

These bacteria are sometimes lumped together under one general term of crenoform organisms, as they all cause equally serious biofouling.[226] This group of organisms can be differentiated by the way the branching of the filaments occur (Fig. 6-15) They utilize the manganous salts better than they do the ferrous salts, and therefore may have manganic oxide deposits on their sheaths, giving rise to black, shiny deposits. They also oxidize organic matter for energy and thus do without the metallic salts.

Sphaerotilus belongs to a different cultural group than the other three, but performs in similar fashion. This organism occurs abundantly in heavily polluted waters rich in organic material such as wastewaters of sugar factories and paper mills. Fig. 6-16 illustrates the sheath and cells, which occur in chains. The sheath may appear to branch, but this is recognized as false branching, as can be seen in Fig. 6-16. The cells emerging from an open end or a break in the sheath wall are called swarm cells. The hollow sheath is of organic nature, but may become encrusted with an accumulation of ferric iron deposits.[227]

The stalked bacteria, the *Gallionella* (Fig. 6-17), were formerly called *Spirophyllum* because of their twisted-ribbon appearance. It was found that this ribbon is really a stalk of almost pure colloidal iron hydroxide excreted by the small oval bacteria cell at one end. These organisms cannot utilize manganese, and their growth is depressed by the presence of soluble organic matter. They are therefore true iron bacteria in the strictest sense of the term. Another peculiarity of these

Fig 6-15   Crenothrix (200×).

organisms is that their optimum growth occurs at about 6 to 10°C; so it is often found in cold well waters or in the wintertime. In the summer months, they give way to *Crenothrix* and *Leptothrix*.

There are other members of the stalked bacteria family not usually thought of as iron bacteria, such as *Siderophacus* and *Nevskia*. The latter often contain globules of fat or sulfur, which tends to confuse the classification. The family of Siderocapsa-

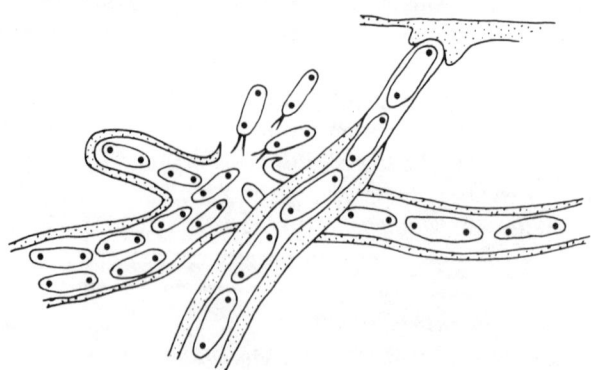

Fig 6-16   Sphaerotilus.

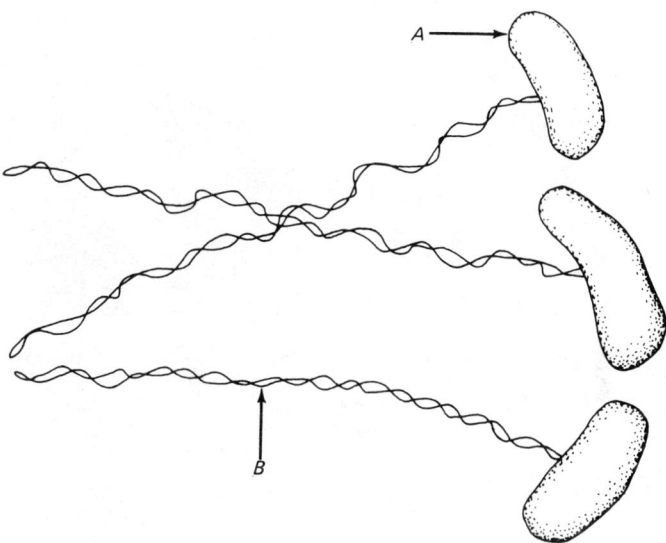

Fig 6-17 Gallionella, showing bacterial cell (A) and colloidal ferric hydroxide (B) deposited from the concave side of the cell, which form flat bands or ribbons extending from the cell. The individual ribbons twist and become entangled with other ribbons.

ceae includes ten genera whose cells are usually surrounded by a thick mucilaginous capsule containing iron or manganese compounds. Only one species of this group of ten genera, *Ferrobacillus ferrooxidans,* has been isolated in North America, but they are all widely distributed in European iron-bearing waters.[228]

Sulfur Bacteria. These bacteria, such as *Beggiatoa* and *Thiobacillus,* obtain the energy necessary for growth by oxidizing the sulfide ion to colloidal sulfur, which they store in their cells. By metabolic action this stored sulfur is eventually oxidized to sulfates. The most common of this group (*Beggiatoa*) is shown in Fig. 6-18. These organisms grow in long filaments in which can be seen granules of free sulfur and often a purple pigment. They grow in a scum on the surface of sulfide bearing waters and are easily detected by microscopic examination. This filamentous group of organisms, if allowed to grow, cause serious taste and odor problems in a distribution system.

Another group of sulfur bacteria that cause severe corrosion, taste, odor, and discolored water are the sulfate-splitting bacteria (*Desulfovibrio desulfuricans*). In anaerobic environments these bacteria convert sulfate and other sulfur compounds to hydrogen sulfide.

Most of the bacteria which inhabit water pipes are attached forms which build colonies resembling drops of colored jelly, encrusting masses varying in size from one-quarter inch up to tubercles as large as 2 inches in diameter. Since these are attached forms, tap samples will not reveal the magnitude of growth inside the

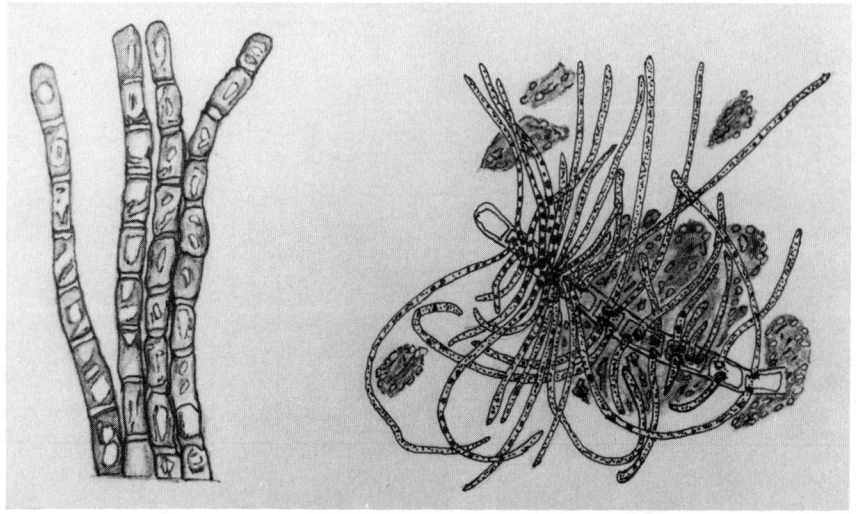

Fig 6-18   Beggiatoa (2000×) and trichomes attached to algal slime (400×).

pipes. For this reason every water distribution system should maintain accessible for routine inspection pieces of pipe in different locations that can be easily removed so that the inside of the pipe may be examined for these attached growths.

It is well known that bacteria growing inside water pipes can cause severe corrosion.[223,229,231] Thomas[232] described how water lines at a steel plant had a life of only a few months. An investigation revealed this corrosion was due to the sulfate splitting bacteria that flourished in the distribution system. This was corrected by chlorination. The action is explained as follows.[231]

The sulfate reducing bacteria splits the $SO_4^=$ ion to form hydrogen sulfide. The hydrogen sulfide either reacts with iron to form ferric sulfide or escapes through the porous ferric hydroxide scale and is oxidized into sulfuric acid by the sulfur bacteria (*Beggiatoa*). This proceeds according to the following equations:

$$SO_4^= + 10H^+ \xrightarrow{\text{sulfate-reducing bacteria}} H_2S + 4H_2O + \text{energy} \qquad (6\text{-}8)$$

and

$$2H_2S + O_2 \xrightarrow{\text{sulfur bacteria}} 2S + 2H_2O \qquad (6\text{-}9)$$

$$2S + 3O_2 + 2H_2O \xrightarrow{\text{sulfur bacteria}} 2H_2SO_4 \qquad (6\text{-}10)$$

When a tubercle is formed, the sulfuric acid is found between the tubercle and the metallic surface of the pipe enclosed by the tubercle. This explains why all such tubercles show evidence of pitting underneath them.

Iron bacteria also form tubercles, and cause both anaerobic conditions within the tubercle and low pH conditions, so that the cycle of this environment plus the catalytic action caused by the iron bacteria keep putting more iron into solution under the tubercle. This cycle continues with the iron bacteria, which lives on the outside of the tubercle, continuing to deposit the iron going into solution from the wall of the pipe, thus enlarging the tubercle.

Corrosion of iron pipe will not proceed when an equilibrium is established between the metal ion concentration in the water and the concentration of electrons in the metal. It is the bacteria that upset this equilibrium by putting more and more metal ion into solution.

Tubercle Formation. Figure 6-19 illustrates the formation of a tubercle. This is one of the most common phenomena of biofouling in a piping system. They seem to form in areas where pipeline velocities are low (<3 ft/sec), or when there are long periods of low velocity. All the evidence indicates that tubercles are of biologic origin,[226,232] apparently the outgrowth of a collection of fairly large colonies of crenoform organisms. These clusters of biologic deposits first start forming on the invert of the pipe. Long threads appear next in these jellylike clusters. The threads begin to acquire a sheath, which thickens by deposition of iron oxide, which the bacteria have taken out of solution from their water environment. Since the tubercles are hollow and spherical in shape, and since all tubercles show evidence of pipe wall corrosion by pitting within the tubercle, this suggests the evolution of CO_2, which would assist in the expansion of the sheaths to a bubblelike form. This would also help to explain the cause of pitting due to a low pH environment. An alternative explanation is that the tubercles form oxygen concentration cells, which are also conducive to pitting. Sometimes these tubercles grow as large as two inches in diameter.

In an attempt to determine the nature of common tubercle deposits, the literature reports iron oxides varying from 25 percent to 90 percent, with several analyses showing appreciable silica content.

This type of biofouling not only promotes severe corrosion but also reduces the carrying capacity of the piping system. Often these tubercles are first discovered when the pipe perforates under the tubercle (see Fig. 6-19). In some cases the tubercles will give off a pigpen odor when broken while still moist. This is a positive indication of organisms undergoing putrefaction. This explains why this type of biofouling can contribute to significant taste and odor problems.

Other Types of Biofouling. Slime-forming organisms can lead to biofouling in a distribution system. This ultimately leads to water quality degradation. Some are fungi (thallophytes) that do not contain chlorophyll. Consequently they must live on organic food or on energy available from inorganic compounds. They are mostly spore-forming and resist the most adverse conditions of moisture, heat, and chemical poisons. These organisms are usually found in waters that are heavily polluted with domestic wastewater.

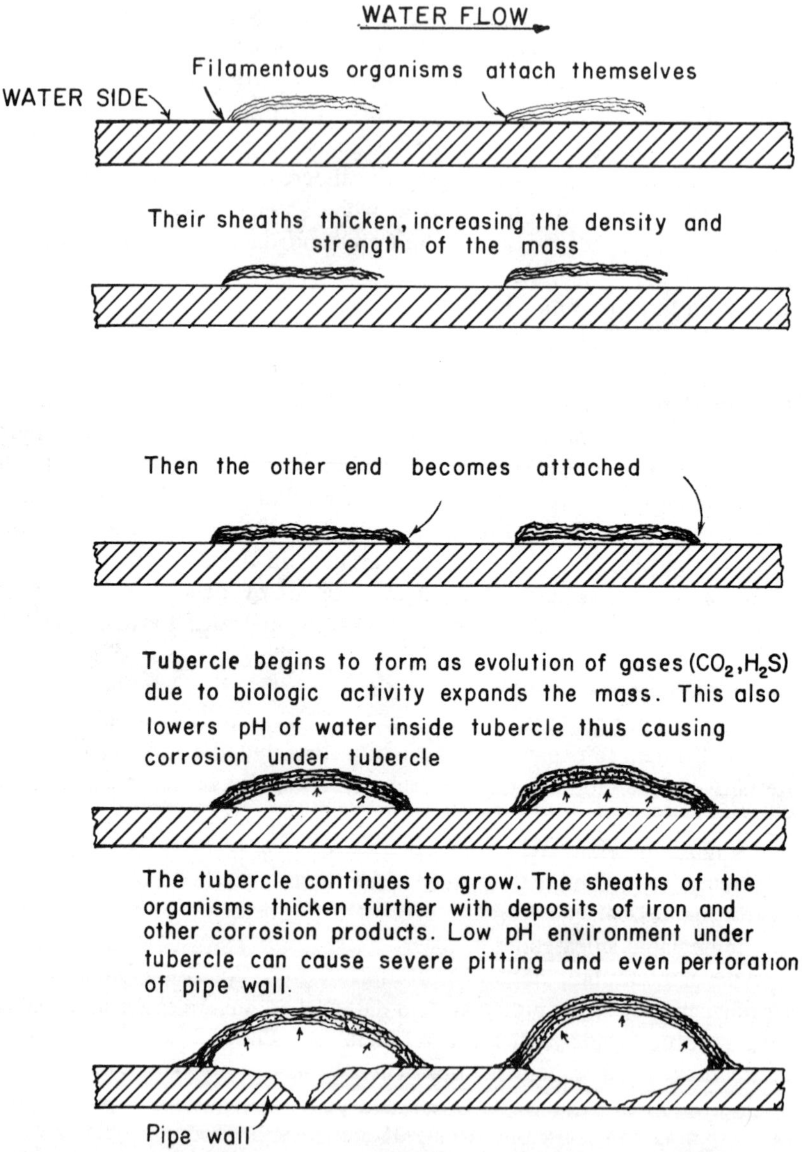

WATER FLOW

Filamentous organisms attach themselves

WATER SIDE

Their sheaths thicken, increasing the density and strength of the mass

Then the other end becomes attached

Tubercle begins to form as evolution of gases (CO_2, H_2S) due to biologic activity expands the mass. This also lowers pH of water inside tubercle thus causing corrosion under tubercle

The tubercle continues to grow. The sheaths of the organisms thicken further with deposits of iron and other corrosion products. Low pH environment under tubercle can cause severe pitting and even perforation of pipe wall.

Pipe wall

Fig 6-19   The evolution of a tubercle.

Bacteria are found in profuse quantities within the zoological slimes that lead to biofouling in a distribution system. They are unicellular plants that sometimes grow in chains or clusters. They are microscopic in size and proliferate rapidly. One organism can produce in a few hours millions of organisms in a colony. They can be classified into three structural types: the coccus, which is spherical;

the bacterium or bacillus, which is rod-shaped; and the spirillum, which is curved. Some types are aerobic, growing only in the presence of dissolved oxygen; others are anaerobic, growing in the absence of dissolved oxygen. Some are capable of producing spores; others surround themselves with a gelatinous sheath or capsule which protects them against heat, dryness, or chemical treatment for long periods.

The predominant organism found in about 50 percent of the slimes examined in the laboratory are short, gram-negative rods. The majority of these belong to the genus *Aerobacter; Aerobacter aerogenes* is the most common species (see Fig. 6-20).

Of the remaining 50 percent of the slimes studied, the predominant organisms are aerobic gram-positive, spore-bearing rods. These bacteria, when properly stained, exhibit a visible capsule. These are of the genus *Bacillus* (see Fig. 6-21). *Bacillus subtilis* and *Bacillus megatherium* have been found in many samples studied. This group develops as mucoid colonies which provide a gelatinous slime binder.

For a more detailed description of the nature, morphology, and structure of bacterial slime growths and the transformation of iron by bacteria in water, see Starkey[233] and Nason.[234]

Coliform Regrowth. The frequency and persistence of coliform "regrowth" in the distribution network has become a major concern to many water utilities. This continuing problem has occurred in what are considered biologically stable systems in spite of the fact that the water leaving the treatment plant is free of coliforms and cross-connections are nonexistent.

American Water Works Service Company, Haddon Heights, New Jersey reported

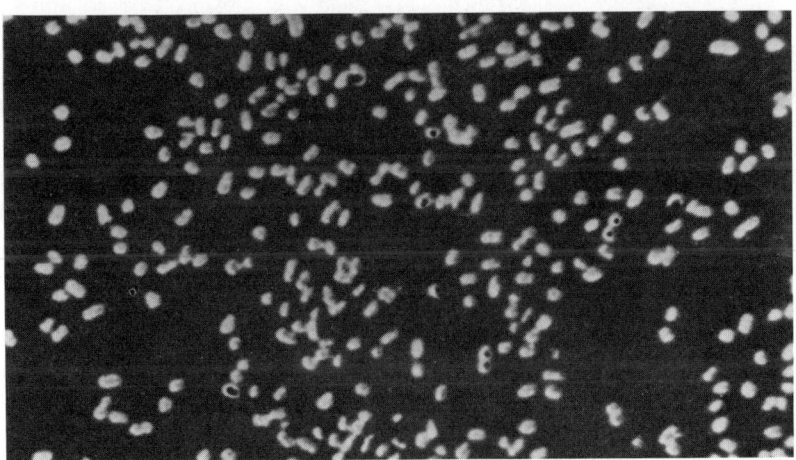

Fig 6-20 Small slime-forming capsulated coccobacilli (approximately 600×) (short gram-negative rods).

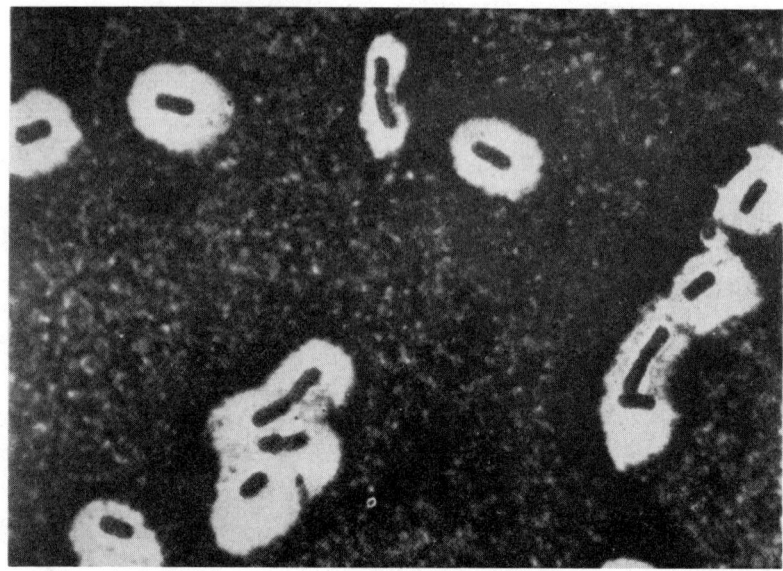

Fig 6-21 Large capsulated bacteria (approximately 550×) (gram-positive spore-forming rods).

in 1984 that six of their Midwestern water utilities had been experiencing low levels of coliforms in their distribution systems during the past four years.[312] In all cases, traditional corrective measures had failed to reduce the levels of bacteria concentration until the chlorine residual was increased to 6 mg/l. These systems suffered from an unexplained appearance of coliforms in distribution systems that habitually maintained free chlorine residuals of 2–3 mg/l.[313]

In one of these systems a zinc phosphate compound was added to the treated water as it entered the distribution system. This treatment seems to have stopped the appearance of coliform organisms. Moser[313] offers the following explanation based upon his belief that the coliforms may be harbored in the gasket material at the pipe joint: the zinc compound added to the water deposits a metallic film at the pipe joints, thereby encapsulating the surviving organisms. This explanation derives from the fact that the distribution system piping involved in the coliform reappearance phenomenon had been poly-pigged prior to the zinc phosphate treatment. This treatment removes tubercles and other scaly deposits that are likely to harbor or encapsulate microorganisms.

Olivieri,[314] who investigated the regrowth problem that has plagued the above systems, believes the answer is the result of a microbiological process. This is the development of a biofilm by the microorganism which protects the organism from the chlorine residual. He further believes that the microorganism develops this film as a consequence of the hostile environment: i.e., free chlorine residual and a starvation level of nutrients. White,[315] who also investigated one of these systems, thinks that age of the pipe, joint material, scaling conditions, and a system free from biofouling are all contributing factors for the survival of a sturdy strain of coliforms. As one noted microbiologist said "when nutrients in the system are restricted to the poverty level as in a clean system the opportunistic organisms that do survive become street tough."

Bacteria Enumeration. Until recently the method for bacteria detection and enumeration (standard plate count) required a protracted length of time—sometimes days. A rapid new method was exhibited at the 1984 Annual AWWA Conference, Dallas, Texas. This is a color-guided test instrument which can perform the enumeration of as few as 100 bacteria per ml within a 2–3 minute test period.[316] This test was developed by the Baylor College of Medicine, Department of Virology and Epidemiology.[317]

Procedure and other details of this test appear on p. 918, *Standard Methods*, 16th ed. This rapid procedure will greatly simplify distribution system quality monitoring.

Control of Biofouling in Distribution Systems. The application of chlorine is imperative to control biofouling in distribution systems, whether it be regrowth of coliform organisms or the growths that cause taste, odor, dirty water, or corrosion. The controversial aspect of this treatment is how to apply it. Some strongly believe that the best method is to push a free chlorine residual to the far reaches of the distribution system. Others feel just as strongly that a combined residual that is predominantly monochloramine should be used.

Free Chlorine vs. Combined Chlorine Residual. The proponents of the combined available residual method cite the following:

1. The combined residual, while less potent, is more persistent and will eventually penetrate farther for a longer time.
2. If a free residual is pushed forward and the system is badly contaminated, this free residual in its travel will gradually be turned back toward the dip and then to the hump of the breakpoint curve and will eventually become all combined available residual. Further, during this process whereby the free residual is being converted to combined residual, tastes will result from the inevitable formation of nitrogen trichloride and dichloramines.
3. This eventual formation of combined residual chlorine is certain to occur in the consumer service pipes long after the distribution pipes have been cleaned.
4. The free residual method reacts much faster than the combined residual, which produces sloughing of the debris in the system caused by the biofouling process. This produces horribly dirty water and objectionable tastes and odors.

Case Histories. The proponents of the free residual method are usually those confronted with a serious situation that needs drastic action and are quick to admit serious consumer complaints while the system is being cleaned.

The ammonia chlorine process was used by Arnold[238] in 1936 to control the growth of filamentous bacteria by intrusion waters in the coast tunnel of the San Francisco water supply. The application of ammonia was discontinued when it was difficult to obtain during the war years, and was never resumed. Ackerman[239] describes controlling tuberculation using chloramine treatment.

Harvill et al.[95,96] used the unique method of utilizing ammonia to produce an induced breakpoint. By varying the ammonia dosage, they were able to control the magnitude of the residual at the dip. The ammonia was applied after the chlorine. This is a variation of the postammoniation technique recommended by proponents of chloramine residuals for distribution systems. This technique was used for about eight years until the system was cleaned, at which time the ammonia was discontinued.[97]

Alexander[243] reported in 1944 upon the successful use of free residual chlorination in cleaning and maintaining a distribution system free from biofouling originally caused by iron- and sulfur-related organisms. Gradually pushing through a 1.0 mg/l free residual over the entire system did not result in complaints of chlorinous tastes. However, the success described by Alexander came about after the system was thoroughly cleaned of massive tenacious films of organic slimes. This cleaning was done by isolating parts of the system and subjecting them to sterilization dosages of 50 to 100 ppm chlorine plus hydraulic and air purging of the pipe lines.

Brown[244] reported the use of free chlorine residuals as high as 1.5 mg/l to clean the Santa Rosa, California, system, but there were numerous complaints about chlorinous odors, which were apparently caused by the formation of nitrogen trichloride. Dechlorination, to limit the residuals going into the system, was tried, after the system had been cleaned, in an attempt to reduce complaints about chlorinous tastes and odors, sometimes described as a "Clorox cocktail" by the newspapers. This dechlorination did reduce these complaints, but soon the black fluffy slime returned to the taps. The use of dechlorination was discontinued in favor of a free chlorine residual in the distribution system in order to restore water quality to the consumer.

Wilson[244] reported the successful use of free residual chlorine, with 0.5 to 0.8 mg/l in the distribution system, which took three years and resulted in many consumer complaints about chlorinous tastes. After the system was pronounced clean, the chlorine residual entering the system was reduced from 2 to 1 mg/l with a noticeable reduction in the number of complaints.

Pomeroy and Montgomery[244] experimented at Torrance, California, between 1940 and 1949, by using both chlorine and ammonia and free residual chlorination. They were unable to clean the system by using chloramine residuals. After the system was rid of the putrefactive condition, free residual chlorination produced a water of excellent quality. While they attributed a large part of the success of cleaning the system to chlorine treatment without ammonia, an intensification of swampy and musty odors, causing a storm of complaints, was experienced when free residual chlorination was substituted for the ammonia-chlorine treatment.

Blair[245] reported in 1954 of the use of free chlorine residuals to clean red water troubles caused by *Crenothrix* that had been plaguing the Palo Alto, California, water system.

Implementing a Clean-up Program. This carefully planned program at Palo Alto attempted to carry a 1.0 mg/l free chlorine residual throughout the entire distribution system. No objectionable tastes and odors were reported whenever the residual was as high as 0.4 mg/l but when it dropped to 0.2 mg/l or less, the complaints were prevalent. After the entire system was cleaned satisfactorily, further complaints were traced to consumer service lines. A phenomenon reported by Wilson[246] was also reported by Blair. In nearly all these instances, the objectionable taste and odor occurred in the morning after the water had stayed in the pipes overnight and all the free chlorine had been consumed. After this water and its chloro products had been discharged from the pipes, the taste and odors disappeared, only to recur when consumer use dropped off and the free chlorine residual again disappeared. The distribution system residual might often show 0.6 mg/l free chlorine residual while the consumer tap would not show even a trace of residual.

The foregoing demonstrated the effectiveness of chlorination in cleaning a distribution system suffering from biofouling organisms. The following is a summary of guidelines for carrying out a program to clean such a system.

1. These systems are usually supplied with underground waters that are more or less deficient in oxygen, so that the first step is to run a dissolved oxygen survey to identify the zones of degradation.

2. Implement a flushing program at the worst places, starting with the zones of least dissolved oxygen.

3. Try to push through the entire system a free chlorine residual of at least 1.0 mg/l.

4. Enlist the aid of the local newspaper in indicating to the consumers the reasons for the program and how they can help by flushing their own service pipes.

5. Keep a continuous record of complaints.

6. Repeat the dissolved oxygen survey as necessary.

7. Utilize portable chlorine residual recorders where possible not only to record the residual variations but to identify as free residual chlorine.

8. Make spot checks of distribution system residuals versus adjacent consumer tap residuals. If these show a wide disparity, institute a special consumer service flushing program at no cost to the consumer for water used.

Chlorine Residuals to Maintain Bacterial Quality. The National Academy of Sciences and the National Research Council issued a joint report in 1955 and a clarification in 1959[247] advising that the establishment of a universal standard for maintaining residual chlorine in water in distribution systems is not desirable at this time, owing to the wide variations in circumstances encountered. This report, pertaining solely to military installations, acknowledges not only the desirability of chlorine residuals to maintain better bacteria control but that loss of the residual could be a warning of possible sabotage. Furthermore, this report recognizes that combined residuals can be maintained within a distribution system much more easily than can free chlorine residuals, but that the combined residuals are such relatively weak disinfectants that it is questionable whether the two types of residual (free and combined) should be considered analagous or even comparable unless the combined residuals used are from ten to twenty times the usual value of free residuals.

When the U.S. Public Health Service standards on water quality were revised in 1942, emphasis was placed on samples of water collected from the distribution system. Water leaving the treatment plant or pumping station could be well within the permissible bacteriological standards but might show considerable deterioration in the distribution system. The result was an increased effort to maintain chlorine residuals in those distribution systems confronted with this deterioration of quality.

Baylis and Kuehn[247] reported that the city of Chicago maintains chlorine residuals throughout their system of approximately 0.4 to 0.5 mg/l some free chlorine residuals and some chloramine.

Crabill[247] reported that:

1. High-level chlorine residuals—0.4—maintain a better quality of water in the distribution systems than do low-level residuals—0.05.

2. The quality of the distributed water is reduced when chlorine residuals are lost.

3. When chlorine residuals are lost, supplemental chlorination will improve water quality.

4. There is a definite relation between water temperature and the persistence of chlorine residuals.

Plowman and Rademacher[248] found that the loss of combined residual chlorine was low when the water was in the 44 to 60°F range, but that during the summer, when the range was 68 to 73°F, there was a rapid loss of residual.

These conclusions indicate a real value in maintaining chlorine residuals throughout the system.

Umbehauer[247] reports that the use of 2.0 mg/l residuals in the far-flung El Paso, Texas, distribution system has been highly successful from a bacteriological standpoint with fewer than the anticipated number of consumer complaints.

The public health aspect of maintaining chlorine residuals in a distribution system for added consumer protection is becoming virtually a necessity. It is well known that a distribution system can become contaminated from cross-connections with non-potable water, or a water main and sewer line break at crossover points, thus causing outbreaks of a waterborne disease. This emphasizes the necessity for maintaining a chlorine residual in the distribution system.

Eliassen's analysis of waterborne disease outbreaks showed that the greatest number of cases among users of public water supplies resulted from contamination of the distribution system.[249] One of the largest outbreaks occurred at Newton, Kansas, with an epidemic of 3000 cases of bacillary dysentery in 1942. In 1952, the Kansas State Board of Health requested that cities using surface supplies maintain a minimum of 0.4 mg/l of free available residual chlorine at all consumer taps.[16] Later the same year, thirty cases of amoebic dysentery broke out among factory workers in South Bend, Indiana, caused by sewage contamination of the distribution system.[250] The 1952 request for distribution system residuals and the chlorination of all supplies was made into an order by the Kansas State Department of Health in 1956.

The investigation of the Cincinnati system reported by Buelow and Walton[240] concluded that the change from combined to free chlorine residual together with an increase in the residual concentration greatly reduced the average monthly coliform counts. Prior to the changeover from combined to free residual, the plant effluent was regulated to maintain 0.85 ± 0.20 mg/l total chlorine; to this 0.2 mg/l ammonia was added. When the changeover was made the chlorine residual was raised to 2.0 mg/l total, about 80 percent was free chlorine. This study emphasized what the Community Water System Survey[129] showed—that a chlorine residual must be maintained throughout the system in order to establish proof that disinfection has been achieved. Also the recurring cases of coliform appearances in the distribution system samples when none were detected in the treatment plant effluent indicate the necessity for continuous disinfection throughout the system. The data collected in the Cincinnati investigation strongly support the need to take bacteriological samples in known problem areas such as reservoirs, dead ends, and the periphery of the system. It has been suggested that chlorine residuals could in some cases be substituted for elaborate bacteriological sampling, provided of course that the residual is properly monitored.

In 1974 the committee on water quality in transmission and distribution systems outlined some basic requirements for a water entering a distribution network.[241] For example, if a water is of low quality with a high nutrient content, it should be carefully flocculated and filtered to produce a product low in aluminum and iron with a turbidity no higher than 0.2JTU. This water should be able to hold a chlorine residual fairly well if the piping system is relatively clean. The goal is to maintain a 0.2–0.3 mg/l total residual in the far reaches of the system. If the residual decays in the system, rechlorination may be necessary.

In 1973 the City of Chicago converted both the Central and South Water Filtration Plants to provide free chlorine residuals throughout the city.[162] In view of the mass of evidence in the literature and case histories of water quality improvement in distribution systems this change had been contemplated for some time. After one year of operation the records indicated that the switchover was a complete success in that there have been fewer consumer complaints, better bacteriological quality, and residuals can be maintained in remote dead-end areas with a modest amount of special flushing techniques.

Snead et al.[242] made a comprehensive report in 1980 for the EPA demonstrating the benefits and the necessity of chlorine residual in distribution systems, with particular emphasis on free chlorine residuals.

The foregoing case histories reflect success with free chlorine residuals. But methods may have to be changed in some instances on account of potential THM formation in these systems. The magnitude of THM formation increases with time in the presence of free chlorine. Chlorine dioxide is an excellent disinfectant for distribution system chlorination because of its slow decay rate in the presence of most zoological slimes. See Chapter 12.

Relay Chlorination. The availability of modern equipment to monitor and record either free or combined chlorine residuals simplifies the task of maintaining effective residuals throughout a distribution system. The Hartford-type installation, applicable to most systems, is described next.[251]

The water is rechlorinated solely by automatic residual control. A chlorine analyzer provides the intelligence to control the chlorination equipment without the necessity of installing expensive flow-sensing devices in the distribution system. This installation provides the means of obtaining residuals in that part of the system where the bacteriological quality of the water is substandard. This also is the part of the system where it is impossible to obtain residuals with chlorination at the treatment plant alone. This single point of rechlorination provides the means of meeting the U.S. Public Health standards. The attempt is to keep 0.8–1.2 mg/l free residual chlorine; however, owing to flow reversals, the residual deteriorates to as low as 0.3 mg/l and climbs to as high as 2.0 mg/l for a few hours each day. This approach has proved successful in many instances. It is particularly useful where open balancing reservoirs deplete the residual in the distribution system.

Practical Considerations. Careful planning of the distribution system and the proper location of balancing reservoirs can largely prevent situations that tend to cause such wide fluctuations in distribution system residuals. As an example, more attention should be given to the prevention of dead ends. This will require a periphery type of grid system to improve the circulation of a distribution system. The literature abounds with dead end water quality problems, but apparently the designers of such systems are totally unaware of the consequences of poor circulation.

Research is also needed to properly determine whether free residual or combined residual chlorine is preferable in a distribution system. As a broad general rule, it is better to have a predominantly monochloramine residual, rather than a free residual, entering a relatively clean system. The chance of developing nitrogen trichloride is nil if the residual entering the system is primarily monochloramine. However, if a system is dirty, then free residual chlorine must be used to do the cleaning as soon as possible. Such applications are bound to produce nitrogen trichloride and dichloramines in quantities sufficient for the consumers to complain of tastes and odors. These obnoxious chlorine compounds will persist until the free chlorine residuals penetrate to all parts of the system, which is accomplished by a continuous flushing program with sufficient high free residuals to do the job.

Another area that needs more attention is that of the corrosion potential of the system residual. As an example, Williams[222] used a combined residual at Brantford, Ontario, to maintain proper water quality and prevent the formation of tastes and odors in the distribution system. Owing to the presence of significant concentrations of organic nitrogen (1–3 mg/l) in the raw water and the necessity of using the free residual chlorination process at the treatment plant, nitrogen trichloride was produced at the treatment plant. If free chlorine residuals were allowed to enter the distribution system, then nitrogen trichloride would develop in the distribution system, causing taste and odor complaints. To prevent the subsequent formation of nitrogen trichloride in the distribution system, postammoniation was practiced.* Sufficient ammonia was added to convert all the free chlorine residual to combined residual, and enough in excess of this to produce a combined residual consisting of 80 percent mono- and 20 percent dichloramine going to the distribution system. If the organic nitrogen content were high, the dichloramine fraction would be higher, and vice versa. The total combined residual going to the distribution system was kept between 0.6 and 0.8 mg/l.

At one time this water passing through the distribution system developed a progressively deteriorating palatability, with the resulting taste described as "bitter irony." A thorough investigation revealed a pickup of zero to 0.8 mg/l ferrous iron and a shift in the chlorine residual to 40 percent mono- and 60 percent dichloramine with the monochloramine deteriorating from about 0.6 mg/l to about 0.15 mg/l. The drop off of chlorine residual in the system was proportional to the pickup of ferrous iron. There was also a drop-off in ammonia content as high as 75 percent in two hours as the water passed through the distribution system. However, samples stored in the laboratory showed complete stability of chlorine residual and ammonia content. The difficulty was finally discovered to be aggressive water. Analysis showed that the heavy prechlorination doses combined with the addition of aluminum sulfate for turbidity removal resulted in a sufficient loss in alkalinity and a lowering of the pH to make the treated water aggressive. This aggressive water contained enough free CO_2 to attack the iron pipe, causing ferrous bicarbonate to go into solution. This resulted in the dechlorination of the monochloramine fraction producing microquantities of ferric hydroxide, which eventually flushed from the system without causing red water troubles. The stability of the dichloramine fraction of the residual is somewhat of an anomaly, but can be ex-

---

* The preferred method to prevent NCl_3 formation and the elimination of dichloramine is complete dechlorination with SO_2 followed by rechlorination. Williams chose postammoniation because it required less capital expenditure for equipment.

plained. Williams points out that when the organic nitrogen increases, so does the dichloramine proportion of the total combined residual. The high organic nitrogen content of the raw water can easily be due to pollution from sewage. Since sewage contains urine and other compounds containing significant quantities of creatinine, the stable dichloramine fraction is probably a chlorcreatinine compound* that exhibits the chracteristics of dichloramine in the amperometric titration procedure. It has been found that such compounds are quite stable but with little or no germicidal power.

Williams confirms the fact that there is no off flavor resulting from the dichloramine fraction. The monochloramine fraction is considerably more susceptible to dechlorination by ferrous iron than is dichloramine, and so the gradual disappearance of monochloramine residual as the water passes through the distribution system is not unexpected.

The problem of the iron pickup and the resulting "bitter irony" taste was solved at Brantford by the addition of lime to the treated water, which restored the alkalinity and raised the pH to normal values in accordance with Langelier's equilibrium index. Now the water has the ability to lay down a slight protective film of calcium carbonate on the metal pipes.

Various experiences, extending over many years, some of which are described in the foregoing text, show that chlorine residuals in a distribution system: control growths in the system; improve the palatability of the water; give added consumer protection against waterborne diseases; protect against accidental and possibly intentional contamination of the system.

Restoring Pipeline Capacity. The application of chlorine to long transmission systems is somewhat akin to the maintenance of chlorine residuals in a distribution system. The principal difference is that there are seldom any consumers supplied by the transmission system, and therefore the public health aspect and the taste and odor control problems are usually not factors. These systems usually terminate at the treatment plant. If not, then the transmission line is simply part of the distribution system. It has been demonstrated that filamentous organisms can create a condition inside pipes to significantly increase the friction loss, thereby reducing the design capacity.

Rogers[253] reported a slime deposit of ⅜ to ½ inch thick attached to the cement lining of a 42- and 48-inch transmission line thirty-four miles long. This deposit was 75 percent ferric oxide and contained *Crenothrix*. The capacity of the line dropped from 30 to 20.4 million gallons in three years. Continuous chlorination of 3 mg/l restored the capacity of the line, and, by maintaining a free residual of chlorine at 0.4 to 0.5 mg/l at the end of the line, prevented growth of the *Crenothrix* and thereby maintained the C factor in the line.

Griswold[254] used chloramine treatment on a 24-inch supply line twelve miles long. After fifteen years of continuous treatment with a terminal residual of 1.0

---

* We now know that the dichloramine fraction in the presence of organic N is a mixture of impotent but stable organochloramines. See Chapter 8.

to 1.5 mg/l combined chlorine, the line showed an average loss of 10 percent in capacity in seven years. The line is now cleaned every five or six years, while prior to chloramine treatment the line had to be cleaned every twenty months; but, in order to achieve the nominal carrying capacity of this line on a continuous basis provided by the chloramine treatment, the line would have to be cleaned every seven months if the treatment were discontinued.

The San Diego aqueduct, which travels some seventy miles from the west portal of the San Jacinto tunnel to the San Vicente reservoir in San Diego originally had a capacity of 104 cfs. This dropped to 94 cfs, and slime growths were suspected. Streicher[255] reported on the success of intermittent chlorination of this line to maintain the original capacity of 104 cfs. After much experimentation it was found that a 2.2 mg/l chlorine dose for two hours twice a week was sufficient.

The experience at Little Rock, Arkansas, described by Jackson and Mayhan,[256] graphically illustrates how much more difficult it is to clean a transmission line after the onset of slime growths than it would be to prevent the growth from forming. A 39-inch transmission line thirty three miles long was put into operation in 1938 with a C factor of 147 and a carrying capacity of 25.32 mgd. In one year this line was reduced in capacity by 20 percent.

The slime deposit of black jellylike organic material varied from a thin film to a mat ¼ inch thick and was heavily interspersed with iron oxide. The dominating organisms were found to be large gram-positive encapsulated bacteria and *Crenothrix*. The water supply originated at Lake Winona, a soft water of only 15 mg/l hardness and an iron content of approximately 0.4 mg/l. Chloramine treatment was begun in 1939 at 1.25 mg/l with a residual of 0.55 mg/l (starch-iodide) at the terminus of the line, which was the influent to the treatment plant. This treatment proved to be only moderately successful in restoring the line capacity. One year later free residual chlorine treatment was begun, and after nine years of a 5 mg/l dose resulting in a 1 mg/l residual at the plant the line was considered to be almost like new. The C factor had risen to 133 with a 22.75 mgd capacity in 1950. At this point, since the line was considered to be relatively clean, intermittent treatment with superdoses of chlorine was substituted for continuous treatment. Doses of 12 mg/l for about 2 or 3 hours per day resulted in residuals of 8 to 10 mg/l at the plant. Dechlorination of these residuals was found to be unnecessary, as they dissipated rapidly in the large settling basins. The intermittent treatment, which consumed only about 250 pounds of chlorine per day, resulted in an annual savings of approximately fifty thousand dollars, thus demonstrating the exorbitant cost of the cure as compared to the cost of prevention.

The first step in any program designed to maintain a transmission line is to anticipate the problem of slime growths. When the line is first put into operation, the coefficient of friction should be determined while it is in the new, clean condition. This will establish an irrefutable reference point to determine when the line needs cleaning, and will provide a means for evaluating any needed chemical treatment. The friction factor should be checked each month of the first year of operation to get a slime growth profile.

The effect of corrosion and tuberculation on the friction coefficient of metal pipelines is well known. However, for many years it was thought that concrete pipe or cement-lined steel pipe would maintain their capacities indefinitely. This has proved to be an incorrect assumption, since it has been abundantly demonstrated that filamentous organisms, freshwater sponges, and other organisms are able to attach themselves in massive quantities to concrete and various types of cement surfaces. The carrying capacity of such lines drops off rapidly without any increase in roughness, simply because of the reduction in the effective diameter of the pipeline. Derby[257] cites the C factor of a 5-foot concrete pipe fifty miles long dropping from 140 to 95 because of massive growths of fresh-water sponges *Asteromyenia plumosa* and *Trochospongilla leidyi.*

Another unknown factor that has caused serious trouble in transmission systems is the possibility of groundwater intrusion into the concrete-lined tunnels carrying the water supply. These waters are often the source of iron and sulfur bacteria infestation. This was the cause of luxuriant growths of *Crenothrix* in the coast tunnel of the San Francisco water supply and the source of sulfur-bearing water in the Santa Barbara supply. The San Francisco coast tunnel capacity decreased 30 percent in three weeks.[238]

Control of such growths in transmission lines requires careful study to determine the proper course of action. This may be cleaning followed by chemical treatment with chlorine, chloramine, and possibly copper sulfate. It may not be possible to accomplish satisfactory cleaning. If chemical treatment is to be used without cleaning, it should be determined whether or not the line velocities are sufficient to flush out the organisms destroyed by the chemical treatment. For long conduits of varying cross sections, chemical treatment may be cheaper and more satisfactory than mechanical cleaning. Intermittent chemical treatment must also be considered where taste and odor problems are not involved. If the organisms cause taste and odor, some program of continuous treatment will probably be necessary to prevent these growths from forming. Open channels are subject to sunlight, which has the ability to destroy free chlorine residuals. Therefore these situations call for some kind of variation of the chemical treatment—possibly a combination of chloramine and copper sulfate, intermittent or continuous. If intermittent, the treatment should be confined to the hours of darkness if it is found that free residual chlorine is most effective.

Aid to Coagulation

Prechlorination to free residual is nearly always effective in improving coagulation or reducing the coagulant dose or both. This phenomenon has been known for some time.[258,259] It is not clearly understood how chlorine acts as a coagulant aid, although it is probably due to its oxidizing effect on organic matter.

Chlorine is also used in the preparation of the ferric ion as a coagulant for use in certain types of color and turbidity removal processes. In these instances chlorine solution from conventional chlorination equipment is added to either a solution of pickle liquor or copperas, both of which are ferrous sulfate. Chlorine converts the ferrous ion to ferric, which hydrolyzes to form ferric hydroxide, a fluffy gelatinous floc. Copperas is a granular free-flowing material easily handled in a dry chemical feeder. The chlorine is added downstream from the solution chamber outlet of the copperas feeder. These systems have special application where the optimum pH for coagulation is in the range of 8 to 9. It requires one part of chlorine to react with 7.8 parts copperas to convert all of the ferrous to ferric iron as follows:

$$6FeSO_4 \cdot 7H_2O + 3Cl_2 \longrightarrow 2Fe_2(SO_4)_3 + 2FeCl_3 + 7H_2O \qquad (6\text{-}11)$$

Chlorinated copperas has been known to form a tough, rapidly settling floc with as little as 0.5 grains per gallon dosage.[260]

Chlorine is also used in the preparation of activated silica, a coagulation aid.

Aid to Filtration

At one time it was thought that the efficiency of filtration depended in part upon the establishment of a very thin film of organic material on the sand grains, and that a chlorine residual would destroy this organic film, thereby decreasing the efficiency of the filter.[261] This notion was soon discarded when it was discovered that the proliferation of microorganisms in the filter media would seriously impair the effectiveness of filtration. Baylis of Chicago insisted that *all* surface supplies should be chlorinated as soon as the water reached the treatment plant, as is recommended for polluted supplies.[262]

Baumann et al. reported in 1963 on a controlled study of prechlorination and its effect on filter efficiency.[263] They found that the prechlorinated filter operated considerably longer than the control filter which was not prechlorinated. Other observations included: prechlorination reduces the depth of suspended solids into the filter sand; less sand was removed during the cleaning process; oxidizable tastes and odors were destroyed; oxidizable organic matter was confined to the surface cake at the sand surface; the bacterial quality of the filter effluent was considerably improved as compared to the unchlorinated control filter.

An adequate free residual chlorine is necessary to maintain filter runs and to prevent the formation of mud balls, which not only impair the filtration efficiency but interfere with proper backwashing. This chlorine residual prevents organic growths that contribute to slime buildup on the filter media. These growths not only reduce the mechanical efficiency of filtration but contribute to the degradation of the bacterial quality of the filtered water.

Certain types of algae are also well known for contributing to the clogging of filters. The most serious offenders are the diatoms, which are present during all seasons of the year—*Asterionella, Fragilaria, Tabellaria,* and *Synedra.*

In Chicago, when the water to be filtered contained approximately 700 organisms per ml., principally *Tabellaria* and *Fragilaria,* the filter runs were only 4.5 hours.

Three days later, when the count was down to 100 per ml the filter runs increased to forty-one hours.[191] In Washington, D.C., filter runs were reduced from an average of fifty hours to less than one hour by the sudden influx of the diatom *Synedra*, which had a concentration in the raw water reaching 4800 cells per ml.[191]

If the filter-clogging blue-green algae—*Anacystis, Rivularia, Anabena*, and *Oscillatoria*—are allowed to grow, they form a loose slimy layer over the sand grains, thereby reducing the flow of water through the filter.

Organic growths on the sand grains or other media within the filter reduce the length of filter runs and cause the formation of mud balls, which seriously impair backwashing efficiency. These same organic growths provide an anaerobic environment, which contributes to the rise of the ammonia nitrogen content in the filter effluent. If this is not counteracted by free residual chlorination, nitrifying bacteria such as *Pseudomonas* may proliferate, causing the conversion of the available nitrogen to nitrites. At this point, with the development of nitrites, the environment becomes most suitable for the proliferation of other bacteria, thereby degrading the bacterial quality of the effluent. This situation can be prevented only by a free residual chlorine. Combined chlorine will not oxidize the nitrites to nitrates; only free chlorine (HOCl) can do this.

A filter that has become fouled by organic growths can often be restored by prechlorination. At first the only residual that will appear in the effluent will be all combined residual—meaning that it is predominantly monochloramine. This suggests the presence of significant amounts of ammonia developing within the filter media because of the organic growths. If the prechlorine dose is gradually increased, the organic material will become more and more oxidized. Within a short period of time—possibly one or two weeks—the total residual of filter effluent should begin to show a trace of free chlorine. When the free chlorine content is 80 to 85 percent of the total residual, it may be assumed that the filter is relatively free from organic growth. Backwashing will become more effective and filter runs will be considerably increased, and the bacterial quality will return to acceptable standards. The application of chlorine ahead of filters that have not been properly cared for by prechlorination will not necessarily achieve the desired results. Some filter media that have been neglected for a long time simply cannot be cleaned by adequate chlorination; these media should be discarded and replaced with new material.

Since free residual chlorination is an aid to both coagulation and filtration, it becomes patently clear that the case for carrying a free chlorine residual throughout the entire treatment process becomes strong indeed. It also indicates the necessity of a special point of application of chlorine just ahead of the filters.

Hydrogen Sulfide Control and Removal

Hydrogen sulfide is probably the most obnoxious and troublesome compound to be dealt with in a potable water supply. It is an almost impossible task to produce a palatable water that is free of taste and odor at all times if hydrogen sulfide is present in the raw water in significant concentrations. Not only does the treatment process require continuous surveillance, but the distribution system and consumers' hot water systems require monitoring.

Hydrogen sulfide occurs mainly in well waters. Occurrence in surface supplies is primarily by groundwater intrusion; however, with the rising pollution of natural waters by sewage and industrial wastes, surface waters may become contaminated with hydrogen sulfide.

Sulfides in well water are probably produced through chemical and bacterial changes under anaerobic conditions far underground. Sulfates may be reduced to sulfides by organic matter under anaerobic conditions, and the resultant metallic sulfide changed to hydrogen sulfide by the action of carbonic acid.

The sulfate reducing bacteria (*Desulfovibrio desulfuricans*) are another source of hydrogen sulfide production. In anaerobic environments these bacteria convert sulfates and other sulfur compounds to H_2S. They have a growth range of pH 5.5 to 8.5 and are found to exist in temperatures of 0 to 100°C, with an optimum range of 24 to 42°C.[264]

Another group of bacteria also plays an important part in sulfur bearing waters. These are the sulfide oxidizing forms. The most prevalent are *Beggiatoa* and *Thiobacillus*. *Beggiatoa* are filamentous white sulfur bacteria that obtain the energy necessary for their growth by oxidizing the sulfide ion to colloidal sulfur, which is then stored in their cells.

Hydrogen sulfide is a flammable and extremely poisonous gas.

Brief exposures (30 minutes or less) to H_2S concentrations as low as 0.1 percent by volume of air may be fatal. The gas is highly soluble in water to the extent of 4000 mg/l at 20°C and one atmosphere. The minimum detectable concentration by taste in water is given as 0.05 mg/l.[264] In aqueous solutions it hydrolyzes as follows:

$$H_2S \rightleftharpoons HS^- + H^+ \quad (6\text{-}12)$$

The hydrosulfide ion (HS^-) further dissociates as follows:

$$HS^- \rightleftharpoons S^= + H^+ \quad (6\text{-}13)$$

At 18°C the hydrolysis constant for Eq. (6-12) is:

$$K_h = 9.1 \times 10^{-8} = \frac{[H^+][HS^-]}{[H_2S]} \quad (6\text{-}14)$$

And for Eq. (6-13)

$$K_h = 1.2 \times 10^{-15} \frac{[H^+][S^=]}{[HS^-]} \quad (6\text{-}15)$$

362 HANDBOOK OF CHLORINATION

Figure 6-22 illustrates the distribution of H_2S, HS^-, and $S^=$ for various pH levels. At pH 7, hydrogen sulfide is approximately 50 percent of the total dissolved sulfides; at pH 5, it is practically 100 percent of the total; at pH 9, it is nearly all hydrosulfide ion. Therefore the existence of hydrogen sulfide in sulfur-bearing waters is pH dependent. This scientific fact has been well documented.[265,268,269,274]

Whenever the equilibrium between the hydrosulfide ion and the hydrogen sulfide in solution is upset, as when H_2S is removed by oxidation, the shift will be to form more H_2S from the remaining dissolved sulfides to reestablish the equilibrium. This stored sulfur gradually disappears by metabolic action, being itself oxidized to sulfate to yield more energy.

The primary objection to waters containing hydrogen sulfide is the offensive rotten-egg taste and odor. The secondary objection is the marked corrosiveness of these waters to both metals and concrete structures. Another serious problem is their ability to promote luxuriant blooms of the various types of filamentous sulfur bacteria that lead to a general degradation of water quality in the system.

The presence of H_2S will turn silverware black, discolor lead base paint, make bathing in tubs or showers extremely unpleasant, and stain all plumbing fixtures. The sulfides in the water react with iron from the mains to form a suspension of iron sulfide that causes a discoloration of the water. This suspension makes laundering almost impossible.

The presence of hydrogen sulfide is noticeable to the extent of 0.5 mg/l even in cold water.[265] If the pH is high, the odor may be slight. Concentrations of H_2S as low as 0.2 mg/l will promote the growth of *Beggiatoa*.[264] The range of hydrogen sulfide concentration in usable sulfur waters is below 10 mg/l. Notable

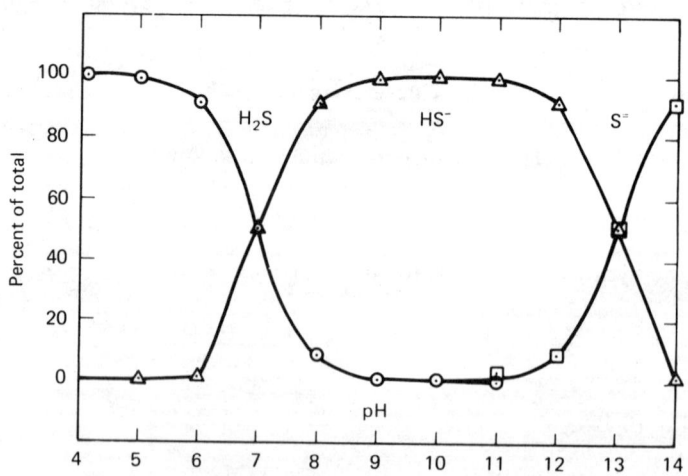

Fig 6-22 Effect of pH on hydrogen sulfide-sulfide equilibrium.

exceptions are the Wadsworth Plant of the Southern California Water Company[266] and the Santa Barbara supply[267] where the H_2S content in the raw waters sometimes reaches 20 mg/l.

The chemistry of the oxidation and removal of hydrogen sulfide is extremely complex. Supposedly the oxidation proceeds to form either elemental sulfur, sulfate, or both, as follows:*

Aeration or oxidation by dissolved oxygen in the water:

$$2S^= + 2O_2 \longrightarrow SO_4^= + S_{\downarrow}^0 \qquad (6\text{-}16)$$

by chlorination

$$H_2S + Cl_2 \longrightarrow 2HCl + S_{\downarrow}^0 \qquad (6\text{-}17)$$

and

$$H_2S + 4Cl_2 + 4H_2O \longrightarrow 8HCl + H_2SO_4 \qquad (6\text{-}18)$$

Theoretically Eq. (6-17) requires 2.1 mg/l chlorine for each mg/l H_2S, and Eq. (6-18) requires 8.5 mg/l for each ppm H_2S. This reaction was amply demonstrated in field experiments by Powell.[274]

The assumption is that if enough chlorine is added to satisfy the natural chlorine demand of the water plus enough to react with the H_2S present, then free sulfur will be formed in Eq. (6-17), and that if more chlorine is added to satisfy the stoichiometric requirements of Eq. (6-18), then all the H_2S will be converted to sulfates. Unfortunately, this is not the case.

When chlorine is added, regardless of the amount, to a sulfur-bearing water, colloidal free sulfur will be formed if enough chlorine is added to more than satisfy the natural chlorine demand of the water. This formation of free sulfur is readily evident as a milky blue turbidity (the Tyndall effect).

Choppin and Faulkenberry[270] state that the oxidation of the alkali sulfides is not simple and direct and may yield as end polysulfides, sulfites, and thiosulfates in addition to elemental sulfur and sulfates. The relevancy of this statement is based upon practical evidence that most sulfur-bearing waters that have been treated for hydrogen sulfide removal will exhibit some magnitude of odor in hot water systems. While the odor is not that of hydrogen sulfide, it is a sulfurous odor. Monscvitz and Ainsworth[267] believe that this odor is the result of the formation of polysulfides (HS_n^-) in the oxidation reaction of hydrogen sulfide. Given time in days, these polysulfides will eventually oxidize to sulfates in the presence of dissolved oxygen in the water. If dissolved oxygen is not present, as occurs in distribution systems, these polysulfides may be reduced back to hydrogen sulfide.

The oxidation of hydrogen sulfide will produce colloidal sulfur which can be removed by filtration. Side reactions will produce polysulfides, which must be

---

* See Chapter 7 for more details on hydrogen sulfide chemistry.

dealt with in order to eliminate threshold odors in the finished water. Monscvitz and Ainsworth suggest the conversion of colloidal sulfur and the remaining polysulfides to sulfates by first adding sulfite (which forms thiosulfate) and by converting the thiosulfate (S_2O_3) to sulfate by rechlorination as follows:

$$S^0 + SO_3^= \rightleftharpoons S_2O_3$$
$$\Updownarrow \qquad \qquad \qquad (6\text{-}19)$$
$$HS_n^-$$

$$S_2O_3 + HOCl \longrightarrow S_4O_6^= + SO_4^= \qquad (6\text{-}20)$$

The reaction of Eq. (6-19) is very rapid. The formation of tetrathionate ($S_4O_6^=$) is not significant since it will convert to sulfate within a short time, with no contribution to any threshold odor. The sulfite ion may be added as sulfur dioxide in an aqueous solution or as sodium metabisulfite. This reaction taken to completion is recognized as the typical dechlorination reaction.

Above pH 9.0, polysulfides do not appear to form. This is probably one reason that any lime-softened sulfur-bearing water will not produce threshold odors resulting from these compounds.

Probably the most important factor in hydrogen sulfide removal is that of contact time. Given time and an oxidizing agent, all dissolved sulfides can be converted to sulfates.

Aeration with contact times of up to three hours have shown hydrogen sulfide removals of 2 to 3 mg/l and 35 to 45 percent removal of dissolved sulfides.[271] Shorter contact times and raw water concentrations of 5 mg/l H_2S reduced to 1 mg/l by aeration, followed by coagulation and filtration, require free residual chlorine of up to 0.35 mg/l for the complete removal of H_2S.[272] Derby[273] reported in 1928 on the satisfactory removal of up to 7 mg/l by conventional lime-softening and chlorination. Concentrations as high as 32 mg/l have been successfully removed by pre- and postchlorination, coagulation, and filtration.[266]

The final concern of the water producer is to protect the quality of the product in the distribution system. This is difficult to accomplish because of the possible growth of sulfur bacteria. While the oxidizing forms that grow in an aeration tower may be beneficial because of their ability to convert the sulfide ion to sulfates, they may find their way to the distribution system and in their life cycle slough off into dead ends and contribute sulfur compounds that under anaerobic conditions will start the entire hydrogen sulfide production cycle over again. For this reason chlorination for the control of these organisms in the distribution system is of utmost importance.

Hydrogen sulfide can be successfully removed by aeration and chlorination, followed by the addition of sulfur dioxide or sulfite for the removal of polysulfides and colloidal sulfur and then rechlorination to convert all the remaining sulfur compounds to sulfates.

The formation of hydrogen sulfide is pH dependent and is encouraged by a lower pH.

Oxidation of sulfur-bearing waters will produce colloidal sulfur that may cause a milky blue turbidity (the Tyndall effect).

As the hydrogen sulfide is removed by oxidation, the equilibrium shifts to form more H_2S from the remaining dissolved sulfides; depending upon the amount of oxidant and contact time, this reaction can continue until all the dissolved sulfides are removed.

When sulfur-bearing waters are oxidized, a combination of sulfur compounds are formed, including colloidal sulfur, polysulfides sulfites, and sulfates.

Aeration, chlorination, coagulation, filtration, and softening have been used successfully for the removal of H_2S.

Chlorination is extremely important not only as an oxidant but also as a germicide to prevent the growth of unwanted sulfur bacteria.

Complete removal of H_2S can best be accomplished by oxidation by chlorine; conversion of resulting colloidal sulfur and polysulfides by metabisulfite or sulfur dioxide to thiosulfates, and then converting to sulfates by rechlorination.

The Iron and Manganese Problem

Nature and Occurrence. Both iron and manganese cause serious problems in potable and industrial water systems. While it is beyond the scope of this text to cover the subject of iron and manganese removal, it is pertinent to discuss the occurrence of these elements in water supplies, their significance, and the role of chlorination as related to compounds of these substances.

These two elements are relatively abundant in the earth's crust. The lithosphere contains approximately 5 percent iron and 0.1 percent manganese.[276] Iron exists in soils and minerals, mainly as insoluble ferric oxide and manganese, as manganese dioxide. Iron also occurs as ferrous carbonate (siderite), which is slightly soluble. Since groundwaters usually contain substantial amounts of carbon dioxide (30 to 50 mg/l), appreciable amounts of ferrous carbonate may be dissolved to form soluble (150 mg/l) ferrous bicarbonate as follows:

$$FeCO_3 + CO_2 + H_2O \longrightarrow Fe(HCO_3)_2 \qquad (6\text{-}21)$$

This is the commonest form in which iron is found in water supplies in troublesome amounts. Iron in natural water supplies may also be present as ferric hydroxide, ferrous sulfate, and colloidal or organic iron.[265]

The insoluble ferric compounds will only go into solution under reducing conditions—in the absence of oxygen. The same holds true for the compounds of manganese.

Manganese most commonly occurs in water supplies as manganese bicarbonate, which is even more soluble than ferrous bicarbonate. Acid mine waters frequently contain manganous sulfate as well as ferrous sulfate; some shallow wells and surface waters contain colloidal or organic manganese.

The occurrence of iron and manganese in water supplies is usually limited to wells and impounded surface supplies. It is rare to find either in flowing streams of normal pH and alkalinity. In waters in the regions of acid mine drainage wastes, particularly along parts of the Allegheny and Monongahela rivers, where the pH is less than 5, there are significant quantities of both iron and manganese compounds in solution.

From all the evidence available it appears that biologic activity is a powerful factor in the dissolving of iron and manganese.[269,277] It is apparent that anaerobic conditions must develop in order for appreciable amounts of iron and manganese to gain entrance to the water suppply.[275] For example, groundwaters that contain appreciable amounts of iron and/or manganese are always devoid of dissolved oxygen and are high in carbon dioxide. High carbon dioxide content indicates that bacterial oxidation of organic matter has been extensive, and the absence of dissolved oxygen shows that anaerobic conditions were developed.

Waters that normally do not contain any iron or manganese while flowing in a stream are certain to contain these compounds after the water has been impounded. The amount that goes into solution depends upon the character of the soil and the amount of plant life. Decomposition of organic matter in the lower strata of the water (hypolimnion) in the reservoir results in the elimination of dissolved oxygen and the production of carbon dioxide, so that the iron and manganese compounds in the flooded soil and rocks are converted to soluble compounds. These soluble compounds rise to the surface in those areas where the fall overturn occurs. At this point they are oxidized and precipitated and then sink to the lower portion of the reservoirs where re-solution occurs in the absence of oxygen. Thus waters near the surface of reservoirs are most likely to be free from iron and manganese. Therefore it is essential that multiple port outlet structures be provided for impounded supplies to allow selection of waters somewhere between the surface water containing algae and the deeper water (devoid of oxygen) that may contain iron and/or manganese. The amount of iron and manganese that gains entrance to either impounded surface supplies or groundwater varies throughout the country. The manganese concentration in the hypolimnion of reservoirs varies from 2.0 mg/l in New Jersey to as much as 20 mg/l or more in the lakes of the Tennessee Valley Authority.[277] The time required for the appearance of significant amounts of iron and/or manganese in man-made lakes also varies with the location. Some areas develop significant concentrations in one year, while others may require ten years.

Significance. While there is no evidence that humans suffer from drinking water containing iron or manganese, these substances contribute to some of the most serious problems ever to confront the waterworks industry. Waters containing ferrous bicarbonate stain everything with which they come in contact a yellowish to reddish brown. Manganese-bearing waters free of iron will produce black stains. Waters containing ferrous bicarbonate usually contain some manganese, and this combination will produce stains varying from dark brown to black. Depending on the concentration of these compounds, consumers' complaints will start first with staining and streaking problems in the laundering process; next, there will be the red water or dirty water complaints; finally, there will appear large visible chunks of material that have sloughed off the distribution piping system. Most industrial wet processes, such as in the textile, pulp and paper, and the beverage industries, cannot tolerate this kind of water.

In addition to the physical phenomena of the presence of iron and manganese in the water, both contribute to and promote the growth of crenoform organisms in the distribution system. These filamentous organisms utilize both iron and manganese in their metabolism, and deposit within the pipelines to form heavy, gelatinous, stringy masses that slough off at intervals, causing a variety of problems for both domestic and industrial consumers. In addition to the unsightly mess that results, the growth of these organisms impairs the hydraulic carrying capacity of the entire system.

Because of the potential damage that these two elements can cause to a water supply system, the U.S. Public Health Service Drinking Water Standards of 1962 have placed a limit of 0.3 mg/l of iron and 0.05 mg/l for manganese in potable water supplies. An AWWA task group[278] suggested limits of 0.05 mg/l for iron and 0.01 mg/l for manganese for an "ideal" quality water for public use. Traditionally manganese was reported as a combination with iron regardless of their ratio. Prior to the 1962 standards the limit was set at 0.3 mg/l for both iron and manganese. While 0.3 mg/l of iron could be tolerated under certain conditions this amount of manganese alone would cause severe difficulties. In view of the new standards, iron and manganese should be reported separately. Manganese in excess of 0.02 mg/l will almost surely cause problems in the distribution system. A large segment of waterworks people believe that all public water supplies should be devoid of manganese. While this is imperative for some industries, most domestic systems can tolerate up to 0.01 mg/l of manganese.

To emphasize the difficulty that may be encountered from these compounds it is pertinent to cite a classic example:

The Metropolitan Water District of Salt Lake City, Utah, has its source of supply water impounded at the Deer Creek reservoir on the Provo River. The pipeline from this reservoir to the distributing reservoir in Salt Lake City is some 40 miles long. The water is chlorinated at the outlet of the Deer Creek reservoir just as it enters the long pipeline. This treatment is for disinfection and prevention of slime growths in the long transmission line. A few years after this system had been in operation, it became coated with a black deposit of iron and manganese. When this coating is about $\frac{1}{32}$ inch thick it breaks off and colors the water. The manganese content is usually a few hundredths of a mg/l and sometimes goes as high as 0.1 mg/l but is continuous throughout the year. The outlet of the reservoir is at the bottom where the dissolved oxygen becomes depleted after spring turnover until it reaches zero. The source of iron and manganese in this water is due to the reducing conditions in the bottom of the reservoir, where both iron and manganese from the soil and biological life readily go into solution. After spring turnover, the iron and manganese rise to the top, where they are oxidized by the available dissolved oxygen in the water. This is followed by precipitation to the lower levels, where they again go into solution, only to become oxidized again by the chlorine and precipitate out in the pipeline. All types of chlorination programs were tried to eliminate this deposition. When chlorine was reduced, matters became worse because this resulted in a buildup of pipeline growths. Therefore the district has had to resort to pipeline cleaning on a regular basis. The solution to this problem is an outlet structure with the ability to withdraw water at selected depths to avoid the areas of iron and manganese concentrations. Unfortunately the present outlet structure is irrevocable.

The Role of Chlorine

Iron Removal. Generally speaking, there are three methods for removing iron: by simple oxidation, by aeration or chlorination, or a combination of both; by precipitation with lime at a pH above 8; or by ion exchange. The major problem in these methods is the removal of the insoluble precipitate from each.

Chlorine, either free or combined, reacts to oxidize ferrous iron as follows:

$$2Fe(HCO_3)_2 + Cl_2 + Ca(HCO_3)_2 \longrightarrow 2Fe(OH)_3\downarrow + CaCl_2 + 6CO_2 \quad (6\text{-}22)$$

The soluble ferrous bicarbonate is oxidized to the insoluble ferric hydroxide, which can be removed by sedimentation and/or filtration, depending on how heavy a floc is produced. While this reaction will take place over a wide range of pH (4–10) the optimum pH is 7.0.[279] The colder the water, the slower are these reactions. This reaction takes a maximum of one hour, and is most rapid at pH 7.0. Each part of iron as Fe oxidized requires 0.64 mg/l chlorine. This reaction consumes 0.9 mg/l alkalinity as calcium carbonate ($CaCO_3$) for each mg/l iron as Fe oxidized.

If the iron present is in the complex organic form, free residual chlorine is more effective than combined in breaking up the iron complex so that oxidation by chlorine can proceed.

It is common practice to preaerate the raw water, which will oxidize some of the iron but, most important, will reduce the carbon dioxide content, causing a rise in pH where further oxidation by chlorine is more effective. While chlorine is usually applied after aeration, it is nearly always desirable to prechlorinate to control the crenoform organisms and to chlorinate again just after aeration for the further oxidation of the iron.

The application of chlorine to iron-bearing waters is imperative regardless of whether or not it is considered a part of the iron removal process, simply to prevent and control the growth of the crenoform organisms, which, if allowed to proliferate, can devastate the entire system and render the iron removal process useless. It should also be emphasized that when iron is present in small quantities (0.3 mg/l) where iron removal is not a factor, chlorine should be used to prevent the growth of the crenoform organisms, which have been known to proliferate in waters containing iron as low as 0.1 mg/l.

Manganese Removal. Chlorine will oxidize soluble manganous manganese to the insoluble manganic form as follows:

$$Cl_2 + MnSO_4 + 4NaOH \longrightarrow MnO_{2\downarrow} + 2NaCl + Na_2SO_4 + 2H_2O \quad (6\text{-}23)$$

The manganese dioxide produced may be removed by filtration as follows: Insoluble MnO_2 plates out on the sand grains. These deposits act as a catalyst to make possible complete extraction of manganese as MnO_2.

The optimum pH for this reaction is between 7 and 8. At pH 8 and with alkalinities on the order of 50 mg/l the time required may be two to three hours. As the pH increases, the time requirement diminishes to the pH values in the softening zone, where oxidation appears to be complete within minutes.[277,279] If the pH approaches 6.0, the time required may be as much as 12 hours. Temperature of the water does not appear to be a significant factor.

It should also be noted that manganese readily precipitates at pH 2 in heavy concentrations of chlorine. Proof of this is the troublesome deposit of manganese often found on the injector throats of chlorination equipment.

Chlorine in the form of free residual is imperative, and is in addition to that required to react with iron, ammonia, hydrogen sulfide, and so on. There should be 1.3 mg/l chlorine for each part of manganese as Mn.[279,280] For each part of Mn oxidized 3.4 mg/l alkalinity as $CaCO_3$ is consumed.

Difficulties from both iron and manganese are much more widespread than is commonly thought. Probably the most troublesome cases are those that are not readily categorized for physical removal of these compounds by an engineered treatment plant. Meyer[281] reported a number of such case histories. The end result is always the same: degradation of water quality in the distribution system; sloughing off of microbiological debris; and red or black water not fit to use. These cases occur throughout the United States. They are most common in underground supplies. The problem usually hits hardest those who can least afford to take proper corrective steps—the small rural community water supply or the individual well owner.

A typical case is a well that consistently contains less than 0.4 to 0.6 mg/l iron and probably 0.1 mg/l manganese. Most wells are capable of supporting troublesome growths of crenoform organisms, and so chlorine is applied to control these growths. Following chlorination there is likely to be a delayed reaction with chlorine and manganese. In such cases deposition of manganese in the remote or dead ends of the distribution system is common, indicating that ten to twelve hours are required for the complete oxidation of manganese. This problem can be overcome by adding a sequestering agent along with the chlorine as follows: Apply the chlorine down the well, ten feet below the pump bowls, as described in Chapter 9. Then add, at the rate of 2 mg/l sodium hexameta-phosphate down the well in the same line as the chlorine solution. This has been found in actual practice to be permissible, probably because there is not enough time for any reaction to set in between the additions of the chlorine and hexameta-phosphate. The latter will prevent the deposition of the manganese in the distribution system.

The procedure to follow for chlorination is: If the water by proper chemical analysis shows more than 0.3 mg/l iron, it should be chlorinated to prevent the growth of crenoform organisms. If there is more than 0.03–0.05 mg/l manganese present, it will probably precipitate out in a delayed reaction. If this occurs, add the hexametaphosphate as described.

This procedure has been known to be successful on iron contents as high as 0.6–0.8 mg/l and higher in some rare cases. Since chlorine is a must on any water over the iron limit, results will dictate whether or not the iron and manganese will have to be removed by a conventional treatment process.*

It should also be added that chlorination may be required in some instances where crenoform organisms proliferate when both iron and manganese are below the U.S. Public Health Service limits of 1962.

Color Removal

The practice of free residual chlorination has been known to bleach true color in waters when color removal by coagulation and filtration has failed. The bleaching effect of chlorine is well known in swimming pool operation. An adequate free chlorine residual produces a polished, sparkling look to the water because of its ability to bleach organic matter.

True color removal by chlorine is most effective in the acid pH zone, between pH 4.0 and 6.8. Usually highly colored, low-turbidity waters are either naturally acid or so lightly buffered that the application of chlorine may be sufficient to reduce the pH. In some cases the simple reduction of pH may cause the color to disappear. Therefore when decolorizing with chlorine, care must be taken so that when the upward readjustment of pH is performed for corrosion control it does not result in color return. This can be predicted by simple laboratory procedure.

Color removal by free residual chlorine is usually instantaneous, and temperature does not seem to be a factor. There is no rule for predicting the optimum dosage. The amount of color removal will vary with local conditions.[279]

Where color is accompanied by turbidity, and coagulation is practiced, prechlorination to a free residual will aid materially in color removal by two means. The

---

* See Chapter 12—the use of chlorine dioxide for iron and manganese removal.

first is through oxidation of part of the color, and the second is through its action as a coagulant aid. Thus two benefits may be expected: better color reduction, and decreased coagulant dosage.

In highly polished nitrified wastewater effluents White has observed that a free chlorine residual can produce a sparkling blue effluent, whereas the same effluent dosed with NH_4OH to produce a combined chlorine residual shows a slight greenish-straw color. This demonstrates the bleaching ability of free chlorine.* So far as is known chloramine or other combined chlorine residuals do not have any bleaching power. If all else fails in an effort to prove that a chlorine residual is in fact a true free available residual and not some complex organic chloramine the bleaching quality of a free residual is certain proof.

Chlorination of High pH Waters

A great many public water supplies are lime-softened resulting in pH values of 10 or more. One of the more notable sources of such supplies is the Missouri River. In the heavily populated areas along its course, the chemical and bacteriological characteristics of this water vary widely, which presents a formidable burden on the treatment plant. It is common practice first to soften the water with lime, which is followed by two-stage flocculation, coagulation, and settling. Chlorine is usually applied just ahead of secondary settling, followed by filtration and postammoniation.

Considerable bacteria reduction occurs in the softening process. Since the chlorine-consuming inorganic compounds of iron, manganese, and sulfur are largely removed by precipitation in the lime-softening process, the remaining chlorine demand is due primarily to nitrogenous compounds still in solution.

At pH 10 and higher, the reaction between free chlorine and the nitrogenous compounds is extremely slow. Temperature is also a factor. As the temperature drops, the chlorine ammonia reaction slows down. Fig. 6-23 illustrates the reaction of chlorine and lime-softened Missouri River water at St. Louis.[282] This water contained 0.9 mg/l ammonia nitrogen, and the temperature was 65°F. Note that after one hour's contact very little chlorine was consumed, and only after twenty-four hours did a typical free residual chlorine curve appear. Fig. 6-24 compares the chlorination of the same river water before and after softening when subjected to a sixty-minute contact period.

The importance of such application of chlorine is in the time factor. Even though the reactions between the chlorine and ammonia do not have sufficient time to go to completion within the treatment plant, there is a considerable benefit in a high free chlorine residual during this period. Chlorine doses of 5 mg/l lower the alkalinity approximately 7 mg/l, which may result in some lowering of the pH and tend to increase the efficiency of the free residual. Postammoniation is desirable to stablize the free chlorine residual for two reasons. At pH values above

---

* When chlorine is used to bleach organic color the required dose may be as high as 15–20 mg/l.

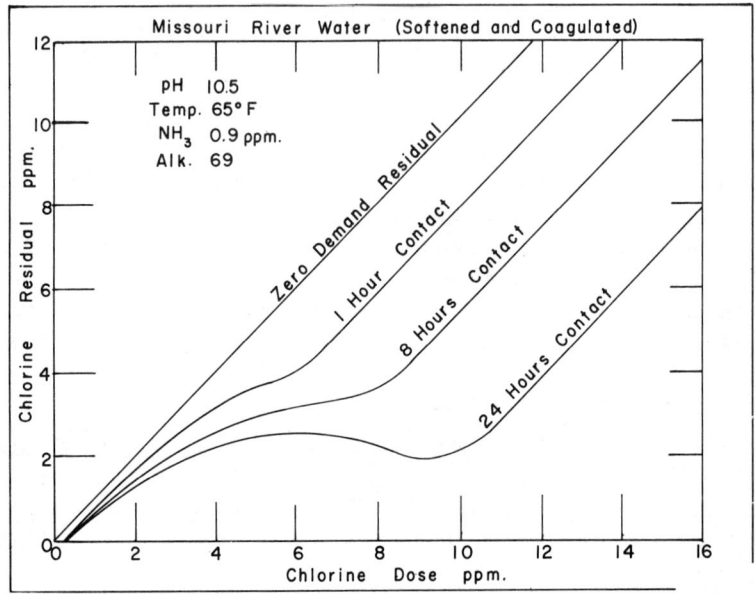

Fig 6-23 Relationship of chlorine and nitrogenous compounds in high-pH waters.

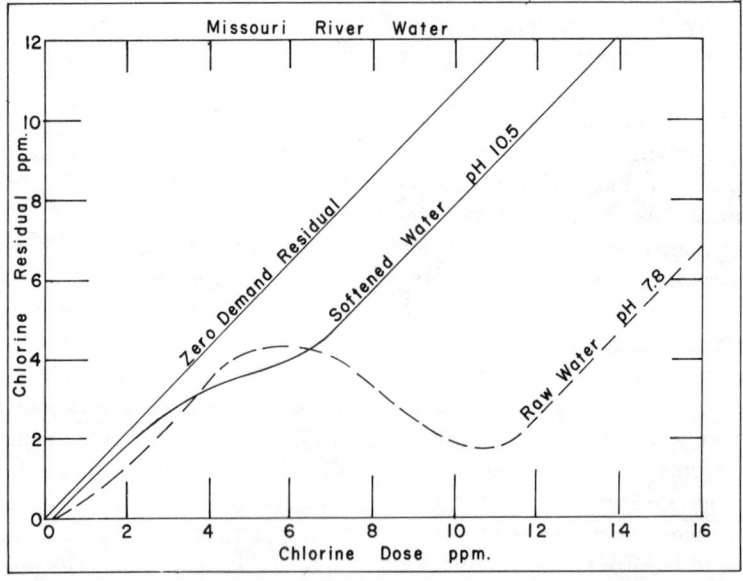

Fig 6-24 Comparison of chlorination of raw water and softened water.

9, the efficiency of monochloramine is known to be superior to free chlorine. (See Chapter 4.) If the free chlorine residual were allowed to go into the distribution system, the continuing slow action of chlorine with the nitrogenous compounds already in the water could eventually, if given the proper amount of time, develop nitrogen trichloride and dichloramine, which could produce objectionable tastes and odors at the consumers' taps. Postammoniation prevents the formation of these compounds. The key to successful chlorination of high pH waters is a long residence time with free chlorine residual followed by postammoniation.

Miscellaneous Applications of Chlorine

Desalting Plants. Current chlorination practice for desalinization plants has gradually evolved so that design engineers are providing pretreatment for the control of marine growths and the destruction of hydrogen sulfide and postchlorination for disinfection.

Prechlorination. Whenever seawater is used for any purpose, it will support and promote the various marine growths: mussels, sponges, bryozoa, and soft forms. Chlorination to control these growths can be on an intermittent basis, as in cooling water treatment. This treatment maintains plant capacity and prevents damage and plugging of tubular heat transfer equipment, thus avoiding costly downtime for repairs and cleaning.

An unexpected development has occurred at several desalinization plants—the appearance of hydrogen sulfide in the raw salt water. Its presence is attributed to decaying vegetation in tropical water supplies. If the supply is from deep wells, it is thought that the organic matter that gives rise to hydrogen sulfide, has been trapped in the coral growth.

The presence of hydrogen sulfide in a desalting plant using an evaporative technique is very costly because it will cause severe corrosion to almost any metal. Sometimes the H_2S comes in slugs, making control difficult. It can be best eliminated by chlorination of the raw water unless the amounts present are excessive (>5.0 mg/l) when chlorination combined with aeration may be more economical.

Postchlorination. The freshwater effluent should be chlorinated as a safety precaution even though the U.S. Public Health Service recognizes the pasteurization effect of the evaporative techniques. At times in the operation of these plants when shutdowns occur, the freshwater system may become contaminated with the raw water. The treatment plant is by definition one huge cross-connection with a contaminated supply. For this reason most engineers include postchlorination in their plant design.

Reflecting Pools. It is common practice to include various kinds of water fountains to enhance the appearance of public recreation centers. These bodies of water require careful control to prevent the growth of any unsightly algae. At the same time the treatment must not give off any obnoxious chemical odors that would be offensive to the public.

These water systems usually contain cascades that provide a certain amount of aeration. It therefore is imperative that chlorination be maintained at a free residual level and that none of the chloramines be allowed to form. It is the chloramines, particularly dichloramine and nitrogen trichloride, which aerate out and cause offensive odors near the pool area.

The best way to treat these waters is to maintain a free chlorine residual of 0.75–1.25 mg/l continuously at a pH of 9.0–10.0 and never to use any algicide that contains nitrogen compounds. If proper free residual chlorination is practiced, algicides will not be required. The chlorination equipment, whether chlorine gas solution feed or hypochlorite, must be controlled by an amperometric residual analyzer calibrated for free chlorine residual. This will assist in preventing the formation of chloramines. The assurance of a *continuous* adequate free chlorine residual will eliminate the necessity of supplementing the chlorine treatment with an algicide.

All recirculated waters will require the continuous addition of caustic, sodium carbonate, or sodium bicarbonate to maintain the elevated pH when using chlorine gas, and most waters will require the addition of these chemicals even if hypochlorite is used. The latter may not add enough caustic to raise the pH to the required level except in waters softened by the lime soda process.

Restoring Wells. Most wells suffer from loss of efficiency with time. As the demand for additional water steadily increases, it becomes just as important to maintain efficiency of existing wells as it is to develop new wells.

There are several causes for the decline in well production: lowering of the ground water table; loss of pump efficiency, owing to worn, corroded, or encrusted parts; plugging of the aquifers by microorganisms and/or scale deposits; biofouling of the well screen area by microorganisms; fouling of the well screen area by deposits of scale and corrosion products or by mud, sand, and silt.

Natural phenomena bring about the deposition of scale and corrosion products and the proliferation of microorganisms that cause biofouling.

Water reaching the underground aquifers as rain picks up carbon dioxide. This is further supplemented by the evolution of carbon dioxide from decaying organic matter in the soil as the rain water penetrates into the earth. This addition of carbon dioxide greatly increases the solvent power of the water, so that the potential scale-forming compounds are readily dissolved. The simple mechanical action of pumping a well causes a draw down of the static water level, which decreases the pressure in the vicinity of the well. This decrease in pressure plus the turbulence in the pump bowl area results in the release of carbon dioxide, which decreases the solubility of the water. Therefore it is at this point that the scale-forming compounds of calcium, magnesium iron and silica are deposited.

The existence of biofouling organisms in underground waters is well known.[283]

The prime offenders are the filamentous iron bacteria: *Crenothrix, Cladothrix, Leptothrix,* and *Gallionella*. Sulfur-bearing waters contribute to the filamentous organism *Begiattoa*. The area of the United States between Lake Michigan and Lake Superior to the north and to the Gulf of Mexico on the south has an overabundance of iron bacteria growing in underground supplies. The area immediately adjacent to Green Bay, Wisconsin, has earned a reputation of harboring the most lush growth of iron bacteria of any place in the United States and perhaps in the world. It is also known that a great many areas in the western part of the United States have underground waters that support troublesome growths of biofouling organisms deep within the aquifers. These areas include, but are not limited to, certain areas of the Rocky Mountains, the alluvial plains of Utah, and the Sacramento and San Joaquin valleys and the Los Angeles basin in California.

Two separate factors are involved in the restoration of well capacity 1) stimulation or redevelopment; 2) cleaning, reconditioning, and replacement of worn parts in the pump and ancillary devices.

Chlorine is used primarily in conjunction with other chemicals and methods involved in the stimulation procedure as a supplement following acid or polyphosphate treatment. It should always be used as a final chemical treatment before putting a well back in service, simply to eliminate contamination that may have resulted from the mechanics of the stimulation procedure. Erickson[284] has described the various methods in considerable detail. The methods used are dependent upon the nature of the problem and the type of aquifer involved.

The successful restoration of well productivity by chlorine alone was reported by Brown[285] in 1942. A group of four wells, with a total initial production of 7090 gpm, had deteriorated in four years to 3350 gpm. One treatment with superchlorine doses restored the total capacity to 6480 gpm, and the second chlorine treatment, a short time later, to 7200 gpm. Houston[286] reported (1946) on the use of heavy doses of chlorine to restore two wells that had lost 50 percent of their productivity. Chlorine treatment of 40 to 50 pounds of chlorine *down the well* was followed by several hundred gallons of muriatic acid to dissolve the scale and encrusted material.

Chlorine with dry ice for stimulation was reported by Suter[287] in 1938. This application to a well suspected of declining production due to biofouling completely restored the well to its original capacity.

The use of dry ice results in the evolution of carbon dioxide. This creates turbulence which mixes the chlorine and builds up a pressure that forces the heavily chlorinated water back into the aquifers. The use of dry ice, while inexpensive, is sometimes difficult to control. It has given way to other more modern methods using compressed air and vibratory explosions.[284]

The application of chemicals to effect a penetration is often done by the surging technique. This consists of pumping for ten to fifteen minutes at a time. As the pump is shut down, the water in the riser column upstream of the check valve surges back down into the well. This is repeated several times, and then the chlorine is allowed to "soak" in the well for twelve to twenty-four hours before being flushed out.

When a well is suspected of loss of productivity due to biofouling, there is no better way to alleviate this condition than with chlorine. It is not always a simple matter to prove that biofouling is the cause. The organisms responsible for these deposits belong to a group not ordinarily isolated in a routine sanitary water analysis. Many of them will not grow on the ordinary culture media employed; some of them are very difficult to cultivate on any media; and they are seldom dispersed in the water in appreciable concentration.

For the destruction of the iron bacteria group (*Crenothrix, Leptothrix, Clonothrix,* and *Gallionella*), a 200 mg/l concentration in the well is effective. The destruction of sulfur bacteria may require as much as 500 mg/l. The chlorine should be applied down the well ten feet or so below the pump bowls and, after the stimulation procedure, be allowed to soak for several hours before flushing. If scale and corrosion deposits are known to exist in the well screen and adjacent areas, the chlorine may be supplemented with acid after first applying an acid inhibitor to protect against corrosion of the metal parts.

When using chlorine for the elimination of biofouling organisms, there is the danger of oxidizing ferrous compounds to ferric hydroxide, which might result in the formation of gelatinous flocs that would settle out in the well only to cause some plugging at a later date. If investigation reveals that this might be a factor, polyphosphates are used along with the chlorine to sequester the formation of ferric precipitates. However, care must be exercised in the use of polyphosphates. In wells with multiple screens in water-bearing strata separated by clay or shales, the action of polyphosphates on these materials will cause the disintegration of the clay and shales, which will infiltrate the sand and gravel strata and tend to reduce the productivity of the well.

The treatment of some biofouling problems occasionally calls for sterner and more costly methods of chemical treatment. Piatek[288] reported (1967) that attempts at Sayreville, New Jersey, to control such fouling in the screens, riser column, and transmission line required such long periods of high dosages of chlorine that a corrosion problem developed. The problem was solved by separating the cleaning procedure into two parts: one for the well and riser column, and the other for the transmission line. Chlorine dioxide was used instead of chlorine. In this case the chlorine dioxide was produced by combining hypochlorite with an aqueous solution of anthium dioxide. This mixture was added along with a polyphosphate. For an 8-inch well containing forty feet of water and 20 feet of air space, a solution of 100 pounds of Calgon in 50 gallons of water was followed by 5 pounds of 70 percent hypochlorite dissolved in 5 gallons of water. This solution was surged once in the well to assure a good mix, and then 15 gallons of anthium dioxide were added. This total mixture was surged for two hours by pumping fifteen minutes and shutting down for fifteen minutes, and repeating. Then it was allowed to stand overnight. This was repeated every two months. Such experiences indicate the necessity of continuous treatment with both chlorine and polyphosphate added down the well at nominal dosages for the continuous destruction of the organisms that cause fouling in the system elsewhere than in the aquifers.

Injection Wells. Reclaimed or other waters are being used more and more to replenish the underground aquifers in those areas that are short of water supply. Injection wells are also used extensively for maintaining oil well production,[289] and for the prevention of sea water intrusion of freshwater aquifers as well as to prevent land subsidence due to withdrawal of oil from underground deposits.

It is the consensus[289,290] that any water used for injection into underground aquifers should be sterilized—free of bacteria. This will provide protection from the growth of filamentous organisms that might plug the aquifers, and will also inhibit the growth in the aquifers of bacteria native to the underground formation that might otherwise proliferate owing to the introduction of a water containing oxygen and nutrients essential to their growth.

It may be appropriate for users of injection water to establish a very high initial chlorine dosage to provide several mg/l free chlorine residual and then taper off to 1.5 mg/l.[290]

Disinfection of Water Mains and Storage Tanks

Chlorine is used exclusively in the disinfection of both water mains and storage tanks. It can be applied as an aqueous solution from chlorine gas in cylinders, as a hypochlorite solution from either granular calcium hypochlorite, or liquid sodium hypochlorite, or as calcium hypochlorite in the tablet form.

When using chlorine gas from cylinders, it is imperative that a conventional chlorinator with injector and booster pump be used. Otherwise, the handling of the chlorine gas can be awkward and hazardous.

Calcium hypochlorite contains 65 percent available chlorine by weight in the granular or tabular form. The tablets, 6 to 8 to the ounce, are designed to dissolve slowly in water. Calcium hypochlorite is dissolved into water to make about a 1 percent (10,000 mg/l) solution, which is injected into the main.

Sodium hypochlorite is packaged in strengths from 5.25 to 16 percent available chlorine. It is available as a liquid in containers varying in size from one-quart bottles to five-gallon carboys, and may also be purchased in bulk for delivery by tank truck in some locations.

Water Mains. The hypochlorite solutions are usually injected into water mains by gasoline- or electric-powered chemical feed pumps or by hand pumps. Special pumps designed to handle only hypochlorite solutions are not mandatory. Ordinary cast iron pumps give many hours of satisfactory operation if they are properly flushed and cleaned after each use period. Hand pumps are also satisfactory.

The methods of applying chlorine to water mains are continuous-feed, slug-dosage, and tablet. These are described in detail in the AWWA committee 8360D report.[291] The AWWA standard for the continuous-feed method calls for chlorination to continue until the entire main contains chlorine solution. The recommended dosage is 50 mg/l, but the requirement is to have no less than a 25 mg/l residual throughout the entire length of the main at the end of twenty-four hours.

For large systems, the slug method is used. This consists of a slug or column of water containing a concentration of at least 300 mg/l to expose all the interior surfaces for a period of at least three hours.

The tablet method is limited to short extensions (up to 2500 feet) and smaller diameter mains (up to 12 inches). Because the preliminary flushing step must be eliminated, this method should be used only when scrupulous cleanliness has been maintained.

Table 6-6 gives the amounts of chlorine required for various size pipes. Table 6-7 shows the number of hypochlorite tablets required.

The confirmation of adequate disinfection is by proper bacteriologic tests made after final flushing.

The recommended procedure for chlorine residual determination is the drop dilution method using acid O-T as the reagent. (See Chapter 5.)

Storage Tanks. The disinfection of elevated tanks, covered ground storage tanks, and ship tanks is achieved most conveniently by the use of portable chlorination equipment. The chlorine solution is added at the rate of 50 mg/l while the tank is being filled.

Portable chlorination equipment must be used carefully and under expert supervision. A typical arrangement, as reported by Tracy,[292] consists of limiting the solution strength to 500 mg/l using a 25 gpm pump with 100 feet of 1-inch chlorine solution hose terminating in a PVC spray nozzle. All workmen must wear oxygen type

Table 6-6 Chlorine Required to Produce 50 mg/l Concentration in 100 feet of Pipe by Diameter

Pipe size (in)	100 percent chlorine (lb)	1 percent chlorine solution (gal)
4	0.027	0.33
6	0.061	0.73
8	0.108	1.30
10	0.170	2.04
12	0.240	2.88
18	0.483	5.80
24	0.875	10.10
36	2.220	26.50

Table 6-7 Number of Hypochlorite Tablets of 5 g Required for Dose of 50 mg/l*

Length of section (ft)	Diameter of pipe (in)					
	2	4	6	8	10	12
13 or less	1	1	2	2	3	5
18	1	1	2	3	5	6
20	1	1	2	3	5	7
30	1	2	3	5	7	10
40	1	2	4	6	9	14

* Based on 3¾ gm available chlorine per tablet.

gas masks, and operators in the tank will require raincoats as well. This operation requires three men, changing nozzle operators every twenty minutes. While it is an uncomfortable job, it is not particularly hazardous. The washing operation is carried out with all the valves closed. All of the chlorine solution will flow to the bottom, and any particulate matter will be carried along or sink to the bottom in the strong chlorine solution. This is then flushed from the reservoir after the washing is completed. Washing reservoirs with hypochlorite solutions is awkward and cumbersome. The use of chlorine solution from portable equipment is not only less expensive but can accomplish the task in a much shorter time.

Summary of Recommendations to Achieve the Objectives of Chlorination

Chlorine Demand. This is a significant water quality parameter and should be monitored continuously as shown in Fig. 6-25. The potential for THM formation is partly a function of chlorine demand. This parameter is a measure of pollution and is related to the presence of nitrogenous compounds. These compounds play an important role in the chemistry of chlorination.

When the ammonia nitrogen content in untreated water reaches 0.15 mg/l and organic nitrogen content of 0.20 mg/l, routine chlorine demand studies as described by Feben and Taras (Chapter 4) should be started. From these studies each water producer can establish chlorine demand constants which will provide a more rational control over the application of chlorine. Furthermore these studies will provide a historical background of raw water quality.

Controlling THMs. For the long-term, the most effective way to control THMs is to exploit pretreatment practices which would improve the quality of the raw water before chlorination. These practices should be patterned after current European and U.K. methods. The objectives of pretreatment should be as follows: 1)

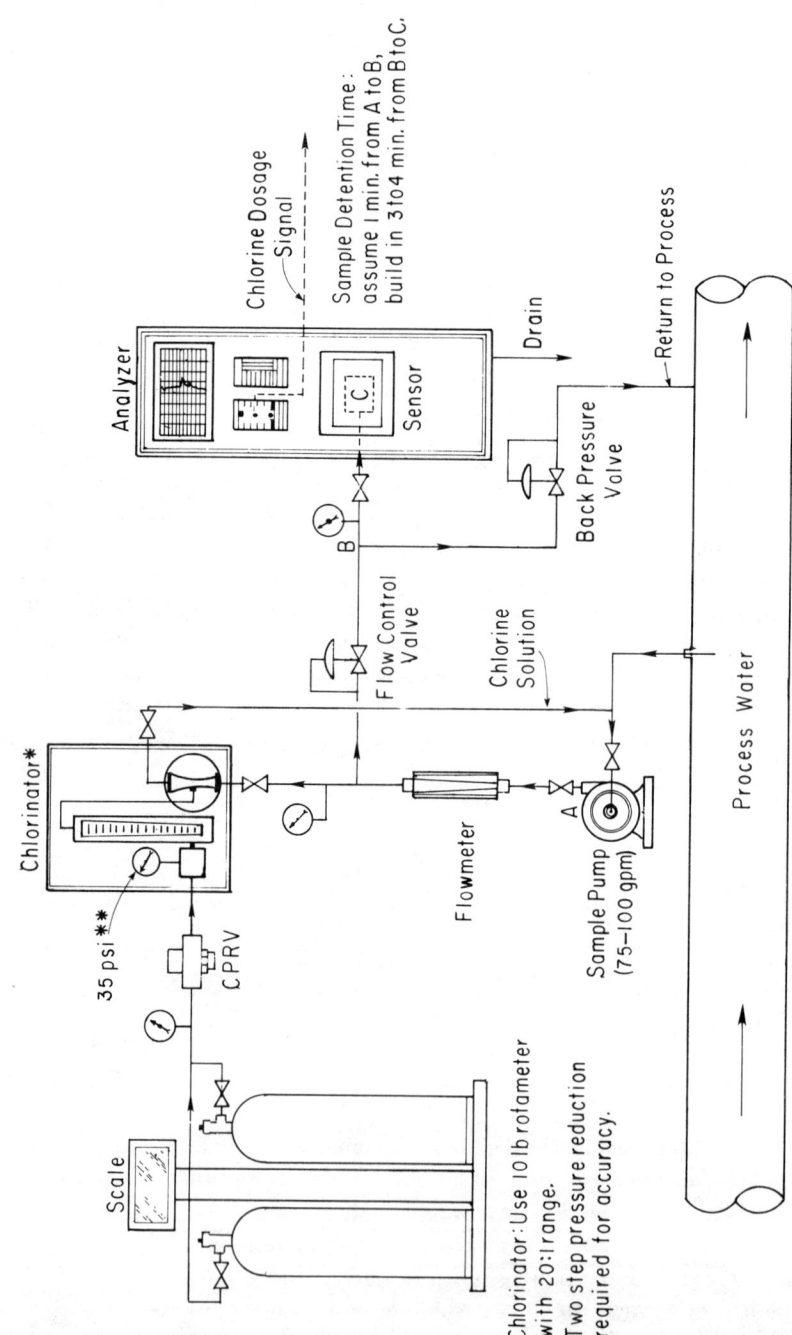

Fig 6-25 Continuous chlorine demand monitoring system.

maximize THM precursor removal, 2) reduce ammonia-N concentration to 0.10 mg/l, 3) reduce organic-N concentration to 0.05 mg/l, 4) limit 15 min. chlorine demand to 0.5 mg/l. These guidelines should improve raw water quality sufficiently to allow the use of the free chlorine residual process without exceeding the EPA MCL for THMs; i.e., 50 µg/l.

Chlorine Residual Analysis. When using the free residual process it is customary to monitor and control on the basis of the free residual fraction. This is not enough. It is suggested that whenever the chlorine residual analyzer is calibrated, a forward amperometric titration for all three fractions (free, mono-, and dichloramine) should be carried out. The best results from the free residual process obtain from a total residual that contains *85 percent free available chlorine* (HOCl).

Nitrogen Trichloride Formation Spells Trouble. Nitrogen trichloride can be generated in copious quantities during the free residual process in some waters. These waters are usually highly polluted with organic nitrogen compounds derived usually from wastewater discharges. Typical conditions for NCl_3 formation at pH levels of 7–9 require a chlorine to ammonia ratio of about 12:1, ammonia-N concentration 0.5 mg/l or greater, and organic N concentration 0.3 mg/l or greater. The higher the concentration of the nitrogen compounds, the greater the production of NCl_3. This compound produces a pungent geranium like odor at low concentrations (0.02 mg/l). However it is practically insoluble in water and is difficult to measure by available methods. Owing to its pervasive and unmistakable odor, and its characteristic eye tearing quality, it is not necessary to attempt detection with analytical methods. Therefore those waters subject to developing significant amounts of nuisance residuals (greater than 20 percent of the total chlorine residual) should be able to generate NCl_3, which is easily scavenged by postaeration. This will also reduce the dichloramine fraction well below the taste and odor producing level. Nitrogen trichloride aerates easily and decomposes rapidly in sunlight. This ability to decay rapidly in the atmosphere, including the hours of darkness, reduces significantly the possibility of it becoming an air pollution problem.

Points of Chlorine Application. If it is practical, use only a single point of application—at the influent of the treatment process—and in a sufficient amount to maintain a free chlorine residual throughout the entire treatment train. When this is not possible, a secondary point of application must be provided. This one is usually just ahead of the filters.

Monitoring Residuals. The chlorine residual should be monitored continuously in vital parts of the treatment process. It should not be limited to the plant effluent. For example, in addition to the analyzer controlling plant effluent residual, an accessory analyzer should record residuals at the influent of the sedimentation basins and filter influent. The sample pumps from each of these locations should operate continuously—one to the analyzer cell and the other to waste via a three-way valve. This three-way valve reverses the flow pattern at a predetermined interval by means of a time clock. For ten minutes, the sample should flow through the analyzer from the sedimentation influent sample pump; the time clock would then actuate the three-way valve, diverting the flow from the other sampling point (filter influent) from waste overflow to the chlorine residual analyzer. This procedure will produce two monitoring traces of chlorine residual in vital parts of the plant. The first would be a guide as to the sufficiency of the prechlorination dosage. The second would indicate whether or not sufficient chlorine was present to provide an adequate residual in the effluent. This situation would also protect the filter media from biofouling.

Residuals should also be monitored in the distribution system. The preferred locations are at trouble spots where there may be degradation of water quality, insufficient chlorine residual, and/or consumer complaints.

Monitoring stations in the distribution system will help determine both the need and location of relay chlorination stations.

Controlling the Chloramine Process. This process carries with it the potential to supply the treated water with nutrients that promote the proliferation of bacteria and other microorganisms which conspire to degrade the quality of any potable water. These nutrients will become available when the chloramine residual is consumed. At this point the chloramine reverts to ammonia nitrogen. If chlorine is added when the residual is depleted, the ammonia nitrogen is immediately converted to chloramine. This is the reason for chlorine relay stations. Chloramination systems must be designed and operated so that measurable chloramine residuals persist continuously throughout the distribution network.

REFERENCES

1. Morris, J. C., "Future of Chlorination," *J. AWWA,* **58,** 1475 (Nov., 1966).
2. Baker, M. N., "The Quest for Pure Water," Amer. Water Works Assoc., New York (1930).
3. Cohen, Lord M. D., "John Snow—Autumn Loiterer?" *Proc. Roy. Soc. Med.,* **62,** 99 (Jan., 1969).
4. "Manual of British Water Supply Practice," Compiled by Inst. Of Water Engrs., Heffer & Son Ltd., Cambridge, England (1950).
5. Whipple, G. C., "Disinfection as a Means of Water Purification," *Proc. AWWA,* 266 (1906).
6. Leal, Dr. J. L., "Sterilization Plant of the Jersey City Water Supply," *Proc. AWWA,* 100 (1909).
7. Anon., "Water Quality and Treatment," 2d ed., Amer. Water Works Assoc., New York (1951).
8. Houston, A. C., "Studies in Water Supply," Macmillan & Co., Ltd., London (1913).
9. Hooker, A. B., "Chloride of Lime Sanitation," (1913).
10. Jennings, C. A., "Significance of the Bubbly Creek Experiment," *J. AWWA* **40,** 1037 (1948).
11. Johnson, G. A., "Hypochlorite Treatment of Public Water Supplies," *Am. Jour. Publ. Hlth.,* 562 (1911).
12. Orchard, W. J., Wallace & Tiernan Company, Pvt. Comm. (1959).
13. Anon., "Strange Fever," *M.D. Magazine,* 221 (Nov., 1969).
14. Laubusch, E. J., "Chlorination and Other Disinfection Processes," Chlorine Inst., New York (1964).

15. Gorman, A. E., and Wolman, A., "Significance of Water-Borne Outbreaks," *J. AWWA,* **31** (1939).
16. Culp, R. L., "History and Present Status of Chlorination Practice in Kansas," *J. AWWA,* **52,** 888 (1960).
17. Darnall, Dr. C. R., "Purification of Water by Anhydrous Chlorine," *Am. J. Pub. Hlth.,* 783 (1911).
18. Tiernan, M. F., "Controlling the Green Goddess," *J. AWWA,* **40,** 1042 (Oct., 1948).
19. Beard, P. J., "The Survival of Typhoid in Nature," *J. AWWA,* **30,** 124 (1938).
20. Craun, G. F., and McCabe, L. J., "Review of the Causes of Waterborne Disease Outbreaks," *J. AWWA,* **45,** 74 (Jan. 1973).
21. Senevirtine, G. "The Wandering of a Vicious Cholera Strain," *San Francisco Chronicle* (Sept. 10, 1973).
22. Dufour, A. P. "Disease Outbreaks Caused by Drinking Water," *J. WPCF,* **54,** 980 (June 1982).
23. Morello, Carol, "Sinister Parasites Play Havoc with Pennsylvania Water Supply," *San Francisco Sunday Examiner and Chronicle* (Mar. 11, 1984).
24. Houston, A. C., "Water Purification," *Municipal Sanitation,* **3,** 4, 148 (Apr. 1932).
25. Laubusch, E. J., "How Safe Is Your Chlorine Residual?" *Pub. Works* (Mar. 1959).
26. Brock, T. D. *Biology of Microorganisms,* 3rd ed. Prentice-Hall, Englewood Cliffs, NJ, 1979.
27. Greenberg, A. E., and Ongerth, H. J., "Salmonellosis in Riverside, California," *J. AWWA,* **58,** 1145 (Sept. 1966).
28. Ross, E. C., Campbell, K. W., and Ongerth, H. J., *"Salmonella typhimurium* Contamination of Riverside, California Supply," *J. AWWA,* **58,** 165 (Feb. 1966).
29. Collaborative Report, "A Waterborne Epidemic of Salmonellosis in Riverside, California, 1965." *Am. Jour. Epdemiol.,* **93,** 33 (1971).
30. Craun, G. F. "Waterborne Outbreaks," *J. WPCF,* **49,** 1268 (June 1977).
31. Lloyd, B., and Morris, R., "Effluent and Water Treatment Before Disinfection," paper presented at the International Symposium: Viruses and Disinfection of Water and Wastewater, Univ. of Surrey, Guildford, U.K., Sept. 1-4, 1982.
32. Dufour, A. P., "Disease Outbreaks Caused by Drinking Water," *J. WPCF,* **55,** 905 (June 1983).
33. Hunter, C. A., and Ensign, P. R., "An Epidemic of Diarrhea in New-Born Nursery Caused by *P. Aeruginosa,"* *Am. J. Pub. Hlth,* **37,** 1166 (1947).
34. Roueche, Berton, "Three Sick Babies," *The New Yorker* (Oct. 5, 1968).
35. Culp, R. L., "Disease Due to Non-Pathogenic Bacteria," *J. AWWA,* **60,** 157 (1968).
36. Scott, J. A., "Schistosomiasis Control in Water Supply Sources," *J. AWWA,* **61,** 352 (1969).
37. Herringer, E. J., "Schistosomiasis Control Is an Engineering Problem," *Pub. Works* (Jan. 1949).
38. Faber, H. A., "How Modern Chlorination Started," *Water and Sew. Works,* **99,** 455 (Nov. 1952).
39. Wolfe, M. S., "Giardiasis," *Pediatric Clinic of North America,* **26,** 295 (1979).
40. Meyer, E. A., and Jarroll, E. L., "Giardiasis," *Am. J. of Epidemiol.,* **111,** 1, (1980).
41. Pelczar, M. J. Jr., and Reid, R. D., *Mirobiology,* 2nd ed., McGraw-Hill, New York, 1965.
42. Rice, E. W., Hoff, J. C., and Schaeffer, F. W., III, "Inactivation of *Giardia* Cysts by Chlorine," *App. and Env. Microbiol.* **43,** 250 (Jan. 1982).
43. Jarrol, E. L., Bingham, A. K., and Meyer, E. A., "Effect of Chlorine on *Giardia lamblia* Cyst Viability," *App. and Env. Microbiol.,* **41,** 483 (Feb. 1981).
44. Garbe, P. L., private communication, Center for Disease Control Atlanta, Georgia, July 5, 1983.
45. Fraser, D. W., et al., "Legionnaires Disease," *New Eng. J. of Medicine,* 297 (Dec. 1977).
46. Herwaldt, L. A., and Fraser, D. W., "Legionellosis: Legionnaires' Disease and Related Diseases," U.S. Dept. of Pub. Hlth. reprint p. 45 (1981).
47. Broome, C. V., and Fraser, D. W., "Epidemiologic Aspects of Legionellosis," *Epidemiologic Reviews,* **1,** 1 (1979).
48. Dufour, A. P., and Jakubowski, W., "Drinking Water and Legionnaires Disease," *J. AWWA,* **74,** 631 (Dec. 1982).
49. Fraser, D. W., and McDade, J. E., "Legionellosis," *Scientific American,* **241,** 82 (Oct. 1979).
50. Chang, S. L., "Viruses, Amoebas, and Nematodes and Public Water Supplies," *J. AWWA,* **53,** 288 (1961).

51. Foot, H. B., Jellison, W. L., Stienhaus, E. E., and Kohls, G. M., "Effect of Chlorination of *Pasteurella tularensis* in Aqueous Suspension," *J. AWWA,* **35,** 7 (July 1943).
52. Olivieri, V., and Cabelli, V., personal communication, Sept. 2, 1982.
53. Morris, R., Finch, P., and Sharp, D. N. "Effect of the Kingsbury Lakes on the Microbiological Quality of the River Tame, U.K. a paper presented at the International Symposium on Viruses and Disinfection of Water and Wastewater, University of Surrey, Guildford, England, Sept., 1–4, 1982.
54. Cabelli, V., "Waterborne Viral Infections," paper presented at the International Symposium on Viruses and Disinfection of Water and Wastewater, University of Surrey, Guildford, U.K., Sept. 1–4, 1982.
55. O'Conner, J. T., Hemphill, L., and Reach, C. D. Jr., "Removal of Virus from Public Water Supplies," EPA report 600752–82–024, Cincinnati, OH, August 1982.
56. Kott, Yehuda, "Effluent Usage and Disposal," paper presented at International Symposium: Viruses and Disinfection of Water and Wastewater, Univ. of Surrey, Guildford, U.K., Sept. 1–4, 1982.
57. Timbury, M. C., *Notes on Medical Virology,* 4th ed., Churchill Livingstone, Edinburgh and London, 1973.
58. *Van Nostrand's Scientific Encyclopedia,* 4th ed., Van Nostrand Reinhold, New York, 1968.
59. Dennis, J. M., "Infectious Hepatitis at New Delhi," *J. AWWA,* **51,** 1288 (1959).
60. Blacklow, N. I. and Cukor, G., "Viral Gastroenteritis Agents," Chapter 90 in E. H. Lennette, A. Balows, W. J. Hausler, Jr. and J. P. Truant (Eds.), *Manual of Clinical Microbiology,* 3rd ed., Amer. Society of Microbiology, Washington, DC, 1980.
61. Wilson, R., et al., "Waterborne Gastroenteritis Due to Norwalk Agent: Clinical and Epidemiological Investigation," *Am. J. Pub. Hlth.,* **72,** 72 (1982).
62. Greenberg, H. B., et al., "Role of Norwalk Virus in Outbreaks of Non-Bacterial Gastroenteritis," *Jour. Inf. Disease,* **139,** 564 (May 1979).
63. Kaplan, J. E., Gary, G. W., Baron, R. C., Singh, N., Schonberger, L. B., Feldman, R., and Greenberg, H. B., "Epidemiology of Norwalk Gastroenteritis and the Role of Norwalk Virus in Outbreaks of Acute Nonbacterial Gastroenteritis," *Annals of Internal Medicine,* **96,** 756–761 (1982).
64. Adler, J. L., and Zickl, R., "Winter Vomiting Disease," *Jour. Inf. Disease,* **119,** 668 (1969).
65. Kaplan, J. E., Goodman, R. A., Schonberger, L. B., Lippy, E. C., and Gary, G. W., "Gastroenteritis Due to Norwalk Virus: An Outbreak Associated With a Municipal Water System," *Jour. Inf. Dis.,* **146,** 190 (Aug. 1982).
66. Taylor, J. W., "Norwalk Related Viral Gastroenteritis Due to Contaminated Drinking Water," *Ann. J. Epidemiol.,* **114,** 584 (1981).
67. Zatotin, B. A., Libiyainen, L. T., Bortnik, F. L., Chernitskaya, E. P., et al., "Waterborne Group Infection of Rotavirus Etiology," *Microbiol. Epidemiol. Immunol.,* **11,** 99–101 (1981).
68. Craun, G. F., "Disease Outbreaks Caused by Drinking Water," *J. WPCF,* **51,** 1751 (June 1979).
69. Crann, G. F., "Disease Outbreaks Caused by Drinking Water," *J. WPCF,* **45,** 1566 (Jan. 1973).
70. Kienle, J. A., "Use of Liquid Chlorine for Sterilizing Water," *Proc. AWWA,* 267 (1913).
71. Cutler, J. W., and Green, F. W., "Operating Experience With a New Residual Recorder Controller," *J. AWWA,* **22,** 755 (1930).
72. Goudey, R. F., "Residual Chlorination On the Los Angeles System," *J. AWWA,* **28,** 1742 (1936).
73. Wolman, A., and Enslow, L. H., "Chlorine Absorption and the Characteristics of Water," *J. Ind. and Eng. Chem.,* **11,** 209 (1919).
74. Howard, N. J., and Thompson, R. E., "Chlorine Studies and Some Observations on Taste Producing Substances in Water, and the Factors Involved in Treatment by the Super- and De-chlorination Method," *J. NEWWA,* **40,** 276 (1926).
75. Watzl, E., "Superchlorination and Dechlorination over Carbon for a Municipal Water Supply," *Ind. and Eng. Chem.,* **21,** 156 (1929).
76. Cox, C. R., "Double Chlorination," *J. AWWA,* **16,** 55 (1926).

77. McAmis, J. W. "Prevention of Phenol Taste with Ammonia," *J. AWWA,* **17,** 3, 341 (Mar. 1927).
78. Williams, D. B., "Control of Free Residual Chlorine by Ammoniation," *J. AWWA,* **55,** 1195 (Sept. 1963).
79. Williams, D. B., "Elimination of Nitrogen Trichloride in Dechlorination Practices," *J. AWWA,* **58,** 248 (Feb. 1966).
80. Raschig, F., "Chloramine," *Verh. Ges. Dent. Naturforsch Aertzl (Germany),* **11,** 120 (1907).
81. Race, Joseph, *Chlorination of Water,* John Wiley & Sons, Inc., New York, 1918.
82. Race, Joseph, "Chlorination and Chloramines," *J. AWWA,* **3,** 63 (Mar. 1918).
83. Race, Joseph, "Discussion of Pre-Ammoniation of Filtered Water," *J. AWWA,* **23,** 411 (Mar. 1931).
84. Adams, B. A. "The Iodoform Taste Acquired by Chlorinated Water," *The Medical Officer,* **869,** 33 (Dec. 1925).
85. Houston, Sir A. C., 19th Ann. Rept: Metro. Water Bd., London, "Chemical and Bacteriological Examination of London Water," 1925.
86. Harold, C. H. H. "Further Investigation into the Sterilization of Water by Chlorine and Some of Its Compounds," *Jour. Royal Army Corps,* **45,** 190, 251, 350, 429 (1925).
87. Adams, B. A. "The Chloramine Treatment of Pure Water," *Med Officer,* **5,** 55 (1926).
88. Lawrence, W. C. "Studies in Water Purification Processes at Cleveland," *J. AWWA,* **23,** 6, 896 (June. 1931).
89. Braidech, M. M. "The Ammonia–Chlorine Process as a Means for Taste Prevention and Effective Sterilization," *Ohio Conf. Water Purif.,* **9,** 67 (1930).
90. Braidech, M. M. "Practical Application of Ammonia–Chlorine Process in Sterilization of Cleveland Water Supply," *J. AWWA,* **22,** 1297 (Sept. 1930).
91. Ellms, J. W., and Lawrence, W. C., "Investigation of Tastes and Odors in the Cleveland Water Supply," *Engr. News-Rec.,* **86,** 1039 (1921).
92. Spaulding, C. H. "Pre-Ammoniation at Springfield, Illinois," *J. AWWA,* **21,** 1085 (Aug. 1929).
93. Harrison, L. B., "Chlorophenol Tastes in Water of High Organic Content," *J. AWWA,* **21,** 542 (Apr. 1929).
94. Committee Report, "Control of Tastes and Odors in Public Water Supplies," *J. AWWA,* **25,** 1490 (Nov. 1933).
95. Harvill, C. R., Morgan, J. H., and Mauzy, H. L., "Practical Application of Ammonia Induced Breakpoint Chlorination," *J. AWWA,* **34,** 275 (1942).
96. Harvill, C. R., Morgan, J. G., Hagar, M. C., and Todd, A. R., "Maintenance of Chlorine Residual in the Distribution System," *J. AWWA,* **34,** 1797 (1942).
97. Harvill, C. R., private communication, 1966.
98. Joint Committee Report, "Chlorine–Ammonia Treatment," *J. AWWA,* **33,** 2079 (Dec. 1941).
99. Anon., "Interim Primary Drinking Water Regulations; Control of Organic Chemical Contaminants in Drinking Water," Environmental Protection Agency, Washington, Washington, DC, *Federal Register,* Part II (Feb. 9, 1978).
100. Morris, J. C. "Aspects of Quantitative Assessment of Germicidal Efficiency," in J. D. Johnson (Ed.), *Disinfection: Water and Wastewater,* Ann Arbor Science, Ann Arbor, MI., 1975, p. 1.
101. Selna, M. W., Miele, R. P., and Baird, R. B., "Disinfection for Water Reuse," a paper presented to the Disinfection Seminar at the Ann. Conf. AWWA, Anaheim, CA, May 8, 1977.
102. Selleck, R. E., Saunier, B. M., and Collins, H. F., "Kinetics of Bacterial Deactivation with Chlorine," *Jour. Env. Engr. Div. ASCE,* **104,** EE6, 1197 (Dec. 1978).
103. Means, E. G., McGuire, M. J., Otsuka, D. J., and Tanaka, T. S., "Impact of Chlorine and Ammonia Application Points on Bactericidal Efficiency of Free Chlorine and Chloramines in Pilot Plant Studies," paper presented at the Ann. Conf. AWWA, Las Vegas, Nevada, June 9, 1983.
104. Griffin, A. E., and Chamberlin, N. S. "Some Chemical Aspects of Break-Point Chlorination," *J. NEWWA,* **55,** 371 (1941).

105. Griffin, A. E. "Chlorine for Ammonia Removal," Fifth Annual Water Conf. Proc. Engrs. Western Penn., p. 27, 1944.
106. Fair, G. M., Morris, J. C., Weill, Ira, and Burden, R. P., "The Behavior of Chlorine as a Water Disinfectant," *J. AWWA* **40**, 1051 (Oct. 1948).
107. Fair, G. M., Morris, J. C., and Chang, S. L. "The Dynamics of Water Chlorination," *J. NEWWA*, **61**, 285 (1947).
108. Morris, J. C., Weil, Ira, and Burden, R. P. "The Formation of Monochloramine and Dichloramine in Water Chlorination," paper presented at 117th meeting, Am. Chem. Soc., Detroit, MI, April 16–20, 1950.
109. Palin, A. T., *Chemical Aspects of Chlorine,* Inst. of Water Engrs. (England), p. 565, 1950.
110. Palin, A. T. "A Study on the Chloro-Derivatives of Ammonia and Related Compounds with Special Reference to Their Formation in the Chlorination of Natural and Polluted Waters," *Water and Water Engineering* (England), p. 151 (Oct. 1950), p. 189 (Nov. 1950), p. 248 (Dec. 1950).
111. Granstrom, M. L. "The Disproportionation of Monochloramine," Ph.D. dissertation in Sanitary Engineering, Harvard Univ., 1954.
112. Williams, D. B., "How to Solve Odor Problems in Water Chlorination Practice," *Water and Sew. Wks.,* **99**, 358 (1952).
113. Williams, D. B., "Control of Free Residual Chlorine by Ammoniation," *J. AWWA,* **55**, 1195 (Sept. 1963).
114. Williams, D. B. "Elimination of Nitrogen Trichloride in Dechlorination Practice," *J. AWWA,* **58**, 148 (Feb. 1966).
115. Palin, A. T. "Determination of Free Chlorine and Combined Chlorine in Water by the Use of Diethyl-*p*-Phenylene Diamine," *J. AWWA,* **49**, 873 (Jul. 1957).
116. Palin, A. T. "Methods for the Determination, in Water, of Free and Combined Available Chlorine, Chlorine Dioxide and Chlorite, Bromine, Iodine, and Ozone, using Diethyl-*p*-Phenylene Diamine (DPD)," *Jour. Inst. Water Engrs.,* **21**, 537 (1967).
117. Palin, A. T. "Analytical Control of Water Disinfection with Special Reference to Differential DPD Methods for Chlorine Dioxide, Bromine, Iodine and Ozone," *Jour. Inst. Water Engrs.,* **28**, 139 (1974).
118. Wei, I. W., and Morris, J. C., "Dynamics of Breakpoint Chlorination," Division of Engineering and Applied Physics, Harvard Univ., Cambridge, MA, May 1973.
119. Saunier, B. M., and Selleck, R. E., "The Kinetics of Breakpoint Chlorination in Continuous Flow Systems," paper presented at the AWWA Ann. Conf., New Orleans, LA, June 22, 1976.
120. Williams, D. B. "The Organic Nitrogen Problem," *J. AWWA,* **43**, 847 (Oct. 1951).
121. Griffin, A. E., "Reactions of Heavy Doses of Chlorine in Water," *J. AWWA,* **31**, 2121 (Dec. 1939).
122. Griffin, A. E. "Observations on Breakpoint Chlorination," *J. AWWA,* **32**, 1187 (Jul. 1940).
123. O'Connell, W. J., Jr., unpublished report "Superchlorination and Ammonia–Chlorine Treatment, Governors Ave. Well Incident," Stanford Univ., Palo Alto, CA, 1939.
124. Clark, N. A., et al., "Human Enteric Viruses in Water: Source, Survival and Removability," *International Conf. Water Pollution Research,* Sept. 1962, Pergamon Press, London.
125. White, G. C., Bean, E. L., and Williams, D. B. "Chlorination and Dechlorination: A Scientific and Practical Approach," *J. AWWA,* **60**, 540 (May 1968).
126. White, G. C., unpublished notes of plant survey, 1959.
127. Sung, R. D., "Effects of Organic Constituents in Wastewater on the Chlorination Process," Ph.D. dissertation, Univ. of Calif., Davis, CA, 1974.
128. White, G. C., "Disinfection: The Last Line of Defense for Potable Water," *J. AWWA,* **67**, 410 (Aug. 1975).
129. "Community Water Supply Study: Analysis of National Survey Finding," Bur. of Water Hygiene, Envir. Health Service USPHS, Dept. of HEW, Washington, DC, July 1970.
130. Anon., "Industrial Pollution of the Lower Mississippi River in Louisiana," U.S. Environmental Protection Agency, Cincinnati, OH, April 1972.

131. Anon., "New Orleans Area Water Supply Study," U.S. Environmental Protection Agency, Cincinnati, OH, draft report released on Nov. 8, 1974.
132. Rook, J. J., "Formation of Haloforms During Chlorination of Natural Waters," *Water Treatment and Examination,* **23** (Part 2), 234 (1974).
133. Bellar, T. A., Lichtenberg, J. J., and Kroner, R. C. "The Occurrence of Organohalides in Chlorinated Drinking Waters," *J. AWWA,* **66,** 703 (Dec. 1974).
134. Symons, J. M., Bellar, T. A., Carswell, J. K., DeMarco, J., Kropp, K. L., Robeck, G. G., Seeger, D. R., Slocum, C. V., Smith, B. L., Stevens, A. A., "National Organics Reconnaissance Survey for Halogenated Organics," *J. AWWA,* **67,** 634 (Nov. 1975).
135. Trussell, R. R., and Umphres, M. D., "The Formation of Trihalomethanes," *J. AWWA,* **70,** 604 (Nov. 1978).
136. Lange, A. L., and Kawczynski, E., "Controlling Organics: The Contra Costa County Water District Experience," *J. AWWA,* **70,** 653 (Nov. 1978).
137. Morris, J. C., Ram, N., Baum, B., and Wajon, E., "Formation and Significance of N-Chloro Compounds in Water Supplies," EPA Report No. 600/2–80–031, Mun. Env. Res. Lab. Cincinnati, OH, July 1980.
138. Symons, J. M., et al. "Ozone, Chlorine Dioxide, and Chloramines as Alternatives to Chlorine for Disinfection of Drinking Water," U.S. EPA, Cincinnati, OH, Nov. 1977.
139. Norman, T. S., Harms, L. L., and Looyenga, R. W., "Use of Chloramines to Prevent THM Formation at Huron, S.D.," *J. AWWA,* **72,** 176 (March 1980).
140. Barrett, R. H., and Trussell, A. R., "Controlling Organics: The Casitas Municipal Water District Experience," *J. AWWA,* **70,** 660 (Nov. 1978).
141. Rook, J. J., "Haloforms in Drinking Water," *J. AWWA,* **68,** 168 (Mar. 1976).
142. Selleck, R. E., private communication, Aug. 1978.
143. Love, O. T., Jr., Carswell, J. K., Miltner, R. J., and Symons, J. M., "Treatment for the Prevention or Removal of Trihalomethanes in Drinking Water," Appendix 3 to Treatment Guide for the Control of Chloroform and Other Trihalomethanes, U.S. EPA, Cincinnati OH, 1976.
144. Anderson, M. C., Butler, R. C., Holdren, F. J., and Kornegay, B. H., "Controlling Trihalomethanes with Powdered Activated Carbons," *J. AWWA,* **73,** 432 (Aug. 1981).
145. Miller, R., and Hartman, D. J., "Feasibility Study of Granulated Activated Carbon Adsorption and On-Site Regeneration," EPA Report, No. 600/52–82–087 Muni. Env. Res. Lab., Cincinnati, OH, Nov. 1982.
146. Oulman, C. S., Snoeyink, V. L., O'Connor, J. T., and Taras, M. J., "Removing Trade Organics from Drinking Water Using Activated Carbon and Polymeric Adsorbents," EPA Report No. 600/52–81–077–078–079, Mun. Env. Res. Lab., Cincinnati, OH, July 1981.
147. Quinn, J. E., and Snoeyink, V. L. "Removal of Total Organic Halogen by Granular Activated Carbon Adsorbers," *J. AWWA,* **72:** 483 (Aug. 1980).
148. Weber, W. J., Jr., and Pirbazari, M., "Effectiveness of Activated Carbon for the Removal of Toxic and/or Carcinogenic Compounds from Water Supplies," EPA Report No. 600/52–81–057, Muni. Env. Res. Lab., Cincinnati, OH, June 1981.
149. Symons, J. M., Stevens, A. A., Clark, R. M., Geldreich, E. E., Love, O. T., Jr., and DeMarco, J., "Removing Trihalomethanes from Water," *Water/Engr. Mgmt.,* p. 50 (July 1981).
150. Staff Report, "The Activated Carbon Dilemma," *Water and Sewage Works,* **125,** 34 (Dec. 1978).
151. J. M. Montgomery Engrs., "Alternative Disinfectants for Trihalomethane Control," Report to the Metropolitan Water District of Southern California, October 1981.
152. Anon., "Mainstream," AWWA, Denver, CO, March 26, 1982.
153. Kennedy/Jenks Engineers, "Trihalomethane Control Investigation," Report to the Alameda County Water District, Fremont, CA, April 1982.
154. Harms, L. L., and Looyenga, R. W., "Preventing Haloform Formation in Drinking Water," EPA Report No. 600/2–80–091 Mun. Env. Res. Lab., Cincinnati, OH, August 1980.
155. Rice, L. M., and Bolding, M. E., "An Alternative Solution to the THM Problem," *Water Engr. and Mgmt.,* p. 59 (May 1981).
156. McGuire, M. J., Shepherd, B. M., and Davis, M. K., "Surface Water Supply Trace Organics

Survey: Maximum Trihalomethane Potential," Metropolitan Water District of Southern Calif. Water Quality Lab. Report, Feb. 1980.
157. Brown, H. A., "Superchlorination at Ottumwa, Iowa," *J. AWWA*, **32**, 1147 (July 1940).
158. Brown, H. A., "Triple Chlorination at Ottumwa, Iowa," *Water and Sewage Works*, **97**, 267 (July 1950).
159. Williams, D. B. "Informal Presentation at AWWA Research Committee meeting on Taste and Odor Control," Chicago, IL, Nov. 12 and 13, 1974.
160. Inhofer, W. R., and DeHooge, F. J., "Free Residual Chlorination of Passaic River Water at Little Falls, New Jersey," paper presented at the AWWA Ann. Conf. Boston, Mass., June 16-21, 1974.
161. American Water Works Association, Special Taste and Odor Research Committee Meeting, Chicago, IL, Nov. 12 and 13, 1974.
162. Willey, B. J., Duke, C. M., and Rasho, J. "Chicago's Switch to Free Chlorine Residuals," *J. AWWA*, **68**, 441 (Aug. 1975).
163. Hoehn, R. C., and Johnson J. D. "An Analysis of Disinfection, Water Quality Control, and Safety Practices in 1978 in the United States Water Utility Industry," AWWA Water-Quality Division, Disinfection Committee, 1978.
164. McCabe, L. J., Symons, J. M., Lee, R. D., and Robeck, G. G., "Survey of Community Water Supply Systems," *J. AWWA*, **62**, 670 (Nov. 1970).
165. White, G. C., "Disinfection: The Last line of Defense for Potable Water," **67**, 410 (Aug. 1975).
166. McDermott, J. H. "Virus Problems and Their Relation to Water Supply," presented at Virginia section AWWA Conference, Roanoke, VA, Oct. 25, 1973.
167. McCabe, L. J., "Significance of Virus Problem," paper presented at AWWA Water Quality Conference, Cincinnati, OH, Dec. 3 and 4, 1973.
168. Engelbrecht, R. S., Foster, D. H., Masarik, M. T., and Sai, S. H. "Detection of New Microbial Indicators of Chlorination Efficiency," paper presented at the AWWA Water Technology Conference, Dallas, TX, Dec. 1-3, 1974.
169. Ptak, D. V., Ginsburg, W., and Willey, B. F., "Identification and Incidence of *Klebsiella* in Chlorinated Water Supplies," *J. AWWA*, **65**, 604 (Sept. 1973).
170. Sorber, C. A., Williams, R. F., Moore, B. E., and Longley, K. E., "Alternative Disinfection Schemes for Reduced Trihalomethane Formation: Vol. 1. Prototype Studies," U.S. EPA Report #600/52-82-037, U.S. Env. Prot. Agency, Cincinnati, OH, Aug. 1982.
171. Ward, N. R., Means, E. G., Olson, B. H., and Wolfe, R. L., "The Inactivation of Total Count and Selective Gram-Negative Bacteria by Inorganic Monochloramines and Dichloramines," paper presented at AWWA Water Qual. Technology Conf., Nashville, TN, 1982.
172. White, G. C., Beebe, R. D., Alford, V. F., and Sanders, H. A., "Wastewater Treatment Plant Disinfection Efficiency as a Function of Chlorine and Ammonia Content," in R. L. Jolley et al. (Eds.), *Water Chlorination: Environmental Impact and Health Effects*, Vol. 4, Book 2, p. 1115, Ann Arbor Science, Ann Arbor, MI, 1983.
173. Buelow, R. W., and Walton, G., "Bacteriological Quality versus Residual Chlorine," *J. AWWA*, **63**, 28 (Jan. 1971).
174. Robeck, G. G., "Substitution of Residual Chlorine Measurement for Distribution Bacteriological Sampling," paper presented at AWWA Ann. Conf., Minneapolis, MN, 1974.
175. Baumann, R. E., and Ludwig, D. D., "Free Available Chlorine Residuals for Small Non-Public Water Supplies," *J. AWWA*, **54**, 1379 (Nov. 1962).
176. Varma, M. M., and Baumann, E. R., "Superchlorination-Dechlorination of Small Water Supplies," State Proj. Rept. Project 353-S, Iowa State Univ. Engr. Exp. Sta., Ames, IA, 1959.
177. Butterfield, C. T., Wattie, E., Megregian, S., and Chambers, C. W., "Influence of pH and Temperature on the Survival of Coliform and Enteric Pathogens When Exposed to Free Chlorine," *Public Health Reports*, **58**, 1837 (1943).
178. Lensen, S. G., Rhian, M., and Stebbins, M. R., "Inactivation of Partially Purified Poliomyelitis Virus in Water by Chlorination, Part II," *Am. J. Pub. Health*, **37**, 869 (1947).

179. Weidenkopf, S. J., "Inactivation of Type I Poliomyelitis Virus with Chlorine," *Virology,* **5,** 56 (1958).
180. Clark, N. A., and Kabler, P. W., "Inactivation of Purified Coxsackie Virus in Water by Chlorine," *Am. J. Hyg.,* **59,** 1159 (1954).
181. Fair, G. M., Morris, J. C., and Chang, S. L., "The Dynamics of Chlorination," *J. NEWWA,* **61,** 285 (1947).
182. Snow, W. B., "Recommended Residuals for Military Water Supplies," *J. AWWA,* **48,** 1510 (Dec. 1956).
183. Brazis, R. A., et al., "Special Report to Department of the Navy, Bureau of Yards, and Docks: Sporicidal Action of Free Available Chlorine," R. A. Taft San. Engr. Center, Cincinnati, OH, 1957.
184. Neefe, J. R., et al., "Inactivation of the Virus of Infectious Hepatitis in Drinking Water," *Am. J. Pub. Health,* **37,** 365 (1947).
185. Baumann, R. E., "Safe Disinfection for Household Water Systems," *Pub. Works* (May 1964).
186. Lin, S. D., "Tastes and Odors in Water Supplies: A Review," *Water and Sew. Wks., Ref. No.,* p. R-141 (1977).
187. Middlebrooks, E. J., "Taste and Odor Control," *Water and Sew. Wks., Ref. No.,* p. R-122 (1965).
188. Spitzer, E. F. (Ed.), *Handbook of Taste and Odor Control Experiences in the U.S. and Canada,* Amer. Water Works Assoc. Denver, CO, 1976.
189. Gerstein, H. H., "Odor Monitor and Threshold Tester," *J. AWWA,* **43,** 373 (1951).
190. Sigworth, E. A., "The Threshold Odor Test," *Water and Sew. Wks., Ref. No.,* p. R-92 (1964).
191. Palmer, C. M., "Algae in Water Supplies," U.S. Dept. of H.E.W. PHS #657, Washington, DC, 1962.
192. Vaughn, J. C., "Tastes and Odors in Water Supplies," *Env. Sci. and Tech.,* **9,** 703 (Sept. 1967).
193. Jenkins, D., Medsker, L. L., and Thomas, J. F., "Odorous Compounds in Natural Waters: Some Sulfur Compounds Associated with Blue-Green Algae," *Env. Sci. and Tech.,* **1,** 9, 731 (Sept. 1967).
194. Silvey, J. K. G., Russell, J. C., Redden, D. R., and McCormick, W. C., "Actinomycetes and Common Tastes and Odors," *J. AWWA,* **42,** 1018 (1950).
195. Erdei, J. F., "Control of Taste and Odor in Missouri River Water," *J. AWWA,* **55,** 1506 (1963).
196. Safferman, R. S., Rosen, A. A., Mashui, C. I., and Morris, M., "Earthy-Smelling Substance from a Blue-Green Alga," *Env. Sci. and Tech.,* **1,** 5, 429 (May 1967).
197. Lalezary, S., Pirbazari, M., McGuire, M. J., and Krassner, S. W., "Trace Taste and Odor Compounds from Water," paper presented at AWWA Ann. Conference, Las Vegas, Nevada, June 8, 1983.
198. Gerber, N. N., and Lechevalier, H. A., "Geosmin, an Earthy Smelling Substance Isolated from Actinomycetes," *Appl. Microbiol.,* **13,** 935 (1965).
199. Safferman, R. S., Rosen, A. A., Mashni, C. I., and Morris, M. E., "Earthy Smelling Substance from a Blue Green Alga," *Env. Sci. and Technol.,* **1,** 429 (1967).
200. Medsker, L. L., Jenkins, D., and Thomas, J. F., "Odorous Compounds in Natural Waters: An Earthy-Smelling Compound Associated with Blue Green Algae and Actinomycetes," *Env. Sci. and Technol.,* **2,** 461 (1968).
201. Dougherty, J. D., and Morris, R. L., "Studies on the Removal of Actinomycetes Musty Tastes & Odors in Water Supplies," *J. AWWA,* **59,** 1320 (Oct. 1967).
202. Baker, R. A., "Threshold Odors of Organic Chemicals," *J. AWWA,* **55,** 913 (1963).
203. Howard, N. J., "Removal of Taste & Odor," *J. AWWA,* **18,** 766 (1926).
204. Howard, N. J., and Thompson, R. E., "Chlorine Studies on Taste Producing Substances," *J. NEWWA,* **40,** 276 (1926).
205. Howard, N. J., and Thompson, R. E., "Progress in Superchlorination at Toronto," *J. AWWA,* **23,** 387 (1931).
206. Bushnell, W. B., "Over-Chlorination for Taste Control," *J. AWWA,* **17,** 653 (1925).
207. Lloyd, J. M., "Superchlorination at Tyler Texas," *J. AWWA,* **31,** 2130 (1939).

208. Lower, J. R., "Superchlorination at Upper Sandusky, Ohio," 19th Ann. Rept. Ohio Conf. Water Purif., p. 69 (1940).
209. Calvert, C. K., "Superchlorination," *J. AWWA,* **32,** 299 (1940).
210. Harlock, R., and Dowlin, R. "Use of Chlorine for Control of Odors caused by Algae," *J. AWWA,* **50,** 29 (1958).
211. Riddick, T. M., "Controlling Taste and Odor and Color with Free Residual Chlorination," *J. AWWA,* **43,** 545 (1951).
212. Adams, C. D., "Control of Tastes and Odors from Industrial Wastes," *J. AWWA,* **38,** 702 (1946).
213. Ettinger, M. B., and Ruchhoft, C. C., "Stepwise Chlorination on Taste and Odor Producing Intensity of Some Phenolic Compounds," *J. AWWA,* **43,** 561 (1951).
214. Burtschell, R. H., Rosen, A. A., Middleton, F. M., and Ettinger, M. B., "Chlorine Derivatives of Phenol Causing Taste and Odor," *J. AWWA,* **50,** 205 (1959).
215. Ryckman, D. W., and Grigoropolous, S. G., "Use of Carbon and Its Derivatives in Taste and Odor Removal," *J. AWWA,* **50,** 1268 (1959).
216. Harrison, L. B., "Super-Chlorination of Phenol Wastes," *J. AWWA,* **17,** 336 (1927).
217. Hale, Frank, "Successful Superchlorination and Dechlorination for Medicinal Taste of a Well Supply," *J. AWWA,* **23,** 373 (1931).
218. Baty, J. B., "Taste and Odor Control by Superchlorination," *Can. Engr.,* **78,** 19 (1940).
219. Brown, H. A., "Superchlorination at Ottumwa, Iowa," *J. AWWA,* **32,** 1147 (1940).
220. Brown, H. A., "Triple Chlorination at Ottumwa, Iowa," *Wat. and Sew. Works,* **97,** 267 (July 1950).
221. Williams, D. B., "A New Method for Odor Control," *J. AWWA,* **41,** 441 (May 1949).
222. Williams, D. B., "Dechlorination Linked to Corrosion," *Water and Sew. Works,* **100,** 104 (Mar. 1953).
223. Wilson, Carl, "Bacteriology of Water Pipes," *J. AWWA,* **37,** 52 (Jan. 1945).
224. Beger, H., "Iron Bacteria in Water Works and Their Practical Significance," *Gas-u.-Wasser,* **80,** 886 (Dec. 11, 1937).
225. Beger, H., "The Biology of Iron Bacteria," *Gas-u.-Wasser,* **86,** 779 (Oct. 23, 1937).
226. O'Connell, W. J., Jr., "Characteristics of Microbiological Deposits in Water Circuits," *Refining,* 66–83 (1941).
227. Peclzar, M. J., Jr., and Reid, R. D., *Microbiology,* 2nd ed., McGraw-Hill, New York, 1965.
228. *Standard Methods for the Examination of Water and Wastewater,* 12 ed., Amer. Publ. Health Assoc., New York, 1965.
229. Reddick, H. G., and Linderman, S. E., "Tuberculation of Mains as Affected by Bacteria," *J. NEWWA,* **46,** 146 (1932).
230. Von Wolzogen Kuhr, C. A. H., and Van Der Vlugt., L. S., "Aerobic and Anaerobic Iron Corrosion in Water Mains," *J. AWWA,* **45,** 33 (1953).
231. Weers, W. A., and Middlebrooks, E. J., "A Review of the Theory and Control of Corrosion," *Water and Sew. Works,* **114,** 156 (May 1967).
232. Thomas, Arba, "Role of Bacteria in Corrosion," *Water Works and Sew.,* **89,** 367 (1942).
233. Starkey, R. L., "Transformation of Iron by Bacteria in Water," *J. AWWA,* **37,** 963 (Oct. 1945).
234. Nason, H. K., "Chemical Methods in Slime and Algae Control" *J. AWWA,* **30,** 437 (Mar. 1938).
235. Love, O. T., Jr., Miltner, R. J., Eilers, R. G., and Fronk-Leist, C. A., "Treatment of Volatile Organic Compounds in Drinking Water," EPA-600/8–83–019, Cincinnati, OH, May 1983.
236. Ridgway, H. F., and Olson, B. H. "Scanning Electron Microscope Evidence for Bacterial Colonization of a Drinking-Water Distribution System," *Appl. and Environ. Microbiol.,* **41,** 274 (Jan. 1981).
237. Ridgway, H. F., Means, E. G., and Olson, B. H., "Iron Bacteria in Drinking-Water Distribution Systems: Elemental Analysis of *Gallionella* Stalks, X-Ray Energy-Dispersive Micro-analysis," *Appl. and Environ. Microbiol.,* **41,** 288 (Jan. 1981).
238. Arnold, G. E., "Crenothrix Chokes Conduits," *Engr. News. Rec.,* **116,** 774 (May 1936) and *Water Wks. and Sew.* **85,** 263 (Apr. 1938).

239. Ackerman, J. W., "Capacity of Cast Iron Main Sustained by Chloramine Treatment," *Water Works and Sew.*, **83**, 159 (May 1936).
240. Buelow, R. W., and Walton, G., "Bacteriological Quality vs. Residual Chlorine," *J. AWWA*, **63**, 28 (Jan. 1971).
241. Victoreen, H. T., "Control of Water Quality in Transmission and Distribution Mains," *J. AWWA*, **66**, 369 (June 1974).
242. Snead, M. C., Olivieri, V. P., Kruse, C. W., and Kawata, K., "Benefits of Maintaining a Chlorine Residual in Water Supply Systems," U.S. EPA Muni. Env. Res. Lab. Report No. EPA 600/2-80-010, June 1980.
243. Alexander, L. J., "Control of Iron and Sulfur Organisms by Super-Chlorination and De-Chlorination," *J. AWWA*, **36**, 1349 (Dec. 1944).
244. Panel Discussion, "Control of Growths in California Distribution Systems," *J. AWWA*, **42**, 849 (Sept. 1950).
245. Blair, G. Y., "Combating Pipeline Growths by Maintaining Chlorine Residuals Throughout a Distribution System," *J. AWWA*, **46**, 681 (July 1954).
246. Wilson, Carl, "Odor and Taste Control as Influenced by Consumer Pipes," *Water and Sew. Works*, **95**, 156 (Oct. 1948).
247. Panel Discussion, "Value and Limitation of Chlorine Residuals in Distribution Systems," *J. AWWA*, **51**, 215 (1959).
248. Plowman, H. L., and Rademacher, J. M., "Persistence of Combined Available Chlorine Residual in Gary-Hobart Distribution System," *J. AWWA*, **50**, 1250 (1958).
249. Eliassen, R., and Cummings, R. H., "Analysis of Water-Borne Outbreaks 1938-45," *J. AWWA*, **40**, 1301 (1948).
250. Offutt, A. C., Poole, B. A., and Fassnacht, G. G., "A Water-Borne Outbreak of Amebiasis," *J. Pub. Health*, **45**, 486 (1955).
251. Minkus, A. J., "Re-Chlorination in the Hartford Distribution System," *J. NEWWA*, **72**, 251 (Sept., 1958).
252. Williams, D. B., Brantford Ont., Pvt. Comm. (1970).
253. Rogers, M. E., "Restoring Pipeline Capacity at Wichita, Kansas," *J. AWWA*, **37**, 713 (1945).
254. Griswold, L. J., "Maintaining Transmission Line Capacity with Chlorine and Ammonia," *Water & Sew. Works*, **96**, 472 (1949).
255. Streicher, Lee, "San Diego Aqueduct Capacity Restored by Chlorination," *Water & Sew. Works*, **100**, 333 (Sept., 1953).
256. Jackson, L. A., and Mayhan, W. A., "Chlorination Maintains Supply Line Capacity," *Water & Sew. Works*, **98**, 248 (June, 1951).
257. Derby, Ray, "Control of Slime Growths in Transmission Lines," *J. AWWA*, **39**, 1107 (1947).
258. Weston, R. S., "The Use of Chlorine to Assist Coagulation," *J. AWWA*, **11**, 446 (1924).
259. Griffin, A. E., "Chlorination a Ten Year Review," *J. NEWWA*, **68**, 97 (1954).
260. Billing, L. C., "Experiences with Chlorinated Copperas as a Coagulant," *Water Wks. and Sew.*, **81**, 73 (1934).
261. Streeter, H. W., and Wright, C. T., "Prechlorination in Relation to the Efficiency of Water Filtration Processes," *J. AWWA*, **23**, 22 (1931).
262. Baylis, J. R., "Improving the Bacterial Quality of Water," *Water Works and Sew.*, **86**, 96 (Mar. 1939).
263. Baumann, R. E., Willrich, T. L., and Ludwig, D. D., "Prechlorination," *Agr. Engr.*, **44**, 138 (Mar. 1963).
264. McKee, J. E., and Wolf, H. W., "Water Quality Criteria," 2d. ed., Calif. State Water Quality Control Board, Sacramento (1963).
265. Nordell, E., "Water Treatment for Industrial and Other Uses," Reinhold, New York (1951).
266. Foxworthy, J. E., and Gray, H. K., "Removal of Hydrogen Sulfide in High Concentrations from Water," *J. AWWA*, **50**, 872 (July, 1958).

267. Monscvitz, J. T., and Ainsworth, L. D., "Hydrogen Polysulfide in Water Systems," Am. Chem. Soc. Mtg., Div. Water, Air, and Waste Chemistry, Minneapolis (Apr. 1969).
268. Black, A. P., and Goodson, J. B., Jr., "The Oxidation of Sulfides by Chlorine in Dilute Aqueous Solutions," *J. AWWA*, **44**, 309 (Apr. 1952).
269. Sawyer, C. N., "Chemistry for Sanitary Engineers," McGraw-Hill, New York (1960).
270. Choppin, A. R., and Faulkenberry, L. C., "The Oxidation of Aqueous Sulfide Solutions by Hypochlorite," *J. Am. Chem. Soc.*, **59**, 2203 (1937).
271. Wells, S. W., "Hydrogen Sulfide Problems in Small Water Systems," *J. AWWA*, **46**, 160 (Feb. 1954).
272. Sammon, L. L., "Removal of Hydrogen Sulfide from a Ground Water Supply," *J. AWWA*, **51**, 1275 (1959).
273. Derby, R. L., "Hydrogen Sulfide Removal and Water Softening at Beverly Hills, Calif.," *J. AWWA*, **20**, 813 (1928).
274. Powell, S. T., and Von Lossberg, L. G., "Hydrogen Sulfide Removal," *J. AWWA*, **40**, 1277 (1948).
275. Fair, G. M., Geyer, J. C., and Okun, D. A., "Water and Wastewater Engineering," Vol. 2, John Wiley & Sons, New York (1968).
276. Robinson, L. R., and Dixon, R. I., "Iron and Manganese Precipitation in Low Alkalinity Ground Waters," *Water & Sew. Wks.*, **115**, 514 (Nov. 1968).
277. Griffin, A. E., "Significance and Removal of Manganese in Water Supplies," *J. AWWA*, **52**, 1326 (Oct. 1960).
278. Bean, E. L., "Progress Report on Water Quality Criteria," *J. AWWA*, **54**, 1313 (Nov. 1962).
279. Griffin, A. E., and Baker, R. J., "The Breakpoint Process for the Free Residual Chlorination," *J. NEWWA*, **73**, 250 (Sept. 1959).
280. Edwards, S. E., and McCall, G. B., "Manganese Removal by Breakpoint Chlorination," *Water and Sew. Works*, **93**, 303 (Aug. 1946).
281. Myers, H. C., "Manganese Deposits in Western Reservoirs and Distribution Systems," *J. AWWA*, **53**, 579 (May 1961).
282. Tuepker, J. L., "Chlorination of High pH Waters," Panel Discussion, Ann. A.W.W.A. Mtg., San Diego, Calif. (May 21, 1969).
283. Griffin, A. E., "Well Rehabilitation by Chlorination," *Water and Sew. Works*, **102**, 277 (June 1955).
284. Erickson, C. R., "Cleaning Methods for Deep Wells and Pumps," *J. AWWA*, **53**, 155 (Feb. 1961).
285. Brown, E. D., "Restoring Well Capacity with Chlorine," *J. AWWA*, **34**, 698 (1942).
286. Huston, W. E., "Restoring Well Capacity with Chlorine," *J. AWWA*, **38**, 761 (1946).
287. Suter, Max, "Cleaning of Wells," *J. AWWA*, **30**, 1130 (1938).
288. Piatek, A., "Preventing Filamentous Scale in Well Water," *Water and Waste Engr.*, 55 (Dec. 1967).
289. Griffin, A. E., "Water Treatment for Water Flooding," *Producers Monthly* (1954).
290. A.W.W.A., Task Group Report, "Experience with Injection Wells for Artificial Ground Water Recharge," *J. AWWA*, **57**, 629 (1965).
291. A.W.W.A. Committee, 8360D Report, "Disinfecting Water Mains," *J. AWWA*, **60**, 1085 (1958).
292. Tracy, Harry, "Tank Disinfection," *J. AWWA*, **43**, 85 (1951).
293. White, G. C., "Chlorination & Dechlorination: A Scientific and Practical Approach," *J. AWWA*, **60**, 540 (May 1968).
294. Rideal, S., "The Influence of Ammonia and Organic Nitrogenous Compounds on Chlorine Disinfection," *J. Royal San Inst.* (England), **31**, 33 (1910).
295. Deberard, H. I., "Chloramine at Denver, Colo., Solves Aftergrowth Problems," *Engr. News-Rec.*, **79**, 210 (1917).
296. Hulbert, Roberts, "Chlorine-Ammonia Treatment Yields Nitrites in Effluent," *Engr. News-Rec.*, **109**, 315 (1933).

297. Shanson, D. C., "Infections and the Gut," Gastroenterology Seminar, St. Stephen's Hospital, Chelsea, London, U.K., Hospital Update, p. 756, June 1983.
298. Markell, E. K., Havens, R. F., and Kuritsubo, R., "Intestinal Parasitic Infections in Homosexual Men at a San Francisco Fair," *Western Jour. of Medicine*, p. 177 (Aug. 1983).
299. Petit, Charles, "Castro District Survey," *San Francisco Chronicle* (Sept. 17, 1983).
300. Kjellstrand, C. M., Eaton, J. W., Yawata, Y., Swofford, H., Kolpin, C. F., Buselmeier, T. J., Von Hartitzsch, B., and Jacob, H. S., "Hemolysis in Dialized Patients caused by Chloramines," paper prepared by the Department of Medicine, Chemistry and Surgery, Univ. Minnesota, Minneapolis, MN, 1974, published in Switzerland in *Nephron*, **13**, 427 (1974).
301. Meyer, M. A., and Klein, E., "Granular Activated Carbon Usage in Chloramine Removal from Dialysis Water," School of Medicine, Nephrology Division, Univ. Louisville, published in "Thoughts and Progress," *Artificial Organs* (Louisville, KY), **7**, 484 (April 1983).
302. Bauer, R. C., and Snoeyink, V. L., "Reactions of Chloramines with Activated Carbon," *J. WPCF*, **45**, 2292 (1973).
303. Snoeyink, V. L., and Suidan, M. T. "Dechlorination by Activated Carbon and Other Reducing Agents," in J. D. Johnson (Ed.), *Disinfection of Water and Wastewater*, Ann Arbor Science Publishers, Inc., Ann Arbor, MI, 1975.
304. Kim, R. B., and Snoeyink, V. L., "The Monochloramine-GAC Reaction in Adsorption Systems," *J. WPCF*, **50**, 122 (Jan. 1978).
305. Blasiola, G. C., "Protecting Aquarium and Pond Fish from the Danger of Chloramines," *Freshwater and Marine Aquarium Magazine*, 1984.
306. Smith, C. E., and Russo, R. C., "Nitrite Induced Methemoglobinemia in Rainbow Trout," *Prog. Fish Cult.*, **37**, 150 (1975).
307. Gratzek, J. B. and Hayter, C., "An Experiment in Filtration for the Freshwater Aquarium," *Pets: Supplies and Marketing* (June 1979).
308. Atkins, P. F., Scherger, D. A., Barnes, R. A., and Evans F. L., "Ammonia Removal by Physical-Chemical Treatment," *J. WPCF*, **45**, 2372 (Dec. 1973).
309. Rittman, B. E., and Snoeyink, V. L., "Achieving Biologically Stable Drinking Water," paper presented at the Ann. Conf. AWWA, Dallas, TX, June 14, 1984.
310. Hack, D. J., "Survey On the Use of Chloramine in Water Supplies of 50 States," paper presented at Chloramination Seminar, Ann. Conf. AWWA, Dallas, TX, June 10, 1984.
311. Barrett, S. E., Davis, M. K., and McGuire, M. J., "Blending Chloraminated Water with Chlorinated Water: Considerations for a Large Water Wholesaler," presented at Ann. Conf. AWWA, Dallas, TX, June 14, 1984.
312. Anon., "Coliforms Resist Treatment in Six Midwestern Systems," *AWWA Mainstream* (Feb. 1984).
313. Moser, R. H., private communication, American Water Works Service Co. at Ann. Conf. AWWA, Dallas, TX, June 12, 1984.
314. Olivieri, V. P., private communication, Ann. Conf. AWWA, Dallas, TX, June 13, 1984.
315. White, G. C., "Investigation of Muncie Indiana Waterworks Disinfection System," Confidential Report to American Water Works Service to Muncie Indiana, August 1980.
316. Anon., Bacteria Detection Instrument Code No. 300-25, Rothmoore Analytical Houston, TX, June 1984.
317. Wallis, C., and Melnick, J. L., "An Instrument for the Immediate Quantification of Bacteria in Potable Waters," Baylor College of Medicine, Houston, TX, 1984.
318. Bingham, A. K., and Meyer, E. A., "Disinfection of *Giardia muris* Cysts in Chloraminated Water," Dept. of Microbiology and Immunology, The Oregon Health Sciences University, Portland, Oregon, Nov. 12, 1981.

7
Chlorination of Wastewater

INTRODUCTION

Uses of Chlorine

The use of chlorine in wastewater treatment falls into the following categories: (1) odor control in wastewater and foul air, (2) prevention of septicity, (3) control of activated sludge bulking, (4) cyanide destruction, and (5) disinfection. Each of these categories will be discussed in detail.

Historical Background

The use of chlorine has generally been associated with the search for a means to control disease in man. Until the late 19th Century the idea persisted that disease was spread by odors and that therefore the control of odors would stop the spread of disease. Hence it is not surprising that chlorine was used as a deodorant long before its value as a germicide was recognized. Although bacteria were discovered about 1680, it was not until about 1880 that investigations revealed that certain bacteria—now described as pathogens—caused specific diseases.

It was subsequently discovered that small concentrations of various chlorine compounds could destroy these organisms. The earliest recorded practice of sewage chlorination on a large scale was in 1854 when the Royal Sewage Commission used chloride of lime to deodorize London sewage. During the next several decades many British patents were issued to control the use of chlorine and chlorine compounds for treating sewage. In 1859 an investigation for the Metropolitan Board of Works, London,[1] by Hofman and Frankland showed that a dosage of 400 pounds of chlorinated lime per million gallons (15 ppm) could delay putrefaction of the raw sewage for four days.

The first known application of chlorine for disinfection was described by William Soper of England in 1879, when he reported the use of chlorinated lime to treat the feces of typhoid patients before disposal into a sewer. The first use of chlorine on a plant scale for disinfection of sewage was made at Hamburg, Germany in 1893. This application was the result of a disastrous waterborne typhoid epidemic there. The first recorded use for this purpose in the United States was in 1894 at Brewster, New York,[2] a small village in the Croton drainage area of the New York City water supply. The chlorine was produced by the Woolf process. This installation was unique in that it made chlorine on the spot by electrolytic decomposition of a brine solution and discharged the resultant hypochlorite solution into the sewage. This plant operated successfully until it was destroyed by fire in 1911.

Meanwhile, the results reported by British and German investigators using chlorinated lime were confirmed by studies made in the United States by Phelps and Carpenter in 1906–07 at Massachusetts Institute of Technology.[3] Plant-scale studies by Phelps and others at Red Bank, New Jersey, Baltimore, and Boston in 1907–8 marked the beginning of effective chlorination practice in the United States. Clark and Gage[4,5] established many of the basic principles of chlorination at the Lawrence Experiment Station prior to 1911. The practical value of chlorinating raw, septic or treated sewage with chlorinated lime was soon confirmed by studies at installations in Philadelphia, Chicago, Providence, and elsewhere. By 1911, eight sewage treatment plants in New Jersey were reported to be using chlorinated lime to protect water supplies, shellfish areas, and bathing beaches.

Until the development of a suitable gas chlorinator in 1913, the adoption of chlorination for disinfection and odor control of sewage was very slow because the chlorinated lime was found to deteriorate in storage and was relatively expensive, messy to handle, and awkward to apply.

Liquid chlorine became available as a commercial product in 1909 at Niagara Falls. The first chlorinator for metering and applying chlorine gas was developed by George Ornstein of the Electro Bleach Gas Company in 1912. In 1913, Wallace & Tiernan marketed the first successful gas chlorinator based on the Ornstein patent, which revived interest in sewage chlorination. Among the first municipalities to employ liquid chlorine for the disinfection of sewage effluents were Altoona, Pennsylvania, and Milwaukee, Wisconsin, in 1914, followed by El Dorado, Kansas, in 1915 and Philadelphia in 1916.

The use of chlorination in wastewater processes has grown tremendously over the years since the development of suitable equipment. In 1958, it was reported that over 2200 plants serving a population of almost thirty-eight million were equipped with chlorination facilities.[6] This represents about 30 percent of all treatment plants in the United States and about 50 percent of the population served by treatment facilities.

Contemporary Chlorination Practices

Chlorination is now established as an integral part of wastewater treatment practice in the United States and Canada. Several hundred technical articles have been written on the subject. It is the consensus that the primary use of chlorine is for disinfection. As early as 1945, Enslow and Symons[7] stated that by definition there are only three processes of sewage treatment: primary treatment, secondary treatment, and disinfection. In recent years tertiary treatment has become an additional process. Sludge disposal is relegated to being a by-product of the process.

Active interest in wastewater disinfection began in the United States about 1945. Up to that time the primary use of chlorine in sewage disposal systems was for odor control, hydrogen sulfide destruction, and prevention of septicity. Most of the sewage treatment plants practicing disinfection during that time belonged to the United States Armed Forces. It was military policy during World War II that sewage effluents at all Army bases in the United States had to be chlorinated. Today, as a result of the 1970 Federal Water Pollution Control Act, almost all wastewater treatment plants are subjected to some disinfection requirement.

In 1961 the first chlorine residual controlled disinfection system for wastewater was installed at Napa, CA. In 1975 chlorine was first used on a plant scale for nitrogen removal from wastewater. Both of these applications proved successful.

The uses of chlorine in wastewater treatment practice may be summarized as follows: disinfecting; controlling odor and preventing septicity; improving grease and scum removal; preventing filter ponding; controlling flies; controlling activated sludge bulking; controlling odor in the sludge-thickening process; controlling waste-activated sludge disposal; controlling foaming; destroying cyanides; destroying phenols; and foul air scrubbing.

Chemistry of Chlorine and Wastewater

The chemistry of the chlorination of potable water, wastewater, and industrial waste is fundamentally the same. The reactions differ only because of the differences in species and amounts of interfering substances, both organic and inorganic. The interfering substances are those that either contribute to excessive consumption of chlorine or impair the bactericidal efficiency of the chlorine residual.

The most important of these interfering substances found in wastewater discharges are: ammonia nitrogen, organic nitrogen, hydrogen sulfide, tannins, cystine, uric acid, humic acid, pickle liquor, cyanides, and phenol.

All of these compounds have been at one time or another isolated and studied for their effect upon the application of chlorine.

Ammonia Nitrogen. With the exception of highly nitrified effluents there is usually an appreciable amount of ammonia nitrogen in all wastewater effluents. The range is on the order of 10 to 40 mg/liter. The ammonium ion exists in equilibrium with ammonia nitrogen and hydrogen and the distribution is dependent upon pH and temperature. The relative distribution can be defined as follows:

$$NH_4^+ \rightleftharpoons NH_3 + H^+; \text{ where } K = 5 \times 10^{-10} \text{ at } 20°C \tag{7-1}$$

where

NH_3 = undissociated ammonia
NH_4^+ = ammonium ion
H^+ = hydrogen ion
K = dissociation constant

According to the dissociation constant, the pH value at which undissociated ammonia and ammonium ion are present in equal proportions (pK) is about pH 9.3 at 20°C. Above pH 9.3 undissociated ammonia (NH_3) predominates; below pH 9.3 ammonium ion (NH_4^+) predominates.

At usual wastewater pH levels the predominant chlorine reaction proceeds as follows:

$$HOCl + NH_4^+ \rightleftharpoons NH_2Cl + H_2O + H^+ \qquad (7\text{-}2)$$

when the chlorine to ammonia nitrogen weight ratio is less than 5:1. If the pH drops below 7, dichloramine ($NHCl_2$) will begin to form, and at a much lower pH nitrogen trichloride will form. However, as soon as the 5:1 weight ratio of chlorine to ammonia nitrogen is exceeded a new set of reactions takes place and are described in Chapter 8.

The ammonia–chlorine reactions in wastewater follow the same pathways as in potable water. This is thoroughly discussed in Chapter 4. It is abundantly clear from the vast numbers of observations of chlorine in wastewater that the speed of reaction is fastest with inorganic compounds. These reactions seem to take precedence over the penetration of bacteria.

The speed of the chlorine–ammonia reaction in wastewater is grossly pH dependent. For example, at 25°C and with a molar ratio of 0.2×10^{-3} mol/l HOCl and 1.0×10^{-3} mol/l $NH_3 = N$, there will be a 99 percent conversion to monochloramine in 0.2 sec at pH 7 and 421 sec at pH 2. The most rapid reaction occurs at pH 8.3 (0.069 sec). This is a significant factor in the location of the injectors with respect to the diffusers when using plant effluent for injector operating water. If the travel time of the chlorine solution is more than 3 min, then about 10 percent of the chlorine feed rate will be consumed in the breakpoint reaction during the travel time to the diffusers. Therefore, the injectors should always be as close to the diffusers as possible.

However, since the chlorination process operates in the neutral pH range, the speed of the chlorine–ammonia reaction to monochloramine is finite but almost instantaneous and takes precedence over bacteria penetration, provided there is proper mixing at the point of application.[8] It can be computed that, in a wastewater containing 30 mg/l ammonia, the molecules of ammonia will outnumber the organisms by a factor of about 10^{13}. This may be a significant factor in the kinetics of wastewater disinfection.

Organic Nitrogen. While the reactions between chlorine and ammonia nitrogen can be predicted with some certainty, the reactions between chlorine and compounds containing organic nitrogen are relatively obscure. Organic nitrogen compounds are present in all wastewaters containing domestic sewage. These discharge significant amounts of proteinacious matter, amino acids, and the various nitrogenous compounds of urine. Studies of some of these compounds have revealed that they are stable over a long period of time, even in the presence of free chlorine.[9] These chlorinated organic nitrogen compounds are classified as N-chloro compounds or organic chloramines. In the procedures for chlorine residual determination, these compounds show up in the dichloramine fraction. They are not considered to have any effective germicidal characteristics. One investigator identified creatinine, a compound of urea, as responsible for the long-lasting ineffective "dichloramine" residuals in swimming pools,[10] and Williams confirmed this at the water treatment plant at Brantford, Ontario. Other investigations have shown that cystine and tannins have the most inhibiting effect upon wastewater disinfection, closely followed by uric acid and humic acid[11] and some of the amino acids.[12] So far as is known the organochloramines do not enter into any reactions with inorganic compounds present in the wastewater.

Tannins. These compounds of organic nitrogen are used in leather-processing works. They exert an extremely high chlorine demand. In one instance, slugs of tanning wastes raised the chlorine demand of a typical primary-treated domestic waste from 12 mg/l to over 60 mg/l. The situation was largely corrected by distributing the waste from the tanning operation over a longer period of time, thus taking advantage of a dilution factor. Tannins should be treated at the source before entry into the collection system.

Sulfites. Inorganic chemicals containing the sulfite ion are used in leather-processing and wine-making. Sulfites will promote the generation of hydrogen sulfide in any sewage collection system, particularly if the wastewater temperature rises above 19°C. Sulfites are easily neutralized by chlorine. This is the reverse of dechlorination. Sulfites should be neutralized at the source.

Pickle Liquor. Pickle liquor from steel-mill operations in significant amounts will definitely affect treatment plant chlorination procedures. This substance ($FeSO_4 \cdot 7H_2O$) is the result of pickling steel products to remove the scale. Ferrous sulfate combines with chlorine to form ferric chloride. Each 7.8 ppm of full-strength pickle liquor will consume approximately 1.0 mg/l of chlorine, depending upon the strength of the former. This reaction, if enough ferric chloride is formed, will have a decided beneficial reaction at any plant practicing prechlorination. The ferric chloride hydrolyzes to ferric hydroxide, a fine coagulant at the pH range found in most wastewater processes. However, if only effluent chlorination is practiced, the floc formed in the chlorine contact chamber or outfall line makes the effluent turbid and of a reddish brown color.

Phenols. Phenols are present in many chemical plant effluents, and will exert a definite additional chlorine demand in wastewater processing. Care must be taken to destroy phenols before they reach the receiving water, because phenols contribute to extremely obnoxious taste and odor problems in water treatment processes.

Theoretically it requires 10 mg/l chlorine to destroy each mg/l of phenol. However, owing to the presence of other compounds usually associated with phenols, this figure can soar to 20 mg/l of chlorine.

Chlorine Demand. The only other specific parameter to cite in connection with the application of chlorine is its relation to the strength of the waste to be treated. It can be said with assurance that the chlorine consumption will increase as the BOD of the waste increases, and will vary as the BOD varies. This is the reason why most domestic wastes have a diurnal variation of chlorine demand from as low as 2:1 up to a high of 5:1. The latter is a result of the effect of infiltration in the collection system, so that the treatment process may be handling practically all runoff water between 2 A.M. and 6 A.M.

A rule-of-thumb estimate of 15 minute chlorine demand for various wastewaters is as follows:

Raw fresh domestic waste	10–15 mg/l
Raw septic domestic waste	15–40 mg/l
Primary effluent	10–16 mg/l
Biofilter effluent (secondary)	4–8 mg/l
Trickling filter effluent	4–10 mg/l
Well oxidized secondary effluent	3–8 mg/l
Multimedia filter effluent	3–6 mg/l
Slow sand filter effluent	2–4 mg/l
Nitrified filtered effluent	2–10 mg/l
Septic tank effluent	30–45 mg/l

These figures can be used to calculate chlorinator capacity for disinfection. Since these are 15 min contact time demands, make the following adjustment to arrive at chlorine dosage: Assume total contact time equal to 45 min. Add 1.5–2.0 mg/l die-away in contact chamber, then add the required residual calculated from the Collins model.

The figures as shown for raw sewage are adequate to size chlorination equipment for odor control or septicity control.

THE ODOR PROBLEM

Significance

If the combined efforts of both the designers and operators of a wastewater collection and treatment system result in obnoxious odors, the public will consider the project a complete failure. There is probably no other type of process in which odor is used as quickly by the general public to measure success or failure. It does not matter if the process produces an effluent which meets the most stringent requirements of the local regulatory body. It must not emit any objectionable odors.

Sources of Odors

Odors are the result of putrefaction of the solids in wastewaters. The amount of these solids is on the order of 0.08 percent by weight. These suspended solids are either in true colloidal suspension or in solution. Proteins make up a large percentage of the organic suspended matter. Removal of the suspended matter in the waste is of vital importance to the operator. Proteins are components of a complex structure containing four elements in varying proportion: carbon, nitrogen, oxygen, and hydrogen. One variety contains a small amount of phosphorous. Proteins may be classed as colloids, and thus contribute to the colloidal nature of domestic wastes. When the oxygen supply in the wastewater is depleted, the microorganisms, enzymes, and inorganic compounds present cause putrefaction to start. Putrefaction can produce a variety of odoriferous compounds—indol, skatole, leucine, tyrosine—all derivatives of amino acids, which are products of the decomposition of proteins.[13]

Many distinctive and unpleasant odors are caused by bacteria and enzymes acting on protein matter. Ammonia, hydrogen sulfide, volatile fatty acids, and mercaptans are also products of putrefaction. Fats, another class of compounds which give off offensive odors, contain but three elements: carbon, hydrogen, and oxygen. They putrefy in two ways: by oxidation, to give a tallowy odor; or hydrolytically, which causes the fats to break down to fatty acids of a volatile nature. These are usually referred to as the "goat acids," the name suggesting the type of odor to expect.

Odor-producing substances, such as amino acids, mercaptans, and goat acids, usually persist only within the confines of the treatment plant. These odors can be greatly reduced by liberal use of landscaping with shrubs and flowers, which have a great capability of both masking and absorbing odors. However, experience has shown that when hydrogen sulfide is controlled, other obnoxious odors are also controlled. The other compounds are difficult to detect and analyze, and so it is assumed that the detection and measurement of hydrogen sulfide will indicate when putrefaction may be occurring and consequently when conditions in the wastewater could produce foul odors.

Identifying the Problem

The easiest odor producing compound to identify and control is hydrogen sulfide. It is generally assumed that when hydrogen sulfide odors are eliminated, the odors from other compounds (cadaverine, indole, mercaptans, and skatole) will also be eliminated or greatly suppressed.

Therefore, evaluation of H₂S emmissions will identify the problem. This can be done by two methods.

Lead Acetate Impregnated Tiles. This method is generally better suited to the confines of the treatment plant. These tiles are placed to form a network around the plant or treatment units as described by Chanin.[14] The tiles are rated on a daily basis and evaluated weekly as shown on Table 7–1.

Air-Sampling H₂S Detectors. A variety of these devices are available. These are generally of three different types. (1) The lead acetate impregnated tape type draws a continuous air stream to the tape. The tape changes color with H₂S concentration. The light transmissibility from the color spot on the tape is automatically recorded as H₂S concentrations. Examples of these units are those made by Research Appliance Corp. and Houston Atlas, Inc. (2) The wet chemistry automatic coulometric type is made by ITT Barton. (3) The solid state electrolytic cell sensors are offered by Texas Analytical Controls, Bio Marine, and others.

The selection of these units for monitoring and/or control of a treatment process should be based primarily upon the effective detection range and the probable maintenance time required for proper operation. For odor control and monitoring purposes in wastewater treatment the H₂S air sampling unit should be capable of detecting concentrations as low as 1 ppm. The lead acetate tape units have proved successful on detection requirements and maintenance time.

One additional piece of instrumentation necessary to solve odor monitoring is

Table 7-1   Lead Acetate Hydrogen Sulfide Detector Tiles—Numerical Rating System

	Observed Color								
	White or Yellow	Very Light Brown		Light brown		Brown		Dark Brown or Black	
Starting Color	24 or 48 hr	24 hr	48 hr	24 hr	48 hr	24 hr	48 hr	24 hr	48 hr
White or yellow	0	1	½	2	1	3	2	4	3
Very light brown	—	¼	¼	1	½	2½	1½	3½	2½
Light brown	—	—	—	¼	¼	2	1	3	2
Brown	—	—	—	—	—	—	—	2	1

Plot on daily basis number corresponding to color change at each station.

Also: Change Per 7 Days = $\frac{7}{\text{Total Days}} \times \Sigma$ (Color Change Numbers For Period)

A weekly number of greater than about 3.5 would indicate an odor problem.

a wind direction and velocity recording unit. This unit provides the information necessary to establish validity of odor complaints.

Hydrogen Sulfide Characteristics

Hydrogen sulfide is a lethal and obnoxious gas generated in wastewater systems and caused by putrefaction. It is the main cause of sewage odors and is detectable in very low concentrations (<1.0 ppm).

1. H_2S is a deadly poisonous gas, and has resulted in many deaths to operating personnel. Many sewer maintenance workers have lost their lives because of this gas. Death can occur within a few minutes' exposure to concentrations as low as 2000 ppm by volume in the atmosphere.*

2. H_2S gives off a "rotten egg" odor that is intolerable.

3. It combines with moisture in the humid air above the water line to form sulfuric acid, which is devastating to concrete structures.[16]

4. It is explosive at a concentration of 4.3 percent by volume in the atmosphere. H_2S is entirely different from other odor-producing compounds. It will not diffuse in air to any appreciable extent, and so it persists for great distances and tends to flow in a laminar configuration. H_2S odors have been known to be obnoxious at distances up to six miles from the point of origin, and in a band no wider than 100–200 feet, with a concentration of dissolved sulfides at the point of origin no greater than 10–12 ppm.[17]

H_2S has another perverse characteristic. It will, over a short period of time, paralyze the olfactory nerve to the extent that an operator working in the area of a nontoxic concentration loses the ability to detect H_2S by odor. This is one of the reasons it is so dangerous in a poorly ventilated area.

At 5 ppm concentration in air (by volume) the odor is moderate and easily detectable. Concentrations of 2 ppm are considered inoffensive and are not hazardous. This is the optimum goal for the control of hydrogen sulfide in wastewater practice. Other effects are as follows:

10 ppm	eye irritation starts
30 ppm	strong unpleasant odor of rotten eggs
100 ppm	coughing; loss of smell in 2–15 minutes
200–300 ppm	red eyes; rapid loss of smell; breathing irritation
500–700 ppm	unconsciousness and possibly death in 30–60 minutes.
700–1,000 ppm	rapid unconsciousness; breathing stops

---

* A specific antidote for the toxicity of H_2S was reported by the Esso Medical Center, Fawley, Southampton, U.K. An ampoule of amyl nitrite is broken into a cloth and the casualty breathes the vapor. When medical aid is available, 10 ml of 3 percent sodium nitrite is injected intravenously over 2–3 minutes. If symptoms recur, a further 5 ml may be given.[15]

1,000–2,000 ppm	instant unconsciousness; death in a few minutes
43,000 ppm	lower explosive limit

Source of Hydrogen Sulfide in Wastewater

Of all the compounds present in wastewater, those containing sulfur have the greatest potential for generating hydrogen sulfide. The putrefaction of organic matter causes the generation of hydrogen sulfide.

The sulfate ion ($SO_4^=$) is the principal sulfur compound in wastewater. When organic matter is present and oxygen is absent, bacteria of the species *Desulfovibrio desulfuricans* will reduce sulfate to sulfide, using oxygen to oxidize organic matter. Letting C represent organic matter, the reaction may be written as follows:

$$SO_4^= + 2C + 2H_2O \xrightarrow{\text{bacteria}} 2HCO_3^- + H_2S \qquad (7\text{-}3)$$

In this reaction 96 grams of sulfate make available 64 grams of oxygen, leaving 32 grams of sulfide. Using an equation showing an average formula for the reacting organic matter would probably indicate that about 42 grams of organic matter would be oxidized. Sulfite, thiosulfate, free sulfur, and other inorganic sulfur compounds sometimes found in wastewater can be similarly reduced to sulfide. The sulfur-bearing organic compounds of the mercaptan group contribute significantly to the odor problem.

Effect of Sulfate Concentration. Most of the sulfate in sewage comes from the water supply. Sulfur in proteins is metabolized by the body and excreted as sulfate. This amounts to only 10–20 mg/l of sulfate added by the sewage. The rate of sulfide generation is affected very little by sulfate concentration if some is present. Variations of sulfate concentrations in the water supply of a given area have no significant effect. Pomeroy reported that in one sewer line the sulfate concentration was reduced to 1 mg/l before there was any significant slowing of sulfide generation.[18] The total amount of sulfide produced in a sewer is not greatly affected by sulfate concentrations as long as there is a substantial surplus.

Other Sulfide Compounds in Wastewater. There are several forms of sulfides found in wastewaters which contribute to the generation of hydrogen sulfide. This mixture is made up of insoluble metallic sulfides, dissolved sulfides, and the secondary sulfide ion, $S^=$. (There are also organic sulfur compounds in which the sulfur is arbitrarily assigned a valence of minus two; these are called organic sulfides. They do not respond to analytical tests used to measure inorganic sulfide, and they do not have the same significance.)

There are also volatile organic sulfur compounds that are very important odor components in wastewater. There are three principal types: mercaptans (thiols), thioethers (also called sulfides), and the disulfides.

There are also nonvolatile sulfur compounds that do not cause the problems associated with the volatile compounds unless they are broken down by biological action to yield inorganic sulfide. These naturally occurring compounds in wastewater are principally the albuminoid proteins.

Hydrogen Sulfide Chemistry

General Discussion. This gas is highly soluble in water: 3618 mg/l at 20°C and one atmosphere. In aqueous solutions it hydrolyzes as follows:

$$H_2S \rightleftharpoons HS^- + H^+ \tag{7-4}$$

The hydrosulfide ion (HS^-) further dissociates as follows:*

$$Hs^- \rightleftharpoons S^= + H^+ \tag{7-5}$$

At 18° the hydrolysis constant for Eq. (7-4) is:

$$K_h = 9.1 \times 10^{-8} = \frac{[H^+][HS^-]}{[H_2S]} \tag{7-6}$$

Figure 7-1 illustrates the distribution of H_2S, HS^-, and $S^=$ for various pH levels. At pH 7 hydrogen sulfide is approximately 50 percent of the total dissolved sulfides;

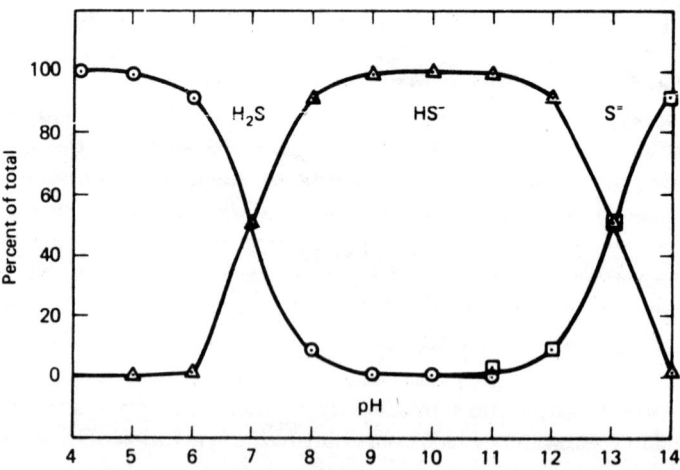

Fig. 7-1 Effect of pH on hydrogen sulfide—sulfide equilibrium.

---

* and for Eq. (7-5):

$$K_h = 1.2 \times 10^{-15} \frac{[H^+][S^=]}{[HS^-]} \tag{7-7}$$

at pH 5 it is practically 100 percent of the total; at pH 9, dissolved sulfides are nearly all sulfide ion. Therefore the existence of hydrogen sulfide in wastewater is pH dependent.

Whenever the equilibrium between the hydrogen sulfide ion and the hydrogen sulfide in solution is upset, as when H_2S is removed by oxidation (Cl_2) or air stripping, the shift will be to form more H_2S from the remaining dissolved sulfides. This shift reestablishes the equilibrium of Eq. (7-4). This stored sulfur gradually disappears by metabolic action, being itself oxidized to sulfate to yield more energy.[19,20]

The chemistry of the oxidation and removal of hydrogen sulfide is extremely complex. Supposedly the oxidation proceeds to form either elemental sulfur, sulfate, or both. These reactions are described in the section "Chlorine for Odor Control" in this chapter.

Categories of Sulfides Important to H_2S Generation. These categories are as follows:[21]

1. *Total sulfides* includes *dissolved* H_2S and HS^- plus the *acid soluble* metallic sulfides present in the suspended matter. The secondary sulfide ion $S^=$ mentioned above is negligible. Copper and silver sulfides are so insoluble they can be ignored.

2. *Dissolved sulfides* is that portion remaining after the suspended solids have been removed by flocculation and settling.

3. *Unionized hydrogen sulfide* may be calculated from the concentration of dissolved sulfide, the pH of the sample, and the practical ionization constant of hydrogen sulfide. (See *Standard Methods*, 15th edition.)

There are several methods outlined in *Standard Methods* for finding the concentration of total and dissolved sulfides. It is important to remember the distinction between these two categories of sulfides in order to understand the problem of odor control. As a general rule, those who live with the problem of H_2S generation in a sewage system run total sulfide determinations first because the procedure is faster. If this concentration is 0.4 mg/l or greater, then a dissolved sulfide test is performed because 0.3 mg/l dissolved sulfides is considered the tolerable limit. Concentrations higher than this will cause odor problems. Below this limit the evolution of H_2S from the wastewater will not cause an odor problem. Experience has shown that in areas of high fluid turbulence the concentration of H_2S in the surrounding atmosphere will be less than 2 ppm. At this concentration there will not be an odor problem.

Hydrogen Sulfide Generation

Causes and Occurrence. In general, it can be said that hydrogen sulfide generation is a result of septic conditions in the wastewater. The onset of this septicity occurs when the oxygen in the sewage or in the enveloping atmosphere is depleted.[22-24]

Factors contributing to the generation of H_2S are as follows: high BOD in the wastewater; high sulfates in the waste (sulfate content of the contributing water supply is not a factor[23]); high sewage temperatures; sluggish and stagnant flow conditions; lack of air cover in force mains; sludge deposits.

Hydrogen sulfide is generated in both free-flowing sewers and force mains. It always forms in a force main provided the detention is in excess of 20 or 30 minutes, but does not necessarily form in free-flowing sewers.

Free-Flowing Sewers:

General Discussion. Pomeroy[23] summarizes the fundamental concept of sulfide generation in sewers as follows: "In free-flowing sewers, sulfides are produced only by slimes on the submerged surface of the sewer and by deposited sludge. In the flowing body of the sewage, sulfides are not generated, but on the contrary, are destroyed by oxygen which is continually being absorbed from the surface."

Sewage flowing in a conduit showing a rapid increase in sulfide concentrations will not necessarily show the same increase when a sample is enclosed in a bottle, because the H_2S generation is produced by sludge deposits in the conduit. Fig. 7-2 illustrates the processes occurring in free-flowing sewers under sulfide buildup conditions. In order to prevent sulfide access to the flowing stream the oxygen concentration in the stream is critical. This range is between 0.1 and 1.0 mg/l oxygen.[21] Before all the sulfides produced can pass into the stream, conditions must be completely anaerobic.

The rate of sulfide generation is roughly proportional to the BOD, which is a measure of the oxygen utilized by microorganisms in oxidizing the sewage to stable end products (CO_2, H_2O, NO_3^-). Since this is a biological phenomenon, it is markedly affected by temperature. Pomeroy combines these two factors into a single factor called "effective BOD." By using 20°C as a standard temperature, and since the biological activity increases 7 percent for each degree C rise (geometrically), the effective BOD may be expressed as follows:

$$\text{Effective BOD (EBOD)} = \text{Standard BOD} \times (1.07)^{t-20} \quad (7-8)$$

Pomeroy also makes the observation that, for a specified temperature and flow condition, there is a limiting sewage strength below which a buildup of sulfide will not occur. In free-flowing sewers, the BOD would have to be greater than 50 ppm to generate sulfide. Below this figure, it is doubtful that sulfides would be generated.

Since sulfide-producing microorganisms can function over a wide range of pH, the latter is not generally a factor. However, raising the pH significantly by adding of caustic or lime is known to greatly diminish the generation of sulfides.

The generation of sulfide in the slime layer of Fig. 7-2 will also be dependent upon the ratio of sulfates to organic matter at a ratio of about 2.3:1. One will

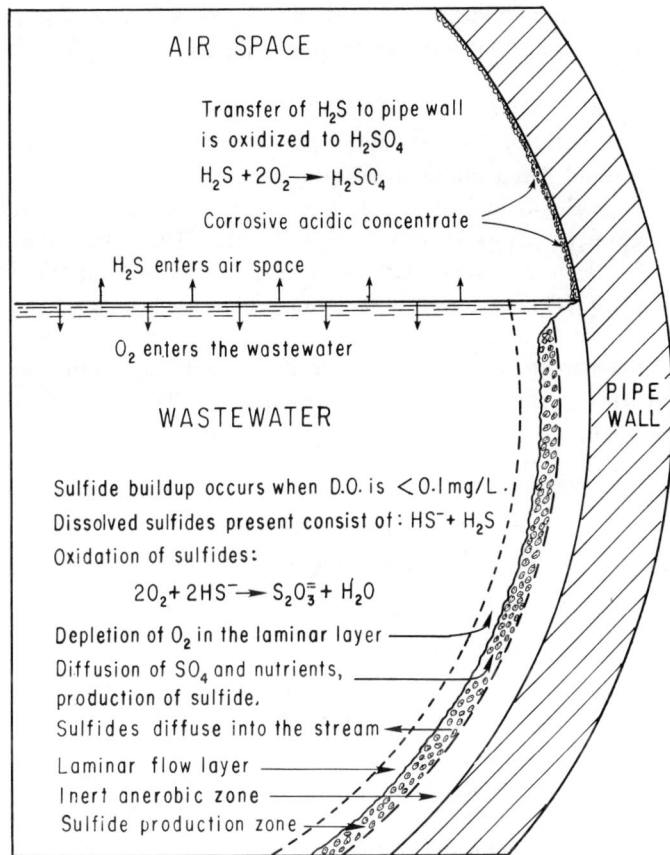

Fig. 7-2 Processes occurring in free-flowing sewers under sulfide buildup conditions.

be used up before the other, and the relatively scarce one will then be the constituent that limits the sulfide production rate. If there is an abundant supply of organic matter, but sulfate is scarce, then sulfide generation will be proportional to sulfate concentration and independent of the amount of organic matter.

It was once thought, because of long detention times in the collection system, that the age of the sewage was responsible for sulfide generation. What matters is not the age but how rapidly the sewage flows. If the velocity is adequate, a considerable degree of purification takes place in long lines, thus preventing a sulfide buildup. At the same time, undue turbulence is to be avoided, since this releases the H_2S and may give rise to odor complaints at manhole locations. On the other hand, high velocities (5 ft/sec and higher) scour the slimes, preventing sulfide buildup. This scouring action is not noticeable until the velocity exceeds 3 ft/sec.[23] Pomeroy states: "For any particular temperature and sewage strength combination, there is a limiting flow velocity above which sulfide buildup will not occur." (This applies only to free-flowing sewers.)

Forecasting Sulfide Buildup. It is the consensus that as little as 0.2 mg/l D.O. will prevent H_2S buildup—the slime layer is the critical factor. It is desirable to keep this in mind when sulfide buildup is a factor.

The Sacramento Regional County Sanitation District consolidated twelve existing wastewater treatment plants into one major facility. This consolidation resulted in 27 miles of collecting sewers with travel times up to 24 hours before reaching the regional plant. This required a long and laborious study of the system in order to provide rational solutions to prevent hydrogen sulfide buildup.[15,26]

In 1946, Pomeroy et al.[23] developed the first equations relating to conditions necessary for H_2S generation in gravity sewers. In 1950, Davey[27] reported on the effect of velocity on H_2S generation. This work was later modified by Pomeroy into what is known as the Z formula:

$$Z = \left(\frac{E_{BOD}}{S^{0.50} Q^{0.33}}\right)\frac{P}{b} \qquad (7\text{-}9)$$

where

Z = function defined by the equation
$E_{BOD} = (BOD)_5 \times (1.07)^{T-20}$ mg/l
T = temperature, °C
S = slope, ft/ft
Q = discharge, cfs
P = wetted perimeter, ft
b = surface width, ft

The value of Z obtained is interpreted as follows:

$Z < 5000$ sulfide is rarely produced
$5000 < Z < 10000$ conditions marginal for buildup
$Z > 10000$ sulfide buildup is common

This formula has been used successfully in predicting generation of sulfide in sewers but there is some uncertainty as to its reliability.[28]

Equations for Partially Filled Pipes. In 1977 Pomeroy and Parkhurst[29] presented a quantitative method for sulfide forecasting. The work by the Sacramento Area Consultants[15,26,30] in their study of the regional collection system concluded that the best predictive equations available were those developed by Pomeroy and

Parkhurst in 1977.[29] The final report confirmed the accuracy of these equations,[30] which are as follows:

$$\frac{d[S]}{dt} = \frac{3.28M'(E_{BOD})}{r} - \frac{2.10m[S](SV)^{0.375}}{d_m} \tag{7-10}$$

where

$\frac{d[S]}{dt}$ = rate of change of total sulfide, mg/l-hr

M' = sulfide flux coefficient

M' is empirically determined to suit conditions. It is usually about 0.4×10^{-3} m/hr when dissolved oxygen is low (<0.5 mg/l) and approaches zero as D.O. concentrations increase. Pomeroy suggests 0.32×10^{-3}.

E_{BOD} = 5 day BOD $\times (1.07)^{T-20}$ mg/l
T = °C
r = hydraulic radius, ft
m (or N) = empirical coefficient for sulfide losses

Pomeroy indicates that m ranges from a conservative value of 0.64 to a less conservative value of 0.96. The Sacramento Calif. County study found the best fit was for $m = 0.64$. This was based upon field study results.[30]

[S] = total sulfide concentration, mg/l
S = slope, ft/ft
V = velocity, ft/sec
d_m = mean hydraulic depth, cross-sectional area divided by surface width.

The negative term of Eq. (7-10) is proportional to the sulfide concentration. When the sulfide losses equal the sulfides generated and the buildup is zero, the total sulfide concentration approaches a limit:

$$\frac{3.28M'(E_{BOD})}{r} = \frac{2.10m(SV)^{0.375}}{d_m} \times [S]_{lim} \tag{7-11}$$

Substituting P/b for d_m/r, where P is the wetted perimeter and b is the surface width of the stream, and using the conservative value of 0.64 for m, Eq. (7-11) becomes:

$$[S]_{lim} = 0.78 \times 10^{-3} \frac{E_{BOD}}{(SV)^{0.375}} \times \frac{P}{b} \tag{7-12}$$

To calculate the sulfide generation in a series of pipe sections with changing hydraulic conditions, the following relationship occurs:

$$t_2 - t_1 = \left(\frac{1.10 d_m}{m(SV)^{0.375}}\right)\left(\log \frac{[S]_{lim} - [S]_1}{[S]_{lim} - [S]_2}\right) \tag{7-13}$$

Where

$t_2 - t_1 = \Delta t =$ flow time (hours) in a given pipe reach with constant slope s, diameter, and flow.
$1.10 =$ meters to feet conversion factor
$[S]_1 =$ sulfide concentration at start of reach, mg/l
$[S]_2 =$ sulfide concentration at end of reach, mg/l
$[S]_{lim} =$ limiting sulfide concentration from Eq. (7-12)
$m =$ empirical coefficient (either 0.64 or 0.96).

By rearranging Eq. (7-13) we get the following expression for $[S]_2$:

$$[S]_2 = [S]_{lim} - \frac{[S]_{lim} - [S]_1}{\log^{-1}\left(\dfrac{m(SV)^{0.375} \Delta t}{1.10 \, d_m}\right)} \tag{7-14}$$

and the value of $[S]_2$ for the first reach becomes $[S]_1$ for the second reach and so on. By calculating $[S]_{lim}$ and t and knowing the hydraulic conditions for each reach of the system, the sulfide buildup can be calculated for a series of reaches.

Force Mains:

General Discussion. It is well known that in very large pipes where the sewage has time to become thoroughly anaerobic, such as in long force mains, a substantial amount of sulfide can be generated in the stream as compared to the amount on the walls. If all the generation were in the stream, then the sulfide concentration discharging from the pipe would be proportional to the detention time but independent of the pipe diameter. This is by no means the case. Pomeroy found[23,31] that a small pipe produces more sulfide in the same time than does a large pipe. A pipe 0.34" in diameter produced 14 mg/l of sulfide per hour; a 96" and 144" complex produced 0.75 mg/l per hour. Pomeroy estimates that in a 12" pipe about 11 percent of the sulfide is produced in the stream, while in a 48" pipe

CHLORINATION OF WASTEWATER 411

about 54 percent is produced in the stream. These figures are only an approximation. Pomeroy developed a mathematical relationship so that the total sulfide production in force mains can be calculated.

Forecasting Sulfide Buildup. The calculation of S_s, which is the weight of sulfide produced in the sewage stream, assuming that the amount is proportional to the volume of the pipe and the effective BOD, C_{EBOD}, can be written:[31]

$$S_s = N \frac{\pi D^2 L}{4} C_{EBOD} \qquad (7\text{-}15)$$

where:

N = a coefficient, lb/day/cu ft/mg/l
D = pipe diameter in ft
L = length of filled pipe in ft

The calculation of S_w, the weight of sulfide produced on the wall, lb/day, may be written[31] for circular pipes:

$$S_w = M \pi DL \, C_{EBOD} \qquad (7\text{-}16)$$

where:

M = coefficient, lb/day/sq ft/mg/l

Adding Eqs. (7-15) and (7-16), the total sulfide production S_T may be expressed:

$$S_T = C_{EBOD} \, \pi \, DL \left(M + \frac{ND}{4} \right) \qquad (7\text{-}17)$$

Substituting P for $N/4M$, Eq. (7-17) becomes

$$S_T = M C_{EBOD} \, \pi \, DL \, (1 + PD) \qquad (7\text{-}18)$$

For calculating the increase in concentration of sulfide in sewage going through a force main, Pomeroy uses the following equation:

$$\Delta C_s = \frac{44.6 \, M \, t \, C_{EBOD}}{D} (1 + PD) \qquad (7\text{-}19)$$

Converting pipe diameters to inches, Eq. (7-19) becomes:

$$\Delta C_s = \frac{535\, M\, t\, C_{\text{EBOD}}}{d}\left(1 + \frac{Pd}{12}\right) \qquad (7\text{-}20)$$

Where C_s is the concentration of sulfide in mg/l and t is the time of passage in minutes. Based on data collected by Pomeroy, it appears that the value of P should be about 0.12. This corresponds to production of about 0.4 mg/l of sulfide per hour in the body of the stream. Substituting this value of P and using K in place of 535 M, Eq. (7-20) becomes:

$$\Delta C_s = Kt\, C_{\text{EBOD}}\, \frac{(1 + 0.01\, d)}{d} \qquad (7\text{-}21)$$

From available field data, Pomeroy concluded that up to a time of ten minutes very little sulfide is likely to be produced and that K would be on the order of 0.0010 in Eq. (7-21). Between ten and sixty minutes, K would be on the order of 0.0020. For detention times of one to five hours, K values would be above 0.0020. Pomeroy suggests $K = 0.0026$, which gives a value of M as 4.85×10^{-6} or say 5×10^{-6}. This he believes to be a conservative figure which will predict more sulfide than will probably be generated.

Chlorine for Odor Control

Historical Background. The widespread utilization of chlorine for odor control has been well documented.[13,22,24,32-39] The task has been accomplished both by up-sewer chlorination and by prechlorination. Probably the most extensive work on up-sewer chlorination has been done by the sanitation departments of Los Angeles and Orange counties in California, where it began in 1930 and 1931.[23]

These two areas, based upon their experiences of the past forty years, consider it far better to design a collection system which will avoid undue buildup of sulfides in the sewage rather than resort to chemical treatment.

Up-sewer chlorination is usually practical only at pumping stations where space can be provided for a safe facility. But these pumping stations are usually adjacent to residential areas, thereby presenting a hazard. Considerably less hazardous is the use of copperas for controlling sulfides. Copperas will, however, not control sulfides to less than about 1 mg/l, whereas complete destruction by chlorine is possible. Two possibilities are: purging the sewer walls of slimes by raising the pH with slugs of caustic, and injecting of air into force mains. If the force main is not designed to handle air injection, this method may create prohibitive pressures from air accumulation. The disadvantages of chlorine is that while it will destroy sulfides, converting them to colloidal sulfur and sulfates, reasonable dosages may not prevent re-formation of sulfides before the sewage reaches the treatment works.[38] These instances indicate the necessity of relay chlorination, such as that practiced at various times by Los Angeles County.

Chemistry. In dilute aqueous solutions, chlorine reacts instantaneously with sulfides to form colloidal sulfur, Eq. (7-22), or sulfates, Eq. (7-23), depending upon the pH, temperature, and ratio of chlorine to sulfides present.[39,42]

$$Cl_2 + H_2O \longrightarrow HOCl + H_2S \longrightarrow S°\downarrow + HCl + H_2O \qquad (7\text{-}22)$$

This reaction requires 2.2 ppm chlorine per ppm of sulfide as hydrogen sulfide. A high pH is required for this reaction to go to completion. At pH 10, the amount oxidized to free sulfur is only slightly more than 50 percent.[42]

The addition of 8.87 ppm chlorine to each part of sulfide theoretically will oxidize the sulfides to sulfates:

$$4H_2O + S^= + 4\ Cl_2 \longrightarrow SO_4^= + 8HCl \qquad (7\text{-}23)$$

Nagano[43] was able to show by a series of tests made in 1950 that the chlorine-sulfide reaction most consistent with experimental evidence is as shown in Eq. (7-23)—that is, chlorine oxidizes sulfur to sulfates in the ratio of 8.87:1. Actually the chemistry of this reaction is so complex that, depending upon the pH and temperature of the sewage, some free sulfur and polysulfides are probably formed along with the sulfates. Given sufficient time in an oxidizing environment, the polysulfides will become sulfates, but in a reducing environment they can easily revert to hydrogen sulfide.

Role of Chlorine. The ability of chlorine to control and eliminate odors cannot be over emphasized. It is the most universal and practical answer to these perennial problems. It controls by:

1. Destruction of bacteria that contribute to the conversion of sulfates into H_2S and organics into CO_2
2. Prevention or limitation of slime layer growth in sewers—the most critical factor in H_2S production
3. Instantly destroying any H_2S present at the point of application
4. Maintaining sewage in a "fresh" condition, which prevents septic action that leads to H_2S generation and other problems associated with septicity

This last is probably the most important attribute of chlorine. Other methods such as air or oxygen injection, hydrogen peroxide, and/or potassium permanganate cannot provide this capability.

Chlorine Requirement. If the wastewater to be treated for hydrogen sulfide control is available, the practical way to determine the optimum chlorine dosage is the chlorine demand test. Use the procedure given in *Standard Methods* and determine the chlorine consumed during a 5-min contact period. The correct dosage will yield a measurable residual at the end of 5 min, which is sufficient time for all the chlorine–sulfide reactions to go to completion.

Prechlorination Facility. Every wastewater treatment plant should have provisions for prechlorination. This provision can be used as a standby system for disinfection, and/or as a supplement for RAS chlorination. In the absence of chlorine demand tests, the equipment capacity should be at least 15 mg/l for peak dry weather flow, assuming a fresh domestic sewage. If the wastewater is to be a mixture of industrial wastes and septic domestic sewage, the optimum chlorine requirement might escalate to 30 or 40 mg/l.

It is not practical to control the chlorine dosage by the chlorine residual method. A chlorine residual is not necessary to control hydrogen sulfide generation. Moreover a chlorine residual analyzer cannot perform on raw sewage.

The ORP method (oxidation–reduction potential) has been tried at many different plants with little success. It requires at least one year of experimentation with the system to determine the ORP control set point. This is a trial and error procedure. These systems have all been abandoned due to the magnitude of operator time required to keep the system on track. This coupled with the difficulty of maintaining clean electrode surfaces makes it an undesirable system.

The preferred method with a good record is based upon the control of chlorination dosage which will prevent generation of H_2S in an aerated sample of chlorinated wastewater. The H_2S that breaks out of an aerated sample of sewage is directly proportional to the dissolved sulfides in the sewage. Proper chlorination control using this air sampling technique is readily accomplished provided that the sample is aerated properly.

One example of adequate aeration is found in a typical sewage sampling device. The wastewater is pumped at a rate of 2–3 gpm to the sampling device where it discharges to waste over a weir. The air suction cone of the H_2S detector is suspended over the weir so that any H_2S being aerated out of the sample flow will be trapped by the detector. One of the most successful installations of this kind utilized an automatic proportional sampler of the raw sewage being prechlorinated for odor control.[44] Chlorination control was maintained so that the *dissolved sulfides* were always slightly less than 0.3 mg/l, which has been established as the critical limit. The air entering the cone of the H_2S detector suction line is pumped by the detector through the analyzer where it comes into contact with a lead acetate saturated tape. The H_2S in the air sample produces a black spot with a density proportional to the amount of H_2S in the atmosphere. The density of the spot is converted into an electric signal fed into a controller. The controller operates the chlorine metering orifice in accordance with a field-observed empirical set point. This system allows the operator to adjust the set point to meter the chlorine in accordance with some acceptable level of H_2S concentration in the atmosphere adjacent to the wastewater surface.

The chlorine control system must be equipped with a bias feature which limits the minimum dosage of chlorine to approximately 8 mg/l. If this is not provided there could be times when H_2S is not being generated and the control system would eventually reduce the chlorine feed rate to near zero. Then when the sudden appearance of H_2S occurred the control system would be unable to react in time to prevent H_2S breakout in the treatment plant area. There are several ways to achieve this minimum dosage control.

It is wise to use two H_2S detectors: one to control the chlorine dosage and the other to monitor plant area odors. The latter analyzer should be a portable type. There are a variety of H_2S analyzer-detectors available for this type of system such as: Research Appliance Corp. (RAC), Houston-Atlas Co., Texas Analytical Instruments and others. Wet chemistry type of analyzers are not recommended.

The control analyzer should be fitted with a 0–3 ppm H_2S range that can be converted to 0–10 ppm in the field. The portable monitor analyzer should have the same capability. Both units should have built-in H_2S concentration recorders. This historical record is of vital importance to the operator.

Local conditions will determine the necessity of a flow pacing signal in addition to the H_2S signal. A flow signal is highly desirable and should be considered. It is well worth the expense even if the flow meter range is limited to 3 to 1. The flow must be measured near to the plant influent. Practically all modern influent pumping stations are designed to use variable speed pumps which follow the diurual flow change of the collection system. However, if a step-rate pump system controls the influent flow then a step-rate control system is an imperative for the chlorine feed rate control—in addition to the H_2S signal.

Automatic chlorine dosing for odor control using an H_2S analyzer to detect the odor potential of the sewage is the surest way to achieve successful control. The analyzer eliminates the need to rely upon human response as a detection device. It is well known that continuous exposure to H_2S in low concentrations will in a brief span of time paralyze the olfactory nerve, completely destroying the human response.

Other Uses of Chlorine in Wastewater

Preventing Septicity:

General Discussion. One of the generally overlooked advantages of prechlorination is the ability of chlorine to prevent septic action by maintaining the waste in a fresh condition, in which state the various types of biologic treatment processes are most efficient. One important result of maintaining a fresh waste is improved clarification. The application of chlorine ahead of the primary clarifier not only improves the settling rate of normal sewage but eliminates septicity in settled sludge, thereby preventing it from rising in the clarifier. This has been reported by many plant operators.[41] One of the most dramatic cases of improved settling occurred when a small domestic waste treatment plant was nearly put out of operation by cannery waste,[45] which was about 40 percent of the total flow, consisting of peaches, pears, and tomatoes. The mixture of the cannery waste and domestic sewage became septic during the time it took to pass through the primary clarifier. Laboratory tests showed that either waste alone would not become septic.

The result of this septicity was a thick, floating scum on the clarifier, pigpen odors, and suspended solids removal of only 36 percent. It was found that the addition of chlorine prevented the septic action which floated the solids to the surface of the clarifier. It was further found that the magnitude of chlorine dosage controlled the length of time the solids would remain settled. By laboratory experiment with Imhoff cones, a settling period of 2 hours was considered adequate. This amounted to a chlorine dose of about 80 percent of the immediate chlorine demand. In practice, this proved sufficient. After the application of chlorine, the thick scum disappeared overnight, the pigpen odor was eliminated, and the suspended solids removal rose to 64 percent.

The comment has often been heard that prechlorination interferes with digester performance, but there is no scientific evidence to support this. From a chemical standpoint, it is highly unlikely because the chlorine would be converted to chloride long before it got to the digester, and the resulting chloride concentration would be so low as to be negligible.

The use of chlorine to prevent and/or control the septicity of sewage has applications other than the extreme case described above. Precisely how much chlorine is required to prevent septicity over a given time span can be determined by using the methylene blue stability test.[46] This can be used as described above for less severe cases of septicity in sedimentation tanks as well as long periphery trunk sewers carrying treated effluents from several plants to a common discharge point.

The problem of septicity is closely allied to the generation of hydrogen sulfide. However septic conditions can occur in situations where hydrogen sulfide formation is not a factor. In the case where solids in a sedimentation basin float to the surface, this is caused by the formation of methane gas in the bottom of the sedimentation tank due to anaerobic conditions. Chlorine added to relieve this type of condition becomes a question of how much is sufficient.

Methylene blue becomes decolorized in the presence of anaerobic conditions caused by reducing bacteria, loss of oxygen, or the formation of hydrogen sulfide. Thus, so long as the blue color persists in a sample containing methylene blue, no septicity or sulfide odor will exist.

Methylene Blue Stability Test. The first step is to determine the 15-min chlorine demand of the sewage to be examined for septicity. This will be a guide to find the optimum dosage for a given situation. If we assume this chlorine consumption in 15 min is 14 mg/l, then it would be appropriate to dose four samples at 3, 5, 10, and 15 mg/l, respectively, provided that preventing septicity for at least 3 hours is required.

Collect sewage samples in 250 ml glass-stoppered bottles so that no air space is left between the liquid and the stopper. Using the same chlorine solution as used in the chlorine demand test, dose the four 250-ml samples as described above. After a 5 sec vigorous mix of the chlorine solution, apply the stopper as directed above and store in the dark at 25–30°C if possible. Since a clean bottle does not resemble the biological conditions in a sewage system, this elevated temperature is an attempt to approach actual conditions.

At the end of one hour add 0.5 ml of a 0.10 percent solution of methylene blue to each 250-ml sample. Be sure to have an equivalent control sample without chlorine added. Observe these samples over a 6–8 hour period and record the status of the blue color in each of the samples as compared to the control. Do this at hourly intervals. If septic conditions develop, the MB color will fade. The optimum chlorine dose will be somewhere between the faded and the nonfaded sample. The time involved in these occurrences is the probable length of time that the "optimum" chlorine dose will keep the sewage fresh and thereby prevent septicity. When the MB color fades that is the onset of septicity. Chlorine can be applied to prevent septicity in raw sewage up to at least 24 hours.[47]

SLUDGE BULKING

Nature of the Problem. The consequences of sludge bulking in the activated sludge process can be disastrous. This phenomenon is responsible for many of the upsets that occur in the activated sludge process and can seriously impair the overall efficiency of the entire plant. Effluents discharged during serious bulking conditions have been known to cause pollution in the receiving waters worse than if the plant had been completely bypassed. Therefore sludge bulking must be prevented at all costs.

Chlorine has been used for at least forty years for control of sludge bulking problems.[41,48] It is a reliable and economical method.

Activated sludge contains large populations of several types of microorganisms. There are two important morphological groups; (1) the floc-formers and (2) the filamentous sludge bulkers.

The floc-forming group of bacteria include the following genera: *Zoogloea, Pseudomonas, Arthrobacter,* and *Alcaligenes.*

The filamentous group (the causative organism of sludge bulking) includes members of the genera *Sphaerotilus, Thiothrix, Microthrix, Parvicella, Beggiatoa,* and others.*

Satisfactory operation of the activated sludge process depends upon controlling the level of the filamentous organism population.

Identifying the Problem.[69,93] There are two tests to perform in order to identify sludge bulking. These are: (1) the measurement of the sludge volume index (SVI), and (2) daily microscopic observations of the RAS organisms.

---
* For illustrations of some of these filamentous organisms, see Chapter 6.

418 HANDBOOK OF CHLORINATION

The most common causes of sludge bulking (poor settling) are: (1) low dissolved oxygen, (2) low food to microorganism ratio (F/M), and (3) nutrient deficiency (N or P or both). However, bulking is a physical phenomenon caused by filamentous bacteria.

Sludge Volume Index. This is the volume in milliliters occupied by one gram of activated sludge after the aerated liquor has settled for 30 minutes. (See p. 130, *Standard Methods,* 15th edition.) The operating personnel should monitor both the D.O. in the aeration tanks and the depth of sludge in these tanks, then calculate the sludge volume index (SVI) of the mixed liquor. When the SVI reaches 150 sludge bulking may begin. Sludges with good settling characteristics in diffused-air aeration systems operating with mixed-liquor suspended-solids concentrations of 800–3500 mg/l will show an SVI range from 35 to 150. Therefore with a fixed and limited capacity return-sludge pump (RAS recirculating pump), the sludge concentration that can be maintained in the mixed liquor (without escaping in the effluent) is reduced as the SVI increases. However, keeping the mixed liquor suspended solids low at high SVI is not important. What is important is the *clarifier*

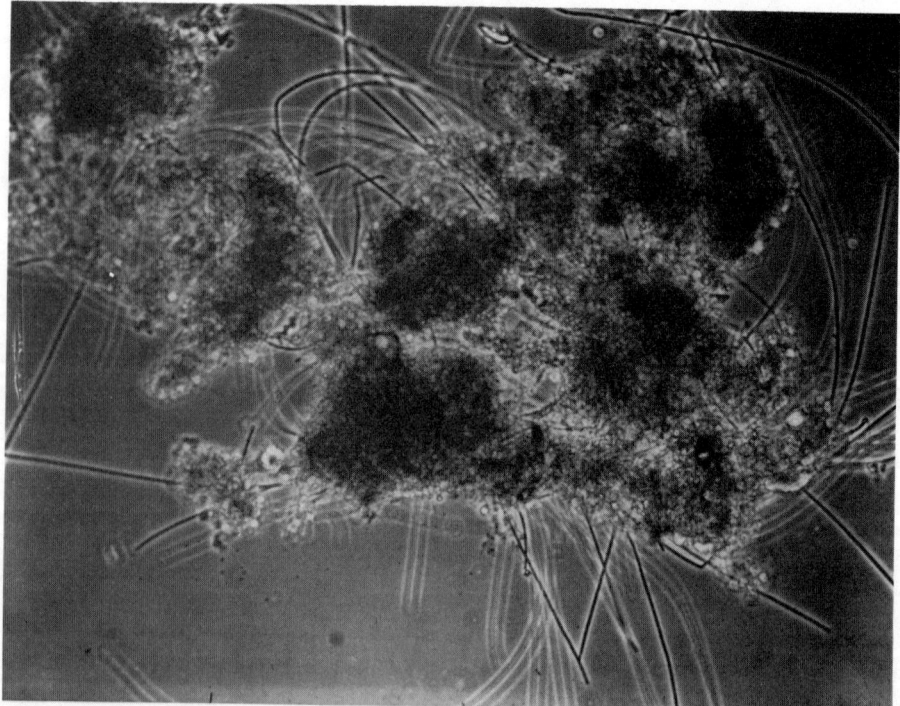

Fig. 7-3 Activated sludge during bulking conditions resulting from healthy filamentous thiothrix (100 × mag.).

flux, i.e., the total pounds per day of suspended solids per square foot of settling area in the clarifier. For example, doubling the MLTSS from 1100 to 2200 with an SVI of 200 would give the same performance if the surface area of the clarifier were also doubled. For acceptable clarifier operation the maximum solids flux goes down as the SVI goes up. Therefore a rising value of SVI is indicative of trouble ahead and prompt action should be taken to bring it under control. The first step in the control program is to set a target SVI. Consensus indicates it should be from 60 to 110, providing the wastewater flow compares to the plant design flow.

Microscopic Observation. Satisfactory operation of the activated sludge process depends upon the population control of the filamentous organisms in the sludge. These organisms interfere with the growth and survival of the floc-formers upon which the process depends. Fortunately the filamentous organisms are more sensitive to chlorine than are the floc-formers and are therefore destroyed first. The floc-formers grow inside the floc, where they are protected from the chlorine by a diffusion barrier. Moreover, chemical reactions between chlorine and compounds inside the floc restrict the penetration of chlorine to the outside shell of the floc.

Figure 7-3 shows a typical healthy bloom of *Thiothrix* in the activated sludge floc. The important element of the photograph is the bridging of the filaments between the floc particles. This is without chlorination showing the onset of bulking.

Figure 7-4 shows the same activated sludge sample during chlorination. There are fewer filaments and the bridging has disappeared. The empty sheaths and the missing cells are the effects of chlorination.

Figure 7-5 illustrates the final phase of bulking control by chlorination. At this point chlorination should be discontinued.

Research on this subject at the San Jose/Santa Clara Water Pollution Control Plant, San Jose, California revealed that besides using a target SVI to regulate the chlorine dose to the RAS, observations of the number and condition of filamentous organisms extending from the activated sludge floc were most helpful.[69] The microbiology laboratory staff at the above plant used a simple but effective filament counting technique, as follows:[69,94]

1. Transfer 50 microliters mixed liquor sample to a glass slide.
2. Cover completely with a 22- by 30-mm cover slip.
3. Using 100× total magnification and starting at the edge of the cover slip, observe consecutive fields across the entire length of the cover slip. (At San Jose, this is 17 fields).
4. The eyepiece is fitted with a single hairline. Count the number of times that any filamentous organism intersects with the hairline.
5. Sum the number of intersections for all fields examined. This is the filament count.

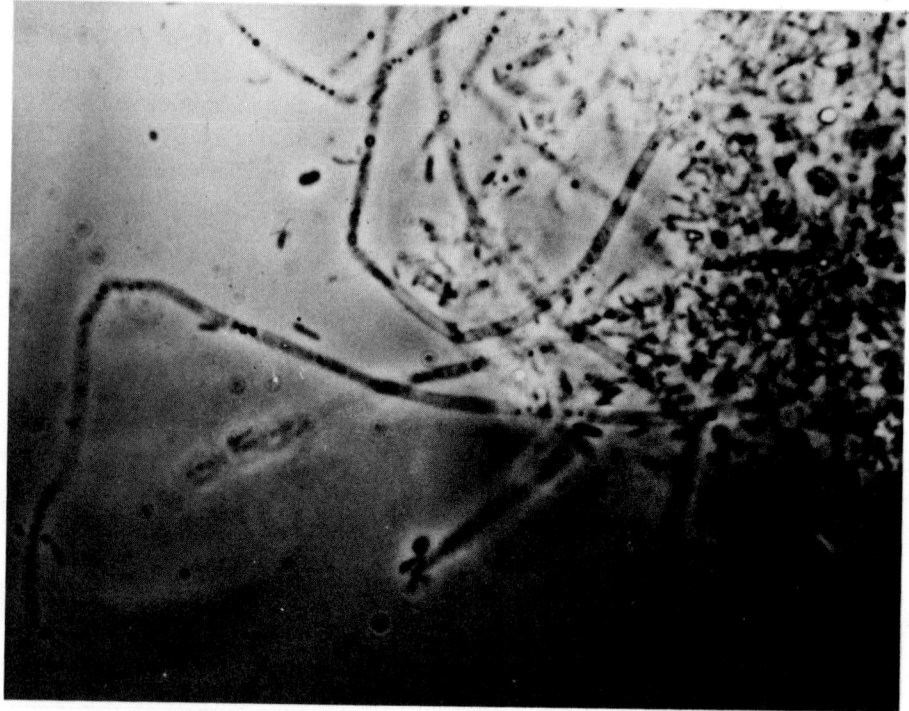

Fig. 7-4 Activated sludge bulking during chlorination. The thiothrix filaments are devoid of sulfur and the cells are deformed (100 × mag.).

Daily observations of the organisms in the activated sludge serves as a basis to check the level of the filamentous organisms. This provides the necessary information required to adjust the chlorine dosage or frequency of RAS passes by the point of chlorination.

Increases in extended filament length do not necessarily precede increases in SVI, nor does a decrease in filament length precede a decrease in SVI. Therefore microscopic enumeration of filamentous bacteria does not allow prediction of changes in the settling characteristics of the sludge. However, microscopic observation of the chlorinated RAS does show the onset of the effectiveness of chlorination, which in turn establishes when to reduce the chlorine dose. The progressive effects of chlorine are to cause cells inside the filament sheaths to become deformed, to pull away from the sheath, to lyse and create empty spaces in the sheaths. Finally, the sheaths become empty and broken. When the filamentous organism is *Thiothrix*, the early effect of chlorine is the disappearance of sulfur inclusions in the sheaths. When dealing with sheathed bacterium such as *Thiothrix*, sludge settleability may be influenced by the empty sheaths even after the organism is no longer viable. Therefore the reduction in SVI may continue for 2 or 3 days after chlorination

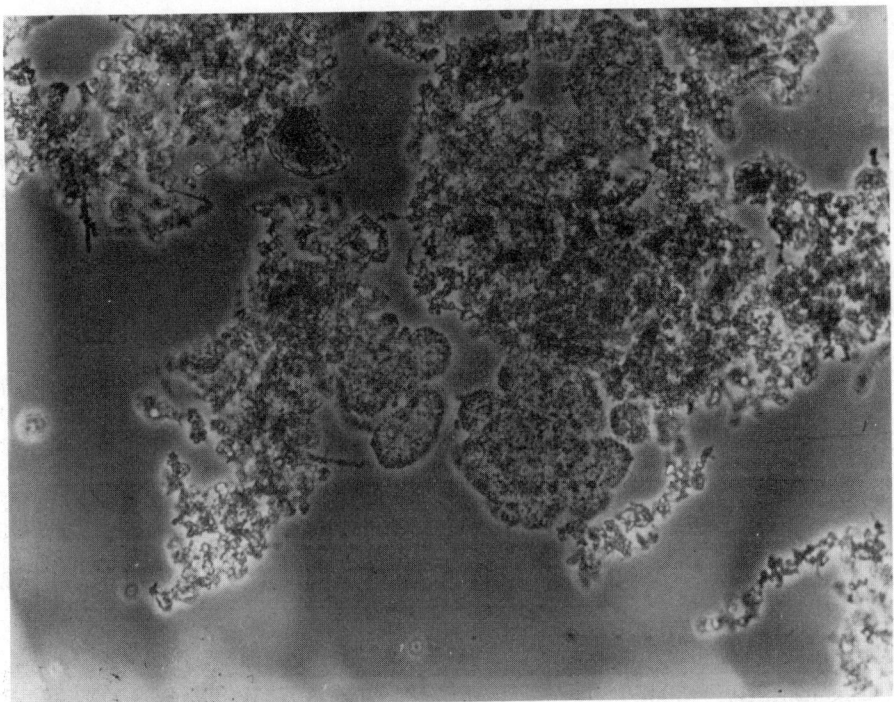

Fig. 7-5 Normal non-bulking activated sludge, showing floc-formers and absence of filamentous organisms (100 × mag.).

has ceased because the sheaths continue to break down and are wasted from the sludge inventory.

Nitrogen Deficiency. There are times when some plants are subjected to heavy seasonal cannery loads during which time ammonia-N is completely consumed by cell synthesis and none remains to produce chloramines. During such an event nitrites are likely to appear in concentrations from 1 to 3 mg/l. Since there is no ammonia-N to form chloramines all of the chlorine will exist as HOCl (free chlorine). Free chlorine reacts immediately with the nitrites to oxidize them to nitrates. Chlorine is consumed in this reaction at a rate of 5 parts chlorine to 1 part NO_2^--N (nitrite ion). Therefore, if RAS chlorination proves ineffective in the presence of good initial mixing it would be prudent to check for both ammonia-N and nitrite in the RAS. (Chloramines do not react with nitrites).

Another situation may occur during these periods of nitrogen deficiency. It is possible that both NH_3-N and NO_2^--N may be absent and RAS chlorination is ineffective. This would most likely be due to the greater reactivity of free chlorine than chloramines. This results in higher chlorine consumption, which must be compensated by increasing the dose. Therefore, during these periods of nitrogen deficiency it may be necessary to add ammonia-N to the RAS system. Either aqueous ammonia (NH_4OH) or anhydrous ammonia (NH_3) may be used.

Control of Sludge Bulking by Chlorination

Historical Background. In 1945 Tapleshay[49,50] began investigating the use of chlorine for the prevention and control of sludge bulking. He theorized that chlorine restores the balanced load between the organic matter and the biological life in the sludge. This investigation of the sludge bulking phenomenon encompassed fifty treatment plants in the United States. He found that the best method of control was when the chlorine dose was correlated to the sludge index (SI = amt. of dry solids in the return sludge) expressed as follows:

$$\text{lb/day } Cl_2 + SI \times F \times W \times 0.0000834 \qquad (7\text{-}24)$$

where:

SI = sludge index (Mohlman)
F = return sludge rate in mgd
W = suspended solids in return sludge in ppm

Chamberlin[41] found in his studies that the usual chlorine dose required to control bulking averaged about 5 mg/l based upon the return sludge flow. This application of chlorine usually resulted in a turbid effluent for the first twelve hours as the reaction began to stabilize. This situation caused some who had tried chlorination to discontinue it and called it a failure. If chlorination had been continued in these cases the operator would have found the turbidity gradually decreased along with the sludge volume index. Usually a decided improvement was noted in three or four days. In these former years some relied solely on intermittent chlorination while others practiced continuous chlorination. However, current knowledge of the problem has provided a more reliable program for control.

Chlorination System:

Operating Plant Example. The following calculations are based upon field data operating in the conventional mode (see Fig. 7-6).

Average daily flow = 10 mgd
Aeration detention time = V/Q = 6 hours
Hydraulic detention time, where R = RAS return rate, mgd = $\dfrac{V}{Q+R}$ = 4.4 hours
Aeration volume (including mixed liquor channel) = 2.5 million gallons

Clarifier area	= 12,500 ft²
Clarifier volume	= 1.3 million gallons
Clarifier overflow rate gpd/ft²	= 800 av.
	= 1050 peak
Mixed liquor total suspended solids (MLTSS)	= 1375 mg/l
Mixed liquor volume suspended solids (MLVSS)	
(MLVSS is about 75–80% of MLTSS)	= 1030 mg/l
Average suspended solids in clarifier (TSS)	= 2000 mg/l
Volume suspended solids in clarifier (VSS)	= 1500 mg/l
Return activated sludge suspended solids (RASTSS)	= 5300 mg/l
RAS return rate = 35 percent of total plant flow	= 3.5 mgd

NOTE: During normal conditions the clarifier is the repository of about 20 percent of the total suspended solids in the activated sludge system. However, during bulking conditions this percentage can escalate to 50 percent.

The food to microorganism ratio is expressed as F/M = 0.30. The food is the lb BOD applied (primary effluent = 135 mg/l) and the microorganisms are the total system volatile suspended solids (TSVSS) including aerators, clarifiers, and mixed-liquor channel.

The mean cell residence time (MCRT) is 6.3 days and the sludge yield is 0.7 lb VSS per lb BOD removed. Therefore the MCRT is related to

$$F/M = \frac{\text{lb BOD applied}}{\text{total system volatile inventory}}$$

The proper RAS rate can be estimated from the following relationship:

$$\frac{\text{RAS flow}}{\text{Primary eff. flow}} = \frac{\text{MLTSS}}{\text{RASTSS} - \text{MLTSS}} \tag{7-25}$$

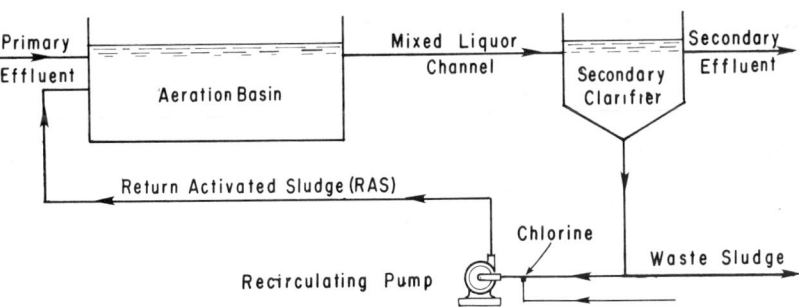

Fig. 7-6 Return activated sludge (RAS) system showing point of chlorination and relevant structures.

424 HANDBOOK OF CHLORINATION

It is best to use a target value for the RASTSS in the above equation, otherwise as the RAS thins out, Eq. (7-25) will call for a higher RAS flow. The objective is to thicken the RAS—more organisms per unit volume. Reducing the RAS flow increases the detention time in the aeration basins and as the RAS flow is reduced the sludge concentration tends to increase due to greater compaction and more settling time in the clarifier.

Substituting the figures from the example above into Eq. (7-25) we get:

$$\frac{\text{RAS flow}}{\text{Primary eff. flow}} = \frac{1375}{5300-1375} = 0.35$$

Therefore RAS flow = 0.35 × 10 = 3.5 mgd.

NOTE: Clarifier suspended solids should be sampled—a "sludge judge" tube may be used—since chlorination is normally practiced under bulking conditions when an abnormally large amount of sludge inventory may be in the clarifiers due to poor settling; i.e., do not assume it is the same as the mixed liquor concentration.

Initial Chlorine Dose. It is recommended to start at a low dose of 4 lb chlorine/day/1000 lb VSS. The VSS is total inventory in the entire system, i.e., aeration basins, clarifiers, and mixed liquor channel. If more chlorine appears to be needed it is prudent to make conservative increases—about 10–20 percent per day.

Sample calculation for chlorine feed rate:
VSS' in aeration basins:

$$\text{MLVSS mg/l} \times 8.34 \times \text{Volume (million gallons)} = \text{MLVSS lb}$$
$$1030 \times 8.34 \times 2.5 = 21,475$$

MLVSS in clarifiers:

$$1500 \times 8.34 \times 1.3 = \underline{16,260}$$
$$\text{Total inventory} \quad 37,735 \text{ lb}$$

So daily feed rate is

$$\frac{37,735}{1000} \times 4 \text{ lb Cl}_2 = 150 \text{ lb/day}$$

The next step is to check the number of exposures of the RAS to the chlorine applied. The minimum number is considered to be 3.

Total sludge returned per day = RAS Volume × 8.34
$$\times \text{RASTSS} \times \text{Volatility} \quad (7\text{-}26)$$

Therefore

Total sludge lb/day = 3.5 mg. × 8.34 lb/mg × 5300 mg/l × 0.75
RASVV = 116000 lb/day

Total inventory = 37,735

$$\frac{116030}{37735} = 3.05 \text{ times per day}$$

The above calculation can be done using TSS if they are used for the RAS and inventory calculations, however, the volatile inventory has already been calculated.

Since the exposure frequency is marginal (3 times/day) it is suggested that if the effects of chlorination are not observed within 48 hours in this example, and providing other parameters are in order the sludge return rate should be increased. This can be done provided the clarifier solids flux loading will allow it.

Clarifier solids flux check: RAS rate (lb/day) must be RASTSS, so

$$\text{RASTSS} = \frac{\text{RASVSS}}{\% \text{ Volatile}} = \frac{116000}{0.75} = 154,700 \text{ lb}$$

Total clarifier area = 12500 ft². Then

$$\text{Clarifier flux} = \frac{154,700}{12,500} = 12.4$$

A safe flux range is 20–25 lb/ft²/day. Therefore the RAS flow in the example above could be safely increased 40–45 percent. In this event the RASTSS concentration will probably decrease slightly. The flux should be rechecked when this reduction in RAS concentration is known.

Final chlorine dosage check at point of injection:

$$\frac{150 \text{ lb/day}}{Q_{RAS} \times 8.34} = \frac{150}{3.5 \times 8.34} = 5.1 \text{ mg/l}$$

This dosage can and does require 6–7 mg/l.

The chlorine must be applied to the sludge at a point where there is effective initial mixing. The suction side of a centrifugal RAS pump is ideal.*

This will provide the necessary initial mixing, which is most important. It assures that the filaments in the total sludge mass will be inhibited, rather than small amounts of the sludge being fully oxidized in passing clumps.

---

* The pump should not be constructed with brass or bronze parts due to possible chlorine corrosion. Fiberglass or all iron pumps are preferred. Moreover, chlorine application should be arranged so that it cannot be applied while a pump is out of service.

Chlorine should be applied continuously until the target SVI is achieved. Intermittent chlorination allows filaments to pass through undamaged and be able to proliferate in those parts of the aeration basin where they have an environmental advantage.[69,93]

Warning of Initial Side Effects. When chlorination of the RAS is first initiated to correct a bulking problem, temporary deterioration in the effluent quality will occur. Therefore the operators of single-stage systems can expect the turbidity to double, the suspended solids to increase by 20 percent, and the BOD by not more than 10–15 percent.

If, however, after RAS chlorination has been initiated a marked increase in BOD occurs in the effluent (40 percent or more), this is a warning signal that chlorine is inhibiting the biological processes and the dose must be reduced.

Two-Stage Sludge Systems. In these systems, such as the San Jose/Santa Clara plant,[69] using high chlorine doses in the first-stage activated sludge system led to a deterioration in the first-stage effluent quality. It allowed this system to operate at high loadings and to achieve adequate removals of the major portion of the organic load to the biological treatment system. Thus the best use of the first stage was to achieve adequate overall removals of organics at the price of some deterioration in effluent quality, while the second stage functioned to polish the effluent before it entered the tertiary filters.

SLUDGE TRANSPORT

Since 1976, Environmental Protection Agency permits have required phase-out programs for ocean disposal of sludge. All ocean dumping of sludge was to stop by 1981. Nearly all of the cities have long since established timetables to cease dumping sludge into rivers, lakes, and ocean seabeds to meet the 1981 deadline. The responses to these actions have caused a variety of changes in the handling of sludge. Of particular interest is the movement of sludge from point of origin to a terminus for further treatment or disposal.

If during the transport operation the sludge becomes septic, hydrogen sulfide will be generated in amounts sufficient to cause intolerable odor and corrosion problems. Pumping sludge in a force main will cause serious problems unless high concentrations of chlorine are applied to prevent septicity. The dose of chlorine will be high enough to cause corrosion of the sludge pumps. Dosages to prevent septicity are usually on the order of 50–125 mg/l.

Work performed by the Sanitation Districts of Los Angeles suggests a different approach. This relatively new discovery is to transport the sludge as a mixture of sludge and raw or treated sewage—whichever is more convenient.[51]

To better understand the phenomenon of sulfide generation in the case of sludge transport, the importance of the Pomeroy i factor must be explained. This is an important variable which concerns rate of H_2S buildup in a sewage collection system:

$$i = \frac{\Delta S_{ox}}{R_R + R_{SO}} \tag{7-27}$$

where

ΔS_{ox} = oxidation rate of sulfides (sulfides removed), mg/hr/l
R_R = rate of D.O. reaction with impurities in process stream, or oxygen removed if no sulfides are present, mg/hr/l
R_{SO} = oxygen uptake rate, mg/hr/l

Determination of the i factor or oxidation capacity of the sewage stream is as follows: Agitate a sample of the sewage in question so that the D.O. reaches about 6 mg/l. To this sample add sulfides to determine the sulfide oxidation capacity. Assume that the total of sulfides removed is 4 mg/hr/l while the total loss of D.O. is 12 mg/hr/l. The i factor is $4/12 = 0.33$. During this test the D.O. and dissolved sulfides are maintained above 0.3 mg/l.

When Los Angeles County sulfide lab ran tests to determine the oxidation capacity of the raw sewage versus raw sewage plus raw sludge they found the i factor went from 0.26 in the raw sewage to 0.43 in the sewage–sludge mixture. This signifies that the oxidation capacity of a sewage stream is increased nearly twofold when raw sludge is added. Therefore whenever sludge is to be transported, the use of a sewage flow to carry the sludge should be investigated.

To determine the feasibility of any sludge transport system the probable condition of the sludge at the terminus must be investigated. Septic conditions and hydrogen sulfide generation must be avoided if at all possible. This can be done conveniently with reasonable chlorine doses. The sludge or sludge and wastewater mixture should be subjected to the methylene blue stability test. (This test is described under the heading "Prevention of Septicity" in this chapter.) Using this test the operator can determine the necessary chlorine dose which will prevent septicity in the flowing system for a given length of time. The time involved should be the length of time required for the flow to travel from the point of origin to the terminus. It is important for the operator to understand that it is *not necessary to carry a chlorine residual to the end of the force main*. The methylene blue test is based upon chlorine dose, not residual.

It is good practice to purge the sludge transport pipeline with caustic on a periodic basis. This will remove biological slimes which might otherwise reduce the oxidation capacity of the system. It will also act to prevent an increase in hydraulic friction head. The recommended dosage is 400 lb 50 percent NaOH per 450 gpm, applied over a 30 min period, about once every 4–6 weeks.

Grease Removal

A great many waste treatment works are troubled by grease. Preaeration, vacuum flotation, and aerochlorination are common methods of removing grease. Mahlie[53] lists the adverse effects of grease as follows: "It blinds screens, accumulates on walls, destroys the paint on steel structures, gives rise to odors, interferes with secondary treatment processes by clogging trickling filters, spray nozzles and sand beds, and interferes with sludge digestion. It also may have an unsightly effect on the receiving waters."

The addition of chlorine for improved grease removal was first reported in 1937 at the Woonsocket, Rhode Island, plant.[54] 2 mg/l of chlorine gas was added to the air used for preaeration for grease removal.

With a contact time of approximately six minutes, increased grease removal over air alone ranged from 189 to 442 percent. About 1938, Keefer and Cromwell[55] tried chlorine dosages from 1 to 10 mg/l with contact times of 5, 10, and 15 min at the Baltimore, Maryland, plant. Chlorine gas was introduced into the diffused air lines. Best results were obtained with a chlorine dose of 5 mg/l and a contact time of five minutes. Increased removals compared to air alone varied from 148 to 800 percent.

These results were so phenomenal that aerochlorination was tried at the two activated sludge plants of Lancaster, Pennsylvania. This treatment is of particular interest because the chlorine was applied as a solution immediately ahead of the air diffusers in the preaeration channel. Chlorine dosages averaged about 2 mg/l with 2–3 min contact. With this treatment, 70–80 percent of the grease was removed, as compared to only 50 percent with air alone. While the data on the use of aerochlorination for grease removal are limited, they do indicate a definite increase in removal over air alone. This is further substantiated by the use of chlorine for grease removal in animal-rendering works. Chlorine appears to break up the emulsions, allowing the grease to float and thus facilitating its collection. Operators have noticed this phenomenon for years. Moreover, the amount of grease recovered is a profitable venture which more than pays for the chlorination system in these rendering works.

Conventional wastewater prechlorination practices have been found to increase the amount of scum removed from the surface of settling tanks. The scum appears to accumulate in a more dense or compact mass, requiring less skimming and resulting in an improved general appearance of the settling basins and clarifiers.

Chlorine for grease removal may be applied as a solution ahead of the primary clarifiers (prechlorination) or ahead of the aeration tanks. It is not necessary to satisfy the chlorine demand of the wastewater. Often as little as 2–5 mg/l will suffice to bring about an improvement in the grease removal.

Aerochlorination seems to be more effective.[54] In this case, the dry chlorine gas is mixed with the diffused air going to the aeration tanks. Conventional vacuum feed solution type chlorination equipment is used; however, air, instead of water, is forced through the injector to operate the equipment. By this method 2–10 mg/l chlorine as chlorine gas and 0.02–0.20 cubic feet of air per gallon of wastewater with a 3–20 min aeration period is the customary practice.

TRICKLING FILTERS AND HIGH-RATE RECIRCULATING BIOFILTERS

Chlorine has proved beneficial in the operation of both trickling filters and high-rate recirculating biofilters. Trickling filters invariably require chlorine for odor control. The worst offenders are those that operate on an intermittent cycle. During the downtime the wastewater is allowed to get septic, giving rise to the formation of odorous gases, which are aerated into the atmosphere through the spray nozzles during the filtering cycle. About 1928 Morris Cohn[40] at Schenectady was one of the first plant superintendents in this country to experiment with the use of chlorine for controlling odors at trickling filters. The chlorine is applied to the dosing tank of the trickling filter unit. Sometimes effective trickling filter odor control can be achieved by the prechlorination point of application, but this should not be depended upon. Proper control is usually obtained by satisfying only a fraction of the immediate chlorine demand (approximately 25 to 30 percent). The chlorine varies from 2 to 6 mg/l.

Chlorinating to less than the chlorine demand is somewhat difficult to control. Effectiveness depends on the number of complaints received; so, rather than risk any complaints, some operators make occasional dissolved oxygen tests at the influent to the dosing tank. The chlorine applied is sufficient if there is some D.O. at this point. If not, the chlorine dose is increased and the secondary effluent is recirculated through the filter until D.O. appears again in the dosing tank.

High-rate recirculating biofilters are generally odor-free, and so chlorine for this purpose is not a consideration. This is a result of continuous operation, which produces a more stable biological environment. However, recirculated filters do suffer from ponding and filter fly nuisance, as does the trickling filter. Chlorine, if used properly, can be effective in controlling these two problems.

Ponding is due to the clogging of the interstices by filamentous growths or excessive accumulation of solids. If the filter is overloaded and the clogging material has penetrated deeply, it may be too late for the chlorine to be of any help. Chlorine is most successful when used in the spring to induce unloading of material accumulated in the filter during the winter. The chlorine attacks the organisms in the zoogleal mass, causing them to loosen so that they can be flushed out of the bed. Removal of this material requires heavy doses of chlorine.

Sufficient chlorine should be applied at the filter influent to produce an amperometric residual of between 4 and 8 mg/l at the nozzles.[55]

Contrary to popular belief, filter ponding on high-rate filters can be prevented and controlled by continuous chlorination to a residual of about 0.5 mg/l.[55] This small amount of chlorine continuously applied seems to create a stable environment which will not allow excessive growths and does not interfere with normal biological processes within the filter.

Some filters never experience ponding, while others experience it as a regular seasonal occurrence in varying degrees. For this reason, operators must find by experience the best way to apply the chlorine. Some find the answers in programming chlorination for a few hours at night one or two days a week. The most difficult to control are the intermittent trickling filters.

The trickling filter and the recirculating filter present ideal places for the propagation of the filter fly, *Psychoda alternata,* which develops rapidly during the warm summer months. Eggs are laid on the surfaces of the zoogleal film in irregular masses containing from 30 to 100 individuals. The eggs hatch into larvae, which penetrate the film, leaving a breathing tube projecting. In this position they feed and are transformed into pupae. When pupation is completed, the shell bursts and the fly emerges. The life cycle may be completed in as little as 12 days. The larvae and pupae are most abundant between 3 and 12 inches below the surface of the filter, although they may be distributed throughout the bed.

There is no entirely effective method for controlling this nuisance. Chlorination has had some success either by limiting the thickness of the zoogleal film or by removing it entirely in the top layer of the filter. The number of larvae and pupae seems to be related to the thickness of the zoogleal film on the filter rock. Best control is achieved in the larvae and pupae stage. Chlorine has no effect on the adult fly. Therefore application of chlorine for the control of *Psychoda* must result in partial removal of the biological growth in the upper layers of filter rock, which is the breeding ground. This should be done only in the fly season. Chlorination is usually programmed in a fashion similar to that of correcting ponding—namely, 2 to 10 ppm O-T residuals at the nozzles to produce some sloughing of the filter rock film.

Chlorination for trickling filter odor control and filter fly control and high rate recirculating filters for ponding control are good operation practices. In many installations the quality of the filter effluent has been improved by this intermediate point of chlorination.

BOD REDUCTION

The ability of chlorine to reduce the BOD permanently is often overlooked in the perception of wastewater treatment. As early as 1859, chlorination of wastewater in the form of chloride of lime was practiced in England for the purpose of delaying putrefaction of the wastewater long enough for the receiving waters to be carried to the open sea. This in effect was chlorination for BOD reduction. It was this application and other applications in the United Kingdom that led to the origin of the five-day BOD parameter. This length of time was chosen because five days happens to be the longest time it takes any major receiving water in the British Isles to flush itself to the nearest open sea.

During periods of critical assimilation capacity* of the receiving waters effluent chlorination has been used successfully to alleviate this situation by its ability to reduce the effluent BOD. This BOD reduction allowed the receiving water to maintain the desired DO levels.

Susag[57] has shown by example how effluent chlorination is more economical than additional treatment facilities to relieve the assimilation capacity of the Mississippi River during critical periods. The Minneapolis–St. Paul Sanitary District treatment plant utilizes the high-rate activated sludge process, which has a predicted BOD removal efficiency of 75 percent. This treatment level was considered adequate for all but about 7 percent of the time (26 days in the year 1980). From this study the cost of effluent chlorination for BOD reduction ranged from 8 to 40 percent of that for additional treatment. Moreover, considering the toxicity limits on chlorine residuals, which requires dechlorination, there is little or no incentive to use chlorine for BOD removal. There are no data on the effect of dechlorination on BOD reduction by chlorination.

BOD reduction by chlorination is by no means confined to effluent chlorination. Chlorination of wastewater for any purpose reduces the BOD. This effect shows most dramatically in prechlorination of a strong sewage (high BOD). Rawn[59] reported a 25 percent BOD reduction as a result of a comprehensive up-sewer chlorination procedure for Los Angeles County Sanitation District.

The full value of this treatment was, however, not only BOD reduction. The sewage arriving at the treatment plant was in a fresh, rather than a septic, state, which had the ultimate effect of trebling the capacity of the plant.

A number of studies of BOD reduction by effluent chlorination have been carried out in the United States,[40,52,57,60,61] The consensus on the effect of chlorination is as follows:

1. For each mg/l of chlorine absorbed there will be a 2 mg/l reduction of BOD.[40,62]
2. The reduction appears to be permanent.
3. The reduction increases with increasing chlorine dose.
4. The unit BOD reduction per mg/l chlorine absorbed is greatest at the dosage producing the least residual, and decreases with increasing chlorine dosage. This is irrespective of total initial BOD.
5. The greatest BOD reduction per pound of chlorine applied up to the first appearance of a chlorine residual occurs in the strongest sewage (highest BOD).
6. The greatest BOD reduction in percent up to the first appearance of a chlorine residual occurs in the most highly treated sewage.

---

* Assimilation capacity of a receiving water is a function of its dilution capacity, D.O. content, and oxygen-consuming organic load of the receiving water itself.

7. Up to approximately 80 percent removals can be expected by chlorination of highly purified sewage (i.e., activated sludge effluents).[52]

The work by Susag indicates that chlorine does not act solely as an oxidant of oxidizable organic matter; if it did so, the percent reduction would remain constant for all incubation periods, which it does not. It is evident that chlorine reacts to reduce and retard BOD, partly by oxidizing organic matter and partly by substitution and addition to unsaturated and saturated compounds to produce compounds that are either inert or resistant to bacterial action.

In summary, the use of chlorine to reduce BOD is not a practical premise for the design of a wastewater treatment system. However, the knowledge that it will reduce BOD is helpful in certain situations that could, for instance, make up-sewer chlorination preferable to some other method of treatment.

NITROGEN REMOVAL

Purpose. Nitrogen in its various forms can deplete dissolved oxygen levels in receiving waters, stimulate aquatic growth, or, above a certain concentration, demonstrate lethal toxicity to aquatic life. Ammonia nitrogen depletes oxygen and produces toxic effects. Toxicity is of special concern and can be eliminated by converting the ammonia-N to N_2 or nitrate.

In 1970 the European Inland Advisory Commission studied the effect of ammonia nitrogen on fresh water fish. A great many European and U.K. surface waters carried high levels of ammonia-N (2–5 mg/l). This study indicated that the allowable limit for unionized ammonia (NH_3-N) (rather than $NH_3 + NH_4^+$—total ammonia) be limited to 0.025 mg/l because this fraction is far more toxic than the ionized fraction (NH_4^+). Unfortunately, it is impossible to directly measure the unionized fraction in the relevant concentration range. NH_3-N probes can only detect as low as 0.03 mg/l. However, unionized ammonia can be *calculated* from easily measured quantities: total ammonia, pH, TDS, or salinity, and temperature. The following expression[58] is used to calculate the unionized ammonia:

$$NH_3\text{-}N = \frac{\text{Total ammonia}}{10^x + 1} \quad (7\text{-}28)$$

where $x = pK_a - pH$.

Example:

$$\begin{aligned}
\text{Total ammonia} &= 3 \text{ mg/l} \\
\text{Temperature} &= 18°C \\
\text{TDS or salinity} &= 15 \text{ g/kg}
\end{aligned}$$

From tables,[63] $pK_a = 9.605$. Then

$$NH_3\text{-}N = \frac{3}{10^{9.605-8.3} + 1}$$

$$= \frac{3}{14.33 + 1} = 0.20$$

The above approach has been incorporated into the discharge requirements by the California Water Quality board with the NH_3-N limit as calculated by Eq. (7-28). Using the limitation of 0.025 mg/l for the receiving water, Eq. (7-28) allows the discharger to calculate the maximum amount of unionized ammonia that could possibly exist in the receiving waters. Using this method there will be no chance of ammonia buildup in the receiving waters.

Current Practices. The EPA "Process Design Manual for Nitrogen Control" (Oct. 1975) describes several methods of nitrogen removal. Air stripping using cooling towers has been tried in many places. The results have always been disappointing. The most reliable method appears to be biological: either fixed growth or suspended growth reactors. These methods produce effluents that range in concentration from an undetectable trace of NH_3-N to about 4 or 5 mg/l depending upon overloading conditions.

Chlorine is used in those cases where the effluent concentration must be limited to about 0.5 mg/l NH_3-N. Occasionally chlorine is used seasonably to assist the biological process.

NITROGEN REMOVAL BY CHLORINE

Introduction. Removal of ammonia nitrogen by chlorine is an extension of the breakpoint reaction. The use of chlorine to remove ammonia nitrogen has been studied on two levels: (1) where chlorine was the sole process, and (2) where chlorine followed a nitrification process. In the first instance ammonia nitrogen concentrations were in the range of 15–25 mg/l, and in the second instance the range was from 0.5 to 2.5 mg/l. The importance of these two ranges of ammonia nitrogen concentration is the contact time requirement. The speed of reaction to complete the breakpoint reaction is a function of the concentration of the reactants, i.e., chlorine and ammonia. The higher the ammonia concentration the faster the reaction, and conversely, the lower the concentration the slower the reaction. Therefore, larger contact chambers are required for the removal of lower concentrations of ammonia-N. These situations are described below.

Rancho Cordova Project. A full-scale demonstration of nitrogen removal by breakpoint chlorination was carried out at the Rancho Cordova secondary treatment plant, Sacramento Co., California, between December 1975 and March 1976.[64,65] During this time the automatic chlorination facility was operated 24 hr. a day 5 days a week. The process flow rates varied from about 0.1 to 1.2 mgd. Influent ammonia nitrogen concentrations were ordinarily in the range of 15 to 25 mg per liter. The breakpoint process succeeded in a constant removal of about 97 percent of ammonia nitrogen.

A number of specific observations and conclusions made as a result of the Rancho Cordova breakpoint chlorination demonstration program are enumerated below.

1. The dosage of chlorine at Rancho Cordova required to reach breakpoint and maintain a controllable free residual in the process stream averaged 10 mg per liter for each 1.0 mg per liter ammonia nitrogen present in the process influent which is the plant effluent.

2. Approximately 70 percent of the breakpoint chlorine dosage was consumed to produce nitrogen gas (N_2) from ammonia (NH_4^+) at pH set points between pH 7 and 8. The oxidation of NH_4^+ to NO_3^- consumed 8 percent to 19 percent of the total chlorine dosed to the system. Overall, about 96 percent of the total chlorine dosage was accounted for in reactions between chlorine and nitrogenous species in specific chemical pathways and free chlorine residual remaining in solution following breakpoint. See Fig. 7-7.

3. Nitrate (NO_3^-) production in breakpoint chlorination was not found to be pH sensitive, with about 1.0 mg per liter NO_3^- (as N) produced from NH_4^+ across a final system pH range of pH 6.5 to 8.5. The production of NO_3^- from NO_2^- was wholly dependent upon influent NO_2^- concentration.

4. Nitrogen trichloride (NCl_3) production was observed to be fairly insensitive to pH across a range of final system pH values from pH 7 to 8. The median value for NCl_3 production was about 0.4 mg/l (as N) when breakpoint effluent was used as the source of chlorine injector water. While the amount of chlorine

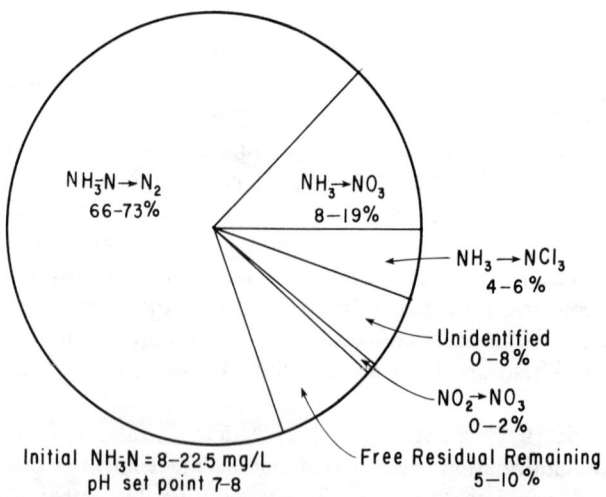

Fig. 7-7 Chlorine consumption during nitrogen removal by breakpoint chlorination.

consumed in the formation of NCl_3 was relatively small (4–6 percent of total dosed), NCl_3 generation affects the minimum ammonia concentration that can be achieved in breakpoint, since its concentration decays slowly in dilute solution and it is converted to ammonia upon the dechlorination with sulfite ion ($SO_3^=$).

5. If the breakpoint process influent is used as injector water, reactions between chlorine and ammonia do occur in the injector water and can consume chlorine in undesirable side reactions. Therefore, injector water should always come from the reacted process effluent. At Rancho Cordova when secondary effluent (process influent) was used as injector water source, the NCl_3 formed in the injector discharge increased the NCl_3 in the process effluent by about 0.2 mg liter.

6. The concentration of organic nitrogen compounds was not affected by breakpoint chlorination.

7. The rate of reaction for breakpoint chlorination was found to vary depending upon the pH control point (final system pH), with fastest rates observed at a set point of pH 7.0. The time to completion was found to be between 60 sec and 90 sec at pH 7.0. The reaction rate slowed considerably at pH set point 6.5, with gradual reductions in rate observed as pH increased from pH 7.3 to pH 8.5.

8. Variations in the amount of mechanical mixing intensity in the zone of breakpoint chemical application had no effect upon overall system chemical consumption and effluent quality. Mechanical mixing, to facilitate a rapid and thorough blending of process chemicals and influent stream, was important in damping free residual oscillations for control purposes.

9. Sodium hydroxide (NaOH) was used throughout the study as an alkalinity supplement. The amount of NaOH required to neutralize all breakpoint acidity (1.53 lb NaOH/lb Cl_2) was essentially identical to that predicted from chemical stoichiometry.

10. The small amount of NCl_3 formed did not present any odor problem, partly because of the closed pipe reactor and partly because of the reliability of the sophisticated control system.

11. Although it was built into the control system, the continuous NH_3-N process influent analyzer signal to the chlorination system was not necessary for the successful operation of the process.

12. The key control parameters, in addition to the plant flow signal, were the chlorine dosage trim signal from the free residual analyzer and the ability of the sodium hydroxide feed system to maintain a set point of the process pH within ± 0.2 pH units.

13. The DPD–FAS titrimetric method for the determination of the chlorine residual species proved most reliable and a time saving method of analysis.

14. The reaction time is so rapid at high NH_3-N concentrations that the nuisance residuals have little chance to form.

Saunier's Research.[66] Saunier made a comprehensive study of the chlorine–ammonia nitrogen chemistry from which he constructed a computer model which predicts reaction times for various concentrations of ammonia nitrogen. This work is discussed in detail in Chapter 4, but repeated here to cover situations where ammonia-N concentrations to be removed are 2.5–3.0 mg/l or less.

These effluents may contain ammonia nitrogen concentrations in the range of 0.5–3.0 mg/l. The chlorine to ammonia nitrogen ratio will be on the order of 10:1 by weight as found for the Rancho Cordova project.

Based upon Saunier's model, to remove 0.5 mg/l ammonia-N will require at least 20 min contact time. If the ammonia-N concentration falls below 0.5 mg/l the contact time will have to be longer. However for an ammonia-N concentration of 2.5 mg/l the contact time required is on the order of 10 min.

The key control parameter in these cases of low ammonia-N concentrations will be the signal generated by the free chlorine analyzer. This analyzer must have the ability to detect free chlorine in the presence of combined chlorine over a range of 1–15 mg/l free chlorine in the presence of about 25 percent combined chlorine. The chlorine dose correction called for by the analyzer must have a correction frequency related to the contact time required to complete the breakpoint reaction. As long as the Cl:N ratio is optimum (10:1) the formation of nitrogen trichloride will not be significant enough to cause a nuisance.

Foul Air Scrubbing

General Discussion. Scrubbing towers are in common usage at wastewater treatment plants. The principal objective is to control odors and thereby prevent an air pollution problem. Equally important is their use to prevent foul air from reaching the plant ventilation system. These towers are imperative when conditions do not allow control of odors at the source. Ventilation air must be free from lethal H_2S gas at all times. Many deaths occur in sewer-related accidents because of the presence of H_2S gas.

Scrubbing towers are also widely used at pumping plants and other collection system structures where uncontrolled sewage odors occur.

There are two general types of scrubbing towers: chemical and biological. The biological tower is dependent upon availability of a treated effluent. These are packed towers that utilize the buildup of a zoological slime on the packing material to oxidize the hydrogen sulfide and other sewage related gases. These towers recirculate the effluent in a fashion similar to a high-rate recirculating filter. This method is usually reserved for large treatment plants. Complete H_2S removal is not always attainable, so if it is required this method requires the use of a backup system of either chlorine or activated carbon.

Chemical tower systems require the use of either chlorine, caustic, potassium permanganate, or activated carbon or some combination of these chemicals. For severe cases in odor-sensitive areas the wet scrubber is followed by an activated carbon tower impregnated with either potassium or sodium hydroxide.[92] The most successful type of wet scrubber is the packed tower shown in Fig. 7-8. The foul

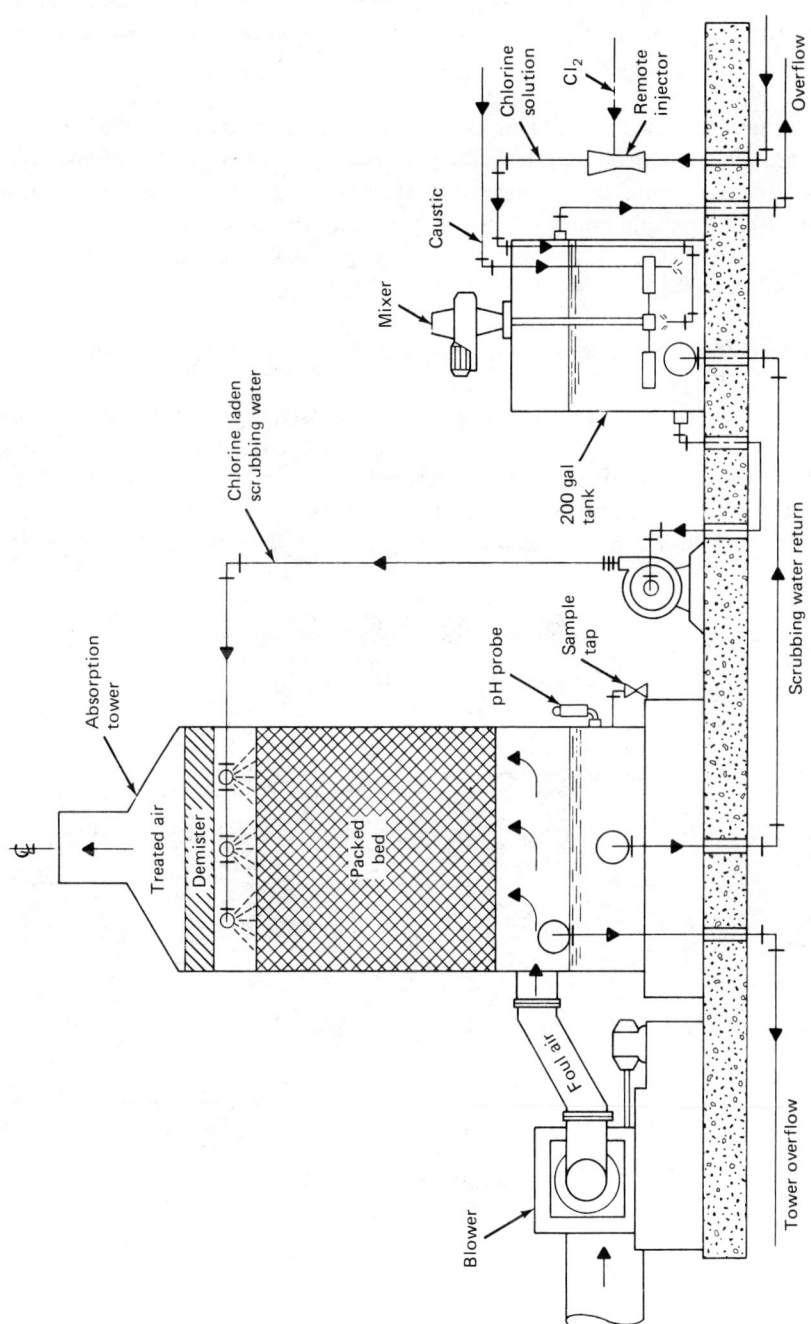

Fig. 7-8 Wet scrubber system for H_2S removal from foul air.

438 HANDBOOK OF CHLORINATION

air enters at the bottom and the scrubbing water enters the top. This is called a countercurrent tower. The packing rings in the tower are designed to provide intimate contact of the recirculated water with the foul air in order to insure maximum efficiency of odor removal.

Current practice consists of using secondary effluent when the scrubber location is at a treatment plant and fresh water for collection system installations. All of these scrubbers operate as a recirculated system using a continuous blow down with a constant makeup water supply. A proper chemical balance must be maintained. A chlorine scrubbing tower is the most practical because it is predictable, reliable and flexible.

A Chlorine Wet Scrubber:

General Discussion. The wet scrubber system (see Figs. 7-8 and 7-9) consists of a tower packed with plastic saddles. These saddles provide the necessary intimate contact between the foul air and the scrubbing water. The scrubbing water is recirculated on a continuous basis with a centrifugal pump with the suction connected to the bottom of the chemical mixing tank. The pump discharges into

Fig. 7-9 Wet chlorine scrubber tower for H_2S removal from ventilation air (courtesy Envirotech Operating Services Fairfield-Suisun WWTP).

the top of the tower as shown in Fig. 7-8. The foul air is blown into the bottom of the tower with a centrifugal fan. Chlorine and caustic are added to the mixing tank. These are the only makeup fluids used in the system. They are usually sufficient to eliminate the necessity for continuous make-up water. Visual inspection supplemented by a total sulfide concentration determination will indicate the need to blow down some or all of the recirculation water.

There are several advantages to a chlorine scrubber as compared to a caustic scrubber. A chlorine scrubber is operated at neutral pH so there is no precipitation of calcium or magnesium ions which occur in caustic towers. There is no corrosion from the chlorine because caustic is added to replace the alkalinity lost in the chlorine–sulfide reaction. The chlorine scrubber provides instantaneous and certain odor removal. Chlorine is far more reliable and flexible than other chemical scrubbers because it is easily controlled by simple monitoring of the residual. The presence of a small residual (0.5–1.0 mg/l) insures the complete destruction of sulfides.

Chlorine–Sulfide Chemistry. Chlorine does not actually remove H_2S, it converts the H_2S and hydrosulfide ion (HS^-) to sulfates. In aqueous solutions hydrogen sulfide hydrolyzes as follows:

$$H_2S \rightleftharpoons HS^- + H^+ \qquad (7\text{-}29)$$

The hydrosulfide ion (HS^-)* further dissociates as follows:

$$HS^- \rightleftharpoons S^= + H^+ \qquad (7\text{-}30)$$

See Fig. 7-1 for the distribution of H_2S, HS^-, and $S^=$ at various pH levels. At pH 7, hydrogen sulfide is approximately 50 percent of the total dissolved sulfides. Whenever the equilibrium between the sulfide ion (HS^-) and the hydrogen sulfide in solution is upset, as when H_2S is removed by oxidation, the shift will be to form more H_2S from the remaining dissolved sulfides to reestablish the equilibrium. Assuming wastewater effluent as the scrubbing water at approximately 22°C (72°F) the solubility of H_2S in water at a pressure of one atmosphere is 3432 mg/l.[21]

There are two important reactions between chlorine and hydrogen sulfide: The first and most significant is the reaction that converts the H_2S to sulfates. This requires 8.32 parts of chlorine to each part of hydrogen sulfide:

$$H_2S + 4Cl_2 + 4H_2O \longrightarrow 8HCl + H_2SO_4 \qquad (7\text{-}31)$$

in the pH range 6.0–9.0. *The optimum pH for this reaction is 6.0. For each part of H_2S removed, 10 parts of alkalinity (as $CaCO_3$) will be consumed.*

---

* This is commonly referred to as the sulfide ion.

The other important reaction is an intermediate one which precipitates free sulfur:

$$H_2S + Cl_2 \longrightarrow 2HCl + S \downarrow \tag{7-32}$$

This reaction requires only 2.10 parts chlorine for each part H_2S. For each part of H_2S removed, 2.6 parts alkalinity (as $CaCO_3$) will be consumed. The pH range is 5.0–9.0. *The optimum pH for this reaction is 9.0.*

In a scrubbing operation the intent is to convert as much of the H_2S as possible to sulfates. But owing to the broad pH range for both of the above reactions it is desirable to maintain the pH in the scrubber system as near to 7.0 as practical. The blowdown from the system will always appear milky due to the presence of colloidal sulfur formed by the intermediate reaction of Eq. (7-32). Maintaining the pH at 7.0 in the system will prevent long-term corrosion.

Loss of alkalinity in the system due to the addition of chlorine must be prevented by the addition of caustic.

In addition to absorbing the hydrogen sulfide and other odor producing compounds the scrubbing water will also absorb some carbon dioxide that may be in the foul air. This will tend to lower the pH and can be accounted for by continuous pH monitoring.

One other facet of chemistry important to the application of chlorine is the presence of ammonia nitrogen in the scrubbing water. If non-nitrified plant effluent is used the NH_3-N concentration will be in the range of 12–22 mg/l. The chlorine dosage of the scrubbing water must be kept at a ratio of not more than 6 parts chlorine to one part NH_3-N. Otherwise chlorine will begin to be consumed by the breakpoint reaction.

Sample Tower Calculation. Assume ventilation air at 20000 SCFM to scrub a possible maximum H_2S concentration of 30 ppm. Fig. 7-9 illustrates this size tower.

$$30 \text{ ppm} \times 20000 \text{ ft}^3/\text{min.} = \frac{600,000}{10^6} = 0.6 \text{ ft}^3/\text{min}$$

Weight of air = 0.075 lb/ft³
S.G. of H_2S = 1.1895

$0.60 \times 0.075 \times 1.1895 = 0.0535$ lb/min = 77.08 lb/day
Chlorine requirement = 8.32×77.08 lb/day = 642 lb/day

NOTE: Each part H_2S requires 8.32 parts chlorine.

Next is the calculation of scrubbing water requirement. Tower company manufacturers recommend a liquid to gas ratio of 1.5–2∶1.

Wt. of ventilation air = 20000 × 0.075 = 1500 lb/min

Use a ratio of 1.75; then

$$1500 \times 1.75 = 2625 \text{ lb/min}$$

$$\frac{2625}{8.34} = 314.75 \text{ gpm}$$

This is the amount of recirculated water required.

Next, calculate chlorine dosage to see how near it will be to the breakpoint.

315 gpm = 0.454 mgd

$$\text{Cl}_2 \text{ dose} = \frac{77.08}{0.454} = 169.8 \text{ lb/mg} = \frac{169.8}{8.34} = 20.36 \text{ mg/l}$$

Therefore the dose, 20.36 mg/l, is far from the breakpoint. Assume NH_3-N concentration in recirculating water is as low as 10 mg/l. The chlorine dose would have to be well in excess of 6 × 10 = 60 mg/l before chlorine starts being consumed by NH_3-N.

Next, calculate the amount of 50 percent caustic required to replace the alkalinity consumed by the chlorine: One gallon of 50 percent sodium hydroxide solution contains 6.38 lb of NaOH and has a sp. gr. of 1.52. Each part of H_2S removed by chlorine consumes 10 parts of alkalinity. Or, said another way, each part of chlorine added consumes about 1.25 mg/l alkalinity in the neutral pH range. (See Chapter 4.) One pound of caustic produces 1.25 mg/l alkalinity.

Therefore, replace the alkalinity consumed by chlorine on a pound for pound basis. This amounts to 642 lb/day. Since there will be some carbon dioxide absorbed in the recirculating process a daily caustic addition should be about 700 lb/day or 700/6.38 = 109.72 or about 110 gallons 50 percent NaOH per day.

Other recommendations: Add chlorine to produce a total chlorine residual of about 1–2 mg/l. Since 700 lb/day chlorine will require a 2-inch injector, adjust the injector throat stem to deliver 30 gpm. This requires a 2-inch disk meter in the injector water line.

Add enough caustic solution to maintain a pH between 7.0 and 7.5.

Monitor pH and chlorine daily, and make a visual inspection of recirculating water overflow to determine its "milkiness." This will establish the frequency of system blowdown. The milkiness results from the inherent formation of colloidal sulfur caused by the reaction in Eq. (7-32).

Monitoring the H_2S in the ventilation air can be done in several ways. The most reliable method is to use a detection device similar to the Research Appliance Corp. portable H_2S detector. Other methods include lead acetate impregnated tiles that turn from white to black when exposed to H_2S. The Sanitation Districts of Los Angeles County use Union Carbide "Gastec" tubes. The tubes are filled with lead acetate impregnated material. Air is injected at one end of the tube and the material discolors as the H_2S is trapped. Concentrations can be read directly on a calibrated scale depending upon the length of discoloration in the tube. Tubes are available in different concentration ranges. Los Angeles uses the 0–120 ppm tubes.[70]

Caustic Towers. Another H_2S scrubbing method is the recirculation of scrubbing water dosed with 50 percent NaOH. The principle of this system is to raise the scrubbing water pH to about 12 so that the H_2S is converted to the HS^- ion. This does not remove the potential of H_2S formation in the scrubbing water. The overflow from this system is discharged into the plant effluent or into the adjacent collection system where it is "removed" by dilution.

The main problem with this system is the water "softening" action of the caustic. Raising the pH precipitates the calcium hardness down to a concentration of 13 mg/l. This requires shutdown and removal by flushing with 5 percent hydrochloric acid. Sanitation Districts of Los Angeles County use these towers for scrubbing foul air from ventilation systems in and around the sewer collection system. They use the following design parameter:[70]

Air flow velocity	350 ft/min
Recirculation rate scrubbing water	25 gpm/1000 cfm
Depth of packing media	8 ft
Removal efficiency at up to 150 mg/l H_2S	90–95 percent

Carbon Columns. These are used in sensitive areas where 100 percent H_2S removal is required when caustic towers are used as the primary H_2S remover. Sometimes they are used to polish the discharge from biological towers. Biological towers cannot be depended upon to achieve 100 percent removal. Therefore supplementary treatment with chlorine or virgin activated carbon is required to assure complete removal. Los Angeles County uses carbon columns to polish off the discharge from caustic scrubbers and seems satisfied with the results.

OTHER METHODS OF H_2S GENERATION CONTROL

Caustic Addition. The Sanitation Districts of Los Angeles County have had many years experience with the intermittent use of caustic to destroy the slime layer in those reaches of free flowing sewers that cause sulfide problems. This treatment is most effective. When the sewage temperature reaches about 20°C in the winter, the treatment is once every two weeks to once a month, and in the summertime, when the sewage temperature reaches 25°C, treatment is once every two weeks.

Treatment consists of a shock dose of 50 percent NaOH at the rate of 620 lb/mg applied over a 30 min period. This is a shock dose of 620 lb 0.1N caustic to 20,880 gallons of sewage. Application of more caustic than the above dose will not improve the effect.

Air-Oxygen Injection:

General Discussion.[67] These two methods are based upon the oxygen uptake of the sewage. To prevent H_2S generation it is necessary to oxidize the sulfides:

$$2HS^- + 2O_2 \longrightarrow S_2O_3^= + H_2O \tag{7-33}$$

This is the predominant reaction in sulfide oxidation. It forms thiosulfate. Approximately one part of oxygen is required to oxidize one part of sulfide as shown by the above equation.

Sewage increases its capacity to oxidize sulfides with age or with the diameter of the pipe. Conversely, the DO diminishes as the size of the line increases.

Typical oxidation rates of sulfides in free flowing sewers can be as high as 5.0 mg/l sulfides oxidized per hour per liter provided the DO can be maintained at 0.3 mg/l or greater. Of this total oxidation only about 0.6 mg/l is from direct chemical action by the DO. The rest is from biological oxidation. Biological oxidation rates vary widely in different sewage. Fresh sewages display lower biological oxidation rates than do older sewages. In large lines the DO is usually less than 0.3 mg/l and as a result the oxidation rate will be considerably less than 0.5 mg/hr/l.

The oxygen uptake rate is:

$$O_2 \text{ uptake} = \frac{[3600 \times 560 \times 10^{-6} \times (SV)^{0.375}] \, O_D}{d_m} \tag{7-34}$$

simplifying:

$$O_2 \text{ uptake} = \frac{[2.016 \, (SV)^{0.375}] \times O_D}{d_m} \tag{7-35}$$

where

O_2 uptake = mg/hr/l
S = slope, ft/ft
V = velocity, ft/sec
O_D = D.O. deficit, mg/l
d_m = mean hydraulic depth, ft

This is the sole means of measuring the sulfide destruction ability of a given sewage. Eq. (7-34) was developed by the Hydrogen Sulfide Laboratory, Los Angeles County Sanitation District.[67]

Oxidation Capacity of Raw Sewage. This factor is known as the i factor, expressed as follows:

$$i = \frac{\Delta S_{OX}}{R_R + R_{SO}} \qquad (7\text{-}36)$$

where

ΔS_{OX} = oxidation rate of sulfides (sulfides removed), mg/hr/l
R_R = oxygen consumed if no sulfides are present, mg/hr/l
R_{SO} = oxygen uptake rate, mg/hr/l

(Determination of the i factor is described in "Sludge Transport," this chapter.) The i factor is a function of the oxygen uptake rate divided by the total oxygen demand when the D.O. is greater than 0.3.

EXAMPLE: Assume S_{OX} = 4 mg/hr/l and $R_R = R_{SO}$ = 12 mg/hr/l. Then $i = 4/12 = 1/3$. If the available DO in the sewage is only 3.0, then only 1 mg sulfides can be oxidized per liter per hour. This is the i factor. Fresh sewage may have an i factor of 0.2, whereas fresh sewage that is aged may have an i factor of 0.6. The i factor varies from sewage to sewage.

It only requires a small amount of DO (0.2) to prevent instream generation of H_2S. This is true regardless of the solids content of the sewage. The slime layer is the critical factor.

U-Tube Aeration. This concept of wastewater aeration has been tried on a limited scale.[25,26] Fig. 7-10 illustrates a typical U-tube installation. Air or oxygen is forced (pumped) or aspirated into the top of the downdrop. The air/wastewater mixture then passes into an enlarged section, *A-B*. In this section increasing head losses will occur; this will be dependent upon the air to water input ratio (by volume). Downleg velocities should be on the order of 1.5 ft/sec to provide enough detention time for optimum oxygen transfer. Section C-D, the upleg, is decreased in size to provide velocities of 4 ft/sec. This will prevent solids from depositing in the bottom of the U-tube. The key variable in U-tube design and operation is the air to water ratio. This ratio should be between 0.05 and 0.10. The critical value is about 0.1 where complete dispersion of air in the wastewater no longer occurs

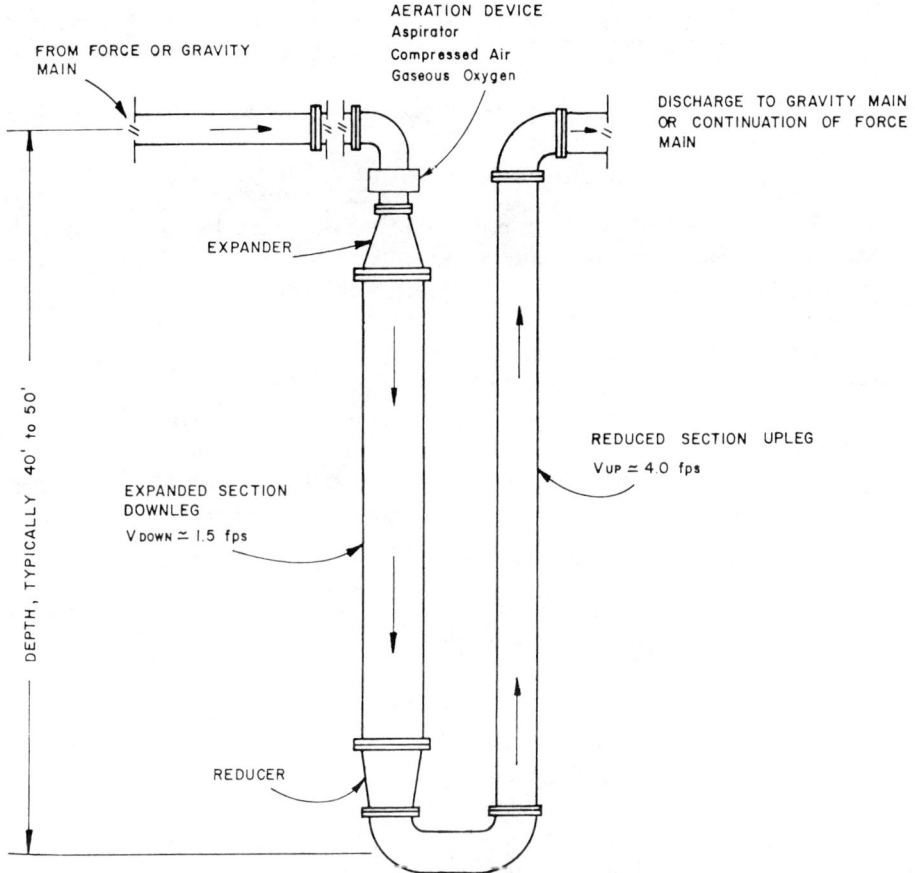

Fig. 7-10 Typical U-tube aeration installation.

and the head loss increases sharply. Oxygen transfer using pure oxygen instead of air is much higher and head losses are less.

According to Pomeroy, the biological oxidation of sulfide in a typical sewage is about 10–15 mg/hr/l after the sewage has been flowing for a few hours. If the sulfide and DO concentrations are not less than 1 mg/l this rate is independent of these concentrations.[28] Hoag et al[25] confirmed this rate in their field studies.

Air and/or oxygen injection does not usually result in complete removal of the sulfides. However, this method can be used to reduce the sulfides to slightly below the critical level of dissolved sulfides, which is considered to be 0.3 mg/l.

Location of the U-tube is a difficult decision to make. Once the location is established it cannot be moved and still remain cost-effective. Some U-tubes have been located near the end of a force main as a means to oxidize the sulfides before the sewage in the force main is released to atmospheric pressure. In these cases the dissolved sulfides must be kept below the critical level at the end of the force main.

Pure oxygen flow rates used in the Sacramento studies were 5–7 SCFM while air was 15–30 SCFM.

Force Mains. Compressed air injection into a force main has been used with varied success. Studies by the Sanitation Districts of Los Angeles County resulted in dividing force mains into two categories, each with its own formula for calculating the quantity of air required for sulfide control.[68]

Category I. This category includes force mains 14 inches in diameter or larger. In this category the oxidation rate within the pipe exceeds the sulfide generation. Therefore only small amounts of air are required to completely remove any sulfide being generated within the force main piping. Moreover, the DO demand of the slime layer attached to the pipe wall is near zero at the air level which controls the sulfide buildup. The mathematical relationship for calculating the quantity of air is as follows:

$$(SCFM)_{control} = \frac{4.0 \times 10^{-3} \times D \times L}{\ln P_{ps}} \qquad (7\text{-}37)$$

where

D = pipe diameter, in.
L = length of force main, ft
$\ln$ = natural log.
P_{ps} = abs. pressure where air is injected, atmospheres
$= \dfrac{34 + \text{Static head} + \text{Friction loss}}{34}$

1 atmosphere = 34 ft

Category II. This category is for force mains smaller than 14 inches in diameter. Since the ratio of sulfide oxidation to sulfide generation is proportional to diameter this ratio will always be lower for smaller pipes than for larger pipes. In this category the capacity of the sewage to oxidize sulfide is less than the capacity of the slime layer to produce sulfides. Therefore the only way to successfully eliminate sulfide buildup in this category is to partially suppress generation within the slime layer by oxygen addition sufficient to create an aerobic zone well within the anaerobic zone of the slime layer. The mathematical expression for determining the quantity of air for force mains smaller than 14 inches in diameter is as follows:

$$(\text{SCFM})_{\text{control}} = \frac{5.35 \times 10^{-3} \times D \times L}{\ln P_{ps}} \qquad (7\text{-}38)$$

When calculating air requirements for this category it is suggested that a 20 percent factor of safety be used if there are any flat slopes that amount to 30–50 percent of total force main length.

Oxygen transfer in a force main is a function of pressure and turbulence. The amount of turbulence and interfacial area determines the rate of oxygen transfer. Around each bubble of air there is a layer of water. The faster the bubble travels (turbulence) the thinner the layer of water, and therefore the greater the O_2 transfer efficiency. This phenomenon explains the effect of the physical slope of a force main. At the lower end of the force main the pressure is higher and the O_2 transfer efficiency will be higher so there will be "overcontrol" at the lower end. Therefore in these cases strive for control in the last 25 percent of the length of this section.

Temperature is an important factor. Sewage temperatures at 20°C and above are certain to produce sulfide generation in a force main.

The addition of air does not add to the friction loss in a rising force main provided there are no negative slopes. The friction effect of air as an additional volume for the pipe to carry is canceled by the "lift" effect of the air. Force mains with negative slopes are usually not amenable to the compressed air treatment. The hydraulics of air addition generally cause difficulties that negate any success derived for sulfide control. It is best to avoid the use of air injection for these situations.

Another important factor relating to sulfide generation in force mains is the pumping cycle, which affects the turnover in the piping system. If there is sufficient DO in the wet well to oxidize the sulfides generated in the force main during the down time, and if complete turnover occurs within one pumping cycle, no sulfides will form.

The most important factor in a compressed air system is the air compressor. It should be sized so that it only operates 50 percent of the time, otherwise maintenance becomes a chronic problem. The compressor capacity should be twice that needed for continuous operation. Arrange to have an adjustable timer to provide an off time up to 15 min. Then adjust the timer to operate the compressor for 7.5 min on and 7.5 min off. If this provides too much oxygen transfer change the timer to 5 min on and 10 min off. The compressor should never be off for more than 15 min.

The following is an example of a compressor selection in an existing force main (Category I), which is shown in Fig. 7-11. The pumps are variable speed. The velocity in the force main will vary from 0.5 to 2.5 ft/sec. Therefore the detention time will vary from 1 hour 40 minutes to about 8 hours.

From Eq. (7-36) we get

$$\text{SCFM} = \frac{4 \times 10^{-3} \times D \times L}{\ln P_{ps}}$$

$$= \frac{4 \times 10^{-3} \times 1.17 \times 15 \times 10^3}{\ln 4.82 = 1.6}$$

$$= 43.88$$

Therefore the compressor should have a nominal capacity of 100 SCFM (3 stage). Operate with timer for 15 min on and 15 min off.

Hydrogen Peroxide. Los Angeles County Sanitation Districts studied the use of hydrogen peroxide for sulfide control and concluded it was an expensive way to put DO in sewage. The Sacramento studies tried H_2O_2 as an alternative and found that the ratio of hydrogen peroxide to dissolved sulfides would have to be at least 3 to 1 to reduce the sulfides to a level of 0.5 mg/l. Therefore the dosage to reach the critical level of dissolved sulfides (0.3 mg/l) might be as high as 4 or 5 to 1. The 1984 cost of 50 percent H_2O_2 in 4000 gallon lots is approximately 37 cents/lb. In estimating the overall cost of H_2O_2, half-life decay due to temperature and impurities within the system should be added to the effective dosage rate.

One other significant factor discovered in the use of H_2O_2 is the lack of effect it has on the BOD, COD, or DO concentrations at the end of force mains except when dosages were 60 mg/l.[25]

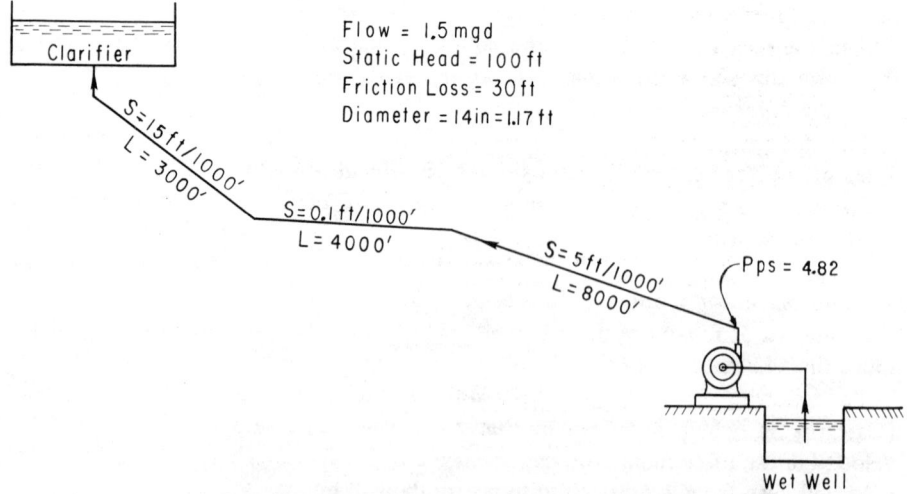

Fig. 7-11 Force main aeration calculation diagram.

INDUSTRIAL WASTES

Cyanide Wastes

Significance. Cyanides are toxic to aquatic life, interfere with normal biological processes of natural purification in streams, present a hazard to agricultural uses of water, and are a menace to public water supplies and bathing. Cyanides exert a toxic action on living organisms, animals, and human beings by reducing or eliminating the utilization of oxygen in a manner similar to asphyxiation. The action at the toxic level is both rapid and fatal.

If the microorganisms in a surface water responsible for proper oxygen balance lose their efficiency or are destroyed, oxygen depletion of the stream ensues. Fish in such a stream stand the chance of being killed either directly by the cyanide content or indirectly by the destruction of the organisms upon which they feed. The toxic threshold level for some species of fish is reported to be 0.05 to 0.10 ppm cyanide as (CN) radical.[71,72] This toxicity increases as the temperature increases and as the dissolved oxygen decreases. The toxic action is apparently related to photosynthesis, for it has been demonstrated that fish in the absence of sunlight can thrive in water containing 1000 ppm cyanide.[40]

Livestock and other animals are likewise endangered when using a stream polluted with cyanides. Lethal doses to animal and man amount to only 4 mg/lb of body weight.[40] A stream used for bathing or a domestic water supply should not contain even a trace of cyanide.

Cyanide wastes definitely impair the biological processes of waste treatment plants, although these processes do reduce the cyanide concentration. The amount of reduction of cyanide concentration in these various processes has never been determined, and therefore it is not possible to establish a maximum allowable limit in the raw sewage entering the plant.

Owing to recent Federal legislation implemented by the Environmental Protection Agency all cyanide wastes discharging into any wastewater treatment collection system must be reduced to cyanates. All cyanide wastes discharging into the environment, whether it be land disposal or surface water must be reduced completely to carbon and nitrogen. The latter situation has been standard practice for many years.

In addition to the cyanides present in metal-finishing wastes, there may be copper and zinc ions, which are also toxic to fish and other aquatic life. The toxic threshold level for some species of fish is reported to be 0.10–0.20 mg/l copper and zinc.

Occurrences. While cyanide compounds are widely used in industry, five processes are responsible for most of the cyanide wastes causing stream pollution and presenting problems in waste treatment plant operation. These processes are: metal plating, case hardening of steel, neutralizing of acid "pickle scum," refining of gold and silver ores, and scrubbing of stack gases from blast and producer gas furnaces.

Metal Plating. The greatest source of cyanide-bearing waste is found in the metal-finishing industry, whose electroplating plants are distributed throughout the country. Most of the cyanide wastes are the rinse waters, spillages, and drippings from the plating solutions of cadmium, copper, silver, gold, and zinc. These plating solutions will vary in concentrations of alkali and "free" cyanide (sodium or potassium), but the various metal baths are similar in concentration. The following shows the contents of a solution for a brass-plating operation:

Copper cyanide	4 oz/gal	30000 mg/l
Zinc cyanide	1¼ oz/gal	9200 mg/l
Sodium cyanide	7½ oz/gal	57000 mg/l
Sodium carbonate	4 oz/gal	30000 mg/l
	Total CN = 39000 mg/l	

Case Hardening of Steel (Nitriding). The process of heat treating steel by cyanide consists of immersing the steel part at a predetermined temperature in a molten bath of a mixture of sodium cyanide, barium chloride, sodium chloride, potassium chloride, and strontium carbonate. When the part has been held for the proper time in this molten bath, it is withdrawn and quenched in a continuous flow of water. The discharge from such a bath may be expected to run as high as 50–100 mg/l of free cyanide.[71]

Neutralizing of Acid Pickle Scum. In the pickling of steel sheet for the removal of mill scale, the metal is usually dipped in dilute sulfuric and hydrochloric acid. As the steel leaves this pickling bath, a coating called pickle smut, consisting of materials not soluble in the acid—iron carbide, silicon, silicon carbide, copper, and copper sulfides—remains on the surface. In addition, ferrous sulfate and ferrous chloride are formed and cling to the surface. If the sheet is then air dried or neutralized with caustic, the ferrous iron is converted to ferric iron, which covers the pickle smut and forms a coating difficult to remove and destructive to the finish of the sheet. It can be removed by agitation in a solution of sodium cyanide (0.2 oz/gal) and sodium hydroxide (0.2 oz/gal). The resulting waste from this solution contains free cyanide in concentrations between 200 and 1000 mg/l.[71]

Refining of Gold and Silver Ores. The present method of gold recovery from ore is the McArthur-Forrest process, first patented in 1887, which is based on the solubility of gold in a cyanide solution. The gold is ultimately precipitated from this solution by zinc dust. The chemical reactions are as follows:

$$4\,Au + 8\,KCN + O_2 + 2\,H_2O = 4\,KAu(CN)_2 + 4KOH \qquad (7\text{-}39)$$

The precipitation of gold by zinc dust is represented by:

$$2\ \text{KAu(CN)}_2 + \text{Zn} \longrightarrow 2\ \text{AU}\downarrow + \text{K}_2\text{Zn(CN)}_2 \qquad (7\text{-}40)$$

The potassium–zinc–cyanide solution remaining after the gold is precipitated in Eq. (7-40) is the waste to be treated.

Alternatively, gold can be dissolved in a sodium cyanide solution:

$$4\ \text{Au} + 8\ \text{NaCN} + \text{O}_2 + 2\ \text{H}_2\text{O} = 4\ \text{NaAu(CN)}_2 + 4\ \text{NaOH} \qquad (7\text{-}41)$$

The gold is then precipitated with zinc dust as follows:

$$4\ \text{NaAu(CN)}_2 + 2\ \text{Zn} = 4\ \text{Au}\downarrow + 2\ \text{Na}_2\text{Zn(CN)}_4 \qquad (7\text{-}42)$$

The recovery of silver from its ore is similar. The reaction is:

$$2\ \text{Ag} + 4\ \text{NaCN} + \text{O} + \text{H}_2\text{O} \longrightarrow 2\ \text{NaAg(CN)}_2 + 2\ \text{NaOH} \qquad (7\text{-}43)$$

The silver is then precipitated by zinc dust or aluminum powder. The remaining cyanide solution must then be treated for the destruction of cyanides before disposal.

Scrubbing of Stack Gases. Cyanide is a by-product of the manufacture of carbides. The cyanide gets into the flue gas and appears in the wastewater from the flue gas scrubber. This wastewater must then be treated for the destruction of the cyanides.

All of these wastes described above are so toxic to man and animal alike that even in remote areas such as Nevada, where there are many gold and silver processing plants, the state department of health requires that the cyanide be completely destroyed, even when these wastes are stored in protected ponds.

Chemistry of the Treatment. There are two types of cyanide wastes: free cyanide (NaCN or KCN), and combined or complex cyanide ($\text{Na}_2\text{Au(CN)}_2$).

Destruction of Free Cyanide to Cyanate. When chlorine in its various forms (represented as Cl_2 below) is applied to a free cyanide such as sodium cyanide, oxidation takes place, producing cyanogen chloride:

$$\text{NaCN} + \text{Cl}_2 \longrightarrow \text{CNCl} + \text{NaCl} \qquad (7\text{-}44)$$

This reaction is practically instantaneous, and is independent of pH.[73] Cyanogen chloride is also a toxic waste, and, since it volatilizes readily (similar to NCl_3), its formation is to be avoided. Cyanogen chloride can be converted to a more stable compound with proper pH control. In the presence of alkali, represented by NaOH, it decomposes to cyanate at pH 8.5 to 9.0 as follows:

$$CNCl + 2\ NaOH \longrightarrow NaCNO + NaCl + H_2O \qquad (7\text{-}45)$$

At a controlled pH of 8.5–9.0, it takes 10–30 minutes for 100 percent completion of this reaction—that is to say, the conversion of free cyanide to cyanate with chlorine. As the pH increases, the reaction time diminishes. At pH 10–11, the time is on the order of 5–7 minutes. If the pH drops to as low as 8.0, cyanogen chloride will begin to form. This is to be avoided.[73]

The theoretical requirements for destruction of free cyanides to cyanates are as follows: 2.73 parts of chlorine for each part of free cyanide as CN; 1.125 parts of caustic as NaOH or 1.225 parts of hydrated lime ($Ca(OH)_2$) per part of chlorine applied.

In actual practice, it requires more chlorine since other chlorine-consuming compounds are usually present.

Usually the waste is of rather high alkalinity, and so the alkali requirements average out less than one part—usually 0.6–0.8 parts of alkali per part of chlorine. If ammonia is present in the waste, causing formation of chloramines, it is not economically desirable to chlorinate beyond the breakpoint to a free chlorine residual, since chloramines will react to destroy the free cyanide. It simply takes a little longer time. For chloramine treatment, allow an additional 15 minutes. The pH control is critical and is to be measured after chlorination.

Either the flow-through or the batch system of treatment is acceptable when it is necessary only to destroy cyanides to cyanates. The flow-through system can be more economical since it is possible to control the reaction between chlorine and cyanide to form cyanate before the reaction between chlorine and cyanate can begin. This economic advantage may not be of primary consideration when it is imperative always to destroy the free cyanide completely.

Destruction of Cyanates to Carbon and Nitrogen. Cyanates are several hundredfold less toxic than are the cyanides, but there are times when it is mandatory to destroy them. The cyanates present as the end product in the destruction of the cyanides at pH 8.5 to 9.0 are not readily decomposed by water or the excess alkali present in the treated waste unless free chlorine is present. The cyanates in the presence of chlorine slowly hydrolyze to form ammonium carbonate and sodium carbonate:[73]

$$3\ Cl_2 + 4\ H_2O + 2\ NaCNO \longrightarrow 3\ Cl_2 + (NH_4)_2\ CO_3 + Na_2CO_3 \qquad (7\text{-}46)$$

Chlorine does not take part chemically in this reaction, but does aid in completing the reaction within one to one and a half hours. It is thought that chlorine acts as a catalyst in Eq. (7-46). The chlorine in Eq. (7-46) must be free available HOCl to complete this reaction. After the hydrolysis of Eq. (7-46) takes place, the chlorine and caustic rapidly oxidize the ammonium carbonate to nitrogen gas, and the carbonates are converted to bicarbonates as follows:

$$3\ Cl_2 + 6\ NaOH + (NH_4)_2\ CO_3 + Na_2CO_3 \longrightarrow \\ 2\ NaHCO_3 + N_2 \uparrow + 6\ NaCl + 6\ H_2O \qquad (7\text{-}47)$$

As part of this reaction, but not shown, small amounts of inert nitrous oxide (N_2O) and volatile nitrogen trichloride (NCl_3) are also formed.

The complete destruction of cyanides and cyanates to carbonates, nitrogen, and nitrous oxide theoretically requires: 6.82 parts of chlorine per part of free cyanide as CN; 1.125 parts of caustic (NaOH) or 1.225 parts of hydrated lime ($Ca(OH)_2$) per part of chlorine applied.

In practice, it requires slightly more chlorine to deal with other chlorine-consuming compounds which may be present. Any ammonia initially present will have to be destroyed by the breakpoint phenomenon in order to insure the formation of free chlorine. This requires approximately 10.0 mg/l chlorine per mg/l of ammonia nitrogen.

The alkali requirements are usually less than one part (0.6–0.8) of alkali per part of chlorine, owing to the frequently high initial alkalinity of these wastes.

Either a flow-through or a batch system can be employed for the complete destruction of cyanides. In either case, the minimum holding time for all reactions to go to completion is one and a half hours. In a flow-through system, short-circuiting becomes critical; therefore it is necessary to monitor the waste at several points, utilizing ORP measurements.

Removal of Metallic Ions. In the electroplating industry, the plating baths and rinse waters contain an excess of cyanide above that required to form the soluble metal cyanide complexes. The excess cyanide is called free cyanide; in the complex metallic form, it is known as the combined cyanide. The latter forms can be those of cadmium: $(Cd(CN)_2 \cdot 2NaCN$; of copper: $Cu(CN) \cdot 2NaCN$; of silver: $AgCN \cdot NaCN$; and of zinc: $Zn\ (CN)_2 \cdot 2\ NaCN$.

The chlorine and alkali requirement for the destruction of cyanides and cyanates in metal plating is based on the total cyanide present = free plus combined.

Chlorination of metallic cyanide wastes differs from that of free cyanides as follows:

1. The metals are more or less converted to insoluble oxides, hydroxides or carbonates.
2. These insoluble compounds require additional facilities for sedimentation and sludge disposal.

3. Additional chlorine is required to oxidize certain metals to a higher form, specifically copper and nickel.
4. When nickel is present, it interferes somewhat with the destruction of cyanides to cyanates.
5. When iron is present as ferrocyanide, it is readily oxidized to ferricyanide, which resists destruction by the alkaline chlorine process.

The chlorination of such plating wastes to the soluble alkali cyanates and the insoluble metallic precipitate is shown by the following reaction, which forms the insoluble zinc carbonate:

$$Zn(CN)_2 \cdot 2NaCN + NaCN + Na_2CO_3 + 10NaOH + 5Cl_2 \longrightarrow \\ ZnCO_3 \downarrow + 5NaCNO + 10NaCl + 5H_2O \tag{7-48}$$

or by forming the insoluble zinc hydroxide:

$$Zn(CN)_2 \cdot 2NaCN + NaCN = 12NaOH + 5Cl_2 \longrightarrow \\ Zn(OH)_2 \downarrow + 5NaCNO + 10NaCl + 5H_2O \tag{7-49}$$

The above reactions are rapid in the presence of a free chlorine residual. If chloramines occur, the reaction will not go to completion, and insoluble metallic cyanides may be found in the sludge. This can be avoided by monitoring to a free chlorine residual or to the proper ORP level. The volume of sludge will be on the order of 1 percent of the initial waste when the total cyanide concentration is on the order of 50–75 mg/l.

When copper or nickel plating rinse waters (which are cyanide free) are mixed with other cyanide wastes and treated, additional chlorine is consumed in converting the cuprous copper and nickelous nickel to higher forms as follows: 0.56 parts of chlorine per part of cuprous copper; 2.25 parts of chlorine per part of nickelous nickel; 1.125 parts of caustic (NaOH) or 1.225 parts of hydrated lime (Ca(OH)$_2$) per part of chlorine applied.

Interference by Nickel. Nickelous nickel forms a cyanide complex which interferes with the destruction of cyanides to cyanates. It is reported that this complex cyanide cannot be completely destroyed to cyanate in a relatively short time, even in the presence of a free chlorine residual.

However, it creates no problem when the cyanides are to be completely destroyed to carbon dioxide, nitrogen, and nitrous oxide.

In a flow-through plant in which the cyanides are to be converted to the cyanates, the presence of nickel may be solved as follows:

1. Separate nickelous waste from the cyanide waste.
2. Apply nickelous waste to inlet of settling basin along with the cyanide waste after chlorination (nickelous waste does not contain cyanide).

3. Allow the nickelous oxide or hydroxide formed by the reaction with the alkali to settle out. Any excess free chlorine in the chlorinated cyanide waste rapidly converts the nickelous nickel to the black nickelic oxide, which hydrolyzes further to the insoluble hydroxide.

Interference by Iron. Photographic process wastes contain potassium and sodium ferricyanides among other significant chemicals deleterious to the quality of receiving waters. Both ferrocyanide and ferricyanide are toxic to some fish in the presence of sunlight at levels as low as 2 mg/l and therefore should not be discharged without treatment to bodies of water used for beneficial purposes.

While the ferrocyanide is readily oxidized to ferricyanide by chlorine, it does not appear that it can be destroyed directly by alkaline chlorination. It is therefore necessary to decomplex the iron cyanide so that it can be destroyed by chlorine. The addition of mercuric chloride to the waste, to decomplex the iron cyanide, has been suggested.[74] This would form mercuric cyanide, which can be destroyed by the alkaline chlorination process, but the mercury would then have to be recovered from the waste.

Process Control. The monitoring and control of cyanide destruction to either cyanates or to carbon dioxide and nitrogen require sophisticated laboratory expertise.

For the flow-through or continuous process treatment, the only satisfactory monitoring technique is the use of ORP control. The ORP measuring cell consists of a measuring electrode and a reference electrode (Fig. 7-12). There are two conveniently spaced ORP plateaus in the destruction of cyanides by the alkaline chlorina-

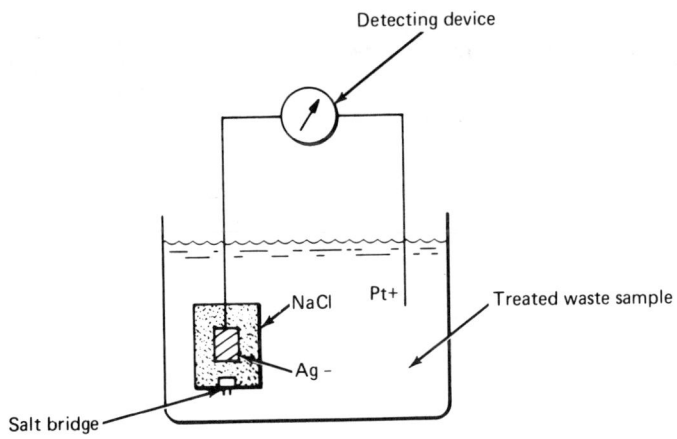

Fig. 7-12 ORP cell for cyanide waste treatment.

456 HANDBOOK OF CHLORINATION

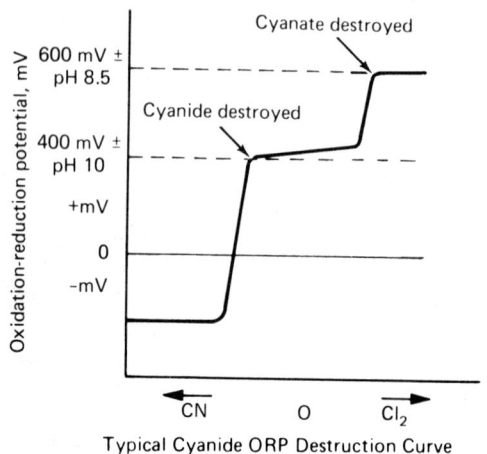

Fig. 7-13 Typical ORP versus cyanide destruction curve.

tion process. Fig. 7-13 shows the first plateau, where cyanide is converted to cyanates at approximately +400 mV. The complete destruction of the cyanates occurs at the second plateau, some +200 mV higher.

These values are relative, depending on the poise of the waste. For example, the first plateau on some wastes might be as low as +325 mV, and the second at +500 mV. Therefore the poise of the waste must be determined in order to establish the set point of the ORP controller.

If the cyanides are to be converted to cyanates only, then it is necessary to run quantitative tests for cyanides and cyanates to determine the optimum chlorine dosage. It is from these tests that the first plateau of the oxidation reduction potential is determined.

The following procedures are suggested:

The modified Liebig titrimetric method, for the determination of total cyanide in raw waste and the total residual cyanide in the treated waste,[75] is applicable where great accuracy is not required and where it is not necessary to quantitatively measure the presence of minute quantities of total residual cyanide.

The pyridine-benzidine method[76] and the pyridine-pyrazolone method[77] can be used for the microanalysis of any residual total cyanide left in the waste after treatment.

A modification of the Herting method[78] is used for determining any residual cyanate left in the waste after treatment.

For the complete destruction of cyanides in a flow-through system, it is necessary to carry a free chlorine residual of 2–5 mg/l at the end of the required contact time. Measuring the ORP at this residual level will determine the poise of the particular waste and thereby establish the set point of the ORP controller.

It is necessary at all times to be able to determine quantitatively the chlorine residual at the selected control points in the treatment system. This is done with the amperometric titrator.[79] For distinguishing between free and combined chlorine, only a properly modified amperometric titration procedure is satisfactory. Heavy metal ions, such as cuprous and silver, as well as high concentrations of cupric ion interfere with the amperometric titration. In the complete destruction of cyanide wastes, dichloramine is always present, and the total combined chlorine may amount to as much as twenty-five times that of the free chlorine. It is possible to titrate free chlorine in the presence of up to 25 mg/l of monochloramine, but the dichloramine must be less than 5 mg/l. The remedy is to apply opposing voltage to make the proper change in the zero point of the ammeter scale on the titrator. This is easily done by use of a 1.5 volt dry cell.[79]

For the batch process control, the following is suggested:

For the destruction of cyanides to cyanates, control to a free chlorine or chloramine residual of 0.2–0.5 mg/l at pH 8.5–9.0 after chlorination.

For the complete destruction of cyanides to carbon dioxide and nitrogen, continued chlorination at a pH not less than 8.5–9.0 is required until a free chlorine residual (0.5–1.0 mg/l) is present after not less than one and a half to two hours retention. This would be at the suction side of the circulating pump and ahead of the point of chlorine application.

When metallic ions are present, the destruction of cyanides to cyanates also requires the appearance of a free chlorine residual.

Proper laboratory analysis will require dechlorination of the sample and usually distillation as well. These procedures are necessary to prepare the sample for the quantitative estimation of total cyanides and cyanates.

In using chlorinated samples for a total cyanide or cyanate analysis, it is essential (1) to remove any remaining cyanogen chloride before dechlorination by increasing the pH of the sample to 10–11 by addition of alkali, preferably NaOH, and (2) remove all residual chlorine by a reducing agent such as sodium sulfite.

Distillation is necessary to obtain the total cyanide in the raw or treated waste when metallic cyanide complexes are present and to separate the cyanide from objectionable turbidity, color, and other substances which interfere chemically with specific tests for cyanides.

Hypochlorite. If the quantities of cyanide to be destroyed are small enough, it is perfectly acceptable to use hypochlorite. The same amount of available chlorine is required. The only difference is that caustic is not required when hypochlorite is used. The following equations demonstrate why equal amounts of available chlorine are required whether it be hypochlorous acid made from chlorine gas and water:

$$Cl_2 + H_2O \longrightarrow HOCl + HCl \qquad (7\text{-}50)$$

or sodium hypochlorite, which is made from chlorine gas, water, and caustic:

$$HOCl + NaOH \longrightarrow NaOCl + H_2O \qquad (7\text{-}51)$$

Therefore, using chlorine water:

$$5\ HOCl + 5HCl + 10\ NaOH + 2NaCN \longrightarrow \\ 2NaHCO_3 + N_2 \uparrow + 10\ NaCl + 9H_2O \qquad (7\text{-}52)$$

or hypochlorite:

$$5\ NaOCl + H_2O + 2\ NaCN \longrightarrow 2\ NaHCO_3 + N_2 + 5\ NaCl \qquad (7\text{-}53)$$

Chlorination Facility. The first chlorination plant for the destruction of cyanides in the United States was built in 1942.[71] Since that time the alkaline chlorination system has been widely accepted. Some of these installations have been well documented.[80-83]

Many plant installations operate mainly on the batch method, but utilize automatic pH control and ORP measurement to control the destruction to cyanates. In this form, the individual waste treatment system is allowed to discharge the cyanates directly to the municipal or other waste collection system. In some localities, wastes from individual plants are collected in a tank truck and taken to a central batch treating plant. Other plants, for example, those in areas of automobile manufacturing, munitions manufacturing, and ore processing, may be the sophisticated completely automated flow-through type.

It is customary to recirculate the cyanide waste through the chlorinator injector wherever possible. If a batch system is indicated, smaller, less expensive equipment may be used, whereas in a flow-through system more units of larger capacity may be required since the chlorinator "sees" the waste only once in the flow-through system. In both systems, the water to power the operation of the injector is always that of the waste itself. Fig. 7-14 illustrates a typical batch system.

Let us assume that the cyanides are to be converted only to cyanates. The total cyanide content of the waste is 1500 mg/l, and the capacity of the batching system is 5000 gal/day. To convert to cyanates requires a chlorinator capable of delivering 4 mg/l Cl_2 per mg/l cyanide = 6000 mg/l total capacity in 5000 gallons.

$6000 \times 8.33 = 49{,}980$ lb/mg mg = million gallons
5000 gal $= 0.005$ mg
$0.005 \times 49{,}980 = 249.9$ lb chlorine required

A 2000 lb/day chlorinator would be the unit of choice. This unit operating at capacity could treat the entire batch in three hours by recirculating all the waste through the injector, as shown in Fig. 7-14. The caustic feeder should be able to

CHLORINATION OF WASTEWATER 459

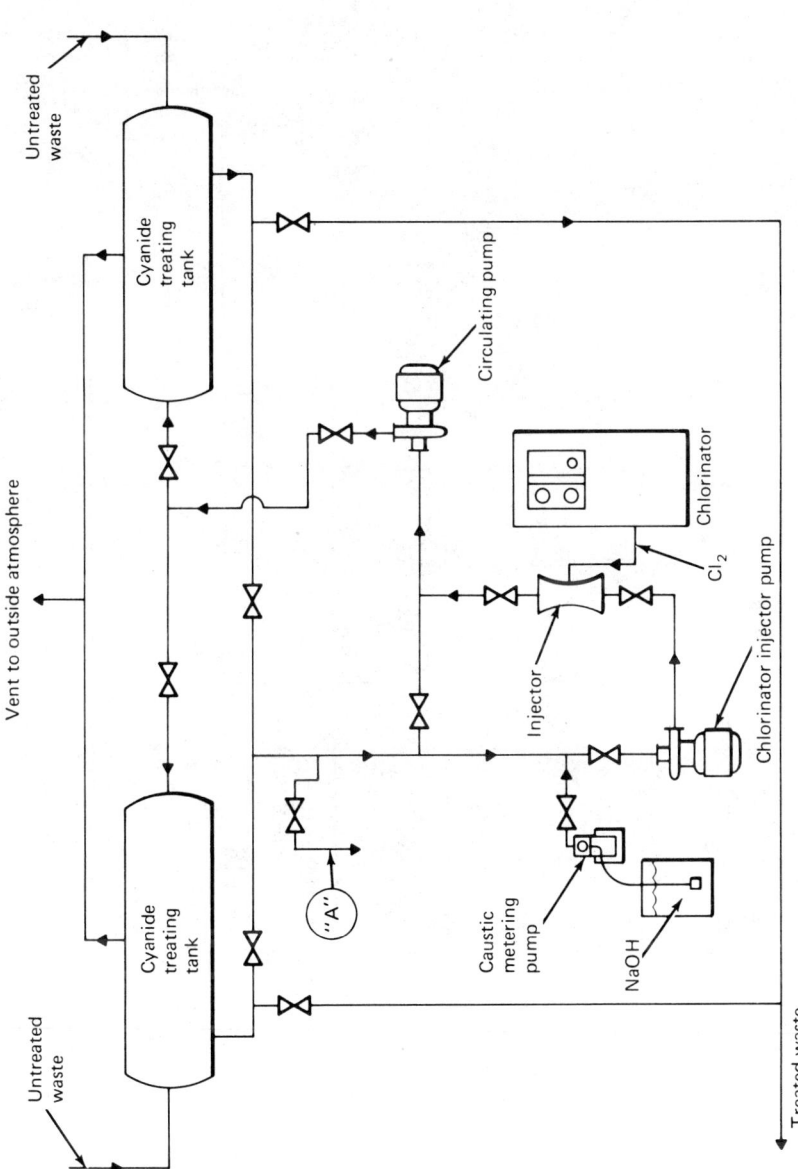

Fig. 7-14 Typical schematic diagram of batch-type chlorination system for the destruction of cyanides. (Free chlorine residual at point "A" denotes complete destruction of cyanides to carbon dioxide, nitrogen, and nitrous oxide.)

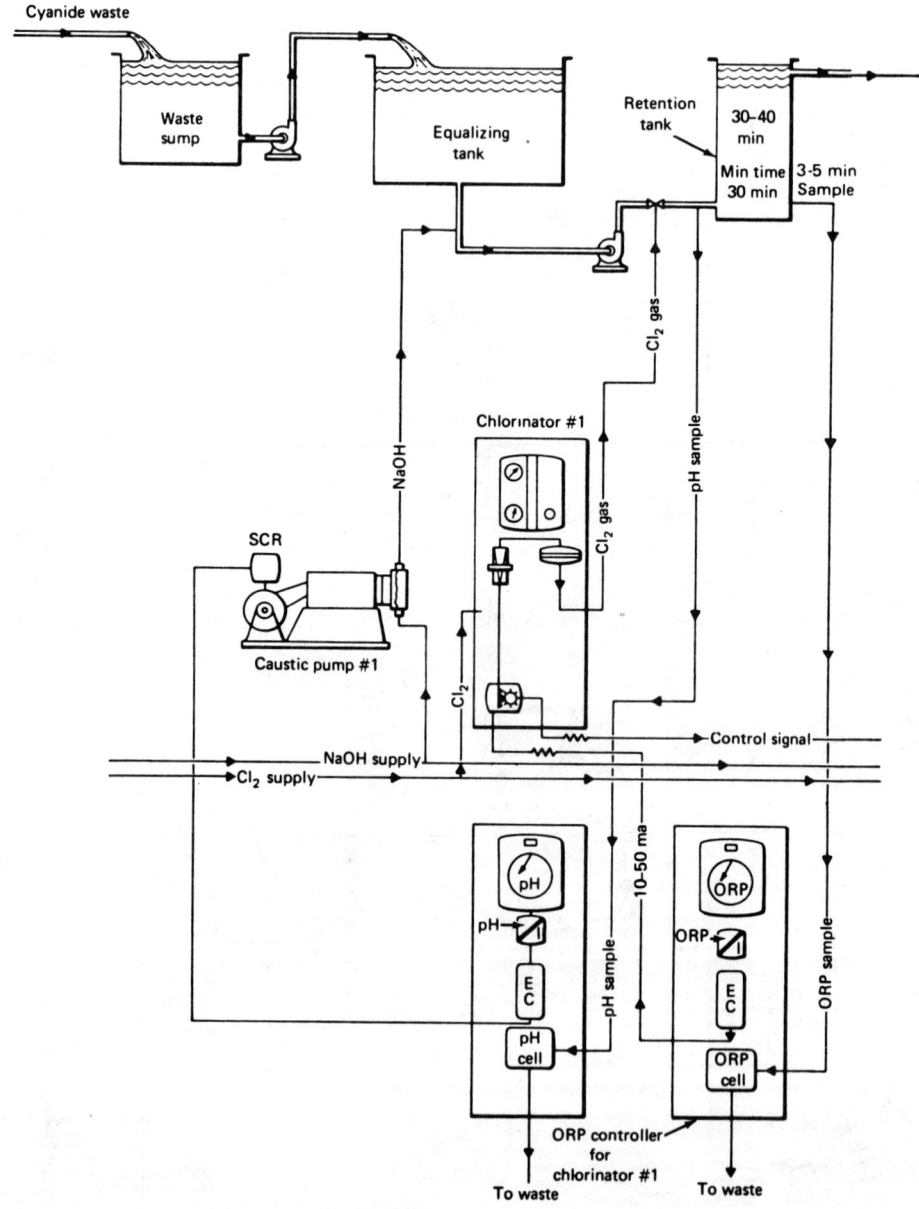

Fig. 7-15 Fully automatic alkaline-chlorine cyanide waste destruction system for continuous flow-through treatment. Chlorinator #1 controlled by ORP controller; this is first-stage destruction to cyanates. Caustic pump #1 controlled by pH controller. Chlorinator #2 controlled by chlorinator #1; this is second-stage destruction to C and N. Caustic pump #2 controlled by chlorinator #2. Chlorinator #3 and caustic pump #3 are backup units operated by alarms on ORP monitoring system.

CHLORINATION OF WASTEWATER 461

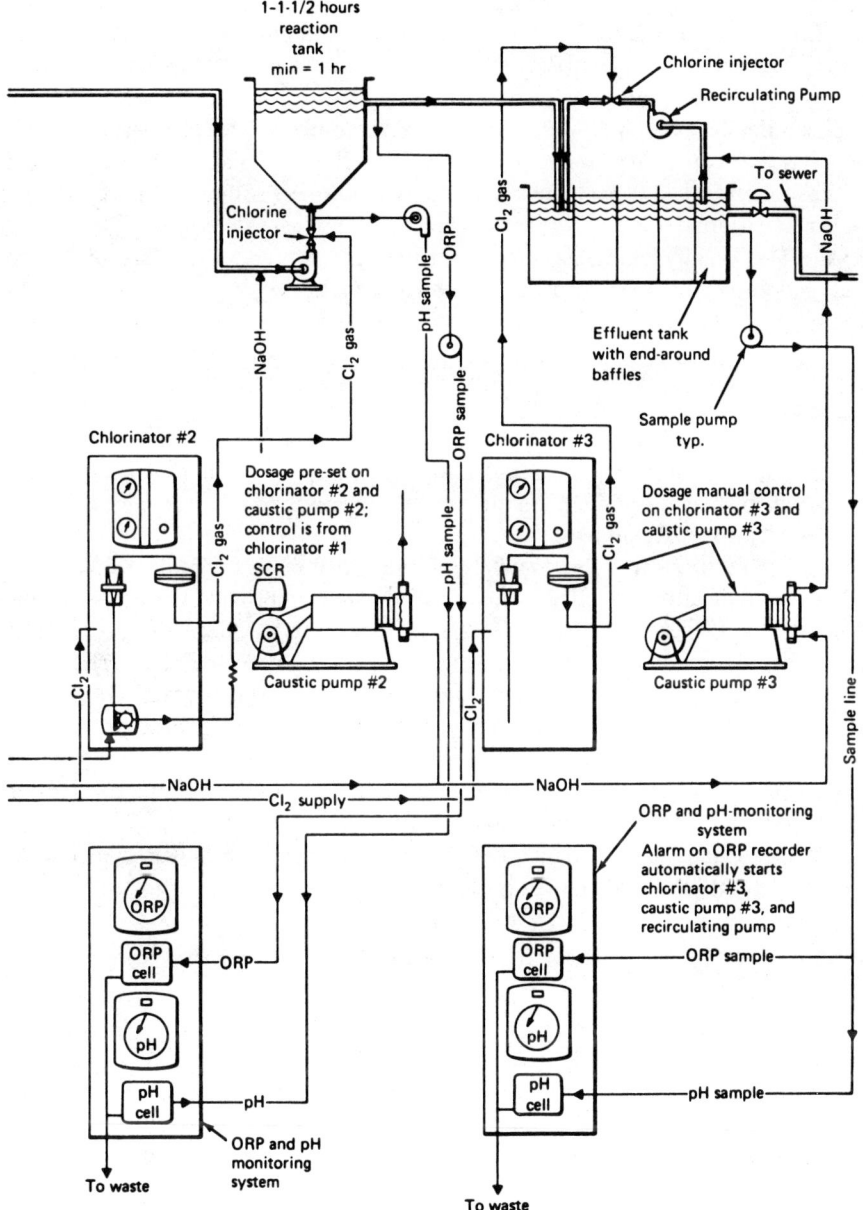

match the 2000 lb rate of the chlorinator; this would amount to a rate of approximately 313 gal/day, so that the caustic feeder should have a capacity of approximately 350 to 400 gpd.

Theoretically it requires 1.125 parts of NaOH to neutralize each part of chlorine. In practice, because these wastes have such high alkalinities and high pH, it usually requires about 0.9–1.0 part caustic per part chlorine. Assume 1.0 part caustic to each part chlorine.

Caustic is usually handled at 50 percent strength. The sp.gr. is 1.52, and therefore each gallon of 50 percent caustic = 6.38 lb NaOH.

$$\frac{2000 \text{ lb Cl}_2 \text{ per day}}{6.38 \text{ lb/gal}} = 313 \text{ gal/day}$$

The system should have an indicating pH meter. Automatic operation or a time clock system is optional. Confirmation of cyanide to cyanate can be achieved by an indicating ORP unit.

Figure 7-15 illustrates a completely automated flow-through system. Here the pH is automatically controlled, using a ratio system to provide the proper amount of caustic to the chlorine being fed. In this case, it has been decided that the complete destruction of cyanide to nitrogen and carbonates is required. The success of this system is based on the ORP control and monitoring system. Once the set point of the ORP controller has been determined by laboratory procedure, described previously, the ORP controller takes over the control and monitoring of the chlorination system. This in conjunction with automatic pH control integrates a completely automatic flow-through system.

Frequent laboratory checks should be made to verify the ORP controller set point as well as the validity of the pH electrodes.

Controlling the Oxidation of Cyanides by the Chlorination Process:

The Potential Curve or Characteristic for the Oxidation of Cyanides. When chlorine in increasing amounts if applied to a cyanide-bearing stream, the pH of which has previously been adjusted to 8.5 to 9.0, the potential of the treated waste increases along a characteristic curve similar to that illustrated by Fig. 7-16. The potential levels plotted vary somewhat with the potential cell; therefore the potentials shown in the figure are only examples of the different levels recorded by a representative cell.

Furthermore, it is important to remember that a change in pH affects the potential. A drop in pH results in an increase in potential, while a rise in pH causes a decrease in potential. It is therefore essential for best control tht the pH of the raw waste be maintained within reasonably close limits prior to chlorination.

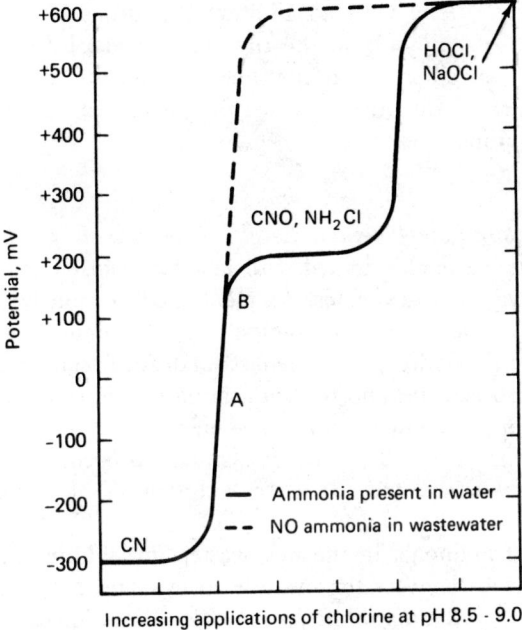

Fig. 7-16 Relationship of potential versus increasing amounts of chlorine in the presence and absence of ammonia in a cyanide destruction system.

The potential system, for purpose of further discussion, is divided into four potential levels: (1) the reduced or cyanide potential level; (2) the cyanate-chloramine potential level; (3) the free chlorine potential level; and (4) the controlling potential level.

The Reduced or Cyanide Potential Level. The reduced potential level is the potential level of the raw waste and the raw waste chlorinated with an insufficient amount of chlorine (less than 2.73 parts of chlorine per part of CN). The constituent most affecting the potential is the cyanide. This reduced potential level can therefore also be called the *cyanide potential level*. In the potential system involved with the chlorination of the waste, it is the *lowest potential level*.

In Fig. 7-16, the cyanide potential level (−300 mV) is maintained with increasing applications of chlorine until practically all of the cyanide has been oxidized to cyanogen chloride and more than 90 percent of the CNCl, within three minutes, has been converted to cyanates, at which time the potential will have risen to zero or +50 mV. Additional retention time, sufficient to complete the hydrolysis of the cyanogen chlorides to cyanates, will result in a higher stabilized potential at the *cyanate potential level* with no additional chlorine. Up to this point, the presence or absence of ammonia in the raw waste make little difference in the end result. However, as chlorination continues beyond the theoretical 2.73 parts of chlorine to 1 part of CN, the presence or absence of ammonia in the raw waste is of extreme importance.

The Cyanate or Chloramine Potential Level. The *cyanate potential level* is the potential level of the waste chlorinated with just sufficient chlorine completely to convert all of the cyanide to cyanates. As mentioned previously, this level is not reached immediately, since a retention period of thirty minutes or more is usually required to complete the hydrolysis of the last small fraction of cyanogen chloride.

As chlorination proceeds beyond the application necessary to reach the *cyanate potential level,* which in practice requires a somewhat greater ratio of chlorine to cyanide than the theoretical 2.73:1 ratio, two situations occur: (1) where ammonia is present in the raw waste, and (2) where ammonia is not present in the raw waste.

In the presence of ammonia in the raw waste, any chlorine applied in excess of that required just to convert the cyanide to cyanates appears as chloramine residual. The *cyanate* and *chloramine potential levels* are identical, and this level is shown on Fig. 7-16 at about +200 mV. Usually, when ammonia is present in the waste, the *cyanate potential level* is the *highest potential level* attained, since it is seldom feasible to convert all of the ammonia to chloramines. However, if enough chlorine is added completely to react with all of the ammonia in addition to the cyanide initially present in the waste, further increases in chlorine application appear as a free chlorine residual, in which case the potential rises to a new maximum at about +600 mV.

When no ammonia is present in the raw waste, any chlorine applied in excess of that required just to convert the cyanide to cyanates appears as a *free chlorine residual,* and the potential rises immediately to a value of about +600 mV, as shown on Fig. 7-16 by the broken curve.

The Free Chlorine Potential Level. It has been mentioned that the *free chlorine potential level* may be attained in two different ways, determined by the presence or absence of ammonia in the waste. The solid curve in Fig. 7-16 illustrates the potential characteristic with ammonia present, and the broken curve shows the characteristic when no ammonia is present in the raw waste. Both curves are identical while the first reaction is being completed—namely, the oxidation of cyanide to cyanogen chloride and the subsequent hydrolysis of this cyanogen chloride to cyanates.

Since oxidation of the cyanates by additional chlorination requires not only the presence of a *free chlorine residual* but also a contact period of perhaps forty-five minutes, the *free chlorine potential level* may be reached momentarily when no ammonia is present by applying more than enough chlorine to complete the cyanate reaction but not enough completely to convert all of this cyanate to nitrogen and carbon dioxide. In this case, the potential will remain at the free chlorine level as the oxidation of the cyanates proceeds until all of the excess free chlorine is used up, at which time the potential will fall to the cyanate level again.

The slowness of this cyanate-free chlorine reaction also accounts for the plateau on the solid curve at +200 mV in the presence of ammonia which represents the *cyanate–chloramine potential level,* since any excess chlorine applied after the cyanates have formed reacts quickly with the ammonia to form chloramines. Because chloramines will not oxidize the cyanates to nitrogen and carbon dioxide, it is necessary to apply enough excess chlorine to (1) combine with all the ammonia, and (2) to react with all the cyanates before the *free chlorine potential level* can be reached and maintained. Thus the length of the cyanate-chloramine plateau at +200 mV is directly associated with the amount of ammonia present in the waste.

The Controlling Potential Level. For purposes of control, it is important to sample the chlorinated waste as soon after the application of chlorine as is possible to obtain a sample which has significance or meaning with respect to the process. In practice, this has been found to be three minutes (the control potential retention period). Point A in Fig. 7-16 represents the midpoint between the *cyanide potential level* and the *cyanate–chloramine potential level*—in this case, −50 mV.

Below this potential, cyanides are likely to be present in increasing small amounts with decreasing small applications of chlorine. Above this potential, a chlorine residual, free or chloramine as the case may be, will be present in increasing small amounts with increasing small applications of chlorine.

With the process under automatic potential control, the potential will at times fluctuate above and below the set point. This raises the possibility of having present small amounts of cyanides in the treated waste at times if the control potential point is set at the no cyanide–no chlorine residual level of −50 mV.

In view of the axiom that cyanides and a chlorine residual cannot be present in solution at the same time, it appears safer to have any possible fluctuation in the control potential occur in the zone in which chlorine residuals in small amounts will always be present after a three-minute retention period. In the representative cell (Fig. 7-16), such a potential control or set point has been set at B +100 mV.

This controlling potential may be computed as follows:

Formula: Controlling potential level in mV. = calibrated cyanide potential in mV plus 80 percent of the potential difference in mV between the calibrated cyanide and cyanate potentials.

Example:

Calibrated cyanide potential = −300 mV
Calibrated cyanate potential = +200 mV
Potential difference = 500 mV
80% of potential difference = 400 mV
Controlling potential level = −300 mV + 400 mV = +100 mV

In some cases, it may be possible to calculate this controlling potential level on the basis of a smaller percentage figure for the potential difference. This applies particularly to plants converting only to cyanates and where the 80 percent figure may produce a continuing chlorine residual which is considered excessive.

In terms of cells other than the representative cell mentioned in this discussion, knowledge of the cyanide and cyanate potential levels for a particular cell can be seen to be of utmost importance. Only from this knowledge is it possible to determine the proper control and alarm points for the individual cells associated with the control system. Normally the calibrated potential levels for individual cells should agree within +50 mV.

If the potential for any cell disagrees beyond this limit, it is suggested that the gold measuring electrode be removed from the cell and placed in a beaker of distilled water (200 ml) containing about two grams of sodium cyanide for about three minutes.

Calibration of Potential Cells. The technique for the calibration at both potential levels is similar, and for each level in potential is as follows:

For the cyanide or reduced potential level, use a cyanide solution buffered to pH 8.4, made as follows: To 10–12 quarts of tap water* in a clean porcelain pail, add 100 g or 0.25 lb sodium bicarbonate ($NaHCO_3$) and 2 g sodium cyanide (NaCN); then dissolve by stirring.

Drain sample pump and potential cell and recirculate this solution through cell assembly. Depending upon the recorder and chart range, it may be necessary to reverse the electrical connections at the cell in order to get this potential on scale. Record the potential until it becomes constant. This potential is the *calibrated cyanide potential* and should be so noted, due regard being made of any reversal of cell connections.

For the cyanate potential level, use a hydrogen peroxide test solution buffered to pH 8.4, since it has the same potential as cyanates. It is made as follows: To 10–12 quarts of tap water in a clean porcelain pail, add 100 g or 0.25 lb sodium bicarbonate ($NaHCO_3$) and sufficient hydrogen peroxide to equal 2 g H_2O_2. Proceed as above, except that it will not be necessary to reverse cell leads. The recorded potential is the *calibrated cyanate or peroxide potential.*

---

* If the tap water contains appreciable amounts of iron or manganese, it is advisable to calibrate the cells with test solutions made up with distilled water.

TREATMENT AND DESTRUCTION OF PHENOLS

Phenolic compounds are toxic to aquatic life, and in minute quantities are deleterious to water supplies for their taste and odor producing capabilities. Wastes from many chemical plants, coke, gas-manufacturing, rubber-making plants, and places where sheep dips are used and livestock disinfected contain these objectionable substances. Phenols are the hydroxyl derivatives of benzene, including various phenols, cresols, and related compounds of higher order.

Phenols can be completely destroyed through oxidation by chlorine, chlorine dioxide, and ozone. Economical considerations eliminate both chlorine dioxide and ozone as a practical method for the destruction of phenols in wastewaters.

Biological oxidation of phenols can be accomplished under rigorously controlled treatment conditions. However, slugs of phenolic compounds discharged to a conventional waste treatment plant (activated sludge or high rate filters) will not have time to biologically adjust so as to oxidize and thereby destroy the phenols. In these instances, phenols will appear in the effluent, which cannot be tolerated. Furthermore, if the effluent from such a plant is chlorinated for disinfection purposes, the level of chlorination will be such that chlorophenols will be formed. This is the worst possible situation, since one part in 10,000 million parts[40] of chlorophenols will impart a detectable off-flavor in a water supply. In addition, chlorophenols are more toxic to aquatic life than are the straight phenols.

Destruction of phenols by chlorooxidation is the preferred chemical method, and is economically practical if the phenolic content of the waste is not in excess of 200 mg/l. The chlorooxidation process must be carried out at pH levels between 7 and 10 and a minimum of 100 mg/l free chlorine residual.

While the ratio of chlorine to phenol to accomplish destruction has been estimated at between 6:1 and 10:1,[83] most work indicates that large excesses of chlorine are required. Three-minute residuals should be on the order of 1000–3000 mg/l, as a result of the enormously high chlorine demand of phenolic wastes.

Chamberlin and Griffin[84] found that a coke plant still waste required 5000 mg/l of chlorine to destroy 100 mg/l of phenols. If the theoretical ratio is valid (10:1), then there were other compounds in this waste which had about 4000 mg/l chlorine demand. It should also be recognized that contact time for the reaction to take place is largely dependent on the amount of excess chlorine available. The minimum chlorine residual in the destruction process should be about 100 mg/l. While this seems unusually high, extremely obnoxious odors will occur during the treatment process if it is not maintained, indicating incomplete oxidation by chlorine.

Some free chlorine residual must be present to avoid these odors. Temperature also has an effect on the speed of the reaction. The rate of phenol destruction increases with increased temperature.[85]

Depending on the amount of excess chlorine residual, the contact time may vary from minutes (five to ten), with residuals of 1000–3000 mg/l, to two to five hours, with residuals of 50–100 mg/l. There is an apparent stepwise reaction with chlorine that first forms trichlorophenols. These compounds are subsequently cleaved by chlorine to yield aliphatic acids as the end product of the phenol destruction.[85] This high residual–contact time relationship brings up the problem of the treated effluent, which, having been treated for phenol destruction, is now a toxic waste because of too high a chlorine residual. This situation makes a holding or contact tank imperative so that the chlorine residual–contact time relationship can be resolved without the necessity of providing dechlorination facilities.

When the phenol content of a waste is in excess of 200 mg/l, consideration of a phenol recovery process is recommended. There are several methods available. Depending on the circumstances, the destruction of phenols may be by a combination of biological treatment and chlorooxidation. Phenol-tolerant microorganisms can completely metabolize phenolic compounds. The removal of these compounds in a biological treatment plant would therefore depend upon establishing such microorganisms in the biological processes. When this has been done, these processes have been known to be able to remove continuously up to 50 mg/l of phenols.

Chlorination Facility. The chlorination facility for phenol destruction is similar to that for cyanide destruction. The waste to be treated is pumped from a holding tank through the chlorinator injector at a constant rate to a chlorine contact or holding tank.

Since phenol destruction will take place only in the pH 7–10 range, the pH of the solution discharge from the chlorine injector must be maintained accordingly. Therefore caustic must be added at the approximate rate of 1.125 parts NaOH to each part chlorine, to replace the alkalinity destroyed by the chlorine. Depending on the buffering action of the waste, this rate may vary from 0.8 to 1.5 parts NaOH for each part chlorine. Therefore the caustic feeder should be capable of feeding approximately 0.235 gallons 50 percent NaOH per pound of chlorine.

The control of the caustic feeder should be by an automatic pH recorder-controller. The control of the chlorinator feed rate can be predetermined by laboratory analysis for the optimum dosage for complete destruction. Since the chlorinator recirculates a constant amount of waste to be treated, the chlorinator control is by manual dosage setting.

TEXTILE WASTES

Wastes resulting from the processing of cotton and wool have a high BOD content (500–1500 mg/l) and are usually highly colored. These wastes are usually subjected to both chemical precipitation and biological types of treatment.[86] Chlorine plays a dual role in the treatment of these wastes: (1) it is used to make a coagulant (ferric chloride); and (2) it bleaches the color out of the treated effluent. One of the most popular coagulants in chemical precipitation of such wastes is the ferric ion.

Since chlorine is required as a bleaching agent, it might well be made available also as an oxidizer to convert the ferrous ion in the form of copperas (ferrous sulfate) to ferric sulfate as follows:

$$6 \text{ FeSO}_4 \cdot 7 \text{ H}_2\text{O} + 3 \text{ Cl}_2 \longrightarrow 2 \text{ Fe}_2(\text{SO}_4)_3 + 2 \text{ FeCl}_3 + 42\text{H}_2\text{O} \qquad (7\text{-}54)$$

The ferric sulfate hydrolyzes to ferric hydroxide, which forms a brown, heavy, gelatinous floc as follows:

$$\text{Fe}_2(\text{SO}_4)_3 + 6 \text{ H}_2\text{O} \longrightarrow 2 \text{ Fe(OH)}_3 + 3 \text{ H}_2\text{SO}_4 \qquad (7\text{-}55)$$

and similarly:

$$\text{FeCl}_3 + 3 \text{ H}_2\text{O} \longrightarrow \text{Fe(OH)}_3 \downarrow + 3 \text{ HCl} \qquad (7\text{-}56)$$

For such a facility, the relatively easy-to-feed copperas ($FeSO_4$), which is a granular material, is fed through a volumetric feeder into a solution mixing chamber. This solution is pumped or ejected into the discharge of the chlorinator injector solution line, where the ferric sulfate is formed. The solution is then delivered to the chemical mixing chamber to provide the floc for the treatment process.

An additional point of application, proportioned by rotameters, is split off from the injector discharge ahead of the copperas addition, to be used as chlorine for bleaching color from the treatment plant effluent or as a disinfectant if combined wastes are involved.

OTHER APPLICATIONS OF CHLORINE IN WASTE TREATMENT

Faber[87] reported in 1947 on the recovery of wool grease in the wool scouring process by the use of hypochlorite. Before raw wool can be processed by the textile manufacturer, it must be cleaned to remove the adhering dirt, vegetable matter, perspiration, and grease. The liquor that results from this cleaning process is an emulsion of grease, organic matter, and fine solids in suspension. It is a brown and thickly turbid liquor, usually covered with a greasy scum, always alkaline and highly putrescible. Each pound of wool scoured will usually produce from 1 to 10 gallons of waste. For this reason the wool waste varies considerably in strength but usually contains approximately 5000 to 10,000 mg/l suspended solids and 5000 to 15,000 mg/l grease, while the BOD range is between 5000 and 6500 mg/l.

The hypochlorite process reacts with the alkalinity present and the soluble soaps to produce a precipitate, so that the wool scouring waste is rapidly separated into its solid and liquid phases as a result of the oxidation of organic matter by the hypochlorite. This process is reported to be able to remove about 98 percent of the grease, of which 70 to 75 percent can be recovered and sold.

Because of this process, practically all of the suspended solids and grease can be removed together with 80–90 percent of the BOD. This represents a genuine accomplishment in the treatment of an industrial waste.

Chlorine is also widely used in rendering plants, to convert inedible poultry, fish, and animal offal and waste products into animal feed and recovered grease. This is accomplished by cooking the inedibles at high temperatures for a period of several hours. The cooked material is pressed to remove the grease, and the pressings are ground. The cooking process gives off odorous vapors, which are captured by condensing in large quantities of water and by burning the noncondensible vapors. About 150 gpm are required for a five-ton cooker. The water which scrubs the stack gases is chlorinated to about 6–8 mg/l, which eliminates the odors and provides a substantial increase in grease removal.[88]

Strong liquid wastes result from the washing of trucks, barrels, and floors. Granstrom[89,90] reported in 1951 that alum and chlorine are the chemicals of choice to deal properly with these wastes. The alum reacts with the bicarbonate alkalinity to form a floc, and the produced CO_2 reduces the pH. The chlorine oxidizes the reduced material and coagulates the protein. The protein coagulation is most effective at pH 4.0–4.5. The optimum dose found by Granstrom was 80 mg/l alum and 260–350 mg/l chlorine. The BOD reduction by chlorine was 35–90 percent for 260–350 mg/l dosages. The reader is cautioned about the wide variation in the strength of these wastes, depending upon the individual operating conditions.

Coagulation and precipitation of proteins by chlorine and a coagulant were reported in 1931 by Halvorson et al.[91] on fish packing house waste. This report also found that the best removal of protein by chlorination was at pH 4. The chlorine dosage, depending upon the protein content of the waste, varied from 100 to 1000 mg/l.

THE SCOTT–DARCEY PROCESS

Although this process is no longer used, it should be preserved for posterity by a brief description here.

L. H. Scott was superintendent of water and sewage treatment at Oklahoma City, and H. J. Darcey was chief engineer of the Oklahoma state department of health. About 1930, they experimented with scrap iron and chlorine to provide a process whereby all the capabilities of chlorine and iron could be used by a single integrated installation. This patented process was utilized by Wallace and Tiernan, who developed equipment to produce ferrous chloride for odor control, ferric chloride for coagulation, and chlorine for disinfection. The Scott–Darcey process was based on the fundamental reactions of chlorine and iron in an aqueous solution:

$$\text{HOCl} + 2\,\text{HCl} + \text{Fe} \longrightarrow \text{FeCl}_2 + \text{HCl} + \text{H}_2\text{O} \tag{7-57}$$

$$\text{FeCl}_2 + \text{Cl}_2 + \text{H}_2\text{O} \longrightarrow \text{FeCl}_3 + \text{H}^+ + \text{Cl}^- + \text{OH}^- \tag{7-58}$$

This process was able to provide odor control, chemical precipitation, and disinfection with one integrated piece of apparatus utilizing such relatively inexpensive raw materials as scrap iron and chlorine. The installation consisted of a reaction tank that was charged with scrap iron. To this tank was added chlorine solution (HOCl). The reaction of the water, chlorine, and iron would stabilize in about twenty-four hours, so that with the proper chlorine/water ratio the solution could be considered predominantly ferrous chloride.

The success of this part of the process depended upon a centrifugal injector, which simultaneously recirculated the solution in the reaction tank and added chlorine gas to the solution under a vacuum. This ferrous chloride solution was used to react directly with H_2S for odor control, producing ferrous sulfide. A coagulant solution was made available by the use of another centrifugal injector, which added the required amount of chlorine to the odor control solution, converting it to ferric chloride and simultaneously discharging it to the point of application.

Chlorine for disinfection was available as a part of the system. One of the valuable side reactions of this process was that ferrous and ferric ion contributed beneficially to conditioning the sludge prior to digestion.

This process lost its popularity during World War II, when scrap iron became extremely difficult to obtain. It has never been revived, but may be in the future because of the great advantages of using the ferrous and ferric ion in waste treatment processes, particularly for its long-lasting effect on the settleable solids which eventually become part of the sludge digestion process. This facet of waste treatment has been overlooked in recent years because of more attractive results in flocculation using chemicals specific for producing a floc but which are relatively poor in enhancing the characteristics of disposal of the sludge formed by these flocculants together with settleable organic matter.

REFERENCES

1. Hofman, A. W., and Frankland, E. "Report on the Deodorization of Sewage," Report to the Metropolitan Board of Works, London, Aug. 12, 1859.
2. Baker, M. N., "Sewage Purification in America," *Engr. News,* 41 (July 13, 1893).
3. Phelps, E. B., and Carpenter, W. T., "The Sterilization of Sewage Filter Effluent," *Technology Quarterly,* **19**, 382 (1906).
4. Clark, H. W., and Gage, S. D., "Experiments Upon the Disinfection of Sewage and the Effluent of Sewage Filters," 43rd Ann. Rept. Mass. St. Bd. Hlth., p. 339, 1911.
5. Clark, H. W., and Gage, S. D., "A Review of 21 Years Experiments Upon the Purification of Sewage at the Lawrence Experiment Station," 40th Ann. Rept. Mass. St. Dept. Hlth., p. 251, 1908.
6. Thoman, J. R., and Jenkins, K. H., "Statistical Summary of Sewage Chlorination Practice in the U.S.," *Sew. and Ind. Wastes,* **30**, 1461 (1958).

7. Enslow, L. H., and Symons, G. E., "Sewage Treatment Processes," *Sewage Works Jour.*, **17**, 5, 984 (Sept. 1945).
8. White, G. C., "Chlorination—How it Works," paper presented at the 5th Annual Symp., Bureau of Eng., Calif. State Dept. of Health, May 1970.
9. Lusher, E. E., "Swimming Bath Water Treatment," *Baths Service* (England), p. 16 (1959).
10. Lomas, P. D. R., "The Combined Residual Chlorine of Swimming Bath Water," *J. Assoc. Pub. Analysts* (England), **5**, 27 (1967).
11. Sung, R. D., "Effects of Organic Constituents in Wastewater on the Chlorination Process," Ph.D. Dissertation Univ. of Calif., Davis, CA, 1974.
12. Feng, T. H. "Behavior of Organic Chloramine in Disinfection," *J. WPCF*, **38**, 614 (1966).
13. Ryan, W. A., "Control of Odors," *Sew. Wks. Jour.*, **5**, 89 (1933).
14. Chanin, G., Elwood, J. R., and Chow, E. H., "Atmospheric Hydrogen Sulphide Studies," *Sew. and Ind. Wastes*, **26**, 1217 (1954); "Correction," ibid., **26**, 1449 (1954).
15. Anon., "Highlights: Wastewater Wisdom Talk," Water Poll. Control Fed., Feb., 1980.
16. Swab, B. H., "Effects of Hydrogen Sulfide on Concrete Structures," *Jour. San. Engr. Div. ASCE*, **1** (Sept. 1961).
17. White, G. C., unpublished field studies, Modesto, CA, 1946.
18. Pomeroy, R. D., private communication, Nov. 7, 1983.
19. McKee, J. E., and Wolf, H. W., "Water Quality Criteria," 2nd ed., Calif. State Water Quality Control Board, Sacramento, CA, 1963.
20. Nordell, E., *Water Treatment for Industrial and Other Uses*, Reinhold, New York, 1951.
21. Anon., "Process Design Manual for Sulfide Control in Sanitary Sewerage Systems," U.S. Env. Prot. Agency, Oct. 1974
22. Heukelekian, H., "Sewage Chlorination for Odor Control," *Water Wks. and Sew.*, **89**, 302 (1941).
23. Pomeroy, R., and Bowlus, F. D., "Progress Report on Sulfide Control Research," *Sewage Wks. Jour.*, **18**, 597 (July, 1946).
24. Heukelekian, H., "Utilization of Chlorine during Septicization of Sewage," *Water and Sew. Wks.*, **95**, 179 (May 1948).
25. Hoag, L. N., McLaren, F. R., and Hall, G. H., "Control of Odors and Corrosion in the Sacramento Regional Wastewater Conveyance System," a report by Sacramento Area Consultants, Sacramento, CA, Sept. 1976.
26. Meyer, W. J., and Hall, G. H., "Prediction of Sulfide Generation and Corrosion in Sewers," a report by J. B. Gilbert and Associates, Sacramento, CA, April 1979.
27. Davey, W. J., "Influence of Velocity on Sulfide Generation in Sewers," *Sew. and Ind. Wastes*, **22**, 1132 (1950).
28. Pomeroy, R. D., private communication, Nov. 1977.
29. Pomeroy, R. D., and Parkhurst, J. D., "The Forecasting of Sulfide Build-Up Rates in Sewers," *Prog. Wat. Tech.*, **9**, 621–628 (1977).
30. Hall, G. H., McLaren, F., and Hyde, W. S., "Analysis and Design for Sulfide and Corrosion Control in the Sacramento Regional Wastewater Collection System," *Calif. Water Poll. Control Bulletin*, p. 40 (Jan 1981).
31. Pomeroy, R., "Generation and Control of Hydrogen Sulfide in Filled Pipes," *Sew. and Ind. Wastes*, **31**, 1082 (Sept. 1959).
32. Hommon, C. C., "Control of Odorous and Destructive Gases in Sewers and Treatment Plants," *Water Wks. and Sew.*, **89**, 277 (1942).
33. Bowlus, F. D., and Banta, P. A., "Control of Sewage Condition by Chlorination," *Water Wks. and Sew.*, **89**, 277 (1932).
34. Anon., "California Operators Symposium—Odor Control," *Sew. Wks. Jour.*, **14**, 883 (1942).
35. Wisely, W. H., "Experiences in Odor Control," *Sew. Wks. Jour.*, **13**, 956 (1941).
36. Wisely, W. H., "Experiences in Odor Control," *Sew. Wks. Jour.*, **13**, 1230 (1941).
37. Nelson, M. K., "Sulfide Odor Control," *J. WPCF*, **35**, 1285 (Oct. 1963).

38. Backmeyer, D. P., and Drautz, K. E., "Miami Tries to Control Sulfides in Force Main," *Wastes Engrg.*, 290 (June 1963).
39. Black, A. P., and Goodson, J. B., Jr., "The Oxidation of Sulfides by Chlorine in Dilute Aqueous Solutions," *J. AWWA*, **44**, 309 (1952).
40. Anon., "Chlorination of Sewage and Industrial Wastes," Manual of Practice No. 4, WPCF, Washington, DC, 1951.
41. Chamberlin, N. S., "Chlorination of Sewage," *Sew. Wks. Jour.*, **20**, 304 (1948).
42. Nordell, E., *Water Treatment for Industrial and Other Uses*, Reinhold, New York, 1961.
43. Nagano, J., "Oxidation of Sulfides during Sewage Chlorination," *Sew. and Ind. Wastes*, **22**, 884 (July 1950).
44. Chanin, G., "Solving Odor Problems by Prechlorination of Flow, Paced by Automatic Continuous Monitoring of H_2S Content of Atmosphere," *Water Wks. and Wastes Engrg.*, 42 (Jan. 1964).
45. Castro, A. J., "The Treatment of Cannery Wastes at Santa Clara, California," *Calif. Sew. Wks. Jour.*, **14**, 2, 22 (1942).
46. Griffin, A. E., "The Regulation of Chlorine Application in Sewage Odor Control Work," *Water and Sew. Wks.*, **80**, 218 (June 1933).
47. White, G. C., unpublished report to Union Sanitary District, 1978.
48. Smith, E. E., "Control of Activated Sludge Bulking by Chlorination," *Water Wks. and Sew.*, **90**, 209 (June 1943).
49. Tapleshay, J. A., "Control of Sludge Index by Chlorination of Return Sludge," *Sew. Wks. Jour.*, **17**, 1210 (1945).
50. Tapleshay, J. A., "Sludge Density Control," *Water Wks. and Sew.*, **93**, 116 (1946).
51. Livingston, J. and Harland, R., private communication, H_2S Lab., L. A. County Sanitation District, L. A., Calif. Nov. 1972.
52. Griffin, A. E., and Chamberlin, N. S., "Exploring the Effect of Heavy Doses of Chlorine in Sewage," *Sew. Wks. Jour.*, **17**, 730 (1945).
53. Mahlie, W. S., "Oil and Grease in Sewage," *Sew. Wks. Jour.*, **12**, 727 (July, 1940).
54. Faber, H. A., "Chlorinated Air Proves Aid in Grease Removal," *Water Wks. and Sew.*, **84**, 171 (1937).
55. Keefer, G. E., and Cromwell, E. C., "Grease Separation Enhanced by Aero-Chlorination," *Water Wks. and Sew.*, **85**, 97 (1938).
56. White, G. C., operating report to H. N. Jenks, Consulting Engineer, Palo Alto, CA, 1943.
57. Susag, R. H., "B. O. D. Reduction by Chlorination," *J. WPCF*, **40**, 434 (1968).
58. Anon. "Water Quality Control Plan San Francisco Bay Basin (2)," Calif. Water Quality Board, San Francisco Bay Region, 1982.
59. Rawn, A. M., "Chlorination in Sewage Treatment," *Water Wks. and Sew.*, **81**, 97 (1934).
60. Baity, H. G., and Bell, F. M., "Reduction of the Biochemical Oxygen Demand of Sewage by Chlorination," *N. Carolina Water and Sew. Wks. Assoc. J.*, **6**, 151 (1928).
61. Enslow, L. H., "Recent Developments in Sewage Chlorination 1926," Proc. 9th Texas Water Wks. Short School, p. 317, 1927.
62. Babbitt, H. E., and Baumann, E. R., *Sewerage and Sewage Treatment*, 8th ed. John Wiley & Sons, New York, 1965.
63. Skarheim, H. P. "Tables of the Fraction of Ammonia in the Undissociated Form for pH 6 to 9, Temperature 0–30°C, TDS 0–3000 mg/l, and Salinity 5–35 g/kg," San. Engr. Res. Lab., Sch. of Pub. Hlth., Univ. California, Berkeley, CA, SERL Report No. 73–5, June 1973.
64. Stone, R. W. "Rancho Cordova Breakpoint Chlorination Demonstration," a report prepared by Sacramento Area Consultants, Sept. 1976.
65. Stone, R. W., Saunier, B. M., Selleck, R. F., and White, G. C., "Pilot Plant and Full-Scale Ammonia Removal Investigations Using Breakpoint Chlorination," paper presented at the 49th Ann. Conf. WPCF, Minneapolis, MN, Oct. 5, 1976.

66. Saunier, B. M. "Kinetics of Breakpoint Chlorination and Disinfection," Ph.D. Dissertation, Univ. Calif., Berkeley, CA, 1976.
67. Livingston, J., private meeting, Los Angeles County Sanitation District, Hydrogen Sulfide Lab., Compton, CA, March 20, 1972.
68. Livingston, J., and Harland, R., "Control of Sulfides in Force Mains Using Compressed Air," in-house report to F. D. Dryden, Sanitation Districts of Los Angeles, CA, Nov. 24, 1971.
69. Beebe, R. D., Jenkins, D., and Daigger, G. T., "Activated Sludge Bulking Control at the San Jose/Santa Clara California Water Pollution Control Plant," paper presented at 55th Ann. Conf. WPCF, St. Louis, MO, Oct. 1982
70. Easley, R. S. and Vidal, B., private communication Sanitation Districts of Los Angeles County, Whittier, CA, Feb. 16, 1984.
71. Dobson, J. G., "The Treatment of Cyanide Wastes by Chlorination," *Sew. Wks. Jour.*, **19**, 1007 (Nov. 1947).
72. McKee, J. E., and Wolf, H. W., "Water Quality Criteria," Sacramento, Calif. St. Water Qual. Control Bd., 2nd ed., 1963.
73. Chamberlin, N. S., and Snyder, H. B., Jr., "Technology of Treating Plating Wastes," Tenth Ann. Wastes Conf., Purdue Univ., May 9–11, 1955.
74. Zehnpfennig, R. G., "Possible Toxic Effects of Photographic Laboratory Wastes Discharged to Surface Water," *Water and Sew Wks.*, **115**, 136 (1968).
75. Ryan, J. A., and Culshaw, G. W., "Use of p-dimethylaminobenzylidene Rhodamine as an Indicator for the Volumetric Estimation of Cyanide," *Analyst* (England), **69**, 370 (1944).
76. Aldridge, W. H., "New Method for the Estimation of Microquantities of Cyanide and Thiocyanate," *Analyst* (England), **69**, 262 (1944).
77. Epstein, J., "Estimation of Microquantities of Cyanide," *Anal Chem.*, **19**, 272 (1947).
78. Herting, Otto, "Analysis of Commercial Cyanides. Estimation of Cyanic Acid," *Zeit. Angew. Chem.*, **14**, 585 (1901).
79. Marks, H. C., and Chamberlin, N. S., "Determination of Residual Chlorine in Metal Finishing Wastes," *Anal. Chem.*, **24**, 1885 (Dec. 1952).
80. Carmichael, D. C., "Cyanide Wastes at Dupont," *Cons. Engr.* (Dec. 1952).
81. Weyermuller, G., and Morris, H. E., "Monsanto Controls Chemical Waste Disposal," *Chem. Processing* (Oct. 1955).
82. Lowder, L. R., "Modifications Improve Treatment of Plating Room Wastes," *Water and Sew. Wks.*, **115**, 580 (1968).
83. Hill, E. A., and Neff, F. J., "Cyanide Waste Oxidized in the Plating Room," *Plating* (Aug 1957).
84. Chamberlin, N. S., and Griffin, A. E., "Chemical Oxidation of Phenolic Wastes with Chlorine," *Sew. and Ind. Wastes*, **24**, 750 (June 1952).
85. Eisenhauer, H. R., "Oxidation of Phenolic Wastes," *J. WPCF*, **36**, 1116 (1964).
86. Chamberlin, N. S., "Application of Chlorine and Treatment of Textile Wastes," *Amer. Dyestuff Reporter* (June 21, 1954).
87. Faber, H. A., "The Hypochlorite Process for Treatment of Wool Scouring Wastes and for Recovery of Wool Grease," *Sew. Wks. Jour.*, **19**, 248 (1947).
88. White, G. C., unpublished chlorination results at Reno Rendering Wks., Reno, Nevada, 1956, 1968, 1969.
89. Granstrom, M. L., "Rendering Plant Waste Treatment Studies," *Sew. Wks. Jour.*, **23**, 1012 (1951).
90. Granstrom, M. L., "Rendering Plant Wastes Treatment Studies # II Pilot Plant Investigation," *Sew. Wks. Jour.*, **24**, 1478 (1952).
91. Halvorson, H. O., Cade, A. R., and Fullen, W. J., "Recovery of Proteins from Packinghouse Waste by Superchlorination," *Sew. Wks. Jour.*, **3**, 488 (1931).
92. Bizzari, R. E., Popeck, J. R., Pickard. D. W., and Drapp, J. E. "H_2S Odor Control on Tampa's Major Sewerage Systems," paper presented at 55th Ann. Conf. WPCF, St. Louis, MO, Oct. 5, 1982.

93. Beebe, R. D., amd Jenkins, D., "Control of Filamentous Bulking at the San Jose/Santa Clara Water Pollution Control Plant," paper presented at 53rd Ann. Conf. Calif. Water Poll. Control Assoc., Long Beach, CA, April 29, 1981.
94. Adams, A., private communication, Envirotech Operating Services, San Mateo, CA, 1980.

8
Disinfection of Wastewater

CONCEPTS OF DISINFECTION

Importance of a Chlorination Disinfection Facility

One of the most useful features of a wastewater disinfection facility is its ability to serve as a monitor for the combined wastewater treatment processes. If all the unit processes are not performing properly the chlorination system will not be able to deliver an effluent that meets the NPDES discharge requirement. This important attribute of disinfection is often overlooked.

Importance of Disinfection

Today the emphasis is on the disinfection of all wastewater effluents in the United States. Sewage disinfection is defined as the process of destroying pathogenic microorganisms in the wastewater stream by physical or chemical means. This is best accomplished by the use of chemical agents such as aqueous solutions of chlorine, chlorine dioxide, hypochlorite, bromine, bromine chloride, ozone, or combinations of these chemicals. Other means which have been used in special conditions and with varying degrees of success are ultraviolet light radiation, gamma radiation, sonics, heat, and silver ions. Present disinfection practices depend almost exclusively on chlorine compounds.

From the viewpoint of health, the disinfection process is the most important stage of wastewater treatment. The objectives of wastewater disinfection are: to prevent the spread of disease and to protect potable water supplies, bathing beaches, receiving waters used for boating and water contact sports, and shellfish growing areas.

Public Health Agency Perspective. The public health agency is deeply committed to the theory of multiple barriers or multiple points of control between a sewage discharge and a water supply intake. These barriers or points of control include wastewater treatment, land confinement, dilution, time, distance, and potable water treatment. Any type of treatment is fallible, so reliance on natural barriers should be maintained as long as possible. Where the natural barriers are eroding, increased emphasis and reliability must be placed onto the artificial barriers of treatment processes. The factors which operate against and deteriorate the effectiveness of natural barriers are: increased population, increased mobility of population, increased recreation, more leisure time, increased sewage discharge, and increased water use. Inequities of rainfall distribution combined with increased water consumption produces increased water recycling of wastewater. This leads to an overall decrease in dilution, time, and distance factors between sewage discharges and potable water intakes.

All of these factors support the efforts of regulatory agencies to improve the quality of wastewater effluents prior to their discharge into the environment. For many areas of use this is the only protection available. *Therefore, disinfection is the last remaining barrier against the transmission of waterborne diseases.*[1]

Legal Concepts. English law has used the doctrine of public nuisance to abate public health hazards. An examination of centuries-old common-law precedents emphasizes that doctrines of common-law nuisance were directed toward abatement of risk rather than abatement of disease. There was no need to show that disease had actually occurred nor was likely to occur. The law operates on common sense public health recognition that conditions conducive to disease should be eliminated.[113] There are many public health hazards today that are the proper subject of judicial scrutiny and judicial action despite conflicting or contradictory evidence on the other side.

It is health risk rather than health harm that is the subject of the action.

Lawyers who practice in the field of environment and public health protection are usually significantly more conservative than the EPA or most state public health agencies. These lawyers advocate maximum control of pathogenic organisms. This philosophy derives from numerous cross-examinations of scientific experts. Such intensive examination over the years has led to the conclusion that the scientific community knows little about the capability of the environment to assimilate the wide variety of pollutants discharged daily.

The legal profession is critical of the EPA position which seeks to justify the "cost-effective" analysis because health risks from discharge of waste into receiving waters are impossible to quantify. Moreover the lawyers cannot accept putting costs and benefits on these risks because there are always costs and other variables that are incapable of quantification.

Contemporary Practices in the USA. Over the last 25–30 years the various regulatory agencies, local, state, and now the EPA (Environmental Protection Agency) have been seeking a set of guide lines for proof of disinfection in various receiving water situations.

The most persistent and aggressive pursuit of wastewater disinfection requirements has been by the California State Department of Public Health, United States. The adequacy of disinfection is evaluated in compliance with a prescribed MPN (most probable number) of coliform organisms as determined by *Standard Methods*[2] for the "confirmed" test procedure. For example, in ocean and saline bay waters used for recreation, 80 percent of the receiving water samples must fall within a coliform MPN of 1000/100 ml. This is approximately equivalent to a median MPN of 230/100 ml.* For other situations, more restrictive median or average coliform concentrations may govern. The discharger is required to disinfect to the degree necessary to maintain a suitable quality in the receiving waters. Often the discharger is given the option of demonstrating compliance by meeting the receiving water quality in the effluent itself. This eliminates the costly process of monitoring several sampling stations in the receiving waters. In most situations the discharger usually is required to apply the disinfection requirements only to the effluent quality. This practice has evolved to a point in the State of California where practically all cases are based on the MPN total coliform in the plant effluent.

The evolution of these requirements is of particlar interest because it relates largely to geographical considerations.

Other states of the union for various reasons have not been as concerned with the protection of receiving waters as has California. The predominant factor which led to the adoption of the aforementioned coliform requirements revolves around the extensive development and use of the California coastline and coastal waters for recreational and shellfish growing purposes. The California coastline is some 720 miles in length. Along this coastline are some of the most beautiful bathing and water sports beaches in the world.

Evolution of Disinfection Requirements. It was the fouling of one of the most beautiful expanses of resort beach areas in California that captured the attention and energies of the California State Department of Public Health (see Fig. 8-1). Beginning about 1920, the Bureau of Sanitary Engineering a division of this agency under the able direction of C. G. Gillespie, entered into a twenty-yr. battle with the City of Los Angeles to clean up their Hyperion outfall discharge which was solely responsible for the fouling of a ten-mile stretch of this magnificent beach. Obviously Los Angeles was the huge stumbling block in the overall California antipollution program envisioned by Mr. Gillespie and his colleagues.

In order to convince the City of Los Angeles that the Hyperion outfall discharge (170 mgd) was polluting this beach area (which currently attracts about one million people on summer weekends) the Bureau of Sanitary Engineering, California State Department of Health, made a year-long study of ten miles of beach in Santa Monica Bay in 1941–1942.[3]

The report of this investigation dated June 26, 1943 led to the immediate quaran-

---

*The MPN of 230/100 ml is used when the statistical procedure utilizes a 5-tube dilution. For a 3-tube dilution the MPN would be 240/100 ml.

Fig. 8-1 Los Angeles County Beach, California, 1974 (Los Angeles Times Photo).

tine of this ten-mile stretch of beach, on the grounds that both the beach and surf waters were polluted with sewage and were therefore dangerous to one's health. After a length of time considered sufficient to correct this hazardous condition of gross pollution, the California State Board of Public Health took the City of Los Angeles to court on a suit based on pollution as determined by the "coliform count" in the bathing waters. Other factors were considered but it was the coliform count that became the most persuasive piece of evidence.

In their law suit against the city, the State Board of Health maintained that 1000 *Escherichia coli*/100 ml as a limiting standard in the surf waters assured safety for the bathers. The State of California won the suit because the judge hearing the case held that the coliform standards used by the Board of Public Health were reasonable after it was demonstrated that there was clear evidence of pollution and physical nuisance in areas where these bacterial limits were exceeded.

This historic case established a statistical coliform concentration baseline which defines the difference between polluted and pollution-free recreational contact waters in the open surf. This has been accepted by the Sanitary engineering profession in the United States as a landmark achievement.

Rationale for Coliform Concentration Requirements. The most comprehensive study of inland recreational waters by Stevenson[4] in 1950–1953 concluded that an MPN coliform concentration of 2300/100 ml may be a threshold quality associated with an increase in the incidence of disease. For those who may have an interest in the comparison between fecal coliform and total coliform, the use of data developed some years later on the fecal coliform content of the same waters and by the use of ratios, a geometric mean fecal coliform content of 400/100 ml was determined to be equivalent to the 2300/100 ml total coliform number.

In theory two different standards could be set for a potable water and for a wastewater discharge. One would be a standard of assured safety where there is no health concern. The other standard would be the threshold of unsafe water (2300/100 ml) and might be used as an indication that quarantine action or abatement action should be undertaken. The California State Department of Health has recommended discharge requirements which, based on experience and judgment, will result in a water quality where there is no health concern. In other words their quality requirements reflect an inherent factor of public health safety, whether it be potable water, wastewater discharging to variable use receiving waters, or water reuse situations. These concepts are described as follows:

1. The limit for surf waters is about 500 times the pollution allowed by the United States Public Health Service standards for potable water (2.2 MPN/100 ml). The 2.2 standard has been universally accepted by water and health experts for many years. Comparing relative ingestion of potable water versus seawater in the course of surf bathing (2–3 ml swim), the figures seem compatible.

2. There is no scientific evidence to indicate that water within this standard causes ill health.

3. The level of indicator organisms is seldom reached or exceeded in saline waters where the cause is not obviously recent waste contamination.

4. A less severe standard might show "approved" areas to lie within visible areas of grease and detritus of waste origin and would therefore seem to be lacking in common sense and decency.

The application of this standard for disinfection first appeared as a required chlorine residual after a specified contact time which would produce the desired quality in the receiving waters. At this time it was thought that disinfection would meet these standards if a 0.5–0.75 mg/liter orthotolidine residual at the end of 30 min. contact time was accomplished. Contact chambers were built to give a theoretical 30 min, detention time at average flow based on the volume of the chamber. Effective mixing and contact chamber short circuiting was never considered. During this period a great many chlorination facilities went into operation (1947–1957), and it became obvious that a residual-contact period requirement often produced effluents of quite different bacterial quality at different plants. After many years of testing and surveillance of these installations, the Bureau of Sanitary Engineering of the California State Department of Health concluded that it was practical, feasible, and superior to prescribe a coliform count directly to the plant effluent, rather than a chlorine residual value as evidence of disinfection.

Current Coliform Requirement in California. The numbers presently in effect are: 80 percent of samples less than 1000/100 ml for coastal bathing waters (equivalent to a median of 230/100 ml); median of 70/100 ml for shellfish growing areas, and a median of 23/100 ml for confined waters used for bathing or other water contact sports assuming the dilution is at least 100 to 1. The requirement for discharge into ephemeral streams or other areas where public exposure to effluents receiving little dilution is for an essentially coliform-free effluent, that is, a median MPN not greater than 2.2/100 ml. There is a subtle implication of the necessity for good operation and adequate treatment to achieve the 23/100 ml requirement. This has been found to be a meetable standard, but it requires a properly operated treatment plant with an effective disinfection system to consistently achieve it. The severe effluent standard of 2.2/100 ml implies the necessity for some type of advanced treatment prior (i.e. filtration) to disinfection to reliably meet this level of disinfection effectiveness, and thereby suggests some virus removal capability for the system beyond that which normally occurs. These are not alternative requirements; only implications of what might be needed to meet the requirement.

There may be cases where the quality of the plant effluent bears no relation to the receiving water quality. For example, a receiving water might have a consistent coliform concentration as high as 2300/100 ml or even greater. If the State Department of Health had decided that disinfection of a wastewater discharging to such a receiving water was necessary, the discharger would not be allowed to simply meet the water quality of the receiving water (2300/100 ml or greater) because this would not be evidence of effluent disinfection.

Administration of Requirements. Administratively, this is how the control system presently operates in California:

1. The discharger applies to the appropriate Regional Water Quality Control Board for permission to discharge wastewater at a given location.
2. The Board then notifies all interested agencies for recommendations.
3. The State Department of Public Health submits its recommendation for the disinfection requirement to the Regional Board.
4. The Board then holds a public hearing to discuss and establish the requirements with the discharger.

The Regional Water Quality Control Board can issue cease and desist action on a discharger for violation of any portion of the requirements including the requirements on disinfection. Further, the Board can and has placed a ban on further connections to the discharger's collection system if the requirements are not met and impose a heavy fine ($10,000) for each day of violation.

Total Coliforms Versus Fecal Coliforms as a Standard. The fecal coliform determination is the latest in a long history of selective tests to separate the strains of coliform bacteria found in wastewater. It has been largely adopted by the United States Environmental Protection Agency for various disinfection requirements: shellfish areas, recreational waters, etc.

Coliforms from the intestines of dogs and cats are mainly *E. coli*. The fecal coliforms from humans and livestock account for about 97 percent of the total. Fish do not have permanent coliform flora in their intestines. The presence of coliforms in fish is evidence of pollution in the water of their habitat. Fecal coliforms are not abundant in soil (as in *E. coli*) since they die off rapidly when deposited in the soil.[5]

Currently, there are no means for distinguishing between fecal coliforms of man and those of other warm blooded animals.[6] Their presence in significant numbers is, however, indicative of fresh pollution. All fecal coliforms in a stream may be accepted as being of fecal origin, whereas an unknown portion of the total coliform bacteria may be of other origins. Determination of fecal coliforms provides a superior indicator of fecal pollution. There is no argument on this point. The question remains: Is the fecal coliform concept a valid parameter for disinfection? The differential fecal coliform test demonstrates the ability to enumerate coliform bacteria originating from fecal sources while suppressing those of soil origin. Consequently, it is a very useful tool in the Sanitary Survey of surface waters where numerous differing sources can contribute to the total coliform content. The California Department of Health has used both the total coliform and fecal coliform enumeration in its surface water studies since 1960. It is their opinion that if a single indicator test for bacteriological quality were to be used, the fecal coliform test would be the best for *fresh water* areas because of its selective ability.[7] While the use of a fecal coliform number as a river water quality objective is appropriate, the use of a fecal coliform standard for a measurement of disinfection would not be appropriate. All available data indicate that the fecal coliform strains are more fragile than the total coliform group and can be more easily destroyed or inactivated by disinfection or natural purification processes. It has been amply demonstrated that on the basis of chlorine demand tests, fecal coliforms can be completely destroyed while significant numbers of total coliforms remain.[5]

There is only sketchy information on the relative resistance of pathogenic agents to chlorination. The data suggests that bacterial agents may be as hardy as coliforms while most viruses are more resistant. The total coliform group is a more conservative indicator of effective disinfection and is more numerous in wastewater than is the fecal coliform group. It is conservatively estimated that the ratio of fecal coliforms to total coliforms in saline waters is 1:70. In San Francisco Bay it has been observed that when the ratio of fecal coliform to total coliforms, is 1:70, that area is generally out of the known pollution areas.[8] However, there is a considerable difference in the fecal to total coliform ratio from saline to fresh water areas. Instead of the 1:70 ratio observed above it is estimated that in wastewater effluents the total coliforms contain about 30 percent fecal coliforms.[9]* This comparison is most significant when disinfection is related to a final coliform count rather than a log reduction in the total coliforms present before disinfection.

Disinfection Efficiency: Bacteria Survival Versus Percent Kill.

The effectiveness of a disinfectant dose for a given contact time is usually expressed as a ratio of logs reduction of initial to final bacteria count, or as a percent destruction of the initial bacteria count. In order to keep the proper perspective when evaluating reports of disinfecting procedures, the minimum acceptable surviving number of total coliforms should not be in excess of 1000/100 ml MPN. For comparing disinfectants of secondary effluents the final MPN should be no more than 230/100 ml, total coliform concentration. Let us see what this means when looking at studies reporting log reduction versus percent destruction.

Assuming a well oxidized secondary effluent, the total coliform concentration before disinfection is probably on the order of 1×10^6 (1,000,000), so a 4-log reduction would produce a final MPN of 100/100 ml, and a 99 percent reduction would yield a final count of 10,000/100 ml and a 99.99 percent reduction would yield 100/100 ml. So for this magnitude of initial count a 99.99 percent kill is the same as a 4-log reduction.

---

* During the past decade, 1975–1985, the consensus has changed to 25 percent.

A well oxidized and filtered tertiary effluent would probably have an effluent coliform count of 50,000/100 ml before disinfection. The coliform requirement for a disinfected tertiary effluent is usually 2.2/100 ml. Therefore, a log reduction greater than 4 is required, and the percent kill must be greater than 99.99. The point to be made is that disinfection efficiencies reported as 99 or 99.9 percent are meaningless. Actually, disinfection efficiency studies based on coliform destruction should specify the range of initial coliform as MPN/100 ml in addition to the log reduction. Then the mathematical model developed by Collins et al. in 1970[10] can be used for verification. This model is as follows

$$y/y_0 = [1 + 0.23\ ct]^{-3} \tag{8-1}$$

where

$y_0 =$ initial coliform MPN/100 ml
$y =$ final coliform MPN/100 ml
$c =$ chlorine residual, mg/liter, at the *end* of contact time t
$t =$ contact time in min.

This model has been verified by several practitioners and researchers since it was first published.

Bacterial Indicator Concepts. The parameter of major importance that would provide proof of disinfection is the resistance of the indicator organism to the disinfectant. Researchers over the years have expressed dissatisfaction with the coliform group as an indicator organism because it is not resistant enough to chlorine to allow any safety factor. For a group of organisms to be an ideal indicator, the following conditions should demonstrate disinfection efficiency:

1. The indicator organism must be more resistant to disinfection than the pathogenic organisms.
2. The indicator must be present in the sample whenever pathogenic organisms are present.
3. The indicator must occur in greater numbers than the pathogens.
4. A simple, rapid, and unambiguous procedure must be capable of enumerating the indicator organisms.
5. The indicator should not regrow or otherwise increase in numbers in the aquatic environment after disinfection.
6. The indicator organism must be randomly distributed in the influent stream.
7. The presence of other organisms must not inhibit the growth of the indicator organism.
8. The indicator organism should be nonpathogenic to man.

The lack of disinfection resistance of the coliform group was clearly demonstrated when the organism *Klebsiella* was found in the City of Chicago's water distribution system.[11] The members of the *Klebsiella* genus can cause severe enteritis in children, and pneumonia and upper respiratory tract infection, septicemia, meningitis, peritonitis, and urinary tract infection in adults. Discovering such a hazardous organism in a distribution system, carrying water which had been coagulated, filtered, and chlorinated sufficiently to carry a small residual in the system, was indeed unsettling. In this instance it was concluded that *Klebsiella* was the organism responsible for most of the positive samples occurring in the routine coliform sampling procedures. The *Klebsiella* group are encapsulated organisms, and once in the distribution system, may be harbored in protective slime and sediment. For these reasons they are more resistant to disinfection than the coliforms. In 1974, Engelbrecht et al.[12] evaluated two promising groups of organisms believed to be resistant to chlorine in the range necessary to inactivate both bacillary pathogens and waterborne viruses. These groups are the acid-fast cultures *Mycobacterium fortuitum* and *M. phlei*, and a yeast *Candida parapsilosis*.

However, it must be recognized that the concept of proof of disinfection for wastewater discharges and water reuse situations is entirely different from that of water to be used for potable purposes. With the situation as it exists in 1976 with respect to the organics in wastewater, it is unlikely that it will be possible to pursue the reclamation of wastewater for potable use. Therefore, proof of disinfection for wastewater and water reuse should always be on the basis of destruction of the indicator organisms. This could conceivably be on the basis of a chlorine residual contact time envelope provided this combination of criteria could be developed for a "consensus" organism.[13,14]

WASTEWATER REUSE

Historical Background. Wastewater reclamation, in the United States, has been practiced since about 1920. One of the first systems to use the total discharge for beneficial use was the activated sludge sewage treatment plant in the Golden Gate Park of San Francisco. This operation began in about 1930. In 1935, there were 62 communities using treated wastewater for crop irrigation and regulations for this use had been in existence for about 15 yr. in California. As of 1970, there are about 600 reclamation projects operating on a continuous basis in the United States. About half of these are in the State of California. The first use, of course, was for crop irrigation. These crops were fiber, fodder, and seed crops, and there was little opportunity for public contact. Use of reclaimed water for landscape irrigation (i.e., irrigation of parks, playgrounds, golf courses, freeways, rights-of-way, and so forth) also has a fairly long history, but it wasn't until the last 10 or 15 yr. that there has been a sudden increase in the number of such installations. Where crop irrigation installations have increased 40 percent, landscape irrigation installations have multiplied nine-fold. There is a trend, then, and a general swing toward the uses of reclaimed water where the public may have more exposure and more contact. Public attention has been drawn to many outstanding reclamation systems in California such as the Santee project which includes swimming in addition to boating and fishing; and to the Contra Costa project of supplying industry cooling water needs from reclaimed wastewater.

Moreover, it is a matter of fiscal responsibility for any agency producing water and treating wastewater to have plans for wastewater reclamation. Otherwise such an agency would not qualify for Federal funds.

California Requirements. In 1968, the California State Board of Public Health adopted standards for the quality of reclaimed water used for crop irrigation, landscape, and for recreational impoundments. It is important to realize that scientists from all fields acknowledged that the establishment of specific quality limits for reclaimed water in terms of BOD, suspended solids, either soluble materials, or other commonly used criteria, was not practical. The monitoring cost would be too great for smaller reclamation operations. The group decided to use as the keystone of the standards, the coliform bacteria concentration in the reclaimed water. The coliform bacteria concentrations which are allowable for different reclamation uses are supplemented in the standards by terms such as "oxidized wastewater," "filtered wastewater," and other descriptive terms which broadly identify the type of reclaimed water that is required without specifically identifying numerous quality limits. Briefly these adopted standards are as follows:[15]

Primary Effluent. It can be used for surface irrigation of processed food crops, orchards, and vineyards; irrigation of fodder, fiber and seed crops. Virtually no public contact or ingestion is possible.

Oxidized Effluent (Secondary). With a median coliform MPN of 23/100 ml it can be used for landscape irrigation, spray irrigation of processed food crops, landscape impoundments, and milk-cow pastures. Public contact is possible but ingestion is very unlikely.

Oxidized Effluent (Secondary). With a median coliform MPN of 2.2/100 ml it can be used for surface irrigation of produce (makes this operation impractical) and restricted recreational impoundments. Public contact and minor ingestion is possible.

Filtered Effluent (Tertiary). With a median coliform MPN 2.2/100 ml it can be used for spray irrigation of produce and unrestricted recreational impoundments. Public contact and minor ingestion are likely.

Other Applications. At present there are many cooling water applications using secondary effluent meeting the 2.2/100 ml MPN coliform requirements. Water reuse is here to stay, at least in California. It is used extensively to recharge groundwater supplies which accomplish two other major objectives: (1) provide a salt water barrier; and (2) prevent land subsidence due to withdrawal of natural gas and oil reserves.

Importance of Disinfection. Disinfection is the most important link in this chain of treatment. Chlorine must be relied upon to do the bulk of the disinfecting and to provide a persisting residual. Ozone may be called on to provide the assurance of viral inactivation.

The Environmental Health Laboratory at the Hebrew University, Jerusalem, Israel has been studying the health risks resulting from spray irrigation of nondisinfected wastewater. Their studies were undertaken to obtain data about the number and types of enteric bacteria dispersed into the air as a result of spray irrigation. Katzenelson and Teltch[47] discovered that coliform bacteria were found in the air at a distance of 400 yd. downwind from the irrigation line, and in one case a *Salmonella* bacterium was isolated 65 yd. from the source of irrigation. These findings inspired an epidemiological survey of the incidence of enteric communicable diseases in 77 agricultural communal settlements practicing spray irrigation with nondisinfected partially treated oxidation pond effluent as compared with those in 130 similar settlements not practicing any form of wastewater irrigation. Katzenelson, Buium, and Shuval[48] reported in 1976 that the incidence of shigellosis, salmonellosis, typhoid fever, and infectious hepatitis was two to four times higher in communities practicing wastewater irrigation. Moreover, it was found that there were no differences in the incidence of enteric diseases between these two communities during the winter nonirrigation season. *The result of these studies has brought about recommendations for strong wastewater-treatment measures including effective bacterial and viricidal inactivation by disinfection to prevent the spread of enteric diseases due to airborne contamination of communities adjacent to spray irrigation projects.*

Viruses

The Virus Hazard. It is of some concern that the frequency of infectious hepatitis has remained at a static level of 50,000–60,000 cases per year in the United States whereas the incidence of typhoid fever has dropped from approximately 2000 cases in 1955 to only 300 cases in 1968.[13] This indicates that although waterborne bacterial infections have been all but eliminated, water utility people may have a more severe task when dealing with waterborne viral infections. The problem of viruses in water supplies has received a great deal of attention in recent years and is well documented.[16-21] In evaluating the virus hazard, two factors should concern the protectors of our water supplies, whether for wastewater disinfection or potable water treatment. These factors are the origin of infectious hepatitis and the origin of the significant increase in gastroenteritis. Gastroenteritis is not a reportable disease as distinguished from relatively well-defined illnesses, e.g., infectious hepatitis, shigellosis, salmonellosis, and typhoid. Yet it can be estimated that the number of gastroenteritis cases occurring per year is hundreds of thousands and possibly millions. (See Chapter 6). During the period 1961–1970 a total of 26,546 cases were definitely attributed to contamined water.[19] Of 52 waterborne-disease outbreaks in the United States in 1971–1972, there were 22 outbreaks of gastroenteritis, amounting to 5615 cases from a total of 6817 cases of waterborne illnesses.[20] The concern here is that these cases may be the result of some yet unidentified viruses. There are more than 100 viruses excreted in human feces that have been reported to be in contaminated water. Any of these could cause a waterborne disease.

Other factors of viral infections that are great cause for concern are included in the evidence researched by McDermott who points out that poliomyelitis virus has hurdled the technical barriers of water treatment of the Paris, France, water supply which consists of coagulation, filtration, and disinfection by both chlorine and ozone.[5]

In the same discussion McDermott refers to the work by Plotkin and Katz that the minimum infective dose by a virus is 1 pfu.[18] Although this statistic may be open to question, the contemplation of this situation and the possibility of 100 or more viruses capable of contributing to a waterborne disease should be of extreme concern to water producers.

All of these concerns are confirmed and substantiated by the 1970 report of the Committee on Environmental Quality Management, ASCE Sanitary Engineering Division.[16] Some of the important conclusions of this report are as follows:

1. There is no doubt that the virus of infectious hepatitis can be transmitted by drinking water and epidemiological opinion uniformly supports this conclusion.

2. Although evidence is scanty, it should also be assumed that the enteric viruses and other possible causative agents of viral gastroenteritis can be transmitted by drinking water.

3. There is no doubt that a positive coliform index means that virus may be present; however, absence of coliform may not mean that virus is absent. The coliform index, therefore, while a good laboratory tool, is not a reliable index for viruses. Greater assurance of the absence of virus would be a turbidity of less than 0.1 Jackson Unit and an HOCl residual of 1 mg/liter after a contact period of 30 min.

4. The evidence available indicates that a risk of hepatitis infection results from the consumption of raw or steamed undepurated shellfish taken from sewage polluted waters and that the Public Health Service Coliform Standard (70 coliforms/100 ml) has been shown by experience to be a reliable indication of risk-free shellfish waters. The Committee of Environmental Quality Management believes that a

high level of protection would be provided by activated sludge treatment and chlorination of the effluent to a level producing an amperometric chlorine residual of 5+ mg/liter after 30 min. contact.*

5. Virus multiplication in polluted water appears not to be a significant possibility, and from the control point of view it can be disregarded.

6. Viruses are present in certain river waters and failure to isolate them results presumably from their low concentrations and the relatively ineffective sampling and concentration procedures employed.

7. Enteroviruses and the virus of infectious hepatitis can survive for prolonged periods under conditions prevailing in drinking-water reservoirs. Long detention times therefore cannot be considered as a safety factor.

8. Enteric viruses differ in resistance to free chlorine. Adenovirus 3 is less resistant than *E. coli* while poliovirus 1 and Coxsackie virus A2 and A9 appear to be more resistant than any of the other enteroviruses studied.

Viruses in Sewage Contaminated Potable Water Supplies. In the United States the enteric virus concentration in raw sewage probably ranges from two or three to more than 1000 infectious virus units/100 ml with peak levels occurring in late summer and early fall.[17] If the coliform bacteria concentration of raw sewage is estimated at 10^7–10^8/100 ml organisms then the enteric virus concentration is perhaps 5–7 orders of magnitude lower.[22,23]

Although available evidence indicates that enteric virus concentrations in drinking water are likely to be very low it is important to be aware of the fact that as little as one virus infectious unit is probably capable of producing an infection in humans.[18] Most enteric virus isolations have been made from heavily polluted surface waters, but Berg and coworkers detected enteric viruses in Missouri River water having fecal coliform concentrations as low as 60/100 ml.[22] While virtually nothing is known about enteric virus levels in United States potable water supplies, monitoring of the potable water supplies of Paris, France, in the 1960s revealed that about 18 percent of the 200 samples examined contained enteric viruses; and the average virus concentration was estimated at one infectious unit/300 liter.[25] The heavily polluted Seine River is one of the major sources of the Paris water supply. Nupen (1974) and co-workers have reported finding enteric viruses in 10 liter samples of drinking water in South Africa.[26]

While there is little quantitative information available on enteric virus levels in sewage-contaminated surface and ground waters, there is plenty of evidence that wastewaters are a primary source of enteric virus contamination of man's environment. With the possible exception of a few poliomyelitis outbreaks, there is no evidence of waterborne outbreaks in the United States caused by other specific viruses. The most prevalent waterborne disease in the United States continues to

---

* Amperometric residual is stipulated to be the total chlorine residual with no reference made to any free chlorine residual fraction.

be gastroenteritis of unknown etiology. Therefore, for reasons described above it is imperative that wastewater-treatment processes and wastewater-reuse systems address themselves to this virus hazard. See Chapter 6.

Virus Inactivation:

General Discussion. To date, the most comprehensive study of virus inactivation on a pilot-plant scale has been the work by the County Sanitation Districts of Los Angeles County. The results of this two-year study are contained in the "Pomona Virus Study—Final Report" prepared for the California State Water Resources Control Board and the U.S. Environmental Protection Agency February 1977. This work has been summarized in a paper by Selna, Miele, and Baird.[27]

The Sanitation Districts of Los Angeles County have been active participants in various water reuse programs since the mid 1950's. Various discharges of disinfected secondary effluent have been spread in percolation basins for the purpose of replenishing groundwater which is used for domestic supplies. Conveyance of this water to the percolation sites is through open flood control channels and in transit is unintentionally used for recreational activities including body contact. About 80 mgd of chlorinated secondary effluent is subject to this unplanned recreational use in flood control channels. These channels are classified as "unrestricted recreational impoundments" by the California State Department of Health, therefore wastewater discharged to such channels must comply with Title 22 of the California Administrative Code. This document contains the effluent quality and treatment system requirements for recreational reuse as decreed by the California State Department of Health. In order to qualify for such use, secondary effluent must be coagulated, settled, filtered, and disinfected to achieve a median total coliform MPN of 2.2/100 ml or less, and to provide an effluent which will protect swimmers against viral illnesses. Since this required treatment is expensive, both from a capital and operational standpoint, the prime objectives of the Pomona study was to investigate alternate methods of tertiary treatment that might be more cost effective than the required Title 22 System and still produce an effluent with the required degree of public health protection.

Treatment System Studies. This work investigated the following four systems:

1. The Title 22 treatment called for by the California State Department of Health was a 40 gpm secondary effluent (NH_3-N, 20 mg/l) treated with alum and ionic polymer followed by flash mixing, flocculation, sedimentation, dual media filtration, and disinfection.

2. A 25 gpm secondary effluent similar in quality to 1 was treated with alum and ionic polymer followed by flash mixing, dual media filtration and disinfection.

3. A 100 gpm secondary effluent also similar in quality to 1 and 2 was treated using two-stage carbon absorption followed by disinfection.

4. 25 gpm of a nitrified secondary effluent (NH_3-N, 0.1 mg/l) similar in quality

to the others, was treated with alum and anionic polymer followed by flash mixing dual filtration and disinfection.

Disinfection Process. Chlorine. Systems 1, 2, and 3 were capable of disinfection by either ozone or chlorine. System 4 was disinfected with chlorine only since nitrification is not considered to enhance the performance of ozone.

Chlorine was applied in systems 1, 2, and 3 to produce two levels of chlorine residual at the end of the contact chamber (i.e., 5 and 10 mg/liter). These residuals required chlorine dosages of about 10 to 15 mg/liter respectively.

In system 4 the free chlorine residual at the outlet of the contact chamber was maintained at 4 mg/liter. This required a chlorine dosage of 10 mg/liter.

All chlorine residuals were measured by the DPD-FAS titrimetric method.

Careful attention was given to the degree of mixing at the point of application. A ⅓ hp flash mixer was installed in a 32 gal confined mixing chamber. This calculates to a G factor of 450. A conventional perforated-type chlorine diffuser was used discharging about three in. from the mixer impeller.

The chlorine contact chambers were designed to have plug flow characteristics. After an optimization study to achieve a coliform MPN of 2.2/100 ml a detention time of 120 min. was selected. A tracer study of the prototype contact chamber revealed a modal time of 98 min. and a minimum time of 58 min. So for the sake of comparison with laboratory studies it might be stated that the bacterial kills and virus inactivation were accomplished with a one-hr contact time and not a two-hr time.

Ozone. The ozone system was arranged to provide a *dosage* level of 10 mg/liter in system 1, and 10–50 mg/liter in system 2, and 6 mg/liter in system 3. In each case the ozone contact time was 18 min. The ozone contactors consisted of six 14-inch diameter PVC columns 18 ft high.

Predisinfection Effluent Quality. The quality of the wastewater at the point of disinfection is of considerable interest. The nonnitrified effluent contained 20 mg/liter NH_3-N, with suspended solids on the order of 1.5 mg/liter and turbidities of 1–1.5 FTU.* The pH was 7.5 and TDS was 580 mg/liter. At this pH the undissociated HOCl in system 4 was about 50 percent.

Results. The virus inactivation results from the Pomona study are illustrated in Figs. 8-2, 8-3, and 8-4. The effectiveness of combined chlorine residual is a real surprise, which advances a totally new concept: *that chloramines do in fact have viricidal efficiency potentially equal to that of free chlorine.* Until the revelation of the Pomona study it was believed that the only viricidal chlorine compound was free chlorine (HOCl). This is not seen to be the case for tertiary effluents.

Moreover, it was found that system 4 using free chlorine increased the predisinfection effluent concentration of chloroform from 0.6 to 198 micrograms per liter, while the other systems using all combined chlorine increased the chloroform

---
* FTU = turbidity measurement using Formazin polymer.

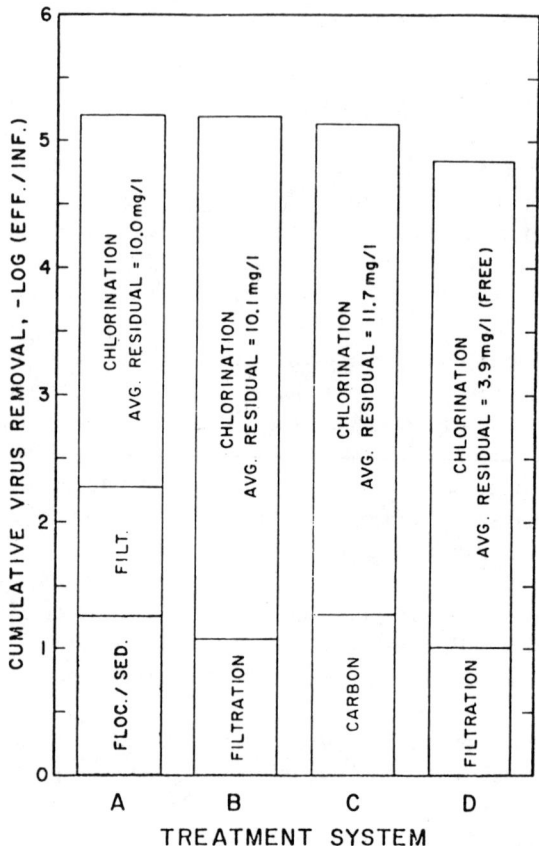

Fig. 8-2 Virus removal at high chlorine residuals.

content on the order of 1.0 to 6.0 micrograms per liter (approximately). This observation plus the demonstrated effectiveness of chloramines as a viricide leads to the conclusion *that there is no benefit from free residual chlorination of tertiary effluents which are required to* meet the Title 22 requirements of the California State Department of Health *where indirect potable reuse is called for.*

A dominant feature of Fig. 8-2 is that the majority of virus removal occurs in the disinfection step. The data presented in this figure indicate that reliance must be placed on the disinfection step rather than the filtration or carbon absorption steps prior to disinfection.

Other Conclusions. Other important conclusions drawn from the Pomona study are:

1. Virus inactivation in tertiary treatment systems employing combined chlorine residuals of 5–10 mg/liter ranged from 4.7 to 5.2 logs. These results were obtained

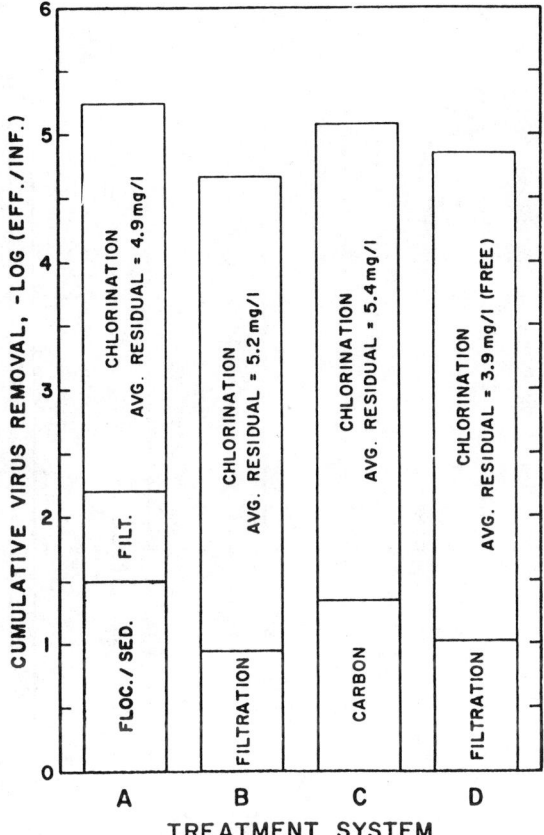

Fig. 8-3 Virus removal at low chlorine residuals.

in poliovirus seeding experiments. The additional virus removal at the 10 mg/liter residual over the 5 mg/liter residual was minimal but the coliform kill was consistently better at the higher residual.

2. The system employing free residual chlorination produced about 4.9 logs virus removal when the average free chlorine residual was 4 mg/liter.

3. All of the chlorination studies indicated consistent capability for attainment of the 2.2/100 ml MPN coliform requirement by all of the systems.

4. In the experiments using ozonation, virus removal ranged from 5.1 to 5.5 logs (see Fig. 8-4); however, attainment of the 2.2/100 ml MPN coliform standard was hampered by water quality variations.

5. Based on the results of the virus experiments it was concluded that system 2 (direct filtration) or system 3 (carbon adsorption) tertiary treatment systems are more cost effective than system 1 (the one required by the California State Dept. of Health).

494 HANDBOOK OF CHLORINATION

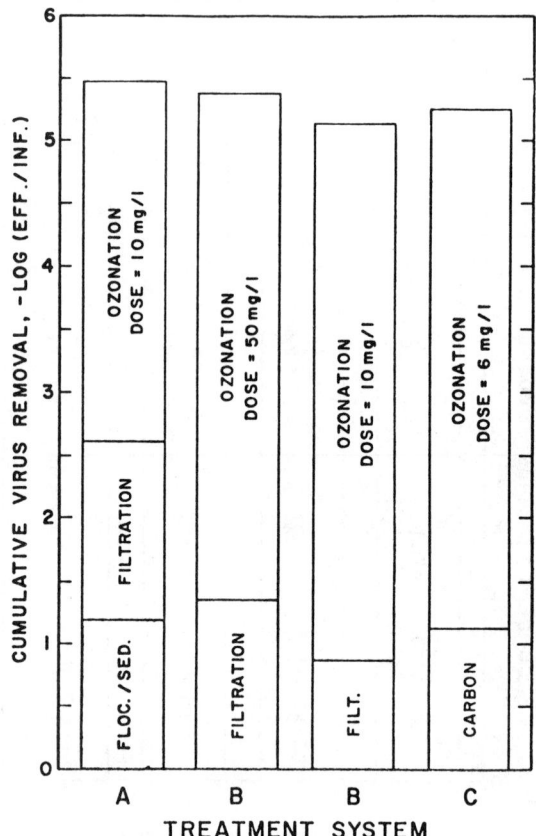

Fig. 8-4 Virus removal at ozonation experiments.

6. The direct filtration system (2) is the lowest cost with chlorination and is estimated at 13.7¢/1000 gal. for total capital and operating cost. This compares to system 1 at 21.5¢/1000 gal. with chlorine, 17.2¢/100 gal. for system 3 with chlorine and 19.9¢/1000 gal. for system 4 with free chlorine. The systems using ozonation were more costly.

Comparison with Other Investigations. It is difficult at best to compare virus inactivation studies owing to evaluation of detection limits, and seeding and analytical procedures; however, it is worthwhile to compare the Pomona study which is an outstanding investigation with other noteworthy virus inactivation studies as follows:

Ludovici et al.[28] performed a large number of pilot-plant experiments using the tertiary effluent from the Tucson, Arizona wastewater treatment plant. This effluent ranged in *pH* from 7.3 to 7.6; NH_3-N from 2.8–5.5 mg/liter; organic N.

1.4–5.6 mg/liter; BOD 1.5–8.5 mg/liter; and COD 15–33 mg/liter. No turbidity information was given, but obviously this is a high quality effluent except for the 5.6 mg/liter organic N which will have some inhibiting effect on the chlorination process. The effluent was seeded variously with polio 1, Coxsackie B1, and Coxsackie B2. The chlorine dosages were 2 and 4 mg/liter which means that all the residuals reported at the end of 30 min. contact time had to be combined chlorine. These were 0.31–1.2 mg/liter for the 2 mg/liter dose and 0.99–2.76 mg/liter for the 4 mg/liter dose in the polio 1 experiments. The mean reduction for the 2 mg/liter dose was 96.19 percent and a mean of 99.3 percent for 4 mg/liter—all at 30 min. contact time. The Coxsackie B1 and B2 viruses were much more susceptible to the combined chlorine residuals where a mean of 99.8 percent reduction was achieved with a 4 mg/liter dose for Coxsackie B1 and a 100 percent reduction in the Coxsackie B2 experiments with the same chlorine dose. Concurrently with these experiments Ludovici et al investigated destruction of total coliforms. With a maximum Y_0 coliform MPN of 19,100/100 ml,* the maximum final coliform MPN during the Coxsackie B2 seeding experiment was 1.1/100 ml with a 4 mg/liter chlorine dose, and a 30 min. contact time. The chlorine residual range was 1.5–2.94 mg/liter. The same results occurred in the polio 1 seeding experiments demonstrating the great enhancement of the chlorination process by a high-quality effluent having a low-initial coliform count.

These virus studies by Ludovici et al.[28] are noteworthy and deserve some comment. For virus inactivation, reporting destruction as a percent of the original number of organisms is a good comparative method, but may not be compatible with what a regulatory agency is likely to concede as appropriate disinfection. We need to have more discussion on a standard of disinfection where there is a public health threat from viruses in certain wastewaters.

Other recent virus inactivation studies include the Louisville, Kentucky, experiments by Pavoni and Tittlebaum.[29] They applied ozone to a 40,000 GPD activated sludge plant effluent seeded with F_2 virus (bacteriophage) at a concentration of 10^{11} plaque units per ml and a rate of 1 ml to 1 liter of sewage. They reported "virtually 100 percent efficiency (inactivation) after a contact time of 5 min. with a total ozone dosage of approximately 15 mg/liter and a residual of 0.15 mg/liter."

Nupen et al. treated a high-quality effluent for water reuse and found that a free chlorine residual beyond the breakpoint, which produced a pH of about 6.0 and a residual of not less than 0.6 mg/liter, inactivated polio virus in 35 min.[26]

One of the major difficulties in the inactivation of viruses is their variable sensitivity to disinfectants. Liu and co-workers tested 16 types of human enteric viruses for resistance to free chlorine in treated Potomac River water. The criteria used was time in minutes required for a 99.99 percent inactivation by a 0.5 mg per

---

* Low initial coliform count (Y_0 in the Collins model) is anything less than 50,000/100 ml. By comparison some raw potable water supplies exceed 3000/100 ml at the intake.

liter free chlorine residual at pH 7.8 and 2°C.[21] These are listed below in descending order:

polio type II (36.5 min.)
Coxsackie B5 (34.5 min.)
E. coli type 29 (18.2 min.)
E. coli type 12 (16.7 min.)
polio type III (16.6 min.)
Coxsackie B3 (15.7 min.)
adenovirus 7a (12.5 min.)
polio type I (12.0 min.)
Coxsackie B1 (8.5 min.)
adenovirus 12 (8.1 min.)
Coxsackie A9 (7.0 min.)
E. coli 7 (6.8 min.)
adenovirus 3 (4.3 min.)
reovirus 2 (4.2 min.)
reovirus 3 (4.0 min.)
reovirus 1 (2.7 min.)

So the picture developed to date by virus inactivation studies is a bit murky.

Future Considerations of Virus Destruction. The results of the experiments described above clearly illustrate the importance of predisinfection processes when treating wastewater effluents. Raw water for potable water supplies is hopefully of much higher quality than the well-oxidized and filtered effluent of a tertiary wastewater plant. Therefore, we have to recognize the existence of two different standards of disinfection related to virus inactivation: one for potable water and the other for wastewater. For the latter, it seems hopeless to expect significant virus destruction unless the effluent is of tertiary quality. As for raw potable water, it is the consensus that a 1.0 mg per liter free chlorine residual at the end of 30 min contact time at a *p*H not to exceed 8.0 will probably destroy all pathogenic viruses.

For tertiary effluents, based on the Pomona study it appears that a *combined* chlorine residual between 5 and 10 mg/liter with a two-hr contact is as good or better than a 4 mg/liter *free* chlorine residual for the same contact time. This is a new concept. Other disinfectants, e.g., chlorine dioxide, might prove to be better viricides than either chlorine or ozone. There is no question of the efficacy of ozone as a viricide, but on a cost-effective basis it seems to be very little better than chlorine with the disadvantage that it may not produce an effluent to meet the coliform limit. All of this gives impetus to the future investigation of chlorine dioxide as a combination viricide and bactericide.

One other item of significance when considering the treatment of waste waters for virus inactivation is the report by Boardman and Sproul[30] that none of the particulate systems investigated—clay, hydrated aluminum oxide, and calcium carbonate—protected the T7 phage from inactivation by chlorine. Therefore, viral absorption on an exposed surface provides negligible protection from disinfection, which raises the question of how much turbidity might be tolerated in a high quality effluent. Total encapsulation of the virus would appear to be the major mechanism by which a particle may be afforded protection due to adsorption.

The Regrowth Phenomenon

Significance of Regrowth. Numerous studies have demonstrated the regrowth phenomenon of coliform and fecal coliform organisms after disinfection. This may be due to the destruction of bacterial predators and may depend upon the presence of certain nutrients in the wastewater of the receiving waters.[7] It has been observed both in wastewaters and in receiving waters downstream from a disinfected sewage discharge. It is the judgment of authorities such as Geldreich that pathogenic bacteria like *Salmonella* and *Shigella,* of the same family as the coliform group, also regrow. If the disinfection process has a comparable effectiveness against such pathogens as with coliforms, a disinfection requirement resulting in a reduction of the numerous coliform bacteria down to a level of 23 or 230/100 ml would virtually assure the absence of the pathogenic bacteria present in sewage in smaller numbers. This would eliminate the pathogen regrowth potential, whereas a more liberal criteria of 200 or 400 fecal coliforms would not. Therefore, the phenomenon of regrowth after disinfection, whether it be a chlorine compound or ozone as the disinfectant, is not of any adverse public health significance.

The Clumping Phenomenon. It has been suggested over the years that "clumping" of organisms in the suspended solids of wastewater effluent contributes to irrational data on the efficiency of disinfection after the breakup of these clumps. Recent investigations by White[31] confirm an extremely small margin of disinfection reliability of a primary effluent. The extension of this observation to storm water overflows leads White to believe that it is not possible to adequately disinfect primary effluents or storm water overflows. This is exemplified by the fact that a primary effluent could be chlorinated with reasonable dosages of about 200 lb/mg to achieve a 230 MPN/100 ml in the effluent at a one-hr plus contact time, but if the sample for this accomplishment were taken in the same place but on the discharge side of the sample pump, the MPN would rise to 30,000/100 ml. This clearly demonstrates the futility of attempting disinfection of primary effluents. The cause of this dramatic increase in MPN is the clumping phenomenon. The sample pump breaks up the clumps thereby releasing great quantities of organisms protected from the disinfectant by the clumps. All of these data lend credence to state regulatory agencies asking for better and more effective predisinfection unit processes such as secondary treatment.

Toxicity of Chlorine Residuals

General Discussion. Chlorine residuals of low concentration whether free or combined are toxic to most fish and other aquatic life depending upon the species and time of exposure. Most oxidants like chlorine and ozone are known to be irritants to both freshwater and saltwater fish.

The resistance of fish to toxic substances varies according to the size of the fish. Fingerlings are very susceptible whereas carp and other large fish are highly resistant to toxic agents. The resistance is also proportional to the size of the fish scales: the larger the scales the greater the resistance. This is why goldfish are hardier than trout.

A review of the literature reveals a general agreement by the investigators as to what constitutes a lethal chlorine residual.

In 1968, Tsai[32] reported that the effects of chlorinated domestic sewage effluents on a fish population may stem not only from indirect degradation of water quality and alteration of stream bottom, but also from direct action on the fish similar to that of industrial discharges. When wastewater is chlorinated other toxic compounds may also be formed. For example, if thiocyanate is present this can be converted to the highly toxic cyanogen chloride by the chlorination process. It is also well known that under certain conditions an array of chloro-organic compounds are formed during the disinfection of wastewater by chlorination, and some of these compounds may also be toxic to aquatic life.[33] Very little is known about how these complex reactions affect aquatic life.

In 1971, Esvelt, Kaufman and Selleck[34] reported in a comprehensive study that the average daily emission of toxicity to the San Francisco Bay system was about 56 percent from municipal sources and 44 percent from direct releases by industry. The studies were performed long before it was thought necessary to be concerned about the concentration of the chlorine residual in the effluent owing to the enormous dilution factor of the receiving waters. The total chlorine residuals encountered (as measured by the back titration procedure with an amperometric endpoint) ranged from 1 to 8 mg/liter. The test fish used throughout the project were golden shiners. Using continuous-flow on-line bioassays they reported a 96 hr TL_{50}* in municipal wastewaters to be 0.2 mg/liter total chlorine residual. Even though the fish bioassay procedure is not necessarily a good measure of toxicity,[35] this investigation revealed the following important conclusions:

1. Toxicity removal by biological treatment was 75 percent while that removed by lime precipitation was 40 percent. Both ion exchange removal of ammonia and sorption of organics when coupled to lime precipitation provided an overall 65 percent removal.
2. MBAS† and NH_3-N represent significant toxicants in municipal wastewaters.

---

* This means that 50 percent of the fish subjected for 96 hours to the specified residual will die. TL = tolerance limit.
† MBAS = Methylene blue active substances.

3. Chlorination increased the toxicity of treated municipal wastewaters in all instances.
4. A chlorinated and dechlorinated effluent (with a slight excess of sulfite ion) was less toxic than either the unchlorinated or chlorinated effluent.
5. Dechlorination completely removed the chlorine-induced toxicity.

Also in 1971, the Michigan Department of Natural Resources conducted four separate studies at different wastewater-treatment plants and reported that for rainbow trout the 96 hr TL_{50} concentration below two plants was 0.023 mg/liter, and that for fathead minnows concentrations less than 0.1 mg/liter were toxic in the plant effluents.[36]

Another recent investigation demonstrated how the use of the orthotolidine method of measuring and control of wastewater chlorine residuals resulted in gross overchlorination beyond disinfection requirements.[37] This practice resulted in a major fish kill in the lower James River (Virginia).

Another case involving fish kills occurred in the Sacramento River (California) below a wastewater-treatment plant discharge. Chlorine-induced toxicity in the wastewater was immediately suspected. Consequently the regulatory agencies conducted bioassays to determine the source of the problem.[38] River water collected above, at, and below the discharge site was used in the static bioassays. King salmon fry were used as the test fish. The test fish were held in the receiving waters 150 ft upstream from the waste plume and in the waste plume 100, 200, and 300 ft downstream from the discharge where very little dilution of the waste discharge with the receiving water occurred. All of the test fish below the discharge died within a captive 14-hr period of exposure while all the fish upstream survived. The total chlorine residuals measured by the back titration amperometric endpoint procedure ranged from 0.2 mg/liter to 0.3 mg/liter during the test period. Therefore, it could only be concluded that the fish kills were caused by the toxicity in the wastewater—presumably the chlorine residual.

Probably the most comprehensive study of the toxicity of chlorine residuals to aquatic life in the receiving waters is the 1975 report by Arthur et al.[39] The disinfection system under investigation by both chlorine and ozone was able to produce an effluent with coliform levels less than 1000/100 ml. All of the studies were in freshwater systems. Both fish and invertebrates were included in the study. The chlorinated effluent was more lethal than the ozonated effluent or the chlorinated-dechlorinated effluent, which confirms the observations and conclusions of previous investigators. The fish were more sensitive than the invertebrates to the chlorinated effluent in the 94-hr tests. The respective 94-hr TL_{50} values of total residual chlorine to fish and invertebrates ranged from 0.08–0.26 and 0.21 to greater than 0.81 mg/liter respectively.

A discussion of the effect of chlorine residuals would not be complete without the consideration of the literature review and analysis of the aquatic life criteria as it relates to treatment of wastewaters provided by Brungs in his 1976 report.[40]

Need for Dechlorination. It was shortly after the Esvelt, Kaufman and Selleck report published in 1971[41] that the Water Quality Control Board, State of California, in cooperation with the Department of Fish and Game began issuing orders for certain wastewater-treatment plants to supplement the chlorination system with dechlorination facilities. The first wastewater-treatment plant to add dechlorination (by sulfur dioxide) in California was the city of Burlingame circa 1972. The orders specified a chlorine residual not to exceed 0.1 mg/liter. For technical reasons explained in Chapter 10 this meant total dechlorination with a slight excess of $SO_3^=$ ion. In the meantime the Sanitary Engineering Research Laboratory, University of California at Richmond, under the direction of Dr. Warren Kaufman, investigated the toxicity of the sulfite ion (a product of overdechlorination with sulfur dioxide) and found none up to more than 10 mg/liter.[42]

As of 1976, there were about thirty chlorination-dechlorination systems operating in California wastewater treatment plants and many more in the design-purchase stage. The California Water Resources Control Board and the Fish and Game Department are in agreement by following the conclusion from the bay studies on toxicity, that a chlorinated and dechlorinated sewage effluent is less toxic to aquatic life than either a chlorinated or unchlorinated effluent (has not been treated by a chlorination process).

Problems leading up to the chlorine toxicity dilemma have been gross overchlorination and poor control systems. It has been argued in some quarters that optimizing the chlorination system and controlling the dechlorination of the effluent to a comfortable 0.5 mg/liter chlorine residual would result in a sewage plume that would lose this small residual quickly in the receiving waters by the effects of chlorine demand and dilution.

It is interesting to note that a study by Stone[43] revealed some interesting facts about the die-away of chlorine residuals in San Francisco Bay. When a chlorinated secondary effluent was diluted with seawater there was no uptake of chlorine residual. The residual depletion was strictly by dilution: the seawater exerted no apparent chlorine demand on the combined chlorine residual which varied from 0.9 to 7.6 mg/liter at the end of 24 min. before dilution with seawater.

Freshwater receiving streams are known to exert a chlorine demand which in most cases could quickly absorb a 0.5 mg/liter chlorine residual. This should be investigated on a case by case basis because dechlorinating to 0.5 mg/liter has many more advantages than the use of excess sulfite ion.

Available Methods of Disinfection

Introduction. The scope of possible methods of wastewater effluent disinfection is great and includes natural processes (predatation and normal die-away), environmental factors (salinity, solar radiation) and methods having certain industrial applications (ultrasonics, heat). Only those methods which appear to have possible general application for wastewater and water reuse disinfection will be explored in detail in this text. In this introduction the general features or characteristics which have influenced overall use of the various disinfection methods will be identified and included in the design coverage except for gamma radiation, and iodine for reasons described below. The 1975 comparative cost analysis where available will also be shown.

Chlorine. Chlorine has and will probably continue to be the dominant disinfectant of wastewaters. It is available in different forms, and the characteristics of these forms greatly influence the system design. Because of the importance of chlorine in wastewater treatment the specific features of gaseous chlorine and chlorine compounds are discussed in detail elsewhere in this text (see Table of Contents).

Liquid-Gas Chlorine. This is the basic chlorine compound. It is used in large volumes by the chemical industry where derivatives of chlorine find use as pesticides for agriculture, plastics, food preservatives, and pharmaceuticals. It is used in large quantities as a bleaching agent for paper and textiles. Only about 4–5 percent of the total annual production in North America is used for sanitary purposes: household bleaches, restaurant sanitizers, potable water treatment, wastewater treatment, swimming pools, cooling waters, and other industrial process water treatment.

Liquid-gas chlorine is manufactured commercially by the electrolysis of a saturated salt solution. The gas collected in the process is moist and must be dried by passing it through a concentrated sulfuric acid solution to remove the moisture. It is then liquified by a combination of compression and cooling and is stored in steel containers from 150 lb cylinders to 90-ton tank cars. Liquid-gas chlorine is the principal form of chlorine used in wastewater disinfection. It is also used in wastewater treatment for odor control, destruction of hydrogen sulfide, prevention of septicity, control of activated sludge bulking, etc.

Hypochlorite. Imported Hypochlorite. "Available chlorine" can be provided either in the form of sodium or calcium hypochlorite. The most popular form is the sodium hypochlorite. Calcium hypochlorite is much too difficult to manage owing to excessive maintenance problems resulting from the deposition of the calcium ion throughout the system. Therefore, calcium hypochlorite could only be considered an emergency alternative.

Sodium hypochlorite is a clear liquid available in concentrations of 5, 10, and, 15 percent by weight trade strength (available chlorine).

Calcium hypochlorite is available either as a dry granular white powder or in tablet form in strengths of either 35 or 65 percent chlorine by weight.

Imported sodium hypochlorite is being used in some large wastewater treatment plants as a measure to avoid the potential hazard of liquid-gas chlorine delivered and stored in containers under vapor pressures of 80–110 psi.

On-Site Hypochlorite Generation. Complete systems are available for the on-site manufacture of hypochlorite solutions by electrolysis, which also avoids the potential hazard of handling the liquid-gas chlorine in pressurized containers. Electrolytic cell systems are available for use with either seawater, brackish water, or concentrated salt brines. The hypochlorite is produced in much the same way the liquid-gas chlorine is manufactured, except that there is no need to separate the chlorine gas and the sodium hydroxide which are the products of the electrolysis. This also eliminates the necessity of the sulfuric acid drying step required in the manufacture of liquid-gas chlorine. In addition to the formation of sodium hypochlorite, hydrogen gas is also a product of the electrolytic action. This is diluted with air and vented to atmosphere in concentrations well below the combustible capability of hydrogen. Equipment for this on-site production includes the electrolytic cells, rectifiers, electric switchgear, brinemaker, brine-treatment unit (where required), water-treatment system for cellwater, cooling equipment, and storage tanks for brine and hypochlorite. For economy the process is operated at a constant rate, and excess hypochlorite solution is stored to meet the high demand periods. Experience with these systems at wastewater treatment plants is relatively limited.

Another method of on-site manufacture of hypochlorite of considerable merit is the use of tank car quantities of chlorine gas supplemented by either calcium hydroxide or sodium hydroxide solution to produce an 8000–9000 mg/liter hypochlorite solution. This system is economically appealing from an equipment cost consideration when the average daily chlorine feed rate exceeds 5–6 tons/day (see Chapter 2).

Chlorine Dioxide. Chlorine dioxide is an unstable gas similar to ozone, so it must be generated at the point of use. It cannot be stored in steel containers as can chlorine. It is generated on-site as an aqueous ClO_2 solution by reacting a solution of sodium chlorite with the aqueous solution of a conventional chlorinator injector discharge. In wastewater treatment it has a distinct advantage in that it does not combine with the ammonia nitrogen normally present. Therefore, in a nitrogen-laden wastewater it is reputed to have a disinfection efficiency for both bacterial and viral destruction comparable to free chlorine. Experience with chlorine dioxide on wastewaters is limited. It is significantly more expensive than chlorine (see Chapter 12).

Bromine, Bromine Chloride, and Iodine. Bromine, bromine chloride, and iodine have been used in various ways as an alternative to chlorine. Bromine and bromine chloride are relatively soluble in water, more so than chlorine; however, bromine is much too hazardous a chemical to handle in the treatment of wastewater. Bromine chloride is much easier to handle because it has a vapor pressure of about 30 psi at room temperature. The materials required to meter bromine chloride are considerably different from those of chlorine so a separate species of equipment is required for feeding and metering this gas (see Chapter 14).

Bromine compounds have an advantage over the various chlorine compounds concerning the toxicity of residuals to aquatic life in the receiving waters: bromine residuals usually die away rapidly compared to chlorine residuals.* However, this characteristic makes bromine or bromine chloride residual control virtually impossible. Moreover, the bromine compounds are more expensive than those of chlorine.

Iodine is a gray-brown crystalline solid which is only slightly soluble in water. It is derived from kelp or oil field brines, and is mined from deposits in South America. It has been used as an effective method of water treatment on an emergency basis. There are so many unknown factors about iodine as a wastewater disinfectant that this, coupled with its high cost and uncertain availability, conspires to eliminate it from consideration as a practical wastewater disinfectant. The higher molecular weight of both bromine and iodine puts them at a distinct competitive disadvantage to chlorine.

Ozone. Ozone is another unstable gas which must be produced at the point of use. It is produced commercially by the reaction of an oxygen-containing gas (air or pure oxygen) in an electric discharge. It is a powerful oxidant and has been used since the early 1900s for odor and color removal as well as disinfection of potable-water supplies in Western Europe and Canada. It has been investigated recently as a process for polishing tertiary effluents for both color removal and disinfection. From these recent investigations it appears that ozone in combination with either chlorine or chlorine dioxide could solve the disinfection problem of both bacterial and viral contamination in tertiary wastewater effluents. This is particularly significant where there is consideration of wastewater reuse (see Chapter 13).

Ultraviolet Radiation. Ultraviolet radiation (UV) from sunlight has a sterilizing effect on microorganisms but most of the radiation from this natural source is screened out by the atmosphere before reaching the earth's surface. Ultraviolet radiation can be produced by special lamps (mercury vapor) and is presently used in special situations for high quality potable water disinfection. The disinfection reaction occurs on the thin film surfaces of water where the microorganisms can be readily exposed to the radiation reaction. With wastewater, lethal action cannot be exerted through more than a few centimeters. Beyond this limiting distance the high absorption of the rays by the water and the suspended solids dissipates the ultraviolet energy. Consequently, the problems of providing effective exposure of sewage effluents containing varying amounts of interfering suspended solids and ordinary turbidity to the UV rays is such that the practical application must be to a very thin sheet of wastewater flow of nearly uniform thickness. The monitoring requirements for proof of disinfection of a UV system are nonexistent.

The UV disinfection concept is applicable when the water or wastewater is of high quality. The process must be carefully monitored and supplemented by termi-

---

* Bromine residuals in wastewater effluents containing organic nitrogen are remarkably stable. This is directly related to the formation of organobromamines, which are similar to organochloramines.

nal chlorination for drinking water. The use of UV in wastewater has possibilities where NPDES requirements are not stricter than 200/100 ml fecal coliforms. See Chapter 14.

Gamma Radiation. In addition to UV radiation, gamma radiation has been investigated recently as a method of wastewater treatment and disinfection. In sufficient dosages, gamma radiation is an effective sterilant and is used as a method of sterilizing surgical instruments. Unlike UV radiation, gamma rays are capable of great penetration. Gamma radiation has the ability to alter organic and inorganic molecules and this effect may benefit tertiary treatment processes. This most convenient source of energy for irradiation is cobalt 60 which is available in virtually unlimited quantity. The cost of radiation energy is high and gamma radiation as a disinfection process for wastewater is not economically or otherwise competitive with other methods.

The proponents of gamma radiation point out that cesium 137 is the major component of nuclear waste material, which is a by-product of nuclear power plants and is therefore available as a source of gamma rays for water purification, and thereby diminishes the amount of chlorine required for disinfection.

Woodbridge[44] claims six and seven orders of magnitude reduction in the concentration of microorganisms have been obtained by irradiation in less than five min. exposure. The reported advantages of this method include reliability, beneficial side effects and no residual effects. Disadvantages are principally associated with safety needs, excessive cost and virtually no operating experience. The application requires considerably more engineering design information than is now presently available. It should not be ruled out, however, as an adjunct to present methods. Because of lack of information as a wastewater disinfectant, nothing more will be said in this text on the subject.

COMPARATIVE COSTS OF DIFFERENT METHODS OF DISINFECTION

Cost comparisons are unfair at best because all things are not equal between the various methods. One has to weigh the advantages and disadvantages of each method; cost is only one of many factors in evaluating the various methods. The method which will predominate: is one which gets the job done easily; is known to have a minimum health and safety risk; is easy to apply, measure, and control; and for which the handling equipment is reliable and easy to operate. The chlorination–dechlorination method fits this description and will certainly remain popular for a very long time.

In 1976 the EPA published a report on the various methods of wastewater disinfection.[45] Tabulated below is the comparative capital and process cost of the various methods studied (see Table 8-1).

Missing from the above tabulation is chlorine dioxide. The capital cost increase over chlorine for a chlorine dioxide installation is small: probably 15 percent.

Table 8-1 Cost Summary

Plant Size (mgd)	1	10	100
Capital Cost	$K	$K	$K
Process			
Chlorine	60	190	840
Chlorine/SO_2	70	220	930
Chlorine/SO_2/aeration[b]	120	360	1,580
Chlorine/carbon	640	2,800	8,400
Ozone/air[a]	190	1,070	6,880
Ozone/oxygen[a]	160	700	4,210
Ultraviolet[a]	70	360	1,780
Bromine chloride	50	130	410
Activated Sludge	1,450	5,790	39,800
Disinfection Cost	¢/Kgal.	¢/Kgal.	¢/Kgal.
Process			
Chlorine	3.49	1.42	0.70
Chlorine/SO_2	4.37	1.75	0.89
Chlorine/SO_2/aeration[b]	7.66	2.39	1.19
Chlorine/carbon	19.00	8.60	3.28
Ozone/air	7.31	4.02	2.84
Ozone/oxygen[a]	7.15	3.49	2.36
Ultraviolet[a]	4.19	2.70	2.27
Bromine chloride	4.52	3.04	2.65
Activated Sludge	55.90	20.20	14.00

[a] Tertiary treatment stage is not included in these costs.
[b] Aeration is not required following dechlorination by SO_2 because a properly designed system will not remove any DO in the effluent. (*author's note*)

The following is the process chemical cost for chlorine dioxide assuming chlorine at 15 ¢/lb and sodium chlorite at 70 ¢/lb; 2 mg/liter dose and 30 min. contact time[46] as shown in Table 8-2 below.

Table 8-2

Design capacity (mgd)	1	10	100	150
Chlorine dioxide (¢/Kgal.)	4	2	1	1

SUMMARY

Many modern wastewater disinfection practices in the United States had their origin in California. In the early 1920's, health officials were alarmed at the possible long-term deleterious effects of raw or poorly treated sewage discharging into the surf waters along the Pacific coast and freshwater streams throughout the state. A comprehensive investigation of the consequences of sewage discharges as related to public health indices resulted in a law suit against the City of Los Angeles.

The presiding judge upheld the State Health Department contention that wherever samples taken from the surf exceeded a 1000 MPN/100 ml statistical coliform concentration, it constituted sewage contamination injurious to public health and that the offending dischargers must be made to provide proper sewage treatment.

During the last thirty-five years, the California State Department of Health has formulated a set of requirements for all receiving waters. These include confined saline waters, estuaries, surface waters, ephemeral streams, and shellfish growing areas. An additional constraint is considered whenever the receiving water is used for bathing or water contact sports. The numbers applied for these situations vary from 230 MPN/100 ml for confined saline waters down to 2.2/100 ml for sewage discharging into ephemeral streams or negative estuaries.

This concept of receiving water quality based upon coliform concentration carries with it the hypothesis that certain degrees of treatment are imperative to achieve the various numbers specified. In other words, the designer should not attempt to depend wholly upon a disinfection system to achieve the desired coliform count in the plant effluent. For example, disinfection of a sewage discharging into a confined saline water or estuary cannot consistently accomplish the required 230/100 ml MPN coliform without secondary treatment. Attempts to "disinfect" raw sewage or stormwater overflows are a waste of chemicals, time, and energy. Similarly tertiary treatment is almost always a necessity to meet a 2.2/100 ml MPN standard. When such a severe requirement is placed upon an effluent, there is a further implication of virus destruction. Virus destruction can only be accomplished on high quality effluents, regardless of the disinfectant used.

Nitrification* of an effluent is no longer considered a necessity when a low coliform count is required. It was once thought that if an effluent was nitrified then the practice of free residual chlorination could be assured which would result in a more reliable and efficient disinfection system. Through the years 1970–1982, it has been found that nitrification to produce a free chlorine residual is not necessarily worth the effort.

Disinfection studies should represent the degree of disinfection based upon the total coliform concentration in the plant effluent after disinfection.

A mathematical model has been developed which relates the coliform concentration before and after disinfection with chlorine residual at the end of a specified contact time. This model clearly demonstrates that the higher the quality of effluent results in lower numbers of coliforms to be destroyed. This translates to higher efficiency of disinfection.

To date there does not seeem to be a better indicator organism for proof of disinfection than the total coliform group.

Wastewater reuse requires a different approach. In these cases the sewage discharge is being used directly for land irrigation, spraying of crops, industrial cooling water, other makeup water requirements, groundwater recharge, prevention of

---

* The only reason for nitrification is to limit the concentration of unionized ammonia in the receiving waters. Ammonia is higly toxic to aquatic life.

salt water intrusion, and prevention of land subsidence due to underground withdrawals. All of these applications depend heavily upon the unit process of disinfection. It becomes the most important link in this chain of treatment for these applications.

Whenever wastewater is used in situations where there is the possibility of human contact or ingestion, the problem of widespread virus infection becomes the most serious concern of public health officials everywhere.

A recent study by the Los Angeles County Sanitation Districts revealed that contrary to previous beliefs, combined chlorine residuals could be nearly as effective as comparative free chlorine residuals in the destruction of viruses, provided that the disinfection facility is properly designed.

All of the data currently available demonstrate conclusively what the Collins mathematical model tells us: that disinfection efficiency is related directly to the quality of the effluent, and that for any quality of effluent the degree of disinfection is directly related to the total chlorine residual and contact time, provided that mixing is rapid and that the contact chamber demonstrates plug flow conditions.

The phenomenon of regrowth which occurs temporarily in some cases downstream from the point of disinfection is not considered significant to public health. The public health practitioners recognize this phenomenon. They firmly believe that all pathogenic organisms are destroyed in the disinfection process.

A source of constant worry to any practitioner of disinfection is the clumping phenomenon. It is theorized that clumps of suspended or colloid-like particles such as are present in raw sewage, primary effluents and poorly treated secondary effluents, may be able to pass through the disinfection system only to break up downstream and spew into the effluent gross amounts of coliforms and possibly pathogens which were sheltered from the disinfectant.

This concept is of grave concern to public health people and is a major reason for assuming that raw sewage and/or primary effluents should not be considered as candidates for the disinfection process.

The toxicity of chlorine residuals to aquatic life is well documented and has given rise to the addition of the dechlorination step to complete the disinfection process, whenever chlorine (or chlorine dioxide) is the disinfectant.

A chlorinated-dechlorinated effluent has been proved to be less toxic than either a chlorinated or a nonchlorinated effluent.

Available methods of disinfection of wastewaters include all of the halogens (Cl_2, I_2, Br_2, $BrCl$, ClO_2), ozone, ultraviolet radiation, gamma radiation, and possibly some combinations with sonics.

Cost effective analyses of these various methods always place the chlorination-dechlorination method in the most favorable position.

From evaluations of the art of disinfection it is clear the process will not provide the desired results unless the other unit processes of the wastewater-treatment system are performing properly. Therefore, a disinfection system is a protective device for public health as well as a sensitive monitor of the entire wastewater treatment process.

CHEMISTRY AND KINETICS OF DISINFECTION BY CHLORINE

Reactions with Wastewater

General Discussion. There are numerous constituents present in wastewater which react immediately with the HOCl from the chlorinator injector discharge or the hypochlorite solution. Consequently, free chlorine (HOCl + OCl⁻) is probably consumed or converted to some form of chloramine in a matter of seconds after mixing with the wastewater stream, owing to the presence of ammonia nitrogen. Very little is known about this specific reaction. Simultaneously with the Cl_2–NH_3 reaction, are the other inorganic reactions of chlorine with reduced substances such as $S^=$, HS^-, $SO_3^=$, NO_2^-, Fe^{++}, Mn^{++}, etc. These substances react with both the free chlorine and combined chlorine (NH_2Cl, $NHCl_2$) to reduce these compounds so that the active chlorine compound is eventually reduced to the stable chloride ion which is nonbactericidal. At this point, there is no measurable residual. The chlorine consumption in the first minute of reaction is probably due to the reactions with inorganic substances. The reactions occurring in the following 2 or 3 min. are probably due to organic chlorine demand. These are much slower reactions. The 10 min. chlorine demand of a fresh domestic sewage may be as low as 5 mg/liter to produce a measurable residual; however, this figure may escalate to 40 mg/liter if the same sewage becomes septic. Ammonia-N does not begin to consume chlorine until the Cl-N dosage ratio exceeds 5:1, i.e. beyond the hump on the breakpoint curve.

The most significant chemical reactions between chlorine and the various chemical constituents in wastewater effluents are those with the various nitrogenous compounds, either inorganic (NH_3-N, NO_2) or organic (proteins and their degradation products). These reactions are described in Chapter 4.

Ammonia Nitrogen. With the exception of highly nitrified effluents there is usually an appreciable amount of ammonia nitrogen in all wastewater effluents. The range is on the order of 10–40 mg/l. The ammonium ion (NH_4^+) exists in equilibrium with the ammonia nitrogen and hydrogen. The distribution is dependent upon pH and temperature. The relative distribution can be defined as follows:

$$NH_4^+ \rightleftharpoons NH_3 + H^+ \quad \text{with } K = 5 \times 10^{-10} \text{ at } 20°C \qquad (8\text{-}2)$$

where

NH_3 = ammonia molecule
H^+ = hydrogen ion
K = dissociation constant

According to the dissociation constant, the pH value at which NH_3 and NH_4^+ are present in equal proportions (pK) is about pH 9.3 at 20°C. Above pH 9.3, NH_3 predominates; below pH 9.3, NH_4^+ predominates.

At usual wastewater pH levels the predominant chlorine reaction proceeds as follows:

$$HOCl + NH_4^+ \rightleftharpoons NH_2Cl + H_2O + H^+ \qquad (8\text{-}3)$$

when the chlorine to ammonia nitrogen weight ratio is 5:1 or less. If the pH drops below 7, dichloramine ($NHCl_2$) will begin to form, and at a much lower pH, nitrogen trichloride will form. However, as soon as the 5:1 weight ratio of chlorine to ammonia nitrogen is exceeded, a new set of reactions takes place. These reactions are described fully in Chapter 4.

The disinfecting power of chlorine in wastewater is greatly enhanced by good mixing at the point of application. White et al.[49,50] found a significant reason for this relationship: Good mixing ensures the maximum formation of monochloramine in Eq. (8-3). If mixing is poor the chlorine species tend to split between monochloramines and organochloramines (organochloramines titrate as dichloramine). However, pure dichloramine does not form when the pH is in the neutral range and the chlorine to ammonia nitrogen weight ratio is 6:1 or less. This is an important consideration because the species described above as organochloramines have little or no germicidal efficiency. The organochloramines derive from the organic nitrogen which is always present in substantial amounts (3–15 mg/l). In highly nitrified and filtered effluents the organic nitrogen is present in amounts from 0.75 to 3.0 mg/l. The lower the organic nitrogen content, the more germicidally efficient the chlorination process will become. The significance of organic nitrogen as it affects the efficiency of disinfection is described below.

Organic Nitrogen. In raw untreated municipal wastewaters organic nitrogen compounds are present as both soluble and particulate. Most occur as insoluble compounds. The soluble compounds are mainly in the form of urea and amino acids. Secondary biological treatment reduces the soluble organic N compounds to a range of approx. 10–14 mg/l. Filtration will remove the remainder leaving about 3 mg/l in the filtered effluent. Filtration preceded by flocculation can reduce this residual to about 0.75 mg/l.

Chlorine is known to react and combine with urea, amino acids, and proteinaceous organic nitrogen compounds to form organochloramines of dubious germicidal efficiency. Very little is known about the kinetics of these reactions, their reversibility, or their germicidal power. Extensive investigations surrounding the presence and identifications of this species of chlorine compounds in swimming pools and potable water have demonstrated that they titrate as dichloramines in the forward titration procedures using either amperometric or DPD methods. How much these compounds interfere with the monochloramine determinations is unknown.[51,52]

Recently White et al.[50] have been able to prove conclusively that the chloramines appearing in the dichloro fraction of a combined residual (organochloramines) have a significantly lower germicidal efficiency than those appearing in the monochloramine fraction.

Observations by White, Selleck, and Collins have indicated that the germicidal efficiency of a combined chlorine residual has a tendency to decrease with time. This decrease has been noticed in some secondary effluents. It is most noticeable in primary effluents when the contact time exceeds 45–60 min. It is thought that this is related to the presence of organic nitrogen. In the cases observed there was a noticeable shift in the chloramine species. While the total residual remained relatively stable between 45 and 60 min the monochloramine concentration decreased and the "dichloramine" increased. Therefore it is theorized that in effluents containing 0.5 mg/l organic nitrogen or more, the monochloramine formed by the presence of ammonia nitrogen in the wastewater slowly hydrolyzes with time to react with organic nitrogen compounds present to form organochloramines, thereby decreasing the overall germicidal efficiency of the remaining residual.

In 1966 Feng[53] discovered a great disparity in the germicidal efficiency between ammonia chloramines and those found in an environment of pure organic nitrogen compounds. He has reported that methionine, an indispensable amino acid for biological growth present in wastewater, forms a measurable chlorine residual with no germicidal power. Feng also investigated the lethal activities of the glycine, taurine, and gelatin chloramines. His work shows that taurine chloramines are as lethally active as ammonium chloramines at pH 9.5, but that their germicidal efficiency falls off as the pH decreases. The glycine chloramines are as germicidally active as monochloramine at pH 4 but are totally inert at pH 7, and the gelatin chloramines are active at pH 9.5 but are inert at pH 7 and 4. There are certain to be other such organic-nitrogenous compounds which contribute to the total chlorine residual which have little or no germicidal effect.

Sung[54] made a controlled laboratory study of fifteen organic compounds representing seven groups to evaluate their individual and combined effect upon the chlorination process. Nine of the fifteen compounds were found to interfere with the germicidal efficiency of the chlorination process. Of these nine compounds five were organic nitrogen compounds. Cystine and uric acid were the severest inhibitors of the nitrogen group. When five of the interfering compounds were mixed together, their combined effect was found to be more pronounced than any of their individual effects, but did not equal the sum of their individual effects. Sung compared the germicidal efficiency of a simulated wastewater with and without the interfering organic compounds. He found that the germicidal efficiency of wastewater containing the interfering compounds by themselves and the resulting chlorine residuals had little or no germicidal effect. The greatest interference was observed to be caused by cystine, tannic acid, humic acids, uric acids and arginine.

Cystine is an amino acid connected by two sulfur groups that is known to

DISINFECTION OF WASTEWATER 511

react with chlorine. Tannic, humic and uric acids are capable of exerting a significant chlorine demand when present in water or wastewater. *Arginine* is a basic amino acid. The reaction between chlorine and arginine is almost instantaneous.

The organic compounds that had little or no interfering effects on the chlorination process were: acetic acid, cellubiose, dextrose, glutamic acid, uracil, and lauric acid.

The significance of the above findings by Sung[54] confirms the theory of interference in the chlorination process by the presence of organic nitrogen. It further points to the fact that present analytical techniques do not provide for separating the chlorine residual fractions into those of equal germicidal efficiency.

It is also interesting to note that Esvelt, Kaufman, and Selleck[55] found that the toxicity of combined chlorine residuals diminished with time. This finding together with those of Sung[54] demonstrates quite convincingly that there are a significant number of organic compounds in wastewater which will react with chlorine to form organic chloramines of little or no germicidal potential, and moreover, these compounds appear to increase in concentration with time. This apparent increase of this chlorine residual fraction with time would also explain the loss of germicidal efficiency of combined chlorine residuals in wastewater with the passage of time, as described elsewhere in this text.*

Effluent Quality as Related to Dosage. The single most important effluent quality parameter as it effects the efficiency of disinfection is the indicator organism concentration in the treated effluent before the application of chlorine. This holds true for non-nitrified effluents based upon observations by White over a period of several years (1970–1983).

Collins Model. The basic concept of adequate wastewater disinfection is expressed in the mathematical model developed by Collins et al.[56] in 1970. This work resulted in the formulation of an equation based upon a comprehensive pilot plant study of a primary effluent. The message portrayed by the equation has been substantiated many times since its publication: if there is good mixing at the point of chlorine application and if there are plug flow conditions in the contact chamber (no short circuiting) one can expect a definitive coliform reduction with a given chlorine residual at the end of a specified contact time. This is the *ct* relationship commonly referred to elsewhere as the chlorine concentration–contact time envelope.

The original model has been subsequently fine-tuned by Collins based on plant-scale studies.[57] This recent work reinforces the practical aspects of his original model, which is represented by the following equation:

$$y = y_o [1 + 0.23 \ ct]^{-3} \qquad (8\text{-}4)$$

where

---

* The loss of germicidal efficiency with time is caused by the reaction of monochloramine with the organochloramines. The monochloramine gradually hydrolyzes to form organochloramine.

y = MPN in chlorinated wastewater at end of time t
y_o = MPN in effluent prior to chlorination
c = total chlorine residual, mg/liter, at the end of contact time
t = contact time, min

The mathematical model of Eq. (8-4) was developed from a pilot plant system that had excellent mixing in a highly turbulent regime and an ideal plug flow contact chamber. Based upon many plant observations, good mixing occurs when the velocity gradient G is approximately 500 and the contact time t_i is not less than 30 min. Contact times longer than one hour should be avoided in effluents containing organic N in concentrations higher than 5 mg/l. Long contact times in these effluents allow the monochloramine fraction of the chlorine residual to hydrolyze and become converted to organochloramines. These organochloramines have low germicidal qualities; therefore the potency of the total chlorine residuals in these effluents show a marked decrease with time.

The Collins model is a valuable tool for sizing chlorination equipment for a new plant, and for evaluating an existing plant. Some examples follow.

Primary Effluent. y_o is usually about 38×10^6/100 ml. Assume discharge into a surf water. In California the y requirement is 1000/100 ml MPN total coliforms. Substituting in Eq. (8-4):

$$\frac{1000}{38 \times 10^6} = [1 + 0.23 \ ct]^{-3} \qquad (8\text{-}5)$$

See key stroke sequence for an HP-21 calculator in the appendix and solve for ct. Assume $t = 30$ minutes and solve for c:

$$ct = 142 \quad \text{and} \quad c = 4.73$$

To allow for the probable immediate (3–5 min) chlorine demand of 6–8 mg/l plus the die-away in the contact chamber (25+ min) of about 1 mg/l, the required chlorine dosage will be about $5 + 8 + 1 = 14$ mg/l.

Secondary Effluent. The median coliform concentration in a well oxidized secondary effluent will be about $2 \times 10^6 = y_o$. Assuming a discharge into a confined body of water the total coliform requirement might be 23/100 ml MPN. In this case ct calculates to 188. This magnitude calls for a contact time longer than 30 min. So c for 45 min is 4.2. To estimate the chlorine dosage assume a 5 mg/l initial chlorine demand and 1.5 mg/l residual decay in the contact chamber. Therefore required dosage will be $4.2 + 1.5 + 5 = 10.7$, say 12 mg/l.

Filtered Effluent. The ease with which these effluents can be treated depends a great deal upon whether or not the filtered effluent has been preceded by coagulation and sedimentation. A conventional water reuse situation would consist of filtration of secondary effluent preceded by coagulation and sedimentation. The y_o of such an effluent would probably range between 3000 and 10,000 coliform per 100 ml. In California, whenever tertiary effluent is required the coliform requirement is usually the same as that for potable water, namely 2.2/100 ml. Assuming y_o = 10,000 and y = 2.2, then c calculates to 2.25 mg/liter for t = 30 min. A tertiary effluent with the predisinfection processes described above would require chlorine dosages on the order of 5–7 mg/liter.

A filtered effluent with chemical coagulation but without sedimentation produces an effluent with coliform concentrations considerably higher, on the order of 50,000/100 ml. In this case using y = 2.2/100 ml, c calculates to 3.96 mg/liter for t = 30 min. This is almost twice the residual required in the preceding case.

These examples clearly demonstrate the effect of effluent quality as it relates to the y_o coliform concentration. The easiest effluents to disinfect that White has investigated to date is a secondary effluent followed by 100-day ponds. In these particular cases, the y_o rarely exceeds 4000/100 ml and the final coliform is usually less than 3/100 ml using chlorine dosages on the order of 3.5–4.0 mg/liter and 15 min contact time. The chlorine residuals at the end of 15 min are usually on the order of 2 mg/liter. This is what an optimized system can do when y_o is a low figure.

In order to provide a touch of conservatism, White always uses the chlorine residual c in the Collins equation as the residual measured at the end of the contact chamber and the contact time t_i as the first appearance of the dye at the end of the contact chamber.

The Collins–Selleck Model.[58,61] This model is a refinement of the Collins model shown in Eq. (8-4). It is described by the following equation:

$$y/y_o = (RT/b)^{-n} \qquad (8\text{-}6)$$

where:

y_o = initial bacterial concentration before chlorination
y = bacterial concentration at end of contact chamber or at time T in minutes
R = chlorine residual at the end of time T, mg/l
T = contact time in minutes
b = the x intercept when y/y_o = 1 or log (y/y_o) = 0 (see Fig. 8-5)*
n = slope of the curve

An easy way to use this equation is to plot the log values on arithmetic paper; log y/y_o on the y-axis and log RT on the x-axis. Examination of the equation shows that when $y = y_o$ there is no kill, so y must be less than y_o to have any

---

* b is sometimes called the lag-time of bacterial kill because kill does not occur until $RT > b$. See Fig. 8-5.

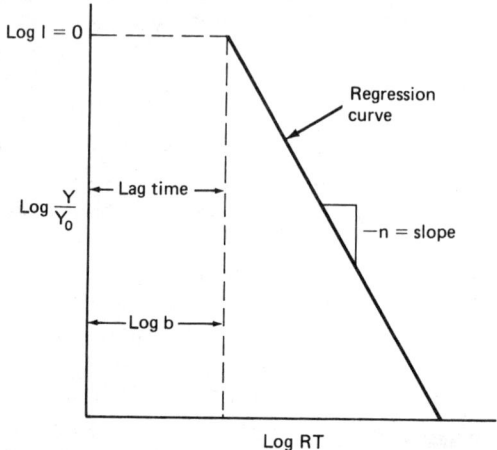

Fig. 8-5 Arithmetic plot of $Y/Y_o = \left(\dfrac{RT}{b}\right)^{-n}$

kill. When there is no kill, $y/y_o = 1$ and $\log 1 = 0$. Therefore the equation begins with 0 on the y-axis.

Now when $RT = b$ then $RT/b = 1$ and $y/y_o = (RT/b)^{-n} = (1)^{-n} = 1.0$ Therefore b is determined when the regression plot intercepts the x-axis. If this equation is plotted on log–log paper the intercept would be 1.0, but on arithmetic paper the intercept is 0 because $\log 1 = 0$. The intercept is where the $\log RT = \log b$. Every point on the regression curve to the right of the zero intercept represents $RT > b$.

When data is not available to plot a bacteria kill regression curve the suggested value for b is 4 when working with total coliforms and 3 for fecal coliforms. When these values are used for b and for $n = -3$ the equation $y/y_o = (RT/b)^{-n}$ is practically identical with the Collins 1970 model which is $y/y_o = (1 + 0.23ct)^{-3}$ where $c = R$ and $t = T$. The insertion of 1 in the latter equation was used to force the regression plot into a straight line at low values of bacterial reduction shown in Fig. 8-6.

The following values of b and n were found in an extensive study by Roberts et al.[59] which was done to compare the feasibility of ClO_2 as a disinfectant to chlorine.

Palo Alto, Calif. secondary effluent:

$$y/y_o = (RT/3.95)^{-2.79} \qquad (8\text{-}7)$$

San Jose, Calif., secondary effluent:

$$y/y_o = (RT/4.06)^{-2.82} \qquad (8\text{-}8)$$

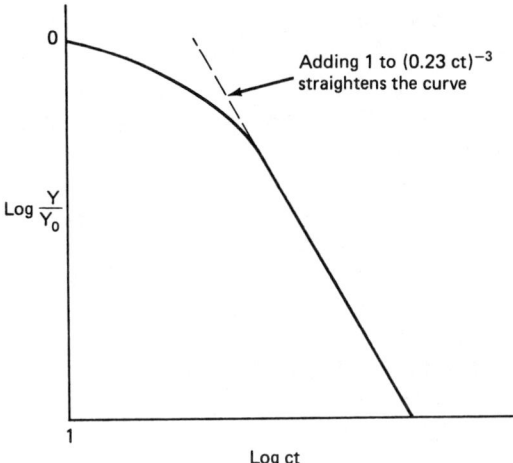

Fig. 8-6 Collius 1970 model: explanation of unity factory.

Nitrified Effluents

Inadvertent Loss of NH_3-N. The chemistry of wastewater chlorination takes on a different aspect when a secondary effluent becomes nitrified. This situation occurs inadvertently in some plants when there is an overload due to cannery wastes. This was a common occurrence in a conventional secondary activated sludge plant that experienced heavy seasonal loads of cannery wastes. During these periods of over loading the ammonia nitrogen in the effluent would disappear completely. When this occurred the organic nitrogen increased from about 6 to 11 mg/l and simultaneously the disinfection efficiency of the chlorination system would fall off suddenly and dramatically.[60] In such a situation one would expect the appearance of the free chlorine species and a subsequent improvement in the disinfection efficiency. This never happened to the amazement of all involved in the project.* Subsequently it was determined that the chlorine applied reacted totally with the organic-N present (13 mg/l) to form organochloramines.

In an attempt to correct for the dramatic decrease in disinfection efficiency, operating personnel increased the chlorine dosage to the maximum capability of the system. While this increase in chlorine dosage represented a two fold increase (20 mg/liter) it was not enough to achieve the NPDES requirement of 230/100 ml MPN total coliform concentration. However when the ammonia-N concentrations exceeded about 1 mg/liter there was never any problem in achieving this disinfection requirement with chlorine dosages below the breakpoint, i.e., less than 10 mg/liter.

---

* During these periods of zero NH_3-N the plant laboratory switched to the forward titration procedure with an amperometric titrator.

To overcome this disappearance of ammonia nitrogen during the canning season, the plant influent was seeded with the secondary digester supernatant which contained approximately 550 mg/liter ammonia nitrogen. There was sufficient quantity of this liquor to dose the plant influent with 10 mg/liter NH_3-N. This remedial action restored the disinfection efficiency at a chlorine dose of 10 mg/l. This plant was later converted to produce a nitrified and filtered effluent.

Tertiary Effluents. Most nitrified effluents are filtered before disinfection. Filtration may in some instances be preceded by sedimentation; if not, the addition of a coagulant is usually provided in lieu of sedimentation. The NPDES requirements for these effluents are usually severe enough that virus removal becomes desirable, hence the need for filtration. Owing to the difficulty in achieving disinfection in the absence of ammonia nitrogen described above, White et al. investigated several plants with nitrified tertiary effluents.[49,50]

All of the plants investigated were required to meet an NPDES total coliform concentration of 2.2/100 ml MPN. All effluents were filtered except one. Some plants had coagulant addition capability ahead of filtration. Some used coagulants and some did not. There was no set pattern and the effluent quality with and without coagulants was undetermined.

The most disturbing element of this investigation revealed that in spite of higher quality effluents (y_o = 30–60,000) none of the plants conformed to the Collins[56] mathematical model. This model has been used with reliable success for predicting required combined chlorine residuals for given contact times to achieve a particular coliform concentration requirement. In general the chlorine dosages required to achieve a 2.2/100 ml coliforms standard were often twice those needed in plants with similar treatment process that *did not nitrify*. This was difficult to believe because the residuals in the nitrified effluents were usually 50–60 percent free chlorine as measured by amperometric titration and DPD–FAS titration.

The San Jose, California water pollution control plant was thoroughly investigated over a 2–3 year period in search of answers to this chlorination anomaly. The chlorine dosage required to achieve a 2.2/100 ml coliform count in the nitrified effluent varied from a low of 17 mg/l to a high of 22 mg/l. The *total chlorine residual* at the end of 49 min contact time required to meet this goal consistently was about 9 mg/l. This residual contained about 50–60 percent free chlorine. About 90–95 percent of the remainder titrated as dichloramine, the rest as monochloramine.*

The Collins model predicts a total chlorine residual of 2 mg/l at 49 min contact time to achieve a 2.2/100 ml coliform level.

A laboratory study using the addition of ammonia nitrogen was made to find the chlorine to ammonia ratio with the best germicidal efficiency. (See Fig. 8-7).

---

* The dichloramine fraction is believed to be germicidally impotent organochloramines derived from the 2–3 mg/l presence of organic nitrogen. Since there was no trace of ammonia-N in these samples there could not be any monochloramine formed. Therefore the appearance of monochloramine is the result of interference from organochloramines in the forward titration procedure.

This turned out to be 6:1, chlorine to ammonia-N by weight. This confirmed the finding of Selleck et al.[61] They showed the most effective germicidal chloramine residual occurs when the chlorine–ammonia-N ratio is on the "breaking" side of the breakpoint curve, as shown in Fig. 8-7 between points A and B. It was further found that a chlorine dosage of 12 mg/l would produce a combined residual of 7 mg/l at the end of 49 min contact time which was sufficient to achieve a 2.2/100 ml coliform count in the effluent. Plant operation with the above chlorine and ammonia-N dosages proved to be consistent with laboratory findings. This treatment consistently produced an effluent meeting the coliform requirement without exceeding the limit of 0.025 mg/l unionized ammonia-N in the receiving waters. The chlorine dose and residual requirement to achieve the disinfection standards have been consistently reduced by 5 and 2 mg/l, respectively.

However, one interesting observation has been made at the San Jose plant. The color of the final effluent is perceptibly different after the addition of ammonia. When ammonia is not added and free chlorine is present in the contact chambers, the effluent is bright and sparkling with a tinge of blue, similar to a well operated swimming pool. When ammonia is added, the effluent retains its "clean" look, but the color is a very light beige that can be characterized as a straw color with a tinge of very light green. This difference in color explains in part why so much more chlorine is consumed when ammonia nitrogen is removed and free chlorine residuals are produced. The additional chlorine is consumed in bleaching the organic color in the nitrified effluent. This is typically consistent with potable water treatment practice where free chlorine residuals are used to bleach colored water. *This also can be taken as proof positive that all of the titrations identifying the free chlorine species were and are in fact free chlorine residuals.* There were so many skeptical responses to the findings described above about the apparent impotence of the free chlorine residuals, one additional step was taken to prove the existence of the free chlorine residuals. This was the use of the Olivieri Biofac method.[63] This method is based upon the fact that a bacterial virus f_2 of the *E. coli* K-13 strain is inactivated in a few seconds by free available chlorine but is affected very little by combined chlorine. The San Jose effluent was subjected to this comparison with the following results: a 10 mg/l chlorine dose was applied to each of two aliquot samples. Sample A was nitrified with only a trace of NH_3-N remaining. Sample B was the same effluent as A except that it was spiked with 1.67 mg/l NH_3-N. Each sample was spiked with the f_2 virus. At the end of 5 minutes contact time sample A measured a total chlorine residual of 7.4 mg/l. Of this total residual 4.5 mg/l was free available chlorine. Sample B measured a total chlorine residual of 8.7 mg/l; there was no free chlorine residual. This sample titrated 7.2 mg/l monochloramine and 1.2 mg/l dichloramine. The control sample without chlorine measured 780 f_2 units. The sample with free chlorine measured zero units and the sample with combined chlorine residual measured 470 f_2 units.

Examination of other plants did provide some clues as to why so much additional chlorine is required to produce a disinfecting residual, and why disinfection can

518 HANDBOOK OF CHLORINATION

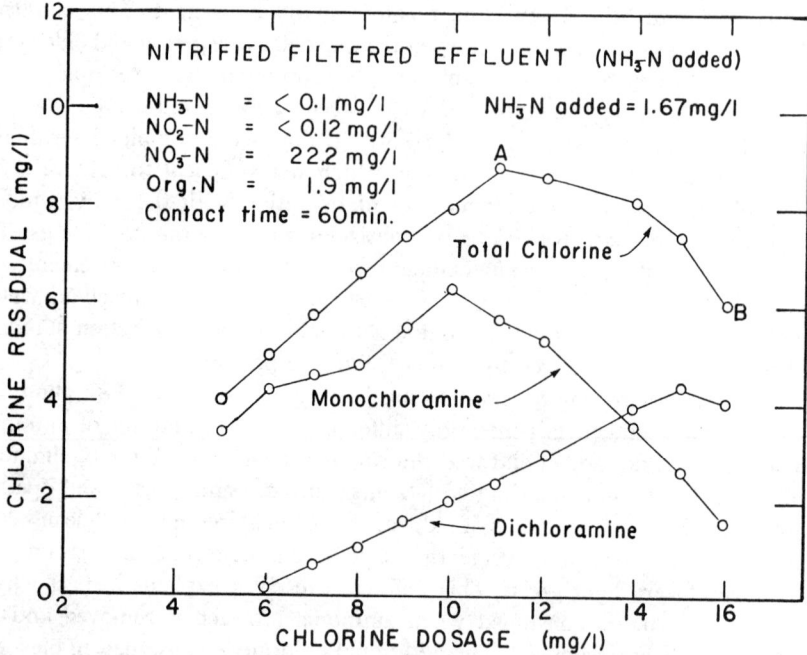

Fig. 8-7 Breakpoint curve: nitrified effluent with ammonia-N added.

be achieved with a lower dose with sufficient ammonia present to form chloramine residuals.

When chlorine is added to a nitrified effluent where the NH_3-N is only a trace, there will usually be an organic N concentration of about 3 mg/l. This is an important factor because only two species of chlorine residual will be produced, i.e., free available chlorine and "dichloramine," which is not true dichloramine but organic chloramines resulting from the presence of organic N. A forward titration reveals some interference by the organochloramines in the monochloramine fraction. Since there is no information available on the kinetics of organochloramine formation in the presence of free chlorine it is difficult to predict with any accuracy the disinfection efficiency of the chlorine–contact time envelope by the Collins model. Organic nitrogen compounds have a wide range of reaction times with chlorine to form the chloramine fraction. Furthermore, the germicidal efficiency of these organic chloramines is most likely to be nil. Given this background it is easy to see that the 9 mg/l chlorine residual at San Jose at the end of 49 minute contact time is not so impressive when half of it is composed or organic chloramines.* When ammonia nitrogen is *absent* there is no formation of a true monochloramine.

---

* The other half is free chlorine.

DISINFECTION OF WASTEWATER 519

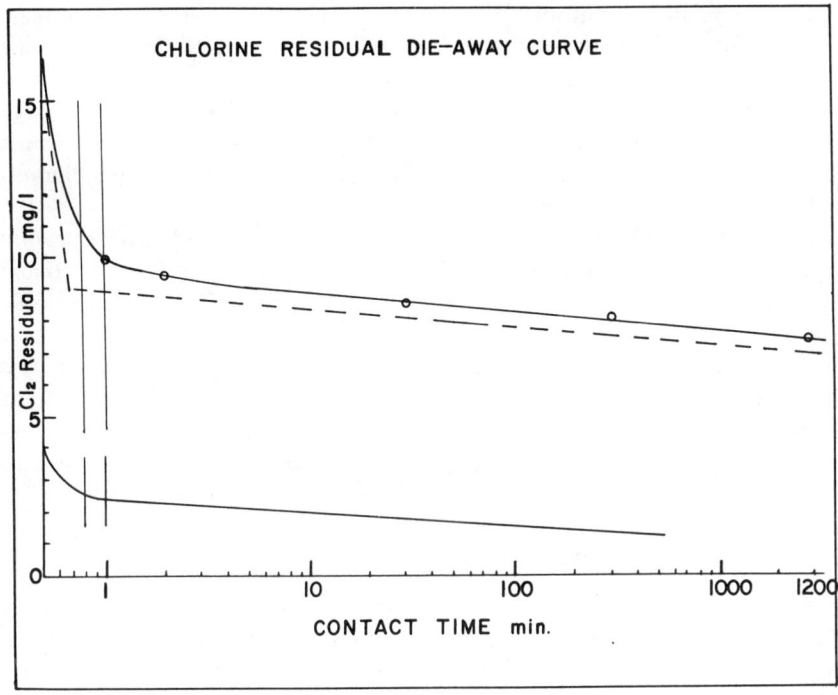

Fig. 8-7a Typical combined chlorine residual die-away curve.

One plant was found which proved to be the exception. This was the Fairfield-Suisin WWTP at Fairfield, California. This plant treats a daily wastewater flow that ranges from 7 to 14 mgd. The secondary effluent is filtered and nitrified before chlorination. Chlorine dosage ranges from 7.5 to 9.25 mg/l. The NPDES disinfection requirement is 2.2/100 ml MPN total coliforms. An investigation revealed the control residual (2–3 min contact time) was 6.0 mg/l TRC. This consisted of 5.4 mg/l free residual (FRC) and 0.60 mg/l "dichloramine" residuals, by forward amperometric titration. The TRC was measured by back-titration with a separate titrator.

The residual at the end of the contact chamber was 2.8 mg/l TRC. This was made up of 2.2 mg/l FRC and 0.60 mg/l "dichloramine." The sample used to measure the residuals at the end of the contact chamber were found to contain 0.04 mg/l NH_3-N and 0.60 mg/l organic N. Of all the nitrified effluents investigated, this plant and one other* used the smallest chlorine dosage to achieve a 2.2/100

---

* The Tapia water reclamation plant of the Las Virgenes Water District, Calabasas, CA in 1973–1974 was using approximately 8 mg/l chlorine dosage and 2 mg/l free chlorine residual at the end of 49 min contact time. Organic N was between 0.9 and 1.4 mg/l. ammonia-N was not detectable. The nitrified effluent was not filtered.[114]

ml MPN coliform concentration. An equally important observation was that the Fairfield plant exhibited the lowest concentration of organic N. Organic N concentration in the other plant effluents (except Las Virgenes) were on the order of 3–5 mg/l. These observations seem to substantiate White's conviction that the organic N content in a nitrified effluent represents a major interference with the germicidal efficiency of free chlorine. It reacts swiftly to form impotent organochloramines which "dechlorinates" the free available chlorine.

Some nitrification plants are able to control the ammonia-N concentration in the effluent to a level of 2–3 mg/l which is sufficient to produce a combined residual containing about 80–85 percent monochloramine in the presence of 3–5 mg/l organic N.* This solves the problem of excessive chlorine consumption and provides more reliable disinfection results. One such plant found that the saving in chlorine (due to allowing 2–3 mg/l NH_3-N to remain) amounted to \$37,000 per year or 4.1 percent of the total plant budget.[64,65] The successes of these plants and the San Jose plant have been duplicated by others. The problem that usually arises is the inability to control the ammonia-N level to the 2–3 mg/l range. Those plants are well advised to stay with complete nitrification and add either ammonium hydroxide (38 percent solution) or anhydrous ammonia.

Nitrification Summary. When the ammonia nitrogen is removed completely and chlorine is applied for disinfection, the free chlorine with its higher oxidizing power is subjected to excessive consumption by the various reducing compounds found in municipal wastewaters. As an example, organic color is bleached by free residual chlorine and is known to consume as much as 10–12 mg/l FRC in a few minutes.

The presence of organic N acts as an equally formidable competitor for the free chlorine because the resulting reaction forms organic chloramines well known to possess little if any germicidal efficiency.

The addition of ammonia-N to these effluents has been shown to result in a big reduction in chlorine consumption together with an increase in disinfection efficiency, provided the chlorine is mixed well with the NH_3-N spiked effluent. *Combined chlorine will not be consumed by organic color.*

Plant-scale investigations have determined that 6:1 Cl to N ratio is the most germicidal. This should be confirmed in the laboratory for each case.

These investigations have also substantiated the findings of Collins et al.,[66] Selleck et al.,[62] and Selna et al.,[67] that if given enough time (about 1 hour) combined chlorine will eventually catch up to free chlorine in bactericidal efficiency. These findings which corroborate one another may be due to a simple phenomenon that chloramines take longer to diffuse through the cell walls of the bacteria than does free chlorine.

---

*The San Jose study proved the importance of mixing at the point of chlorine application. Poor mixing resulted in gross formation of organochloramines compared to monochloramine in the ammonia-N spiked effluent.

One other effect that is present when free chlorine is the primary disinfectant is the likelihood of bacterial encapsulation by calcium ions present in the wastewater. Collins et al.[66] reported a grossly poorer bacterial kill with free chlorine in a plant using the lime precipitation process and it was believed that encapsulation could have been the cause. This phenomenon was first reported by Ingols[68] while studying the effect of calcium ion on the disinfection efficiency of swimming pool treatment methods.

FORMATION OF ORGANOCHLORINE COMPOUNDS

General Discussion. These compounds enter surface waters from many sources, both point and nonpoint. Little is known about the chemical stability, biological degradation, distribution, and ecosystem behavior of the majority of these compounds. Most of the surface water pollutants identified thus far are present in microgram per liter concentrations or less. At this concentration level, the highly chlorinated pesticides and hydrocarbons probably represent a more serious problem with respect to potable water treatment than those of the chloro-organics formed during wastewater disinfection. These organochlorine species of compounds are formed by the chloramine reaction in chlorinated wastewater effluents which discharge into surface waters.[69]

The formation of a wide variety of organochlorine compounds as a result of wastewater chlorination practices has been well documented by Jolley[69,70] and others.

The importance of the formation of these halogenated-organic compounds is the possible public health risk when they appear in potable water. The carcinogenic nature of several of these compounds has been demonstrated in the laboratory, using high concentrations and laboratory test animals. However, no direct cause-and-effect link with cancer in man has yet been established, and probably never will be.

The EPA is making a concerted effort to accumulate as much data as possible on the public health risk of these compounds. Their primary concern is the formation of the trihalomethanes (chloroform and bromoform) as a result of chlorination practices.[71] These compounds are known carcinogens so it is prudent to pursue a course of practice which would reduce or eliminate the formation of these compounds.

It is interesting and important to observe that there is no evidence that chlorination practices using chloramines will form any trihalomethanes; therefore, sewage discharges not treated beyond the breakpoint will not form trihalomethanes. *This makes the practice of free residual chlorination of wastewater effluents highly speculative.*

The Rancho Cordova Investigation. A recent comprehensive study by Stone[72] covered the halomethane formation in the following situations: activated sludge effluent containing 20 mg/liter ammonia nitrogen chlorinated to the breakpoint with dosages of 200 mg/liter or more and resulting in substantial free chlorine residuals (7–12 mg/liter); same effluent but with chlorination for disinfection (3–4 mg/liter residual at 30 min contact); same effluent except nitrified and filtered and chlorinated; a primary effluent dosed with 11 mg/liter chlorine; and an oxidation pond (nitrifying) effluent downstream from the activated sludge effluent.

The results of this study reveal a rapid formation of chloroform resulting from breakpoint chlorination as was expected. At a pH of 7, 88 µg/liter formed in 2 min and 123 µg/liter at 5 min.

The activated sludge effluent chlorinated for disinfection purposes formed only 4 µg/liter of chloroform in 4 hours.

The highly polished activated sludge effluent dosed with chlorine at 11 mg/liter led to the formation of three times the amount of halomethane compounds than formed by disinfection chlorination of the activated sludge effluent. This is undoubtedly a result of the formation of free chlorine residuals in the highly polished effluent. In this environment of free residuals there is a steady growth of halomethane formation.

Chlorination of the primary effluent to 14 mg/liter caused no change in chloroform concentration, a slight decrease in dichloroethane, and an appreciable increase in dibromochloromethane. Overall, a steady increase in total halomethane concentration was noted to a level of approximately 0.210 µm/liter* at 4 hours contact time. This is almost the same as that measured for the polished activated sludge effluent at the same contact time, even though differences in the form of chlorine residuals exist. The residual in the primary effluent is all combined chlorine.

Chlorination of the oxidation pond effluent to a dosage of 6.7 mg/liter yielded no change in THM concentration from unchlorinated levels through a 4-hour chlorine contact period even though chlorine residuals of 2.5–3.5 mg/liter were maintained throughout that time. Carbon tetrachloride and dibromochloromethane concentrations in the pond effluent were unaffected by chlorination. The low-ammonia nitrogen concentration of the nitrifying pond effluent insured that the chlorine residual existing throughout the 4-hour contact period was a mixture of both free and combined residuals.

A sample of the chlorine injector water was analyzed for THM concentration to determine if the high-chlorine concentrations (1000–3000 mg/liter) present in the injector water could cause THM formation prior to the injection of the chlorine solution into the process liquid. The injector water source was activated sludge effluent which had not been subjected to any previous application of chlorine. At the time of this particular sampling, the injector water chlorine concentration was on the order of 1800 mg/liter at a pH of 2.5. The residence time in the injector water system from the time of chlorine addition to the point of application was about 20 sec. Approximately 0.23 µm/liter* of THM were observed to be

---

* µm = micromoles.

formed in the injector water under the above conditions. However, in the overall system, dilution of the injector water into the process stream results normally in a 200 or 300 to 1 dilution so that the net impact of THM formation due to the chlorination system itself is negligible. It is likely that the low pH value and the short contact time in the injector water system may be responsible for keeping THM concentrations below that observed for breakpoint chlorination of the process stream.

The above observations of THM formation in sewage treatment processes were also compared with data on municipal tap water and raw surface water. Analysis of potable water taken from a municipal water system (surface water) in northern California showed THM levels almost identical to those reported as a median value of the NORS report,[73] with measured values of 25 μg/liter chloroform and 4.3 μg/liter bromodichloromethane. The total halomethane concentration was observed to be 0.235 μm/liter for the northern California potable water sample as compared to 0.218 for the median value of the NORS report. Analysis of samples collected above and below the wastewater discharge of the Rancho Cordova plant into the American River demonstrated that with the dilution factor of 250:1, current analytical techniques would be unable to detect the contribution of THM concentrations from this discharge to background THM levels in the American River.

The Occoquan Watershed Survey. A yearlong study of watershed runoff in this water service area located in northern Virginia near Washington, D.C. was reported by Hoehn et al. in 1976.[74] This report involves a system owned and operated by the Fairfax County Water Authority which provides potable water to more than 600,000 inhabitants in this rapidly urbanizing area. The Occoquan watershed collection system receives both agricultural and urban runoff as well as treated sewage from eleven plants in the area. The purpose of the report was to determine the effects of various factors in a given runoff area which might contribute to the formation of trihalomethane in the potable water supply at the downstream end of this runoff system. The findings of this comprehensive study support other evidence that chloroform concentrations appearing in surface waters do not necessarily have their origin in chlorinated wastewater discharges.

For example, the upstream control at Catharpin on Bull Run had a mean concentration of 2.2 μg/liter chloroform in all samples; while at Bull Run 2.3 miles below the last of eleven chlorinated sewage treatment plant discharges the samples showed a mean concentration of 3.2 μg/liter of chloroform. However, while the intake of the water treatment plant which is downstream from the Bull Run sampling point showed only 3.0 μg/liter chloroform (mean) the treated potable water showed a mean of 232.6 μg/liter chloroform. *The latter must be attributed to the practice of free residual chlorination at the treatment plant.*

Pomona, California Virus Study. This comprehensive investigation of various tertiary effluent treatment processes has been described above. The chloro-organic compounds investigated confirm other findings concerning the formation of chloroform due to chlorination procedures. It was found that free residual chlorination of a tertiary effluent increased the average chloroform concentration 330-fold (0.6–198 μg/liter) compared to a sixfold increase during combined chlorine residual chlorination. The tertiary effluent before chlorination contained chloroform concentrations on the order of 0.6 to 1.5 μg/liter.

Montgomery Maryland Simulation Study. A planned construction of a major advanced wastewater treatment (AWT) facility in the vicinity of Dickerson, Maryland has raised some concern about the discharge into the Potomac River at a point approximately 22 miles upstream from the Washington, D.C. water supply intake and 8 miles upstream from the proposed Leesburg, Virginia, water supply intake[75] This AWT facility is designed to remove ammonia nitrogen by breakpoint chlorination. This will be followed by activated carbon adsorption and dechlorination. The effluent disinfection requirement is 2.2 MPN total coliform per 100 ml or less. A major consideration in the selection of this process was its projected ability to obtain the desired disinfection with free chlorine residuals which also should provide virus inactivation.

This study clearly demonstrated that *breakpoint chlorination results in the formation of appreciable amounts of chloroform and bromodichloromethane.* And that excessive amounts of free chlorine resulting from pursuing the breakpoint produces increased concentrations of organochlorine compounds none of which are desirable in potable water.

The results of this study confirm other investigations, particularly the comprehensive work by Stone.[72]

CHLORINATION FACILITY DESIGN

Factors Unique to a Wastewater Chlorination System

General Discussion. There are several design considerations unique to a wastewater facility. These will be described here separately. Those features common to both potable water and wastewater are described in Chapter 9.

1. The chlorinator capacity is usually ten times that of a potable water facility of the same hydraulic capacity.

2. The injector system is supplied with treated effluent and therefore must be pumped. This requires standby power for reliability. Moreover the chlorine–ammonia reaction time must be considered in the design of the solution lines to prevent unnecessary chlorine consumption.

3. Special considerations are necessary for the design of diffusers to provide maximum dispersion and prevent off-gassing.

4. The chlorine residual analyzers are different from potable water analyzers

because they must handle a lower quality sample and are always arranged to measure total chlorine residual.

5. The control scheme is committed to a specific contact time and a prescribed NPDES discharge requirement.

6. Most chlorination facilities are followed by a dechlorination system. This affects the chlorination facility design.

FACTORS AFFECTING SYSTEM EFFICIENCY

Introduction. The strict requirements set forth by the California regulatory agencies for wastewater disinfection focused on the performance of every wastewater chlorination system. Therefore, having been involved in every aspect of a great many wastewater chlorination systems, White decided in 1972 it was time to evaluate the performance of these systems and to determine the elements of an optimum system.[76] Over a six-month period in 1972, White made a personal investigation requiring at least three and sometimes four visits to 34 plants. From 1972 to 1976 inclusive, another 12 plants were investigated to see if the findings of the 1972 survey could be corroborated. The plant effluents investigated in the 1972 survey included eleven primary, fifteen activated sludge, four secondary using high-rate recirculating biofilters, and four secondary followed by oxidation ponds. The second group included both secondary and tertiary effluents. From 1976 to 1982 White spent many days investigating the perceived anomalies of the disinfection process when applied to nitrified effluents.

Elements of an Optimum System. These investigations revealed that regardless of the treatment process, the most effective chlorination systems were those that included the following elements:

1. A proper and workable automatic chlorine residual control system
2. Adequate initial mixing
3. Sufficient contact time (not less than 30 min at peak dry weather flow) in a contact basin that has a minimum of short circuiting, i.e., 80–90 percent plug flow
4. Competent and dedicated personnel
5. Laboratory facilities sufficient to provide proper support to operating personnel
6. Reliability

These factors are examined below in this text. The chlorine demand, pH, temperature, and initial coliform concentration (before chlorination) of the wastewater are beyond the operator's control but must be reckoned with. This becomes the duty of the operating personnel, but they have to have an optimum facility to accomplish the job.

Optimized Mobile Pilot Plant Study[77-79]

Description of Plant and Study. In 1979, the California Department of Health Services completed a comprehensive design optimization study[79] on eight secondary and tertiary wastewater treatment plants. The major part of the project consisted of concurrent studies on a mobile optimized chlorination pilot plant and the existing full-scale system at each of the eight study sites. The trailer-mounted pilot plant contained optimum design features, including rapid initial mixing, reliable automatic chlorine residual control, and plug-flow-type contact tanks providing up to two hours detention time. Disinfection efficiency was measured by total coliform bacteria and iodometric chlorine residual tests. Concurrent fish bioassays on the same effluents were performed by the Department of Fish and Game. In addition to the above concurrent studies, special pilot studies were made of initial mixing, residence time distribution, effects of chlorine residual and contact time, and dechlorination.

Results and Conclusions of Study

1. At all locations the optimized pilot plant used significantly less chlorine than the full-scale plant. This amounted to an overall average savings in chlorine of 46.7 percent or 8.9 mg/l. The higher chlorine dosages required by the full scale plants were due variously to inadequate contact time, unreliable chlorine control system and/or inadequate operation and maintenance. The reduction in chlorine dosages was accompanied by an overall average reduction in effluent chlorine residual of 3.3 mg/l, which implies significant savings in sulfur dioxide used for dechlorination.

2. All plants studied except one met their disinfection requirements.

3. Overall, the pilot plant effluents were 43 percent less toxic to test fish than the full scale plants.

4. The chlorine dosages and residuals required to disinfect the various effluents to the same total coliform level differed fivefold in pilot plant studies. The dosages required to achieve an effluent MPN of 2.2/100 ml total coliforms with 60 min contact time varied from 4–22 mg/l in different effluents. The corresponding control residuals ranged from 2.8 to 11.5 mg/l. Poorer quality effluents and those containing industrial wastes required higher chlorine dosages and residuals than good quality domestic effluents.

5. Nitrified effluents required higher chlorine dosages for disinfection than some of the non-nitrified effluents. This phenomenon is explained elsewhere in this chapter.

6. The Collins–Selleck mathematical model $N/N_0 = (RT/b)^n$ in good quality effluents adequately described the relationship between the chlorine residual–contact time envelope (RT) and total coliform survival.

7. At contact times longer than 60 minutes a noticeable decrease in the bacterial

kill rate was observed. This phenomenon is partially explained under the subheading "Contact Time" in this chapter.

8. Tracer tests were necessary to assess adequately the performance of chlorine contact tanks. The use of the length to width ratio alone was found not to be sufficient.

9. Dechlorination by sulfur dioxide removed all chlorine-induced toxicity from the effluents.

10. The use of sulfur dioxide as a dechlorinating agent requires automatic dosage control based upon flow pacing, continuous chlorine residual measurement and rapid initial mixing. No additional contact tanks are necessary.

11. Operator attendance is mandatory to keep the chlorine control system in proper working order. This includes daily cleaning and calibration of the chlorine residual analyzers. Therefore, improved operator training is necessary to maintain and properly operate the disinfection system.

12. The pilot plant chlorine control system performed significantly better than the full-scale systems because of the following factors: (a) very short loop time, 60 sec, (b) adequate initial mixing, $G = 500$; (c) constant flow rate; (d) instrument compatibility; and (e) good operation and maintenance.

The pilot plant control system was able to maintain the control residual within the desired ± 0.5 mg/l range. It was observed that poorly designed and/or operated chlorine control systems were responsible for a major portion of the excessive chlorine dosage used at most of the full-scale plants studied. The pilot chlorine contact tank also performed better than most of the full scale tanks due to better plug-flow characteristics and longer minimum contact time.

SIZING THE CHLORINATOR

Influence of Plant Flow Meter. The first step in the design of a chlorination facility is the determination of the maximum chlorine needed for the given situation, including prechlorination, intermediate, and postchlorination applications. While this text is specific for disinfection (postchlorination), total capacity must include the requirements for the other purposes stated above. Chlorinators should be divided into two groups, one group for prechlorination and intermediate points of application, the other group for postchlorination (disinfection). Equipment should be arranged so that the first group can provide standby service for the disinfection equipment. A third group may be desirable if both pre- and postchlorination is to be continuous. The third group should be arranged for both intermediate chlorination and standby for either pre- or postchlorination.

Chlorinator capacity is a function of the flow signal and the chlorine demand of the wastewater. Attention is called here to the term "flow signal." Since it is imperative that the chlorine feed rate be controlled in proportion to the flow, the chlorinator capacity must be sized to the capacity of the primary flow meter. While it is true that chlorinators have overriding dosage control, this feature must be reserved for variations in the chlorine demand of the wastewater. For example, at a proposed plant, the primary flow meter to be used has a range of 0–10 mgd, which is the ultimate design capacity of the facility. Furthermore, the chlorine demand is to be a maximum of 150 lb/million gal. The chlorinator capacity for that plant must be $10 \times 150 = 1500$ lb/day regardless of the initial low flows that may be expected for the first several years. This capacity is required because the chlorinator needs the full range of the primary meter flow signal so that it can also provide a dosage range to meet the diurnal chlorine demand variation.

This concept applies to both pre- and postchlorination, but not to intermediate chlorination because the latter chlorine feed rate is usually on a manual control basis.

In the above example the chlorinator would be ordered as a 2000 lb unit fitted with a 1500 lb rotameter and a 2000 lb spare rotameter.

Effect of Chlorine Demand. The required chlorine dosage for prechlorination is difficult to predict with any degree of accuracy in the absence of some historical evidence. If the sewage is moderately fresh, the dosage may be as low as 10–12 mg/liter. If sulfite wastes such as from a tannery or a winery are a factor, the chlorine demand might be as high as 40–50 mg/liter. If the sewage is septic upon arrival, the demand will most likely be from 30 to 40 mg/liter.

Intermediate points of application of chlorine require approximately the following equipment capacities.*

1. Secondary sedimentation tanks (10 mg/liter)
2. Return activated sludge—control of bulking (5 mg/liter)
3. Ahead of biofilters or trickling filters (10–15 mg/liter)
4. Sludge thickener line—odor control (50 mg/liter)

The use of chlorine at these points of application is rarely on a continuous basis, so the prechlorination equipment can be used for these purposes. If prechlorination is to be used continuously for odor control, additional equipment should be supplied for the intermediate points of application. This group of equipment should then be sized so as to provide standby service for either pre- or postchlorination.

Industrial wastes have a profound effect on the disinfection requirements of domestic wastes. This is further complicated by the seasonal nature of some industrial wastes and extent to which these wastes are pretreated before being discharged into the domestic collection system.

For domestic wastes with not more than 1–2 percent industrial waste, the following are minimum capacity guidelines for the designer:†

---

* All dosages are based upon treatment plant flows (PDWF) while sludge thickener dosage is for this specific flow rate.
† These dosages are based upon NPDES requirements of 200/100 ml MPN fecal coliforms or 1000/100 ml MPN total coliforms.

Primary effluent	150–200 lb/mg
Secondary effluent	50–75 lb/mg
Secondary plus ponds	50 lb/mg
Tertiary (not nitrified)	30–50 lb/mg

Effluents that require a free chlorine residual require about 10 parts of chlorine for each part of ammonia nitrogen.

If industrial wastes are present to the extent of 10 to 25 percent of the total wastewater flow, it may be necessary to increase the chlorine requirement for primary effluents by a factor of two and perhaps more depending on the nature of the wastes. This may apply to secondary effluents as well, depending on how effectively the treatment process can cope with the industrial waste. The more uncertain the factor of industrial waste, the more uncertain becomes the required quantity of chlorine for disinfection. In this case, laboratory determination of chlorine demand will be necessary.

CHLORINE CONTROL SCHEMES*

Flow Pacing. Chlorinators for both pre- and postchlorination should be provided with flow proportional control. The prechlorinator should be controlled from the influent flowmeter and the postchlorinator from an effluent flowmeter. Never attempt to control the postchlorinator from the influent meter or vice versa; it will not work because of the lag time between the two measuring points. As discussed previously, the flow proportional signal should be used to control the chlorine metering orifice in the chlorinator. This can be done either pneumatically or electrically. If it is to be pneumatic, the signal must be linear (3–15 psi). If it is to be done electrically, there are a variety of linear signals. The one most preferred is the analog milliamp signal, with outputs of 1–5, 4–20, and 10–50 mA. The most common is 4–20 mA.

While there is local dosage control provided on the chlorinator cabinet for individual chlorinator control, a ratio relay (0.4–4.0 range) should be installed in the flow meter electrical signal circuit. This instrument will provide the operator with the means to make precise dosage adjustments without interfering with the range and accuracy of the flow pacing signal.

Chlorine Residual Control:

General Discussion. In addition to the flow proportioning control, residual control for disinfection is necessary because of the diurnal variations in chlorine demand, and the necessity for close regulation of the disinfection process to meet NPDES discharge requirements.

The key factors for a successful residual control system is the ability to obtain

---

* For more details on control strategy illustrated by instrumentation schematics, see Chapter 9.

a homogeneous sample downstream from the chlorine diffuser within 30 seconds after the application of chlorine at average dry weather daily flow. At peak flow this time would be reduced to about 15 sec. These time values are based upon a mixing time of not more than 3–5 sec. It will then take the sample another 45–60 sec to reach the cell in the analyzer since it has to pass through a filter and the internal piping within the analyzer; therefore, the earliest the sample can reach the measuring cell is about 1½–2* min after the chlorine has been mixed with the wastewater. If the chlorine has been well mixed, the control residual will lie on the flat part of the residual die-away curve shown in Fig. 8–7(a). Another mandatory requirement of a well-designed system is to utilize the remote injector concept by installing the injector as close to the point of application as is physically possible. The average loop time should be on the order of 2 min, with a maximum not to exceed 5 min.

Residual control is synonymous with dosage control. The analyzer transmits a signal which calls for either more or less chlorine. The control strategy can be either the floating type or it can be a separate PID loop. There are a variety of arrangements which are described below.

Compound-Loop Control. This has been the predominant method of residual control for wastewater treatment since it was first introduced in 1961. This system was unique to Wallace and Tiernan equipment. In 1984, Wallace and Tiernan announced the elimination of this control method in favor of a dual signal system.

This method utilizes two signals that are sent to different components in the chlorinator. The chlorinator combines these two separate pieces of intelligence to produce a given chlorine flow rate. The wastewater flow meter sends a pacing signal to the motorized chlorine orifice valve. This signal changes the area of the valve opening. The signal from the residual analyzer is sent to the vacuum regulating valve in the chlorinator. The change in this valve position changes the differential "head" across the above described metering orifice. The flow through an orifice varies with area and differential pressure across the orifice ($Q = A\sqrt{2gh}$). This method has many advantages, which are described in Chapter 9. The principal advantage is its uncommonly wide range capability of at least 100 to 1. One other important feature is the ability of the residual signal to control when the flow signal becomes unreliable at low flows. See Fig 9-42 page 669.

Multiplied Signal Control. This method has been in widespread use as a process instrumentation technique for many years. When used for chlorination and dechlorination control the wastewater flow signal from the treatment plant flow meter and the chlorine residual signal are sent to a multiplier. The output of the multiplier controls the chlorine metering orifice positioner. This method is limited to the practical range of the chlorine metering valve and rotameter, which is significantly less than the range of a compound-loop system with equivalent accuracy described above and in Chapter 9.

* At average dry weather flow.

Dual Signal Control. This is a variation of the multiplied signal. It was introduced by Wallace and Tiernan in 1978. The Wallace and Tiernan electric positioner is equipped with a manually controlled dosage potentiometer. The use of the positioner potentiometer is the basis for automatic residual control. The residual analyzer is fitted with a potentiometer. The potentiometer signal is integrated by the circuitry in the positioner so that the positioner responds to two separate signals—one from the flow meter and one from the potentiometer in the analyzer. An override switch on the front of the chlorinator permits manual dosage adjustment.[81]

Straight Residual Control. This method operates without benefit of a process flow signal. This has proved to be a successful and reliable method.[82] The success of this system depends upon three factors: (1) the residual signal must be transmitted to the chlorine orifice positioner (via a PID controller); (2) the positioner travel time from zero to maximum must be on the order of 15 sec; (3) the loop time should be limited to approximately 45 sec at peak flow conditions. See Chapter 9 page 664 Fig 9-41 for a typical potable water system schematic.

INJECTOR SYSTEMS

General Description. The injector system is the heart of the entire chlorination facility. If this system is inoperative, no other part of the system can function. The various parts of this system include: (1) the operating water supply to the injector; (2) the injector; (3) the injector vacuum line from the chlorinator; (4) the injector discharge system described as the chlorine solution line; (5) the diffuser at the point of application; and (6) the necessary gages required to monitor and troubleshoot the system. Design details are fully presented in Chapter 9. The following text serves to cover situations that are peculiar to wastewater installations which may be different from potable water systems.

Water Supply. Every injector system is designed to create sufficient vacuum (16–22 in. Hg) to move the chlorine from the supply containers into and through the chlorinator to the injector suction port, where it is dissolved into the injector water supply. The injector must then be able to deliver the chlorine solution to the chlorine diffuser. The injector water supply pressure must be sufficient to create the required vacuum and overcome the friction loss in the solution piping and diffuser. Additionally this pressure must overcome any positive static head between the injector centerline and the hydraulic gradient at the diffuser. The amount of water required must be sufficient to limit the chlorine solution strength to 3500 mg/l. A broad rule of thumb is 40 gal of water/day/lb of chlorine, more or less, depending upon the local conditions.

Plant effluent is universally used for the injector water supply. Non-nitrified effluents contain 15–25 mg/l ammonia nitrogen. If the breakpoint reaction between ammonia-N and chlorine is allowed to proceed in the chlorine solution line, 10 parts of chlorine will be consumed for each part ammonia-N before the chlorine solution reaches the point of application. If proper precautions are not taken the chlorine wasted in this reaction could amount to about 10 percent of the total chlorine used. This reaction is both time and pH dependent. At pH 3 the reaction will go to 90 percent completion in about 4 minutes. The usual pH of chlorine solution derived from wastewater is between 2 and 3, depending upon the chlorination rate. Therefore chlorine solution lines should be designed to limit the solution travel time to not more than 4 minutes. If this is not practical the injector should be moved nearer to the point of application.

Hydraulic Considerations:

Positive Head. Static pressure at the discharge of the injector should be limited to no more than 4–5 psi, for the following reasons: (1) Injector operating pressures and water quantity escalate rapidly as the back-pressure on the injector rises beyond 5 psi, particularly when the chlorine feed rate is in excess of 500 lb/day.* (2) Head loss through the diffuser is used as a chlorine mixing device. Allow 8 to 10 ft loss through the diffuser for this function. To meet this requirement injectors should be located as close to the diffuser as possible and at an elevation above the hydraulic gradient at the diffuser location. Injectors may be installed in any direction, i.e., upright, horizontal, or upside down.

When these conditions cannot be met and the static head at the injector is 10–15 psi or more, a chlorine solution pump should be installed adjacent to the injector discharge as shown in Fig. 9-53 on page 693 Chapter 9. For wastewater installations the Fybrock Div. of Robt. D. Norton Co., Fairfield, NJ can furnish Fiberglas pumps with a good experience record for pumping chlorine solution. The installation must be arranged so there is always a positive pressure (approx. 2–3 psi) on the discharge of the injector.

Negative Head. An open-channel diffuser can be a great nuisance and even a hazard if it is not properly designed. Anytime a negative head exists in the chlorine solution line, molecular chlorine will break out of the chlorine solution and cause serious chlorine gas emission at the diffuser. Therefore, if the sewage level at the diffuser is below the injector throat, the hydraulic gradient from the injector to the diffuser must be calculated to provide a reasonable injector back-pressure. Assume this back-pressure is 5 ft (approx. 2 psi). Then the friction loss through the diffuser holes plus the line losses minus the difference in elevation between sewage level and injector throat must equal approximately 5 ft of head. This is one of the reasons for having a compound gage on the solution line at the discharge of each injector. The minimum recommended size for diffuser holes is 3/8 in.†

---

* For chlorinators 500 lb/day and smaller this back-pressure limitation is not a factor.
† This is for wastewater systems diffusers.

The recommended velocity through the holes to provide initial mixing is 22–26 ft/sec. All open-channel diffusers should be constructed for easy removal. They may become plugged or the hole configuration might have to be changed. This requires a flanged connection so that the entire diffuser can be lifted bodily from the channel. Diffusers are available with projecting pins that fit into slotted wall brackets for easy removal.

Figure 8-8 illustrates a problem in diffuser design for an open channel. Assume a 200 gpm flow through a 3-in. injector and 8000 lb/day chlorine so as to limit the chlorine solution strength to 3500 mg/liter chlorine. Use 4-in. chlorine solution piping to the diffuser. Applying Bernoulli's energy equation we get

$$P_1 + E_1 + \frac{V_1^2}{2g} - \text{losses} = P_2 + E_2 + \frac{V_2^2}{2g} \tag{8-9}$$

Ignoring velocity head differences from Fig. 8-8 we get the following:

$$5 + 240 - \text{losses} = 0 + 220$$

$$-\text{losses} = -25 \text{ ft}$$

The equivalent length of the chlorine line is assumed to be 150 ft; this is made up of 75-ft linear pipe lengths plus 5 elbows at 15 ft each. Therefore, the friction losses in the solution piping will be about 5 ft. So, to prevent a negative head condition at the diffuser by maintaining 5 ft of head at the injector discharge, the diffuser will have to consume $25 - 5 = 20$ ft of head loss.

The next step is to choose a diffuser hole size. The minimum acceptable size is ⅜ in. Use the equation

$$Q = CA\sqrt{2gh} \tag{8-10}$$

where

$C = $ orifice coefficient $= 0.75$
$A = $ area of orifice ft^2
$h = $ head loss, ft
$Q = $ discharge, cfs

Using $A = 0.0008$ and $h = 20$, Q calculates to 0.0215 cfs = 9.7 gpm. Therefore, $200/9.7 = 20.6$; so use 20 or 21 holes.

Diffusers should always be designed on the basis of one diffuser for each injector, unless more than one injector is to be operated simultaneously.

Remote Injectors. Fig. 8-9 shows the installation of a 2-in. remote injector. Larger injectors (3 in. and 4 in.) should be installed in the horizontal position when

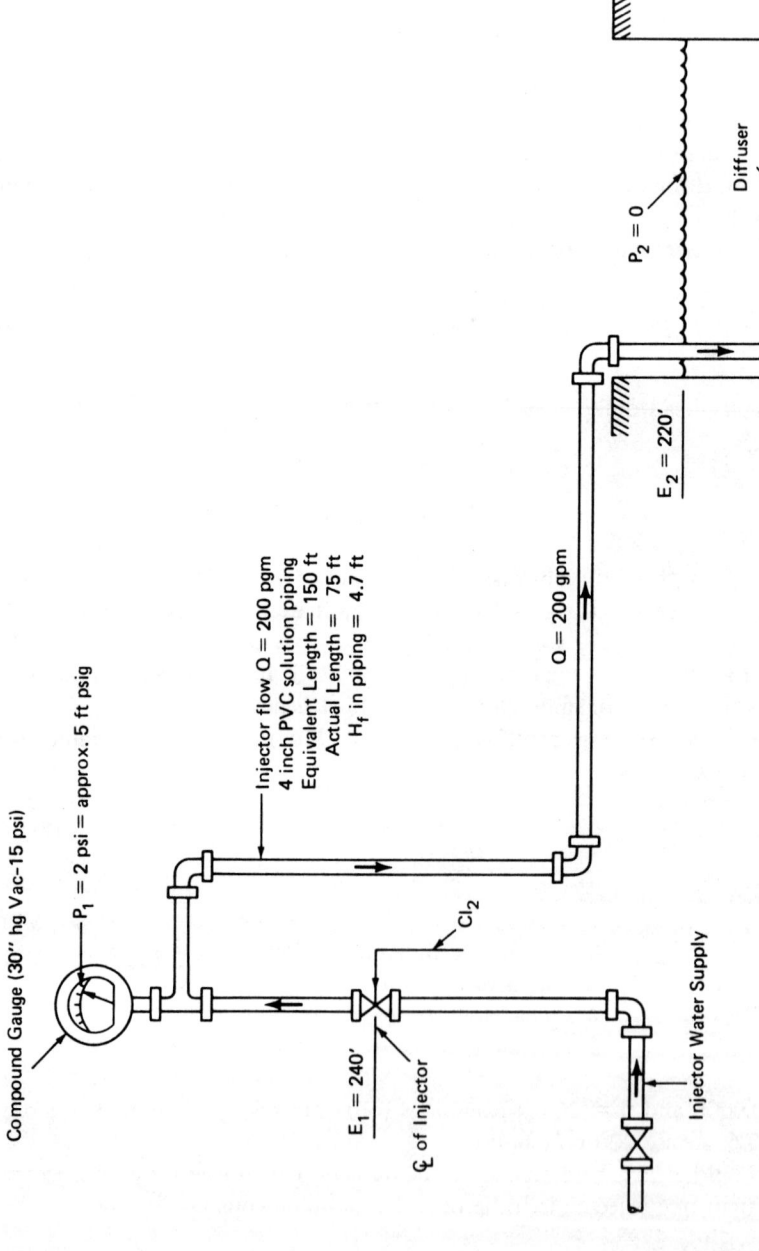

Fig. 8-8 Example of diffuser design for open channel application under negative head conditions.

Fig. 8-9 Remote injector utilizing injector water to provide sample to chlorine residual analyzer.

possible. This low profile position reduces injector back-pressure and allows for easy disassembly of the injector. The vacuum line between the chlorinator and the injector should be sized so that the total pressure drop in this line does not exceed 1.5 in. Hg at maximum chlorine flow. (See Ch. 9 for sample calculation.) A shutoff valve should be provided at each end of the vacuum line. A vacuum gage should be installed adjacent to, but downstream from, the shutoff valve at the injector end.

Monitoring Devices:

Flow Meters. Adjustable throat injectors (i.e., the 2, 3, or 4 in.) should be provided with a flow meter of reasonable accuracy over a 3:1 range. A direct-reading rotameter utilizing a bypass orifice type installation is satisfactory. However an in-line propeller meter is more expensive but makes for a neater piping arrangement.

Flow meters are necessary for the operator to achieve optimum injector adjustment for the local conditions to insure maximum efficiency, proper chlorine solution strength, and adequate back-pressure.

Gages and Alarms. Gages should be provided to show the operating water pressure for each injector in the system. Each injector should be provided with a chlorine solution pressure gage mounted immediately downstream from the injector discharge. These must be compound gages reading to 30 in. Hg vacuum and to 15 psi pressure and equipped with a silver diaphragm protector. These gages are necessary for the proper analysis of the injector operation. They provide the operator with the back-pressure readings for various quantities of water passing through the injector. In some situations this is critical information that cannot be obtained in any other way. It also constitutes proof of proper or improper diffuser design and/or chlorine solution line design.

Every installation should be provided with a loss of water pressure alarm to indicate a failure in the water supply. Overpressure switches are not necessary for the injector system.

Chlorine Solution Lines. The piping downstream from the injector is the chlorine solution line. It is permissible to manifold the injector discharge from two or more chlorinators into one point of application,* but a solution line to the point of disinfection should not be manifolded to any other point of application. The most desirable arrangement is for each injector (in a multiple injector system) to have its own solution line and diffuser. For pipe materials, see "Materials of Construction." Chlorine solution flow proportioning systems are not worth the expense and most end in failure. Some waters, when carrying a chlorine solution, release a tremendous amount of carbon dioxide and other gases. The bubbles of gas in the solution passing through the rotameter can cause sufficient vibration to severely limit the accuracy of the reading. Glass tube rotameters should not be used because this vibration can cause the rotameter float to shatter the glass tube. The only rotameters satisfactory for chlorine solution lines are the straight-through metal (Hastelloy C) or PVC tube type with dial indication.

Diffusers. There are three categories of diffusers in wastewater treatment: prechlorination, intermediate chlorination, and disinfection. These diffusers are required to perform different tasks, so they must be discussed separately.

Prechlorination. The choice of chlorine solution diffuser for the best results at this point of application is a challenge to the designer. Usually the diffuser will be in a closed conduit flowing partially full, thereby conforming to open-channel flow. There is practically no mixing capability in an open channel, and, owing to the large amount of debris in the raw wastewater influent, mechanical mixers are totally impractical. Therefore the designer must select the diffuser with the best mixing characteristics. Fig. 8-10 shows one of the most satisfactory types of

---

* Provided the solution line discharges into a multiple diffuser system so that there is one diffuser per injector.

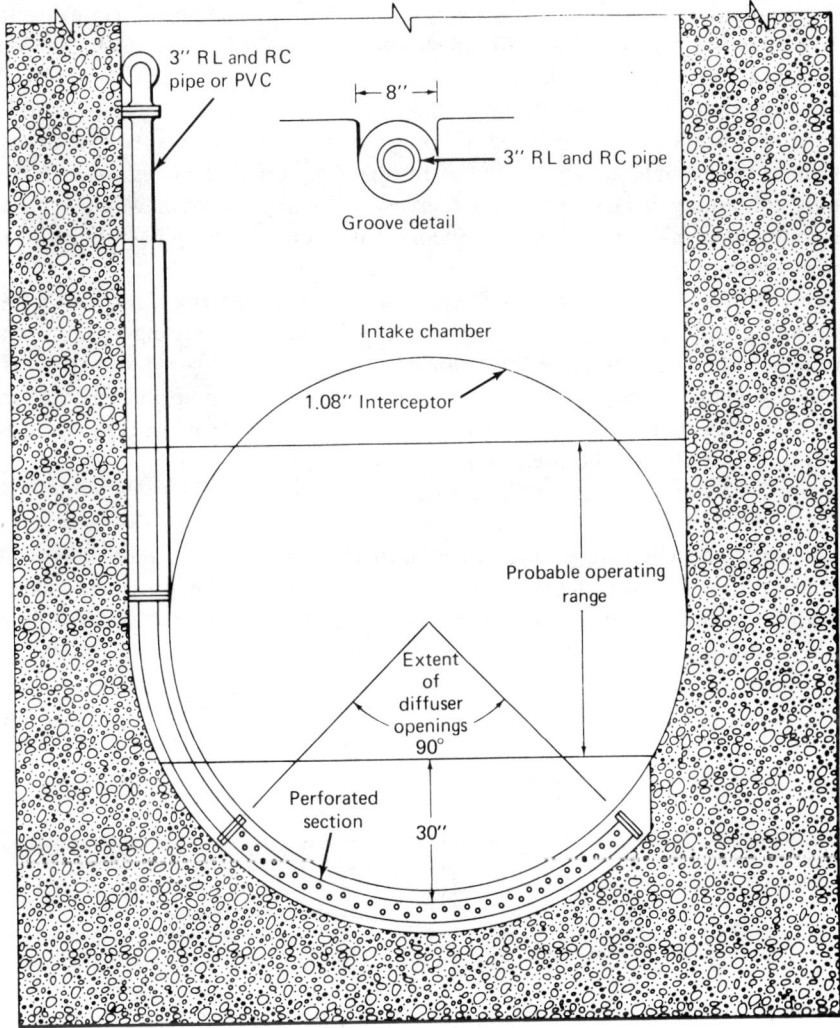

Fig. 8-10 Perforated diffuser shaped to fit into recess of circular conduit especially for prechlorination.

prechlorination diffuser.* This type can be made of either PVC or hard-rubber-lined and rubber-covered steel pipe. The curved and perforated sections are heat-treated to conform to the radius of the conduit and are laid in a recess in the conduit. The pipe is perforated, so that chlorine solution is discharged mostly below the minimum water surface level. This diffuser eliminates the problem of

---

* One such diffuser has been in operation without removal or maintenance problems for more than 35 years.

538 HANDBOOK OF CHLORINATION

the accumulation of debris which is always prone to collect on any object projecting into the flowing sewage. The diffuser should be designed to be easily removed. This may be necessary to clean the slot of silt and other debris as well as to inspect for plugging of perforations. Experience indicates that over a period of three or four years chlorine diffusers do plug up in wastewater plants. The deposits that cause the trouble appear to be the result of highly chlorinated organic compounds containing a large portion of grease, which provides a waxlike binder. The above type of diffuser must be incorporated into the original design of the influent pipeline.

Figure 8-11 shows the perforated type of diffuser laid on the invert of the pipe and running axially with the conduit. This type is used for existing systems when there is no other way to apply the chlorine solution.

Hanging spray nozzle diffusers have been used for prechlorination diffusers. Some have been successful but again experience has shown that the type of diffuser shown in Fig. 8-10 is to be preferred.

Figure 8-12 illustrates the diffuser in a closed conduit, when the influent line to the plant is a force main. Although it is well known that the best mixing occurs when the chlorine is discharged into the center of the pipe, this is not practical because of the debris that will accumulate on the projected diffuser. Therefore the diffuser must end flush with the inside wall of the conduit, but it should be designed for a velocity of 22–26 ft/sec in order to project the chlorine

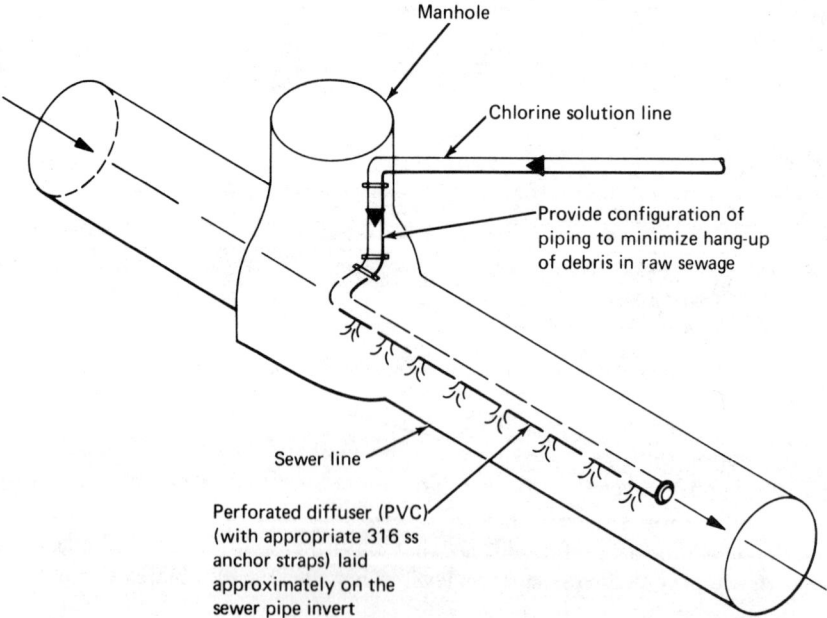

Fig. 8-11 Chlorine diffuser laid on pipe invert (prechlorination only).

DISINFECTION OF WASTEWATER 539

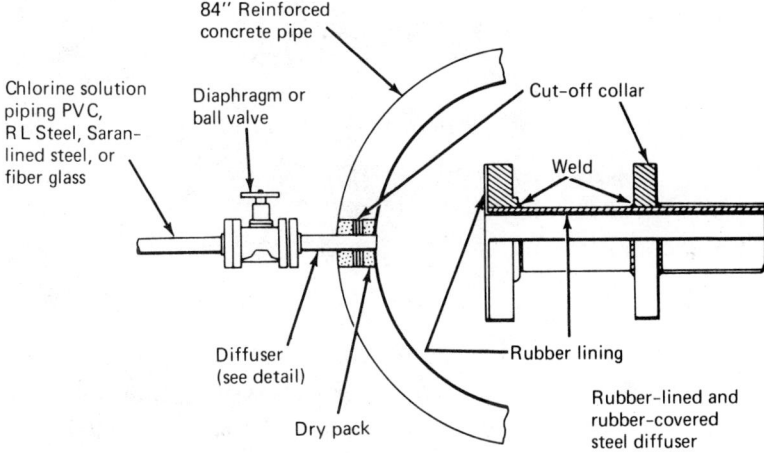

Fig. 8-12 Prechlorination diffuser flush with pipe wall to prevent debris fouling (orifice velocity 20–25 ft/sec).

solution toward the center of the conduit. Successful results have been obtained by following these guidelines. This velocity across the diffuser orifice will result in an 8–10-ft pressure drop. Therefore the chlorine solution line and injector water supply will have to be designed to accommodate this additional injector backpressure.

Intermediate Points of Application. The mixing and dispersion of chlorine solution into return activated sludge, supernatant liquor, secondary sedimentation, and recirculating biofilters do not present the difficulties associated with pre- and postchlorination.

For example, return activated sludge, supernatant liquor, and other points of intermediate chlorination are usually into a closed pipe. For these situations a pipeline diffuser is used (see Figs. 8-13 and 8-14). This type of diffuser is designed

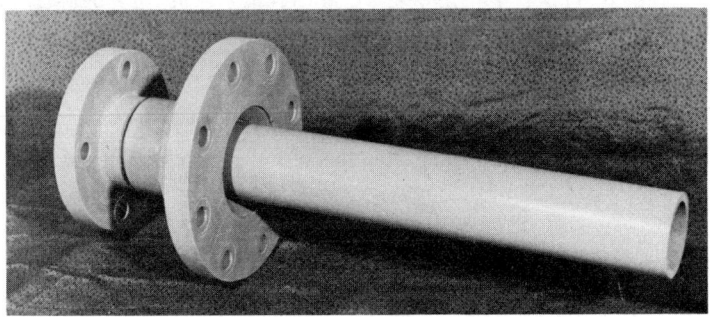

Fig. 8-13 Chlorine solution diffuser (courtesy Chlorine Specialties Inc.).

540 HANDBOOK OF CHLORINATION

Cl₂ Sol. Line Size	Diffuser No.
1½"	1½ × 3 × D
2"	2 × 3 × D
2½"	2½ × 4 × D
3"	3 × 4 × D
4"	4 × 6 × D
6"	6 × 8 × D

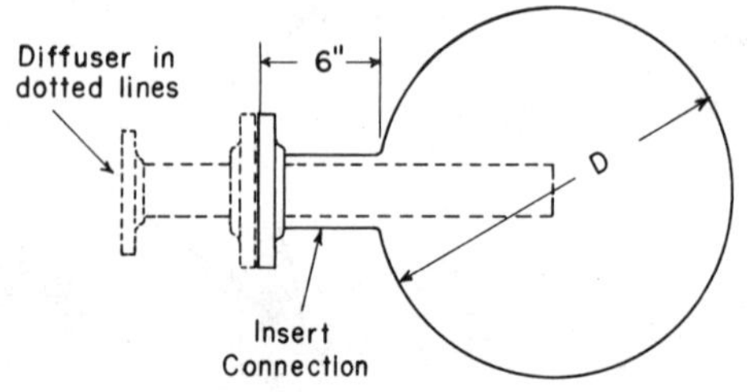

Fig. 8-14 Dimension of pipeline diffuser.

to deliver the chlorine solution to the center of the pipe. This provides a complete mix in 10 diameters of the pipe. Return activated sludge chlorine should be applied in the recirculation pump suction.

Secondary sedimentation tanks are usually accommodated by the perforated type of horizontal diffuser in an open distributing box ahead of such structures.

Disinfection (Postchlorination). At this location there is never a problem of debris accumulation so the designer has a wide array of design choices. However these diffusers are of vital importance to one of the most critical factors affecting disinfection efficiency—initial mixing. Therefore the design and configuration of these diffusers is described below under "Mixing."

Materials of Construction. The materials used for the injector system are those employed in good water works practice for the water system up to the inlet of the injector assembly. From this point forward, a corrosive chlorine solution will be encountered which requires special materials. The chlorine solution lines can be either Sch 80 PVC, Fluoroflex-K (Kynar), rubber-lined steel, Saran-lined steel, Kynar-lined, or certain types of fiber cast pipe.

Valves on solution lines can be either diaphragm or ball type. Diaphragm types are usually flanged, rubber-lined, or PVC-lined cast iron Saunders-type valves. The ball type PVC valve is preferable up to 2½-in. size, although the ball-type valve is available in much larger sizes. The diaphragm type should be considered for sizes 3-in. and larger.

The injector vacuum line between the chlorinator and the injector is the most controversial of all the piping used for the chlorination facility as to material selection. This line carries moist chlorine gas under a vacuum. It is preferable to use Sch 80 PVC or Kynar pipe, although some types of fiber cast pipe are also suitable. Ball type PVC valves should be used instead of diaphragm type valves on this line. There is no known corrosion resistant metal pipe available for this use. Saran and Saran-lined steel pipe have been used for this purpose; glass pipe has been found to be impractical. The only time that this particular chlorine carrying line becomes a consideration is in the case of remote injector installations. It should be noted that when the injector is located in the chlorinator room, the manufacturer supplies Sch 80 PVC for the interconnecting piping.

The diffusers and the piping leading to the diffusers are customarily made up of Sch 80 PVC pipe and fittings. If the specifications for underwater piping require steel construction for additional strnegth, all the underwater piping and diffusers must be made of rubber-lined and rubber-covered steel pipe. The diffuser holes must also be rubber covered. This results in extremely expensive construction costs that are rarely worth the expense.

Nuts and bolts for assembly of the underwater portion should be 316 stainless steel. All other bolts should be galvanized or cadmium plated steel. Pins and slotted wall brackets for diffusers are available in PVC.

As described previously, all gages on the solution and vacuum lines should be silver diaphragm protected. This is a standard item with the chlorinator manufacturers. Vacuum line gages are for vacuum only; solution line gages must be compound gages as described in the section on Gages.

Initial Mixing

General Discussion. The question of optimum mixing of chlorine solution into an ammonia laden wastewater has been a formidable puzzle for designers. The principal competing reactions in a chlorine–wastewater scheme is the HOCl with ammonia-N versus HOCl and the microorganisms.

However, how long HOCl persists during the initial mix remains a controversial question among many investigators. Under normal conditions of wastewater disinfection the reaction between HOCl and NH_3-N is so rapid it is probably not measurable. However, it can be estimated. Assume the following condition: $ph = 7.0$, temperature $= 25°C$, NH_3 in wastewater $= 14$ mg/l as N, and chlorine dosage $= 10.5$ mg/l. Using Professor Morris' rate constants the HOCl–NH_3-N reaction is 99 percent complete in 0.2 seconds. At pH 4 it is 147 seconds. The peak rate occurs at pH 8.3. Therefore it is the opinion of both White and Collins that under normal conditions there is no measurable effect of HOCl. Moreover, since the molecules of NH_3 outnumber organisms by a factor approximating 10^{13}, it is practically certain that all of the disinfection occurring in the initial mixing phase is by monochloramine.*

There is one other competing reaction that has a profound effect upon disinfection efficiency. This is the chlorine–organic N reaction. This was observed in a completely nitrified effluent[50] which was converted from free chlorine residual to a combined residual by adding NH_4OH in a CL:N ratio of 6:1. With little or no mixing the

---

* An exception to this is the Pentech and Fischer and Porter Systems addition system, where chlorine gas instead of chlorine solution is used. In this case molecular chlorine has a definite effect in the initial mixing phase.

final residual consisted of 45 percent monochloramine and 55 percent organochloramine. When there was thorough rapid mixing the total residual was composed of 85 percent monochloramine and 15 percent organochloramine. The total coliform concentration at the end of 50 min contact time was 1300/100 ml for little or no mixing and 30/100 ml for thorough mixing. The reason for this phenomenon has yet to be explained. It is believed that where poor mixing is involved a hydrolysis reaction takes place whereby the monochloramine reacts with the organic N to form organochloramines.

The difficulty with initial mixing is the lack of information that would allow a valid description of what to expect from different "levels" of mixing.

Historical Background. The earliest known mention of the importance of initial mixing in the chlorination of wastewater was by Rudolfs and Gehm.[84] They reported in 1936 on a comprehensive study of wastewater disinfection by chlorination. As a result of this study, they advised that "maximum bacterial kill can be obtained . . . provided good mixing is allowed. . . . However, they made no attempt to define "good mixing."

The first known documentation of chlorine dispersion in a wastewater effluent occurred at the City and County of San Francisco, Richmond–Sunset WWTP in 1940.[83] The chlorine diffuser consisted of a series of hanging nozzles across a 4-ft-wide channel. The nozzles were slotted, which produced a thin layer of chlorine solution in the shape of a 90° fan. The nozzles were positioned about 18 in. below the wastewater surface in a 3-ft-deep channel. Residuals were measured by starch–iodide titration. The diffusers were located upstream from a Palmer–Bowlus flume. The residuals were measured about 50 ft downstream from the diffuser. The results demonstrated quite conclusively that there was gross stratification of the chlorine solution due to laminar flow and the absence of turbulence. Ten feet beyond the point of residual measurement the effluent discharged into an outlet structure where there was a high degree of turbulence. The chlorine residuals became uniform throughout this structure as a result of this turbulence.

In 1948 Eliassen, Heller, and Krieger[88] presented a statistical approach to sewage chlorination. This study examined the effect of initial mixing upon bacterial destruction. They compared three levels of mixing intensity: (1) no mixing; (2) 10 swirls with glass rod at approx. 100 rpm; (3) 15 seconds in a Hamilton-Beach egg-beater. They found a significant difference in bacterial destruction by chlorine between the unmixed and the mixed samples. However they found no improvement in the samples mixed by the egg-beater. These findings tend to support recent findings which are discussed later in this text.

Heukelekian and Day[90] reported in 1951 on a study of 5 sewage treatment plants using chlorine for disinfection. They found that in those plants not provided with special initial mixing compartments, gross segregation of chlorine concentration will occur. This is the result of low-velocity (0.5–0.75 ft/sec) laminar flow. Fig. 8-15 illustrates the data collected by Heukelekian and Day demonstrating

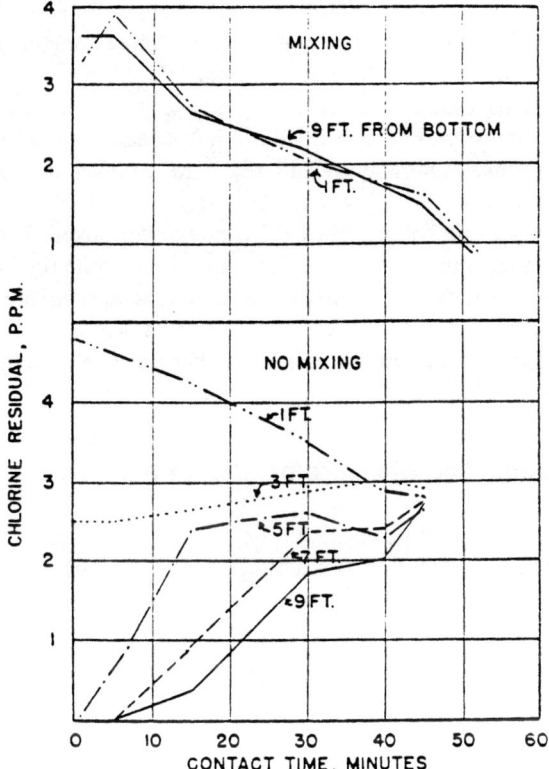

Fig. 8-15 Segregation in a chlorine contact tank: mixing vs no mixing.

the gross differences in the homogeneity of chlorine residuals with and without mixing in a chlorine contact chamber.

About 1968 Collins began a pilot plant study at the University of California at Berkeley in an attempt to delineate the kinetics of wastewater disinfection with chlorine.[56,87] This work demonstrated conclusively that initial mixing had a profound effect upon the kinetics of the disinfection reaction.

Other research work in the 1970s at Johns Hopkins University by Krusé et al.[89] and at the University of California at Berkeley by Stenquist and Kaufman,[93] Collins and Selleck,[58] and Selleck, Saunier, and Collins,[62] as well as field studies by Longley,[91,92] all presented evidence that initial mixing has a singular effect on the efficiency of disinfection. These findings were further confirmed by the California State Department of Health mobile pilot plant study reported by Sepp and Bao in 1980.[79] A variety of field investigations between 1970 and 1980 covering some 40 treatment plants has convinced White that initial mixing—its quality and intensity—is an important factor in wastewater disinfection. However, the intensity and quality of initial mixing is difficult to define.

544 HANDBOOK OF CHLORINATION

Calmer et al.[94] reported in 1984 the results of a combined literature search and field study of the initial mixing phenomenon. They concluded that there was little or no benefit to coliform destruction where the mixing intensity was greater than $G = 500$. This indicates a wide disparity in beliefs of what the optimum level of mixing intensity should be. Additional research needs to be performed on operating systems, preferably at full scale, to establish the appropriate level of mixing intensity.

The work by Vrale and Jorden[61] in 1971 is often overlooked. It has an important but almost hidden message. Their work indicated that, while the G factor (velocity gradient) is a useful measure of mixing intensity, it should not be used as the sole criterion, and that the type of mixing may have a profound effect. The implication is that jet mixing is probably superior to other types of mixing for two streams of flowing water.

Velocity Gradient: The Measure of Mixing Intensity:

General Discussion. The fluid mechanics of mixing has been represented by the velocity gradient for many years.[85] It is a dimensionless number expressed as follows:

$$G = \sqrt{\frac{P}{\mu V}}$$

where

G = mean velocity gradient, sec^{-1}
P = power requirement, ft lb/sec
V = mixing chamber volume, ft^3
μ = absolute fluid viscosity, lb-sec/ft^2

When water flows through a pipe or along any solid surface or through another fluid moving at a lesser velocity or in an opposite direction *the motion is resisted by drag.* The nature of this drag is fluid shear, and the cause of this phenomenon is the viscosity of the fluid. The work of shear results in *energy dissipation* as heat and is described as "head loss" or friction loss. The velocity gradient is related directly to the total shear per unit volume per unit of time. The G number provides a somewhat better understanding of turbulence and problems involving internal energy dissipation *as it relates to head loss which in turn relates to mixing.*

In chlorination of wastewater, intensity of mixing is a key factor in disinfection efficiency. Defining the role of the G number as a parameter for disinfection efficiency as it is related to the intensity of mixing is important, because experience has demonstrated that it is not the sole factor. The type of mixing is also important (e.g., a jet system or a turbine mixer, etc.).

Current Practices. Various consulting engineering firms in California use G numbers varying from 500 to 1000. The Sanitary Districts of Los Angeles County use a different approach. They rely on turbine mixers rated at 1 hp/1000 gallons contained in the mixing compartment. White used to believe that the G number should be approximately 1000 for superior mixing, and that the intensity of mixing required was partly dependent upon the magnitude of coliform destruction required.

Recent operating experience (1978–1982) indicates acceptable values of G can be significantly less than 1000 regardless of disinfection requirements. At the San Jose/Santa Clara WWTP in San Jose California turbine mixers provide a G number of 475 for a consistent total coliform concentration in the filtered effluent of 2.2/100 ml at 60 min t_i contact time and 12 mg/l chlorine dose (2 mg/l NH_3-N is added before chlorination to the filtered-nitrified effluent). Another treatment plant discovered there was sufficient mixing at the point of application due to hydraulic turbulence and jet action of the diffuser that the action of the mechanical mixers was unnecessary, so they were turned off.

Recent Developments. The most compelling example was the performance of a jet-type chlorine diffuser devised by Egan[95,96] in 1978 at the Los Angeles–Glendale water reclamation plant (filtered secondary effluent), Los Angeles, California. Egan retrofitted a perforated horizontal diffuser shown in Fig. 8-16 to a jet diffuser shown in Fig. 8-17. The jet system alone contributes a G number of 189. Between

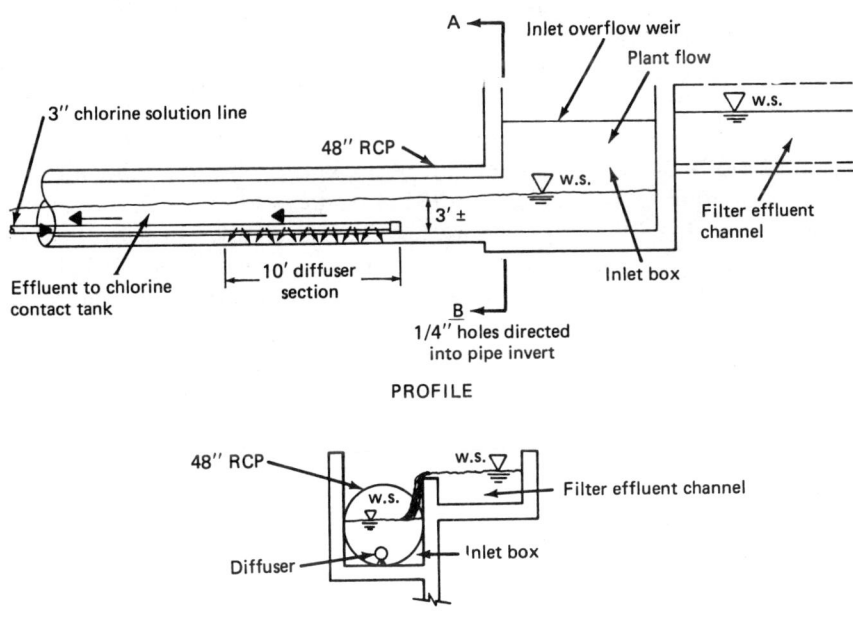

Fig. 8-16 Original Los Angeles-Glendale diffuser.

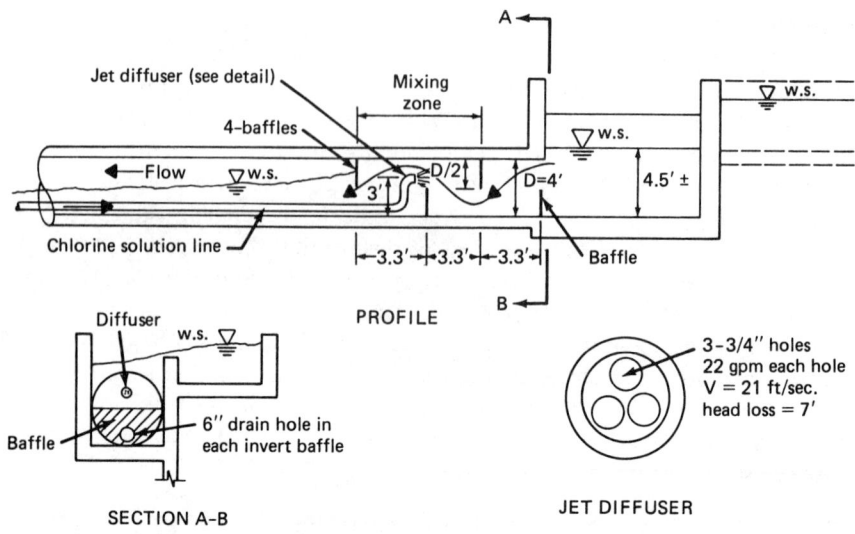

Fig. 8-17 Egan retrofit of Los Angeles-Glendale diffuser.[95,96]

baffles B and C, $G = 205$, and the countercurrent effect produces $G = 232$. For a total $G = 626$.* This is known as the Egan Effect. The calculations are based upon reasonable hydraulic assumptions. The countercurrent value of G is based upon the energy dissipation caused by the opposing flows of chlorine solution and wastewater. The total G value is not surprising because the efficiency of the Egan mixing scheme resulted in a consistent compliance with the NPDES requirement of 2.2 MPN/100 ml total coliforms at a chlorine dosage of only 4.8 mg/l (reduced from 10–12 mg/l), which resulted in contact chamber effluent residuals of 0.4–0.8 mg/l after 3 hours detention. The original diffuser (Fig. 8-16) did not produce consistent compliance with the coliform requirement at elevated chlorine dosages of 10–12 mg/l. The disinfection efficiency of the jet diffuser and baffles corresponds well to the Collins model.

A dye study was performed which demonstrated the ability of the mixing system to produce a homogenous mixture of chlorine solution and wastewater. Therefore it follows that intensity of initial mixing may not need to be as high as $G = 1000$ regardless of disinfection requirements.

The subject of mixing intensity as it relates to disinfection efficiency needs more research and laboratory study of the fluid mechanics of mixing chlorine with wastewater. The objective would be to quantify intensity versus homogeneity of the mixture in a given time frame.

---

* See Appendix for calculations.

Diffusers as Mixers:

General Discussion. The notion that a chlorine solution diffuser could be designed to provide adequate initial mixing using the velocity gradient as the principal parameter was originated by Egan at the Los Angeles–Glendale Reclamation Plant circa 1976.[96] The validity of such a notion is supported by the results outlined above.

The jet diffuser shown in Fig. 8-17 has a 3-port nozzle through which 67 gpm chlorine solution flows. At 22 gpm per ¾ inch port the calculated head loss is 4.03 ft. This calculates to $G = 189$, which in itself is not impressive.

The most prevalent situation for diffuser location is in an open-channel-type structure. There are some situations where the diffuser can be located in a surcharged conduit. However, designing a diffuser for both of these situations requires the use of the same principles of fluid mechanics plus the phenomenon of the Egan Effect.

Grid-Type Open-Channel Diffuser. First, choose a structure location upstream but adjacent to the contact chamber where the average daily flow is on the order of 1.0–1.25 ft/sec.

Second, select a target G number, such as 500.

Third, assume a head loss due to the jet action of the diffuser is from 7–10 ft.

Fourth assume injector is adjacent to mixing chamber.

A typical grid diffuser is shown in Fig. 8-18. The following are the calculations for this hypothetical diffuser.

EXAMPLE. Assume wastewater flow 10 mgd, and flow in channel to contact chamber = 0.75 ft/sec. Also assume mixing occurs in 3 sec. At 0.75 ft/sec this horizontal area of mixing is $L_m = 2.25$ ft.

From the velocity gradient formula:

$$G = \sqrt{\frac{P}{2.5 \times 10^{-5} \times V}} \quad (8\text{-}11)$$

$$P = \frac{Qh}{3960 \times \text{eff.}}$$

where

Q = total flow through diffuser, gpm
h = loss in diffuser orifice, ft
eff. = 1.0

Assume plant flow is 10 mgd and chlorine dosage is 15 mg/l. This requires about 1250 lb/day chlorine; therefore a 2-inch adjustable injector could be used.

548 HANDBOOK OF CHLORINATION

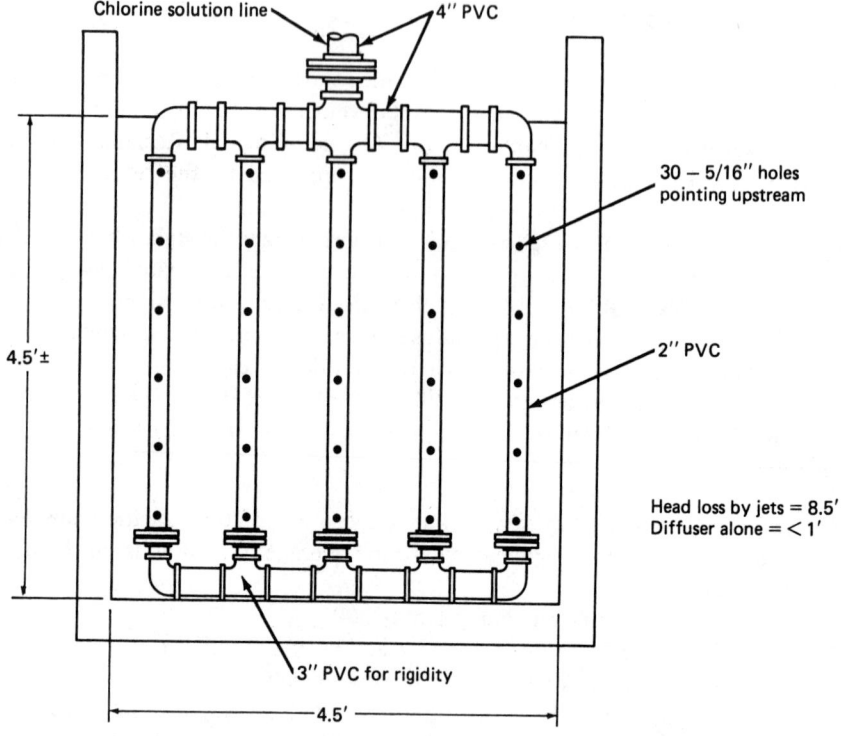

Fig. 8-18 Open-channel grid diffuser based upon the Egan concept.

However a mixer-diffuser needs to use as much water as is practical, so a 3-inch injector is preferable. Use a $\frac{5}{16}$-inch-diam. hole and 4 gpm per hole. This requires an injector operating water supply of 120 gpm. This size hole produces a head loss $h = 8.5$ ft. Therefore the mixing power generated by the diffuser in Fig. 8-18 is

$$P = \frac{120 \times 8.5}{2960 \times 1.0} = 0.26$$

$$V = 3 \times 3 \times 2.25 = 20.25$$

$$G = \sqrt{\frac{550 \times 0.26}{2.5 \times 10^{-5} \times 20.25}} = 548.18$$

Therefore a chlorine solution diffuser that exerts no more than 10 ft of friction loss in the chlorine solution line is well within the limits of acceptability for injector operation. The major difference between the previous concept of diffusers and the one shown in Fig. 8-18 is double the amount of injector water flow.

DISINFECTION OF WASTEWATER 549

Restricting the flow in the channel by the diffuser plus the countercurrent effect (perforations directed upstream) will add about 200 to the G number. Therefore it appears that achievement of $G = 500+$ is practical and much less energy intensive than the use of mechanical mixers.

Mechanical Mixers. Mechanical mixers have been widely used in potable water treatment and more recently for mixing chlorine solution in wastewater. Ideally the mixing device should be able to homogenize the chlorine solution and the wastewater in a fraction of a second. Mixing Equipment Company of Rochester, New York claims that it is practical to provide complete mixing within 1–3 sec. (regardless of flow) depending upon how much power input is allowable.[98] Such a mixer is illustrated in Fig. 8-19.

This arbitrary mixing time is for either closed conduit or open channel flow using propeller mixers. Based upon a comparison of existing installations with various types of mixing, a complete mix in 3 sec would be rated as adequate. Fig. 8-20 illustrates the use of a mechanical mixer in open channel flow. Sizing procedures for these mechanical mixers are proprietary secrets but for general guidelines Mixing Equipment Co. advises that they size mixers for dispersing chlorine solution in a flowing stream on the basis of 0.3–0.6 hp/mgd flow. The mixing basins are sized for theoretical retention times of 5–15 sec. This information is helpful for the design of an optimum velocity gradient.

Fig. 8-19 Propeller mixer (courtesy Link Belt Co.).

550 HANDBOOK OF CHLORINATION

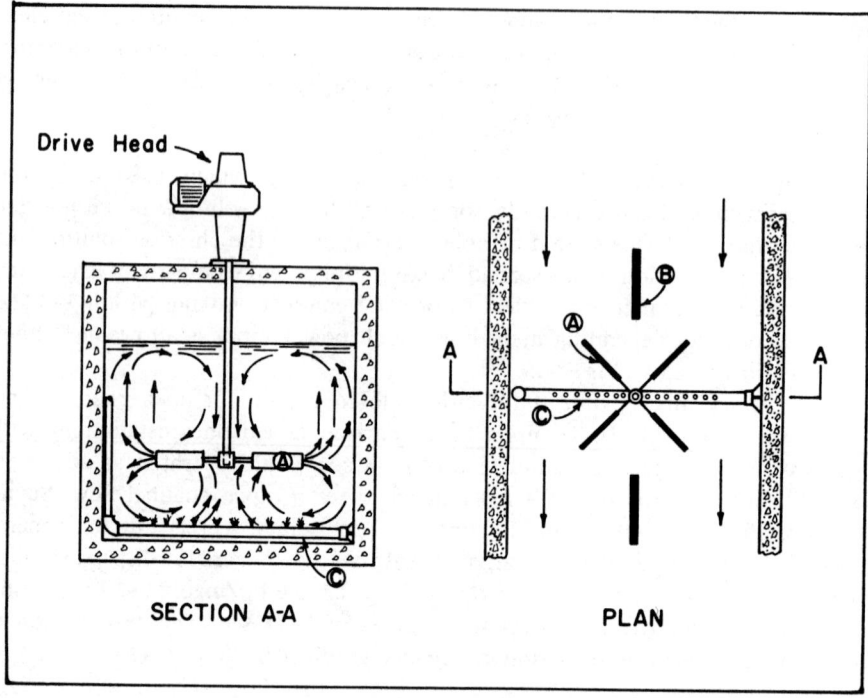

Fig. 8-20 Propeller mixer in open channel flow showing location of baffles. A = radial flow impeller, B = baffles, and C = chlorine diffuser (courtesy Mixing Equipment Co. Rochester, N.Y.).

Let us take a condition of 5 mgd flow with 10 sec theoretical detention time in the mixing chamber, assuming a 3 hp mixer as the design parameter, which is the upper limit of design quoted by the manufacturers. So G, the mean velocity gradient, is

$$G = \sqrt{\frac{3 \times 550}{2.735 \times 10^{-5} \times 77.37}} = 948$$

Hydraulic Mixers:

Closed Conduits. Mechanical mixers for closed conduits running full are not necessary, provided certain design criteria are followed. White found by hydraulic laboratory dye studies and a plant-scale field demonstration that if the chlorine were injected at the center of the flow, the chlorine solution would be completely mixed in ten diameters of pipe length, provided the Reynolds number was 2000 or greater.[98] This rule of thumb is valid for pipe sizes up to 30 inches. The dosing and sampling arrangement is illustrated in Fig. 8-21. If the pipe is larger than

DISINFECTION OF WASTEWATER 551

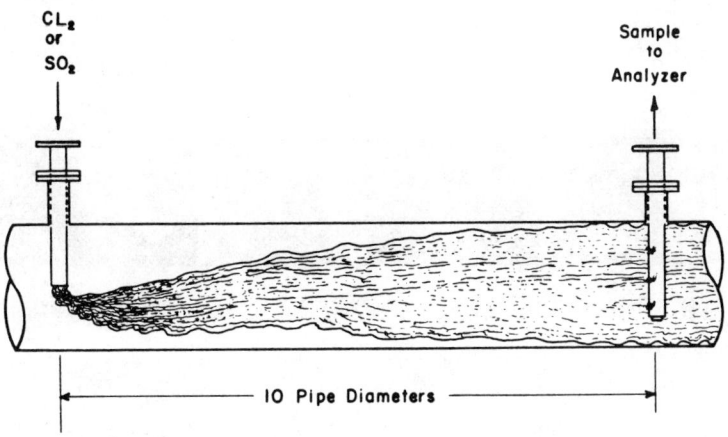

Fig. 8-21 Pipeline as a mixing device.

30 in. in diameter, use the across-the-pipe diffuser illustrated in Chapter 9. Perforate the diffuser across the middle one-third of the diameter. Make a similar device for the analyzer sampling probe. Whenever possible use the chlorinator injector pump to provide the analyzer sample flow. This serves to shorten appreciably the loop time for residual control systems. This arrangement can be used when it is possible to utilize a short run of surcharged pipe immediately upstream from the contact chamber. The designers should always be alert and innovative to seize the opportunity to capitalize on hydraulic turbulence for mixing chlorine and wastewater.

Static Mixers. These can be used to advantage where 10 diameters of pipe length are not available. Fig. 8-22 illustrates the concept of a static mixer. The mixer elements are made to fit inside pipe in lengths equal to $1.5D$. The manufacturer's rule of thumb for the number of elements required is based upon local conditions of velocity. For flows 2 ft/sec or greater two elements are recommended. Flows that go as low as 0.5 ft/sec, four elements are required to produce proper mixing. The G factor can be calculated for each situation but it is not the principal criterion.

Hydraulic Jump. This device was first used as a chemical mixing device in a potable water treatment plant in 1927.[99] Fig. 8-23 illustrates a typical hydraulic jump. The diffuser should be located as shown. This is the preferred location even though it may cause some chlorine odors in the turbulent zone. Placing the diffuser in the turbulent zone does not provide an equivalent mix. Moreover, since the turbulent zone moves laterally with flow changes it is almost impossible to choose an optimum location. The hydraulic jump should be designed to provide at least 9 in. of cover over the diffuser. The diffuser perforations should be pointing

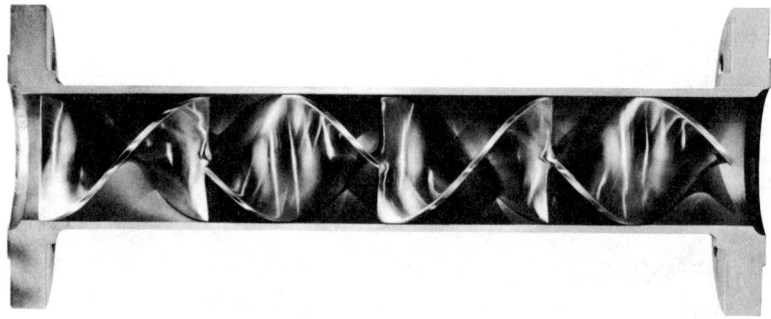

Fig. 8-22 Static mixer (courtesy Kenics Corp., Danvers, MA).

upstream and the velocity through these holes should be approximately 22 ft/sec. If properly designed, a hydraulic jump can be an excellent mixing device, but it requires about a 2 ft fall as shown to provide the necessary turbulence.

The Parshall flume or the Palmer–Bowlus flume should not be considered as mixing devices. They do not generate sufficient turbulence.

A partially closed sluice gate or other type of gate valve placed immediately downstream from the chlorine diffuser can develop a turbulent regime sufficient to achieve adequate mixing.

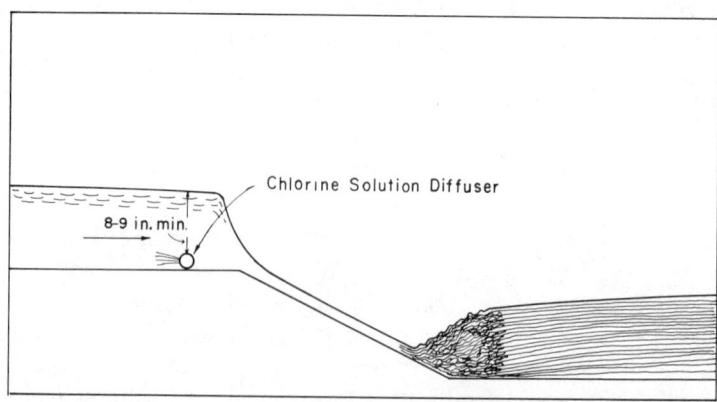

Fig. 8-23 The hydraulic jump as a mixing device.

This type of mixing device has performed as well as or better than mechanical mixers in documented cases.[94] The chlorine diffuser should be placed immediately upstream from the sluice gate. If a 36 inch sluice gate is throttled to provide a 6 inch head loss, this will produce a G value of 614 assuming a 10 mgd flow and a 3 second mixing time. If the diffuser is designed for G = 200 the total G will be more than adequate for good mixing.

The Pentech System. This system was made available for wastewater chlorination by the Pentech Div. of Houdaille Ind. Inc., Cedar Falls, Iowa in the mid-1970s. The system is illustrated in Fig. 8-24. It utilizes either the direct injection of chlorine gas under a vacuum or chlorine solution from the discharge of a conventional chlorinator. When chlorine gas is injected directly, the pump shown becomes the "injector pump" and the jet nozzle assembly takes the place of the injector.

The entire effluent stream is forced through the reactor tube. This provides plug flow in a highly turbulent regime. This system provides almost instantaneous dispersion of chlorine throughout the mass of effluent. Pentech claims this mixing system will generate a G number as high as 10000 and that total mixing occurs in about 0.2 sec with the ability to cope with variable flow rates and to provide a system hydraulic gradient which does not introduce a significant head loss.[100]

Pentech recommends a minimum energy dissipation rate to be maintained so that the turbulent mixing zone will have a mixing rate t^{-1} (sec^{-1}) of somewhere between 10 sec^{-1} and 5 sec^{-1} depending upon the size of the system and with corresponding mixing residence times of 1.0 sec or less. These limits require mixing in times far less than those described above for other systems. In accordance

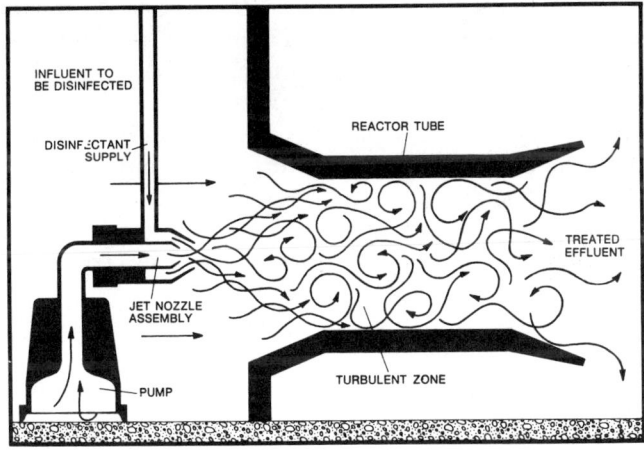

Fig. 8-24 Pentech injector mixer (courtesy Pentech-Hondaille)

with fluid dynamic principles, the mixing rate t^{-1} is directly related to the specific turbulent energy dissipation rate in the turbulent mixing zone and is inversely related to the square of the scalar macroscale, L_s, of the turbulence structure of the mixing zone as follows:

$$t^{-1} = K(e/L_s^2)^{1/3} \qquad (8\text{-}12)$$

where

e = specific turbulent energy dissipation rate
L_s = scalar macroscale
K = constant, which is 0.489 for the cgs system of measurement

The specific turbulent energy dissipation rate e is further defined as:

$$e = \frac{P}{pV} \qquad (8\text{-}13)$$

where

P = net power lost to fluid
p = fluid density
V = fluid volume

The scalar macroscale L_s, for a system such as illustrated in Fig. 8-24 may be approximated as about $0.131D$ for purposes of calculation for equipment design, where D is the mixing parallel diameter (cm).

It is also useful to define a mixing number θt^{-1}, which is the product of the mixing residence time and the mixing rate. This characterizes the product stream inhomogeneity. Mixing efficiency numbers from about 1.5 to 15 or greater should be applied to achieve superior disinfection results.

Moreover, for a flow-through system with continuous mixing, the specific energy requirement (the energy dissipated per unit throughput of product stream or the work done in mixing the product stream) should be at least about 0.2 hp/mgd of treated effluent. For a given level of mixing, the energy requirement will increase with increasing values of L_s, but will generally be in the range of from about 0.2 hp/mgd to 3 hp/mgd of effluent to be treated.

The average residence time θ for the mixer illustrated in Fig. 8-24 may be readily determined from the volume of the turbulent mixing cone. The effluent and disinfectant may generally be assumed to be mixed to within acceptable disinfectant concentration gradient limits upon reaching a point adjacent to the base of the mixing cone at its intersection with the mixing parallel. The volume V of the mixing cone thus defined may be calculated as follows:

$$V = \frac{D^3}{24 \tan(\alpha/2)} \qquad (8\text{-}14)$$

where

D = diameter (cm) of the intercepting conduit at the point of intersection with the mixing cone
α = included angle of the mixing cone

By way of example, a 5 mgd flow is to be treated utilizing 5 hp input and the jet mixer shown in Fig. 8-24. The mixer parallel diameter is 20 in. × 2.54 = 50.8 cm and $\alpha = 28°$. Therefore the mixing rate t^{-1}, the residence time θ, and the energy requirement are calculated as follows:

$$V = \frac{D^3}{24 \tan(\alpha/2)} = \frac{(50.8)^3}{24 \tan 14°} = 68{,}827 \text{ cm}^3 = 2.43 \text{ ft}^3$$

$$\theta = \frac{2.43 \text{ ft}^3}{5 \text{ mgd} \times 1.55 \text{ cfs/mgd}} = 0.31 \text{ sec}$$

$$e = \frac{P}{\rho V} = \frac{(5)(550)(30.48)(454.5)(980)}{(1)(68{,}827)} = 542{,}436 \text{ cm}^2/\text{sec}^3$$

$$L_s = (0.131)(20)(2.54) = 6.65 \text{ cm}$$

$$t^1 = 0.489 \left[\frac{542{,}436 \text{ cm}^2}{44.22 \text{ cm}^2 \text{ sec}^3} \right]^{1/3} = 11.28 \text{ sec}^{-1}$$

$$e = \frac{5 \text{ hp}}{5 \text{ mgd}} = 1 \text{ hp/mgd}$$

Unfortunately this system has not been rigorously evaluated so no conclusions can be drawn about its practicality or cost-effectiveness.

Fischer and Porter System.[101] This is what might be termed a modified Pentech system that requires a specially designed mixing chamber. The complete system is illustrated in Fig. 8-25. In operation a small portion of the wastewater to be chlorinated is pumped to the ejector-diffuser assembly by a submersible pump. This flow creates a vacuum which pulls chlorine gas into the diffuser portion of the assembly where it is discharged under pressure into the turbulent mixing zone in the fiberglas reactor tube. This tube is mounted in a baffle wall in the mixing chamber which seals the channel and directs all the remaining wastewater flow through the tube where it is mixed with the chlorine. Rapid kills are claimed

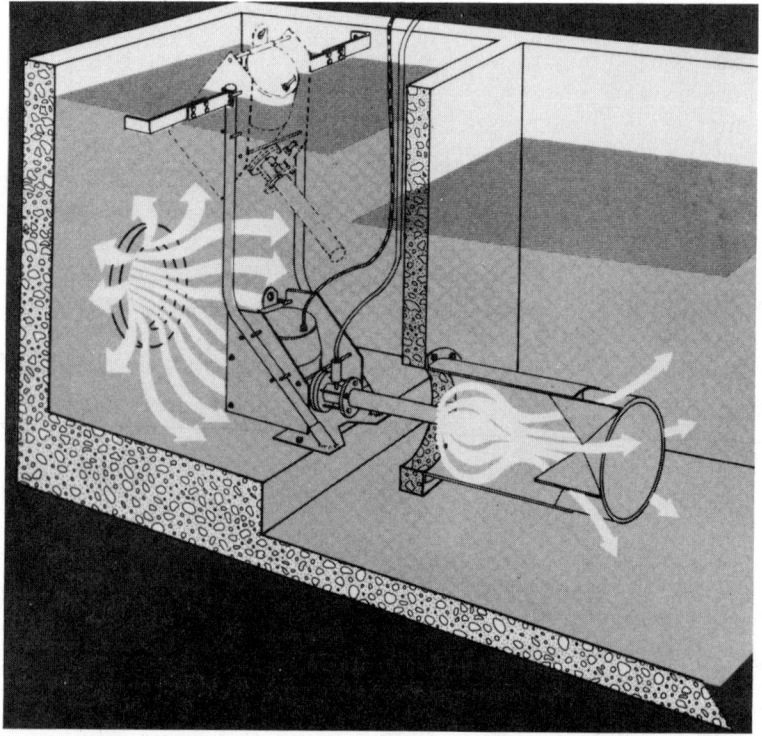

Fig. 8-25 Molecular chlorine-wastewater mixing device (courtesy Fischer and Porter Co.).

because the chlorine is in its molecular form. There is no current documentation to substantiate a claim for better kills by this method than for the chlorine solution method.

Although a constant-volume pump is used to provide the power to maintain the chlorine gas flow, the wastewater flow through the reactor tube can vary within the usual diurnal ranges encountered in most wastewater plants.

The chlorination equipment can be controlled by the usual conventional methods of flow pacing and residual control.

These systems are designed to produce a turbulence of $G = $ approx. 1650. They are most popular for treatment plants in the flow range of 0.5–3 mgd, but can be designed for much larger treatment plants.

CONTACT CHAMBERS

Function. The chlorine contact chamber must be designed to provide the optimum distribution of residence time for contact between the disinfectant and the microorganisms to be destroyed. There are many considerations required to develop this efficiency. Therefore, the chlorine contact facility should be considered as an integral part of the overall wastewater-treatment process which means it becomes a unit process such as sedimentation, aeration, sludge digestion, etc.

A chlorine contact chamber is not a mixing chamber, and should not be used for that purpose. The applied chlorine should be thoroughly mixed with the wastewater prior to entry into the contact chamber.

Distribution of Residence Time:

General Theory. The distribution of residence time may differ appreciably in chambers of different geometrical configuration, although chamber volumes and flow rates are identical. The ideal reactor is one which can provide equal residence time for all the molecules in the chamber. This is described as 100 percent plug flow.

Figure 8-26 illustrates the dye tracer response in an ideal plug-flow chamber. This ideal situation compares to three other rectangular chambers less than ideal.

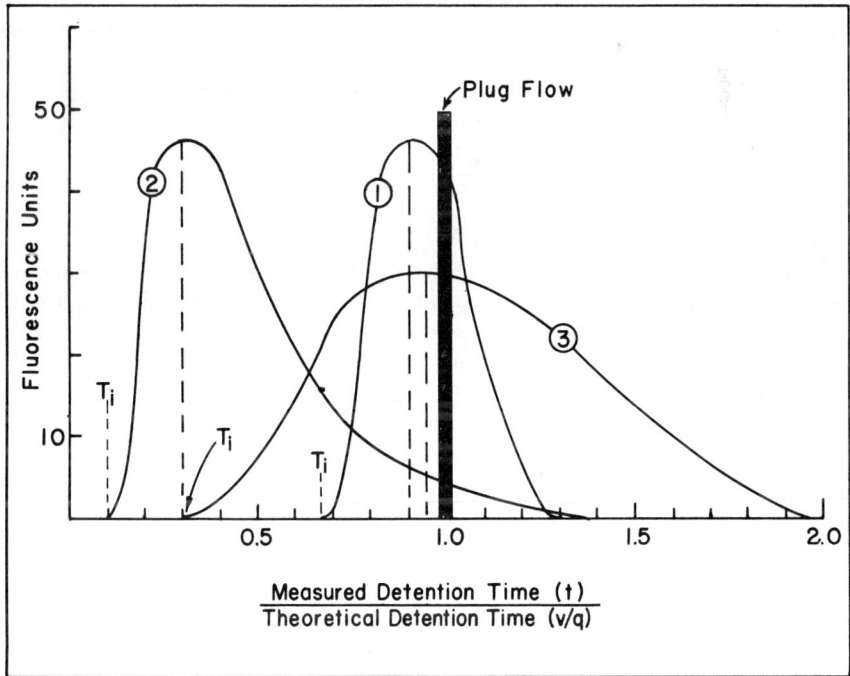

Fig. 8-26 Dye tracer studies of distribution residence times for various chlorine contact chambers configurations.

Curve No. 1 illustrates the results of proper longitudinal baffling, with a modal time of 0.9 and approximately 95 percent plug-flow characteristics. Curve No. 2 is typical for what may be expected of circular tanks and rectangular tanks with poor baffling arrangements. Curve No. 3 is an actual dye response curve representing a seemingly well-baffled tank which exhibits excessive short-circuiting (back-mixing). This tank is considered properly baffled since the peak dye concentration occurs at the theoretical detention time (V/Q). However, the mere achievement of unity modal time does not satisfy all the requirements of the flow pattern in a contact chamber. In this case the tank was cross-baffled, which is to be avoided if possible, as this configuration promotes short circuiting, which in turn significantly diminishes the plug-flow characteristics.

Effects of Short-circuiting. Ignoring initial mixing effects, the batch-type reactor (beaker) produces results equivalent to any reactor that obtains 100 percent plug-flow characteristics. The effects of the residence time distribution on disinfection efficiency are illustrated in Fig. 8-27, which compares the destruction of coliform bacteria in a batch reactor with that of a continuous flow system afflicted with short-circuiting characteristics. The chlorine residuals are approximately equal in the two reactors. Fig. 8-27 clearly demonstrates the gross effect of residence time distribution upon disinfection efficiency between the ideal plug-flow chamber (batch reactor) and the chamber with gross short-circuiting characteristics. It also demonstrates that the design criterion for residence time based upon chamber volume divided by flow rate (V/Q) is meaningless. For example, there exists a difference of four orders of magnitude in the coliform kill in the two reactors for a contact time t of 37 min.

Analysis of Residence Time. The conventional method of analyzing residence time is by the dye tracer dispersion technique. Since chlorine residuals normally carried in wastewater contact chambers do not affect the fluorescent quality of the dye it is permissible to conduct a dye study during chlorination. The procedure consists of injecting a dye like Rhodamine WT at the entrance to the contact chamber and measuring the dye concentration with a fluorometer at the exit of the chamber. The best place to introduce the dye is into the chlorine solution discharge line or the injector water supply. However this requires shutting off the flow of chlorine as the low pH of the chlorine solution plus the high concentration of the chlorine will significantly affect the fluorescent quality of the dye. This point of application is desirable because it provides a reliable indicator of diffuser efficiency as a mixer.

The shape of the dye concentration–time curve provides the necessary data to analyze the hydraulic flow pattern of any contact chamber. The conventional parameters used to describe the performance of a contact basin are presented in Table 8-3.

Each parameter shown in Table 8-3 can be used to predict contact basin perfor-

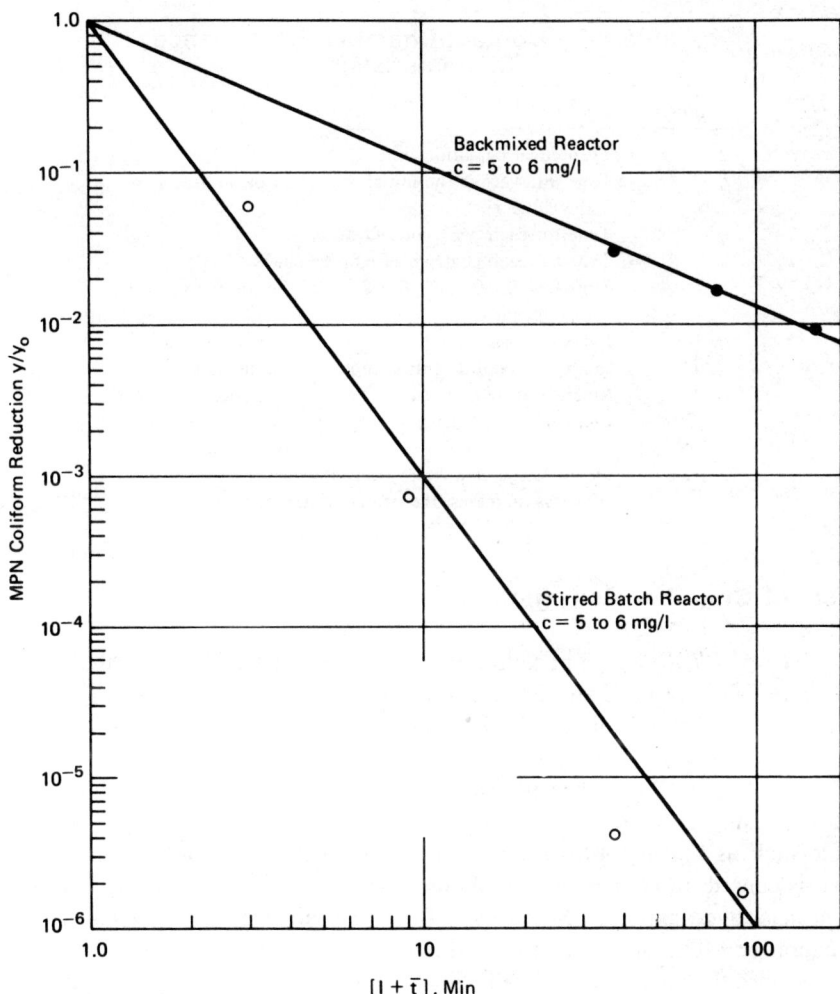

Fig. 8-27 Comparison of reactor (contact chamber) performance.

t = contact time
c = amperometric chlorine residual (total) at the end of the contact time t, mg/liter
Y = final coliform bacteria MPN/100 ml
Y_o = initial coliform bacteria MPN/100 ml.

mance. Fig. 8-28 illustrates a typical dye dispersion curve. This curve represents a long, narrow, and shallow gunite-lined channel with an L to W ratio of 21:1, an unbaffled pipe entrance, and a sharp-crested weir at the outlet.[103] The importance of the shape of the dye dispersion curve is discussed below (see "Contact Chamber Evaluation").

Table 8-3 Contact Chamber Performance Characteristics

Parameter	Definition	
T	Theoretical detention time	
t_i	Time interval for the initial indication of the tracer in the effluent	
t_p	Time to reach peak concentration	
t_g	Time to reach centroid of effluent curve	
t_{10}, t_{50}, t_{90}	Time for 10, 50, and 90 percent of the tracer to pass at the effluent end	
t_i/T	Index of short-circuiting	Ideally, all these parameters will approach 1.0 under perfect plug-flow conditions
t_p/T	Index of modal detention time	
t_g/T	Index of average detention time	
t_{50}/T	Index of mean detention time	
T_{90}/t_{10}	Morril Dispersion Index; indicates degree of mixing; as t_{90}/t_{10} increases, the degree of mixing increases	

Effect of Chamber Configuration:

Length to Width Ratio. The field study by Sepp and Bao[61,79] indicated that the L/W ratio is not in itself an accurate descriptor of plug-flow characteristics. Their pilot tank had a much higher L/W ratio than the plant-scale tanks but did not have a correspondingly smaller dispersion number. Studies by Marske and Boyle, and Trussell and Chao, have indicated the dispersion number usually decreases with increasing L/W ratio, but the correlation is poor. Other factors must play a role such as the depth-to-width ratio (H/W), extent of dead space, and eddy currents caused by poor plug-flow characteristics at the baffle turning points.

The field investigation by Marske and Boyle[103] evaluated seven different chamber configurations. The ones with longitudinal baffles proved to be the most efficient. The one with the most practical value is shown in Fig. 8-29. This is a longitudinally baffled chamber with a flow length to width ratio of 72:1. This chamber provides 95 percent plug-flow conditions and exhibits a modal time of 0.7. While it would be difficult to improve upon the percentage of plug-flow conditions, the modal time of this tank could be increased to 0.9, as has been demonstrated by others, with some slight changes such as elimination of the square corners in the tank. This work by Marske and Boyle substantiates the claims that long, narrow channels and/or conduits make the best chlorine contact chambers.

Depth to Width Ratio (H/W). The contact chamber analysis by Trussell and Chao[104] shows that depth can have an effect on the dispersion index but not nearly to the same extent as the length to width ratio. Based-upon the results of the plant-scale versus the mobile pilot plant study by Sepp and Bao,[79] the data indicate

DISINFECTION OF WASTEWATER 561

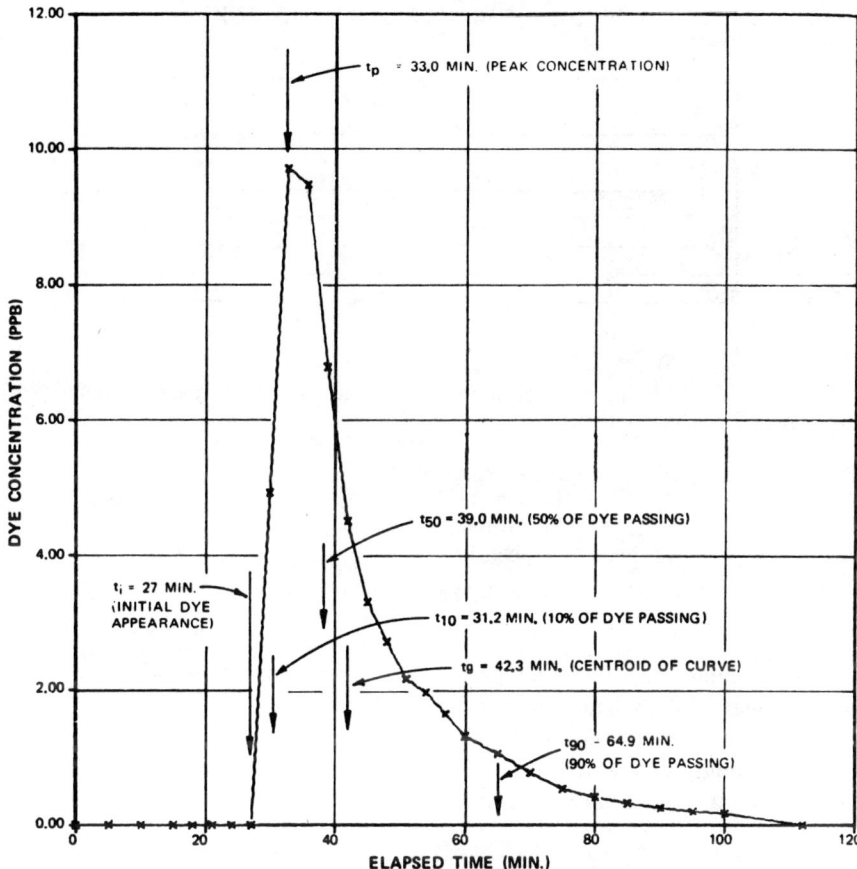

Fig. 8-28 Typical dye dispersion curve.

Theoretical detention time $(V/Q) = 50$ min.
Modal time $= 0.7$–0.8
$t_i/T = 0.36$–0.54 (depending on wind direction)
Morrill index $= 2.1$
Chemical engineering dispersion index $= 0.08$
Percent plug flow $= 60$–70 percent (depending
                       upon wind direction).

(From Marske and Boyle.[18])

the H/W ratio should be 1.0 or less. Therefore a compromise is a square cross-section at peak flow (maximum water surface) and a slightly rectangular section at lower flows.[105]

Circular Chambers. Circular chambers, least of all circular clarifiers, are not acceptable as chlorine contact chambers unless they are specially designed with

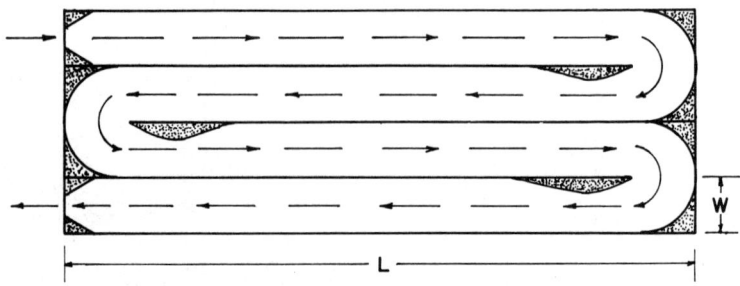

Fig. 8-29 Chlorine contact chamber with longitudinal baffling and optimum plug flow characteristics.

Flow length: Width = 72:1
L:W = 18:1
Modal time = 0.70
% plug flow = 95

(From Marske & Boyle.)

an outer annular ring. The poor performance of a conventional circular clarifier versus a longitudinally baffled serpentine flow and the special annular ring clarifier is clearly shown in the Marske and Boyle investigation.[103] Any other circular chambers would be expected to perform just as poorly as a conventional circular clarifier. (See also "Modified Chambers" below).

Outfalls. When available these structures provide almost perfect plug-flow conditions when flowing full or partly full. Fig. 8-30 illustrates the plug flow characteristics of a 4000 ft outfall line discharging into San Francisco Bay. This represents a dye-tracer study by Kennedy Engineers for the City of Richmond, California. In order for an outfall to qualify as a contact chamber, an effluent sampling site must be available for chlorine residual monitoring and dechlorination control.

Other Physical Characteristics Affecting Residence Time. Marske and Boyle[103] also studied the following physical characteristics used in the design of a contact basin: (1) the depth of the basin as it relates to the effect of surface wind; (2) the effect of outlet weir configuration; and (3) the effect of turbulence. The effects of these physical characteristics are summarized below.

Wind. Two tracer tests conducted on a very long, narrow contact channel approximately 3 ft deep, indicated that wind may cause surface currents resulting in

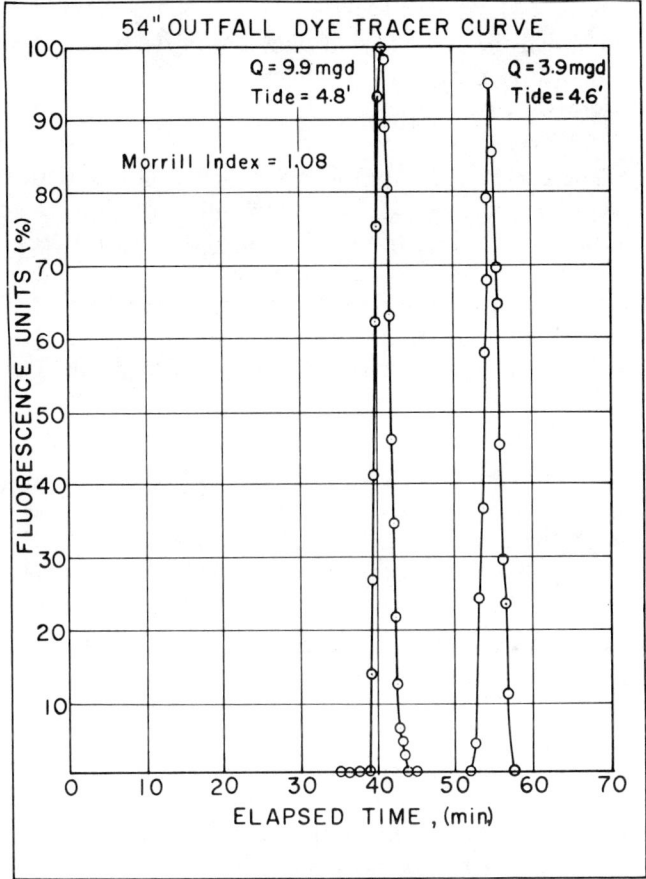

Fig. 8-30 A free-flowing outfall dye dispersion curve.

some short-circuiting. In the tests, a down stream wind resulted in a relative time ratio, t_i/T, of 0.36 whereas an upstream wind resulted in a t_i/T of 0.54.*

Consequently, due to wind effects, a unidirectional shallow basin or trench may not provide the plug-flow distribution of residence times that would be expected from the geometrical configuration. In this case, the designer would be well advised to replace this type of chamber with a closed conduit.

Weirs. Tracer tests, performed on a 3.5-ft-wide Cipolleti weir and on an 18-ft-wide sharp-crested weir, indicated that all hydraulic performance parameters were

---

* The term t_i is the time interval between the injection of the tracer and its first appearance in the effluent, and T is the contact time, calculated from V/Q, where V is the basin volume and Q is the flow rate.

improved when the sharp-crested weir was employed. The value of t_i/T increased from 0.19 to 0.27 and the percentage of plug flow in the basin increased from 38 to 58 percent. Consequently, it is recommended that contact chamber overflow weirs extend across the entire width of the final channel of the contact chamber.

Turbulence. There is no necessity to provide turbulence such as is required in the mixing phase. Turbulence in a contact chamber will not improve coliform destruction; it may in fact be detrimental to the disinfection process. Turbulence during the contact period may reduce the monochloramine residual by aeration, and turbulence may cause back-mixing. The monochloramine residual is the most potent disinfecting compound of any combined residual occurring in wastewater discharges, so it should not be subjected to any loss by aeration in the contact chamber.

Effects of Baffles:

General. Longitudinal baffles are superior to horizontal baffles. The latter cause many times more back-mixing in the chamber than do longitudinal baffles.

Special Structural Design Situations. Sometimes it is impossible to design a contact chamber with longitudinal baffling because of seismic design considerations. In these cases the horizontal baffling is usually integrated in the design as structural members. In these instances all of the corners should be eliminated as shown in Fig. 8-31 to minimize dead spaces and short-circuiting. Fig 8-32 is a photograph of Fig. 8-31 after construction. This illustrates the special smooth concrete surface which is more amenable to cleanliness of this chamber.

Whenever the effluent MPN coliform requirement is 2.2/100 ml the specifications for the concrete in the contact basins should call for a porcelainlike smooth finish which is free from any kind of pits or recesses, because these are breeding areas for the bacteria. Moreover, such situations call for special attention to frequent cleaning. Actually contact chambers designed with either longitudinal or horizontal baffling for effluents which are to achieve a coliform MPN of 2.2/100 ml must be kept as free from slime and algae deposits as a well-kept swimming pool.

Modification Baffles. Hart[106] and Hart and Vogiatzis[107] have shown that significant improvements can be made by using diffusion baffles. These are illustrated in Fig. 8-33. These modifications have been able to change the t_i/T from 0.08 to 0.50 and the dispersion index from 0.056 to 0.004. In terms of disinfection efficiency, the modified unit showed a 24 percent improvement,[107] while chlorine consumption was reduced by 10 percent for equivalent performance.

One type of modification proved to be effective is a V-notch launder type weir at the chamber exit.

DISINFECTION OF WASTEWATER 565

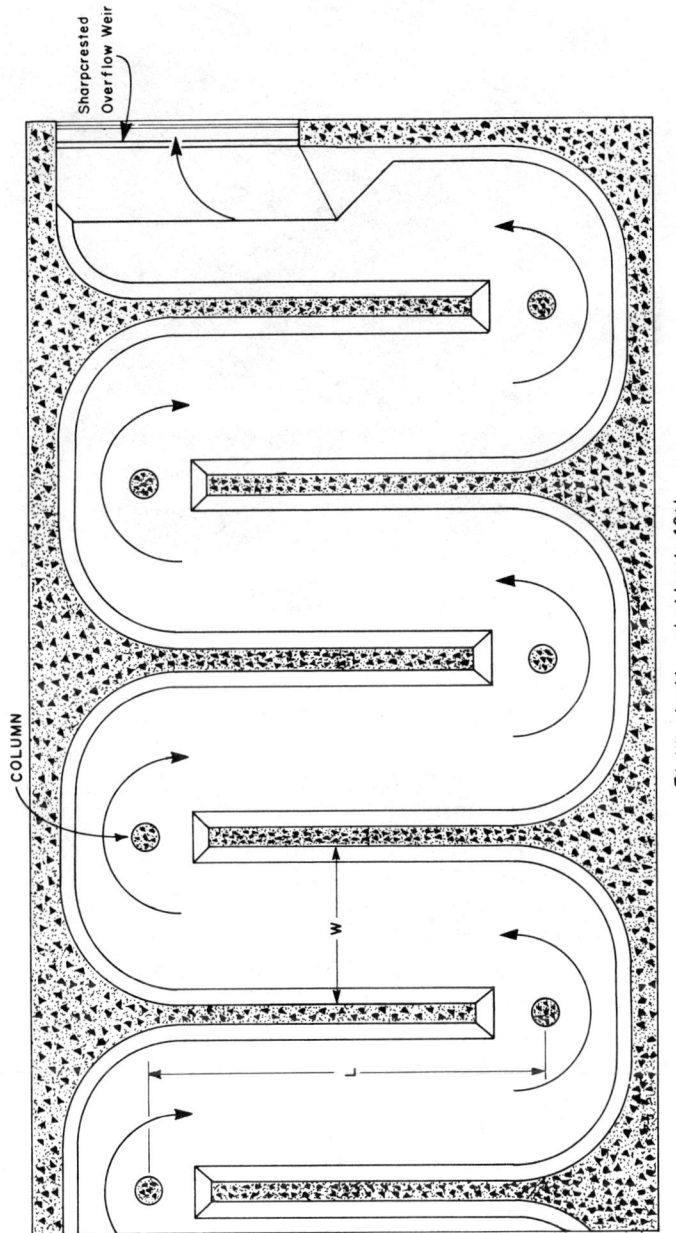

Fig. 8-31 A well baffled contact chamber with horizontal baffles from contract drawings.

Fig. 8-32 Contact chamber shown in Fig. 8-31 after construction.

Provisions for Cleaning. It is known that 50 percent of the suspended solids remaining in the effluent will precipitate in the contact chamber. This is the result of velocities that range from a minimum of about 0.06 ft/sec to a maximum of about 0.4 ft/sec. Therefore the contact chamber must be provided with a means for easy cleaning on a regular basis.

Contact Chamber Evaluation:

General. All contact chambers should be evaluated in-situ as soon after being put into operation as is possible. The information developed must be based on the statistical analysis of the dye dispersion curve of the basin at various flow rates. This describes the hydraulic performance of the contact chamber as related to disinfection and considers the shape of the entire curve rather than the central tendency values.

The purpose of evaluating a contact chamber is to determine the capability of a given chamber to provide disinfection sufficient to meet the NPDES requirements. The most important factor in evaluating a chamber is the time it takes for the initial appearance of the dye at the chamber exit. This time (t_i) must not be less than 30 min, and should be closer to 60 min. The next most important index is the modal time, represented by t_p in Table 8-3. This figure should be close to 90 percent of the theoretical contact time $T = Q/V$.

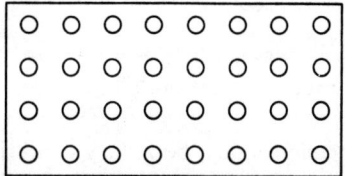

Typical Baffle

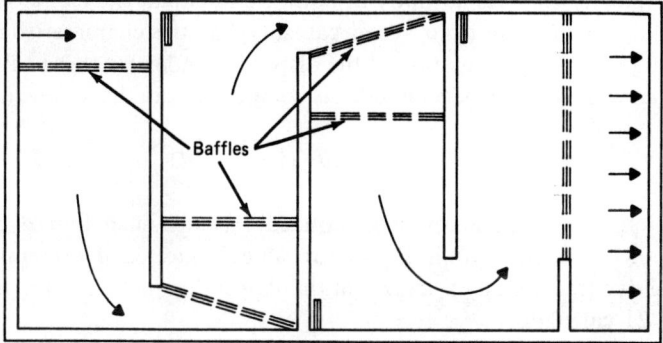

Contact Chamber with Diffusion Baffles

Fig. 8-33 Use of diffusion baffles to improve dispersion index in existing chlorine contact chambers[106] (courtesy *Journal WPCF*).

Finally, the overall performance can be characterized by either the dispersion index d or by the Morrill index. Both of these methods are described below.

Dispersion Index. The chemical engineering dispersion index d was introduced by Thirumurthi.[108] This index is calculated from the variance of the dye dispersion curve, which relates dye concentration with time. As conditions approach ideal plug flow the value of d approaches zero. The algebraic calculations are delineated in Refs. 103 and 109. The dispersion index d calculated by this method includes all points on the dye dispersion curve illustrated in Fig. 8-28. The index described by this mathematical approach demonstrates a strong statistical probability of correctly describing the contact chamber efficiency, which is related to the dispersion index. The dispersion index is

$$d = \frac{D}{uL} \tag{8-15}$$

where

$D =$ the longitudinal dispersion coefficient, ft²/sec
$u = L/T,$ average chamber velocity, ft/sec

L = reactor length, ft
σ^2 = Variance of the tracer curve, which is a measure of the spread of the curve.*
It is the square of the standard deviation.[109]

In computing the dispersion index, boundary conditions must be observed (i.e., whether the chamber is a closed or open vessel). Outfalls would be characterized as open vessels. Most chlorine contact chambers fall into the closed vessel category where the velocity changes abruptly at both the inlet and the exit of the chamber. It is well known that the open vessel category has the characteristics nearest to those of plug flow, so computation of the dispersion index for these will be ignored.

For closed vessels the dispersion index is computed from the following expression:

$$\sigma^2 = 2d - 2d^2 (1 - \epsilon^{-1}/d) \qquad (8\text{-}16)$$

When there is a small amount of reduced variance such that d calculates to less than 0.05 the second term in Eq. (8-16) can be neglected. The complete solution of a typical dye tracer study is shown in the Appendix, including the key sequence for an HP 21 calculator.

Use of the dispersion index has a major advantage over other parameters for contact chamber evaluation. It can be used in mathematical models to predict both hydraulic performance and disinfection efficiency. The dispersion number represents an expression of the entire shape of the curve. This is in contrast to other dispersion parameters which only consider one or two points of the traces.

Morrill Index. Another method of analysis which is considered practically equal in statistical confidence to the dispersion index is the Morrill Index.[110] This is a mathematical representation of dispersion such that ideal or 100 percent plug flow produces a Morrill Index of 1.00. This analysis utilizes only two points in the dye dispersion curve shown in Fig. 8-20. The Morrill Index is characterized by the expression t_{90}/t_{10}. This means that the time in minutes required for the passing of 90 percent of the dye (65 min) divided by the time in minutes required for the passing of 10 percent of the dye (31 min) equals the Morril dispersion index, or $65/31 = 2.10$. This compares with the longitudinally baffled chamber with the ideal length to width ratio of 72 : 1 and a Morrill Index of 1.48.

Therefore, it would be safe to declare that any Morrill Index from minimum to maximum flow rates on the order of 1.5 to 2.5, respectively, could be considered as acceptable conditions for disinfection purposes. A sample calculation of the Morrill Index is shown in the Appendix.

---

* Tracer curves often exhibit a long tail caused by recycling of the tracer from dead spaces and backmixing.

Dispersion in Long Narrow Structures:

Closed Conduits and Round Pipes. Under the heading "Outfalls," above, these structures flowing full or partly full are considered to have nearly ideal plug-flow characteristics. Therefore very little actual data has been documented on this type of contact chamber. It is believed that for a long, straight outfall the dye curve shown in Fig. 8-30 would be representative. Trussell and Chao[104] and Trussell and Pollock[80] have discussed this case and offer the following empirical equation as suitable where turbulent flow intensity is $N_R = 10000$ or greater:*

$$d = 895000 f^{3.6}(D/L)^{0.859} \quad (8\text{-}17)$$

where

D = pipe diameter, ft
L = length of pipe, ft
f = Darcy–Weisbach friction coefficient

Experience has demonstrated that a 30–45 minute residence time at peak dry weather flow at a velocity of 1.5–2.0 ft/sec will produce nearly ideal plug flow.

Sample calculation of dispersion index.

1. Given:

$Q = 5.2$ mgd $= 8.06$ cfs $= 484$ cfm
$L = 0.6$ miles
$t =$ contact time $= 30$ min

2. Calculate diameter for 30 min contact time

$$t = \frac{(\text{area})(\text{length})}{\text{flow rate}} = \frac{\frac{\pi D^2}{4} L}{Q} \quad (8\text{-}17a)$$

$$D = \sqrt{\frac{4tQ}{\pi L}} = \sqrt{\frac{4 \times 30 \times 484}{\pi(3170)}}$$

$D = 2.41$ ft, use standard size: 30 in. $= 2.5$ ft

---

* Empirical equation is from F. Sjenitzer, "How Much Do Products Mix in a Pipeline?" *The Pipeline Engineer,* D-31 (Dec. 1958).

3. Solve for Reynolds Number

$$N_R = VD/v \qquad (8\text{-}17b)$$

where:

V = pipeline velocity, ft/sec
D = pipe diameter, ft
v = kinematic viscosity, ft²/sec at 70°F

$$V = Q/A = \frac{8.06 \text{ cfs}}{\frac{3.14 \times 6.25}{4}} = 1.64 \text{ ft/sec}$$

$$N_R = \frac{1.64 \times 2.5}{1.06 \times 10^{-5}} = 386{,}792$$

4. Select friction factor as follows: From Fig. 11 in the Appendix select e as 0.003 for average concrete pipe, 2.5 ft in diameter. This intersects the es/d ordinate at approx. 0.0015 for $\epsilon = 0.003$. Then from Moody's diagram (Fig. 10, Appendix) select f from ϵ/D ordinate = 0.0015, where it intersects Reynolds number 3.9×10^5. This is $f = 0.0225$

5. Calculate dispersion number from Eq. (8-17):

$$d = 89{,}500 \, (f)^{3.6}(D/L)^{0.859}$$
$$= (89{,}500) \, (0.0225)^{3.6} \times (2/3170)^{0.859}$$
$$= 0.0002$$

6. Conclusion: Perfect plug flow.

Open Channels. Trussell and Chao[104] have shown that the dispersion index can be estimated by the following expression:

$$d = \frac{24.4 \, n \, (R_H)^{5/6}}{L} \qquad (8\text{-}18)$$

where

n = manning's coefficient
R_H = hydraulic radius, ft
L = length of channel, ft

By manipulation such as inserting coefficients of channel geometry and making simplications according to good practice, Trussell has presented the following as an abbreviated form of Eq. (8-18):

DISINFECTION OF WASTEWATER

$$d = 0.14/\beta, \text{ where } \beta = \text{channel } L/W \tag{8-19}$$

This expression should be used with some caution because it is for long, straight channels without structures causing eddy currents or other hydraulic disturbances. Trussell has plotted field data collected from three sources[79,80,104] shown in Fig. 8-34. This suggests the use of a nonideality coefficient as a measure of effectiveness of a design that would approach the predicted performance of a straight channel:

$$k = d\beta/0.14 \tag{8-20}$$

The data in Fig. 8-34 show coefficients of nonideality ranging from 1.33 to 15 and dispersion numbers ranging from 0.0054 to 0.25. Ten of the 22 plants studied by the three sources had a nonideality coefficient of approx. 3-4. This suggests that such performance is a reasonable expectation for a rational design. Seven of the 22 plants had dispersion numbers near 0.01 or less, therefore low dispersion conditions are attainable.

Sample calculation using Eq. (8-18). Channel flow is 20 cfs and velocity is 1.5 ft/sec. Assume Manning's coefficient is $n = 0.015$ and the channel is 3.65 ft wide and water depth is also 3.65 ft; then $R_H = 10.95$ ft; $L = 800$ ft,

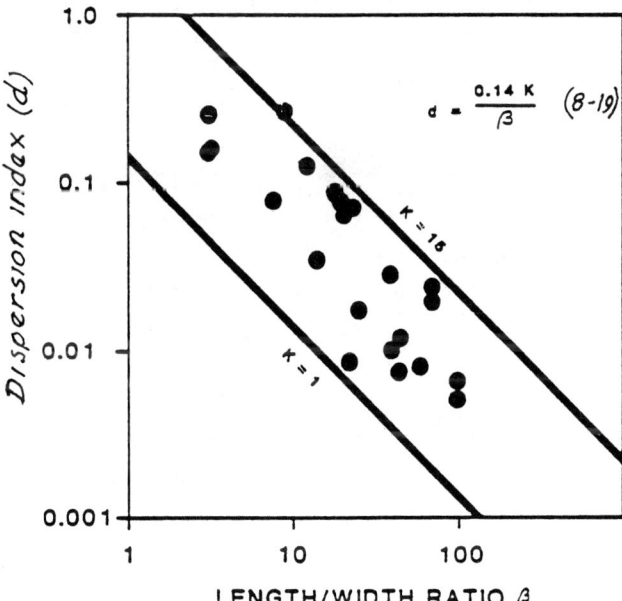

Fig. 8-34 Dispersion index versus length to width ratio of long channels. (From Trussell and Pollock.[80])

$$d = \frac{24.4 \times 0.015 \times (10.95)^{5/6}}{L = 800 \text{ ft}}$$
$$= 0.00334$$

This is equivalent to ideal plug-flow characteristics.

Optimum Design Considerations:

Sanitation Districts of Los Angeles County, Example[111.] The optimum design of a contact chamber can be achieved if sufficient information is available. The factors which govern the design are as follows: (1) initial coliform concentration prior to disinfection, (y_0); 2) final coliform concentration in the disinfected effluent (y); (3) the 2–3 min. chlorine demand; or the chlorine dosage required to produce the desired residual; and (4) the residual–contact time envelope (ct) to achieve the final coliform concentration (y). When this information is known for a given effluent, the contact chamber can be designed for minimum annual cost and chlorine contact time.[19]

A differential equation based on the annual cost (capital amortization plus operation and maintenance) can be developed. To find the minimal cost this equation is differentiated with respect to t and the result is set equal to zero. The answer represents the point where annual cost with respect to a change in contact time is a minimum. This equation takes the form of:

$$\frac{\partial AC}{\partial t} = a - b\frac{K}{t^2} = 0 \qquad (8\text{-}21)$$

Therefore

$$t = \sqrt{\frac{b}{a}K} \qquad (8\text{-}22)$$

where

AC = annual cost
$K = ct$
c = chlorine residual, mg/liter at time t
t = contact time, min

The AC (annual cost) consists of capital amortization plus operation and maintenance. This cost is made up of the following factors:

1. CRF (capital recovery factory). This consists of the cost of the chlorine contact chamber, the dechlorination station, the chlorine, and the sulfur dioxide, and the cost of labor. This is usually taken at i (interest) = 7 percent and n = 20 yr.

2. Chlorine Contact Tanks = y/gal. × z gpm × t (min.)
3. Cost of dechlorination station
4. Cost of chlorine
5. Cost of sulfur dioxide

From this a differential equation can be formulated which will be as follows.

AC = CRF (dollars for chlorine contact tank cost) + dechlorination station cost + annual chlorine cost (based on dosage required (+ annual sulfur dioxide cost (based on maximum residual required).

When the above term AC is factored out it will declare numerically the relations between contact time and chlorine residual at the end of this contact time which is related to the values of a and b. These are a function of items 1–5 described above. The solution to this equation provides the minimum contact time for a given flow of wastewater to produce the coliform concentration limit (y) for a given situation.

Referring to Eq. 8-22; $K = ct$; this is the chlorine residual contact time envelope. For a given situation the upper and lower limits for this factor should be selected. From this, contact time t should be tabulated together with chlorine residual. Let us assume that the lower limit is 500 at a given peak flow to accomplish the desired disinfection, then the upper limit based on an average flow would be

$$\frac{Q_1 = \text{peak flow}}{Q_2 = \text{ave flow}} \times 500 = K$$

for contact chamber design conditions. Therefore, c and t for this value of K can be computed and these will be the design parameters for contact time and chlorine residual.

Kennedy–Jenks Engineers Example.[105] This method requires knowledge of the wastewater quality, the same as the above example.

1. The ct envelope versus the log kill of coliform organism (y/y_0) is drawn. If there are known seasonal variations of y_0 an upper and lower ct envelope may be constructed, one for the lower limit of y_0 and one for the upper limit. The term y is the coliform concentration at the end of time t and is the disinfection requirement prescribed for the effluent.

2. The next step is to conduct some laboratory chlorine demand studies in order to establish the chlorine dose required to produce the desired residual at several different contact times.

3. The information developed in steps 1 and 2 will provide the necessary information to find the cost of chlorine and sulfur dioxide.

4. The next step is to determine the contact chamber construction cost for various selected contact times.

5. All of the above information is then tabulated against contact times. Then the cost of each function for a given time t is totaled and converted to a "present worth figure."

6. The last step is to draw a curve or series of curves using the present worth figure as the ordinates and contact time as the abcissa. The series of curves may be one each from the two ct envelopes, one dry weather flow, and one for wet weather flow and another for the present worth figure of the chemicals. The lowest point on the curve to be used will represent the optimum contact time for a given known wastewater quality and disinfection requirement.

Optimum Contact Time.[79] The following approach to produce an adequately disinfected effluent at minimum cost determines the minimum contact time. Using Eq. (8-23) and simple economic analysis, a differential equation for partial cost of chlorination can be written from which the most cost effective contact time can be calculated:

$$N_1/N_o = \left(\frac{RT}{b}\right)^n \tag{8-23}$$

where

N_o = average (median or geometric mean) concentration of coliform bacteria immediately upstream from chlorine application
N_1 = NPDES total coliform concentration in effluent
R = chlorine residual (mg/l) after contact time T
T = contact time, minutes
n = slope of the curve
b = the x-intercept when N_1/N_o = 1.0 mg/l, minutes

The symbols used in the following analysis are:

CRF = capital recovery factor (amortization)
$S = N_1/N_o$ required bacterial destruction
T = mean residence time in contact chamber, min
C_c = unit cost of chlorine, \$/lb
C_s = unit cost of sulfur dioxide, \$/lb
C_t = unit cost of contact chamber, \$/gal
D = total chlorine demand, mg/l
CE = DE = cost of chlorination equipment or dechlorination equipment
Q = mean effluent flow, 10^6 gal/day

Annual cost of chlorine is:

$$365Q(R + D)8.34 \times C_c = 3044Q(R + D)C_c \tag{8-24}$$

Annual cost of sulfur dioxide is:

DISINFECTION OF WASTEWATER

$$365Q(R)8.34 \times C_s = 3044QRC_s \tag{8-25}$$

Total cost of chlorine contact chamber is:

$$\frac{10^6}{24 \times 60} QTC_t = 694QTC_t \tag{8-26}$$

Total partial annual cost is:

$$\text{CRF}(694QTC_t + \text{CE} + \text{DE}) + 3044Q[(R+D)C_c + RC_s] \tag{8-27}$$

From $S = (RT/b)^n$,

$$R = bS^{1/n}/T \tag{8-28}$$

Substituting this into Eq. (8-27), we get:

$$\text{Total cost} = \text{CRF}(694QTC_t + \text{CE} + \text{DE}) + \\ 3044[(bS^{1/n}T^{-1} + D)C_c + bS^{1/n}T^{-1}C_s] \tag{8-29}$$

Differentiate Eq. (8-29) with respect to T, to optimize detention time, and then equate to zero:

$$\frac{d(\text{total cost})}{dT} = \text{CRF}694QC_t - 3044Qbs^{1/n}T^{-2}(C_c + C_s) = 0 \tag{8-30}$$

Therefore optimum residence time is:

$$T = 2.1 \frac{[bS^{1/n}(C_c + C_s)]^{1/2}}{\text{CRF} \times C_t} \tag{8-31}$$

where $n = -3$.

Sample calculation using Eq. (8-31): Assume CRF = 0.095 at 8 percent interest for a 20 year period; CE = \$60,000 for a 10 mgd plant; cost of chlorine, \$0.10/lb; cost of sulfur dioxide, \$0.15/lb; cost of contact chamber, \$0.35/gal. Assume also that

$N_0 = 2.4 \times 10^6$ MPN/100 ml;
$N_1 = 240$ MPN/100 ml; $\quad D = 5.0$ mg/l; $\quad S = 1 \times 10^{-4}$
$b = 6 \quad$ and $\quad n = -3$

Then for an optimized system:

$$T = 2.1 \left[\frac{6.0(10^{-4})^{-1/3}(0.10 + 0.15)}{0.095 \times 0.35} \right]^{1/2}$$

$$= 2.1 \left[\frac{6.0(10)^{4/3}(0.10 + 0.15)}{0.095 \times 0.35} \right]^{1/2}$$

$$= 2.1\sqrt{964.49}$$

$$= 65.22 \text{ min}$$

Summary of Initial Mixing and Contact Chamber Requirements

Initial Mixing. This event should occur in the most turbulent zone of the effluent entrance to the contact chamber where the velocity is on the order of 2–3 ft/sec. If mixing is to be achieved by a grid-type diffuser, the diffuser should be located in this turbulent zone.

The mixing device should be able to generate a G number of about 500. The objective is to provide complete mixing of the chlorine solution with the wastewater flow in not more than 3 sec. A variety of solutions to the problem of initial mixing are available to the designer. The use of a closed conduit for the mixing zone should not be overlooked.

Contact Chambers. The most important parameter in the evaluation of these structures is t_i. This is the time it takes for the first appearance of dye at the exit of the chamber. This should never be less than 30 min.

The design objective is to achieve a dispersion index no larger than 0.02, and preferably 0.01. This can be done using longitudinal baffles where the L/W ratio is greater than 40, provided the turning zones are designed to eliminate eddy currents and the H/W ratio is 1.0 or less. Where possible and convenient, longitudinal baffles with an L/W ratio of 70 are preferred.

The exit structure of the contact chamber should be provided with either a diffuser baffle or a launder type weir. These embellishments must be able to cope with the change in H due to variations in hydraulic flow.

In general the design of a contact chamber should provide a distribution of residence times approaching ideal plug-flow characteristics. This cannot be overemphasized. It is apparent, therefore, that a long, straight pipeline will fulfill these conditions.

Existing contact chambers can be retrofitted to achieve significant improvement in disinfection efficiency.

RELIABILITY PROVISIONS FOR CHLORINATION FACILITY

The need for continuous and dependable disinfection has been stressed. Chlorination system failure can be due to a number of causes, so the design of the system must include provisions to either prevent failure or allow immediate corrective action to be taken. Although assured reliability is essential, design provisions for this are often slighted.

Chlorine Supply. As with any chemical feed process, one of the most frequent interruptions in treatment is caused by the exhaustion of the chlorine supply. Five features are essential in order to maintain continuous chlorine feed: (1) an adequate reserve supply of chlorine sufficient to meet normal needs and bridge delivery delays and other possible contingencies; (2) chlorine container scales; (3) a manifolded chlorine header system; (4) an automatic device for switching to a full chlorine container when the one in use becomes empty; and (5) an alarm system to alert operating personnel of imminent loss of chlorine supply. These five features are discussed elsewhere in this text. Without them it is not possible to assure uninterrupted chlorine feed even with full-time operator attendance and no equipment breakdowns.

The chlorine header system is needed for two reasons: (1) to provide a connected on-line chlorine supply which is adequate to assure uninterrupted flow of feed for whatever period the system may be unattended; and (2) to allow switchover to a full cylinder without interruption of feed.

Power Failure. Power outage usually results in water supply failure, which in turn automatically shuts down the chlorination system. A range of special provisions can be employed to assure reliability of power and water supply, depending upon the particular situation. A standby power source and duplicate pumps are the measures most often taken here.

Standby Equipment. The design of the chlorine feed system should provide for continued operation in cases of equipment failure. Where both pre- and postchlorination are to be practiced, separate chlorination systems should be provided for each plus a standby system. If prechlorination is not continuous, it may be possible to use the prechlorination system as the standby for disinfection. The units, piping, and accessories should be designed with this application in mind. If prechlorination must be carried out continuously, or if no prechlorination is to be done, a standby system capable of replacing the postchlorination system during repairs, maintenance, or emergencies should be provided. Standby equipment of sufficient capacity should be available to replace the largest unit during shutdowns. This includes standby pumps for the injector water supply.

Spare Parts. In addition to standby equipment, the equipment manufacturer should be consulted regarding vulnerable components. These components should be a part of the plant's inventory of spare parts.

Water Supply. As mentioned above, during a power failure, the injector water system will be shut down unless there is an alternative supply that does not require power, such as an elevated tank. Standby equipment to provide injector water in the event of a power failure would consist of an engine-driven injector supply pump. Every injector water supply system should have such a standby pumping unit. There is no way to operate the chlorination system without an adequate water supply.

Chlorine Residual Analyzers. Every system using an analyzer for chlorination control should be backed up by an effluent monitor analyzer that can be switched over to the control function in the event of control analyzer failure. Provisions should also be made for standby sample pumps. An adequate supply of all necessary chemicals should be on hand at all times.

Records and Reports

Reliable, continuing records should be kept to establish proof of performance, and to justify decisions, expenditures, and recommendations. Records also serve as a source of information for plant operation, modification, and maintenance.

Adequate records of disinfection are also important from the standpoint of regulatory agencies. Records will facilitate public health surveillance and enable the regulatory agencies to assess compliance with state regulations.

There are two classes of records which should be maintained at a wastewater treatment plant using chlorination: (1) descriptive, planning, and inventory records related to the physical plant; and (2) performance records.

Physical Facilities. The following records, referring to the chlorination–dechlorination unit, should be available for reference at the plant:

1. Design engineer's report, including basis of design, equipment capacities, population served, design flow, reliability features, and other data
2. Contract and "as built" plans and specifications
3. Shop drawings and operating instructions for all equipment
4. Cost of each equipment item
5. Detail plans of all piping and electrical wiring
6. A complete record of each piece of equipment, including name of manufacturer, identifying number, rated capacity, and dates of purchase and installation
7. Supply of chlorine and dechlorinating agent, including reserves and availability estimate

Records of Operation. Daily records should be kept of the following:

1. Chlorine: daily quantities used, both for pre- and postchlorination; chlorine residuals, including method used (preferably continuous recording of residuals); dosage

2. Wastewater flow: preferably continuous recording; total daily flow treated; maximum, minimum, and average daily flow.
4. Results of bacteriological analyses: coliform bacteria and other required tests.
5. Daily inspection: operational problems, equipment breakdowns, periods of chlorinator outage, diversions to emergency disposal, and all corrective and preventive action taken.

Records of Equipment Inspection, Maintenance, and Repair. In addition, the following important records and plans should be kept at each facility:

1. Routine equipment maintenance schedule and record
2. Annual equipment inspection and maintenance record
3. Plan for prearranged repair service
4. Emergency plan for chlorinator failure
5. Emergency plan for accidental chlorine or sulfur dioxide release

Reports. Monthly operating reports should be prepared, containing information on chemical usage, wastewater flows, laboratory analyses, and significant operational problems.

REFERENCES

1. H. F. Collins, private communication, Chief, Environmental Health Division, California Health Services, June 1983.
2. *Standard Methods for the Examination of Water and Wastewater,* 14th ed., American Public Health Assoc., 1975.
3. "Report on a Pollution Survey of Santa Monica Bay Beaches in 1942," California State Board of Health, June 26, 1943.
4. Stevenson, A. H. "Studies of Bathing Water Quality and Health." *Am. J. Public Health,* 43, 529 (1953).
5. California State Department of Health Symposium, "Fecal Coliform Bacteria in Water and Wastewater," Berkeley, CA, May 21, 1968.
6. Unz, R. F. "Fecal Coliforms and Fecal Streptococci in the Bacteriology of Water Quality," *Water and Sewage Works,* 115, 238 (1968).
7. Jopling, W., "Statement in Support of California Recommended Disinfection Requirements," paper presented by the California State Department of Public Health at a joint meeting of Arizona, California, Nevada, and EPA regarding waste discharge requirements and water quality objectives for the Colorado River, Las Vegas, Nev., Oct. 28, 1975.
8. "National Symposium on Estuarine Pollution," Stanford Univ., Palo Alto, CA, 1967.
9. Jopling, W., and Young, C., private correspondence, California State Dept. of Health, Berkeley, CA, 1976.
10. Collins, H. F., Selleck, R. E., and White, G. C., "Problems in Obtaining Adequate Sewage Disinfection," *ASCE J. San. Eng. Div.,* 97, 549 (Oct. 1971).
11. Ptak, D. V., Ginsburg, W., and Willey, B. F., "Identification and Incidence of *Klebsiella* in Chlorinated Water Supplies," *J. AWWA,* 65, 604 (Sept. 1973).
12. Engelbrecht, R. S., Foster, D. H., Masarik, M. T., and Sai, S. H., "Detection of New Microbial Indicators of Chlorination Efficiency," paper presented at the AWWA Water Technology Conference, Dallas, TX, (December 1–3, 1974).

13. White, G. C., "Disinfection: The Last Line of Defense for Potable Water," *J. AWWA*, **67**, 410 (Aug. 1975).
14. White, G. C., "Disinfection Committee Report," paper presented at the AWWA Annual Conference, Minneapolis, MN, June 1975.
15. Jopling, W. F., "Water Re-use Standards for the State of California," paper presented at the annual WPCF Conference, Anaheim, CA, May 9, 1969.
16. Committee on Environmental Quality Management, "Engineering Evaluation of Virus Hazard in Water," *ASCE J. San. Engr. Div.*, **96**, 111 (Feb. 1970).
17. Sobsey, M. D., "Enteric Viruses and Drinking Water Supplies," *J. AWWA*, **67**, 414 (Aug. 1975).
18. Katz, M. and Plotkin, S. A. "Minimal Infective Dose of Attenuated Polio-virus for Man," *Am. J. Public Health*, **57**, 1837 (1967).
19. McDermott, J. H., "Virus Problems and Their Relation to Water Supply," paper presented at Virginia Sect. Meeting, AWWA, Roanoke, VA, Oct. 25, 1973.
20. McCabe, L. J. "Significance of Virus Problems," paper presented at AWWA Water Qual. Conf., Cincinnati, OH, Dec. 3 and 4, 1973.
21. "Effect of Chlorination on Human Enteric Viruses in Partially Treated Water from Potomac Estuary," Proc. Congr. Hearings, Proc. Serv. No. 92–94. Washington, DC, 1973.
22. Clarke, N. A., and Kabler, P. W., "Human Enteric Viruses in Sewage," *Health Lab Sci.*, **1**, 44 (1964).
23. Geldereich, E. E., and Clarke, N. A., "The Coliform Test: A Criterion for the Viral Safety of Water," in V. Snoeyink and V. Griffin (eds.), *Proc. 13th Water Qual. Conf.*, Univ. of Ill., Urbana, IL, 1971.
24. Berg, G. "Reassessment of the Virus Problem in Sewage and in Surface and Renovated Waters," *Prog. Water Technol.*, **3**, 87–94 (1973).
25. Coin, L., et al., "Modern Microbiological Virological Aspects of Water Pollution," *Ad. Water Pollution Research*, Proc. 2nd International Conf., Pergamon Press, New York, (1966), pp. 1–10.
26. Nupen, E. M., Bateman, B. W., and McKenny, N. C., "The Reduction of Virus by the Various Unit Processes Used in the Reclamation of Sewage in Potable Waters," paper presented at the Virus Symposium, Austin, TX, April 1974.
27. Selna, M. W., Miele, R. P., and Baird, R. B., "Disinfection for Water Reuse," paper presented for the Disinfection Seminar at the Ann. Conf. AWWA, Anaheim, CA, May 8, 1977.
28. Ludovici, P. P., Philips, R. A., and Veter, W. S., "Comparative Inactivation of Bacteria and Viruses in Tertiary Treated Wastewater by Chlorination," in *Disinfection: Water and Wastewater*, J. D. Johnson (ed.), Ann Arbor Science, Ann Arbor, MI, 1975, p. 359.
29. Pavoni, V. L. and Tittlebaum, M. D. "Virus Inactivation in Secondary Wastewater Treatment Plant Effluent Using Ozone," Water Resources Symp. No. 7, in J. F. Malina Jr. and B. P. Sagik (eds.), *Viruses in Water and Wastewater Systems*, Univ. of Texas, Austin, TX, April 1974.
30. Boardman, G. D. and Sproul, O. V. "Protection of Viruses During Disinfection by Absorption to Particulate Matter," paper presented at the 48th Annual Conf. WPCF, Miami, FL, Oct. 1975.
31. White, G. C., "Disinfection Facility Evaluation City of San Francisco," unpublished report, 1976.
32. Tsai, C. F., "Effects of Clorinated Sewage Effluents on Fishes in Upper Patuxent River, Maryland," *Chesapeake Sci.*, **9**, 83 (June 1968).
33. Jolley, R. J., "Chlorine-Containing Organic Constituents in Chlorinated Effluents," *J. WPCF*, **47**, 601 (Mar. 1975).
34. Esvelt, L. A., Kaufman, W. J., and Selleck, R. E., "Toxicity Assessment of Treated Municipal Wastewaters," paper presented at the 44th Annual Conf. of the WPCF, San Francisco, CA, Oct. 4–8, 1971.
35. Esvelt, L. A., private correspondence, Oct. 1971.
36. "Chlorinated Municipal Waste Toxicities to Rainbow Trout and Fathead Minnows," Water Pollution Control Research Series No. 18050 GZZ 10/71, Bur. of Water Mngmt., Mich. Dept. of Nat. Resources for the EPA, Oct. 1971.

37. Bellanca, M. A., and Bailey, D. S., "A Case History of Some Effects of Chlorinated Effluents on the Aquatic Ecosystem in the Lower James River in Virginia," paper presented at the 48th Annual Conf. WPCF, Miami Beach, FL, Oct. 5–10, 1975.
38. Collins, H. F., and Deaner, D. G., "Sewage Chlorination Versus Toxicity—A Dilemma?" *ASCE J. Environ. Eng. Div.,* 761 (Dec. 1973).
39. Arthur, J. W., Andrew, R. W., Mattson, V. R., Olson, D. T., Glass, G. E., Halligan, B. J., and Walbridge, C. T., "Comparative Toxicity of Sewage Effluent Disinfection to Freshwater Aquatic Life," EPA Report 600/3–75–012, Research Lab., Duluth, MN, Nov. 1975.
40. Brungs, W. A. "Effects of Wastewater and Cooling Water Chlorination on Aquatic Life," EPA Report 600/3–76–098, Research Lab., Duluth MN, Aug. 1976.
41. Esvelt, L. A., Kaufman, W. J., and Selleck, R. E., "Toxicity Removal from Municipal Wastewaters," SERL No. 71-8, Sanitary Engineering Research Lab., Univ. of Calif., Berkeley, Oct. 1971.
42. Kaufman, W. J., private correspondence, 1972.
43. Stone, R. W., Kaufman, W. J., and Horne, A. J. "Long-Term Effects of Toxicity and Biostimulants on the Waters of Central San Francisco Bay," SERL Report 73-1, Univ. of Calif., Richmond, CA, 1973.
44. Woodbridge, D. D., and Cooper, P. C., "Reduction of Chlorination by Irradiation," paper presented at the AWWA Disinfection Seminar, Anaheim, CA, May 8, 1977.
45. A Task Force Report, "Disinfection of Wastewater." U.S. Environmental Prot. Agency No. 430/9-75-012, Washington, DC, March 1976.
46. Love, O. T. Jr., Carswell, N. K., and Symons, J. M. "Comparison of Practical Alternative Treatment Schemes for Reduction of Trihalomethanes in Drinking Water," paper presented at the IOI Workshop on Ozone and Chlorine Dioxide, Cincinnati, OH, Nov. 17–19, 1976.
47. Katzenelson, E., and Teltch, B., "Dispersion of Enteric Bacteria by Spray Irrigation," *J. WPCF,* **48,** 710 (April 1976).
48. Katzenelson, E., Buium, I., and Shuval, H. I. "Risk of Communicable Disease Infection Associated With Wastewater Irrigation in Agricultural Settlements," *Science,* **194,** 944 (Nov. 26, 1976).
49. White, G. C., Beebe, R. D., Alford, V. F., and Sanders, H. A. "Wastewater Treatment Plant Disinfection Efficiency as a Function of Chlorine and Ammonia Content," in Jolley, R. L. et al. (Eds.), *Water Chlorination: Environmental Impact and Health Effects,* Vol. 4, Book 2, Ann Arbor Science, Ann Arbor, MI, 1983.
50. White, G. C., Beebe, R. D., Alford, V. F., and Sanders, H. A., "Problems of Disinfecting Nitrified Effluents," Municipal Wastewater Disinfection, Proc. of Second Ann. Nat. Symposium, Orlando, FL, EPA Report 600/9–83–009, July 1983.
51. Lomas, P. D. R., "The Combined Residual Chlorine of Swimming Bath Water," *Jour. Assoc. Pub. Analysts* (England), **5,** 27 (1967).
52. Fair, G. M., Morris, J. C., Chang, S. L., Weil, Ira, and Burden, R. P., "The Behavior of Chlorine as a Water Disinfectant," *J. AWWA,* **40,** 1051 (1948).
53. Feng, T. H., "Behavior of Organic Chloramine in Disinfection," *J. WPCF,* **38,** 614 (1966).
54. Sung, R. D., "Effects of Organic Constituents in Wastewater on the Chlorination Process," Ph.D. Dissertation, Univ. of Calif., Davis, CA, 1974.
55. Esvelt, L. A. Kaufman, W. J., and Selleck, R. E. "Toxicity Assessment of Treated Municipal Wastewaters," paper presented at the 44th Ann. Conf. of WPCF, San Francisco, CA, Oct. 4–8, 1971.
56. Collins, H. F., Selleck, R. F., and White, G. C., "Problems in Obtaining Adequate Sewage Disinfection," *ASCE J. San. Engr. Div.,* **97,** SA 5, Proc. #8430 (Oct. 1971).
57. Collins, H. F., White, G. C., and Sepp, E., "Interim Manual For Wastewater Chlorination and Dechlorination Practices," California State Dept. of Health, Feb. 1974.
58. Collins, H. F., and Selleck, R. E., "Process Kinetics of Wastewater Chlorination," SERL Report No. 72–5, Sanitary Engineering Research Laboratory, Univ. of Calif., Berkeley, Nov. 1972.
59. Roberts, P. V., Aieta, E. M., Berg, J. D., and Chow, B. M., "Chlorine Dioxide for Wastewater

Disinfection: A Feasibility Evaluation," Tech. Report No. 251, Civil Engr. Dept., Stanford Univ., Palo Alto, CA Oct 1980.
60. White, G. C., unpublished field studies, San Jose–Santa Clara Water Pollution Control Plant, San Jose, CA, 1973–1976.
61. Vrale, L., and Jordan, R. M., "Rapid Mixing in Water Treatment," *J. AWWA*, **63**, 52 (Jan. 1971).
62. Selleck, R. E., Saunier, B. M., and Collins, H. F., "Kinetics of Bacterial Deactivation with Chlorine," *J. Env. Eng. Div. ASCE*, p. 1197 (Dec. 1978).
63. Snead, M. C., Olivieri, V. P., and Dennis, W. H., "Biological Evaluation of Methods for the Determination of Free Available Chlorine," in W. J. Cooper (Ed.), *Chemistry in Water Reuse*, Vol. 1, Chapter 19, Ann Arbor Science, Ann Arbor, MI, 1981.
64. Bhupinder, S. D., and Baker, R. A., "Controlling Nitrification to Reduce Energy Usage and Treatment Costs," report to Central Contra Costa San. Dist., Walnut Creek, CA, 1981.
65. Bhupinder, S. D., and Baker, R. A., "Role of Ammonia-N in Secondary Effluent Chlorination," *J. WPCF*, **55**, 454 (May 1983).
66. Collins, H. F., Selleck, R. E., and Saunier, B. M. "Optimization of the Wastewater Chlorination Process," paper presented at the National Conference on Environment and Res., sponsored by EE Div. ASCE, Seattle, WA, July 12, 1976.
67. Selna, M. W., Miele, R. P., and Baird, R. B., "Disinfection for Water Reuse," paper presented in the Disinfection Seminar at the AWWA Annual Conf. Anaheim, CA, May 8, 1977.
68. Ingols, R. S., private communication, Georgia Inst. of Tech., Atlanta, GA, July 1977.
69. Jolley, R. L., and Pitt, W. W., Jr., "Chloro-Organics in Surface Water Sources for Potable Water," paper presented for the Disinfection Symposium at the Ann. AWWA Conf., Anaheim, CA, May 8, 1977.
70. Jolley, R. L., "Chlorine Containing Organic Constituents in Chlorinated Effluents," *J. WPCF*, **47**, 601 (1975).
71. Symons, J. M., "Interim Treatment Guide for the Control of Chloroform and Other Trihalomethanes," report by the EPA Env. Res. Lab., Cincinnati, OH, June 1976.
72. Stone, R. W., "The Formation of Halogenated Organic Compounds in Wastewater Chlorination," unpublished report by Brown and Caldwell Cons. Engrs., Walnut Creek, CA, 1977.
73. Symons, J. M., Bellar, T. A., Carswell, J. K., Demarco, J., Kropp, K. L., Robeck, G. G., Seeger, D. R., Slocum, C. J., Smith, B. L., and Stevens, A. A., "National Organics Reconnaissance Survey for Halogenated Organics in Drinking Water," *J. AWWA*, **67**, 634 (Nov. 1975).
74. Hoehn, R. C., Randall, C. W., Bell, F. A., Jr., and Shaffer, P. T. B. "Trihalomethanes and Viruses in a Water Supply." A paper presented at the ASCE National Conf. on Env. Eng. Res. Dev. and Design, Seattle, WA (July 12–14, 1976).
75. Breidenbach, A. W., "Interim Report on Montgomery Simulation: Study of Formation and Removal of Volatile Chlorinated Organics," EPA MERL Cincinnati, OH, (July 8, 1975).
76. White, G. C., "Disinfection Practices in the San Francisco Bay Area," *J. WPCF*, **46**, 84 (Jan. 1974).
77. Sepp, E., and White, G. C., "Manual for Wastewater Chlorination and Dechlorination Practices," State of California, Dept. of Health Services, Sanitary Engineering Section, March, 1981.
78. Sepp, Endel, "Optimization of Chlorine Disinfection Efficiency," *J. Environ. Engrng. Div. ASCE*, **107**, EE1 (Feb. 1981).
79. Sepp, E., and Bao, P., "Design Optimization of the Chlorination Process, Vol. I, Comparison of Optimized Pilot System With Existing Full-Scale Systems," Wastewater Research Div., Mun. Env. Res. Lab., Cincinnati, OH, and Calif. Dept. of Health Services, Berkeley, CA, 1980.
80. Trussell, R. R., and Pollock, T., "Design of Chlorination Facilities for Wastewater Disinfection," paper presented at Preconference Workshop, Ann. Conf. WPCF, Atlanta, GA, Oct. 2, 1983.
81. Anon., "Wallace and Tiernan V-Notch Chlorinators," Wallace and Tiernan–Pennwalt, Belleville, NJ, Cat. No. 25.052, Rev. 8–79.
82. Finger, R., private communication, Renton WWTP, Renton, WA, April 1984.

83. Fraschina, K., "Chlorine Residual Distribution in Open Channel Flow," Richmond-Sunset Wastewater Treatment Plant, San Francisco, CA, 1940.
84. Rudolfs, W., and Gehm, H., "Sewage Chlorination Studies," Bulletin No. 601, New Jersey Agric, Exper. Sta., New Brunswick, NJ, March 1936.
85. Camp, T. R., and Stein, P. C., "Velocity Gradients and Internal Work in Fluid Motion," in *Civil Engineering Classics: Outstanding Papers by Thomas R. Camp*, ASCE, New York, 1973, p. 203.
86. Longley, K. E., "Engineering Design of a Disinfection System," paper presented at National Conf. Environmental Engineering ASCE, Minneapolis, MN, July 14–16, 1982.
87. Collins, H. F., "Process Kinetics of Wastewater Chlorination," Ph.D. Dissertation in Engineering, Univ. of California, Berkeley, CA, 1971.
88. Eliassen, R., Heller, A. N., and Krieger, H. L., "A Statistical Approach to Chlorination," *Sew. Wks. J.*, **20**, 1008 (1948).
89. Krusé, C., Kawata, K., Olivieri, V., and Longley, K., "Improvement in Terminal Disinfection of Sewage Effluents," *Water and Sew. Wks.*, **120**, 57 (June 1973).
90. Heukelekian, H., and Day, R. V., "Disinfection of Sewage With Chlorine; III Factors Affecting Coliforms Remaining and Correlation of Orthotolidine and Amperometric Chlorine Residuals," *Sew. and Ind. Wastes*, **23**, 155 (Feb. 1951).
91. Longley, K. E., "Mixing and Chlorine Disinfection," paper presented at the Ann. WPCF Conf., Minneapolis, MN, Oct. 4, 1976.
92. Longley, K. E., "Turbulence Factors in Chlorine Disinfection of Wastewater," *Water Research*, **12**, 813 (Nov. 1978).
93. Stenquist, R. J., and Kaufman, W. J., "Initial Mixing in Coagulation Processes," Univ. of California, Berkeley, CA, SERL Report 72-2, Feb., 1972.
94. Calmer, J. C., Sanchez-Adams, R. M., and Tobin, L. D., "Chlorine Mixing Energy Requirements for Coliform Disinfection of Non-nitrified Secondary Effluents," paper presented at Ann. Conf. WPCF, New Orleans, LA, Oct. 1984.
95. Egan, J. T., "Chlorine Solution Diffusers and Mixers in Effluent Channels," *Calif. Water Poll. Control Assoc. Bulletin*, p. 38 (April 1978).
96. Egan, J. T., private communication, City of Colorado Springs, CO, Nov. 1983.
97. Mixing Equipment Co., private communication, Rochester, NY, 1973.
98. White, G. C., Bean, E. L., and Williams, D. B., "Chlorination and Dechlorination, A Scientific and Practical Approach," *J. AWWA*, **58**, 540 (1968).
99. Levy, A. G., and Ellms, J. W., "Hydraulic Jump as a Mixing Device," *J. AWWA*, **17**, 1 (Jan. 1927).
100. Mandt, M. G., private communication, Pentech Div., Houdaille Ind. Inc., Cedar Rapids, IA, March 1977.
101. Anon., Enhanced Disinfection System Fischer and Porter Specification 71 ED 1000, June 1981.
102. Deaner, D. G., "Effect of Chlorine on Fluorescent Dyes," *J. WPCF*, **45**, 507 (March 1973).
103. Marske, D. M., and Boyle, V. D., "Chlorine Contact Chamber Design—A Field Evaluation," *Water and Sewage Wks.* **120**, 70 (Jan. 1973).
104. Trussell, R. R., and Chao, J. L., "Rational Design of Chlorine Contact Facilities," *J. WPCF*, **49**, 659 (1977).
105. Calmer, J. C., and Adams, R. M. "Design Guide Chlorination–Dechlorination Contact Facilities," in-house report, Kennedy/Jenks Engineers, San Francisco, CA, July, 1977.
106. Hart, F. L., "Improved Hydraulic Performance of Chlorine Contact Chambers," *J. WPCF*, **51**, 2868 (Dec. 1979).
107. Hart, F. L., and Vogiatzis, Z., "Performance of Modified Chlorine Contact Chamber," *J. Env. Engr. Div. ASCE*, **108**, EE3, 549 (June 1982).
108. Thirumurthi, D., "A Breakthrough in the Tracer Studies of Sedimentation Tanks," *J. WPCF*, **41**, Part 2, R405 (Nov. 1969).

109. Levenspiel, O. *Chemical Reaction Engineering,* John Wiley & Sons, Inc., New York, 1962, p. 250; 2nd ed., 1972, p. 260.
110. Morrill, A. B., "Sedimentation Basin Research and Design," *J. AWWA,* **24,** 1442 (1932).
111. Stahl, J. F., "Chlorine Contact Chamber Design, Pomona Water Renovation Plant," in-house report for Sanitary Districts of Los Angeles County, CA, Whittier, CA, ca 1972.
112. Egan, J. T., private communication, City of Colorado Springs, CO, May 4, 1984.
113. Karaganis, J. V., "Sewage Treatment: The Present Situation Won't Do," *Water and Sew. Wks.,* **125,** Editorial, (Nov. 1978).
114. Hedenland, L. D., private communication, Las Virgenes Municipal Water District, Calabasas CA, Dec. 10, 1974.

9
Chlorine Facilities Design

PREFACE

The conventional chlorination facility for use in potable water and wastewater treatment consists of three principal parts: chlorine supply, metering system, and injector system. In addition there is ancillary equipment: safety equipment, metering and control instrumentation and chlorine residual analyzers.

CHLORINE SUPPLY SYSTEM

Chlorine is packaged in special steel containers of various sizes, as follows:

1. 100- and 150-lb cylinders
2. Ton containers
3. Single-unit tank cars
4. Multiple-unit tank cars (TMU) containing fifteen 1-ton cylinders
5. Tank trucks of 15–20 tons capacity
6. Stationary storage tanks

The selection of the various sizes is discussed in this chapter.

100- and 150-lb Cylinders. The pertinent dimensions and tare weight of these containers are shown in Fig. 9-1.

Minor variations in these dimensions depend upon the cylinder age and the manufacturer. The 150-lb cylinder is so popular that the 100-lb size may be considered obsolete. Some packagers have available a 35-lb cylinder, which is suitable for laboratory or test work on a small scale.

The packager fills these cylinders with liquid chlorine to approximately 85 percent of total volume; the remaining 15 percent is occupied by the chlorine gas. There must be strict adherence to these figures in order to prevent hydrostatic rupture of the cylinder in the event of abnormally high ambient temperatures. As the temperature rises, the liquid chlorine expands. Theoretically, the cylinder could get hot enough to completely fill the remaining 15 percent occupied by gas and

Chlorine Cylinder

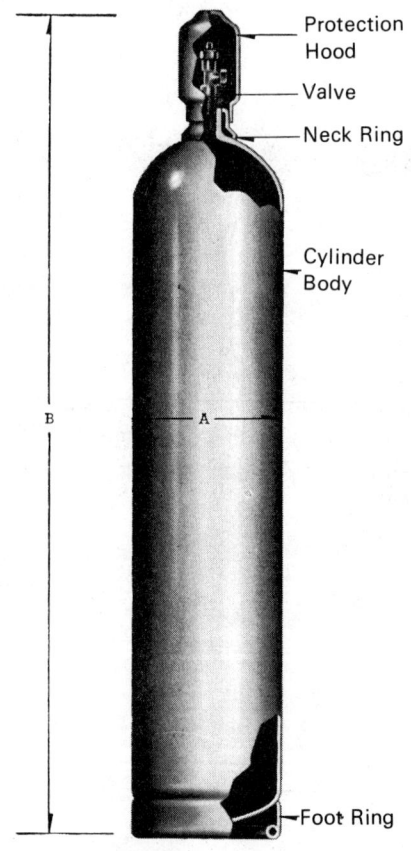

Net Cylinder Contents	Approx. Tare, Lbs.*	Dimensions, Inches	
		A	B
100 Lbs.	73	8 1/4	54 1/2
150 Lbs.	92	10 1/4	54 1/2

*Stamped tare weight on cylinder shoulder does not include valve protection hood.

Fig. 9-1. Chlorine cylinder dimensions (courtesy PPG Industries, Chemical Division).

therefore rupture the cylinder. However, the outlet valve on these cylinders is fitted with a fusible plug, the core of which will melt at approximately 158°F, thus preventing rupture of the cylinder during instances of abnormally high temperatures. When the plug melts, the liquid chlorine discharged through the core opening (1/8" diameter) cools so rapidly that it freezes, momentarily halting the flow of liquid chlorine. By this time the danger of cylinder rupture is over, and a trained operator wearing air or oxygen breathing equipment can apply the chlorine safety kit, stop the leak, and remove the cylinder from the area.

The gross weight of a full 150-lb cylinder varies from 250 to 285 lb. Therefore these cylinders are best handled by a two-wheel cylinder handtruck. *Never use slings, or try to pick up cylinders with hoisting equipment attached to the protective cap or valve.* The water volume of a 150-lb cylinder is 14.4 gallons.

The most important design considerations are as follows:

1. Direct sunlight must never reach the cylinder.
2. The maximum withdrawal rate should be limited to 40 lb/day/cylinder.
3. Minimum allowable room temperature is 50°F.
4. Heat must never be applied directly to the cylinder.
5. Sufficient space should be allowed in the supply area for at least one spare cylinder for each one in service.

Provisions should be made so that the operator can determine the amount of chlorine left in the supply system. This is most effectively done by weighing scales.

Figure 9-2 illustrates a new concept in scales for weighing 100- or 150-lb chlorine cylinders. This is Capital Controls Co. Series 1350 LO-Line Digital scale. It has a low-profile platform which transfers the cylinder weight to a strain gage transducer. The electronics package within the scale system provides a digital readout in pounds of chlorine remaining (or kilos). This system is unique because it can also provide a 4–20 mA output signal to a timer, which can then convert loss of weight at the scale into a chlorine flow rate. This replaces the usual chlorine gas-flow transmitters. The weighing accuracy is claimed to be ±0.5 percent of capacity (±0.75 lb for a 150-lb cylinder).

There are individual weighing surfaces for each cylinder which can be arranged to handle two banks of cylinders. There can be from one to six cylinders per bank. This is equivalent to a maximum of 900 lb chlorine exclusive of tare weight. There is a 3-digit readout and a summator readout for each bank of cylinders plus a mechanism to adjust out the tare weight for each cylinder.

Automatic switchover devices are readily adapted to this scale regardless of the number of cylinders in each bank.

Ideally the chlorine supply system should always be at a lower temperature than the chlorinator when withdrawal is from the gas phase. This reduces the possibility of reliquefaction at the chlorinator. The distance between the chlorine supply and the chlorinator should always be as short as conveniently possible.

588 HANDBOOK OF CHLORINATION

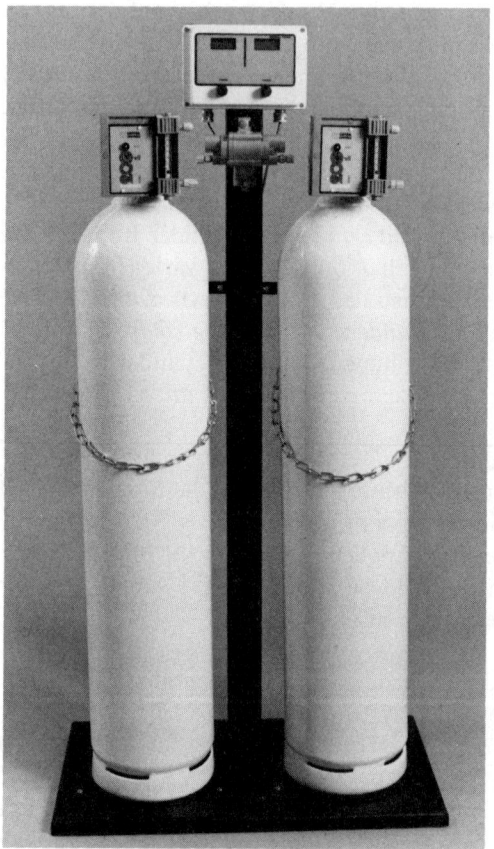

Fig. 9-2. Two cylinder digital scale with output signal (courtesy Capital Controls Co.).

The greatest difficulty in operating such installations is caused by reliquefaction of the chlorine gas. This occurs mostly at the first point of pressure reduction; when it occurs, impurities in the chlorine gas are deposited at this point, which is the chlorine inlet pressure-reducing valve of the chlorinator. This phenomenon is a result of the hot gas in the cylinder passing very slowly (inches per hour) through the piping between the cylinder and chlorinator and cooling during the night. This cooling causes the gas to reliquefy. The amount and frequency of trouble depends upon how much cooling takes place. This is a function of the difference in ambient temperatures between day and night, the volume of gas between cylinder and chlorinator, and the velocity of gas flow. The lower the feed rate and the greater the distance between cylinder and chlorinator, the greater the chance of reliquefaction, other things being equal. Reliquefaction will not occur if the cylinder is kept cooler than the chlorinator. This brings up the necessity for some type of insulation, which will be discussed below in this chapter.

The minimum allowable temperature for the chlorine storage area is about 50°F. Below this temperature the flow of chlorine becomes sluggish and erratic, particularly for smaller installations, from 1 to 20 lb/day. Therefore all 150-lb cylinder installations utilizing one or two cylinders should be housed in adequately insulated areas with provision for heating to 65°F.

Heat should never be applied directly to a chlorine cylinder. Steel will ignite spontaneously at about 483°F in the presence of chlorine. For example, it is possible to "burn" a hole in a chlorine cylinder by directing the rays of an infrared lamp onto the cylinder at the liquid level. With heat on one side, the liquid chlorine on the other side acts as a catalyst and will support the burning of the steel until the chlorine has become exhausted.

Sufficient space should be allowed for the scale and storage for one spare cylinder for each one connected in service. The distance between the cylinders should be sufficient so that standard four-foot flexible connections can be used. Each cylinder hookup should consist of an auxiliary cylinder valve and flexible connection.*

Filters and traps ahead of all chlorinator control apparatus are highly desirable, to prevent the impurities inherent in chlorine from reaching the chlorinator control mechanisms. Most small chlorinators have a built-in strainer of some type, which should be preceded by a convenient inlet trap.

If the storage area for in-service cylinders is properly designed, external chlorine pressure-reducing valves are not required. If the location of these cylinders is remote, an external reducing valve should be installed as close as possible to the cylinders.** If the chlorine supply line is longer than ten or fifteen feet and is subject to much variation in temperature due to poor insulation from ambient temperature changes, an external chlorine pressure-reducing valve should be installed close to the cylinders to avoid reliquefaction. If the vapor pressure in the chlorine cylinder is 100 psi, the gas between the cylinder and the chlorinator will reliquefy if the temperature drops below 80°F. If a reducing valve is utilized to reduce the pressure from 100 to 40 psi, the temperature would have to drop to below 32°F before reliquefaction will occur. (See Chapter 1.)

Automatic Switchover. The ability to switch from an empty cylinder to a full cylinder automatically increases the reliability of any chlorination system. Automatic switchover devices can perform in pressure systems or vacuum systems.

For systems up to 500 lb/day the vacuum system is ideal. A typical system is illustrated in Fig. 9-3. The vacuum regulator check units shown adjacent to the cylinder may be located remotely from the chlorinator, e.g., for separate housing location of cylinders.

The system functions as follows: Gas under pressure enters the vacuum regulator

---

* Cylinder mounted chlorinators do not require flex connections nor auxiliary cylinder valves. Nevertheless they are still subject to reliquefaction.
** When remote vacuum systems are used, external pressure reducing valves may be required to provide a two-step pressure reduction owing to ambient conditions.

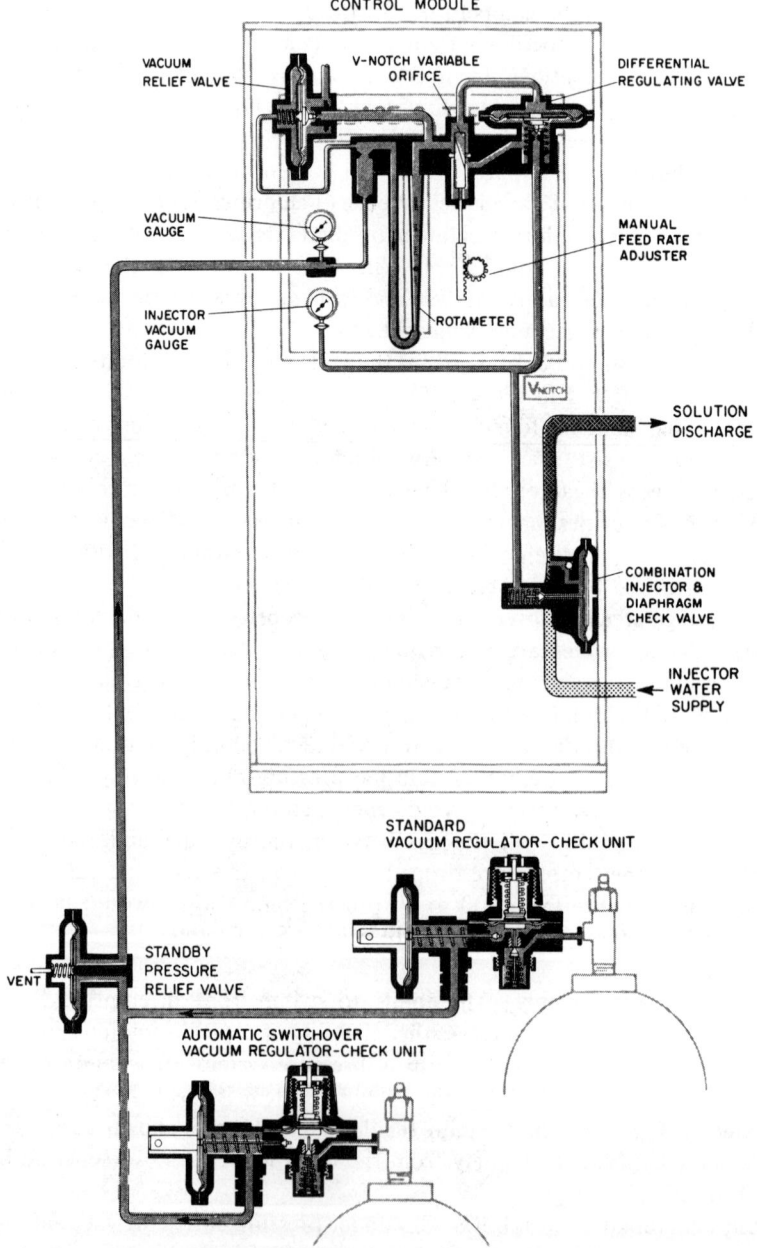

Fig. 9-3. Automatic cylinder switchover system—vacuum differential type (courtesy Wallace and Tiernan Div. Pennwalt Corp.)

CHLORINE FACILITIES DESIGN 591

check unit. As the gas flows through the two valves the cylinder pressure is reduced to the vacuum in the system created by the chlorinator injector. The valves will not open unless the minimum operating vacuum is produced (approx. 10 in. Hg). Flow of gas will be indicated by a depressed red indicator button. If the first valve allows gas to flow when a vacuum is not present, the second valve will remain closed and contain the pressure in the vacuum regulator check unit, which is designed to contain full container pressure. These vacuum regulator check units are fitted with mechanical latches called "detentes." One valve remains open until the cylinder or cylinders connected to this valve are exhausted. This situation allows the operating vacuum to rise significantly higher than normal ($\approx$3–5 in. Hg). This rise in vacuum level provides sufficient force to unlatch the second unit which then takes over the gas supply function.

For safety reasons a standby pressure relief valve must be installed as shown in Fig. 9-3. This valve must be vented to atmosphere well above ground level.

Positive pressure systems are described later in this chapter.

Ton Containers:

General Discussion. Unlike the 150-lb cylinders, either gas or liquid may be withdrawn from ton containers; consequently each container has two outlet valves. Also unlike the small cylinders, the ton containers have six fusible plugs—three in each of the dished heads. Fig. 9-4 illustrates the ton container. They are transported either by truck (Fig. 9-5) or by multiple-unit tank cars (TMU). A truck can carry a maximum of fourteen containers; a TMU, fifteen, as shown in Fig. 9-6. *The water volume of a ton container is 192 gallons.*

The gross weight of these containers (3500 lb) dictates that proper handling equipment must be used. The container is designed for use in the horizontal position. Each cylinder must be positioned so that the outlet valves line up in the vertical before being connected to the supply system. In other words, the container must

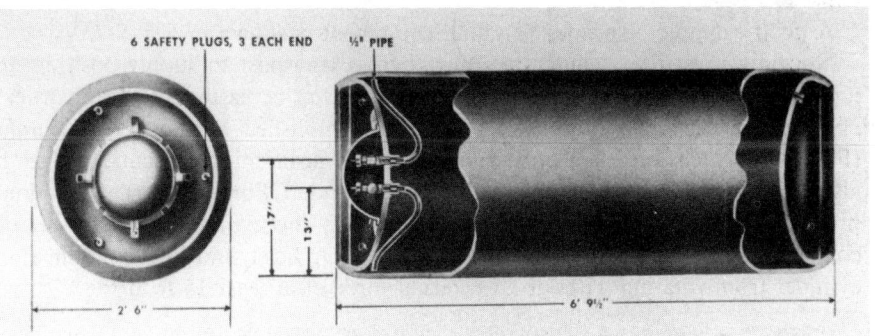

Fig. 9-4. Ton container for chlorine (courtesy Chlorine Institute).

592 HANDBOOK OF CHLORINATION

Fig. 9-5. Ton container truck and trailer rig with boom (courtesy PPG Industries, Chemical Division).

be positioned so that one eductor tube is in the gas position and the other in the liquid when the container is full.

Handling Equipment. Proper handling equipment includes the following:

1. Two-ton capacity electric hoist
2. Lifting bar
3. Cylinder trunnions
4. Monorail for hoist
5. Cinch straps*

A good example of proper handling equipment is shown in Fig. 9-7.

Not only must the cylinders be moved from transport to supply position; it is most important that each cylinder to be connected be easily rotated in order to align the outlet valves vertically. This is accomplished by a pair of trunnions (Fig. 9-8), which serve not only as a method of positioning the outlet valves but also for the spacing and support for each cylinder. Further, if the trunnion is properly designed, it should function to contain the cylinder in the event of a collision with an incoming cylinder on the traveling hoist, and to prevent an empty cylinder from rotating without an external force of at least 15 ft lb.**

---
* Where seismic forces are of concern, plastic cinch straps should be used to prevent the cylinder from becoming dislodged from the trunnions.[70]
** A recent earthquake with the epicenter at Livermore CA and a 5.8–6.2 Richter force failed to move banks of ton containers at two wastewater treatment plants nearby. Plant container alignment was 90° different from each other.

CHLORINE FACILITIES DESIGN 593

Fig. 9-6. TMU car (courtesy Chlorine Institute).

One of the critical design dimensions for ton container installations is the distance from the bottom of the monorail to the floor of the container room. The monorail must be high enough to pick up a cylinder off the truck and also high enough to lift one cylinder over another that is on the floor. Usually the governing distance

Fig. 9-7. Typical chlorine storage with trunnions for rotation and spacing of ton containers (courtesy Chlorine Specialties, Inc.).

Fig. 9-8. Trunnion for ton container (courtesy Chlorine Specialties Inc.)

is the height of the truck bed above the container floor. Fig. 9-9 illustrates how to arrive at the minimum distance of the monorail above the floor.

Gas Withdrawal. Space requirements depend upon whether the cylinders are for liquid or gas withdrawal, the rate of withdrawal, the quantity price break for the number of cylinders delivered at one time, and the length of time containers can be used without incurring demurrage charges.

At room temperature the maximum gas withdrawal rate from a ton cylinder is approximately 400 lb/day. If the maximum chlorinator capacity is 400 lb/day, then one cylinder in service will suffice. If it is a 500 lb/day unit, two ton cylinders must be in service simultaneously. Theoretically, gas withdrawal can be used up to any capacity if enough cylinders are connected to the supply header. Switching from gas withdrawal to liquid withdrawal utilizing an evaporator is necessary when a continuous rate of 1500 lb/day is reached. The one exception to this rule of thumb is in intermittent operating installations, as employed on cooling water circuits. At such installations it is customary practice to withdraw rates up to 1000 lb/day for thirty minutes if the temperature of the storage area never goes below 50°F. In warmer areas 1500 lb/day gas withdrawal is safe up to one hour from a single ton cylinder. This assumes a temperature-pressure restoration period of no gas withdrawal of at least twice the length of time as the withdrawal period.

For up to 1500 lb/day maximum continuous withdrawal, space should be provided for four cylinders in simultaneous service, four standby cylinders, and four empty spaces for the next delivery. Beyond this rate, an evaporator should be installed. However, there is one exception, as follows:

In hot climates where the "in shade" summer temperatures exceed 95 to 100°F on a consistent basis it is desirable to consider the use of an evaporator. Normally those climates experiencing summer temperatures of 100°F usually experience winter temperatures of 15°F. Therefore, the evaporator concept takes care of both extremes of climatic temperature. Liquid withdrawal of chlorine to an evaporator from a supply system is least affected by ambient temperature. The only precaution is to prevent direct sunlight on the cylinders. For winter operation no special

CHLORINE FACILITIES DESIGN 595

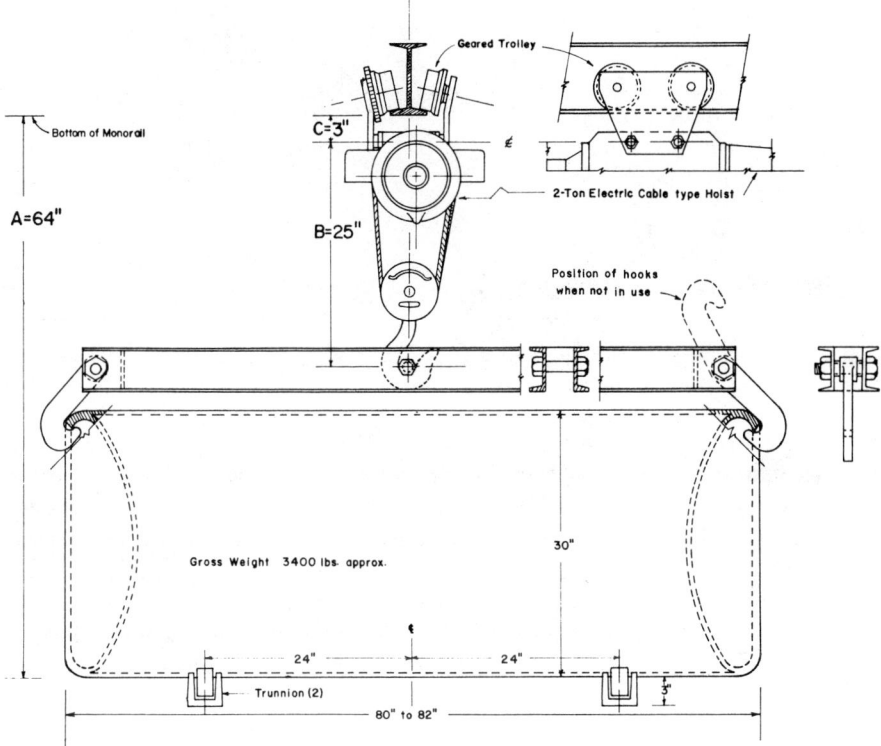

Fig. 9-9. Monorail height location. To determine monorail height above grade add truck bed height to distance "A" as illustrated. Check dimensions "B" and "C" with hoist manufacturer (courtesy Chlorine Specialties Inc.)

precautions are required (such as artificial heating) because the gas temperature at the outlet of the evaporator will usually *exceed* 100°F regardless of room temperature and the temperature of the gas in the vacuum line to the injector will never fall below the critical temperature (35°F) where chlorine hydrate occurs.

Evaporators are available in capacities of 4000 lb/day, 6000 lb/day, and 8000 lb/day. In a pinch, one cylinder can discharge liquid to an evaporator at a rate as high as 12,000 lb/day. This means that an evaporator can be used to conserve space for cylinder storage if necessary. The optimum storage requirements should be based on the quantity discount price break that is offered by the local chlorine supplier. This usually occurs at a quantity of five, thus dictating space for five in service, for five empties, and a vacant space for the incoming five, or a total space for fifteen ton cylinders.

If the gas phase is used, the same consideration must be given to the design of ton cylinder storage space as for 150-lb cylinders. All ton cylinder installations using gas withdrawal should be equipped with a special filter installed as close

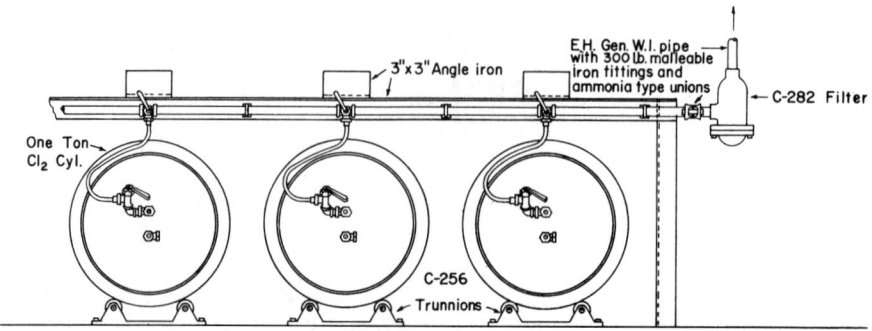

Fig. 9-10. Gas withdrawal system for ton containers.

as possible to the last ton container (Fig. 9-10). If the header between the last cylinder and the chlorinator is subject to temperature variations of 20°F or more over a twenty-four-hour period, then it is desirable to install an external pressure-reducing valve. This valve prevents liquefaction of the chlorine in the header system and the chlorinator mechanism caused by wide variations of the ambient temperature. This valve should be installed immediately downstream from the filter shown in Fig. 9-10. This filter is a combination sedimentation trap and filter. The filter medium is spun fiber glass, which is held in place by a stainless steel insert and screen assembly.

Liquid Withdrawals. The chlorine header system for liquid withdrawal is somewhat different from that for gas withdrawal. The piping and support system are the same, except that the flexible connections to the auxiliary header valve are connected to the bottom cylinder outlet valve (the top valve is for gas withdrawal). Because of the evaporator in the system, the filter is located immediately downstream of the evaporator outlet. The filter must always be installed in the gas phase. It is not possible to filter out chlorine impurities in the liquid phase, because the impurities are in solution. This places the filter just upstream from the chlorine pressure-reducing valve, which now becomes an automatic shut-off valve as well.*

A complete liquid withdrawal system is illustrated in Fig. 9-11 which also shows proper spacing of the evaporator-chlorinator system.

It is to be noted that some engineers prefer to use duplicate header systems from the containers to the evaporators. In this way one header can be taken out of service for the required periodic cleaning without interruption of the entire facility. Most systems rely on a single header; however, duplicate headers have advantages from an operator's viewpoint.

---

* This pressure-reducing and shut-off valve is always a part of the evaporator system. It is electrically interlocked with the evaporator water bath temperature, and automatically shuts off the chlorine supply in the event the water bath temperature falls below 150°F.

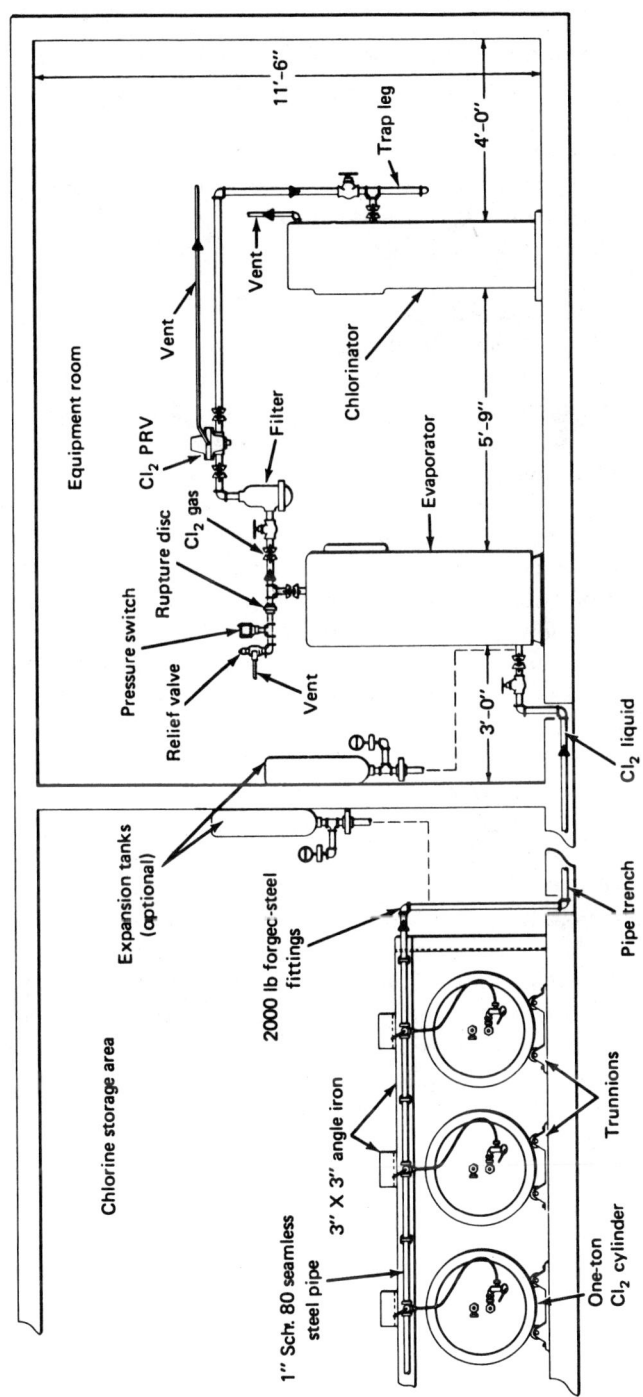

Fig. 9-11. Liquid withdrawal system showing chlorinator and evaporator spacing and ancillary equipment.

When contemplating the use of liquid withdrawal versus gas withdrawal, the following advantages of a liquid withdrawal system should be considered:

1. The danger of reliquefaction of chlorine between the containers and the chlorinator is all but eliminated.

2. Fewer cylinders need to be connected at one time. Liquid withdrawal rates of a ton cylinder can be as high as 10,000 lb/twenty-four hours.

3. The evaporators, although insulated, do give off some heat in the equipment room.

Liquid-withdrawal systems do not have the critical design problems with regard to temperature considerations as do the gas-withdrawal systems, except for inadvertent trapping of liquid in the header system which constitutes a temperature-pressure hazard. If liquid chlorine is trapped between two shut-off valves and the ambient temperature rises a few degrees, the liquid chlorine will try to expand. Since it cannot expand it will exert a pressure in accordance with the vapor pressure temperature curve shown in the Chlorine Institute Manual. For example, suppose a container connection full of liquid chlorine is trapped by yoke shut-off valves and at each end a ton container or a tank car (at an ambient temperature of 80°F) is removed, then the vapor pressure would be 100 psi. Now if the connection full of liquid were allowed to reach a temperature of 90°F by placing it in the sun, the hydrostatic pressure of the contained liquid would probably exceed the tensile strength of the already fatigued flexible connection resulting in a rupture of the flexible connection.

A note of caution about manifolding ton containers withdrawing liquid: always be sure that the temperature of the cylinders is about the same; never connect "hot" cylinders to the manifold simultaneously with cylinders already in use.

The cylinders can be placed in carport-type open structures with only a sunshield when the system is operating entirely on liquid withdrawal (Fig. 9-7). If the cylinder storage area is remote from the chlorine control system, it is desirable to install an expansion tank in the header system using a frangible disk and pressure alarm in series with the expansion tank. (See Fig. 9-12). This device is necessary to protect the system against the condition where an operator might close both the outlet valve on the header and the inlet valve to the evaporator, thereby trapping liquid in the header. Any subsequent ambient temperature rise would result in a pressure rise in the liquid chlorine sufficient to rupture the header piping. This is a direct result of the hydrostatic pressure due to the expansion of the liquid chlorine (or SO_2) which is greater than that of the steel pipe or copper flexible connection as a function of a rise in ambient temperature.

Since chlorine headers must be cleaned occasionally, it is desirable in most cases to install duplicate headers between the cylinder area and the chlorination equipment.

In either liquid or gas withdrawal systems using ton containers or tank cars, a sediment trap should always be provided where the gas line enters the chlorinator. Fig. 9-11 illustrates such a trap. This minor item saves much maintenance time

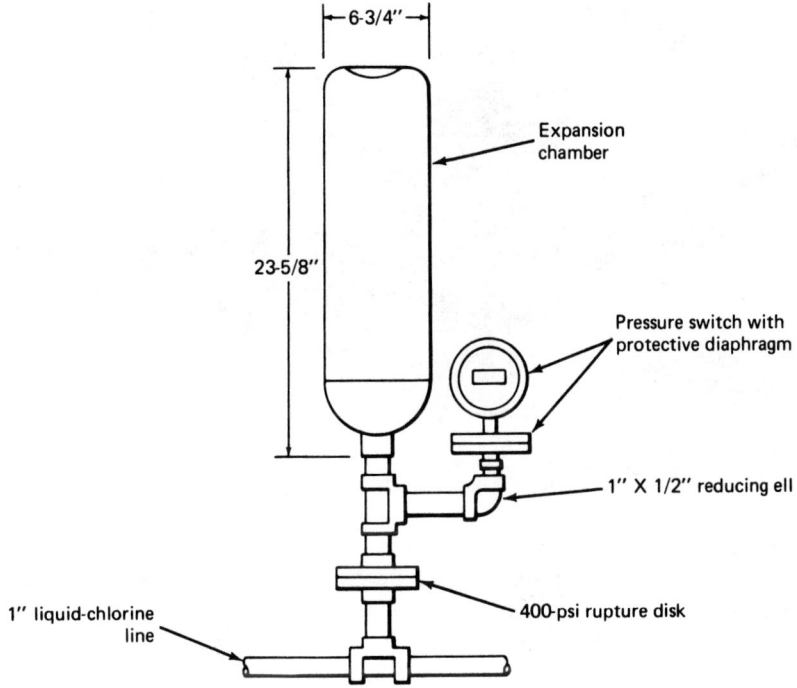

Fig. 9-12. Liquid chlorine expansion chamber. Connection is to liquid phase only. All fittings are 2000 lb CWP forged steel. All piping is seamless carbon steel Sch 80. Teflon tape should be used at all threaded joints.

on chlorination equipment. The chlorine gas line nearly always enters the chlorinator from the vicinity of the ceiling. The necessary downdrop to each chlorinator is a potential collection place for chlorine sludge caused by the inherent impurities in the chlorine. These deposits are the result of the reliquefaction phenomenon. A trap 12 inches long and capped will catch most of this debris and prevent it from entering the control mechanism of the chlorinator.

Platform Scales. These are available for weighing one or more cylinders, up to a maximum of five. They are available in dial indicating and/or dial indicating plus recording models.

Weighing Devices Using a Hydrostatic Load Cell with Remote Dial Readout. These are popular for installations using ton cylinders. Fig. 9-13 illustrates a load cell unit for weighing one ton cylinders. This unit features a remote digital readout at a much lower cost than beam- or pipe-lever-type scales. This type of scale is also available with electronic type load cells. These systems are capable of providing a variety of chlorine inventory information.

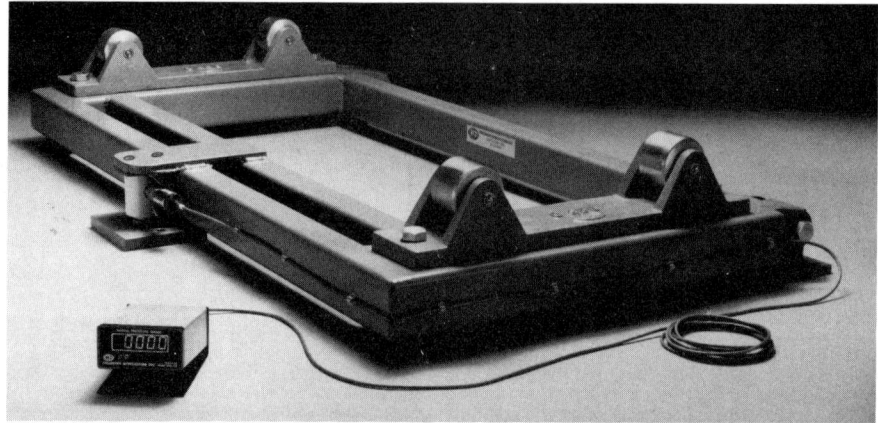

Fig. 9-13. Ton cylinder load cell type scale with digital readout (courtesy Chlorine Specialties, Inc).

External Chlorine Pressure Reducing Valve. Gas withdrawal systems which operate under pressure as opposed to the remote vacuum systems are subject to chlorine reliquefaction due to the ambient diurnal temperature changes. To prevent, reliquefaction, which is highly undesirable, a pressure reducing valve should be installed in the gas header. It should be located adjacent to the last cylinder in the header system. These valves are available in capacities of 500, 2000, and 10,000 lb/day.

Chlorine Gas Filter. Every installation using ton containers for gas withdrawal should have a chlorine gas filter as close as possible to the last cylinder, and always upstream from the external reducing valve or remote vacuum regulating valve.

Expansion Tanks. These tanks are necessary only when there is danger of liquid chlorine becoming trapped in the supply line. Then if there were a significant ambient temperature rise, hydrostatic pressure from the liquid chlorine trying to expand would rupture the pipe. One thing is certain, however: spring-loaded relief valves discharging to atmosphere should never be used on liquid chlorine lines. The chance of such a valve reseating properly is remote.

Figure 9-12 illustrates an expansion tank recommended by the Chlorine Institute. This tank utilizes a frangible disk that will rupture at some desired pressure between 300 and 400 psi, thereby allowing the liquid trapped in the system to enter the expansion tank.* This immediately produces some vapor pressure in the expansion tank, which actuates the pressure switch and sounds the alarm.

Systems employing long liquid lines and/or chlorine storage tanks usually have electrically operated valves at both ends of the lines. When there is a power failure

---

* If for any reason the header piping is subjected to a vacuum the rupture disk will be damaged. Therefore these disks must be specified for vacuum operation on the pressure side.

or a pressure drop, these valves close. An expansion tank is required on such lines—equal in volume to at least 20 percent of the line volume between the automatic valves. The chlorine pressure-reducing and shut-off valve on the discharge side of the evaporator will also close on power failure, trapping the liquid in the evaporator. This will cause an immediate pressure rise in the evaporator, since the liquid chlorine in the evaporator will tend to reach the temperature of the water in the surrounding water bath. The vapor pressure in the chlorine chamber of the evaporator will rise until there is equilibrium in the temperature gradient between the liquid chlorine and its surrounding water bath. This would be close to 160°F, which is equivalent to a vapor pressure of 315 psi.

This condition is exaggerated when two or more evaporators are manifolded together and operating at maximum capacity at the time of power failure. In such cases, the vapor pressure could reach 425 psi.

To alleviate this condition, one or more expansion tanks must be connected into the system on the liquid inlet line to each evaporator just downstream from the automatic electric shut-off valve on the liquid supply header. Fig. 9-11 illustrates the proper arrangement of an expansion tank to take care of any liquid which might get trapped in the evaporator; it also illustrates the conventional location of an expansion tank to handle any liquid which might get trapped in the supply header system.*

For installations where the supply system is close coupled to the evaporators, there is no appreciable danger from closing valves at the outlet of the supply system simultaneously with the closing of the inlet to the evaporator; therefore, expansion tanks are not required. It is also deemed necessary to provide pressure relief of the evaporator between the inlet shut-off valve and the downstream shut-off valve in some instances.

Operators should be thoroughly and properly instructed never to close valves in this system so as to trap the pressure in any part of the liquid portion. The basic rule is never to shut off the outlet of the supply container unless the system is being secured. In these instances, the pressure on the entire system is then allowed to go to zero gage pressure.

Chapter 11 outlines operating procedures for securing a liquid supply system so as to eliminate the necessity for relying on expansion tanks for those systems that do not use automatic electric shut-off valves on the liquid lines.

Gages. If the system is utilizing gas withdrawal and it is close coupled, gages in the chlorine gas header can be omitted, since the gage on the chlorinator will suffice. However, if this gage fails, it does not leave the operator with any means to determine whether or not the system is "live." An extra gage is always useful.

If the system utilizes an external pressure-reducing valve, there should be one gage upstream from this unit and one downstream.

For a liquid withdrawal system, the evaporator has a gage showing the system

---

* Fig. 9-11 also illustrates the 1975 mandatory evaporator relief system (see p. 637). The expansion chamber concept is the safest because it prevents a chlorine emission.

602 HANDBOOK OF CHLORINATION

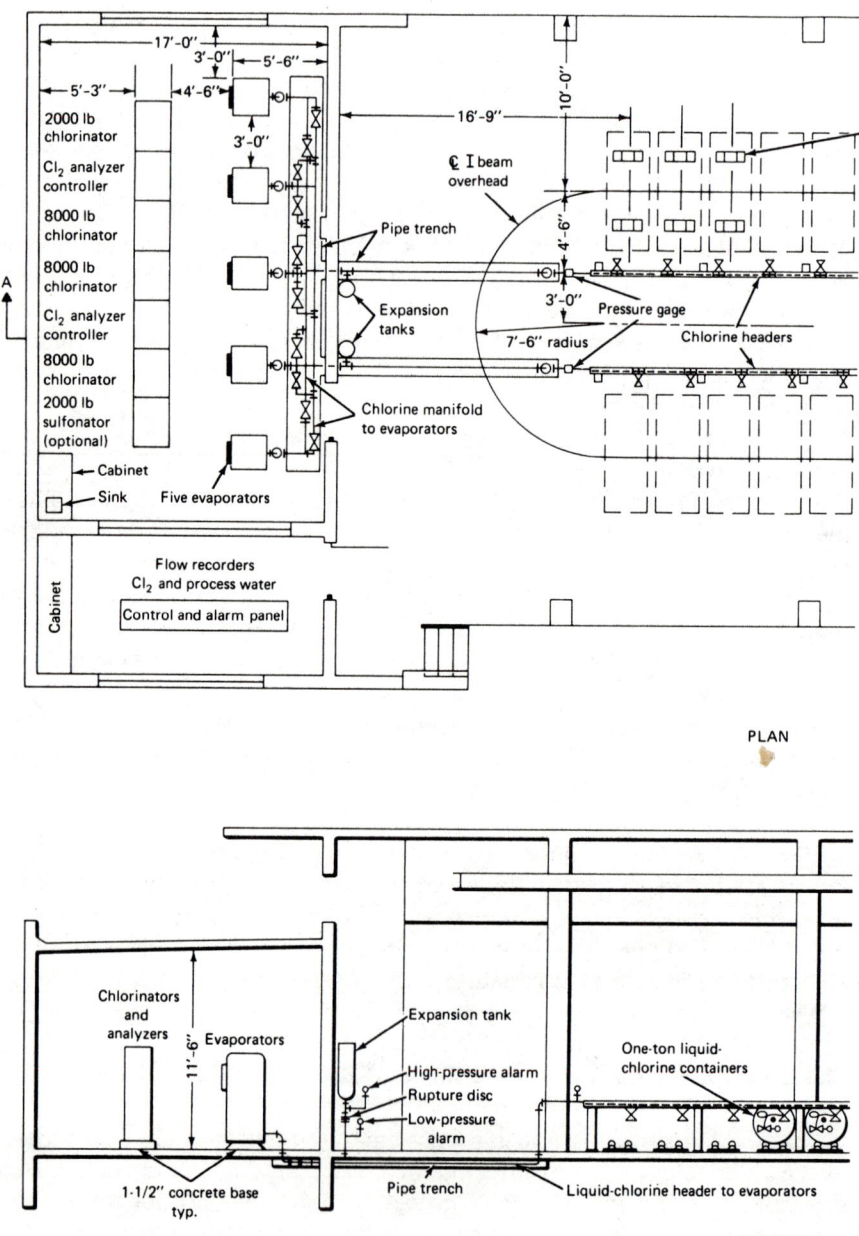

Fig. 9-14. Chlorine storage and handling facilities, equipment and space requirements for 24000 lb/day capacity.

CHLORINE FACILITIES DESIGN 603

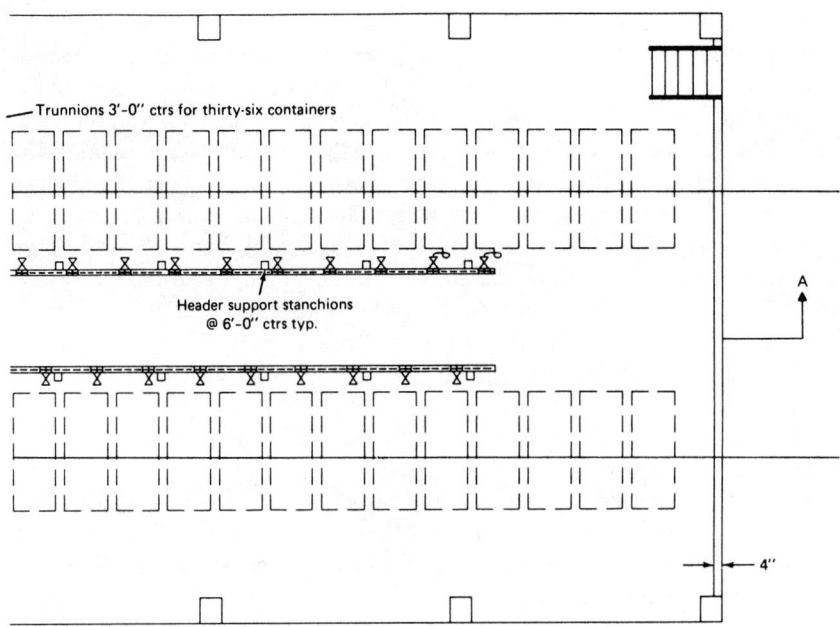

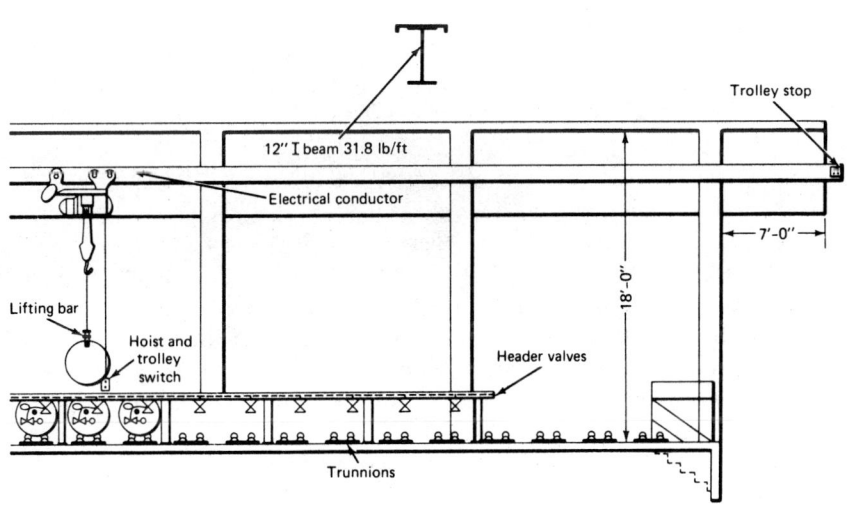

pressure. It is useful to have another gage on the liquid line between the containers and the inlet shut-off valve to the evaporator. A gage downstream from the pressure-reducing and shut-off valve should be considered, unless the system is a one-chlorinator installation that is close coupled. In the latter case, the chlorinator gage could be sufficient. However, when chlorinators and evaporators are manifolded together, it is always desirable to have gages downstream from each reducing valve. Similarly, when these systems are interconnected, it is also desirable to have gages upstream from each reducing valve. When more than one evaporator is involved, there should be a gage on the liquid chlorine inlet line to each evaporator upstream of the evaporator inlet shut-off valve and downstream from the evaporator isolating valve. These gages are extremely helpful to the operator, for safety reasons as well as for troubleshooting.

Alarms. Pressure switches that sound an alarm are helpful, and should be considered on most installations. Two kinds of pressure alarm devices are available. One is an adjustable mercoid type switch that has adjustable contacts which close for either rising or falling pressure—the two functions are not now available in the same housing. The other is a combination indicating gage and pressure switch. The U.S. Gauge Co. Model 3050 is adjustable to sound an alarm at both high and low pressures. Its contacts are adjustable over the entire range of the gage.

All liquid withdrawal systems should have a high-pressure alarm, located on the liquid supply header between the chlorine containers and the evaporators, which will be activated when the pressure reaches 150 psi.

Gas withdrawal systems do not need high-pressure alarms.

A low-pressure alarm is highly desirable. It warns the operator of imminent exhaustion of the chlorine supply. In some cases it can be used in lieu of weighing devices.

For example, consider that five ton containers are in service simultaneously in a chlorination system with a usage rate of 4000 lb/day. If the low-pressure alarm is set at 30 psi on the liquid supply line, there would be approximately 30 to 40 lb of chlorine in each cylinder and 30 to 40 lb in the evaporator and piping system, or a total of approximately 150 to 200 lb of chlorine remaining in the system at 30 psi. At 4000 lb/day, the usage is 167 lb/hour. Thus, when the alarm sounds, the operator has about one hour to prepare for putting another bank of cylinders in service. The situation is the same whether the withdrawal is liquid or gas.

Fig. 9-14, a typical ton container handling system for a chlorination facility with a maximum capacity of 24,000 lb/day, indicates space requirements and the arrangement of handling facilities.

Container Withdrawal Rates[2]:

General Discussion. The sustained maximum chlorine withdrawal rate will vary from plant to plant, depending primarily upon the temperature surrounding the

Table 9-1 Vapor Pressure vs. Container Liquid Temperature

Temperature, °F	Chlorine	Sulfur Dioxide, psi	Ammonia
0	14		16
10	21		24
20	28	2	34
30	37	7	45
40	47	12	59
50	59	18	75
60	71	26	93
70	86	34	114
80	102	45	138
90			
100			

container, characteristics of the chlorinator installed, and the size of the container.

Chlorinators require a minimum gas pressure for proper operation, again depending upon the type of chlorinator. Most vacuum type units require at least 10 psi but are more reliable at 15 psi inlet pressure. The gas pressure at the point of withdrawal is a function of the chlorine liquid temperature in the container. This relationship is shown in Table 9-1. The table is derived from graphs that can be found in the Appendix.

For a given chlorinator, the temperature at which the minimum required gas pressure is reached is termed the "threshold temperature." For example, Table 9-1 shows the pressure needed for a typical vacuum-operated chlorinator occurs at about 0°F (14 psi). Therefore 0°F is the threshold temperature for that type of chlorinator. The liquid chlorine in the cylinder or container must be kept at 0°F to maintain the gas pressure required for proper operation.

When gas is withdrawn from the container, the liquid in the container is chilled. To maintain the liquid at the threshold temperature, heat must pass from the air surrounding the cylinder through the walls of the container and into the liquid. Therefore the maximum sustained rate of withdrawal is directly dependent upon the rate at which this heat transfer takes place. Several factors influence the heat transfer rate, including ambient temperature, air circulation, humidity of the air, amount of liquid remaining in the container, and size and type of container. In general the important factors are: (1) ambient air temperature and (2) size and type of cylinder. Ambient air temperature is simply the air temperature surrounding the container.

Container Withdrawal Factor. The size and construction of the cylinder can be related to the ambient temperature with a number called a "withdrawal factor." Table 9-2 lists several withdrawal factors.

Table 9-2 Withdrawal Factors

Gas	Withdrawal Factor
Chlorine	
150 lb cylinder	1.0
1 ton container	8.0
Sulfur dioxide	
150-lb cylinder	0.75
1-ton container	6.0
Ammonia	
150-lb cylinder	0.4
250-lb cylinder	0.8 estimated
800-lb container	3.2 estimated

Sample Calculation. The equation used to calculate the maximum withdrawal rate from a container is as follows:

$$(\text{Temp. of Room} - \text{Threshold Temp.}) \times \text{Withdrawal Factor} = \text{Max. Withdrawal rate, lb/day}$$

Assume room temperature or ambient temperature surrounding the container is 70°F, the withdrawal temperature is 10°F. The withdrawal factor is 8 for a ton container. Then we have $(70 - 10) \times 8 = 480$ lb/day. The rule of thumb is often quoted as 400 lb/day, which is a more practical figure. The example shown in the calculation would cause excessive moisture condensation on the cylinder, resulting in copious quantities of water accumulating on the floor of the container area. Visible moisture in the chlorine or chlorinator areas should be *avoided*.

Automatic Switchover, Ton Containers:

General Discussion. Unattended chlorination stations must have special provisions to preclude the possibility of chlorine supply failure. Traditionally chlorine supply systems have been provided with scales for determining the amount of chlorine remaining in the connected cylinders. If this is the sole means of accounting for the chlorine supply, it requires frequent checking and operator judgment to determine when the cylinder will become empty. This diminishes facility reliability. The operator may find that he is faced with repeated interruptions in chlorination.

To solve the problem of chlorine supply reliability a great many stations have adopted the automatic switchover concept in lieu of scales. There are two types: (1) the pressure system and (2) the vacuum system.

The Pressure System. This system has been in use since about 1966. It is illustrated in Fig. 9-15. This arrangement of pressure reducing valves, gage, valves, and piping

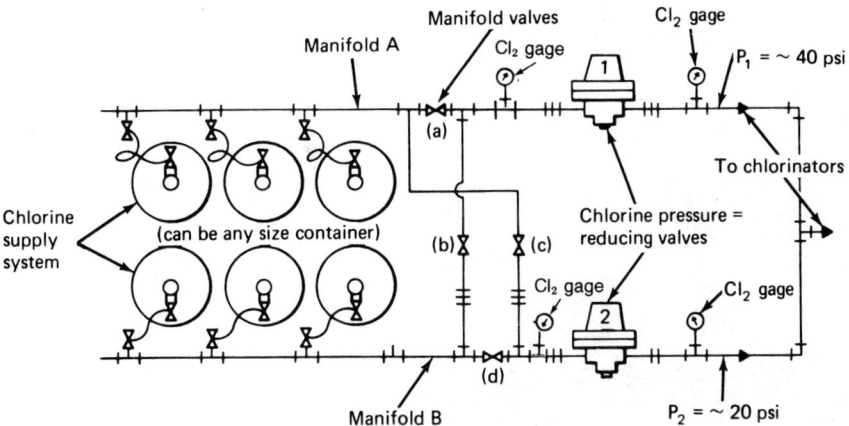

Fig. 9-15. Automatic chlorine cylinder switchover (gas phase only). By reversing manifold valves a,b,c, and d, chlorine can be routed from manifold A to CPRV #2 and manifold B to CPRV #1.

usually requires less capital outlay than required for weighing devices. Use of this system greatly improves the overall reliability of the chlorination facility.

The system consists of two independent manifolds, each equipped with a manually adjusted chlorine pressure-reducing valve (CPRV) and the necessary gages. CPRV-1 is set 10–20 lb higher than CPRV-2. Therefore, only gas will flow through regulator 1, and only from manifold A. When all the liquid chlorine in the cylinders connected to manifold A vaporizes, and the gas pressure drops to the set limits, CPRV-2 will open, allowing the full cylinders on manifold B to supply the chlorine. With the higher pressure setting on CPRV-1, gas from manifold B cannot flow back into the cylinders connected to manifold A. The lower pressure reading on each gage is evidence to the operator that chlorine is now coming from manifold B. After the empty cylinders connected to manifold A have been replaced by full ones, the manual valves ahead of the regualtors are reversed so that the gas from manifold B now passes through CPRV-1 (higher pressure), which will again cause automatic switchover when the cylinders are empty.

The Vacuum System. The Wallace and Tiernan version of this system utilizes the dual arrangement of the vacuum-regulator check unit used in the remote vacuum arrangement for ton containers.[3] The switchover system is illustrated in Fig. 9-16. For automatic switchover capability two vacuum regulators are fitted with mechanical detents. One regulator feeds gas until the container to which it is connected is nearly exhausted. The resulting rise of vacuum to higher than normal provides sufficient force to unlatch the detent in the second regulator which then takes over the gas supply function. The original supply continues to feed with the new supply, insuring exhaustion of gas from the original supply container.

608 HANDBOOK OF CHLORINATION

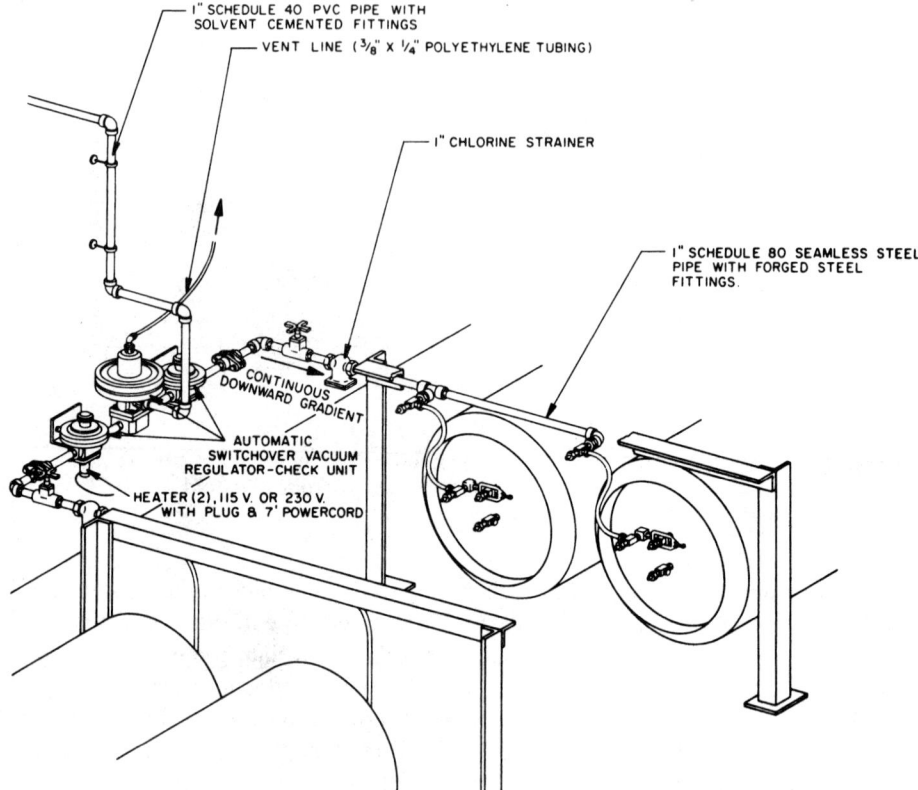

Fig. 9-16. Automatic chlorine cylinder switchover: Vacuum type, gas phase (courtesy Wallace and Tiernan Div. Pennwalt Corp.).

Referring to Fig. 9-16, it is imperative that the vacuum-regulator check unit be vented to outside atmosphere well above ground level where chlorine fumes cannot cause injury to personnel or damage to foliage.* Avoid areas routinely used by personnel, near windows, recirculated or ventilation air ducts.

This automatic switchover unit remains in the closed position and will not open until a vacuum is produced on the downstream side by the chlorinator injector. The entire unit is designed to withstand full container pressure. If the first valve passes gas without a vacuum, the second valve will remain closed and contain the cylinder pressure within the unit. In the extremely unlikely event that the

---

*When multiple chlorinators are involved serious consideration should be given to the use of a mini-absorption tank for these and other similar vents.

second valve passes gas (without a vacuum) the built in pressure relief valve will allow this gas to pass out the vent to atmosphere.

Liquid Chlorine. Automatic switchover for ton containers operating from the liquid phase is shown in Fig. 9-17. The pressure switch is in the common liquid line if multiple evaporators are in use. This arrangement can be used with some modification for tank cars. However, the preferred method for tank cars is the reserve tank.

When the pressure is on the liquid side of the evaporator it is presumed to be set at 20–25 psi. This will insure that all of liquid has been exhausted and only vapor remains in the supply system. The electrically operated valves in the supply line should open slowly enough to accommodate the near empty situation downstream from these shut-off valves.

Another method used is a sonic liquid chlorine level sensor as used in the reserve tank system. This system depends upon the operator to strip the exhausted supply system of any liquid and the remaining vapor. As shown in Fig. 9-18 this stripping

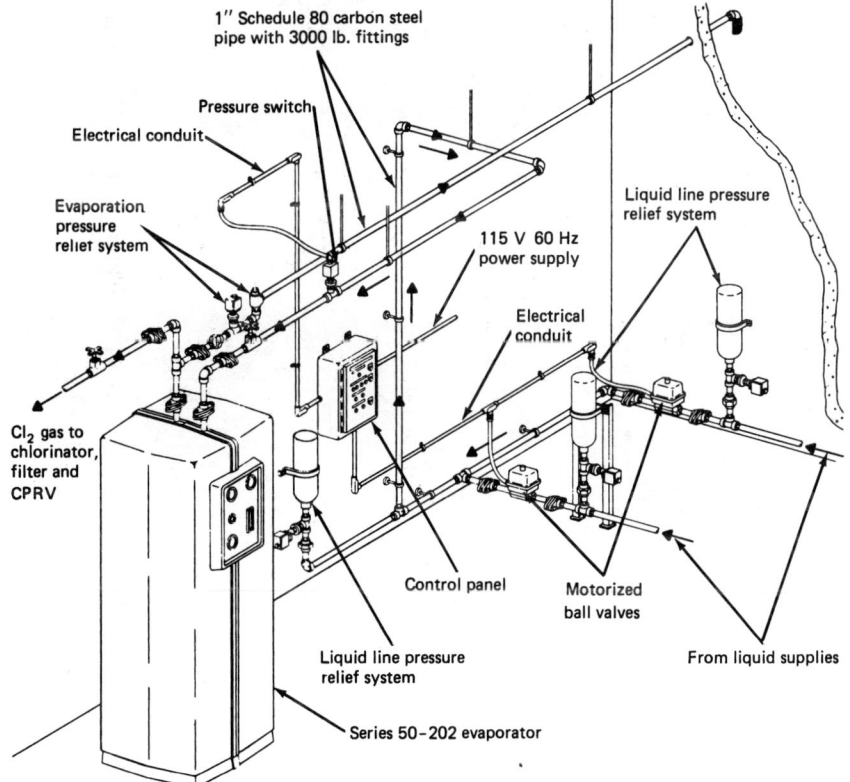

Fig. 9-17. Automatic chlorine supply switchover system for liquid withdrawal (courtesy Wallace and Tiernan Div. Pennwalt Corp.).

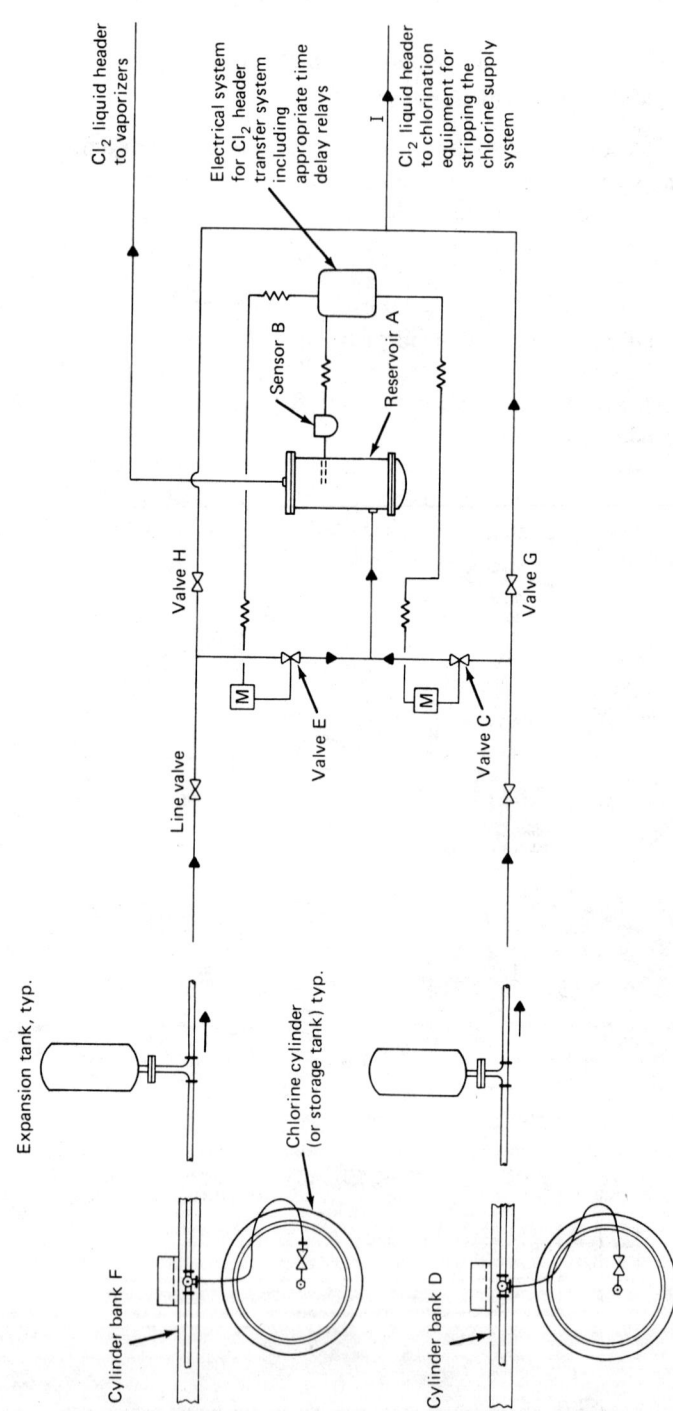

Fig. 9-18. Automatic liquid chlorine supply switchover system using a sonic level sensor.

line is a separate header which should be connected to the alternate top liquid connection on the evaporators.

This system has been used for many years in the pulp and paper industry.

Selection of Container Size. The first step in the design of a chlorination facility is to decide on the size of the chlorine containers to be used. This depends primarily on the average daily consumption of chlorine. It becomes a problem of economics and logistics, although a few mitigating circumstances might take precedence over these factors, such as length of haul and accessibility to the chlorine storage area.

For the purpose of economic evaluation, the following are the relative costs of chlorine in the various size containers. These prices fluctuate considerably, depending upon the area, but will usually remain in relative ratios.

Size Container	Cents/lb Cl_2
100- to 150-lb cylinder	15
Ton containers	7½
55 to 90 ton tank car	3¾
Storage tank	3½

Another cost factor to be considered is the demurrage charge on the containers. Depending upon the chlorine packager, this may be a flat charge for any container kept longer than thirty, sixty, or ninety days, or it may be reflected in a price break for ordering multiple quantities at one time plus a demurrage charge on each container kept longer than ninety days.

Other things being equal, and considering the practical aspects of an installation, 150-lb cylinders are the containers of choice when the average daily consumption of chlorine does not exceed fifty pounds. These cylinders are readily handled by one man using a special two-wheel hand truck.

When the daily consumption is in excess of fifty pounds, ton containers are to be considered. More space must be allotted and special handling equipment must be provided for these containers.

Tank cars should be considered when the average daily consumption of chlorine reaches two tons per day, which would require a weekly delivery of fourteen ton cylinders. This is about the maximum quantity which can be handled by truck rigs for one load.

The use of a storage tank may be considered for any installation using tank cars, particularly if the average daily consumption is four tons per day or more.

Designer's Check List for Chlorine Supply System: Ton Cylinders. The following is a guide to remind the designer to check the chlorine supply layout for all necessary accessories:

1. Cylinder weighing scales or load cells
2. Trunnions for ton containers
3. Ton cylinder lifting bar
4. Chlorine gas filter
5. External chlorine pressure-reducing valve
6. Liquid chlorine expansion tank
7. Appropriate gas and liquid supply pressure gages
8. High- and low-pressure indicating switches for alarms. (High-pressure alarm is used only on liquid systems.)
9. Condensate traps at inlet to chlorinators
10. Appropriate header valves and shut-off valves

Single Unit Tank Cars:

General Description. Chlorine is available in five sizes of single-unit cars: 16, 30,* 55, 85, and 90 tons. Each car is equipped with an outlet dome that is made identical for all tank cars approved by the Chlorine Institute. This dome contains two liquid outlet valves that are in line with the longitudinal axis of the car and two gas outlet valves on an axis at right angles to the liquid valves. In the very center of the dome is a safety relief valve that will expel gas to atmosphere under overpressure conditions. In each liquid outlet line, there is installed a safety check valve described as an excess flow valve. In case the car is in an accident and the valves are sheared off, the check valves will jam in a shut tight position due to the momentary velocity of the liquid exiting through the sheared off valve. This is why any tank-car withdrawal system should be equipped with a rotameter to see that the unloading rate is less than that which will cause the check valve to jam closed. This rate is usually 7000 lb/hr.**

Tank car dimensions are shown in Table 9-3.

Figure 9-19 illustrates the pertinent features of tank car construction. These cars are provided with a cork or a combination fiber glass and foam insulation from 4 to 6 inches thick, covered with an outer steel jacket, to minimize vaporization and pressure buildup in transit.

The only opening in the tank, at the center on top of the car, is closed by the dome plate, which is sealed by a lead gasket and secured with a ring of bolts.

The mechanism for chlorine withdrawal from a tank car is shown in Fig. 9-20. Four forged steel angle valves are mounted on the cover plate which is secured to the dome; a fifth is a safety relief valve. These valves have Monel seats and stems, and are protected by a false dome with a hinged cover. Openings around the false dome, protected by close-fitting covers, permit access for connection to the valves. The two angle valves on the longitudinal axis of the car are for liquid withdrawal. An eduction pipe extends from each liquid discharge valve to the

---
* The 16 and 30-ton cars are being phased out of service in most areas.
** Excess flow valve rates vary for car size.

CHLORINE FACILITIES DESIGN

Table 9-3 Dimensions of Tank Cars

	Length Over Strikers*	Overall Height†	Height to Valve Outlet†	Extreme Width‡
TMU	42' 4"–47' 0"	6' 8"–7' 6"	—	9' 6"–10' 1"
16-Ton	32' 2"–33' 3"	10' 5"–12' 0"	9' 3¼"–10' 0"	9' 2"–9' 6½"
30-Ton	33' 10"–35' 11½"	12' 4½"–13' 7"	11' 3"–11' 9"	9' 3"–9' 10"
55-Ton	29' 9"–43' 0"	14' 3"–15' 1"	12' 6"–13' 4"	9' 3"–10' 7½"
85-Ton	43' 7"–50' 0"	14' 11"–15' 1"	13' 2"–13' 4"	10' 5½"–10' 6½"
90-Ton	45' 8"–47' 2"	14' 11"–15' 1"	13' 2"–13' 4"	10' 5½"–10' 6½"

* Add 2' 6" for length over center line of coupler knuckles.
† Heights are for empty cars, and are measured from top of rail. Heights for loaded cars may be 4" less.
‡ Width over grab irons.
Note: Height to manway platform is 6 to 10" less than height to center line of valve.

bottom of the car. Just beneath each of these valves in the eduction pipe are excess flow valves, as shown in Fig. 9-21. These protective devices are designed to close automatically in case of a major break in the angle valve or in the chlorine unloading pipeline. These valves seat only when the clorine flow reaches approximately 7000 lb/hour. This flow rate shut-ff varies with car size.

For the layout of a single-unit car unloading site, the designer should consult the *Chlorine Institute Manual*[1] which quotes from the ICC regulations for chlorine unloading.

A dead-end siding should be provided for chlorine unloading. Tracks should be level. The car should be protected by a locked derail at least one car length away from the end of a car hooked up for unloading. If the car must be on an open siding, both ends should be protected. If a switch is involved, it should have a lock; the keys for the switch and the derail should only be in the hands of the person responsible for unloading.

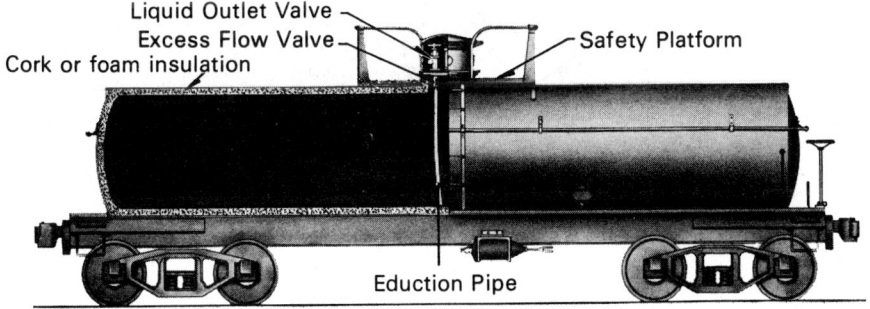

Fig. 9-19. Single unit chlorine tank car (courtesy PPG Industries, Chemical Division).

614 HANDBOOK OF CHLORINATION

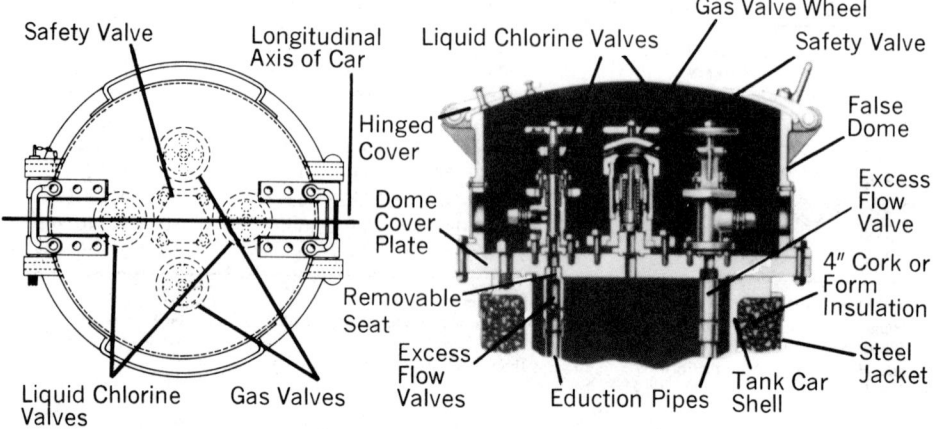

Fig. 9-20. Chlorine tank car dome (courtesy PPG Industries, Chemical Division).

An operating platform must be provided at the unloading point to provide easy access to the dome for connecting and disconnecting the loading lines and for operation of the valves. This platform should be of such height that it is suitable for working on cars of any of the sizes listed in Table 9-3.

Flexible Connections. Two types of flexible connections for use between the car dome and the piping system are available: the annealed copper loop, and the flexible

Fig. 9-21. Tank car excess flow valve (courtesy PPG Industries, Chemical Division).

reinforced metal hose. The connections are fitted with 2-bolt ammonia-type unions with a lead gasket joint. It is customary and desirable to connect a pressure gage to one of the gas valves on the car dome. An L-shaped pipe assembly is used to permit mounting the gage where it can be seen from the unloading platform.

One flexible connection is used to join one of the liquid withdrawal valves to the liquid header pipping, and the other is used to connect one of the gas withdrawal valves to the air padding system. This gas connection can also be used to connect the gas withdrawal valve to the gas header piping.

Flexible connections should be plugged at both ends with corks immediately upon removal from the tank car and stored in a heated compartment. This minimizes corrosion and extends the active life of the flexible connections.

Header System. Two completely separate headers are required between the tank car (or storage tank) and the evaporators: one for liquid and one for gas. It is at times necessary to withdraw gas from the supply system. In the event of a leak in the tank car discharge system or header piping the ability to withdraw gas can be used to reduce the tank car pressure. Accordingly, a field test was made with a 55-ton car approximately half full at the beginning of the test.[4] Three 8000-lb chlorinators operating simultaneously were used for this test. Withdrawal of approximately 2030 lb of chlorine reduced the car pressure from 98 psi to 37 psi in 3¾ hours. Ambient temperature was 55°F. The pressure drop caused a liquid chlorine cooling effect of 47°F. This represents 43 lb liquid chlorine per degree F drop. However, from 37 psi to 20 psi required only 30 min and 208 lb of chlorine to produce and additional cooling of 24°F, or 9 lb chlorine per degree F.

The presence of two header systems is also desirable for standby service. Tank car consumers using a duplicate header system have reported frequent use of the second or standby header for routine maintenance. Each of these headers should be equipped with a pressure gage immediately downstream from the shut-off valve on the unloading platform.

Air Padding System. This is required for unloading the tank car and for pressurizing cars in cold climates. Chlorine is usually shipped cold. Assuming a median temperature of 40°F, the vapor pressure will be 45 psig. Use only clean dry air free of oil and foreign matter. Air must be dried to at least a −40°F dewpoint measured at atmospheric pressure.

The air padding system should incorporate the following elements.

1. Oil-free air compressor with after-cooler and receiver
2. Air dryer
3. Air filter (oil absorber type)
4. Dewpoint indicator
5. Dewpoint alarm
6. Vapor pressure alarm system

Table 9-4. Tank Car Capacity

	55-ton Car	90-ton Car
Water capacity, gal	10564	17298
Total car volume, ft^3	1410	2335
Liquid Cl$_2$ volume 60°F, ft^3	1240	2035
Gas volume space, ft^3	170	300

Table 9-4 itemizes the volume characteristics of a 55 ton and 90 ton car.

Table 9-5 gives the amount of padding air required to pressurize a 55-ton and a 90-ton car to various pressures.

The air flow required to unload a tank car at a given rate can be calculated by the following formula:

$$\text{SCFM Air} = \frac{\text{lb/day Chlorine} \times (P_{car} - 25)}{1.8 \times 10^6} \qquad (9\text{-}1)$$

Selection of an oil-free air compressor is the objective of a reliable air padding system. An example is the Corken Pump Co. of Oklahoma City, OK. Any selected compressor should be supplemented with a Deltech filter for compressed air. The compressor should be equipped with an ASME code receiver of appropriate size with an air-cooled after-cooler mounted on the receiver.

Suitable air dryers are available from Deltech Engineering Co., Century Park, Newcastle, DE 19720 and Lectrodryer, Box 4599, Pittsburgh, PA 15205. The air dryer can be the heat reactivated type capable of operating up to 150 psi. The air flow capacity of the dryer should be 1.5 times the compressor capacity discharging at a maximum dewpoint of −40°F. This part of the system should be equipped with a dewpoint indicator and a high humidity alarm.

A critical part of the air padding system is the check valve which prevents chlorine vapor from entering the system. Installations using ordinary ball check valves have invariably suffered from chlorine vapor getting into the air pad system. Spring-loaded check valves have proved to be somewhat more reliable, but they too have caused corrosion problems.[5]

Table 9-5. Dry Air Required to Pressurize Tank Car*

Car Size	Volume of Free Air (ft^3) Required to Achieve Car Pressure Shown (PSI)					
	60	75	90	105	120	135
55-ton car	170	340	510	680	850	1020
90-ton car	300	600	900	1200	1500	1800

* Assume median Liquid Cl$_2$ temp. = 40°F, and car pressure 45 psig.

* This is the only instance where a check valve is optional in the chlorine supply system. Check valves at other locations (flex connections, header piping etc.) should never be used. In a chlorine atmosphere they are unreliable due to corrosion products. Moreover they would be hazardous to operating personnel.

The Chlorine Institute Pamphlet No. 4, "Consumer Air Padding of Chlorine Single Unit Tank Cars," illustrates a power operated shut-off valve instead of a check valve. This is the preferred method. Fig. 9-22 illustrates a power operated shut-off valve system used for many years by the Pennwalt Corporation, Tacoma, WA.[6]

Each air padding system should have its own separate and independent air supply. When both chlorine and sulfur dioxide are present the air padding system must never be used interchangeably. Each liquid–gas supply must have its own air-padding system.

Whenever air padding is deemed unnecessary, then nitrogen purging facilities should be considered mandatory. This consists of a sufficient number of nitrogen cylinders which contain nitrogen gas at 2000 psi and a common pressure regulator with an outlet pressure regulated at 150 psi. This system should be arranged to allow the operator to purge the entire header system back into the containers. This system is most advantageous in the handling of liquid chlorine.

Absorption Tanks. In addition to purging and air padding systems, tank-car systems should be provided with a liquid chlorine absorption tank. This tank should be able to absorb all the liquid chlorine in the header piping and evaporators. The absorption tank should be made of reinforced Fiberglas or rubber-lined steel. It should have a vent, a special connection for caustic, and one for makeup water. There should also be a sampling tap to determine the effective absorption capacity of the solution. The piping of the liquid chlorine to the tank must have a barometric loop (see reserve tank system). The downdrop of this loop and the sparger (diffuser) should be made of Kynar pipe. The size of the absorption tank is based upon the stoichiometric combination of caustic (NaOH) and chlorine which is 1.13 lb of caustic for 1.0 lb chlorine. The absorption tank must have a minimum depth of 8 ft and the caustic solution is usually kept between 0.5 to 1.0 lb/gal.

EXAMPLE. 1000 lb chlorine requires 1130 lb caustic for neutralization. One gallon of 50 percent caustic contains 6.38 lb caustic. Therefore 1130/6.38 = 177 gal of 50 percent caustic is required for each charge of the absorption tank to neutralize 1000 lb of liquid chlorine. So to provide a tank having about 1.0 lb/gal caustic, the absorption tank should have a capacity of approximately 1200 gal.

A PLAN TO DEAL WITH A MAJOR CHLORINE LEAK

Definition of a Major Leak. How to deal successfully with a major leak is a formidable task. The two most discussed events are: a direct hit by an aircraft and or a planned act of sabotage. The latter is usually dismissed on the basis that proper security measures can provide the necessary deterrence. The air crash scenario is usually dismissed as an improbability. However an air crash accompanied

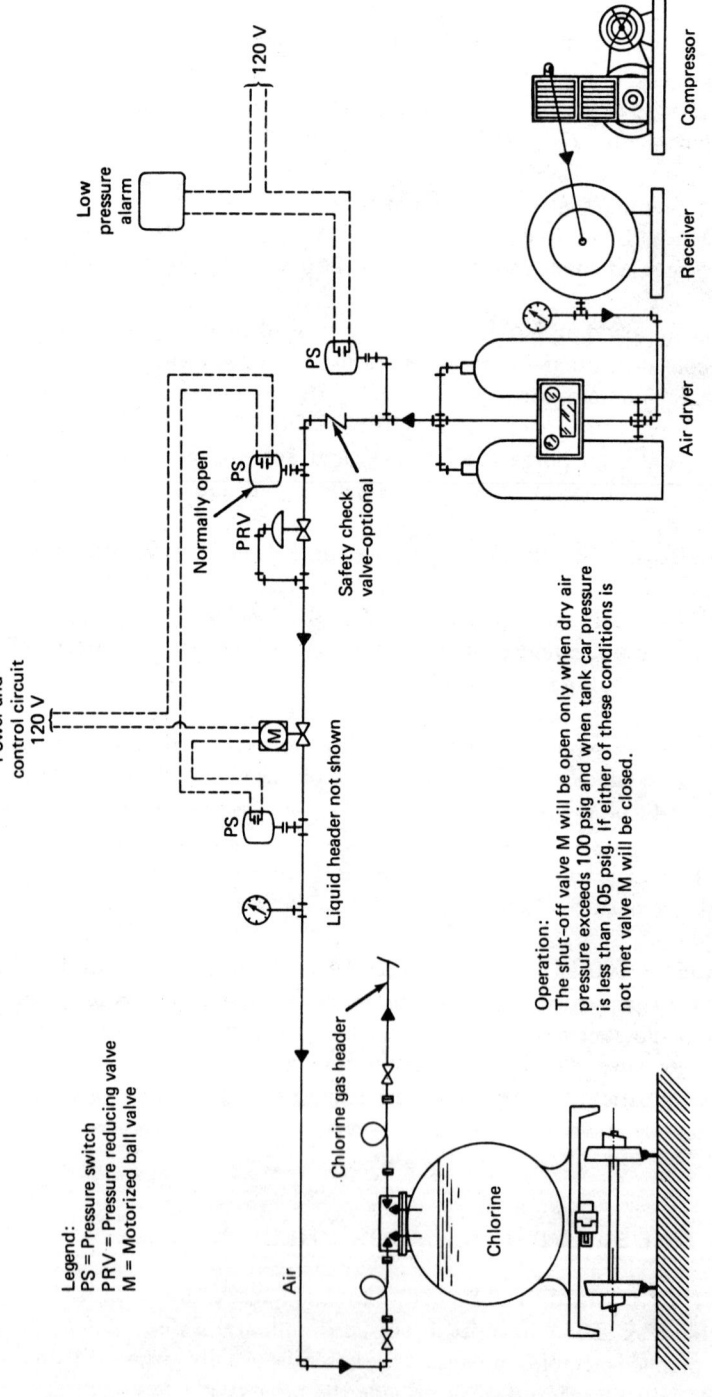

Fig. 9-22. Power operated shut-off valve, air padding system.

by exploding fuel results in the following consequences. The aircraft impact or subsequent explosion would probably rupture the chlorine container(s). The ensuing fire would instantly vaporize the liquid chlorine and the chlorine vapor would rise quickly with the heat of the fire. This sequence of events would serve to greatly diminish or eliminate chlorine exposure in the surrounding area. This was clearly demonstrated when a freight train derailment severely damaged a 90-ton chlorine tank car. The car was ruptured by the couplers from an adjoining butane tanker. All 90 tons of chlorine were released. The butane tank car exploded and all of its contents were consumed by fire. The heat from this fire vaporized the liquid chlorine which disappeared into the upper atmosphere due to the rising hot air from the butane fire. A subsequent investigation revealed that no one in the surrounding area or at the scene of the accident was found who had experienced any exposure to chlorine (See Ch. 1).

The consensus definition of the most probable major leak is a guillotine break in the liquid chlorine header between the chlorine supply system and the chlorine evaporators. The only safe and practical way to deal with this kind of a leak is to contain it in the room and neutralize released chlorine on the spot.

Containment with Controlled Release to Outside Air. It is assumed that the liquid chlorine header will break in the chlorinator-evaporator room and the spill will amount to no more than 2000 lb. of liquid chlorine. This would comprise the contents of all the header piping and evaporators.*

To dispose of or neutralize this amount of chlorine safely, the room where the spill occurs must be able to contain the chlorine vapor. The normal ventilation system must automatically close as soon as a leak is detected. This will prevent any chlorine release to the outside air.

In a containment situation where the room is filled with chlorine it should be noted that a Chemtrec Emergency Response team will not enter any confined area to secure a leak until the chlorine concentration has diminished to a level compatible with breathing apparatus capability.

Therefore to neutralize a major leak it is imperative to install a chlorine neutralizing scrubbing system described below.

Leak Scenario. The 1 in liquid chlorine header to a multiple evaporator installation breaks and releases 2000 lb of liquid chlorine. Ambient temperature is 80°F and liquid line pressure is 90 psi at the time of the break. This will put two one inch orifices into action; one from the supply side and the other from the evaporators (back flow). Under the conditions described liquid chlorine will flow at approx. 0.43 ft.^{3}/sec from each orifice. This amounts to $0.43 \times 2 \times 88$ lb/ft^3 = 75.68 lb/sec. However a liquid chlorine leak is not that simple. At the moment of pipe rupture about 20 percent of the 2000 lb. of liquid will flash to vapor (400 lb)

---

* Assume each evaporator will spill 125 lb. of chlorine. This would be a worst case operating condition.

and the room temperature will drop about 30°F or 50°F. This event will consume about 5–6 seconds leaving 1600 lb (approx. 18 ft^3) which will spill on the floor in the next 20–22 seconds.

If the floor space between the 40 ft. long wall and the evaporators has a grated trough (gutter) and if the floor is sloped properly the remaining liquid can be captured in a confined space. This arrangement will diminish the vaporization rate of the remaining liquid chlorine. The liquid chlorine collection trough should be about 1 ft. deep and 6 inches wide to minimize surface area. According to Howerton[81] the chlorine in the trough will vaporize at a rate of 15 to 20 lb/hr/ft^2. The above dimensions provide a 20 ft^2 surface area so the remaining liquid chlorine will require about 4–5 hours to vaporize. Vaporization will be intermittent because of the cyclic action of the chlorine going from stages of freezing to thawing. This vaporization is dependent upon how long it takes for the ambient temperature to be restored in the chlorinator-evaporator room. Therefore the total time involved to neutralize the leak described above may take up to 5 hours.

Scrubber System:

System Description. The venturi-ejector fume scrubber is the method of choice for a major leak. This is the method used by modern-day chlorine/caustic plants to neutralize chlorine releases during plant upset conditions.[82] A typical system is shown in Fig. 9-23.

Typically a scrubber is designed to recirculate the room air until all of the chlorine spill has been absorbed by the continuous passage of room air through a caustic* solution. The air recirculating rate should be sufficient to provide one complete room air turnover every two minutes. The scrubber system depends upon a chlorine detector to close the normal ventilation system and to activate the scrubber blower. The room air is then diverted to the scrubber suction where it travels to the inlet of the venturi throat. Simultaneously with the blower activation the caustic recirculating pump starts and delivers caustic solution to the inlet of the venturi. The room air is mixed with the caustic solution in the throat of the injector. This is similar to a chlorinator injector operation. The air flow travels through the entire vertical distance of the caustic tank and discharges to the mist eliminator. The scrubbed air is then returned to the chlorinator-evaporator room.

Theory of Operation. The caustic tank is sized to hold a sufficient quantity of 7 percent sodium hydroxide to convert the chlorine gas leak to sodium hypochlorite. The rule of thumb for absorption towers is 1.0 lb/gal of caustic. The stoichiometric combination of NaOH and chlorine is 1.13 lb caustic per lb of chlorine. Therefore a 2000 lb chlorine leak would require a theoretical 2260 lb of caustic. This translates to a 2260 gallon caustic tank. However the scrubber manufacturers recommend a 3600 gallon caustic tank for a 2000 lb spill.[83] This factor of safety is necessary to allow for dead spaces in the tank since mechanical mixers are not practical.

---

* Sulfite should not be considered as a substitute for caustic. Molecular chlorine will depress the pH of the solution which will release a cloud of SO_2 vapor.

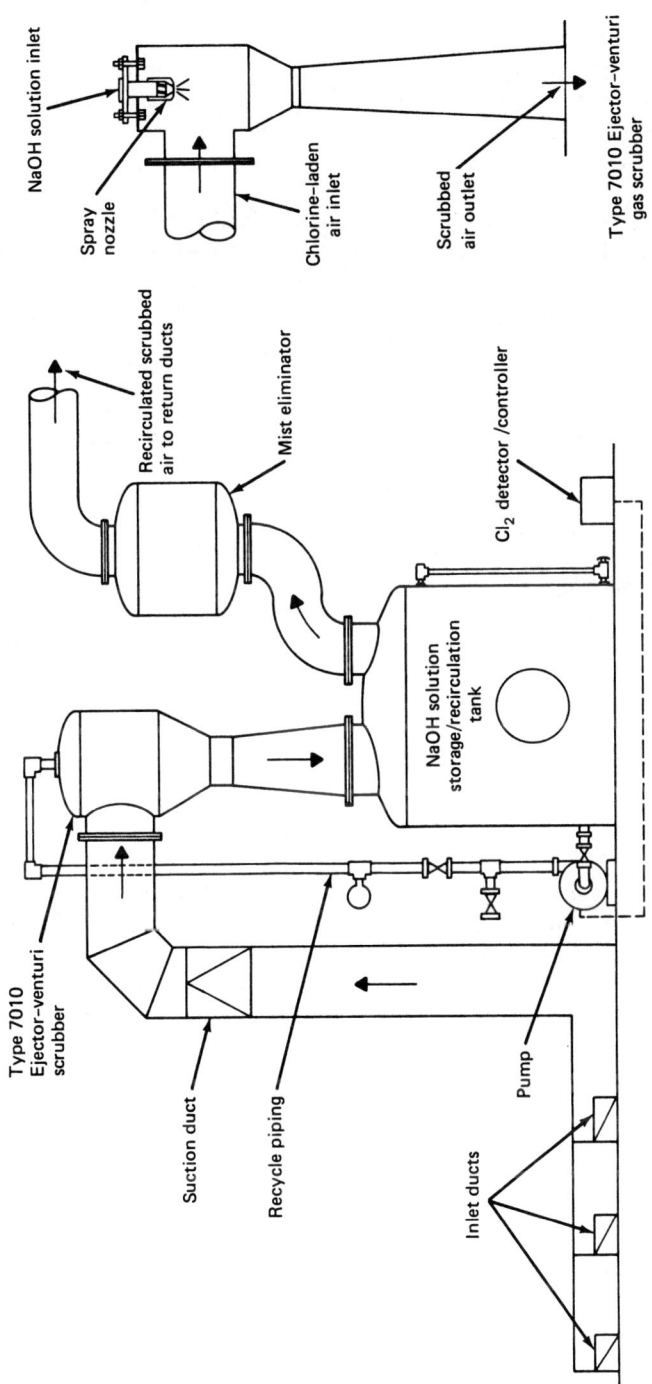

Fig. 9-23. Chlorine scrubber system (courtesy Ametek-Schutte and Koerting Div., Durham, N.C.)

The scrubber operates until the chlorine concentration in the room air is reduced to 1 ppm. Either or both the room air and caustic solution can be monitored during the scrubbing period. Instrumentation is included with the system to provide shutoff of the scrubber and reactivation of the normal room ventilation system.

Reliability. This system has a long history of reliability because it is not dependent upon the solubility of chlorine in water due to the caustic recirculating system nor is it dependent upon scrubbing water as in wet scrubbers. If the leak is greater than expected additional caustic can be added to displace the caustic converted to hypochlorite.

Cost. The equipment required for a 2000 lb. leak as described above is approximately $35,000. For a lesser leak; 1000 lb. the capital cost would be about $25,000.

Spent Caustic Disposal. The spent caustic will be a sodium hypochlorite solution about 7000 mg/L. This is easily disposed of at a water or wastewater plant provided it can be metered in small quantities over a given period of time. The hypochlorite can be easily destroyed by catalytic decomposition using nickel and iron as catalysts.* The use of sulfites to dechlorinate the hypochlorite solution is not recommended. The heat of reaction between the sulfite ion and hypochlorite is far too great at these concentrations.

Chlorine Tank Car Supply Monitoring Systems

General Discussion. Single-unit tank cars pose a problem in determining when exhaustion of the liquid chlorine is imminent. Reliance on pressure drop is not the total answer, and where air padding is used on a continuous basis (cold climate), temperature drop monitoring is not reliable. Track scales are the only direct means, but these constitute an exorbitant expense, about $50,000 to $75,000. Scales for storage tanks are much less expensive (about $5,000).

A convenient method for solving the supply monitoring problem would be to meter the flow of liquid chlorine exiting the car. Unless the car is padded to a pressure of 175–200 psi it is not possible to measure with any accuracy the flow of liquid chlorine due to its physical characteristics. As the liquid flows past a point of slight pressure drop, as in a flow meter, the liquid flashes to vapor causing a large change in density. This makes calibration of a flow meter practically impossible. *A magnetic flow meter will not respond to liquid chlorine.*

Tank cars are well insulated so that they maintain vapor pressure equilibrium until all of the liquid has been discharged.

EXAMPLE. A 90-ton car displays a vapor pressure of 70 psi. When all the liquid is gone there will remain in the car approx. 2500 lb of chlorine gas provided

---
* If available seawater will destroy the hypochlorite solution in a few hours owing to the presence of heavy metal ions.

the car has not been air padded during use. This is a significant amount of chlorine that should be used by the consumer.

Chlorine Pressure Switch. A chlorine pressure switch installed in the liquid header immediately upstream from the evaporators can be used for monitoring or automatic switchover service. A pressure switch set to alarm at 20 psi will advise the operator that approximately 1050 lb of chlorine is left in the car. This may be sufficient time to switch to a new car. The other alternative would be to set the pressure switch to 20 psi and use the alarm circuit to activate the motorized valves in an automatic switchover system. The time between the occurrences described above is totally dependent upon withdrawal rate and to some extent upon ambient temperature. Therefore a reliable and accurate pressure switch is essential for liquid chlorine supply monitoring. The US Gauge Co. Model 3050 is a combination indicating gage and pressure switch with adjustable contacts for both high and low pressures over the entire range of the gage, 0–160 psi.

Three Detection Methods for Liquid Chlorine Exhaustion. The liquid chlorine exhaustion point has been detected in three ways: (1) A cylindrical vessel has been used which is mounted vertically and equipped with a liquid level sensor which alarms when the level falls some predetermined distance in the cylinder. (2) All of the chlorinators have been equipped with chlorine flow recorders and a totalizer that will alarm at 500 or 1000 lb chlorine remaining. (3) The liquid chlorine temperature has been continuously monitored adjacent to the point where the flexible connection to the tank car joins the liquid header. As the liquid chlorine approaches the point of exhaustion it attempts to vaporize as it exits the tank car. This premature vaporization causes a sudden drop in the temperature of the liquid as it enters the header piping at the unloading platform. A temperature sensor equipped with an alarm will notify the operator of imminent liquid exhaustion. The reliability of these methods is dependent upon the withdrawal rates of liquid chlorine at the moment of exhaustion.

Reserve Tank for Liquid Chlorine Supply. A system used for more than half a century in the pulp and paper industry is the reserve tank concept. This system was developed by the late Brian Shera of Pennwalt Corp., formerly Pennsylvania Salt Co., Tacoma, WA.

This is a liquid chlorine flow-through tank, as shown in Fig. 9-24 which provides an active reserve when the car has been emptied of liquid. The tank may also be installed vertically if preferred. These tanks have been made in sizes from 1000 to 16,000 lb of chlorine. The capacity should provide approximately 1–3 hr reserve of chlorine at peak demand. Liquid chlorine enters and leaves the tank through pipes extending to the bottom of the tank. Venting of gas is necessary to provide a minimum of 20 percent gas volume as a safety factor against excessive filling of the vessel with liquid. This is accomplished by a dip pipe of an appropriate

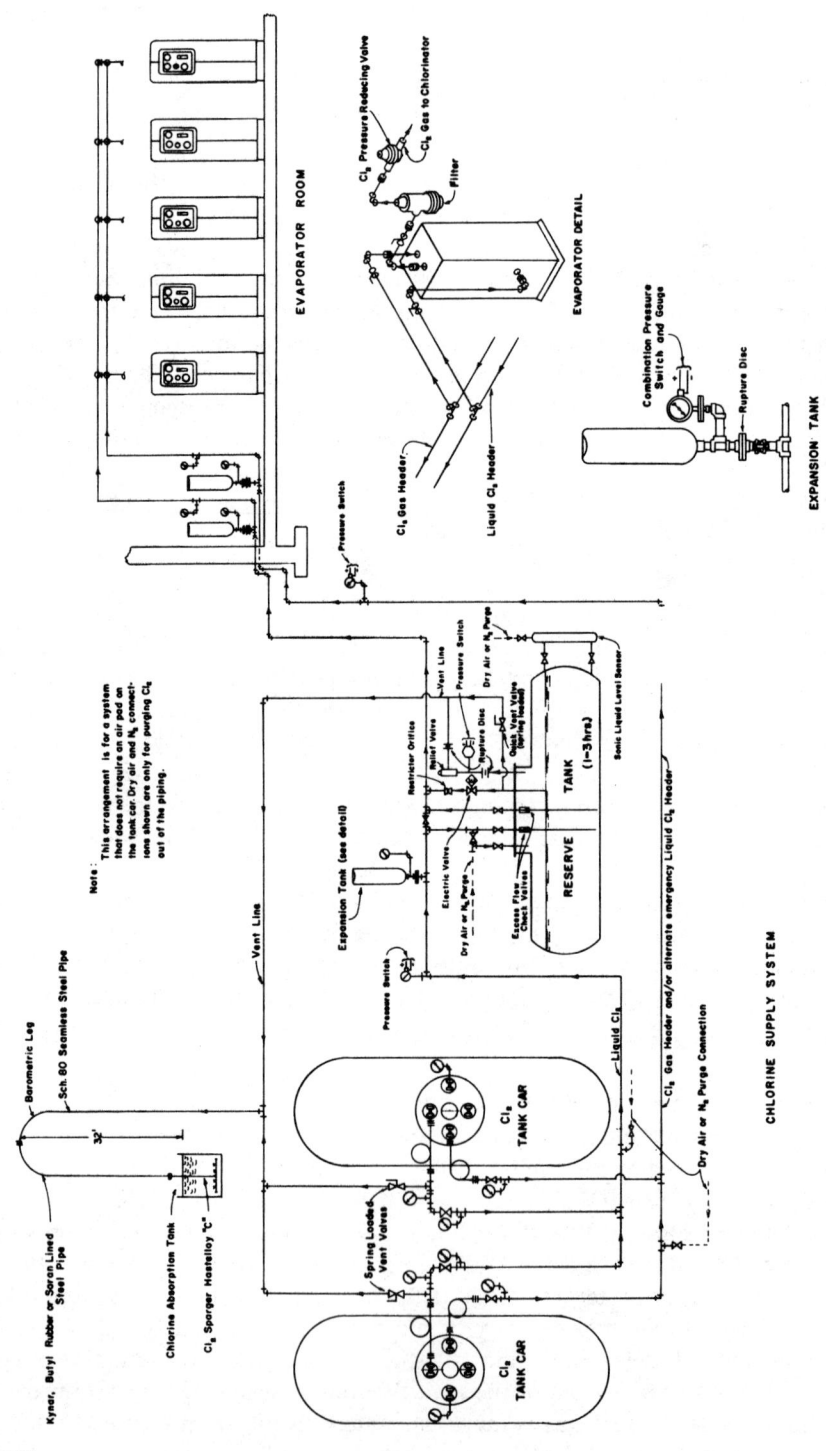

Fig. 9-24. Reserve tank system for chlorine supply system.

length to provide the gas volume. The vent piping is fitted with an electric shutoff valve and a restrictor orifice. Upon leaving the restrictor orifice the vent pipe joins the liquid lines going to the vaporizers. This restrictor orifice allows a minor flow of gas that will flashover from the small amount of liquid being bled from the tank thus keeping the liquid level constant in the reserve tank so long as chlorine is flowing from the tank car. This vent also allows the liquid level to rise to its normal height when the reserve tank is being filled. When the liquid chlorine flow from the tank terminates, the chlorine level in the reserve tank will drop and this will be detected by the ultrasonic level sensor. The sensor sends a signal through a control box which will engage an alarm and also close the electric shutoff valve in the vent line. The operator now has two choices: shutoff the liquid flow from the reserve tank and bleed the remaining gas in the tank car to the process; or immediately switch to a full car. When the time comes to switch to a new car, the reset button on the electric valve is actuated so as to open the vent line and allow refilling of the reserve tank. The reserve-tank piping and valving accessories are equipped with both automatic and manual pressure relieving devices all discharging to the absorption tank. This prevents any overpressuring of the reserve tank. The pressure relief line has a rupture disc followed by a pressure switch alarm. This is followed by a spring loaded relief valve protected from moisture entry by a rupture disc which is installed backwards for ruptures at low pressure. In addition, there is a spring-loaded manual quick vent valve to the absorption tank. Fig. 9-24 illustrates all of the appropriate expansion tanks and pressure switches required to give the operator the necessary operating information.

The following are the recommendations of the Chlorine Institute for chlorine storage tanks as set forth in Pamphlet No. 5, "Facilities and Operating Procedures for Chlorine Storage," 2nd Ed., Jan. 18, 1962.

- Minimum design volume, 25.6 ft^3 for each 2000 lb of chlorine to be stored
- Minimum working pressure, 225 psig plus ⅛-in corrosion allowance
- Design and fabrication, compliance with the ASME-UPV Code, 300°F, 70 percent weld efficiency, spot X-ray or 100 percent X-ray longitudinal and circumferential seams
- Nozzle necks, Sch 160, seamless steel pipe
- Piping, Sch 80 seamless steel
- Flanges, 300 lb weld-neck, slip-on, or screwed
- Fittings, buttweld extra heavy
- Fittings, screwed, 3000 lb
- Valves, 300 lb, flanged, forged steel body
- Valves, 300 lb, screwed, forged steel body

The purge connections shown are essential for the inspection and repair of the system. They also prevent the emission of chlorine to the atmosphere and the entrance of moisture to the piping and tank system, thus accelerating corrosion and deteriorating the entire system. This is the only way to provide a completely closed system which prevents penetration of atmospheric moisture.

Fully Automatic Tank Car Valve Shut-off System. (CVA)*:

Historical Background. The Clorox Co. of California spent several years and large sums of money developing this valve shut-off system—only for tank cars.[85] The Clorox Company has a nationwide operation of producing sodium hypochlorite from chlorine in tank cars. They have a great many of these installations in the U.S.A. In 1979 they experienced a serious leak at their Oakland, California plant. The leak occurred after a full tank car had been connected. A small leak developed in a 2 bolt flange near the car valve. The operator tried to stop the leak by tightening the bolts without shutting the car valve. As he tightened one bolt, the nut was stripped off the other bolt, partially opening the flange. This caused a significant leak of liquid chlorine. However, the size of the leak was not sufficient to activate the excess flow valve in the tank car. The leak required evacuation of people working in the immediate area and the rerouting of traffic on a heavily traveled freeway. This incident motivated the Clorox Company to seek a permanent solution to this type of problem, hence the development of this unique emergency tank car security system. It is probably the most significant contribution towards safety in chlorine handling in the last 50 years. This system incorporates the latest technology in mechanical, electrical and instrumentation engineering. Reliability of the system has been established by rigorous and lengthy field testing and operation.

System Description. Three separate components make up the entire system. They are: a DC motor operated valve actuator, an instrument control module and leak detector units.

The motor operated valve actuator is the heart of the system. It is a unified piece of machinery which is shaped to fit inside the circular tank car dome. The actuator fits over the tank car outlet valve so that there is no need to make any modifications or adjustments of the tank car valve. The gear train operated by the DC motor is designed to exert a 50 ft-lb torque on the valve stem. The torque is designed to be self limiting which prevents the actuator from shearing the valve stem or damaging the valve in any other manner.

In addition to the valve actuator the Clorox Company developed a Chlorine Detector Logic Panel (CDLP) to increase the sensitivity and reliability of the chlorine leak detector. The CDLP electronic package incorporates the logic to *close the valve only.* Closing time is 18 seconds. The tank car valve must be opened manually using a handwheel attachment on the top of the CVA.

The CVA will close under the following conditions:

---

* Car valve actuator.

1. Chlorine detection by Fischer & Porter Detectachlor
2. Manual chlorine alarm
3. Manual or automatic fire alarm
4. Low air pressure
5. 120V power failure
6. Low battery power
7. Disconnected service cord to CVA
8. Manual trip at the Chlorine Detector Logic Panel or at CVA.

Power failure will not impair the system owing to the separate self contained DC power source which is under continuous recharge. Unfortunately this valve shut-off system is not available commercially at this time.

Stationary Storage:

General Discussion. The use of stationary chlorine storage at both water and wastewater treatment plants has become popular due to the increased availability of bulk chlorine deliveries by truck. Fig. 9-25 illustrates a semi-trailer chlorine tank which is a part of the chlorine tanker fleet operated by the Metropolitan Water District of Southern California, Los Angeles. These trailers are loaded with chorine at the Stauffer Chemical Co. plant located at Henderson, Nevada (near Las Vegas). They are hauled some 300 miles to various treatment plants in the

Fig. 9-25. Chlorine semi-trailer tanker (courtesy Metropolitan Water District of Southern California).

628 HANDBOOK OF CHLORINATION

Los Angeles Basin area. This practice began in the early 1960s. The chlorine tank illustrated in Fig. 9-25 has a capacity of 17 tons. Newer trailers have a capacity of 19 tons. The trailers are off-loaded into 25-ton storage tanks. These chlorine tank trailers are made by the Evans Tank Car Co. of Lubbock, Texas. General American Transportation Corporation of Chicago and American Car and Foundry (ACF Industries) of New York also make chlorine tank trailers of equivalent capacity.

Unloading System. When the decision has been made to utilize a storage tank to be loaded from either a tank car or tank truck a special unloading system has to be designed. A typical storage tank and tank-car unloading system is shown in Figs. 9-26 and 9-27. The essential features are:

1. Unloading platform
2. Storage tank and sun shield
3. Weighing device
4. Air padding system
5. Eductor

Fig. 9-26. Typical stationary chlorine storage system (courtesy East Bay Municipal Utility District).

CHLORINE FACILITIES DESIGN 629

6. Chlorine gas and liquid headers
7. Gages
8. Pressure switches and alarms
9. Expansion tanks
10. Flexible connections

The storage tank capacity for truck tankers usually is 25 tons, but the storage tank capacity for cars should be commensurate with the size of tank cars likely to be delivered. Cars usually have a capacity of 55 or 90 tons of liquid chlorine. The volume of liquid chlorine increases considerably with increasing temperature. Therefore, to provide adequate room for this expansion, the Chlorine Institute recommends that the weight of chlorine in a tank must never exceed 125 percent of the weight of water at 60°F that the tank will hold. On this basis, each ton of chlorine will require 192.2 US gallons of tank capacity.

The storage tank should be designed in accordance with the Chlorine Institute recommendations, as set forth in their pamphlet No. 5, "Facilities and Operating Procedures for Chlorine Storage." Tanks should be designed for 120 percent of the maximum expected working pressure, or not less than 225 psi. To allow for

Fig. 9-27. Typical stationary chlorine storage system (courtesy East Bay Municipal Utility District).

corrosion, the wall must be ⅛ inch thicker than required by the design formula code.

The tank should be designed so that it can accept a tank car dome assembly. This consists of the four outlet valves, safety valve, eductor tubes with excess flow valves, and cover assembly. It is best to purchase this entire assembly from one of the chlorine tank car manufacturers, such as American Car and Foundry, Union Tank Car Company, or General American Transportation Company.

Weighing Device.   This is imperative. Weighing can best be accomplished by the use of either a lever scale system or load cells. These systems cost about the same, but the load cell is more popular since it lends itself to remote readout and/or recording using a 4–20 mA output signal.

Air Padding System.   This requirement is mandatory for stationary tanks. It is used for both tank car unloading (transfer) and process. It is necessary for purging the tank to allow inspection. For details see "Single Unit Tank Cars" in this chapter.

Ejector System.   Some kind of system must be provided to evacuate the tank for both inspection and repairs. This can best be accomplished by using a standard 2" chlorinator injector. The inlet of this aspirator type of injector is connected to a supply of water at about 50 psi. The throat of the injector (suction side) is connected to the gas withdrawal line on the tank.

Piping and Header System.   Fig. 9-28 illustrates a typical piping system for a chlorine storage system. It is important to point out the necessity of two header systems: one for gas withdrawal, and one for liquid withdrawal. Both of these enter the chlorination system at the liquid inlet side of the evaporator. Gages should be provided as shown.

The flexible connections from the tank car to the tank may be of either annealed copper tubing or flexible metal hose.

Copper tubing, with a two-foot diameter expansion loop and silver-soldered copper nipples on each end, has been the most widely used. On each end an ammonia type union is threaded onto the copper nipple.

In recent years flexible metal hose has been available and seems well suited to this application. It should be made of corrugated Monel with Monel wire braid and Monel nipples that are helium arc welded to the hose. The ammonia unions are threaded to each end. The total length of these flexible connections need be only ten feet.

Expansion tanks, equivalent to 20 percent of the header volume, should be used between the storage tank and the inlet to the evaporator.

As an added precaution to prevent overfilling of the tank in the event of a failure in the weighing system, dip tubes can be threaded into the dome section

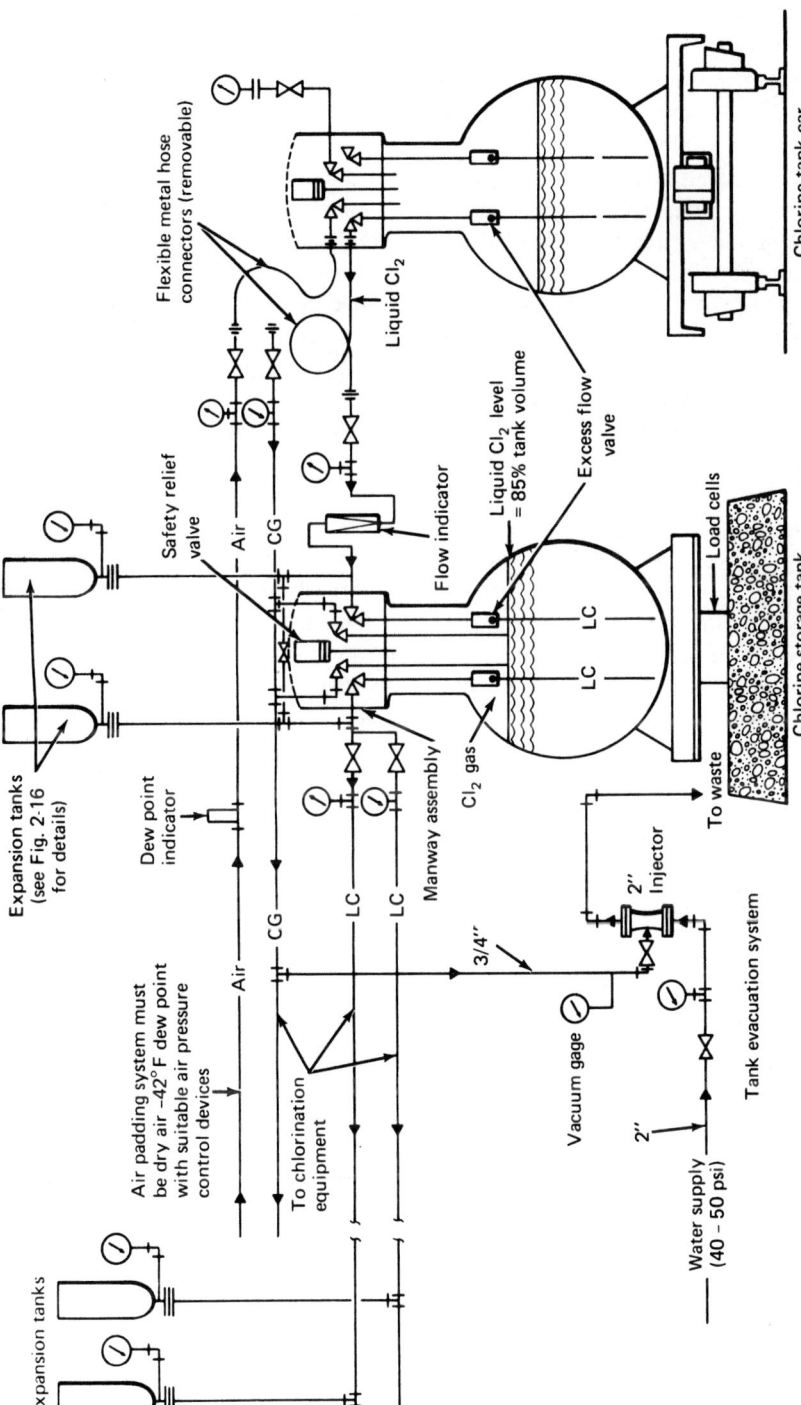

Fig. 9-28. Bulk chlorine storage and tank car unloading system. LC refers to liquid chlorine; CG refers to chlorine gas. The liquid chlorine header is in duplicate. A separate gas header is provided so that any car that arrives with too high a pressure (>125 psi) may be relieved to a lower pressure. (Dip tubes must extend far enough into the tank so that liquid contents cannot exceed 85% of the total tank volume.)

just under each gas outlet valve. Fig. 9-28 shows how these tubes come into operation. The storage tank is being filled from the tank car via line L. Gas is being withdrawn to the chlorination system via line G, which joins the chlorine liquid line at the evaporator. Whenever the tank is filled "too full," liquid will be discharged through line G to the evaporators. The operator will immediately notice the cooling in line G, signifying that the liquid transferred from the car has reached the bottom of the dip tubes. The operator should then stop the unloading process.

The length of the tubes can be arbitrarily calculated on the basis of the maximum ambient temperature to be expected in the vicinity of the tank. For example, at 100°F the liquid chlorine of a "full" tank would occupy 92 percent of the volume. This corresponds to 85 percent at 70°F, which is the recommendation of the Chlorine Institute. The depth of liquid to fill 92 percent of the tank can be calculated. The distance from the liquid to the inlet of the gas withdrawal valve determines the length of the dip tubes.

In some areas of the United States and Canada, chlorine is available in tank trucks (Fig. 9-29). It is usually imperative that the user transfer the contents of the truck to a storage tank rather than tie up the truck for the length of time of consumption of the contents. Storage tanks for tank truck operations follow the same design requirements as those loaded from tank cars, except for size. The capacity of most tank trucks ranges from 15 to 20 tons. Thus, the storage tank capacity need not usually exceed 25 tons. Depending on the length of haul and his chlorine consumption, the user should consider duplicate tanks.

Materials of Construction:

Chemical Reactions. Liquid chlorine is always packaged in steel containers. Liquid chlorine which is absolutely free of moisture will not react with iron or steel

Fig. 9-29. Bulk chlorine tank truck (courtesy Evans Tank Co. Lubbock, Texas).

which is similarly free of moisture. From a practical standpoint, it is not possible to package chlorine free of moisture; therefore a very small amount of ferric chloride develops in each container of chlorine. This moisture is unavoidable because there is always some moisture in the atmosphere and because containers are cleaned by being flushed with water. Although packagers are meticulous about drying the containers after flushing, first with steam and then with hot, dry air, enough moisture still remains to form some ferric chloride, a corrosion product of chlorine and iron or steel. This corrosion process is very slow. The design of containers and pipe lines should provide extra wall thickness to take care of this factor.

The two most significant chemical reactions of liquid chlorine and the materials of construction are:

1. Liquid chlorine* will spontaneously ignite and support combustion of carbon steel at 483°F. (See Chapter 1.)
2. It will attack and dissolve PVC at ambient temperatures.

The lessons to be learned here are:

1. Never apply heat directly to a chlorine container. An infrared lamp applied to the liquid portion of the cylinder can cause the burning of a hole through a steel cylinder.
2. Never use PVC or simiiar plastic materials anywhere in the liquid chlorine system or anywhere where chlorine gas is under pressure that is related to the vapor phase of the liquid–gas supply system. This usually means between the chlorine containers and the inlet pressure-reducing valve of the chlorinator.

Supply System. The chlorine supply system should consist of steel and cast iron products. The *supply system* is defined as that part of the system that begins at the chlorine containers and terminates at the inlet to the chlorinator.** From the chlorinator and beyond, the materials of construction are entirely different and are discussed elsewhere.

The supply system piping must be Sch 80 black seamless steel and fittings must be 2000 lb forged steel. *Do not use bushings* (they cannot meet the 2000 lb criterion). Use reducing fittings instead.

All unions should be ammonia-type with a lead gasket joint. *Never use a ground joint union.* Filter bodies and reducing valve bodies are usually cast iron. Expansion tanks should be of welded steel construction, but can be a standard 100- or 150-lb chlorine cylinder. Valves for the chlorine supply system should be Chlorine Institute approved. Two types of valves are used, one is for main line shutoff purposes and the other for isolating cylinders (header valves). Header valves are

---

* Chlorine cannot be in the liquid phase at 483°F.
** There is an exception: when the remote vacuum system is used, PVC pipe and fittings may be used downstream from the vacuum regulator.

identical to the outlet valves of ton containers; bronze bodies with monel seat and stem. Main line valves can be either the ball-type or rising stem-type. The ball-type is more popular because it utilizes a lever which not only indicates at a glance the position of the valve, but also makes it easier to operate the valve.

All gauges on the supply system must be equipped with a protector diaphragm. The diaphragm should be of silver and the diaphragm housing can be either Hastelloy "C" or silver cladded steel. Shutoff valves should not be used ahead of gauges. Gauges require a minimum of maintenance so that when replacement is needed the entire supply line should be drained of pressure before replacing or removing the gauge. The value of a shutoff valve for this purpose is lost, because a valve in a chlorine supply system loses its reliability if it is not operated on a frequent basis.

In assembling the piping system, either welded or threaded construction can be used; welded is preferable. If threaded construction is used, the contractor must be cautioned to use sharp dies and all threaded pipe must be cleaned with solvent before assembly. Pipe dope should not be allowed; instead use teflon tape for thread lubricant.

Valves Liquid and Gas Service. Valves on the chlorine supply side of the installation should be approved by the Chlorine Institute. The auxiliary header and container valves are identical to cylinder valves—but without fusible plugs—of bronze bodies with Monel stems and seats. Line valves are of the ball type or the rising stem type. The ball type is more popular, as it uses a handle indicating immediately whether the valve is in the open or closed position. The valves are of steel body with Teflon seat and Monel ball.

Construction. Extreme care must be taken during the fabrication of the chlorine supply system. Sharp dies must be used to cut the threads, and all joints must be cleaned of oil, dirt, and debris before assembly. The most satisfactory thread compound is Teflon tape; litharge and glycerine are usually detrimental to the quality of the pipe work.

For that portion of the system using PVC pipe, socket weld joints should be installed where possible. Every thread cut on a piece of PVC represents a potential weakness in this spot, since apparently adverse internal stresses are created during the thread cutting process. Threaded joints will crack at the threads if subjected to even mild vibration or shock. The older the joint, the weaker it gets.

Evaporators:

General Discussion. When the rate of chlorine withdrawal exceeds 1500 lb/day an evaporator should be installed. This changes the supply system to liquid withdrawal, which has different characteristics than gas-withdrawal systems.

Evaporators are available in capacities from 400 lb/day to 10000 lb/day. In a pinch, one cylinder can discharge liquid to an evaporator at a rate as high as 12,000 lb/day. This means that an evaporator can be used to conserve space for cylinder storage if necessary. The optimum storage requirements should be based upon the quantity-discount price break that is offered by the local chlorine supplier. This usually occurs at a quantity of five, thus dictating space for five in service, five empties, and a vacant space for the incoming five, or a total space for fifteen cylinders.

Electric Heater Type. The most widely used evaporator is the electric heater type. These are available from Capital Controls, Fischer and Porter, and Wallace and Tiernan Div. of Pennwalt. These units are equipped with G.E. Calrod heating elements of various sizes depending upon the vaporization requirement. It takes approximately 65,000 Btu to vaporize 8000 lb of chlorine. However, the evaporator must have a wide margin of safety to allow for the partial filling of the chlorine vessel with impurities inherent in the manufacture of chlorine. Therefore, to provide a sufficient safety factor, the chlorine gas which is vaporized must contain at least 20°F of superheat to prevent misting or liquid chlorine fallout in the gas discharge piping.

Misting occurs when the evaporator is pushed beyond its capacity. This is detrimental to the chlorinator because the little globules of mist contain the various impurities inherent in the production of chlorine. These impurities will plate out at the various stages of pressure reduction. This is another reason why a chlorine gas filter should always be installed immediately downstream from the evaporator. The important details to examine when comparing evaporators are as follows: volume of chlorine container vessel (extra volume is required for inherent sludge deposits); surface area in contact with water bath; volume of vapor space as determined by depth of penetration of gas discharge pipe into the container vessel; magnitude of hydrostatic pressure test; and allowable working pressure.

Beatty's Mixture. For more than 15 years the Metropolitan Water District of Southern California has been using a liquid mixture of water and corrosion inhibitor instead of a continuous flow of fresh water for the evaporator water bath. The mixture consists of 1 pint of closed-system corrosion inhibitor No. B-239 to 10 gallons of demineralized water. Each 8000 lb/day evaporator installation is supplied with 5 gallons of this solution. Operators add the solutions to the evaporator water bath on a biweekly basis from the 5 gallon containers. In 1982 MWD reported[7] that use of this fluid for evaporator heating has eliminated scaling problems with immersion heaters, and has eliminated the need for cathodic protection of water bath and chlorine container vessels.

Hot Water Type. Evaporators are also available for use with recirculated hot water. One type utilizes the intermittent recirculating flow of hot water pumped in a circuit between a heat exchanger and the evaporator water bath. A temperature probe actuates the recirculating pump to maintain the water bath between 170 and 180°F. In another hot water arrangement, a treatment plant utilizes a closed-loop hot water system, in which water at 200°F is intermittently pumped through a coil in the evaporator hot water bath at approximately 10 gpm. This arrangement requires an independent water bath makeup system.

Steam Type. Vaporizers are available that use live steam instead of recirculated hot water. These units are available from Whitlock Mfg. Co. in capacities as high as 5000 lb/hr of chlorine. Custom-made evaporators are also available from specialty manufacturers in Seattle, Washington and Toronto, Canada. These vaporizers can be furnished for either steam or water heated by a special propane-fueled heater.*

*Chlorine Pressure Reducing and Automatic Shutoff Valve.*** Integral with the evaporator is the electrically interlocked chlorine pressure shutoff valve on the discharge line of the evaporator. The circuit that operates this valve, whether an air solenoid for a pneumatically operated valve or an electric motor operator, is connected to the low-temperature alarm circuit. The alarm circuit sounds and deenergizes the CPRV circuit when the water bath temperature drops to 150°F. This protects the chlorinator against possibly receiving severely damaging liquid chlorine from the evaporator.

Cathodic Protection and Insulation. All evaporators using a water bath should be equipped with a cathodic protection system to protect both the water bath tank and the outside of the chlorine container from aggressive water corrosion. This system is provided with an indicating ammeter on the evaporator instrument panel to verify cathodic protection.

The outside of the water bath should be insulated with a ½-inch covering of urethane foam.

Accessories. Accessories that are consistent with good practice and should be standard equipment but which are still considered by the manufacturers as optional are: a gas temperature gage and automatic water level control of the water bath. Standard accessories include a gas pressure gage, water level indicator, and low-temperature alarm switch. A high-temperature alarm switch is optional.

Electrical Requirements: Electric Heater Type. 1. A three-wire 240 or 480 V circuit for the heater elements in the evaporator water bath. The load requirement is 12 kW for 6000 lb/day and 18 kW for 8000 lb/day.

2. A two-wire 120 V circuit is needed for the following functions:

---

* The Metropolitan Water District of Southern California uses one of these evaporators as part of their 30000 lb/day mobile chlorination system.
**This valve is an imperative for remote vacuum systems.

a. Air solenoid or electric operator on chlorine pressure reducing and shutoff valve downstream from the evaporator, interlocked with low-temperature alarm
b. Low-temperature alarm
c. High-temperature alarm (optional)
d. Solenoid valve to makeup line to water bath
e. Water level pressure switch

3. Alarms for each evaporator should include low temperature of the water bath, and low water bath level.

Evaporator Pressure Relief System:

Historical Background. Until as recently as 1974 there was no mandatory requirement to provide any pressure relief device for the liquid chlorine vessel in an evaporator. However, the manufacturers of chlorine evaporators have always been conscious of the possibility of liquid vessel ruptures. The prevailing belief was and still is that any pressure relief system creates more of a hazard than it might prevent. Therefore, evaporators have always been designed with enough strength to hold any vapor pressure that could conceivably be encountered. Furthermore, the overall system design mitigates against any possibility that might allow the liquid vessel to get "skin" full of liquid chlorine. As a final precaution, the classic design of an evaporator is to have the connections to the liquid vessel made with lead gasketed "ammonia-type" unions. These unions act like relief valves under extremely high pressures. However, since the liquid chlorine vessel is fabricated according to the ASME Boiler and Pressure Vessel Code, it is subject to rigid inspection regardless of who the manufacturer may be and must therefore be certified accordingly.

As of December 1975,[8] all pressure vessels manufactured in accordance with Division 1 Section VIII of the ASME code must be protected from overpressuring by means of a safety device. This safety device need not be provided by the vessel manufacturer, but must be provided prior to placing the vessel in service. This latest code defines the general requirements of pressure relief devices. The relief valve must be able to relieve the pressure in the liquid chlorine vessel when this pressure exceeds 110 percent of the rated working pressure of the vessel.

The Chlorine Institute specifies in Pamphlet No. 9 that chlorine vaporizing equipment must have a pressure relief device.[9] This can be either a rupture disc or a spring loaded relief valve or both; preferably discharging to an adsorption system. When both are used the section between should be equipped with a vent or pressure alarm.

The State of California safety orders for "Unfired Pressure Vessels" indicates that in addition to compliance with the ASME Code the following control is also required.

467. Controls: (a) Any pressure vessel not specifically covered or exempted elsewhere in these orders shall be protected by one or more safety valves or rupture discs set to open at not more than the allowable working pressure of the vessel* and by such other controlling and indicating devices as are necessary to insure safe operation.

Current Practice. Both Fischer and Porter[10] and Wallace and Tiernan, Div. of Pennwalt Corp.[11] provide as optional equipment a relief valve system for all their various types of evaporators. Both illustrate the location of the relief valve on the gas and *not the liquid* phase of the evaporator connections. This current arrangement is shown in Fig. 9-30. It should be noted that all rupture disks can be damaged if subjected to a vacuum. They must be specified to withstand 25 in. Hg vacuum on the pressure side.

Relief Valves. The Fischer and Porter relief valve Model 71P1412 has been manufactured for over 20 yr. This valve opens at about 275 psig and seats tightly at about 200 psig. Fischer and Porter report a high-confidence level for this valve.

The Wallace and Tiernan valve carries Part No. U25470. It opens at 560 psig and closes at 550 psig. This valve is purchased from Crosby Valve Co. and other suppliers such as Dresser Industries and Ferris Valve Co.[12]

Safety Considerations. The discharge of the relief valve system (vent) brings up serious questions about the hazards of chlorine leaking to the atmosphere at high pressures. Whenever either of these valves open to relieve pressure they become subject to atmospheric corrosion. Maintenance routine should therefore require that the valve be overhauled after each opening or closing cycle. Moreover, this valve must be protected at all times from chlorine vapor by installing a rupture disc as recommended by the manufacturer (upstream from the valve). This keeps the valve clean and dry during the periods of nonuse.

Owing to the potential hazard of a chlorine leak it is always advisable but not mandatory to discharge the vent from this system into a chlorine absorption tank. Therefore, all of these systems should be designed so that at a later date a barometric loop and absorption tank can be conveniently added to the relief valve vent. (See section describing "Reserve Tank.")

*Absorption Tank for Relief Valve Vent System.*** In addition to a suitable size absorption tank (containing NaOH), the discharge piping between the relief valve discharge and the absorption tank must contain a barometric loop. (See Fig. 9-24). The barometric loop is mandatory with an absorption tank because it prevents the almost certain intrusion of moisture into the chlorine gas supply piping.

---

* Both Fischer and Porter and Wallace and Tiernan use a working pressure design of approximately 500 psi at 212°F and a hydrostatic pressure test of 1450 psi at 125°F.[12,13]
** This tank should also be considered for remote vacuum regulator vents and external CPRV vents.

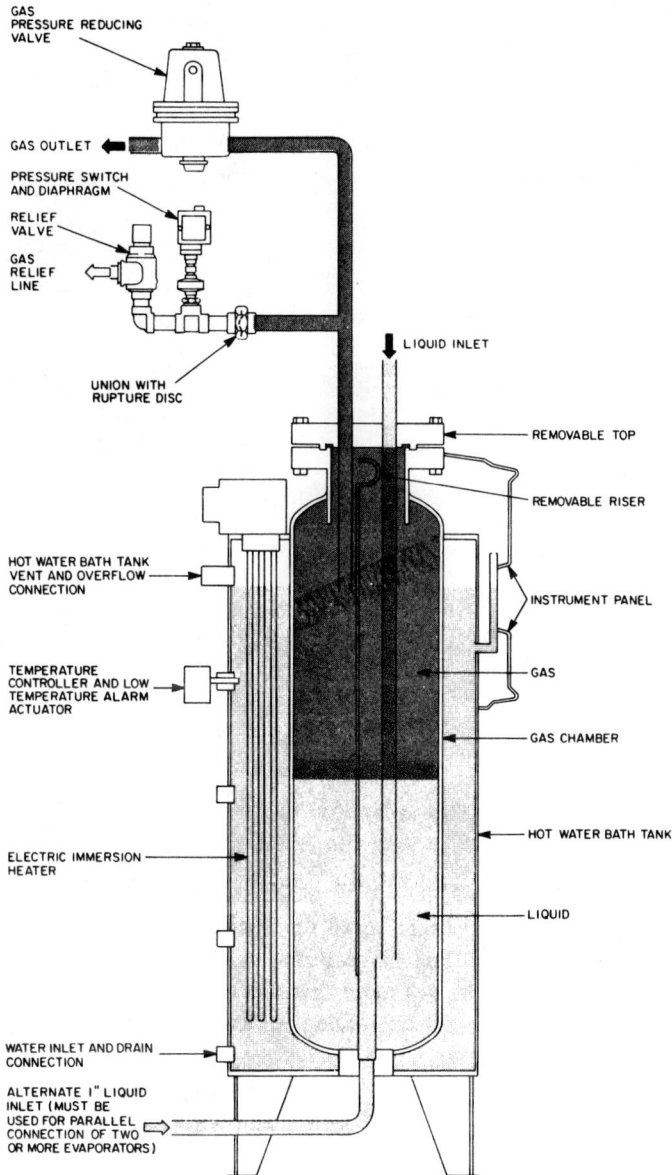

Fig. 9-30. Evaporator showing relief system (courtesy Wallace and Tiernan Div. Pennwalt Corp.).

The absorption tank should be capable of neutralizing a maximum of 150 lb liquid chlorine from each evaporator during an overpressure crisis situation. This is about the amount of liquid chlorine which would be contained in each evaporator if the liquid vessel were full of liquid chlorine. While this is an improbable situation it provides a generous safety factor to the size of the absorption system. To this amount of chlorine should be added the amount in the liquid piping and expansion tanks* between the supply tanks and the evaporators.

Monitoring Relief System. As shown in Fig. 9-30 there will be a pressure switch to monitor a critical overpressure situation sufficient to rupture the frangible disk. However, it is recommended that if this relief system is vented to the atmosphere and not to an absorption chamber the vent should be combined (but separated) in such a way that this vent and the usual chlorinator and chlorine pressure reducing valve vents be monitored continuously by a chlorine leak detector.

SAFETY EQUIPMENT AND ACCESSORIES

Breathing Apparatus. There are two types of breathing apparatus, the canister-type gas mask and the oxygen or air-type breathing unit. The canister-type gas mask is limited in effectiveness; it is appropriate for changing chlorine cylinders or normal maintenance work. It is not satisfactory for use in repairing a leak. Therefore, either of the following types of equipment should be furnished: the air-type breathing unit (with 30-minute air supply, as manufactured by Mine Safety Appliance Company or Scott Aviation Company) or the oxygen breathing apparatus (as manufactured by MSA). The latter is similar to a canister type. When the seal on the unit is broken, the unit manufactures its own oxygen which lasts for 45 minutes. These canisters must be discarded after the seal is broken.

Chlorine Container Emergency Kits. Every chlorination station should have at least one chlorine container emergency kit. These kits are available for 150-lb cylinders, ton containers, tank cars, tanker trucks, and stationary storage tanks. These kits are available from Chlorine Specialties Inc., San Francisco, CA and Indian Springs Mfg. Co., Baldwinsville, NY. The kits are designed to seal off a leaking fusible plug, a leaking outlet valve, a tank car relief valve, or a moderate-size rupture in the container shell. Emergency Kit C for tank cars, tanker trucks and storage tanks** is illustrated in Fig. 9-31. Kit A is for 150 lb cylinders and Kit B is for ton containers.

---

* It is reasonable to expect that the rupture discs on the expansion tanks might also be overpressured at the same time.
** Provided the storage tanks and trucks are fitted with a Chlorine Institute tank car dome assembly.

CHLORINE FACILITIES DESIGN 641

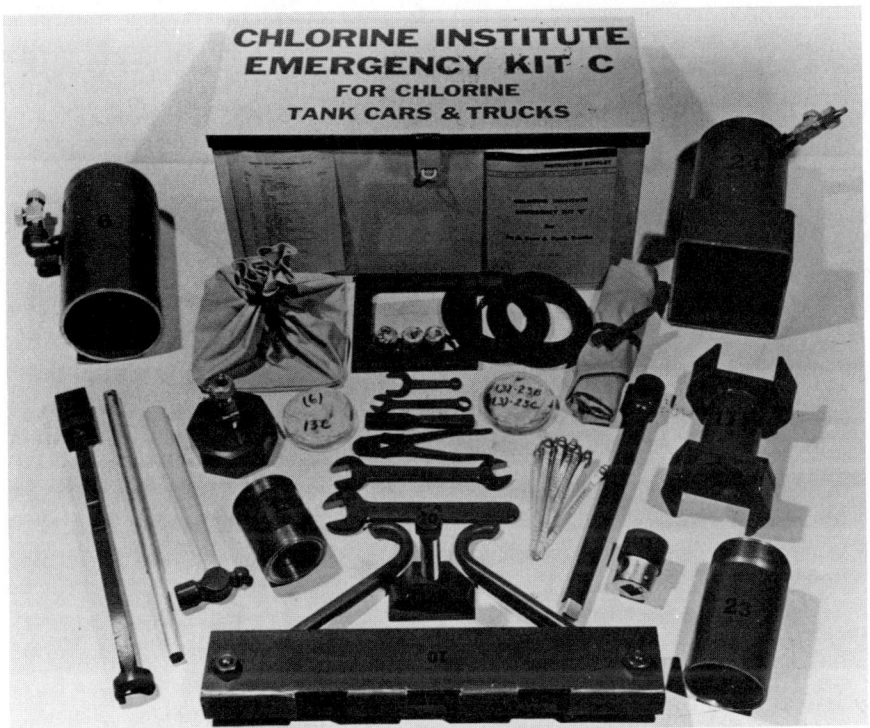

Fig. 9-31. Ton container emergency kit (courtesy Chlorine Institute N.Y. and Chlorine Specialties San Francisco).

Leak Detectors:

General Discussion. There are two categories of leak detectors: (1) continuous monitors of the working environment ambient air, and (2) hand-held personnel leak locators. There should be at least one continuous detector for every chlorination station to prevent hazardous situations for both personnel and the surrounding population. Furthermore, these leak monitors are necessary to meet OSHA requirements for maximum contamination levels of chlorine (1 ppm) in the working area.

The second category of detectors are for searching or checking for leaks in the fabricated components of the chlorine supply system which is under chlorine equilibrium vapor pressure.

Since the continuous monitors are for area leaks, more than one detector per installation may be required. For example, one each should be provided for tank car and/or storage tanks area, ton cylinder room, evaporator–chlorinator room,

642 HANDBOOK OF CHLORINATION

and at the discharge of chlorine vent lines. Remote injector locations do not warrant a detector. Most leak detectors are designed as single-point samplers, therefore multiple sampling point units should be used for detection at multiple sources. Fig. 9-32 illustrates the single-point system and Fig. 9-33 shows the central system module capable of monitoring multiple sensors for a variety of sampling points.

Capital Controls. The Advance Series 1030 detector uses a totally solid-state gas sensor.[14] It consists of an electronic enclosure which contains the alarm and warning circuits and a remote sensor. These two components are required for each area sampling point as shown in Fig. 9-32. The sensor is a solid-state voltammetric type based upon proprietary electrochemical gas-sensing electrodes. Each sensor contains a sensing electrode and a reference electrode immersed in an electrolyte bound in an absorbant matrix. No chemicals are required, thus eliminating any need for regular maintenance. The Series 1030 is designed to permit a separation

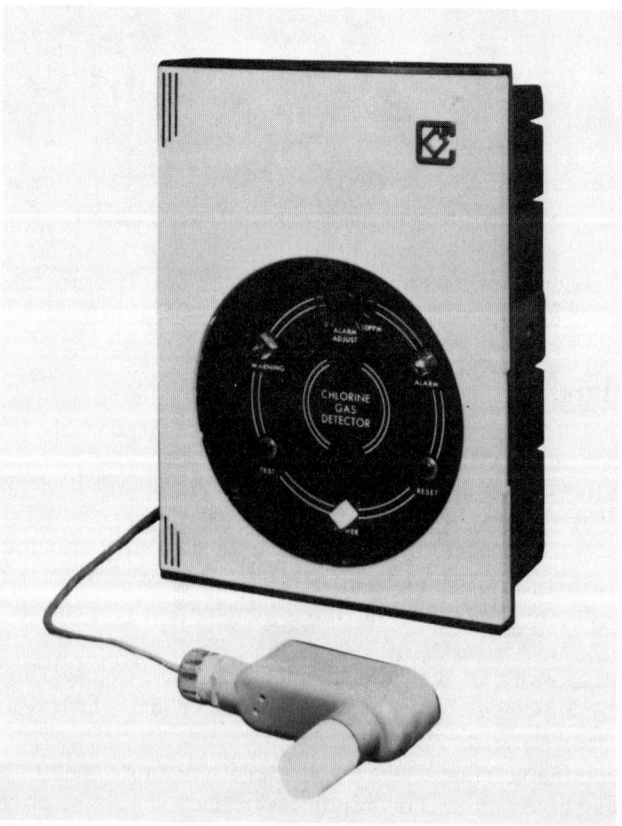

Fig. 9-32. Chlorine leak detector and sensor (courtesy Capital Controls Co.).

CHLORINE FACILITIES DESIGN 643

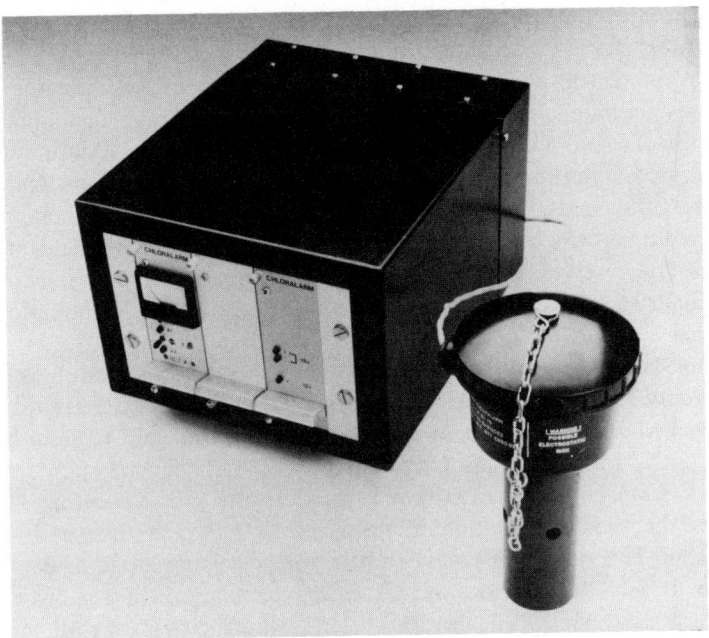

Fig. 9-33. Draeger Safety-Chloralarm (courtesy Draeger International and ICI, U.K.).

up to 3000 feet between the sensor and the electronics enclosure. The electronic circuitry is arranged to provide a dual monitoring system such that separate alarm and warning circuits provide redundant chlorine detection with backup alert in the event of either a slow accumulation of vapor or a major leak.

The Series 1030 meets the OSHA maximum concentration level of 1.0 ppm chlorine allowed in working environments. It is field adjustable in the alarm mode from 0 to 10 ppm of chlorine in the air sample. The instrument response time is 30 seconds. It is responsive to chlorine concentrations less than 1 ppm. The detection of chlorine gas is accomplished by a voltage applied across the two electrodes embedded in the sensor cell. The voltage level is selected for the gas of interest. When chlorine is present an electric current flows between the electrodes. This event actuates the warning and/or alarm circuits. The sensor recovery time after chlorine exposure is 5 minutes.

Draeger Safety Chloralarm. This detector system was designed by ICI, the major producer of chlorine in the United Kingdom. The licensee in the USA is Draeger Safety, a division of National Draeger Inc.[15] The Chloralarm detector was designed specifically for multiple sample source detection. Fig. 9-33 illustrates a single source system with one sensor. Large systems are housed in a free standing cabinet with a viewing panel behind a lockable front door. Eight, sixteen, or twenty-four channel systems are available with a built-in power unit.

The Chloralarm is a typical amperometric wet chemistry system used primarily for large chlorination process industries and other large consumers. The remote cable-connected sensors can be located almost any distance from the central control module. Two platinum electrodes are immersed in an electrolyte containing a small concentration of calcium bromide. When the air sample contains chlorine an electric current flows through the cell. This current triggers the alarm system. The response time is about 30 seconds. The sensor is designed to be installed outdoors in adverse weather conditions which also serves to protect it from mechanical damage. The large electrolyte reservoir eliminates the need for frequent replenishment. Under normal industrial conditions the sensor should operate without attention, for at least one year.

The Chloralarm has a unique built-in system check. Each module is provided with a switch whereby the entire function of that module and its associate sensor may be checked without having to visit the remote sensor and without having to expose the sensor to chlorine gas. Operation of the system check switch disconnects the 600 mV supply and provides a 2 V pulse to the remote sensor. This pulse electrolyzes the bromide solution in the electrolyte, which releases bromine at the electrodes. The system responds as if it were receiving an equivalent amount of chlorine. If the Chloralarm indicates an alarm condition it demonstrates that the sensor, sensor cable, cell electronics, and alarm system are all in working order.

The control module contains a meter scaled 0–5, 0–10, 0–25, and 0–50 ppm together with potentiometers for setting two alarm levels (per sensor) plus toggle switches for displaying the alarm levels.

Enterra.* This company markets a line of toxic gas sensors.** The chlorine gas detector is their Chlor-Guard Model No. 5152.[16] It incorporates a gas diffusion sensor which does not require any maintenance. This sensor generates a current directly proportional to the concentration of chlorine. Response time is 90 percent in 15 seconds, and the sensor may be located 500 ft from the alarm module.

The alarm module contains alarm lights, horn, and alarm relays. When a low level of chlorine gas is sensed the yellow warning light will flash, and where a higher level of chlorine concentration is detected the red lamp will glow steadily. The concentrations of chlorine gas at which these alarms occur are adjustable on-site from 1 to 10 ppm.

The Enterra toxic gas sensors are classified as voltammetric, diffusion-limited electrochemical devices. Each sensor contains a noble metal sensing electrode and a reference electrode immersed in a supporting acidic electrolyte. A diffusion membrane isolates the sensing electrode from the ambient air. In operation a small voltage is applied across the two electrodes, the voltage level being a function of the gas of interest. When the sensor is exposed to chlorine in the ambient air,

---

* Formerly Exidyne.
** See schematic Fig. 10-2, sulfur dioxide detector.

diffusion begins. This allows the gas into the sensor through the diffusion membrane. When the gas comes in contact with the polarized sensing electrode, an oxidation reaction occurs at the counter (reference) electrode. This reaction generates a current in the system which is proportional to the concentration of chlorine gas at the diffusion membrane–sensor interface. The electrode reactions for chlorine are:[17]

Sensing electrode:

$$Cl_2 + 2e^- \longrightarrow 2\ Cl^- \tag{9-2}$$

Counter electrode:

$$Pb + H_2SO_4 \longrightarrow PbSO_4 + 2H^+ + 2e^- \tag{9-3}$$

Fischer and Porter. Their model series 17CA 1000 Chloralert is a low-concentration ambient-air chlorine detector.[18] The detection level is fixed at 1 ppm chlorine by volume (3 mg/m^3). This instrument will sense and alarm at chlorine levels below the threshold of olfactory detection to conform to OSHA maximum tolerable limit.

The Chloralert uses a glycerine-based electrolyte containing a small amount of potassium bromide. Dual platinum electrodes are partially immersed in the KBr* electrolyte. A low voltage is applied across the electrodes which intensifies the electrochemical response.

An internal blower draws an air sample into the unit where it passes through a flow indicator to permit adjustment of the sampling rate. From the flowmeter the air sample flows to the electrode sensing cell where an electric current is generated by the presence of minute concentrations of chlorine. This current is sensed and monitored by a self contained electronic circuit and is used to trigger the alarm system. The instrument response time is a matter of 2–3 seconds, and it desensitizes equally fast once the leak disappears.

Once the Chloralert is actuated, the circuitry is specifically designed to demand the alarm to be acknowledged with the spring-loaded reset switch. If the air sample still contains chlorine above the 1 ppm level, the alarm contacts will transfer to their normal (non-alarm) position while the reset switch is depressed and will immediately return to the alarm position when the reset switch is released. The test switch, when depressed together with the reset switch, substitutes a simulated sensing cell current, thus allowing a sample check to be made of the circuit performance.

This instrument is designed for single sample point application and is limited to 25 feet from the source of air sampling. The sensing cell is designed for electrolyte replacement once per year.

Wallace and Tiernan. The Wallace and Tiernan Series 50-125 Chlorine Detector consists of a dual electrode sensor and an electronic package which controls and

---

* When chlorine comes in contact with this electrolyte Br_2 is released by the oxidation action of chlorine. The free bromine triggers the alarm circuit.

monitors the sensor.[19] The sensor unit is provided with a 5-liter electrolyte reservoir. This administers a drip rate flow of electrolyte which wets the surface of the dual electrodes. The reservoir has enough capacity for about 3 months continuous operation. The electrolyte contains a small amount of potassium iodide (KI) in solution with glycerine and distilled water. A 0.4 voltage is applied across the dual platinum electrodes. This sharpens the sensitivity and reduces the electrode response time, which is a matter of 2–3 seconds. This detector is available in three sensitivities: 0.5, 1.0, and 3.0 ppm chlorine in air.

When the concentration of chlorine in the air sample reaches or exceeds the detection level, the red alarm light will remain on until the chlorine concentration recedes below the detector sensitivity. When the alarm condition has been eliminated, the sensor should be cleaned and wetted with electrolyte again by using the wash bottle supplied with the unit. This speeds up the return of the detector to normal (non-alarm) operating condition. The detector circuitry can be tested by actuating the "PRESS TO TEST" button on the front panel. The red alarm light should go on. The sensor electrodes must be coated with the electrolyte—otherwise the detector will not alarm for chlorine. Therefore the drip rate of the electrolyte must be checked by the operator on a frequent basis. The drip rate will vary with temperature and barometric pressure changes. The electrolyte in the supply reservoir is protected from ambient air contamination by an activated carbon filter. Whenever the detector is exposed to a high concentration of chlorine, the filter should be replaced.

This detector is designed for single-source air sampling. It is equipped with a built-in motor driven fan which draws the air sample into the sensor cell. The air sample source should be limited to 25 ft from the detector location and not lower than 4 ft from the sensor elevation.

Leak Locators. American Gas and Chemical Co., Northvale, NJ markets two products that assist the operator in locating chlorine, ammonia or sulfur dioxide leaks. These products are identified by number for each specific gas.

For chlorine, CDP-100 is a chlorine-sensitive spray which reacts chemically with minute chlorine leaks causing a visible color change from white to yellow at the point of leak. It is easily removed with a damp cloth.

For personnel protection and leak searching there is a rechargeable, hand-held detector model CGT-701. This instrument can locate chlorine leaks as low as 1 ppm. It uses a solid-state sensor which requires no maintenance and cannot be poisoned.

Expansion Tanks. These tanks are mandatory for liquid chlorine supply systems. These chambers are necessary when there is danger of liquid chlorine becoming trapped in a length of pipe. If this situation is followed by a significant ambient temperature rise then hydrostatic pressure will develop. This may be sufficient to rupture the pipe because the coefficient of expansion of steel is much less than that of liquid chlorine. The recommended type of expansion tank with rupture disc and chlorine pressure switch as recommended by the Chlorine Institute is illustrated in Fig. 9-12. The frangible disc ruptures at pressures between 300 and 400 psi, thereby allowing the liquid trapped in the system to enter the expansion tank. This immediately produces some vapor pressure in the expansion tank, which actuates the pressure switch and sounds the alarm. Spring loaded relief valves discharging to the atmosphere should never be used on liquid chlorine lines. The chance of such a valve reseating properly is remote. Therefore, the discharge of any and all such valves should be to an absorption system.*

FLOW OF CHLORINE IN PIPES UNDER PRESSURE

The fluid mechanics of liquid and gaseous chlorine are entirely different, and so they will be discussed separately.

Liquid Chlorine. The flow of liquid chlorine is restricted to the piping between the supply containers and the evaporators. This flow of liquid will not be similar to that of other liquids. For this reason, the friction loss of flowing liquid chlorine cannot be accurately predicted, as has been confirmed by field tests with rates of liquid chlorine flow up to 18,000 lb/day in approximately 1000 ft equivalent length of 1″ Sch. 80 pipe.[20] This is due to the phenomenon of the liquid chlorine "flashing" to vapor whenever there is a sudden change of the flow rate in the system. Most chemical processes function at some fairly precise optimum capacity, but a chlorinator facility is designed to have a wide range of flow capability—at least twenty to one.

Therefore, when there is a sudden change in the flow rate, the liquid in the evaporators suddenly vaporizes rapidly, causing an abrupt pressure drop in the system, due to the change in demand. The vaporization process will extend back into the pipeline leading from the storage containers and create pockets of gas impeding the flow of the liquid. This vapor flashing occurs first at points of highest friction loss, such as entrance and exits of valves and fittings.

Finally, after the vaporization becomes stabilized in the evaporators, the pressure will rise in the evaporators, causing the flashing to cease in the pipeline. Then the system performs normally until there is another abrupt change in the liquid chlorine flow rate. The longer the pipeline, the longer the system takes to stabilize. In a 500 ft line, the pressure will begin to restore itself in about five to ten minutes, depending on the rate of flow change, and will take two or more hours for complete restoration of pressure. This phenomenon is also related to temperature. If the chlorine pressure in the storage tank is at 80 psig, then the vapor temperature would be only 68°F. A long liquid line on a warm day or one exposed to the sun might be considerably warmer than 68°F, which would tend to warm the chlorine above its vapor pressure and to cause flashing in the line. The line would,

---
* Exception: See "Evaporater Relief System" p. 637.

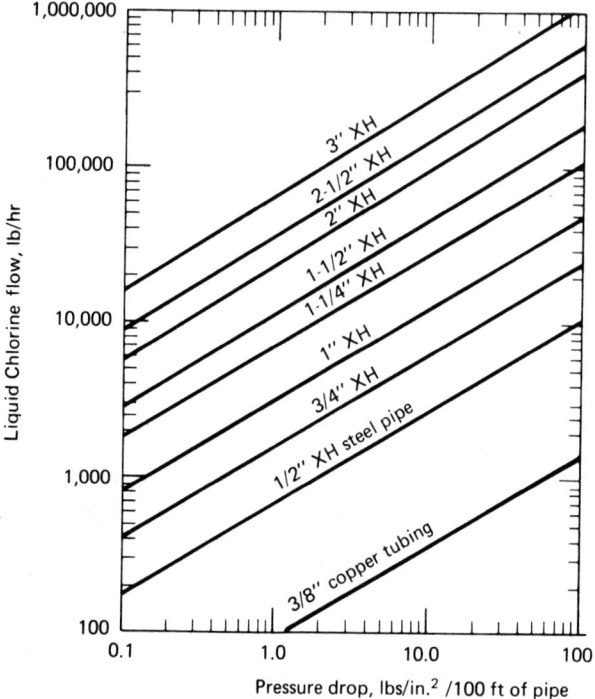

Fig. 9-34. Friction loss in liquid chlorine piping. (Clean new pipe; vaporization in the pipes will cause much higher drops.) (courtesy Hooker Chemical Co.)

however, be cooled by the flow of liquid chlorine, and the flashing would cease when sufficiently cooled below the vapor pressure. Therefore, depending upon the ambient temperature, the pressure in the chlorine containers, and the rate of change of liquid flow in the system, many combinations of pressure fluctuations will be noted. Sudden pressure drops of 10 to 15 psi in a 1" line 500 to 1000 ft long, with a maximum flow of 24,000 lb/day would not be unusual. Flows of 24,000 lb/day and less are considered "low" by chlorine manufacturers. This is only 1000 lb/hr, compared with the excess flow rate valve restricting withdrawal rates from tank cars at 7000 lb/hr.

Fig. 9-34 shows the estimated pressure drop in various size pipes carrying liquid chlorine. On the basis of sudden changes in flow under various kinds of climatic conditions and container pressures, it is recommended that the pipe size be limited to a maximum pressure drop of 0.25 psi per 100 ft for lines 500 ft and longer and of 0.5 psi per 100 ft for shorter lines. Thus, the following size lines should be limited to the capacities shown in Fig. 9-34:

CHLORINE FACILITIES DESIGN 649

	Maximum Liquid Cl$_2$ Flow lb/Day Length of Line	
Line Size	Less than 500 ft	500–1500 ft
¾"	24,000	17,000
1"	48,000	33,600
1¼"	100,000	72,000
1½"	168,000	115,000

Actually, ¾" pipe should not be used for lines longer than about 100 ft. Furthermore, most manufacturers have standardized on 1" inlets and outlets for the various chlorine piping accessories. While the flows shown in Fig. 9-34 may seem enormous for such small lines, it should be noted that the velocity of liquid chlorine in a 1" pipe at 48,000 lb/day is only slightly more than 1 ft/sec.

Example:

$$V \text{ ft/sec} = \frac{Q \text{ cu ft/sec}}{A \text{ sq ft}} \tag{9-4}$$

The inside diameter of a 1" Sch. 80 pipe is 0.957"

$$A \text{ sq ft} = \left(\frac{0.957}{12}\right)^2 \cdot \frac{\pi}{4} = 0.00499 \text{ sq ft}$$

$Q = 48,000$ lb chlorine per day, so

$$\frac{48,000}{1440 \text{ min} \times 60 \text{ sec}} = 0.5556 \text{ lb/sec}$$

Liquid chlorine at 68°F weighs 87.8 lb/cu ft; so

$$Q = \frac{0.5556}{87.8} = 0.00632 \text{ cu ft/sec}$$

Therefore

$$V = \frac{0.00632}{0.00499} = 1.27 \text{ ft/sec}$$

Therefore the figures given in Fig. 9-34 are conservative for liquid chlorine pipe systems. The designer is cautioned to be meticulous about arriving at the proper equivalent length of pipe, accounting for elbows, sudden enlargements, ammonia unions, tees for pressure gages, switches, and line valves. Most of these values are given in the Appendix to the present book.

The flashing phenomenon of liquid chlorine is one reason the flow cannot be properly measured. The usual differential pressure method for measuring the flow of liquids, including the variable area meters (rotameters) cannot be properly calibrated since the basic principle of their operation depends on a pressure drop which initiates the flashing phenomenon. At this point, there are both gas and liquid, with entirely different flow characteristics, passing through the measuring device simultaneously. Under proper conditions, a rotameter will give at best a rough approximation. The conductivity factor of liquid chlorine is so low that magnetic meters cannot be used either.

Gaseous Chlorine. Since the flow of chlorine gas follows the laws of fluid dynamics, the friction loss in such a system can be predicted with reasonable accuracy.

In pipelines carrying gas, the designer should have two special concerns: the piping system from the evaporators to the chlorinators, and the line carrying the gas from the chlorinator to the injector under a vacuum. When remote injectors are used, this calculation of line size is critical because the head loss tolerance is low—1½ to 2" Hg vacuum. The chlorinator is relying on the injector for its operating energy, so that little should be wasted in friction loss. However, on the pressure side between the evaporators and chlorinator, a 10 psi pressure drop can be tolerated. A critical situation develops when the velocity of the gas under pressure is so great that it undergoes sufficient cooling to cause condensate or ice formation on the outside of the pipe.* Therefore if the velocity is kept below 35 to 40 ft/sec, this phenomenon will probably not occur and the pressure drop will be easily tolerable.

Most manufacturers arrange for one evaporator to serve each chlorinator independently. The openings out of the evaporator are 1"; the chlorine filters and regulating valves are also made for 1" pipe. Therefore the pipe size for any 8000 lb/day chlorinator and evaporator should always be 1". Let us investigate the velocity and possible head losses that will occur at maximum output of the chlorinator.

Gas leaving the evaporator will be at approximately 100°F. Let us assume a gage pressure of 85 psi, indicating that the gas has a certain amount of superheat, which is normal for a properly operating evaporator.

The density (ρ) of the gas = 127 lb/100 cu ft (See Fig. 1, Appendix)

$$Q = 8000 \text{ lb/day} = 0.0926 \text{ lb/sec}$$

Converting to cu ft/sec:

---

* As the gas cools it becomes denser and the friction loss increases.

CHLORINE FACILITIES DESIGN 651

$$\frac{0.0926 \text{ lb/sec}}{1.27 \text{ lb/cu ft}} = 0.0729 \text{ cu ft/sec}$$

Now

$$V = \frac{Q}{A}, \quad A \text{ for a } 1'' \text{ Sch. 80 pipe} = 0.00499 \text{ sq ft}$$

so

$$V = \frac{0.0729 \text{ ft}^3/\text{sec}}{0.00499 \text{ ft}^2} = 14.6 \text{ ft/sec}$$

This tolerable velocity exists between the evaporator and the external chlorine pressure-reducing and shut-off valve. (See Fig. 9-35.) Downstream of the CPRV,

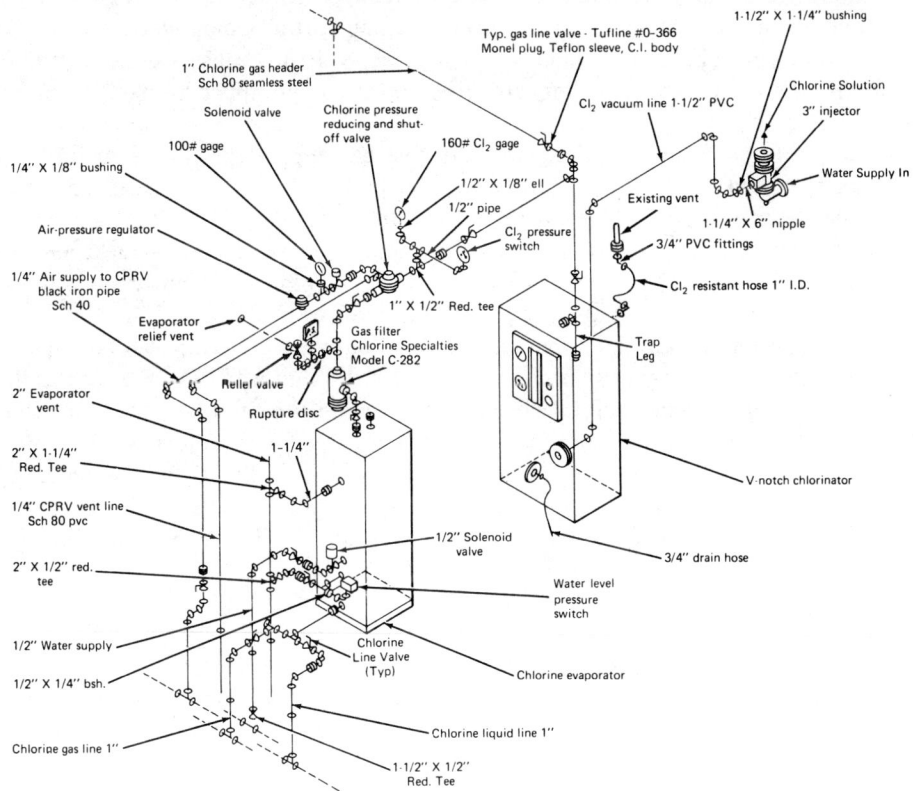

Fig. 9-35. Piping schematic evaporator, filter, CPRV, and chlorinator. Chlorine gas and liquid line fittings are 2000 lb forged steel. All unions on chlorine gas and liquid piping are ammonia type with lead gasket.

conditions change abruptly. Usually the CPRV is adjusted to give a downstream pressure of 40 to 60 psi. Let us now see what the velocity is at the reduced pressure of 40 psi. The temperature of the gas will drop about 25°F owing to this pressure reduction, so that the density of the chlorine gas will be $\rho = 0.7$ lbs/cu ft. Therefore

$$Q = 8000 \text{ lb/day} = 0.0926 \text{ lb/sec} = \frac{0.0926}{0.7} = 0.1325 \text{ ft}^3/\text{sec}$$

and

$$V = \frac{0.1325}{0.00499} = 26.51 \text{ ft/sec}$$

As the velocity increases, the pressure drop due to friction losses increases exponentially. This further reduction in pressure reduces the density of the gas, thereby initiating further increase in the velocity, which further compounds the friction loss problem. The above velocity, 26.5 ft/sec, will not cause excessive friction losses in a conventional layout, but does approach the upper limit of maximum allowable velocity.

The rule of thumb for designing chlorine gas supply systems is to limit the velocity at maximum flow to 50 ft/sec and preferably not more than 35 ft/sec.

Whenever evaporators are manifolded together, the designer must provide piping large enough to stay below this velocity limit. In these instances the rotameter and chlorine pressure-reducing valve will have 1" pipe size connections, as will the chlorinator, but the piping required may have to be 2".

As an example, consider the case of manifolding three evaporators, capacity 8000 lb/day, discharging to a common header through a CPRV, thence through a transmitting rotameter, for recording and totaling the chlorine flow, and thence to each of three chlorinators, as illustrated in Fig. 9-36. Note that the CPRV has a 1" inlet and outlet, but has a nominal capacity of 32,000 lb/day. Similarly, a 1" straight-through metal tube rotameter can easily handle 24,000 lb/day of chlorine. It is permissible to allow the high velocity of chlorine gas through these two devices, but the rest of the piping must be at least 1½".

The gas will cool from the inlet temperature at the CPRV from 90 to 100°F down to an average of about 65 to 70°F between the CPRV and the chlorinator. The CPRV will be adjusted to give a downstream pressure of approximately 35 psi at 24,000 lb/day flow. This gives a chlorine gas density of approximately 0.65 lb/cu ft.

Using 1½" pipe, we get:

$$Q = 24,000 \text{ lb/day} = 0.2778 \text{ lb/sec} = \frac{0.2778}{0.65} = 0.4265 \text{ ft}^3/\text{sec}$$

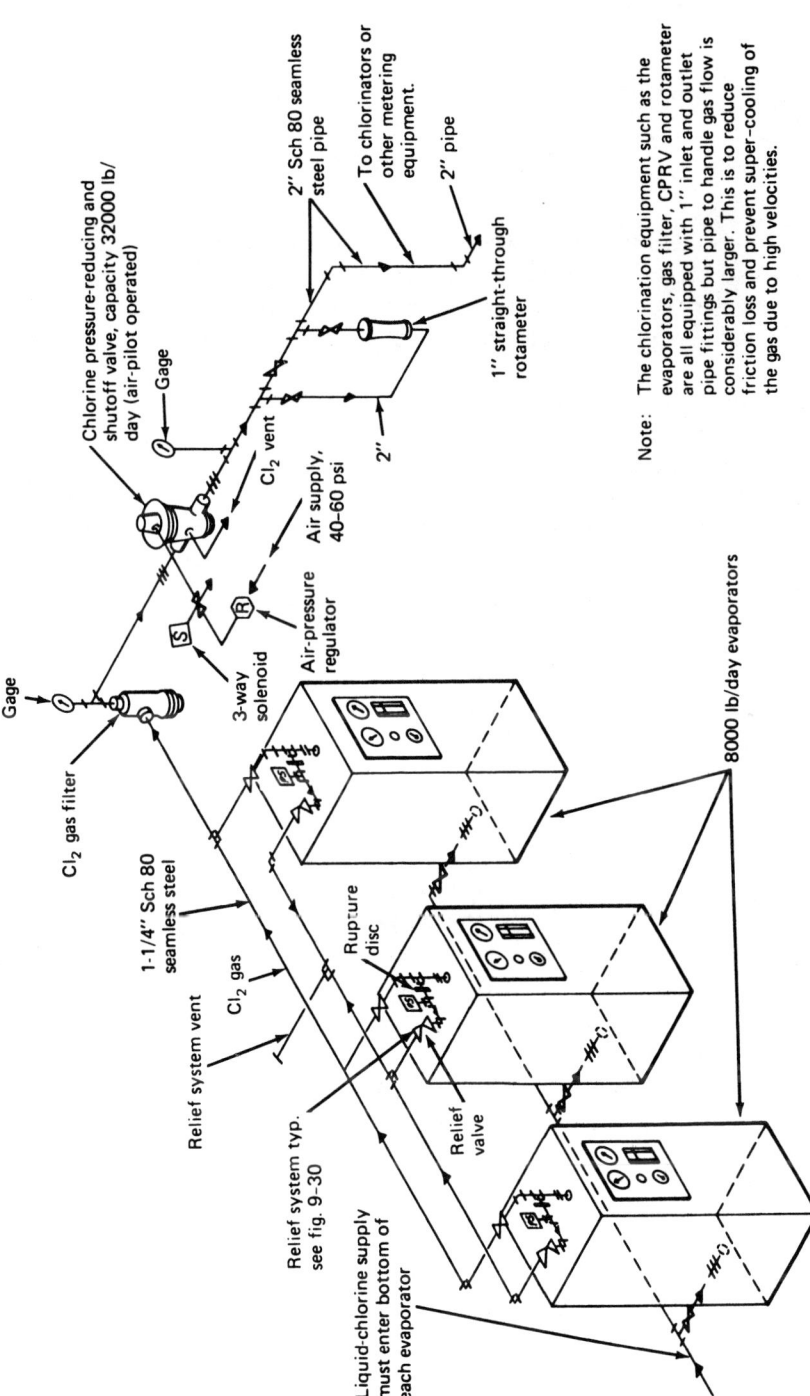

Fig. 9-36. Manifold chlorine evaporators. The chlorination equipment, such as the evaporators, gas filter, CPRV, and rotameter, is all equipped with 1-in. inlet and outlet pipe fittings, but pipe to handle gas flow is considerably larger. This is to reduce friction loss and prevent super-cooling of the gas owing to high velocities.

and

$$V = \frac{0.4265}{0.01225} = 34.82 \text{ ft/sec}$$

This is close to the upper limit of velocity, which starts a vicious cycle: as the velocity increases, the gas becomes cooler; as the gas cools, it becomes denser, thus increasing the friction loss. As the friction loss increases, it becomes more and more difficult to regulate the density through the rotameter in order to maintain its calibration.

The equivalent length of pipe between the CPRV and chlorinator may be as high as 60 ft, even though the two units are no more than 10 to 15 ft in pipe length apart, depending upon the number and kinds of fittings in the piping layout.

Another method of analysis is to use Reynolds number (N_r), a dimensionless quantity that relates velocity, viscosity, and size of pipe. It can be expressed in the following ways:[21]

$$N_r = \frac{pVD}{u'} = \frac{6.32\ W}{ud} \tag{9-5}$$

where

V = velocity, ft/hr
D = diameter, ft
W = mass flow, lb/hr
d = diameter, in
p = density, lb/cu ft
u = viscosity, cp
u' = viscosity, lb/hr-ft
$u' = 2.42u$

Taking the case of 24,000 lb/day flow in a 1" pipe, Reynolds number is calculated as follows:

$$N_r = \frac{6.32 \times 1000}{0.0127 \times 0.957}$$

The temperature of the gas at this velocity in a 1" pipe will drop to about 40°F; hence $u = 0.0127$ cp and therefore

$$N_r = 654,600$$

Not let us make the same calculation for the case where the maximum recommended allowable velocity in 1" pipe is two evaporators manifolded together = 16,000 lb/day flow, therefore

$$N_r = \frac{6.32 \times 666.7}{0.0129 \times 0.957}$$

The temperature of the gas at this velocity in a 1" pipe will be about 50°F; hence $u = 0.0129$ cp, and therefore

$$N_r = 364{,}175$$

Now for the 24,000 lb flow in a 1½" pipe which shows a velocity under the 50 ft/sec maximum limit, Reynolds number will be

$$N_r = \frac{6.32 \times 1000}{0.0131 \times 1.500} = 321{,}630$$

The temperature of the gas at this velocity is estimated to be approximately 60°F.

It would appear that the size of the pipe must be such that the Reynolds number is kept below 365,000 for any piping system between the CPRV and the chlorinator.

CHLORINATORS

Historical Background (See Also Chapter 6):

Direct Gas Feed. The first gas chlorinators developed circa 1906–10 were of the direct gas feed type. No matter how ingenious or competent the design, these chlorinators suffered disastrously from flooding by water at the point of application. This occurred in spite of a negative head at the point of application. Maintenance due to corrosion was excessive to say the least.

The Suck-Back Phenomenon. This occurrence is related to the affinity chlorine has for water. If the chlorine feed is shut off or if the feed rate is low enough (<0.25 lb/day), chlorine gas in the feed line adjacent to the diffuser will gradually be absorbed by the water. If this condition persists a vacuum begins to build up in the feed line, which will eventually pull the water from the point of application all the way back to the chlorine cylinder—in spite of the vapor pressure in the cylinder! This is the suck-back phenomenon that plagued all direct-feed chlorinators until special corrosion-resistant back-pressure valves were designed to prevent this occurrence.

The Bell-Jar Era. In an all-out effort to solve the two major problems of chlorine gas metering, the suck-back phenomenon and impurities in the gas, C. F. Wallace developed the vacuum solution-feed bell-jar line of chlorinators. This line of chlorinators enjoyed great success from circa 1922 to about 1960. Fig. 6-2 (Ch. 6) illustrates a flow-paced automatic unit—maximum capacity 500 lb/day. The operation of the chlorine pressure reducing valve was visible inside the bell jar. The vacuum created by the injector was transmitted to the bell jar, where chlorine impurities present were spewed into the glass jar and could be easily retrieved for analysis. The evidence provided by operators of these chlorinators resulted in the production of a special "clean" grade of chlorine for water supplies, wastewater, and cooling waters. The suck-back phenomenon was resolved by the incorporation of the vacuum relief system which automatically allowed air to enter the chlorinator, thereby breaking any vacuum that might result from suck-back. All solution-feed chlorinators today have vacuum relief systems. The two salient points that were responsible for the success of bell-jar chlorinators were: (1) they were made entirely of corrosion-resistant materials, and (2) the opration was completely visible. Their major disadvantage was feed-rate capacity limitation. For feed rates in excess of 500 lb/day the design became complicated, which resulted in excessive production cost.

Development of the Spring-Loaded Diaphragm Concept:

General Discussion. The famous bell-jar line of chlorinators pioneered by Wallace and Tiernan was replaced virtually overnight by the introduction of PVC injection molding. Once again the original spring diaphragm concept was revived and its success was made possible by the research into plastics and corrosion-resistant diaphragms, particularly by DuPont. The very first series of chlorinators marketed by Wallace and Tiernan were based upon the spring-loaded diaphragm principle (ca 1913). Owing to severe corrosion problems this concept was dropped as soon as the visible vacuum models by C. F. Wallace were introduced (ca 1920). When PVC injection molding and teflon diaphragms became available the chlorinator design approach reverted to the original spring-loaded diaphragm principle, utilizing the technology of the newly arrived plastics industry (ca 1955).

This design concept, pioneered by Fischer and Porter, was a much needed stimulus for the chlorinator industry. This was a small specialty manufacturing industry at that time, but once the ingredient of quality competition was introduced the results for the consumer were dramatic. To counter the impact of the Fischer and Porter "all plastic" line, Wallace and Tiernan introduced their line of "plastic" chlorinators but with a revolutionary idea of chlorine feed-rate control. Wallace and Tiernan introduced the V-notch orifice concept, which they had been keeping under wraps for at least 10 years. This concept of chlorine metering control proved to be unique. From this point forward Wallace and Tiernan developed unbelievably accurate and flexible control systems. Their success spurred their competitors to greater achievements, some of which are described below.

Theory of Operation. These chlorinators consisted of the following components:

1. Inlet chlorine pressure-reducing valve
2. Indicating meter (rotameter)
3. Chlorine metering orifice
4. Manual feed rate adjuster
5. Vacuum differential-regulating valve
6. Pressure-vacuum relief valve
7. Injector

Figure 9-37 is a flow diagram of a Wallace and Tiernan V-800 chlorinator illustrating a "pressure" type system as opposed to the remote vacuum arrangement. The chlorine gas enters the system through the pressure-vacuum regulating valve at which point the inlet chlorine pressure from the supply system is reduced to some constant level of negative (vacuum) pressure (the level of vacuum varies with each manufacturer). The gas then passes through the metering orifice (V-notch) to the differential vacuum regulator and then to the injector vacuum line.

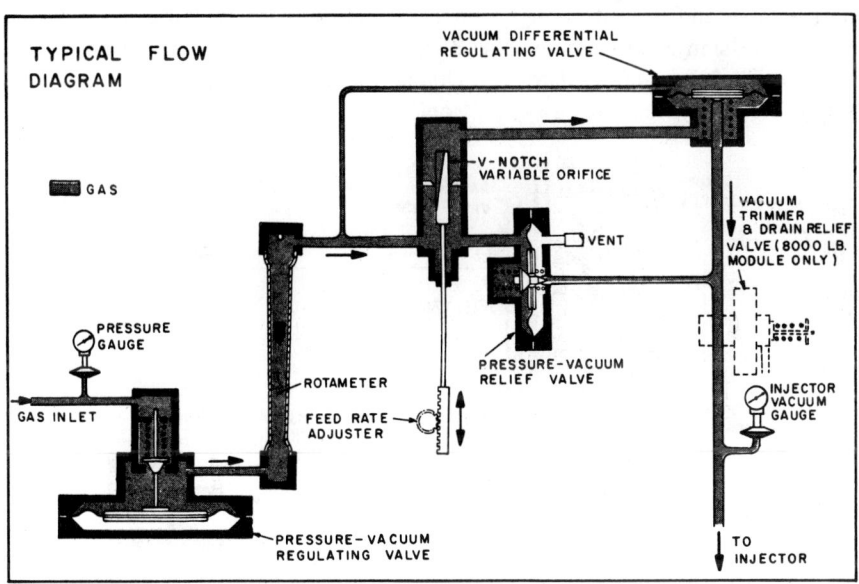

Fig. 9-37. Flow diagram V-800 series vacuum solution feed chlorinator (courtesy Wallace and Tiernan Div. Pennwalt Corp.).

The chlorine flow through the V-notch orifice is based upon the classic flow formula $Q = AV$, where A is the area of the orifice opening (position of V-notch orifice positioner) and V is the velocity of the gas through the orifice. This gas velocity is best expressed in terms of the differential pressure across the orifice to produce a given velocity. This is equal to $C\sqrt{2gh}$, where h equals the differential vacuum across the differential vacuum regulator and C is the velocity coefficient of the V-notch orifice. The pressure–vacuum regulating valve illustrated in Fig. 9-37 is designed to maintain a constant pressure (usually a slight vacuum) upstream from the metering orifice and control device. The vacuum differential regulating valve is designed to maintain a constant pressure drop (h) across the metering orifice (V-notch orifice).

Sonic Flow Concept. The most recent development in chlorinator design which affects accuracy and control modes is the concept of sonic flow. Previously described concepts assumed that the gas flow through the metering orifice (feed-rate valve) was a function of the differential pressure across that valve (h). This is true for a wide range of differential pressures; however, if the velocity through the valve is increased to the speed of sound in the gas flow at that point, a different set of conditions is encountered. Once the sonic velocity is reached, the flow through the valve is no longer a function of the pressure drop (h) across the valve. Under these conditions, gas flow is directly proportional to the area of opening in the control valve and is entirely independent of the downstream pressure, which is a function of the injector vacuum. Therefore, when sonic flow conditions are attained, the differential vacuum regulator is no longer a necessary component of chlorinator design. Fig. 9-38 illustrates a typical sonic flow design by Capital Controls. Sonic

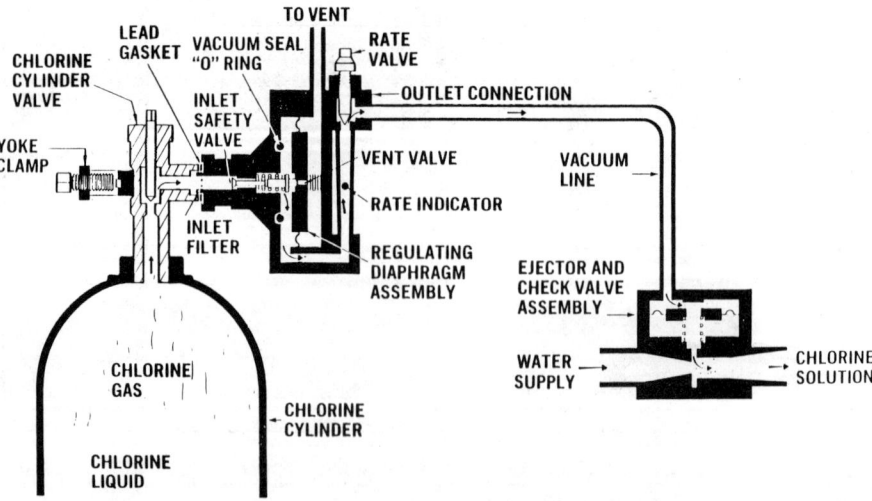

Fig. 9-38. Flow diagram sonic flow type chlorinator (courtesy Capital Controls Co.).

flow chlorinators are limited to 2000 lb/day capacity owing to the additional energy required by the injector systems to achieve sonic velocities across the chlorine feed-rate valve.

Remote Vacuum. This is a concept whereby the chlorine supply system operates under a vacuum from the cylinder to the inlet of the injector. Originally the advantage of this arrangement was to increase the margin of safety in the handling of chlorine. However, this is a minor benefit, because analysis of chlorine emission accidents indicates that the supply piping is rarely a factor in these occurrences. The major benefit is the use of the remote vacuum concept for automatic cylinder switchover. Remote vacuum systems are available up to 8000 lb/day as shown in Fig. 9-39.[22] This is of doubtful value because the system is applicable only with the vapor phase. An 8000 lb/day chlorinator requires an evaporator, therefore the vapor under pressure travels only a short distance where space is at a premium for mounting the vacuum regulating valve and pressure-check, and pressure relief

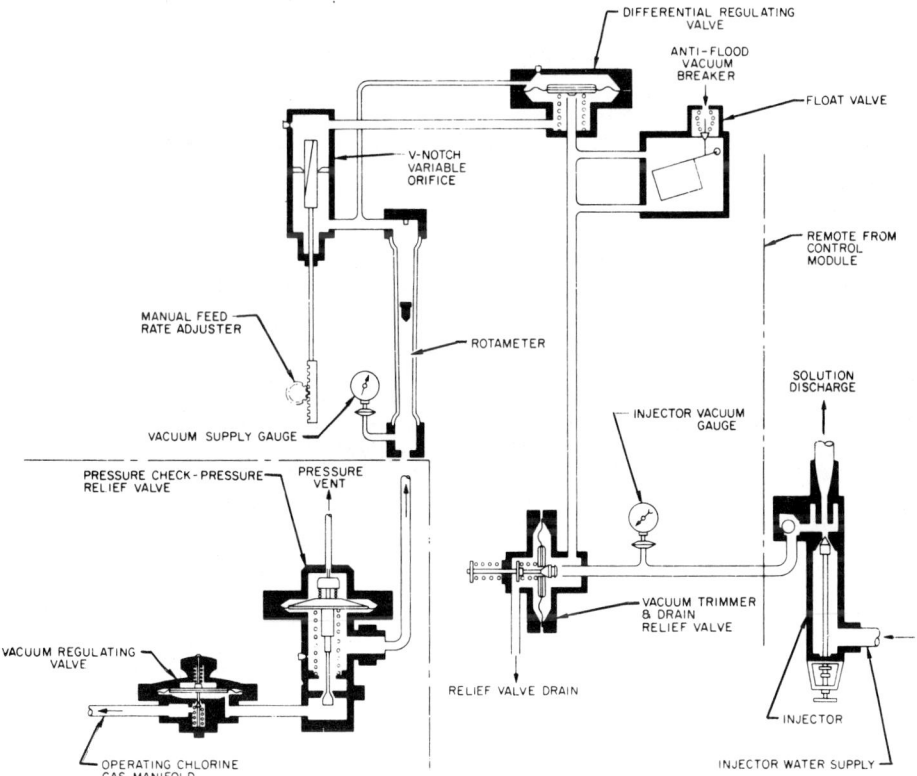

Fig. 9-39. Flow diagram remote vacuum system for 8000 lb/day chlorinator (courtesy Wallace and Tiernan Div. Pennwalt Corp.).

660 HANDBOOK OF CHLORINATION

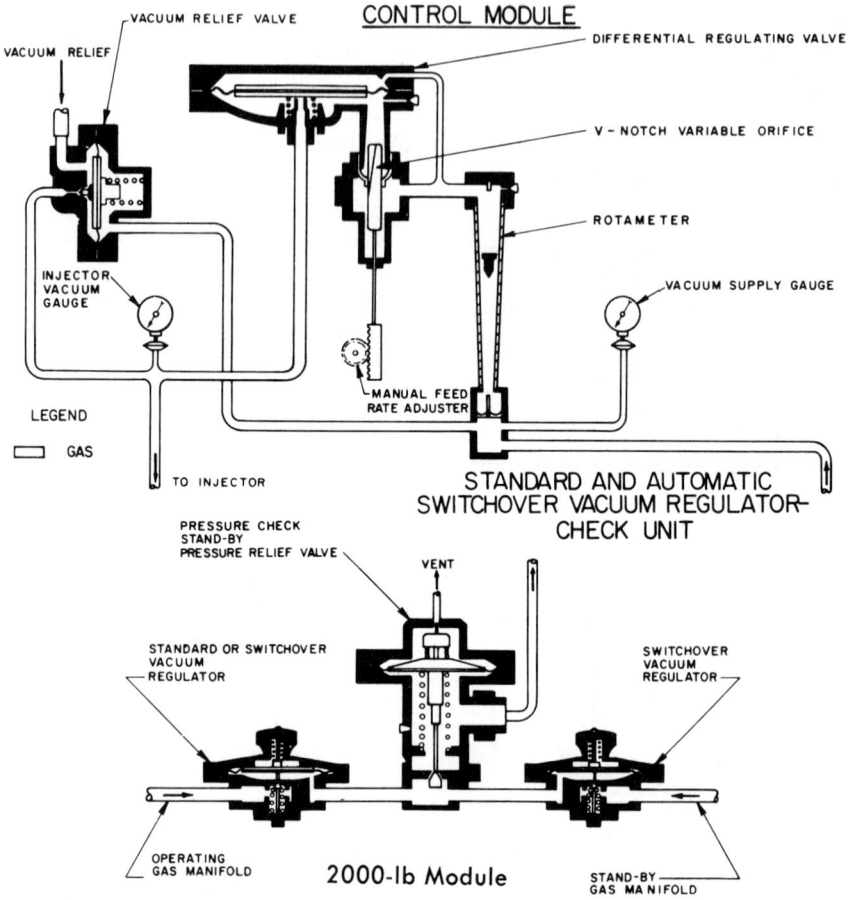

Fig. 9-40. Flow diagram of combination remote vacuum and automatic switchover system; 2000 lb/day chlorinator (courtesy Wallace and Tiernan Div. Pennwalt Corp.)

valve. Therefore when liquid is used instead of vapor the chlorinator should be arranged as shown in Figs. 9-11 and 9-37 with the pressure–vacuum regulating valve located within the chlorinator cabinet.* The 8000 lb/day unit (Wallace and Tiernan) utilizes a combination vacuum-trimmer and drain-relief valve in combination with an anti-flood vacuum breaker instead of a conventional vacuum relief valve. The 500 lb and 2000 lb Wallace and Tiernan chlorinators use a conventional vacuum relief valve with the remote vacuum arrangement as shown in Fig. 9-40.[22]

It should be noted that although remote vacuum capability is available up to 8000 lb/day, automatic switchover by the remote vacuum principle is limited to 2000 lb/day capacity. *Pressure sensing for automatic switchover is available for any capacity.*

---
* This arrangement is currently available only on special order.

Control Strategies:*

General Discussion. All chlorinators are arranged for basic manual control. This control is accomplished by opening and closing a valve characterized for a nominal range of 10:1. Most manufacturers provide rotameters calibrated over a 20:1 range. Chlorinator models are usually separated into capacity categories of 100, 500, 2000, and 8000 lb/day. Some manufacturers stretch the capacity of the 8000 lb/day size to 10,000 lb/day, depending upon certain site conditions.

For each of the models described above there is a variety of metering tubes and other internal accessories available to cover the entire capacity range of each category. For example, for a 2000 lb chlorinator, the available rotameter tube sizes are: 50, 75, 100, 150, 250, 500, 1000, 1500, and 2000 lb/day.

Flow Pacing. This is accomplished by the use of an analog signal transmitted from the process flow meter to the motor drive on the shaft that positions the chlorine metering orifice. This valve is designed and characterized so that the flow of chlorine varies *directly* as the process flow changes. Therefore the signal transmitted by the process flow meter must be linear with process flow. This signal can be either electric (4–20 MA) or pneumatic (3–15 psi). In the Wallace and Tiernan chlorinators the chlorine orifice valve is commonly referred to as the V-notch plug. The Fischer and Porter valve is known as the Chloromatic valve. Depending upon local conditions an optional item for flow paced control is the use of a ratio relay. This is desirable when the primary meter is oversized for future conditions.

> EXAMPLE. A treatment plant is designed for 50 mgd in the year 2000. Current peak daily demand is only 20 mgd. Install a ratio relay with a range of 0.4–4.0 in order to get the full range of the chlorinator.

When considering flow pacing by itself there is tradeoff between range and accuracy. A chlorine dosage that is flow paced only will only be as accurate as the transmitted flow signal. The accuracy of this signal depends upon the primary metering element. Designers should specify the accuracy they want over a specified range, i.e., 3:1, 5:1, 10:1. Many makers of primary elements are reluctant to claim ±2 percent accuracy over a range that exceeds 4:1. This is a crucial factor in precise chlorine dosage control. It is discussed in further detail below.

Direct Residual Control. For new systems where retrofitting is not a problem, this method has considerable appeal. The key elements that make up a successful operating system are: (1) selection of proper injector water pump, (2) careful design of chlorine diffuser and residual sample tap, and (3) a short loop time.

The appropriate way to describe the direct residual control concept is by an example. A 48 inch pipe carries an average daily flow of 12 mgd and summer peaks of 24 mgd. Low flow during the early morning hours is 2 mgd. This represents

---
* See Appendix for control function definitions.

662 HANDBOOK OF CHLORINATION

a 12:1 flow range and a velocity range of 0.31–2.96 ft/sec. The first step is the diffuser design to provide proper initial mixing at the point of chlorine injection. Assume chlorine dosage will be a maximum of 2.5 mg/l. This amounts to 500 lb/day at 24 mgd flow. This is the upper limit for a fixed-throat injector, so choose a 2 inch adjustable throat injector capable of passing at least 70 gpm. This is for mixing reasons, as will be shown. Use an across-the-pipe diffuser as shown in Fig. 9-40a with two cluster jets of three holes each pointing upstream. Using the Egan concept of upstream jet energy for mixing, design chlorine diffuser perforations for 25 ft/sec velocity. This requires six ½" holes at 12 gpm per hole for a total amount of injector water of 72 gpm. This diffuser will develop a head loss of 11.75 ft.

The next step is to calculate the G factor

$$G = \sqrt{\frac{550 \times P}{\mu V}} \tag{9-6}$$

$$P = \frac{Qh}{3960} = \frac{72 \times 11.75}{3960} = 0.21 \tag{9-7}$$

In order to calculate G the volume of the initial mixing area affected by the energy of the jet will have to be decided. Experience in wastewater chlorination indicates that mixing will occur in 2–3 seconds. The longer the time selected for mixing, the lower the G value will be. Assume 3 sec at average flow, 12 mgd, where $V = 1.48$ ft/sec. This means the mixing volume is 55.77 ft³.

$$G = \sqrt{\frac{550 \times 0.21}{2.35 \times 10^{-5} \times 55.77}} = 296.86$$

The next step is to calculate G for the dissipation in energy caused by the counterflow—this is called the Egan effect. Referring to Fig. 9-40a it is assumed that the obstruction of the 3-inch diffuser with the two cluster jets and the impinging effect of the cluster jets will dissipate some of the energy in the 12 mgd counterflow. The diffuser restriction increases the velocity of the mainstream from 1.48 ft/sec to 1.65 ft/sec and $V^2/2g = 0.04$ ft. Assume a head loss of 0.04 ft in the mainstream at this point.

$$P = \frac{8343 \text{ gpm} \times 8.34 \times 0.04}{60 \times 550} = 0.08 \text{ hp}$$

and

$$G = \sqrt{\frac{550 \times 0.08}{2.35 \times 10^{-5} \times 55.77}} = 183.23$$

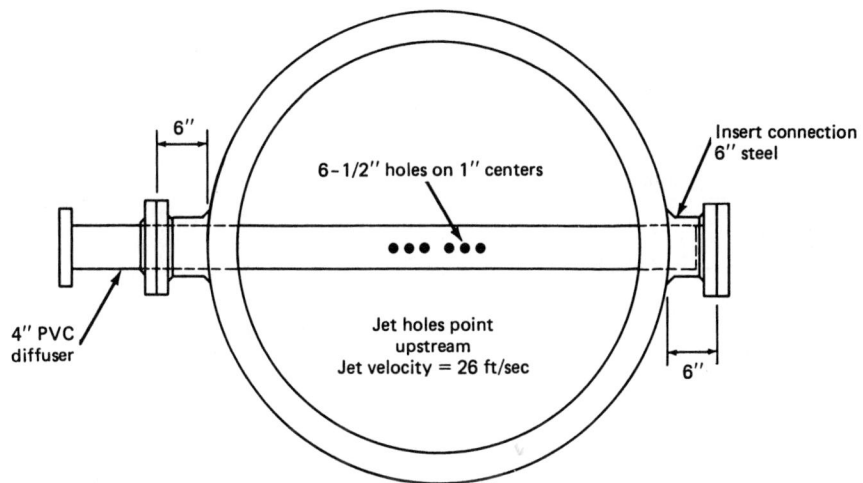

Fig. 9-40a. Egan type cluster-jet counterflow diffuser.

Therefore it is reasonable to believe that the total energy dissipated can be translated into a G factor equal to the jets 297 plus the counterflow 183 for a total $G = 480$. This amount of turbulence is sufficient for adequate mixing.

Sampling Tap. The downstream sampling tap which is also the injector pump suction should be a duplicate of the chlorine diffusers with the holes pointing upstream, except that the number of holes must be increased to reduce head loss to one ft. This arrangement will insure that the sample received by the analyzer will be consistently homogeneous. Moreover, using the injector pump in this fashion will tend to short-circuit the main line flow in such a way as to prevent abnormally low velocities during low flow periods. See Fig. 9-41.

Loop Time. Estimating the loop time (system response) is the next step. Referring to Fig. 9-41, the loop-time circuit is defined as A-B-C-D-E. The travel time through all of these segments will be constant by design except from A-B and C-D. These segments of the loop will vary with the velocity of chlorine in the vacuum line and of water in the 48 inch pipe from low flow to peak flow conditions.

From Fig. 9-41 use the following distances:

$$A\text{-}B = 300 \text{ ft}$$
$$B\text{-}C = 30 \text{ ft}$$
$$C\text{-}D = 10 \text{ diam.} = 40 \text{ ft}$$
$$D\text{-}D_1 = 10 \text{ ft}$$
$$D_1\text{-}E = 15 \text{ ft}$$

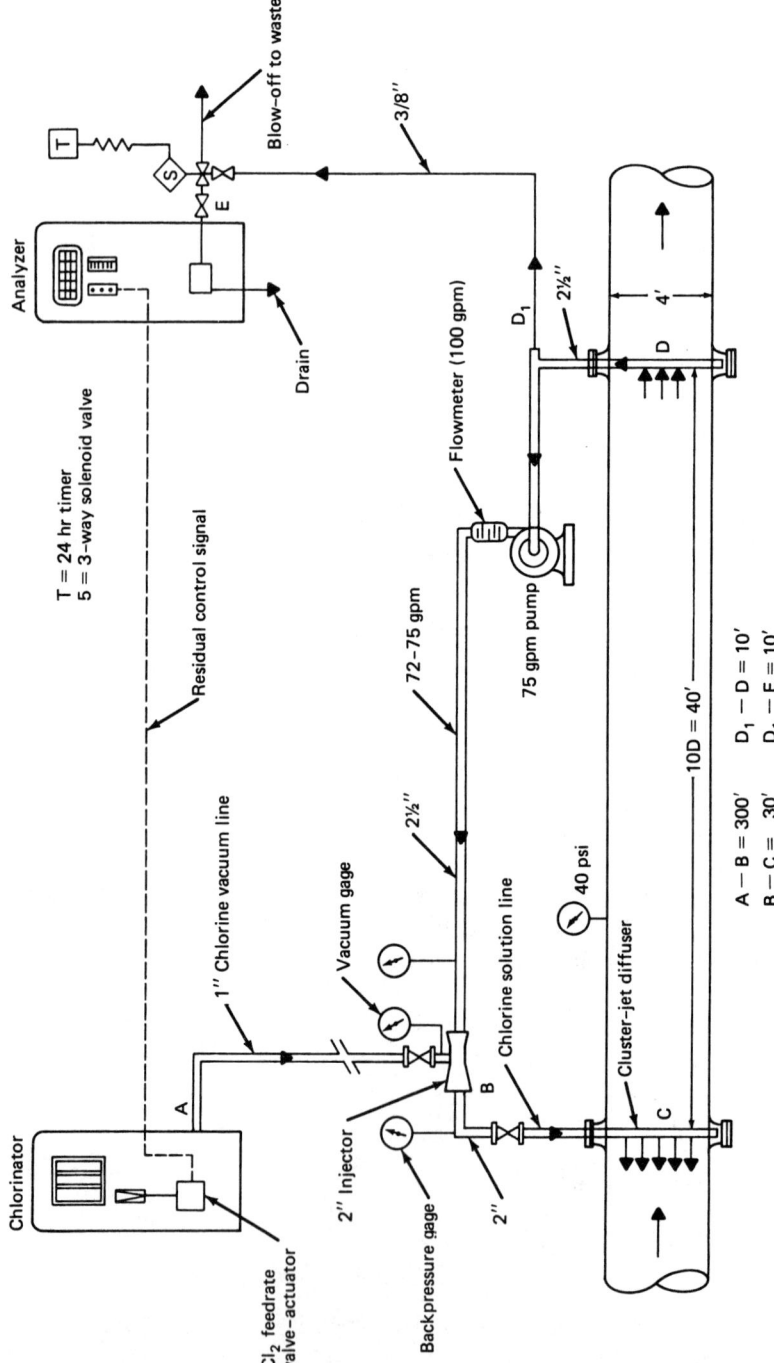

Fig. 9-41. Direct residual control system using injector water pump to provide analyzer sample which minimizes control loop dead time.

Injector pump flow by pump selection and control valve will be 72–75 gpm. Sample flow to the analyzer pump suction will be 1 gpm and pressure in 48 inch pipe is estimated at 40 psi.

The next step is to calculate the flow in each of the segments:

1. *A-B*, chlorine gas flow at 500 lb/day max. Try a 1 inch PVC vacuum line:

$$Q = 500 \text{ lb/day} = 0.0115 \text{ lb/sec}$$

Density of Cl_2 gas at injector vacuum of 20 in. Hg = 0.06 lb/ft³

$$Q = \frac{0.0115}{0.06} = 0.1917 \text{ cfs}$$

and

$$V = \frac{0.1917}{0.005} = 38.34 \text{ ft/sec}$$

Now calculate for friction loss to see if 1 inch pipe is large enough:

$$P = \frac{11.89 \times L \times f \times (W)^2}{10^9 \times P \times d^5} \tag{9-8}$$

where

L = equiv. length of vacuum line, ft
f = Darcy's friction factor
p = density of chlorine gas, lb/ft³
d = diameter of pipe, inches

The next step is to find the friction factor f by calculating the Reynolds Number:

$$N_R = \frac{6.32 \times W}{u \times d \times 24}$$

where

u = viscosity of chlorine gas = 0.0133.

$$N_R = \frac{6.32 \times 500}{0.0133 \times 0.957 \times 24} = 10345$$

From Fig. 8 (Appendix) the friction factor $f = 0.07$. Therefore

$$P = \frac{11.89 \times 300 \times 0.07 \times (500)^2}{10^9 \times 0.06 \times (0.957)^5}$$

$$= 1.3 \text{ in Hg.}$$

so 1" line is big enough. Therefore at 500 lb/day the gas velocity will be about 38 ft/sec and at low flow, 40 lb/day, the vacuum will be higher and the gas less dense.

$$Q = 40 \text{ lb/day} = 0.00046 \text{ lb/sec}$$

$$Q = \frac{0.00046}{p} = \frac{0.00046 \text{ lb/sec}}{0.03 \text{ lb/ft}^3} = 0.01543 \text{ cfs}$$

so

$$V = \frac{0.01543}{0.005} = 3.06 \text{ ft/sec}$$

Therefore the travel time for the chlorine in the vacuum line will vary from 100 sec at 40 lb/day rate to 8 sec at 500 lb/day. However, the response time following a dosage change occurs at a rate of 50 percent change in the first 25 percent of the time required to achieve the total dosage change of, say, 100 sec at 40 lb/day chlorine feed rate. For example, if the controller called for a 5 lb/day rate change at the 40 lb/day feed rate, the response time in the vacuum line to change the feed rate to 42.5 lb/day would be about 25 percent of 100 sec or 25 sec. This demonstrates the necessity to keep the chlorinator module as close to the injector as possible. *It also demonstrates that remote vacuum and automatic switchover systems should be designed so that the vacuum line between the chlorine supply system and the chlorinator be kept to a minimum,* i.e., not more than ± 15 ft.

2. *B-C.* This is the chlorine solution line and diffuser segment. The injector should be as close as physically possible to the diffuser inlet. Assume a distance of 10 ft. At 72–75 gpm the velocity in a 2 inch pipe will be about 8 ft/sec and will cause a head loss of about 1.5 ft. The response time for this segment will be negligible—about 1 sec.

3. *C-D.* This is the 40 ft length of 48 inch pipe between the chlorine diffuser and the pump suction which is also the sampling tap. At peak flow rate (24 mgd) $V = 2.96$ ft/sec and travel time to sample tap $= 14$ sec; at low flow (2 mgd) $V = 0.25$ ft/sec, so travel time will be 160 sec to the sample tap.

4. *D-D_1.* This is the segment where the sample is carried in the pump suction which is a 3 inch pipe. Assume this is 10 ft. At 75 gpm, $V = 3.4$ ft/sec, so travel time is a constant 3 sec.

5. *D_1-E.* This segment is the sample line to the analyzer from the injector

pump suction. The flow requirement to the analyzer is only 1 gpm maximum. Therefore use a ⅜ in. PVC pipe (brass or copper not allowed). The velocity in this pipe will be 1.67 ft./sec at 1.0 gpm. Assume this pipe is 10 ft long. Travel time will be a constant 6 sec.

Table 9-7 summarizes the response time elements. This represents a 9:1 change in response rate for a 12:1 flow treatment range. It is obvious that a great improvement can be made if the chlorinator could be moved to shorten the vacuum line to a reasonable distance of 10–20 ft. However, this wide range of response can be handled by a proper control system. It is recommended, however, that the upper limit of response time for low flow conditions should never exceed 5 min.

Table 9-7 Loop Time

	Loop Segment	Low Flow (2 mgd), seconds	Peak Flow (24 mgd), seconds
A-B	Cl_2 vacuum Line	100	8
B-C	Cl_2 solution Line	1	1
C-D	48 in. pipeline	160	14
D-D_1	Sample line, pump suction	3	3
D_1-E	Sample line to analyzer	6	6
	Total loop time	270	32*

*Response time for the sample within the analyzer is about 15–20 seconds.

An embellishment would be to use an alternative sample point for flows less than 5 mg/day, located 20 ft downstream from the diffuser. This is usually unsatisfactory because the sample tends to be stratified with various levels of chlorine residual. Moreover, since this is an exercise in direct residual control the above concept requires a flow meter to switch the valves on the two pump suction points.

The control system shown in Fig. 9-41 requires the electric valve operator for the chlorinator feed rate to be able to change the feed rate from zero to maximum in at least 15 seconds. The analyzer is required to have a 4–20 mA transmitter which will provide a linear signal from zero to full scale residual over a specified range of 0–1, 2, 5, 10, or 20 mg/l chlorine residual. This signal is transmitted to an adjustable residual set point controller with proportional, integral (reset), and derivative control functions. This controller drives the chlorine feed-rate valve. Potable water chlorination systems have been known to operate well on a 0.1 mg/l dead band where the residual span is 0–2 mg/l. One wastewater plant has operated successfully with a dead band of 0.2 mg/l where the residual span is 0–5 mg/l.[23]

A most important requirement in the operation of continuous residual analyzers is the necessity to purge the sample line at frequent intervals in order to scour away the biological slimes that are prone to build up in the sample lines. Referring to Fig. 9-41 there is shown a sample blow-off line controlled by a timer. The purging cycle should be once every 24 hours for 2–3 min. The orifice on the downstream side of the ⅜ inch solenoid valve should be a ⅜ × ¼ reducer ending in about 6 inches of ¼ inch pipe.

The injector water pump for the system shown will be required to have a TDH of approximately 60 psi to provide 100 psi inlet pressure to the injector.

Compound Loop Control. This control strategy was first introduced in 1960 by Wallace and Tiernan Inc. (U.S. Patent No. 2,929,393). The control scheme is based upon two separate but independent signals, so that each can change the chlorine feed rate separately. One signal represents process flow changes and the other registers chlorine residual changes. The first chlorine residual control for a wastewater plant utilized this system.[79] The installation was at the Napa Sanitary District treatment plant, Napa, CA in 1961. Since that time there have been several hundred such installations. However, in 1984 Wallace and Tiernan elected to drop this system in favor of the remote vacuum method. Since the compound loop system is being used extensively and since it can be made available on special order the following discussion is deemed worthwhile.

The difference between this method and other systems using two signals is as follows: each signal controls a different component within the chlorinator, and each performs an independent chlorine feed-rate change by a different function of gas flow mechanics. Other systems using dual signals integrate the two signals by external instrumentation which transmits a single multiplied signal to the chlorine feed-rate valve. (Exception: the Fischer and Porter Chloromatic feed-rate valve contains built-in multiplier instrumentation.)

As illustrated by Fig. 9-42 the process flow signal is transmitted to the motor-operated chlorine feed-rate valve and the chlorine residual signal goes to the vacuum differential regulating valve. This concept is based upon the fundamental law of fluid mechanics: $Q = AV$, where Q = chlorine flow, A = area of the chlorine orifice opening, and V = velocity through the orifice. The velocity through the orifice follows the following law of fluid mechanics: $V = \sqrt{2gh}$, where h is the head loss required to initiate a gas flow change. In this method h is the vacuum differential across the chlorine orifice which is controlled by the vacuum differential regulating valve.

Chlorine Feed-Rate Valve. This component is described by Wallace and Tiernan as the V-notch orifice positioner. The motor drive is capable of changing the feed rate from zero to maximum in 15 seconds. There is an internal feedback loop which controls and verifies valve travel for any signal from 4 to 20 mA. This valve is controlled by the signal from the primary flow meter.

Vacuum Differential Regulating Valve. This valve changes the chlorine feed rate by changing the vacuum differential across the chlorine rate valve orifice. On Wallace and Tiernan equipment the vacuum range is from 8 to 88 in. H_2O, zero to maximum feed rate. This vacuum level can be controlled by a vacuum

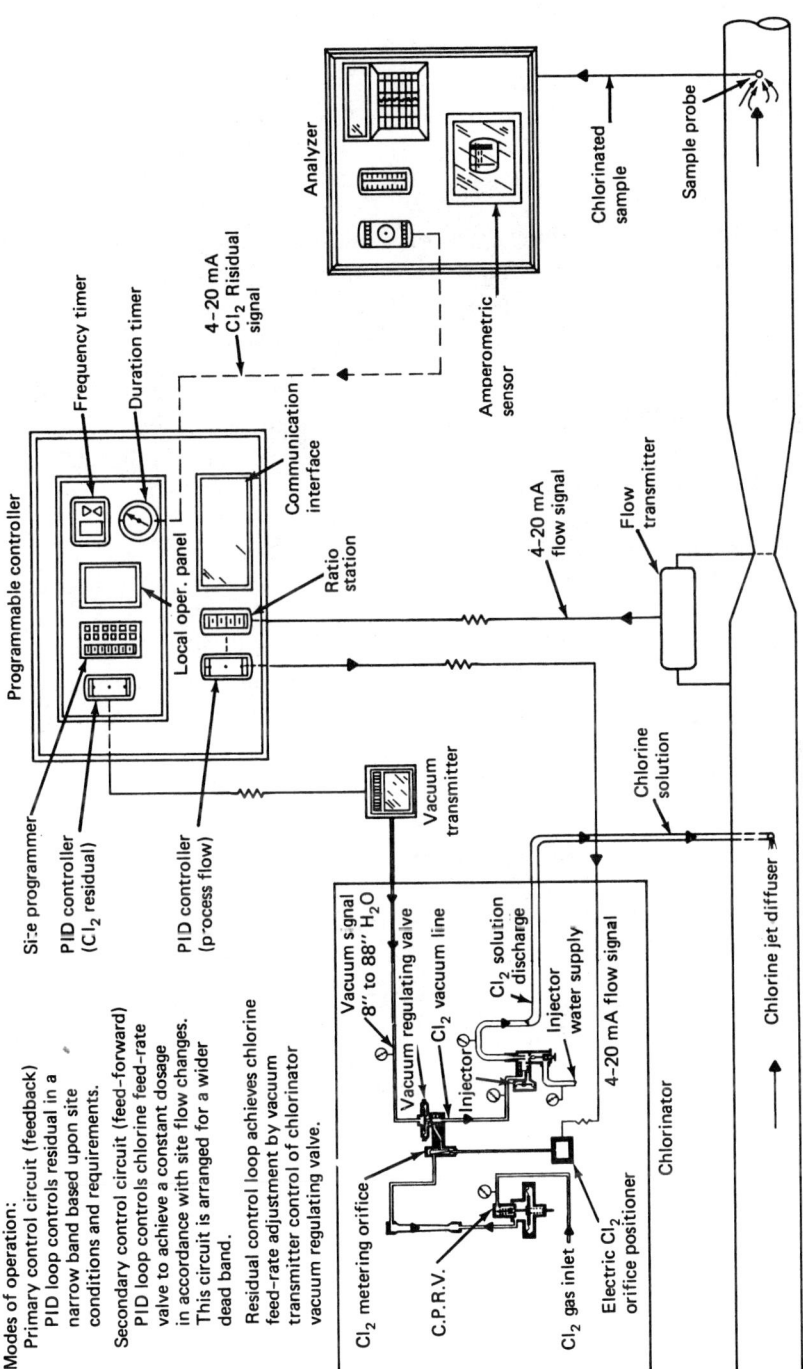

Fig. 9-42. Compound-loop residual control system using a programmable controller.

transmitter capable of changing the vacuum over the complete range in about five seconds. This vacuum level can also be controlled by a motorized vacuum valve which has a much slower vacuum rate change, approximately 15 minutes from 8 to 88 in. H_2O. This provides floating control by a cycle and duration timer.

Table 9-8 illustrates the mathematics of variable vacuum control as a function of chlorine feed rate. This is informative because it is recommended procedure to adjust the vacuum level of the vacuum regulator in the mid-range or 50 percent of valve travel. This is shown as 28 inches H_2O vacuum.

Theory of Operation. Fig. 9-42 illustrates a potable water chlorination system which was retrofitted from flow pacing control to compound loop control using a programmable controller to perform the intelligence functions. The programmable controller for each chlorinator installation requires the control strategy to be designed for specific field conditions. This particular system described by Fig. 9-42 had the following characteristics:

1. The chlorinator was located at the outlet of a reservoir where reverse flow occurred on a daily basis.

2. Water flow meter in a 60 inch concrete pipe had a range far exceeding any practical use, 0–200 cfs. The actual flow range in the high demand season was from 10 to 120 cfs—a 12:1 range.

3. The piping configuration was such that upstream supplies entering this pipeline occasionally contributed some measurable chlorine residual.

Table 9.8 Arithmetic of Variable Vacuum Control
Wallace and Tiernan Equipment

Chlorine Feed Rate, %	Chlorine Flow as a Factor	Vacuum Differential = Cl_2 × Flow	Chlorine Flow as a Factor = Cl_2 Flow^2	Vacuum Signal to Chlorinator, in. H_2O (Variable Vacuum + Fixed Vacuum = Total)		Vacuum Gage in. H_2O
				Flow Factor × Range	Fixed Vacuum Zero Signal	
100	1.00	100	1.00	80	8	88
90	0.9	81	0.81	64.8	8	72.8
80	0.8	64	0.64	51.2	8	59.2
70	0.7	49	0.49	39.2	8	47.2
60	0.6	36	0.36	28.8	8	36.8
50	0.5	25	0.25	20.0	8	28.0
40	0.4	16	0.16	12.8	8	20.8
30	0.3	9	0.09	7.2	8	15.2
20	0.2	4	0.04	3.2	8	11.2
14.3	0.143	2	0.02	1.6	8	9.6
*						
10	0.1	1	0.01	0.8	8	8.8
0	0	0	0	0.0	8	8.0

* Practical limit of usable range = 7:1

CHLORINE FACILITIES DESIGN 671

Owing to the above characteristics it was decided to use the PC in the following manner:

1. Primary control was a PID feedback loop between the chlorine residual analyzer and the vacuum transmitter. This loop controlled the chlorine flow through the vacuum differential regulator. This control loop was designed with a narrow dead band.

2. Secondary control was achieved by a second feed forward PID loop which controlled the chlorine feed-rate valve to provide a constant dosage (lb Cl/cfs) all in accordance with water flow information from the flow meter. This loop was designed to operate on a wider dead-band. This loop was only for abrupt flow change corrections. If an alarm condition occurred for X seconds or minutes the feed-rate valve motor would open or close for N seconds.

3. If the chlorine residual level entered the PID alarm zones for X seconds the mathematical constant that converts the water flow rate to chlorine dosage (lb Cl/cfs) units is changed causing a greater or lesser dosage when the feedforward loop operates. This dosage computation change protects the system from wide variations of chlorine demand due in part to chlorine residuals in the system upstream from the chlorine diffuser.

4. The reverse flow sensor system shuts down the chlorinator for a specific length of time to allow for flow oscillation to cease. The chlorinator is restarted when steady outflow occurs. Reverse flow detection can be performed by the use of an auxiliary analyzer or a thermal probe sensing system. See "Reservoir Outlets" in this chapter.

5. The loop time for the chlorination system was determined in the field. It was 2 min 25 sec at 47 cfs. Deducting the travel time from C to D at 47 cfs the loop time constant calculated to 1.62 min. From this loop time characteristic the correction cycle frequency exhibited a variation from 2.0 min at 120 cfs to 5.5 min at 10 cfs and up to 11 min at flow reversal conditions.

6. The next step was to determine the required length of correction duration for an arbitrary dead band setting of 0.25 mg/l on the residual control PID loop. Using a 1000 lb chlorine rotameter and allowing each correction to be one-half the dead band or 0.125 mg/l, the chlorine feed-rate change required for each correction from 5 to 120 cfs flow is shown in Table 9-9. Since the chlorine feed-rate valve response is a constant at about 50 lb/sec it was decided to use a wider dead band for flow pacing control so that at about 30–40 cfs flow the chlorine feed-rate adjustment would be taken over by the residual control loop, where the feed-rate change can be as low as 10 lb per correction. Table 9-9 itemizes the chlorine feed-rate change per correction based upon one-half the dead band change in dosage = 0.125 mg/l.

Using the loop time information developed on site, the PC can be programmed with mathematical constants so that the chlorine feed rate change will respond quickly and accurately to wide flow fluctuations and chlorine demand changes. The use of dual PID loops provides greater flexibility over the wider range of

Table 9-9 Chlorine Feed-Rate Change per Correction

Water Flow		Cycle (min)	Cl_2 Flow Change, lb/day
cfs	mgd		
120	77.5	2.0	80.60
110	71.1	2.0	73.94
100	64.6	2.0	67.18
90	58.2	2.5	60.53
80	51.7	2.5	53.77
70	45.2	2.5	47.01
60	38.8	2.5	40.35
50	32.3	2.5	33.54
40	25.8	3.0	26.83
30	19.4	3.0	20.18
20	12.9	3.5	13.42
10	6.5	5.5	6.76
5	3.2	9.5	3.33

conditions. Assuming that the PID flow pacing loop has a ±3 percent accuracy over a 4:1 flow range and the residual control loop has at least a 7:1 range, a worst case situation is provided with at least a 28:1 range—far in excess of normal necessity. Other features of this type of dual signal control include the ability to isolate and operate manually each of loops either locally or remotely in case of analyzer failure.

The most important feature is the automatic takeover by the residual control loop when the process flow drops below the accurate range of the flow meter. More than 20 years of operating experience has demonstrated that this method of dual signal feedback control is a favorite with operating personnel because it is easy to adjust, calibrate, maintain, and understand.

When restarting the system or when recalibrating, the operator manually positions the vacuum valve to read 28 in. H_2O vacuum. This is followed by adjusting the chlorine feed-rate valve to give the appropriate chlorine dosage. The PC allows site-specific corrections and adjustments to account for loop time and historical flow patterns. Dual adjustable dead bands provide more flexibility and greater reliability of the control system.

Dual Signal Control Systems:

Introduction. This type of control strategy employs the use of two separate signals multiplied together to produce a single output which controls the chlorine feed-rate valve within the chlorinator. This is the method used by the three most active chlorinator manufacturers. This strategy is described in various ways, including compound loop control. It is not, however, the same as the compound loop control system described above.

In order to compare the systems offered by the different manufacturers *a common chlorination problem in potable water treatment was presented to them*. The strategies described for each manufacturer were based upon the following field conditions:

A 60 inch reinforced concrete pipe, served by a balancing reservoir, has an existing chlorine diffuser and a residual sample tap already in place. A venturi meter with a dp transmitter which generates a 4–20 mA output signal is located nearby. The flowmeter range is 0–200 cfs; however, historical records show that maximum flow has never exceeded 120 cfs. Low flows are on the order of 5 cfs due to almost flow reversal conditions.

The loop time at 47 cfs has been established as 2 min 25 sec. The distance from the chlorine diffuser to the residual sample point is 116 ft. Chlorine dosage is approximately 0.75 mg/l. The flow signal is not considered reliable under 50 cfs. Assuming one correction per loop time, frequency of correction will vary from 2 min up to 7 min.

Capital Controls Co. This firm announced a new, scientifically advanced chlorinator control system in 1984.[24] Fig. 9-43 shows a simplified schematic of this system. It is described by the manufacturer as a compound-loop control system, but it should not be confused with the Wallace and Tiernan compound loop system described above. It is a dual signal system which controls only one component in the chlorinator—the feed-rate valve.

Controller. The entire control system is shown in Fig. 9-44. It is packaged in a cabinet so that the operator can perform all the one-time adjustments, system calibration and residual set point adjustments from one convenient location.

The process water flow signal is modified by a dosage control adjustment (ratio station). It is a one-time set based upon flowmeter range and chlorine feed-rate valve sizing. The second signal, the residual signal, is compared to the set point. If the residual is at the set point no corrective action occurs. When corrective action is required an error signal is transmitted to the deviation gain. The deviation gain is a site-specific field adjustment (one-time set) which reflects the characteristics of the system. The deviation gain provides a wide choice of speed response to the amount of the error signal. For example, it permits a 10 percent difference of signal from set point to be altered from one percent of valve travel to 100 percent of valve travel.

The valve correction signal is transmitted simultaneously to the proportional multiplier circuit and the integral/timing circuit. These two circuits provide independent functions. The proportional circuit permits a rapid valve response while the integral circuit provides slow fine-tuning.

The proportional multiplication function is the same as proportional band. It is field adjustable, site specific, one-time set to fit local requirements. The integral/timing circuit sets an internal time delay (lag) in sampling the residual error signal. The adjustment is set on site and must have a time period equal to or greater than the loop time. (For a description of loop time, see Fig. 9-41.)

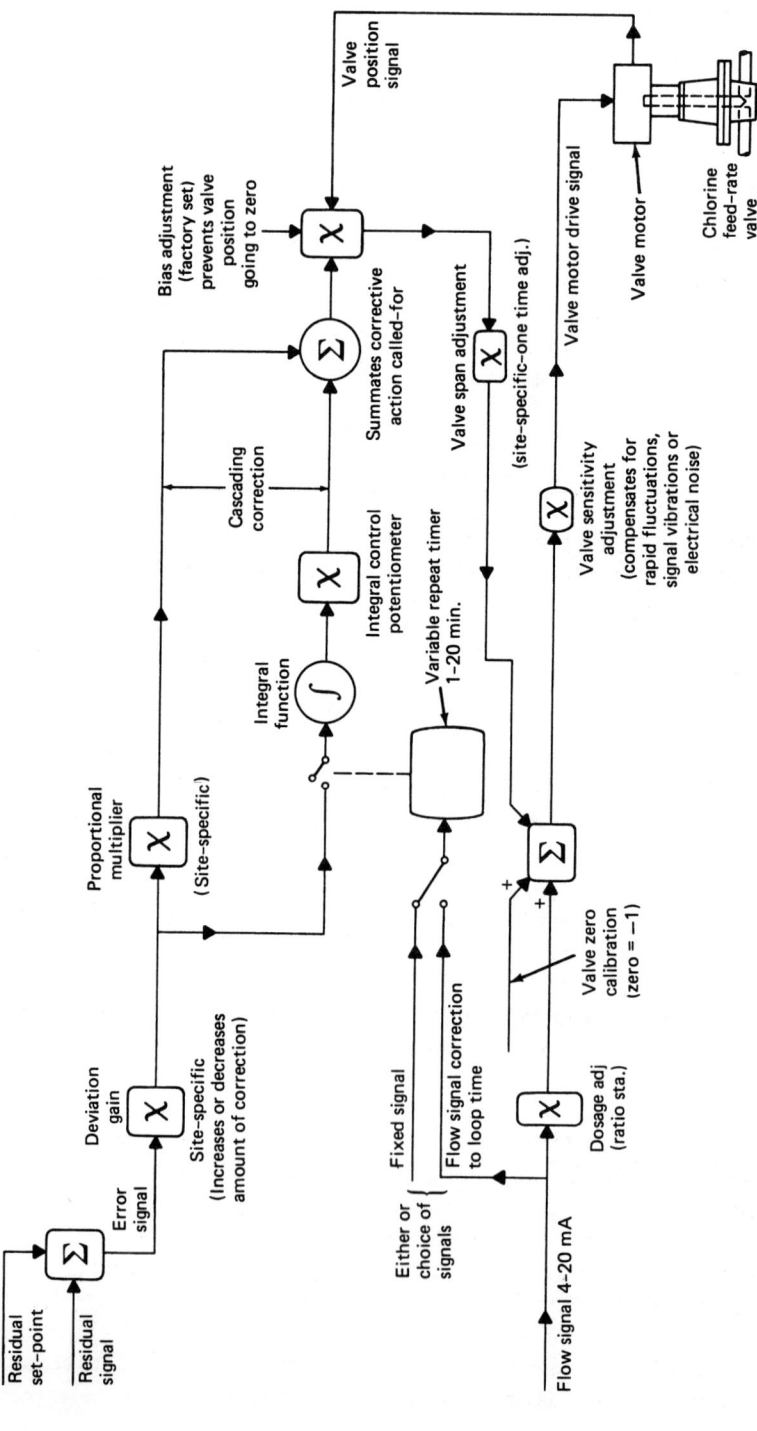

Fig. 9-43. Dual-signal residual control system (courtesy Capital Controls Co.).

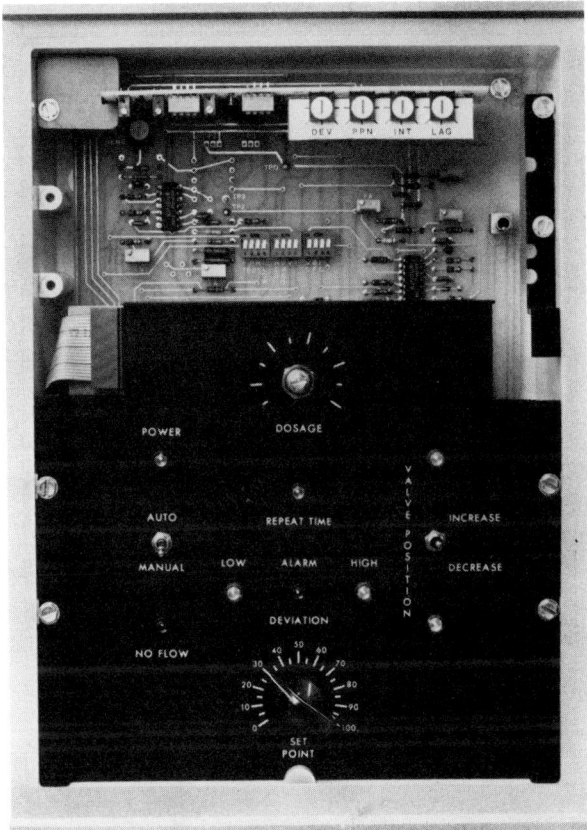

Fig. 9-44. Dual-signal control system electronic controller package (courtesy Capital Controls Co.).

The integral/timing circuit allows the valve to remain relatively dormant until the chlorine residual analyzer has time to measure a true sample. The lag adjustment varies frequency of correction. The length of correction is dependent upon the time taken by the rate valve to move to a new position. There is no change in output until a new correction signal is received after the lag-time cycle repeats. The lag time cycle frequency is arranged to be proportional to flow or set at a predetermined fixed value.

The integration circuit is the same as reset. It integrates or sums the error signal for a two second period after the lag-time cycle ends. The signal is held until the next lag-time cycle occurs and then recorrects itself. The integral function is the reset adjustment. It is field set for fine tuning at the time of startup.

At this point in the circuitry the signals from the proportional and integral loops are summed and the result is sent to a multiplier. This multiplier also receives a signal from the chlorine feed-rate valve feedback potentiometer.

A factory offset (bias) is added to permit the residual control function to override the flow proportioning function. This provides added protection and greater system reliability, particularly during low water flows or when the water flowmeter signal is erratic or unreliable.

The offset principle is described in the following mathematical model:

$$(R + C)\left(\frac{V}{100} + \theta\right) + (-F) + (-C) \times (\theta) = 0 \qquad (9\text{--}9)$$

$$V = \left[\frac{F - (R)(\theta)}{R + C}\right] 100 \qquad (9\text{--}10)$$

where

R = residual control signal voltage
C = factory set constant voltage (bias)
θ = constant ratio
F = flow signal voltage
V = valve position (percent)

Schematic of Mathematical Model

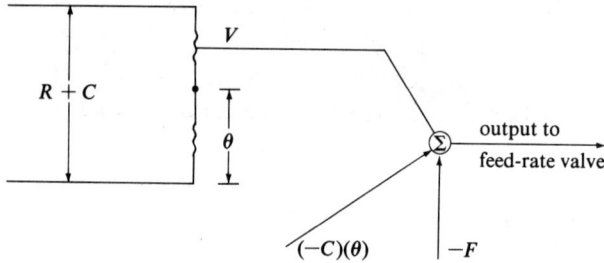

Equation (9–9) represents the summation of the input signals from chlorine residual, process flow, and the offset (bias)—which prevents the feed-rate valve from complete closure. (Chlorinators cannot be designed for zero flow conditions, except by interrupting the injector operation.) Eq. (9–10) represents the feed-rate valve position as percent open. This equation also shows how the residual control system can operate alone as a controller by modifying the bias signal to the chlorine feed-rate valve. The circuitry also allows the flowmeter control to act without residual control.

The multiplied signal has a valve span adjustment which is site-specific and usually a one-time set. This adjustment serves as a ratio station to trim a grossly over or undersized valve. The valve span is field adjustable.

Switches in the controller allow the operator to select any of these operating modes: manual, flow paced, flow paced–residual, residual alone, and dechlorination.

Chlorine Feed-Rate Valve. This is a constant-speed-motor driven linearized valve which requires approximately 70 seconds to travel from zero to 100 percent feed-rate position in 12½ revolutions. The valve can move physically through 16 turns. The valve also drives a ten-revolution geared feedback potentiometer. This allows fine positional resolution. One contact drives the motor upscale and another drives it downscale. The system operation is as follows: the controller sends out the appropriate contact closure, then the signal from the feedback potentiometer is compared with the control signal at point A. When these two signals equalize, the controller opens its output contacts and the motor holds its position.

The flow control signal works in the conventional feedforward manner—linearly with flow. The residual control signal, however, is applied to the rate valve feedback potentiometer and serves to vary the sensitivity of the valve. A large voltage across the potentiometer means the valve does not have to move very far to satisfy its feedback requirements. A lower voltage across the feedback potentiometer means the valve will have to physically move a greater distance (revolutions) in order to satisfy the same required voltage change. Changing the voltage to the feedback potentiometer allows the feed-rate valve to open or close completely independently of the flow signal.

This system has had insufficient field testing to allow any comment on its reliability. However, the advanced technology used in the circuitry makes this system appear to be tailor made for chlorination installations.

Fischer and Porter. Their dual signal system is built upon two principal components: (1) a 16-turn Chloromatic valve located within the chlorinator, and (2) a microprocessor-based controller located near the chlorinator for easy operator access.

The Chloromatic valve is a wide-range chlorine feed-rate valve with a characterized plug, which is arranged to accept a single signal input for automatic control. There are 3200 different valve positions in which the valve motor may stop. The schematic arrangement of the dual signal system is shown on Fig. 9-45.

The 53MC microprocessor controller is a single PID loop instrument which can be programmed as a special purpose controller including auxiliary instrument functions that may be part of the control loop.[71] This instrument is based upon a standard PID controller algorithm which can be configured through a built-in side keyboard. It features a modern gas discharge display of process information, both in vertical bar chart form and precise digital displays. Unusual control problems can be solved by using the FAPTRAM programming language. The programmable controller has a pressure sensitive keyboard which contains 13 functional keys used for configuration and programming plus 12 numerical keys for data entry. Three special keys are provided for tuning the constants in the PID loop.

Owing to the grossly oversized venturi meter stated in the field conditions, Fischer and Porter recommend two DP transmitters across the venturi to increase the reliable range. Assuming that reliability of the venturi meter will be acceptable

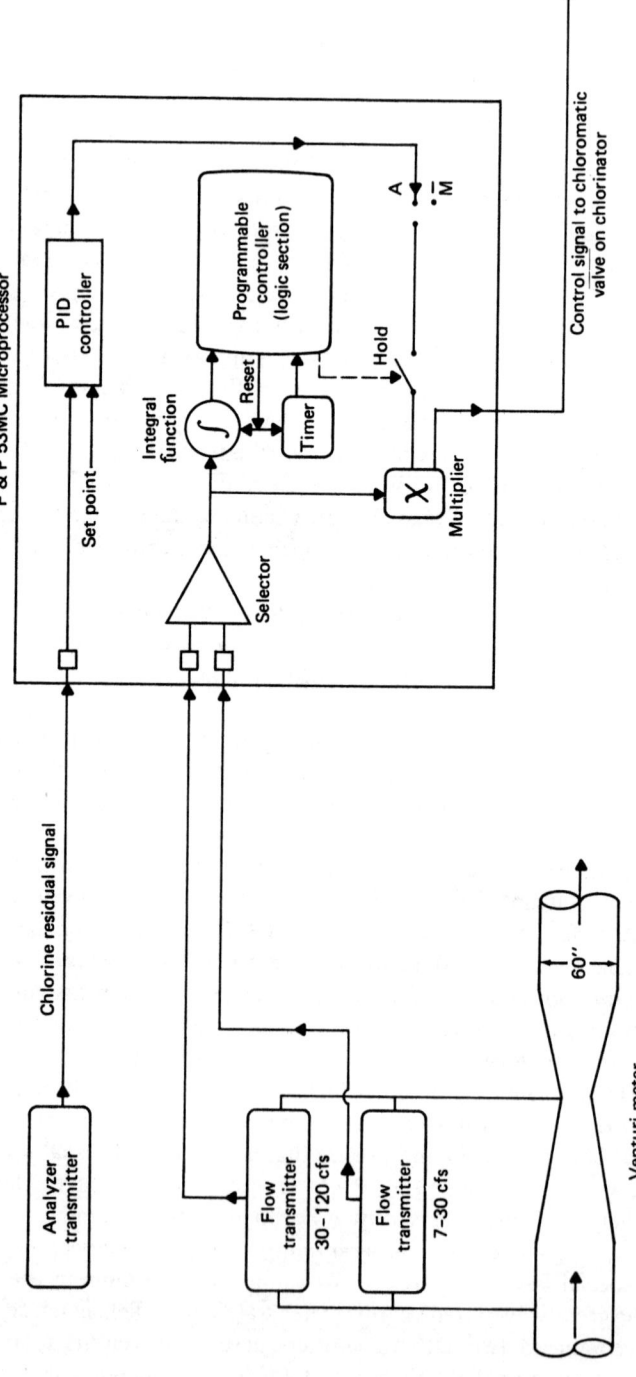

Fig. 9-45. Dual-signal residual control system (courtesy Fischer and Porter Co.).

over a 4:1 range with maximum flow at 120 cfs, the first DP would measure from 120 down to 30 cfs, while the second would pick up at 30 cfs and go down to approximately 7 cfs. An accurate flow signal at low flows will greatly improve the quality of the controller program.

The two flow inputs and the residual signal serve as the inputs to the controller. The analyzer signal is sent to the microprocessor PID controller section. This is the feedback loop.

The controller selects either flow signal depending upon the flow range and passes it on to be totalized in the integrator section. This portion of the microprocessor integrates the water flow over a given period of time, which results in an average flow signal. This quantity determines both the frequency and amount of controller correction.[72] This provides the solution to the problems that usually occur at low flows.

From the integrator section the signal goes to the programmable controller. This segment of the instrument creates a reset signal for both the integrator and the timer. The timer is adjustable to control the on time of the controller.

Both the residual control signal and the selected flow signal are then sent to the multiplier, which furnish a single output to the Chloromatic valve. The multiplier has a built-in ratio station which can be fine tuned at the controller.

The dosage rate can be called up by the data panel for on-site operator information. The dead band will not exceed one percent, which is beyond detection on the chlorinator rotameter.

The programmable controller provides the operator with the means to fine tune the proportional band gain, the reset action to reduce the error signal, and the derivative action to adjust to rapid rates of change in residual measurements. This includes on time for the sample and hold logic and the off or hold time for the logic based upon totalized flow.

Wallace and Tiernan. Their dual system is based upon the dosage control element in the chlorine feed-rate valve positioner. These positioners have always used an adjustable potentiometer to achieve manual dosage control. To achieve automatic dosage control a motor-driven potentiometer is installed in the residual analyzer. This 10-turn potentiometer is positioned by a 2-phase reversible motor. A residual signal from the high-off-low switch in the analyzer causes the motor to drive the potentiometer in one direction or the other. The analyzer provides set point and dead band adjustment. These adjustments are one-time set to fit field conditions.

The schematic shown in Fig. 9-46 illustrates the operation of this system. It is described as a floating control feedback loop. The programmable controller integrates the feedforward flow pacing signal with the feedback residual signal.

The programmable controller should contain sufficient options to allow the operator to adjust contact closure frequency to fit the system loop time over the process flow range, and to vary closure duration for the same conditions. Other options

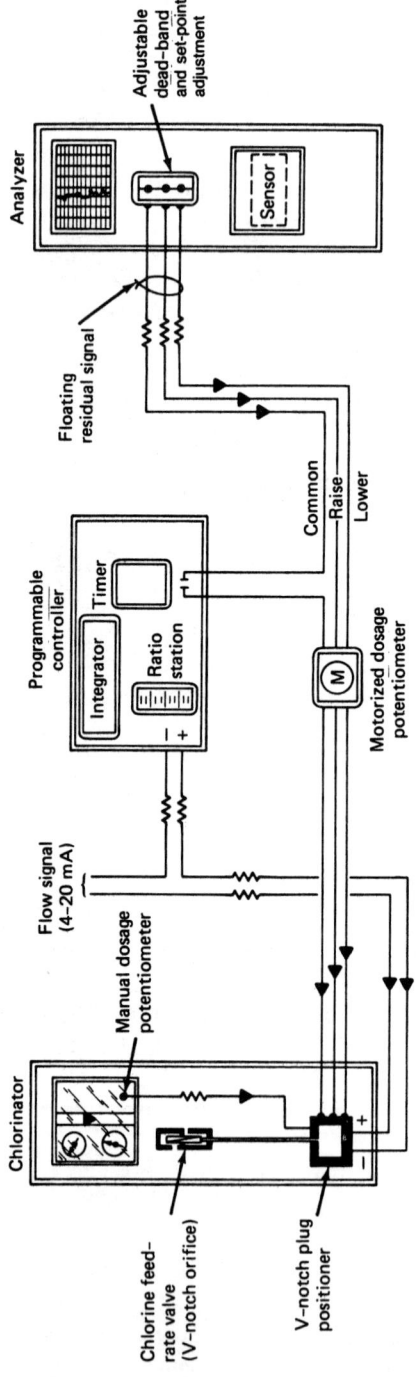

Fig. 9-46. Dual-signal residual control system (courtesy Wallace and Tiernan Div. Pennwalt Corp.).

might include the ability to "freeze" the chlorine orifice positioner during a power failure and/or interrupt the chlorine feed-rate during reverse flow conditions.

INJECTOR SYSTEMS

General Description. The power developed by the injector allows the chlorine to flow from the supply containers through the chlorinator, which is the metering system, and then through the injector vacuum line to the injector inlet. At the injector, the chlorine dissolves in the injector water to form a mixture of hypochlorous acid (HOCl) and molecular chlorine (Cl_2). This is the chlorine solution which flows in the solution lines to the diffuser at the point of application. This method of feeding chlorine gas is unique to water treatment processes (potable water, wastewater, and industrial waters). In the beginning, chlorinators metered and controlled directly into the process flow. This practice proved to be impractical over the long term. There were and are some exceptions. The difficulties with the direct feed system resulted from feed-rate sensitivity due to ambient temperature changes and low solubility of chlorine in water at atmospheric pressure. The vacuum system invented by G. F. Wallace overcame these difficulties and provided several advantages:

1. It is the easiest method of dissolving chlorine in water.
2. Chlorine is easily handled when in solution.
3. Since the injector creates a vacuum, it allows chlorine to be metered under a vacuum. This is the most accurate way of metering, since constant density is maintained and is not affected by ambient temperature changes.
4. Operation under a vacuum is safer than operating under pressure.
5. A metering system can be easily designed to stop automatically if the vacuum should fail.
6. Available vacuum can be used for automatic switchover of containers.
7. Injector vacuum level can be used to provide two alarm signals: (1) loss of chlorine supply, (2) injector malfunction or loss of water pressure.
8. Injector vacuum can be used to generate a variable chlorine feed rate as a function of chlorine residual (compound loop control).

System Description. The injector system is the heart of the entire chlorination facility. If this system is inoperable, no other part of the system can function. The various parts of this system include: (1) the operating water supply to the injector; (2) the injector; (3) the injector vacuum line from the chlorinator; (4) the injector discharge system, described as the chlorine solution line; and (5) the diffuser at the point of application.

Operating Water Supply:

Water Quantity. Injectors are designed to develop about 25 in. Hg vacuum at the injector inlet when the chlorinator is at maximum chlorine feed rate and the

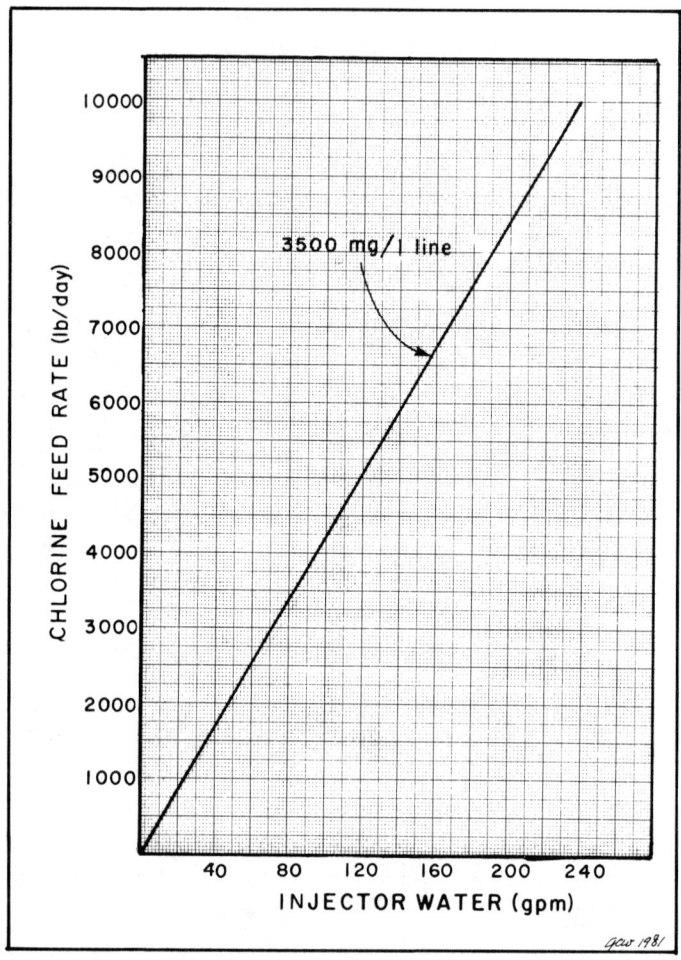

Fig. 9-47. Injector chlorine solution curve for limiting chlorine concentration of 3500 mg/L.

chlorine solution is discharging at atmospheric pressure, i.e., no back-pressure. The water flow through the injector must be sufficient to limit the chlorine solution strength to 3500 mg/l. Above this strength molecular chlorine appears in significant amounts, which breaks out of solution causing off-gassing at the point of application if open to the atmosphere, and gas binding in solution lines under low negative heads (see Table 4-1). Either condition is intolerable. Fig. 9-47 shows the minimum amount of water required to limit the chlorine solution to a maximum strength of 3500 mg/l for various gas flow rates. A rule of thumb is: approx. 40 gal. water/lb/day chlorine, more or less depending upon local conditions.

Water Pressure vs. Back-Pressure. A second factor is the pressure at the point of application of the chlorine. This is known as injector back-pressure. A higher back-pressure requires a higher injector inlet pressure and more operating water to make the injector function properly. The injector also has minimum operating water requirements. Chlorination equipment manufacturers use injector operating curves that specify how much water at which pressure is required for a given amount of chlorine to be applied against a given back-pressure. It is the designer's responsibility to make a hydraulic analysis of the chlorine solution line between the injector and the point of application to establish the amount of back-pressure to be expected. With this information and the maximum chlorine feed rate desired, the chlorinator manufacturer can then advise the necessary water supply inlet pressure required at the injector and the optimum injector operating water quantity. The back pressure should never be allowed to drop below 2 psi.

In spite of manufacturers' injector curves it is unwise to operate any injector on less than 50 psi inlet pressure unless there are extenuating circumstances.

Flow Meters. Injectors are usually available in four sizes: 1" fixed-throat type, 2", 3", and 4" adjustable-throat type. The fixed-throat injector assembly utilizes interchangeable throats and tailways with various-size openings for different conditions of chlorine feed rate and hydraulic conditions (back-pressure). Therefore the water flow through a fixed-throat injector is predictable. This is not the case with the adjustable-throat injector. For this reason, when adjustable-throat injectors are used, it is mandatory that the operator be provided with an indicating flow meter (for each injector) with a 3 to 1 operating range and a head loss at maximum flow not to exceed 5–6 ft.

Pressure-Regulating Valves. In cases where more than ample pressure is available to operate the chlorinator there is no need to use pressure-regulating valves upstream from the injector except in cases of abnormally high pressure—that is, > 150 psi.* The reason for not requiring pressure-reducing valves is the hydraulic characteristics of an injector. It will consume all of the pressure available on the upstream side of the injector.

Hydraulic Gradient Analysis:

General Discussion. It is necessary to consult manufacturers' injector efficiency curves to determine injector operating water pressure and quantity. The reader is cautioned that some manufacturers may have different sets of curves for different modes of vacuum operation, i.e., remote vacuum, variable vacuum control, and/ or sonic flow vs. non-sonic flow. Whenever the chlorinator feed rate is varied via the differential vacuum regulator, higher internal operating vacuum levels must

---

* The exception is when booster pumps are used or where cavitation of the injector tailway occurs.

684 HANDBOOK OF CHLORINATION

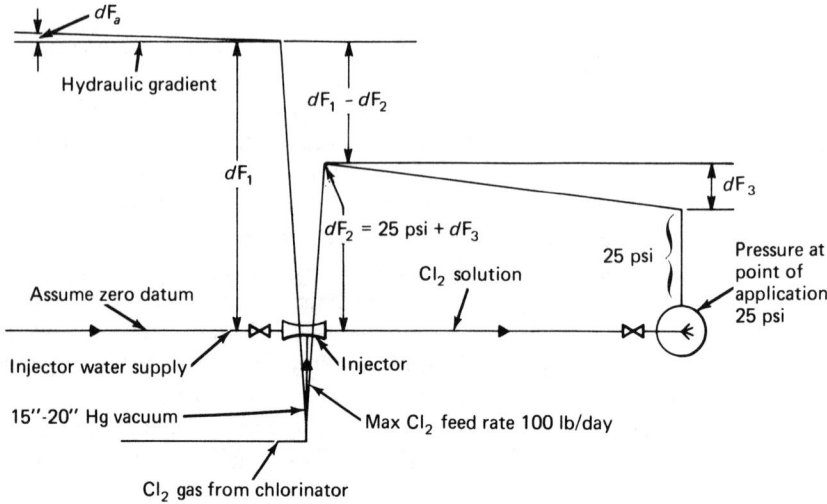

Fig. 9-48. Typical injector hydraulics, where dF_a = friction loss in injector water supply line to the injector; dF_1 = required operating pressure (consult manufacturer's curves); dF_1–dF_2 = pressure drop across injector required for satisfactory operation; dF_3 = friction loss in chlorine solution line from injector to point of application. The chlorinator injector "back pressure" is dF_2. dF_1 is also dependent on the amount of chlorine to be injected as well as on control mode, i.e., use of variable orifice positioner or differential regulator.

be used; therefore the injector requires more power, and hence more water pressure for the same back-pressure. This is the case with the Wallace and Tiernan compound loop control system shown in Fig. 9-42. When remote vacuum is used injector vacuum levels need to be increased about 5 in. H_2O.

Injectors can be installed in either the vertical or horizontal position. The only precaution required is to change the position of the check valve inlet block on the 3″ and 4″ injectors so that the ball-check will operate as it does in the vertical position.

Figure 9-48 is a typical injector hydrualic gradient diagram where dF_1 = IOP = injector operating water pressure, dF_2 = BP = injector back-pressure, or hydraulic gradient immediately downstream from the injector (outlet of tailway). Injector water quantity Q is dependent upon three factors: (1) maximum chlorine gas feed rate; (2) size of injector; (3) back-pressure. However if the water pressure available to operate the injector is greater than required by the gradient analysis shown in Fig. 9-48, this water pressure will determine the amount of injector operating water for the case of the one-inch fixed-throat injectors. The throat size adjustment provided for in the 2-, 3-, and 4-inch injectors allows the operator to adjust the inlet water flow to conform to the hydraulic gradient analysis.

CHLORINE FACILITIES DESIGN   685

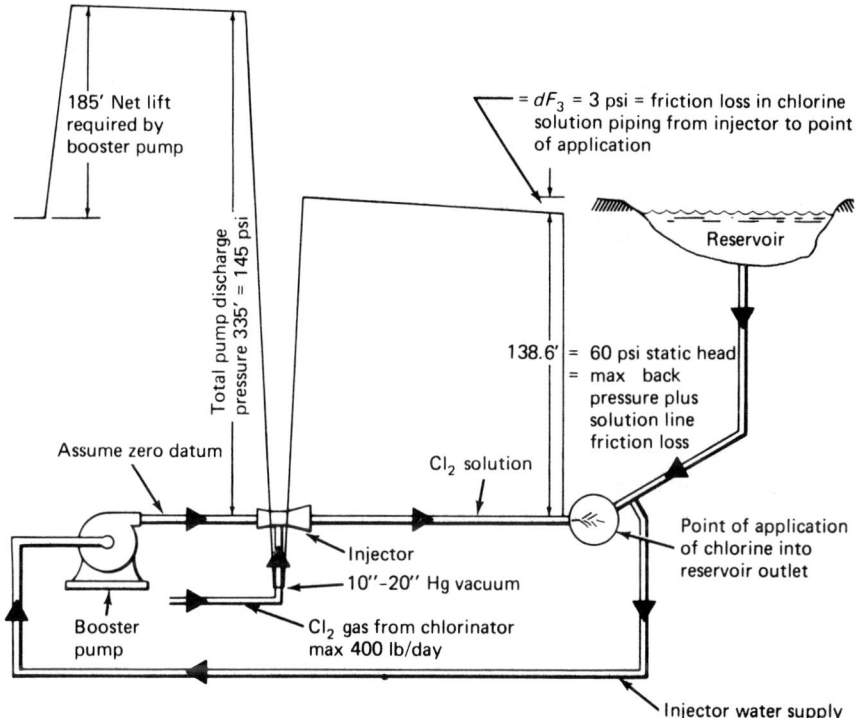

Fig. 9-49. Hydraulic gradient injector system. Chlorine feed rate = 400 lb/day. Manufacturer's injector curves for this feed rate at 65 psi back pressure requires minimum injector water $Q = 20$ gpm. For the pump, the following amounts are practical in terms of safety factors: $Q = 40$ gpm for a turbine and $Q = 25$ gpm for a centrifugal.

Example I. Fig. 9-49 illustrates the hydraulic gradient of the injector system described as follows:

Water is to be chlorinated at the outlet of an impounding reservoir. The location of the equipment will be such that the maximum static head at this point is 140 ft (60 psi). This then is the maximum back pressure at any time. The peak water flow to be treated is estimated at two and a half times the average daily consumption of 4 mgd = 10 mgd. Chlorine demand tests show that during summer months the demand is as high as 2.8 mg/l. Therefore the chlorinator should be sized to feed at least 4 mg/l in order to provide an adequate free chlorine residual. The capacity of the chlorinator must be 4 mg/l × 8.33 lb/mg × 10 mgd = 330 lb/day. The selection would be a chlorinator with a maximum capacity of 400 lb/day—one of the breaking points on overall size of chlorinators. Consultation with the manufacturers' injector curves reveals that a fixed throat injector will be used and that the most efficient size for feeding 330 lb/day maximum versus a 60 psi back-pressure will require an inlet pressure of 155 psi, which will produce a flow through the injector, $Q = 20$ gpm. Therefore select a booster pump with $95 \times 231 = 220$ ft total net head lift and Q of 20 gpm plus a 25 percent (25 gpm) safety factor if the pump is to be centrifugal or 100 percent (40 gpm) if it is to be a turbine type. Additional allowance for friction loss in the solution line should be added to the pressure at the point of application. Assume that the solution line has an equivalent length of 150 ft. A chlorine solution flow of 20 gpm* in a 1½" pipe will have a pressure drop of 4 ft per hundred. This will then raise the downstream back-pressure to approximately 63 psi. Since the pump chosen should have some factor of safety, we will select a pump to deliver 40 gpm at a total net head of 250 ft for a turbine and 25 gpm at a total net head of 250 ft for a centrifugal. Because of the high head, a turbine pump is preferred in this instance.

Example II. Installations in which the point of application is below the injector require that the chlorine solution be sized so as to provide an artificial back-pressure of approximately 2 psi just downstream from the injector. This example involves a 2000 lb/day chlorinator using a 2" adjustable throat injector. The point of application is 25 ft below the injector, and the water surface is 15 ft below the injector. Therefore the static back-pressure is -15 ft. The minimum amount of injector water will be 50 gpm. In negative head situations, there will be greater opportunity for developing chlorine fumes at the point of application owing to breakout of molecular chlorine under a negative pressure. Therefore add 15 percent to the injector water flow to decrease the chlorine concentration—try 65 gpm. The injector curves indicate a requirement of 25 psi with a back-pressure of 2 psi to meter 2000 lb chlorine per day.

Next choose a chlorine solution line small enough to produce 2 psi back pressure at the outlet of the injector. See Fig. 9-50 for the hydraulic gradient. Analysis of the solution line shows an equivalent length of 96 ft. Therefore choose a 2" pipe size. At 65 gpm the friction loss in 96 ft of pipe is approximately 12 ft. Therefore provide a diffuser with an 8 ft head loss. The total loss of 20 ft meets the requirement to provide a 2 psi back-pressure at the injector to prevent fuming at the diffuser. In all cases diffuser head losses of 8–10 ft are highly desirable because of the jet velocity attainable. This results in superior initial mixing. (See Chapter 8.)

Remote Injectors. Installation of injectors remote from the chlorinator location is more the rule than the exception. This is especially true in the wastewater application, which has already been described in Chapter 8. In potable water treatment there are two major advantages in having the injector adjacent to the diffuser: (1) it shortens the loop time in residual control application because gas velocities five times that of chlorine solution velocities can be tolerated; and (2) head-loss inherent in solution lines can be transferred and added to the diffuser loss to provide better mixing. See the "Diffuser" section this chapter.

---

* The flow through the injector is limited by the discharge pressure of the pump. In the case of the turbine pump the excess flow recirculates in the by-pass. (See Fig. 9-52.)

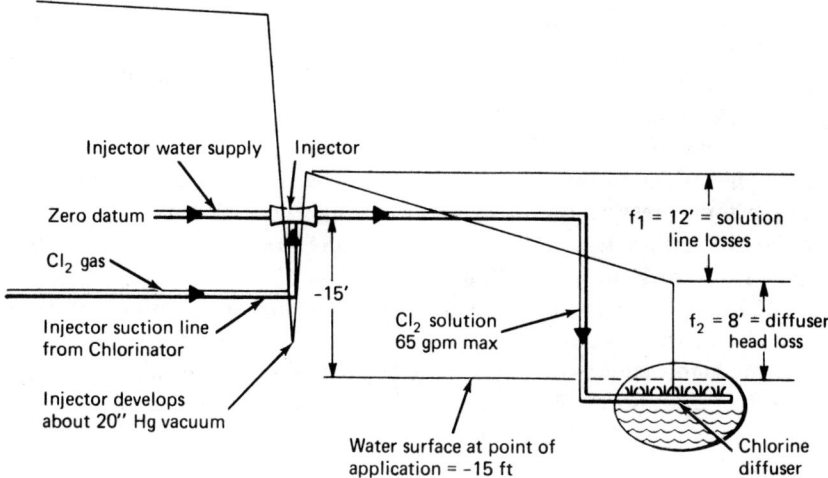

Fig. 9-50. Hydraulic gradient of injector system to prevent negative head at the diffuser.

Vacuum Line Design. The design of a long vacuum line (> 100 ft) should include the following features. The vacuum line (regardless of length) between the chlorinator and the injector should be sized so that the total pressure drop in this line is not more than one and one-half in. Hg. A shutoff valve should be provided at each end of this vacuum line. A vacuum gage should be installed adjacent to, but downstream from, the shutoff valve at the injector end. The injector should always be installed in a horizontal position to achieve a low profile (better hydraulics) and to allow easy disassembly of the injector.

The optimum pipe size for vacuum lines between the metering equipment (chlorinators or sulfonators) and the injector is subject to a great deal of scrutiny. White has investigated the hydraulic characteristics of three remote injector systems varying in distances from 750 to 8740 ft. The most comprehensive study was of a chlorination system with 3-2000 lb/day chlorinators remotely located from a single 3-in. injector connected by one 3-in. PVC vacuum line 8740 ft long.[25] Friction loss data were plotted showing the friction loss factor f as a function of Reynolds Number, N_r (see Fig. 8, Appendix). The response time (lag time) between the chlorinators and the injector was determined by 30-sec. interval amperometric titrations of the chlorinated effluent immediately downstream from the chlorine diffuser, which was within 25 ft of the injector. As a check on these field observations Fischer and Porter Co. Engineering Department set up a laboratory experiment using a 400 ft ¾-in. vacuum line to a remote injector and a variable chlorine feed rate of 150–500 lb/day. The results of this experiment when extrapolated agreed closely with the field results by White and Stone described above.[25]

Several important conclusions were drawn from these observations: 1) for any given system the lag time appears to be almost constant regardless of magnitude of change of the gas feed rate; 2) the higher the vacuum level in the vacuum line the shorter the lag time because the lower density of the gas provides a higher velocity; 3) lag time is independent of total volume of the vacuum line; 4) friction factor f varies significantly with Reynolds Number (see Fig. 8 Appendix); and 5) when the total pressure drop is greater than about 1.5 in. Hg there is a noticeable decay in the vacuum level of the entire line (see Fig. 9-51). Referring to this figure, the total pressure drop was about 1.5 in. Hg at flows less than 500 lb/day. Above this flow (i.e., 500–5500 lb/day), the pressure drop was practically constant. It varied from 2.88–3.06 in. of Hg, while the vacuum level varied from 5.25 to 22.25 in. Hg. Over this range of vacuum level and at a constant temperature of 68°F the density of gas varied from 1.55 lb/ft³ at the low vacuum to 0.048 lb/ft³ at the high vacuum—this is a range of 32–1. From these values it is obvious that the flow conditions of this vacuum line are highly unstable. The instability of this sytem is a result of the high-total pressure drop.

These observations bring up the question of the design procedure for an optimum size vacuum line for a given length and amount of gas flow. It would seem prudent to design for the severest conditions. Instead of designing for the minimum vacuum

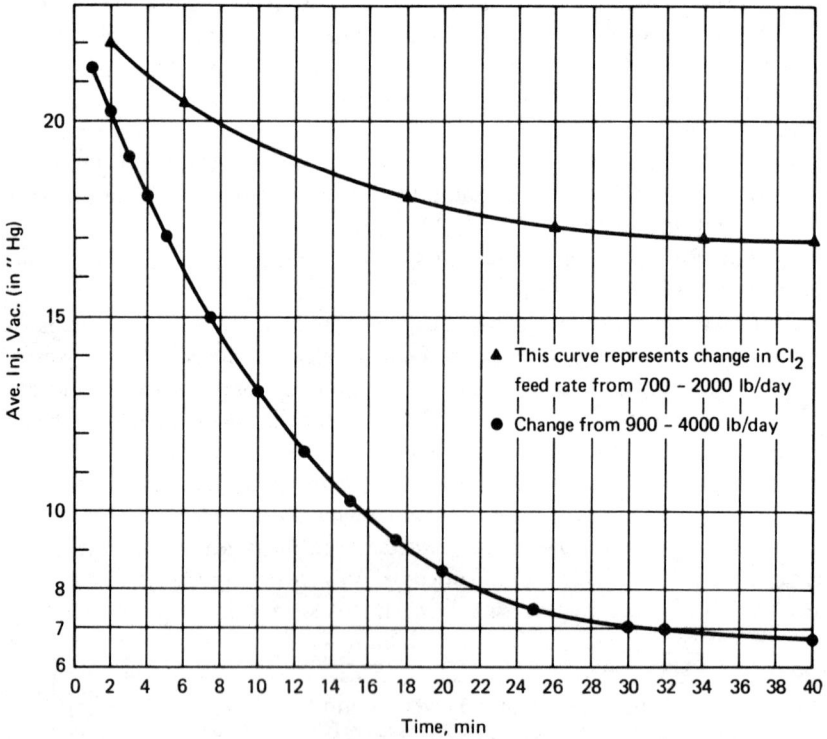

Fig. 9-51. Vacuum level decay in long vacuum lines due to excessive friction loss.

conditions allowable for the proper operation of chlorinators and sulfonators (about 8–10 in. Hg), the maximum injector vacuum level should be assumed (22–23 in. Hg). At this level the *total* pressure drop in the system, regardless of pipe length, should be limited to 1.50–1.75 in. Hg. One such system designed to include these factors has been in operation long enough to draw the following conclusion: it is a stable system with rapid response and therefore close to the optimum design. The maximum vacuum decay level from minimum to maximum feed rate is only about 3 in. Hg (i.e., from 24–21 in. Hg, vacuum).

The design procedure is to first choose a line size. This particular case was for a dechlorination system—3300 ft from sulfonator to injector, 8000 lb/day maximum feed rate. Let us try a 4-in. Sch 80 PVC pipe, $d = 3.826$ in. Using the following equation corrected to pressure drop in inches of Hg:

$$\Delta P = \frac{11.89 \times L \times f \times W^2}{10^9 \times \rho \times d^5} \qquad (9\text{–}11)$$

where

ΔP = total pressure drop, in. Hg
L = length of line in ft
f = friction factor from Fig. 8, Appendix
W = lb/day Cl_2 or SO_2
ρ = density of gas, lb/ft^3*
d = inside pipe diameter, in.

To find f from Fig. 8, Appendix, it is necessary to calculate the Reynolds Number of the system:

$$N_r = \frac{6.32 \times \omega}{\mu \times d} \qquad (9\text{–}12)$$

where

ω = lb/hr gas flow
μ = viscosity of gas in cp*
d = inside pipe diameter in in.

$$N = \frac{6.32 \times 8000}{0.0133 \times 3.826 \times 24}$$

$N = 41{,}399$

---

* For gas density values, see Fig. 2, Appendix.
* For gas viscosity values see Fig. 3, Appendix.

So from Fig. 8, Appendix,

$$f = 0.027; \therefore \Delta P = \frac{11.89 \times 3300 \times 0.027 \times (8000)^2}{10^9 \times .05 \times (3.826)^5}$$

$\rho = 0.05$ lb/ft^3 at 22 in. Hg. vac. and 68°F.

Answer: $\Delta P = 1.65$ in. Hg.

At 6000 lb/day flow ΔP is about 0.16 in. Hg. Now to calculate the lag time in this system it would be appropriate to calculate it for 4000 lb/day gas flow, but at 25 in. Hg vacuum. So 4000 lb/day is 0.05 lb/sec. The gas density at 25 in. Hg vacuum and 68°F is 0.03 lb/ft^3 so;

$$Q = \frac{0.05 \text{ lb/sec}}{0.03 \text{ lb/ft}^3} = 1.54 \text{ cfs}$$

$$V = \frac{1.54}{0.07986} = 19.32 \text{ ft/sec}$$

Therefore, the lag time for any change in gas feed rate is on the order of 3 min. To show that the lag time would be nearly the same at 8000 lb/day assuming a 22 in. Hg vacuum level, 8000 lb/day = 0.09 lb/sec and the density = 0.048.

$$Q = \frac{0.09 \text{ lb/sec}}{0.048 \text{ lb/ft}^3} = 1.88 \text{ cfs}$$

$$V = \frac{1.88}{0.07986} = 23.54 \text{ ft/sec}$$

So at double the gas flow the lag time decreased to about 2.35 min. Therefore, as the flow of gas changes, the density changes in a direction which provides an almost constant system lag time regardless of flow change.

It is to be noted that the physical properties of chlorine and sulfur dioxide gas under a vacuum are so similar that calculations for either are interchangeable; so for simplicity, long vacuum lines can be designed for either gas based on the physical characteristics of chlorine.

Booster Pumps:

Turbine vs. Centrifugal. Chlorinator or injector water booster pumps are as important as the injector or the chlorinator. If the booster pump is not large enough, the chlorinator will not operate. Therefore never undersize or try to economize while attempting to select a pump. If anything, oversize. It is always a good idea

CHLORINE FACILITIES DESIGN 691

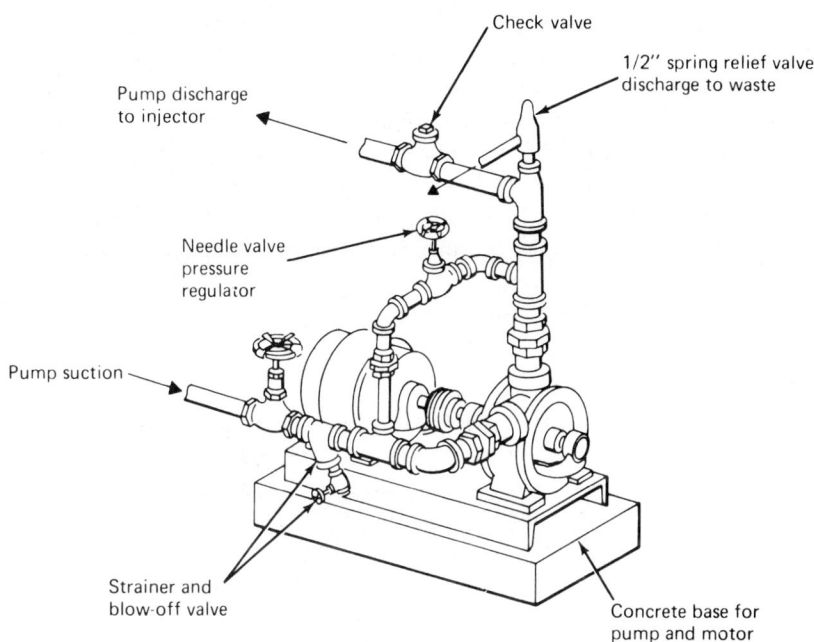

Fig. 9-52. Turbine type booster pump showing by-pass piping. (courtesy Wallace and Tiernan Div. Pennwalt Corp.).

to add 10 percent to the estimated back pressure, and be sure that the amount of chlorine to be fed is a little more than calculated.

Estimating the injector operating water depends on whether the pump to be used is a centrifugal or a turbine.

If the pump selected is a centrifugal, it will make a difference whether it is a unibuilt type or has an outboard bearing. If it is the latter, a higher wear factor for Q is allowable than if it is unibuilt. Picking a Q in excess of 15 percent of what the fixed throat injector will pass at the discharge pressure of the pump will overload the motor thrust bearing and cause undue wear. A centrifugal pump with an outboard bearing can easily handle the injector Q plus 25 percent. With adjustable throat injectors this is not a problem, because the injector Q is easily adjusted in the field.

A turbine pump, which is usually the one of choice for all fixed injector throat chlorinators, should be arranged with an adjustable pressure bypass assembly, as shown in Fig. 9-52. In these cases, the pump is selected to deliver 2 Q. In other words, if the injector requirement is 3 gpm, select a pump to deliver 6 gpm. The regulating bypass assembly allows the excess water to flow back into the suction. This is done by the needle valve in the bypass. As the pump wears over a period of time, and the pressure drops off, the bypass is closed off more to compensate for this wear.

With water that has a quantity of sand, it is more desirable to use a centrifugal pump, because sand will ruin a turbine pump in a very short time.

Most of the chlorinator booster pump installations require a turbine pump because of high head and low Q requirements. For example, all chlorinators in the capacity range of 3 to 400 lb/day use a fixed throat injector. The largest fixed throat injector will pass about 22 gpm. Most injector water requirements are for less than 15 gpm.

Centrifugal pump selection usually begins when the injector operating water approaches 15 gpm.

High Back-Pressure Conditions. There are situations of high static pressures that are beyond the ability of high-head multistage 3500 rpm turbine pumps and or high-head centrifugal pumps. When conditions exist that are beyond the capabilities of these pumps, the chlorine solution has to be pumped. It is fair to say that the upper limit for a conventional turbine pump application would be a condition for a fixed-throat injector where the maximum chlorine feed rate is 400 lb/day and the maximum back-pressure 140 psi. When the chlorine feed rate and back-pressure combination exceeds 100 lb/day and 75 psi, respectively, pumping chlorine solution should be investigated. When a 2-inch injector is required, pumping chlorine solution should be investigated when the combination of chlorine feed rate and back-pressure is 1000 lb/day and 60 psi, respectively. For 3-inch and 4-inch injectors the chlorine solution should be pumped when the back-pressure reaches 15–20 psi.

When it becomes necessary to pump the chlorine solution the pump system should be arranged as shown in Fig. 9-53. The two basic requirements for this system are: (1) the regulating valve should be adjusted so that there is always approximately 2–5 psi back-pressure on the injector (a negative head must be avoided); and (2) the pump should be sized to deliver the required injector water plus an additional 30–40 percent dilution water.

This system operates on the principal that the spring-loaded regulating valve creates a pressure drop when the pump is operating. The pressure on the downstream side of this valve becomes the injector "back-pressure." Therefore the pressure drop across the regulating valve plus the friction loss in the system equals the TDH* requirement of the pump. If the chlorinator utilizes a fixed-throat injector, the largest-size throat should be used so as to insure minimum chlorine solution concentration.

Two types of pumps are available for this service. The type that has been in service for the longest time for pumping chlorine solution is the titanium pump line of the Duriron Co.[25] The City of San Francisco Water Dept. has reported excellent service by these pumps for more than 20 years. In recent years the Duriron

---

*TDH = total dynamic head.

CHLORINE FACILITIES DESIGN

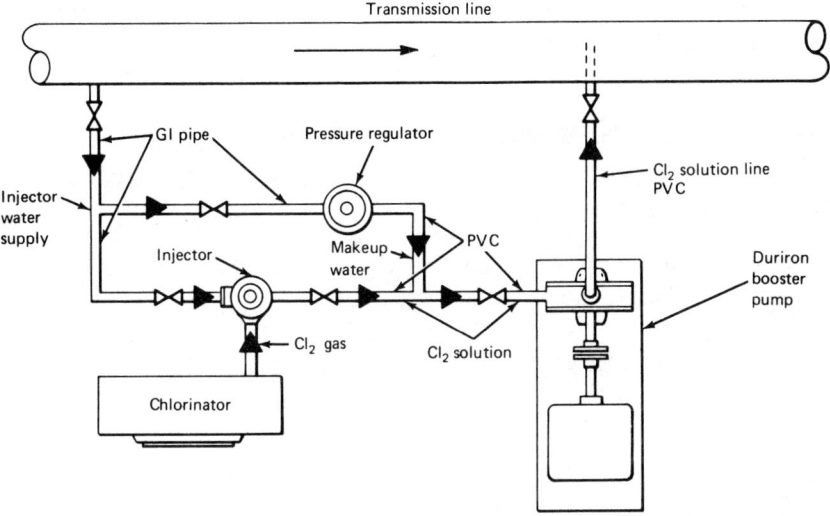

Fig. 9-53. Schematic arrangement showing booster pump downstream of injector. All pipe and valves downstream of injector and pressure regulator must be corrosion resistant, i.e., PVC, rubber-lined steel, Saran-lined steel, or fiber glass.

Co. has made a line of reinforced Fiberglas pumps for this service.[27] The Fybroc Division of the Met Pro Corporation also have a line of Fiberglas pumps available for this service.[28]

Example. Assume a transmission line with a maximum static head of 150 psi and a chlorine requirement of 1000 lb/day. Using Wallace and Tiernan Injector Performance Curve 25.100.190.021 shows that the best a 2-inch injector can do with 150 psi initial water pressure is 80 psi back-pressure at 80 gpm. Therefore the chlorine solution pump must be able to boost 80 gpm injector water plus about 25 gpm makeup water to 155 psi plus 10 psi friction loss or 165 psi. So $165 - 80 = 85$ psi $\times 2.31 = 196$ ft TDH. This situation calls for a pump to deliver 105–110 gpm at a TDH of 205–210 ft.

Designers Check List for Injector System. The injector system and chlorine solution lines usually include all of the following components:

1. Injector water-pressure gage
2. Injector vacuum gage for remote injector installations
3. Injector vacuum line shut-off valve at remote injector loction
4. Chlorine solution pressure gage located immediately downstream from the injector to indicate injector back-pressure (not required on fixed throat injector installations)
5. Injector water-pressure switch for low-water-pressure alarm

694 HANDBOOK OF CHLORINATION

6. Injector water flow meters for multiple chlorinator installations of chlorinators using 2", 3", and 4" injectors
7. Chlorinator–sulfonator built-in vacuum switch and alarm for both high and low vacuum
8. Compound back-pressure gage for injector discharge

CHLORINATION STATIONS

Principal Considerations. The design and layout of any chlorination station must be based upon the following considerations:

1. Chlorine container selection
2. Chlorinator capacity
3. Points of application
4. Injector requirement
5. Methods of control
6. Alarms
7. Safety equipment
8. Chlorine residual testing facility
9. Chlorine storage inventory system
10. Housing requirements—space, ventilation, lighting, and heating
11. Reliability provisions

All of these items are addressed in detail elsewhere in this text.

Individual Deep-Well Station. This is probably the most widely used type of chlorinator station. The chlorine may be applied either down the well or at the well pump discharge. Fig. 9-54 shows the application down the well. Whenever possible, this is the preferred method for a variety of reasons.

1. No injector booster pump is required.
2. Chlorine solution lines may be used simultaneously to feed a sequestering agent to prevent deposition of iron or manganese in the distribution system.
3. Chlorine solution line provides means for intermittent purging and cleaning of aquifier with heavy doses of chlorine.
4. It provides longer contact time.

Special attention should be given to the point of application and the injector hydraulics. The chlorine solution line must terminate at least ten feet below the pump bowls; the chlorine solution strength should be limited to about 100 ppm; and the alkalinity of the water should be at least 100 ppm.[29] These factors prevent chlorine corrosion of the pump and well casing.

The injector water line must be taken off the pump discharge line downstream from the check valve. Therefore, when the pump shuts down, the pipeline from the check valve drains back into the well casing. This creates a negative head on

CHLORINE FACILITIES DESIGN   695

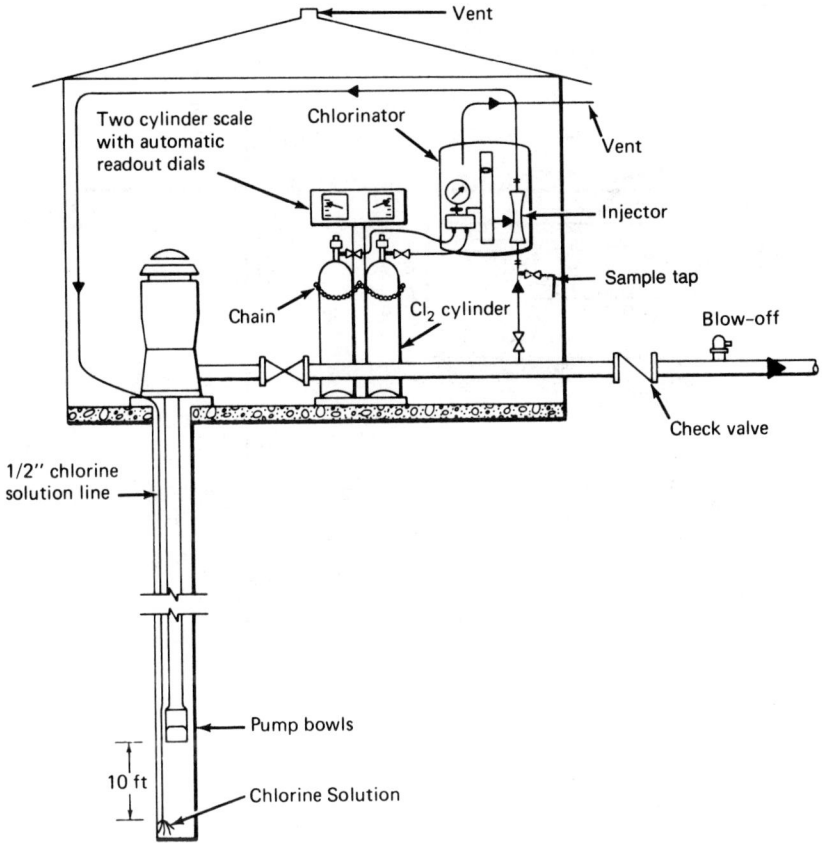

Fig. 9-54. Typical chlorinator installation with chlorine applied down well.

the upstream side of the injector, which soon equalizes to become the same as the negative head in the chlorine solution line. This equalizing of negative head on both sides of the injector allows the chlorinator to automatically shut down without any siphon action from the chlorine solution line. If the injector water line is taken off downstream from the pump discharge check valve, the chlorinator will continue to operate regardless of the well pump. Putting a solenoid valve in the injector water line at this point is useless, because the chlorinator would continue to operate without injector water, owing to the siphon action of the negative head on the chlorine solution line.

When it is not convenient to put the solution line down the well the alternative method is to utilize a chlorinator injector booster pump and apply the chlorine at the well discharge, as shown in Fig. 9-55. In this case the booster pump suction and the chlorine solution discharge must be downstream from the well pump check valve. Then, when the well pump shuts down, the injector pump also shuts

696 HANDBOOK OF CHLORINATION

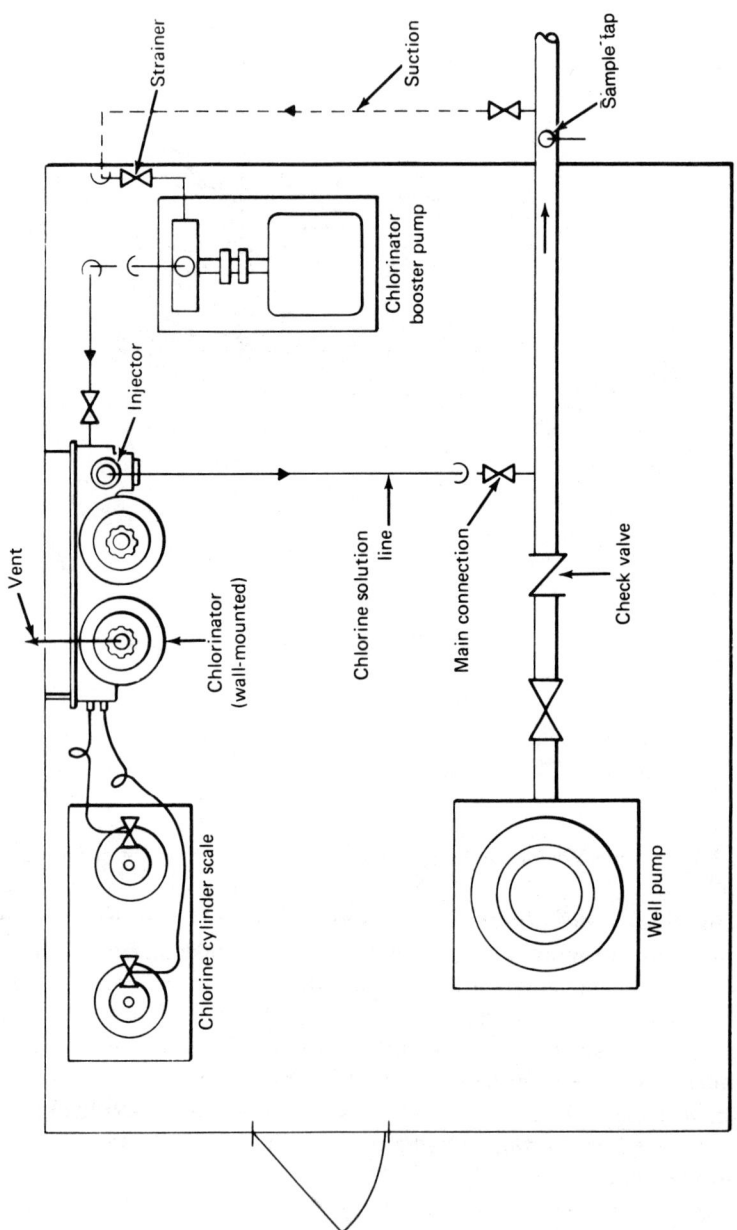

Fig. 9-55. Typical chlorinator installation with point of application in well pump discharge.

down and the pressure on both sides of the injector equalizes, thereby automatically stopping the chlorinator. This is known as semiautomatic operation.

Multiple-Well or -Pump Stations. If there is more than one well station discharging into a common line, it is often preferable to use one chlorinator and one point of application in the common well discharge line. For such a situation the chlorinator control could be either flow-proportional from a primary flow meter or step-rate from an adjustable-rate controller.

For purposes of illustration, the method of step-rate control will be used. This method is practical at a pumping station when the pump controls are at one location or near one another. For widely scattered well stations, the electrical wiring becomes expensive and cumbersome. Fig. 9-56 shows a schematic of a pump station with three pumps of different capacities arranged to provide a separate rate of chlorination for the various combinations of pumps. The usefulness of step-rate control diminishes as the number of rates exceeds four or five, simply because the electrical interlocking gets overly complicated. However, if a primary metering device is not going to be provided at a multiple-pump station, the only method that can be used for chlorinator pacing is the step-rate type. With this in mind, the designer should select pumps of similar capacities, so that the control is limited to only four or five rates.

Relay Stations and Transmission Lines. Probably the only simple manual control chlorination stations are those which treat a constant flow of water that is being transferred from one reservoir system to another, with no user takeoffs between. These examples of constant flow are hydraulic rarities. There is usually some fluctuation in flow worth accounting for, particularly if the objective of chlorination is to provide a chlorine residual in the entire transmission line for purposes of maintaining the "C" factor or water quality control.

When it is necessary to maintain chlorine residuals in a distribution system, relay stations are often used. Because of problems of logistics, it is not necessarily convenient to install a primary metering device to pace the chlorinator; it is generally more desirable to rely entirely on straight residual control. The schematic for such a system is shown in Fig. 9-41 and 9-57. The key to a successful direct residual control system is to design a short loop time, 60–90 sec. at average flow, and utilize the chlorinator injector pump to supply the sample to the residual analyzer. Both the chlorine solution diffuser and sample take-off should incorporate the design principles described for Fig. 9-41 and 9-57.

This method is suitable if the flow changes in the distribution system reflect the usual diurnal domestic consumption. The chlorine residual controller can react fast enough to keep up with these fluctuations. Such a control cannot be used alone if the change in flow is abrupt, as reflected in a step-rate change caused by a nearby pumping station.

698 HANDBOOK OF CHLORINATION

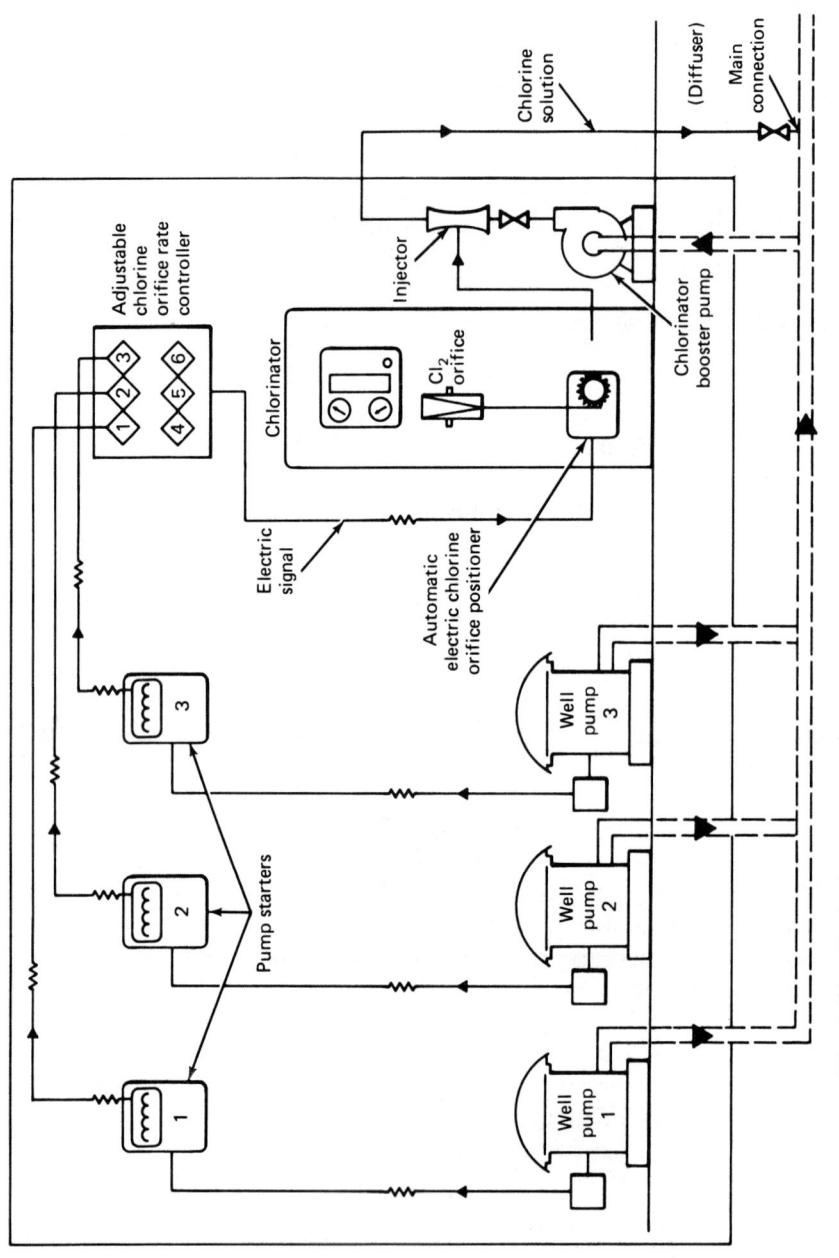

Fig. 9-56. Typical step-rate control utilizing electrically operated chlorine orifice positioner.

CHLORINE FACILITIES DESIGN 699

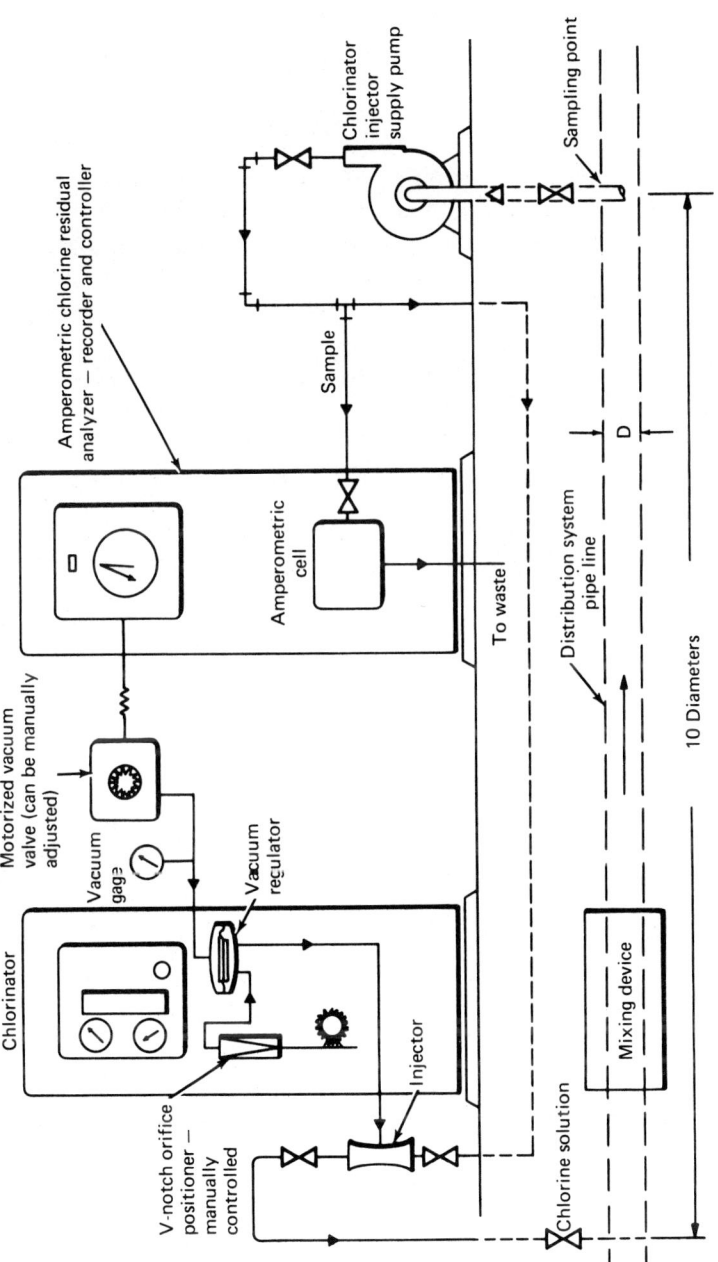

Fig. 9-57. Distribution system relay station with direct residual control. See also Fig. 9-41.

Reservoir Outlets:

General Discussion. It is common engineering practice in distribution system design to use balancing reservoirs that float on the distribution system hydraulic gradient. This is a convenient way to balance the flow in a large system. It also helps to solve the problem of pressure equalization by using reservoirs as pressure breaks in hilly terrain. The flow into and out of these reservoirs is almost always by a common pipe. Reversal of flow does occur in many of these systems where the outflow is a function of distribution system demand. During daytime hours the reservoir is supplying water for normal and peak demands but during low flow conditions in the early morning hours the system may be designed to fill the reservoir, thereby causing at least a once-daily flow reversal.

It is also common practice to chlorinate the outflow of a surface reservoir. This in itself is not an easy task. However, to provide adequate and reliable chlorination for a reverse flow condition is even more challenging.

The usual method of chlorination control solely by flow pacing from a primary meter is a risky practice because of abnormally wide daily flow ranges experienced by these reservoir systems. Systems designed for flow reversal have much wider flow ranges because the flow has to reach zero before it can reverse. Therefore in every one of these systems the range is infinite, i.e., zero to any maximum value.

Chlorination control for reservoirs will be discussed below in two categories: (1) outflow only and (2) unpredictable reversal of flow. However, it is necessary first to discuss primary meters and reverse flow detection.

Primary Flowmeters. There is a wide variety of flowmeters available, but none have the accuracy to provide the necessary chlorine dosage control over the flow ranges experienced at the majority of these systems. The designer or operator should accept the flow meter as most accurate and acceptable for chlorinator control in the range of 4:1. Therefore these chlorination systems should be supplemented with residual control. Retrofitting existing systems using existing flow meters must add residual control. The principal flowmeter accessory for chlorination control is an analog flow signal transmitter, preferably 4–20 mA or 10–50 mA. A pneumatic signal is acceptable but not preferred for a variety of reasons. The reasoning behind establishing an "accurate" range (4:1) for the flowmeter is to allow this signal to blend properly with the residual signal. This is described below.

Flow-Reversal Detection. This is a difficult task in any hydraulic system because the basic requirement is the ability and assurance of being able to detect zero flow. Reliable devices such as the thermal probe system[30] are expensive. Laminar-type flow switches that flex with the water flow are inherently insensitive to flows less than about 0.5 ft/sec. Other devices that operate on a turbine-type movement are more sensitive.

The need for a flow reversal switching device in a chlorination system is a site condition requirement. The user may find it desirable to shut the chlorination system down during this period, or be able to actuate valves to direct reverse flow to a compartmented reservoir.

Any chlorination system dealing with a reverse flow condition must have added protection of adjustable time delay relays. These relays will protect the chlorination control system from chatter due to the oscillating prism of water usually encountered when the flow makes the transition from forward to reverse. Appropriate switch gear can lock in the forward or reverse function long enough to allow the system to stabilize in either direction.

Reversal Detection by Chlorine Residual Analyzers. This type of reservoir chlorination system requires two chlorine residual analyzers arranged as shown in Fig. 9-58. In order to explain the operation of the analyzer control system the function of each principle element of the system will be described.

1. *Chlorine diffuser.* This is the jet injection type for a pressurized conduit. For pipes 36 inches in diameter or less, terminate the nozzle $D/6$ inches from inside wall. For conduits larger than 36 inches, use an across-the-pipe diffuser perforated across the middle third of the diameter (see Fig. 9-40a). The injection nozzle should be fitted with an orifice to provide an exit velocity of 22–26 ft/sec. Use the same velocity for the across-the-pipe diffuser. This diffuser design will eliminate the need for a mechanical mixer provided that the sampling taps are approximately 10 diameters from the chlorine diffuser.

2. *Injector water supply.* Design piping and injector pump to provide as much water as the injector will allow regardless of chlorine flow rate. This will enhance mixing and help to maintain a reasonably constant loop time over a wide water flow range.

3. *Injector location.* The injector should be placed as close to the point of application as is practical.

4. *Sample taps.* The sample taps for each of the analyzers should terminate at the center of the pipe for diameters up to and including 36 inches. For larger pipes the sampling taps should be made the same as the chlorine diffusers. The sample piping for Analyzer No. 2 should be designed for about 3 gpm at 5 ft/sec to provide a scouring velocity and to minimize loop time. Diffuser-type taps should have sufficient holes to limit the entrance loss to 1 ft.

5. *Injector water pump.* This pump serves three functions: (1) it operates the injector; (2) it provides excess injector water to insure good mixing and a more constant loop time; (3) it provides sample water to both analyzers.

6. *Analyzer No. 1.* This analyzer controls the chlorine residual in the outflow water to the distribution system. It is assumed there will always be a chlorine residual in the sample line to this analyzer.

7. *Analyzer No. 2.* This analyzer operates on the biased sample principle. This arrangement is necessary in order that the analyzer may stay in calibration, i.e.,

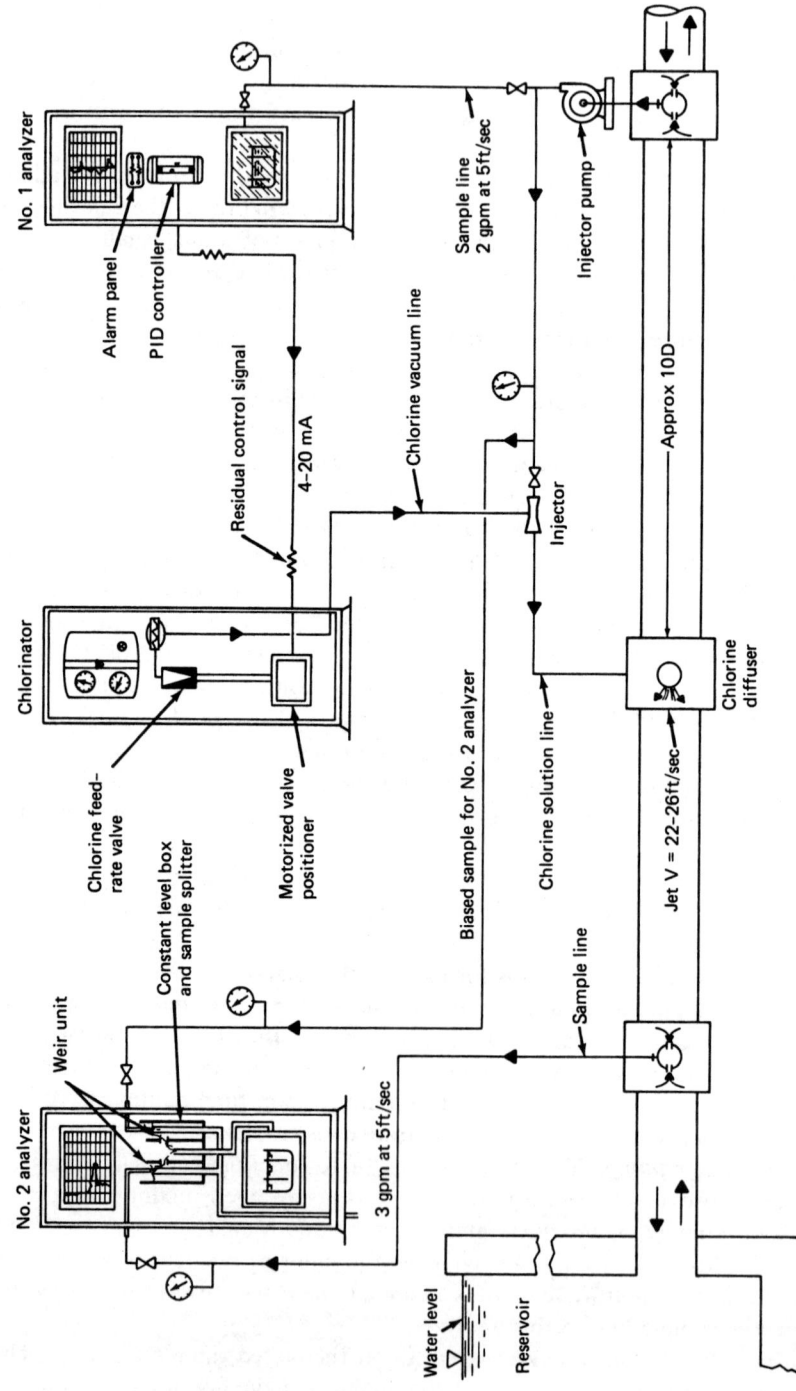

Fig. 9-58. Flow reversal detection and direct residual control using two residual analyzers (unpredictable reversal).

it must not be exposed to zero chlorine residual for any length of time. This is the analyzer which detects the flow reversal, as described below.

This system controls the chlorinator feed rate in the outflow mode by direct residual control, as described in "Control Strategies," earlier in this chapter. Assume the residual control band is set between 0.4 and 0.6 mg/l. In the forward direction both analyzers will be receiving a sample in this band range, except that Analyzer No. 2 cell will receive a sample diluted 1:1 with unchlorinated reservoir water. Therefore this analyzer will actually be "seeing" a residual in the 0.2–0.3 range. This is sufficient to maintain its calibration equivalent to Analyzer No. 1.

When the reservoir flow begins to reverse to the inflow mode nothing much will happen for several minutes. Then there will be evidence of residual decay in the distribution system and Analyzer No. 1 will call for more chlorine; this will not be seen by Analyzer No. 1 but will appear as a residual increase on Analyzer No. 2. When this residual (which is diluted) reaches an arbitrary level—say 0.6 mg/l on Analyzer No. 2—an adjustable residual alarm switch is tripped actuating a time delay relay which "freezes" the chlorine feed rate at that dosage level and locks out the residual control signal from Analyzer No. 1.

When the reversal begins to change to the outflow mode the chlorinated portion of the reservoir inflow water will pass by Analyzer No. 2 without incident and begins to register on Analyzer No. 1. By field observation of this occurrence the operator will be able to determine where to set the low-residual alarm switch on Anlyzer No. 2. This probably will be about 0.1 mg/l. When this low-level alarm switch is tripped it actuates a time delay relay which restores the automatic chlorinator controls to analyzer No. 1. The key to success of this system or any direct residual control system is a short loop time, 60–75 sec at average flow, and good mixing by short circuiting the chlorine solution to the sample tap by utilizing the injector water pump as a sample pump.

Water Treatment Plants:

General Discussion. A chlorination station for a conventional treatment plant should be designed for precise control, flexibility, reliability, and safety. It should also be provided with the necessary monitoring equipment and alarms for critical functions plus the required safety equipment for dealing with chlorine emissions. The principle considerations are: points of application, chlorinator capacity, control system, contact time, monitoring and control analyzers, injector system and chlorine solution lines, safety equipment, and standby equipment.

Points of Application. This is the first consideration because it prefaces the number of chlorinators required. In spite of the recent concern about the THM formation resulting from prechlorination this remains one of the crucial points of application. This is described in Chapter 6. Normally this point is located to allow 2 or 3 minutes contact time ahead of the flash mixers. Although the plant may be designed to use the free residual process and carry a free residual through the entire plant to the clear well, an intermediate point of application should be provided just ahead of the filters. This point may be used intermittently on a manual basis for shock-type treatment to clean up any filter biofouling that may occur.

The other point of chlorination would be for final disinfection or postchlorination for distribution system residuals. In some plants that use the free residual process, this point of application is used for dechlorination to trim the residual entering the distribution system.

Chlorinator Capacity. Somewhere in the plant there will be a flowmeter with a given range. The range of this meter should be the basis for the chlorinator capacity. For example, the engineers may be talking about a 10 mgd plant but will be providing a 20 mgd flowmeter. To simplify chlorinator selection, provide a pre- and postchlorinator with the capacity to treat 20 mgd—the full range of the flowmeter. Then if one or the other is not in continuous use it can be used for pre- and/or intermediate chlorination or as a standby. In any event, one chlorinator must be available for standby service.

The actual capacity of the equipment must be calculated on the basis of the 30 min chlorine demand of the water to be treated. Chapter 6 delineates chlorine dosages for the various points of application.

Control System. If prechlorination is to be continuous it should be flow-paced controlled and monitored by a residual analyzer after a 15–20 minute contact time. Prefilter chlorination should be manually controlled only. Postchlorination should be under a combination of flow-paced and residual control. Flow pacing for prechlorination should be from an influent meter and postchlorination from an effluent meter. Flow pacing prechlorination by an effluent meter or postchlorination by an influent meter should never be attempted. The time lag and intermediate uses of plant water for backwashing, etc., destroys the control function of flow pacing.

Mixing. * If the chlorine diffusers are designed in accordance with the recommendations of the "Diffuser Design" section of this chapter, there will be no need for additional mechanical mixers. See Fig. 9-40a.

Contact Time. The only contact time that may be critical in a conventional treatment plant is postchlorination. Usually the clearwell is used as a contact chamber. Otherwise the transmission line may be used to provide the necessary contact time before the water reaches the first consumer.

If the treatment plant is relying upon a free residual for disinfection, 10 minutes contact time is sufficient. If chloramine treatment is used the contact time should be increased to 50–60 minutes.

---

* See also diffusers as a mixing device in Chapter 8.

CHLORINE FACILITIES DESIGN 705

Monitoring and Control Analyzers. Chlorine residual analyzers should be used to monitor all prechlorination activities. Also analyzers should be used to monitor and control the residual in the water entering the distribution system. Large water systems will also find it useful to monitor chlorine residuals at strategic points or trouble spots in the distribution system.

Safety Equipment. All treatment plant chlorination stations should be equipped with the safety equipment described under that section.

Injector System and Chlorine Solution Lines. Experience has demonstrated it is preferable to locate the injectors adjacent or near to the diffusers. The use of remote injectors eliminates the temptation to manifold the chlorine solution lines in the chlorinator room. Manifolding solution lines and using rotameters to split the chlorine flow downstream from the injector is excessively costly and reduces the chlorination system to an uncontrollable hazard resulting in an operator's nightmare. Locating the diffuser adjacent to the point of application requires that there be no more than one diffuser per injector and no more than one injector per diffuser. In some cases the vacuum lines to the various injectors may be manifolded at the chlorinators to achieve redundancy—but splitting the chlorine solution discharge from the injector should not be permitted.

CHLORINE RESIDUAL ANALYZERS

Historical Background. There are two methods by which the chlorine residual can be continuously analyzed: colorimetric and amperometric.

The colorimetric method, using orthotolidine reagent, has been tried and discarded many times. Wallace and Tiernan Company began investigating the possibilities of automatic chlorine residual control using orthotolidine about 1927. Units were developed that could continuously record the O-T residual and at the same time adjust the chlorinator feed rate. The first of these was installed in 1929 at Little Falls, NJ; the second, in 1930 at Rahway, NJ.[31] A modification of these units was installed about 1930 in Los Angeles,[32] and operated for many years. One of these units is shown in Fig. 9-59. The latter units differed in that the O-T residual analyzer had a fixed color disk for a given O-T residual of 0.3 ppm. If the color in the analyzer did not match that of the disk, the error signal between the two drove the control valve on the "trimming chlorinator" until the two colors matched and the error signal became zero. The residual record then was one of deviation. Some years later Caldwell[33] developed an O-T recorder based on Harrington's work,[34] using neutral orthotolidine. The colorimetric recorders fell into disuse primarily because the photoelectric cells were not able to differentiate accurately enough between the relatively small changes in the color representing a residual change. Also, after it was realized how important free chlorine residual was, compared to total chlorine, another method had to be found, since it was not practically

706 HANDBOOK OF CHLORINATION

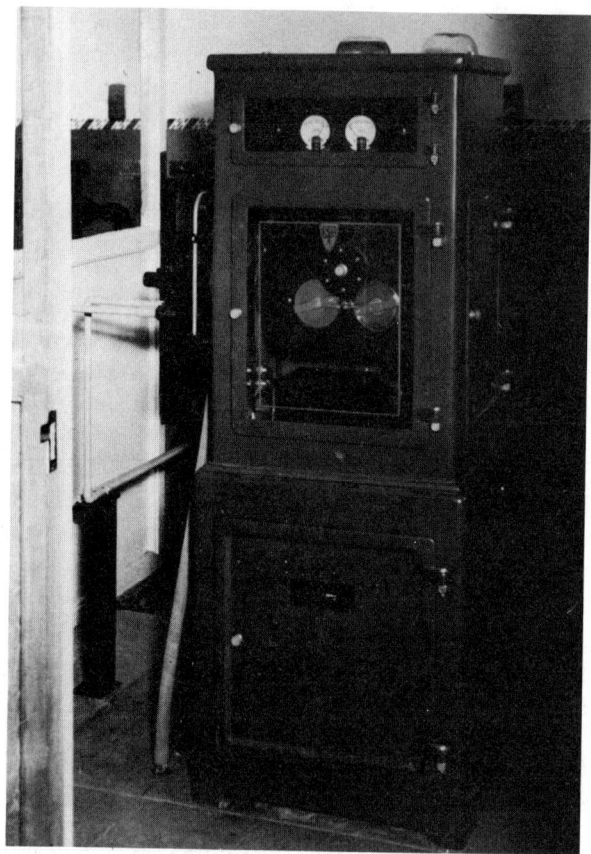

Fig. 9-59. Colorimetric O-T analyzer used for chlorine residual control by City of Los Angeles Dept. of Water and Power from 1930 to 1950s at McClay Highline Station.

possible to do this with orthotolidine.[35] In addition, awkward mechanical operation helped make the system impractical.

Circa 1955 Fischer and Porter Co. developed a sophisticated O-T analyzer which survived for a few years until their amperometric analyzer was developed. Hach Chemical Co. also marketed an O-T analyzer about this same time which was unable to survive competition from the amperometric analyzers. However, Hach did introduce the first successful DPD colorimetric analyzer in 1980. This unit is described below.

The search for a better method of chlorine residual determination resulted in the development by H. C. Marks of Wallace and Tiernan Company, about 1942,[36] of the amperometric titrator, in part inspired by Griffin's work in 1939–1940 on the breakpoint phenomenon. The first amperometric chlorine residual recorder was placed on the market by Wallace and Tiernan Company about 1948. Among the first installations were those at Wyandotte, MI,[37] Allentown, PA,[38] and Brantford, Ontario, Canada. Results from these early installations proved so dramatic that continuous residual recording became popular in the treatment of public water supplies. Operating experience from these installations demonstrated two important characteristics of the amperometric cell. These were predictable changes in cell output current due to changes in sample temperature and pH. Changes in pH were corrected by additions of pH 4 buffer solution, and temperature was corrected by either manual adjustment or automatically by the installation of a thermistor.[39] Most current models come equipped with a thermistor.

During this period Wallace and Tiernan company made a polarographic cell for the City of Chicago Water Department which was arranged to measure combined (chloramine) chlorine only. It did not require chemicals but it used a large sample, 1 liter per minute. Twenty years later, when Chicago built the Central Water Filtration plant, analyzers were required to measure free chlorine only. The same type of analyzer was used except that the applied voltage across the electrodes was changed. This made the same cell specific for free chlorine.

Since the 1970s two notable contributions were made. The first was the development by Morrow[40] for Fischer and Porter Co. of the Chlor-Trol analyzer, which was capable of measuring free chlorine in a sample containing significant chloramine concentrations. The second was the development of the selective membrane electrode by Johnson et al.[41] By this time many companies were actively engaged in developing instrumentation for continuous chlorine residual measurement. These participants included Foxboro, Orion, Uniloc (Rosemount), Delta Scientific, Enterra, Chlortect, and IBM. Other accomplishments have been modifications that have extended the use of chlorine residual monitoring in wastewater, and finally the step from monitoring to automatic residual control.

The step from residual recording to residual control took nearly ten years to gain acceptance in the United States. Residual control was being used in England for the London water supply as early as 1950 and for controlling superchlorination and dechlorination stations of underground supplies where the only other treatment was microstraining.

The use of continuous residual analyzers for wastewater chlorination began circa 1960. The immediate purpose was to control the chlorination system. Prior to 1960 there was limited use of analyzers on highly polished tertiary effluents. The pathway to successful analyzer operation on wastewater effluents has been a rough one. One of the first successful installations on secondary treated effluents was made at the Napa Sanitary District, Napa, CA in 1961, where the unit installed was the same as the version used for potable water, except that a motorized filter was installed on the analyzer cell sample line.

When this unit was tried for operation on primary treated effluent, various problems were encountered which were related to the quality and characteristics of the wastewater. The high grease content in the primary effluent is particularly bothersome in that it coats the electrodes. The insulation effect of the grease causes a severe distortion of the calibration, resulting in erroneous readings. This became an unwarranted maintenance problem.

The amount of buffer solution required to lower the pH of most primary effluents is very high, and an appreciable excess of iodide is also necessary to make the reaction with combined residual chlorine proceed as fast as possible. This made the chemical cost of operation a most significant factor.

Experience has shown that chlorine residual analyzers are not reliable for use with primary effluents owing to the poor quality of the sample, and the inability to maintain a sample flow through the cell. Another factor discouraging to operating personnel is the time required to keep the electrode surfaces clean enough to maintain system calibration. Recalibrations for primary effluent samples need to be made at least twice a day.

For many years chlorine residual control was used solely as an adjunct to flow-paced control. Now there are many installations controlled solely by chlorine residual.

The advent of the programmable controller has stimulated interest in the use of residual control. This type of instrumentation allows the precise formulation of a residual control system to any site-specific case.

General Discussion and Definitions:

Amperometric. Continuous analyzers which use two electrodes for measurement are usually categorized as *amperometric*. This was true when referring to the original Fischer and Porter and Wallace and Tiernan analyzers. However, it is no longer true with the array of analyzers currently available.

Electrode reactions are characterized by the transfer of electrons between the electrode and substances in the sample solution. The electrode reaction which initiates the electron transfer results in the flow of current between the electrodes. When the electrode reactions occur spontaneously upon short-circuiting the two electrodes, the system is described as a *galvanic cell.* If, however, the reactions are forced to occur by the imposition of an external electromotive force the cell is referred to as an *electrolysis cell,* i.e., the generation of chlorine. In the above definitions the term "reactions" refers to the chemical changes of the solution which might occur. In the measurement of chlorine residuals the current flows between the electrodes primarily as a characteristic of the species involved. The dual electrode analyzers like those made by Capital Controls, Fischer and Porter, and Wallace and Tiernan are galvanic cells—amperometric type. These cells respond in a linear fashion only to the elemental halogens, i.e., bromine, chlorine, and iodine. They do respond to most chloramines, but not in a linear fashion.

Voltammetry. When the galvanic cell is subjected to an external applied voltage to provide the measurement of a particular chemical species in the presence of other species the system is described as *voltammetry*. In this case the voltage applied provides qualitative information on the electroactive substance present and the current measured provides the quantitative information of the species selected by the applied voltage. This technique is used to isolate the free chlorine species that ocurs in the presence of chloramines. This is a variation of the *amperometric* measurement technique. The use of the term *amperometric* has come into use to indicate concentration measurement based upon the measurement of the current flowing between two electrodes at a constant potential at the indicator electrode.

Polarography. This is a broader term used to describe analytical measurement of trace materials in solution, metallic ions, etc.[42] This method utilizes the principles of voltammetry, i.e., current–voltage curves, but with a polarographic cell. A polarographic cell is an electrolytic cell consisting of a nonpolarizable reference electrode, a readily polarizable electrode, and an electrolyte solution containing electro-oxidizable or electro-reducible material.

At this point the difference between a polarized and a depolarized electrode should be understood. An electrode is said to be polarized when it adopts the externally impressed potential (voltage) on it with little or no change in the rate of the electrode reaction, i.e., no change in current. Therefore if only one electrode in the cell is polarized, its potential changes in the same amount as the change in applied voltage.

At the other extreme an ideally depolarized electrode is one that retains a constant potential regardless of the magnitude of the current flowing between the electrodes. Therefore the nonpolarizable electrode potential is not altered by the changes in the externally applied voltage.

When an increasing electromotive force (applied voltage) is impressed across the electrodes of a polarographic cell, and if the resulting current is plotted as a function of the applied voltage, a curve is obtained as shown in Fig. 9-60. The extension of the wave curve along the current axis is directly related to the concentration of the material in the solution sample and its inflection point (0.8 V) is at the applied voltage which is characteristic for this particular species of material. Examples of this type of cell for measuring species of chlorine residual are described below.

Potentiometry. This method is used for measuring trace materials in solution by means of an electrode containing two sensing elements. The method is based upon the constant current technique of species separation which measures the potential at each element. It is not widely used in the measurement of chlorine residuals.

The potentiometric method for chlorine residual is only applicable to the measurement of total residual chlorine because it measures the iodine species released from the potassium iodide electrolyte oxidized by total residual chlorine. A platinum (redox) sensing element develops a potential which depends upon the relative concentration of the iodide and iodine species in the sample. The second element,

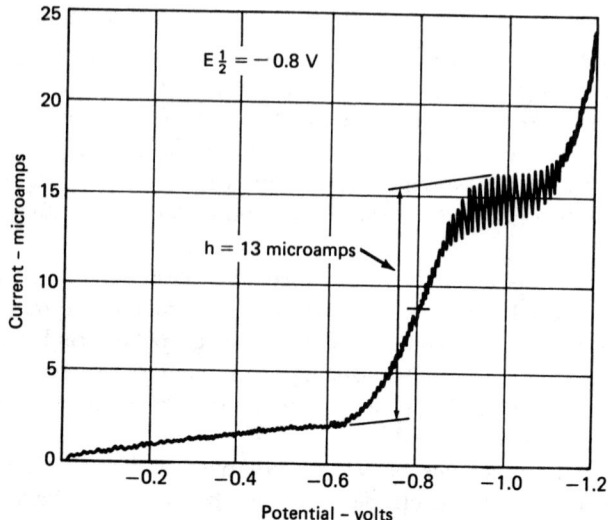

Fig. 9-60. Polarographic curve.

the iodide element, develops a potential that depends upon the iodide ion concentration in the sample. The electrode then in turn measures the difference between these two potentials developed at the two sensing elements. The net potential represents the iodine concentration directly related to the total residual chlorine as measured in the amperometric method.

Designer's Evaluation Check List. The following limitations should be examined when selecting or evaluating an analyzer for site-specific conditions.

Species Measurement. The most important function of any analyzer is its ability to measure any given species of chlorine residual without interference from other species present in the sample. For example, how dependable can a given analyzer be for measuring free chlorine residual in the presence of some chloramine residual? What proportion of chloramine residual will interfere with the accuracy of the free residual measurement? This is part of analyzer accuracy limitation.

Accuracy. This is the agreement between the amount of a component measured by the analytical method and the amount actually present. It is usually expressed as percent of full scale over a specified range.

Precision. This is referred to as reproducibility of measurement when repeated on a homogeneous sample under controlled conditions. Normally this should be expressed in percent of the full scale reading over a specified range.

Sensitivity. This is described as the least concentration of chlorine species detectable in any given sample.

Range. This is most important. Some analyzers are limited to a detectable range of only 0–2.0 mg/l. Analyzers capable of 0–10 or 0–20 mg/l are preferable to those with a limited range.

Response Time. This characteristic is of prime importance if the analyzer is to be used for residual control. The manufacturer usually expresses this time as that necessary to provide a percentage response to a step change in residual concentration after the sample has entered the analyzer.

For residual control purposes the 90 percent response time should not exceed 10 seconds and initial response should occur in less than 10 seconds.

Temperature Compensation. All analyzers should be equipped with thermistors to compensate for temperature variations in the sample. These variations affect the response time and current output of the electrodes. Colorimetric analyzers experience major changes in response time due to changes in both sample and ambient temperatures.

Wallace and Tiernan, Division of Pennwalt Corp.:

General Discussion. This family of continuous analyzers was the first to be marketed for measuring chlorine residuals. These analyzers can be divided into three categories: potable water, wastewater, and cooling water. The first two are amperometric. The third type (cooling water) is polarographic. They are described below.

Amperometric Cell. Theory of Operation. Fig. 9-61 illustrates a typical arrangement of the analyzer cell. The measuring electrode is copper and the reference electrode is either platinum or gold. The electrodes are mounted coaxially so that the sample water flows continuously through the cell. The cell produces a small (0–150 microamp) DC current proportional to the amount of free halogen (chlorine, bromine, iodine, chlorine dioxide) in the sample. However, since the cell is capable of measuring an oxidation reaction the cell will also produce a current due to the presence of chloramines in the sample. This response is not linear, but acts to interfere with free residual measurements. The limiting range of these cells is 0–20 mg/l, but they are available in ranges of: 0–1, 0–2, 0–5, 0–10, and 0–20 mg/l. The current response is linear over the entire range of 0–20 mg/l. Speed of response is 10 seconds or less.

The basic principle of this cell is that a current will flow between the two electrodes proportional to the free halogen concentration in the flowing sample. To measure this current the electrodes are short circuited.

A minor flow of current will always be present even in the absence of a free

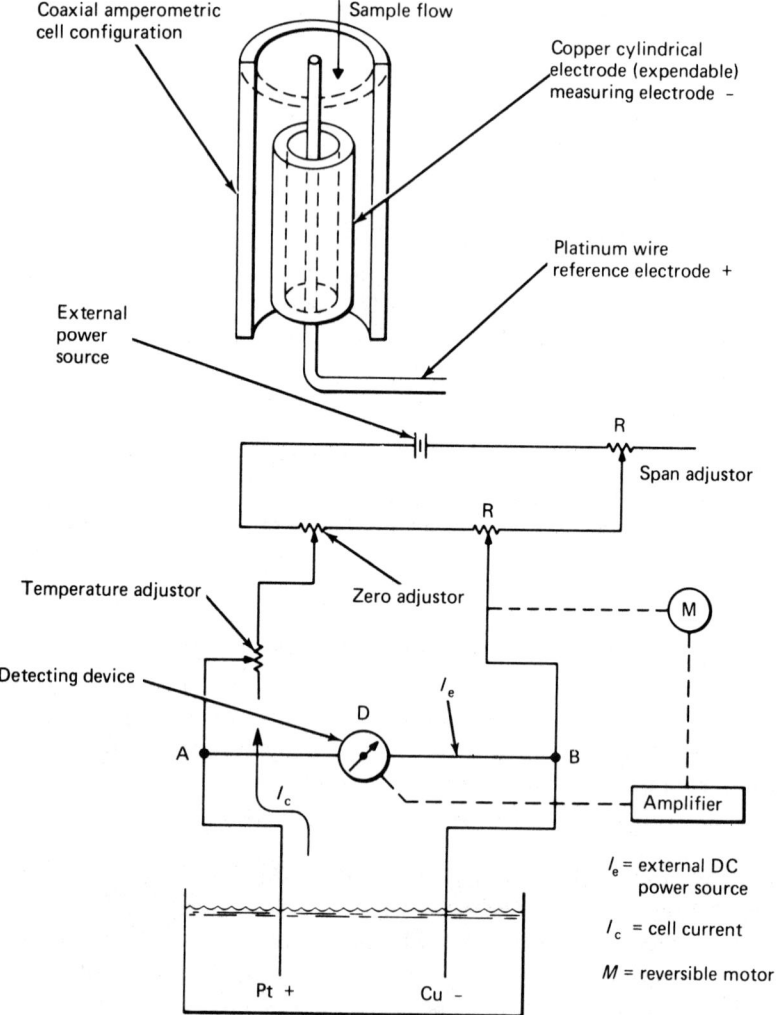

Fig. 9-61. Details of amperometric cell for chlorine residual analyzer (courtesy Wallace and Tiernan Division, Pennwalt Corp.).

halogen; therefore the current detecting device must be able to delete this current. This is shown as the zero adjustment in Fig. 9-61.

There are two ways of measuring the current produced by an amperometric cell. One way, illustrated above, measures the current flow directly. The other way is to pass the current through a known resistance and measure the potential drop. This is known as the IR drop method (see "Capital Controls").

Referring to Fig. 9-61, an external DC power source is used for comparing

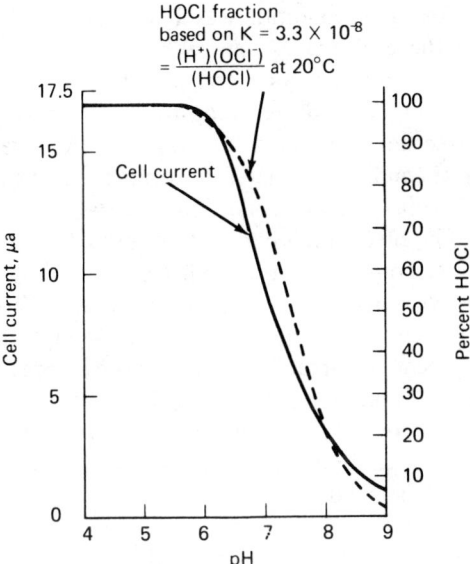

Fig. 9-62. Effect of pH on amperometric cell current. (All values are for 1.0 mg/l free chlorine residual.)

the current produced by the free halogen concentration in the cell. Current flows from A to B. As this current increases, the detecting device D indicates this and generates a signal which is amplified and drives a reversible motor which positions a slider R. This movement increases the current from the external power source, which flows from B to A. As soon as the increase from the external source is sufficient to balance the potential between A and B, the motor stops and the recording or indicating mechanism comes to rest (null position) at the new value of an increased current generated by the free halogen in the amperometric cell.

*p*H *Effect.* Current flow in amperometric cells are grossly affected by changes in *p*H. Fig. 9-62 illustrates the magnitude of this effect upon a 1.0 mg/l free chlorine residual. This is called the *p*H coefficient. It is dealt with in either one of two ways by Wallace and Tiernan analyzers: addition of either carbon dioxide gas, or liquid sodium acetate—acetic acid (*p*H 4 buffer) to the sample.[43] The cell current is most stable in the *p*H range of 4 to 4.5. The amount of buffer required depends upon the alkalinity of the sample. The analyzer is designed to meter either type of buffer additive.

Temperature Effect. All amperometric cells respond to changes in sample temperature. The change in cell current due to temperature change in the sample for a fixed residual is mathematically predictable, and therefore can be automatically compensated by the installation of a thermistor[39] in the circuitry as shown (temperature adjuster).

Diffusion Phenomenon. This condition was first recognized by Marks and Campbell in 1950.[44] When the cell current approaches zero or remains at zero in the absence of free halogen residual, a concentration of electrons begin to accumulate at the measuring electrode, thereby repelling the negative ions. This impedes the speed at which the free halogen can diffuse through the layers of electrons that accumulate under these conditions. Therefore, sample flushing action and electrode bombardement tend to alleviate this condition.

Electrode Fouling. The effect of this condition was discovered early in the development of these analyzers. To prevent the sample cell from being fouled by biological slime, organic debris, or chemical deposition (alum floc, ferri-floc) provision is made to continuously recirculate a quantity of abrasive grit which bombards both electrodes and the adjacent cell area. This keeps the electrodes clean and maintains the electrode surface at optimum molecular equilibrium.

Sample Flow Rate. Current flow in the cell will vary with sample flow rate. It must be observed that in calibrating an analyzer with a sample containing free halogen, the zero check and adjustment is made by stopping the sample flow. This demonstrates the effect of sample flow rate upon cell current. The sample hydraulics must be such that the flow through the cell will be constant. Variable pressures in the sample source must be controlled by a suitable external pressure regulating valve.

Zero Current Effect. It has been observed that there is no practical means of calibrating the amperometric cell when free halogen is absent from the sample. Consider the zero check described above, i.e., completing the zero adjustment with the sample flow stopped while the sample contains a significant concentration of free halogen. When a sample not containing any free halogen is passed through the cell the analyzer will always indicate a slightly negative reading—below zero.

Potable Water Practice. The analyzer for this purpose is described in Catalog No. 50.245. It can measure either free or combined chlorine. This analyzer has been successful in monitoring and controlling the free chlorine residual process for many years. The operating experience that has been accumulated indicates clearly that the treated water total chlorine residual should contain at least 85 percent free chlorine. When this analyzer is arranged to control or monitor free residual in the presence of combined residual, interference from the combined chlorine will be about 10 percent of the total residual chlorine at sample pH 4.5.[45]

When total residual chlorine is to be measured (free and combined) it is necessary to add potassium iodide along with the buffer solution. Total residual chlorine in the sample at pH 4.0–4.5 will release free iodine in proportion to the total chlorine residual. Therefore the amperometric cell will generate a current in proportion to the free iodine liberated from the potassium iodide.

The Wallace and Tiernan cell requires a sample flow of 360 ml/min. At this rate 15 g KI/l/mg/l total chlorine residual is required. Normally the KI comes in tablet form at 3 g/tablet. The tablets are dissolved into the buffer solution.

Wastewater Practice. The analyzer for wastwater is considerably different from the potable water model. It is described in Wallace and Tiernan Catalog No. 50.215. This unit is a modification of the potable water unit. It is designed to process low quality wastewater effluent. A sample dilution system (with fresh water) is provided to reduce chemical cost (pH 4 buffer and KI). A motorized filter is installed in the incoming sample line to eliminate the sample debris encountered in treated sewage. This analyzer requires the same amount of KI per mg/l of total chlorine. Allowance must be made for dilution water added.

Polarographic Cell. This cell uses a reference electrode and a measuring electrode illustrated in Fig. 9-63 and is described in Wallace and Tiernan Catalog No. 50.236.

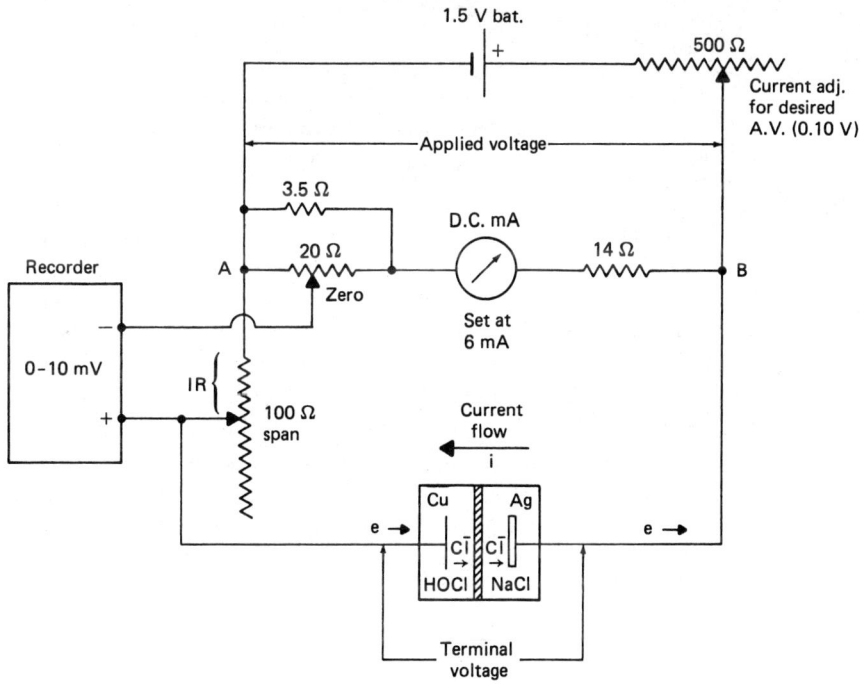

Electron is from Cu Ag (opposite to current flow)
Terminal voltage = A.V. − IR drop across span potentiometer
Applied voltage = Total IR drop across AB
Applied voltage = 0.006 × (14Ω + 3Ω) = 0.102 V approx.

Fig. 9-63. Polarographic cell (courtesy Wallace and Tiernan Div. Pennwalt Corp.).

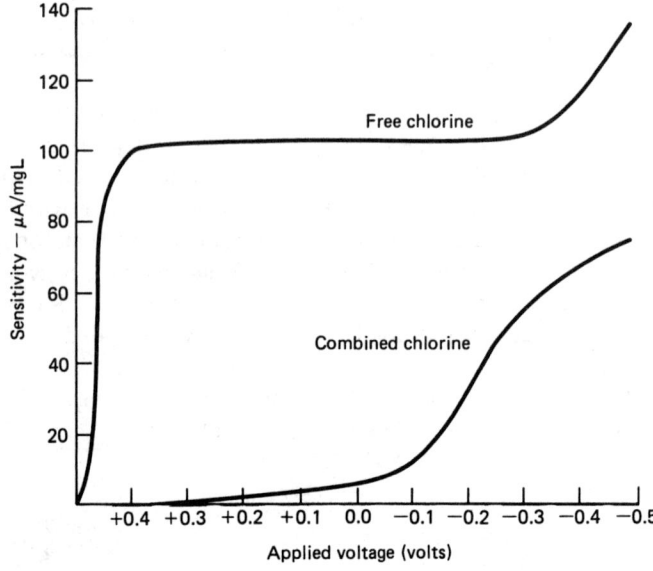

Fig. 9-64. Polarographic curve showing free chlorine versus chloramines.[58]

The measuring electrode is the only electrode that is in the sample. This cell is capable of measuring either free chlorine or free iodine and it does not require any pH buffering or potassium iodide. The use of applied voltage to the electrodes allows the selection of either chlorine residual species. Fig. 9-64 demonstrates that the free chlorine voltage current measurement curve obtains at +0.3 V applied voltage while combined chlorine obtains at −0.07 V (approx.). In this cell the reference electrode is silver–silver chloride and the measuring electrode is platinum. The range of this analyzer is limited to 0–2 mg/l free chlorine. It is not affected by pH changes in the range of 5–8.5, nor is it affected by the presence of combined chlorine residual. Since no chemicals are added to reduce pH, a true sample is analyzed. Changes in salinity and the presence of such cooling water treatment compounds as chromates, phosphates, and defoamers do not interfere with its accuracy. A thermistor compensates for sample temperature changes between 35°F and 120°F. Grit is replaced once a month and electrolyte tablets (NaCl), once every 3 months. This unit was designed specifically for condenser cooling water systems.

Capital Controls:

General Discussion. Two amperometric analyzers are offered. The flow diagram is shown in Fig. 9-65.[46]

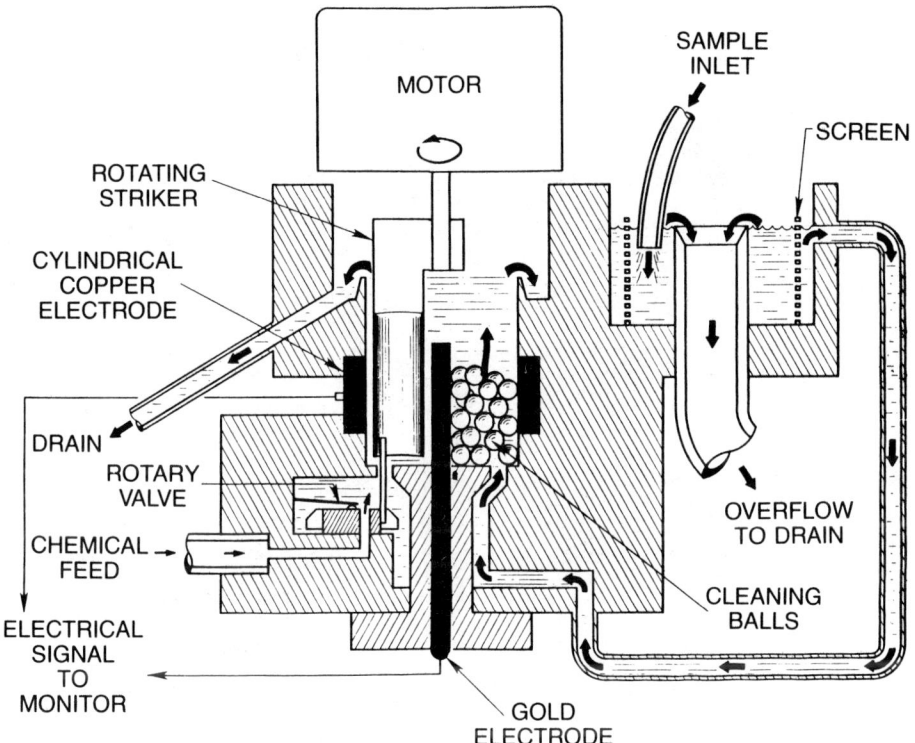

Fig. 9-65. Amperometric analyzer flow diagram (courtesy Capital Controls Co.).

No. 871. This is for the measurement of free chlorine. Analyzer includes a single reagent feed system for feeding pH buffer solution.

No. 872. This unit measures total residual chlorine. It is furnished with two separate reagent feed systems. One for pH 4 buffer solution and the other for potassium iodide solution.

Both the buffer solution and potassium iodide solution are stored in separate containers (7 days capacity) and fed from constant head reservoirs. Each solution is fed through a rotary valve located on either side of the sensing cell. These valves are actuated by an extension of the striker used to clean the sensing cell. The valves move with each contact of the striker. This provides a precise amount of solution to be fed during each valve revolution.

The sample flow to the filtering chamber should be about 500 ml/min, but only 100 ml/min passes over the overflow weir in the filtering chamber to the cell. This is the amount of sample that needs to be treated with reagents. This sample flow rate is controlled by fixed differential weir levels between the inlet chamber and the cell chamber.

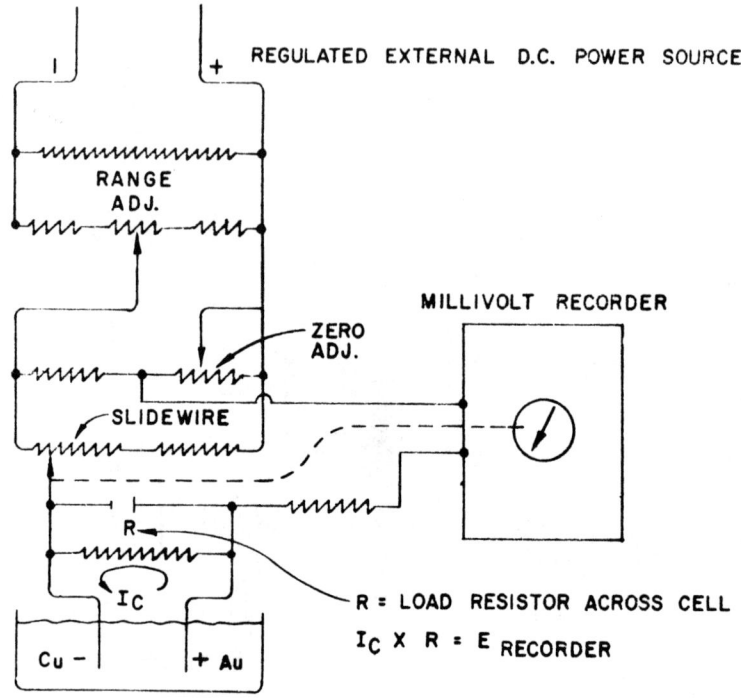

Fig. 9-66. Amperometric cell current measurement by the IR drop circuit.

The surfaces of both electrodes are kept clean by the continuous action of PVC balls agitated by a motor driven rotating striker.

Residual Measurement. The cell circuitry for these analyzers differs somewhat from the previously described Wallace and Tiernan analyzers. The cell current is measured indirectly. Fig. 9-66 illustrates a typical *IR* drop circuit which is used in the Capital Control analyzers. The cell current passes through a resistance and the voltage (potential) drop across this resistance is measured. The major difference between this method and the current measurement method is that the external power source is used solely for zero adjustment. This is a necessary feature, since some current will flow in all waters even if a free halogen residual is zero, therefore this natural current must be neutralized.

The current output of these analyzers is 50 millivolts at full scale. These units are offered in ranges of 0–0.5, 0–1.0, 0–2.0, 0–3.0, 0–5.0, 0–10, and 0–20 mg/l.
Sensitivity is 0.01 mg/l.
Speed of Response. About 4 seconds elapses from the time the sample enters the analyzer to the first signal indication of changed residual. Full-scale residual change requires 1.5–2.0 minutes to reach correct indication—depending upon analyzer range.[46]

Delta Scientific (Xertex):

Historical Background. The 1984 models marketed under the above name derives from the research and development work by Johnson, Edwards, and Keeslar.[47,48] The importance of developing an instrument to measure undissociated HOCl was reported by White[49] in 1974. The purpose of such an instrument was to equate more precisely the germicidal power of free chlorine in terms of residual measurement. All of the existing methods measure HOCl plus OCl^-. It is well documented that the germicidal efficiency of the HOCl fraction is 80–100 times that of the OCl^- ion (see Chapter 4). Therefore measurement of HOCl alone instead of HOCl + OCl^- would provide a more realistic relationship between "Free chlorine" and germicidal efficiency.

The Johnson cell was designed as a direct and selective amperometric membrane electrode for the exclusive measurement of undissociated HOCl in the presence of OCl^-, combined chlorine, and other interferences found in the control of disinfection by chlorine in natural waters.[48] The sensor was described as a polarographic cell. It consisted of a silver anode immersed in a 1.7 M KCl solution and a gold cathode. The gold measuring electrode was separated from the "HOCl porous" membrane. The silver/silver chloride half cell acted as the reference electrode. The membrane is used to establish a diffusion-limiting barrier next to the cathode (gold electrode). Any oxidizable compounds able to diffuse through the membrane are reduced, which generates a flow of current within the cell. Membrane selectivity was further provided by a positive voltage applied to the cathode. This positive voltage makes it difficult for OCl^- and chloramines to be reduced (i.e., obtain electrons). Research by Johnson et al.[48] revealed that interference by OCl^- ion was 0.0 percent, monochloramine 1.3 percent, dichloramine 3.0 percent, and nitrogen trichloride 725 percent. Interference by NCl_3 would be disastrous to any attempt to measure free chlorine after breakpoint in some waters.

On the basis of the above research and several years of development, Delta Scientific marketed three models as a division of National Sonics, then a part of Envirotech. The first model was based upon the Johnson cell, designed to "measure disinfecting power"—more specifically, it was intended to measure undissociated HOCl in the presence of OCl^- and chloramines. This was the Series 8224/8324 HOCl analyzer.[50] The second model, Series 8225/8325, was designed to measure all chloramine species and hypochlorous acid. This analyzer was intended for the wastewater market. During this period of marketing Delta Scientific altered their stance significantly because of the wastewater market potential. In 1978 they announced their series 8000 Polarographic Wet Chemistry Analyzers:[51] Model PWC-82125 Module (for measuring Total Residual Chlorine), together with a Model 8224 or 8324 Analyzer Transmitter.

Owing to difficulties with the original idea of measuring only HOCl for free chlorine, the sample conditioning module adds pH 5 buffer solution to the sample which converts HOCl + OCl^- to all HOCl, which is measured by the Delta polarographic sensor.

To measure total residual chlorine the sample conditioning module adds acified potassium iodide solution to the sample, converting all of the chlorine residual to free iodine, which is measured by the Delta Scientific polarographic sensor.

Sometime later Delta Scientific made another change by offering a model which was specific for ammonia chloramine species.[52] This was designated the CAC Model 82125 (combined ammonia chloramines). This model contained an electrolyte solution of potassium iodide. The current response between the reference electrode, the electrolyte, and the chloramines diffusing through the membrane at the measuring electrode was proportional to the concentration of molecular triiodide ion

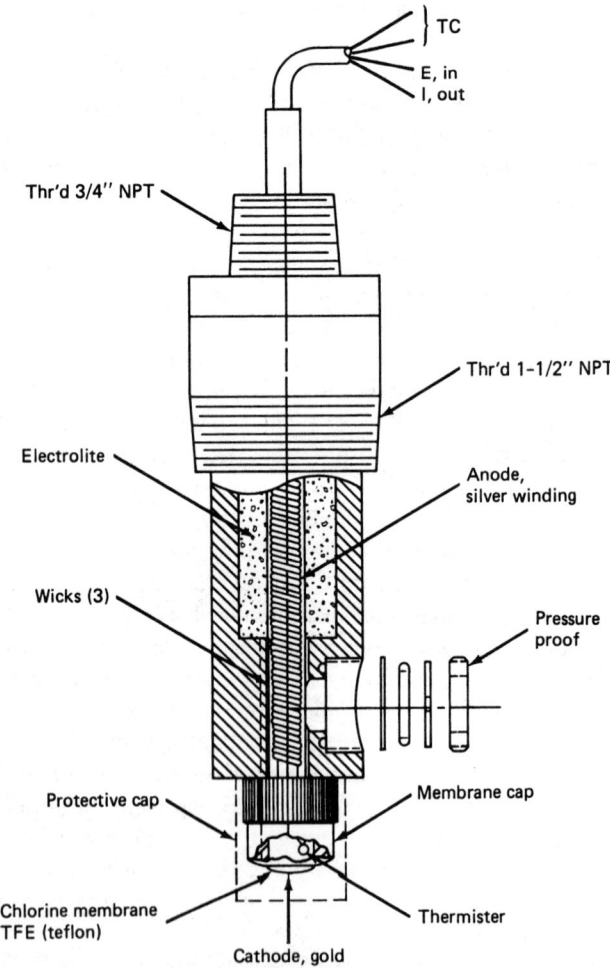

Fig. 9-67. Chlorine sensor polarographic membrane cell (courtesy Delta Analytical Div. Xertex Corp.).

(I_3^-) produced upon oxidation of iodide by the membrane-permeable ammonia chloramine species. The practical value of this analyzer has never been made clear.

During 1984 Delta Scientific announced a new 900 Series of analyzers. These are digital microprocessor based analyzers which measure sensor current and temperature and sensor output, and accurately compute the chlorine residual species using a polynomial equation. These instruments are described below. To better understand these instrument descriptions the following terminology is used by Delta Scientific: FAC = free available chlorine = undissociated HOCl; TFC = total free chlorine = HOCl + OCl$^-$; TRC = total residual chlorine = HOCl + OCl$^-$ + ammonia chloramine + organochloramines; CAC = combined available chlorine = HOCl + OCl$^-$ + ammonia based chloramines (i.e., mono- and dichloramine).

Delta Scientific Model No. 924. This instrument measures either FAC or TFC.[53] Therefore it is specific for free available chlorine and at the same time is subject to interference by both NCl$_3$ and combined chlorine. NCl$_3$ has a gross effect while combined chlorine has a much lesser effect.

If the pH is separately measured it can be used to measure either FAC or TFC manually. When it is equipped with a pH probe the instrument measures or computes either FAC or TFC, temperature, and pH. It provides these results for display and instrument output. Chlorine residual values can be selected for 0–1, 0–5, or 0–10 ppm full scale. The analog outputs for chlorine are 0–10 mV, and temperature is 0–10 mV corresponding to 0–50°C.

This analyzer does not require the addition of any reagents. The heart of this instrument is the chlorine sensor (see Fig. 9–67). This is a classic polarographic cell. It consists of a semipermeable membrane, a silver anode immersed in a salt solution, and a gold cathode. A constant voltage—appropriate for the free chlorine plateau—is applied to the gold cathode. The membrane establishes a diffusion-limiting barrier next to the gold cathode where the free chlorine species, which diffuse through the membrane, are reduced. Residual chlorine diffusing through the membrane reacts electrochemically with the salt solution and causes a current to flow between the gold cathode and the silver/silver chloride reference electrode (anode). The current is directly proportional to the chlorine concentration:

$$I = \left(\frac{nFAP}{b}\right)C = KC \qquad (9\text{-}13)$$

where

I = current, mA
n = number of electrons transferred in the reduction reaction, dimensionless
F = Faraday's constant, mA-sec/mmole
A = area of the electrode, mm^2

P = permeability of the membrane to the chemical species diffusing through the membrane, mm^2/sec
b = membrane thickness, mm
C = concentration of the chemical species being reduced, mmole/mm^3
K = cell constant, mA-mm^3/mmole

The terms n, F, A, and b are constant for any given membrane and chemical species.

The microprocessor operates as follows:
1. It measures the chlorine sensor output and temperature and computes FAC.
2. It measures pH sensor output and computes temperature corrected pH.
3. It computes TFC from the temperature-corrected FAC and pH, because the concentration of HOCl is also temperature dependent (see Chapter 4).

An internal battery is provided to maintain the polarizing voltage on the gold cathode during power failures. This is a rechargeable battery which is kept charged during normal operation and has a life measured in years.

The speed of response is 90 percent in 20 sec.
Linearity over 0–5 and 0–10 ppm is unknown.
Stability is ±1 percent full scale for 24 hours.
Electrolyte life is approximately 6–9 months.
The electrolyte is 60 ml buffered salt solution.
Detection limit is unknown. Precision is unknown.

This analyzer is designed on the basis of a probe application, i.e., no sample lines required. The sensor can be located up to 50 ft from the analyzer. Sample movement must be not less than 1 ft/sec at the gold cathode.

Delta Scientific Model No. 925 Chlorine CAC/TRC.[37] This instrument was designed for wastewater applications and uses a special sensor.[38] The sensor used in this analyzer is physically the same as the Model 924 with the following exceptions: the electrolyte is a buffered potassium iodide solution instead of a salt solution, and the applied voltage to the gold cathode is different. This difference in applied votage allows the sensor to be specific for combined chlorine. This is a typical voltammetric application. Under the above conditions, the analyzer measures CAC, for whatever that is worth.* This mode of measurement does not require the addition of any reagents.

To measure TRC this model analyzer requires the use of the PWC Sample Conditioning Unit. This unit treats the sample with a buffered potassium iodide solution, which converts TRC to free iodine. Delta Scientific advises it is imperative that the NH_3–Cl_2 ratio to be at least 4 to 1 if the sensor is to work directly in the wastewater stream.[55]

The model 952 analyzer has the same performance characteristics as the model 924.

---

* Probable widest application would be wastewater effluents.

Fischer and Porter Co.:

General Discussion. Three basic models of analyzers are available for measuring the various chlorine residual species under a variety of conditions. These analyzers incorporate conventional solid two-electrode amperometry. The electrochemical cell consists of a gold (or platinum) measuring electrode and a copper reference electrode. The halogen reaction occurs at the gold cathode. The copper reference electrode is the sacrificial anode. The copper reference electrode is nonpolarizable such that when a potential is applied *across* the two electrodes, the potential at the measuring electrode is the only one that is affected.

Anachlor™ Series 17B4200C. This model is used to measure either free or total chlorine.[56] Free chlorine is measured by buffering the sample to pH 4 and the addition of potassium bromide. At this pH free chlorine will oxidize the bromide ion to free bromine (Br_2). The galvanic cell measures the current which represents the concentration of free bromine which is proportional to the free chlorine present:

$$2Br^- + HOCl + H^+ \longrightarrow Br_2^0 + Cl^- + H_2O \tag{9-14}$$

and the bromine then reacts at the measuring electrode:

$$Br_2^0 + 2e^- \longrightarrow 2Br^- \tag{9-15}$$

In effect the bromine depolarizes the measuring electrode (gold), allowing copper to go into solution (at the electrode) as electrons flow to replenish the polarizing layer.* The pH of the sample is depressed to 4 to convert all the OCl^- to $HOCl$. This condition provides maximum current flow which is not affected by pH variation from 3 to 5. pH 4 is optimum for measuring free halogens. In the event that chloramines are present with free chlorine they will also generate a current at the measuring electrode. This interference is on the order of 10 percent of the total combined chlorine residual present. Chloramines will not, however, oxidize any of the bromide ions to free bromine.

The depolarizing current produced by the reactions occurring at the electrode passes through the load resistor. The potential drop across this resistor is directed through the thermistor and millivolt-to-current converter to provide a temperature compensated 4–20 mA DC output signal.

The gold measuring electrode is mechanically connected to the shaft of the motor above the cell which rotates the electrode at 1550 rpm. This helps to maintain ideal, reproducible electrolytic conditions and allows the cell output to be indepen-

---
* Free chlorine has always shown a tendency to cause a film-type coating to occur at the surface of the noble metal (Au,Pt) measuring electrodes. This produces a poisoning effect. Free iodine and free bromine do not exhibit this characteristic.

dent of any slight variations in the sample flow rate. Thirty nylon balls are placed in the annular space between the electrodes where their constant motion, maintained by the rotating measuring electrode keeps both electrode surfaces free from foreign matter.

This analyzer is available in selected ranges up to 0–20 mg/l. Accuracy is ±2 percent of range and sensitivity is 0.01 mg/l. Response time is 30–40 seconds upon introduction of sample into sample inlet tank. Cell response is less than 5 seconds full scale.[56] Buffer solution and potassium bromide is pumped into the sample by a peristaltic pump which operates at 1 rpm. Reagent consumption is 0.5 l/day. Sample flow rate through cell is approximately 100 ml/min. Sample to analyzer can be 0.5–3 gpm depending upon local conditions.

Total chlorine residual is measured by changing the buffer additive reagent from potassium bromide to potassium iodide. TRC reacts to oxidize the iodide ion to free iodine at pH 4. Free iodine generates a current at the measuring electrode in proportion to the TRC.

Anachlor™ Series 17B4200-2. This model is a modification of the preceding analyzer. This analyzer was designed for controlling the free residual process which was dependent upon the breakpoint phenomenon.[57,58] This means that the analyzer must be able to distinguish between free and combined chlorine residuals. Morrow and Roop[58] described how they prevented combined chlorine residual interference with the free residual. First they found that chloramine will not oxidize the bromide ion at pH 4. Next they examined the separation of the voltammetric curves for free chlorine and chloramines by appropriate applied voltage to the two electrodes. Finally they found a suitable applied voltage that gave sufficient separation of these curves to avoid interference by combined chlorine residual.* This analyzer proved able to perform as described on a wastewater nitrogen removal project at Rancho Cordova California in 1975.[80]

Since this analyzer is identical to the Anachlor Series 17B4200C (described above), except for the external voltage applied to the electrodes, it has the same operating characteristics and requirements.

Chlor-Trol™ Free Residual Chlorine Analyzer Series 17K/1000. This analyzer measures free chlorine without the use of any reagents and is supplied with automatic temperature compensation.[59] The amperometric cell consists of two concentrically mounted stationary copper electrodes. A small DC voltage is applied to the electrodes. This causes polarization of the electrodes thereby preventing current flow in the circuit. The presence of free chlorine in the flowing sample depolarizes the electrodes and permits a current to flow in the circuit. This depolarization current is directly proportional to the free chlorine residual concentration in the sample. This current is measured by a series-connected microammeter located within the indicator. The instrumentation contains provisions for zero and span adjustment. Both high and low alarm contacts are also included.

---

* See Fig. 9-64.

The range of this analyzer is limited to 0–2 mg/l free chlorine. Accuracy is 0.1 mg/l, and sensitivity is 0.02 mg/l changes in residual. This unit is specific for free residual.* Changes of ±10 percent of sample flow does not affect the cell output. Sample flow through the cell is 750 ml/min (0.2 gpm). Variations in sample water pH between 5.0 and 9.0 have no effect on cell performance. Sample water must be free from solids which will not pass through a 170 mesh screen (88 microns) Maximum sample pressure is 40 psi.

Amperometric Probe. In 1984 Fischer and Porter was granted a patent (No. 4,441,974) covering a nutating chlorine residual analyzer probe.[84] This probe is unique in the field of chlorine residual measurements. It is a membrane type probe designed for field mounting indoors or outdoors, in open process tanks or channels. It can measure either free or total chlorine residuals without any chemical addition to the sample. The internal chemistry of the probe enables an accurate residual measurement unaffected by pH variations in the process water. It is constantly in motion (nutation). This maintains the integrity of the process sample at the membrane/sample interface preventing depletion of the chlorine species being measured in the region immediately adjacent to the membrane.

The probe illustrated in Fig. 9-68 includes a platinum measuring electrode, an oxidizable silver counter-electrode and an electrolyte which in combination with the electrodes defines an electro-chemical cell. The output current depends upon the amount of chlorine residual passing into the cell through a diffusion membrane permeable to the chlorine species being analyzed. In order to maintain analytical sensitivity of a membrane probe instrument it is highly desirable to simulate rapid sample flow past the membrane. To accomplish this, the probe is vertically supported through a flexible coupling. It contains an internal motor with an unbalanced rotor connected to the shaft of the probe. Rotation of the rotor causes the probe to nutate about its vertical axis. This simulates the effect of rapid sample flow. The nutate or wobble rate is 700–800 cycles per minute.

The internal chemistry of the probe makes use of the reaction between the various chlorine species and bromide salts. Free chlorine reacts with bromide salts to produce free bromine at pH 4.0–4.5. Combined chlorine and bromide salts also react, but at pH levels below 2.0 to produce free bromine. The pH of these reactions is controlled within the probe.

Free chlorine measurement. Where this is the chlorine species to be measured the electrolyte within the probe is a concentrated solution of alkali bromide (KBr) buffered to pH 4 with an acetic acid/acetate buffer.

Combined chlorine measurement. This measurement does not include free chlorine, only combined chlorine. In this arrangement the saturated solution of potassium bromide is buffered to a pH value approaching zero. It has been found that

---

* If chloramine residual develops in the free residual process and it is a small constant amount (10 percent of total residual) it can be calibrated out of the free residual reading.

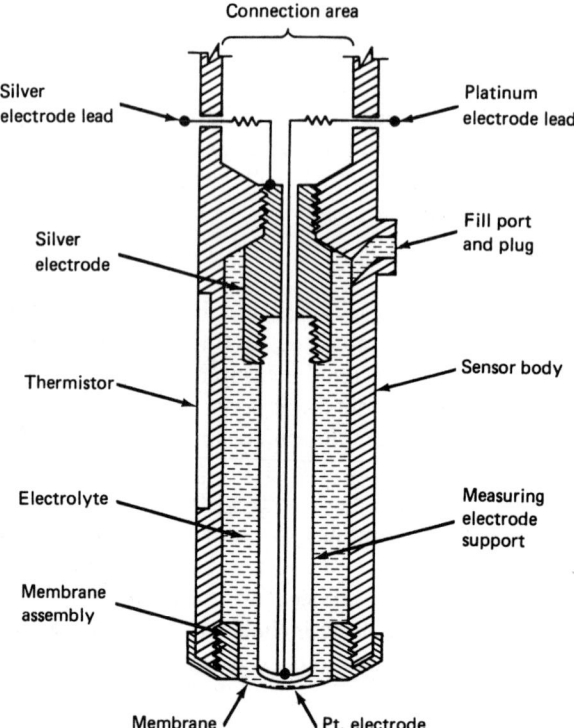

Fig. 9-68. Oscillating probe for a new scientifically advanced amperometric cell (courtesy Fischer and Porter Co.).

at pH levels below 2 the bromine analog signal represents only the combined chlorine species present.

Bromide was chosen rather than the conventional iodide because it avoids the precipitation of insoluble iodide salts within the probe which interferes with the electrode reactions. Furthermore iodide solutions at acidic pH levels undergo autooxidation, spontaneously generating free iodine in the solution giving rise to false positive probe currents.

This probe is basically an amperometric device in spite of the fact that a +0.4 Vdc from an external source is applied to the noble electrode. It generates a measurable electric current as a function of the concentration of the species measured. As soon as the chlorine species penetrates the membrane it is reduced to chloride and bromine is generated in proportion to the chlorine residual. Bromine does not contaminate the electrolyte within the probe because it is quickly reduced at the noble electrode back to bromide with a corresponding oxidation of the silver counter-electrode.

CHLORINE FACILITIES DESIGN 727

and
$$Br_2 + 2e \longrightarrow 2Br^-$$

$$Ag^0 \longrightarrow Ag^+ + e.$$

Measurement of the resulting current flow shown by these reactions yields a signal proportional to bromine concentration and therefore to chlorine concentration.

An electronic package is associated with the probe. The analyzer displays the residual chlorine concentration on an integral meter and transmits a 4–20 mA signal for remote instrumentation requirements.

The probe has a chlorine concentration range up to 20 mg/l. While no chemicals are required, the manufacturer recommends the electrolyte be changed once every two months. One year of field testing has demonstrated maintenance of the probe is negligible—10 min. every two months to empty and flush the electrolyte reservoir and refill it with the buffered bromide salt solution.

EPCO: Chlortect® Chlorine Monitor.[60] This is one of two analyzers based upon the design of the "Flux Monitor" by Marinenko of the U.S. National Bureau of Standards.[61] The other is marketed by IBM (see below). This instrument was designed purposely to detect extremely low levels of chlorine residual in order to determine precisely the toxic effects of total chlorine residuals upon various species of fish and other aquatic life.

This instrument measures total chlorine residual only over a range of 2 ppm. The least amount detectable is claimed to be 2 ppb. The digital display reads from zero to 1000 ppb. Internal calibration is selectable for 1, 10, 100, and 500 ppb.

Two reagents are required: potassium iodide and pH 4 buffer. These reagents and chlorinated sample are pumped by their separate peristaltic pumps. There are two cells in the analyzer; one is coulometric and the other is amperometric. The amperometric cell consists of a platinum measuring electrode (cathode) and a nonpolarizable reference electrode (Ag/AgCl) immersed in a salt solution. The coulometric cell consists of a platinum anode and a nonpolarizable reference electrode (Ag/AgCl). The operation of the analyzer utilizes three well established electroanalytical techniques: coulometry, electrolysis, and amperometry. The coulometric cell provides the instrument with internal calibration by electrolytic on-site generation of a known amount of iodine.* This is accomplished by passing a fixed amount of current through the cell (see Fig. 9-69). The magnitude of the current depends upon the calibration range selected and is determined by Faraday's Law of Electrolysis.

The iodine generated by the stoichiometric reaction between the total chlorine

---

* A chlorine demand-free sample of deionized water must be available for calibration procedure as shown in Fig. 9-69.

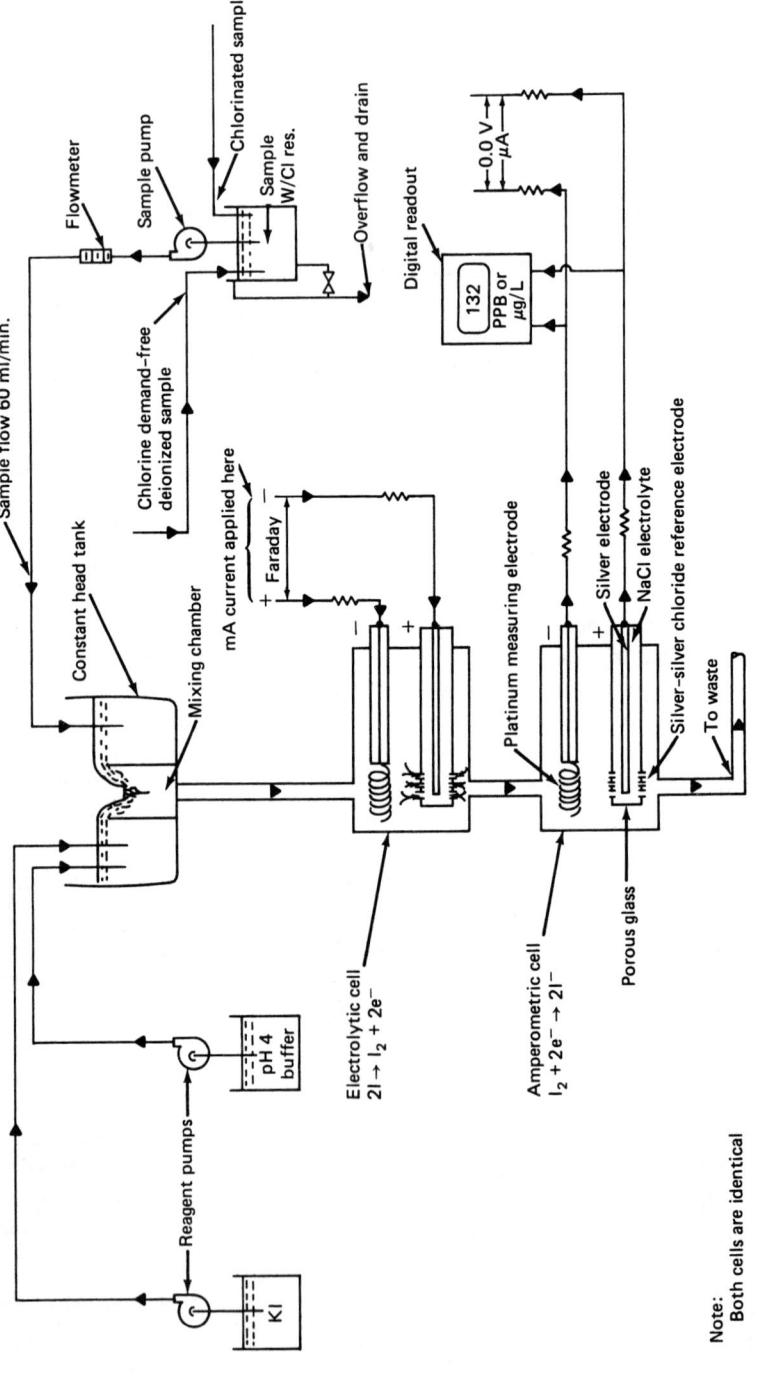

Fig. 9-69. High resolution amperometric cell with internal colometric calibrating circuit utilizing an electrolytic cell (courtesy of EPCO and IBM).

residual in the sample and the potassium iodide added to the sample is measured by the amperometric cell as shown in Eq. 9-16.

$$2 H^+ + OCl^- + 2I^- \xrightarrow{pH\ 4} I_2 + H_2O + Cl^- \qquad (9\text{-}16)$$

$$I_2 + 2e^- \longrightarrow 2I^- \qquad (9\text{-}17)$$

The current produced by the reduction of I_2 to I^- in the amperometric cell is directly proportional to the concentration of I_2, which in turn corresponds to the concentration of the total residual chlorine. This amount is displayed directly in ppb or ppm on the digital panel meter.

If any sample preparation is required due to the presence of undesirable foreign matter, the user must assume this responsibility.

IBM EC/250 Series Chlorine Analyzer.[62,63] This analyzer is based upon the same principles described above for the EPCO analyzer. These two analyzers have the same schematic arrangement of operation. There are some minor differences. For example model 2A incorporates a heavy duty sample pump and a "self cleaning" filter. Calibration ranges are 10, 100, and 500 ppb; detection limit = 2 ppb; linear range = 10 ppm; accuracy, ±10 percent for linear range; reproducibility ±3 percent; reagent flow rate 1 ml/min; sample flow rate 60 ml/min; and response time 90 percent in 3 min. Optional features are: automatic temperature control, automatic calibration, battery power adapter, high/low alarm and 12 V DC power pack.

Orion Research. This manufacturer supplies two chlorine monitors for the continuous analysis of total chlorine residual—only. Model 1570 is for clean water[64] and Model 1770 is for "dirty water."[65] These analyzers use the potentiometric method for the determination of total chlorine residual. This is the only known potentiometric analyzer commercially available at this time (1985).

This method of analysis is based upon detection of the concentration of iodine in the sample using a special electrode. A pH buffer is added to the sample to lower the pH to between 3.0 and 4.0. KI is then added to the sample to convert all the residual chlorine to iodine. The electrode contains two sensing elements. A platinum electrode develops a potential that depends upon the relative concentration of iodine and iodide in solution:

$$E_1 = E_0 + (S/2)\,(\log\,[I_2]/[I^-]^2) \qquad (9\text{-}18)$$

$$= E_0 + (S/2)\,\log\,[I_2] - S\,\log\,[I^-] \qquad (9\text{-}19)$$

where

E_1 = potential developed by the platinum sensing element, mV

E_0 = a cell constant, mV

S = monovalent electrode slope (58 mV/decade at 20°C)

$[I_2]$ = iodine concentration, moles/l

$[I^-]$ = iodide concentration, moles/l

A second electrode, the iodide element, develops a potential that depends upon the iodide ion concentration in solution:

$$E_2 = E_0' - S \log [I^-] \qquad (9\text{-}20)$$

where:

E_2 = potential developed by the iodide sensing element, mV
E_0' = a cell constant, mV

The electrode thus measures the difference between potentials developed at the two sensing elements:

$$\begin{aligned} E = E_1 - E_2 &= (E_0 - E_0') + (S/2) \log [I_2] \\ &= E_0'' + (S/2) \log [I_2] \end{aligned} \qquad (9\text{-}21)$$

where

E_0'' = potential measured by the analyzer, mV

$E_0'' = E_0 - E_0'$ = a cell constant, mV

The net potential measured by the electrode is converted by analog electronics to read directly as residual chlorine in mg/l. The electrode is calibrated with a standard of known equivalent residual chlorine concentration.

The preceding equations demonstrate that the output of the electrode is proportional to the log of the iodine concentration, and thus proportional to the log of the total residual chlorine concentration. This enables the electrodes to measure the residual chlorine concentration over a 4 decade range from 0.001 mg/l to 10 mg/l.

The only difference between the two models is in the quality of the sample water that is acceptable. Model 1770 is for so-called dirty water and is described here. The sample stream enters the inlet block and the flow is directed at high velocity parallel to the inlet screen. Considerably less than 1 percent of this flow is taken through the screen into the monitor. The surface of the screen is subjected to considerable shear force, which keeps the screen surface clean. This design allows the analyzer to be used on unfiltered samples. The sample then flows to the flow cell block. The reagent is passed through a purification column to remove any background iodine, which might have formed during storage, and then is delivered to the block by a separate reagent pump. Sample and reagent are combined within the flow cell block, where they are mixed and agitated by a stream of air. The reagent, which contains sufficient acid to give a resulting pH between 3.0 and 4.0 (depending upon the alkalinity of the sample), also contains iodide. The turbulent mixture is directed out of the flow cell block through a mixing loop. Under these conditions, any total residual chlorine reacts completely to form iodine; the resulting iodine concentration is equal to the sample residual chlorine concentration before reaction. The sample then returns to the flow cell block and is directed past a sensing electrode. The turbulence in the mixing loop and agitation against the electrode not only promotes mixing but also helps to minimize fouling of the electrode by sample debris.

Electrode response is affected by sample temperature so automatic temperature compensation circuitry is incorporated in the analyzer.

Specifications for the analyzer claim a sensitivity of 0.001 mg/l, an accuracy and a precision of ±10 percent and ±5 percent, respectively. Note that for precision and accuracy the specifications are stated as a percentage of the actual reading, whereas manufacturers of other residual chlorine analyzers state their specifications for precision and accuracy in terms of percentage of the full-scale reading. The analyzer response time is 2 minutes to obtain a full response from a change in residual chlorine concentration in the sample. Minimum sample flow is 0.5 gal. per minute and minimum allowable water pressure is 20 psi. Maximum allowable is 100 psi.

The manufacturer recommends weekly checks for leaks, replenishment of reagent, and analyzer recalibration monthly. The analyzer is equipped with a sample tap and valve for calibration grab samples. There are no operating data on how often this analyzer must be calibrated for wastewater practice.

Uniloc, Div. Rosemount Inc.[66] The Model 852 Chlorine Monitor has been replaced by Model 853. The sensor is Model 450. This analyzer is for measuring free chlorine only. Uniloc does not provide any control instrumentation, but the monitor can provide either current or voltage analog outputs which are compatible with any recorder, controller, or signal converter.

The sensor measures HOCl plus OCl$^-$ but it is blind to combined chlorine. They do not make a unit that measures either combined or total residual chlorine.[67] The analyzer operates on the polarographic principle. A constant 200 mV voltage is applied to a gold cathode and a silver anode. The anode is immersed in an electrolyte chamber of salt solution (NaCl) so that the anode becomes the familiar silver–silver chloride reference electrode. The free chlorine species (HOCl + OCl$^-$) migrate through the membrane between the sample and the electrode chamber. This results in a sensor output current proportional to the free chlorine residual in the sample. The membrane is supposed to have selective diffusion characteristics

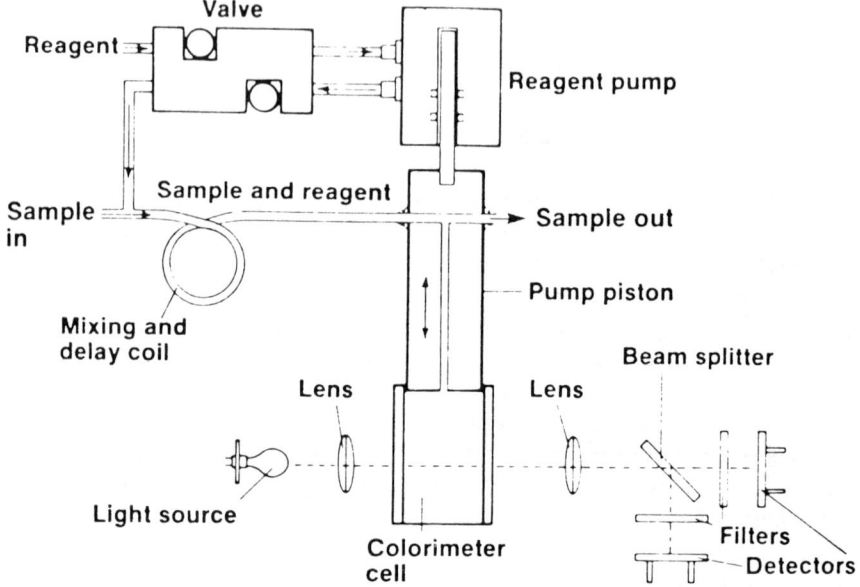

Fig. 9-70. Colometric analyzer showing fluid path and mechanical operation (courtesy Hach Co.).

which allows passage of free residual chlorine species but not combined chlorine. The sensor is temperature compensated.

A four-position range switch permits selection of 0 to 1, 0 to 5, 0 to 10, or 0 to 20 mg/l residual chlorine. High and low alarms with individual adjustable set points are available over the entire range by means of calibrated potentiometers. The alarm set-point can also be displayed on the digital or analog display. Deadband can be adjusted up to 15 percent of full scale.

This analyzer does not require any reagents. The response time in the 0–1 mg/l range is 5 minutes; 0 to 5 mg/l range is 1 minute and 0–10 mg/l range is 30 seconds. Sensitivity is 0.1 ppm at constant pH and ±1°C.

Hach Colorimetric.[68,69] The Hach company is the only supplier of continuous colorimetric analyzers in North America today (1985). These analyzers use the DPD colorimetric method of analysis. The 31100 model is used to analyze free chlorine over a range of 0–2 mg/l. The 61100 model covers a range of 0–5 mg/l. The 31300 model is designed to analyze total chlorine residual.

Figure 9-70 shows the basic mechanical operation and fluid path of sample and reagent through these analyzers. The reagent pump and sample pump pistons are synchronized precisely so that the reagent is added simultaneously with the sample. As the sample piston rises, drawing the sample in, the reagent piston discharges reagent into the sample stream. A mixing and delay coil provides time

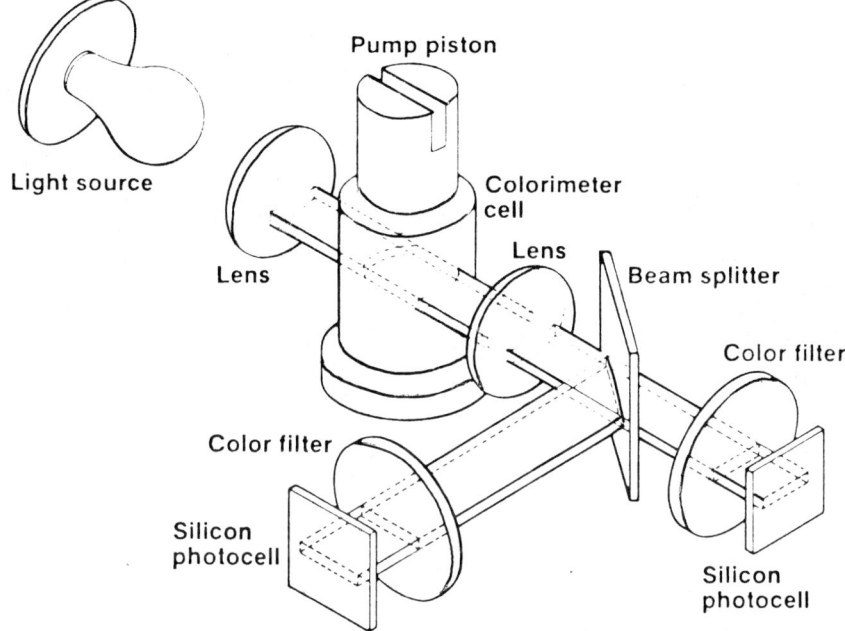

Fig. 9-71. DPD colorimetric analyzer, optical design (courtesy Hach Co.).

for the sample and reagent to completely mix and the colorimetric reaction to progress to completion before the sample enters the colorimetric cell. After analysis (the other half of the pump cycle) the sample pump discharges while the reagent pump draws in the next portion. Valves (not shown in Fig. 9-70) open and close during the parts of the cycle when the pistons are stationary.

Figure 9-71 shows the optical design of the analyzers. The light beam passes through a lens before entering the cylindrical pump/sample cell. After passing through the cell, the light encounters a beam splitter, which directs part of the beam through a filter to a reference detector located at 90° to the light path. The other, main part of the light beam passes through a filter to a sample detector. The signals from the two detectors are fed to the electronics package for processing. The single-beam/dual-wavelength capability provides a correction for any changes in light source or sample cell condition, and can compensate for moderate changes in sample turbidity and color by the reference wavelength used in the analyzer. Although these features increase the reliability when the analyzer is used to determine residual chlorine concentrations in wastewaters that may contain suspended solids and other contaminants, sample pretreatment is still required for removal of debris and large solids which may foul or plug the analyzer.

A block diagram of the analyzer electrical system is shown in Fig. 9-72. Light, after passing through the sample, strikes the reference and sample silicon photocells.

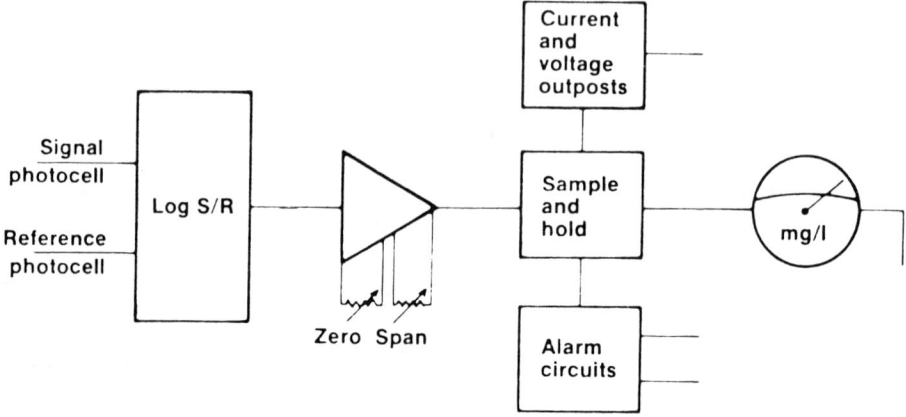

Fig. 9-72. DPD colorimetric analyzer, block diagram of electrical system (courtesy Hach Co.).

This produces a current, which goes to a log-ratio converter. The converter sends out a signal equal to the log of the ratio between the two currents. The log-ratio signal travels to an amplifier with gain and offset controls and then to a sample-and-hold circuit. A microswitch is activated when the sample piston is at the top of the cylinder/sample cell (allowing an unobstructed light path). This advises the sample-and-hold circuit to put the amplifier output into storage. The meter and other external readouts indicate this value, which is stored until the next cycle.

Free available chlorine DPD colorimetric analyzers are available with one of four factory-preset ranges. The widest range is 0–5.0 mg/l. Total residual chlorine DPD colorimetric analyzers are available with one of three factory-preset ranges. The widest range is 0–2.0 mg/l. The analyzers have no automatic temperature compensation. The effect of varying sample temperatures or ambient temperature can be significant.[52] Hach offers an optional sample heater which will maintain the sample temperature within ±1.5°C. This should be included as an integral part of these analyzers.

A major drawback of these analyzers is the response time to variations in sample concentrations. Unlike other analyzers, the concentration is determined using discrete samples. One complete sample cycle requires time for entry of the sample into the analyzer, addition of reagent, color development, and expulsion of the sample from the analyzer. The supplier's specifications state the following response times:

Free available chlorine	Initial response in 30 seconds, 97% in 2 minutes
Total residual chlorine	Initial response in 5 minutes, 95% response in 7 minutes

This response time is unacceptable for either residual control or the dechlorination process. These analyzers are relatively insensitive to drift, as demonstrated by the following supplier specification:

> A drift rate of less than 0.5 percent of full scale over 24 hours when measuring 0 mg/l residual chlorine with a constant sample temperature of 25°C and a constant ambient temperature of 30°C.

The sensitivity claimed by Hach is 0.02 mg/l, precision is ±0.5 percent of full scale at constant ambient and sample temperature, and accuracy is ±5 percent for the same conditions.

It is unlikely that these analyzers could be used for anything other than monitoring.

HOUSING

General. Many important design provisions for chlorination housing relate to the safe use of chlorine and the protection of those working with it. Consequently, many chlorine room design provisions are required elements of state standards.

Chlorinator and sulfonator rooms should be at or above ground level. Container storage should be planned so that it is separate from chlorinators and accessories. It is logical to locate the chlorination room near the point(s) of application to minimize the length of chlorine lines. Other general site considerations include a location which permits ease of access to facilitate container transport and handling, adequate drainage, and separation from other work areas.

Separation. Proper design standards require either a completely separate chlorination building or a room completely separate from the remainder of the building with access only through an outside door. There should be no apertures of any type from the chlorination room to other parts of a common building through which chlorine gas could enter other work areas.

Fire Hazard. The building should be designed and constructed to protect all elements of the chlorine system from fire hazards. If flammable materials are stored in the same building, a fire wall should separate the two areas. Fire-resistive construction is recommended. Water should be readily available for cooling cylinders in case of fire.

Space Requirements. Modern chlorination equipment is available in modules so that the chlorinators and accessory equipment can be arranged in a panel-like array. There should be about four feet between the front of a module and the nearest wall and about two feet on the sides and rear. Fig. 9-11 illustrates space requirements for chlorinator–evaporator installations and Fig. 9-14 illustrates space requirements for a ton container supply area.

The smallest area used for the installation of a chlorinator, weighing scales, and a spare cylinder* of chlorine should not be less than 6 feet by 6 feet.

There should be adequate room provided to allow ready access to all equipment for maintenance and repair. There should be sufficient clearance to allow safe handling of equipment containers. Absolute minimum clearance around and in back of equipment is 2 feet.

Some general minimum space guidelines are as follows:

1. Plants with one chlorinator feeding less than 200 pounds per day should have at least 64 square feet.
2. Plants using two chlorinators with a total feed rate of up to 400 pounds per day should have at least 160 square feet.
3. For each chlorinator–evaporator unit, 160 square feet should be provided.

Ventilation. Adequate forced air ventilation is required for all chlorine equipment rooms. An exception to this would be small chlorinator installations (<100 pounds per day) located in separate buildings if the windows and doors can provide the proper cross-circulation. For a small building, windows in opposite walls, a door with a louver near the floor, and a rotating-type vent in the ceiling usually provide the necessary cross-ventilation.

Factors to be considered in the design of a ventilation system are: air turnover rate, exhaust system type and location, intake location and type, electrical controls, and temperature control.

A forced air system should be capable of providing one complete air change in 2–5 minutes. Since chlorine gas is 2½ times heavier than air, it is logical to provide air inlet openings for ventilation fans at or near floor level. For small installations it is common to employ an exterior exhaust fan with the intake duct extending to the chlorine room floor. A wall-type exhaust fan is an acceptable alternative. The exhaust system should be completely separate from any other ventilation system. For larger installations a blower-type fan is needed. The use of free-moving, gravity-operated louvers may be advantageous in colder climates for conserving room heat when the blowers are not in operation; however, venting systems should not have covers. The fan discharge should be located so as to not contaminate the air supply of any other room or nearby habitations. It is mandatory that the ventilation discharge be located at a high enough elevation to assure atmospheric dilution, e.g., at the roof of a single-story building.

Air inlets should be so located as to provide cross-ventilation. To prevent a fan from developing a vacuum in the room and thereby making it difficult to open the doors, louvers should be provided above the entrance door and opposite the fan suction. In some cases, it may be necessary to provide temperature control on the air supply so that the chlorination system is not adversely affected. A

---

* This space is for chlorination systems using a 150 lb cylinder.

signal light indicating fan operation should be provided at each entrance when the fan can be controlled from more than one point.

Wind Socks. All installations should locate at least one wind sock on the chlorine supply structure. This is very valuable in the event of a leak.

Doors. Exit doors from the chlorination room should be equipped with emergency hardware and open outward. Some design guides recommend two means of exit from each room or building in which chlorine is stored, handled, or used; however, this would not appear to be essential in most cases.

Inspection Window. A means should be provided which permits viewing of the chlorinator and other equipment in the chlorination room without entering the room. A clear glass, gas-tight window which is installed in an exterior door or interior wall of the room is recommended. Door windows appear to be a logical provision even with a separate wall inspection window.

Heating. The chlorinator room should be provided with a means for heating and controlling room air temperatures above 55°F. A minimum room temperature of 60°F has been recommended as a good practice. Ideally, the heating system should be able to reliably maintain a uniform moderate temperature throughout the chlorination room.

Hot water heating is generally preferred because of safety considerations and the uniformity of temperature which this method of heating provides, without the extremes which might be experienced with failure of a steam heating system. Electric heating is suitable and forced air heating would be appropriate if an independent system is provided for the chlorination room or building. Central hot air heating is not acceptable since gas could escape through the heating system.

Chlorine vapor leaving a container will condense if the piping temperature is significantly lower than the temperature of the container. Design should provide a higher temperature in the chlorinator room than in the container room. This applies to systems using the gas phase from the containers. Elimination of unnecessary windows may aid in maintaining uniform building temperatures.

If container storage and chlorination equipment are in separate rooms, the temperature of the chlorine container should not be allowed to drop below 50°F *if evaporators are not used.*

Drains. It is generally desirable to keep the plant floor drain system separate from that of the chlorinator. Drainage from a chlorinator drain relief valve may contain chlorine. Consequently, hose, plastic pipe, or tile drains are recommended. The discharge should be delivered to a point beyond a water-sealed trap or disposed of separately where there is ample dilution.

Scale pits are generally designed with floor drains having a water-sealed trap.

In actual practice, most traps probably do not contain enough water to form a seal and it would be preferable to provide a straight pipe drain outside to grade.

Vents. Chlorinators, external chlorine pressure reducing and shutoff valves, remote vacuum systems, and automatic switchover systems have vents. Chlorinators have a pressure vacuum relief system which should be carried to the outside atmosphere without traps to a safe area, with one vent for each chlorinator. In the case of a malfunction, the operator can easily determine which chlorinator is malfunctioning if the vents are separate. The ends of the vent lines should point down, be covered with copper wire screen to exclude insects, and should not be more than 25 feet above the chlorinator. The line should have a slight downward pitch from the high point (directly above the chlorinator) to drain any condensate away from the chlorinator. It is acceptable to run the vent vertically (but no more than 25 feet) above the chlorinator to the roof, with a 180° return bend at the exit.

External chlorine pressure-reducing and shutoff valves should be checked for vents. When supplied, these vents should drain away from the valves. In other words, these valves should be located high enough so that the individual drains will have a continuous downgrade to the outside atmosphere.

Manufacturer's instructions should be followed for vents required with remote vacuum and automatic switchover systems. Evaporators* have water bath vapor vents which can be manifolded together and discharged to the atmosphere without traps.

Electrical. Controls for fans and lights should operate automatically when the door is opened and there should be provisions to activate these manually from outside the room. Switches for fans and lights should be outside of the room at the entrance. A signal light indicating fan operation should be provided at each entrance when the fan can be controlled from more than one point.

RELIABILITY PROVISIONS

The need for continuous and dependable disinfection has been stressed. The chlorination system can fail due to a number of causes and, therefore, the design of the system must include the necessary provisions to either prevent failures or allow immediate corrective action to be taken. Although assured reliability is essential, design provisions for this are often slighted.

Chlorine Supply. As a chemical feed process, one of the most frequent interruptions in treatment is caused by the exhaustion of the chlorine supply. Five features are essential to maintain continuous chlorine feed: (1) an adequate reserve supply

---
* Evaporator relief system vents require special consideration for local conditions.

of chlorine sufficient to meet normal needs and bridge delivery delays and other possible contingencies; (2) chlorine container scales; (3) a manifolded chlorine header system; (4) an automatic device for switching to a full chlorine container when the one in use becomes empty; and (5) an alarm system to alert operating personnel of imminent loss of chlorine supply. These five features are discussed elsewhere in this text. Without them it is not possible to assure uninterrupted chlorine feed even with full-time operator attendance and no equipment breakdowns.

The chlorine header system is needed both to provide a connected on-line chlorine supply which is adequate to assure uninterrupted flow of chlorine for whatever period that the system may be unattended and to allow switchover to a full cylinder without interruption of chlorine.

Power Failure. Power outage usually results in water supply failure which in turn automatically shuts down the chlorination system. A range of special provisions can be employed to assure reliability of power and water supply depending upon the particular situation. As discussed previously, these may be in the form of a standby power source and pumps.

Standby Equipment. The design of the chlorine feed system should provide for continued operation in cases of equipment failure. Where both pre- and postchlorination are to be practiced, separate chlorination systems should be provided for each plus a standby system. If prechlorination is not to be continuously used, it may be possible to use this system as the standby system for disinfection. The units, piping, and accessories should be designed with this application in mind. If prechlorination must be carried out continuously or if no prechlorination is to be done, a standby system, capable of replacing the postchlorination system during repairs, maintenance or emergencies should be provided. Standby equipment of sufficient capacity should be available to replace the largest unit during shutdowns. This includes standby pumps for the injector water supply.

In addition to standby equipment, the equipment manufacturer should be consulted regarding vulnerable components. These components should be a part of the plant's inventory of spare parts.

Water Supply. As mentioned above, during a power failure the injector water system will be shut down unless there is an alternate supply that does not require power, such as an elevated tank. Standby equipment to provide injector water in the event of a power failure would consist of an engine driven injector supply pump. Every injector water supply system should have such a standby pumping unit. There is no way to operate the chlorination system without an adequate water supply.

Chlorine Residual Analyzers. Every system using an analyzer for chlorination control should be backed up by an effluent monitor analyzer that can be switched over to the control function in the event of control analyzer failure. Similarly, backup capability should be provided for analyzers controlling and monitoring the dechlorination process. Provisions should be made for standby sample pumps. Sample lines should be piped to facilitate flushing or purging to remove biological slimes.

AMMONIATION FACILITY

AMMONIA—NH_3

Useful Ammonia Compounds. Ammonia is commercially available in four forms: anhydrous ammonia, NH_3, commonly stored and transported as a liquid in pressure vessels; aqua ammonia, NH_4OH, most commonly a 20–30 percent solution of ammonia in deionized or softened water; white crystalline ammonium chloride, NH_4Cl, very soluble in water; and ammonium sulfate, $(NH_4)_2SO_4$, a gray-green crystalline solid which is hygroscopic. The latter two must be dissolved in water to be useful in waterworks practice.

Source, Availability and Uses. Ammonia, NH_3 does not appear free in nature. It occurs in compounds as the NH_4^+ ion. However, the primary source for the nitrogen atom is in the atmosphere. All ways to make ammonia consist of combining one atom of nitrogen with three atoms of hydrogen.

Ammonia is extremely important commercially. Its most notable uses are for refrigeration and fertilizers. Other uses include explosives, petroleum refining, rubber, textiles, chemicals, pulp and paper, and metallurgy. It is a byproduct of the destructive distillation of coal and coke, but most of the ammonia produced in the United States is by the Haber process. This process produces ammonia by direct synthesis. Gaseous nitrogen and hydrogen are mixed in correct proportion and heated under pressure (up to 1000 atm.) at 400–600°C and passed over a catalyst. The pure nitrogen for the process is obtained from liquid air, while natural gas (methane, CH_3) is the usual hydrogen source. This is why oil companies are usually the important suppliers of ammonia. For example, Union Oil Co. is the main supplier on the West Coast of U.S.A. owing to their participation in the Alaska Pipeline. They have an ammonia manufacturing plant in Kenai, Alaska. From here it is shipped to West Coast ports and exported to Far East countries.

About 30 billion tons of ammonia are produced annually in the U.S.A. Its production is exceeded only by sulfuric acid and oxygen.

Physical and Chemical Characteristics. Ammonia is a colorless gas with a very pungent, irritating odor. However, when it is released into the atmosphere it creates a rather dense white fog. This is due to its immediate reaction with the moisture in the atmosphere. Ammonia is highly soluble in water.

CHLORINE FACILITIES DESIGN

At normal temperatures and pressures anhydrous ammonia is a gas. It is easily liquified by pressurizing in a container and it is commonly stored and transported as a liquid. When the liquid reverts to a gas a great amount of heat is absorbed. This is why it is extensively used for refrigeration.

At atmospheric pressure liquid ammonia has a density of 42.6 lb/ft^3, approximately two-thirds that of water. The vapor pressure of ammonia at 70°F is 114 psi compared to chlorine at 85 psi (see Fig. 12, Appendix). The important properties of ammonia are shown in Table 9-10.

Table 9-10 Physical Properties of Ammonia[70]

Molecular symbol	NH$_3$
Molecular weight	17.031
Boiling point at 1 atmosphere*	−28°F (33.3°C)
Freezing point at 1 atmosphere	−107.9°F (−77.7°C)
Critical temperature	271.4°F (133.0°C)
Critical pressure	1657 psia (114.2 bars)
Latent heat at −28°F (−33.3°C) and 1 atmosphere	589.3 Btu/lb (13.71 × 10^5 J/kg)
Relative density of vapor compared to dry air at 32°F (0.0°C) and 1 atmosphere	0.5970
Vapor density at −28°F (33.3°C) and 1 atmosphere	0.0555 lb/ft^3 (0.8899 kg/m^3)
Specific gravity of liquid at −28°F (−33.3°C) compared to water at 4°C	0.6819
Liquid density at −28°F (−33.3°C) and 1 atmosphere	42.57 lbs/ft^3 (681.9 kg/m^3)
Specific volume of vapor at 32°F (0.0°C) and 1 atmosphere	20.78 ft^3 (1.297 m^3/kg)
Heat of solution at 0% conc. by wt.	347.4 Btu/lb (8.081 × 10^5 J/kg)
Heat of solution at 28% conc. by wt.	214.9 Btu/lb (4.999 × 10^5 J/kg)

* atmosphere = 760 mm Hg = 1.01325 bars.

Characteristics which Affect Use in Water Treatment:

Weight. Molecular weight of ammonia (17.03) is half that of chlorine (35.5). Therefore a 2000 lb/day chlorinator can only feed 950 lb/day ammonia.

Ammonia gas is lighter than air, so leaking vapor will rise quickly; in contrast, chlorine, which is heavier than air, tends to settle toward the ground.

Heat of Vaporization. Nearly 5 times that of chlorine. Therefore, a 10,000 lb/day chlorine evaporator can vaporize only 2,000 lb/day ammonia.

Solubility in Water. Almost 50 times that of chlorine. This simplifies the requirements for mixing at the point of application and reduces significantly the quantity of injector water required. This minimizes the burden when injector water must be softened to a hardness not exceeding the solubility of calcium carbonate (35 mg/l).

Reaction with Water. 1 mg/l NH_3 will *increase the alkalinity* of water by 2.9 mg/l (as $CaCO_3$), whereas 1 mg/l chlorine reduces the alkalinity 1.4 mg/l.

Corrosivity. Dry ammonia, either liquid or gas is not corrosive to any metals. Moist ammonia will not corrode iron or steel but will react (to corrode) with copper, brass, zinc and most copper alloys. Never use galvanized pipe. Valves, flex connections, and other hardware used in chlorine supply systems contain pure copper and/or copper alloys—therefore they are not interchangeable with comparable ammonia supply hardware.

Water Softening Reaction. This is the one and only disadvantage of ammonia use in water treatment. Ammonia reacts with the hardness in water to produce a softening effect comparable to excess lime softening. This occurs at the interface between ammonia vapor and water: at the inlet to the injector throat and at the exit of the direct feed gas diffuser. At these two interfaces a scale will develop directly in proportion to the calcium and magnesium hardness. Stoppage occurs quickly due to this scale. This is why all injector water should be softened to the maximum solubility of calcium carbonate (35 mg/l). For the direct feed diffuser, the one supplier of this equipment solved the hardness problem with a diffuser design modification many years ago.

Physiological Effects. Persons having chronic respiratory disease or persons who have shown evidence of undue sensitivity to ammonia should not be employed where they will be exposed to ammonia.

Ammonia is not a cumulative metabolic poison; however, ammonia in the ambient air has an intense irritating effect upon the mucous membranes of the eyes, nose, throat, and lungs. High levels of ammonia can produce corrosive action on these tissues which can cause laryngeal and bronchial spasms, as well as edema which will obstruct the breathing passages. Conscious people are protected by its pungent odor, but unconscious people are not. Table 9-11 lists human physiological response to various concentrations of ammonia in air. It should be noted here that individuals differ in their sensitivity to ammonia. Some are highly reactive to low concentrations while others show a significant tolerance to the irritative effects.

SUPPLY SYSTEM: ANHYDROUS AMMONIA

Cylinders. Two sizes, 100 lb and 150 lb, are usually available, but not as common as they were. Cylinders are equipped with a dip tube so that when they are placed horizontally, liquid can be withdrawn.

Table 9-11 Physiological Response to Ammonia[70]

	Concentration, ppm
Least perceptible odor	5
Readily detectable odor	20–50
No discomfort or impairment of health for prolonged exposure	50–100
General discomfort and eye tearing; no lasting effect on short exposure	150–200
Severe irritation of eyes, ears, nose, and throat; no lasting effect on short exposure	400–700
Coughing, bronchial spasms	1,700
Dangerous, less than ½ hour exposure may be fatal	2,000–3,000
Serious edema, strangulation, asphyxia, rapidly fatal	5,000–10,000
Immediately fatal	10,000

800 lb Container. These are still available but are not as popular as bulk storage tanks. They are similar to chlorine ton cylinders. They are equipped with dip tubes for vapor or liquid withdrawal and are the same physical size as chlorine ton containers. Therefore a chlorine ton container lifting bar and trunnions will fit these cylinders.

Storage Tanks:

General Consideration. Bulk storage tanks sized to fit the consumer's needs are the most popular means of on-site ammonia storage. For example, in the Southern California area the Union Oil Co., Chemical Division, has a large fleet of 8000 gallon tank trucks used to service customer storage tanks on a routine basis. USS Agrichem (U.S. Steel Co.) and Chevron provide similar service in other areas.

Consumers in the water treatment industry are advised to purchase refrigeration grade ammonia instead of commercial grade because it is moisture free. Commercial grade ammonia purposely contains a small amount of moisture to reduce stress fractures (fatigue). This prolongs tanker life.

Design Factors. It is common practice in the industry for the supplier to build or furnish the consumer with a storage tank. Certain features should be a part of a proper storage tank. The shell should be designed for at least 250 lb working pressure.[70] There should be one liquid outlet with an excess flow check valve inside the tank and two vapor outlets, all with stainless steel trim shutoff angle valves (similar to a chlorine tank car). In the vicinity of these valves there should be a safety relief valve with a vent line which discharges 2–3 feet above "roof" height. It should end in a double 90 degree elbow to prevent rain from entering the vent and it should have a moisture trap-leg adjacent to the valve discharge. This moisture trap will eliminate moisture interference with the relief valve.

Filling Density. The allowable filling density for uninsulated stationary storage tanks is 82 percent by volume and 56 percent by weight. Therefore an integral part of the storage tank must be a device to measure the weight or volume of ammonia at any time.

Inventory Control. The preferred method of determining the status of ammonia supply is by weight. Therefore the contents of the tank should be weighed by either a set of industrial lever-type scales or a load cell. A load cell with a 4–20 mA signal for remote display is preferable to scales.

Withdrawal Rates. This is an important factor when selecting the size of a storage tank. Assume that the maximum withdrawal rate will never exceed 1000 lb/day and assume that this much vapor will be needed at an ambient temperature of 50°F. How big will the storage tank have to be? The formula for withdrawal rates for ammonia is as follows:[2]

$$[\text{Room temp}(°F) - \text{Liquid temp}(°F)] \times \text{withdrawal factor} = 1000 \text{ lb/day}$$

Liquid temperature is the temperature of the ammonia at minimum allowable inlet pressure to the ammoniator (10 psi). This pressure is required to actuate the remote vacuum regulator. At 10 psi the liquid temp is $-5°F$. Let the withdrawal factor be W:

$$[50 - (-5)] W = 1000 \text{ lb/day}$$

$$W = \frac{1000}{55} = 18.2$$

An 800 lb cylinder (192 gal) has a withdrawal factor of 3.2:[2]

$$\frac{W}{192} = \frac{18.2}{3.2}$$

$$= 1094 \text{ gal}$$

Therefore an 1100 gallon tank when full could provide a 1000 lb/day vapor withdrawal rate at 50°F ambient temperature.

Evaporators. Ammonia evaporators are identical to chlorine evaporators* but they must not be used interchangeably because this practice could produce an

---
*Except that inlet and outlet valving have 316 SS trim.

explosive mixture of chlorine and ammonia. The major difference between chlorine and ammonia evaporators is capacity. A 10,000 lb/day chlorine evaporator can only vaporize 2,000 lb/day ammonia.

Evaporators for ammonia are being used in water treatment practice where the ammonia is supplied in 800 lb cylinders. Evaporators can and should be avoided by the use of custom-made storage tanks.

Leak Detectors. As of 1985 continuous leak detector monitors similar to chlorine detectors are not available. However American Gas and Chemical Co. of Northvale, NJ market the following products for locating ammonia leaks at the source.

Their aerosol powder, ADP-219, can be sprayed onto areas of suspected leaks (pipe joints, fittings, valves, etc.) as a yellow powder coating which changes to a dark blue in the presence of a small leak. The coating can easily be removed with a damp cloth.

Another aerosol spray, ADS-100, generates white smoke upon contact with ammonia fumes. This spray tends to neutralize the escaping ammonia, making it less hazardous, and produces a visible fog which grows more intense as it approaches the leak.

Their CG Tracer is a small, hand-held device sensitive to both ammonia and combustible gases. It can quickly locate an ammonia leak.

Potential Hazard of a Major Leak. There are two physical properties of anhydrous ammonia that reduce significantly the hazard of a major leak. These factors can be graphically illustrated by the events of a recent major leak. A Chevron tanker truck overturned in Richmond, California, resulting in a 6 inch gash in the tank shell. The Chemtrec emergency response team arrived to see a dense white fog rising into the atmosphere above the overturned truck.[73] Fire hoses were sprayed into the fog (ammonia plus atmospheric moisture) which disappeared immediately upon contact with the water. This is the result of the high solubility of ammonia in water. Ammonia is so much lighter than air that it rises quickly above and beyond people at ground level, which is greatly different from chlorine. After the fog was washed into the storm drains, water was sprayed into the gash on the tank. By this time the vaporization of the liquid due to the leak, cooled the liquid sufficiently so that the water formed an ice cover which sealed off the liquid ammonia long enough to right the truck and move it to a safe place where it was emptied without further incident. The high latent heat of vaporization of liquid ammonia is directly responsible for the rapid formation of the ice cover. This is the characteristic which makes ammonia a universal refrigerant. From the above accident description it can be seen that ammonia is a relatively safe gas. Furthermore, the levels of concentration required to produce a dangerous environment are many more times those for chlorine or sulfur dioxide.

Materials of Construction. The supply system under pressure should be Sch. 80 seamless steel pipe with 3000 lb forged steel fittings. Welded joints are preferred.

Use bell reducer-type fittings. Bushings should never be used. Unions should be 2-bolt flanged with a lead gasket joint. Valves should be steel with 316 stainless steel trim.

All ammonia piping under a vacuum should be Sch. 80 PVC with solvent weld joints.

AMMONIATORS

Direct Feed.[71] This type of ammoniator is available from only one manufacturer as of 1984 (Wallace and Tiernan). Metering capacities up to 1000 lb/day are available. All sizes come equipped with a special diffuser designed to prevent interference from scale formation and back-flooding of the equipment. The pressure at the point of application *must not exceed 15 psi.*

The manufacturer advises that there were over 200 direct-feed ammoniators in operation in 1984. Because of their stainless steel construction and the characteristics of ammonia there are no corrosion problems, hence they are rugged, low-maintenance devices. White recently observed a pair of these units in operation for nearly 50 years.

The direct feed unit can be furnished for either manual control or automatic flow-paced control. The automatic version costs about three times as much as the manual unit. Operating experience with the automatic type has proved to be satisfactory.

Solution Feed[72]:

General Description. Solution feed ammoniators are available from the top three chlorinator manufacturers. They are identical in design to chlorinators and sulfonators except for minor differences in materials of construction to conform to the chemical characteristics of ammonia vapor and aqueous solution. They are available up to capacities of 950 lb/day and are usually arranged for remote vacuum operation. Automatic switchover components are optional. The major design difference is in the sizing of the injectors, which is described below.

Control Strategies. Ammonia should be applied at a fixed ratio to chlorine for the best and most consistent results. Hence the ammoniator should be controlled by the same signals which control the chlorinator. The precise ratio of ammonia to chlorine can be achieved by rotameter feed-rate selection and use of the manual dosage adjustment on the ammoniator.

Injector System:

Water Requirements. The major disadvantage of a solution-feed ammoniator is the softening reaction caused by the formation of ammonium hydroxide:

$$NH_3 + H_2O \longrightarrow NH_4OH \tag{9-22}$$

The ammonium hydroxide precipitates the calcium and magnesium hardness down to the solubility limit of $CaCO_3$, which is 35 mg/l. Therefore the injector water should be softened to this level of hardness to prevent the injector from plugging due to hardness scaling.* Owing to the high solubility of ammonia in water, much less water is required for solution feed ammoniators than for chlorinators. For example, one manufacturer requires only 4.25 gpm with 50 psi injector water for 240 lb/day ammonia versus 6 psi back-pressure and only 2.5 gpm versus 2 psi back-pressure.[73] Therefore the manufacturer should be consulted early to determine injector water requirements, particularly if softening is to be involved.

The above figures are based upon the use of 1-inch fixed-throat injectors which are limited in ammonia capacity to 240 lb/day. Multiple injectors are required to achieve a maximum capacity of 950 lb/day. In any case, duplicate injectors and duplicate diffusers are recommended for solution feed ammoniators.

Vacuum Lines. Sizing of these lines is critical owing to the use of remote vacuum and remote injectors. The following analysis will serve as an example of the necessary calculations using these formulae:

Head loss in vacuum line:

$$\Delta P \text{ (in. Hg)} = \frac{11.89 \times L \times f \times W^2}{10^9 \times p \times d^5} \tag{9-23}$$

where

L = equivalent pipe length, ft

f = friction factor (see Figs. 10 and 11, Appendix)

W = lb/day ammonia

p = density of NH_3, lb/ft³

d = pipe diameter, inches

Reynolds number:

$$N_R = \frac{6.32 \times W}{\mu \times d \times 24} \tag{9-24}$$

---

* Capital Controls announced (1984) a special solution-feed ammonia injector with a flexible liner timed to flex by external water pressure on a programmed basis. The flexible liner is intended to shred the hardness scale buildup.

where

$$W = \text{lb/day ammonia}$$
$$\mu = \text{viscosity of } NH_3, \text{ centipoise}$$
$$d = \text{pipe diameter, inches}$$

Use the Reynolds number to select the friction factor from Fig. 8, Appendix. Calculate the density of ammonia at 20 in. Hg vacuum and 50°F:

$$PV = NRT \qquad (9\text{-}25)$$

where

$P = -20$ in. Hg or $(30 - 20)/30 = 10/30$ atm. $= 0.33$ atm.

$T = 50°F$ or $283°K$

$R = 0.08205$

Hence

$$0.33V = (0.08205)(283)$$
$$V = 70.36 \text{ l/mole}$$

$$17.03 \text{ g/mole} \times \frac{1 \text{ mole}}{70.36} \times \frac{1 \text{ lb}}{454 \text{ gm}} \times \frac{28.3 \text{ l}}{\text{ft}^3} = 0.015 \text{ lb/ft}^3$$

Therefore

$$p = 0.015 \text{ lb/ft}^3 \text{ at 20 in. Hg and } 50°F$$

Solve for Reynolds number using viscosity of NH_3 at 0.0095 centipoises at 50°F. Try 1.25 inch pipe and use max. capacity of the ammoniator (950 lb/day):

$$N_R = \frac{6.32 \times 950}{0.0095 \times 1.25 \times 24}$$
$$= 21{,}066$$

f from Fig. 8 (Appendix) is 0.03.

Now solve for total head loss in 300 ft of pipe using Eq. (9-23):

$$\Delta P = \frac{11.89 \times 300 \times 0.03 \times (950)^2}{10^9 \times 0.015 \times (1.25)^5}$$
$$= 2.11 \text{ in. Hg}$$

Therefore a 1.25-inch diameter PVC pipe is borderline, so use 100 ft of 1.5-inch and 200 ft of 1.25-inch.

Diffusers and Solution Lines:

Mixing. The solution lines should be sized so as to not exceed 1.5 ft total head loss. The diffusers should be designed with holes that will create a head loss of 8–11 ft. This will translate to a G factor of 250–300, which is sufficient turbulence for a quick mix. The orifice jets must point upstream.

Back-Pressure. The injector should be located as close to the diffuser as possible, which in nearly all instances calls for a remote injector. The centerline of the injector throat should be above the hydraulic gradient of the plant in order to keep the back pressure as low as possible. If it is not practical to keep the back pressure below 10 psi (to minimize injector water consumption) the ammonia solution should be pumped with a small turbine pump. Optimum back pressure is about 2 psi.

Contact Time. The reaction of ammonia nitrogen with chlorine to form chloramines is practically instantaneous, so contact time is not a factor—rapid dispersion of the ammonia solution is the principal factor.

Flushing. Diffusers should not only be provided in duplicate, they should be arranged so that they can be flushed with acid or chlorine solution to dissolve any formation of carbonate scale.

ADVANTAGES OF SOLUTION FEED ANHYDROUS AMMONIA

A great many chloramine systems in the U.S.A. use ammonium hydroxide solution. When faced with a decision to select aqua or anhydrous ammonia, Engineering Science, consulting engineers, Pasadena, CA conducted a survey of both forms and concluded that anhydrous ammonia was more advantageous for the following reasons:[74]

- Much smaller space requirement for storage
- Simplicity of metering and automatic control
- Simplicity of operation and maintenance, since solution feed ammoniators are virtually identical to chlorinators
- Lower cost

AQUEOUS AMMONIA (NH$_4$OH)

Source and Availability. While there may be a single source of ammonia locally there will be several distributors of ammonium hydroxide solution. The ammonia market is so competitive—domestic and foreign—that ammonia in either form is ubiquitous.

Local distributors have a fleet of trucks and will provide storage tanks upon request. Aqua ammonia costs about 14 percent more than anhydrous ammonia; however, storage tanks are less expensive than the pressurized tanks required for anhydrous ammonia.

Shipping can be made by 8000 gallon tank cars, 4000 gallon tank trucks, 375 and 750 gal. drums, or 30 gal. carboys.

Physical and Chemical Characteristics. Commercial strength is approximately 30 percent. The density of Grade A, 29.4 percent solution, is 0.8974 at 60°F compared to water at 1.

This strength solution corrodes copper, aluminum alloys, and galvanized surfaces. When this solution is dispersed into the process water by the diffuser a water softening reaction will occur at the perforation in the diffuser. This will promote the deposit of calcium scale in proportion to the hardness of the water. Provisions should be made to allow the diffusers to be cleaned with acid or chlorine solution.

Potential Hazards. Ammonium hydroxide is a caustic solution and can be hazardous to persons in contact with the solution. Usual precautions required for caustic solutions should be observed.

If the solution is being applied at an installation using hypochlorite solution, steps must be taken to prevent aqua ammonia solution from being unloaded into the hypochlorite storage tank or vice versa. In such an event an explosive mixture of nitrogen trichloride is likely to form with disastrous results. If not, the evolution of nitrogen trichloride is certain to cause a hazardous air pollution problem.

SUPPLY SYSTEM

Storage Tanks. The local supplier should be consulted on this matter. Make certain the supplier incorporates all of the necessary piping to facilitate unloading and that the tank has a proper sight glass. Tank and appurtenances should be made of mild steel. The words "AQUA AMMONIA—NH_4OH CAUSTIC" in large letters should be highly visible on each side of the tank.

METERING AND CONTROL SYSTEM

Design Considerations. The most difficult part of the aqua ammonia facitity to design properly is the metering and control system. It is most exacting because ammonia N has to be applied at a constant ratio to chlorine. Depending upon local conditions this ratio will be approximately 4 to 1 chlorine to ammonia. This means the feed rates will be 3–5 times lower than chlorine. Furthermore, since the aqua ammonia will be about 30 percent NH_3 or 300,000 mg/l compared, to

10,000 mg/l anhydrous ammonia solution in a solution feed ammoniator, the designer is faced with the problem of precise control of small quantities of solution. For example, a water supply uses 50 lb/day chlorine and the ammonia-N requirement is 12.5 lb/day (4:1 ratio); then at 2.25 lb NH_3-N per gallon of 30 percent solution an aqua ammonia feed rate of only 0.23 gallons per hour will be required. Obviously aqua ammonia is not suitable for small water supplies unless it is diluted with softened water.

Diaphragm Pumps. These are widely used in waterworks practice when they fit the requirements. If these pumps can fit the metering requirement the control system should be patterned after those described in Chapter 2 for hypochlorite.

The designer should strive to eliminate pulsing activity of the diaphragm pumps in order to salvage the range and accuracy of flow metering equipment required to control these pumps.

Diffusers and Solution Lines. Reasonable care should be used in the design of the solution lines in order to minimize transit time and thereby decrease system dead time. This will make the system more compatible with the chlorinator control system.

Owing to the fact that there are no real back-pressure (up to 125 psi) problems when using diaphragm pumps, the designer is free to put the required head loss into the diffuser (8–12 ft) to provide good mixing with the process water.

All diffusers should be in duplicate so that they may be taken out of service for cleaning to remove scale deposits caused by hardness in the water.

RELIABILITY PROVISIONS

This is not a factor in water treatment. Ammonia application interruptions for short periods will do little harm. THM formation may escalate during these periods. Health effects of such an event are unknown and are probably insignificant. Disappearance of ammonia in the treatment process may result in off-flavors at the consumer's tap. This may or may not result in consumer complaints.

REFERENCES

1. Anon., *Chlorine Institute Manual,* 4th ed. pp. 6, 10, 12, 13, and 14, New York, 1969.
2. Baker, R. J., "Maximum Withdrawal Rates from Chlorine, Sulfur Dioxide and Ammonia Cylinders," AWWA Publication, OP-Flow, p. 4, April 1980.
3. Anon., Vacuum Regulator Check Unit—3000 lb/da. capacity, Wallace and Tiernan Div. Pennwalt Corp., Belleville, NJ, Book No. WBB50.177, Feb. 1982.
4. Whiteman, T. J. "Chlorine System Test," in-house report, East Bay Municipal Utility District WPCP, Oakland, CA, Feb. 14, 1980.
5. White, G. C., survey of five tank car and bulk storage air padding systems, April 1982.
6. Anon., "Stop-Valve System for Chlorine Tank Car Padding," Pennwalt Corporation Dwg. No. T50W016 Tacoma, WA, July 9, 1968.

7. Beatty, A., private communication, Metropolitan Water District of Southern California, Los Angeles, CA, March 1982.
8. Albers, R. G., "Manufacturers' Data Reports," private communication issued by Pressed Steel Tank Co., Milwaukee, WI, Dec. 22, 1975.
9. "Chlorine Vaporizing Equipment," Chlorine Inst. Pamphlet No. 9, 2nd ed., New York, 1970.
10. "Technical Information for Handling Chlorine, Sulfur Dioxide, and Ammonia from Supply to Point of Application," Fischer and Porter Instr. Bull. 70-9001 Revision 1, Publ. No. 22155, Warminster, PA, 1977.
11. "Evaporator Series 50.202," Wallace and Tiernan Div. Pennwalt Corp., Belleville, NJ, Rev. Jan. 1977.
12. Walker, T. B., personal communication, Wallace and Tiernan Div., Pennwalt Corp., June 1977.
13. Nagel, Wm., private communication, Fischer and Porter Co., July 1977.
14. Anon., "Advance Series 1030 Chlorine Gas Detector," Capitol Controls, Colmar, PA, July 1982.
15. Anon., "Draeger Safety Chloralarm," National Draeger Inc., Pittsburgh, PA, Jan. 1983.
16. Anon., "Chlor-Guard Model 5152 Chlorine Gas Detector," Exidyne Inc., Exton, PA, May 1983.
17. Becker, J., private communication, Exidyne Inc., Exton, PA, June 1984.
18. Anon., "Chloralert Chlorine Detector Series 17CA 1000," Fischer and Porter Co., March 1978.
19. Anon., "Series 50-125 Chlorine Detector Instruction Book," Wallace and Tiernan–Pennwalt, Belleville, NJ, June 1980.
20. White, G. C., and Cariss, F., unpublished field data, East Bay Municipal Utility District, Oakland, CA, 1965.
21. Houston, R., "Friction in Pipes," *Product Engr.*, 191 (Aug. 1957).
22. Anon., "Wallace and Tiernan V-Notch Chlorinators" Wallace and Tiernan–Pennwalt Corp. Belleville, NJ, Catalog No. 25.052, Rev. Aug. 1982.
23. Finger, R. E., private communication, Renton, WA, Wastewater Treatment Plant, Metropolitan Seattle, WA, 1984.
24. Connell, G. F., private communication, Capital Controls Co., Colmar, PA, July 27, 1984.
25. White, G. C., and Stone, R. W., "Factors Affecting the Feed Rate Response Time in a Long Injector Vacuum Line for Chlorinators and Sulfonators," unpublished in-house report for Brown and Caldwell consulting engineers, Walnut Creek, CA, Mar. 1974.
26. Anon., The Duriron Co. Bulletin No. P-10-101 r, Dayton, OH, 1983.
27. Anon., The Duriron Co. Bulletin No. P-17-101a, Dayton, OH, 1979.
28. Anon., MetPro Corp. Fybroc Div. Bulletin 15B1, Hatfield, PA, 1982.
29. Rossum, J. R., private communication, California Water Service Co., San Jose, CA, 1962.
30. Murphy, D., private communication, FCI Fluid Components, Inc. Bull. SF-3/79, San Marcos, CA, Jan. 1984.
31. Cutler, J. W., and Green, F. W., "Operating Experiences with a New Automatic Residual Control Recorder Controller," *J. AWWA*, **22**, 755 (1932).
32. Goudey, R. F., "Residual Chlorination on the Los Angeles System," *J. AWWA*, **28**, 1742 (1936).
33. Caldwell, D. H., "Automatic Chlorine Residual Indicator and Recorder," *J. AWWA*, **36**, 771 (1944).
34. Harrington, J. H., "Photo-Cell Control of Water Chlorination," *J. AWWA*, **32**, 859 (1940).
35. Baker, R. J., and Griffin, A. E., "Development of Instrumentation in Chlorination," *J. AWWA*, **50**, 489 (1958).
36. Marks, H. C., Bannister, G. L., Glass, J. R., and Herrigel, E., "Amperometric Methods of Control of Water Chlorination," *Anal. Chem.*, **19**, 200 (1947).
37. Hazey, F. J., "Amperometric Chlorine Residual Recording," *J. AWWA*, **43**, 292 (1951).
38. Krum, H. J., "Residual Chlorine Recording Experiences," *Water Wks and Sew.*, **98**, 376 (Sept. 1951).
39. Clark, G. C., "Amperometric Techniques for Chlorine Residual," *Inst. and Automation* (April, 1954).

40. Morrow, J. J., U.S. Patent No. 3.4.3, 199 (Nov. 26, 1968).
41. Johnson, J. D., Edwards, J. W., and Keeslar, F., "Chlorine Residual Measurement Cell: The HOCl Membrane Electrode," *J. AWWA,* **70,** 341 (June 1978).
42. Kolthoff, I. M., and Lingane, J. J., *Polarography,* Vols. I and II, Interscience Publishers, New York, 1965.
43. Anon., "Residual Chlorine Analyzer for Water Treatment," Wallace and Tiernan Div. Pennwalt Corp., Belleville, NJ, Cat. No. 50.245, Rev. Jan. 1982.
44. Marks, H. C., and Campbell, G. A., "Dual Electrode Measuring Cell," in-house report, Wallace and Tiernan Inc., Belleville, NJ, 1950.
45. Huebner, W. B., private communication, Wallace and Tiernan–Pennwalt, Belleville, NJ, Aug. 28, 1984.
46. Anon., "Advance Series 870 Chlorine Residual Analyzers," Capital Controls Co., Colmar, PA, Bulletin No. A1. 1870.6, 1982.
47. Johnson, J. D., and Edwards, J. W., "An Amperometric Membrane Halogen Analyzer," *Proc. Div. Envir. Chem. ACS,* **14,** 169 (April 1974).
48. Johnson, J. D., Edwards, J. W., and Keeslar, F., "Chlorine Residual Measurement Cell: The HOCl Membrane Electrode," *J. AWWA,* **70,** 341 (June 1978).
49. White, G. C., "Disinfection: Present and Future," *J. AWWA,* **66,** 689 (Dec. 1974).
50. Anon., "Directions for Series 8224/8324 and 8225/8325 Continuous Automatic Chlorine Monitor/Controller," Delta Scientific, National Sonics Div. Environtech Corp., Lindenhurst, NY, March 1976.
51. Anon., "Polarographic Wet Chemistry Analysis," Series 8000 Analyzers. Delta Scientific Products National Sonics Div., Envirotech Corp., Lindenhurst, NY, 1978.
52. Stanley, W., and Nossel, R., "Measurement of Residual Chlorine Compounds in Wastewater With Amperometric Membrane Electrodes," Jolley, R. L. et al. (Eds.), *Chemistry and Water Treatment of Water Chlorination,* Vol. 4, Book 1, p. 699. Ann Arbor Science (Butterworth Group) Ann Arbor, MI, 1983.
53. Anon., "Model 924 Chlorine Measurement System," Product Data Sheet 924, Delta Analytical (Xertex Corp), Hauppage, NY, 1983.
54. Anon., "Model 925 Chlorine Measurement System," Product Data Sheet 925, Delta Analytical (Xertex Corp), Hauppage, NY, 1983.
55. Obear, P. H., private communication, Xertex Corp. Huappage, NY, July 8, 1984.
56. Anon., "Instruction Bulletin for Series 17B4200 Rev. 4 Anachlor™ Residual Chlorine Analyzer Transmitter," Fischer and Porter Co., Warminster, PA, Nov. 1979.
57. Anon. "Instruction Bulletin for Series 17B4200-2 Rev 2 Anachlor™ Residual Chlorine Analyzer Transmitter," Fischer and Porter Co., Warminster, PA, Dec. 1978.
58. Morrow, J. J., and Roop, R. N., "Advances in Chlorine Residual Analysis," *J. AWWA,* **67,** 184 (April 1975).
59. Anon., "CHLOR-TROL™ Free Residual Chlorine Analyzer, Series 17K1000," Fischer and Porter Co., Warminster, PA, June 1975.
60. Anon., "The Chlortect® Chlorine Monitor for the Accurate Measurement of Chlorine Residuals," U.S. Patent No. 3,966,413, EPCO, Danbury, CT, Oct. 1980.
61. Marinenko, G., Huggett, R. J., and Friend, D. G., "An Instrument with Internal Calibration for Monitoring Chlorine Residuals in Natural Waters," *Jour. Fisheries Research Board Canada,* **33,** 822 (April 1976).
62. Anon., "EC/250 Chlorine Analyzer User's Manual," IBM Instruments Inc., Danbury, CT, 1982.
63. Kutt, J. C., and Vohra, S. K., "A Simple Approach to Chlorine Analysis," *Amer. Lab.* (Dec. 1983).
64. Anon., "Instruction Manual, Model 1570 Chlorine Monitor," Orion Research Inc., Cambridge, MA, 1982.
65. Anon., "Instruction Manual, Model 1770 Chlorine Monitor," Orion Research Inc., Cambridge, MA, 1981.

754 HANDBOOK OF CHLORINATION

66. Anon., "Instruction Manual, Model 853 Residual Chlorine Analyzer/Transmitter and Model 450 Sensor Assembly," Uniloc Div. of Rosemount Inc., Irvine, CA, Dec. 23, 1980.
67. Hoffman, F., private communication, Uniloc Div. of Rosemount Inc., Irvine, CA, May 13, 1981.
68. Anon., "Pump-Colorimeter Colorimetric Free and Total Chlorine," Models 31100, 31300, and 61100, Hach Company, Loveland, CO, Bulletin 1071, Feb. 1984.
69. Anon., "Free and Total Chlorine—Pump Colorimeter Colorimetric Models 31100, 31300, and 61100," Hach Company, Loveland, CO, Feb. 1984.
70. "Load Hugger 5000 lb. Capacity Bulletin," Lift All Products, Manheim, PA, (1974); and Liftex Slings, Inc. Bulletin A76-TD, Libertyville, IL, Dec. 1976.
71. Anon., "No. 53 MC Controller," Product bulletin, Fischer and Porter Co., Warminster, PA, June 1982.
72. Hayes, T. H., private communication, Fischer and Porter Co., Warminster, PA, Aug. 15, 1984.
73. Sloat, R., private communication, Chief, Chemtrec Emergency Response Team, Dow Chemical Co., Pittsburg, CA, Jan. 1984.
74. Anon. "Anhydrous Ammonia," CGA Pamphlet G-2, 6th ed. Compressed Gas Association Inc. New York, NY 1977.
75. Anon. "Direct Feed Ammoniator," Wallace and Tiernan-Pennwalt, Belleville N.J., Catalog 60.215, 1980.
76. Anon. "V-Notch Ammoniator Series V-800 Remote Vacuum Arrangement." Wallace and Tiernan-Pennwalt, Belleville, N.J. Catalog 60.222, Aug. 1977.
77. Baker, R. J. and Rudolph, G. C., Private communication Wallace and Tiernan-Pennwalt Belleville, N.J., March 1982.
78. Bentwood, R. W., Reichenberger, J. C., Suggs, D. and Clements, E. V., "Chloramine Disinfection for THM Control" Paper presented at National Conference on Environmental Engineering, ASCE, Univ. Sou. Calif. June 26, 1984.
79. Goodwin, H. E., "Automated Chlorination System" Pub. Wks. **93**, 74 (1962).
80. Stone, R. W. "Rancho Cordova Breakpoint Demonstration" A report prepared by Sacramento Area Consultants. (Sept. 1976).
81. Howerton, A. E. "Estimated Area Affected by a Chlorine Release," Chlorine Institute Report 71, (March 1969).
82. Horowitz, N. C. "Selecting Materials for Chlorine Gas Neutralization," Chem. Engineering, p. 105 April 6, 1975.
83. Anon. "Emergency Chlorine Scrubbing Systems for Water Purification Plants," Ametek-Schutte and Koerting Div., Durham, N.C. Bulletin 7S/21B (1977).
84. Dailey, Leo, private communication, Fischer and Porter Co., Warminster, PA, Jan. 29, 1985.
85. Hankins, R. A., private communication, Clorox Co. Tech Center, Pleasanton, CA, Jan. 31, 1985.
86. Roop, R. N., private communication, Fischer and Porter Co., Warminster, PA, Aug. 16, 1976.

10
Dechlorination

INTRODUCTION

General Discussion. Dechlorination is the practice of removing all or a specified fraction of the total combined chlorine residual. In potable water practice dechlorination is used to reduce the residual to a specified level at a point where the water enters the distribution system. In some cases where taste and odor control is a severe problem, control is achieved by complete dechlorination, followed by rechlorination. This removes the taste producing nuisance residuals and prevents the formation of NCl_3 in the distribution system. Dechlorination of wastewater and power plant cooling water is required to eliminate chlorine residual toxicity which is harmful to the aquatic life in the receiving waters.

The most practical method of dechlorination is by sulfur dioxide and/or aqueous solutions of sulfite compounds. Other methods used are: granular activated carbon and hydrogen peroxide.

Ammonia has been described as a dechlorinating agent. This is a misnomer. The addition of ammonia converts the free chlorine to monochloramine. When this method of "dechlorination" is used, its sole purpose is to prevent the formation of NCl_3 in the distribution system. NCl_3 *will not form if* HOCl *is absent.* Nitrite (NO_2^-) has been used to dechlorinate the HOCl fraction of a total chlorine residual. This is a special case where the combined residual would interfere with the accuracy of an analyzer arranged to record free only. In this case the analyzer is arranged to measure total residual. At frequent intervals nitrite is automatically injected into the sample. The nitrite consumes the HOCl fraction, which is instantly reflected on the recorder chart. The dip on the chart is a record of the HOCl fraction which the operator needs to know in order to control the free residual process. This is a special case.

Wastewater Research. In the late 1970s the Sanitary Districts of Los Angeles County were commissioned by the EPA to make a comprehensive study of wastewater dechlorination. This investigation and study included a literature search and review, a pilot plant study and a full-scale evaluation in the field. The field survey involved the canvassing of 55 operating plants in California by mail, telephone, and site visits to selected facilities. This project was reported in 1980 by Chen and Gan.[1] The important conclusions of this report are given below.

1. The sulfur dioxide process was the most cost effective of the three methods studied. To dechlorinate completely a 5 mg/l total chlorine residual the sulfur dioxide method total cost was 2 cents/1000 gal; the holding pond method cost was 4.5 cents and carbon adsorption cost was 13.5 cents. These were the only methods studied.

2. The feed-forward method of control, with flow as the primary signal and residual as the secondary signal, was the most commonly used system for the sulfur dioxide method.

3. Adding an overdose of SO_2 was essential to achieve consistent dechlorination. This provided the sulfur dioxide method with generally good reliability.

4. The residual chlorine analyzer was the weakest link in the SO_2 system. Therefore it is the most important component. The inherent weakness of amperometric type analyzers is their inability to maintain calibration stability for long enough periods of time (~24h) and to remain in calibration at zero residual.

5. Depletion of dissolved oxygen or reduction in pH was not observed in the pilot plant studies at a sulfur dioxide to residual chlorine dosage ratio of 2 to 1.

6. No significant physical-chemical degradation was found in the effluent after dechlorination by sulfur dioxide.

7. Bacterial regrowth in the ten minute sampler after dechlorination was observed to be a one to two order of magnitude increase in total coliform density in all three processes.

8. The regrowth, which was predominantly in the total coliform group, seemed to originate from contamination by the existing microorganism communities in the dechlorinated effluent rather than from the reactivation of injured bacteria cells. Fecal streptococci in the effluent remained relatively unchanged after dechlorination.

SULFUR DIOXIDE

Properties of Sulfur Dioxide:

Physical Characteristics of SO_2. Sulfur dioxide is a colorless gas with a characteristic pungent odor. It may be cooled and compressed to a colorless liquid. When liquid sulfur dioxide in a closed container is in equilibrium with the SO_2 gas, the pressure within the container varys with the ambient temperature. This is similar to the characteristics of chlorine. The relationship for both sulfur dioxide and chlorine between temperature and vapor pressure is shown in Fig. 6 in the Appendix. While sulfur dioxide is similar in many ways to chlorine, the most notable difference is in the vapor pressure. At 70°F, SO_2 vapor pressure is approximately 35 psi, while chlorine is approximately 90 psi. The next most important difference is its

Table 10-1

Characteristic	SO_2	Cl_2
Molecular weight	64.06	70.91
Latent heat of vaporization at 32°F, Btu/lb	161.8	109.1
Liquid density at 60°F, lb/ft³	87.2	88.8
Solubility in water at 60°F, g/l	120.0	7.0
Specific gravity of liquid 32°F (water = 1.0)	1.486	1.468
Vapor density at 32°F and 1 atm, lb/ft³	0.1827	0.2006
Vapor density compared to dry air at 32°F and 1 atm, lb/ft³	2.264	2.482
Specific volume of vapor at 32°F and 1 atm, ft³/lb	5.47	4.98
Critical temperature, °F	314.8	291.2
Critical pressure, psia	1141.5	1118.4

solubility in water. The solubility of sulfur dioxide is 120 g/l while chlorine is only 7 g/l, at 60°F.

The important characteristics of both sulfur dioxide and chlorine are shown in Table 10-1 (from *International Critical Tables*).

Sulfur dioxide is not flammable or explosive in either the gaseous or liquid state. Like chlorine, *dry* sulfur dioxide is not corrosive to ordinary metals, but in the presence of any moisture it is extremely corrosive. Therefore, with one known exception the same materials of construction are used for handling both sulfur dioxide and chlorine. The exception is the use of 316 SS for sulfur dioxide diffusers and valve trim for valves in the SO_2 vapor and liquid lines (monel trim is used for chlorine valves). The ability of 316 SS to withstand the corrosivity of SO_2 aqueous solution is helpful where PVC does not have the rigidity or structural strength. Although chlorinators and sulfonators may seem identical in construction they should never be used interchangeably. If evaporators used for chlorine are to be used for SO_2 they must be thoroughly cleaned. *All of the corrosion products from handling chlorine must be removed from the container vessel before use with sulfur dioxide.*

Physiological Effects. Fatal accidents from sulfur dioxide vapor are rare because persons in the lethal concentration of SO_2 cannot breathe and are compelled to seek the open air.[2] Low concentrations cause a sensation of suffocation, coughing, sneezing, and tearing of the eyes. Following this low level exposure the victim will have the sensation of a heavy chest cold. However, unlike exposure to chlorine, the victim will experience a rapid recovery from these symptoms after a few minutes in the open air free of SO_2. Liquid sulfur dioxide may cause severe injury to the skin and eyes due to the freezing action caused by the rapid evaporation of the liquid.

Physiological responses to various concentrations of SO_2 vapor are as follows:[2]

Effect	SO_2 in Air, ppm
Least detectable odor	3–5
Least amount causing immediate eye irritation	20
Least amount causing throat irritation	8–12
Least amount causing immediate coughing	20
Max. conc. allowable for prolonged exposure	10
Max. conc. allowable for ½ to 1 hr exposure	50–100
Dangerous for short exposure	400–500

First aid treatment for victims of SO_2 exposure is generally the same as for chlorine exposure. Remove all clothing; keep patient warm with blankets; do not excite or exercise the patient. Inhalation of ammonium carbonate fumes relieves respiratory irritation, and coughing is relieved by drinking a mixture of 3–5 drops of chloroform in a full glass of water. If this is not available have the patient sip 80 proof whisky or vodka from a teaspoon or straw. Two ounces of either should be sufficient for complete relief from coughing. Depending upon the severity of the exposure the patient should use copious amounts of water for flushing the eyes. This is particularly useful and necessary if liquid SO_2 has touched the victim's skin. The Compressed Gas Association should be consulted for the first aid supplies to be kept on hand. For skin burns, a 5 percent freshly made solution of tannic acid powder should be used to moisten sterile bandages placed over the skin burn area (8 level tablespoonfuls of tannic acid powder in an 8 oz glass of boiled water).

Operating Significance of SO_2 Characteristics. 1. *Vapor Density.* The densities of SO_2 and chlorine gases are so nearly the same that a chlorine-indicating rotameter can be used to meter sulfur dioxide without appreciable consequences. To convert chlorine feed rate to sulfur dioxide multiply by 0.95. For example a 1000 lb/day chlorine rotameter will become 950 lb/day sulfur dioxide.

Since the vapor density and viscosity of sulfur dioxide are so close to those of chlorine, long vacuum line calculations can be sized using the same variables and formulas as for chlorine vacuum lines.

2. *Vapor Pressure.* At 70°F the vapor pressure of SO_2 is approximately 35 psi, while chlorine's is 90 psi. This low vapor pressure causes problems in the SO_2 supply system. One major effect is the withdrawal rate. Assuming that 10 psi is the minimum operating pressure for a sulfonator and that the SO_2 ton container is at 70°F, the withdrawal rate will be:[3]

$$\text{Cylinder pressure} - \text{min. operating pressure} \times \text{Withdrawal factor} = \text{maximum withdrawal rate, lb/day}$$

The withdrawal factor for a ton cylinder of $SO_2 = 6.0$, therefore:

$$35 - 10 \times 6 = 150 \text{ lb/day}$$

This compares to a chlorine withdrawal rate of 440 lb/day, where the operating pressure at the chlorinator is 35 psi rather than 10 psi for SO_2.

The other major effect of the low vapor pressure is reliquefaction. This will be a chronic problem unless the pressure in the cylinder is raised artificially, or the SO_2 is transferred to an evaporator as a liquid. The ways to overcome the reliquefaction phenomenon are discussed later in this chapter.

3. *Solubility in Water.* The solubility of sulfur dioxide in water at 60°F is 120,000 mg/l as compared to 7,000 mg/l for chlorine. This makes the handling of SO_2 solutions easier than for chlorine. For example, negative pressures or excess turbulence at open channel diffusers must be avoided in the case of chlorine. Not so for sulfur dioxide. Consider the case of using a weir as a mixing device for a SO_2 solution. In the case of chlorine the diffuser should be placed downstream from the weir—in the zone of turbulence—to minimize off-gassing. For the SO_2 solution the diffuser is placed on the upstream side of the weir and the full benefit of the weir as a mixing device is realized.

Reactions with Chlorine. The various reactions involved in the sulfur dioxide dechlorination process are shown below. The first reaction is the formation of sulfurous acid in the sulfonator injector:

$$SO_2 + H_2O \longrightarrow H_2SO_3 \tag{10-1}$$

The sulfurous acid reacts with the various chlorine residual species as follows:

$$HOCl + H_2SO_3 \longrightarrow HCl + H_2SO_4 \tag{10-2}$$

$$NH_2Cl + H_2SO_3 + H_2O \longrightarrow NH_4Cl + H_2SO_4 \tag{10-3}$$

$$NHCl_2 + 2H_2SO_3 + 2H_2O \longrightarrow NH_4Cl + HCl + 2H_2SO_4 \tag{10-4}$$

$$NCl_3 + 3H_2SO_3 + 3H_2O \longrightarrow NH_4Cl + 2HCl + 3H_2SO_4 \tag{10-5}$$

These equations illustrate that all of the chlorine species can be dechlorinated with SO_2. The equations also illustrate that acids are produced which affect alkalinity and possibly pH. For each part of chlorine removed 2.8 mg/l alkalinity as $CaCO_3$ is consumed.

The stoichiometric relationship requires 0.9 parts SO_2 to remove 1.0 part of chlorine residual. In practice this ratio can be as high as 1.05 parts SO_2 to each part chlorine.

For practical purposes the reactions described by Eqs. (10-1) through (10-5) are complete in a matter of seconds (<10).

The great majority of potable waters have sufficient alkalinity buffering power (>40 mg/l) so that there is no cause for concern about lowering the pH with SO_2 addition (2–3 mg/l max.). The same is true for wastewaters. The alkalinity of chlorinated wastewater effluents is usually in the range of 150–200 mg/l as

CaCO₃. Dechlorination of a 12 mg/l chlorine residual will consume only 34 mg/l alkalinity as CaCO₃.

Reaction with Dissolved Oxygen. There has been some apprehension about the possibility that excess SO₂ might consume a significant amount of dissolved oxygen in the receiving waters downstream from a dechlorinated wastewater discharge. The reaction between the excess sulfite ion ($SO_3^=$) from dechlorination and the dissolved oxygen is:

$$SO_2 + H_2O + O_2 \longrightarrow H_2SO_4 \qquad (10\text{-}6)$$

This means it will require 4 parts of SO₂ to remove one part dissolved oxygen. Owing to the speed of reaction of Eq. (10-6) there would never be sufficient time available to complete the reaction at the optimum reactant concentrations. Plant-scale experience has confirmed this. The study by Chen and Gan[1] demonstrated no effect upon pH (7.2) due to an excess sulfite level of 2.6 mg/l in the plant effluent. Moreover the D.O. increased from 4.8 to 6.1 mg/l after dechlorination with sulfite. This was probably due to a high degree of turbulence caused by the mechanical mixers. There have not been any reports of sulfur compounds used for dechlorination having any effect upon dissolved oxygen consumption or pH change in the receiving waters or in the dechlorinated effluents.

Sulfite Compounds:

General Discussion. Sulfite compounds are used in solution and are primarily for smaller installations where feed rates for sulfur dioxide (<100 lb/day) are not practical. Sulfites are also used on large installations where storage of sulfur dioxide might be considered a hazard. These solutions are applied with metering pumps. Feed rate control and process monitoring of these systems need more attention and require more complex instrumentation than sulfur dioxide systems.

There are four sulfur compounds to be considered as alternative chemicals to sulfur dioxide for dechlorination: sodium sulfite, sodium bisulfite, sodium metabisulfite, and sodium thiosulfate. Sodium sulfite is available only as a white powder or crystals. It is extremely difficult to handle in the dry form because it is hygroscopic, therefore it is never used as a dechlorinating agent. Sodium thiosulfate is used in solution, but almost entirely as a laboratory chemical. It is not a satisfactory dechlorinating agent for treatment plant use because its reaction with chlorine is slow and not amenable to metering control situations. Therefore only sodium bisulfite and sodium metabisulfite are practical alternative chemicals to sulfur dioxide.

Sodium Bisulfite (NaHSO₃). This is a white powder or granular material, generally purchased as a solution in strengths up to 44 percent. It can be handled wet in stainless steel, PVC, or Fiberglas (tanks). It is usually metered as a dechlorinating agent by diaphragm-type pumps.

The reaction between bisulfite and chlorine residual is as follows:

$$NaHSO_3 + Cl_2 + H_2O \longrightarrow NaHSO_4 + 2HCl \qquad (10\text{-}7)$$

Each part of chlorine residual removed requires 1.46 parts of sodium bisulfite. The usual solution strength is 38 percent, which has a specific gravity of 1.3. The manufacturer's data sheet[4] advises that at this specific gravity a 38 percent solution contains 3.5 lb of sulfite per gallon. This calculates to 2.17 lb sulfur dioxide.

For each part chlorine removed, 1.38 parts of alkalinity as $CaCO_3$ will be consumed.

Sodium Metabisulfite ($Na_2S_2O_5$). This is a cream-colored powder readily soluble in water and available in solutions of various strengths.

The reaction between metabisulfite and chlorine residual is as follows:

$$Na_2S_2O_5 + 2Cl_2 + 3H_2O \longrightarrow 2NaHSO_4 + 4HCl \qquad (10\text{-}8)$$

Therefore each part of chlorine residual consumed requires 1.34 parts of sodium metabisulfite, and for each part of chlorine removed 1.38 parts of alkalinity as $CaCO_3$ will be consumed.

HYDROGEN PEROXIDE

General Discussion. Hydrogen peroxide has been used with only limited success as a dechlorinating agent. It reacts with free available chlorine (FAC) as follows:

$$Cl_2 + H_2O_2 \longrightarrow 2HCl + O_2 \qquad (10\text{-}9)$$

Therefore 0.48 mg/l of hydrogen peroxide will remove 1 mg/l free chlorine (HOCl).

In the 1970s, Dupont and FMC companies explored the use of hydrogen peroxide as a dechlorinating agent, as a hydrogen sulfide control chemical, and as a supplementary source of oxygen in wastewater treatment. It failed as a dechlorinating agent because it does not remove chloramines. Moreover, it has been discovered that hydrogen peroxide enters into side reactions that reduce its efficiency as an oxidizing agent. Furthermore, it is highly unstable and is subject to deterioration by various types of contamination which occur in packaging, transport, and on-site handling. Hydrogen peroxide is also expensive compared to other chemicals. In the San Francisco Bay Area a 50 percent solution was selling for 37 cents/lb in 4000 gallon deliveries (1984).

ACTIVATED CARBON

General Discussion. Activated carbon is packaged in two forms for water and wastewater use. The powdered form (PAC) is metered by weight and carried to the point of application as a water slurry. This is used as a pretreatment process to suppress taste and odor. It is often applied near the point of prechlorination. Operating experience indicates that only about 10 percent of the applied carbon reacts with the chlorine as a dechlorination reaction.[5] Granular activated carbon (GAC) is the other form of carbon used in water treatment processes. In this form it is used as a filter medium in conventional filter basins. It is used in vertical towers for the adsorption of organics and other undesirable compounds. PAC has not been used intentionally as a dechlorinating agent, but GAC has been used for many years.[6] It is the method of choice in the beverage industry. GAC has been used extensively as a filter medium in potable water treatment plants and has proved effective and reliable as a dechlorination agent.[7] It also produces a most palatable water.

GAC has not been successful as a dechlorinating agent in wastewater treatment. This is probably due to one or more of the following reasons: GAC may not have the ability to remove the organochloramines that form when significant concentrations of organic nitrogen is present; it may not be possible to design an effective carbon bed with the knowledge currently available; or it may be that the time required for the overall dechlorination reaction in wastewaters is much too long.

Chemistry of GAC Dechlorination

Definition of Carbon Mode of Action. The terms adsorption, saturation, and absorption are used frequently when describing the mode of action of GAC (granular activated carbon).

Adsorption is a phenomenon consisting of the adhesion in an extremely thin layer of the molecules of gases, dissolved substances, or of liquids to the surfaces of solid bodies with which they are in contact.

Absorption is a process of soaking up like a sponge. This is in sharp contrast to adsorption.

Saturation is a volumetric expression. It is used in GAC reactions to explain what occurs when the attractive force of the carbon is depleted.

Free Chlorine. GAC reacts with free available chlorine as shown

$$C^* + HOCl \longrightarrow CO^* + H^+Cl^- \qquad (10\text{-}10)$$

Where C^* and CO^* indicate active carbon and a surface oxide on carbon, respectively.[8] If significant amounts of HOCl are allowed to react with the carbon some of the oxygen attached to the surface may be emitted as CO or CO_2 gas. The stoichiometric reaction will be as follows:

$$C + 2Cl_2 + 2H_2O \longrightarrow 4HCl + CO_2 \qquad (10\text{-}11)$$

In this reaction 1.0 part chlorine will destroy 0.00845 parts of carbon. Since the ultimate reaction is chlorine conversion of the GAC to carbon dioxide no regeneration of carbon is required—it must be physically replaced because it is destroyed.

Chloramines. Studies by Bauer and Snoeyink demonstrated that pure chloramines can be dechlorinated by GAC as follows:
Monochloramine:

$$2NH_2Cl + H_2O + C^* \longrightarrow NH_3 + HCl + CO^* \qquad (10\text{-}12)$$

Dichloramine:

$$2NHCl_2 + H_2O + C^* \longrightarrow N_2 + 4HCl + CO^* \qquad (10\text{-}13)$$

The reaction time used in these observations was 20 hours. During these reactions the carbon apparently accumulates surface oxides which partially oxidize the NH_2Cl and $NHCl_2$ nitrogen to N_2.

Chloramines are adsorbed by the GAC to the depletion of the attractive force, which is described above as saturation. Therefore, from the moment of saturation the chloramines will break through and the carbon will have to be regenerated. Although both free chlorine and chloramines are adsorbed by the carbon, HOCl is adsorbed to the concentration reaction at equilibrium to form CO_2 and the chloramines are adsorbed to the point of depletion of the attractive force (saturation). Chloramines deplete the carbon while HOCl destroys the carbon.[10]

GAC Filter Bed Design. In water treatment practice, water that passes through a GAC filter bed is considered completely dechlorinated unless depletion of the carbon occurs. Therefore the water so treated must be rechlorinated if a residual is required in the finished water.

A detailed study of the carbon–chlorine reaction[11] indicates the following relationship between flow rates, bed depths, concentration of influent and effluent chlorine, as well as the granular carbon itself:

$$\log \frac{C_I}{C_E} = \frac{B \times \text{bed depth (ft)}}{\text{filtration rate (gpm/sq ft)}} = \frac{B}{V} \qquad (10\text{-}14)$$

in which C_I is the concentration of chlorine in influent (ppm), C_E is the concentration of chlorine in effluent (ppm), B is the efficiency constant for each carbon, and V is the flow rate (gpm/cu ft).

Hager and Flentje[7] report that a granular carbon medium in a 1 mgd filter at 2.5 gpm/sq ft and a 2.5 ft bed can process 700,000,000 gallons of 4 ppm free chlorine residual, and with 2 ppm chlorine a similar bed would last approximately 6 years at 1 mgd rate. The dechlorination reaction proceeds concurrently with the absorption of contaminants. Long-chain organic molecules, such as those of detergents, seem to reduce the dechlorination efficiency somewhat, but many common impurities and phenols have little effect upon the dechlorination reaction. A rise in temperature and a lowering of pH favor the reaction, and it has been found that chlorine–ammonia and other nitrogenous compounds of chlorine tend to react much slower than free chlorine.[7] Residual chlorine concentrations can be more easily maintained in the backwash water when combined chlorine residual, instead of free chlorine, are present.

Dechlorination by activated carbon for a conventional water treatment plant has certain limitations that should be recognized, and should not be relied upon as the sole means of controlling the finished water residual. The exceptions are in the beverage industry, where the raw water is of superior quality and the dechlorination must be complete.

In August 1977 the Cincinnati Water Works entered into a cooperative agreement with the United States EPA to pursue a feasibility study of municipal water treatment using GAC adsorption and on-site carbon regeneration.[12] This project demonstrated that the GAC removed all of the free available chlorine and all but a trace of the combined chlorine. These removals permit the growth of bacteria within the carbon bed with potential for carryover into the distribution system. Therefore a remedy for such a situation would be to construct a postchlorination facility followed by additional clear-well capacity to provide sufficient chlorine contact time for disinfection before the treated water enters the distribution system.

DECHLORINATION BY STORAGE

Holding lagoons are a possibility as long as time and space are available at a reasonable cost. FAC is vulnerable to destruction by sunlight. The rate of destruction is a function of cloud cover and latitude. Observations at water treatment plants where free residual chlorination is practiced, and the sedimentation tanks and filters are not covered, indicate that FAC dechlorination by sunlight is about 0.75–1.25 mg/l per hour between 10 A.M. and 2 P.M., Lat. 35°N, June through August. A similar phenomenon has been observed in swimming pools from California to Florida.

Chloramines do not show this type of decay. They are far more stable in sunlight than FAC. There are no reliable observations on chloramine residual decay due to UV destruction. It is believed at most to be 0.2 mg/l per sunlight hour between 10 A.M. to 2 P.M. at Lat. 30–40°N.

FERROUS SULFATE

Although not considered as a dechlorinating agent, any ferrous salt in solution will react with free chlorine, monochloramine, and—much more slowly—with dichloramine to form a ferric salt that eventually hydrolyzes to form ferric hydroxide, which is an excellent coagulant. Strockbine[13] reported the successful use of the chemical along with prechlorination to produce an excellent floc for coagulation, which at the same time did control (by dechlorination) the heavy doses required for prechlorination.

It should also be emphasized that metal pipe corrosion produces ferrous ion in a distribution system which acts to dechlorinate the distribution system residual. On a stoichiometric basis, 65.33 lb/mg of $FeSO_4 \cdot 7H_2O$ (ferrous sulfate) will be required to dechlorinate 1 ppm of chlorine. This represents a ratio of 1 part chlorine for the conversion of 7.8 parts of ferrous sulfate by weight.

OTHER METHODS OF DECHLORINATION

Aeration and holding lagoons have been mentioned as possible means of dechlorination. Aeration has no effect upon the removal of undissociated HOCl. At concentrations of FAC at high pH (>8.5) where the OCl^- dominates, an observer might perceive a rather strong chlorine odor. This is probably due to the slight release of chlorine monoxide. There is no loss of HOCl.

Chloramines will be lost in aeration, but the loss is minimal—up to a maximum of 10 or 15 percent for monochloramine and 20 percent for dichloramine. However, nitrogen trichloride is completely removed with only a slight amount of aeration.

CURRENT PRACTICES

Potable Water. The earliest reported use of dechlorination as a potable water treatment process was by Howard and Thompson[14] in 1926. Their report was based upon an exhaustive study on taste and odor control for the City of Toronto water supply. This study began in 1922. In the years following the discovery of the breakpoint phenomenon and the implementation of the free residual process, dechlorination became an important tool for the treatment plant operator. The free residual concept required that a prechlorination dose be sufficient to maintain an adequate residual throughout the entire plant. At some convenient place the plant effluent residual was trimmed by dechlorination to the appropriate level before entering the distribution system.[15] When the free residual is subject to decay by sunlight when passing through sedimentation basins and filters, the prechlorination dose can be adjusted automatically by a time clock to provide different dosage levels for night time and day time operation. This reduces the operating range burden for the dechlorination system. Sulfur dioxide is the preferred chemical because the equipment and control modes are the same as for chlorination systems. In most cases the prechlorinator is flow paced with automatic night–day dosage change and the sulfonator operates by direct residual control without flow pacing.

Chlorination followed by dechlorination has been used in the British Isles for more than thirty years. Its principal use is for disinfection of underground supplies where contact time is short. Using the chlorine residual–contact time concept, the chlorine residual can be elevated to the desired amount for the available contact time which may be only 5–10 minutes at best. The residual at the end of this time may be as high as 3–5 mg/l, depending upon the local conditions. This residual is trimmed to the desired distribution system level by dechlorination.

This concept of the residual–contact time product has been used as an interim solution for *Giardia lamblia* destruction pending construction of filtration facilities at surface supplies contaminated by the intrusion of this pathogenic cyst. Cysts usually require long free chlorine contact time to perform adequate disinfection. These cysts require at least 4 mg/l residual for one hour contact time to be destroyed, or 8 mg/l for 30 minutes. Obviously either of these residuals would have to be trimmed by dechlorination before entering the distribution system.

Dechlorination has been found beneficial for waters that are burdened with high concentrations of ammonia-N and organic N. When the free residual process is practiced in these waters terminal dechlorination may be partial or complete. This depends upon the ratio of free chlorine to total combined chlorine residual just prior to the point of dechlorination. If at this point the free residual chlorine is 85 percent or more of the total, it is usually satisfactory to dechlorinate to a desired free chlorine residual because the amount of nuisance residuals—those made up of other than free chlorine—is so small that it will not be a factor in the palatability of the finished water or cause any problems in the distribution system, other things being equal.

When the ratio of free to total residual drops below 85 percent, this usually indicates the presence of significant amounts of organic nitrogen in the raw water interfering with the free residual process. At this point the combined chlorine residual contains what appears to be dichloramine but may be some type of poly-N-chlor compounds as a result of the reaction of chlorine and organic nitrogen. This is an undesirable situation since these nuisance residuals seem only to contribute to off flavors in the water. Furthermore, it is at this point that nitrogen trichloride is most likely to form. Even in trace quantities, this compound gives off an obnoxious odor. If the resulting combined residual is allowed to proceed into the distribution system, the remaining free chlorine continues to react with the combined residual to produce more nitrogen trichloride. Therefore in such waters it is desirable to go to complete dechlorination. This removes all of the nuisance residuals and prevents further formation of either dichloramine or nitrogen trichloride, both of which produce off flavors. This water, to which has been added a slight excess of sulfur dioxide, is then rechlorinated. Since dechlorination by sulfur dioxide will not remove any of the ammonia, there may be sufficient ammonia remaining after dechlorination to form a predominantly monochloramine residual upon rechlorination. This depends entirely on the amount of ammonia in the raw water that has not been destroyed by the free residual process. Usually postammoniation for such waters is required to produce the proper monochloramine residual. If this is done, the water leaving the treatment plant will not contain any nuisance residual fractions that will adversely affect the quality of water at the consumer's tap.

Other special applications requiring dechlorination are: ahead of demineralizers, boiler makeup water, certain food plant operations, and the beverage industry.[6] In these cases the dechlorination process is arranged to remove all remaining chlorine residual.

Wastewater. The widespread use of chlorine in wastewater discharges and cooling water treatment at steam power stations came under close scrutiny in the middle 1960s. Esvelt et al.[16] reported in 1971 on the toxicity of wastewater effluents being discharged into San Francisco Bay. This study covered a span of about 3 years and proved that chlorinated effluents were more toxic to aquatic life than unchlorinated effluents.

These studies also showed that a dechlorinated effluent is less toxic than either the chlorinated or the unchlorinated effluent. These studies used continuous-flow on-line bioassays with golden shiners (*Notemigonus chrysoleucas*). In these studies the fish were captive in tanks of various chlorine residuals for a minimum of 96 hr.

In 1973 Brungs[17] turned out a comprehensive report in which the toxicity of chlorine to aquatic life is quantified. This paper reviews and discusses the uses of chlorine and chlorine chemistry and emphasizes toxicity studies in the field and laboratory. The literature review comprises more than 150 references. The following are the finite conclusions of Brungs' report:

1. In areas receiving wastes treated continuously with chlorine, total residual chlorine should not exceed 0.01 mg/l for the protection of more resistant organisms only, or exceed 0.002 mg/l for the protection of most aquatic organisms.

2. In areas receiving intermittently chlorinated wastes (power plants), total residual chlorine should not exceed 0.2 mg/l for a period of 2 hr/day for more resistant species of fish, or exceed 0.04 mg/l for a period of 2 hr/day for trout or salmon. If free chlorine persists, total residual chlorine should not exceed 0.01 mg/l for a period of 30 min/day for areas with populations of trout and salmon.

To the chlorination–dechlorination practitioner the above restrictions translate to a zero chlorine residual at all times. This is the interpretation by the California Fish and Game Commission as well as the California Water Quality Resources

Board who have jurisdiction over these matters. Zero residual is the consensus among other state regulatory boards; however some exceptions do exist.

Therefore the basic design of wastewater dechlorination systems are designed to produce a zero chlorine residual in the effluent before it leaves the plant. There are always exceptions. In some cases the outfall may be long enough to consume up to 0.5 mg/l chlorine residual before reaching the receiving waters. In such a system a closed loop residual control system using a 0.5 mg/l set point can be used. The major problem with this kind of installation is the ability to produce a monitoring system to satisfy the regulatory agency. Most complaints by regulatory personnel in California are about noncompliance with maintaining a continuous zero residual in the effluents.

Reclaimed Water. Water conservation is receiving increased attention in the arid states. By its own nature it must always have a persisting residual to prevent biofouling in the distribution system. Therefore these effluents do not require dechlorination unless they are discharged into the environment containing aquatic life.

DECHLORINATION FACILITY DESIGN

Factors Affecting System Efficiency. The principal factors affecting a successful operating installation are the same as for the chlorination system with one exception. This exception is monitoring for zero residual. The other principal factors are: the control system, adequate mixing, sufficient contact time, and trained and dedicated personnel. All of these will be discussed in detail below.

Sulfur Dioxide Supply System:

General Discussion. Sulfur dioxide is available in ton containers, tank trucks or tank cars. Owing to the low vapor pressure of SO_2 special precautions must be taken when using ton containers to prevent reliquefaction. Tank truck deliveries are by far the most popular modes of delivery in California. This type of delivery requires the user to provide a storage tank. Some sulfur dioxide suppliers provide a storage tank as a ploy to have a continuing contract with the users. Tank cars are designed the same as those for chlorine, therefore guidelines used for chlorine tank car layouts can be used for SO_2 cars.

Ton Containers: Gas Withdrawal. Owing to its low vapor pressure, SO_2 gas reliquefies quite easily. This causes operating problems with the control equipment. This low vapor pressure is also a limiting factor in the movement of the sulfur dioxide from the supply system to the control equipment.

Withdrawing from the gas phase usually requires the application of heat to the cylinders. Gas-phase systems operate best at around 90–100°F, and it is acceptable practice to apply heat directly to SO_2 ton cylinders provided there is control to limit the heating to 100°F. The maximum gas withdrawal rate at 70°F from a ton container is about 180 lb/day without reliquefaction. At 100°F it is about twice as much.

Whenever heat is applied to SO_2 cylinders, a pressure-reducing valve must be installed immediately downstream from the cylinders to prevent reliquefaction in the SO_2 header system. Otherwise the benefits of the heat are lost. The use of liquid withdrawal, which requires evaporators, is preferred over gas withdrawal except for rates of 150 lb/day or less.

Ton Containers: Liquid Withdrawal. When SO_2 is to be withdrawn as a liquid it is highly desirable to have a nitrogen padding system. This requires use of nitrogen tanks manifolded to the header system. These tanks are pressurized to 2000 psi, therefore they must discharge into a pressure regulating valve capable of reducing the nitrogen pressure to 150 psi. This arrangement can also be used to purge the piping system as required; however, it must not be used to purge the chlorine system. These two systems must be completely separate.

Air padding can also be used, but it must be dried first. This complicates the installation. This is why nitrogen is preferred for ton containers. The nitrogen—or air—pad should be capable of elevating cylinder vapor pressure to 100 psi.

Materials of Construction. All of the piping, pipe fittings, pressure gages, pressure switches, and expansion tanks shall be of the same material as for chlorine. The only exceptions are valves. Line valves and auxiliary valves must have 316 SS trim, rather than monel which is used for chlorine.

Accessory Equipment. Sulfur dioxide ton containers are identical to chlorine containers except for the color—SO_2 tanks are usually blue, while chlorine tanks are silver colored. Therefore all of the accessory equipment such as weighing devices, trunnions, lifting bar, filters, and pressure reducing valves (with 316 SS trim) are the same as for chlorine. (See Chapter 9).

Safety Equipment. The breathing apparatus and the container emergency kits are also the same as for chlorine.

Automatic Cylinder Switchover. Use the same system as for chlorine. (See Chapter 9.)

Remote Vacuum. Same as for chlorine.

Storage Tanks. Owing to the widespread use of sulfur dioxide in industry and agriculture a large proportion of SO_2 production is delivered by road tankers of 17–20 ton capacity. These suppliers have established the practice in many states of supplying stationary storage tanks for the users. These tanks do not conform to what might be described as "Chlorine Institute approved." They do not have: excess flow valves, safety relief valve, multiple vapor or liquid outlet valves, or any emergency device. Therefore the user is advised to install a stationary tank equipped with a standard chlorine tank car dome assembly modified for SO_2 handling. This means replacing all monel trim valves including excess flow valves with 316 SS trim. This approach insures greater safety and reliability. Moreover, the user can then utilize the chlorine tank car emergency kit. All piping and header systems, including expansion tanks, gages, alarms, etc., should conform to the guidelines set forth in Chapter 9 for chlorine.

Flexible Connections. Storage tanks eliminate the need for flexible connections. Flexible connections used for chlorine are still being used for SO_2 ton containers and tank cars without incident.

Leak Detectors: SO_2:

Introduction. There are two basic categories of sulfur dioxide detectors. They are ambient and source detectors. Ambient detectors are for air pollution monitoring; they require greater sensitivity than source detectors.

Ambient SO_2 detectors are usually made by manufacturers who also make NO_x

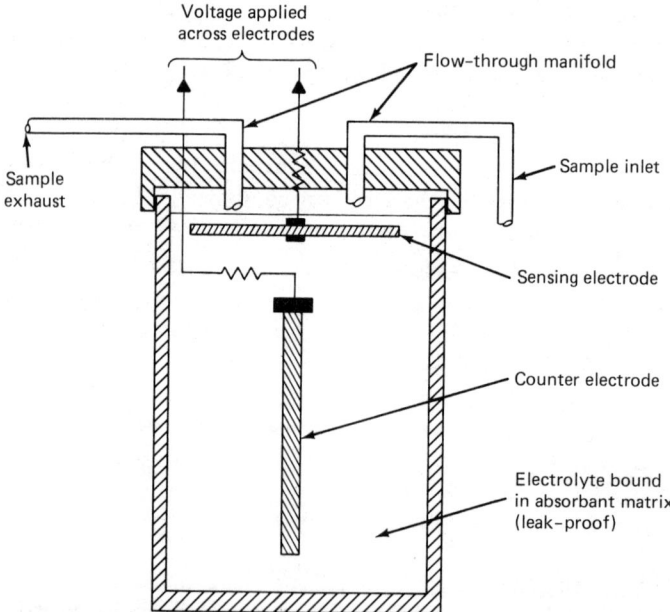

Fig. 10-1. Leak-proof electrochemical voltammetric sulfer dioxide sensor (courtesy of Interscan).

detectors. An example of an ambient air SO_2 detector is the pulsed fluorescent analyzer by Thermo Electron Corporation, Waltham, MA.

Source detectors are used principally for monitoring industrial stack gases. The detectors for SO_2 leak detection at water and wastewater plants are required to detect concentrations of 5 ppm. The two most commonly used on the Pacific Coast are described here.

Stack gas monitors must be able to detect in the 1 ppm range. Therefore these monitors are in a different category from water and wastewater SO_2 leak detectors.

Interscan Corporation. This company founded in 1974 markets analyzer-detectors for toxic gas monitoring. These analyzers are equipped with a voltammetric sensor (U.S. Patent Number 4,017,373). (See Fig. 10-1.) It is an electrochemical gas detector operating under diffusion controlled conditions. The Model No. 1247 has an SO_2 operating range of 0–10 ppm.[18]

Gas molecules from the sample are adsorbed onto an electrocatalytic sensing electrode, after passing through a diffusion medium. These gas molecules are reacted electrochemically at an appropriate applied voltage to the sensing electrode. This reaction generates an electric current directly proportional to the gas concentration. This current is then converted to a voltage for local readout, or remote recorder.

The generated diffusion-limited current, I_{LIM}, is directly proportional to the gas concentration as follows:

$$I_{LIM} = \frac{nFADC}{L}$$

where

I_{LIM} is the limited diffusion current in amperes,
F is the Faraday constant (96,500 Coulombs),
A is the area of reaction interface, cm^2,
n is the number of electrons/mole of reactant,
L is the diffusion path length,
C is the gas concentration, moles/cm^3,
D is the gas diffusion constant, representing the product of the permeability and solubility coefficient of the gas in the diffusion medium.

An external applied voltage maintains a constant potential on the sensing electrode relative to a nonpolarizable reference counterelectrode. This electrode therefore can sustain a current flow without suffering a change in potential; thus it acts as a reference electrode. This eliminates the need for a third (reference) electrode and a feedback circuit, as would be required for sensors using a polarizable air counterelectrode.

The sensor principal used by Interscan allows gas detectors to be made specific for each toxic gas. Sensor selection and design for specificity is accomplished by careful consideration of the following factors: (1) selection of an appropriate sensing electrode potential within the voltammetric curve plateau for a given toxic gas; (2) choice of an electrocatalyst for the sensing electrode; (3) the reaction kinetics of the interfering gas; and (4) the solubility of the gas in the thin liquid film at the electrode.

Interscan claims an accuracy of ± 2 percent of full scale, limited only by the accuracy of the calibration standard.

Chemical reagents are not required. The sensor electrolyte (Fig. 10-1) is immobilized, similar to dry batteries. The bound electrolyte enables the analyzer to be operated in any position. The sensor has a leakproof reservoir. This provides longer sensor life and ability to withstand pressurization, and eliminates the possibility of reference electrode contamination in a hostile environment.

*Enterra.** This company markets a line of toxic gas sensors which includes sulfur dioxide, chlorine, and hydrogen sulfide. These instruments are voltammetric sensors based upon proprietary electrochemical gas sensing electrodes. Each sensor contains a noble metal sensing electrode and a counter electrode immersed in a supporting acidic electrolyte. A diffusion membrane isolates the sensing electrode from the ambient air. This type of sensor is classified as a voltammetric diffusion limited electrochemical device. Fig. 10-2 illustrates the geometry of the sensor.

The life of the sensor depends primarily upon the amount of lead used in the counter electrode and the loss of water from the sensor through evaporation. Life expectancy is on the order of 2 years operating on low gas concentrations.

The standard range of the Enterra unit has a range of 0 to 25 ppm SO_2. These detectors are factory calibrated and the alarm points are set at 5 ppm and 10 ppm.[19] At 5 ppm the DANGER lamp will begin to flash and the SPDT relay will operate. At 10 ppm the ALARM lamp will light, the audible horn will sound, and the DPDT will operate. This is consistent with the physiological effects of SO_2 vapor, i.e., the maximum concentration allowable for prolonged exposure is 10 ppm.[2]

Flow of Sulfur Dioxide in Pipes. Due to the physical and chemical similarity of chlorine and sulfur dioxide the hydraulics of SO_2 vapor or liquid flow in pipes can be analyzed using the same factors as those used for chlorine, i.e., viscosity, density, Reynolds number, and friction factor. This holds true for both pressure and vacuum calculations.

Sulfonators. Except for minor modifications in materials of construction, sulfonators are identical to chlorinators. See Chapter 9.

Evaporators. Evaporators for liquid sulfur dioxide are identical to those used for liquid chlorine. However, because of the difference in latent heat of vaporization,

---

*Formerly Exidyne.

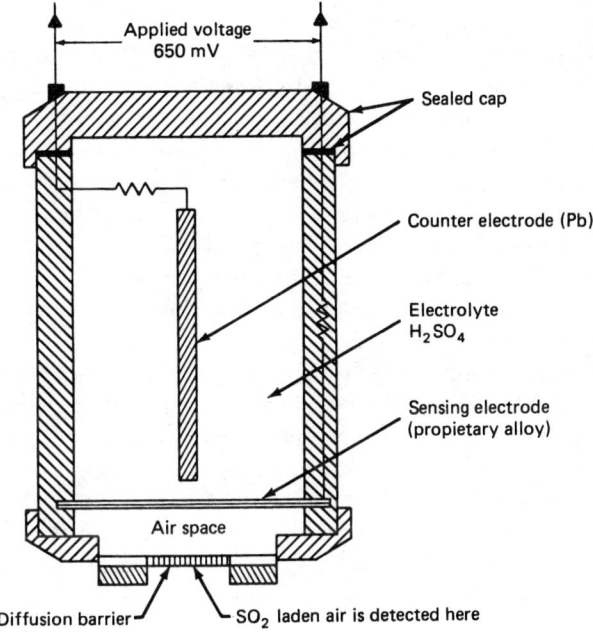

Fig. 10-2. Voltammetric diffusion limited sulfur dioxide sensor (courtesy of Enterra).

an 8000 lb/day chlorine evaporator will only vaporize about 5600 lb/day sulfur dioxide. This is based upon latent heat of vaporization of SO_2 at 164.5 and Cl_2 at 115.

Dechlorination Control Strategies:

General Discussion. *Potable Water.* Dechlorination practices in potable water are generally concerned with trimming the chlorine residual. Therefore conventional closed loop control strategies may be used. See Chapter 9. Complete dechlorination of potable water followed by rechlorination is rare.

Wastewater. There are three major reasons why the strategies for dechlorination control have to be different from those used for potable water: (1) there is no analytical system currently available to measure the concentration of the sulfite ion; (2) dechlorination must be complete (zero); and (3) chlorine residual analyzers are not capable of maintaining their calibration in a zero residual environment.

The following strategies are primarily for wastewater systems—zero control.

Feed-Forward. Because of the above reasons this method became the most popular method in the decade 1975–1985. This system requires an effluent flow signal and a chlorine residual signal measured at the end of the contact chamber. The two signals are combined to generate a control signal for the sulfur dioxide control equipment. This scheme is illustrated by Fig. 10-3. The principal advantage of this system is its simplicity and use of conventional equipment. The analyzer is always measuring residuals well above zero, hence it is not subject to electrode polarization and subsequent calibration drift. The effluent flow signal is easily obtained from the effluent flowmeter.

This system suffers from inherent weaknesses. The primary weakness is the lack of a mechanism for continuously monitoring system performance, i.e., the measurement of the residual in the dechlorinated effluent.

For this control system to function properly, the measured effluent flow signal, the SO_2 feed rate, and the measured chlorine residual must all be accurate. Moreover the selected reaction ratio between the chlorine residual and sulfur dioxide must remain constant. Should any of these measured values drift out of calibration, or the reaction ratio vary due to competing reactions within the effluent, the desired dechlorination level will not be fulfilled. It is for this reason that this method of control must rely upon the addition of a slight excess of SO_2. Therefore it is not possible to document system performance on a continuous basis. This has been the principal complaint of the water quality surveillance staff of the California Water Quality Resources Board, San Francisco Bay Region.

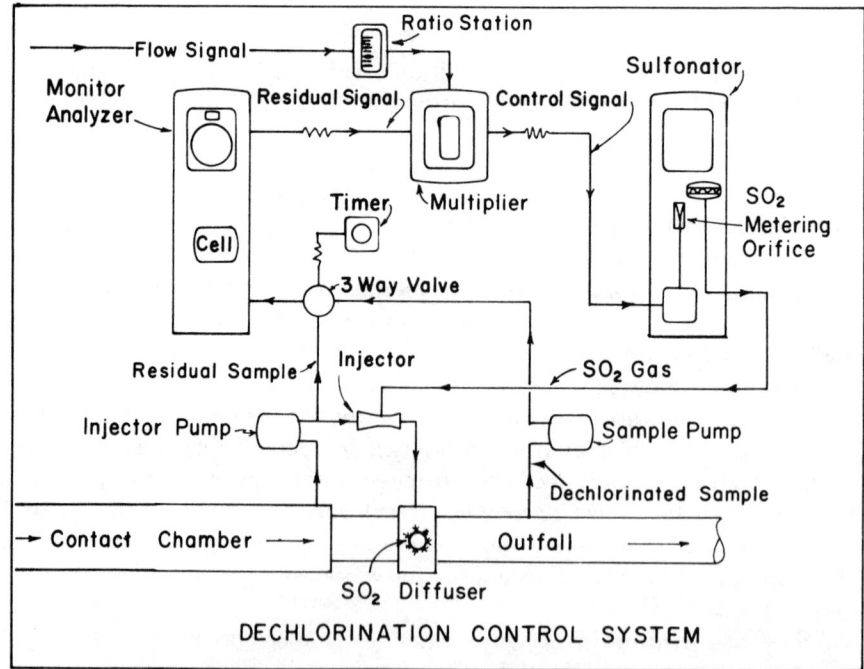

Fig. 10-3. Feed forward control of a sulfur dioxide dechlorination system.

Two-Step System: Feed-Forward and Feedback. This is a variation of the system described above. It is intended to use the standby sulfonation equipment to provide a two-level dechlorination system. These systems have been operating since 1976. This system is shown in Fig. 10-4. Sulfonator No. 1 controls the chlorine residual in the effluent from the chlorine contact basin to 0.5 mg/l or thereabout using a conventional feedback loop. The standby sulfonator (sulfonator No. 2) provides the necessary SO_2 required to destroy the remaining chlorine residual. In practice, sulfonator No. 2 is usually set to provide 0.3–0.5 mg/l excess of SO_2.

The operation is briefly as follows. The lead sulfonator is controlled by a compound loop control thusly: the chlorine residual analyzer sends a dosage control signal to the sulfonator differential regulating valve and an effluent flow meter sends via a ratio control station a flow signal to the sulfonator metering orifice positioner. The ratio station is imperative because it allows adjustment of wastewater flow and chlorine demand characteristics.

The No. 2 sulfonator is controlled by feed forward control. The same flow signal utilized by the lead sulfonator is sent to another ratio station and thence to a signal multiplier where it combines with the chlorine residual from the effluent chlorine residual analyzer. The signal from the multiplier controls the sulfonator metering orifice positioner.

Proper operation of this system shown in Fig. 10-4 requires proper selection of metering orifices for the two sulfonators. The following example will illustrate this point.

Assume the maximum wastewater flow to be 10 mgd and maximum expected chlorine residual to be dechlorinated is 6 mg/liter. The lead sulfonator will have to provide 1.0 mg/liter SO_2 × 5.5 mg/liter CL_2 × 10 mgd = 459 lb/day so choose a 500 lb/day rotameter–orifice combination for sulfonator No. 1. To dechlorinate the remaining chlorine residual requires 1.0 mg/liter SO_2 × 0.5 mg/liter Cl_2 × 10 mgd = 42 lb/day, so provide sulfonator No. 2 with a 50 lb/day orifice meter. This arrangement will provide the best flexibility of a system based upon the situation described above.

A variation of the above can be accomplished with only one sulfonator, as follows. The sulfonator injector discharge could be split to provide a dechlorinating solution for application immediately downstream from the primary or main point of the sulfur dioxide application. The sulfonator would be controlled to a 0.5 mg/liter chlorine residual by a compound loop control system. The second addition of sulfur dioxide would be a small portion (about 10 percent) of the total sulfonator injector discharge sufficient to dechlorinate the remaining chlorine residual. To control this system properly requires two flow measuring devices; one upstream from the sulfonator injector and one in the split to the second point of application. The line to the second point would also have to be equipped with a corrosion resistant control valve possibly supplemented by an orifice-type restrictor. The solution lines to both points of application would have to be carefully analyzed so as to provide the necessary hydraulic conditions to allow precise control of the injector discharge split.

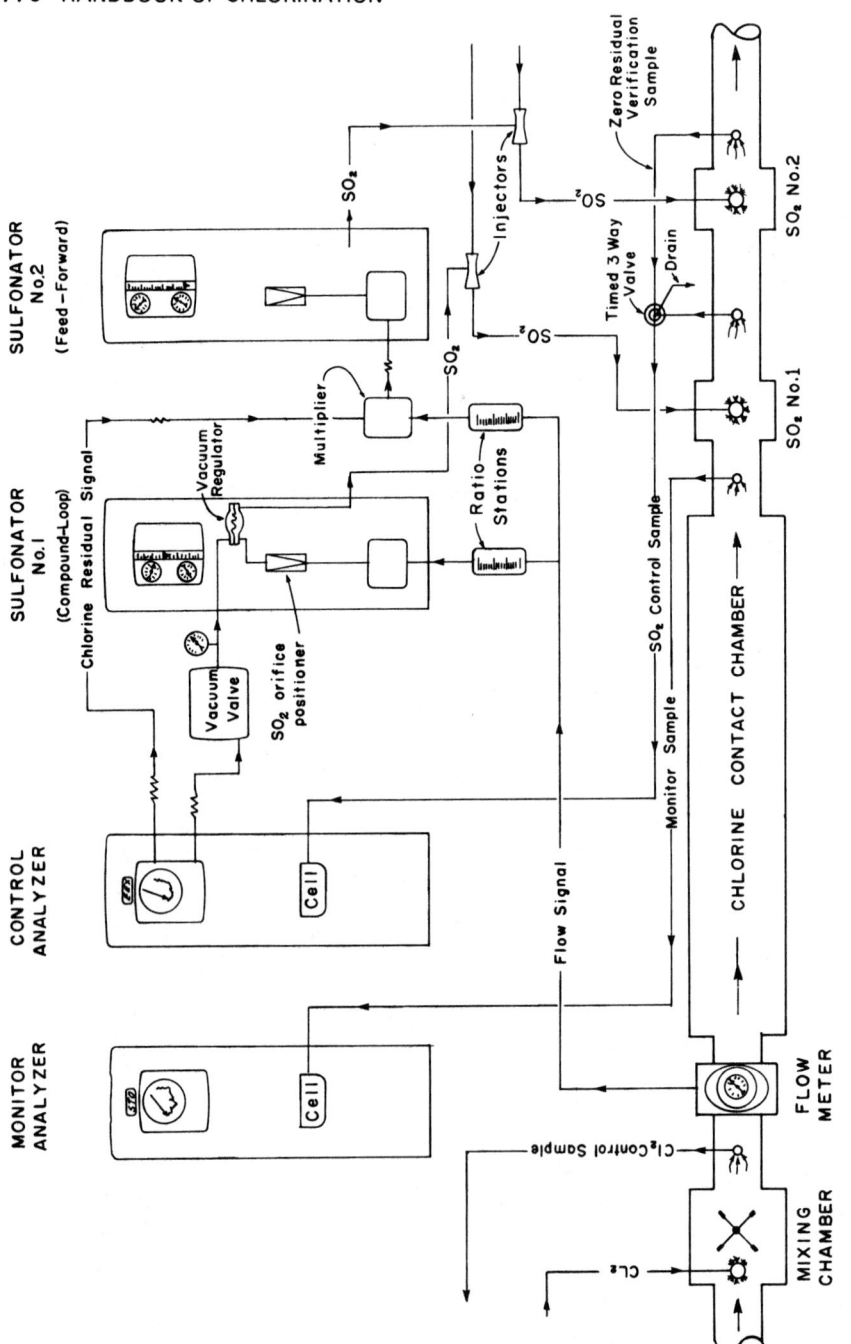

Fig. 10-4. Two-step dechlorination system utilizing compound-loop control.

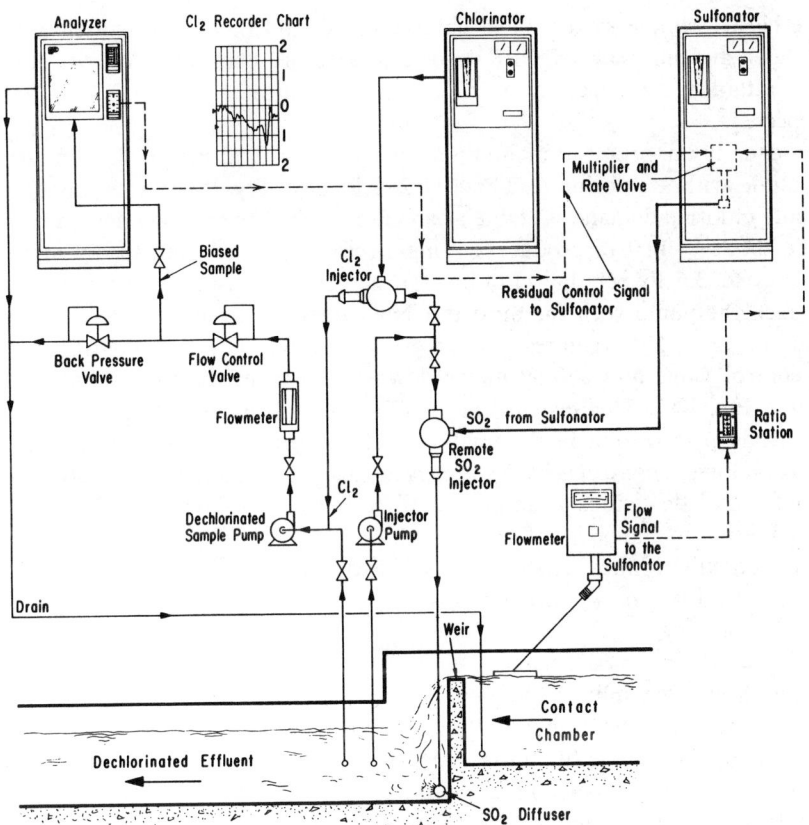

Fig. 10-5. Compound-loop control dechlorination system using a chlorine biased control sample.

Biased Sample. This system has been used successfully for several years. It is illustrated in Fig. 10-5. This is a scheme whereby the dechlorinated sample of the sulfonator control analyzer is biased with a constant artificial chlorine dosage. This results in a total residual of approximately 2 mg/liter in the biased sample when there is zero residual in the dechlorinated sample. The residual analyzer–recorder receiving this biased sample is arranged as shown to record chlorine residuals from 0 to +5 mg/liter and 0 to −5 mg/liter. Whenever the chlorine residual appears in the minus area of the chart this signifies that there is a chlorine residual in the dechlorinated effluent. The operators are immediately alerted to this situation by an alarm signal from the residual recorder. This system is preferred over the other systems described above because of the feedback loop that controls the sulfonator to a 2 mg/l SO_2 concentration in the biased sample to the analyzer. The success of this system is largely dependent upon the proper hardware selection in the biased sample system. The chlorinator should be equipped with a 10 lb/day meter and an external CPRV so that accurate control at 2–3 lb/day rate can be attained. An accuracy of ±2 percent of indicated rate (±0.05 mg/l) can be expected.

It is equally important to maintain a constant measured sample flow in quantity compatible with a chlorine feed rate of 2–3 lb/day. Try 75 gpm and assume the 5 minute chlorine demand is stable at about 2 mg/l. The chlorine dosage required will be about 4 mg/l to produce a 2 mg/l residual. At 75 gpm (0.108 mgd) this calculates to 3.6 lb per day chlorine. To achieve a constant sample flow of 75 gpm, pick the pump with the most efficient hydraulics to provide 30 psi pressure at the analyzer. This requires a back pressure valve. For flow control install a flow control valve and a flow meter downstream from the pump discharge, as shown in Fig. 10-5. This system has the following advantages: There is no danger of electrode polarization in the analyzer cell, so that the analyzer maintains its calibration longer; control is by the preferred closed-loop method; and both chlorine and sulfur dioxide residuals can be read directly from the analyzer recorder chart.

By adhering to principles of bias system components selection described above, this method will produce reliable results. The deviation experienced on one operating installation was on the order ±0.10 mg/l.

Zero Residual: Renton System. The concept of using a dechlorinated sample biased with an even split of the chlorinated sample was implemented by William Miks[20] at the Palo Alto Water Pollution Control plant in 1979. He diverted part of the chlorinated sample (used for chlorination control) to the fresh water dilution compartment of a Wallace and Tiernan analyzer. This sample was mixed 1 to 1 with the dechlorinated sample which entered the other half of the sample dilution tank. This was convenient at Palo Alto because the chlorine control analyzer and the sulfonator control analyzer were located adjacent to each other. This method of "zero control" is still operating successfully (1985). The sulfonator control analyzer is calibrated to provide about 0.3 mg/l negative residual.

A more sophisticated method using the above concept was developed by Kennedy/Jeuks Engineers in 1980.[21] This method was presented at the WPCF annual conference in 1983.[22] The application of sulfur dioxide is controlled solely by residual measurement.

The Renton system is based upon the fundamental concept of the Kennedy/Jenks system. It has operated almost flawlessly for two years.[23] Flow pacing is not used. The control system provides for a slightly negative chlorine residual (−0.20 mg/l) to insure reliability. This system is illustrated in Fig. 10-6. It is a retrofitted system which utilizes both Fischer and Porter and Wallace and Tiernan equipment. The success of this system is due in part to the short loop dead time—60 seconds at 42 mgd average daily flow.

The dechlorinated effluent sample is biased with the chlorinated effluent sample to provide a sample that can be measured by a conventional residual analyzer.

DECHLORINATION 779

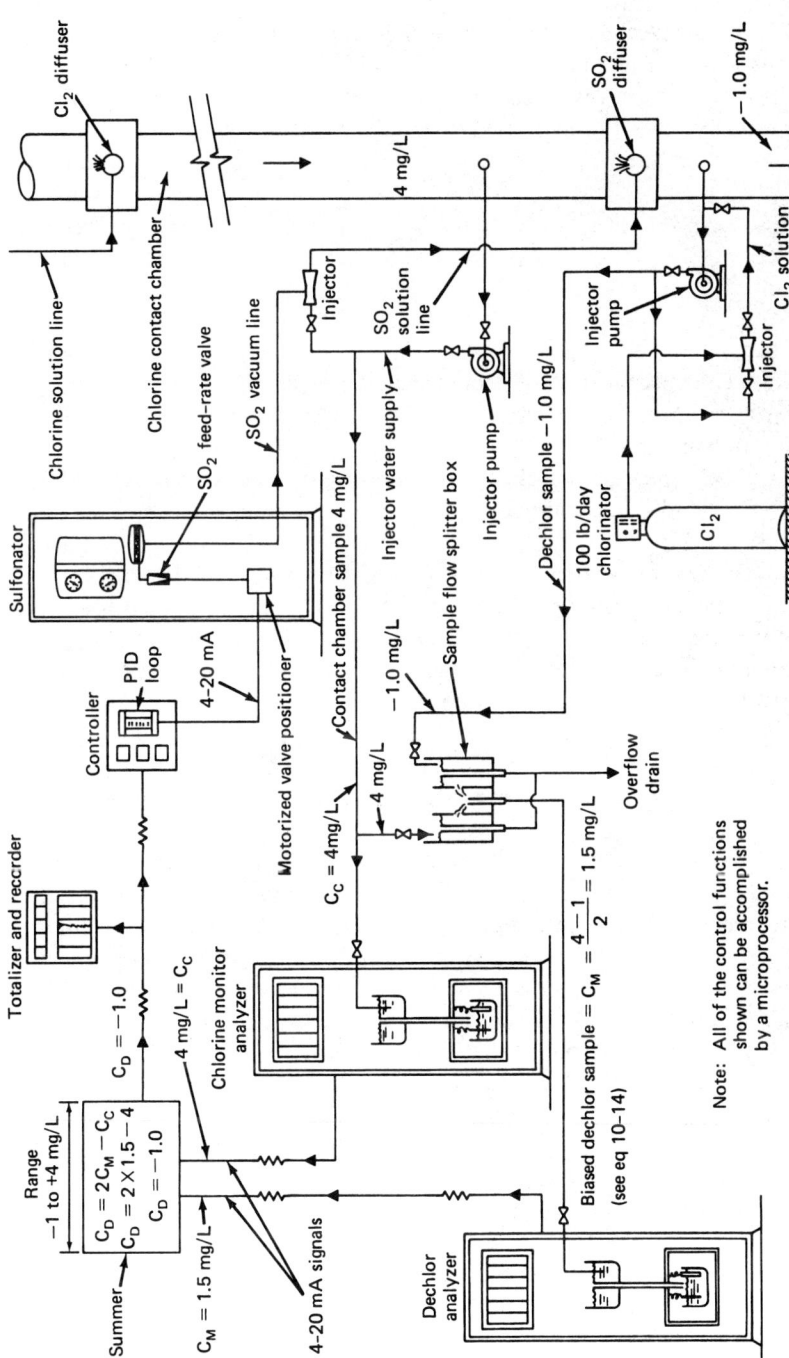

Fig. 10-6. Zero residual dechlorination system (after Renton Washington WWTP[23]).

The chlorinated and dechlorinated sample flows are proportioned to a 1:1 ratio by using constant-head tanks.

The chlorine residual in the biased sample of the dechlorinated effluent is calculated electronically from the measured residuals in the chlorinated and dechlorinated samples and the fixed mixing ratio of the two samples based upon the following equation:

$$C_D = 2C_M - C_C \qquad (10\text{-}15)$$

where

C_D = chlorine residual of dechlorinated sample
C_C = chlorine residual of chlorinated sample at end of contact chamber
C_M = chlorine residual of 1:1 mixture of chlorinated and dechlorinated sample

Each analyzer generates a signal proportional to a range of 0–5 mg/l. These two signals are sent to a "summer" (add or subtract). The output of this summer goes to a controller which provides a PID loop to control the metering orifice on the sulfonator. The motor drive mechanism which operates the SO_2 metering valve is capable of changing the feed-rate from zero to maximum in 15 seconds. The deadband on the controller is equivalent to 0.1 mg/l residual. The speed of response of the SO_2 control valve in the sulfonator and the 60-second loop dead time at 42-mgd average daily flow are largely responsible for the success of this system.

EXAMPLE. A hypothetical example shows how the system works. Refer to Fig. 10-6. Assume the dechlorination system is not in operation and the chlorine residual exiting the contact basin is 4 mg/l, so $2C_M = 8$ or $C_M = 4$. Therefore the signals going to the summer are 8 and 4 and the signal going to the controller is +4, calling for SO_2. The operator starts the sulfonator and it overdoses to produce 1 mg/l of sulfite ion so that postresidual is −1 mg/l. From Eq. (10-15), $2C_M = +3$. Going to the summer are a +3 and a −4 (to be subtracted), which produce a −1 signal to the controller. The 4–20 mA output of the summer is calibrated so that a 4 mA signal represents a −1 mg/l chlorine residual signal and 20 mA represents +4 mg/l. This corresponds to the 0–5 mg/l range specified for the analyzers. The controller reduces the SO_2 feed rate to produce a zero residual in the dechlorinated sample, hence $C_M = (4 + 0)/2 = 2$ and $2C_M = 4$. Therefore the two signals going to the summer are 4 and the output of the summer is a zero signal.

Biofouling. Operating personnel have found it imperative to control biofouling in the dechlorinated sample line. They have found that a 100 lb/day chlorine feed rate for 10 minutes daily has effectively controlled the microbiological growth in the dechlorinated sample line. This has also eliminated the need for extensive daily analyzer cleaning and has resulted in much better reliability. Analyzer cleaning has been reduced to a weekly basis.

Reliability Provisions. Two levels of backup are provided, since this system depends upon the reliable operation of two chlorine residual analyzers. These are shown in the schematic Fig. 10-7. They are illustrated separately from the zero residual scheme to avoid any confusion.

The first level of backup is a feed-forward control loop. The signal that drives the SO_2 rate valve positioner is switched to the plant effluent flow signal. Then

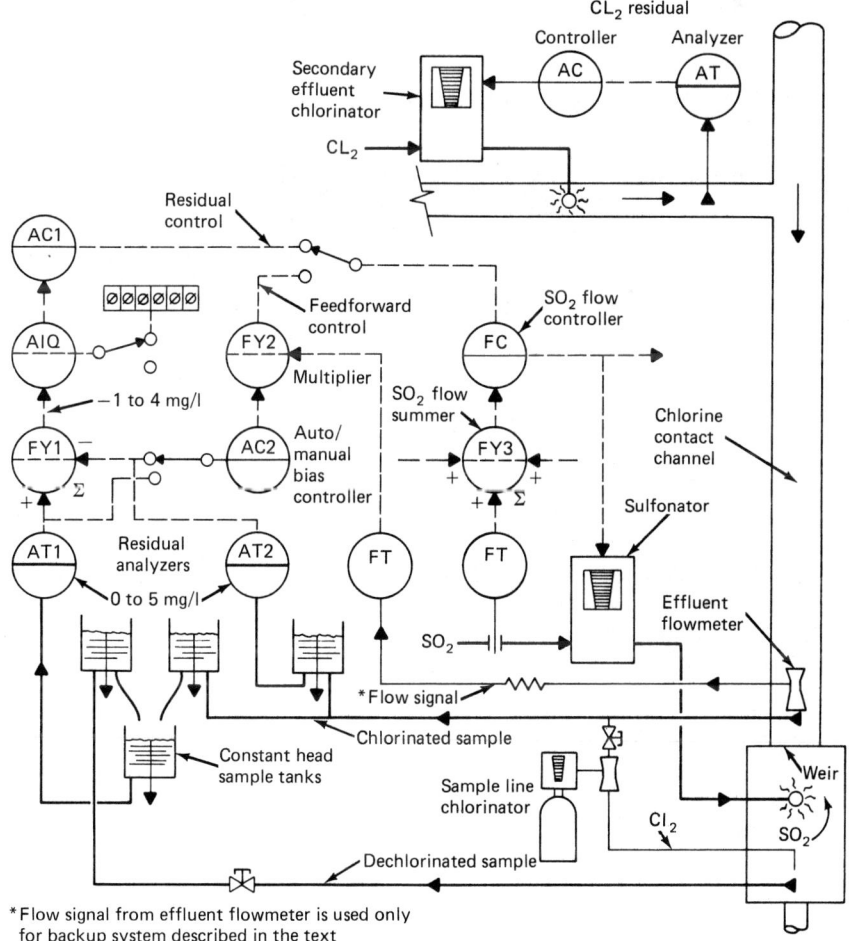

Fig. 10-7. Renton zero residual dechlorination system showing backup systems arrangement.[23]

either analyzer output (using 100 percent presulfonation sample) can be combined with the plant flow signal to generate a sulfur dioxide feed rate signal to the SO_2 rate valve positioner.

In the event that the presulfonation analyzer malfunctions, the postsulfonation analyzer signal can be used by shutting off the dechlorinated sample to the constant head tank. This converts the postsulfonation analyzer to a presulfonation analyzer. This output signal of the converted analyzer is then switched to the auto/manual bias controller by means of a selector switch.

The second level of backup is comparable to the first level but does not require the use of either analyzer. With the loss of both analyzers a simulated analyzer output signal can be plugged into the bias auto/manual station shown in Fig. 10-7. This signal is manually set and is based upon the control setting of the analyzer output that controls the chlorinator feed rate.

EXAMPLE. Chlorinator control analyzer set point is 3 mg/l and operating experience indicates chlorine residual die-away in contact chamber at peak daily flow is 0.5 mg/l. Operator manually sets simulated analyzer output signal at 2.5 mg/l to insure complete dechlorination. This is somewhat similar to the Miks system except no analyzer is involved.

While neither of these backup control systems provides a measurement of the dechlorinated effluent residual, they assure that the dechlorination requirement can be met.

A third analyzer was considered for additional backup for the two on-line analyzers. However, operating experience proved that repair of a malfunctioning analyzer would be quicker than bringing the standby analyzer on line.

ORP Control. This method has been used with some success for monitoring a dechlorination process where the chlorine must be completely removed, as in the case of dechlorination of process water going to demineralizers (chlorine residual will eventually damage ion exchange resins). The practical value of ORP control of chlorination or dechlorination is in qualitative and not quantitative control.

The effectiveness of this control method has a bearing on the type of chlorine residual present. In the case of combined chlorine residual alone, the response between zero residual and some residual is rapid, but in the case of free chlorine residual in the presence of combined residual, up to 30 minutes or more is required to return to the base potential level when all of the residual is removed by a reducing agent such as sulfur dioxide. This is known as the poisoning effect of free residual chlorine on the electrode system. This is known to occur with both platinum and gold to silver–silver chloride electrode systems.

Operating installations have demonstrated that if only combined chlorine is present, a control point for complete dechlorination at about +300 mV would give rapid response. If only free chlorine is present, or if only free chlorine is to be removed in the presence of combined chlorine, a control point at +550 mV would give rapid response. Where both are present and complete dechlorination is required, it would be necessary to add enough ammonia to the sample water only of the ORP cell to convert all of the free chlorine to combined residual to prevent poisoning of the electrode. The control setpoint then would be selected for combined chlorine residual.

Therefore, the ORP concept can be used for controlling certain types of dechlorination processes. This method might have some practical value for controlling dechlorination of cooling water; however, it is doubtful that it could succeed as a control strategy in wastewater dechlorination.

ORP control is the method of choice in the control of cyanide destruction. See Chapter 7.

Orion Sulfur Dioxide Electrode. This specific ion electrode was first announced about 1973.[24] This is a gas-detecting electrode which senses the level of dissolved sulfur dioxide in aqueous solutions. Nevertheless it is used as a specific ion electrode. This electrode can measure sulfur dioxide in aqueous solutions, and sulfite and bisulfite after conversion to sulfur dioxide by acidifying the sample to pH 1.5. The range of detection of this electrode is linear on a semi-log plot over three decades of concentration between $2 \times 10^{-6}M$ and $2 \times 10^{-3}M$ (0.128–128 mg/l).

This is a laboratory instrument. The sample solutions must be prepared before detection can begin and the electrode must be calibrated several times a day. SO_2 concentrations are obtained from the calibration curves.

There is no known application of this electrode on a continuous flowing sample. As of 1984 it has not been made available for continuous monitoring of dechlorinated wastewater effluents. Nor has any other techniques for measuring sulfite ion been made available for this purpose.

Monitoring. As can be seen from the description of the control strategies above, there is no way to provide continuous monitoring of the dechlorinated "residual" for feed-forward control systems. The most popular method in California, where all wastewater residual requirements are for zero residual, is the intermittent recording of the dechlorinated sample for a brief interval (5–7 min) each hour.

This method requires a three-way valve on the sampling line inlet to the chlorine residual analyzer which controls the sulfonator feed-rate.* The dechlorinated sample pump operates continuously. A time clock operates the three-way valve at the analyzer which allows the dechlorinated sample to be diverted from waste to the amperometric cell for a brief interval each hour. This is usually sufficient to provide convincing evidence of a continuous zero chlorine residual. However, the surveillance people are not completely satisfied with this system.

Sulfonator control during the intermittent sampling period of the dechlorinated sample is accomplished as follows:

---

* This system is illustrated on Fig. 10-3.

During the brief verification interval to determine zero residual, a sample-and-hold instrument would continue to generate a signal equal to the last known value of chlorine residual. This allows the system to respond to flow changes. At the end of the verification, the interval timer would again select a sample from the wastewater before dechlorination. However, another timer would delay resumption of normal control (release of sample and hold timer) until the analyzer had been purged of the dechlorinated sample.

Another method tried at a few treatment plants was the use of two analyzers arranged to switch sequentially from the chlorinated sample to the dechlorinated sample at an interval controlled by a time clock. It was thought that the analyzer receiving the dechlorinated sample could maintain its calibration for a period of one or two hours. This system was a failure. It was discovered that the analyzer would drift out of calibration within a few minutes after the dechlorinated sample started through the cell. It was also demonstrated that the higher the excess SO_2 residual the quicker the cell lost its calibration. The maintenance and operating personnel found it impossible to solve the calibration problem.*

To summarize, the bias sample and the zero residual–Renton system described above are the only methods used to date which can provide continuous monitoring of the dechlorinated sample.

Injector Systems

Water Requirement. Calculations of injector water requirements are made on the same basis as those for chlorine. Use the manufacturer's injector curves for chlorine for calculating injector operating requirements for SO_2.

Materials of Construction. The SO_2 vacuum line and solution line require the exact same materials as for chlorine with one exception: the sulfur dioxide solution diffuser can be made of 316 S.S.

Diffusers. Diffuser location is less critical for SO_2 solution than for chlorine because SO_2 gas is several times more soluble in water than chlorine. Therefore the danger of off-gassing due to turbulence at or adjacent to the diffuser is nonexistent with SO_2 application, whereas it is a serious problem with chlorine. For example, the SO_2 diffuser can be placed immediately adjacent to but upstream from a weir and utilize the turbulence on the downstream side for immediate mixing without fear of releasing SO_2 vapor due to the turbulence.

Mixing. The application of sulfur dioxide to wastewater differs considerably from the application of chlorine. The reaction is inorganic in nature. The $SO_3^=$ ion reacts with the chlorine residual to convert the active chlorine to chloride

---

*White recently found an exception at the Fairfield-Suisun WWTP where the analyzer measures a sample containing a 0.5 mg/l excess SO_2 and maintains its calibration within 0.1 mg/l over a 24 hr period. The analyzer is calibrated daily.

and the sulfite ion ($SO_3^=$) to sulfate ion ($SO_4^=$). This reaction takes precedence over any side reaction that might be encountered in wastewater. This reaction with chlorine occurs in a matter of seconds. Probably 15–20 seconds at the most. Mixing of the sulfur dioxide solution to achieve dechlorination is much less demanding than the mixing requirements for chlorine to achieve disinfection.

Mechanical mixers are not required if the structure is amenable to the required diffuser design. Examples for closed surcharged conduits and open channels for chlorine application are illustrated and described in Chapters 8 and 9. It is recommended that SO_2 solution diffusers be designed after the Egan concept of nozzle contraflow to produce a velocity gradient of 250–300.

Contact Time. Since the reaction time for the complete conversion of the total chlorine residual to chloride by the sulfite ion is on the order of seconds, there is no need for any contact chamber between the SO_2 diffuser and the receiving waters. However, common sense dictates that the structure downstream from the diffuser must be suitable for obtaining a dechlorinated sample about 5 feet downstream from the estimated mixing zone.

Sample Lines. The dechlorinated sample line must be kept free from biological slime growths. These lines carrying dechlorinated effluent will be subject to a rapid development of organic debris and slimes throughout the transport system. Any such biofouling introduces errors in the analyzer system whether it be for control or for monitoring.

Two methods are capable of removing and/or controlling these growths: intermittent purging with either chlorine or caustic. Chlorine is the logical choice. It should be applied into the sample line collection pipe. See Fig. 10-6. This should be made available for every chlorine residual analyzer.

All sample lines should be sized to provide line velocities up to 5 ft per second. This acts to keep the line scoured and acts to decrease the dead time in the control loop.

The sampling point for coliform bacteria population measurement for proof of disinfection should be 5–10 ft upstream from the SO_2 diffuser. Otherwise effects of coliform regrowth after dechlorination will distort proof of disinfection.

Housing. The designer should use the same guidelines for housing SO_2 equipment as those used for chlorine. These design provisions are based upon safe operating conditions for personnel, which include but are not limited to space, ventilation, protection from fire, location of wind socks, door hardware, inspection windows, heating, electrical controls, drains, SO_2 vents, and evaporator vents. These considerations are described in detail in Chapter 9.

Separation. Chlorine gas and sulfur dioxide gas should never be allowed to come in contact with each other. Therefore it is imperative that the supply system piping should be so located and arranged that such an occurrence could never happen.

The most positive way to accomplish this is to have completely separate areas for the supply systems and house the sulfonators and evaporators in a separate room from the chlorination equipment.

Reliability. The dechlorination system can fail for the same reasons which cause failure of the chlorination system. Therefore the reliability provisions described in Chapter 9 for chlorine are generally applicable to sulfonator systems.

DECHLORINATION FACILITY DESIGN: BISULFITE SOLUTION

General Discussion. Sodium bisulfite solution is the preferred chemical if for any reason sulfur dioxide is not being used. In large systems this is usually because of potential hazards of SO_2 vapor. For small systems it would be due to the difficulties that surround the metering of small quantities of SO_2 vapor—less than 10 lb/day.

Design Considerations. The components and their arrangement and selection are almost precisely the same as for hypochlorite facilities. These details are discussed in Chapter 2. In general there are three types of systems: (1) pumped, (2) gravity, and (3) eductor. Each of these systems provide the energy necessary to move the solution from the point of storage to the point of application.

In practically all cases the pumped system is preferred. The pumped system has two variations: first is the positive-displacement-type metering pump system which is controlled solely by the pump, and second is the centrifugal-type pump system which is controlled by a modulating control valve on the downstream side of the pump. These systems are discussed in Chapter 2.

Designer's Check List:

Positive-Displacement Pump System. The following are the vital components for this system:

1. Diaphragm pumps with back-pressure valves and accumulators on both the suction and discharge to dampen pulsation effect
2. Magnetic flowmeter or rotameter to continuously monitor bisulfite feed rate with a remote recorder
3. A flow pacing signal from treatment plant effluent flowmeter and a 3 or 4 to 1 ratio relay
4. A monitoring system that will provide zero residual control by the use of chlorine residual analyzers.

These systems are discussed in this chapter.

Centrifugal Pump System. The components for this system are slightly different from the diaphragm pump system:

1. A centrifugal-type pump
2. A modulating control valve which controls the pump discharge and meters the bisulfite solution
3. A rotameter type flowmeter with an indicating transmitter and remote flow recorder to monitor the flow of bisulfite
4. A flow pacing signal from the treatment plant effluent flowmeter and a 3 or 4 to 1 ratio relay
5. A monitoring system with the same features as for the diaphragm pump system

Chemical Storage System. Regardless of the type of system a storage vessel will be required. These tanks should be constructed of reinforced Fiberglas and should be equipped with a remote type of level sensing instrumentation, atmospheric vent, drain, pump suction, and sample tap for laboratory use in checking solution strength.

Most important is the piping configuration and safety valving used to load the tank. The objective is to prevent any other chemical from being loaded into this tank. Preferably this tank should be a discreet distance from the hypochlorite tank or other chemical tanks. Mixing bisulfite with hypochlorite produces a violent temperature reaction. Other mixtures may do the same or release a cloud of SO_2 vapor.

Design Calculation. The important mathematical relationship is: each gallon of 38 percent sodium bisulfite contains the equivalent of 2.17 lb sulfur dioxide.

REFERENCES

1. Chen, Ching-Lin, and Gan, H. B., "Wastewater Dechlorination State-Of-The-Art Field Survey and Pilot Studies," Municipal Environ. Res. Lab., EPA-600/S2-81-169, Cincinnati, OH, Oct. 1981.
2. Anon., "Hazards of Sulfur Dioxide and Caustic Soda, A Committee Report," *J. AWWA,* **31,** 489 (Mar. 1939).
3. Baker, R. J., "Maximum Withdrawal Rates from Chlorine, Sulfur Dioxide and Ammonia Cylinders," *AWWA Op Flow,* p. 4, Apr. 1980.
4. Anon., "Sodium Bisulfite 38% Solution Data Sheet," Dupont Industrial Chemicals Dept., Wilmington, DE, Sept. 1975.
5. Vaughn, J. C., private communication, Southside Filter Plant, Chicago, IL, June 1968.
6. Baker, R. J., "Dechlorination Ahead of Demineralizers and Similar Applications," paper presented at 23rd Int. Water Conf., Pittsburgh, PA., October 24, 1962.
7. Hager, D. G., and Flentje, M. E., "Removal of Organic Contaminants By Granular Carbon Filtration," *J. AWWA,* **57,** 1440 (Nov. 1965).
8. Suidan, M. T., Snoeyink, V. L., and Schmitz, R. A., "Reduction of Aqueous HOCl with Activated Carbon," *J. Env. Engr. Div. ASCE,* **103,** 677 (Aug. 1977).

9. Bauer, R. C. and Snoeyink, V. L., "Reactions of Chloramines With Active Carbon," *J. WPCF*, **45,** 2290 (Nov. 1973).
10. Baker, R. J., private communication, Wallace and Tiernan, Div. of Pennwalt Corp., Belleville, NJ, Nov. 1980.
11. Magee, V., "The Application of Granular Activated Carbon for Dechlorination of Water Supplies," *Proc. Soc. Water Tr. and Exam.* (England), **5,** 17 (1956).
12. Miller, R., and Hartman, D. J., "Feasibility Study of Granular Activated Carbon Adsorption and On-Site Regeneration," Municipal Envir. Res. Lab., EPA-600/S2-82-087, Cincinnati, OH, Nov. 1982.
13. Strockbine, W., "Superchlorination with Dechlorination by Ferrous Sulphate," *J. AWWA*, **32,** 1176 (1940).
14. Howard, N. J., and Thompson, R. E., "Chlorine Studies and Some Observations on Taste Producing Substances in Water and the Factors Involved in Treatment by the Super- and De-chlorination Method," *J. NEWWA*, **40,** 276 (1926).
15. White, G. C., "Chlorination and Dechlorination: A Scientific and Practical Approach," *J. AWWA*, **60,** 540 (May 1968).
16. Esvelt, L. A., Kaufman, W. J., and Selleck, R. E., "Toxicity Removal from Municipal Wastewater," SERL Report No. 71-7, Sanitary Engineering Research Lab., Univ. of Calif., Berkeley, CA, Oct. 1971.
17. Brungs, W. A. "Effects of Residual Chlorine on Aquatic Life," *J. WPCF*, **45,** 2180 (Oct. 1973).
18. Anon., "Toxic Gas Monitoring Instrumentation," Interscan Corporation, Chatsworth, CA, 1984.
19. Anon., TOX-ALARM Sulfur Dioxide Detector Model No. 5183 Instruction Manual, Exidyne Eastern Div., Exton, PA, Feb. 1983.
20. Miks W. D., personal communication, Water Pollution Control Plant, Palo Alto, CA, 1979.
21. Calmer, J. C., private communication, Kennedy/Jenks Engrs., San Francisco, CA, 1980.
22. Adams, R. M., Calmer, J. C. and Nourse, J. C., "Dechlorination Process Control Using a Biased Dechlorinated Sample," paper presented at 55th Ann. Conf. Calif. Water Poll. Control Assoc., Palm Springs, CA, May 4–6, 1983.
23. Finger, R. E., Harrington, D., and Paxton, L. A., "Development of an On-Line Zero Chlorine Residual Measurement and Control System," paper presented at Water Pollution Control Federation 56th Ann. Conf. Atlanta, Georgia, Oct. 4, 1983.
24. Anon., "Instruction Manual, Sulfur Dioxide Electrode Model No. 95-64," Orion Research Inc., Cambridge, MA, 1973.

11
Operation and Maintenance of Chlorination and Dechlorination Equipment

OPERATION: CHLORINATION EQUIPMENT

PHYSICAL FACILITIES: INVENTORY

The operator should have available for reference the design engineer's report delineating the design criteria. This information is vital to the overall operation of the system because it specifies the capabilities of the equipment. Of equal importance is the manufacturer's instruction books, one for each specific piece of equipment. It is imperative that the operator take the time to familiarize himself with the content of these books. Since the manufacturers always provide supervision of installation the operator should make it a point to have the manufacturer's representative clarify any specific features of the equipment not adequately described in the instruction books.

If the chlorination system is part of a treatment plant complex the following items should be available to the operator:

1. Operation and Maintenance manual
2. Contract plans and specifications
3. A complete record of each piece of equipment including name of manufacturer, manufacturer's representative (if not purchased directly from the manufacturer), identifying number (model), serial number, rated capacity, dates of purchase and date of installation

SUPPLY SYSTEM PREPARATION

Introduction. The start-up of a new system or one that has been out of service for a considerable length of time is the most critical period of operation. Various factors contribute to this situation. One common cause is the failure of the pipe fitters to clean up the pipe after threading and/or the use of improper thread lubricant. Another more serious situation is the accumulation of moisture in the system. This occurs in those instances where the system stands idle during construction for a long period of time with the supply system open to the atmosphere.

In one instance severe damage to a new system resulted from the accumulation of approximately 3 inches of water in the bottom of the evaporator chlorine container vessel. Therefore certain precautions must be taken to ensure that the entire supply system is clean, dry, and gastight.

The recommended precautions are usually part of the contract specifications which requires the contractor to clean and dry the chlorine supply system and test it for leaks before any chlorine is allowed in the system. The procedure for cleaning, drying, and testing is outlined in the maintenance section of this chapter. The discussion on operation will assume that all of chlorine supply piping between the containers and chlorinators, and the evaporators are clean, dry, and gastight.

Small Systems. Installations that do not fall into the above category, such as those that involve less than 100 ft of pipe and do not utilize evaporators, are required only to thoroughly clean the pipe threads (before assembly) with trichlorethylene, or other suitable chlorinated solvent. Never use hydrocarbons or alcohols. The residual solvent may react adversely with the chlorine.

Following this, some suitable thread lubricant should be carefully applied to each joint. Teflon tape is satisfactory for pipes up to 1 inch in diameter. Other lubricants can be: John Crane Plastic Lead Seal No. 2; a mixture of linseed oil and graphite; a mixture of linseed oil and white lead; or a mixture of litharge and glycerine. The latter must be mixed to a heavy, syruplike consistency from litharge powder and glycerine. Since it hardens or "sets up" quite rapidly, it must not be made up too far in advance of its use. While this mixture does harden in a short time it should never be relied upon as a joint sealant. There is no substitute for good threads in a chlorine or sulfur dioxide piping system.

Obviously if the supply piping is assembled with welded, forged steel fittings, thread cleaning and lubrication is unnecessary.

STARTUP (GAS SYSTEM)

Four Important Steps. The most critical period in the operation cycle occurs when a system is put into operation for the first time or after a prolonged shutdown.

1. The very first step is to start the injector water system and make certain that the hydraulic conditions are satisfactory. This is done by checking the injector supply pressure gage and the injector vacuum gage. If the conditions are satisfactory, the vacuum gage should show a reading above 10 in. Hg.*

If the chlorinator is not equipped with a vacuum gage, remove the tubing at

---

* There are some exceptions to this figure, to be explained later.

the injector vacuum inlet and place a hand over the opening while the injector water is running. If the injector is performing properly, the suction will be felt instantly on the portion of the hand over the opening. This will be a strong suction requiring some effort to remove the hand. If the suction is feeble, the hydraulic conditions are not proper and should be investigated further.

When the injector system is operating properly, the chlorine gas may be turned on, but before this is done the chlorine metering orifice should be in a partially open position, approximately 25 percent of maximum chlorine feed rate. To accomplish this, the metering valve control knob should be rotated to approximate this feed rate. If the chlorinator is arranged for some type of automatic operation, it must be disconnected from this mode of action and placed in manual control position.

2. The next step is to verify that all of the tubing, manifold, and auxiliary valve connections are correct and that all union joints are properly gasketed.

3. Next, check to see that all chlorine valves on the supply line to the chlorinator are closed.

4. Then "crack" open the chlorine container valve and, with a bottle of ammonia, check all the joints between this valve and the next one downstream to make sure that there are no leaks. If there are none, fully open the next valve downstream and proceed as before. Continue until chlorine gas pressure is showing on the chlorinator gage.

If there are no leaks, then the chlorinator is ready for further testing—that is, for rangeability, automatic control, and so on.

Procedure When a Leak Occurs. If there is a leak, the first step is to close the cylinder valve and relieve the entire supply system of chlorine pressure. NEVER ATTEMPT TO REPAIR A LEAK BY TIGHTENING A PACKING GLAND, PIPE FITTING, OR ANYTHING ELSE WHEN GAS OR LIQUID PRESSURE EXISTS. THIS LEADS TO DISASTER. If the cylinder valve is only cracked open, it can be closed quickly, which is the purpose of this procedure. Next open all other valves wide, increase the chlorine feed rate to maximum. This will reduce the pressure to zero very quickly. Repair any leaks which result from a poor fit at a gasketed joint. This might require, in addition to a new gasket, refacing the cylinder valve seat with a flat file. When this is accomplished, the system is ready to be started again.

A gross leak, such as that caused by a missing gasket or hole in a fitting, is another matter. Under these conditions, the operator should close the main cylinder valve and retreat to a safe area where the oxygen breathing apparatus is located, don this equipment, and return to perform the gas evacuation procedure through the chlorinator.

In small installations, where only one flexibile connection is involved, the amount of gas which might leak out will be gone a few minutes after the cylinder valve is closed, and so the operator can return to correct the leak without the need of a gas mask in most cases.

Checking the Chlorine Supply System. On many occasions there can be an unforeseen obstruction in the supply system between the chlorine cylinders and the chlorinator. Therefore, after all the leaks have been corrected, verify that the chlorinator will reach its maximum capacity as specified. This is the most important criterion of any chlorinator installation. If it will not perform to capacity, the automatic mode of operation will be unreliable and inaccurate.

If the injector system is providing a stable 15 in. Hg vacuum and the chlorine feed rate does not reach the maximum reading on the rotameter scale, then the difficulty is most likely in the supply system—such as a restriction due to a buildup of impurities from the chlorine gas. Location of the obstruction can be done as follows: If the obstruction is upstream from the chlorinator inlet pressure reducing valve, the gas pressure on the chlorinator gage will fall off rapidly when an attempt is made to reach maximum feed rate. If the obstruction is in the chlorinator inlet pressure-reducing valve, there will be no change in the chlorine pressure gage at the chlorinator. To verify that the inlet CPRV is restricted, close the chlorine gas inlet valve to the chlorinator. If the gage pressure does not fall off rapidly then the obstruction is in the inlet pressure-reducing valve in the chlorinator.

Checking the Vacuum Relief System. It is easy to verify whether the fault is due to a weak spring in the vacuum relief device. While the chlorinator is operating, disconnect the vent line and plug the opening of the vacuum relief device. If the spring is weak it is allowing air into the chlorinator. When the vent is plugged (with a thumb) the rotameter indicator will immediately jump to a higher position, showing an increase in chlorine feed rate. Solution: Replace spring in the vacuum relief device.

Insufficient Vacuum: Checking the Injector System. If the chlorinator vacuum is not sufficient (less than 10 in. Hg) the difficulty is usually traceable to the injector system, which includes the injector operating water supply, the chlorine vacuum line, the chlorine solution line, and the diffuser system. The difficulty will consist of one or more of the following:

1. Vacuum leak in chlorinator or vacuum line to injector
2. Insufficient water pressure
3. Excessive friction loss in chlorine solution line caused by inadequate pipe size
4. Too much head loss in diffuser due to insufficient holes, or holes too small
5. Too much static head downstream from the injector
6. A restriction in the injector throat and/or tailpiece

A compound gage with a protector diaphragm installed immediately downstream from the injector discharge will verify items 3, 4, and 5. The manufacturer's injector curves must be consulted to see if the back-pressure on the injector as indicated by the above gage is within the limits for the existing operating conditions. This information together with a hydraulic analysis will identify the problem.

Injector Hydraulics; Effect of Air and Gas Binding. Another factor that contributes to excessive friction loss in the chlorine solution line is the phenomenon of air and gas binding. This problem is most likely to occur in solution lines longer than 100 ft that terminate in a closed conduit under some pressure.

There are two situations in which air and gas binding will reflect a vacuum of less than 10 inches Hg on startup under satisfactory hydraulic conditions:

1. In those installations where the injector is more than 300 feet from the chlorinator, the vacuum line between the chlorinator and the injector will become filled with air during startup. If the line is long, this large amount of air reduces the injector vacuum, sometimes to less than 5 in. Hg. The situation gradually corrects itself as chlorine gas displaces the air. The problem is caused by the fact that the air will not go into solution as does chlorine, and this greatly impairs the efficiency of the injector.

2. The same thing occurs in long solution lines. The air taken into the system via the vacuum relief system will reflect a deterioration of the injector vacuum. But when the chlorine is turned on and all the air is purged from the solution lines, the injector vacuum will return to normal.

Another phenomenon results from gas binding in the chlorine solution line. It should be explained, however, that the effect of air or gas binding in the solution line is to increase the back pressure at the injector, thereby reducing its efficiency. At the injector there is a direct reaction by chlorine and water to produce a release of carbon dioxide in proportion to the alkalinity of the water. Other gases are also released, owing to the raction between chlorine and the ammonia ion (NH_3) in the water. This is particularly pronounced in wastewater practice, where the effluent is used for injector water. These gases are forms of nitrogen resulting from the breakpoint reaction by chlorine. They have about the same solubility as carbon dioxide and so are expected to be released under these conditions. Copious quantities of such gases have been observed in various glass metering devices on chlorine solution lines. This phenomenon is characterized by a surging in the chlorine solution line. Elimination of this situation calls for patience on the part of the operator. The injector should be adjusted to give the maximum amount of operating water at the least back pressure. This will provide maximum flushing capability at maximum injector efficiency. Each trial adjustment should be allowed to proceed for several hours before changing to another adjustment. The predominant symptom of this gas binding condition is a fluctuation on the vacuum gage from 5 to 15 in. Hg at a frequency of about once every two seconds or less. At the same time, the chlorine feed rate may oscillate as much as 10 to 20 percent of scale, or even more, depending on how near to maximum the feed rate is when the fluctuation is greatest. During the experimental adjustment of the injector, try to keep the chlorine feed rate at about 50 percent of scale. The line will flush itself faster at this feed rate.

The best way of avoiding such a situation is to keep the chlorine solution lines as short as possible, locating the injectors remotely, keeping the solution lines at such an elevation as to reduce the injector back pressure, and to add 15 to 30 percent more injector water pressure for any solution line in excess of 100 feet if the injector is larger than 2 inches. Injector systems utilizing the smaller sizes—that is, one-inch and two-inch—do not display the magnitude of surging noted in the three-inch and four-inch systems.

Vacuum Leaks. One other sure sign of trouble, either from injector hydraulics or a vacuum leak, is the behavior of the vacuum gage when the chlorine feed rate is changed from minimum to maximum. If the vacuum at the lower feed rate is 20–22 in. Hg and if it deteriorates significantly (to 13–15 in. Hg) with increasing feed rates, this signifies trouble in the system and should be investigated.

The chlorinator itself can be a source of trouble if there is a vacuum or an air leak within the unit. If there is a sizeable vacuum leak it can be discovered by shutting off the injector water suddenly and applying ammonia on all the joints. This sudden removal of vacuum within the chlorinator momentarily puts it under a slight positive pressure, and chlorine will be expelled into the atmosphere. Very small vacuum leaks will not show up in this procedure.

The obvious points to check are the various tubing connections, which may have loosened in shipment. Next, the O-rings that seal the metering tube may be defective or have been damaged during installation. Next, check the vacuum relief device, which sometimes arrives with a defective spring or a scored seat and so can let air into the control unit, thereby impairing the vacuum. This leak can be determined while the chlorinator is in operation, as follows: Put the chlorinator on manual control at about 50 percent feed rate, disconnect the vacuum relief vent line, and place your thumb over the opening. If there is a leak, the rotameter will "jump" to a higher reading.

STARTUP (LIQUID SYSTEM)

Evaporator:

General Discussion. The procedure for start-up on a system using liquid chlorine is generally similar to that using gas, but one big difference is in the role of the evaporator.

The evaporator is an extension of the chlorine container system. Whatever happens in the container is reflected in the evaporator. The danger existing in a liquid system is the possibility of trapping chlorine liquid in a pipeline. If this occurs and there is a significant ambient temperature rise, the liquid chlorine will expand and rupture the pipe if the expansion of the liquid stresses the pipe beyond its ultimate tensile strength. For this reason, the liquid line between the evaporator and the chlorine supply system should always remain open while the evaporator is operating.

The first step preparatory to starting up a liquid system is to verify that the system is dry.* This can be done by heating the water in the evaporator and passing dry air ($-40°F$ dew point) through the evaporator cylinder and all the chlorine lines between the containers and chlorinators. This may take several hours. It is essential to do this, because evaporators may sit idle on the job site for many months, accumulating moisture while the construction is being completed. This moisture collects in the bottom of the chlorine cylinder. If the moisture is not removed, the resulting formation of ferric chloride will pass through the chlorine control mechanism with disastrous results. Whenever this occurs, the entire chlorine supply system must be flushed with water and thoroughly dried with steam and hot, dry air. In addition, the chlorination equipment must be dismantled and cleaned. Care in cleaning all threaded pipe joints, to remove cutting oils and other foreign matter, and the use of Teflon tape instead of other joint lubricating or sealing compounds, will prevent the production of chlorine-induced impurities which cause severe maintenance problems.

When the operator is convinced the chlorine supply system is clean and dry, the next step is to start the evaporators. This is done by filling the water bath and adjusting the control devices so that the water level rides between the limits established by the manufacturer.

Most evaporators are heated by electric immersion heaters; however, recirculated hot water or live steam is sometimes used to heat the water bath. Never energize the electric immersion heaters until there is the proper amount of water in the evaporator. Usually this is provided for in the electric controls by an overriding circuit.

When the water bath temperature reaches 150°F, the chlorine pressure-reducing and shut-off valve will open, and the system is ready to operate; however, the operator must wait until the water bath reaches its maximum temperature and then adjust the controls so it operates between 170 and 180°F.

Now the system is ready to start. First, start the injector water system and proceed exactly as in the start-up procedure, using gas; then follow the same procedure to check for chlorine leaks.

Always start a liquid system first with gas. If the system is started on liquid and there is a leak, many times more gas is released at the leak: one volume of liquid chlorine is equivalent to 456.8 volumes of gas. Furthermore, it takes that much longer to empty the system of liquid to repair the leak.

---

* See the "Maintenance" section in this chapter for further details on dealing with this problem.

Once the system has been checked with gas for leaks, it is then ready to be switched to liquid and ready for control adjustments and calibration.

Cathodic Protection System. After the water bath is filled to the proper level add ¼ lb. (max.) sodium sulfate (Na_2SO_4) to increase the conductivity of the water.[1] The instrument panel ammeter should be observed after the sodium sulfate has been dissolved in the water bath. When a new evaporator is placed into operation for the first time the ammeter reading may be very low or zero. However, if any movement above zero is attained by turning the current knob clockwise, the cathodic protection system is satisfactory. A higher reading is not attained initially because of the insulating effect of the paint finish on the outside of the liquid chlorine container vessel. As this paint finish deteriorates with time a higher ammeter reading will be observed.

If the ammeter continues to read zero, check the cathodic system electrical connections. If after some months' operation, when the ammeter needle has moved into the midrange of scale the needle decreases substantially, *check the anodes for their condition.* Whenever the ammeter reads in the high range (250 mA or more), turn the current knob counterclockwise to bring the needle near to the midpoint of the scale. A reading of 50–250 mA, or in the green area, is satisfactory—except at initial operation, when a low current flow is both normal and satisfactory. See the "Maintenance" section of this chapter for routine inspection of the cathodic protection system.

Superheat. The proper operation of an evaporator depends upon the water bath temperature that will provide 20°F superheat to the gas being vaporized. This condition is dependent upon the vapor pressure of the liquid. For example, read the liquid vapor pressure on the evaporator instrument panel and determine the corresponding liquid temperature from the chlorine vapor pressure curve. Then read the chlorine gas temperature of the gas exiting the evaporator. This gas temperature should be kept 20°F higher than the corresponding liquid temperature obtained from the chlorine vapor pressure curve. When it is no longer possible to achieve 20°F of superheat, the chlorine container vessel in the evaporator must be cleaned or the immersion heaters must be replaced.

Excessive Chlorine Pressure. Another condition that may be encountered in the start-up or routine operation of a liquid system is that of abnormally high pressure in ton containers or tank cars. In a tank car, high pressure is caused by air padding; but in ton containers, it results from exposure to heat, such as sunlight. Tank cars are insulated. The usual pressure range of ton containers is 85 to 100 psi; tank cars usually arrive with some air padding so as to show 125 psi. Whatever the incoming pressure, the evaporator efficiency will be greatly improved if the tank car or container pressure is reduced to 80 to 85 psi by operating from the gas phase until the pressure drops to this range.

The most significant case is that of the tank car, because they are not only often aid-padded but are filled with liquid chlorine at very low temperatures. This requires more work for the evaporator.

Assume that the temperature of the liquid in a tank car is 60°F and that the temperature of the evaporator water bath is 180°F. Fig. 6 in the Appendix shows that the vapor pressure of liquid chlorine at 60°F is approximately 70 psi without any air padding, and the thermal difference through the chlorine cylinder wall of the evaporator will be 120°F. With this temperature difference, the evaporator can transfer quantities of BTUs and thereby evaporate more chlorine. However, if the tank car happens to be air-padded to the maximum allowable pressure in accordance with the *Chlorine Institute Pamphlet #54*, the chlorine pressure due to air padding can be 190 psi for liquid chlorine at 60°F. The pressure in the evaporator cylinder then becomes 190 psi, instead of 70 psi, owing to the air padding.

Again according to the vapor pressure curve, 190 psi vapor pressure is equivalent to 120°F. Therefore the evaporator "thinks" that the chlorine is at 120°F, so that there is now only a 60°F temperature differential between the water bath and the liquid chlorine, or half as much at the lower pressure; consequently there will be only half as many BTUs transferred, and the evaporator can evaporate only half as much chlorine. If the evaporator is forced beyond this 50 percent capability, liquid chlorine will flash over into the gas phase and cause damage to the chlorinator equipment.

Tank cars should be delivered to chlorinator stations without air padding; but if one arrives padded, it is easy to remove the air padding simply by operating on the gas phase until the pressure drops to around 70 psi. Almost all chlorination equipment can operate to maximum capacity with 20 psi chlorine pressure at the chlorinator. This means that padding is required only when tank car liquid temperature is in the 10°F range, and then only a small amount of padding is required.

In hot ton containers, it is not uncommon for the pressure to be as high as 180 psi (116°F) as a result of sitting in the sun on a hot day. The pressure can be quickly reduced by operating on the gas phase if the rate of withdrawal is somewhat in excess of 500 lb/day per cylinder.

TO STOP OR SECURE AN INSTALLATION

Four Steps. 1. The first step is to shut off the chlorine supply system. If it is to be for a short period, any auxiliary valve near the chlorinator may be used for this purpose. If it is to be for a long time or for major repairs, it is best to shut down at the main container valve. In this way all the chlorine in the system can be removed.

2. When the chlorine pressure gage reaches zero, and if any disassembly of

798 HANDBOOK OF CHLORINATION

equipment is involved, remove the plastic plug usually located in the chlorine pressure-reducing valve assembly (in the chlorinator) to evacuate all the residual chlorine in the machine while the injector system is still running. If it is a small chlorinator, break the flexible connection at the cylinder. The injector vacuum will pull air through the flexible connection, thereby purging the entire system of chlorine. This also introduces moisture into the system from the atmosphere, but this is inevitable in the maintenance and operation of this equipment.

3. After the chlorine has been purged to the satisfaction of the operator, the injector system may be shut down, thus securing the entire installation.

4. At this point, reconnect either the flexible connection or the plastic plug and then proceed with repairs as required.

TO SECURE AN EVAPORATOR

If the chlorination system is to be shut down temporarily for repairs or adjustments on any part downstream of the evaporator, it is necessary only to close the evaporator outlet (gas) valve. (See Fig. 11-1)

Always leave the liquid system intact. Think of the evaporator liquid, the liquid in the chlorine header piping, and the liquid in the storage system as one entity. Never separate one from the other by closing a valve.

In Fig. 11-1 valve *A* represents the shutoff of the storage system; valve *B*, the inlet to the evaporator; valve *C*, the outlet of the evaporator; and valve *D*, the

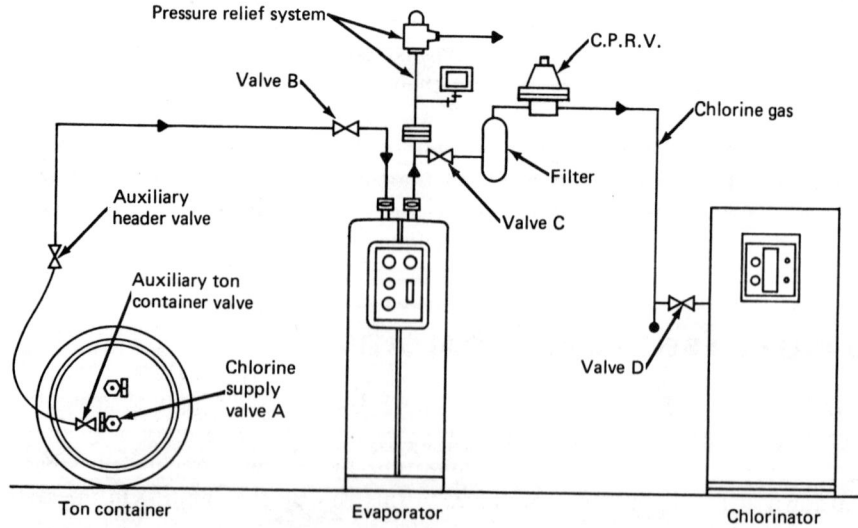

Fig. 11-1. Method used to secure an evaporator.

OPERATION AND MAINTENANCE OF CHLOR-DECHLOR EQUIPMENT 799

inlet supply to the chlorinator. Valves *C* and *D* can be closed at any time and still leave the evaporator operating without any danger. In these instances, there will be a temperature rise in the liquid chlorine (because evaporation has ceased), and this tends to push the liquid back into the containers, thereby equalizing the pressure system.

Whenever it becomes necessary to close valve *B*, do not close valve *C* or *D* until the evaporator has been emptied of chlorine, indicated on the evaporator pressure gage.

If it becomes necessary to close valve *A*, allow the system to empty itself from valve *A* until the pressure at the evaporator reads zero before closing any other valve.

If there is any likelihood that valves *A* and *B* might be closed inadvertently without draining the liquid, then an expansion tank should be installed between these two points.

Fig. 11-2 shows an installation with a long liquid line between the containers and evaporators. This line has valves at either end which automatically close upon power failure, and is protected by an expansion tank. In a power failure, the automatic gas valve downstream of the evaporator also closes, trapping liquid in the evaporator, which heats up to the temperature of the water bath in the evaporator, since no gas is being evaporated.

If the water bath temperature is as high as 180°F, the vapor pressure in the evaporator will rise to approximately 400 psi. This pressure will be exerted on the system between valve *B* and the chlorine pressure-reducing valve. This portion

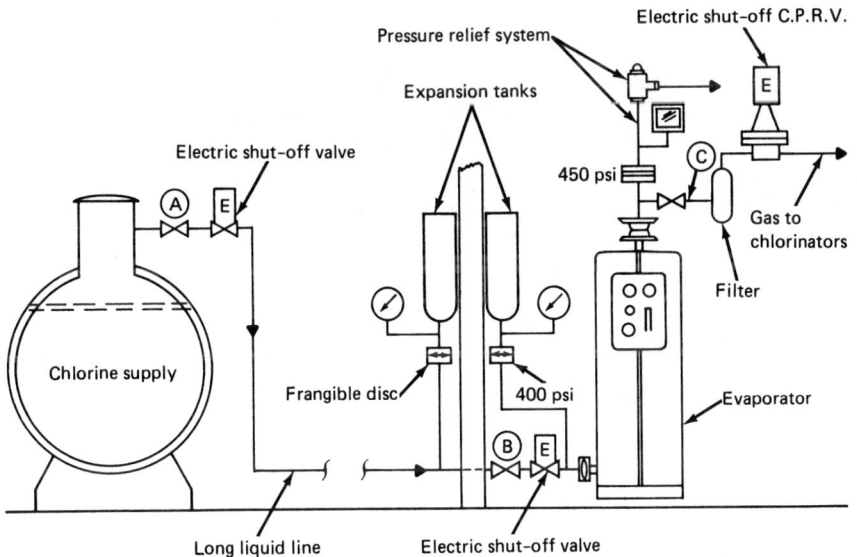

Fig. 11-2. Chlorine supply system with valves that close when power fails.

of the system should be able to accommodate this pressure. However, for extreme caution, this portion of the system can be provided with an alarm system. Connect the incoming liquid chlorine line downstream of the inlet shut-off valve to a frangible disk and expansion tank arrangement so that the disk ruptures at approximately 250 psi. The liquid will flow into the expansion tank connected to an alarm pressure switch set for any significant pressure—for example, 10 to 20 psi. The volume of the expansion tank will be sufficient to lower the pressure of this portion of the system to that of ambient temperature conditions.

CHLORINE SUPPLY SYSTEM

100- and 150-lb Cylinders. Installations using these small cylinders experience the most difficulty from the temperature related properties of chlorine gas.

Never connect a cylinder that has stood in the hot sun for any length of time without allowing it to cool to the temperature of the chlorinator room. This usually takes overnight. If it is connected while hot, the gas between the cylinder and the chlorinator will cool rapidly at night, causing liquefaction of chlorine at the inlet pressure-reducing valve. Impurities in the chlorine gas precipitate out during this liquefaction process and quickly plug up the small ports in the chlorinator.

If a hot cylinder must be connected immediately to the chlorinator, it should be artificially cooled with water. This is accomplished by wrapping the cylinder with absorbent material, such as a blanket or burlap, and soaking with a continuous small stream of water. As a guideline to operators, particularly those in hot climates, when the chlorine pressure gage reaches 120 psi, trouble can be expected. The chlorinator room should be sufficiently insulated, or the cylinder should be artificially cooled, to prevent the chlorine cylinder pressure from exceeding 120 psi. This temperature–pressure problem is accentuated even further in areas where the ambient temperature has a wide fluctuation from low to high in a 24-hour period. This situation exists in desert areas, such as the Far West and Southwest. Insulation against this variation is the proper solution. This problem is also accentuated at those installations where the feed rate is well below the maximum draw-off rate for the chlorine cylinder. The maximum withdrawal rate for a 150-lb cylinder is 40 lb/day at 68°F. Higher rates will produce cooling of the liquid chlorine in the cylinder, thereby reducing the vapor pressure. At a multicylinder installation where the chlorine feed rate is in excess of 40 lb/day, an easy way to cool off a hot 150-lb cylinder is to connect this cylinder by itself to the chlorination equipment and withdraw the gas at rate well in excess of 40 lb/day. As an example, withdrawing chlorine from a 100- or 150-lb cylinder at 100 lb/day will reduce the temperature and consequently the pressure by at least 20°F in 30 min. This temperature–pressure reduction will stabilize at the ambient temperature when other cylinders are connected so that the withdrawal rate is normal again.

Because of this temperature–pressure property of chlorine the sun should never be allowed to shine on any chlorine cylinder discharging chlorine in the gas phase.

Moreover there is a potential danger of a serious leak by allowing cylinders to be exposed to direct sunlight. All chlorine cylinders are equipped with fusible plugs which melt at 158–165°F to prevent cylinder rupture in case of fire. Although the instances are rare, there have been cases of fusible plugs melting from the sun's heat.

Ton Containers. When these cylinders are used for gas withdrawal they behave in the same way as do the 100- and 150-lb cylinders. The maximum rate of gas withdrawal is 400 lb/day at 68°F. These cylinders can be cooled by water as described above or by exceeding the vapor withdrawal rate. When multiple cylinders are used for gas withdrawal it is customary to install a chlorine pressure-reducing valve in the chlorine header adjacent and close to the cylinders. Reducing the chlorine pressure to about 40 psi will eliminate liquefaction in the header and in the chlorinator. Reducing the pressure is like cooling the gas. The liquefaction will occur in the reducing valve, so to minimize fouling of this valve by chlorine impurities which drop out during liquefaction this valve should always be preceded by a filter.

Ton containers have two outlet valves and six fusible plugs. One outlet valve is for gas withdrawal and one is for liquid withdrawal.

It is imperative to position these cylinders so that the two outlet valves line up in a vertical plane. In this position the top valve is for gas withdrawal and the bottom valve is for liquid withdrawal. If these two outlet valves are not aligned vertically, serious difficulties will be encountered. A gas-phase withdrawal system may withdraw the liquid chlorine, which will severely damage the chlorination equipment. A liquid-withdrawal system will withdraw gas; this will immediately reflect in a severe loss of chlorine supply.

When manifolding ton containers that are being used for liquid withdrawal, certain precautions must be taken. Never manifold a hot ton cylinder with others that are cooler. Cool the hot cylinder by the same procedure described for cooling the 100- and 150-lb cylinders. If this practice is not followed, the possibility arises of filling the partially empty, cool ton cylinders by one or more of the new hot cylinders. Filling a cylinder to liquid capacity could result in the eventual hydrostatic rupture of the filled cylinder(s). Such a situation would obviously cause a disastrous chlorine accident.* Therefore, always be sure that the new cylinder being connected is at the same temperature as the cylinders already connected before turning them into the system. This will be revealed by the vapor pressure of the cylinders. The operator is cautioned to connect the new cylinder(s) in such a way that he can verify and compare the vapor pressure between the new cylinders and those already in use. If there is a significant difference in these pressures, the new cylinders will have to be allowed to reach the temperature of the cylinders on the line by remaining in the storage area overnight before being connected.

---

* This is why the Chlorine Institute recommends against manifolding ton containers that are operating in the liquid phase. However such manifolding is common practice throughout the world.

Otherwise, the new cylinders will have to be artificially cooled as described for 100- and 150-lb cylinders.

The best way to operate a ton container supply system is to use a group of cylinders until empty. When this occurs, secure the empty cylinders and activate a new group of full cylinders until empty and so on. It is best never to turn a new cylinder into a supply system of partly full cylinders. The maximum liquid withdrawal rate from a ton cylinder is 12,000 lb/day.

Tank Cars and Storage Tanks. The most important operating consideration for this type of supply system is proper instruction. A qualified representative from the chlorine supplier should be engaged to demonstrate the use of safety equipment and to outline safety precautions and preventive maintenance procedures. The most critical time for these large systems is the original startup, or startup after a prolonged shutdown. Always start these systems on the gas phase first. A vapor leak is less lethal than a liquid leak. A tank car or storage tank installation capable of delivering either vapor or liquid should have two separate chlorine headers all the way to the evaporators, one for gas and one for liquid.

When starting liquid withdrawal, the operator must be careful to open the tank car (or storage tank) liquid valve very slowly and allow the piping system to fill up completely. Otherwise if the valve is opened quickly the sudden flow of liquid chlorine will jam shut the emergency ball check valve that is in the liquid withdrawal tubes inside the tank dome assembly.

Chlorine Leak Detector:

Unit Description. Most leak detectors for chlorine are of the amperometric type. However, there are others that respond to discoloration of impregnated strips or compounds, while some use a gas diffusion electrode.

The amperometric type utilizes dual platinum electrodes for a sensor. The sensor is continuously wetted with an electrolyte solution (high conductivity) which is in constant and continuous contact with the air sample of choice. The air sample is carried to the detector with either a built-in fan or a remote blower unit depending upon the local conditions. Chlorine in the air sample is detected in seconds by the sensor. These detectors can detect 1 ppm or 3 ppm chlorine by volume in the air and can be altered in the field to alarm at 0.5 ppm or 3 ppm.

The detector has local indication by a red alarm light. Additional relay contacts are provided for external alarms and other equipment such as ventilating fans and emergency chlorine shutoff valves.

The electrolyte is a glycerine-based potassium iodide solution diluted with distilled water. The wetting action of this solution keeps the platinum electrodes of the sensor clean. When the air in contact with the sensor contains chlorine it reacts with the potassium iodide by releasing free iodine initiating an electric current across the platinum electrodes. Through an electronic circuit board the detector activates the alarms.

The electrolyte tank is protected at all times from chlorine contamination by an activated carbon air filter. This tank is also equipped with a level indicator.

Preparation for Initial Operation. [5] Remove the sensor which contains the platinum electrodes. Be careful not to touch or disturb the electrodes. Body oil from the fingers or hands can insulate the electrodes. Clean the sensor with detergent (Tide) using a soft brush then rinse the sensor with distilled water. Check to see that the glass orifice is clean and dry and avoid air bubbles in the sensor system because they will block the flow of the electrolyte to the sensor. Reinstall the sensor.

Connect the drain tubing to provide for disposal of spent electrolyte into a suitable drain or container. The size of the electrolyte reservoir is usually about 5–6 quarts.

Now fill the reservoir with the electrolyte furnished with the unit. Follow the directions on the container. Do not leave any air space in the reservoir. Air space results in a faster drip rate, which increases unnecessary consumption of electrolyte.

Then check the electrolyte flow to the sensor. The sensor and its electrodes must be completely wetted by the electrolyte. If not, use a cotton swab to spread the electrolyte around the sensor. Spraying the sensor with electrolyte or distilled water by means of a plastic wash bottle will also help wet the sensor.

The initial drip rate should be one drop every 5 seconds to 1 minute. After a few hours, depending upon the amount of air space in the reservoir, this rate should decrease to one drop every 2–5 minutes. The drip rate will vary with temperature and barometric changes.

The next step is to test the response of the detector to chlorine. First turn on the power. Next mix 3 or 4 drops of pH 4 buffer solution with a similar amount of houschood bleach (Clorox) in a small beaker. This mixture will generate a small amount of chlorine gas. Hold the beaker one inch below the suction inlet pipe. The detector should respond in seconds. If not, then consult the manufacturers troubleshooting guide.

After this initial alarm—or any future alarm—flush the sensor with distilled water using the plastic wash bottle supplied with the detector.

Leak detectors should be checked on a routine basis at least once a week.

CHLORINE CONTROL AND METERING SYSTEM

Injector System. The control and metering system will not function without an adequate injector system. The injector system provides the power to pull gas from the containers through the chlorinator and then dissolves this gas into the injector water supply to provide a chlorine solution at the point of application. All injector systems should be operated with sufficient water to provide a solution concentration not to exceed 3500 mg/liter of chlorine. This usually figures to be 40 gal/day/lb of chlorine.

The injector system should be evaluated as follows. Turn on the water supply and allow the hydraulics of the system to stabilize. With the chlorine supply secured, the injector system should be indicating a vacuum in excess of 20 in. Hg. Otherwise, something is wrong. Assuming the vacuum valve is operating properly, open the chlorine supply to the chlorinator and adjust the chlorine feed rate to about 25 percent of full scale. If the system is adequate, the admission of the chlorine to the injector may only lower the vacuum to about 15 in. Hg or higher. As long as this vacuum does not deteriorate rapidly, it can be assumed the injector system is in order. Let the system remain at this condition momentarily then change the chlorine feed rate to 100 percent of full scale. If the injector system is adequate, the chlorine feed rate response will be instantaneous and the injector vacuum gauge will not deteriorate much below 15 in. Hg. Large installations where 100 percent of full scale is 2000 lb/day or more, and where the injector back pressure is greater than 3–5 psi, the injector vacuum might deteriorate to 10 in. Hg or less upon feed rate change from 25–100 percent of full scale on the rotameter feed rate indicator. Chlorinators as a general rule will not operate properly at injector vacuum values much less than 10 in. Hg. This figure should be checked with the manufacturer.

Booster Pumps. Pumps used to supply chlorinator injector operating water can be of either a centrifugal or a turbine type.

The centrifugal pump is used at installations where the total dynamic head is comparatively low (110–140 ft) and the volume is high (60–220 gpm). Centrifugal pumps are the pumps of choice for use at treatment plants, both potable water and wastewater. They are also used in instances where more than one injector is being operated from a single pump. The centrifugal pump has longer wear characteristics, requires less maintenance, and has no operating adjustments.

The turbine-type pump is an entirely different matter. It is used at installations where the chlorinator utilizes a 1-inch fixed-throat injector. The maximum capacity of these chlorinators is 500 lb/day and the injector requirements vary from 1 to 25 gpm. The turbine pump is capable of operating these chlorinators over a wide range of injector back-pressures up to about 100 psi. These situations occur on a water transmission main or on the discharge of a well or a well field. A turbine pump is rarely ever used at a wastewater plant for chlorinator operation. By design the turbine is a positive-displacement type of pump and therefore it must be equipped with an adjustable bypass assembly for regulating the discharge pressure.

This type of pump is selected to deliver almost twice the required amount of water passing through the injector, so that when it is new the pump will bypass from the discharge line to the suction line through an adjustable needle valve. As the pump wears, the operator must throttle down on the needle valve in order to maintain proper pressure at the injector.

Upon startup the needle valve is adjusted to give a discharge pressure that will just barely pull the maximum capacity of the chlorinator. This pressure is noted by the operator, who should then throttle the needle valve enough to raise the discharge pressure another 5 psi. This will provide an operating safety factor. As the pump wears with time the operator will find it necessary to throttle the needle valve in order to maintain chlorinator performance.

Chlorinators. If the chlorinator feed rate adjustment is manual, there are no critical adjustments to be made. The unit is adjusted manually to give the desired residual. No further adjustment or calibration is necessary. It is important, however, to choose a rotameter of the proper and most useful range. For example, if the desired fed rate is 15 lb/day, a 30 lb/day rotameter would be a wiser selection than a 100 lb/day size.

If the chlorinator is to be paced by a flow signal the rotameter must be selected to fit the range of the flow meter. For example, suppose the meter range is 0–30 mgd and the desired dosage is 16 lb chlorine per m.g.; then the maximum chlorine feed rate will be 480 lb/day, so use a 500 lb/day rotameter.

When the chlorine feed rate is paced by a flow signal, the chlorinator must be adjusted for zero and span. These adjustments are carefully outlined in the manufacturer's instruction book. This adjustment is usually a combination of signal input and mechanical linkage, depending on the type of signal—pneumatic or electric.

The important point when attempting adjustments to the chlorinator control mechanism is that these adjustments can only be made properly when there is a flow of chlorine through the machine. Furthermore, a chlorinator is not designed for zero flow conditions while the injector is in operation. Thus, the zero adjustment is contingent upon a mechanical linkage. As long as the injector is operating and even though the chlorine metering orifice is adjusted to read zero on the rotameter scale, some chlorine will be passing through the chlorinator. This is easily verified. Shut off the inlet gas valve to the chlorinator and watch the chlorine gas gauge pressure on the chlorinator instrument panel gradually creep towards zero.

The most important aid to chlorinator adjustment and calibration is the use of signal simulators to simulate input signals from flow and dosage measurement devices to the chlorinator. The signal simulators used should have the capability of providing input signals over the entire range of the chlorinator.

Special adjustments are required for chlorinators operating from a variable vacuum signal as the sole means of controlling the chlorine feed rate. This mode is commonly used for straight residual control. The special adjusting procedure is outlined in the manufacturer's instruction manual and has to do with the zero adjustment of the chlorine inlet reducing valve. The accuracy of this adjustment, as well as the overall accuracy of this method of control, is greatly improved by the use of an external pressure reducing valve installed in the supply system upstream from the chlorinator inlet. This valve takes the burden off the pressure reducing valve in the chlorinator. Without that additional valve, the chlorinator inlet pressure reducing valve would be the primary instrument of control in a variable vacuum system.

Chlorine Residual Analyzer:

General Discussion. This unit has zero, span, and temperature adjustment. If a thermistor is used, it automatically compensates for temperature changes in the sample.

The zero adjustment is a simple one. The sample flow to the analyzer is shut off, or one lead to the cell is disconnected (depending on the manufacturer). Under these conditions the indicator or pen on the recorder should go to zero. If not, the potentiometer marked zero is rotated until the pen reads zero. The span adjustment depends upon the calibration.

The amperometric analyzer is generally calibrated by the use of an amperometric titrator. Other titrimetric methods can also be used such as starch–iodide or DPD ferrous ammonium sulfate. These methods are outlined in the 15th edition of *Standard Methods.* In wastewater applications the back-titration method using the amperometric titrator is preferred. The DPD ferrous ammonium sulfate and starch–iodide titrations are equally acceptable in wastewater practice. Colorimetric methods are not reliable for wastewater calibrations.

Advanced wastewater treatment plants that nitrify obtain free residual chlorine together with combined. In these cases it is desirable to differentiate the free from the combined; therefore, the forward titration procedure must be used.

Whenever extensive calibration of the analyzer is required or desired it is desirable to have a signal simulator available. This instrument provides checkpoints throughout the range of the analyzer cell output.

There is a significant difference between the operation of an analyzer for potable water treatment and the operation of one for wastewater.

Potable Water. There are many installations interested only in free chlorine. Measuring free chlorine can be done with or without chemicals. If the pH is constant, as it is in some instances, no buffer solution is required.* No KI is required because free chlorine is measured directly by the analyzer. In most of these cases there is not enough combined chlorine to cause any error in the free reading. If, however, the pH varies (which causes a variation in cell output) then pH 4 buffer solution or carbon dioxide must be added to the sample to stabilize the pH at 4.

Water systems that add ammonia to produce chloramines or those that treat water with a natural ammonia content are concerned with measuring total chlorine residual. In these instances it is imperative that the sample pH be lowered to 4 by buffer solution and that potassium iodide be added to the buffer solution in sufficient quantity to convert the total residual to free iodine. The amount of KI

---

*Constant pH is a rarity, moreover maximum cell current occurs at pH 4 therefore it is prudent to add pH 4 buffer.

to be added to the buffer solution depends upon the sample flow through the analyzer and the magnitude of the residual. The cell flow rates vary from one manufacturer to another (i.e., the Capital Controls cell flow is 100 ml/min, the Fischer and Porter cell flow is 150 ml/min, and the Wallace and Tiernan cell flow is 360 ml/min). The Wallace and Tiernan cell requires 15 grams of KI per liter of buffer per mg/l total chlorine residual.

Wastewater. All wastewater analyzers have (or should have) a sample inlet-line disk-type strainer (Cuno). The drain valve on this strainer should always be wide open for continuous flushing. The sample pump or sample supply should be able to deliver about 12 gpm sample at 25 psi at the inlet to the filter. Under these conditions there will be about 9–10 gpm discharging to waste through the filter leaving sufficient capacity to supply the cell with an adequate continuous sample.

Chlorine residual analyzers for wastewater applications need many more calibration checks than those measuring residuals in potable waters. Furthermore, wastewaters generally have a significantly higher alkalinity than potable water, therefore pH control of the sample through the cell is critical. Similarly, the amount of KI added is also critical because of higher residuals required in wastewaters.

The sample pH must be kept at 4.5–5.0, preferably 4.5. Multicolored pH papers are satisfactory for this determination. Sufficient potassium iodide must be added to the pH 4 buffer solution to allow the chemical reaction between the KI and the chloramine residual to go to completion. The chlorine residual reacts with the potassium iodide to release free iodine in proportion to the total residual. The analyzer cell measures the free iodine. A rule of thumb for KI addition is 15 grams per liter of buffer solution per 150 ml/min cell flow of undiluted sample per mg/l of total chlorine residual. The Wallace and Tiernan wastewater analyzer is arranged for freshwater dilution of the sample. Dilution is usually on a 1:1 ratio.

The operator is advised to consult the manufacturer's instruction book to make sure of the proper calibration and operating procedures with respect to the buffer solution, KI addition, etc. This is extremely important in the operation of a wastewater analyzer.

An easy way for the operator to determine whether or not sufficient KI is being added is to add about 1 ml of KI solution (from the amperometric titrator test solutions) to the sample head box or diversion box and observe the chlorine residual indicator on the analyzer. If it does not move, the KI addition is sufficient. If it deflects up scale the KI addition is insufficient.

Owing to the importance of the addition of a buffer solution and KI, it is obvious that the operation of the reagent pumps is critical. Therefore, the operator must check the sample pH daily.

Once the analyzer has been put into operation, it is not necessary to respan the instrument each day. It is more desirable to plot a graph of the indicated versus the titrated sample readings. The curves of each of these groups of reading should indicate a span adjustment either upward or downward. This procedure should be done at first on a weekly basis, inasmuch as most wastewater systems require respanning the analyzer once a week.

Daily inspection of the sample system to the cell is necessary to assure a continuous sample. The noble metal electrode in the cell may be either gold or platinum, depending upon the manufacturer. This should be kept free of grease deposits by wiping with a soft cloth. However the sacrificial copper electrode should never be disturbed.

White has found that some wastewater residual control systems are operating without the benefit of KI addition to the buffer solution. This is proof that the amperometric cell is responsive to the combined chlorine residual. This is considered unreliable because the cell response to chloramine is not linear as it is with free iodine or free chlorine. However, these systems not using KI show remarkably consistent control, somewhat superior to some systems using KI for measuring total chlorine residual. This phenomenon requires further investigation.

Zero adjustment of analyzers used to monitor the dechlorinated sample will always expose a calibration problem that is best ignored. These analyzers are used primarily to monitor and control the sulfonation process. However they are often arranged to monitor the dechlorinated effluent for five or ten minutes every hour. When the dechlorinated sample reaches the cell the analyzer indicator or pen will drop below zero in spite of the fact that zero and span have already been adjusted for a sample containing chlorine residual. This phenomenon is probably due to the presence of excess SO_2 being applied. It does not signify a negative residual. It does, however, demonstrate that when a dechlorinated sample passes through the cell it changes the molecular surface of the electrodes so that they are slightly different (electrochemically) than when the zero calibration check is made with a sample containing a chlorine residual. This phenomenon does not necessarily affect the span adjustment of the analyzer.

Amperometric Titrator. Operating personnel should practice the titration procedure initially two or three times per day, to establish confidence in the reliability of their results. The proper degree of confidence can come only after several titrations. Daily titrations after that are in order.

Before starting any titrations (or when standing idle) it is helpful to keep the electrodes sensitized. This is done by immersing them in the sample jar containing tap water with about a 5 mg/l free chlorine residual. If the titrator is being used in the back-titration mode for wastewater then the tap water should contain small concentrations of free iodine. This precaution provides continuous sensitivity of the electrodes, which results in greater repeatability and a higher degree of reliability.

In recent years there has been a demand of sorts to be able to read chlorine residuals in receiving waters as low as 0.005 mg/l. It has been shown that the sensitivity of the Wallace and Tiernan titrator can be increased from 0.1 mg/liter to 0.001 mg/liter by the following procedure:

1. Remove the ammeter from the titrator and substitute a Rochester converter which will deliver a milliamp output signal.
2. Carefully encase the sample jar in aluminum foil and ground it.
3. Dilute the phenylarsene oxide four times. This modification can measure 0.001 mg/liter chlorine residual.

Chlorine Flow Transmitters. The only satisfactory way to measure the flow of chlorine gas flowing through a chlorinator is by a differential transmitter. To accomplish this, an orifice is placed in the chlorinator gas discharge line. The orifice should be sized so that the vacuum differential will be 12 in. of water at maximum Cl_2 flow. The orifice is specific for a given size rotameter. The transmitter consists of a spring-loaded diaphragm assembly which must be mounted on top of the chlorinator in order to minimize access of moist chlorine gas to the diaphragm assembly. Since the relationship of the chlorine flow to the vacuum differential is a square root function, the diaphragm movement is converted to a linear function by a characterized spring. This makes the diaphragm movement directly proportional to gas flow. The diaphragm controls the armature in a differential transformer which puts out an AC signal. This signal is converted to a 4–20 mA DC signal in a separate unit identified as a demodulator. Snap-in resistors are available to change the output signal to fit various categories of receivers.

It is first necessary to verify the differential across the orifice at 100 percent of chlorine flow. This can be done by connecting a water manometer across the tubing leads to the transmitter. This requires disconnecting the leads to the transmitter. The manometer should read between 10 and 14 in. H_2O differential. If not, check the size of the flow measuring orifice with the manufacturer's instructions.

Assuming the orifice is the correct size, adjust the zero setting of the demodulator to produce a milliampere current equal to the recorder (receiver) zero signal. This most likely will be 4 mA. Turn on the chlorinator and operate it at 100 percent flow. Then adjust the coarse span and the fine span in the demodulator to obtain a milliampere signal equal to the receiver–recorder 100 percent signal. This will probably be 20 mA. Recheck zero and span and adjust as required.

When not working on the electronic equipment inside, keep the demodulator cover closed at all times to prevent moisture and chlorine fumes from entering the enclosure.

The chlorine flow transmitting system is now ready to operate.

Alarms. Various types of alarm units may or may not require adjustment.

The chlorine high- and/or low-pressure supply alarms require only simple mechanical adjustments.

The high- and low-vacuum alarm switches on the chlorinator are usually preadjusted at the factory and need no further adjustment.

Other alarms may be high- and low-limit switches on the chlorine orifice positioner that will require field adjustment. These alarms may also be installed in the chlorine flow recorders. Adjustments are made at the factory, but can be changed in the field.

Safety Equipment. See "Maintenance: Chlorination Equipment," later in this chapter.

OPERATION: DECHLORINATION EQUIPMENT

PHYSICAL FACILITIES

The operator should have available for reference the same material suggested for the chlorination equipment.

SUPPLY SYSTEM PREPARATION

The same precautions must be taken for the sulfur dioxide system as those described earlier in this chapter for the chlorination equipment.

The procedure for cleaning and drying the piping and the evaporators is also the same as it is for chlorine. (The procedure for cleaning and drying will be found in the "Maintenance" section of this chapter.)

SULFUR DIOXIDE SUPPLY SYSTEM

General Discussion. The vapor characteristics of sulfur dioxide make it prone to reliquefaction—an undesirable situation when one is attempting to meter a gas. Unless the container is artificially padded with a propellant, the cylinder pressure at room temperature will be about 35 psi. See Fig. 6 in the Appendix for the sulfur dioxide vapor pressure curve. If the cylinders are stored outdoors in a carport-type structure, the SO_2 gas will surely reliquefy in the header between the container and sulfonator. One way to relieve this situation is by the application of heat on the cylinders followed by a pressure-reducing valve installed immediately downstream from the cylinders. A better alternative is to install an evaporator and artificially raise the vapor pressure of the cylinders by an air-pad system or a bottle(s) of nitrogen. The design considerations to provide an operable system have been described previously.

Sulfur dioxide is available in ton containers, tanker trucks, and tank cars. Tanker trucks are common throughout the U.S.A. Suppliers using tanker trucks are usually able to furnish a storage tank for the user.

Ton Containers. If the gas phase of these containers is to be used, then heat must be applied to the containers. Strip heaters or hot-air blower-type heaters that are arranged to automatically shut off at 100°F are required. Care must be taken that the heat applied to SO_2 cylinders is not directed to adjacent chlorine cylinders. Sulfur dioxide gas-phase systems do not operate satisfactorily at ambient temperatures much below 85°F. They operate best at 100°F with an external reducing valve in the SO_2 header adjacent to the cylinders. The installation of an evaporator for these systems is the most practical solution.

Although manufacturers of sulfur dioxide may claim that gas withdrawal rates from ton containers can be as high as 500 lb/day, the operator is best served if the withdrawal rate is kept to 175–200 lb/day. The reliquefaction problem is kept to a minimum at this rate.

Stationary Storage Tanks. Owing to the fact that sulfur dioxide tankers are available and that the pressure of SO_2 vapor is so low (35 psi) the use of storage tanks is popular. The maximum delivery in road tankers is 20 tons. It takes about 2–3 hours to unload such a tanker. After the storage tank has received the tanker load, it should be air padded to about 60 psi. This is a convenient operating pressure. At this figure, the gas leaving the evaporator should be reduced to a sulfonator inlet pressure of about 25 psi. The sulfur dioxide storage tank should be designed so that the suction line on the tanker's compressor can be connected to the gas phase of the storage tank. This speeds up the unloading operation by utilizing the SO_2 tank vapor pressure.

Safety Equipment. The safety equipment for sulfur dioxide operations is the same as that for chlorine.*

SYSTEM STARTUP

The operator is referred to the two "Startup" sections under "Operation: Chlorination Equipment" in this chapter. The same guidelines apply.

Superheat. In the use of an SO_2 evaporator, which is a highly desirable piece of ancillary equipment, the same amount of superheat is required as with chlorine to assure satisfactory operation of this unit. Therefore the operator should review the discussion of this topic in the "Evaporator" section under the heading "Chlorination Equipment: Operation," earlier in this chapter.

Sulfur Dioxide Supply Pressure. As compared to chlorine, the higher the SO_2 vapor pressure the better it is for the operation of the facility. There is rarely any likelihood that "high" SO_2 vapor pressure would ever be an operating problem. If such a situation would occur, the same procedures outlined for chlorine would be applicable.

---

* Except the leak detector. SO_2 leak detectors operate on a different principle.

TO STOP OR SECURE AN INSTALLATION

The operator should follow the same procedure described under this heading "Chlorination Equipment: Operation." The identical guidelines apply for sulfur dioxide.

SULFUR DIOXIDE CONTROL AND METERING SYSTEM

Injector System. The only difference between the use of an injector system for sulfur dioxide and chlorine is that SO_2 gas is much more soluble in water than is chlorine. Therefore, if the operation of a sulfur dioxide injector system is fashioned the same as that for chlorine, the operator will be well on the side of safety. *The use of 3500 mg/liter as the maximum strength of sulfurous* acid solution in the injector discharge will prevent breakout of molecular SO_2 at the point of application.

Booster Pumps. The same applies here as for the chlorination system.

Sulfonators. The same operating procedures are required for the sulfonators as has been previously described for chlorinators. This includes adjustments and calibration. However, since a great many sulfonator control systems are based on feed forward control, a slightly different set of operational conditions faces the operator. This concerns the function of the Ratio Station installed on the flow pacing signal which is ahead of the flow-residual signal multiplier (see Chapter 9).

The Ratio Station is a precise dosage control instrument particularly valuable in "feed forward systems." Fig. 11-3 provides the operator with dosage values for a given sulfonator orifice meter. The one shown is a 4000 lb/day meter used with a 200 mgd (maximum) plant flow meter. When the ratio station is set at unity (1.0) the SO_2 dosage will be 2.4 mg/liter. At a ratio setting of 2.0 the dosage will be doubled and so on.

EXAMPLE: Operating with a 1.0 ratio station setting, a 4000 lb/day SO_2 meter controlled from a 0–200 mgd flow meter the SO_2 dosage would be

$$\frac{4000 \text{ lb/day}}{200 \text{ mgd} \times 8.34 \text{ lb/gm}} = 2.4 \text{ mg/liter}$$

for any flow from 0 to 200 mgd.

Figure 11-4 provides the operator with the necessary information to set the system to dechlorinate a given chlorine residual to zero for any plant flow rate. The set of curves shown on Fig. 11-4 are for a 4000 lb/day SO_2 rotameter. The solid line represents a sulfur dioxide to chlorine ratio of 1:1 (by wt). The

OPERATION AND MAINTENANCE OF CHLOR-DECHLOR EQUIPMENT 813

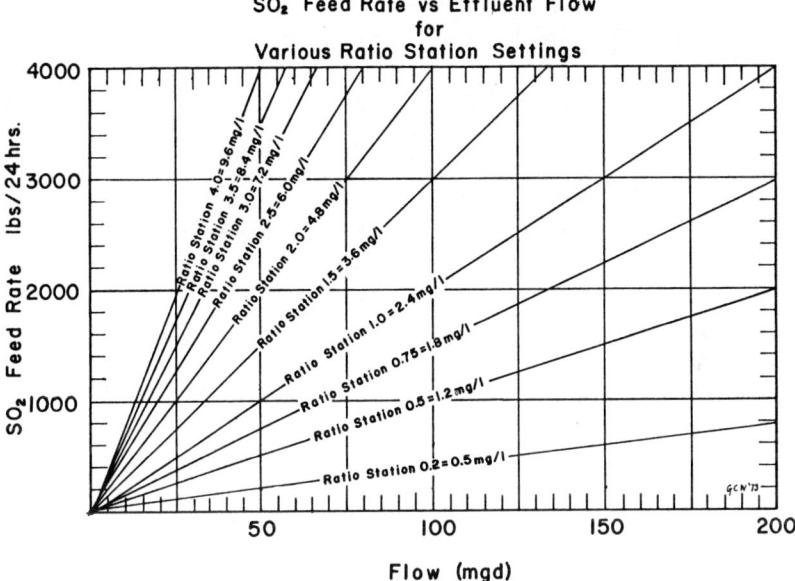

Fig. 11-3. Sulfur dioxide feed rate versus effluent flow for various ratio station settings.

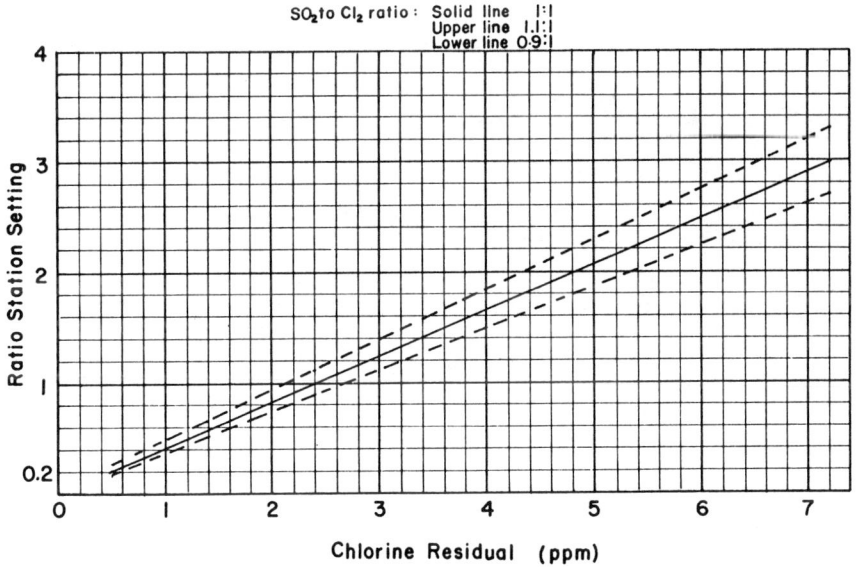

Fig. 11-4. Ratio station settings to achieve zero chlorine residual using a 4000 lb/day SO_2 rotameter.

lower dotted line represents the stoichiometric ratio of 0.9:1. The upper dotted line is the probable upper limit of SO_2 to Cl_2 as 1.1:1.0.

EXAMPLE: If the chlorine residual analyzer is showing 4 mg/liter residual, the *proper setting* of the Ratio Station to feed enough SO_2 to produce a zero residual in the dechlorinated effluent will be about 1.65 using the solid line as a reference. If the selected Ratio Station setting to produce a zero residual falls above or below the dotted lines of Fig. 11-4 for a 4 mg/liter chlorine residual (i.e., is greater than 1.8 or less than 1.5), *something is wrong*. Either the analyzer has drifted out of calibration or the analytical technique of measuring the chlorine residual is in error. Therefore, the laboratory personnel should make further analyses of both chlorinated and dechlorinated effluent samples to verify the calibration of the analyzer. Fig. 11-4 gives Ratio Station settings values for chlorine residuals regardless of plant flow between 0–200 mgd. The same approach would be used regardless of the effluent plant flow meter range.

Analyzers. At present (1983), the chlorine residual analyzer is the only instrument available for the control of the sulfur dioxide system. This is well documented in other sections of this text. If instrumentation ever becomes available to measure the concentration of sulfurous acid (H_2SO_3) or the sulfite ion ($SO_3^=$), the operating requirements are certain to be substantially different from those of the chlorine residual analyzers.

MAINTENANCE: CHLORINATION EQUIPMENT

CHLORINE SUPPLY SYSTEM

General Discussion. The primary objective of maintenance of any chlorination facility is the prevention of leaks, primarily those occurring under pressure. Therefore any joint in the supply system between the chlorine supply and the chlorine pressure-reducing valve in the chlorinator is a potential source of leak. Beyond the inlet chlorine pressure-reducing valve, the chlorine gas is under a vacuum. The chlorine pressure gage connection inside the chlorinator is also a source of leaks. The exception to this occurs with remote vacuum systems where the pressure in the supply system is converted to a vacuum adjacent to the chlorine cylinders. (See Chapter 9). This eliminates chlorine under cylinder pressure in the piping between the cylinders and the chlorinators, or between the evaporators and the chlorinators. It does not eliminate the potential of leaks from flexible connections, header valves, cylinder valves, gasket failures, or fusible plugs on containers. These are the major sources of leaks. Serious leaks in the chlorine header piping proper are rare.

Chlorine gas leaks are insidious in nature. The usual method of detection (26°BE ammonia—not household grade) will not detect the very small leak, the one which cannot be detected by odor. These leaks can proceed unnoticed for weeks before reaching a critical stage. American Gas and Chemical Co. Ltd. markets products for detecting small leaks of chlorine and ammonia. Their literature[2] describes a product (CDP-100) which is sprayed on critical joints. It reacts chemically with imperceptible chlorine leaks, causing a visible color change from white to yellow at the leak point. The product is easily removed with a damp cloth. The sensitivity of this detector is claimed to be 0.00001 Std. cc/sec.

Locating Leaks. The operator is advised to look for two different signs of a chlorine leak at a gas joint under pressure as follows: discoloration and moisture.

Discoloration at Joint. Most chlorine gas header accessories are cadmium-plated over copper tubing and bronze or brass fittings. If a leak is in progress (even those undetectable by ammonia or by odor), the cadmium plating will disappear and the base metal will take on a reddish color, signifying dezincified brass or corroded copper. Some green copper chloride scum will also appear around the edges of the corroded metal.

Moisture Formation. Small droplets of liquid may appear on the underside of a chlorine liquid or gas joint. This is the most insidious of all leaks, since it may go undetected on 1″ and ¾″ steel header systems for a long period. Usually these lines are painted with a good grade of durable paint so that discoloration and products of corrosion may not be readily visible. It is best to paint these lines bright yellow. Then if a minute leak occurs, the first evidence will be a moist spot of brown rust, which is highly visible.

While these small leaks are in progress, corrosion is taking place at the threads of the joint and proceeds until a massive leak suddenly occurs. A continuous method of monitoring these leaks is by means of a chlorine gas leak detector, capable of detecting leaks as small as 0.5 ppm chlorine concentration in the air. These leak detectors monitor the air in either the chlorine container room or the chlorine equipment room, or in both, by drawing in a continuous sample of air.

Maintenance Steps to Prevent Leaks. 1. Each time a gasketed chlorine joint is broken, replace the gasket with a new one.
  2. On a threaded pipe joint, wire-brush the threads and use Teflon tape for thread lubricant.
  3. Replace all chlorine gas header valve packings at least once a year. Try to obtain the Teflon-type packing where recommended. Failure of valve packing is one of the most common causes of chlorine valve leaks.

Other Sources of Leaks. Potential sources of gas and liquid leaks are the diaphragm protectors used on chlorine gas gages and pressure switches. The silver diaphragms in these protectors fail in time owing to metal fatigue. The lives of these diaphragms vary widely, depending upon pressures carried and frequency of cycling, from seven to ten years.

When the silver diaphragm fails, the glycerine or fluorocarbon on the top side of the diaphragm diffuses back into the chlorine header system. The chlorine reacts with the glycerine to produce a large, gooey mass that resembles scrambled eggs and is difficult to remove. If this goes undetected, the chlorine will quickly corrode the bourdon tube in the gage (or pressure switch), thereby causing a massive leak. These diaphragm failures are most likely to go undetected in large systems on the liquid chlorine phase, because the liquid chlorine carries away the small amount of glycerine into the evaporator, where the result may not be observed for some time.

In view of the overall consequences, the preventive maintenance program should consider the replacement of all gages and pressure switches with diaphragm protectors once every five years.

Repairing Leaks. The most important advice to an operator is this: *Never attempt to repair a leak on the chlorine supply system when it is under the chlorine container pressure.* Tightening packing glands, union joints, or flexible connections, etc., while the system is under pressure can only lead to disaster.

If the outlet valve of a container, tank car, or storage tank is leaking, the following steps should be taken in order. (1) Call the chlorine supplier to send expert help. (2) Don air or oxygen breathing apparatus (not a canister-type mask). (3) Ready the container emergency kit for action. (4) First try gingerly to tighten the valve packing; if this does not stop the leak, apply the emergency container kit and await the arrival of the chlorine packager representative.

Leaks at threaded joints of a flexible connection or at the union connection of an outlet valve require a thorough cleaning with a steel wire brush. These joints must be regasketed. Small and apparently insignificant leaks that can later result in devastating leaks often occur at the chlorine cylinder outlet valve. To prevent this situation, the operator should always carry a 1-in. flat file so that he can reface any chlorine cylinder outlet valve. This allows the proper seating surface between the cylinder valve and the auxiliary cylinder valve. This maintenance procedure has proved to be well worth the effort.

Never reuse a gasket that has been in a joint after the joint has been disassembled.

The question of lead gaskets versus asbestos or paper composition gaskets is a continuing controversy. The paper- or asbestos-type gaskets tend to cover up gross deficiencies in chlorine cylinder outlet valve maintenance. This gasket is favored by the chlorine cylinder packagers. On the other hand, once a lead gasket has passed the leak test of a joint, it will never result in a chlorine leak at that particular joint providing it is not disturbed by disassembly and reassembly. In contrast, a paper composition gasket can develop a serious leak by failure during operation. This type of gasket should be avoided because a gasket failure during operation can result in a much more serious leak than one by a lead gasket. Furthermore, the failure of a lead gasket has never proved to be the cause of a chlorine leak once that system was connected without leaks. A lead gasket joint will leak from the very beginning if it is not properly seated. This is the reason a lead gasket joint should be regasketed each time it is disassembled.

The use of yoke-type connections between flexible connections and cylinder valves or header valves minimizes the problem of leaks at these joints. Lead gaskets should be used at these joints. Tank car flexible connections use the 2-bolt ammonia-type union with a replaceable lead gasket for attaching to the car outlet valve and the supply header. These gaskets must also be replaced after each disconnect.

Flexible Connections. The most vulnerable parts of a chlorine supply system are the flexible connections. These are usually made of 2000-lb annealed copper tubing, cadmium plated. Whenever these connections are exposed to the moisture in the atmosphere during a cylinder change, corrosion sets in. Internal corrosion of these flexible connectors is a function of the number of times the tubing is exposed to the atmosphere. A positive way to determine the reliability of a flexible connection is to bend it slightly. If it "screeches," discard it immediately. This phenomenon signifies that internal corrosion is excessive and the tubing is liable to rupture prematurely by crystallization. Flexible connections for 100-lb, 150-lb, and 1-ton cylinders should be discarded on a regular basis. For the smaller cylinders discard after ten cylinder changes and for the ton containers once per year.

Flexible copper connections on tank cars have heavier walls, are subject to much less flexure, and can withstand more usage—up to thirty tank cars before replacing. These connectors are gradually being replaced by flexible metal hose connections that show an even longer safe life than the copper connectors. Whenever a tank car flexible connection is disconnected and not in use, it should be plugged at both ends with cork or wooden plugs and then stored in a warm place. This will practically eliminate internal corrosion. All unions on chlorine header systems should be the ammonia type (2-bolt) with a recessed gasketed joint. It is desirable to replace the gasket joints every five years, even if the joint has not been taken apart. *Ground joint unions should never be used.*

Safety Equipment. Safety equipment refers to breathing apparatus, chlorine container emergency kits, leak detectors, and automatic ventilating systems.

Chlorine container emergency kits for each size of container should be ready for instant use. Practice using these devices should be a part of the maintenance schedule. The same applies to the emergency breathing apparatus. Special safety drills at prescribed intervals should be made available for all operating personnel on both the use and handling of container kits and breathing apparatus. Demonstration of the chlorine leak detector response to chlorine should be a part of this drill. This will also familiarize operating personnel with the chlorine leak alarm system.

LEAK DETECTOR: AMPEROMETRIC TYPE[5]

The following is a maintenance check list to provide reliability of operation of the detector.

Weekly Check List. Check the following items:

1. Verify there is electrolyte in the reservoir.
2. Check the drip rate from the sensor (2–10 minutes per drop).
3. Check detector response to chlorine (use test kit).
4. If sensor is not clean and completely wet, flush it with distilled water.

Monthly Check List. Check the following items:

1. Verify circuitry response to test pushbutton.
2. Drop rate. It should be not less than one drop every ten minutes or more than one drop every minute.
3. See that sensor is clean and covered with electrolyte (no dry spots).
4. Check reservoir level in sight tube to see that it is not empty. Only fill reservoir when required. When not required electrolyte consumption will increase.
5. Check drain container and dispose of spent electrolyte if necessary.
6. Verify that fan is operational and that the blade turns frequently.
7. Check for leaks around the grommet, O-ring, sight tube joints, etc.
8. Check detector response to chlorine.

Annual Check List. Check the following items:

1. Gasket behind circuit board. Replace if ripped, cracked or badly worn.
2. O-ring around vent cap unit. Replace.
3. O-ring around glass orifice. Replace.
4. Grommet (sensor mounting). Replace if leaking or badly worn.
5. Sight tube. Replace if yellowing or cracked.
6. Check fan wiring for loose or broken connections.
7. See that drain tube connection is not loose, cracked, or leaking.
8. Check sensor for loose or broken connections
9. Replace mounting bolts if they are rusted or in any way corroded.
10. Carbon filter: replace if carbon is wet.

CLEANING AND TESTING CHEMICAL PIPING

General Discussion. The chlorine supply system (including evaporator if used) should be cleaned after the passage of each 250 tons chlorine, or before the startup of a new system. The difficulty usually encountered with new systems is a result of the new piping and equipment remaining idle and open to the atmosphere for a significant length of time. Moisture collects in these systems during this construction period and if they are not cleaned and dried before chlorine or sulfur dioxide is turned into the system, the resulting corrosion and distribution of corrosion products throughout the facility erupts into an unholy mess.

There are two methods for cleaning steel pipe used for chlorine or sulfur dioxide service. One is a chemical detergent method applied to an assembled system on-site. The other is the pickling system, where the pickling agent is either chlorine or sulfur dioxide depending upon the service.

In either case valves and gages should be removed and/or isolated from the system. Reserve tanks can be cleaned along with the piping, but it is believed to be better practice to clean these two items separately.

Chemical Detergent Method:
1. Degrease with caustic, trisodium phosphate, sodium metasilicate, and a surfactant. Circulate this mixture of chemicals through the piping for two hours at 180–190°F.
2. Flush the system until pH is neutral.
3. Descale with inhibited hydrochloric acid at 150–160°F for 2–4 hours.
4. Flush system with citric flush.
5. Passivate system with a passivating agent.
6. Blow dry with nitrogen.

Pickling Method. Depending upon the condition of the system it may be necessary to swab or flush the system to remove all vestiges of oil or grease deposited during the threading and coupling procedure during field fabrication. If this is necessary use trichlorethylene or some other chlorinated solvent. If the piping is assembled with welded forged steel fittings, flush with a strong caustic solution. When solvent is used it should be followed by a thorough flushing with cold water.

After the flushing procedure the system should be completely assembled and pressurized with air to 300 psi, soap tested, and all leaks repaired. During the pressure test all equipment such as pressure gages, pressure reducing valves, pressure switches, rupture disks, chlorinators and sulfonators should be isolated to prevent any possible overpressure damage.

A hydrostatic test may be substituted for air if it is more convenient.

The system should then be pickled in place by charging it with chlorine or sulfur dioxide (as the case may be) to normal operating pressure and then checked for leaks using 26° Baumé ammonia solution. The pipe must be evacuated to repair any of the leaks. This is imperative if welding is required. After the leaks have been corrected the system is again recharged with either chlorine or sulfur dioxide to operating pressure and held for a 24-hour pickling period. At the end of the pickling period the gas is evacuated and the system is flushed with fresh water at a minimum velocity of 2½ ft/sec for a time required to pass a quantity equal to being flushed or until the water is clean. The system is then drained and flushed with steam, followed by purging with dry air at 140°F. This flushing cycle shall then continue until the air existing in the system is at zero relative humidity at 150 psi. Zero relative humidity occurs when the dew point is $-40°F$.

During the air flushing cycle it is sometimes desirable that all parts of the system be tapped sufficiently to remove any remaining bits of scale rust or other debris. The pickling process removes all of the mill scale and oxides of iron, which have a great tendency to cause maintenance problems in the entire system.

After the air drying cycle is completed the equipment, valves, and other ancillary equipment removed or isolated is reassembled into the system. The system is then subjected to a 300 psi test with either dry air or nitrogen.

Both of the procedures for cleaning as described above should be supervised by a technician who has had extensive knowledge in these systems. It is this technician who will delineate which pieces of the system have to be isolated and which ones have to be removed.

TROUBLESHOOTING THE SUPPLY SYSTEM

The operator should acquaint himself with the length of time it takes the chlorine supply system to empty itself at a given chlorine feed rate. With the chlorinator at a certain optimum feed rate, close the chlorine cylinder or header valves and note how long it takes the pressure gage on the chlorinator to reach zero. Any appreciable or noticeable sluggishness in the supply system to reach zero pressure condition signifies restrictions caused by impurities somewhere in the system between the cylinders and the chlorinator. This can occur in a single 150-lb cylinder hook up when there is stoppage either in the auxiliary cylinder valve or in the four-foot flexible connection between the cylinder and chlorinator.

Sometimes the header system or flexible connections become so clogged that they cannot be emptied through the chlorinator. These cases require extreme caution. The operator must then close off the outlet valve on each cylinder in operation, thereby securing the chlorine supply. Then he should don proper breathing apparatus and break one of the joints in the header system upstream of the stoppage point. It is best to break the joint nearest the auxiliary cylinder valve; this will allow the trapped gas in the header system to be released to the room ventilation system. The operator should then leave the area for enough time to allow the ventilation system to remove the gas.

This procedure places full dependence on the tight closing of the chlorine cylinder outlet valves. If one or more of the cylinder valves happen to be leaking, the header system should be reconnected as quickly as possible and immediate arrangements made to bypass the header system. (This is why large systems should have duplicate header systems.) It may not be possible to by-pass the header system safely. In such a case, steps must be taken to utilize the appropriate Chlorine Institute Emergency Kit to seal off the leaking cylinders. Separate kits are available for 150-lb. cylinders, ton containers, and tank cars. The tank car kit fits all sizes of tank cars.

When the cylinders have been secured and removed, then the header system can be "unplugged," as described above. The importance of keeping the header system in an operative condition at all times cannot be overemphasized.

Another situation that occurs occasionally is a source of great dismay to the operator, as well as being destructive to the chlorination equipment. This occurs in a gas system using ton cylinders when the feed tubes in the ton container develop a crack below the liquid level. These tubes are shown in Fig. 9-4. The crack usually develops where the tube is threaded into the cylinder outlet valve, and allows liquid instead of gas to enter the system. This is evidenced by icing or extreme cooling and the collection of moisture on the header system and possible severe icing on the rotameter tube in the chlorinator. The liquid chlorine is trying to vaporize and, so in doing, is using up heat, thus causing the local cooling effect.

Liquid chlorine will quickly damage the plastic parts of the chlorinator. Once the operator recognizes the difficulty he should immediately secure the cylinder and allow the header system to drain itself of liquid chlorine through the chlorinator; then he should reduce the chlorine feed rate to allow more vaporization to protect the chlorinator equipment. Then he should disconnect and rotate the defective cylinder 180°. This will put the other feed tube in position. It is rare that both tubes will be defective in the same cylinder.

CHLORINE STORAGE TANKS

The most important maintenance here is on the valves in the dome of the tank. There should be on hand at all times a complete set of spare dome valves, including the safety valve. The safety valve should be replaced every two years, but the two liquid and two gas valves in the dome should be serviced annually and sent to a company specializing in valve repair. After cleaning and repair, the valves should be hydrostatically tested for 500 psi.

It is recommended that the storage tank be hydrostatically tested every three years. To simplify this procedure, it is desirable to have two extra 3- or 4-inch flanged connections at either end of the tank and on the crown. One of these is for filling the tank with water; the other is for the overflow. If special flange connections are not available, the safety valve can be removed and used as the water inlet. The two gas valves should be manifolded together for the overflow.

The procedure is to fill the tank with a fire hose as quickly as possible and to keep adding water until the overlow water becomes clear. When this is accomplished, either remove all the water with air padding or pump it out. Then go into the tank to inspect it and remove any remaining scale or debris by scraping.

Never enter a storage tank that has not been properly flushed out. Otherwise, remaining chlorine will combine with body perspiration to form strong hydrochloric acid, which can cause painful burning of the skin.

The hydrostatic test is made after the tank has been washed, cleaned, and inspected. It is then filled a second time for this test.

Prior to filling the tank with water, the tank must first be evacuated of all gas by pulling a vacuum on the tank with the eductor especially provided for this purpose. After the tank is allowed to stay in a vacuum condition for three or four hours, it is ready to be filled with water. At this point, there is still plenty of residue in the tank that, if allowed to remain, would gas off for a long time.

EVAPORATORS

Liquid Chlorine Vessel. This piece of equipment is subject to filling with a gooey mass of sludge that accumulates from inherent impurities in liquid chlorine. The amount of sludge that accumulates in the bottom of the liquid chlorine vessel is primarily a function of the amount of chlorine passing through the evaporator. In general, evaporators should be inspected for sludge once a year or after passing 250 tons of chlorine, whichever comes first. In addition to this, the operator should keep a close watch on the evaporator superheat. If this cannot be maintained at 20°F, it signifies that the evaporator is losing capacity due to sludge buildup in the liquid container or that there is an immersion heater failure.

Immersion Heaters. These heating elements are the heart of an evaporator. Scale deposits from hardness in the water-bath water can and do build up on these elements. These deposits can become so severe that annual replacement may be necessary. This phenomenon should be watched carefully during the first year of operation—by routine physical inspection every 90 days.

Cathodic Protection System. The anodes should be inspected annually. Under normal conditions the anodes will be found nearly expended and should be replaced.

Before new anodes are installed, water should be drained from the system. Then the water-bath tank and chlorine vessel should be inspected for corrosion. If the cathodic protection system has been operated in the recommended range, little or no corrosion will be observed and long service may be expected.

After the new anodes are in place, fill the water-bath with fresh water and add ¼ lb sodium sulfate to provide proper conductivity for the water.

Cleaning the Liquid Chlorine Vessel. Cleaning the evaporator liquid container consists of dismantling and removing the chlorine vessel and flushing it with cold water until the flushing water is clean. The inside should be visually inspected for pitting. If the pitting is severe, the vessel should be replaced.

After all the flushing water has been removed, reassemble the evaporator, fill the water bath, and heat to 180°F. Then attach the aspirator so that a vacuum can be exerted on the inside of the vessel.[6] The vacuum should be about 25 in. Hg and should continue for 24 hr with the water-bath at 180°F to remove all moisture from the inside of the chlorine vessel.

CHLORINE GAS FILTER

This unit should be inspected every six months. The condition of the filter element will give a clue to the condition of the header piping system and evaporator. The filter elements should be replaced at each inspection, and the sediment trap washed in cold water and dried before reassembly. Most chlorine impurities consist of ferric chloride, which dissolves readily in water. The complex chlorinated hydrocarbons and other gunk will not dissolve in water. If these accumulations appear in the filter or the external chlorine pressure-reducing valve, they can usually be removed by either trichloroethylene or isopropyl alcohol.

CHLORINE PRESSURE-REDUCING VALVE: EXTERNAL

This unit consists of a spring-loaded stem, actuated by a three-layer silver diaphragm assembly. The stem throttles the chlorine gas going through the seat assembly. This is the point of pressure drop, and this is where deposits of the chlorine impurities are most likely to occur. The filter placed upstream from the CPRV should remove most of these impurities. It is inevitable that some will deposit on the stem and seat assembly of the CPRV. Deposits that cannot be removed with a soft cloth can be removed with isopropyl alcohol, trichlorethylene, or Chlorothene-Nu. The latter is an inhibited 1,1,1-trichloroethane.

The spring opposing the silver diaphragm action will suffer fatigue and should be replaced every two or three years. Every five years the entire assembly should be dismantled and the diaphragm inspected for fatigue. The diaphragm as well as the lead gaskets on either side of it should be replaced at this time.

AUTOMATIC SWITCHOVER SYSTEM

Pressure Type. This system utilizes two chlorine pressure-reducing valves and is described in Chapter 9. For reliable operation these valves should be preceded by a gas filter to minimize valve maintenance. The area of vulnerability is the stem and seat assembly. These parts should be inspected and cleaned at least once every three months. Another area of vulnerability is the spring that is used to oppose the spring-loaded diaphragm. This spring operates entirely in a chlorine gas atmosphere and therefore suffers a form of corrosion fatigue. It should be replaced every two years. The spring that loads the diaphragm operates in ambient air and does not suffer from corrosion from normal operation. A ruptured diaphragm or a substantial gas leak in the vicinity of the valve will damage this spring.

Vacuum Type. This type of system utilizes two vacuum regulator units that contain a series of spring-loaded stem and seat assemblies. These units are designed to reduce cylinder pressure (85–100 psi) to about 60 in. H_2O vacuum. This is a much greater pressure drop than for the pressure unit described above. The effect of this greater pressure drop is to precipitate more impurities on the seat and needle assembly. Therefore it is imperative to have a chlorine trap and filter assembly located upstream from the switchover unit. This assembly should be inspected on a regular basis. At the same time the seat and stem assemblies should be examined for deposits and cleaned according to manufacturers instructions. All of the springs should be replaced every two years. They are subject to metal corrosion fatigue since they operate in a chlorine vapor atmosphere.

REMOTE VACUUM SYSTEM

This system is designed to reduce the chlorine or SO_2 pressure in the cylinder, at the cylinder, to a vacuum as described above. The components include a vacuum regulator check unit and a standby pressure relief device. This system should also be protected by a trap and filter unit. It requires the same attention as the automatic vacuum switchover system.

CHLORINATORS

Modern chlorinators consist of a series of spring-loaded diaphragm units that form the basis for control of the chorine gas through the installation, with certain vacuum values at various points. Therefore it is essential that all the joints be vacuum-tight for proper operation.

Most of the maintenance problems occur from metal fatigue of the springs in each of the diaphragm assemblies and from improper stem and seat closure in these diaphragms caused by impurities in the chlorine gas. All springs should be replaced every two years; the stem and seat units should be inspected and cleaned annually.

The rotameter tube and float assembly should be removed and cleaned periodically at least every six months. The chlorine metering orifice should be dismantled for inspection after six months' operation, because impurities deposited here will give a clue as to the condition of the rest of the system.

Since the injector system is a vital part of the chlorinator, it should be inspected occasionally. By using a water pressure gage on the water supply and if an injector vacuum gage is on the chlorinator, the difficulty with the injector system can be readily determined. If there is sufficient water pressure but the injector is not producing sufficient vacuum, only a couple of things can be wrong: the inlet to the injector throat may be partially plugged by ordinary debris or a tiny bit of gravel; or the discharge line to the point of application may have some stoppage, causing high back pressure. These conditions are readily determined by inspection.

One other common condition unknown to novice operators is the deposition of iron and manganese on the injector throat. This problem is most prevalent on chlorinators with a fixed throat injector. Manganese deposits a slick black coating, which reduces the friction loss through the throat to such an extent that it prevents the formation of a vacuum at the throat, thereby making the chlorinator inoperative. Iron deposits will eventually reduce the size of the throat until it becomes so small that is will not pass sufficient water to produce a proper vacuum. This phenomenon is the result of the accumulation of sufficient amounts of iron and manganese in the injector water supply that become oxidized to their insoluble states at the point where the chlorine concentration is the highest—at the injector throat. Iron has a tendency also to slightly delayed reaction, and will sometimes leave deposits in the injector tailway, causing a restriction having the same net effect.

This situation is easily remedied by proper maintenance. The operator should keep on hand a spare injector throat and tailway and make an exchange at regular intervals consistent with proper operation. The throat and tailway are quickly cleaned by submerging the entire unit in muriatic acid. After a few minutes' soaking, remove and rinse in clean water; it is then ready for reuse.

Injectors are equipped with a device to prevent back flow of water into the chlorinator. These are usually spring-loaded diaphragm assemblies. The springs operate in a moist chlorine atmosphere, and so they should be replaced routinely every two years.

BOOSTER PUMPS

Chlorinator booster pumps are as vital to the installation as is the power supply, the chlorine supply, or the chlorinator itself.

Two types of pumps are employed: turbine, or centrifugal. The turbine is the more common on installations of 400 lb/day or less, because the requirements are for high pressure and low volume. These units are furnished with a built-in adjustable bypass assembly that allows for wear on the pump impeller, which is critical for a turbine pump. The pump is usually designed to deliver twice the required amount of injector water.

When the pump is new, half of the pump discharge is bypassed to the suction. Adjusting the needle valve in the bypass will assure this function. As the pump wears, the discharge pressure falls off; then the operator must restore the proper pressure by readjusting the needle valve. When the needle valve is operating in an almost fully closed position, it is time to disassemble the pump and replace the impeller.

If sand is present, a turbine pump will wear out quickly. The sand must be removed or the pump suction relocated to a point in the distribution system where sand is not present.

Centrifugal pumps are maintained in the usual way. However, they differ from turbine pumps in that there is no adjustable bypass assembly and no pressure-relief valve. Sand is not so critical in the operation of this type of pump. The best choice in a centrifugal pump is one that has an outboard bearing. The Unibuilt close coupled type of pump will require more maintenance, as the inboard bearing at the motor is subject to frequent failure.

CONTROL DEVICES

Electric. Most control devices have a combination of plug-in transistorized components which give long and reliable service. The operator should, however, be able to obtain replacements for any of these components in the event of failure. Devices with moving parts, such as potentiometers or safety devices dependent upon mercury switches, are subject to wear, and will require eventual replacement. The operator should have spares of these items also. Reversible motors used in these control devices, such as electric chlorine orifice positioners and motorized vacuum valves, are also subject to failure, even though they do give long and reliable service. The operator should check the source and availability of these motors. The manufacturers of these special duty motors may run into production problems, resulting in delayed delivery schedules. Most items which are components for electric control and alarm devices are off-the-shelf items and are readily available.

Pneumatic. Air control devices are subject more to maintenance problems caused by dirty or wet control air than they are to the wearing of moving parts. However, the moving parts, such as belloframs and other types of diaphragm assemblies, are subject to failure due to wear, and spares should be carried. Flapper valves and seat assemblies can be cleaned and reseated many times before replacement is required. The various springs required in a pneumatic system have a long life—as much as five or six years—because they operate in a normal atmospheric environment.

CHLORINE RESIDUAL ANALYZER

General Discussion. There are two principal categories of analyzers: potable water and wastewater. These two are available to measure either free or total residual chlorine. The only analyzer available to measure free chlorine in wastewater (*in presence of chloramine*) is the Fischer and Porter unit with an impressed voltage applied across the electrodes.

The typical amperometric continuous analyzer consists of a cell assembly that accepts the sample from the sample piping and/or pumping system, plus electronic circuitry that interprets and records (or indicates) the intelligence from the cell. Wastewater analyzers differ from potable water analyzers in the way the sample is delivered to the cell. Wastewater analyzers are equipped with a special filter on the sample line.

Sample Line. The sample line to the analyzer is probably the most neglected component of a chlorine residual measuring system. It is subject to biofouling whether by potable water or wastewater. This exerts a chlorine demand which gives a false reading at the analyzer. Usually the sample flow velocity is far below what would be considered a scouring velocity, and in wastewater systems, where grease and other organics are involved, the problem can be acute. Therefore it is good practice to flush the sample line with caustic* or chlorine on a routine basis. The frequency of flushing will have to be determined by experience and trial and error.

Amperometric Cell. This unit consists of a two-electrode system, one platinum or gold, and the other copper. The copper electrode loses metal ions in accordance with the amount of chlorine residual it is measuring, and so will have to be replaced in time.

The cell is equipped with some type of electrode bombardment system which keeps the electrodes clean and in molecular balance. A common type is grit, which has to be replaced periodically in the cell.

If the cell is operating on potable water, little maintenance is required, depending upon how clean the water is. Some water, particularly raw or untreated, will require the installation of a manually operated filter on the sample line.

If the cell is operating on wastewater, it is imperative that the sample line be equipped with a motorized filter. This operation requires more maintenance on the sampling system, because of the presence of suspended solids, fibrous materials, and grease, which are present in varying amounts, depending upon the type of wastewater treatment process.

A continuous and adequate flow of sample to the cell is the most important single factor in this type of operation. The maintenance and inspection program will vary widely from plant to plant.

Electrodes. The electrodes should be cleaned by flushing with fresh water. Do not disturb the sacrificial electrode. A typical amperometric analyzer consists of a copper sacrifical electrode and a noble metal (gold or platinum) permanent electrode.

The copper electrode needs to be replaced depending upon the amount of chlorine residual passing through the cell. Replacement periods might vary from 6 months to 3 years. A wastewater analyzer measuring residuals on the order of 5 mg/l will use up one copper electrode about every 9 months. It is permissible to clean the platinum electrode with a dampened soft cloth and mild abrasive.

If the output signal is erratic and the cell refuses to remain in calibration, then both electrodes should be cleaned with the recommended abrasives.

---

* Some operators at wastewater plants have installed a point of chlorination into the dechlorinated sample line which is chlorinated heavily for 10 min./day.

It is imperative that maintenance personnel become familiar with the manufacturer's instructions before cleaning these electrodes and for placing the analyzer back into operation.

Recorder or Indicator. This unit consists of plug-in transistorized components that are subject to the same maintenance problems as are other electric control devices. The most common failures are in the potentiometers for the various adjustments, the plug-in amplifier, the chart motor, and the relay mechanism on the control circuits.

Inspections. The following is a suggested routine for daily, weekly, and quarterly maintenance and inspection for a wastewater analyzer:

Daily

1. Check sample flow as follows:
   a. Make certain that pressure on inlet line of sample is adequate—about 25 psi just upstream of motorized filter.
   b. See that drain on motorized filter is in the wide open position and that the filter is continuously flushing.
   c. Check the electrode bombardment system to see that the grit or other devices are in proper motion (swirling around the platinum electrode).
2. Check the buffer pumps. Count the number of drops per minute to assure the proper pH of the sample entering the cell.
3. Check to see that the sample flow through cell is adequate.
4. With pH papers, check sample pH as it leaves the cell. It should be between 4.5 and 5.0—never > 5.0.
5. Check supply in buffer reservoirs.

Weekly

1. Shut off control signal (if any) to chlorination equipment.
2. Flush entire sample line up to and including motorized filter with high pressure plant water (50 to 75 psi).
3. Remove platinum or gold electrode assembly and wash thoroughly.
4. Add grit to cell upon restarting.
5. Recalibrate cell (span adjustment and zero adjustment), using back titration method with amperometric titrator. Be certain that buffer pumps are in the off position during zero adjustment.
6. The reagent additive system will not require any regular maintenance unless the pumping system fails to perform properly.

Quarterly

1. Clean entire cell assembly with soft, wet rag, to remove copper oxide from copper electrode. This will result in erratic operation for about twelve hours. After this time, molecular equilibrium is reached and stability of the system returns. This requires a second calibration with a cleaned copper electrode within twenty-four hours after starting.
2. Disassemble outer case of motorized filter strainer assembly; flush and clean.

Since the filter is so important on a wastewater system, it is desirable to have available a complete unit with motor drive for replacement or cannibalization.

AMPEROMETRIC TITRATOR

The best maintenance of this unit is frequent use—at least once per day. When not in use, the electrodes should be allowed to soak in 1 to 2 ppm free residual chlorine water. If the titrator is being used for wastewater analysis or for total chlorine residual only, then the electrodes should be allowed to soak in a 2 to 5 ppm free iodine solution. This keeps the electrodes sensitized to either free chlorine or free iodine, as these are the only elements measured by the titrator—just the free halogens.

The platinum electrode or electrodes should be periodically cleaned with a mild abrasive household cleanzer, applied with the fingers and then removed by washing in cold water. All electrode contact surfaces should be inspected and cleaned regularly.

If these steps are followed, the titrator will show sharp responses with clear-cut end points and repeatable results.

ELECTRICAL SWITCHGEAR

All electrical switchgear is normally housed in a room away from the chlorination equipment and chlorine supply. Such pieces of switchgear as the control and alarm devices should be in weatherproof enclosures so that any massive chlorine leak will not adversely affect the copper contact surfaces and exposed terminal strips. All switchgear manufactured by chlorinator manufacturers is protected in such a way; ancillary items manufactured by others should be similarly enclosed. In humid climates it is wise to place dessicators in the various types of switchgear housing if condensation is significant.

Good housekeeping is the best watchword throughout the chlorination facility for proper maintenance. The operator must not tolerate chlorine leaks or water leaks of any magnitude; he must not allow any equipment to become dirty either internally or externally. Any sign of corrosion should be immediately investigated.

If these guidelines are followed, the equipment will perform properly and be free from the hazards of chlorine gas.

MAINTENANCE: DECHLORINATION EQUIPMENT

SULFUR DIOXIDE SUPPLY SYSTEM

General Discussion. The characteristics of sulfur dioxide are so similar to those of chlorine that almost everything said about maintenance for chlorine is applicable to the sulfur dioxide system. However, there are some exceptions.

The life of the diaphragm protectors on the vapor and liquid pressure gages in an SO_2 system is much longer than for chlorine because of the low vapor pressure of SO_2 (35 psi).

The corrosivity due to the entrance of moisture is almost the same as that for chlorine, but this action is different on different metals. For example, in a chlorine system, line valves and header valves use monel trim, but these are not interchangeable for sulfur dioxide. Valves for the latter should have 316 SS trim.

Moisture Problems. Sulfur dioxide producers claim that the moisture content does not exceed 40 ppm.[3] However, White[4] has found that this can escalate to more than 100 ppm when a packager transfers the SO_2 from the producer's tank car into the packager's truck tanker. However, certain packagers have a much better and less corrosive product than others who may be somewhat careless in their transference of an otherwise acceptable product in a tank car to their tanker trucks. Recent experience in the San Francisco Bay Area indicate that some of the supplies of sulfur dioxide are more corrosive than chlorine by a factor of 10 to 1. This suggests lack of moisture control during the SO_2 transfer procedure. Attempts to quantify and monitor the moisture content by the user proved fruitless. However, a preventive maintenance procedure was worked out by one of the wastewater plant operators.[4] It is described below.

Preventive Maintenance. The following procedure requires an evaporator. It was the failure of three SO_2 evaporators which inspired this procedure. It will boil off any excess moisture that may accumulate in the supply system.

Once every day or once every three days, depending upon the SO_2 usage, the evaporator under consideration should be shut down by closing the inlet liquid line. The sulfonator vacuum relief line should be plugged and the sulfonator be allowed to operate so as to create a vacuum within the SO_2 evaporator container vessel. The moisture that may collect in the supply system will always float upon the liquid SO_2 in the evaporator container vessel. Therefore withdrawing all of the liquid SO_2 in the evaporator is step number 1. Step number 2 is the boiling off of the moisture in the container vessel by pulling a vacuum through the sulfonator while maintaining a temperature of 160–180°F in the SO_2 evaporator water bath.

The amount of moisture that collects in the container will depend upon the moisture content in the SO_2 and the rate of SO_2 withdrawals. Therefore each installation will require a special procedure of boil-off time and frequency of boil-off. In most cases moisture will be visible in the sulfonator rotameter tube, and when this moisture disappears the boil-off procedure can be terminated. This procedure requires redundancy in the SO_2 evaporator equipment to allow intermittent evaporator shutdown periods.

Those installations not using evaporators will probably not experience corrosion as severe as those with evaporators. This is directly related to the increased corrosive activity of the heated SO_2 caused by the elevated temperature of the moisture in the evaporators.

However, those systems not using evaporators suffer a malaise of considerable importance. Owing to the low vapor pressure of SO_2 it reliquefies readily in the supply system. This phenomenon interferes severely with the ability to meter the SO_2 with any degree of accuracy.

The best system for handling SO_2 dechlorination procedures is to use a storage tank that can be pressurized to 75–100 psi and deliver the SO_2 to conventional vaporizing equipment. Boiling off the moisture described above should reduce the corrosive maintenance problem to within the limits currently experienced by the usual chlorination system.

The above moisture removal procedure eliminated corrosion failure of the evaporators at the plant where three sulfur dioxide evaporators failed in less than six months. The chlorine evaporators at this same plant have been in operation for at least twenty years. This illustrates the magnitude of the problem and the vast difference between the packaging of chlorine versus sulfur dioxide.

The above procedure was devised after a careful examination was made of the lower section of the SO_2 container vessel in all three evaporators that failed. There were a series of pin-holes distributed around the circumference of the vessel right at the SO_2 liquid–vapor interface.

LEAK DETECTION

One of the major maintenance differences between chlorine and sulfur dioxide is that of continuous leak detection equipment. While the spot check use of a solution of NH_4OH to locate leaks is the same for SO_2 as it is for chlorine, the continuous leak detectors are vastly different. The procedure for detecting SO_2 in the air is much more involved and therefore the equipment is more complex than a chlorine leak detector. The best advice is to adhere to the manufacturers procedure for calibration and recommendations for maintenance.

SULFUR DIOXIDE CONTROL AND METERING SYSTEM

Everything that has been said for chlorine under this section applies to sulfur dioxide. The only precaution to the operator is this: never interchange chlorination equipment, particularly evaporators, for sulfur dioxide duty. While chlorination equipment and sulfonation equipment are, as presently manufactured, identical for all practical purposes, they should never be used alternately for chlorine and then sulfur dioxide. This is because a violent heat of reaction occurs when sulfur dioxide comes in contact with corrosion products of chlorine in an evaporator or chlorinator and vice versa. This reaction is most noticeable in an evaporator.

REFERENCES

1. Anon., "Installation, Operation and Maintenance Instructions," Book No. WAA 50.206, Wallace and Tiernan Div. of Pennwalt Corp., Belleville, NJ, 1979.
2. Anon., "Chlorine Leaks," American Gas and Chemical Co. Northvale, NJ, July 1983.
3. Anon., "Sulfur Dioxide Technical Handbook," Cities Service Co. Atlanta, GA, 5th ed., 1979.
4. White, G. C., "Equipment Corrosion from Sulfur Dioxide," report to a client, Aug. 1980.
5. Anon., "Series 50–125 Chlorine Detector," Wallace and Tiernan Div. of Pennwalt Corp., Belleville, NJ, June 1980.
6. Anon., "Operation and Maintenance Instructions, Chlorine Evaporator," Book WAA 50.206, Wallace and Tiernan Div. of Pennwalt Corp., Belleville, NJ, 1979.

12
Chlorine Dioxide

HISTORICAL BACKGROUND

INTRODUCTION

The major use of chlorine dioxide today is for the bleaching of wood pulp in the production of paper. In this process chlorine dioxide is generated from sodium chlorate using sulfur dioxide as the reducing agent (chlorate → chlorite). This method is not used for chlorine dioxide generation in small quantities.

In the early 1940s the Mathieson Chemical Co. (now Olin Corp.) made sodium chlorite available, produced from sodium chlorate. The on-site production of chlorine dioxide from chlorine and sodium chlorite followed shortly thereafter. The men from Mathieson who pioneered this process were J. F. Synan, J. D. MacMahon, and J. P. Vincent.[27] Until a few years ago the only supplier in the U.S.A. was the Olin Corp. However, in the late 1970s Kuhlman, a large chemical supplier in France, has been able to supply it to the North American markets at competitive prices.

Chlorine dioxide generated on-site from chlorine and sodium chlorite has been used to bleach flour for about thirty years. Chlorine dioxide has also been used successfully in treating reclaimed water used to transport fruit and vegetables in food processing plants. In potable water treatment it has successfully eliminated chlorinous tastes and odors, particularly those caused by phenol pollution. It has been found to be superior to chlorine in the removal of iron and manganese. It is slightly more potent than chlorine as a bactericide and viricide. Many distribution system problems related to water quality degradation have been solved with chlorine dioxide. More recently (ca 1972) it was discovered that chlorine dioxide does not promote formation of trihalomethanes (THMs) and is at the same time effective in reducing THM precursors.

NORTH AMERICAN PRACTICE

In 1938 Granstrom and Lee[1] reported on a survey of 52 U.S. plants using chlorine dioxide for taste and odor control. In 1978 a report by Public Technology Inc.,[2] advised that 84 plants were using chlorine dioxide. Only one of these, at Hamilton, Ohio, used the chemical solely for disinfection. The principal use in North America is for the control of taste and odor caused by phenolic compounds in the raw water supply. Most of the plants in the 1978 survey had been using chlorine dioxide for at least 15 years and had capacities of less than 5 mgd. Seventy-seven of the installations were located in Georgia, Ohio, Michigan, and Pennsylvania. A Wallace and Tiernan survey in 1981 recorded 260 installations in North America. Most of these were being used for destruction of phenolic tastes, and iron and manganese removal. In another survey,[2] Canada was reported to have ten installations, all of them in the province of Ontario. These installations used chlorine dioxide for the control of tastes and odors caused by industrial pollution of the receiving waters.

In 1972 the U.S. Environmental Protection Agency reported on the industrial pollution of the Lower Mississippi River in Louisiana.[3] The report disclosed the presence of chloroform (a known carcinogen) in the New Orleans potable water supply. It was believed to have formed as a byproduct of chlorination. This disclosure triggered further investigations confirming the formation of undesirable chloroorganics due to chlorination. A search to find alternatives to chlorination began in an attempt to reduce or eliminate the formation of trihalomethanes (THMs) caused by the then current practices of chlorination. The work by Miltner[4] revealed that chlorine dioxide does not form THMs in drinking water, and moreover it reacts to reduce the precursor concentration in the water. This revelation revived interest in the use of chlorine dioxide for potable water disinfection, and several installations have been made since that time, the most notable being the one located on the shores of Lake Mead, Nevada. This is a 400 mgd plant serving the Las Vegas area, owned and operated by the Las Vegas Valley Water District.

EUROPEAN PRACTICE

In the last few years two factors have changed the European approach to water treatment: one is the restriction of ammonia nitrogen content in the finished water and the desire to reduce THM formation. On December 19, 1978, the Ministers of the European Economic Community signed an agreement limiting the level of ammonia nitrogen (as NH_4) to 0.05 mg/liter as a standard and never at any time to exceed 0.5 mg/liter. The details of how compliance is resolved between the standard and maximum levels is not clear. However the water producers are changing their process train to meet the 0.05 mg/liter level. Standards for THM levels have not been set. It is likely that the E.E.C. will probably go along with the U.S. EPA recommendations. None of the above applies to the United Kingdom. They are not concerned with either ammonia-N or THM concentrations.

In Europe, where many surface water supplies suffered from industrial pollution, particularly phenol spills, a 1977 survey[7] reported several thousand chlorine dioxide installations at water treatment plants. The countries where it is used extensively included Switzerland, the Federal Republic of Germany, France, Belgium, and the Netherlands. Chlorine dioxide popularity in this area resulted from its superior power to oxidize phenols and chlorophenols without imparting off-flavors and without adding a chlorinous taste. It is being used for pretreatment to remove iron and manganese because of its superiority over chlorine. It is not uncommon to find the following train of processes in a French water treatment plant: ClO_2 pretreatment at raw water pump station (0.5–2.0 mg/liter); flocculation and sedimentation; filtration with sand and activated carbon; ozonation (1–2 mg/liter); clear well storage for 30 min; breakpoint chlorination to remove ammonia-N, if any; another storage reservoir; and then postchlorination of finished water for disinfection and distribution system protection (0.3–0.5 mg/liter dosage). A variation of this train of treatment might include dechlorination ahead of postchlorination to minimize THM formation, followed by chlorine dioxide for postchlorination. Activated carbon is the preferred dechlorinating agent because if sulfur dioxide is used there can be a significant return of ammonia-N.

Before the concern over trihalomethanes, chlorine was used in pretreatment for the removal of ammonia nitrogen. In the 1976 period, however, plants which had been using chlorine dioxide for the sole purpose of destroying the chlorophenols expanded their use of chlorine dioxide as described above. All three of the Paris suburban plants, located on the Seine, the Marne, and the Oise rivers, are using chlorine dioxide for both pre- and post-treatment (1980).

At Toulouse, on the Garonne River in southwestern France, chlorine dioxide is used in pretreatment and presterilization, taste and odor control and color removal. Prevention of THM formation was a key factor in many plants in France and other Western European countries, with the result that CIFEC of Paris, manufacturer of on-site chlorine dioxide generating equipment, installed 75 units in 1977 and a like number in 1978 and 1979, and the number is still growing. Practically all of these installations are switchovers from chlorine. Typical ClO_2 dosages in France vary from 1 to 2.5 mg/liter. Chlorine dioxide has been used in French water supplies for 17 years. The chief of water quality control for the city of Paris is enthusiastic about the benefits of chlorine dioxide and is not in the least concerned about the health hazards of the chlorite or chlorate ions resulting from the application of ClO_2.

Operating personnel at the Paris plants have pointed out that chlorine dioxide produces a more palatable water than the combination of chlorine and ozone. Whenever the chlorine dioxide is shut down the change in finished water taste is immediately noticeable. This is usually followed by consumer complaints.

At the Taillefer plant in Brussels, Belgium, chlorine dioxide is used in treating the Meuse River waters to oxidize the organic complexes of iron and manganese, following "presterilization" by chlorine. In Germany and Switzerland chlorine dioxide is used mainly for post-treatment to provide a stable residual in the distribution system. A typical dosage is about 0.3 mg/liter. Whenever this dosage is insufficient the process treatment train is adjusted and revised to reduce the oxidant demand.

GREAT BRITAIN PRACTICE

The water supplies of Great Britain do not suffer from industrial pollution to the same extent as those in Western Europe. Only a few chlorine dioxide installations existed in the late 1970s. Chlorine dioxide has apparently been successful in controlling water quality problems in distribution systems for the Severn–Trent Water Authority[9] and the Huntington Treatment Plant of the Liverpool Corporation.[10] The latter turned to chlorine dioxide when the ammonia nitrogen content in the water escalated. This condition caused free chlorine residuals far in excess of the plant's dechlorination equipment capacity. As chlorine dioxide does not react with ammonia nitrogen, it was possible to produce a proper quality water with chlorine dioxide dosages of less than 0.5 mg/liter, whereas chlorine dosages as high as 4.5 mg/liter were required to produce sufficient free chlorine residuals to maintain proper water quality in the system.

The remaining few chlorine dioxide installations in Britain are being used primarily for the control of chlorophenolic tastes and other tastes and odors that are usually aggravated by chlorine. There are also some installations used for color removal. Although the THM problem aroused greater interest in the process, there is no official move as yet to require abandonment of the free residual process practiced almost universally since 1950.[11]

PHYSICAL AND CHEMICAL PROPERTIES

The properties of chlorine dioxide, like those of chlorine, must necessarily be described in two separate categories: (1) gas and (2) aqueous solutions. Chlorine dioxide is almost never used commercially as a gas because of its explosiveness. For potable water and wastewater treatment processes, it is only used in aqueous solutions.

Chlorine dioxide was discovered by Humphrey Davy in 1814. He prepared the gas by pouring a strong solution of sulfuric acid on potassium chlorate. Subsequently Millon replaced sulfuric acid with hydrochloric. This basic reaction is still followed today in commercial production using sodium chlorate instead of potassium chlorate:

$$NaClO_3 + 2HCl \longrightarrow ClO_2 + Cl + NaCl + H_2O \qquad (12\text{-}1)$$

Chlorine dioxide gas has an intense greenish yellow color with a distinctive odor similar to chlorine but more irritating and more toxic. The odor is evident at 14–17 ppm in air, and at 45 ppm it is irritating.* The gas is soluble in water

---

* Studies on guinea pigs showed that 350 ppm was rapidly fatal, 150 ppm caused death in 44 min, whereas 45 ppm and less did not cause death after several hours exposure.[13]

at room temperature. At 25°C it is about 23 times as concentrated in aqueous solution as in the gas phase at atmospheric pressure (760 mm Hg) with which it is in equilibrium. Therefore it is about ten times more soluble than chlorine.[12] Water solubility of chlorine dioxide depends upon temperature and pressure. The solubility of chlorine is about 7 g/l at 20°C and atmospheric pressure. Therefore according to Gordon et al.[12] chlorine dioxide solubility in water is about 70 g/l at the same conditions. In waterworks practice chlorine dioxide solution is made under partial pressure conditions (vacuum). These solutions have been known to reach concentrations of 40 g/l.

Chlorine dioxide vapor is a deeper shade of green than chlorine. The odor, while somewhat similar to that of chlorine and chlorine monoxide, is detectably different to experienced personnel. A gas at ordinary temperatures, it can be compressed to a liquid. It has a density of 2.4 (air = 1), a boiling point of 11°C, and a melting point of −59°C.[15] Exposure to chlorine dioxide gas has, in addition to the usual toxic effects of chlorine gas, an effect of producing violent headaches and general fatigue lasting for several days. Chlorine dioxide is an unstable gas which is explosive at temperatures higher than −40°C;[13] therefore, it must be generated at the point of use. The major use of chlorine dioxide is for bleaching pulp used in the manufacture of paper. Chlorine dioxide is almost never used commercially as a gas because of its explosiveness. Concentrations of the vapor (about 11 percent) in air may give mild explosions or "puffs," whereas concentrations over about 14 percent in air will sustain a decomposition wave set off by an electric spark.

Chlorine dioxide explodes when its temperature is being raised, when it is being exposed to light, or when allowed to come into contact with organic substances. This tendency to explode at the slightest change in environment is even greater when it is in the compressed liquid state. It has been observed that the mere transfer from one container to another can cause an explosion. Furthermore, these explosions are of the same magnitude as those of hydrogen–oxygen mixtures.[16] Dilution of ClO_2 gas with inert gases reduces the explosion hazard. Industrially, ClO_2 is handled by mixing with air so that the chlorine dioxide content is between 8 and 12 percent.[15]

Exposure of the gas to light results in photochemical decomposition.[17] At first exposure to light, a large quantity of red liquid forms on the walls of the containing vessel. Continued exposure renders the liquid colorless. The products of this type of decomposition are chlorine heptoxide (Cl_2O_7), chlorine monoxide (ClO), chlorine (Cl_2), and possibly oxygen.

One of the important physical properties of chlorine dioxide is its high solubility in water, particularly chilled water. Since it is about ten times more soluble than chlorine this difference in solubility provides a practical means of separating the two gases in the manufacturing process. Paradoxically, it is extremely volatile and can be easily removed from dilute aqueous solutions by a minimum of aeration.

Although chlorine dioxide is soluble in water it does not react chemically with water as does chlorine. For example, it is easily expelled from solution in water by blowing a small amount of air through the solution. This phenomenon together with the difference in solubility of ClO_2 and Cl_2 at certain temperatures is the basis upon which these two gases can be separated in some manufacturing processes. For this reason ClO_2 is not stable while in solution in an open vessel. Moreover, the strength of solution deteriorates rapidly under these circumstances. This inherent instability of chlorine dioxide solutions requires that solution lines be designed so *that there is no possibility of ClO_2 gas coming out of solution.*

It is significant to note that as the chlorine dioxide leaves the solution, the yellowish green color of the solution fades and finally becomes colorless. The remaining solution will contain negligible amounts of chloric acid ($HClO_3$), indicating a very weak acid reaction with water. Aqueous solutions of ClO_2 are also subject to some photodecomposition. This reaction is a function of both time and intensity of the ultraviolet light source, and the products are chloric and hydrochloric acid.[18] The rate of photodecomposition from an ultraviolet light source is low as compared to decomposition from its volatility, described above.

Aqueous solutions of ClO_2 will retain their strength for several months if properly stored in the dark.

The vapor pressure of ClO_2 is sufficient to be troublesome. Even the weakest of solutions gives off an objectionable odor. For example, a 1 mg/liter solution will provide a vapor pressure of 10 mg/liter in the air above the solution at equilibrium. A concentration this low gives an objectionable odor.[13]

Chlorine in dilute aqueous solutions has virtually no vapor pressure nor odor since it reacts with the water to form chloride ion, hypochlorous acid, and hypochlorite ion. In strong acid solutions, however, chlorine dioxide is much more soluble than is chlorine.[13]

The high volatility of chlorine dioxide in an aqueous solution gives it a characteristic which is most desirable in the field of odor control of foul air. This is often associated with wastewater treatment and certain odoriferous industrial processes which contribute to air pollution.

METHODS OF GENERATING CHLORINE DIOXIDE

INTRODUCTION

The methods used for generating on-site ClO_2 are sharply divided into the classifications large and small. For large production units such as the bleaching of paper pulp and textiles, ClO_2 is generated from sodium chlorate ($NaClO_3$). For small production units such as potable water, wastewater or industrial processes, ClO_2 is generated from sodium chlorite ($NaClO_2$). Since sodium chlorite is commercially produced by the reaction of ClO_2 with H_2O_2 in an NaOH solution, it becomes apparent that the sodium chlorate route is the most economical for large-scale operations.

CHLORINE DIOXIDE

For large-scale production, four different processes are currently used in North America. They are generally named for the company which developed them or the reducing agent used. They are as follows:

1. The Mathieson or SO_2 process.
2. The Solvay or methanol process.
3. The Hooker R-2 process.
4. The Hooker SVP® process.

Most of the chlorine dioxide produced in North America is by the first two processes.[13] Increasing tonnage requirements call for changes in the process which will reduce capital and operating costs. To this end the chloride reduction processes (Olin chloride reduction, Hooker R-2 and Electric Reduction Co. ER-2) are becoming a major factor.[13]

THE MATHIESON OR SULFUR DIOXIDE REDUCTION PROCESS

This process consists of blending a 45 percent solution of sodium chlorate with 66°Bé sulfuric acid in the top of a reaction vessel.[19] Air containing 10 percent SO_2 is blown into a diffuser at the bottom of this vessel and chlorine dioxide plus air is extracted at the top of the vessel.

The basic reaction is:

$$2\ NaClO_3 + H_2SO_4 + SO_2 \longrightarrow 2\ ClO_2 + H_2SO_4 + Na_2SO_4 \qquad (12\text{-}2)$$

Side reactions also take place, including:

$$2\ NaClO_3 + 5\ SO_2 + 4\ H_2O \longrightarrow Cl_2 + 3H_2SO_4 + Na_2SO_4 \qquad (12\text{-}3)$$

Figure 12-1 illustrates the Mathieson process. This process produces chlorine dioxide in an air or nitrogen mixture and is not contaminated by the presence of chlorine. The exit gases flow through a scrubber (against a portion of the chlorate feed) to remove any sulfur dioxide leaving the generator.

THE SOLVAY OR METHANOL PROCESS

Methanol is used as the reducing agent in this process. The overall reaction is written as follows:

$$2NaClO_3 + CH_3OH + H_2SO_4 \longrightarrow 2ClO_2 + HCHO + Na_2SO_4 + 2H_2O \qquad (12\text{-}4)$$

An intermediate reaction is:

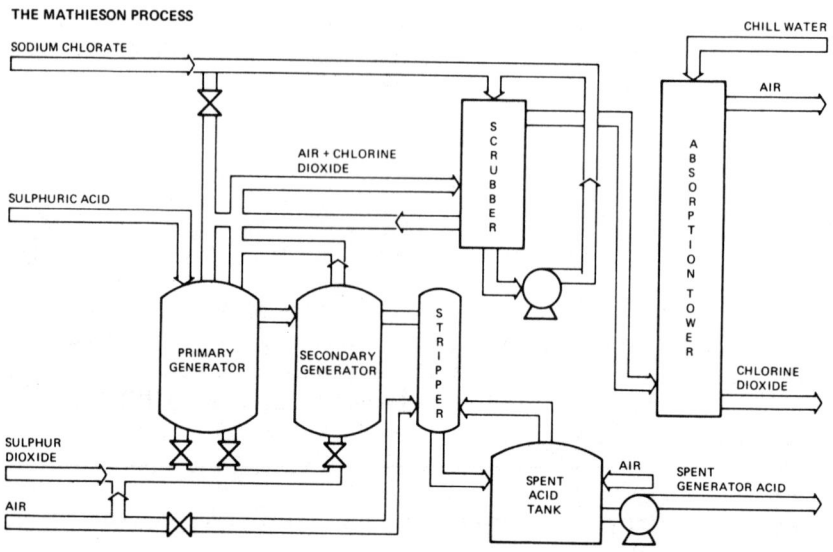

Fig. 12-1. Mathieson process for generating chlorine dioxide.

$$2NaClO_3 + CH_3OH + 2H_2SO_4 \longrightarrow ClO_2 + \tfrac{1}{2} Cl_2 + CO_2$$
$$+ 2NaHSO_4 + 3H_2O \quad (12\text{-}5)$$

The reaction efficiency may be determined by analyzing the exit gas or the chlorine dioxide solution for relative quantities of ClO_2 and Cl_2. Since Cl_2 should not be present as a reaction product, its level is an indirect measure of the reaction efficiency. This procedure may not be applied to the Mathieson system, however, since any SO_2 escaping will destroy the Cl_2 and give erroneous results.

The operation of the Solvay process is quite similar to the Mathieson process, except that the methanol is added to the incoming stream of the 45 percent sodium chlorate solution immediately ahead of the generator. In this process methanol and 75 percent H_2SO_4 are fed to the secondary generator. This drives the reaction to completion. As a result the acidity is increased. The ClO_2 is recovered by the same technique used in the Mathieson process. The reaction rates in the Solvay process are considerably slower than in the Mathieson system, and therefore high operating temperatures are required. The primary generator is run at about 60°C and the secondary at 63°C.[20]

CHLORIDE REDUCTION

This process consists of blending 66°Bé sulfuric acid with a 20 percent sodium chlorate solution and a 10 percent sodium chloride solution in a reaction vessel.[20] Air is blown into the bottom of the vessel driving off the gases formed which are diluted by the incoming air. The gas is composed of 1 part chlorine dioxide, ½ part chlorine, and air. This goes to an absorption tower with water at 42°F which absorbs the ClO₂ and produces a solution of about 8 gm/liter. The chlorine gas is extracted from the top of the absorption tower and is sent to another absorber. The basic reaction of this process is:

$$2NaClO_3 + 2NaCl + 2H_2SO_4 \longrightarrow 2ClO_2 + Cl_2 + 2Na_2SO_4 + 2H_2O \quad (12\text{-}6)$$

JASZKA-CIP PROCESS

There are additional routes to chlorine dioxide generation which are tailored to substantially lower production rates lower than those described above. The Canadian International Paper Process has fallen into disuse because of its somewhat lower efficiency.[20] However, for substantially lower production rates this process is an attractive alternative. In this process the sulfuric acid required is generated on site by the overreduction of sodium chlorate.

The reaction is:

$$2NaClO_3 + SO_2 \longrightarrow 2ClO_2 + Na_2SO_4 \quad (12\text{-}7)$$

In this process, SO₂ is introduced into the base of a packed column and flows countercurrent to a stream of concentrated sodium chlorate solution. The flow of chlorate solution is relatively low in contrast to the volume of gas; therefore, the process may be started and stopped quickly.

Jaszka of Hooker (U.S. Patent No. 3,950,500; 1976) made a significant improvement in this process by introducing chlorine with the sulfur dioxide in the packed tower. This reduces the chlorate requirement to oxidize the sulfur dioxide to sulfuric acid. Therefore, increased efficiencies are obtained.

It is of interest to note that the Societe Universelle de Prodicts Chimiques et d'Appareiliages of Paris has described a chlorine dioxide generator based on the sodium chloride reduction of sodium chlorate in sulfuric acid solution.[20] This unit has a capacity of up to 1000 lb/day ClO₂ which makes it very interesting for wastewater or potable water applications.

LABORATORY PREPARATION

There are several methods available for laboratory generation of chlorine dioxide. The most popular is described in *Standard Methods*[21] and illustrated in Fig. 12-2. This method was used by Hood[22] and Roberts et al.[23,24] This method utilizes the reaction of sulfuric acid with sodium chlorite. A compressed air supply is used to transfer the chlorine dioxide gas to a final flask which represents the stock solution. This flask must be in an ice bath, otherwise the strength of the

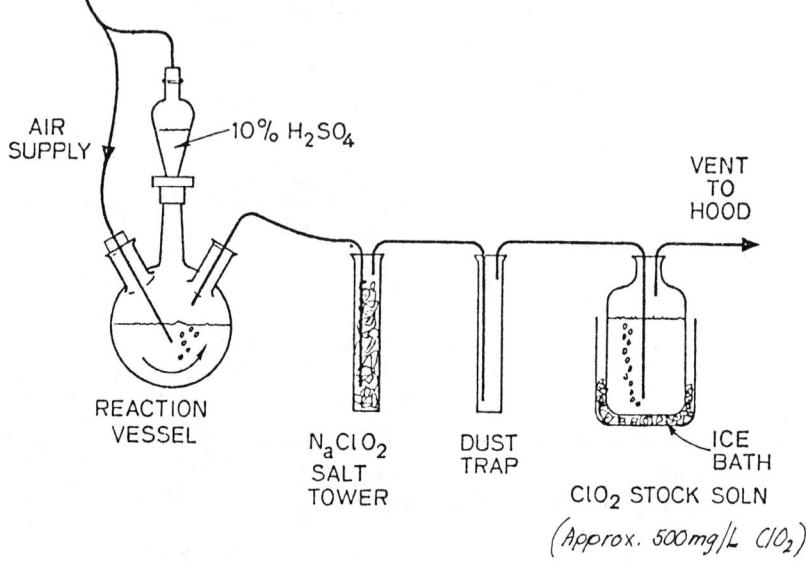

Fig. 12-2. Laboratory chlorine dioxide generator using acid activator of sodium chlorite.[23]

solution will vary considerably. The reaction should be performed in subdued light to prevent decomposition of the stock solution.

Several stoichiometric relations have been reported for acid activation of a chlorite solution.[12] The two which are most widely accepted for sulfuric acid are:

$$4NaClO_2 + 2H_2SO_4 = 2Na_2SO_4 + 2ClO_2 + HCl + HClO_3 + H_2O \quad (12\text{-}8)$$

and

$$10NaClO_2 + 3H_2SO_4 = 8ClO_2 + 5Na_2SO_4 + 2HCl + 4H_2O \quad (12\text{-}9)$$

The first reaction, Eq. (12-8), is catalyzed by the chloride ion, which is also a product of the reaction. It has been reported not only that the reaction is accelerated by the chloride ion, but also that the stoichiometry is altered to that of the second reaction, Eq. (12-9).[12] Some observers have reported that chlorine is formed during this reaction and is one of the products along with the chlorine dioxide, but most of the evidence appears to be against the formation of significant amounts of chlorine.[12]

When chlorine dioxide is generated from sodium chlorite by acid activation, chlorine dioxide is not the only end product. The final composition is dependent upon several variables, among which are: sodium chlorite concentration; purity of sodium chlorite used; acid concentration; pH of reaction mixture; reaction time; temperature of reaction.

The Stanford group[23] performed a mass balance on chlorine in their laboratory reactor which generated the chlorine dioxide. The concentration of sodium chlorite in the reaction vessel was $0.083M$, whereas the concentration in commercial reactors is typically $2.0M$. Reaction time was 3 hours to assure that the reaction was complete. The following equation shows the initial composition of the reaction mixture equated to the final composition (pH 1.75):

$$0.083ClO_2^- + 0.0004\ ClO_3^- + 0.0177Cl^-$$
$$= 0.0037ClO_2^- + 0.0206ClO_3^- + 0.0424Cl^-$$
$$+ 0.0380ClO_2 + 0.0007\ Cl_2 \qquad (12\text{-}10)$$

Collecting terms and relating all species to a reaction involving four moles of chlorite—analagous to the reaction in Eq. (12-8)—the yield is:

$$4ClO_2^- = 1.92\ ClO_2 + 1.02\ ClO_3^- + 1.25\ Cl^- + 0.04\ Cl_2 \qquad (12\text{-}11)$$

This stoichiometry is essentially that of Eq. (12-8).

It can be seen from the above that the yield from the laboratory generator using sulfuric acid as the chlorite activator is less than 50 percent.

COMMERCIAL ACID–CHLORITE GENERATORS

Rio Linda Chemical Co. The Stanford group[23] evaluated a full-scale continuous generator using sulfuric acid as the chlorite activator. The recovered mass of chlorine atoms in their investigation represented 104 percent, well within the limits of experimental error. Using 4 moles of chlorite ion the stoichiometry of this generator was found to be as follows:

$$4ClO_2^- = 2.63ClO_2 + 0.53ClO_3^- + 1.01Cl^- + 0.05Cl_2 \qquad (12\text{-}12)$$

This result can be compared with the laboratory reaction, Eq. (12-11), or on the basis of 10 moles of reacting ClO_2^- as follows:

$$10ClO_2^- = 6.58ClO_2 + 1.32ClO_3^- + 2.52Cl^- + 0.12HOCl \qquad (12\text{-}13)$$

Table 12-1 shows that only 78 percent of the added chlorite is participating in the reaction, and of that 78 percent, only 66 percent is converted to chlorine dioxide. This gives an overall yield of 51 percent; that is, only 51 percent of the added chlorite is converted to chlorine dioxide in the sulfuric acid–chlorite commercial generator. Therefore under the best of conditions the maximum yield of chlorine dioxide to be expected from the sulfuric acid activator of sodium chlorite is in the range of 50–55 percent.

Table 12-1 Sulfuric Acid Activated Chlorite

Reactant	Chlorine Atoms (moles/liter)
ClO_2	0.00
ClO_2^-	0.0214
ClO_3^-	0.0003
Cl_2	0.00
Cl^-	0.0021
Total	0.0238
Final Composition (pH 2.1)	
ClO_2	0.0109
ClO_2^-	0.0048
ClO_3^-	0.0025
Cl_2	0.0001
Cl^-	0.0063
Total	0.0246

The Stanford group[23] then used the Rio Linda generator to evaluate the hydrochloric acid activation of sodium chlorite to produce chlorine dioxide. A mass balance on chlorine atoms was again performed for this generator. The results of this mass balance are shown in Table 12-2.

The stoichiometry expected from hydrochloric acid activation of sodium chlorite is as follows:

$$5NaClO_2 + 5HCl = 4ClO_2 + 5NaCl + HCl + 2H_2O \qquad (12\text{-}14)$$

Table 12-2 shows that 97 percent of the original chlorite has reacted. This compares to 77 percent reported for sulfuric acid activation. It also shows that 78.2 percent of the reacted chlorite is converted to chlorine dioxide. In three field experiments the average conversion of chlorite to chlorine dioxide was 78.5 percent. Therefore it is reasonable to expect consistent reactor yields of 75 percent. This is less yield than the chlorine/chlorite method, because only 80 percent of the theoretical quantity of chlorine dioxide is produced from the same amount of sodium chlorite as when chlorine is used.

One of the most important aspects of both acid methods is the purity of the final composition of the product. The amount of free chlorine formed is negligible in both systems. This is important in two ways: (1) the absence of free chlorine in the chlorine dioxide solution means that no THMs will be formed by the addition of this solution to a process stream; and (2) no free chlorine means that all chlorine residual species (free and combined) will be absent in the process stream, eliminating the necessity of chlorine species differentiation in analytical procedures for chlorine

Table 12-2 Hydrochloric Acid Activated Chlorite

Reactants	Chlorine Atoms (moles/liter)
ClO_2	0.00
ClO_2^-	0.0110
ClO_3^-	0.0000
Cl_2	0.0000
Cl^-	0.0260
Total	0.370
Final Composition (pH 2.2)	
ClO_2	0.0086
ClO_2^-	0.0003
ClO_3^-	0.0003
Cl_2 and $HOCl$	0.0001
Cl^-	0.0277
Total	0.0369

residuals. This overcomes one of the major difficulties in the use of chlorine dioxide.

Combining sodium chlorite with an acid has long been a popular method for generating pure chlorine dioxide gas in the laboratory for bench scale studies. This method is also being used at water treatment plants in Italy and Switzerland. Three water treatment plants in Zurich are using the method. The system at the Lengg plant in Zurich is described by Valenta and Gahler.[25] This system was put into operation in 1974. It is a batch-type system as compared to a continuous generator. The Swiss system receives 32 percent hydrochloric acid and 24.2 percent sodium chlorite solution. These chemicals are put into separate storage tanks. These chemicals are then transferred to separate dilution tanks. The hydrochloric acid is diluted to 9 percent and the sodium chlorite to 7.5 percent. The dilution water must be softened. According to Valenta and Gahler[25] this acid–chlorite system requires 25 percent more sodium chlorite to produce a pound of chlorine dioxide than the chlorine–chlorite method. The latter is described below.

Fischer and Porter Co. Fischer and Porter Co. began marketing an on-site continuous generator in Europe in 1978. This unit is based upon the Zurich system in that it uses hydrochloric acid activation of sodium chlorite. The Fischer and Porter system[26] is designated as Series T70G 1000 and is currently limited to about 300 lb/day chlorine dioxide. This system differs from the Zurich system in three respects: (1) it is a continuous process, whereas the Swiss process makes chlorine dioxide on a batch basis; (2) it uses an injector system which delivers a diluted chlorine dioxide solution to the point of application, whereas the Swiss system pumps the concentrated chlorine dioxide solution; and (3) the Fischer and

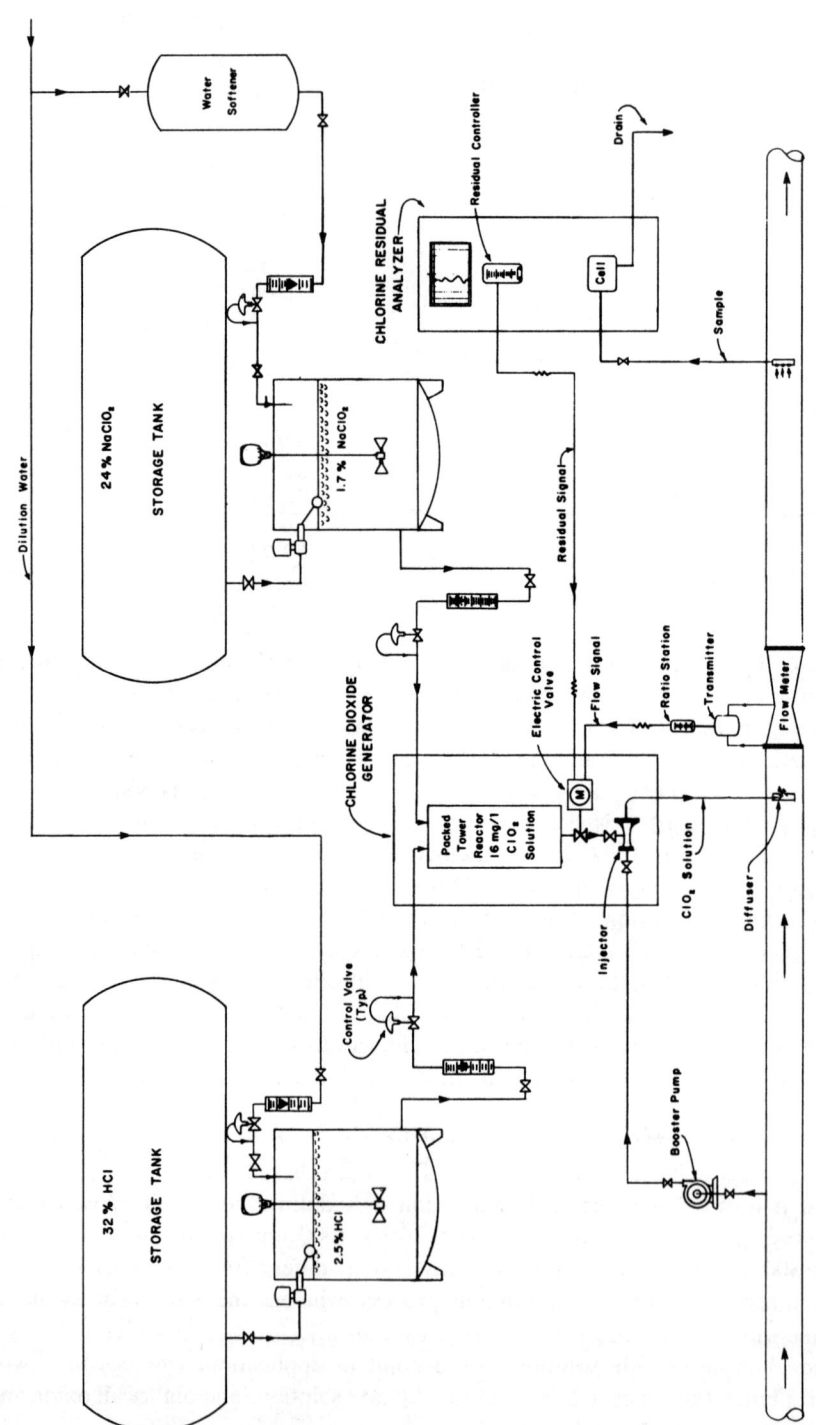

Fig. 12-3. Acid-chlorine method of chlorine dioxide on site generation.

Porter system dilutes 32 percent hydrochloric acid to 2.5 percent, and 24 percent sodium chlorite solution to 1.7 percent versus 9 percent and 7.5 percent, respectively, for the Swiss batch system.

The Fischer and Porter acid–chlorite system schematic is shown in Fig. 12-3. It is not an exact replica of their Series 70G 1000 system because of certain proprietary considerations. A 94 percent yield is claimed by Fischer and Porter for this system. Valenta and Gahler in the shakedown of their batch system at the Lenng plant claim yields of 91–96 percent. However, they list free chlorine production of 4.0–7.5 percent. This compares to Fischer and Porter's claim that their system does not generate any free chlorine. The latter claim is in agreement with the results of the investigations by the Stanford group.[23]

CHLORINE–CHLORITE SYSTEMS

General Discussion. The chlorine–chlorite process reaction can be described in its simplest form as follows:

$$2NaClO_2 + Cl_2 \longrightarrow 2ClO_2 + 2NaCl \tag{12-15}$$

Thus, 1.34 lb pure sodium chlorite ($NaClO_2$) will react with 0.5 lb chlorine to produce 1.0 lb chlorine dioxide. The technical sodium chlorite used in this process is only about 80 percent pure, so that in practice 1.68 lb chlorite will be required for the above reaction if a 100 percent yield is to be achieved. If the chlorine and chlorite are reacted stoichiometrically, the resulting pH is close to 7. However, the production of chlorine dioxide proceeds more favorably at a much lower pH. In order to fully utilize the more expensive of the two ingredients, excess chlorine is used, which lowers the pH and drives the reaction further toward completion. While this increases the yield it is detrimental to the purity of the final solution because of the excess chlorine. Excess chlorine enters into a competing reaction which has the effect of destroying the chlorine dioxide once it has been formed:

$$HOCl + ClO_2 \longrightarrow ClO_3^- + HCl \tag{12-16}$$

In this reaction chlorine dioxide is oxidized to chlorate ion. If more than a small excess of chlorine is present, the negative effect becomes sufficient to offset the benefits that the excess chlorine should have provided. Some researchers have claimed maximum yield when the pH is on the order of 3.5–4.0.[1,31] In practice it has been found that a lower pH is more desirable.[29]

The goal of chlorine dioxide generation is to produce a solution of the highest purity with the greatest yield.

Purity. Purity is defined as the ratio of chlorine dioxide to the total of all oxidative chlorine compounds as determined by iodometric titrations:

$$P = \frac{ClO_2}{ClO_2 + Cl_2 + NaClO_2} \times 100 \text{ percent}$$

Yield. Yield is defined as the amount of chlorine dioxide generated as a percent of the amount of chlorine dioxide that could theoretically have been generated according to Eq. (12-15).

The Stanford group[23] were able to achieve chlorine dioxide yields between 93 and 98 percent when chlorine was about 4 percent in excess of the stoichiometric requirement based upon chlorite. The reaction time in the laboratory reactor was far in excess of that used in continuous commercial generators. This method of generation has the potential of a significantly better yield than the acid-chlorite method.

Effect of Sodium Chlorite Alkalinity. The pH of commercial grade sodium chlorite solutions (25–38 percent) is about 12. They contain about 2–3 percent sodium carbonate and 2 percent sodium hydroxide.* The alkalinity contained in the sodium chlorite solution must be neutralized before the pH can be lowered to permit optimum conditions of the chlorine–chlorite reaction. It will take 5 parts of chlorine to neutralize each percent sodium carbonate in the sodium chlorite solution. This means that in actual operation of this method the system must be able to commit itself to providing a small excess of chlorine to overcome the sodium chlorite alkalinity ($NaCO_3$ + $NaOH$).

Effect of Injector Water Alkalinity. It requires about 5 percent excess chlorine to neutralize each 100 mg/l alkalinity in the injector water.

It is apparent that the chlorine–chlorite needs "acid assistance" to provide optimum conditions for the stoichiometric reaction between chlorine and chlorite. Therefore the operator will eventually have to choose between excess chlorine or supplementary acid.

Analysis of Solutions:

General Discussion. It is necessary to analyze the chlorine dioxide solution at the discharge of the reactor in order to monitor the operation of a generating system. The concentration of this solution may be in the range of 5,000–40,000 mg/l ClO_2. In this range chlorine dioxide is extremely volatile and will vaporize easily from an aqueous solution. Therefore in sampling this solution for analysis precautions must be taken to minimize contact of the sample with the air. This can be accomplished as follows: Place a flexible sample line that reaches the bottom of the sample container and allow several volumes of the sample to overflow the

---

*The AWWA specification for aqueous sodium chlorite solution proposes a limit (by weight) of 6.5 percent sodium carbonate and 3.0 percent for sodium hydroxide.[70]

container, then slowly remove the sample line and cap the sample container so that there is no air space in the container. The sample container should be fully protected from any sunlight to avoid photolytic decomposition of the ClO_2.

The analysis of these solutions is a complicated procedure compared to the analysis of chlorine species. The Stanford group[23,32] examined two methods to be used in their study. These were: the DPD–ferrous titrimetric and the amperometric method. They encountered continuing problems with the DPD–ferrous technique, which they abandoned in favor of the amperometric method.

The Stanford group also measured the total chlorine dioxide concentration with a spectrophotometer using a calibrated curve based upon ultraviolet radiation absorbance by ClO_2. Using this method, a reaction product sample was diluted so that the ClO_2 concentration would be in the range of 5–60 mg/l.[23]

Chemical Reactions.[32] The chemical basis for the analytical procedure, which consists of successive titrations of combinations of chlorine and chlorine dioxide species are pH dependent in accordance with the following reactions with the iodide ion:

$$Cl_2 + 2I^- = I_2 + 2Cl^- \tag{12-17}$$

pH: 7, 2, <0.1

$$2ClO_2 + 2I^- = I_2 = 2ClO_2^- \tag{12-18}$$

pH: 7

$$2ClO_2 + 10I^- + 8H^+ = 5I_2 + 2Cl^- + 4H_2O \tag{12-19}$$

pH: 2, <0.1

$$ClO_2^- + 4I^- + 4H^+ = 2I_2 + Cl^- + 2H_2O \tag{12-20}$$

pH: 2, <0.1

$$ClO_3^- + 6I^- + 6H^+ = 3I_2 + Cl^- + 3H_2O \tag{12-21}$$

Reaction Steps. The reduction of chlorine dioxide to chloride requires the transfer of five electrons. This reaction requires two steps. In the first step, which is carried out at pH 7, one electron is transferred to the chlorine dioxide molecule to form chlorite, Eq. (12-18). This represents one-fifth of the chlorine dioxide reduction.

When the pH is lowered to 2, the second step reduction occurs by the transfer of four more electrons to the chlorite molecule. This reduces the chlorite molecule to chloride, which completes the reduction of chlorine dioxide, Eq. (12-20). The overall reduction of chlorine dioxide to chloride is given by Eq. (12-19). Any chlorite present in a sample in which the pH is lowered to 2 will also be reduced to chloride, Eq. (12-20).

Analytical Procedure[32]:

Glassware. The glassware used for the analysis of chlorine dioxide solution should be stored separately from other laboratory glassware and not be used for any other purposes. Chlorine dioxide reacts with glass to form a hydrophobic surface-coating on the glassware. To satisfy any ClO_2 demand, all glassware should be aged in a strong solution of chlorine dioxide solution (200–500 mg/l) for 24 hours, and rinsed only with water between uses.

Reagents

1. Standard PAO solution 0.00564N
2. Phosphate buffer solution pH 7
3. Potassium iodide granules, reagent grade.
4. Saturated sodium phosphate (Na_2HPO_4) solution: Prepare a saturated solution of $Na_2HPO_4 \cdot 12H_2O$ with hot de-ionized distilled water
5. Potassium bromide (KBr) solution, 5 percent: Dissolve 50 g of KBr and dilute to 1 l; store in a brown glass stoppered bottle; make fresh weekly
6. Concentrated hydrochloric acid (12M HCl)
7. Hydrochloric acid solution (2.5M HCl): Dilute concentrated acid
8. Purge gas: Nitrogen gas is required to purge chlorine dioxide from samples prior to analysis for the remaining species; gas must be free from contaminants and passed through a 5 percent KI scrub solution; discard the solution at the first sign of color

Sample. A convenient volume for titration is 200–300 ml for steps 1–4. The sample volume for step 5 should be 15 ml. This can be adjusted to provide an appropriate dilution, but the ratio of sample to HCl must be maintained.

Procedure
   Step 1.
      a. Add 1 ml pH 7 buffer solution to 200 ml titrator sample jar.
      b. Add 200 ml of sample to jar with minimum air contact.
      c. Add 1 g KI granules while sample is being stirred.
      d. Titrate with PAO to endpoint.
      e. Record ml PAO/ml sample.
This is *Reading A.*
   Step 2.
      a. Continuing with the sample used in Step 1, add 2 ml of 2.5M HCl.
      b. Allow mixture to react in the dark for 5 min.
      c. Titrate with PAO to endpoint.
      d. Record ml PAO/ml sample.

This is *Reading B*.
Step 3.
a. Add 1 ml pH 7 buffer solution to the purge vessel.
b. Add sample.
c. Purge with nitrogen gas for 15 min. Use a gas-dispersion tube for proper gas–liquid contact.
d. Add 1 g KI granules while stirring sample.
e. Transfer sample from purge vessel to titrator sample jar.
f. Titrate with PAO to endpoint.
g. Record ml PAO/ml sample.

This is *Reading C*.
Step 4.
a. Continuing with sample from step 3, add 2 ml of 2.5M HCl.
b. Allow this mixture to react in the dark for 5 min.
c. Titrate with PAO to endpoint.
d. Record ml PAO/ml sample.

This is *Reading D*.
Step 5.
a. Add 1 ml of KBr and 10 ml of 12M HCl to 50 ml reaction flask and mix.
b. Carefully add 15 ml of sample with a minimum of air contact. Mix and stopper flask immediately. Allow mixture to react in the dark for 20 min.
c. Add 1 g KI granules and shake flask.
d. Transfer reacted mixture from the reaction flask to titrator sample jar containing 25 ml of saturated NA_2HPO_4 solution. Rinse reaction flask thoroughly with de-ionized distilled water and add rinse water to titrator sample jar. Final volume for titration should be about 200–250 ml.
e. Titrate with PAO to endpoint.
f. Repeat steps 5a–5e using distilled de-ionized water instead of sample in order to determine blank value.
g. Record ml PAO for sample − mL PAO for blank/ml sample.

This is *Reading E*.

Titration Summary

1. $Cl_2 + \frac{1}{5} ClO_2$ = A
2. $\frac{4}{5} ClO_2 + ClO_2^-$ = B
3. Cl_2 (only the part not volatized by the nitrogen purge). = C
4. ClO_2^- = D
5. $Cl_2 + ClO_2 + ClO_2^- + ClO_3^-$ = E

Arithmetic Calculations. All calculations are based upon the equivalents of PAO titrant required to react with the equivalents of oxidants present. The equivalent

Table 12-3 Equivalent Weights for Calculating Concentrations on a Mass Basis

pH	Species	Molecular Weight (mg/mol)	Electrons Transferred	Equivalent Weight (mg/eq)
7	ClO_2	67,450	1	67,450
2, 0.1	ClO_2	67,450	5	13,490
7, 2, 0.1	Cl_2	70,900	2	35,450
2, 0.1	ClO_2^-	67,450	4	16,863
0.1	ClO_3^-	83,450	6	13,908

weights of the chlorine dioxide species are all pH dependent. Table 12-3 tabulates the equivalent weights.

In the following equations N is the normality (eq/l) of the titrant (PAO = $0.00564N$). A–E are ml PAO/ml sample.

1. ClO_2, mg/l = $5/4 \times (B - D) \times 0.00564 \times 13490$
2. Chlorine, mg/l = $\dfrac{A - (B - D)}{4} \times 0.00564 \times 35450$
3. Chlorite, mg/l = $D \times 0.00564 \times 16683$
4. Chlorate mg/l = $[E - (A + B)] \times 0.00564 \times 13908$

Chloride. To complete the analysis of the chlorine dioxide solution the concentration of chloride ion should be determined. The procedure to use is the mercuric nitrate method (408B) as detailed in *Standard Methods,* with the following exception: The chlorine dioxide solution sample should be diluted to yield a final concentration of approximately 10 mg/l. To 100 ml of this solution add 4.0 ml of indicator–acidifier reagent instead of 1.0 as specified in *Standard Methods.* Titrate the sample as specified with $0.141N$ mercuric nitrate. This is Reading F. Then,

$$\text{Chloride, mg/l} = \frac{F(\text{ml}) \times 0.141 \times 35450}{\text{sample vol. (ml)}}$$

Wallace and Tiernan Analytical Procedure. The Wallace and Tiernan group preferred the iodometric titration method to determine the amounts of chlorine dioxide, chlorine, and sodium chlorite in the generator discharge. The species determination is done by performing three titrations using aliquots of the same sample of concentrated solution exiting the generator. One sample is acidified, one neutral, and one causticized followed by neutralization to pH 7.[29,33] Some precautions are recommended by Wallace and Tiernan: In the first titration it is essential that the potassium iodide be added immediately to the sample. This prevents disproportionation of the chlorine dioxide to chlorite. The KI must likewise be added before acidification. This prevents loss from volatilization. In the third titration pH 12 was found to be better than pH 11. A reaction time of one hour in the dark is recommended before reduction to pH 7 in preparation for titration. With these precautions results of the titrations agreed with spectrophotometric analysis of chlorine dioxide. The following is a summary of the titrations:

pH 2 I_2 observed $= Cl_2 + 0.4ClO_2 + 0.5ClO_2^-$
pH 7 I_2 observed $= Cl_2 + 2ClO_2$
pH 12 then pH 7 I_2 observed $= Cl_2$

The spectrophotometric method of analysis of the generator solution discharge is an alternative to the iodometric method. It has the advantages of being simple and rapid. However, most samples should be diluted before analysis. With the precautions cited above the iodometric titrations will agree with spectrophotometric analysis when it is assumed the extinction coefficient is 1150 and that chlorine dioxide has maximum absorption at about 3600 A.[12]
Since

$$ClO_2 \text{ conc. (moles/liter)} = \frac{A = \text{max. absorption}}{E = \text{extinction coeff.}}$$

it is apparent that the lower the number for E gives a higher value for concentration. According to Kieffer and Gordon[12] the average of seven published values of the extinction coefficient is 1196.

A commonly used field test of the chlorine dioxide yield is to operate the system with the chlorinator turned off and measure the chlorite concentration in the discharge. Then with the chlorinator operating, determine the chlorine dioxide concentration in the discharge. The yield is then calculated by percent conversion. Since chlorine and chlorite are mutual interferences in spectrophotometric analysis[33] purity cannot be determined by this method.

The Olin System. The Mathieson Chemical Corp. (now Olin) first made sodium chlorite available in 1940. It was first used in water treatment in 1944.[27] Prior to the introduction of sodium chlorite, chlorine dioxide had always been generated by the reduction of sodium chlorate. This method is not amenable to the small quantities used in water or wastewater treatment. The sodium chlorate reduction method is only applicable when chlorine dioxide is produced on a large scale (100–500 tons/day). These quantities are used in the bleaching of wood pulp used in paper making.

The availability of sodium chlorite made possible the production of chlorine dioxide from existing chlorination equipment. Wallace and Tiernan Co. (now a Division of Pennwalt Corp.) pioneered this process in 1944 in cooperation with Mathieson Chemical Co.[34] The chlorination equipment was modified so as to merge the chlorine solution discharge from a conventional solution-feed gas chlorinator with a solution of sodium chlorite from a diaphragm metering pump. Both of these chemical solutions were fed to the base of a glass reaction tower filled with porcelain Raschig rings. As this mixture flowed upward through the tower, chlorine dioxide would form. The contact time was only about 1 minute. When the chlorine dioxide was forming properly the mixture in the tower turned a greenish yellow color. Absence of this color indicated that no chlorine dioxide was forming. Wallace and Tiernan advised that the solution strength should not be less than 500 mg/l. Under ideal conditions this would provide a metering range of ClO_2 of about 6:1. To obtain the best possible yield of ClO_2 from this system requires the use of excess chlorine.[1,28] This insures the complete activation of the chlorite in the time available in the reaction tower. At molar ratios of less than 1:2 of chlorine to chlorite, the activation will not be complete. Since chlorite costs about 85¢/lb this is an undesirable situation. Therefore two conditions have to be met for the maximum yield from this system in the reaction time available. (1) A chlorine to chlorite ratio of 1:1 by weight, and (2) proper pH (3.5–4). At chlorine concentrations of 500 mg/l this pH range will be satisfied in most waters used for the chlorinator injector supply. This method will produce a chlorine dioxide solution containing about 30 percent ClO_2 and 70 percent $HOCl$. At pH 4 the concentration of molecular chlorine will be negligible. This system is still being used in water treatment plants for the destruction of phenols, control of taste and odors, and reclaimed water in food processing plants.

Olin Water Services of Kansas City. This firm provides a special service for promoting the sale of their sodium chlorite product. They can provide a complete generating unit to retrofit an existing chlorinator to produce chlorine dioxide from chlorine gas. Such a unit is illustrated in Fig. 12-4. Olin Water Services makes this equipment available to the customer with the proviso that the sodium chlorite be purchased from Olin. They can also furnish their Model 350 Dioxolin complete for generating ClO_2 from hypochlorite. This system is equipped with three metering pumps, one for hypochlorite, one for sodium chlorite, and one for sulfuric acid to acidify the two reacting solutions to the optimum pH. It is rated at 192 lb/day capacity ClO_2 production.

The Olin system provides the customer with a 25 percent solution of sodium chlorite which has been cut back from the hot (85°F) 50 percent solution. This is delivered in 55 gal drums and eliminates the handling problems of the solid sodium chlorite and the hot solution. The Olin system is also provided with an optional acid injection point to cover situations where the existing or proposed chlorine gas injection system is liable not to be able to meet the conditions of optimum pH. Owing to the short reaction time and limited $HOCl$ concentration in the chlorinator injector discharge, the best ClO_2 yield of this method is on the order of one part ClO_2 to 2.5 parts of chlorine. Since this system does not put out a "pure" solution of chlorine dioxide, proof of the results of the process

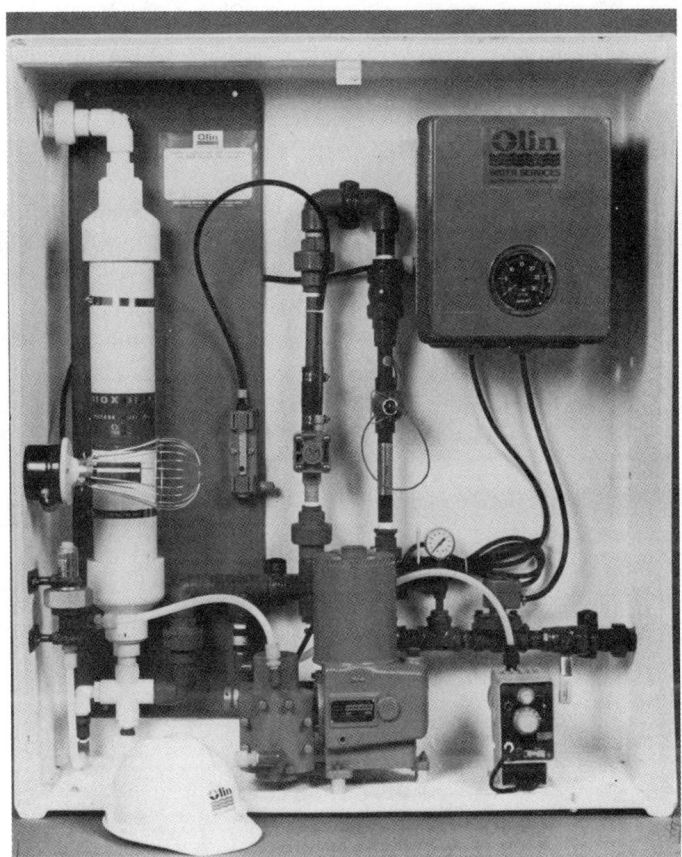

Fig. 12-4. Olin Water Services model 150 Dioxolin chlorine dioxide generator (courtesy Olin Water Services Co.).

is greatly complicated by the difficulty of isolating the true chlorine dioxide residual from the various species of chlorine residuals forming as a result of the predominant species (HOCl) in the solution discharging from the ClO_2 generating tower.

There are limitations to the Olin once-through generating system. It is not particularly adaptable to potable water or wastewater treatment processes where the chemical demand varies over a wide diurnal range (on the order of 6–8:1, based on both chemical demand and flow change). It is most suited to situations of constant demand such as cooling tower circuits and food-plant process waters.

The Wallace and Tiernan System. This method of chlorine dioxide generation is a modification of the chlorine/chlorite process.[29,30] In addition to the usual mixing of chlorine solution with a solution of sodium chlorite, a small amount of hydrochloric acid is added. During the research of this method it was found that suppressing the formation of chlorate ion by avoiding excess chlorine and independent regulation of pH provides maximum yield.[29] Additionally, the presence of the chloride ion acts as a catalyst in the reaction to produce chlorine dioxide:

$$2NaClO_2 + HOCl + HCl \longrightarrow 2ClO_2 + 2NaCl + H_2O \quad (12\text{-}22)$$

Operating data indicate that a 93 percent yield is attainable (see definition of yield). For design purposes yield should be calculated at 90 percent. This system will produce a chlorine dioxide purity of 95 percent. This is accomplished with a ratio of 0.5, chlorine to chlorite, and 0.1 mole of H^+ ion added as hydrochloric acid. This mole ratio and acid added was found to be optimum. For example, the yield at 0.55 ratio was equal, but nearly 7 percent excess chlorine (unreacted) resulted. Excess chlorine should be avoided because it causes a competing reaction that has the effect of destroying the chlorine dioxide which has already formed:[23]

$$HOCl + ClO_2 \longrightarrow ClO_3^- + HCl \quad (12\text{-}23)$$

Also excess chlorine in the chlorine dioxide solution should be avoided if THM formation in the process stream is of concern.

Figure 12-5 illustrates the relation of unreacted chlorite versus mole ratio of

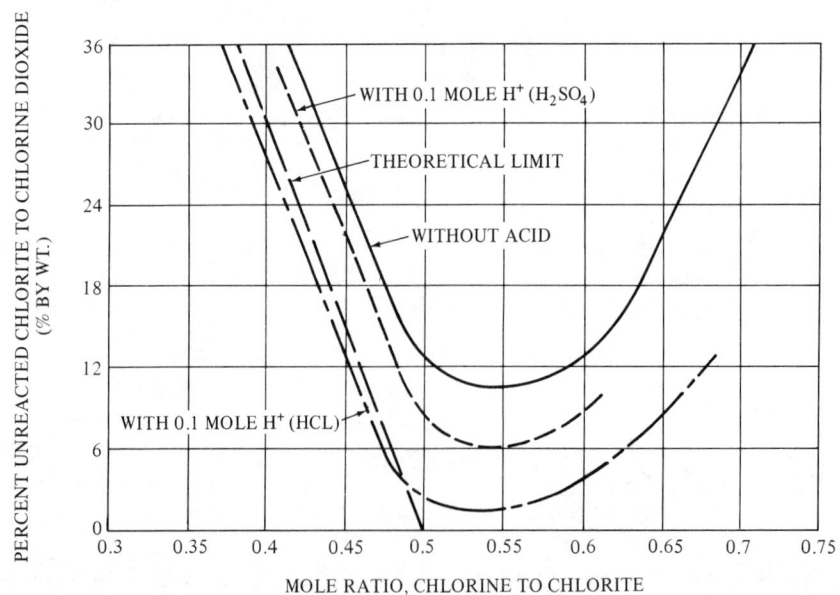

Fig. 12-5. Percent unreacted chlorite versus mole ratio. (courtesy Wallace and Tiernan Div. Pennwalt Corp.).

chlorine to chlorite. The mole ratio should be chosen to produce the smallest percentage of unreacted chlorite.

Typical chemical requirements for this system are as follows: 10 lb of chlorine, 15.2 gallons of 16.7 percent sodium chlorite solution, and 15.2 gallons of 0.8 percent hydrochloric acid solution for each 17 lb of chlorine dioxide produced.

The first installation of this Wallace and Tiernan system was at the Tempe, Arizona water treatment plant in 1980. This system is currently available in sizes up to 510 lb/day ClO_2. Larger sizes are available as custom sized systems. They are available with or without acid. A 10 percent excess of chlorine is recommended for units without acid. Fig. 12-6 illustrates the modified chlorine/chlorite generator using acid. The Wallace and Tiernan system is readily adaptable to both flow pacing and residual control.

Fischer and Porter System. Fischer and Porter Co. designed a ClO_2 generating system in 1980 which was based upon a slightly different approach than other chlorine/chlorite systems. This system is based upon the mixing of chlorine and chlorite solutions just before entering the reactor vessel. Special attention was given to the mixing of these chemicals followed by a minimum of 7 minute reaction time in the reactor. Pilot tests on this system were most encouraging. The solution purity ranged from 95 to 98 percent. The first prototype of this system was installed at the Las Vegas Valley Water District 400 mgd treatment plant located near Boulder City, Nevada on the shores of Lake Mead. Operating experience has revealed that it is necessary to use a 5 percent excess of chlorine to achieve maximum chlorite conversion.[38] It was also found that chlorine solution concentrations as low as 500 mg/liter is sufficient to keep the solution pH below 4 with Lake Mead water. Furthermore, when the pH reaches or exceeds 4 the yield drops off about 2 percent. At this plant a loss of one percent yield amounts to about $6000 per yr. in chemicals. The yield from this installation is reported to be as high as 97 percent.[38] The purity of the solution is usually on the order of 95 percent.

Monitoring the efficiency of the chlorine dioxide produced is quite often accomplished by back calculation. This procedure consists of shutting off the chlorine to the reaction tower and then measuring the chlorite in the tower discharge. This is followed by turning the chlorine on and measuring the chlorine dioxide in the solution discharge from the reacting tower. The yield is then calculated as a percent conversion.

The operating personnel advise that the isolation, differentiation and measurement of the chlorite and chlorate ions in the treated water is a major problem. The EPA requires that all chlorine dioxide installations treating potable water must be monitored for these species and that the sum of the two must not exceed 1.0 mg/liter. This analytical problem is one of the major disadvantages of chlorine dioxide use.

The system described above is readily adaptable to both flow pacing and residual control.

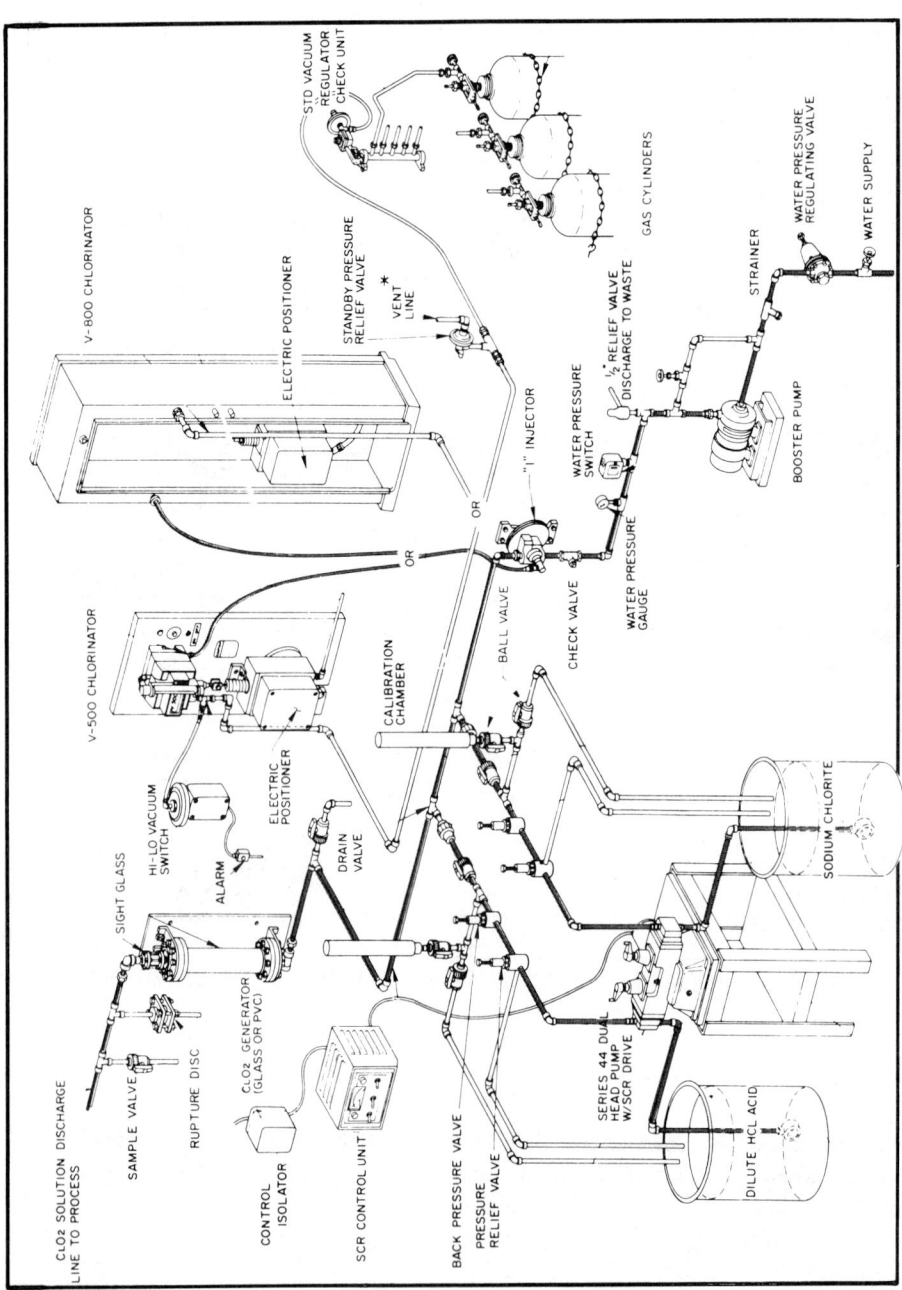

Fig. 12-6. Chlorine dioxide generating system with supplemental acid feed. (courtesy Wal-

CHLORINE DIOXIDE 859

The CIFEC System. This system utilizes a chlorine solution enrichment loop and a longer reaction time in the reaction tower than does the Olin method. The manufacturer claims that the chlorine dioxide solution discharge has a purity of 95 percent.[35] See Fig. 12-7. The recirculating pump powers the chlorinator injector and delivers the chlorine dioxide solution to the point of application. The backpressure at the point of application plus the friction loss in the solution line must not exceed 30 psi. Back-pressures greater than this require a chlorine dioxide solution pump or another injector located downstream from the generator. Each installation is hydraulically customized. This usually means a recirculating pump with specific characteristics for each installation. The service water pressure must be at least 2 psi greater than the pressure at the exit of the generator. The continuous recirculation of the injector discharge is designed to maintain an HOCl concentration between 4000 and 5000 mg/l. In most waters this would result in a pH somewhat less than 2 in the chlorine solution. Under these conditions the chlorine solution would contain about 65 percent molecular chlorine and 35 percent HOCl. This would make it imperative to have at least 5 psi pressure in the chlorine–chlorine dioxide solution lines at all times. Otherwise the molecular chlorine or the ClO_2 might come out of the solution as a gas. Fig. 12-7 illustrates the completely automatic CIFEC generator. This system was designed for a varying water flow and chemical demand situations which characterize both potable water and wastewater. The size of the reactor vessel varies with the capacity of the system. Currently these systems are available in capacities from 1–10 lb/day up to 50–1000 lb/day ClO_2.[35] These units are equipped with alarms to warn the operator of any malfunction and to automatically shut the system down in case of loss of chlorine, chlorite, or makeup water or in case of power failure.

This equipment is adaptable to flow pacing plus residual control. Any 4–20 mA analog signal from both a flow meter with ratio station and a chlorine residual analyzer with a 4–20 mA transmitter fed into a multiplier can be properly characterized through controllers to regulate the three variables: chlorine, chlorite, and makeup water.

The manufacturers of this equipment advised in 1982 that there were more than 200 of these units installed in Western Europe and 7 in the USA. The "Bioxyde de Chlore" system can be purchased separately and retrofitted to existing chlorination equipment.

The Rio Linda Chemical Co. System.[36] This recently patented system appeared on the market about 1980. It is different because it utilizes a vacuum injection system for the sodium chlorite. Instead of injecting the sodium chlorite solution into the discharge of the chlorinator injector, it is discharged into the vacuum line of the chlorinator immediately upstream from the injector (see Fig. 12-8). Therefore the sodium chlorite solution reacts directly with the chlorine gas under a vacuum before being diluted with the injector operating water. Rio Linda Chemical Co. claim that the reaction between the chlorite and chlorine

860 HANDBOOK OF CHLORINATION

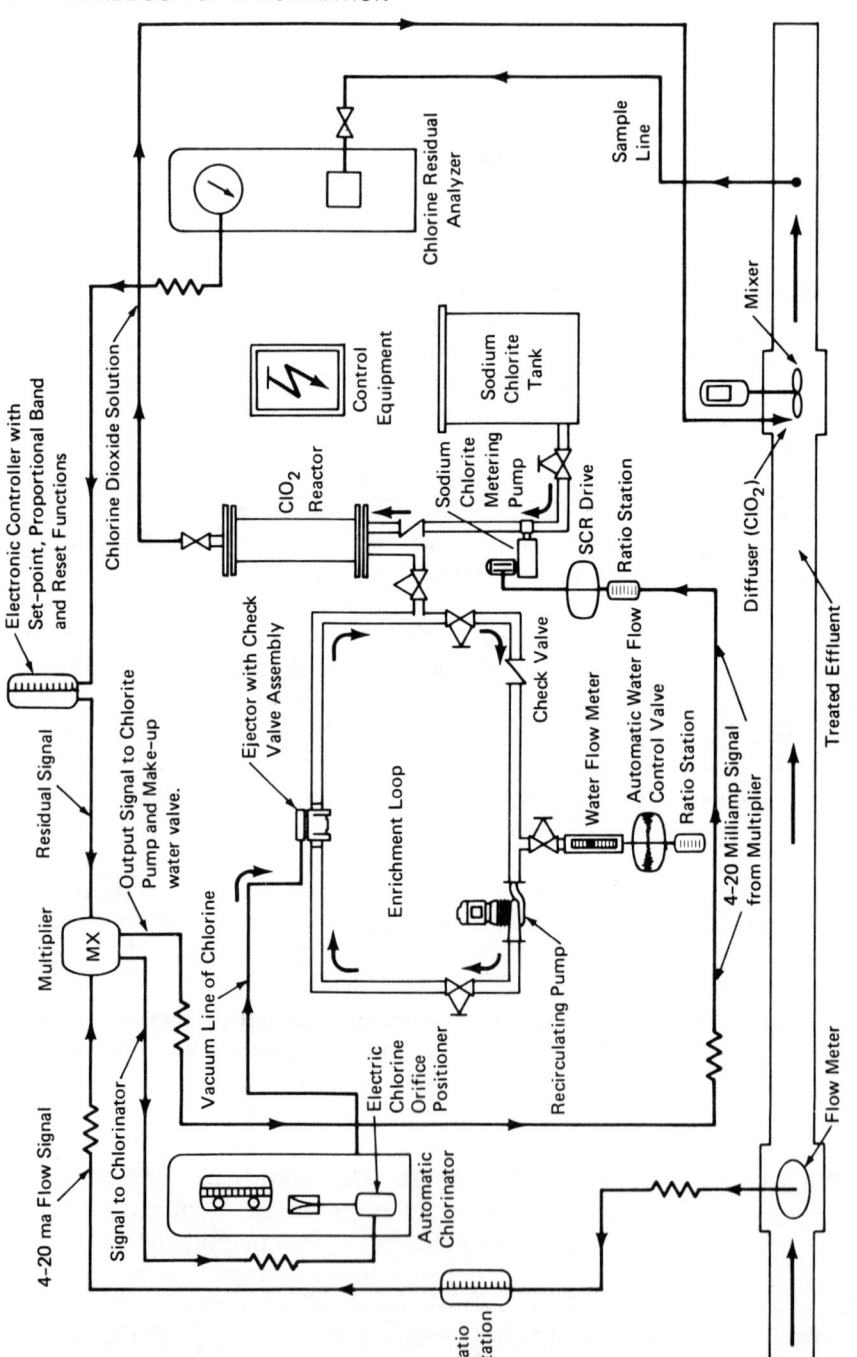

Fig. 12-7. CIFEC automatic residual control chlorine dioxide generating system. (courtesy CIFEC Paris, France).

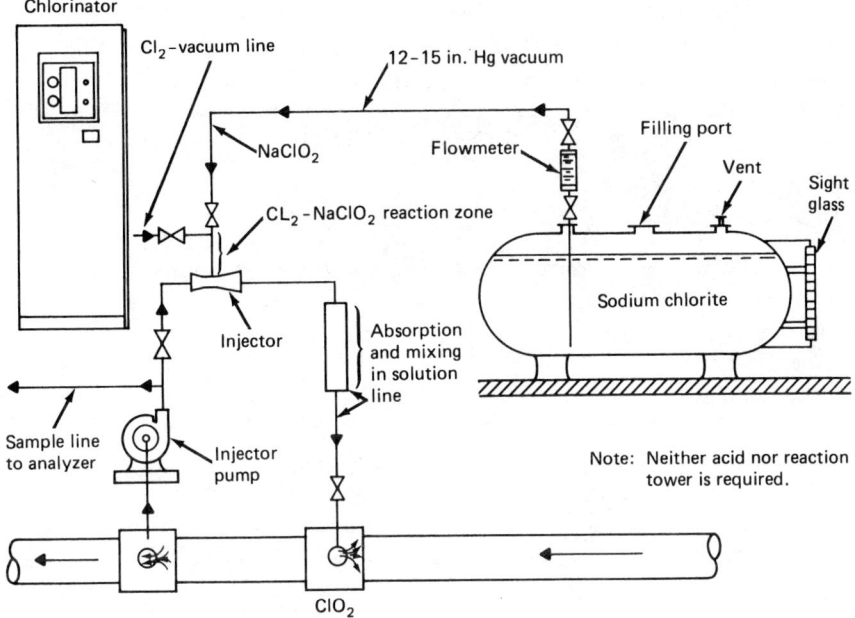

Fig. 12-8. Vacuum type chlorine dioxide generator system (courtesy Rio Linda Chemical Co.).

gas is 100 times faster than the reaction between chlorite and chlorine solution. They also claim yields of 95–98 percent and will guarantee in writing a 95 percent yield. The reaction proceeds as follows:

$$2NaClO_2 + Cl_2 \text{ (gas)} \longrightarrow 2ClO_2 \text{ (gas)} + 2NaCl \qquad (12\text{-}24)$$

This system is different from other chlorine/chlorite systems in another respect. The feed rate of the sodium chlorite solution is controlled by the vacuum produced by the chlorinator injector. Vacuum levels of 12 in. Hg or greater provide satisfactory operation of the chlorite feed system.

This system is not required to provide a controlled water supply to produce an optimum reaction pH in the discharge of the injector. Therefore the only requirement is to supply injector operating water in sufficient quantity and pressure to develop the maximum capacity of the chlorinator for a given back-pressure at the local conditions.

The operating range of this system should be considered 10:1, at least until more field experience is available for automatic control units.

The fear of possible chlorine dioxide explosions occurring at the point of chlorine–chlorite mixing has been considered. Rio Linda Chemical Co. claim that explosions will not occur when chlorine gas and sodium chlorite are mixed under a vacuum.[37]

If the vacuum fails the flow of chlorine gas and sodium chlorite ceases and so does the reaction that produces chlorine dioxide.

Most existing chlorination equipment can be retrofitted to use this system for chlorine dioxide production. However, certain precautions must be taken and the chlorinator operating conditions need to be investigated on a case by case basis.

CHEMISTRY OF CHLORINE DIOXIDE

The chemistry of chlorine dioxide in water treatment is not completely understood. In many instances it has been thought of as a more powerful oxidant than chlorine, but this is not so in the narrow pH ranges encountered in water or wastewater treatment practices.

Although chlorine dioxide does not belong to the family of "available chlorine" compounds (those chlorine compounds that hydrolyze to form hypochlorous acid), nevertheless, the oxidizing power of chlorine dioxide is referred to as having an "available chlorine" content of 263 percent, calculated as follows:

The chlorine in chlorine dioxide is 52.6 percent by weight. Since the chlorine atom undergoes five valence changes in the process of oxidation to the chloride ion:

$$ClO_2 + 5e^- = Cl^- + 2O^= \qquad (12\text{-}25)$$

the equivalent available chlorine content is $52.6 \times 5 = 263$ percent. In effect this indicates that chlorine dioxide theoretically has 2.63 times the oxidizing power of chlorine. This can be substantiated by the reactions in the liberation of iodine from iodide in the acid starch–iodide analytical procedure:

Chlorine dioxide:

$$ClO_2 + 5I^- + 4H^+ \longrightarrow Cl^- + 2\tfrac{1}{2}I_2 + 2H_2O \qquad (12\text{-}26)$$

Chlorine:

$$HOCl + 2I^- \longrightarrow OH^- + Cl^- + I_2 \qquad (12\text{-}27)$$

The implication that chlorine dioxide displays an oxidizing power five times that of chlorine is very misleading because its oxidizing power is pH dependent. In the neutral pH range where all of the potable water and wastewater reactions usually occur chlorine dioxide is reduced to chlorite; but at pH 2 it is reduced to chloride and undergoes five valence changes Eq. (12-26). However, when chlorine dioxide is reduced to chlorite there is only one valence change. This is the reason for the one-fifth factor used in calculating ClO_2 concentrations by the iodometric analytical procedure.

$$ClO_2 \text{ (aq)} + e^- \longrightarrow ClO_2^- \qquad (12\text{-}28)$$

Therefore at pH 7, chlorine dioxide demonstrates only one-fifth of its potential oxidizing power.

At one time it was thought that when chlorine dioxide destroyed phenols it utilized its full oxidizing potential of 263 percent "available chlorine."[15] According to Masschelein[33] this is not so. For aqueous phenol solutions ranging from 10^{-5} to $10^{-4} M$ in the presence of an excess chlorine dioxide at pH 6.9–7.0, the first attack is accomplished in 30 minutes. Chlorine dioxide consumption increases rapidly to 2 moles per mole of phenol oxidized. The consumed chlorine is found in the form of chlorite at least at the rate of 90 percent. At this pH chlorite does not react significantly with phenols.

The oxidizing power of chlorine dioxide can be compared to chlorine based upon equivalent weights. The equivalent weight equals the molecular weight divided by the number of electrons transferred:

$$Cl_2 + 2e^- \longrightarrow 2Cl^- \tag{12-29}$$

Therefore the equivalent weight for the chlorine reaction is $70.90/2 = 35.45$. And for chlorine dioxide it is $67.45/1 = 67.45$ [see Eq. (12-28)]. So $67.45/35.45 = 1.9$. This calculation indicates that 1.9 mg of ClO_2 is equivalent in oxidizing power to 1.0 mg of chlorine.

The redox couple expressed by Eq. (12-29) is 0.954 V.[39] The redox couple for the hypochlorous acid to chloride reaction is 1.49 V.

Although chlorine dioxide is extremely soluble in water it does not react chemically with water. There is some doubt as to whether or not it disproportionates according to the following equation:

$$2NaClO_2 + H_2O \longrightarrow 2ClO_3^- + 2ClO_2^- + 2H^+ \tag{12-30}$$

If this reaction occurs at all in either acidic or neutral solutions, there is agreement in the literature that it is very slow. However, according to Feuss[40] ClO_2 can be forced to disproportionate to chlorate and chlorite by raising the pH to 11 or 12 with caustic:

$$2ClO_2 + 2OH^- \longrightarrow ClO_3^- + ClO_2^- + H_2O \tag{12-31}$$

Ammonia. One chemical property of chlorine dioxide that is unique in water treatment is that *it will not react with ammonia nitrogen,* but it will oxidize nitrites to nitrates.

Sulfur Dioxide. At neutral pH or above, sulfurous acid (H_2SO_3) reduces chlorine dioxide to chlorite. At low pH it is reduced to chloride.[43] Therefore chlorine dioxide can be dechlorinated by sulfur dioxide or the sulfite ion.

Bromides. Chlorine dioxide does not oxidize bromides to HOBr or Br_2 as does chlorine.[12]

Ozone. When using chlorine dioxide as a final treatment preceded by ozonation, care must be taken to minimize or destroy the ozone residual because chlorine dioxide is oxidized by ozone to Cl_2O_6.[12] The hydroperoxides formed during ozonation act to dechlorinate both ClO_2 and HOCl.

Nitrogen Trichloride. This compound has the ability to oxidize chlorite to produce chlorine dioxide.[42]

Hypochlorous Acid. When chlorine is present as HOCl it will oxidize chlorine dioxide to chlorate:

$$2\ ClO_2 + HOCl + H_2O \longrightarrow 2ClO_3^- + Cl^- + 3H^+ \qquad (12\text{-}32)$$

This reaction is much faster in a neutral solution than in an acidic solution.[12] However, hypochlorous acid reacts slowly with the chlorite ion in a neutral solution:[41]

$$HOCl + ClO_2^- + OH^- \longrightarrow ClO_3^- + Cl^- + H_2O \qquad (12\text{-}33)$$

FORMATION OF HALOGENATED ORGANICS

It has been well documented that chlorine dioxide does not form THMs (trihalomethanes).[7] It does, however, have the ability to significantly reduce the precursors of THMs (30–40 percent reduction).

Chlorine dioxide reduces to chlorite within normal pH ranges encountered in potable water and wastewater treatment when it reacts with organics.[33]

When phenols are encountered and the ClO_2 dosage is insufficient to destroy the phenols, then the reaction will produce chloroquinones, chlorohydroquinones, and chlorophenols. These products will impart a strong medicinal taste to the water.

Compounds which can be halogenated by chlorine dioxide are those with carbon–carbon double bonds.[33] The Stanford group[23] compared the formation of TOX in two wastewater effluents between chlorine dioxide and chlorine. Their results showed insignificant amounts of TOX increase in 24 hours by ClO_2. However, with chlorine the increase was significant. This suggests a low level liability of TOX formation with the use of chlorine dioxide.

GERMICIDAL EFFICIENCY

Since the introduction of chlorine dioxide (1944) for the treatment of water supplies, several investigations have been made to determine its germicidal efficiency. All of these investigations were carried out as a comparison to chlorine. Most of these comparisons were performed in sterile, chlorine-demand-free systems. Therefore chlorine dioxide was being compared to free chlorine in these instances. More recently several investigations of chlorine dioxide versus chlorine in ammonia-laden wastewater have been made. These and others are described below.

In 1947, Ridenour and Ingols[44] concluded that chlorine dioxide was at least as effective as chlorine, and in contrast to chlorine the bactericidal efficiency of chlorine dioxide was relatively unaffected by pH values between 6 and 10. Ridenour and Armbruster[45] found in 1949 that less than 0.1 mg/liter chlorine dioxide destroyed the common water pathogens *Eberthella typhosa, Shigella dysenteriae,* and *Salmonella paratyphi B* at temperatures between 5°C and 20°C at pH values above 7 with a 5-min. contact period. An increase in pH brought about an increase in germicidal efficiency.

Ridenour and Ingols[46] reported in 1949 that *chlorine dioxide was clearly superior to chlorine in the destruction of spores on an equal OTA residual basis.* Using a 5-min. contact period, a 99.9 percent reduction of *B. subtilus* in a chlorine demand-free chlorine suspension required a 1.0 mg/liter ClO_2 residual, as compared with a 3.5 mg/liter free chlorine residual. They believed that in the case of the spores, the cell wall was penetrated by the use of the five valence changes in the oxidation of chlorine dioxide, but that vegetative cells do not utilize this phenomenon.

Bedulvich et al.[47] concluded in 1953 that the bactericidal efficiency of chlorine dioxide towards *E. coli, Salmonella typhosa,* and *Salmonella paratyphi* was as great or greater than that of chlorine.

In 1953 Hettche and Ehlbeck[48] investigated the action of chlorine dioxide on poliomyelitis virus and reported that it is more effective than either ozone or chlorine. This superior effectiveness may be explained by a chemical characteristic of chlorine dioxide reported by Ingols and Ridenour in 1948.[49] They found that chlorine dioxide (but not chlorine) reacted with peptone in amounts that followed the laws of adsorption. Since viruses have a protein coat, it may be assumed that chlorine dioxide will be adsorbed by the virus coat. This would cause higher local concentrations on the surface of the virus than would be expected from the measured residual and may account for the claimed effectiveness of the viricidal effect of chlorine dioxide.[10] This theory appears to be compatible with the findings of Benarde et al.[50] described below.

In 1965 a critical comparison of ClO_2 and chlorine was made by Bernarde, Israel, Olivieri, and Granstrom.[51] This work was made possible by the availability of proper analytical techniques and the detailed physiochemical findings of Granstrom and Lee[1] in 1958. Bernarde et al. used an improved method of generating ClO_2 solutions of much greater purity than those produced by the methods of earlier investigators. They used a spectrophotometer for the quantitative analysis of stock solutions, residuals, and dosages. These studies were carried out using sterile unchlorinated sewage effluent to which a known cell density of *E. coli* had been added. The effluent was poised at pH 8.5 and had a BOD of 160 mg/liter.

It is significant to note that a 5 mg/liter dose of chlorine resulted in a 90 percent

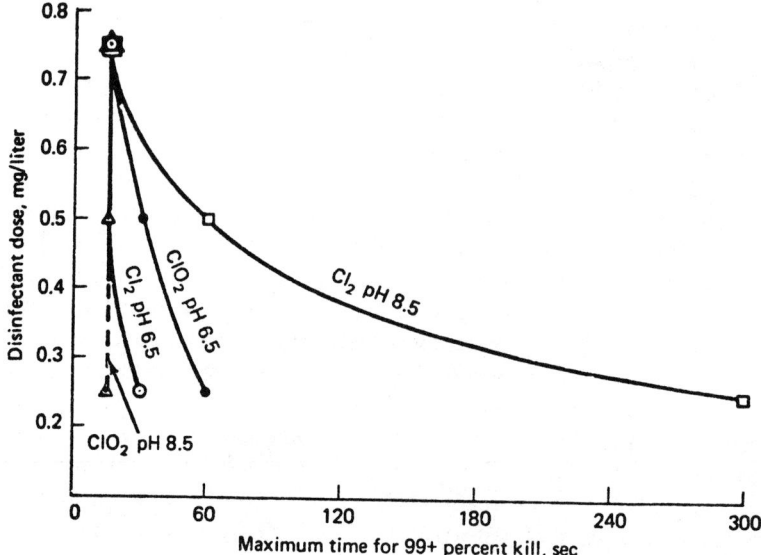

Fig. 12-9. Relative germicidal efficiency of chlorine and chlorine dioxide[51] (Dosage in mg/L vs. time required for a 99+ percent destruction of E. coli at the pH indicated.).

kill after 5 min, compared to 2 mg/liter dose of chlorine dioxide. This is surprising because chlorine dioxide exhibits a higher chlorine "demand" in wastewater than does chlorine. In this particular instance the comparison of the chlorine dioxide efficiency was being compared to combined chlorine, since the medium was sewage effluent containing ammonia nitrogen.[52]

It was found that at pH 4.0, 6.45, and 8.42 the chlorine dioxide molecule remained unaltered and intact, and that it must therefore be the bactericidal compound. This confirms the hypothesis that chlorine dioxide does not react with water to hydrolyze, as does chlorine.

Another important achievement of this investigation was establishing the disinfecting ability of ClO_2 and its relative efficiency as a function of pH. Fig. 12-9 illustrates the relative efficiency of both ClO_2 and chlorine with respect to pH. At pH 6.5, chlorine appears to be more efficient at the lower dosages. Both compounds are equally efficient at an initial dose of 0.75 mg/liter. Increasing the pH to 8.5 shows a dramatic change in efficiency. As would be expected, the chlorine efficiency drops, because at this pH level only 8.72 percent of the residual is HOCl while at pH 6.5 the residual is 89.2 percent HOCl. As the chlorine efficiency drops with the increase in pH, the chlorine dioxide efficiency increases. A 99+ percent destruction of *E. coli* in 15 sec with a 0.25 mg/liter dose of ClO_2 is noted at pH 8.5, as compared with a 0.75 mg/liter dose required for the same destruction by chlorine.

Additional work about the same time by Bernarde et al.[50] investigated the mechanism of kill by chlorine dioxide. This work so far indicates that ClO_2 does not react sufficiently with amino acids to alter their characteristic structures, thus eliminating the possibility of a reaction within the cell. They found, however, that in some unknown fashion ClO_2 abruptly inhibits protein synthesis, which is probably the mechanism by which it destroys these vegetative organisms.

In 1978, Longley et al.[53,54] compared the disinfection efficiencies of chlorine and chlorine dioxide in a sidestream from the effluent of a 90 mgd wastewater treatment plant. The flow of the sidestream through the contactor was about 0.9 mgd. The contact chamber was designed to give rapid mixing and plug flow at minimal head loss. The test organisms were fecal coliforms and coliphage. Chlorine dioxide and total chlorine residuals were determined by the amperometric method. Free available chlorine was determined by the FACTS method (Syringaldazine). This investigation revealed that a 5.0 mg/liter chlorine dioxide dose resulted in a 5 log reduction of fecal coliforms. This proved to be greater by more than one log reduction using the same dose of chlorine at 30 min contact time. Moreover, the inactivation of fecal coliforms by chlorine dioxide in 3 min exceeded the inactivation by chlorine at 30 min for the same dosage.

Even more dramatic results were observed for the viricidal efficiency of chlorine dioxide relative to chlorine under similar test conditions. The observed chlorine dioxide coliphage inactivation of 3 log reduction at 3 min contact time exceeded the comparable chlorine inactivation by nearly a 2 log reduction. It was also observed that the chlorine dioxide demand was greater than the chlorine demand when both were expressed in terms of chlorine. Although not mentioned in the report the wastewater examined was laden with ammonia nitrogen so that the chlorine species was always combined chlorine.

The most comprehensive investigation of chlorine dioxide as a disinfectant was done by the Stanford group between 1977 and 1981. The program was directed by Professor Paul V. Roberts. The graduate students participating in this project were Aieta, Berg, and Chow.[23,24,55-57] They used the natural coliform populations in the effluents from three different wastewater treatment plants. All three produced conventional activated sludge effluents. Two plants filtered the secondary effluent. However the focus of this investigation was on the nonfiltered secondary effluents.

All of the coliform studies were based upon the natural population in the secondary effluent. The animal virus used was Poliovirus I, LCS strain grown in Buffalo Green Monkey kidney cells. A coliphage isolated from the secondary effluent was grown on host *E. Coli* B to a titer of about 10^9 MPN/ml was used in the experiments.

Chlorine dioxide and chlorine residuals were measured by the iodometric backtitration method using an amperometric titrator in accordance with *Standard Methods*. The measurement of chlorine residual was made at pH 4 to insure quantitative recovery of all forms of chlorine residual. The measurement for chlorine dioxide residual was made at pH 7. At this pH there is no interference from any chlorite formed from the reduction of chlorine dioxide. It is necessary however to perform a blank titration at pH 7 and to correct the chlorine dioxide residual measurement accordingly. This is to account for the slower speed of reaction between the PAO and I_2 at pH 7, which would give additional time for the I_2 to react with the iodine demand of the sample. Blank corrections for chlorine residuals at pH 4 are not required. All doses and residuals throughout this investigation have been reported as mg/l chlorine and mg/l chlorine dioxide.

Several disinfection experiments were run on different days. Dosages were compared for 2, 5, and 10 mg/l chlorine dioxide and chlorine. Contact times selected were 5, 15, and 30 min. In order to compare the bactericidal efficiency of chlorine and chlorine dioxide and to determine the relationships of dosages and contact times to bactericidal efficiency, the logarithm of the organisms' survival ratio was calculated for each experiment. Results are shown on Fig. 12-10.

Chlorine dioxide demonstrated a more rapid coliform inactivation than chlorine at the shortest contact time (5 min). The mean log survival ratio in one group of experiments was −4.40 for chlorine dioxide compared to chlorine at −3.44. The dosage for each was 10 mg/l. However at 30 min contact time the ratio* was −5.81 for chlorine dioxide compared to −5.53 for chlorine. These experiments demonstrate that for contact times between 30 min and 1 hour in a secondary effluent, chlorine dioxide is only slightly more efficient as a bactericide than chlorine. It should be noted that in this instance chlorine dioxide is being compared to combined chlorine as a disinfectant. Another aspect to consider is the comparative consumption of each disinfectant. At 30 min contact time chlorine dioxide consumption was 5.13 mg/liter compared to 1.92 mg/l of chlorine. This resulted in residuals of 4.87 mg/liter for chlorine. The only advantage of chlorine dioxide over chlorine for wastewater application would be that less chemical would be required for dechlorination. One other significant point might be made for chlorine dioxide in a potable water situation where the magnitude of [$N(0)$] is low. Chlorine dioxide demonstrates a more rapid kill rate than chlorine at shorter contact times.

In the experiments with a nitrified and filtered effluent chlorine dioxide was shown to be superior to chlorine, except at long contact times.[23] The advantage of chlorine dioxide was found to be due to filtration and not nitrification. The presence or absence of NH_3-N has no effect on the germicidal efficiency of chlorine dioxide.

Selected experiments were conducted to evaluate the viricidal effectiveness of chlorine dioxide as compared to chlorine (combined).[23] Palo Alto secondary effluent (non-nitrified) was dosed with 5 mg/l chlorine dioxide and chlorine. The sample used contained in-situ coliphage and an inoculum of Poliovirus I (10^6 PFU/100

---

* Log survival ratio is:

$$\log_{10}[N(t)/N(0)] = \log_{10}[N(t)] - \log_{10}[N(0)]$$

where

$N(t)$ = number of organisms surviving at time t
$N(0)$ = number of organisms at time zero

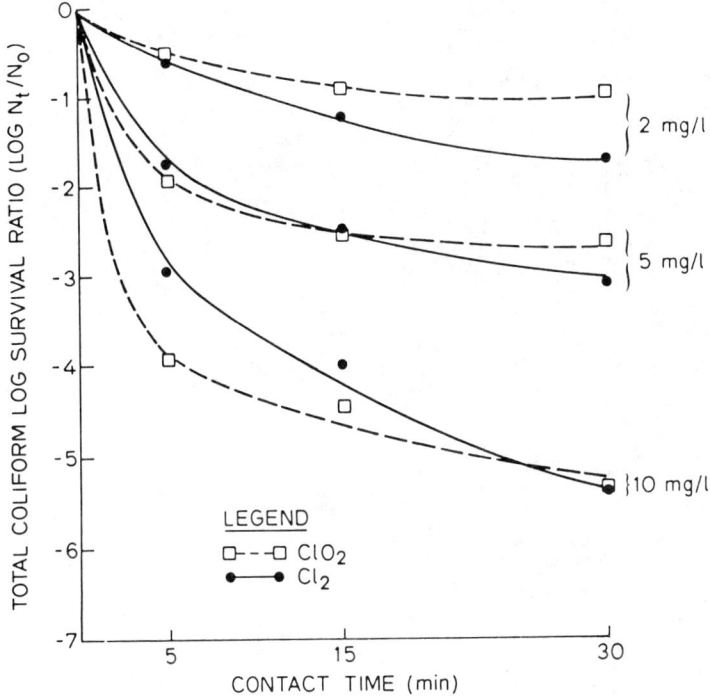

Fig. 12-10. Comparison of germicidal efficiency between chlorine dioxide and chlorine[23] showing coliform survival in a non-nitrified secondary affluent.

ml). Samples were taken at 2, 5, and 10 min contact times and analyzed for coliphage (Kott method), Poliovirus (standard plaque assay), and for fecal coliforms (MF method). The results are shown in Fig. 12-11. These experiments demonstrate that chlorine dioxide is a far superior viricide than is combined chlorine. This agrees with work by other investigators. The reaction of combined chlorine with coliphage also agrees with previous work by Olivieri.[58]

HEALTH HAZARDS FROM BYPRODUCTS

Interest in the use of chlorine dioxide for potable water treatment was revived in the 1970s because of the discovery that chlorine dioxide does not enter into any reactions that cause the formation of THMs. This factor made chlorine dioxide a viable candidate for an alternative to chlorine.[7] However the United States EPA expressed doubts about the toxicological effects of the presence of chlorite ion which is an end product of ClO_2 application. They substantiated these doubts by imposing a 1 mg/liter dosage limit of ClO_2 together with a monitoring requirement of both the chlorite ion (ClO_2^-) and the chlorate ion (ClO_3^-) in the treated waters.

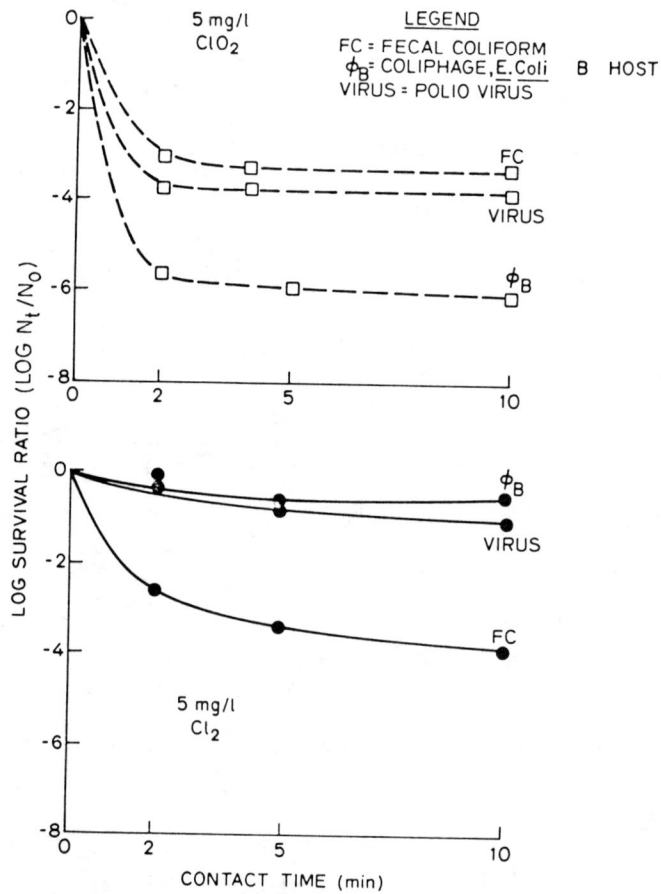

Fig. 12-11. Comparison of viricidal efficiency between chlorine dioxide and chlorine using an in situ coliphage and an inoculum of poliovirus I in a non-nitrified secondary effluent.[23]

In a limited study the EPA estimates that each part of ClO_2 consumed in a water treatment reaction reverts as follows: ClO_2^- 50 percent, ClO_3^- 25 percent, and Cl^- 25 percent. As the pH drops, ClO_2^- to Cl^- will become more dominant.

About 1978 the EPA began a comprehensive study of the effects of *large doses* of chlorine dioxide on *rats*. These efforts confirmed previous work that demonstrated the effect of the chlorite ion in drinking water. This is significant because it was shown that the chlorite ion is capable of oxidizing hemoglobin. This results in a condition known as methemoglobinemia.[5] Other studies have revealed that hemolytic anemia is produced at much lower levels of chlorite ion concentration than those required to produce methemoglobinemia.[6] However, the users of chlorine dioxide may find comfort in the results of a recent study on adult males.[59,60]

This controlled study using a group of healthy adult males demonstrated no detrimental physiological responses caused by daily ingestion of 5 mg/l of chlorite over a period of 12 weeks. The study was sponsored by the U.S. EPA Health Effects Research Laboratory; project officer was R. J. Bull.

A recent epidemiologic study of 198 persons in a rural village was conducted by the U.S. EPA Health Effects Research Laboratory.[61] The study exposed the participants for 3 months to water disinfected with chlorine dioxide. Chlorite ion concentrations in the distribution system ranged from 2.43 to 7.03 mg/liter. Statistical analysis of the data failed to identify any significant exposure-related effects.

A survey of the literature reveals a wide disagreement over the potential risks of using chlorine dioxide in drinking water. Despite the lack of sufficient convincing data on these risks it has been recommended that the chlorite ion residuals should be nil, based upon the possible threat to infants who have a decreased ability to reduce methomoglobin.[62,63] Based on these reports the Norwegian Health Authority has recommended the total absence of chlorite ion in drinking water.[8,64] Similarly, in Germany the applied dose of ClO_2 is unofficially limited to 0.3 mg/l to prevent possible adverse health effects.[64] In France there is extensive use of chlorine dioxide. It has been used to treat the water supplies of Paris since 1966. The Chief of Water Quality Control for the City of Paris advises that there has been no question about health effects from the use of chlorine dioxide, and if need be there would be no hesitation to go to dosages as high as 1.5–2.0 mg/liter.[65,66] There are many chlorine dioxide installations in Switzerland used primarily for taste and odor control (destruction of phenols). There is no official limit placed upon ClO_2 dosage.

It is clear from the above that it is too soon to be certain about the health risks resulting from the use of chlorine dioxide in potable water. More epidemiological data must be gathered in order to reach some reasonable agreement on this controversy.

USES IN WATER TREATMENT

Control of Phenolic Tastes and Odors. The first known use of chlorine dioxide as a water treatment process in the United States was at the No. 2 Niagara Falls, New York, Water Treatment Plant in January, 1944.[27,67,68] This early work was sponsored by the Mathieson Chemical Company, which had first made solid sodium chlorite available in 1940. This chemical made possible a simple method of generating chlorine dioxide at the point of use. In September, 1944, the process was installed at the Western New York Water Company plant at Woodlawn, New York. Other plants which adopted the process early were those at Greenwood, South Carolina; and Tonawanda, North Tonawanda, Lockport, and Port Colbourne, New York. All of these plants experienced serious trouble from phenolic tastes and odors as a result of industrial waste pollution of the raw water. There were also varying degrees of taste and odors from algae and vegetation decomposition in the summer. Prior to the use of chlorine dioxide, all of these plants were experimenting with superdoses of chlorine. However, the chlorination equipment was not very sophisticated, and a method of accurate measurement of a free chlorine residual was not known: so it is not possible to properly evaluate the chlorination process used at that time. A few years later, a survey of these plants[12] led to the conclusion that chlorine dioxide could successfully assist chlorination in the control of tastes resulting from the raw water being contaminated with phenolic substances and algae. Chlorine dioxide seemed also to assist in reducing consumer complaints where it was necessary to carry free chlorine residuals in the distribution system. The success of this early work inspired further investigation into the properties of chlorine dioxide. From this work, it was found that chlorine dioxide is specific for the control of phenolic and chlorophenolic tastes and odors.[49,69,71] Several theories have been advanced to explain this mechanism. Todd[72] thought that superoxidation by high doses of chlorine converted the phenols to an insoluble pentachlorphenol, which then settled out in the sedimentation process. Faber[73] and Adams[74] theorized that during the process of chlorination three substances were formed in a stepwise process: orthochlorphenol, parachlorphenol, and trichlorphenol.

Of these three products, orthochlorphenol produces the greatest magnitude of taste, while parachlorphenol produces about one-fourth as much, and trichlorphenol is practically tasteless. They theorized that with the use of chlorine alone it was probable that a mixture of these substances was formed depending on the relative amounts of free and combined chlorine residual present. When chlorine dioxide was used, its superoxidizing power converted this mixture entirely to the tasteless compound of trichlorphenol. Following these observers, a careful study of phenolic tastes was made by Ingols and Ridenour[75] in 1948. They concluded that chlorophenolic taste resulting from chlorination of phenol-bearing waters was due to a quinone-like substance—probably dichlorquinone—and that the further oxidation of this substance by either chlorine or chlorine dioxide effected a rupture of the benzene ring and formed maleic acid, which is tasteless, rather than a substitution in the ring as in pentachlorphenol.

Even though Ettinger and Ruchhoft[75] proved conclusively that chlorophenolic tastes could be eliminated by a chlorooxidation process, it was not until 1959 that Burtschell et al.,[76] with modern analytical techniques, explained the mechanism of this oxidation process. They confirmed that progressive chlorination proceeds as previously suggested by Adams and Faber. The technique of Burtschell et al. revealed that chlorination proceeds to produce 2-chlorophenol, 4-chlorophenol, 2,4-dichlorophenol (2,4DCP), 2,4,6-trichlorophenol (2,4,6 TCP), and the more highly chlorinated 4,4-dichloroquinone. All of these products may contribute to the intensity of the taste and odor. At the point of maximum intensity, the major contributor is the previously overlooked compound 2,6-DCP (dichlorophenol). This compound is destroyed, thereby removing the objectionable medicinal taste, when the weight ratio of chlorine to phenol is 4:1.[76] The chlorine must be either free available (HOCl) or chlorine dioxide to accomplish this destruction.

There are four reasons why chlorine dioxide is the chemical of choice in dealing with tastes and odors arising from phenolic compounds:

1. ClO_2 will not react with ammonia or nitrogenous compounds, including the simple amino acids, as will chlorine.[77] It is always available, as would be HOCl if the latter did not react to form chloramines from ammonia and other nitrogenous compounds.
2. ClO_2 reacts completely to destroy the taste-producing phenolic compounds many times faster than does free available chlorine. This is undoubtedly a result of its oxidizing capacity of two and a half times that of HOCl.
3. In all cases studied so far, it has been reported that ClO_2 will always destroy any chlorophenol taste caused by prechlorination.
4. The efficiency of ClO_2 is not impaired (as is HOCl) by a high pH environment.

It has been found in actual practice that chlorine dioxide is most effective when applied as a posttreatment process following prechlorination.[27,78,79] This added safety factor of ClO_2 as a posttreatment process allows higher prechlorination doses to accomplish proper algae control, disinfection, and so on, without fear of producing a water subject to consumer complaints from tastes and odors. This can also result in lower ClO_2 dosages as the higher prechlorination dosage satisfies more of the chlorine demand that would otherwise consume higher amounts of ClO_2.

For the control of phenolic tastes and odors, it has been found that the sodium chlorite requirement may vary from 0.3 to 10 lb/mg.[1] While a chlorine dioxide residual is tasteless in itself, some observers have noted that when this residual exceeds 0.8 to 1.0 ppm it may impart a slight metallic taste.[80]

To determine the correct ClO_2 dosage, each water with a taste and odor problem must be studied by itself—each as a special case. A great many anomalies have been reported, and it must be concluded that these are a result of complex kinetic interrelationships in the series of reactions leading to the formation and ultimate destruction of the taste producing substance.

Algae and Decaying Vegetation. Chlorine dioxide is usually associated with correcting only phenolic and chlorophenolic tastes and odors; however, in many cases, it has been found successfully to control a variety of tastes occurring from both algae and decaying vegetation.[79,81-85] In some of these cases, it seems certain that actinomycetes were the offending organisms and that tastes resulting can be controlled.

Ringer and Campbell[83] used chlorine dioxide in preference to copper sulfate as an algicide in open reservoirs. They believed that the effectiveness of chlorine dioxide in this application was due to the changes ClO_2 makes in the benzene ring (for which it has a definite affinity) to render it tasteless and odorless. They thought that since the pyrrole rings of chlorophyll closely resemble the benzene rings, the effect of chlorine dioxide might be duplicated on the pyrrole rings. If, then, the chlorine dioxide oxidizes the chlorophyll, plant metabolism would cease, owing to the interruption of protein synthesis. This results in damage to the plant to the point at which loss of water from the protoplasm brings about hypertonic shrinkage (plasmolysis). This process is irreversible, causing the plant to die, and is documented by photographs. Furthermore, the essential oils remaining from the destroyed organisms were rendered tasteless and odorless. This particular application was carried out largely during the night hours, and apparently the reaction between the chlorine dioxide and the algae was so swift that decomposition of ClO_2 by sunlight was not a factor.

In the survey by Granstrom and Lee[1] of fifty-two treatment plants reporting the use of chlorine dioxide for taste and odor problems, only seven had algae problems. From perusal of the literature describing case histories of chlorine dioxide for taste and odor control of algae and decaying vegetation, it is not clear whether or not proper facilities utilizing prechlorination and automated post dechlorination would achieve the same results.

Chlorine dioxide has been successful in controlling musty tastes,[84] fishy tastes, odors from mallomonas,[81] and similar problems from *Anabena, Asterionella, Synura, Vorticella,* and others,[81,82] undoubtedly including actinomycetes. In nearly all these cases, the increased cost of using chlorine dioxide has been offset to a certain extent by the reduction in the amount of activated carbon required. In some situations, chlorine dioxide is more effective at high pH levels (10–12) but more costly; more chlorine dioxide is required to accomplish the disproportionation reaction, which slowly converts some of the chlorine dioxide to chlorite.

Manganese Removal. Manganese in minute quantities, a common ingredient of impounded water and of a great many well waters, can cause offensive black water in distribution systems. The difficulties encountered are well known: staining of clothes and plumbing fixtures, black water, and incrustation of mains, often giving rise to tremendous amounts of debris at the consumers' taps.

Some of the difficulty can be attributed to the action of free available chlorine, which reacts so slowly on manganese that twenty-four hours after treatment the manganese will become oxidized and precipitate out of solution. This occurs in the distribution system. Chlorine dioxide reacts much more rapidly with the manganese compounds than does chlorine, and so it is often used, instead of chlorine, in conjunction with manganese removal. As with the other applications of chlorine dioxide, chlorine, because of its lower cost and other considerations, is usually applied at the head of the removal process to satisfy the initial chlorine demand and to start the oxidation process. Chlorine dioxide, because of its cost, is usually limited to applications where the total manganese content is less than 1.0 mg/l. Best results for complete oxidation of manganese by chlorine dioxide occur when the pH is higher than 7.[76]

The overall reaction is as follows:

$$2ClO_2 + MnSO_4 + 4NaOH \longrightarrow MnO_2 + NaClO_2 + Na_2SO_4 + 2H_2O \quad (12\text{-}34)$$

Therefore 2.45 parts of ClO_2 are required to remove 1.0 part Mn and 2.45 parts ClO_2 are equivalent to 4.12 parts of 80 percent $NaClO_2$ (sodium chlorite) from which the ClO_2 is generated plus 2.45/2 × 0.90 = 1.36 parts of chlorine* So 4.12 parts of sodium chlorite are required to oxidize 1.0 part of manganese. With the 1983 cost of sodium chlorite at $2.28 per lb,[88] 4.12 × 8.34 × 2.28 = $78.34 for sodium chlorite. The required chlorine will cost 1.36 × 8.34 × 0.20 = $2.24. Thus it will cost 78.34 + 2.24 = $80.58 to remove each part of manganese from each *million gallons* of water, excluding the oxidation of other inorganic compounds, such as iron and hydrogen sulfide likely to be present. It also does not include any natural chlorine dioxide demand that the water may have.

Iron Removal. Similarly, chlorine dioxide can be used effectively for the oxidation of iron from the moderately soluble ferrous state to the ferric ion, resulting in the formation of ferric hydroxide, which forms a heavy gelatinous brown floc, which can be removed by sedimentation followed by filtration. Here too, the optimum pH conditions are always higher than 7, preferably 8 or 9, and must be determined by laboratory procedures. The rapidity with which chlorine dioxide reacts in the oxidation of iron is not as significant as in the case of manganese. Chlorine is the chemical of choice (except for chlorine dioxide) in this case, unless the water being treated has a high ammonia content or other nitrogenous compounds that would seriously interfere with the formation of free available chlorine. This would be an economic consideration unless there were side reactions causing taste and odor problems treatable only by chlorine dioxide.

The oxidation of iron by chlorine dioxide is as follows:

$$ClO_2 + FeO + NaOH + H_2O \longrightarrow Fe(OH)_3 \downarrow + NaClO_2 \quad (12\text{-}35)$$

Therefore 1.2 parts of ClO_2 are required to remove 1.0 part iron as Fe, and 1.2 parts ClO_2 are equivalent to 2.02 parts of 80 percent $NaClO_2$ from which the ClO_2 is generated, plus a like amount of chlorine. So 2.02 parts of sodium chlorite are required to oxidize 1.0 part of iron. Again assuming a cost of $2.28 per pound of sodium chlorite and twenty cents per pound for chlorine, the cost to remove each 1.0 mg/liter of iron as Fe from one million gallons of water would be (2.02 × 2.28 × 8.34) + [(2.02/2) × 0.9 × 8.34 × 0.20] = $40.28.

Usually the amounts of iron and manganese involved in the removal processes are so small that the consumption of alkalinity caused by the above reactions will be insignificant. Manganese removals are usually less than 1.0 mg/liter and

---

* This assumes a 90 percent yield of ClO_2 from $NaClO_2$.

those for iron less than 5.0 mg/liter. However, if there is a significant reduction in alkalinity, the following chemicals can be used to restore the alkalinity (the amounts shown are based upon the removal of 1 mg/liter as a stoichiometric value):

	Manganese	Iron
Commercial grade hydrated lime, $Ca(OH)_2$	3.0 mg/l	1.50 mg/l
Soda Ash, Na_2CO_3	7.75 mg/l	3.80 mg/l
Caustic, NaOH	2.92 mg/l	1.43 mg/l

Color Control. Chlorine dioxide has well-known bleaching powers. It is used extensively to bleach the pulp used in the manufacture of paper.[87] However, the role of chlorine dioxide in color control of treated potable water is not well known. Free available chlorine has been known to bleach certain organic colors that cannot be otherwise removed. Palin[77] experimented with both chlorine and chlorine dioxide as bleaching agents in an attempt to remove peat color, often a problem in moorland water supplies in the British Isles. He found chlorine to be somewhat more effective than chlorine dioxide. In a water with initial color of 48 degrees Hazen, treated with 10 mg/l of both chemicals up to a contact time of 22 hours, chlorine removed 37 degrees Hazen of color, while chlorine dioxide removed only 30 degrees Hazen.

There is no doubt that chlorine dioxide will have some bleaching effect, depending upon the organic composition of the color in the water. It may or may not be more effective than free chlorine.

The water quality personnel at the Choisy Le Roi water treatment plant, Paris, point out that the color of the treated water is noticeably better when chlorine dioxide is being used.

WATER REUSE

Food Processing. Chlorination of process water in canneries and frozen food packaging became an established practice about 1946 in the U.S.A. It was so successful in the potato dehydration plants that its use spread rapidly to fruit and vegetable canning operations.[88] Hall and Blundell[89] reported on experimental projects for three plants—a corn cannery, a pea cannery, and a freezing plant. They observed that free residual chlorination prevented slime growths and eliminated objectionable plant odors. Moreover free residuals of 10 mg/liter did not produce any off-flavor in the final products. They concluded that in-plant chlorination was the best way to control slime growths, which usually occur where there is regular or intermittent contact with water, such as in washers, elevators, inspection tables, gutters, and floors. There are many hundreds of these installations operating in the United States today. However, some important changes have been made.

In the beginning the cannery water supply was a once-through system. However, after some 10 years of breakpoint chlorination, canneries attempted to conserve water, primarily to cut down on the cost of treating the wastewater. Thus a recycling system for food processing plants was evolved. It was soon discovered that the ammonia nitrogen concentration increased significantly in the recycled water, so that it became difficult to control the breakpoint process unless multiple points of application were used. It was found impractical to pursue the free residual process using breakpoint chlorination because of excessive chlorine consumption combined with the generation of intolerable amounts of nitrogen trichloride that pervaded the working area.* Since it is well known that chlorine dioxide does not react with ammonia nitrogen, the Green Giant Co., Le Sueur, Minnesota decided in 1957 to try chlorine dioxide.[90] Their Blue Earth plant chlorination system was retrofitted to generate chlorine dioxide. The prime objective was to determine the feasibility of water conservation through water recycle while maintaining acceptable bacteria counts in the complete system. The experiment was a success. The chlorine dioxide treatment of the once-used water was highly effective in the control of bacteria growth and biofouling in pea and corn canneries. The persistent residuals and the lack of reaction with ammonia nitrogen make possible one point rather than the multiple points of application necessary with rechlorination of used water. Chlorine dioxide residuals in the reuse water did not result in the generation of offensive odors in the plant nor were there any off flavors produced in the product.

The success at this plant rapidly spread to other food processing plants. Since 1957 a majority of these processing plants have retrofitted their existing chlorination systems or purchased new chlorine dioxide systems. These plants include all types of food processing. The success of chlorine dioxide used in the recycle systems is the result of the chemical characteristics of chlorine dioxide: (1) it will not react with ammonia nitrogen, and (2) the residuals persist over a long period of time, so that a single point of application is sufficient. In most cases the chlorine dioxide dosages vary from 2 to 8 mg/l. When chlorine was used on a once-through system the dosages were on the order of 10 mg/l. When recycling was used most plants found it difficult to control biofouling with this dosage applied at two or more points of application. Moreover, it was soon discovered that when breakpoint chlorination was practiced on once-used water, nitrogen chloride was formed. This produced extremely offensive odors in the plant and in some cases off-flavors were produced in the product.

Reclaimed Wastewater. Great interest was centered upon the improvement of wastewater disinfection techniques during the 1970s in the U.S.A. The properties of chlorine dioxide make it an interesting possibility for special situations. The

---

* Each part increase of NH_3-N concentration requires a minimum of 10 parts increase of chlorine to achieve breakpoint.

first critical study was by Bernarde et al.[51] in 1965. This was a laboratory investigation using sterile solutions seeded with *E. coli*. A comparison with chlorine indicated that chlorine dioxide was considerably superior to chlorine, particularly at short contact times.

There is very little field experience with the application of chlorine dioxide as a disinfectant for either secondary or tertiary effluents. There are no known installations (other than pilot plants) on wastewater in the U.S.A. at this time. The most reliable source of information concerning chlorine dioxide as a wastewater disinfectant is the work done by the Stanford group.[23] This has been described above. Their findings, which dealt with wastewater effluents and the native population of organisms, indicate that chlorine dioxide is only slightly superior to chlorine as a bactericide but is a much superior viricide. Owing to the high cost of generating chlorine dioxide (about $3.00/lb.) its primary attraction would be as a viricide. This is of considerable interest to health agencies where wastewater reclamation is considered. In these situations one other factor favors chlorine dioxide over chlorine. This is the greater persistence of ClO_2 residuals over chlorine in closed distribution systems.

Many laboratory investigations have shown that chlorine dioxide has a higher rate of kill in a shorter contact time than does chlorine. This is misleading. The kills in the short contact times are about a magnitude of 2–3 logs. In wastewater reuse situations 5–6 logs kill are required because the treated water for these cases must be able to meet an MPN coliform count of 2.2/100 ml. To achieve this, contact times of 30–45 min are required. At these longer contact times chlorine performs almost as well as chlorine dioxide.

The findings of the Stanford group have been confirmed by a lengthy investigation made in France. This investigation covered a period of 2–3 years. It examined the disinfection efficiency of various disinfectants on effluents of varying quality; i.e., primary, secondary, and filtered secondary.[91]

The Stanford group developed a mathematical model similar to the Collins Model[14] for chlorine dioxide. This expression is based upon numerous samples taken on different days from the effluent of a modern conventional secondary (activated sludge) treatment plant. It is as follows:

$$y/y_0 = (0.54ct)^{-2.90} \qquad (12\text{-}36)$$

where y is the MPN/100 ml coliform requirement in the plant effluent, y_0 is the coliform concentration in the plant effluent before chlorination, t is contact time in minutes, and C is chlorine dioxide residual at time t in mg/liter.

This compares to the model for chlorine as follows:

$$y/y_0 = (0.19ct)^{-3.15} \qquad (12\text{-}37)$$

This compares to the Collins model:

$$y/y_0 = [1 + 0.23ct]^{-3} \qquad (12\text{-}38)$$

The chlorine dioxide model is the first of its kind for wastewater. It should be helpful where chlorine dioxide is considered for the treatment of high quality effluents.

One final but important point concerns the consumption of chlorine dioxide during the contact period. On the same dosage and contact time it was found that more ClO_2 was used than chlorine, thereby resulting in a lower ClO_2 residual. This would require less dechlorinating chemical for chlorine dioxide. However, as described above, chlorine dioxide is only likely to be a candidate for a wastewater reuse situation where dechlorination would probably not be practiced.

DECHLORINATION

Dechlorination is an important consideration in wastewater treatment and in the chlorination of steam power plant cooling water. Chlorine dioxide reacts stoichiometrically with both sulfur dioxide and sulfite solutions similarly as does chlorine.

The overall reactions can be expressed as follows. In sulfur dioxide solution:

$$SO_2 + H_2O \longrightarrow H_2SO_3 \qquad (12\text{-}39)$$

$$5H_2SO_3 + 2ClO_2 + H_2O \longrightarrow 5H_2SO_4 + 2HCl \qquad (12\text{-}40)$$

Similarly with sulfite compounds:

$$5Na_2SO_3 + 2ClO_2 + H_2O \longrightarrow 5Na_2SO_4 + 2HCl \qquad (12\text{-}41)$$

From the above equations it can be seen that it will require 2.5 mg/liter SO_2 for each mg/liter of chlorine dioxide residual (expressed as ClO_2) to dechlorinate a sample to zero residual. This is the stoichiometric relationship. In practice it would be well to design on the basis of 2.7 to 1. It is expected that chlorine dioxide residuals would be much lower than those currently used in chlorination practice. This is due to two phenomena. First, chlorine dioxide consumption (demand) is usually greater than that of chlorine, i.e., at the same dosage chlorine dioxide achieves the same or better results than chlorine with lower residuals. Second, chlorine dioxide residuals die away at a slightly faster rate than chlorine residuals in open bodies of water. However, in closed systems such as a waterworks distribution system they appear to persist longer than chlorine. This would also be true if a closed-conduit outfall were used as a chlorine contact chamber in a wastewater system.

CHLORINE DIOXIDE FACILITY DESIGN

INTRODUCTION

There are several equipment configurations that the designer can select. Owing to the high cost of sodium chlorite (about $2.50/lb) the choice of system and equipment should be based upon a system that will produce the highest yield of chlorine dioxide from the sodium chlorite. The other consideration is the purity of the solution. If controlling the THMs in a potable water supply is the objective, then a 95 percent chlorine dioxide solution should be a design criterion. If THM control is not a factor, as could be the case in iron and manganese removal, then the use of excess chlorine to drive the reaction to completion may be acceptable.

Unfortunately, operating experience of the various systems is somewhat limited in the U.S.A. Most of the operating systems are in Western Europe. The dominant system there is the CIFEC system, which uses a chlorine solution enrichment loop.

CHLORINE–CHLORITE SYSTEM

General Discussion. This system is the most popular because it uses conventional chlorination equipment (gas). There are four variations of this system which are described below. Each system consists of a chlorine gas metering device (chlorinator), a water-operated injector (eductor), a sodium chlorite metering system, and a chlorine–chlorite reaction chamber (generating tower).

The chlorine supply system considerations are the same for chlorine dioxide generation as for chlorine, but in some cases the chlorine injector requirement can be quite different.

The sodium chlorite metering system requires special consideration because of the necessity to maintain precise control of the chlorite to chlorine ratio in order to achieve the maximum yield. Storage requirements for both liquid and solid sodium chlorite also deserve special attention.

Control strategies must be examined closely to determine if the system is practical from the operator's point of view.

Chlorine and Chlorite Mixed Under Pressure. This is the oldest system in use. It was used when sodium chlorite was first introduced to make chlorine dioxide in small quantities. The success of this system depends upon the ability of the chlorinator to produce a chlorine solution with a $pH < 4$, preferably 3.5. The sodium chlorite in this system is mixed into the chlorine solution under pressure adjacent to the injector discharge. The design of the injector system is critical.

All of the major chlorinator manufacturers furnish rotameters calibrated over a 20:1 range. They do not state, however, what the accuracy is over this range. Flow metering manufacturers are more realistic. They claim only a 10:1 range.

CHLORINE DIOXIDE 881

When generating chlorine dioxide by mixing sodium chlorite solution with the chlorinator injector discharge, the question becomes one of the chlorine concentration range in the chlorine solution produced by the chlorinator. The maximum allowable strength has been arbitrarily chosen as 3500 mg/l chlorine. Many years ago it was found that injector efficiency began to deteriorate rapidly as this concentration was exceeded. This was due largely to the rapid increase of molecular chlorine in the chlorine solution. The molecular chlorine has the tendency to create substantial gas pockets in the solution lines, causing excessive back pressure on the injector. At 3500 mg/l chlorine solution at pH 2, the molecular chlorine concentration is 50 percent and the hypochlorous acid is 50 percent. At 5000 mg/l the molecular chlorine is about 65 percent.

The lower limit of solution strength for most waters is 500 mg/l chlorine. This concentration is necessary to maintain a pH low enough (3–4) to allow the reaction between the chlorine and the chlorite to go to completion, thereby preserving the maximum yield of chlorine dioxide.

Therefore the range of any chlorinator is limited to about 7 to 1. On this assumption let us design a 2000 lb chlorinator injector system for the production of chlorine dioxide:

The minimum allowable chlorine feed rate will be $2000/7 = 285$ lb/day. The next question becomes: What is the maximum allowable injector water flow that will provide a minimum concentration of 500 mg/liter chlorine solution at 285 lb/day chlorine feed rate?

Solution: 500 mg/l × 8.34 = 4170 lb/mg.

Using injector water flow in units of mg/day:

$$Q(\text{inj}) \times 4170 = 285 \text{ lb } Cl_2/\text{day}$$

$$Q(\text{inj}) = 0.06835 \text{ mg/day}$$

Converting to gpm:

$$Q(\text{inj}) = \frac{0.06835 \times 10^6}{1440 \text{ (minutes/day)}}$$

$$= 47.46 \text{ gpm}$$

The next question becomes: Will 48 gpm be sufficient to feed 2000 lb/day chlorine? An examination of one manufacturer's 2-in. variable-throat injector curves shows that it will be sufficient under certain conditions. The injector water flow of 48 gpm will produce a 3500 mg/liter chlorine solution, and at 50 psi injector inlet pressure, will be able to feed 2000 lb/day chlorine with injector back pressure of 7 psi. The back-pressure is critical. Therefore, the injector discharge system should be designed to limit the back pressure to 5 psi. A higher back-pressure could be tolerated by the injector by raising the inlet pressure. This is a risky procedure unless the injector water flow is also increased. If the water flow is maintained at 48 gpm and if the inlet pressure is increased to 60 or 70 psi there is danger of cavitation in the injector. Cavitation can damage the tailway in the injector. However the most significant effect is the effect of a profusion of air bubbles in the chlorine dioxide solution. These air bubbles trap chlorine dioxide gas making it extremely difficult for operating personnel to monitor the yield. The bubbles produced by cavitation can cause gas binding in the ClO_2 solution line. To overcome this hydraulic phenomenon it is best to either pump the chlorine dioxide solution to the point of application or to use a liquid–liquid eductor.

Whenever an existing chlorinator is to be retrofitted for chlorine dioxide production, the injector water supply and the solution line must be examined carefully using the guidelines described above.

A variation of the above system is to supplement the chlorine and sodium chlorite with hydrochloric acid. The acid is added downstream from the point of sodium chlorite addition in the chlorine injector discharge line.[30] Wallace and Tiernan (Div. of Pennwalt Corp.) have investigated this system and report the best yield with the highest purity solution occurs when the chlorine to chlorite ratio is 0.5 and the hydrochloric acid added produces 0.1 mole H^+ ion.[29] Operating quantities for this system which is described in their bulletin 85-200 are as follows: 10 lb chlorine, 15.2 gallons of 16.7 percent sodium chlorite solution, and 15.2 gallons of 0.8 percent hydrochloric acid solution for each 17 lb of chlorine dioxide.

The addition of the acid, which assures the proper chlorine–chlorite mixture pH at all times, eliminates the necessity of designing the injector supply system to provide minimum chlorine solution concentration of 500 mg/liter. The acid provides greater system flexibility, which results in a higher operating range of 10:1 versus 7:1 without acid. Moreover, there is no danger of injector cavitation due to the freedom of using the amounts of injector operating water necessary to avoid cavitation.

The Fischer and Porter system is still another variation of the chlorine–chlorite pressure system. Their design is based upon a 7 min contact time in the chlorine dioxide generating tower. When the operating range is greater than 7:1 it is necessary to use multiple units of varying size chlorinators. Each chlorinator is equipped with its own generating tower sized according to the capacity of the chlorinator.

Still another variation of this type system is a system widely accepted in Western Europe. This is the CIFEC unit of Paris, France (see Fig. 12-7). This system recirculates the chlorine solution from the chlorinator injector in what is called an enrichment loop. This insures an optimum pH for the chlorine–chlorite reaction. The sizing of the injector and the recirculating pump is critical. This is proprietary information. Existing chlorinators can be modified in the field to operate with the CIFEC generating system.

All of the above systems have limitations on the maximum allowable pressure at the discharge of the generator. If this pressure must be exceeded then the chlorine dioxide solution will have to be pumped or injected into the point of application by a liquid–liquid eductor.

All of the above systems must apply the sodium chlorite solution under pressure. This can be done in two ways: positive displacement pumps and air pressure.

1. Positive Displacement Pumps. This is the most widely used system because diaphragm pumps lend themselves well to the situation. They are easily adaptable to automatic control by either or both of two ways: (a) variable motor speed, or (b) automated stroke adjustment. However there are specific accessories that should be included to provide the operator with a system that is convenient and reliable.

First, the pump discharge should be metered. The best choice is a magnetic flow meter.

Second, the pulsating flow from the pumps must be dampened to smooth out the pulses. Therefore accumulators should be installed on both the suction and discharge of each pump.

The additional cost for these accessories is well justified because they will allow the operator to prevent the waste of sodium chlorite ($3.00/lb.).

2. Air Pressure. For some installations where fairly large sodium chlorite supply tanks are involved, air padding is a practical solution. Use of a 10 psi air pad system, or whatever pressure is necessary to deliver the chlorite to the point of application, allows the use of rotameters for flow measurement and conventional flow control valves.

Liquid–liquid eductors are not applicable because of the dilution factor and the hardness limitation of the eductor water.

Chlorine and Chlorite Mixed in a Vacuum. The use of the chlorinator vacuum to mix the chlorine and sodium chlorite is a relatively new concept. It was first introduced in the U.S.A. about 1980. There are two patented systems currently available. Both can be used to retrofit existing chlorinators.

The Rio Linda Chemical Co. This system is arranged as shown in Fig. 12-8 to mix the chlorite with chlorine gas in the chlorinator vacuum line adjacent to the injector.

The sodium chlorite piping and the piping configuration at the chlorine–chlorite mixing point is designed in accordance with the results from field testing. Therefore this is proprietary information.

The use of vacuum mixing simplifies all of the design considerations necessary in the pressure systems. The chlorinator injector system is sized in the same fashion as it would be for just chlorine alone. There is no need for concern about dilution of the chlorine dioxide solution due to too much injector operating water.

There are no special limitations on injector back pressure conditions. The same guidelines for conventional chlorinator installations apply to the Rio Linda generating system. Existing chlorinators are easily retrofitted because when a chlorinator is converted to chlorine dioxide, the chlorine required for equivalent dosages of ClO_2 is half that of chlorine alone. This reduction in the amount of chlorine required means that any existing chlorinator installation will have a more than adequate injector system.

Another important feature of this system is the ease with which sodium chlorite can be fed into the chlorinator vacuum line. Field experience has shown that the available vacuum created by the injector is more than adequate for feeding the sodium chlorite solution.

Rotameters if selected properly are satisfactory for measuring the sodium chlorite feed rate if the system is manually controlled. Flow paced automatic control requires a control valve similar to the Fischer and Porter Chloromatic valve, actuated by a 4–20 mA signal that will maintain the proper chlorine to chlorite ratio. In some cases chlorite flow measurement by a magnetic meter may be more desirable than an indicating rotameter. Rotameters with transmitters are also a consideration.

Fischer and Porter Co. This system was patented in 1981.[93] It is arranged so that a conventional water operated injector provides the necessary energy to draw chlorine and sodium chlorite into the bottom of the reactor tower. The reactor was designed to extend the contact time between the chlorite and the chlorine. Owing to the fact that the reaction in the tower proceeds under a vacuum, chlorine exists mostly in the molecular form. This situation enhances significantly the completion of the chlorite–chlorine reaction to produce a high-purity chlorine dioxide solution. Operation of this system at solution strengths as high as 29,000 mg/l ClO_2 has not shown any evidence of explosion or even puffs due to degassing.[93]

The reacting tower is placed upstream from the injector in order to overcome the many drawbacks of conventional injector practice. The tower is typically PVC and is packed with rings, saddles or static mixers. It functions to mix the gas–liquid phases and to provide the necessary reaction time.

The chlorine gas is fed into the bottom of the tower through a vacuum regulator, flowmeter and a Chloromatic rate valve. A back-flow preventer valve is provided to prevent the reactor contents from flowing back into the chlorine supply system. The chlorinator used in this system is a conventional Fischer and Porter Model 1700 which uses the principle of sonic flow. The injector is required to maintain a vacuum at the inlet to the injector of 12 in. Hg. If for some reason this vacuum fails, the chlorine vacuum-regulating valve will automatically close. The control system will then automatically shut down the chlorite feed pump and the water supply to the tower.

The flow of diluting water and sodium chlorite to the reaction tower is independent of the negative pressure in the tower. The flow of sodium chlorite is controlled by a Chloromatic rate valve. The water flow control to the tower is not critical, as it serves only to adjust the reactant concentration within the optimum range.

This system is adaptable to fully automatic control; i.e., flow paced and residual.

In practice the ClO_2 generated in the reactor should be maintained between 5,000 and 25,000 mg/l. At concentrations lower than 5,000 mg/l the reaction is slow. At concentrations higher than 25,000 mg/l there is a potential for forming liquid ClO_2, which is a dangerous chemical. The system is designed to provide a 20:1 operating range. This means that dilution water, sodium chlorite, and chlorine gas must be automatically controlled over this range. The sodium chlorite solution should be commercially available 25 percent strength.

There are no installations of this system in the U.S.A. or Canada. This is due primarily to lack of interest in the use of ClO_2. At present it is being marketed in France by Fischer and Porter Co.

Chlorite Storage Tanks. The recommended strength of sodium chlorite solution is 25 percent. The on-site storage tanks can be made of either Fiberglas or rotationally molded high-density crosslinked polyethylene.

These tanks should be fitted with an atmosphereric vent, drain, and liquid level indicator in addition to the filling line and sampling tap. It is always desirable to consult the supplier of sodium chlorite to determine the piping and valving details for the filling line.

The 25 percent sodium chlorite solution is very stable so tank size would be predicated on practical delivery times. There is no fire hazard when using sodium chlorite solutions.

ACID–CHLORITE SYSTEM

The acid–chlorite method using hydrochloric acid is the preferred system. Referring to Fig. 12-3, the various components for a complete system are shown. These are listed as follows:

1. A softened water system for diluting both the acid and the chlorite
2. Flow meters and control valves for each dilution water
3. A control system to provide the proper dilution of the acid and chlorite
4. Flow meters and control valves for the measurement and control of the acid and chlorite into the reaction tower
5. Packed tower reactor to provide sufficient time to complete the conversion of the chlorite to a chlorine dioxide solution
6. An injector system (with or without booster pump) to provide motive power to draw the ClO_2 solution out of the tower into the point of application
7. A flow control valve located on the outlet of the tower, which should be able to operate automatically from a flow paced signal, with residual control optional
8. The flow control signal, which should be equipped with a ratio station
9. A continuous chlorine residual analyzer—a preferred item, but optional
10. Acid and chlorite storage tanks
11. Day tanks for both acid and chlorite

The configuration of the storage tanks and day tanks should be such that the chemical flow can be by gravity all the way to the tower reactor. If this is not possible, then the chemicals will have to be transferred by pumps or by an air pad on the storage tanks. It is highly desirable to have gravity flow from the day tanks to the tower reactor. Otherwise diaphragm metering pumps will be required. Metering and control using diaphragm pumps is much more complex than gravity flow.

Sodium chlorite storage tanks should be made of the same materials described under "Chlorine–Chlorite System" in this chapter. Piping for $NaClO_2$ and ClO_2 should be Sch. 80 PVC.

Acid piping and storage tanks can be constructed from the same material as for $NaClO_2$ and ClO_2. This will provide uniformity in the complete installation.

ANALYTICAL MEASUREMENTS

Chlorine Dioxide Solutions. Procedures for analyzing the ClO_2 concentration in the solution discharge from the generating tower are described under the heading "Chlorine–Chlorite Systems" in this chapter. This covers all the necessary analyses required to calculate the yield and purity of the solutions.

Chlorine Dioxide Residuals. Chlorine dioxide residuals that are a result of ClO_2 solutions at 95 percent purity will contain very little chlorine. Assume the remaining 5 percent consists of equal parts Cl_2 and ClO_2^-. That would amount to 0.0375 mg/l Cl_2 in a dose of 1.5 mg/l of ClO_2. However the residual would be on the order of 0.5 mg/l ClO_2. Assuming two-thirds of the chlorine was lost in satisfying demand, now the chlorine residual level is down to 0.0127 mg/l. This is at the minimum detectable level.

Therefore all residuals should be measured as ClO_2 and assume that the measurable residual is all ClO_2. The preferred method is described on pages 307–308 Section 410B, *Standard Methods,* 15th ed. as follows:

Titration for total available chlorine:

- Add $6N$ NaOH to raise sample pH to 12.
- After 10 min add $6N$ H_2SO_4 to reduce to pH 7.
- Add 1 ml KI solution.
- Titrate with standard PAO solution to the amperometric end point.
- This is reading *B*.

Titration for total available chlorine and one-fifth of available ClO_2:

- Adjust sample to pH 7 with pH 7 phosphate buffer solution.
- Add 1 ml KI solution.
- Titrate with PAO solution to the amperometric endpoint
- This is reading *C*.

Titration for total available chlorine, ClO_2, and chlorite:

- Add 1 ml KI solution to sample.
- Add sufficient $6N$ H_2SO_4 to lower sample pH to 2.
- After 10 min add sufficient $6N$ NaOH to raise pH to 7.
- Titrate with PAO solution to the amperometric end point.
- This is reading D.

Calculations. Indiviual titrations can be converted into concentrations of chlorine by the following:

$$\text{mg/l chlorine} = \frac{\text{ml PAO} \times 200}{\text{ml sample}} \qquad (12\text{-}41)$$

where ml PAO is amount used for the individual titration for B, C, and D.
Chlorine and chlorine dioxide fractions are calculated as follows:

$$\text{mg/l chlorine dioxide as } ClO_2 = 1.9(C - B)$$

$$\text{mg/l chlorine dioxide as chlorine} = 5(C - B)$$

$$\text{mg/l chlorite as chlorine} = (4B - 5C + D)$$

Continuous Residual Analyzers. These analyzers will respond to concentrations of ClO_2 over a wide range of concentrations. The sample going to the measuring cell should be buffered to pH 4. Given the short sample time in the cell, very little if any interference from ClO_2^- is expected. Experience might prove that the analyzer could operate reliably without the pH 4 buffer.

SUMMARY

ADVANTAGES OF CHLORINE DIOXIDE

Chlorine dioxide is at least as effective a bactericide as chlorine. In many cases it has proved to be superior.

It is far superior as a viricide to chlorine, which makes it a promising candidate for water reuse disinfection.

It does not react with ammonia nitrogen.

It does not react with oxidizable material to form trihalomethanes.

It reacts to destroy up to 30 percent of the THM precursors. In some cases this would allow post–free residual chlorination without the formation of undesirable concentrations of THMs.

The efficiency of chlorine dioxide disinfection is relatively unaffected at pH levels between 6 and 10, but is more effective in the alkaline region, making it attractive for use with high-pH lime-softened water.

It is specific for the destruction of phenols, which cause taste and odor problems in potable water supplies. In most cases of phenol contamination, chlorine cannot by itself destroy the phenols. This results in objectionable chlorophenol tastes and odors. Chlorine dioxide has the ability to destroy these chlorophenols.

Chlorine dioxide does not react with water as other halogens do.

Chlorine dioxide has a long experience record in the removal of iron and manganese. It is superior to chlorine, particularly when the iron and manganese occur in complexed compounds.

It has been used successfully in the control of water quality degradation and taste and odors in distribution systems. Case histories have shown that the chlorine dioxide residuals will persist longer than chlorine in distribution systems and is more effective than chlorine at much lower residuals.

DISADVANTAGES OF CHLORINE DIOXIDE

At the current cost of sodium chlorite ($2.80/lb, 1983) it is probably not cost-effective for most applications in both potable water and/or reclaimed water. The chemical cost is 5 times that for chlorine.

Chlorine dioxide cannot be transported so it must be generated on-site.

One of the more serious disadvantages of chlorine dioxide is the possibility that it may adversely affect the health of humans. This notion has not been expressed yet as a serious possibility by the users in Western Europe.

Chlorine dioxide prepared by some processes may contain significant amounts of free chlorine which could defeat the objective of using ClO_2 to avoid the formation of THMs.

If the dosage exceeds more than 6 lb/million gallons of sodium chlorite a pronounced metallic taste in treated potable water will occur.

One of the most serious difficulties with chlorine dioxide is the lack of reliable and practical techniques for the routine evaluation and differentiation of chlorine dioxide, chlorine, chlorites, and chlorates in both the generator discharge and the treated water. This is not the case where chlorine is used.

REFERENCES

1. Granstrom, M. L., and Lee, G. F., "Generation and Use of Chlorine Dioxide in Water Treatment," *J. AWWA*, **50**, 1453 (1958).
2. Miller, G. W., et al., "An Assessment of Ozone on Chlorine Dioxide Technologies for Treatment of Municipal Water Supplies," Executive Summary, USEPA 60018-78-018, Cincinnati, OH, Oct. 1978.
3. Industrial Pollution of the Lower Mississippi River in Louisiana, U.S. Environmental Protection Agency, Cincinnati, OH, Apr. 1972.
4. Miltner, R. J., "The Effect of Chlorine Dioxide on Trihalomethanes in Drinking Water," Master of Science Thesis, Univ. of Cincinnati, Cincinnati, OH, 1976.
5. Bull, R. J., "Health Effects of Drinking Water Disinfectants and Disinfectant By-Products," *Env. Sci. and Tech.*, **16**, 554A (1982).

6. Heffernan, W. P., Guion, C., and Bull, R. J., *J. Env. Pathol. Toxicol.*, **2**, 1487 (1979).
7. Symons, J. M., et al., "Ozone, Chlorine Dioxide, and Chloramines as Alternatives to Chlorine for Disinfection of Drinking Water," Environmental Protection Agency, Cincinnati, OH, Nov. 1977.
8. Love, O. T., et al., "Treatment for the Prevention or Removal of Trihalomethanes in Drinking Water," Appendix 3 to Treatment Guide for the Control of Chloroform and Other Trihalomethanes, Env. Prot. Agency, Cincinnati, OH, 1976.
9. Kennett, C. A., "Experience With the Use of Chlorine Dioxide for Distribution Problems," Water Research Centre Conf., Medmenham Laboratory, Medmenham, Bucks, England, 1978.
10. Dowling, L. T., "Chlorine Dioxide in Potable Water Treatment," *Water Treatment and Exam.*, **23**, 190 (1973).
11. McConnell, G., "Halo-Organics in Water Supplies," *Jour. Inst. of Water Engrs. and Scientists*, **30**, 431 (Nov., 1976).
12. Gordon, G., Kieffer, R. G., and Rosenblatt, D. H., "The Chemistry of Chlorine Dioxide," S. J. Lippard (Ed.), in *Progress in Inorganic Chemistry*, Vol. XV, J. Wiley & Sons, New York, 1972, pp. 202–286.
13. *Kirk-Othmer Encyclopedia of Chemical Technology*, 2nd Ed. Vol. 5, Interscience, New York, 1964.
14. White, G. C., *Disinfection of Wastewater and Water for Reuse*, Van Nostrand Reinhold, New York, 1978.
15. Sconce, J. S., *Chlorine: Its Manufacture, Properties and Use*, Reinhold, New York, 1962.
16. Kesting, E. E., "The Manufacture and Properties of Chlorine Dioxide," *TAPPI*, **36**, 166 (1953).
17. Booth, H., and Bowen, E. J., "Action of Light on ClO_2 Gas," *J. Chem. Soc. London*, **127**, 510 (1925).
18. Bowen, E. J., and Cheung, W. M., "Photodecomposition of Chlorine Dioxide Solution," *J. Chem. Soc. London*, p. 1200, Part I (1932).
19. Shera, B. L., consulting engineer, private communication, Tacoma, WA, 1974.
20. Gall, R. J. "Chlorine Dioxide: An Overview of its Preparation, Properties, and Uses," paper presented at the IOI Meeting, Cincinnati, OH, Nov. 17–19, 1976.
21. *Standard Methods for the Examination of Water and Wastewater*, 15th ed., APHA, AWWA, WPCF, Washington, DC, 1980.
22. Hood, N. J., "A Laboratory Evaluation of the DPD and Leuco Crystal Violet Methods for the Analysis of Residual Chlorine Dioxide in Water," Master of Science Thesis, Virginia Polytechnic Institute, Blacksburg, VA, Oct. 1977.
23. Roberts, P. V., Aieta, E. M., Berg, J. D., and Chow, B., "Chlorine Dioxide for Wastewater Disinfection: A Feasibility Evaluation," Technical Report No. 251, Civ. Engr. Dept., Stanford Univ., Palo Alto, CA, Oct. 1980.
24. Aieta, E. M., Berg, J. D., and Roberts, P. V., "Comparison of Chlorine Dioxide and Chlorine in the Disinfection of Wastewater," paper presented at the WPCF Ann. Conf., Anaheim, CA, Oct. 1–6, 1978.
25. Valenta, J., and Gahler, W., "Chlordioxanlage," *Gaswasser, Abwasser*, **9**, 566 (1975).
26. W. H. Nagel, Private communication, Fischer and Porter Co. Warminster, PA, March 3, 1978.
27. Synan, J. F., MacMahon, J. D., and Vincent, G. P., "Chlorine Dioxide, A Development in the Treatment of Potable Water," *Water Wks. and Sew.*, **91**, 423 (1944).
28. "Chlorine Dioxide," Cat. File 60.300 Rev. 7–75, and 60.310 Rev. 9–76, Wallace and Tiernan Div. Pennwalt Corp., Belleville, NJ, 1976.
29. Jordan, R. W., Kosinski, A. J., and Baker, R. J., "Improved Method for Generating Chlorine Dioxide," TA 1053-C, Wallace and Tiernan Div. Pennwalt Corp., Belleville, NJ, April 1980.
30. Anon. "Chlorine Dioxide Systems," Cat. File 85.200, Wallace and Tiernan Div. Pennwalt Corp., Belleville, NJ, 1980.
32. Aieta, E. M., Roberts, P. V., and Hernandez M., "Determination of Chlorine Dioxide, Chlorine, Chlorite and Chlorate in Water," *J. AWWA*, **76**, 64 (Jan. 1984).

33. Masschelein, W. J., "Chlorine Dioxide Chemistry and Impact of Oxychlorine Compounds," Ann Arbor Science, Ann Arbor, MI, 1979.
34. Anon., "Chlorine Dioxide," Cat. File 60.300, Wallace and Tiernan Inc., Revised Aug. 1966.
35. Anon., "Bioxyde de Chlore Générateur," CIFEC, Paris, France Cat. 87, July 1980.
36. Anon., "A New Chlorine–Chlorite Process," Rio Linda Chemical Co. Inc., Rio Linda, CA, 1982.
37. Aieta, M., engineering consultant, personal communication, Nov. 1982.
38. Rexing, D. "Las Vegas Valley Water District Chlorine Dioxide Installation," paper presented at the California-Nevada Section meeting, Long Beach, CA, April 16, 1982.
39. Troitskaya, N. V., Mischenko, K. P., and Flis, I. E., *Zh. Fiz. Khim,* **33,** 1614 (1959).
40. Feuss, J. F., "Problems in the Determination of Chlorine Dioxide Residuals," *J. AWWA,* **56,** 607 (1964).
41. Emmeneger, F., and Gordon G., *Inorganic Chemistry,* **6,** 633 (1967).
42. Booth, G. M., Wallace and Tiernan Co., Inc. U.S. Patent No. 2,454,124 (1949).
43. Halpern, J., and Taube, H., *J. Amer. Chem. Soc.,* **74,** 375 (1952).
44. Ridenour, G. M., and Ingols, R. S. "Bactericidal Properties of Chlorine Dioxide," *J. AWWA,* **39,** 561 (1947).
45. Ridenour, G. M., and Armbruster, E. H. "Bactericidal Effects of Chlorine Dioxide," *J. AWWA,* **41,** 537 (1949).
46. Ridenour, G. M., Ingols, R. S., and Armbruster, E. H. "Sporicidal Properties of Chlorine Dioxide," *Water Works Sewage,* **96,** 276 (1949).
47. Malpas, J. F., "Disinfection of Water Using Chlorine Dioxide." *Water Treat. Exam.,* **22,** 209 (1973).
48. Hettche, O., and Ehlbeck, H. W. S., "Epidemiology and Prophylaxes of Poliomyelitis in Respect of the Role of Water in Transfer," *Arch. Hyg. Berlin,* **137,** 440 (1953).
49. Ingols, R. S., and Ridenour, G. M., "Chemical Properties of Chlorine Dioxide," *J. AWWA,* **40,** 1207 (1948).
50. Bernarde, Melvin A., Snow, W. B., Olivieri, Vincent P., and Davidson, B., "Kinetics and Mechanism of Bacterial Disinfection by Chlorine Dioxide," *Appl. Microbiol.,* **15,** 2,257 (1967).
51. Bernarde, M. A., Israel, B. M., Olivieri, V. P., and Granstrom, M. L., "Efficiency of Chlorine Dioxide as a Bactericide," *Appl. Microbiol.,* **13,** 776 (Sept. 1965).
52. Bernarde, M. A., Snow, W. B., and Olivieri, V. P., "Chlorine Dioxide Disinfection Temperature Effects," *J. Appl. Bact.,* **30,** 159 (April 1967).
53. Longley, K. E., Moore, B. E., and Sorber, C. A., "Relative Wastewater Disinfection Efficiencies of Chlorine and Chlorine Dioxide in a Gravity Flow Contactor," paper presented at the Annual WPCF Workshop, Houston, TX, Oct. 7, 1979.
54. Longley, K. E., Moore, B. E., and Sorber, C. A., "Comparison of Chlorine and Chlorine Dioxide as Wastewater Disinfectants," paper presented at the Ann. Conf. Water Pollution Control Federation, Anaheim, CA, Oct. 3, 1978.
55. Aieta, E. M., Chow, B., and Roberts, P. V., "Chlorine Dioxide: Analytical Measurement and Pilot Plant Evaluation," paper presented at the National Symposium of Wastewater Disinfection, Cincinnati, OH, Sept. 18–20, 1978.
56. Berg, J. D., Aieta, E. M., and Roberts, P. V., "Effectiveness of Chlorine Dioxide as a Wastewater Disinfectant," paper presented at The National Symposium on Wastewater Disinfection, Cincinnati, OH, Sept. 18–20, 1978.
57. Aieta, E. M., and Roberts, P. V., "Disinfection With Chlorine and Chlorine Dioxide," *ASCE J. Environ. Engr.,* **109,** 783 (Aug. 1983).
58. Snead, M. C., and Olivieri, V. P. "Biological Evaluation of Methods for the Determination of Free Available Chlorine," Chapter 19, p. 401, W. J. Cooper (Ed.), in *Chemistry in Water Reuse,* Vol. 1, Ann Arbor Science Publisher, Ann Arbor, MI, 1981.
59. Bianchine, J. R., Lubbers, J. R., Chahuan, S., and Miller, J., "Study of Chlorine Dioxide in man," EPA report No. 600/1-81-068 NTIS PB82-109356, Dec. 1981.

60. Lubbers, J. R., Chauhan, S., and Bianchine, J. R., "Controlled Clinical Evaluations of Chlorine Dioxide, Chlorite and Chlorate in Man," *Fund. and Appl. Toxicology,* **1,** 334 (1981).
61. Michael, G. E., Miday, R. K., Bercz, J. P., Miller, R. G., Greathouse, D. G., Kraemer, D. F., and Lucas, J. B., "Chlorine Dioxide Water Disinfection. A Prospective Epidemiology Study," *Arch. of Environ. Hlth.,* **36,** 20 (Jan./Feb. 1981).
62. Musil, J., Kontek, Z., Chalupa, J., and Schmidt, P., "Toxicological Aspects of Chlorine Dioxide Application for the Treatment of Water Containing Phenol," *Sbornik Vsoke Skoly Chem-Technol. V. Praz,* **8,** 327 (1963).
63. Berndt, H., "Studies on Water Treatment and Disinfection with Chlorine Dioxide," *Arch. Hyg. Bacteriology,* **148,** 10 (1965).
64. Myhrstad, J. A., and Samdal, J. E., "Behavior and Determination of Chlorine Dioxide," *J. AWWA,* **61,** 205 (1969).
65. Vilagines, R., Chief Water Quality Control, City of Paris, private communication, 1982.
66. Vilagines, R., Montiel, A., Derreumaux, A., and Lambert, M., "A Comparative Study of Halomethane Formation During Drinking Water Treatment by Chlorine or its Derivatives (ClO_2) in a Slow Sand Filtration Treatment Plant and in Wastewater Treatment Plants," Paper presented for the Disinfection Seminar at the Annual AWWA Conf., Anaheim, CA, May 8, 1977.
67. Synan, J. F., MacMahon, J. D., and Vincent, G. P., "A Variety of Water Problems Solved by Chlorine Dioxide Treatment," *J. AWWA,* **37,** 869 (1945).
68. Aston, R. N., "Chlorine Dioxide Use in Plants on the Niagara River," *J. AWWA,* **39,** 687 (1947).
69. Vincent, G. P., MacMahon, J. D., and Synan, J. F., "Use of Chlorine Dioxide in Water Treatment," *Am. J. Pub Health,* p. 1045 (Sept. 1946).
70. Huebner, W. B., private communication, Wallace and Tiernan Div. of Pennwalt, Belleville, NJ, Jan. 31, 1983.
71. Ingols, R. S., and Ridenour, G. M., "The Elimination of Phenolic Tastes by Chloro-Oxidation," *Water and Sew. Wks.,* **95,** 187 (1948).
72. Todd, A. R., "Maintenance of Chlorine Residual in the Distribution System," *J. AWWA,* **34,** 1805 (1942).
73. Faber, H. A., "A Theory of Chlorine Dioxide Taste and Odor Reduction," *J. AWWA,* **39,** 691 (1947).
74. Adams, B. A., "Substances Producing Taste in Chlorinated Water," *Water and Sew. Wks.,* **33,** 109 (1931).
75. Ettinger, M. B., and Ruchhoft, C. C., "Stepwise Chlorination on Taste and Odor Producing Intensity of some Phenolic Compounds," *J. AWWA,* **43,** 561 (1951).
76. Burttschell, R. H., Rosen, A. A., Middleton, F. M., and Ettinger, M. B., "Chlorine Derivatives of Phenol Causing Taste and Odor," *J. AWWA,* **51,** 205 (1959).
77. Palin, A. T., "Chlorine Dioxide in Water Treatment," *J. Inst. Water Engrs.* (England), p. 61 (Feb., 1948).
78. McCarthy, J. A., "Chlorine Dioxide for the Treatment of Water Supplies," *J. NEWWA,* **59,** 252 (1945).
79. Mounsey, R. J., and Hagar, C., "Taste and Odor Control with Chlorine Dioxide," *J. AWWA,* **38,** 1051 (1946).
80. Griffin, A. E., private communication, 1959.
81. Coote, R., "Chlorine Dioxide Treatment at Valparaiso, Indiana," *Water and Sew. Wks.,* **97,** 13 (Jan. 1950).
82. Harlock, R., and Dowlin, R., "Chlorine and Chlorine Dioxide for Control of Algae Odors," *Water and Sew. Wks.,* **100,** 74 (Feb., 1953).
83. Ringer, W. C., and Campbell, S. J., "Use of Chlorine Dioxide for Algae Control at Philadelphia," *J. AWWA,* **47,** 740 (1955).
84. Malpas, J. F., "Use of Chlorine Dioxide in Water Treatment," *Eff. and Water Treatment Jour.* (England), p. 370 (July 1965).

85. Atkinson, J. W., *Water and Water Engr.* (England), **66,** 146 (1962).
86. Griffin, A. E., "Significance and Removal of Manganese in Water Supplies," *J. AWWA,* **52,** 1326 (Oct. 1960).
87. "Pollution Control Symposium," *Chemical Eng.,* **58,** 135 (May 1951).
88. Barker, P. J., private communication, Rio Linda Chem. Co., Rio Linda, CA, Jan. 1983.
89. Hall, J. E., and Blundell, G. C., "The Use of Breakpoint Chlorination and Sterilized Water in Canning and Freezing Plants," NCA Convention, Atlantic City, NJ, 1948.
90. Synan, J. F., and Malley, H. A., "Chlorine Dioxide: An Alternative to Chlorine for Disinfection in Water Systems," paper presented to the Engineering Panel of the Campden Food Preservation Research Assoc., Chipping Campden, Glos., England, Oct. 21, 1975.
91. Saunier, B. M., Michelon, B., and Jamody, M., "Essais de Disinfection des Eaux Usées Urbaines de Ville de Montpellier," report edited by Agence de Bassin Rhône Méditerranée Corse et Français Ministere L'Environnement, Sept. 1982.
92. Ratigan, B. J., Philadelphia, PA, and Fischer and Porter Co., Warminster, PA, "Chlorine Dioxide Generating System," U.S. Patent No. 4,250,144, Feb. 10, 1981.
93. Kingsmore, L., Private communication, Fischer and Porter Co., Cerritos, CA, Jan. 19, 1983.

13
Ozone

HISTORICAL BACKGROUND

Ozone (O_3) has been used for more than sixty yr. for water treatment on the European continent. More than a thousand municipal water-treatment plants use ozone as part of their chemical treatment.[1] Most of these are in Western Europe, particularly France and Switzerland, but usage is spreading to other countries. Most of these installations are primarily for taste and odor control, and color removal; they are almost without exception backed up by chlorination in spite of the fact that ozone is an admirable disinfectant. The first experiments with ozone were performed in France in 1886 by de Merintence.[2] Ozone was first used on a full-scale basis at Nice, France in 1906 for the treatment of the municipal water supply. This water comes from the river Var which originates close by in the Alps. The quality of this water is usually excellent and only deteriorates during flood periods following heavy rainfalls. The resulting pollution consists mainly of suspended materials which are easily removed by sand filtration. Ozone is used to destroy coliform organisms and the grassy or earthy tastes which accompany periods of high runoff.[3]

The Nice plant is still in use. Enlargements made in 1922 and 1951 increased its capacity to about 25 mgd. Today (1976) the largest ozone installations are used for water treatment in the Paris area. Three separate plants treat water from the Oise, Marne, and Seine Rivers. The combined capacity of these treatment plants is 360 mgd and the ozone production (from air) is approximately 5 tons/day.

In Switzerland, ozone has been used during the past 25 yr. for treating spring waters, ground waters, and surface waters. Following a massive phenol discharge in 1957 at St. Gall, disinfection has been by ozone instead of chlorine. As of 1977, there are about 150 installations using ozone for treatment of potable water and industrial processes. The largest has a capacity of 1750 lb/day. This facility is at the Lengg Lake plant in Zurich. The next largest systems are 650 lb/day in Geneva and St. Gall.[4]

The first large Russian filtration-ozonation plant was built in 1911 in St. Petersburg (Leningrad) having a total capacity of 12 mgd. Owing to difficulties in operation and maintenance this unit was shut down in 1922.[2] Subsequently, large ozonation installations have been included in water-treatment plants for Moscow (317 mgd at 4 mg/liter O_3), Kiev (106 mgd at 5 mg/liter O_3), and Gorski (82 mgd at 2.0 mg/liter O_3). Other plants (under construction in 1976) using ozone as a disinfectant are those of Singapore, Malaya; Lodz, Poland; and Chiba, Japan. None of these designs are for ozone usage capacities in excess of 5 mg/liter with the median design at 3.4 mg/liter.[5]

CHEMISTRY OF OZONATION

Introduction. The role of ozone in wastewater treatment may be classified as both an oxidant and a germicidal compound. These are the same properties exhibited by aqueous chlorine; therefore, there is tendency to view the two substances as competitors. However, the competition is not exact nor universal, for the ways in which the functions are accomplished are somewhat different.[6] It should be emphasized that ozone and an aqueous chlorine solution can act in complementary fashion, each performing some tasks more usefully than the other. *Therefore, there are three distinct faces of ozone: (1) as a bactericide; (2) as a viricide; and (3) as a powerful chemical oxidant.*

Ozone as a Bactericide. In recent years there have been a number of laboratory and pilot-scale studies to evaluate the effectiveness of ozone in disinfecting wastewater. This work has also provided information on the mechanisms by which ozonation occurs.

The potent germicidal properties of ozone have been attributed to its high oxidation potential. Research studies indicate that disinfection by ozone is a direct result of bacterial cell wall disintegration. This is known as the "lysis phenomenon." This mechanism of disinfection by ozone is indeed different from that by chlorine. Although the exact chemical action of chlorine is uncertain, it is generally agreed that the chlorine residual in an aqueous solution diffuses through the cell wall of the microorganism and attacks the enzyme group, the destruction of which results in the death of the microorganism.

Ozone has long been recognized as an excellent disinfecting agent, but reliable quantitative studies of the fundamental germicidal activity of ozone are so few that our knowledge of its real potency is meager compared to what we know about chlorine and other disinfectants. This is partly because of the superior oxidizing power of ozone. It is most difficult to experimentally obtain the necessary time-dependent relations with extremely small ozone concentrations.[6]

Venosa pointed out in his comprehensive review of the literature dealing with the germicidal efficiency of ozone, that there exists much controversy, contradiction, confusion, and nonfactual subjective judgment on the use of ozone. One of the most serious failures by the various investigators has been their inability to distinguish between the concentration of ozone applied and the residual ozone necessary for effective disinfection. It must be recognized that the same principle for controlling chlorination, which is by residual, should also be applied to the control of ozone.

In 1976, Morris[6] presented an excellent summary of what is currently known about the germicidal efficiency of ozone. First he laid to rest the fallacy that ozone displays an "all-or-nothing" effect on bacterial kill. This was the result of the interpretation of the work done by Fetner and Ingols in 1956,[8] and has been often quoted to substantiate this effect. Morris has emphasized that this so-called all-or-nothing effect is neither real nor significant.[6] The effect appears simply because of the inability or failure of investigators to space the dosage concentrations of ozone reagent close enough. Ozone is so strong a germicide that concentrations of only a few micrograms per liter are needed to measure germicidal action. The spacing of the concentrations used by Fetner and Ingols was about 0.1 mg/liter or 100 µg/liter, a large enough gap to go from zero kill at a dosage just equal to demand, all the way to a very rapid kill at a concentration equal to demand plus 0.1 mg/liter. This example is just one of many which confronts the researcher in attempting to evaluate the potency of disinfectants. To overcome this situation Morris[9] developed the concept of the lethality coefficient for a given disinfectant. He treated all of the significant developed data on ozone, beginning with the work by Kessel et al. in 1943 to develop this lethality coefficient:

$$\Lambda = 4.6/Ct_{99} \qquad (13\text{-}1)$$

Where

C = residual concentration in mg/liter.
t_{99} = time in min. for 99 percent microorganism destruction (2-log destruction).

The weighted mean results of these evaluations are shown in Table 13-1. The values are considered by Morris to be valid only within a *factor of two*. The ranges of values do not warrant any greater confidence than this. Comparison of the values in this table with similar values obtained for chlorine (HOCl) are shown in Table 13-2. These values of the lethality coefficient are computed from the 1967 tabulation by Morris.[9] These tabulations by Morris clearly illustrate that ozone is a more powerful germicide against all classes of organisms listed, by factors of 10–100. The relative sensitivities of the various types of organisms are the same as for HOCl. As would be expected, bacteria are the most sensitive and of the usual types of vegetative bacteria examined, all exhibit about the same sensitivity. Viruses are more resistant by about a factor of ten, but not enough forms have been tested with ozone to determine the range of resistances of the difficult viruses. Cysts and spores, as with aqueous chlorine, are about a factor of ten times more resistant than viruses.

Table 13-1 Parameters For Disinfection by Ozone[6]
(pH 7; 10–15 °C)

Organism	Λ[a]	$C_{99:10}$[b]
Escherichia coli	500	0.001
Streptococcus faecalis	300	.0015
Polio virus	50	.01
Endamoeba histolytica	5	.1
Bacillus megatherium (spores)	15	.03
Mycobacterium tuberculosam	100	.005

[a] Λ = specific lethality coefficient = $\ln 100 \div Ct_{99}$
[b] $C_{99:10}$ = concentration in mg/liter for 99 percent destruction or inactivation in 10 min.

Not much experimentation has been done with the activity of ozone at various pH levels. The germicidal efficiency of ozone does not seem to be affected significantly within the pH range of 6–8.5. Even less is known about the effect of treated wastewater temperature on germicidal efficiency. One thing is certain however, the higher the water temperature the lower the efficiency of ozone mass transfer, which translates to lower germicidal efficiency.

For oxidizing compounds used as disinfectants, their superior oxidizing characteristics might provide them with high-lethality coefficients; however, these same characteristics might also cause them to have higher consumption rates if placed in an environment such as wastewater abounding in compounds which react rapidly with oxidizing agents. Therefore, the more reactive the compound the fewer the "miles per gallon."

Ozone as a Viricide. Selna, Miele, and Baird[10] of the Sanitation Districts, Los Angeles County, made a comprehensive study of water reuse disinfection for unrestricted recreational purposes. According to the California Department of Public Health guidelines, in order to qualify for such use a well oxidized secondary effluent must be coagulated, settled, filtered, and disinfected to achieve a median total coliform MPN of 2.2/100 ml, or less. The required treatment is expensive

Table 13-2 Values of Λ At 5 °C
[(mg/liter)$^{-1}$ (min.)$^{-1}$]

Agent	Enteric Bacteria	Amoebic cysts	Viruses	Spores
O_3	500	0.5	5	2
HOCl as Cl_2	20	0.05	1.0 up	0.05
OCl$^-$ as Cl_2	0.2	0.0005	<0.02	<0.0005
NH_2Cl as Cl_2	0.1	0.02	0.005	0.001

both from a capital and operational standpoint; therefore, the Sanitation Districts investigated less costly tertiary treatment alternatives to the required system during a two year study at the Pomona research facility.

One of the objectives of this study was to provide an effluent which would protect swimmers against viral illnesses. The pilot systems employed were from 25 to 100 GPM. Four treatment systems were evaluated as follows:

1. Coagulation, sedimentation, filtration, and disinfection.
2. Coagulation, filtration, and disinfection.
3. Two-stage carbon adsorption and disinfection.
4. Nitrification, filtration, and disinfection.

Ozonation was tested as an alternative to chlorination in systems 1, 2, and 3. System 4 was an investigation of free residual chlorination with a 2-hr contact time. All ozone contact times were designed for 18 min. Ozone dosages varied from 10 mg/liter for System 1; both 10 and 50 mg/liter for System 2, and 6 mg/liter for System 3 (Carbon adsorption). Ammonia nitrogen in effluents 1, 2, and 3 were approximately 20 mg/liter and suspended solids (predisinfection) about 1.5 mg/liter. The total COD was on the order of 20–25 mg/liter. Cumulative virus removal was best in System 1 (i.e., 5.5-log removal with an ozone dose of 10 mg/liter.) System 2 with an ozone dose of 50 mg/liter was only about 5.4 logs removal. System 2 with a lowered ozone dose of 10 mg/liter did almost as well in virus destruction as with 50 mg/liter. System 3 using carbon adsorption and a 6 mg/liter dose provided a 5.25-log removal. These results compared generally with chlorine (free or combined) of 4.6–5.25 logs removal depending upon the system (see Chapter 8).

This investigation defines with confidence the viricidal capabilities of ozone under full-scale treatment plant conditions. It also reveals that the final ozonated effluent coliform concentrations did not routinely meet the required MPN standard of 2.2 MPN/100 ml. This continues to confirm the fact that ozone is a superior viricide but is not a reliable bactericide.

Ozone as an Oxidant. Ozone has a wide array of attributes attractive to its use in potable water treatment such as taste and odor control, color removal, and iron and manganese removal. These oxidizing powers are quite valuable in the polishing of low-quality supplies including water reuse situations.

Ozone oxidizes inorganic substances completely and rapidly, e.g., sulfides to sulfates, ferrous iron to ferric, manganous ion to manganese dioxide or permanganate, and nitrites to nitrate. Of even greater importance is ozone's capability of breaking down organic complexes of both iron and manganese which usually defy the usual procedures of iron and manganese removal from potable waters.

Oxidation of organic materials by ozone is more selective and incomplete at the concentrations and pH values of aqueous ozonation. Unsaturated and aromatic compounds are oxidized and split at the classical double bonds, producing carboxylic acids and ketones as products.[11] Because of the high reactivity of ozone, oxidation of organic matter in the aqueous environment whether it be potable water or wastewater will consume ozone in varying amounts. Therefore, one of the most significant parameters for evaluating ozone is the determination of the immediate ozone demand. Oxidation of the (organic) material is usually incomplete. It is estimated that the reduction in TOC may be only 10–20 percent, although decreases in COD and BOD are generally greater, ranging up to 50 percent COD reduction as with Montreal water.[12] There are also instances where COD appeared to increase, resulting from conversion to more readily oxidized compounds.

Ozone exerts a powerful and effective bleaching action on the organic compounds which contribute to the color in wastewater and potable water. The ability of ozone to attack these compounds which contribute to the color, some of which are the humates and fulvates, makes ozone a fine wastewater polishing agent.

The ability of ozone to destroy taste forming phenolic compounds is probably its most important contribution to the field of potable water treatment. Moreover, this ability appears to be capable of destroying other taste forming compounds of unknown origin. There are two major mechanisms by which ozone may react with organic material.[13] The first of these is a direct additive attack in which ozonides and ultimately peroxides are formed together with a splitting of the organic molecule. The other mechanism is an accompaniment to the decomposition of ozone. This decomposition proceeds by way of the formation of the free radicals OH, HO_2 and HO_3 as described above. These free radicals, especially OH, are highly reactive against all sorts of organic material and may lead to autooxidation of a wide variety of organic matter, particularly those present in wastewater effluents. The free radical autooxidation mechanism may well be involved in the disappearance of residual ozone after the initial rapid demand has been satisfied.[6]

Physical and Chemical Properties. Ozone (molecular weight 48), an allotropic form of oxygen, is an unstable blue gas with a characteristic pungent odor to which it owes its name. It is derived from the Greek word "ozein," meaning "to smell." Discovered by van Marum in 1785, it is produced commercially from dry air or oxygen formed by the corona discharge of high-voltage (4000–30,000 V) electricity. Ozone is also formed photochemically in the earth's atmosphere. It is one of the hazardous elements of smog, and its concentration in the atmosphere is an indicator of the smog intensity. Ozone content in the atmosphere in excess of 0.25 ppm is generally considered injurious to the health of man. Ozone levels of 1.0 ppm in the atmosphere are extremely hazardous to health. Ozone weighs approximately 0.135 lb/ft^3 at one atm. It is a powerful oxidizing agent. Its oxidation potential is -2.07 V referred to the hydrogen electrode at 25°C and at unity H-ion activity. Only fluorine has a more electronegative oxidation potential.[14] Ozone is extremely corrosive, so that materials of construction must be very carefully chosen. Porcelain and glass do not react with ozone. PVC is used, but it is suspected that a reaction takes place, resulting in the loss of ozone.[15] *It completely disintegrates rubber and attacks all plant life.*

The solubility of ozone in water is a limiting factor that greatly affects the process of ozonation. At 20°C the solubility of ozone is only 570 mg/liter.[17] While ozone is more soluble than oxygen, chlorine is twelve times more soluble. In pure aqueous solution, ozone is thought to decompose as follows.[1]

$$O_3 + H_2O \longrightarrow HO_3^+ + OH^- \tag{13-2}$$

$$HO_3^+ + OH^- \longleftarrow 2HO_2 \tag{13-3}$$

$$O_3 + HO_2 \longrightarrow HO + 2O_2 \tag{13-4}$$

$$HO + HO_2 \longrightarrow H_2O + O_2 \tag{13-5}$$

The free radicals (HO_2 and HO) that form when ozone decomposes in aqueous solutions have great oxidizing power, and in addition to disappearing rapidly, (Eq. 13-4), may react with impurities, e.g., metal salts, organic matter, hydrogen and hydroxide ions present in solution. These free radicals formed by the decomposition of ozone in water are apparently the principal reacting species. Ozone, while it exists, does not lose its oxidizing capacity in an aqueous solution.

Inorganic Reactions. Ozone reacts rapidly to oxidize ferrous and manganous ions into their insoluble (ferric and manganic) ions resulting in either a floc that precipitates, or a scum that clings to the water surface. Sulfides and sulfites are readily oxidized to sulfates, and nitrites to nitrate. The oxidation of iodides to iodine is the basis of the usual analytical determination of ozone. Bromides and chlorides are similarly oxidized to bromine (Br_2) and chlorine (Cl_2), respectively, and these reactions are slow and dependent upon the concentration of reactants.

The ammonium ion (NH_4^+) is apparently not attacked under the conditions normally found in wastewater treatment, so there is no waste of ozone oxidizing capacity or side reactions with the ammonia nitrogen in wastewater. However, at a 12 : 1 molar ratio of consumed ozone to ammonia, the ammonia will be completely oxidized to nitrate so long as the pH remains alkaline.[16]

Ozone reacts to oxidize ferrous and manganous compounds to form insoluble ferric and manganic compounds. When iron and manganese are present in complex organic compounds, as is often encountered in groundwaters, ozonation requires supplemental catalytic action by filtration or absorption (GAC) to execute successful iron and manganese removal. When iron and manganese react with ozone a brown scum will form on the surface of the treated water. Since it is light and fluffy it is difficult to remove. Therefore ozone must be applied upstream from the coagulation process.

Ozone reacts effectively to destroy cyanides. When complex cyanides are present, ozone is more effective than chlorine. Inorganic cyanide reactions are as follows:

$$CN^- + O_3 \longrightarrow CNO^- + O_2 \qquad (13\text{-}6)$$

$$2CNO^- + 3O_3 + H_2O \longrightarrow N_2 + 2HCO_3 + 3O_2 \qquad (13\text{-}7)$$

4.5 mg/l ozone is required to destroy 1 mg/l cyanide as CN^-.[70]

Effect of Ozone Residuals on Chlorine and Chlorine Dioxide. French practice when using either chlorine or chlorine dioxide or both in the treatment train along with ozone is to apply the chlorine compounds where there are no ozone residuals. This is because ozone reacts to destroy both HOCl and ClO_2. If the chlorite ion is present ozone converts it to chlorate.[71] Moreover, the hydroperoxides formed during ozonation will also act to dechlorinate both HOCl and ClO_2.

A recent study by Haag and Hoigné[65] has shown that ozone reacts with chlorine (HOCl/OCl$^-$) by the following reactions:

$$O_3 + OCl^- \longrightarrow O_2 + [Cl-O-O^-] \longrightarrow 2O_2 + Cl^- \ (77\%) \qquad (13\text{-}8)$$

$$2O_3 + OCl^- \longrightarrow 2O_2 + ClO_3^- \ (23\%) \qquad (13\text{-}9)$$

That is, when a solution of hypochlorite is exhaustively ozonated, 77 percent of the chlorine is found as chloride ion, while 23 percent is found as the chlorate ion. Overall:

$$1.2 O_3 + OCl^- \longrightarrow 2O_2 + 0.77Cl^- + 0.23ClO_3^- \qquad (13\text{-}10)$$

The above equations are written for the OCl$^-$ ion instead of HOCl because the undissociated form does not react.

Ozone was found to react with monochloramine according to the following equation:

$$NH_2Cl + 3O_3 \longrightarrow 2H^+ + NO_3^- + Cl^- + 3O_2 \qquad (13\text{-}11)$$

Approximately 4 moles of ozone are lost per mole NH_2Cl. No chlorate is found in the reaction.

Ozonation of Seawater[65]. When seawater is ozonated the following reactions occur

$$O_3 + Br^- \longrightarrow O_2 + OBr^- \qquad (13\text{-}12)$$

$$O_3 + OBr^- \longrightarrow [O_2 + BrOO^-] \longrightarrow 2O_2 + Br^- \qquad (13\text{-}13)$$

These reactions form a chain reaction which catalytically destroys ozone. As the concentration of Br$^-$ in seawater is 65–70 mg/l, in the reaction with O_3 at concentrations for disinfection, the dominant mode of O_3 consumption is with Br^-. This is no disadvantage, because an equivalent amount of HOBr is formed as O_3 is lost, and HOBr is an excellent disinfectant. These reactions will occur only if the contactor is designed for optimum plug flow conditions with a transfer efficiency of at least 80–85 percent.

Organic Reactions. Ozone reacts readily with unsaturated organic compounds, adding all three oxygen atoms at a double or triple bond. The resulting compounds are called ozonides. Decomposition of ozonides results in a rupture at the position of the double bond, causing the formation of aldehydes, ketones, and acids. Ozone readily destroys phenolic compounds* and is capable of bleaching organic color found in some waters. These last two characteristics are responsible for the popularity of ozone in the treatment of low quality surface waters in Western Europe.

Glaze et al.[18] has reported on the ability of ozone to destroy humic acid which is the precursor of THM (trihalomethane) formation.† Guirguis et al.[19] reported that ozone makes organic compounds more adsorbable by carbon. Prengle et al.[20] reports that, with time and proper dosage, ozone plus UV light can reduce malathion to carbon dioxide and water after forming three or four intermediate compounds such as alcohol, aldehydes, and oxalic acid after a one hr contact time. Not only was the pesticide (malathion) destroyed, as indicated by gas chromatographic analysis, but also the total organic content of the water. Likewise Richard[21] revealed from his studies that ozone can degrade the pesticides parathion and marathion to phosphoric acid.

Toxicity of Ozone. During the International Ozone Institute meeting in Cincinnati, Ohio (Nov. 1976), several speakers presented information on some of the known toxic properties of ozone. Falk and Moyer[22] reported on a review of scientific literature dealing with reactions of ozone with organic materials as they might occur under treatment conditions. All organic substances considered were taken from a list of compounds that had been found in drinking water by the EPA. Ozonolysis of some of these compounds forms a variety of hydroperoxides which are known to be mutagenic. Ozonolysis of pesticides can produce epoxides, some of which have been shown to be carcinogenic.

Hartemann, Block and Maugras from France[23] reported on the preliminary results of their studies that the toxicity induced by the ozonation of organic materials may be lower than those induced by chlorination which was attributable to the chlorine residual.

Kinman et al.[24] found that ozonated wastewater was more toxic than unozonated wastewater.

Spanggord and McClurg[25] of Stanford Research Institute (SRI) found that a

---

* 2.0 mg/l O_3 is required to destroy 1 mg/l phenols.[70]
† This illustrates the value of preozonation in an ozone–chlorination combination.

very high concentration of ozone produces mutagenic compounds when reacted with ethanol.

Simmon and Eckford,[26] also of SRI, found an increase in mutagenesis after ozonation of ethanol, benzidine, and nitrilotriacetic acid, but 26 other compounds studied were not found to be mutagenic after ozonation.

The public health significance of all these findings is not known as of this writing. The French, who have been ozonating for at least 50 years, do not express any concern, even though they are continuing their toxicity investigations.

CURRENT PRACTICES: POTABLE WATER

United States. Owing to the popularity of chlorination there are very few ozone installations in the U.S.A. The first plant to use ozone in the United States is believed to be the one at Whiting, Indiana. This installation dates from 1940. It is used primarily for the destruction of phenolic tastes and odors. It is still in use today. Ozone application is followed by chlorination for disinfection, and to provide a persisting residual in the distribution system. As a result of applying the ozone ahead of chlorination, trihalomethanes were held to less than 1 μg/l.[27] The other installations were at Strasburg, Pennsylvania, which began operating in 1973, and at Monroe and Bay City, Michigan, and at Saratoga, Wyoming, all of which began in 1978.

In Philadelphia, Pennsylvania, the Belmont Plant, with 36 mgd capacity, was commissioned in 1949 and operated successfully until 1959, when it was taken out of service. The ozone process was discontinued in favor of chlorination when the plant capacity was increased and the contact basins provided contact exceeding twenty hours, making free residual chlorination much more economical than ozone.[2]

There are a variety of reasons why there has been limited use of ozone in the U.S.A. This is mostly due to the availability of superior quality of water for potable use as compared to the waters of Western Europe. The inadequacies of ozone—mainly, the interference by manganese and iron—are more pronounced in the United States. These factors, coupled with lower cost, greater flexibility, and better reliability of chlorination are the contributing reasons for its limited use in the United States.

Ozone can be more attractive for the treatment of low quality potable water supplies where taste, odors, and color are dominant factors. These situations have been far more abundant in Western Europe than in North America. Ozone has long been considered the superpolishing agent for potable waters (i.e., color, taste, and odor removal, plus disinfection).

In the 1980s, however, there is a more compelling situation in North America which may provide the needed impetus to establish ozone as a desirable supplement in the treatment train for potable water: water quality at the source is gradually deteriorating in the United States. Ozone is being investigated as a pretreatment tool providing significant benefits in the overall treatment process of potable water.

The Hackensack Water Co., Harrington Park, NJ, conducted two separate pilot plant studies from 1979 to 1983.[28] The purpose of these studies was to evaluate preozonation in removing color, turbidity, iron and manganese along with the use of conventional coagulants followed by dual media filtration, post-chloramination, and pH adjustment. Their findings were to be used in the design of a 100 mgd addition to an existing 50 mgd plant. The results of these studies revealed that preozonation and direct filtration will reduce present alum usage by 60 percent, present chlorine usage by 75 percent, and present caustic usage by at least 50 percent. In addition, the sludge produced in the present treatment process will be reduced by at least 50 percent and the finished water will contain less dissolved solids, such as sulfates, sodium, and chlorides. Projected chemical savings for 100 mgd are expected to be $244,000. The ozonator capacity is only 2500 lb/day (2mg/l). Therefore all of these benefits have proved in this case that the additional capital cost for the ozone system is well worth it.

The City of Los Angeles Dept. of Water and Power is building a 100 mgd water treatment plant that includes preozonation for improved coagulation, better solids removal, and taste and odor control. Post-treatment will be conventional chlorination. The ozone generation system will have a maximum capacity of 7510 lb/day.

Many other studies incorporating the use of ozone are being pursued in the field of water reclamation and wastewater reuse. Ozone is an attractive process for pretreatment of tertiary effluents designed for recharging the ground water.

Pilot plant studies in 1982 on two difficult-to-treat Florida waters were reported by Elefritz et al.[89] in 1984. The results of these studies were used to design modifications for the existing treatment facilities. The major innovation was the use of preozonation.

The waters studied contain about everything troublesome to the treatment process: they are biologically unstable, which encourages luxuriant algae growths; they contain hydrogen sulfide, iron, and organic color; and they require softening. The decision to study these waters was a direct result of excessive THM formation caused primarily by the free chlorine residual process. Preozonation replaced prechlorination, and chlorine was moved to postchlorination for disinfection and water quality control in the distribution system.

In both studies ozone was applied ahead of the softening process and immediately after pH control, by recarbonation. The results from preozonation (1–3 mg/l at each point) were remarkable. Chlorine dosage at one plant was reduced from 25 mg/l to 3 mg/l and THM formation was reduced from a potential of 930 μg/l to 80 μg/l. The turnkey cost of the ozonation system (Emery, Ind.), was $530,000 for 500 lb/day capacity. In spite of this seemingly high cost the system has proved cost-effective. There have been many intangible benefits from preozonation: the water is more palatable, the color has improved significantly, and it is biologically more stable. The latter will have a pronounced effect on water quality in the distribution system.

White believes these Florida experiences with low-quality water may help to change the U.S.A. philosophy of pretreatment benefits. He has long contended that the 15 minute chlorine demand of any raw water should be a surrogate quality parameter as important as THM potential. More data are needed to make a chlorine demand commitment, but he believes it should be about 2 mg/l.

Canada. The province of Quebec has been the most active in the use of ozone. As of 1978* there were 23 water treatment systems using ozone in Canada. All but one which is in Ontario, are in Quebec. This is undoubtedly a result of the close ties between Quebec and France. The first Canadian installation was in 1956 at the Ste-Therese Quebec filtration plant. During the next 13 years, ozonated water increased from 4 to 137 mgd (U.S. gallons). The largest plants include: City of Laval, 61 mgd (3 systems); City of Quebec, 58 mgd; City of Chomeday, 47 mgd; and the newest and largest system, the Montreal filter plant, at 300 mgd for the first phase.[30,31] The ozone equipment for the first phase has a 7500 lb/day capacity. This calculates to a dosage of 3 mg/l. This plant also has a postchlorination system for disinfection and to maintain a distribution system residual. The Canadian plants report energy consumption for ozone treatment mostly in the range of 10–15 kWh/lb of O_3. Contact times vary from 2 to 20 minutes, depending upon the quality of the water and temperature. Nadeau and Pigeon[30] reported that it requires 12 kg/hr to produce a 0.4 mg/l ozone residual while treating 54 mgd (U.S. gallons). This calculates to a dosage of 1.4 mg/l ozone. Any water with such a relatively low oxidant demand must be of high quality.

The main use for ozone in Canada is to eliminate the seasonal taste and odor problems and to provide backup disinfection of surface waters.

France. As of 1982 there were about 600 French water plants using ozone. Practically all of the plants are supplied with surface waters. The primary purpose cited for ozone use were taste and odor control, destruction of phenols, organics removal, viral inactivation and bacterial destruction. Some of these plants use ozone for color removal and iron and manganese removal. Many plants report that ozone increases the efficiency of turbidity removal. Ozone dosages range from 0.15 mg/l to 10 mg/l. Contact time is usually 5–10 minutes. Average power consumption for 33 plants checked was 14 kWh/lb of ozone generated.

In 22 of the 63 plants surveyed, ozone was the only oxidant used. Chlorine was being used for final disinfection in 26 plants, while chlorine dioxide was being used at 13 of 63 plants. In the five years since the survey a great many chlorinator installations have been retrofitted to generate chlorine dioxide as a step to reduce THM formation. Before this conversion to chlorine dioxide, chlorine was being used for ammonia-N removal. This required prechlorination dosages as high as

---

* In the middle 1970s the U.S. government funded several surveys of potable water treatment practices in Canada, Great Britain, and Western Europe. The findings of these surveys on the use of ozone are contained in this text and are referenced accordingly.

16 mg/l. However, when these plants convert to chlorine dioxide it is necessary to revise the treatment process to remove ammonia-N in order to meet the allowable limit required for drinking water. This is usually accomplished by biological nitrification.

The French water treatment process is similar to the conventional treatment train used in the U.S.A. This consists of prechlorination followed by flash mixing, coagulation, sedimentation, filtration, ozonation and terminal disinfection with either chlorine or chlorine dioxide. The only marked difference is the ozonation step. This allows lower dosages of the terminal disinfectant. Chlorine dosages are about 1.0 mg/l and chlorine dioxide 0.6 mg/l in this step.

Switzerland. Switzerland has had about 25 years of experience in the use of ozone for the treatment of spring supplies, groundwater and surface waters. About 150 water plants use ozone. Many are small plants. Overall there is a variety of plant sizes.

Several are less than 1 mgd, with a gradation in sizes up to the largest, which is the Lengg plant in Zurich with a capacity of 60 mgd. Ozonator capacity is 260 lb/day (4.33 mg/l). As of 1974 the unit processes were: prechlorination, rapid sand filtration (without chemical flocculation), slow sand filtration, and ozonation followed by postchlorination. The plant was to be expanded and at that time chemical precipitation followed by sedimentation before rapid sand filters and GAC between ozonation and postchlorination[32] were planned.

Ozone is used for the usual variety of purposes such as bacterial and viral destruction, taste and odor removal, and organics removal. Ozone dosages range from 0.3 to 1.5 mg/l. For final or terminal disinfection it is estimated that 80 percent of the Swiss plants that use ozone also use chlorine dioxide as a final disinfectant. Some plants use GAC directly following ozonation. The average power consumption at five different plants was reported to be approximately 15 kWh/lb ozone for generation and contacting.

Federal Republic of Germany. The approach to water treatment in Western Germany is substantially different from other European countries. For example, it is not uncommon to find a fully chemically treated river water pumped back into the underground aquifers only to be pumped out of the ground as drinking water.[33] One of the major problems for Germany is the Rhine River. It is grossly polluted along its entire length. It is the source of water supply for many communities. German water specialists believe that the most serious problem with the Rhine is the high level of dissolved TOC, which ranges from 6 to 9 mg/l. Bernhardt[34] at Siegburg claims that if the TOC concentration exceeds 2 mg/l it is not possible to maintain sufficient free residual to reduce the standard plate count below the German standard, which is 20 organisms per ml at 22°C. When such a situation occurs additional treatment must be incorporated within the plant to provide a higher quality water. Two parameters which must be adhered to is that the treated water must have dissolved organics less than 2 mg/l TOC and an oxidant demand (chlorine demand) less than 0.5 mg/l. In other words if the treated water consumes more than 0.5 mg/l chlorine to produce a stable residual in the distribution system, then the water treatment process must be modified to reduce the chlorine demand.

Ozone usage in West Germany is more varied than in any other country. The purposes of ozonation, dosage levels, methods of O_3 diffusion (contacting), and the assortment of equipment in use does not fall into any consistent pattern as is the case in other countries. Power consumption also varies. It seems to be higher than most plants in other countries. The lowest power consumption found was at Duisburg, 7 kWh/lb O_3.

Some 136 municipal water systems in Western Germany are using ozone. The oldest installations date from about 1955 at the Dusseldorf water plants. Of 31 answering the 1978 questionnaire, 24 use ozone for the reduction of organics, 13 for taste removal, 8 for viral inactivation, 7 for odor removal, and 5 for color removal. Ozone dosages ranged from 0.15 mg/l at the Diez-Lahn plant to 5.7 mg/l at the Osterode plant. Chlorine dioxide was used in 9 of the 31 plants for final disinfection, 10 used chlorine, 9 used ozone as the only treatment step, and 3 used ozone as the only oxidant. Ten plants were using GAC as an absorbent following the ozone step.

Owing to the conviction that the most urgent problem in treating water from the Rhine was too great a concentration of dissolved organics, Kuhn et al.[35] developed a treatment train to deal with the problem. In this process the water is first allowed to filter through a berm into a small catchment area which became the raw water supply. The first treatment step was ozonation. This was followed by chemical coagulation, flocculation, sedimentation, and GAC filtration. The final step was disinfection by either chlorine dioxide or chlorine. This complicated process produced a high-quality water with a chlorine demand less than 0.5 mg/l. The ozone step ahead of the GAC filters increased the TOC removal from 1 mg/l to 3 mg/l when the ozone dose was 3 mg/l.

While in pursuit of higher organic removal in potable water supplies the German investigators were exploring better methods of treating wastewater discharges into the Rhine, hoping to decrease the dissolved organics at the source.

Austria. In Austria 42 plants were using ozone in 1978. A few plants were supplementing the ozone treatment with chlorine. The high quality of the water available throughout this country is reflected by the low ozone dosages used (0.06–1.2 mg/l). Some plants used ozone for color and organics removal. Power consumption for ozone generation ranged from 7 to 25 kWh/lb of O_3.

The Netherlands. Most ozone systems in the Netherlands were less than five years old in 1977. The Dutch philosophy exhibits a preference for physical and biological processes instead of chemical treatment. They rely heavily on large storage reservoirs which receive river water from infiltration systems. Their goal is to have several months' storage capacity. The flow sheet for the new plant at Amsterdam scheduled for completion in 1979 included pumping river water into a lake, mixing one-to-one with groundwater consisting of 80 percent infiltration water, coagulation with ferric chloride, followed by ozonation, powdered activated carbon, rapid sand filtration, slow sand filtration, and final chlorination.[34]

A new plant for Rotterdam was planned to take water from the Biesbosch reservoirs which are equipped with destratification aerators to control the growth of plankton. After several months detention in the reservoirs the water was to be coagulated and flocculated followed by sedimentation, ozonation, mixed media filtration, GAC, and final chlorination. Many plants were planning to switch final disinfection from chlorine to chlorine dioxide. Several undesirable experiences with biological regrowth problems when ozone was the last step resulted in its application well ahead of final chlorination.

Primary use of ozone in the Netherlands is for color removal and taste and odor control. Some plants use ozone for bacterial and viral destruction but rarely as a last step. Ozone dosages vary from 0.25 to 5 mg/l. Power consumption was reported to be about 8–9 kWh/lb O_3. Most plants practiced off-gas destruction. Of the seven plants reporting, four plants used chlorine for final disinfection. One plant used ozone as the last step, and in the other two ozone is followed by sand filtration as the last step.

Great Britain. The British rely heavily on physical and biological processes for water treatment and avoid chemical treatment whenever possible. The physical processes include microstraining and slow sand filters. They also look to open storage as a desirable treatment step. The Metropolitan Water Board of London are strong proponents of huge storage reservoirs equipped with artificial destratification for plankton control.

There are eighteen plants in Great Britain known to be using ozone, which is primarily for color removal. Most of them have been using ozone only a few years. The EPA survey team sent out fifteen questionnaires to British plants using ozone; six responses were received.[29] Each of the six plants responding indicated that ozone was used only for color removal. One of the largest of these plants is the Watchgate treatment plant which can supply up to 140 mgd to the city of Manchester. This plant, which uses ozone for color removal, began operating about 1975.[36] Disinfection is by chlorination. The chlorine residual (0.8 mg/l) is automatically controlled before entering the distribution system by the use of sulfur dioxide. Most of the plants using ozone have been in use for only a few years. While the treatment processes vary from plant to plant, ozone is applied after a filtration step in each of the six plants responding. In four of the plants microstraining is the filtration method.

Ozone is also used for other purposes besides the main objective which is color removal. These are taste and odor removal, iron and manganese removal, viral inactivation, and microorganisms destruction.

Table 13-3 Notable Ozone Installations: Water Treatment

Location	O_3 lb/day	Capacity mg/l
Moscow, USSR	10560	4.2
Kiev, USSR	4224	3.5
Manchester, U.K.	2640	2.5
Lodz, Poland	2112	3.3
Rotterdam, Netherlands	1200	4.8
Zurich, Switzerland	1560	2.8
Lake Constance, Germany	3000	2.1
Choisy-Le-Roi, Paris	6600	3.5
Neuilly-Sur-Marne, Paris	6400	4.8
Mery-Sur-Oise, Paris	3600	6.2

Chlorine is used as a final disinfectant in each of the six plants. In 3 of the 6 plants, chlorination follows ozonation, with no filtration step in between. Chlorine dioxide and ozone are not used together in any British plant. Chlorine dioxide is often used as a distinct unit process added stepwise (sometimes with chlorine) to insure a better persisting residual.

Ozone dosages reported by four plants averaged 2.37 mg/l. Power consumption for ozone generation, air preparation, contacting, and off-gas treatment averaged 13.5 kWh/lb O_3 for the plants responding.

Other Installations. Some notable installations around the world are shown in Table 13-3.

There are also installations at Singapore, Malaya and Chiba, Japan. Only one of the above installations has an ozone capacity greater than 5 mg/l which is the rule of thumb by designers of these systems. This is probably the influence of the major supplier of ozone generators (Trailigaz of Paris) which has had long and successful experience with ozonation of potable water.

THE OZONE–BIOLOGICAL ACTIVATED CARBON PROCESS

Historical Background. In the years following World War II the Germans began to study the use of activated carbon as a multi-purpose tool in the processing of potable water. At first it was used as a dechlorinating agent. In those times it was standard practice to use chlorine for ammonia nitrogen removal in the heavily polluted waters of the Ruhr and Rhine Rivers. This resulted in residuals much higher than desired in the finished waters, hence the necessity for dechlorination.

Another important factor in the German scheme of things is their philosophy of a drinking water supply. They are convinced that groundwater is the most important raw material for drinking water. In the overcrowded regions along the Rhine and the Ruhr there was no way to avoid the use of surface water as the raw material for their drinking water supplies. Because of their successful experiences with groundwaters, treatment procedures were investigated which could produce a finished water with a 2 mg/l TOC or less and a chlorine demand of 0.5 mg/l or less.[35] In addition to the employment of water passage through the ground such as sand bank filtration or percolation systems, activated carbon filters and ozone were thoroughly investigated as unit processes. Combinations of ozone and GAC were installed in water plants near Dusseldorf, West Germany, in the late 1950s, but the synergistic interaction of the two processes was not fully recognized until ten years later.

Process Description. This is also known as the Mulheim Process.[37] This process produces a surface water by physiochemical procedures practically equivalent to groundwater. This scheme has been used at the Dohne plant in Mulheim, Germany, since April 1977. The treatment process is as follows: Ruhr River water is pumped from a side channel into a mixing chamber where chlorine used to be applied. Instead, ozone is applied here at an average dose of about 1 mg/l (depending upon turbidity). Coagulants are also added here. The water then passes to a pulsating type flocculating plant. The clarified water leaving the flocculating system passes through the ozonation contact basin, which has a mean retention time of 5 minutes. Ozone dosage here is 3 mg/l. The effluent from the ozone contactor goes to a double-layer group of pressure filters followed by an activated carbon filter. The water then flows into 15 injection wells and two filter basins. After the water is pumped from the injection wells to the distribution system chlorine is added.

Significant Features. Preozonation leads to a transformation of the organic matter in the raw water. Ozone probably converts the long-chain, large-molecule organics which are nonbiodegradable to smaller, more biodegradable organics, and at the same time charges the water with dissolved oxygen. This effect of ozonation is the cornerstone of success of the Mulheim process. The water now has an abundance of oxygen together with biodegradable organic compounds which together promote the growth of aerobic bacteria. The best place for this regrowth to take place is in a GAC filter. The organics there were both pore adsorbed and surface adsorbed on the granular activated carbon. The aerobic conditions within the carbon filters build up a biological system which achieves a high degree of nitrification. Bacterial regrowth after the water leaves the GAC filter is eliminated by passing the water through the ground or a slow sand filter.

Other Users. Users of this process or modifications thereof are: The Lengg plant, Zurich, Switzerland; Bremen, West Germany on the Weser River; and La Chapelle at Rouen, France. The Rouen plant (14 mgd)[38] deserves special attention because it was a new plant using a variation of the Mulheim process. The raw water is pumped from 100 foot wells located adjacent to the heavily polluted Seine River to an intake structure where it receives a preozonation dose of 0.7 mg/l and 3 minutes contact time.

In addition to the effect of preozonation in the Mulheim process, ozone at Rouen was effective for iron and manganese removal. Preozonation is followed by direct sand filtration. The sand filters remove the oxidized iron and manganese plus some flocculated organics rendered insoluble by ozonation. Initial nitrification of the ammonia present takes place in the sand filter. There is no chemical addition other than ozone upstream from the sand filters. The granular activated carbon filters are located underneath the sand filters. The carbon bed is only 30 inches deep and has a theoretical contact time of 9 minutes. Both the sand and carbon filters are flat rectangular units rather than a columnar configuration. The carbon filters perform the removal of organics by adsorption and ammonia removal by the aerobic biological cultures found on the surface of the carbon. Postozonation for disinfection is 1.4 mg/l and is applied sequentially to two different diffuser chambers. The contact time is 12 minutes in order to be certain of viral destruction. The dosage of ozone is consistent with French practice. The ozonators at the Rouen plant are automatically controlled by ozone residual monitors to provide an ozone residual of 0.4 mg/l at the end of the contact chamber. The finished water is stored in two underground reservoirs, each having a capacity of 350,000 gallons. This stored water is treated with sufficient chlorine to provide a residual at the customer's tap of about 0.5 mg/l. The ammonia removal through the plant is about 86 percent. Therefore, since the final ammonia content is about 0.4 mg/l, the chlorine residual in the distribution system will be all combined residual. This will not promote the formation of THMs. At this time it is not possible to compare the overall performance of the Rouen plant with the Mulheim process because the organics concentrations are not reported in the same terms. The carbon filters operated 26 months at Rouen without any necessity to be regenerated. This was similar to the Mulheim experience. It is thought that the aerobic biological activity on the surface of the granular carbon does in some way effect a "regeneration" process.

Uses in the United States. GAC in the United States has been used for at least half a century in the food and beverage industry. It has been used primarily for taste and odor control in the treatment of potable water. The largest use has been with powdered activated carbon rather than the granular form used as filters.

A proposed requirement by the U.S. EPA of GAC filtration in the potable water treatment train as an effective way to meet interim Federal water quality standards led to several pilot plant studies funded by the EPA in 1978–79. Some water utilities also made independent studies. More questions seemed to be raised than answered about both powdered and granular activated carbon.[39]

PREOZONATION: POTABLE WATER

General Discussion. Until recently (1980) European ozonation practice has placed the application as the final treatment step. The work done by Trussell[47] on reuse of waters inspired a similar investigation to determine the effects of preozonation on the treatment of Seine River water.[69] The pilot plant was located at the Choisy-le-Roi water treatment plant, which uses the Seine River as a source of raw water. The treated water serves the city of Paris. This pilot plant study verified the ability of ozone to significantly increase turbidity removal, particle numbers, COD, and UV absorbance, all at very low O_3 dosages: 0.8–1.2 mg/l. Floc settling velocities were observed. It is believed that this, coupled with an increase in floc size by preozonation, was the reason for the increased removal of particle volume. It was also observed that the net effects of preozonation appears to depend strongly upon the quality of the water being treated. The Seine River quality varies widely over a year's period.

The many benefits of preozonation as demonstrated by pilot plant studies greatly influenced the decision of the designers to select this process for the new treatment plants at Hackensack, NJ and Los Angeles, CA. These plants are described in this chapter.

USE OF OZONE IN WASTEWATER TREATMENT

General Discussion. The first reported investigations of ozone as a wastewater treatment process was by the United States Army ca 1955.[40] Their studies indicated that it was effective and economical despite the fact that ozone consumption might have to be as high as 100 mg/l. Their laboratory results demonstrated that ozone could be used successfully for the sterilization of sewage seeded with *B. anthracis, B. subtilis,* influenza virus, and for inactivation of *Clostridium botulinum* toxin.

One of the first pilot plant investigations of ozone for disinfecting sewage was carried out at the Eastern Sewage works in the London Borough of Redbridge, England. This work was reported in 1967.[41] The pilot plant received the equivalent of a secondary effluent and provided for microstraining (35 μ), prechlorination, ozonation, coagulation, and rapid sand filtration for a flow of about 35 gal/min. The ozone dose was kept between 20 and 25 mg/liter to keep the color at or below 10° Hazen. When chlorine was used, the maximum dose used was 20 mg/liter. With these dosages and on a microstrained secondary effluent containing 1.7×10^6 MPN total coliforms/100 ml; ozone alone produced a 4.5-log reduction down to 90/100 ml, while chlorine alone achieved a 6-log reduction down to 1 coliform/100 ml. (No data were available on contact time). However, ozone performed well as a polishing agent. It achieved good color removal and was observed to break down the detergents.

In the United States, the Sanitation Districts of Los Angeles County made a comprehensive study on the use of ozone at their Pomona, California Water Renovation Plant. This study spanned a period of several years in the early 1970s,[10,42-44] Disinfection by ozone was compared to chlorine on the basis of both bactericidal and viricidal efficiency. Owing to the fact that treated wastewater from five activated sludge plants discharge into two rivers that contain undiluted secondary effluent during the swimming season, the disinfection requirement by the California State Department of Health specifies a seven-day median total coliform of 2.2 or less per 100 ml. Therefore this study examined this requirement as well as the EPA standard of 200/100 ml fecal coliforms and a variety of effluents. Using a well oxidized secondary effluent it required 50 mg/l dose of ozone to meet the 2.2/100 ml total coliform and only 10 mg/l dose to achieve 200/100 ml fecal coliforms. They observed the best kill at 3 min contact time.

There is little or no current interest in wastewater disinfection in Europe or the British Isles. Therefore there are no operating ozone installations in these areas for disinfection of wastewater effluents.

About 1974, the EPA, under the able and intelligent guidance of A. D. Venosa, Director, Wastewater Disinfection Research Program, organized several projects to examine the reliability and practicality of ozone as a wastewater disinfectant.[41] The most notable of these early investigations was carried out at Fort Southworth, Kentucky; Grandville, Michigan; Cleveland, Ohio; and New York City. The results were encouraging. The ozone dosage required to meet a fecal coliform standard of 200/100 ml MPN was on the order of 15–20 mg/l.

Venosa made some important observations resulting from these projects. He has pointed out that the most serious practical problem was inadequate process control. When ozone was compared to chlorine (Grandville, MI[45]), chlorine produced the most uniform results because the dosage required to achieve consistent disinfection was easily and reliably controlled by continuous residual control equipment. This was not possible with ozone, therefore manual control was the only means of dosage regulation.

Venosa further observed the excellent nonlinear correlation between the ozone concentration in the exhaust gas from the contactor and the fecal coliform concentration in the final effluent. This suggested that reliable disinfection could be achieved by the use of exhaust gas monitoring as a control strategy.[46] These findings are described in "Wastewater Disinfection Research" in this chapter.

Operating Installations. There are only a limited number of ozone installations at wastewater treatment plans. The following is a listing of known installations in operative use in 1984 with as much detail data currently available:

The system at Frankfort, Kentucky is apparently successful, having survived

Location	Plant Size, mgd	O_3 Capacity, lb/day	Max. Dose, mg/l
Frankfort, KY	24	5	3
Southport Plant, Indianapolis, IN	125	8400	8
Brookings, SD	6	250	5
Vail, CO			2
Kennewick, WA	7.5	250	4
Cleveland Westerly AWTP, Northeast Ohio Regional Sewer Dist.	100	3900	4.7
Meander AWTP, Mahoning Co., OH	8	200	3

the usual startup problems. They are experiencing difficulty with automatic control components.

The Indianpolis Southport system operates only in summer when disinfection is required.

The system at Brookings, SD is experiencing serious difficulties due to a poor contactor design.

Vail, Colorado is operating successfully. The operators are pleased with its reliability. The dosage is low for wastewater—about 1.5 mg/l. This is understandable because the NPDES requirement is 6000/100 ml MPN total coliform.

The Kennewick, Washington system is operating with limited success due to high nitrite concentration in the effluent, which consumes a significant amount of ozone. There is also a contactor efficiency problem. These difficulties combined with a capacity of only 4 mg/l conspire to limit the success of this installation.

The Cleveland Westerly AWTP ozone installation was originally designed as a disinfection system in 1974.[83,88] This was the result of a long-range economic evaluation by the design engineers. This perception changed in the intervening years during the construction period.* As of 1984 the Westerly AWT plant is in the startup stage. The generators are by Emery Industries, Emerzone #EG-625. There are six units on line with one standby. The ozone production per generator is 556 lb/day at 1.5 weight percent ozone using air as the carrier gas. Activation of the plant was planned for late 1984.

The new facilities at the Westerly plant designed by Engineering-Science include conventional headworks, clarification and phosphorous removal with lime coagulation, recarbonation, preozonation, dual media sand filtration, granular activated carbon absorption, disinfection by chlorination, followed by dechlorination. This process train flow using ozone ahead of the filters, which are followed by 17 ft GAC columns, was rigorously tested by pilot plant studies.[88] Results of this pilot

---

* Ozone application was changed from disinfection to preozonation.

study were similar to those reported by Trussell et al.[47,48] on preozonation of a tertiary effluent.

The Meander AWTP in Mahoning Co. Ohio uses ozone for disinfection. The equipment was furnished by W. R. Grace Co., who sold their ozone line of equipment to Union Carbide Co. before the contract was completed. This system demonstrated remarkable disinfection efficiency in the initial stages of operation. When running on oxygen feed, fecal coliform counts of zero were obtained, and 10 or below when using air for carrier gas.[88] The current NPDES requirement is 200/100 mL MPN fecal coliforms.

Preozonation: A Tertiary Process:

General Discussion. Trussell et al.[47,48] reported in 1975 on a comprehensive study of an alternative tertiary treatment process which would produce an effluent to meet the strictest requirements of the California State Department of Health. The conventional process usually proposed by the State for similar situations requires coagulation–flocculation, followed by sedimentation, filtration, and disinfection by chlorine. The alternative process reported on here is the one suggested by the consultants directing this project.* This process consists of preozonation of a secondary effluent followed by flocculation, coagulation, sedimentation, filtration, and disinfection with either ozone or chlorine. The investigation demonstrated the many benefits or preozonation followed by chlorine for final disinfection. Moreover, it proved to be less costly than the conventional method.

This alternative process was chosen because it was believed it would produce an effluent to meet the following objectives:

1. Reduced incremental salt increase
2. Greater virus removal efficiency
3. Lower tertiary treatment costs
4. Reduced organics
5. Increased color and turbidity removal
6. Disinfection to 2.2/100 ml coliforms

The ozone pretreatment step at 10 mg/l and 10 min contact time consistently removed approximately 50 percent of the turbidity and suspended solids, and 70 percent of the color from the secondary effluent. COD was usually decreased but not reproducibly. Coliform removals were usually 2–3 logs in the preozonation step; however these removals were inconsistent.

Virus Destruction. The number of seeded viruses was generally reduced by 3 logs or more following pretreatment with 10 mg/l ozone. Results from preozonation

---

* James M. Montgomery Consulting Engineers, Pasadena, CA.

are shown in Table 13-4. Virus concentrations are reported as PFU/1000 gal based upon expected levels of enteric viruses in the secondary effluent and observed percent removals of seeded viruses tested in the ozonation process at the small-scale prototype plant. All the other data are median values.

Table 13-4 Preozonation Summary of Results (Dose = 10 mg/l; Contact time = 10 min)

Parameter	Unit	Before O_3	After O_3	% Removal
Turbidity	TU	5.5	2.8	50
Color	scu	46	14	70
SS	mg/l	12	5	60
COD	mg/l	87	44	50
Coliform	MPN/100 ml	3.8×10^5	1.9×10^3	99.5
Virus	PFU/1,000 gal	450,000	450	99.9

One other significant achievement of preozonation was the effect on the alum dose. The results of coagulation, flocculation and filtration are shown in Table 13-5. The alum dosages used were 10 and 20 mg/l. There was only a slight improvement in results using 20 mg/l alum. Without preozonation alum dosages of 90 mg/l were required to achieve comparable turbidities.[47]

Final Disinfection. Both chlorine and ozone were compared as final disinfectants for bacterial destruction. Fig. 13-1 shows this comparison. As shown, reozonation of the final effluent with ozone dosages as high as 10 mg/liter failed to meet the 2.2/100 ml coliform objective. The chlorine residual in the contact basin was all "combined" because the NH_3-H content of this effluent was about 35 mg/liter. In the full scale plant the effluent will be nitrified so free chlorine residuals will be formed. In the chlorination step for fecal disinfection a 2 mg/l dose removed approximately 2 logs of coliforms with combined residual after 30 min contact

Table 13-5 Coagulation/Flocculation/Filtration Results (Alum dose = 10–20 mg/l)

Parameter	Unit	After O^3	After C/F/F	% Removal
Turbidity	TU	2.8	1.2	60
Color	scu	14	11	20
SS	mg/l	5	3	40
COD	mg/l	44	35	20
Coliform	MPN/100 ml	1.9×10^3	9.8×10^2	50
Virus	PFU/1,000 gal	450	45	90

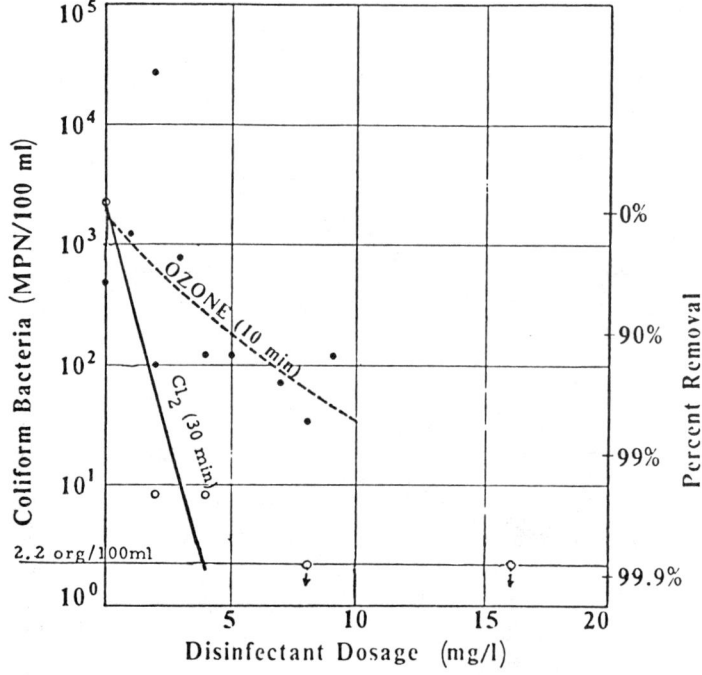

Fig. 13-1. Comparison of bacteria destruction: ozone versus chlorine.[47]

time. This dosage did not consistently achieve the coliform requirement of 2.2/100 ml. The suggested dosage for the full-scale plant is 4 mg/l, which resulted in virus removals greater than 2 logs. The final effluent is expected to have the following quality: turbidity, 0.5–1.5 TU; color, 5–10 SCU; suspended solids, 1–2 mg/l; coliform bacterial, 2.2 organisms/100 ml.[48] As described above, preozonation gave a 3–4 log removal of seeded viruses. Since ozone failed in the final disinfection process, reozonation for virus destruction was omitted from the investigation.[49] However, Fig. 13-2 illustrates the shape of the curves comparing combined chlorine with ozone for seeded virus destruction with a one- or two-run experiment. Therefore, the 5-log removal shown for ozone is not to be considered reliable.[49]

Projected Full-Scale Plant. Table 13-6 illustrates the probable effectiveness of ozone as a pretreatment process for wastewater reuse or tertiary effluents.

WASTEWATER DISINFECTION RESEARCH

Application of Ozone:

Solubility of Ozone. Ozone is a gas; therefore, its solubility in a liquid is governed by Henry's Law: *The weight of any gas that dissolves in a given volume of a liquid,*

OZONE 917

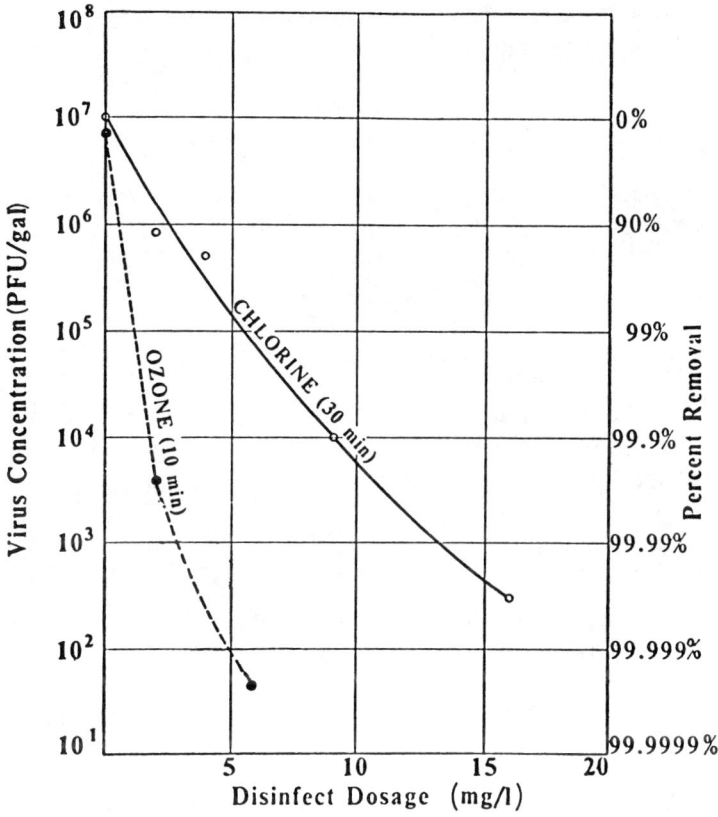

Fig. 13-2. Comparison of virus destruction: ozone versus chlorine.[47]

Table 13-6 Projected Quality Parameters of Full Scale Plant Effluent[47]

Parameter	Unit	Before Treatment	After Treatment	% Removal
Turbidity	TU	3–5	1.0	80
Color	SCU	20–40	5.10	75
SS	mg/liter	10–15	1–3	75
COD	mg/liter	40–60	10–20	60
Coliform[a]	MPN/100 ml	10^4–15^5	≤2.2	99.998
Viruses[b]	PFU/1000 gal	450000	0.5–0.05	98.98

[a] The secondary coliform MPN is usually much higher than is shown in this table. The MPN for this investigation varied from 4.9×10^4 to 3.5×10^6 total coliforms.
[b] Seeded viruses.

at constant temperature, is directly proportional to the pressure that the gas exerts above the liquid. In equation form:

$$Y = HX \quad (13\text{-}14)$$

where

Y = partial pressure of the gas above the liquid, mm Hg

X = concentration of the gas in the liquid at equilibrium with the gas above the liquid, mol gas/total mols of gas plus liquid

H = Henry's law constant (varies with temperature), mm Hg/mol fraction.

The terms in Eq. (13-14) are difficult to understand in practical terms. However, by converting them to units of concentration or mg/l, Henry's law is more easily understood. The terms then become:

Y = concentration of gas above the liquid in equilibrium with the gas dissolved in the liquid, mg/l

X = concentration of gas in the liquid in equilibrium with the gas above the liquid, mg/l

$$H = \frac{\text{mg gas/liter gas}}{\text{mg gas/liter liquid}} \quad (13\text{-}15)$$

Henry's Law simply expresses the concentration of gas above the liquid which must exist in order for a given concentration of gas to be dissolved in the liquid. The lower the value of H, the more soluble the gas.

The solubilities of ozone and oxygen are compared in Table 13-7. Henry's constants were taken from the *International Critical Tables* and converted to units of concentration.[50]

Table 13-7 shows that ozone is about 13 times more soluble in water than oxygen at standard temperature and pressure (H = 20.4 for O_2 versus 1.56 for O_3).* This means that only 1.56 mg/l ozone in air is required to maintain 1.0 mg/l ozone in water, while 20 mg/l oxygen in air is required to maintain 1.0 mg/l oxygen in water under equilibrium conditions at STP. The efficiency of production of ozone in air above about 1.0 weight percent (12.9 mg/l at STP) decreases substantially. Consequently, 8.3 mg/l is the maximum concentration that can be expected to dissolve in the water at that concentration in air, assuming 100 percent mass transfer efficiency and an ozone demand free water. Thus, even though ozone is more soluble than oxygen, when using air as the carrier gas, less ozone will

---

* Chlorine is twelve times more soluble than ozone.

Table 13-7 Solubility of Ozone and Oxygen in Water According to Henry's Law[a]

Temperature	Oxygen (air)			Ozone, 1.0 Weight %[b]		
	H, mg O_2/l air / mg O_2/l water	Y, mg O_2/l air	X, mg O_2/l water	H, mg O_3/l air / mg O_3/l liquid	Y, mg O_3/l air	X, mg O_3/l water
0	20.4	299	14.6	1.56	12.9	8.3
10	25.4	289	11.4	1.86	12.5	6.7
20	29.9	279	9.3	2.59	13.1	4.7
30	34.2	270	7.9	3.80	11.7	3.1

[a] Henry's constant H is a function of temperature, not concentration. Therefore the values of H for oxygen are the same whether the oxygen is 21 percent (i.e., in air) or 100 percent.
[b] 12 mg/l O_3 per liter carrier gas = 1.0 wt. percent at 20°C and atmospheric pressure, or lb O_3 per 100 lb carrier gas.

dissolve on an absolute basis owing to the lower concentration of ozone in the air (i.e., partial pressure). This emphasizes the necessity to achieve the maximum contactor efficiency (TE), because of the difficulty of maintaining high partial pressures of ozone above the process liquid (potable water or wastewater).

Transfer Efficiency Concept. The transfer efficiency (TE) of a given gas–liquid contactor (mixing chamber) is an inherent property of the contactor and is a function of the gas* flow rate relative to the liquid flow rate. Transfer efficiency is defined as follows:

$$\text{TE} = \frac{100\,(Y_1 - Y_2)}{Y_1} \tag{13-16}$$

where

$Y_1 = $ mg O_3/l inlet carrier gas
$Y_2 = $ mg O_3/l exhaust gas from contactor.

TE is the fraction of ozone in the gas that has been transferred to the liquid expressed as percent.

Applied Dose. This is defined as follows:

$$D = Y_1 \left(\frac{Q_G}{Q_L} \right) \tag{13-17}$$

where

$Q_G = $ gas flow rate, l/min
$Q_L = $ liquid flow rate, l/min

The applied dose multiplied by the fraction transferred to the liquid is the *absorbed dose*.

Absorbed Dose. The transfer of ozone, also called the absorbed dose, is defined as follows:

$$\begin{aligned} T &= Y_1 \left(\frac{Q_G}{Q_L} \right) \left(\frac{Y_1 - Y_2}{Y_1} \right) \\ &= \frac{Q_G}{Q_L} (Y_1 - Y_2) \end{aligned} \tag{13-18}$$

---

* Gas in this sense means the gas flow discharge from the ozone generator. It can be derived from prepared air or pure oxygen delivered to the generator.

where T is the amount of ozone transferred to the liquid mg/l. Eq. (13-17) indicates the applied dose can be varied by changing either Y_1 (inlet carrier gas concentration) or the Q_G/Q_L ratio. Operating experience has shown that TE (transfer efficiency) is more sensitive to shifts in the gas flow rate than ozone concentration in the carrier gas.[50] This is important, since an increase in the gas flow (Q_G) relative to the liquid flow may not result in a linear increase in absorbed ozone dose.

Application of Henry's Law and the concept of ozone transfer efficiency is useful for designing and optimizing ozone contactors.

Control of Disinfection Efficiency:

Factors Affecting Disinfection Efficiency. It is necessary to review these factors in order to understand an acceptable control strategy. First it is desirable to understand the proper terminology and to differentiate between applied and absorbed ozone dose:[51]

$$D = \text{applied dose, mg/l to the liquid} = Y_1 Q_G/Q_L \tag{13-19}$$

where

Y_1 = ozone concentration in carrier gas, mg/l gas
Q_G = carrier gas flow rate l gas/min
Q_L = liquid flow rate l liq./min (this is process water flow through the contactor chamber)

When the ozone dose is increased by changing the Q_G/Q_L ratio, the transfer efficiency (TE) of the contactor decreases more rapidly than when the dose is increased by changing Y_1.[52,53] Therefore it is desirable to maintain a constant Q_G/Q_L ratio by flow pacing and change dosage by changing Y_1.

Absorbed Ozone Dose (T). As already described, this is the arithmetical difference between the ozone concentration in the inlet carrier gas and the ozone concentration in the exhaust gas stream, multiplied by the Q_G/Q_L ratio:*

$$T = (Y_1 - Y_2) Q_G/Q_L \tag{13-29}$$

where

T = absorbed ozone dose, mg/l in the liquid†
Y_2 = concentration of ozone in the exhaust gas stream leaving the contactor, mg O_3/l gas.

---

* Liquid in process flow.
† Technically, T is not the absorbed dose but rather the transferred dose, for it consists of both consumed ozone and unconsumed residual.[52]

Venosa and Meckes[36,52,53] have shown that total and fecal coliform levels in a given effluent can be predicted if the demand properties of the effluent (TCOD, NO_2, TSS, and TOC) and the absorbed ozone dose are known.

Exhaust Gas Monitoring. Using this as an operational control strategy to achieve optimum disinfection efficiency has the following advantages: (1) true ozone is being measured, not total oxidant, which is what residual methods measure; (2) the reaction is instantaneous; (3) measurement technology is convenient and reliable; (4) it is easily automated; (5) it is not subject to interferences; (6) it is not affected by sudden changes in effluent quality (ozone demand); and (7) it is useful over a wide range of secondary effluent quality.

The control arrangement, which is a variation of the compound-loop principle, is as follows:

1. Carrier gas flow to contactor is flow paced by a signal from process flow meter.
2. The signal from the exhaust gas O_3 concentration monitor, changes the generator power input which changes the O_3 concentration in the inlet carrier gas.

This method of control has been rigorously tested and was reported by Venosa[52] in 1983. The key to the success of this system depends upon the ability of the process flow metering device to maintain the carrier gas flow to the process flow constant over the range of process flow variation. The process flow meter signal should be equipped with a 0.4–4.0 ratio relay and/or a microprocessor.*

Residual Control:

General Discussion. Experience to date (1984) indicates that residual control is doubtful as a control strategy for ozone application. The overriding factor that contributes to this uncertainty is the rapid die-away phenomenon of an ozone residual. Analyzers cannot track fast enough to cope with the speed of residual decay. The other factor is the measurement of free ozone versus total oxidants. There is some doubt as to the ability of the galvanic analyzer to measure free ozone residuals. There is little doubt, however, about its ability to measure total oxidants (free ozone + ozonides + hyperperoxides). However, monitoring total oxidants by the galvanic analyzer is a possibility. See "Residual Recording" in this chapter.

Continuous Analyzers. There are two continuous analyzers available for the measurement of ozone. One is the Fischer and Porter Series 17L-2000, which is specific

---

* The transfer efficiency decreases when the gas to liquid flow ratio changes. This upsets the overall control system.

for free ozone. It can also be arranged to measure total residual oxidants. The other one is the Delta Scientific membrane call analyzer. Neither of these analyzers have been given sufficient opportunity to demonstrate that residual control of ozone systems is practical.

Another factor always present where oxidants are involved is the presence of substances that interfere with the analytical procedure. Copper ions poison the amperometric cell, while manganese ions interfere with some analytical procedures used in the calibration of the analyzers.

Residual control may never receive the necessary consideration to prove its practicality if the exhaust gas monitoring system continues to be as reliable as reported by the Venosa group. However, continuous residual recording for monitoring total oxidant residual may have practical value. It may be an important tool to be used for increasing the reliability of the ozone process.

Forecasting Coliform Concentrations in an Ozonated Effluent:

Effluent Quality. In order for ozone to be effective in wastewater disinfection the effluent must be of good quality. There is a significant difference between a filtered secondary effluent and a nonfiltered secondary. Other than the absorbed ozone dose the total COD and nitrites are the most significant factors. In the case of filtration it is the reduction of COD and not suspended solids which has the greatest effect. The work by Venosa et al.[51] and Gan et al.[42,43] confirm these observations.

Filtered Effluent. Venosa et al.[51] developed a mathematical model which can be used to predict either the total or fecal coliform concentrations in the ozonated effluent in accordance with the following equations

$$\log_{10} TC = 3.95 + 0.030 TCOD + 0.50 NO_2^- \text{-}N - 3.05 \log_{10} T \quad (13\text{-}21)$$

$$\log_{10} FC = 3.34 + 0.029 TCOD + 0.48 NO_2^- \text{-}N - 3.4 \log_{10} T \quad (13\text{-}22)$$

where

TC = total coliforms/100 ml
FC = fecal coliforms/100 ml
T = absorbed ozone dose, mg/l
$TCOD$ = total chemical oxygen demand, mg/l
$NO_2^-\text{-}N$ = nitrite nitrogen, mg/l

Unfiltered Secondary Effluent. Venosa et al. compared a poor quality unfiltered effluent with the filtered effluent and developed the following equations to predict the final coliform concentrations in the ozonated effluent.

It is important to note that their analysis of the data indicated additional factors were found to influence the correlation of effluent coliform densities.

$$\log_{10} TC = 0.96 \log_{10} TC_i - 0.89 + 0.012 TCOD + 0.60\, NO_2^-N \\ + 0.013 TSS - 4.024 \log_{10} T - 0.57 R^{1/2} \quad (13\text{-}23)$$

where

TSS = total suspended solids mg/l
TC_i = initial coliform concentration/100 ml
R = ozone residual mg/l

In this case of total coliforms the results were affected by both the initial coliform density and the ozone residual.

$$\log_{10} FC = 4.06 + 0.020 TCOD = 0.37 NO_2^--N + 0.012 TSS \\ - 3.94 \log_{10} T - 0.59 R^{1/2} \quad (13\text{-}24)$$

where

TSS = total suspended solids, mg/l
R = ozone residual, mg/l

In Eq. (13-24) the final fecal coliforms were not affected by the initial fecal coliform concentration. However both the final fecal and total coliforms were affected by the total suspended solids (TSS) and the ozone residual (R).

High-Level Disinfection With Ozone:

Marlborough, MA Study. Owing to a current trend in certain coastal states to provide protection for shellfish growing areas and similar high quality receiving waters in other areas, standards have been imposed requiring high level disinfection of wastewater effluents. In 1977 Marlborough, MA built a pilot plant to receive secondary effluent from the Easterly Wastewater Plant. This pilot plant was arranged to ozonate either filtered or unfiltered secondary, filtered or unfiltered nitrified effluent, or a combination of these. The results of this study were reported by Stover and Jarnis in 1979.[54] They explored the efficiency of ozone to achieve final total coliforms of 70/100 ml, which is the level considered necessary to protect shellfish, and 2.2/100 ml. The latter level is required for water reuse situations and other sensitive areas in arid states where effluents flow in ephemeral streams where bathing can take place. (See Chapter 8 for situations where the 2.2 coliform level is mandatory in California).

This study was aimed at determining the relationship between effluent quality,

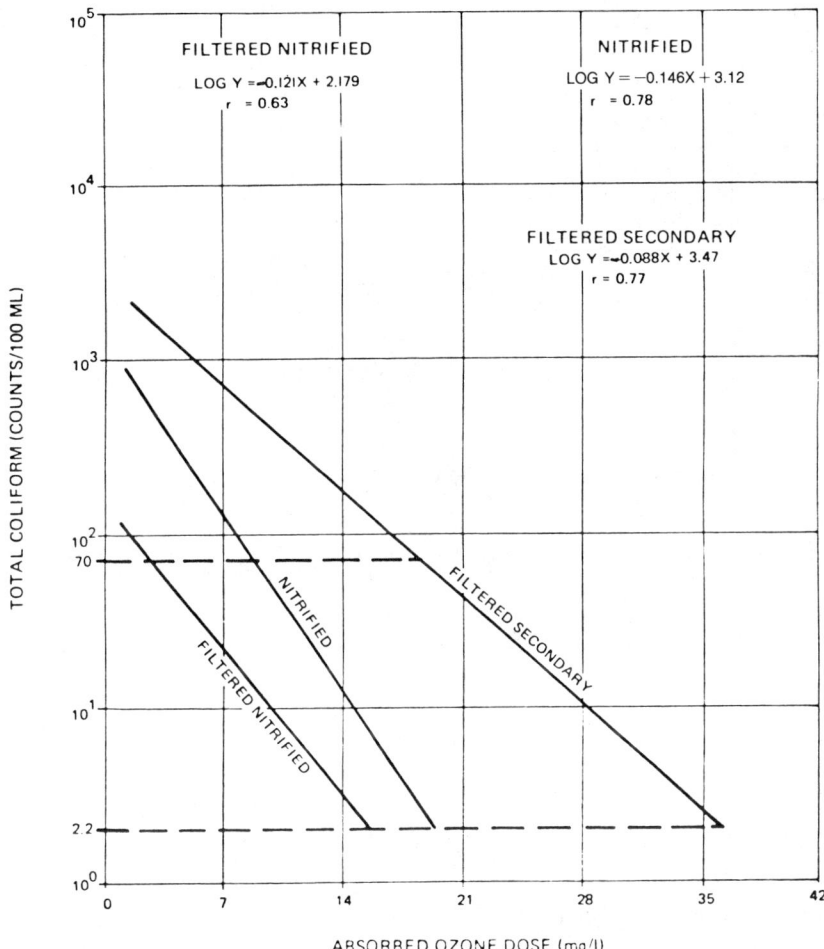

Fig. 13-3. Effluent total coliforms versus absorbed ozone dose[54] (courtesy *Journal WPCF*).

absorbed ozone, and total residual oxidants. The latter were measured by the back-titration method used to measure chlorine residuals with an amperometric endpoint.

Absorbed ozone concentrations of 2–35 mg/l were applied to the different effluents at contactor hydraulic reaction times of 1–10 minutes, yielding total effluent residual oxidant concentrations from 0.4 to 8.0 mg/l. The results showing the relationship of absorbed ozone dose and total residual oxidants to the final coliform count are presented in Figs. 13-3 and 13-4. The data for these plots are shown in Table 13-8. Fecal coliforms were never detected in the ozonated effluents at high-level disinfection.

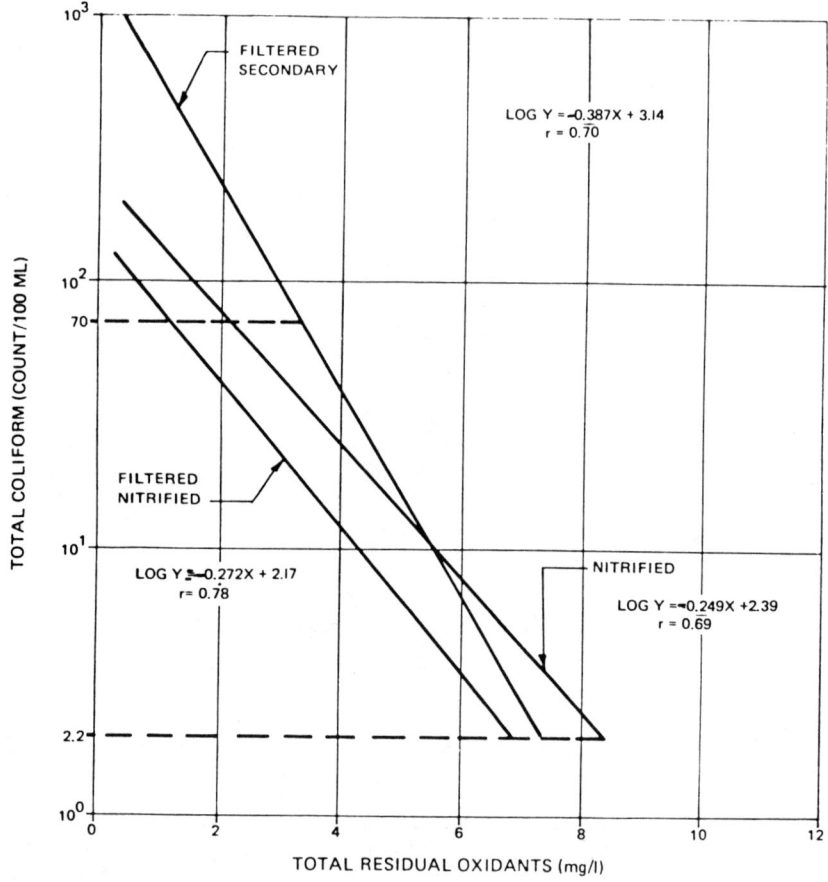

Fig. 13-4. Effluent total coliforms versus total residual oxidants.[54] (courtesy Journal WPCF).

The log total coliform reduction versus the log absorbed dose was plotted for all data points for both the nitrified effluents. Because the two nitrified effluents investigated were essentially of the same chemical water quality, with the exception of initial coliform concentration, the ozone requirement to achieve a certain log reduction would be expected to be the same for both effluents. The same plot was made for the combined secondary effluents. Then the required absorbed ozone dose was calculated from these plots to achieve a 70/100 ml total coliform. For the nitrified and secondary effluents the doses are 5.2 and 14.3 mg/l, respectively.

This study has shown that the absorbed ozone dose gives good correlation with coliform reduction. It also demonstrated that the absorbed ozone dose is significantly dependent upon effluent quality.

High-Level Disinfection with Ozone and UV. The EPA carried out a study to determine whether the combination of UV radiation and ozone would

Table 13-8. Absorbed Dose and Residual Requirements for Disinfection (Figs. 13-3 and 13-4)

Water Quality	Absorbed Dose mg/l	Residual mg/l
High-level disinfection (less than 2.2 total coliforms/100 ml)		
Filtered nitrified	15	6.7
Nitrified	19	8.2
Filtered secondary and secondary	36	7.2
70 Total Coliforms/100 ml		
Filtered nitrified	3	1.2
Nitrified	9	2.2
Filtered secondary and secondary	18	3.3

be sufficiently better than either disinfectant alone to justify the use of both techniques for wastewater disinfection.[55] This investigation revealed that the amount of applied ozone needed to achieve a fecal coliform limitation of 14/100 ml (approximately 60/100 ml total coliform) could be reduced as much as 80 percent if UV light either preceded or followed ozone addition. The ozone dose required to achieve the target coliform count alone was approximately 14 mg/l in all test runs. When UV was applied after ozone, the ozone dose could be reduced to about 3 mg/l. Simultaneous addition of UV and ozone in the same column resulted in a less than additive effect. Apparently UV reduces ozone to molecular oxygen. The sequential application of ozone first, followed by UV radiation exposure was slightly better than UV light before ozone.

The sequential addition of ozone and UV was found to be more cost-effective than either ozone or UV alone for plant sizes greater than 10 mgd. For smaller plants, UV alone appears to be a better choice than ozone.

This investigation was not intended to quantify the optimum UV dose. The purpose was to discover if UV did or did not have an effect upon ozonation. This is the reason the UV dose is not mentioned.

COMMERCIAL OZONE GENERATORS

Introduction. Ozone must be generated at the point of use because it is subject to rapid decomposition, reverting back to oxygen. Ozone can be generated from atmospheric air or from pure, commercially produced oxygen. The process may be by electrical discharge or by photochemical action using ultraviolet light. The latter produces the ozone layer in the upper atmosphere. The most practical method for producing ozone is by electrical discharge.

When ozone is produced by electrical discharge in the corona, the gas stream feeding the generator must be dried to at least −40°F dewpoint, because water vapor lowers ozone production efficiency significantly. When air is used instead of oxygen as the carrier gas, air preparation is critical. Dessicant dryers regenerated by heat should have the hot air blower equipped with 5 micron filters.[56]

Theory of Generation. Ozone is produced from the oxygen in air or pure oxygen when a high-voltage alternating current is imposed across a discharge gap in the presence of either of these gases. Ozone is generated by electric discharge as follows.[57] The key to the process is in the field between the electrodes of stray electrons left over from previous discharge or from background radiation. These electrons become excited and accelerated within the high-energy field between the electrodes. The alternating current causes changing polarity, which in turn causes the negatively charged electrons to be attracted first to one electrode and then to the other, like bouncing balls. As the velocity of the electrons increases, they attain enough energy to split some O_2 molecules in two, which in turn combine with other O_2 molecules to form O_3. The visible effect of the incomplete oxygen molecule breakdown in the air gap between the highly charged electrodes is known as the *corona glow*. Factors which must be considered in the design of electric discharge generators are; voltage, frequency, dielectric material property and thickness, discharge gap, and absolute pressure within the discharge gap.

Under optimum conditions, ozone production depends upon the following relationships.[58]

$$V \sim pg \qquad (13\text{-}25)$$

and

$$(Y/A) \sim f\epsilon V^2/d \qquad (13\text{-}26)$$

Where

Y/A = ozone yield per unit area of electrode surface under optimum conditions
V = Voltage across the discharge gap (peak V)
p = gas pressure in the discharge gap (psia)
g = width of discharge gap (in.)
f = frequency of applied voltage (Hz)
ϵ = dielectric constant
d = dielectric thickness (in)

When studying the above relationships it is obvious that the ozone generator manufacturers are confronted with formidable problems. First of all, the basic method is inherently inefficient. Commercially available corona discharge generators produce ozone in the exiting gas in concentrations varying from 0.5 to 4.0 percent by weight of the carrier gas. Therefore, when making ozone from oxygen most of the oxygen passing through the generator is unchanged. So for economic reasons this oxygen must be used elsewhere in the overall scheme or be dried out and recycled through the generator.

The yield of a generator is related to the square of the voltage, yet high voltages increase the possibility of electrode failure caused by dielectric or electrode puncture. Higher voltages also result in higher pressures (the discharge gap being set), and this means higher operating temperatures. High operating temperatures increase the rate of ozone decomposition.

Dielectric thickness appears in the denominator of the yield relationship which indicates that a thin dielectric is desirable. Thin dielectrics, however, are more susceptible to failure by puncture. It is evident that the problem is to attain maximum yield while simultaneously economizing on maintenance and replacement costs. The solutions to these problems are difficult to achieve. Some of the methods to increase ozonator efficiencies are as follows: High frequency is less damaging to dielectric surfaces than high voltage. Frequencies as high as 2000 Hz are now in use, emphasis is placed upon this parameter. Glass has been found to be a practical dielectric material. It is cheap and readily available. Ingenious generator designs optimize heat removal. In addition, solid-state circuitry, and acid-resistant materials reduce power requirement cost and increase reliability.

So in a modern ozone generator with the latest manifestations of the ingenuity of the designers, only about 10 percent of the energy applied results in the production of ozone.

The largest portion of this energy loss is by the heat generated. The remainder of the loss is through light and sound. The decomposition of ozone back to oxygen is greatly accelerated with increasing temperature, so all high concentration ozonators must employ a method of heat removal.

Assuming a clean dry oxygen-rich gas is being fed to an ozone generator and an efficient method of heat removal is available, production of ozone per unit area of electrode surface under optimum conditions is a function of:

1. Peak voltage across discharge gap
2. Absolute gas pressure in gap
3. Width of discharge gap
4. Frequency of applied voltage
5. Dielectric constant
6. Thickness of the dielectric

Therefore, to optimize the ozone yield, the following conditions should exist.

1. The combination of gas pressure and gap width should be arranged so the voltage may be kept relatively low for reasonable operating pressures.*

---

* Lower voltage protects the dielectric and/or the electrode surfaces from high voltage failure.

2. A thin material with a high dielectric constant should be used.
3. High frequency current (300–2000 Hz) should be used because high frequency increases the ozone yield and is less damaging to the dielectric than high voltage and prolongs the life of the equipment.
4. An efficient heat removal system is essential.

The most significant factor is the efficiency of heat removal. While the voltage and/or frequency can be continually increased to produce more and more ozone, the additional heat produced must be removed as efficiently as possible. Otherwise if the temperature rises, the additional ozone produced will only decompose back to oxygen.

Types of Ozone Generator. Currently there are three types of ozone generators in use (1977). These are: the Lowther Plate; the Otto Plate; and the Tube. All three are illustrated in Fig. 13-5.

Otto Plate. This is the first of the ozone generator designs developed by Otto in 1905. It is known as the "Otto partial injection system." This design is still being used extensively in Western Europe. This ozonator is made up of a series of sections arranged in the following sequence: (1) a water-cooled cast aluminum block which acts as the grounded electrode; (2) a glass plate dielectric; (3) followed by an air gap and another glass dielectric; (4) and then a high voltage stainless steel electrode. A complete unit includes the mirror image of the dielectrics and their corresponding air gaps together with a water cooled, grounded electrode. One of the inherent disadvantages of the Otto-type generator is that the air blown into the ozonator discharge gap area is limited to low pressure.

Cooled Tube. This generator is composed of a number of tubular units. The outer electrodes are stainless steel tubes fastened into stainless steel tube spacers and surrounded by cooling water. Centered inside the stainless steel tubes are tubular glass dielectrics whose inner surfaces are coated with a conductor which acts as the second electrode. The outer electrodes are arranged in parallel and are sealed into a cooling water system. This group of water-cooled tubular units is then enclosed in a gas-tight vessel so that either air or oxygen may be fed to the unit at one end and ozone collected at the other. The glass tube is sealed so that the feed gas passes only through the discharge gap. Fig. 13-6 illustrates a large tube-type ozonation system. Fig. 13-7 illustrates a small tube-type system.

*Lowther Plate.** This is the most recent species of ozone generating equipment. It is air cooled and utilizes either atmospheric air or pure oxygen. This generator

---
*This type of generating system was marketed for several years in the U.S.A. by W. R. Grace Co. Union Carbide Co. purchased this system from Grace (ca 1976) as a feature to supplement their Unox process. In 1984 the Union Carbide Ozone–Unox system is being marketed by Lotepro Inc., a division of the German Linde Co.

OZONE 931

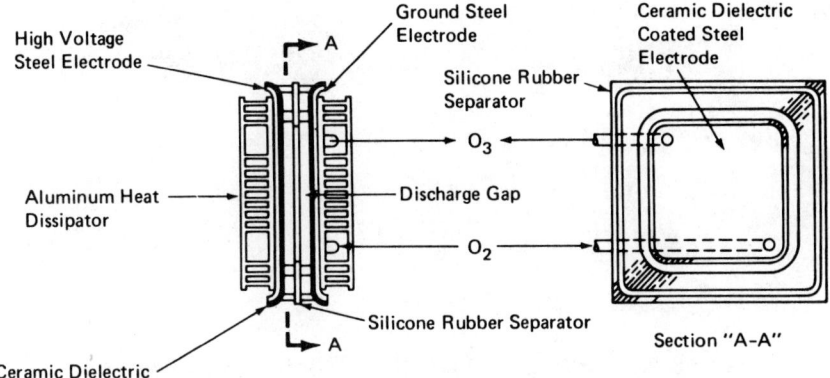

Lowther Plate Generator Unit

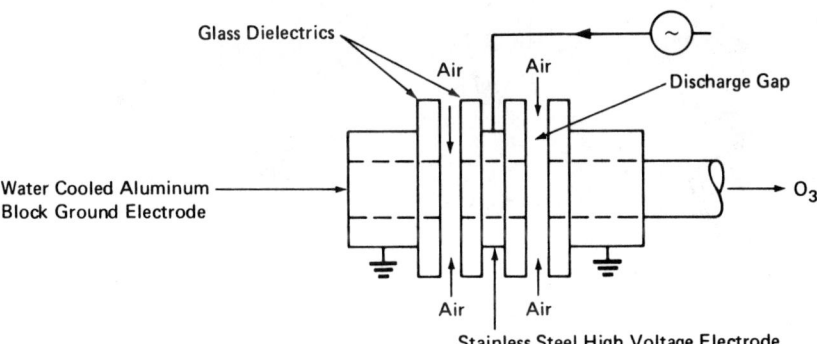

Otto Plate Type Generator Unit

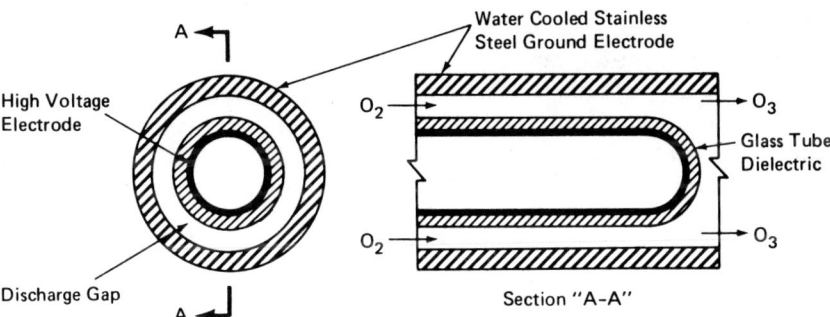

Tube Type Generator Unit

Fig. 13-5. Schematic illustration of commercial ozone generators.

Fig. 13-6. Typical tube type ozone generator (Trailigaz) [courtesy Compagnie Giverale Des Eaux, Paris France].

is made up of a gas-tight "sandwich" consisting of an aluminum heat dissipator; a steel electrode coated with a ceramic dielectric; a silicone rubber spacer which provides the precise amount of discharge gap and; a second ceramic-coated steel electrode with an air (or oxygen) inlet and ozone outlet. The ozone exits through a second aluminum heat dissipator.

These sandwich-type units are pressed together in a frame and are manifolded for either air or oxygen inlet flow. Cooling is accomplished by a fan which moves ambient air across the heat dissipators.

The manufacturers using this type of a generator cite the following characteristics.[58]

1. Uses air cooling.
2. Thin dielectrics provide greater ozone yield which in turn enhances heat removal.
3. The small discharge gap allows higher operating pressures.
4. Low operating voltages provide longer dielectric life.
5. High frequency operation allows higher ozone production without resorting to higher voltages.
6. Heat removal is more efficient.

Fig. 13-7. Two Emerzone® Model EG 250 tube type ozonators capacity 250 lb/day. (courtesy Emery Co. Cincinnati, Ohio).

7. Provides a high-yield efficiency which results in smaller space requirements for the equipment.
8. Power requirements are less than for generators based on older technology.

The operating characteristics of the different types of generator are shown in Table 13-9. The power requirements for the different type of generators are shown in the following table.

Ozone Generating Power Requirements[a]

Type	Power Required (kWh/lb)	
	Air	Oxygen
Otto	10.2	—
Tube	7.5–10.0	3.75–5.0
Lowther	6.8–8.8	2.5–3.5

[a] For 1.0 percent by wt ozone, = 12 mg/l O_3 per liter carrier gas.

Ozonator Manufacturers:

United States. The foremost manufacturers in this country are: Emery Industries, Cincinnati, OH, and PCI, Stanford, CT. These companies can provide a complete

Table 13-9 Typical Ozonator Operating Characteristics

Type	Feed	Dew Point of Feed, °F	Cooling	Pressure, psig	Discharge gap, in.	Voltage, kV, peak	Frequency, Hz	Dielectric Thickness, in.
Otto	air	−40	water	~0	0.125	7.5–20	50–500	0.12–0.19
Tube	air, oxygen	−60	water	3–15	0.10	15–19	60	0.10
Lowther	air, oxygen	−40	air	1–12	0.05	8–10	2000	0.02

range of ozone generators from the smallest to the largest that would likely be used for potable water or wastewater. Emery has had a long and varied experience in the use of ozone and the manufacture of ozone generators. Their principal activity is the commercial production of organic chemicals which requires the use of ozone. The generating units furnished by both Emery and PCI use the cooled tube design. There are some lesser manufacturers with considerable experience who provide generators for special situations. These include Cochrane Co. and Crane Co. Then there are companies who make small generators (less than 50 lb/day), such as Grissin of New Jersey.

Foreign. The dominant names in Europe are Trailigaz and Degremont in France, Kerag in Switzerland, and Gebruder-Hermann in Germany. These are all tube-cooled systems. In Japan the Suni-sonic uses the Lowther plate and Mitsubishi uses the water-cooled tube system. It is not known if the German Linde Co. will manufacture the Union Carbide Lowther plate system in Germany or the United States.

Comparison of Systems. The older Otto plate design, developed early in the 20th Century, is the least efficient and is gradually being replaced by cooled tube systems. The most recent development is the Lowther Plate, hence it is the most efficient. However it has not yet made a significant penetration of the market. The most popular system is the cooled tube design. All of these generators produce an air carrier gas containing 0.5–3.0 percent by weight of ozone. When using oxygen as the carrier gas twice as much ozone is produced compared to air.

OZONE FACILITY DESIGN

General Considerations. The key elements of an ozone system are: air supply pretreatment (unless oxygen is used); ozone generator; ozone absorption chamber (contactor); continuous ozone production measurement; ozone metering and control instrumentation; continuous monitoring of ozone concentration in exhaust gas;

ozone destructor in exhaust gas; ambient ozone monitor in the ozone system area. Fig. 13-8 illustrates a Trailigaz ozonation system (capacity 7656 lb/day).

All installations should be provided with flow pacing capability so that a constant gas-to-process-flow ratio can be maintained. For installations dedicated to disinfection a programmable controller should be provided to allow dosage trim (changing O_3 conc. in gas stream) to maintain constant ozone absorption rate under conditions of variable ozone demand.

The ozone contacting system (absorption chamber) must be capable of dispersing the ozone into the process stream so that the bubbles of ozone will stand "shoulder-to-shoulder" within the regular distribution of microorganisms. A standby power supply system (or alternative source) should be made a part of every ozonator installation owing to its complete dependency upon electricity.

The ozone generator should be supplied so that there is total redundancy. Recent surveys of operating installations indicate that redundancy and an adequate supply of spare parts are necessary.[56,59]

Three Ozone Systems:

Once-Through Oxygen System. This is the simplest and most cost-effective system for wastewater disinfection, and by far, the one for greatest commercial application.[60] The process flow is G-D-E-F, as shown in Fig. 13-9. Oxygen is produced on-site by a cryogenic or pressure swing adsorption process. Dry oxygen is fed to the ozone generator and the off-gas from the contacting unit exits through the ozone decomposer and the resulting oxygen is used to feed an oxygen activated sludge process.

Oxygen Recycle. In this process (G-D-E-F-A-B-C, Fig. 13-9) the oxygen-rich off-gas is recycled to the compressor A and the refrigerant dryer B for cleaning and drying before returning to the ozone generator. Oxygen lost from this recycle loop by ozone conversion and reaction with the wastewater plus the dissolved oxygen added to the wastewater is made up from the oxygen source. Additionally, recycled gas is purged, and makeup oxygen added to maintain oxygen concentration in the recycle loop. This purging procedure is necessary because of the oxygen dilution by nitrogen gas, which is stripped from the ozone mixing chamber, and the carbon dioxide produced by ozone oxidation of organic matter.

Once-Through Air. This process is (A-B-C-D-E-F- discharge to ambient) as shown in Fig. 13-9. The choice favoring this system is purely economic: power costs; ozone generator efficiency; air versus oxygen costs; and plant size.

If an oxygen activated sludge system is not part of the treatment process, the best method is either recycled oxygen or once-through air.

General Purpose Facility. When ozone is used in a water treatment plant for purposes other than disinfection, the Europeans have found from experience

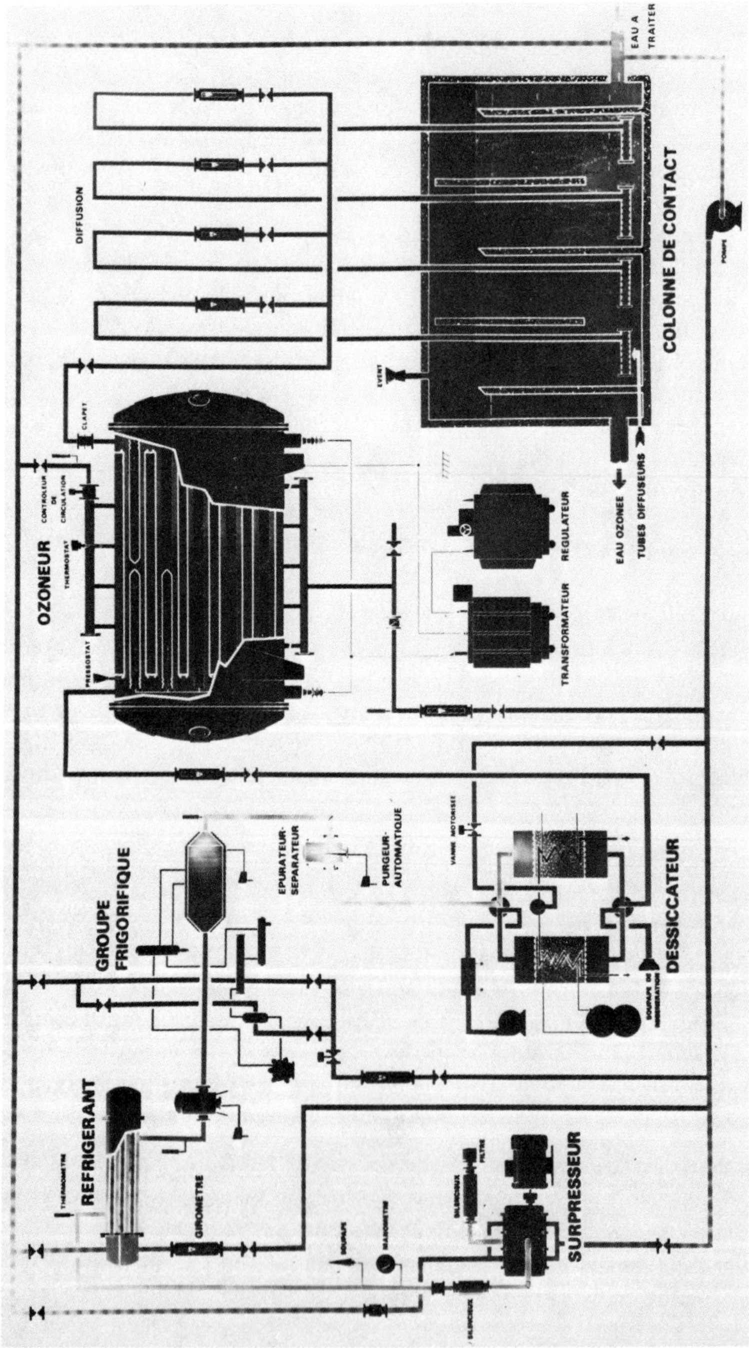

Fig. 13-8. Ozonation system (7656 lb/day) schematic for the Choisy-Le-Roi treatment plant, City of Paris Water Supply (courtesy Compagnie Giverale Des Eanx.).

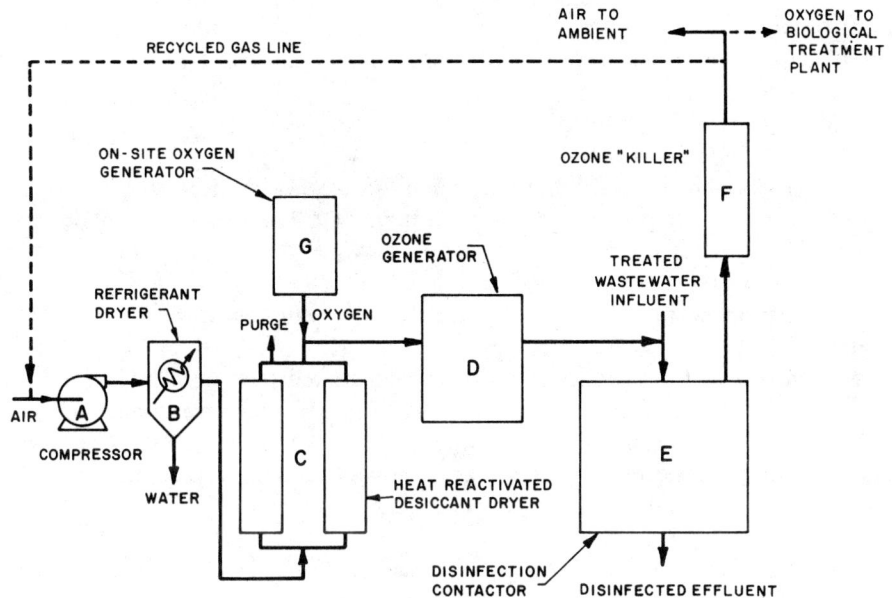

Fig. 13-9. Flow diagram of various ozone disinfecting systems (courtesy Union Carbide Co.).

with their surface waters that ozone-generating capacity of 5 mg/l is ample. In France, for example, for treatment of heavily polluted sources, such as the rivers adjacent to Paris, the current criterion for satisfactory disinfection is 0.4 mg/l residual ozone for a minimum contact period of 4–5 min. This practice is termed "full ozonation."* This practice is also acceptable when ozone is used as a pretreatment process followed by chlorination for final disinfection of a highly polished wastewater effluent.

Wastewater Disinfection. Application of ozone to a secondary effluent presents an entirely different chemical situation when compared to surface water supply treatment. In order for the ozone to be germicidally effective it first must overcome the initial immediate ozone demand (2–3 min.). Owing to the high reactivity of ozone the demand can be extremely high. Moreover this demand is variable and unpredictable over any 24-hour period. Therefore ozone application must be ultimately controlled to reflect both process flow and ozone demand changes. Scientific advances in the field of control instrumentation now make this possible. See the section "Control System" in this chapter.

---

* A 5 mg/l ozone dose has been found to be ample to achieve "full ozonation."

Disinfection Guidelines. There are several levels of disinfection for wastewater effluents. These levels are reported in one of the following ways:

1. MPN total coliforms surviving
2. MPN fecal coliforms surviving
3. Log reduction of initial coliform (total or fecal) concentration after disinfection
4. Percent kill—99 percent = 2 log reduction, 99.999 percent = 5 log reduction, etc.

Current practice for disinfection guidelines is established either by the EPA requirement, whch is 200/100 ml fecal coliforms MPN, or by the state regulatory agency. Many states have disinfection requirements more stringent than the EPA 200/100 ml fecal coliforms. States with shellfish growing areas limit the total coliform concentration in the effluent to 70/100 ml. California uses several levels of total coliform concentration depending upon the situation. (240, 23, and 2.2). See Chapter 8. To convert fecal coliform concentration to total coliforms multiply by 4.

Therefore, to make a rational design of an ozone disinfection system it is necessary to know what the peak coliform concentration will be in the effluent before disinfection. Then, knowing the disinfection requirement the log reduction or percent kill can be calculated arithmetically.

Ozone Demand and Absorption. This terminology is a carryover from chlorination practices. In chlorination it is easy to establish because it is the arithmetic difference between the chlorine dose and the chlorine residual after a specified contact time. This is difficult to determine for ozone because of the following factors: (1) there is no consensus on the optimum contact time; (2) ozone residual measurements are not reliable; and (3) ozone is so reactive there is little if any true ozone residual after 3–5 minutes in the contactor.

Therefore recent work by Venosa et al. on EPA funded projects has demonstrated that the best correlation for reliable disinfection by ozone is to continuously measure the ozone concentration in the exhaust gas. Knowing this and the amount of ozone dosage, control is achieved by maintaining absorption in the contactor at at constant level for a given disinfection requirement. The amount of absorption is directly related to disinfection efficiency.[46,50,52]

Forecasting Ozonator Capacity:

Potable Water. Historically, ozonators for water treatment plants in Europe have been designed to provide a maximum ozone dosage of 5 mg/l. This is not a coincidence. This was slightly more than the maximum absorbance of ozone in these treated waters. This practice was during the time when ozone was universally applied as the final step. For the plants that serve the City of Paris, the criterion for adequate disinfection has been 0.4 mg/l total residual oxidants after a minimum contact period of 4–5 minutes. This practice is termed "full ozonation." The dosage to achieve this criterion is usually never more than 4 mg/l.

Wastewater: Secondary Effluents. Venosa et al.[46] studied the effect of various ozone dosages upon the relationship between coliform destruction (fecal and/or total) and absorbed ozone dose (Fig. 13-10). The same relationship was studied for ozone concentrations in the exhaust gas (Fig. 13-11). Using these data it is possible to estimate the required applied ozone dose for a given NPDES requirement.

EXAMPLE. Assume NPDES requirement is 200/100 mL MPN fecal coliforms; $\log_{10} 200 = 2.3$. Examination of Fig. 13-10 shows that at log = 2.3 the absorbed ozone dose (ozone transferred) will be about 5 mg/l. A similar examination of Fig. 13-11 shows that ozone concentration in the exhaust gas will be about 0.7 mg/l.

The absorbed dose (T) equals the applied ozone dose (D) times the fraction that is transferred:

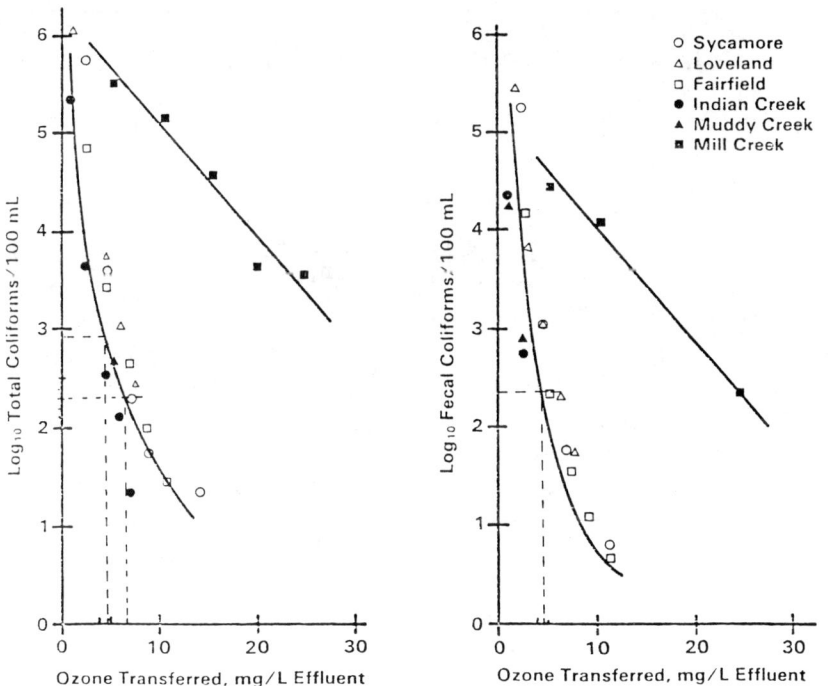

Fig. 13-10. Ozone absorbed (transferred) versus total coliforims and fecal coliforims.[46] (courtesy Journal WPCF).

940 HANDBOOK OF CHLORINATION

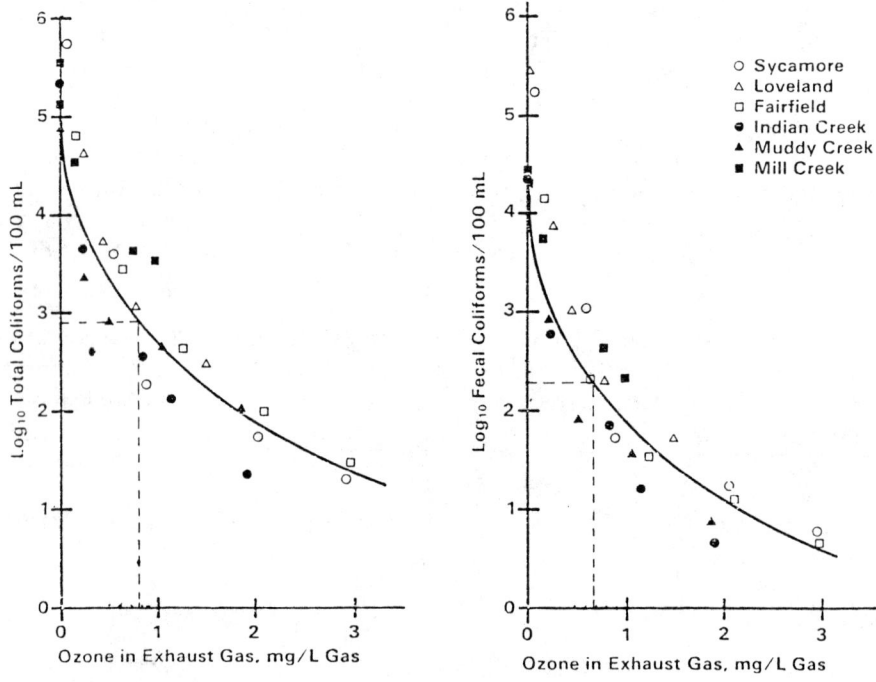

Fig. 13-11. Ozone in exhaust gas versus total coliforims and fecal coliforims[46] (courtesy *Journal WPCF*).

$$T = Y_1 \left(\frac{Q_G}{Q_L}\right) \left(\frac{Y_1 - Y_2}{Y_1}\right) \quad (13\text{-}27)$$

where:

Y_1 = concentration of ozone in carrier gas, mg/l
Q_G = carrier gas flow rate, l/min
Q_L = liquid flow rate, l/min
Y_2 = concentration of ozone in exhaust gas, mg/l

From Fig. 13-10, $T = 5$ mg/l and from Fig. 13-11 $Y_2 = 0.7$ mg/l. Solve for Y_1, in Eq. (13-27):

$$5 = Y_1 \left(\frac{Q_G}{Q_L}\right) \left(\frac{Y_1 - Y_2}{Y_1}\right)$$

$$= \frac{Q_G}{Q_L}(Y_1 - 0.7)$$

Assume $\dfrac{Q_G}{Q_L} = 0.5$; then:

$$Y_1 = \dfrac{5.35}{0.5} = 10.7 \text{ mg/l}$$

Say 11 mg/l.

To the above values for Y_1 a safety factor of 30 percent should be added to provide for unexpected levels of ozone demand or for disappointing contactor efficiency. Therefore the ozonator capacity should be $11 \times 1.3 = 14.3$, say 15 mg/l = 125 lb/mg.

As a check on the above calculation, suppose the NPDES requirement is to be in equivalent total coliforms. This would be approximately four times the fecal coliforms or 800 total coliforms; $\log_{10} 800 = 2.9$. Inspection of Figs. 13-10 and 13-11 shows that the ozone absorbed and ozone in the exhaust gas is practically identical to the amounts for fecal coliforms.

Preozonation: Tertiary Effluents. In the United States, pilot plant operation has been the basis for determination of ozonator capacity. The optimum ozone dosage is dependent upon water quality of each specific source. There are many variables to be considered which affect ozone consumption: color, turbidity, nitrates, COD, Fe, Mn, and a variety of organics.

Ozone Dosage Metering and Measurement. The most effective way for continuous measurement of ozone feed rate in the carrier gas is by ultraviolet absorption using a UV photometer. Modern technology has replaced the previous crude methods of inferential measurement. This technology is provided by the Dasibi[61] ozone photometer. It is a fully automatic instrument designed for continuous monitoring of ozone generating systems. This measurement technique is absolute, based upon the Beer–Lambert law, and is a direct measurement of ozone concentration. The photometer response is a function of: optical path length; ozone–UV absorption coefficient; and ozone concentration. No explosive gas reagents or wet chemicals are required.

The Dasibi unit has a digital display panel which reads out the instantaneous ozone flow from the generator. This is accomplished by the use of a minicomputer which calculates the product of the carrier gas flow times the ozone concentration. The metering unit is connected by a feedback loop which controls the carrier gas feed compressor. This provides a constant ozone concentration in the carrier gas for any given ozone generator power setting.

Exhaust Gas Monitoring System. The ozone concentration in the exhaust gas is crucial to ozone disinfection efficiency, therefore a simple reliable method is needed to accurately measure this concentration. This measurement should be made with a UV photometer as described above for ozone concentration in the carrier gas.

Photometer Calibration. The ozone concentration in the inlet and exhaust gases should be determined iodometrically by the method of Birdsall, Jenkins, and Spadinger.[51,62] This procedure will be required to calibrate the UV photometers used to continuously measure the ozone concentration in these gas streams.

Contacting System:

General Discussion. The contacting system is a structure of some kind within the treatment process where the ozone in the carrier gas is transferred to the process flow. It is sometimes also called the absorption chamber or the mixing chamber. The ozone contactor is the most important part of any ozonation system. Difficulties with contactors are the result of improper design. Most of the ozone system failures are caused by poor contactor design. The solubility of ozone in water is a limiting factor that greatly affects the ozonation process. Moreover ozone does not enter into any appreciable reactions with water—such as going into a solution to form another compound composed of water and an ozone derivative. At a water temperature of 20°C it is only soluble to the extent of 0.57 g/l.[14] Therefore special attention must be given to the contacting system.

An ideal mixing device will produce large bubble-area to gas-volume ratios. This allows maximum diffusion area by producing many small bubbles of ozone. The mixer should also provide enough turbulence to reduce film thickness which offers resistance to ozone transfer. It should also provide for large ozone concentration differences between the gas and liquid phases.

Masschelein et al.[63] have investigated many different types of contactors trying to determine the system that would produce the best transfer efficiency over a wide range of conditions. The two systems with the best overall performance of those studied appear to be the liquid-circulating aeration turbine and the water recirculation system. These gave good dispersion of ozone with ozone losses of about 6–8.5 percent at dosages of 1.5 mg/liter.

One of the large ozonator manufacturers preferred the diffuser system in a baffled contact chamber. This contactor is illustrated in Fig. 13-12. The diffusers are sintered alumina disks and 90 percent transfer efficiency is claimed.

Venosa et al.[51] studied the performance of a stirred turbine reactor (STR) shown in Fig. 13-13 and a three column bubble diffuser (BD) shown in Fig. 13-14. After a rigorous field test and statistical analysis of the transfer efficiency over a variety of conditions it was concluded that the BD system had the best TE.

While 90 percent TE is commonly claimed by ozonator manufacturers, the various studies by the Venosa group strongly indicate that 80 percent efficiency is a more prudent figure to use when estimating the capacity of a proposed ozonator installation.

Ozone Destruction Unit. The exhaust gas from the contactor system must not be vented to atmosphere because this constitutes an air pollution problem. The ozone decomposer or destruct unit is usually furnished with ozonator. This unit

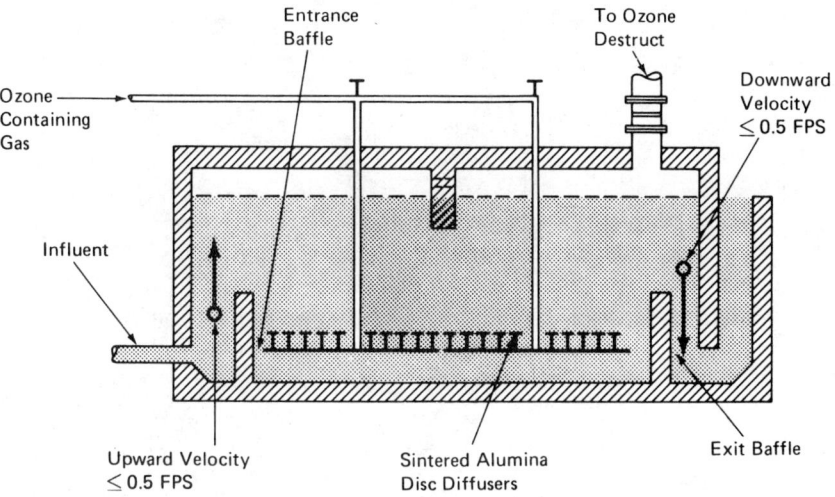

Fig. 13-12. Ozone contactor (courtesy Union Carbide Co.).

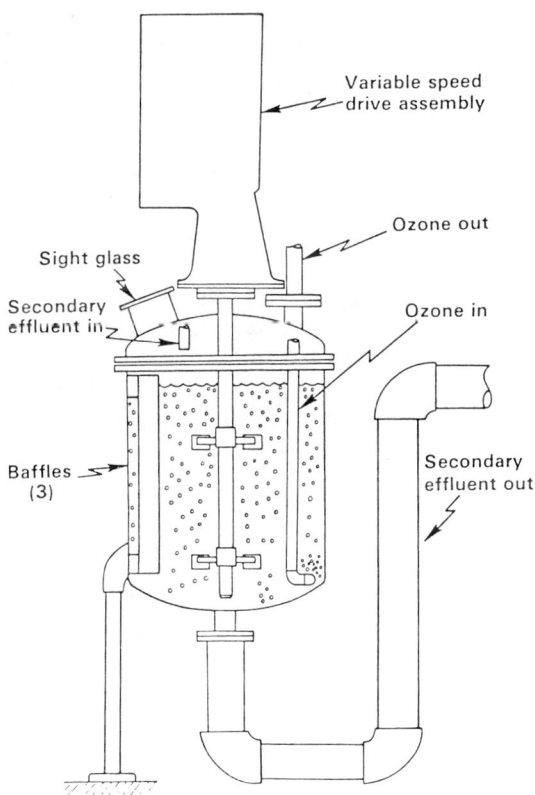

Fig. 13-13. Stirred turbine reactor.[51]

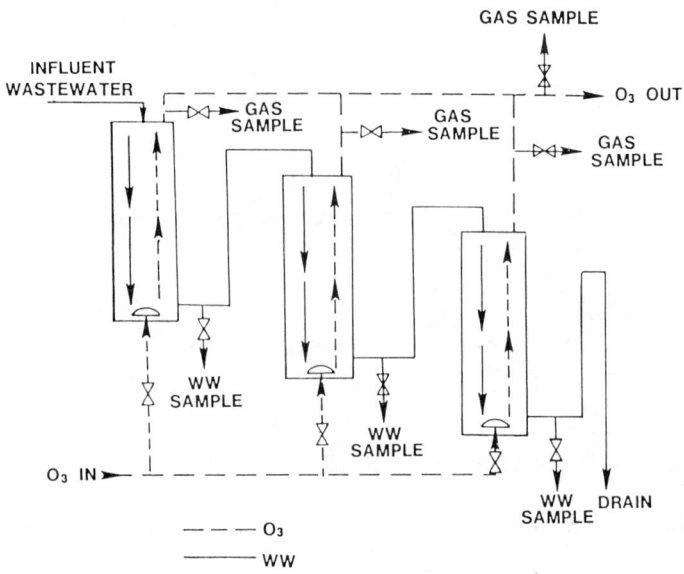

Fig. 13-14. Bubble diffuser reactor.[51]

converts the ozone to oxygen by passing the ozone through a metallic ozide catalytic decomposer. Some units destruct solely by heat, heat–catalyst destruct, and/or activated carbon. The ozone in the decomposer exit gas should be less than 0.1 percent by volume. The exit gas should be monitored for safety on a regular basis.

Contact Time. Owing to the speed of ozone reactions, the contact time taken by itself is not as important as contact time for chlorine reactions. The worldwide consensus for the optimum contact time appears to be 10 minutes. If the contactor is designed with the optimum hydraulic characteristics, i.e., 95 percent plug flow, then maximum transfer efficiency will occur well within the 10 minute contact time at peak flow.

Safety Considerations. The air space in all mixing chambers and contacting systems should be monitored for hydrocarbons in order to alert operating personnel to explosive conditions.

Ambient air in the ozone generator room should be continuously monitored for ozone leaks. If the ozone generator is air cooled the ambient conditions surrounding the generator can be purged into the ventilation system. This air can be used to heat the building, as there is normally a 15°F temperature rise in the cooling air.

MEASUREMENT OF OZONE RESIDUALS

Introduction. Analytical procedures for the quantitative measurement of ozone residuals are burdened with many problems. Ozone cannot be measured successfully if the data are collected in a slovenly manner. Ozone is such a dynamic oxidizing agent that the life of a measurable residual is short. The longer the time between collection and analysis the greater the error in measurement will be.

When ozone is applied to water or wastewater the following occurs: (1) ozone reacts with organic and inorganic compounds and other material present in the substrate; (2) some ozone is decomposed directly to oxygen; and (3) some is lost to the atmosphere through incomplete gas transfer. The analytical procedure must quantify the ozone molecules remaining as a residual.

The nature of inorganic chemistry is such that any analytical procedure is subject to chemical interferences. In the case of ozone this dilemma is aggravated by the wide range of compounds normally found in wastewaters which will interfere and produce significant errors in the quantitative analysis of ozone residuals. These are the ferric ion, manganese dioxide, nitrites, and peroxides. As if this were not enough, the ozonation process itself produces oxidants which interfere with the quantitative selectivity of analysis of a true ozone residual. These interferring oxidants are nongermicidal—if they were germicidal they could not be considered as interferences. They are broadly classified as ozonides and hydroperoxides. There is a parallel case with the chlorination process which involves the dichloramine fraction of a total chlorine residual. It has been found that a combined chlorine residual in wastewater usually contains a small fraction of organic chloramines which have little or no germicidal effect. In the forward titration procedure these compounds seem to appear in the dichloramine fraction and are attributed to the organic nitrogen always present in wastewater effluents. Therefore, it is reasonable to assume that whatever the analytical procedure may be for measuring ozone residuals in aqueous solutions it will probably not be specific for ozone.

EPA Investigation. In 1980 the EPA, under the direction of A. D. Venosa, arranged the evaluation of several analytical methods for the measurement of ozone residuals in aqueous solution. This was done by Gordon and Grunwell.[64]

The following methods were examined:

1. Standard iodometric
2. Amperometric
3. Arsenic III back-titration
4. DPD
5. Indigo dye
6. Arsenic III direct oxidation
7. Delta electrode
8. Direct measurement of UV absorption

Methods 5–8 are noniodometric.

This study concluded that for free ozone measurement no iodometric method can be recommended. The methods of choice according to Gordon and Grunwell are: indigo dye,[66] and Arsenic III direct oxidation.

The main reason for not accepting any of the iodometric methods is their inherent inability to distinguish between free ozone and other oxidants. Therefore these methods all measure total oxidants. This is not a calamity. The elaborate investigation by Stover and Jarnis[54] proved this. They used a modified amperometric procedure which did in fact measure total oxidants. They compared the oxidant residuals with absorbed ozone dose as it related to disinfection efficiency and found suitable agreement. This method is described below.

Modified Amperometric Method[67] This method measures total oxidants, which are probably free ozone plus ozonides and hyperperoxides. From a practical point of view this is acceptable when relating ozone dosages to a broad range of disinfection requirements.

The procedure involves the addition of an excess of potassium iodide so that all of the ozone is reduced immediately, therefore minimizing loss of ozone residual due to its various routes of decomposition. Moreover, the iodine released by the reaction between KI and O_3 is more stable in aqueous solution than is ozone. Another feature which improves the sensitivity of this procedure is that for each 0.2 mg/liter ozone, a full 1.0 mg/liter of iodine is released. One molecule of ozone gives one molecule of iodine. The ratio of molecular weights would be:

$$I_2/O_3 = 253.82/48 = 5.29/1.0 \qquad (13\text{-}28)$$

So for a titrator with a sensitivity of 0.01 mg/liter iodine, the sensitivity for ozone would be 0.002 mg/liter. Titrators with this order of sensitivity are commercially available.

Amperometric titration is not free of analytical problems, so the analyst should be aware that certain ions in solution may poison the electrodes. In all of the copper–noble metal electrode combinations, 0.3 mg/liter of copper and 0.1 mg/liter of aluminum are likely to cause analytical errors. Loss of ozone due to agitation by the titrator mixing device is avoided by carefully adding the sample with ozone residual to the sample jar containing the necessary amount of KI. Therefore, the ozone reacts to liberate iodine before any aeration can take place. The liberated iodine is then titrated with $0.00564N$ phenylarsenoxide reagent (PAO).* According to Dailey and Morrow:[68]

$$O_3 \text{ mg/l} = \text{ml PAO} \times 0.675 \qquad (13\text{-}29)$$

---

* Sodium thiosulfate ($Na_2S_2O_3$) has been recommended as a titrant for this and other residual procedures. Phenylarseneoxide is preferred because of greater stability and speed of reaction.

When using this method the operator must be sure to keep the titrator cell unit sensitized to iodine. When not in use the titrator electrodes should be allowed to soak in a weak iodine solution in the sample jar.

This method has the advantage of speed, convenience, repeatability, and accuracy for total oxidants which closely resemble free ozone.

Standard Methods. The iodometric procedure described in 423A of *Standard Methods* is too time consuming and cumbersome to be considered practical. This method requires the ozone in aqueous solution to be washed out into a reaction container containing KI. It is described here for the reader's convenience.

The procedure consists of passing a pure air or nitrogen stream through the sample and then through an absorber containing a potassium iodine solution. This solution is then treated with sulfuric acid to reach a pH of 2.0 and titrated with $0.005N$ sodium thiosulfate titrant until the yellow color of the liberated iodine is almost discharged. This is followed by the addition of a starch indicator solution which imparts a blue color. The titration is then continued carefully but rapidly until the blue color just disappears. This is the classical iodometric endpoint. The endpoint can also be determined amperometrically described below using phenylarseneoxide as the titrant.

The chemical relationship which describes the ozone–iodometric analysis is as follows:

$$O_3 + 2I^- + H_2O \longrightarrow O_2 + I_2 + 2(OH)^- \qquad (13\text{-}30)$$

The iodine released in Eq. (13-30) is titrated against either thiosulfate or phenylarseneoxide to provide a quantitative answer.

The precision given for the iodometric method is ± 1.0 percent. This is probably close for high concentration of ozone in solution, but it is not correct for low levels of ozone concentration. The starch–iodide procedure begins to show significant error when the released iodine is less than about 2.0 mg/liter. This concentration of iodine in a neutral solution is equivalent to an ozone concentration of about 0.4 mg/liter. Therefore, the starch–iodide procedure has a large error when the ozone concentration is less than about 1.0 mg/liter.

Continuous Residual Recording:

Introduction. Development of dependable instrumentation to provide reliable and continuous analysis of the true ozone residual would be a singular achievement and at the same time enhance the case for ozone.

Significant progress has been made by Fischer and Porter Co. to provide an on-line analyzer for this task.[68,73] Additionally, Johnson et al. developed a membrane cell analyzer specific for ozone.[74,75] This analyzer is now the Delta Scientific ozone analyzer marketed by Xertex Co. of New York.

Amperometric Cell. Fischer and Porter Co. have investigated two types of amperometric cells; the galvanic analyzer and the Ozotrol analyzer.[68] The series 17L 2000 ozone analyzer is a flow through continuous measuring cell. It uses a gold measuring electrode and a copper reference electrode without any applied voltage across the electrodes and does not require any chemicals. This is known as a *galvanic analyzer*. This unit will respond only to free ozone residuals when it is used without chemicals. This unit can be modified by adding a 5 percent potassium iodide solution which will respond to total ozone residuals (i.e., free ozone plus ozonides and hydroperoxides). This analyzer, the 17L 2000, standard or modified version is the one recommended for tertiary effluents. Either the standard or modified unit is subject to interferences from nitrites, ferric ion, and manganese dioxide to the same extent as residual chlorine analyzers. The standard free ozone unit is also subject to interference from aluminum ion (>0.1 mg/liter) and copper ion (>0.3 mg/liter). These interferences, if significant, can be dealt with by the addition of a sample conditioning chemical. Additionally, residual chlorine directly interferes with this analyzer, either the standard or modified.

The other unit investigated by Fischer and Porter[68] utilizes dual copper electrodes with an applied voltage across the electrodes. This is referred to as the Ozotrol analyzer which is a modification of their 17K 1000 Chlor-trol unit which measures free chlorine residual without any chemicals. The use of two copper electrodes with applied voltage produces current-voltage curves similar to those obtained with the gold-copper combination. Current output is directly proportional to ozone concentration. It responds only to free ozone and cannot be modified to measure total ozone residuals. This unit is not suitable for tertiary effluents. It is applicable only to special situations of clean waters such as is found in the water-bottling industry. If there is any iron in the water the ozone residual present would cause the electrodes to become fouled by the ozone precipitation of the iron.

In the decade following the introduction of the galvanic analyzer (17L 2000) by Fischer and Porter in 1973, field experience has shown the unit to be accurate and reliable under selected conditions.[76] Pilot plant studies have also demonstrated that attempting to correlate free ozone measurements using the iodometric procedure is grossly unreliable. It is best to use the modified version of the 17L 2000 Fischer and Porter galvanic analyzer and to measure *total oxidants* instead of free ozone. Stover and Jarnis[54] got good correlation between total residual oxidants and coliform destruction. Venosa et al.[46] also observed good correlation between total residual oxidants and coliform destruction. In these instances the modified amperometric method was used to measure *total residual oxidants*.[72] However, it should be emphasized that the best observed correlation between ozone application and disinfection is that between the absorbed ozone dose and coliform destruction. Continuous residual recording of total residual oxidants should be standard procedure as a backup device to monitor the ozonation process.

In order to obtain an accurate and reliable installation, special attention must be given to the analyzer sample line. The line should be as short as physically possible, and as small as possible in order to get the sample to the analyzer cell as quickly as possible (1 min). Under no circumstances shall the sample line undergo a negative head. Moreover, the sample should go directly to the cell from the contactor effluent without being exposed to any atmosphere air surface.

Membrane Cell. An amperometric membrane cell has been investigated by Johnson and Dunn[75] as a sensitive and selective analytical technique for the determination of free ozone residuals. This work is an outgrowth of the investigation by Johnson et al.[77] during the development of a similar type cell for the measurements of undissociated HOCl.

The advantage claimed for the membrane cell is greater selectivity of oxidants to be measured and elimination of electrode fouling by the use of a microporous fluorocarbon membrane. Interfering substances are supposedly unable to pass through the membrane; therefore, they are neither measured as a false concentration nor able to foul the noble metal electrode surface. This cell is specific for free ozone residual. It will not measure the combined ozone residuals containing ozonides and hydroperoxides. Moreover, chlorine residuals can be excluded from the ozone residual by changing the applied voltage across the electrodes.

The configuration of the membrane cell is such that it is better described as a probe. It is similar to the dissolved oxygen probe, but with a new and improved membrane material and a positive applied voltage. It is available from Delta Scientific Corp. A brief description of the cell follows.

A 1.6 molar solution of potassium chloride which is saturated with silver chloride fills a cavity in the base of the probe and acts as an electrolyte. This solution is in contact with a solid silver reference electrode. As voltage is applied across this silver–silver chloride interface the silver electrode is oxidized, losing one electron, and combines with the chloride ion to form a silver chloride coating on the electrode surface. This reaction supplies the electrons necessary to reduce the ozone and maintains a constant reference potential. The released electrons flow through the gold (positive) measuring electrode to the ozone in spite of the opposing applied voltage force supplied by the battery. At the gold electrode, ozone residual in the sample passes through the membrane and is reduced to oxygen.

$$O_3 + 2H^+ + 2e^- \longrightarrow O_2 + H_2O \qquad (13\text{-}29)$$

This reaction "consumes" two electrons per molecule of ozone reduced. This flow of electrons through the gold electrode is measured and recorded as an electric current. The selectivity of this method is dependent upon the microporous membrane to allow only small diameter molecules to reach the gold electrode and by the high positive opposing voltage which prevents the weaker oxidants from reaching the gold measuring electrode.

This measuring system has not had enough field application to determine whether or not it is a reliable method for measuring ozone residuals in either secondary or tertiary effluents. One of these units is installed at the plant in Frankfort, KY. Operating experience is not available at this time.

Gordon and Grunwell[64] reported in their study that the Delta Scientific probe was unstable and unpredictable. It had to be recalibrated for each run. Therefore more field experience will be required to complete the Delta probe adaptation to measuring ozone residuals.

OPERATION AND MAINTENANCE

A Study of Operating Installations. A nine month study was made by Junkins et al.[59] of existing ozone facilities to analyze operation and maintenance factors affecting the efficiency and reliability of the ozone process. Seven wastewater and five water treatment plants were visited and the operating personnel interviewed. Their findings have been summarized by Venosa.[52] The following is a condensed summary of this study by Roy F. Weston Inc.[59]

Ozone Generators. This is a high maintenance item. Excessive heat buildup in an ozonator room causes heaters to shut down the units. Many generator failures are due to blown fuses and cell failure. System electronics are too complicated for average WWTP personnel to perform routine maintenance and repair work. Ozone leakage is a definite hazard. System instrumentation must be continually calibrated.

Air Drying System. This is one of the most sensitive components of air ozonation system. Improper dewpoint setting or control causes dielectric failure. Dewpoint indicators are often unreliable, or not provided, and dessicant dryers are not always dependable. When using air as the carrier gas it is imperative to supply dry air to the ozone generator. If not, the generators will suffer frequent failures.

Contactors. Most of the adverse comments pertained to poor efficiency which was the result of improper design. This has been the single most severe operation problem. Other complaints were centered around ozone leaks due to improper seals in the contactors.

Ozone Destruct Unit. These units are not necessarily reliable so therefore require careful attention.

Some operators have reported corrosion problems in piping and valve components in the ozone analyzers.

COMPARATIVE COSTS

General Considerations. To evaluate ozone on cost alone is an unfair evaluation of its capabilities. The cost-effectiveness of ozone as a bactericide for secondary effluent as compared to chlorine will always demonstrate the superiority of chlorine. The only chance ozone has as a secondary effluent bactericide is when it is part of the UNOX process. However, ozone for potable water treatment and wastewater reuse is a different matter. There are situations where cost alone is not the dominant factor. Ozone must be evaluated on its capabilities as a powerful oxidant in addition to its ability to destroy viruses and bacteria. Ozonation is an excellent process for polishing low-grade potable waters and wastewater for reuse.[47,69]

The reality of the cost disparity between ozone and chlorine systems is revealed by the following figures showing the cost of each system in the simplest form. Comparative cost of 500 lb/day systems, manual control: chlorinator, $3300; ozonator (air supply), $300,240.[79] Chemical cost, 500 lb/day: chlorine, $60/day; ozone $600/day.

Gumerman et al.[78] prepared a comprehensive cost study for both chlorine and ozone. Operation and maintenance costs for 500 lb/day systems are; for chlorine (ton containers) $11,000/year; for ozone $66,000/year.

Ozone is grossly energy-intensive. Power requirements for generating 500 lb/day ozone are approximately 1,300,000 kWh/yr.[78] Chlorine energy requirement (ton cylinders) is approximately 1100 kWh/yr for 500 lb/day consumption.

Therefore if ozonation is selected as a cost-effective process, it is because it can perform as no other process or combination of processes can. This is why it was chosen as a preozonation process for both Hackensack, NJ Water Company and the Los Angeles, CA Aqueduct Filtration Plant. The following are some vital statistics:

Hackensack NJ. This is a 2500 lb/day ozone system which cost $4,000,000 and requires approx. 40,000 ft² floor space, is provided with standby power,[86] and is followed by postchlorination. Ozonators are by Trailigaz, France.

Los Angeles Aqueduct Filtration Plant. This is a 7510 lb/day ozone system which delivers a maximum dose of 1.5 mg/l (600 mgd). The equipment is by Brown-Boverie of Switzerland but provided by Air Products Co., Allentown, PA. The ozone system cost $6,830,000 and occupies 1800 sq ft. The contactor occupies 14,700 sq ft.[87] The disinfection system consists of four 8000 lb/day chlorinators.

Cost Studies. Several ozone feasibility studies have been made which compare ozone to chlorine specifically for wastewater disinfection.[80-83]

EPA Summary. Some of these studies and others have been summarized in the EPA Task Force Report, March 1976.[84] These findings are shown in Table 13-10 for the ozonation process as compared with the chlorination/dechlorination process shown in Table 13-11.

The costs outlined above should be considered as tentative, because the parameters of disinfection are not specific. The following cost comparisons are specific and therefore are of more practical value for the designer.

Table 13-10 Ozone Disinfection Cost

A. Ozone Generated From Air			
Plant size (mgd)	1	10	100
Capital cost	190,000	1,070,000	6,880,000
Disinfection cost ¢/1000 gal.	7.31	4.02	2.84
B. Ozone Generated From Oxygen			
Capital cost[a]	160,000	700,000	4,210,000
Disinfection cost ¢/1000 gal.	7.15	3.49	2.36

[a] The reduced cost here represents the elimination of drying atmospheric air for the process.

Sacramento, CA. In 1975, the Sacramento Regional County Sanitation District of California made an ozone feasibility study.[80] The disinfection requirement for the plant discharge is a total coliform MPN of 23/100 ml. The PWWF is 240 mgd. The effluent from this plant is assumed to be a well-oxidized secondary effluent resulting from the activated sludge process. There is some cannery waste during the dry weather flow season of July to September. For this situation ozone equipment manufacturers recommended design capacity of 20 mg/liter. However, the design engineers thought this would be much too low in view of recent operating experiences by the County Sanitation Districts of Los Angeles at their Pomona water reclamation plant. These studies indicated that a 40 mg/liter dosage would be more appropriate to achieve the disinfection requirement of 23/100 ml MPN total coliforms. The results of this report are shown in Table 13-12.

Power cost is based on 0.9 ¢/kWh.

This analysis shows the capital cost of an ozonation system is between seven and eleven times as expensive as the chlorination/dechlorination system, and the annual operating cost for ozone is somewhere between 1.2 and 2.3 times as expensive as the chlorination/dechlorination method. On an average annual cost basis, the ozonation system is between three and five times as expensive as the chlorination/dechlorination system.[80]

Richmond, CA. Another analysis for a less stringent disinfection requirement was made for the City of Richmond, CA, discharging a well oxidized activated

Table 13-11 Chlorination Disinfection Cost

Plant size	1	10	100
Capital cost	60,000	190,000	840,000
Disinfection cost ¢/1000 gal.	3.49	1.42	0.70

Dechlorination With Sulfur Dioxide

Plant size	1	10	1000
Capital cost	11,000	29,000	94,000
Disinfection cost ¢/1000 gal.	0.88	0.35	0.19

Table 13-12 Cost Comparison of Ozonation and Chlorination/Dechlorination 240 mgd Design Capacity Current Cost Basis—ENR 2300 ($)

Type of Cost	Chlorine 10 mg/liter Dosage	Ozone 20 mg/liter Dosage	40 mg/liter Dosage
Total capital cost	2,150,000	14,250,000	23,450,000
Average annual cost			
capital[a]	173,000	1,148,000	1,890,000
operating[b]	373,000	439,000	858,000
total	546,000	1,587,000	2,747,000
Average unit cost[c]			
(¢/1000 gal.)	1.2	3.5	6.0
Average annual local cost[d]	395,000	583,000	1,094,000

[a] Capital recovery at 7% for 30 years.
[b] Based on 125 mgd average annual flow.
[c] Based on total average annual cost.
[d] Sum of operating cost and local amortized capital cost share based on 87.5% Federal grant for capital facilities.

sludge effluent into San Francisco Bay.[82] The effluent standards to be met are total coliform MPN of 70/100 ml with no more than 10 percent of these samples exceeding 230/100 ml.

Little reliable data are available to predict the ozone dosage required to achieve this particular disinfection objective. However, the following design parameters were selected based upon proposals from the ozone equipment suppliers:

- Ozone dosage ADWF 15 mg/liter.
- Ozone dosage PWWF 6 mg/liter.
- Contact time at PWWF 10 min.

The system chosen, as recommended by the manufacturers, was based upon the use of high-purity oxygen recycled to the generators. Oxygen was to be supplied locally by the Airco plant at $50–$60/ton. Owing to certain cost differentials between the two ozone equipment suppliers, separate estimates are shown in Table 13-13. A separate calculation was made for a chlorination/dechlorination system which is included in the tabulation. The analysis shown in Table 13-13 is based upon the present worth concept.

Table 13-13 Present Worth Analysis: Chlorination/Dechlorination Versus Ozone

Cost Element	Chlor./Dechlor.	PCI Ozone	Union Carbide Ozone
Total Capital	$990,000	1,380,000	1,570,000
Total O and M	2,600,000	4,620,000	4,870,000
Total Project	3,590,000	6,000,000	6,440,000
Annual O and M Cost	93,000	130,000	149,000

Table 13-14 Annual Operating Cost

Operation of Maintenance	Chlorination	Ozonation
Disinfection 50 mgd flow	77,000[a]	91,000[b]
Deodorizing at headworks	16,000	1,000
Control of carbon column biofouling	16,500	5,000
Total O&M cost	109,500	97,000

[a] Includes cost of chemical, electrical power and maintenace.
[b] Includes cost of electric power and maintenance.

From the above two analyses using the most optimistic dosage and contact times proposed by the ozone equipment manufacturers, disinfection by ozone is substantially more costly than by chlorination/dechlorination.

Cleveland, OH.* However, a report prepared in 1974 for the Cleveland Regional Sewer District[83] for the Westerly Advanced Wastewater Treatment Facility resulted in a recommendation for ozone rather than chlorination.† This was based upon an ozone disinfection dosage of 6 mg/liter. Disinfection was not defined for MPN total coliforms. The comparative cost of the two systems is summarized in the following tables.

It is clear from Tables 13-14 and 13-15 that the cost of disinfection by ozone is considerably greater than by chlorination. However, in this case, ozonation was selected on the basis of longer equipment life (20–40 yr); the relatively inexpensive cost of deodorizing the screening and degritting facilities; and the control of the biological slimes in the carbon columns. That these additional factors might influence the choice of ozonation are at the best tenuous. More operating data concerning these parameters must be developed.

Other long range economic evaluations have been made attempting to justify the use of an ozonation disinfection system instead of an existing chlorination process requiring the addition of a dechlorination facility. One such system involved the conversion of a primary treatment plant to provide secondary treatment. The Unox process was chosen for this conversion. Moreover, the existing chlorination system needed to be supplemented with dechlorination capability. In spite of all these factors the long range economic factors as presented by the engineers demonstrated that the chlorination/dechlorination process was the one of choice.[85]

Houston, TX. Consultants for the city of Houston made a different type of disinfectant evaluation. They compared ozone, liquid-gas chlorine and liquid hypochorite

---
* This is the Westerly Advanced Wastewater Treatment Plant owned by the Northeast Ohio Regional Sewer District and not the City of Cleveland.
† During the time between this report and the start of construction this recommendation was changed. Now (1984) the role of ozonation has been moved to preozonation and chlorination has been selected for disinfection.

Table 13-15 Total Annual Cost of Disinfection

Itemized Costs	Chlorination	Ozonation
Capital cost	1,593,000	2,457,000
Amortized annual cost	136,000	205,000
Oper. and maintenance cost	109,500	97,000
Total annual cost	245,000	302,000

for their new 400 mgd (peak flow) wastewater-treatment plant.[81] Then evaluations were prepared to determine capital cost and annual operation costs for each disinfection system. The present worth method was used to relate capital cost to annual cost over a 20-yr. design period. Parenthetically, it should be mentioned that the high-purity oxygen activated sludge treatment process (UNOX) was found to be the most cost effective process.

The design dosage of 5 mg/liter chlorine was assumed to provide a 1.0 mg/liter residual at the end of 20 min. contact time at peak flows. It was further assumed that a 5 mg/liter ozone dose would be sufficient to meet or exceed the state standards for disinfection. This was also expected of chlorine, and no claim was made that either one would be more effective in eliminating total coliform organisms. In developing the costs of chlorine (in tank cars) versus ozone, the capital cost assessed chlorine at $4,640,000 for the contact chamber, and $1,000,000 for ozone. This calculates to an ozone contact time of 4.3 min. at peak flow.

Table 13-16 tabulates the Houston analysis; the ozone figures are based upon an arithmetical average of three equipment suppliers.*

Table 13-16 Present Worth Analysis of Chlorine Disinfection Versus Ozone[a]

Cost Items	Chlorine	Ozone
Capital cost	4,840,000	7,137,000
Present worth of interest	2,411,000	3,552,000
Present worth of annual costs	1,226,000	919,000
Net present worth	8,477,000	11,608,000

[a] A planning period of 20 yr. and an interest rate of 6.125 percent was used.

SUMMARY

Ozone is a highly unstable compound which has to be generaged at the point of use. It is energy-intensive and has a low solubility which adversely affects the

---

* Houston chose commercial grade sodium hypochlorite for disinfection for safety reasons. The only Houston installations using liquid chlorine are those supplied by 150 lb cylinders.

reliability and efficiency of the process. The equipment required to generate ozone and control the process is enormously expensive when compared to other chemical processes used for potable water and wastewater treatment.

Ozone has many attributes that make it singularly effective as a pretreatment process for potable water. It can destroy phenol tastes and remove certain organic colors from potable water supplies. It is an excellent viricide, but is not a reliable bactericide. It does not contribute to the formation of trihalomethanes. On the contrary it is known to reduce THM precursors. It is most effective in improving coagulation, which in turn increases the length of filter runs. All of this reduces significantly the chemical and energy cost of treating potable water. Ozone as a pretreatment process also improves carbon absorption efficiencies. It is also capable of iron and manganese removal, particularly the organic complexes of these elements.

Both surface and groundwater supplies are being subjected to ever increasing contamination by both domestic and industrial wastes. Simply put, the waters of the United States are threatened with continuous degradation. Therefore many water utilities are turning to ozone as a polishing agent in order to meet water quality standards. These standards are becoming more stringent with the passage of time.

Ozone is also most effective as a pretreatment process when reclamation of wastewater effluent is considered. Ozone combined with filtration of these waters followed by disinfection with chlorine assures virus destruction.

The use of ozone as an alternative to chlorine for the disinfection of potable water and/or wastewater has not met with any success. There are three major reasons for its failure as a replacement for chlorine: (1) more expensive, (2) residual control is unreliable, (3) does not produce a persisting residual.

The interest in ozone as a disinfectant for wastewater has all but disappeared. However, there will always be special situations where it is used for this purpose.

Commercial ozone generators have been around for a long time. In spite of years of operating experience ozone systems experience failures and require a significant amount of maintenance. Ozone is energy intensive, therefore each installation must be provided with standby generating power.

Recent manufacturing developments have progressed to the point where flow pacing the ozone feed rate is reliable. Now instrumentation is available where the concentration of ozone in both the carrier gas and exhaust gas can be measured, recorded, and controlled. Heretofore ozone feed rate measurement was inferential.

The design of ozone contacting systems which are responsible for the efficiency of the ozone process has been thoroughly investigated in the last decade. It is now possible to reliably achieve 90+ percent transfer efficiency with modern technology.

Cost studies published to date nearly always compare ozone to chlorine or chlorination and dechlorination. This is unfair to ozone in most cases, but more important, ozone and chlorine cannot be compared. *Therefore the selection of ozone should not be based upon cost, but upon its merits.*

Technology has not progressed enough to solve the dilemma of *residual measurements* of ozone. The most promising approach appears to be based upon total residual oxidants and not pure ozone. More research is required in this area provided it is found necessary to even measure these residuals. This is not necessary when preozonation is practiced or where ozone is used as a polishing agent, as in European practice.

Ozone's popularity in Europe is due primarily to two factors: (1) Surface waters in France and Germany have long suffered from industrial wastes which produced foul-tasting water; ozone alleviated this. (2) The French government, understanding this benefit, made every effort to encourage and support the manufacture and sale of ozonation equipment.

It is gradually becoming more popular in the United States.

REFERENCES

1. McCarthy, J. J., and Smith, C. H., "A Review of Ozone and Its Application to Domestic Wastewater Treatment," *J. AWWA,* **66,** 718 (Dec. 1974).
2. Dyachov, A. V., "Recent Advances in Water Disinfection," paper presented at the Annual Conf. of the Int. Water Poll. Research, Amsterdam, Neth., Sept. 1976.
3. Richards, W. N., and Shaw, B., "Developments in the Microbiology and Disinfection of Water Supplies," *J. Inst. Water Eng. Sci.,* **30,** 191 (June 1976).
4. Schalekamp, M., "Experience in Switzerland with Ozone, Particularly the Neutralization of Hygienically Undesirable Elements Present in Water," paper presented at the Annual AWWA Conf., Anaheim, CA, May 1977.
5. Stone, B. G., and Trussell, R., "Application of Ozone for Viral Disinfection," paper presented at the Annual Calif. Sect. Mtg. AWWA, San Diego, CA, Oct. 30, 1975.
6. Morris, J. C., "The Role of Ozone in Water Treatment," paper presented at the Annual Conference AWWA, New Orleans, LA, June 24, 1976.
7. Venosa, A. D., "Comparative Disinfection of Wastewater Effluent with Chlorine, Bromine Chloride, and Ozone," paper presented at the Forum on Disinfection with Ozone, Chicago, IL, June 2–4, 1976.
8. Fetner, R. H., and Ingols, R. S., "A Comparison of the Bactericidal Activity of Ozone and Chlorine against *Esch. coli* at 1°C," *J. Gen. Microbiol.,* **15,** 381 (1956).
9. Morris, J. C., "Aspects of the Quantitative Assessment of Germicidal Efficiency," in Chapter 1 J. D. Johnson (Ed.), *Disinfection, Water and Wastewater,* Ann Arbor Science, Ann Arbor, MI, 1975.
10. Selna, M. W., Miele, R. P., and Baird, R. B., "Disinfection for Water Reuse," paper presented at the Disinfection Seminar at the Annual AWWA Conf., Anaheim, CA, May 8, 1977.
11. Bailey, P. S., "Reactivity of Ozone with Various Organic Functional Groups Important to Water Purification," First Int. Symposium on Ozone for Water and Wastewater Treatment, Proc. IOI Waterbury, CT, 1975.
12. Dellah, A., "Study of Ozone Reactions Involved in Water Treatment and the Present Chlorination Controversy," Proc. 2nd Int'l. Symposium on Ozone Technology, Montreal, PQ, Canada, May 11–14, 1975.
13. Hoigné, J., and Bader, H., "Identification and Kinetic Properties of the Oxidizing Decomposition

Products of Ozone in Water and its Impact on Water Purification," Proc. 2nd. Int'l Symposium on Ozone Technology, Montreal, PQ, Canada, May 11–14, 1975.
14. Hann, V. A. and Manley, T. C., "Ozone," *Encyclopedia of Chemical Technology*, pp. 735–753, Interscience, New York, 1952.
15. O'Donovan, D. C., "Treatment With Ozone," *J. AWWA*, **57**, 1167 (1965).
16. Singer, P. C., and Zilli, W. B., "Ozonation of Ammonia in Wastewater," *Water Research*, **9**, 127 (1975).
17. Kinman, R. N., "Ozone in Water Disinfection," p. 123 in Evans, F. L., *Ozone in Water and Wastewater Treatment*," Ann Arbor Science, Ann Arbor, MI, 1972.
18. Glaze, W. H., Rawley R., and Lin, S., "By-Products of Organic Compounds in the Presence of Ozone and UV Light," paper presented at the IOI Meeting, Cincinnati, OH, Nov. 17–19, 1976.
19. Guirguis, W. A., Srivasta, P., Meister, T., Prober, R., and Hanna, Y., "Ozone Reactions with Organic Material in Sewage Non-Sorbable by Activated Carbon," paper presented at the IOI Meeting, Cincinnati, OH, Nov. 17–19, 1976.
20. Prengle, H. W., Jr., Mauk, C. E., and Payne, J. E. "Ozone–UV Oxidation of Pesticides in Aqueous Solution," paper presented at the IOI Meeting Cincinnatti, OH, Nov. 17–19, 1976.
21. Richard, Y., "Organic Materials Produced Upon Ozonation of Water," paper presented at the IOI Meeting, Cincinnati, OH, Nov. 17–19, 1976.
22. Falk, H. L., and Moyer, J. E., "Ozone as a Disinfectant of Water," paper presented at the IOI Meeting, Cincinnati, OH, Nov. 17–19, 1976.
23. Hartemann, P., Block, J. C., and Maugras, M., "Biochemical Aspects of the Toxicity Involved by the Ozone Oxidation Products in Water," paper presented at the IOI Meeting, Cincinnati, OH, Nov. 17–19, 1976.
24. Kinman, R., Rickabaugh, J., Elia, V., McGinnis, K., Cody, T., Clark, S., and Christian, R., "Effect of Ozone on Hospital Wastewater Cytotoxicity," paper presented at the IOI Meeting, Cincinnati, OH, Nov. 17–19, 1976.
25. Spanggord, R. J., and McClurg, B. J., "Ozonation Methods and Ozone Chemistry for Selected Organic Compounds in Water," paper presented at the IOI Meeting in Cincinnati, OH, Nov. 17–19, 1976.
26. Simmon, V. F., and Eckford, S. L., "Methods for Evaluating the Mutagenic Activity of Ozonated Chemicals," paper presented at the IOI Meeting, Cincinnati, OH, Nov. 17–19, 1976.
27. White, G. C., "Other Methods of Disinfection," Ch. 2 in *The Quest for Pure Water*, Vol II, 2nd ed., AWWA, Denver, CO, 1981.
28. Hoven, D. L., Schwartz, B. J., and Weng, Cheng-Nan, "Ozone: an Economical Choice for Hackensack Water Company's Drinking Water," paper presented at Ann. Conf. AWWA, St. Louis, MO, June 1983.
29. Miller, G. W., Rice, R. G., Robson, C. M., Scullin, R. L., Kuhn, W., Wolf, H., and Carswell, J. K. "An Assessment of Ozone and Chlorine Dioxide Technologies for Treatment of Municipal Water Supplies," *Executive Summary*, EPA 600/8–78–018. Cincinnati, OH, Oct. 1978.
30. Nadeau, M., and Pigeon, J. C., "The Evolution of Ozonation in the Province of Quebec, Canada," Proceedings First Int. Symposium on Ozone for Water and Wastewater Treatment, Washington, DC, Dec. 2–4, 1973, p. 157.
31. Dellah, A., "Ozonation at Charles-J. Des Baillets Water Treatment Plant, Montreal, Canada," Proceedings IOI Symposium, Montreal, Canada, May 11–14, 1975, p. 715.
32. Leduc, P., "Ozone Improves Taste, Odor and Color of Water," *Water and Sewage Works*, **125**, 49 (Dec. 1978).
33. Rook, J. J. "Developments in Europe," *J. AWWA*, **68**, 279 (June 1976).
34. Symons, J. M., "Trip Report for European Travel," EPA, Cincinnati, OH, Oct. 29, 1974.
35. Kühn, W., Sontheimer, H., and Kurz, R., "Use of Ozone and Chlorine in Water Works in the Federal Republic of Germany," *J. AWWA*, **70**, 326 (June 1978).
36. Anon., "Ozone Treatment Licks Color Problem," *Water and Sewage Works*, **122**, 52 (April 1975).

37. Sontheimer, H., Heilker, E., Jekel, M. R., Nolte, H., and Vollmer, F. H., "The Mulheim Process," *J. AWWA,* **70**, 393 (July 1978).
38. Rice, R. G., Gomella, C., and Miller, G. W., "Rouen, France Water Treatment Plant," *Civil Engineering—ASCE,* **76** (May 1978).
39. Staff Report, "The Activated Carbon Dilemma," *Water and Sew. Wks.,* **125**, 34 (Dec. 1978).
40. Venosa, A. D., "Ozone as a Water and Wastewater Disinfectant: A Literature Review," in E. L. Evans, II (Ed.), *Ozone in Water and Wastewater Treatment,* Ann Arbor Science Publishers Inc., Ann Arbor, MI, 1972.
41. White, G. C., *Disinfection of Wastewater and Water for Reuse,* Van Nostrand Reinhold, New York, 1978.
42. Ghan, H. B., Chen, C. L., and Miele, R. P., "The Significance of Water Quality on Wastewater Disinfection with Ozone," paper presented at the Forum on Disinfection with Ozone, Chicago, IL, June 2–4, 1976.
43. Ghan, H. B., Chen, C. L., Miele, R. P., and Kugelman, I. J., "Wastewater Disinfection with Ozone," paper presented at the CWPCA Annual Conference, Los Angeles, CA, April 1975.
44. Parkhurst, J. D., "Pomona Virus Study," Sanitation Districts of Los Angeles, Whittier, CA, Feb. 1977.
45. Ward, R. W., Giffin, R. D., DeGraeve, G. M., and Stone, R. A., "Disinfection Efficiency and Residual Toxicity of Several Wastewater Disinfectants," EPA Report, Grant No. S-802292, Cincinnati, OH, 1976.
46. Venosa, A. D., and Meckes, M. C., "Control of Ozone Disinfection by Exhaust Gas Monitoring," *J. WPCF,* **55**, 1163 (Sept. 1983).
47. Trussell, R., Nowak, T., Ismail, F., Jopling, W., and Cooper, R., "Ozone as a Pretreatment for Coagulation, Filtration and Disinfection," paper presented at the CWPCA Annual Conf., Los Angeles, CA, April 1975.
48. Trussell, R. R., et al., "Chino Basin Municipal Water District Prototype Plant Study," J. M. Montgomery Engineers Pasadena, CA, Mar. 1975.
49. Trussell, R., private communication, J. M. Montgomery Engrs., Pasadena, CA, 1977.
50. Venosa, A. D., and Opatken, E. J., "Ozone Disinfection—State of the Art," paper presented at Pre-Conf. Workshop, 52nd Ann. Conf. WPCF, Houston, TX, Oct. 7, 1979.
51. Venosa, A. D., Meckes, M. C., Opatken, E. J., and Evans, J. W., "Disinfection of Filtered and Unfiltered Secondary Effluents in Two Ozone Contactors," paper presented at 52nd Ann. WPCF Conf. Houston, TX, Oct. 7–11, 1979.
52. Venosa, A. D., "Effectiveness of Ozone as a Municipal Wastewater Disinfectant," paper presented at Pre-Conf. Workshop 56th Ann. WPCF Conf. Atlanta, GA, Oct. 1, 1983.
53. Venosa, A. D., et al., "Comparative Efficiencies of Ozone Utilization and Microorganism Reduction in Different Ozone Contactors," in A. D. Venosa (Ed.), *Progress in Wastewater Disinfection Technology,* EPA-600/9-79-018 U.S. Environ. Prot. Agency, Cincinnati, OH, 1979.
54. Stover, E. L., and Jarnis, R. W., "Obtaining High Level Wastewater Disinfection With Ozone," *J. WPCF,* **53**, 1637 (Nov. 1981).
55. Venosa, A. D., Petrasek, A. C., Brown, D., Sparks, H. L., and Allen, D. M., "Disinfection of Secondary Effluents with Ozone and U. V.," *J. WPCF,* **56**, 137 (Feb. 1984).
56. Varas, A. J., Novak F., and Fitzgerald, J., "Making Ozonation Reliable and Economical," paper presented at Ann. Conf. AWWA, Las Vegas, Nev., June 1983.
57. Ogden, M., "Ozonation Today," *Ind. Water Eng.,* **7**, 36 (June 1970).
58. Rosen, H. M., "Use of Ozone and Oxygen in Advanced Wastewater Treatment," *J. WPCF,* **45**, 521 (Dec. 1973).
59. Junkins, R., Eckhoff, T., and Watkin, A., "Operation and Maintenance Evaluation of Ozone and Ultraviolet Disinfection," paper presented at Env. Engr. Div. ASCE National Conf. on Env. Engrng., Minneapolis, MN, July 14–16, 1982.
60. Rosen, H. M., "Wastewater Ozonation: A Process Whose Time Has Come," *Civ. Engineering ASCE,* 65 (March 1976).

61. Dasibi Model 1003-AH, "Ambient Air Quality Ozone Photometer," Glendale, CA, 1975.
62. Birdsall, G. M., Jenkins, A. C., and Spadinger, E., "Iodometric Determination of Ozone," *Anal. Chem.*, **24**, 662 (1952).
63. Masschelein, W., Fransolet, G., and Genot, J. "Techniques for Dispersing and Dissolving Ozone in Water, Parts I and II," *Water and Sewage Works*, **122**, 57 (Dec. 1975) and **123**, 34 (Jan. 1976).
64. Gordon, G., and Grunwell, J., "A Comparison of Analytical Methods for Residual Ozone," paper presented at Ann. Conf. AWWA St. Louis, MO, June 1983.
65. Haag, W. R., and Hoigné, J., "Kinetics and Products of the Reactions of Ozone with Various Forms of Chlorine and Bromine in Water," paper presented at Ann. Conf. AWWA St. Louis, MO, June 1983.
66. Bader, H., and Hoigné, J., "Determination of Ozone in Water by the Indigo Method," *Water Research*, **15**, 449 (1981).
67. Kinman, R. N. "Analysis of Ozone: Fundamental Principles," First Int. Symposium on Ozone for Wastewater Treatment, Wash., DC, Dec. 1973, p. 56.
68. Dailey, L. and Morrow, J. J., "On Stream Analysis of Ozone Residual," paper presented at the First IOI Conf., Wash., DC, 1970.
69. Saunier, B. M., Selleck, R. E., and Trussell, R. R., "Preozonation as a Coagulant Aid in Drinking Water Treatment," *J. AWWA*, **75**, 239 (May 1983).
70. Anon., "Emerzone Systems Catalog," Emery Industries Cincinnati, OH, Oct. 13, 1980.
71. Lambert, M., personal communication, CIFEC, Paris, France, 1982.
72. Venosa, A. D., Optaken, E. J., and Meckes, M. C., "Comparison of Ozone Contactors for Municipal Wastewater Effluent Disinfection," EPA-600/2-79-098, August, 1979.
73. "Dissolved Ozone Analyzer," Specification 17L 2000, Fischer and Porter Co., Warminster, PA, 1973.
74. Singer, P. C., and Johnson, J. D., "Reactions of Ozone in Aqueous Systems: Water Treatment and Analytical Implications," paper presented at the IOI Meeting Cincinnati, OH, Nov. 17–19, 1976.
75. Johnson, J. D., and Dunn, J. F., "Ozone Amperometric Membrane Electrode," Publication 415, Dept. Env. Sciences and Eng., Sch. of Pub. Hlth., Univ. of North Carolina, Chapel Hill, NC, 1976.
76. Hayes, T. J., private communication, Fischer and Porter Co., Warminster, PA, Feb. 16, 1984.
77. Johnson, J. D., Edwards, J. W., and Keeslar, F., "Real Free Chlorine Residual Probe: HOCl Amperometric Membrane Electrode," paper presented at the AWWA Conf. Minneapolis, MN, June 1975.
78. Gumerman, R. C., Culp, R. L., and Hansen, S. P., "Estimating Costs for Water Treatment as a Function of Size and Treatment Efficiency," EPA-600/2-78-182, August 1978.
79. Schwan, M. C., private communication, Emery Industries Cincinnati, OH, Ann. Conf. WPCF, Las Vegas, NV, Sept. 29, 1980.
80. Hoag, L. N., and Salo, J. E., "Ozonation Feasibility Study," Sacramento Regional Wastewater Management Program, Sacramento, CA, June 1975.
81. Matson, J. V., and Coneway, C. R., "Economics of Disinfection," paper presented at the IOI Forum on Ozone Disinfection, Chicago, IL, June 2–4, 1976.
82. Calmer, J., and Adams, R. M., "Evaluation of Disinfection by Ozone," In-House Report, Kennedy Engineers, San Francisco, CA, 1977.
83. "Feasibility Study of Ozone Disinfection of Wastewater Effluent for Westerly Advanced Wastewater Treatment Facility, Cleveland Sewer District." Engineering Science Ltd., Cleveland, OH, March 1974.
84. "Disinfection of Wastewater: EPA Task Force Report No. 430/9–75–012," Cincinnati, OH, March 1976.
85. Consoer, Townsend, and Assoc. Cons. Engrs. Chicago, IL, East Bay Mun. Util. Report, "Disinfection and Chlorine Residual Removal," 1974.

86. Weng, Cheng-Nan, private communication, Buck, Seifert and Jost, Englewood Cliffs, NJ, Jan. 25, 1984.
87. Hoover, M., private communication, Brown & Caldwell Engrs, Pasadena, CA, Feb. 27, 1984.
88. Rownd, W. H., private communication, Mgr. Engineering–Science, Cleveland, OH, March 13, 1984.
89. Elefritz, R. A., Porter, D. W., and Morris, S. F., "The Application of Ozone in Softening Processes for Cost-Effective THM Control: Two Case Histories," paper presented at the AWWA Southeast Ann. Conf., Jekyll Island, GA, April 29, 1984.

14
Bromine, Bromine Chloride, Iodine, and UV Radiation

BROMINE (BR$_2$)

Preface. All bromine species used in water and wastewater treatment revert to bromides after being consumed in the oxidation process. This in itself is not an issue. However when a potable water treatment plant chlorinates water containing bromides, they are oxidized to hypobromous acid and bromamines if any ammonia nitrogen is present. These compounds react with natural precursors in the water to form yet another series of trihalomethanes which are considered to be carcinogenic.

Therefore the use of bromine as an alternative or a supplement to chlorination is not considered practical nor acceptable from an environmental point of view. The consensus is that there are more than enough bromides occurring naturally in the environment to be dealt with, so that adding more bromides to our waterways only compounds the problem.

The following text on bromine is an attempt to describe what is known about its properties, characteristics, occurrence, water chemistry, and how it is used.

Occurrence. Bromine was discovered in seawater, in 1826, by Antoine J. Balard. It derives its name from its offensive odor: the Greek word bromos is stench. It does not occur in nature as a free element. It exists primarily in the bromide form, and is found widely distributed in relatively small proportions. Bromides available for extracting bromine occur in the ocean, salt lakes, brines or salt deposits left after these waters evaporated during earlier geological periods, and from the mineral bromyrite. The bromide content of seawater is about 70 mg/liter. The total bromine content in the earth's crust is estimated at 10^{15} to 10^{16} tons.[1]

The Dead Sea in Israel is one of the richest sources of bromine in the world, containing nearly 0.4 percent at the surface and up to 0.6 percent at deeper levels. In Western Europe the most significant source is in the salt deposits at Strassfurt, Germany. Principal sources in the United States are the brine wells in Arkansas (Arkansas produces about three fourths of the national bromine output), Ohio,

Michigan, and West Virginia with a bromide content ranging from 0.2 to 0.4 percent. Those in Michigan underlie a large area of the Great Lakes region and occur in various sandstone strata at depths of 700–8000 ft. Here the bromide contents vary from 0.05 to 0.3 percent and are generally higher in the deeper levels.

Bromine Production. The recovery of bromine from seawater was first achieved on a commercial scale in 1924 by the Ethyl Corporation.[2] This process involved treatment of the seawater with chlorine and analine. The first successful bromine plant was put into operation at Kure Beach, North Carolina, in 1933, and was capable of extracting 3000 tons of bromine per year. In this plant, the process consisted of adjusting the pH of seawater to 3.5 with sulfuric acid, followed by the application of chlorine. The bromine, liberated by the chlorine, was removed as a dilute bromine gas with a current of air and absorbed in a sodium carbonate solution, from which it was recovered by acidification and stripping with steam. The critical part of this type of bromine extraction is the control of the pH at 3.5.

Oxidation of bromide to bromine can be accomplished either chemically or electrochemically. The electrochemical methods are no longer significant for commercial production. Chemical oxidation can be effected by either chlorine compounds, or oxygen containing compounds such as manganese dioxide, bromate, or chlorate.

The extraction of bromine from bromide compounds requires four steps: (1) oxidation of bromide to elemental bromine (Br_2); (2) separation of the bromine from solution; (3) condensation and isolation of the bromine vapor; (4) purification. Current bromine production methods are based on the modified Kubierschky steaming-out process and the H.H. Dow blowing-out process.

Kubierschky Process. In this process the raw brine is preheated to about 90°C, treated with chlorine in a packed tower, then placed in the steaming-out tower into which steam and additional chlorine are injected. The outgoing brine is neutralized with caustic and used to preheat the raw brine. From the top of the steaming-out tower, the halogen and steam vapor passes into a condenser and then into a gravity separator. Vent gases from the separator return to the chlorination system, the upper water layer containing Br_2 and Cl_2 is returned to the steaming-out tower, and the lower layer containing crude bromine passes on to a stripping tower. From the stripping column, bromine is purified in a fractionating column which produces a 99.8 percent pure liquid bromine as the final product.

The H.H. Dow Process. This process utilizes air instead of steam for the "blowing-out" step in the extraction of bromine. It is a more economical extracting agent than steam, especially when the bromine source is as dilute as in seawater. In the process, the halogens are absorbed from the air in a sodium carbonate solution, or by sulfur dioxide reduction.[1]

$$Br_2 \, (Cl_2) + SO_2 + 2H_2O \longrightarrow 2HBr(2HCl) + H_2SO_4 \tag{14-1}$$

Bromine can then be separated by chlorinating the mixed acids in the blowing-out tower. The theoretical yield is 2.2 tons of bromine per ton of chlorine.[3]

From 1973 to 1975, the estimated total annual bromine production in the United States was about 220,000 tons.[4]

In the years following World War I, the demand for bromine was for pharmaceutical bromides, the organic chemical industry, and photography. However, the biggest boon to the bromine industry was the discovery of tetraethyl lead as an antiknock ingredient in gasoline to accommodate the more powerful high-compression automobile engines. But this ingredient posed a serious problem: deposits of lead in the engine. It was found that a mixture of ethylene dibromide and ethylene dichloride added to the tetraethyl lead was an excellent scavenger which prevented lead deposition in the engine. These lead halides were sufficiently volatile to be expelled in the engine exhausts. It is estimated that about 70 percent of the 1973–1975 production of bromine was used to make ethylene dibromide for gasoline. However, future air pollution controls are almost certain to ban the use of tetraethyl lead in gasoline, thereby wiping out the major market for bromine production. This might be a boon to the water pollution control industry. This will be discussed later in this chapter. Bromine at a lower price becomes a most interesting disinfectant, particularly for water reuse situations.

Physical and Chemical Properties. Bromine is a dark brownish red, heavy, mobile liquid. It gives off, even at ordinary temperatures, a heavy, brownish red vapor with a sharp, penetrating, suffocating odor. The vapor is extremely irritating to the mucous membranes of the eyes, nasal passages, and throat, and is extremely corrosive to most metals. Liquid bromine is likewise corrosive and destructive to organic tissues. In contact with the skin, it produces painful burns which are slow to heal.

Bromine (Br_2; atomic number, 35; molecular weight, 159.832; specific gravity, 3.12) weighs 26.0 lb/gal, and has a boiling point of 58.78°C. Of the metals used to handle bromine, lead is the most versatile.[3] Bromine reacts with lead to form a dense superficial coating of lead bromide, which, if not disturbed, prevents further attack. This is similar to the reaction of chlorine and silver. Tantalum is completely resistant to bromine; wet or dry, at temperatures up to 300°F.

Nickel and its alloy, monel, resist dry bromine and are especially useful as a material for shipping containers. Other nickel alloys, including the hastelloys, are less suitable. Iron, steel, cast iron, stainless steel, and copper are attacked by bromines, either wet or dry. Silver withstands dry bromine.

Bromine handled in lead, nickel, or monel containers should be dry (less than

0.003 percent mositure)[3] and should be protected from ordinary air, from which it can readily absorb enough moisture to make it severely corrosive to these materials.

Bromine is 3 times as soluble as chlorine (i.e., 3.13 g/100 ml water at 30°C). This is an important characteristic when considering the physical aspects of applying bromine to a process stream. Dispersion and diffusion is made "easier" and diffuser design to prevent off-gassing is less of a problem than with chlorine.

Chemistry of Bromine in Water and Wastewater. Bromine is unique in being the only nonmetallic element which is liquid at ordinary temperatures. It reacts with ammonia compounds in solution to form bromamines and displays the breakpoint phenomenon similar to chlorine.

Bromine in water hydrolyzes:

$$Br_2 + H_2O \rightleftharpoons HOBr + H^+ + Br^- \qquad (14\text{-}2)$$

for which the equilibrium constant is 5.8×10^{-9}.

Depending upon the pH, the proportion of dissociation of hypobromous acid (HOBr) and hypobromite is:

$$\frac{[OBr^-][H^+]}{[HOBr]} = K = 2 \times 10^{-9} \qquad (14\text{-}3)$$

Like chlorine, bromine reacts with ammonia forming bromamine. Both Galal and Morris[5] and Johnson and Overby[6] have identified and studied the rate reactions of the compounds NH_2Br, $NHBr_2$, and NBr_3 using the ultraviolet absorption spectrophotometry technique. They reported rapid formation of all bromamine species; however, once the bromamines have formed, a series of decomposition reactions take place. The major chemical difference between the bromamine species and the chloramine species is that the formation of the bromamine species is reversible from monobromamine through dibromamine to tribromamine and back again by fast reactions simply by changing the pH of the solution.

Bromine displays a breakpoint similar to chlorine, and it is the decomposition of the dibromamine which is the basis for this reaction. Tribromamine is the major species of combined residual bromine present beyond the breakpoint. In the pH range of 7–8 it decomposes in accordance with the following equation:

$$2NBr_3 + 3H_2O \longrightarrow N_2 + 3HOBr + 3Br^- + 3H^+ \qquad (14\text{-}4)$$

La Pointe, Inman, and Johnson[7] have shown that the breakpoint occurs when the bromine to ammonia nitrogen molar ratio is 1.5. This is precisely the stoichiometric amount of bromine required to oxidize all of the ammonia to nitrogen gas.

For wastewater disinfection, it is of considerable practical significance that the predominant species of bromine compounds is dibromamine over a pH range of 7–8.5. This is because dibromamine has a germicidal efficiency almost equal to that of free chlorine. Dibromamine is very active and usually displays a rapid decomposition, reverting to the bromide ion. At this point the bromine residual is extinguished. This feature of bromine reactions does not alway occur in wastewater application, in spite of the presence of excess ammonia-N. To date this aberration is thought to be due to the formation of stable organic bromamines resulting from the presence of organic N.

Free bromine residuals (hypobromous acid, HOBr) which would occur in highly nitrified effluents, do not decompose nearly as rapidly as the bromamines. Their persistence increases with decrease in halogen demand of the water being treated.

Reactions with Chlorine. The reactions with chlorine and the bromide ion are the only ones of particular interest. It is important to realize that free chlorine* (HOCl) has the ability to oxidize bromide ions to form hypobromous acid (HOBr) in the pH range of 7–9. This phenomenon was the basis of a patent issued to Marks and Standkov.[8] It was put to use in the application of chlorine to recirculated condenser cooling water at gasoline refineries in the 1950s. Bromide salts were added to the cooling water in the atmospheric tower basin. This was followed by intermittent chlorination. The only chemical consumption was that of chlorine. The chlorine added converted the bromides to hypobromous acid. When the bromine residuals disappeared they reverted to the bromide ion. Thus the intermittent chlorination kept repeating the process to control algae in the towers and tower basins and slime in the condenser tubes.

However, chloramine residuals will not oxidize the bromide ion at pH 4 but free chlorine will. This is the basis of a free chlorine residual analyzer developed by Fischer and Porter in 1968.[9] This analyzer does not suffer from any significant chloramine interference when high concentrations of combined residual are present with free chlorine. A bromide salt is added along with pH 4 buffer to the sample. The free chlorine oxidizes the bromide ion to free bromine† which is measured by the analyzer cell.

Other applications of this phenomenon are described below.

Use of Bromine in Water Processes:

Potable Water. The use of free bromine (Br_2) in potable water is probably nonexistent.

The only known use in municipal potable water treatment was at Irvington, California about 1938. It was discontinued after a reasonable trial period because

---

* It was thought at the time that combined chlorine could oxidize bromide ions, which it cannot do. This was discovered during the investigations of THM formation by chlorination.
† Free bromine is Br_2.

it did not solve the distribution system problem of water quality degradation. The bromine applied reacted so quickly and completely with the zoological slimes on the walls of pipes that it was impossible to obtain a residual downstream from the point of application. It also imparted a high intensity medicinal taste to the water.[2]

As early as 1955, significant efforts were being made to produce solid or dry granular disinfectants using the best attributes of bromine. U.S. Patent 781,730 was issued to the Diversey Corporation of Chicago, Illinois for the invention of a stable dry product composed of hypochlorite and alkali metal bromide. This product was claimed to have extraordinary disinfectant properties when placed in aqueous solution due to the formation of the hypochlorite–hyprobromite mixture.

In 1967 and 1969 patents were issued to Jack F. Mills et al. of the Dow Chemical Co.[10,11] The invention described in these patents represents a process for treating water with elemental bromine obtained from the polybromide form of an anion exchange resin. An effective method for the preparation of this resin is to pass an essentially saturated solution of bromine in aqueous sodium bromide slowly up through a bed of quarternary ammonium anion exchange resin. The resulting polybromide resin in wet form contains about 48 percent bromine. Development of the polybromide resin system as a practical means for disinfection has been concentrated on units capable of treating small quantities of water—potable water for household use and swimming pools.[12] The Everpure Co. of Chicago has developed disposable cartridges containing bromine impregnated resin to feed predetermined amounts of bromine into water for disinfection.[13] The polybromide resin is sealed permanently into the cartridge to prevent its escape into the water system. Polybromide resins with bromine loadings of 25 percent have a very low acute oral toxicity. Direct contact with undiluted 25 percent resin is only moderately irritating to the skin but is capable of producing uncomfortable irritation upon direct contact with the eyes.

The disposable cartridge-type brominator has been installed and operated aboard offshore oil well drilling rigs, some remote land stations, and on ocean going vessels which use seawater distilling systems as a source for their potable water supply.

Wastewater. The use of bromine in wastewater or water reuse situations is unknown in the United States or Canada.

Cooling Water. About 1983 the electric power industry began an investigation using bromine in condenser cooling water treatment. The objective was to determine whether or not dechlorination could be avoided using bromine, owing to the expected rapid decay of bromine residuals. The system adopted by some of the steam generating plants amounted to pumping a bromide salt solution into the chlorine solution discharge of existing conventional chlorination equipment. The chlorine oxidizes the bromide ion in the salt solution to free bromine which goes into solution as hypobromous acid, the active ingredient. Whether or not the resulting bromine residuals will decay fast enough to meet the NPDES discharge requirements is unknown at this time.

Swimming Pools. The most widespread use of bromine as a disinfectant began in Illinois during World War II. The Illinois State Department of Health, under the direction of C. W. Klassen, began an investigation of the use of liquid bromine in the elemental form as a substitute for chlorine when the latter became scarce during the war years. After a trial run in a few swimming pools for a couple of years, it was concluded by bacteriological results that it was doing an efficient job of disinfection. Permission to use elemental bromine was granted to a number of additional pools in 1947, and before the end of that year there were about twenty-five outdoor and thirty indoor pools where bromine was applied instead of the difficult-to-obtain chlorine. However, in order to avoid the hazards of handling liquid bromine, bromine in stick form was introduced about 1958. The stick is a compound of both bromine and chlorine known as bromo-chloro-dimethyl hydantoin (Dihalo). The use of this chemical has been documented by Brown et al.[14] It is thought that this compound, which hydrolyzes to form hypobromous acid, has an additional capability in that it also releases some HOCl by hydrolysis, which reacts with the reduced bromides to form more hypobromous acid. This form of bromine can produce satisfactory results for indoor and very small outdoor swimming pools. The current cost of the bromine sticks is $1.60/lb (1975); therefore its use on large outdoor pools would be decidedly uneconomical.

Br_2 Facility Design. Current practice is limited to the use of bromine in stick form or resin impregnation systems described above for swimming pool and potable water treatment. Neither of these methods is practical for wastewater or water reuse situations. The hazards of handling and the difficulties in metering molecular bromine (liquid or vapor) make it impractical to use. These difficulties stem from the unique characteristic of bromine which allows it to be in the liquid form at room temperature and pressure. This is one of the major reasons why BrCl is being pursued as a practical means for the bromination of wastewater. BrCl has a vapor pressure of 30 psi at room temperature (20°C). This simplifies the packaging and handling problems.

Owing to the undesirable handling characteristics of molecular bromine there can be no valid recommendations for a facility to handle molecular bromine.

BROMIDES: ON-SITE GENERATION OF Br_2

System Description. The on-site oxidation of bromide salts in an aqueous solution to form free bromine (Br_2) is an old concept. In 1948, a patent was issued to Marks and Strandkov of Wallace and Tiernan Co. which involved on-site production of either iodine or bromine by the oxidation of their respective salts by either free or combined chlorine at pH 7–8.[8] Chemically the iodine release

BROMINE, BROMINE CHLORIDE, IODINE, AND UV RADIATION 969

was predictable and quantitative, but the bromine release was not quantitative or predictable at this pH. The iodine system failed because control was difficult. However, in recent years there has been a renewed interest in the on-site generation of bromine from bromide for disinfection of reclaimed water. A patent was issued in 1973 to A. Derreumaux (France) for such a system.[15] This system is different from the one proposed by Marks in that the bromide salt solution is oxidized by a chlorine solution at a carefully controlled pH of 1–2. It conforms to all the established kinetics of bromine chemistry. Free bromine is produced in the Derreumaux process by injecting a bromide salt solution of known concentration into the injector discharge of a conventional chlorinator. The reaction between the chlorine and bromide ion to convert the bromide to hypobromous acid takes place in a reactor where the pH for the reaction is kept below 2. The reaction proceeds as follows:

$$Cl_2 + 2NaBr \longrightarrow 2NaCl + Br_2 \qquad (14\text{-}5)$$

Thus 2.90 lb of pure sodium bromide (NaBr) will react with 1.0 lb chlorine to produce 2.25 lb bromine (Br_2).

The patented Derreumaux system is illustrated in Fig. 14-1. It consists of conventional chlorination equipment, a reaction vessel, and a metering pump for the addition of the bromide salt. The success of this system is its ability to consistently produce an almost pure solution of hypobromous acid at a controlled pH of less than 2. This insures maximum quantitative conversion of the bromide ion in accordance with Eq. (14-5). In cases where the chlorinator injector water is plant effluent

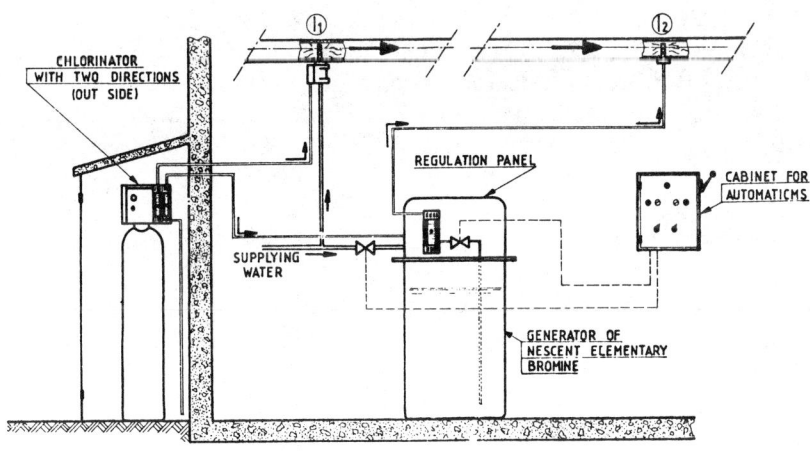

Fig. 14-1. The Derreumaux bromine generating system (courtesy CIFIC Co.).

(i.e., high NH_3 concentration), there will not be any chlorine lost in side reactions with ammonia nitrogen at this low pH and with the reaction time involved. After the bromide is oxidized to free bromine it will remain as Br_2. Therefore, at this controlled low pH in the reactor zone, the free bromine will not be wasted on side reactions with the ammonia nitrogen present.

The most important feature of this process is the sequential addition of chlorine followed by bromine. In practice the chlorine is added first to satisfy the immediate halogen demand (3–5 min. contact time) with a conventional compound loop residual control system. Then bromine is added on a flow proportional basis relying on manual control of the bromine dosage.

This procedure serves three purposes: (1) it reduces significantly the amount of bromine required; (2) it improves the disinfection efficiency[16]; (3) it solves the problem of trying to cope with bromine demand changes normally found in wastewater effluents. Field experience has shown that owing to the rapid die-away of bromamine residuals it is not practical to control the dosage by residual control.[17] In a comparative situation it was shown that chlorine is far more able to meet the diurnal variations in halogen demand, particularly those caused by the usual unpredictable and uncontrollable plant upsets. This is to be expected because of the much greater reactivity of bromine with the various constituents in wastewater.

This method has been used in France for both wastewater and reclaimed water treatment.

The presumed advantages of bromine in wastewater are considered to be: (1) the presence of ammonia-N in wastewater converts the bromine applied to the dibromamine, which is nearly as germicidal as free chlorine; (2) dibromamine, the dominant species, is so highly reactive with wastewater constituents that residual should die away rapidly;* (3) owing to these two characteristics contact times may be reduced from current chlorination practices of 45–60 minutes to perhaps 20–30 minutes thus reducing debromination requirements if necessary.

However, the process must be cost effective. Most of the bromide salts being marketed today in the United States and Canada are very high grade and are used primarily in photographic emulsions so their cost is relatively high—in excess of 70¢/lb for available bromine.

Dow Chemical produces a low-grade calcium bromide used primarily as a well pack fluid. It is marketed in two forms: (1) a 53–54 percent calcium bromide solution (specific gravity of 1.7–1.72 g/ml); (2) an 80 percent calcium bromide flake. The bromide ion content of the solution is about 42–43 percent and in the solid form is about 64 percent. Assuming the following reaction:

$$Cl_2 + CaBr_2 \longrightarrow Br_2 + CaCl_2 \qquad (14\text{-}6)$$

---

* Owing to the presence of significant concentrations of organic N (3–10 mg/l) in wastewaters organic bromamines will form as their chlorine counterparts always do. This species persists as a measurable residual for extended periods of time (several hours).

it would take approximately 2.82 lb of pure calcium bromide plus 1 lb. of chlorine to make 2.25 lb of bromine.

Tank-truck lots of the solution currently cost 28.5¢/lb in 3500 lb lots FOB Ludington, Michigan (1977 prices).

Current Practices in the U.S.A. The only known installations are those used by the electric power industry. Several power plants are experimenting with this method of on-site generation of bromine with the hope that they can escape the need to dechlorinate the cooling water discharge. These discharges usually contain chlorine residuals in the range of 0.5–2.0 mg/l total residual chlorine. The hope is based upon the rapid die-away phenomenon of bromine residuals. The success or failure will depend upon site-specific conditions. This will vary depending upon the water quality at each location. The two most significant water quality parameters in addition to halogen demand are: (1) ammonia-N and (2) organic N concentration. If sufficient ammonia-N is present to convert all of the residual to bromamine, and if there is *insufficient* organic N (and contact time) to form organobromamines, the residuals should die away in a few minutes.

From the power industry's point of view it is much more appealing to retrofit an existing chlorination system to on-site generation of bromine than it is to install a dechlorination facility. From an environmental point of view it would be more appropriate to relax the intermittent chlorine residual discharge requirements than to force a situation where an alternative will contribute to adding more bromides into the environment.

Comparison with Other Methods:

Advantages. This system of on-site generation of free bromine has the following advantages over the application of molecular bromine and bromine chloride.

1. It can be adapted to conventional chlorination equipment and controls.
2. Conventional metering equipment (diaphragm pumps) can be used for feeding the bromide salt solution.
3. It is readily adaptable to conventional compound loop control utilizing existing chlorine residual analyzers.
4. The equipment specific for this process has been proved satisfactory by sufficient field experience.
5. It eliminates hazards of handling liquid molecular bromine or bromine chloride.
6. It eliminates equipment problems of corrosion and the chemical problem of dissociation encountered in the use of bromine chloride and molecular bromine.

Disadvantages

1. The major disadvantage of this process is chemical cost. The least expensive high-grade bromide salt costs about $0.70/lb in 500-lb lots (1976). The low-grade $CaBr_2$ may be the answer.
2. The availability of bromine compounds are limited. However, since there has been a mandatory elimination of leaded gasoline for automobile engines in the United States on account of air pollution, this eliminates the production requirements of bromine compounds which are used as lead scavengers in gasoline. In 1973, 70 percent of the total United States bromine production was for this purpose.
3. The reaction of bromamines in wastewater effluents with organics and the subsequent formation of undesirable bromoorganics is a potential hazard.
4. The end product of bromination—bromide ion—can be a potential hazard to the environment.
5. There is insufficient field experience for proper evaluation.

BROMINE CHLORIDE, BrCl

Physical and Chemical Properties. Bromine chloride is classified as an interhalogen compound because it is formed from two different halogens. These compounds resemble the halogens themselves in their physical and chemical properties except where differences in electronegativity are noted. Bromine chloride at equilibrium is a fuming dark-red liquid below 5°C. It can be withdrawn as a liquid from storage vessels equipped with dip tubes under its own pressure (30 psig 25°C). Liquid BrCl can be vaporized and metered as a vapor in equipment similar to that used for chlorine.

Bromine chloride is an extremely corrosive compound in the presence of low concentrations of moisture. Although it is less corrosive than bromine[18] great care must be exercised in the selection of materials for metering equipment. Like chlorine and bromine, it may be stored in steel containers. This is based on the assumption that BrCl or Br_2 is packaged in an environment where the air is dry (i.e., a dew point not higher than −40°F).

When in contact with skin and other tissues liquid BrCl, like Br_2, causes severe burns. Low concentrations of the vapor are extremely irritating to the eyes and respiratory tract.

Bromine chloride exists in equilibrium with bromine and chlorine in both gas and liquid phases as follows:[6]

$$2 \text{ BrCl} \rightleftharpoons Br_2 + Cl_2 \qquad (14\text{-}7)$$

There is little information on the equilibrium in the liquid state. A recent study by Mills[19] and a previous investigation by Cole and Elverum[20] indicates less dissociation in the liquid (20 percent) than in the vapor state. The equilibrium constant for the vapor phase dissociation of BrCl is close to 0.34 which corresponds to a degree of dissociation of 40.3 percent at 25°C.[1,20]

The density of BrCl is 2.34 g/cc at 20°C.

Figure 14-2 compares the vapor pressure temperature curves for bromine chloride, bromine, and chlorine.

The solubility of BrCl in water is 8.5 g/100 cc at 20°C. This is 2.5 times the solubility of bromine and 11 times that of chlorine.[1]

Bromine chloride forms a yellow crystalline hydrate, BrCl · 7.34 H_2O at 18°C and 1 atm. This compares to the formation of chlorine hydrate (Cl_2 · $8H_2O$) which forms at 9.6°C. This is a significant characteristic because the formation of these hydrates causes operational problems in metering equipment.

Preparation of Bromine Chloride. Bromine chloride is prepared by adding an equivalent amount of chlorine (as a gas or liquid) to bromine until the mixture has increased in weight by 44.3 percent.

$$Br_2 + Cl_2 \rightleftharpoons 2\,Br\,Cl \qquad (14\text{-}8)$$

Bromine chloride may also be prepared by the reaction of bromine in an aqueous hydrochloric acid solution. In the laboratory it can be prepared by oxidizing a bromide salt in a solution containing hydrochloric acid. This produces the following reaction:

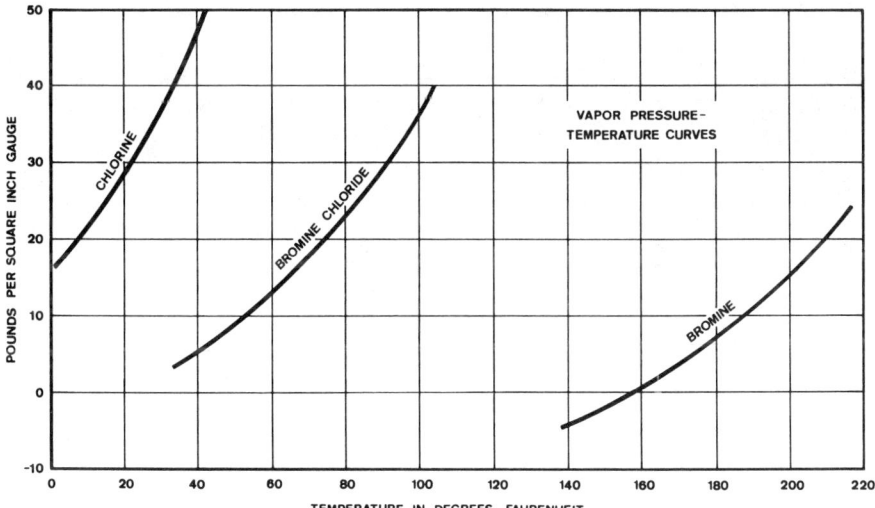

Fig. 14-2. Vapor pressure vs temperature curves from bromine, bromine chloride and chlorine (courtesy Capital Controls Co.).

$$KBrO_3 + 2KBr + 6 HCl \longrightarrow 3BrCl + 3KCl + 3H_2O \tag{14-9}$$

Chemistry of Bromine Chloride in Water. Bromine chloride vapor appears to hydrolyze exclusively to hypobromous acid and hydrochloric acid.

$$BrCl + H_2O \rightleftharpoons HOBr + HCl \tag{14-10}$$

Whereas bromine vapor (or liquid) hydrolyzes to hypobromous acid and hydrogen bromide.

$$Br_2 + H_2O \rightleftharpoons HOBr + HBr \tag{14-11}$$

The formation of HBr represents a significant loss in the disinfecting potential of the expensive bromine molecules. In the hydrolysis reaction of BrCl, any HBr formed by the dissociation of elemental bromine is presumed to be oxidized quickly to HOBr by the HOCl remaining in solution:

$$HBr + HOCl \longrightarrow HOBr + HCl \tag{14-12}$$

However, in the case of wastewaters, the ammonia nitrogen present would immediately convert any HOCl in solution to chloramines in near neutral pH environments. Chloramines cannot oxidize any HBr formed to hypobromous acid (HOBr) at these pH levels.[21] Therefore, the problem of dissociation of bromine chloride is significant in the presence of ammonium ions.

It is of practical significance that the hydrolysis constant for BrCl in water is 2.94×10^{-5} at 0°C, compared with the same constant for chlorine which is 1.45×10^{-4} at 0°C. It is paradoxical that BrCl is several times more soluble in water than chlorine, yet it hydrolyzes by a factor of 10 times slower. The hydrolysis constant of molecular bromine is 0.7×10^{-9} at 0°C which is significantly different from bromine chloride.

Bromine chloride combines with ammonia the same as does molecular bromine to form bromamines. At the usual pH levels encountered in wastewaters (7–8.5), the dominant species will be dibromamine.[24] Typical reactions are as follows:

$$NH_3 + HOBr \rightleftharpoons NH_2Br + H_2O \tag{14-13}$$

$$NH_2Br + HOBr \rightleftharpoons NHBr_2 + H_2O \tag{14-14}$$

$$NHBr_2 + HOBr \rightleftharpoons NBr_3 + H_2O \tag{14-15}$$

Bromine chloride is presumed to have a higher speed of reactivity than bromine. In wastewater, the formation of bromamines is probably much faster than the formation of chloramines. Of greater practical significance is the rapid dieaway of the bromamine residuals. The half-life of bromamine residuals in secondary effluents is about 10 min. If organobromamines are formed the half-life is much longer. In highly nitrified effluents the predominant species would be HOBr. This species of bromine residual persists for a significant length of time in a low halogen demand environment.*

Bromine Chloride Facility Design:

Current Practice. The first attempt at plant scale use of BrCl was at Grandville, Michigan[22] in 1974 and 1975. This was an EPA sponsored project intended to evaluate several wastewater disinfectants. It was not a strict exercise in disinfection because the coliform reduction ratio was too low. The MPN of total coliforms before disinfection ranged between 8×10^4/100 ml and 3×10^6/100 ml. The disinfection requirement was to achieve 10^3/100 ml MPN.† According to the model developed by Collins et al.[23] the required chlorine residual-contact time for the destruction of the lower range of coliform (8×10^4) is 7 and for the higher range (3×10^6) is 23.5. Therefore, if mixing at the point of application is adequate and if 30 min. contact time is available and assuming 80–90 percent plug flow conditions in the contact chamber, the predicted chlorine residual requirement would range from 0.23–0.8 mg/liter to meet the 1000/100 ml total coliform disinfection requirement. For comparable quality effluents this level of chlorine residual is considered very low.

The chlorine dosage at the beginning of this project was 2.9 mg/liter resulting in a 2.0 mg/liter residual at the end of 30 min. This demonstrates a chlorine demand of 0.9 mg/liter in 30 min which appears to be unreasonably low for such an effluent.

At the middle of the project the chlorine dosage was lowered to 2.3 mg/liter which produced a 1.0 mg/liter residual at the end of 30 min. This residual compares to the mathematical model prediction of 0.8 mg/liter for a total coliform concentration before chlorination (y_o) of 3×10^6/100 ml. This is the high range of y_o for the above project.

The BrCl dosage was lowered from 3.6 mg/liter to 3.0 mg/liter early in the project and about the middle of the project it was lowered to 2.0 mg/liter where it remained for the rest of the project. Because of the rapid die-away of the BrCl residual, this process could only be controlled by dosage, while the chlorination process was controlled by residual.

The performance of BrCl was disappointing. Chlorine succeeded in meeting the requirements on the Grandville project 90 percent of the time, while BrCl

---

* An example of this is the practice of power plant condenser cooling water chlorination, where seawater is used. If the ammonia nitrogen content is less than 0.2 mg/liter and the chlorine dose is 1 to 2 mg/liter the chlorine applied immediately converts (stoichiometrically) an equivalent amount of the bromide ions present into HOBr. The resulting bromine residual persists in the condenser discharge plume nearly as long as would comparable chlorine residuals.
† Good disinfection is usually considered as a 4- to 5-log reduction of initial total coliform content.

met the requirements only 80 percent of the time. This was largely the result of equipment failures. The reasons are described later in this chapter.

BrCl System Evaluation. The Potomac Electric Power Company studied and compared two parallel systems at the Morgantown Steam Electric Station.[25] This is an 1100-MWe, fossil-fuel, two-unit generating facility using low-salinity (7000 mg/l) estuarine water for once-through condenser cooling. One system used chlorine and the other used bromine chloride. Both disinfectants were applied at levels of 0.5 mg/l or less. Two properties of bromine chloride were examined: decay rate of bromine residuals and biofouling control effectiveness. The BrCl metering and control unit is shown in Fig. 14-3.

On an equal weight basis, the continuous low-level application of BrCl resulted in equally good control of biofouling in the condensers. On an equimolar basis, therefore, BrCl would be nearly twice as effective but would cost about 2½ times as much as chlorine at current market prices.

The flaw in this investigation stems from the fact that the chlorine applied converted the bromide ion* to HOBr, which reacted with the ammonia-N present (0.25–0.50 mg/l) to produce bromamine. This would explain the nearly equal effectiveness of the two halogens on an equal weight basis.

The BrCl and chlorine residuals dissipated in three decay phases. A very rapid residual loss occurred in the first phase, which lasted less than 1 min. Quasi first-order decay kinetics characterized the second phase, which lasted about 10 min. In the third phase the decay was relatively slow, more than 45 min. The BrCl-induced oxidants dissipated faster than chlorine-induced oxidants. The effluent of the chlorobrominated cooling water contained only two-thirds to one-half the amount of oxidant present in the chlorinated cooling water, even though the BrCl and chlorine were applied at equimolar concentrations.[25]

The report on this parallel study of BrCl and chlorine recommended the consideration of BrCl as an alternative to chlorine for biofouling control. This recommendation was based upon the ability of BrCl to control biofouling at low-level concentrations (less than 1 mg/l) and its relatively rapid residual dissipation compared to chlorine. Whether or not the use of BrCl will eliminate the need for debromination to meet NPDES discharge requirements has not yet been determined.

Metering and Control Equipment. A limited field experience quickly revealed that a new species of equipment would have to be designed to solve the complex problems of construction materials which could stand up to the corrosivity of bromine chloride.

At the Grandville, Michigan project[22] the equipment chosen was a modified Wallace and Tiernan pressure gas-feed chlorinator fitted with an injector. This model (No. 20–055) was chosen because it was constructed of materials thought to be able to withstand bromine chloride. Many problems concerning corrosion

---

*There were sufficient bromide ions in the estuarine water to effect this conversion.

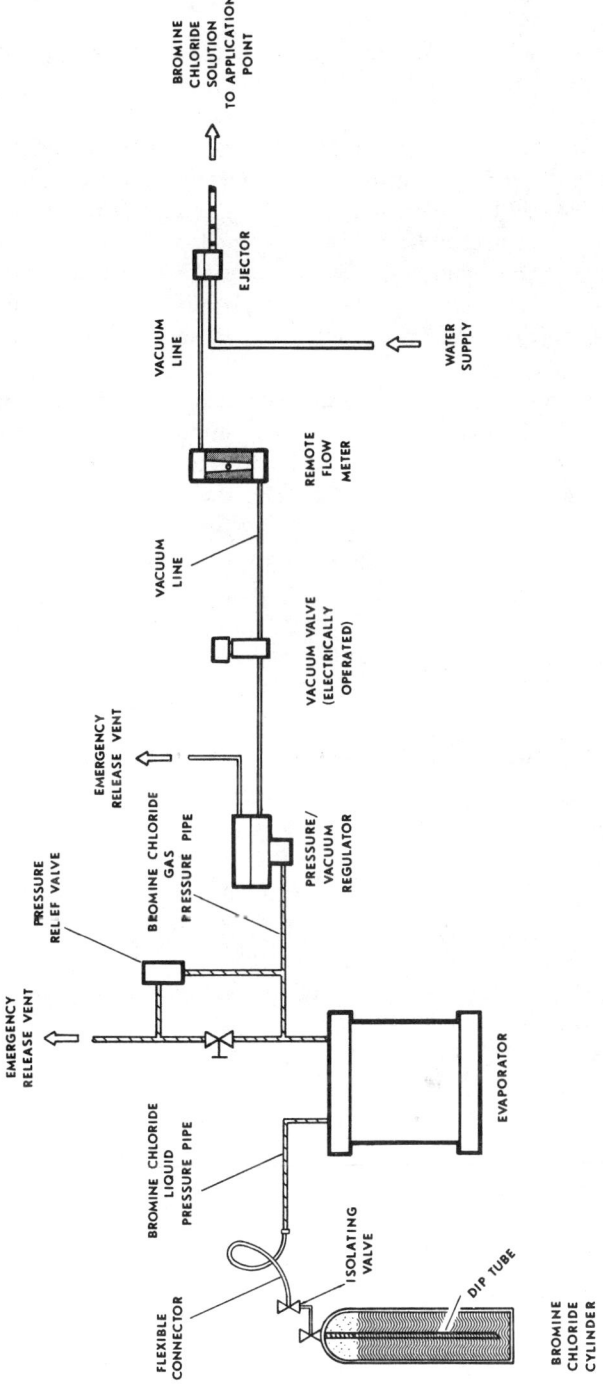

Fig. 14-3. Bromine chloride metering and control equipment (courtesy Capital Controls Co.).

and dissociation of the vapor were encountered, requiring various modifications to the equipment.

At about the same time the Public Service Electric and Gas Co. of New Jersey entered into a project at one of their power stations to evaluate the difference between chlorine and bromine chloride using conventional chlorination equipment for both gases on a side by side installation. It took only a few hours operation to observe that conventional chlorination equipment is not suitable for bromine chloride.[26] Most of the plastic and rubber materials were found to be entirely unsuitable because of permeation and/or direct chemical attack, leading to nearly complete failure of this material. A successful experimental model was constructed on the direct gas-fed approach using slightly modified current production components such as a stainless steel indicating rotameter, chlorine line valves, and chlorine pressure reducing valves. Kynar pipe and valves were used adjacent to the injector where moist BrCl vapor might be encountered. The "suck-back" phenomenon was dealt with by using a modified chlorine check valve adjacent to but not part of the injector assembly.

Figure 14-3 illustrates a system subsequently developed by Capitol Controls, Colmar, PA. This system is essentially the same concept as the Public Service Co. experimental model described by Cole.[26] Fig. 14-3 represents the system provided by Capital Controls Co. for the Potomac Electric Power Co. investigation.

Injectors. The injectors for use with BrCl may require modifications. The one used on the project described by Cole[26] was a 3-in. Wallace and Tiernan model A452. This unit was about 10 years old and did not require any modification. The one used on the Grandville project did require some modification but that used by Capital Controls on the Potomac project was a standard chlorine injector.

Automatic Controls. The system shown in Fig. 14-3 is readily adaptable to automatic control by simply automating the control valve.

Conventional amperometric analyzers may be calibrated to record bromine residuals, but because of the evanescent characteristics of bromine residuals, automatic residual control might not be practical. Field experience is too limited to make a judgment on this facet.

Solution Lines and Diffusers. The material and the design for this part of the system can be of the same material used for chlorine.

Storage and Handling. As in all liquid-vapor exchange chemical systems, the problem of reliquefaction is of importance. This problem increases with decreasing vapor pressure equilibrium of the liquified gas. Sulfur dioxide reliquefies much more easily than chlorine because of its low vapor pressure (i.e., 35 psi at 68°F). This compares with a vapor pressure of 30 psi at 68°F for BrCl. In the case of

BrCl it is more significant because dissociation occurs during reliquefaction and all of the chemical advantages of bromine chloride are lost.

Reliquefaction can be prevented by either raising the temperature of the liquid-vapor container (bromine cylinder) or by external pressure padding with compressed air or nitrogen.

Padding the container to about 50 psi and withdrawing the liquid to an evaporator similar to the practice of handling sulfur dioxide is a preferred procedure. This requires that the bromine chloride containers be fitted with dip tubes and outlet valves similar to chlorine or sulfur dioxide ton containers. One connection would be for pressurization of the vapor and the other for liquid withdrawal. If this were the case then the container storage and handling system would be the same as for chlorine or SO_2.

Evaporators. It is generally conceded that the most practical way to handle bromine chloride is to withdraw the liquid and vaporize it under controlled conditions.

Evaporators used for both chlorine and sulfur dioxide can be used without modification for bromine chloride. The piping arrangement and accessories would be the same as for chlorine. The downstream reducing valve would be set at 25 psi and the containers pressurized to 50 psi. The evaporator should be large enough to provide 30°F superheat.

Materials of Construction. Although bromine chloride is much less corrosive than bromine,[18] great care must be taken in the selection of materials throughout the design of this facility. The material for handling the liquid and/or the vapor under pressure is the same as for chlorine. Therefore, Sch. 80 seamless steel pipe can be used between the containers and evaporator, and between the evaporators and metering system.[26]

Bromine chloride vapor or liquid cannot be handled in PVC or ABS plastics, even at very low pressures, as can chlorine and sulfur dioxide. The preferred piping material for moist BrCl vapor is Kynar, a vinylidenefluoride resin. Similarly, Kynar valves should be used when required with this piping material.

Safety Equipment. Chlorine leak detectors can be readily adapted to detect BrCl in the atmosphere.

Chlorine container kits can be used on BrCl containers if the BrCl is packaged in containers of similar design.

Comparison With Other Methods:

Advantages

1. Use of BrCl eliminates the hazards of handling Br_2.
2. It is presumed to have greater germicidal efficiency than Br_2 and chloramines.

3. At the 1976 price of 20¢/lb it is cheaper than on-site generated bromine.
4. Contact time should be less than for chlorine so contact chambers would be smaller.

Disadvantages

1. Equipment for metering and injection of BrCl is still in the experimental design stage.
2. There is limited field experience.
3. If used alone, residual control will be difficult due to rapid die-away of bromine residuals.
4. It is subject to reliquefaction which causes dissociation of the vapor. This defeats the purpose of bromine chloride.

Costs. The following is the estimated cost for wastewater disinfection using bromine chloride as determined by the 1976 EPA Task Force Report.[27]

Plant Size (mgd)	1	10	100
Capital Cost ($)	47,000	129,000	414,000
Disinfection Cost (¢/1000 gal.)	4.52	3.04	2.65

GERMICIDAL EFFICIENCY (Br_2 AND BrCl)

A review of the literature from 1945–1976 reveals a considerable lack of agreement on the germicidal efficiency of bromine compounds; i.e., free bromine (Br_2), hypobromous acid (HOBr), monobromamine (NH_2Br), and dibromamine (NHB_2). The major difficulty in sorting out the literature was to determine which species was in fact being investigated. This has to do with bromine chemistry. The publication edited by Johnson[28] is the most informative on the germicidal chemistry of bromine. The most informative investigations of the germicidal efficiency of bromine are those by Wyss and Stockton,[29] Marks and Strandkov,[30] Johannesson,[31] McKee et al.,[32] Koski et al.,[33] Schaffer and Mills,[34] Sollo et al.,[35] and Johnson and Sun.[24]

Although this work was done in the laboratory under ideal conditions, the following pertinent conclusions can be made with respect to wastewater disinfection.

In a wastewater effluent containing an ammonia nitrogen content of 10–30 mg/liter and a pH from 7–8 the dominant species of bromine will be dibromamine.[24]

Dibromamine appears to be almost as germicidal as hypobromous acid (HOBr) but not as germicidal as free bromine (Br_2) which only exists below pH 7.

Therefore, the germicidal efficiency of dibromamine is almost comparable to free chlorine (HOCl) as it has been demonstrated that hypobromous acid and hypochlorous acid have about the same germicidal efficiency.[6,24,28] This is significant in wastewater treatment because the application of bromine to wastewater will result in the formation of dibromamine which is *theoretically* many times more germicidal than chloramines. The difficulty in resolving these comparisons is the total lack of field experience. As described above the Grandville, Michigan work[22] revealed disappointment in the germicidal efficiency of bromine. This is probably due to the fact that bromine is such a powerful oxidant, much of it is lost in side reactions with organic matter in the sewage.

Koski and Stuart[33] using the AOAC (Association of Official Agricultural Chemists) method for evaluating disinfectants, reported that liquid bromine at 1.0 ppm is as effective as the 0.6 ppm of available chlorine control, with *E. coli* as the test organism, but that 2.0 ppm liquid bromine is necessary to provide activity equivalent to the 0.6 ppm available chlorine control when *S. faecalis* is the test organism. McKee et al.[32] found that bromine in settled sewage at fifteen-min. disinfectant contact times was less effective than either chlorine or iodine. It is to be presumed that this phenomenon is a result of bromine becoming irretrievably bound up with the organic matter in sewage, or forming bromamines that decompose rapidly into components that do not have any disinfecting power. The interesting point here is that the residual bromine, which is a combination of HOBr and bromamine, appears to be as colicidal as chlorine, but the difficulty is that not much of the bromine becomes a stable residual. This is based on milliequivalents per liter of the halogen. On a dosage basis, however, chlorine is far superior to either bromine or iodine because much of the bromine or iodine is utilized in satisfying the halogen demand. Consequently, they are the least effective of the halogens where sewage is to be disinfected.[32] Eight mg/liter of chlorine will achieve disinfection comparable to 45 mg/liter of bromine or iodine. This is why the literature on this subject stresses the necessity of first satisfying the halogen demand of the sewage with the old reliable less-expensive chlorine. Kott in 1969[36] and subsequent investigations has demonstrated the superior germicidal efficiency of a combination of chlorine and bromine. First, chlorine is applied to satisfy the immediate halogen demand, followed by the application of bromine.

This corresponds to the work of Sollo et al.[16,35] and to the field experiences of CIFEC.[37] The latter reports that a 3 mg/liter dose of chlorine in a wastewater effluent at Bormes, France, containing 90 mg/liter ammonia nitrogen followed by an 8 mg/liter dose of bromine gave a 100 percent reduction of 10^6 coliforms with only 10 min. contact.

According to Mills, bromine chloride is superior in germicidal efficiency to molecular bromine. This notion is derived from the fact that the chemical reactivity of BrCl is much faster than molecular bromine. So far there is too little information to substantiate this supposition.

TOXICITY OF BROMINE RESIDUALS

The acute toxicity tests with chlorobrominated wastewater effluent reported by Ward et al.[22] on the Grandville, Michigan project indicates that bromine residuals are not as lethal to fish as those of chlorine compounds. This supports the finding of Mills[29] who concluded that chlorobromination produced a less toxic effluent because bromamines are less stable than chloramines and thus do not persist as long. This would be true regardless of how the active bromine compounds were applied. The active ingredient would be mostly dibromamine at the pH normally encountered in wastewater treatment. However, if organobromamines are formed due to the presence of certain organic N compounds, the total bromine residual may persist for hours in a wastewater effluent.

BROMO-ORGANIC COMPOUNDS

There is continuing concern over the use of halogens as disinfectants of potable water, wastewater, and water for reuse. This is a direct result of the recent (1975) detection of undesirable halogenated organic compounds in the nation's drinking water.

In addition to the chloro-organic compounds, a variety of bromo-organic compounds were also detected. These included bromodichloromethane, dibromochloromethane, and bromoform. The source of these compounds, generally referred to as precursors, is uncertain, but appears to be distributed by nature as a natural phenomenon. The precursors of these compounds have been found in soil and in the chlorophyll of blue-green algae. It was recently suggested that the occurrence of halogenated bromine compounds was the result of fallout from automobile exhausts. It was postulated that since bromine is used as a lead scavenger in leaded gasoline, these bromine compounds were deposited as a residue on paved roads and therefore reached our water supplies by natural runoff. This possibility is probably overemphasized because bromides are found almost everywhere in nature. However, phasing out of leaded gasoline will certainly curtail this source as a precursor.

An example of how the ubiquitous bromide ion in our environment can effect a potable water supply, was the Contra Costa County Water District experience.[38] This water utility serves 160,000 people in the cities of Concord, Clayton, Port Costa, Walnut Creek, Pleasant Hill, and Martinez, California. The principal source of water supply is from the Sacramento–San Joaquin Delta. During two back-to-back drought years, salt water intrusion from the tidal effect in San Francisco Bay increased the bromide concentration in the water going to the treatment plant. Chlorination of this water converted the bromides to bromine species. These bromine compounds reacted with precursors in the water to form trihalomethanes far in excess of the EPA minimum contaminant level of 100 $\mu g/l$. This forced the water district to take corrective measures. After studying several treatment changes in an effort to reduce the THM precursors, none of which was entirely successful, the district had to switch their chlorination practice to chloramines. This brought the TTHM levels well below the EPA minimum contaminant level (MCl) for THMs.

It is important to recognize that all of the bromine species are better oxidizing agents than their analogous chlorine species. For example, the oxidation of cellulose by hypobromous acid is much faster than by hypochlorous acid, and the oxidation of glucose to gluconic acid by hypobromite is 1300 times faster than by hypochlorite. Therefore, it is not surprising that bromine is a lesser halogenating agent than chlorine simply because it is a more potent oxidizing agent.

There are many organic reducing agents in wastewaters. These include organic alcohols, aldehydes, amines, and mercaptans. They are completely oxidized by bromine chloride resulting in the formation of inorganic chloride and bromide salts as the major byproducts.

The carbon–bromine bond is less stable than the carbon–chlorine bond and there is a possibility that in addition to metabolism by hydroxylation, well established for chlorobiphenyls, a reductive debromination may be a degradation pathway of brominated aromatics. In general, compounds which are more readily brominated are also more susceptible to either hydrolytic or photochemical degradation. This reduces the incidence of bromo-organic compound formation. Moreover, the chemical bond strengths show that bromine bonds are weaker than chlorine bonds. Thus bromine compounds are generally more unstable than those of chlorine and are therefore easier to chemically degrade to innocuous inorganic compounds. The fact that bromine residuals die away after a few minutes contact time in potable water does not mean that a similar die-away pattern will occur in highly polluted water or wastewater effluents. There is evidence that the presence of certain species of organic-N compounds react with the bromine species to form organobromamines that have a very slow decay rate similar to organochloramines. Therefore much more investigative work would have to be done to be able to predict with reasonable accuracy the characteristics of bromine residual die-away in various situations.

MEASUREMENT OF BROMINE RESIDUALS

General Discussion. Regardless of the method of bromine application, whether it be generated on-site, injected as molecular bromine, as an aqueous solution of either bromine or bromine chloride, the resulting residuals respond the same to all analytical procedures.

At the Grandville, Michigan project[22] the bromine residuals were determined by a spectrophotometric procedure developed by Mills of Dow Chemical Co.[39] Other researchers used the DPD FAS titrimetric method[35] and Johnson[6,24,28] used a thiosulfate-iodide amperometric titration which measured the oxidation of iodide to iodine at pH 7 by the bromine residuals. Larsen and Sollo[40] used the bromcresol purple colorimetric procedure.

There are no known analytical methods available to an operator or laboratory technician capable of distinguishing between the three bromine species: hypobromous acid (HOBr), monobromamine (NH_2Br), and dibromamine ($NHBr_2$). Therefore the only measurement available is total bromine residual.

When chlorine and bromine are used sequentially, as in water reclamation, there may be a mixture of chloramine and bromamine owing to the presence of ammonia-N in the wastewater. Total residual measured either as bromine or chlorine is the preferred choice, because most instances would involve the use of a continuous residual analyzer. This instrument can only be calibrated for total residual in the event of a mixture of chlorine and bromine residuals. This can be done using either the amperometric method or the DPD–ferrous method.

Differentiating between total chlorine and total bromine in the same sample can be done by Palin's latest DPD method[41] described below.

Amperometric Method. The amperometric method is well suited. Since the amperometric cell has a linear sensitivity to all the free halogens, and since free and combined bromine compounds will release iodine quantitatively at pH 4, the procedure for total available bromine is determined on the titrator by using the procedure given for total residual chlorine—using buffer solution pH 4 and potassium iodide solution. Then, titrating for bromine, the pipette reading in milliliters must be multiplied by 2.25 to express the answer in ppm of bromine.

DPD Differentiation Method.[41] The following procedure assumes free chlorine, combined chlorine, and free and combined bromine residuals. Hereafter the term "residual bromine" is taken to mean free bromine plus bromamines.

The reagents required are: (1) standard ferrous ammonia sulphate (FAS) solution (1 ml = 0.100 mg available chlorine); (2) DPD No. 1 powder (a combined buffer-indicator reagent); (3) potassium iodide crystals.

Free Chlorine. To a 100 ml sample add 0.5 g DPD No. 1 powder, mix rapidly to dissolve and titrate immediately with the FAS solution. This is reading A. (If free chlorine is not present this step may be omitted.)

Combined Available Chlorine. Add to the previous sample several crystals (approximately 1.0 g) of potassium iodide, mix to dissolve and after standing for two min. continue titration with FAS solution. This is reading C.

Residual Bromine. To a *second* 100 ml sample add 2 ml of 10 percent wt/v glycine solution and mix.* Then add approximately 0.5 g DPD No. 1 powder, mix and titrate with the standard FAS solution. This is reading Br.

Calculations. For a 100 ml sample, 1 ml FAS solution = 1 mg/liter available chlorine.

Residual Bromine = Br
Free Available Chlorine = $A - Br$
Combined chlorine = $C - A$

---

* If it has already been determined that free chlorine is absent this step may be omitted.

If it is desired to report the residual bromine in terms of bromine, multiply the Br reading result by 2.25. (This is the ratio of the molecular wt of bromine to chlorine.)

EFFECT OF SEAWATER CHLORINATION

Chlorination of seawater and estuarine saline waters reacts to convert the excess bromides into the bromine species. If ammonia-N is not present the species will be hypobromous acid. If ammonia-N is present mono- and dibromamine will be present. Therefore all of the remaining residual after chlorination will be bromine residual. This chemistry is explained in Chapter 4. Measuring these residuals is usually carried out for total residual and reported as chlorine residuals.

IODINE

Physical and Chemical Characteristics. Iodine, a nonmetallic element with an atomic weight of 126.92—the heaviest of the halogen group—was discovered by Curtois in 1811. It is the only halogen which is solid at room temperature, and it can change spontaneously into the vapor state without first passing through a liquid phase. It is isolated as a shining blackish gray crystalline solid of specific gravity 4.93 with a peculiar chlorine-like odor. It melts at 113.6°C to a black mobile liquid and boils at 184°C under atmospheric pressure to produce the characteristic violet-colored vapor.[42]

Occurrence and Production. Iodine is always found combined, as in the iodides. It is prepared from kelp and from crude Chile saltpeter. This saltpeter ($NaNO_3$) contains about 0.2 percent sodium iodate ($NaIO_3$). Historically, the United States has depended upon imports for almost all of its iodine needs. Originally from Chile, it is produced as a coproduct from natural nitrate production. In recent times, however, Japan has become the dominant factor in iodine world trade. The 1974–1976 United States consumption of iodine is on the order of 6–7 million lb. During 1975, United States imports totaled 5.3 million lb, of which 93 percent was from Japan and the remainder from Chile.[43] Chile's production appears to be stabilized at a maximum of 4.4 million lb; this is from natural sources.

The small United States output in 1975 came entirely from the Dow Chemical Co. in Midland, Michigan which recovered iodine and several other chemical products from a mixed-salts-type of natural well brine. There are some reportedly high-grade brines in northern Oklahoma which may be exploited by several companies. In the fall of 1975, Houston Chemical Co. broke ground for its 2 million lb/yr iodine plant at Woodward, Oklahoma.

Uses. Iodine and its compounds are used as catalysts in the chemical industry (production of synthetic rubber) food products, pharmaceutical preparations, stabilizers (as nylon precursors), antiseptics, medicine (treatment of cretinism and goiter), inks and colorants, and industrial and household disinfectants.

It was first used in medical practice for treatment of goiter by Prout in 1816.[44] The first specific reference to the use of iodine in wounds was in 1839. Iodine was officially recognized by the *Pharmacopoeia* of the United States in 1830 as tincture of iodine. The first official United States tincture was a 5 percent solution of iodine in diluted alcohol.

Use in Water Treatment. The first known field use for iodine in water treatment was in World War I by Vergnoux,[45] who reported rapid sterilization of water for troops. Investigations on the use of iodine as an alternative method of disinfecting water supplies for United States troops were made in the years following World War I. Hitchens (1922)[46] recommended the use of 5 ml of a 7 percent tincture of iodine for a Lyster bag (about 140 liters or 37 gallons), and stated that, even with raw Potomac River water, such a treatment rendered the water safe for drinking in thirty minutes. This was followed by Dunham's[46] recommendation (1930) to increase the dosage to 10 ml of a 7 percent tincture of iodine per Lyster bag, or 2 to 3 drops per canteen. Pond and Willard[47] verified in 1937 that 2 drops per liter of a 7 percent tincture of iodine is sufficient to render any water innocuous within fifteen minutes, and that three drops per liter did not give any better results.

During World War II, a series of studies at Harvard University by Chang, Morris, et al.[48,49] led to the development of globaline tablets for disinfecting small or individual supplies for the U.S. Army. Each tablet contained 20 mg tetraglycine hydroperiodide, of which 40 percent is I_2 and 20 percent iodide, combined with 85 mg acid pyrophosphate. One tablet imparts 8 mg I_2 to a liter of water. Chang[46] lists the following advantages of the globaline tablet over the tincture: (1) Iodide is present in half the amount of I_2. (2) There is a definite dosage. (3) I_2 loss is insignificant. (4) The presence of an acid salt counteracts high pH. (5) The treated water is more palatable (less taste and odor). (6) The tablets are convenient for field use. The globaline tablets superseded the Halazone tablets for troop use, because one globaline tablet per quart or liter destroys the cysts of *Entamoeba histolytica* in ten minutes, whereas six Halazone tablets are required to accomplish the same kill.[46] Likewise, the human enteric pathogens (bacterial, viral, and protozoan) are satisfactorily destroyed in the same time[48,49,50] under ordinary conditions. If the water is very cold, deeply colored, or highly turbid, the dosage should be increased to two tablets per liter and the contact time extended to twenty minutes.

Iodination of water supplies has been limited largely to emergency treatment by the military. The use of iodine as a disinfectant for water has been recognized for a long time, but has never generated enough interest to displace the popular use of chlorine. As compared to chlorine, the very high cost of iodine and its possible physiologic effects are the main reasons for its limited use. The military did not give their unqualified approval to the use of iodine as a disinfectant for water until a six-month study was made at a naval installation in the Marshall Islands in 1949–50. The use of iodine in the drinking water revealed no untoward effects from its ingestion by a large group of service men[51] whose average intake was 12 mg/day for sixteen weeks and 19.2 mg/day for ten weeks.

More recently (1965) Black et al.[52,53] have completed a demonstration project, with a grant from the U.S. Public Health Service, to study the effects of prolonged use of iodine at three correctional institutions in Florida. Iodination of these water supplies was carried out over a period of nineteen months, serving approximately seven hundred persons under carefully planned chemical, medical, and bacteriological controls. Iodine proved to be a satisfactory disinfectant and in doses up to 1.0 ppm did not produce any discernible color, taste, or odor in the water. There was no evidence from the medical investigation that this level of iodine has any adverse effect upon the general health or thyroid function.

Montezuma's Revenge. The most popular application is in the tablet form or the tincture of iodine solution for emergency treatment of small supplies on a batch basis such as water supplies for troops on bivouac, or travelers in foreign countries where "traveler's diarrhea" is a common occurrence. This type of emergency treatment has proved to be reliable and effective.

The U.S. Public Health Service lists the iodine dosage for emergency water treatment as: 5 drops of a 2 percent tincture for 30 minutes *per quart* (or one liter) for clear water and 10 drops for turbid water. The potential for unpleasant tastes with iodine is about the same as with chlorine. Iodine does not react with nitrogenous compounds as chlorine does, so the resulting taste in polluted waters will be different.

Chemistry of Iodination. There are two methods of dosing a water supply with iodine. The most practical method is the use of iodine crystals. The discussion that follows will be limited to this method. The other possible method is the use of the "iodine bank" scheme. This procedure requires the presence of an excess of iodide. Elemental iodine is released from the iodide quantitatively by the application of an oxidant such as chlorine. This is the procedure used in swimming pools and has no apparent practical application in water supply treatment. This method utilizes the only advantage of iodine over chlorine—that is, that iodine will not react with nitrogenous compounds to form iodomines as chlorine does to form the objectionable chloramines.

Therefore the following chemistry of weak solutions of elemental iodine is based on the use of iodine crystals without the presence of added iodide. The work by Chang[50] (1958) is the basis of this chemistry.

Four different substances must be considered in the reactions of elemental iodine with water:

Elemental iodine I_2
Hypoiodus acid   HIO
Periodide or tri-iodide ion I_3^-
Iodate ion IO_3^-

First is the hydrolysis of I_2 to form hypoiodus acid:

$$I_2 + H_2O \rightleftharpoons HIO + H^+ + I^- \tag{14-16}$$

$$\frac{[HIO][H]^+[I]^-}{[I_2]} = K_h \tag{14-17}$$

$$K_h = 3 \times 10^{-13} \text{ at } 25°C$$

Fig. 14-4 shows the I_2 hydrolysis data for concentrations of iodine in the 0.1–5 mg/l range at 18°C and pH 6.0–8.0, the probable range of use for small water supplies. Table 14-1 shows the effect of pH on the hydrolysis of iodine with a total iodine residual of 0.5 ppm.

Second is the possible dissociation of hypoiodous acid with variation in pH to

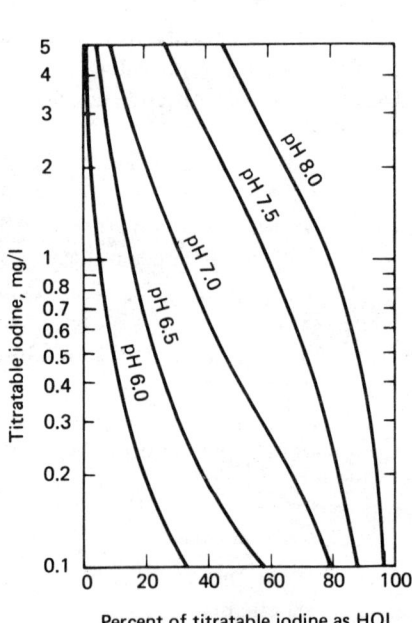

Fig. 14-4. Percent of titrable iodine as I_2 and HIO in water at 18°C and pH as indicated.[50]

Table 14-1

pH	Content of Residual %		
	I_2	HIO	IO^-
5	99	1	0
6	90	10	0
7	52	48	0
8	12	88	0.005

form hypoiodite ion. It is of interest to compare this phenomenon with that of hypochlorous acid in the case of chlorine and how it effects the germicidal efficiency of that compound.

$$HIO \rightleftharpoons H^+ + IO^- \tag{14-18}$$

$$\frac{[H^+][IO^-]}{[HIO]} = K_a = 4.5 \times 10^{-13} \tag{14-19}$$

$$\frac{[HIO]}{[IO^-]} = \frac{[H^+]}{K_a} \tag{14-20}$$

From Eq. (14-20) it is possible to calculate the ratio of undissociated acid to hypoiodite ion at any pH. The dissociation constant of hypoiodous acid reveals that it is only slightly stronger as an acid than pure water. The formation of hypoiodite ion (IO^-) is insignificant, as can be seen from Table 14-1. Compare the formation of hypoiodite ion with that of the hypochlorite ion for chlorine in Table 14-2.

Both elemental iodine and hypoiodous acid are effective germicides. At pH 8, the effective chlorine germicide HOCl is present at 21.5 percent, while the ineffective hypochlorite ion is present at 78.5 percent. With iodine, however, the ineffective hypoiodite ion is present only at 0.005 percent with elemental iodine at 12 percent and hypoiodous acid at 88 percent. Therefore the formation of the ineffective hypoiodite ion can be ignored.

Table 14-2

pH	Content of Residual % at 20°C		
	Cl_2	HOCl	OCl^-
5	0.0	99.74	0.26
6	0.0	97.45	2.55
7	0.0	79.29	20.71
8	0.0	27.69	72.31

Third is the possibility of the formation of the bactericidally ineffective tri-iodide ion I_3^-. Chang reports that when using either iodine crystals, iodine tablets, or tincture of iodine, the formation of tri-iodide ion can be ignored. The amount of iodide formed from hydrolysis of I_2 at the 1 to 2 ppm level and at pH values of 6.5 to 8.0 would be so small that the tri-iodide ion I_3^- formed in the water should not be much more than 1 percent of the total titrable iodine.

Fourth is the formation of the iodate ion, which has no apparent cysticidal or viricidal activity. Iodate formation in solutions of elemental iodine (without added iodide) proceeds as follows:

$$I_2 + H_2O \rightleftharpoons HIO + H^+ + I^- \tag{14-21}$$

$$3HIO + 3(OH)^- \rightleftharpoons (IO_3)^- + 2I^- + 3H_2O \tag{14-22}$$

These equations show that alkaline pH values shift the reactions to the right and the acidic pH values to the left. Iodate formation is not likely to be a problem interfering with the disinfection efficiency of iodine until the pH rises above 8.0 and the contact time runs over thirty minutes. Since most of the disinfection action of iodination is accomplished in the first fifteen to twenty minutes, the small losses of titrable iodine to iodate need not be a problem in treating waters at pH levels below 8.4.

Germicidal Efficiency. It is the consensus of many investigators[46,47,52,54-58] that elemental iodine I_2 and hypoiodous acid HIO are the two most powerful disinfecting agents among the titrable iodine species. The relative amounts of titrable iodine existing as I_2 and HIO in an aqueous solution of elemental iodine depends primarily on the strength and pH of the solution and, to a much lesser extent, upon the temperature. The comparative disinfection efficiency of I_2 and HIO varies considerably, depending on the type of organism. Wyss reported[54] that against *B metiens* spores I_2 is three to six times greater than HIO, but Chang[46] reported that in the destruction of cysts of *E. histolytica,* I_2 is two to three times more effective than HIO. On the other hand, Clarke et al.[56] found that hypoiodous acid destroys viruses at a rate considerably faster than that achieved by I_2; HIO has also been claimed to be three to four times more active than I_2 in destroying *E. coli.*[46] The difference in effectiveness of hypoiodous acid and elemental iodine on the type of organism suggests a different mode of action, depending on whether or not the organism is protected by a membrane such as a spore or a cyst. The superiority of I_2 over HIO in the destruction of spores and cysts is apparently due to the greater penetrating power of I_2. In the case of vegetative bacteria, which have a physiologically active cell membrane, the oxidizing power of the compound becomes the important factor. This is supported by the fact that the normal redox potential (E_0 at 25°C) of HIO is 990 mV and that of I_2 536 mV.[46] In virus destruction, which probably proceeds as a function of the oxidizing power of the compound, it can be supported further when comparing the viricidal efficiency of hypoiodous acid and that of hypochlorous acid. HOCl has a normal redox potential of 1490 mV; its viricidal efficiency is five times greater than that of HIO[56], while that of HIO is at least forty times greater than that of I_2.[46] These differences support the belief that virus destruction is a result of the reaction of the disinfectant with the protein shell and not with the nucleic acid core; if the latter were involved, the protein shell would act as a protective shield, and the destructive process would not be much different from that of spores and cysts.

In order to relate these differences in the germicidal efficiency of both I_2 and HIO, Chang[46] has prepared Fig. 14-5 which shows the relation to titrable iodine and contact time for the 99.9 percent destruction of cysts, virus, and bacteria by I_2 and HIO at 18°C. In practice, these parameters allow a good margin of safety. Satisfactory bactericidal results can be obtained with 0.1 to 0.2 ppm in less than four minutes. Virus destruction by I_2 lies beyond the practical limit of 1 to 2 ppm, but for HIO it is well within these limits. The I_2 cysticidal curve lies close to the upper limit of practical range: 1.0 ppm in thirty minutes to 2 ppm in thirteen minutes.

Compared with free chlorine residuals, iodine is definitely inferior. Free chlorine is five times more cysticidal than HIO, two hundred times more viricidal, and two times more cysticidal than I_2.

The 1960 report by McKee et al.[32] demonstrates that iodine is an effective disinfectant in wastewater; however, wasted side reactions due to the halogen demand of wastewater and its heavier molecular weight *requires 45 mg/liter iodine* to achieve the same level of disinfection as *8 mg/liter of chlorine*.

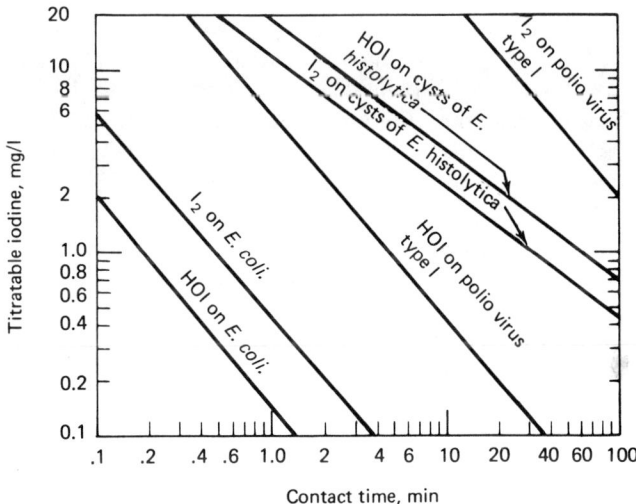

Fig. 14-5. Time versus concentration relationship in the destruction (99.9 percent) of cysts, viruses and bacteria by I_2 and HIO at 18°C.[46]

Limitations of Iodination. Iodination can be used where chlorination or more complete treatment of water is either not possible or impractical. Such use suggests that this process would probably be limited to remote rural areas or undeveloped countries where little or no adequate supervision and expertise are available. Further, it is the consensus that on a prolonged treatment basis, because of possible undesirable physiological effects, dosages of 1–2 ppm with occasional dosages of 4–5 ppm are the practical limits. This would require an investigative procedure to determine the desirability of iodination.

Comparison with Chlorination. The cost of iodination is an important factor against its use. Iodine crystals were available at approximately $2.50 per lb FOB New York from the Chilean Nitrate Sales Corporation in 150-lb containers. This compares with chlorine at 15¢ per lb. However, in a remote location, transport of the iodine crystals and the installation of an electrically driven pump are about the only considerations for applying iodine. The apparatus for a hypochlorite installation is not too much different.

Free chlorine (HOCl) is five times more viricidal and four times more cysticidal than HIO, and is two hundred times more viricidal and two times more cysticidal than I_2. This obvious superiority of chlorination in disinfecting power, together with its low cost and pharmacological inertness, makes chlorination the method of choice over iodination in water treatment. The emphasis here is on free rather than combined chlorine.

Iodination Facility. The application of iodine for the treatment of water is best accomplished by the use of a saturator. Iodine is the least soluble halogen in water, and its solubility is significantly dependent on temperature,[52] as shown in Fig. 14-6. For the usual water supply, the strength of the iodine solution will be in the range of about 200–300 ppm. At these concentrations, 316 SS is satisfactory for piping and metering pumps. Not all plastics are suitable: nearly all clear plastic tubing will discolor rather quickly, and some will disintegrate in a matter of months. Fig. 14-7 illustrates an adequate arrangement for the application of iodine into a water supply. The system consists of a saturator for dissolving the iodine crystals. The temperature of the water will determine the strength of the solution to be injected by the metering pumps. Since the accuracy of dosage is critical, it is strongly recommended that only metering pumps be used instead of injection devices that produce a vacuum by means of water pressure. The latter devices cannot provide proper dependability of accurate feed rates compared to a positive displacement metering pump. Furthermore, the metering pumps should be paced from a flow-measuring device through either an automatic stroke positioner or a variable speed SCR drive if the water flow is variable.

Wastewater. Iodine has never been tried on a plant-scale basis as a wastewater disinfectant, but only in the laboratory for purposes of investigating its germicidal

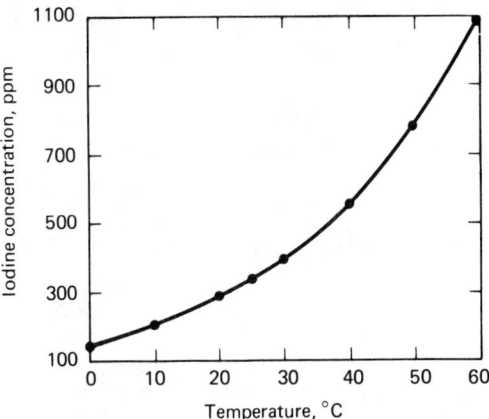

Fig. 14-6. Solubility of iodine in water (concentration as a function of temperature).

efficiency. The cost and nonavailability of iodine mitigate heavily against its use as a disinfectant for wastewater.

Cost. The 1976 price for crude iodine crystals in 100-lb drums was $2.60/lb and USP granular iodine was $4.00–$5.00/lb in 100-lb drums. This compares to chlorine at 10–15¢/lb and bromine chloride at 20¢/lb.

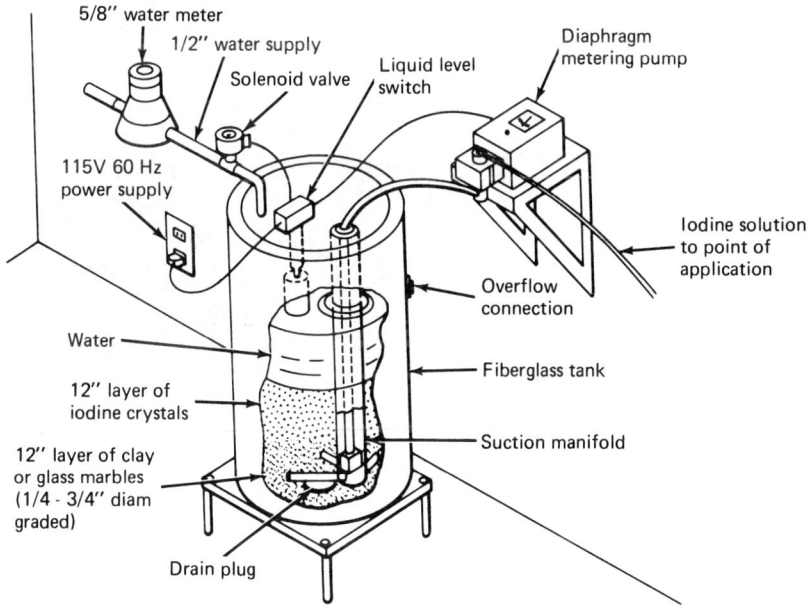

Fig. 14-7. Iodination installation.

Determination of Iodine Residuals. The amperometric titration method is the most practical method to use. Moreover, it is more precise because the electrodes can be sensitized in a few minutes with a dilute solution of iodine. This gives a sharp endpoint.

The DPD–ferrous titrimetric method is equally satisfactory for precise measurements.

ULTRAVIOLET RADIATION

Introduction:

Physics of Radiation. Radiation results from the emission of energy that can be converted into thermal, chemical, or mechanical energy. Radiations are usually divided into two main groups—corpuscular and electromagnetic wave radiations. UV is an example of electromagnetic radiations, and consequently has no matter associated with it. It is similar in this respect to X-rays and gamma rays. It differs from corpuscular radiations exemplified by electrons, protons, neutrons, and certain other subatomic particles. Electromagnetic radiations are believed to be made up of waves that comprise the wide range of radiations from electric waves, radio waves, infrared or heat rays, and from visible light to UV light, roentgen rays, gamma rays, and secondary cosmic rays. The entire group of such radiations is known as the electromagnetic spectrum. This group of rays travel at the speed of light; however, they differ widely in wave length from 10^{11} cm for electric waves down to 10^{-11} for secondary cosmic rays.[59]

It has been well established that bacteria, viruses, and fungi can be destroyed by UV whether suspended in air, suspended in liquid, or deposited on surfaces. The killing or inactivation of microorganisms as a result of exposure to UV light is expressed as the product of the intensity, I, of UV energy, and the time of exposure, t. This leads to the postulation that the same rate of kill can be obtained using either high-intensity UV energy for a short time, or low-intensity UV for a proportionally longer time so that the product of it is equal in both cases.

Germicidal Lamps. UV radiation is generated by special sources of mercury vapor known commonly as germicidal lamps. The principle of all germicidal lamps is the same: that of electron flow between electrodes through ionized mercury vapor. The arc in a fluorescent lamp operates on the same basic principle and produces the same type of UV energy. The difference between the two is that the bulb of the fluorescent lamp is coated with a *phosphor* compound which converts UV to visible light. The germicidal lamp is made of special glass which is not coated, so it transmits UV rays generated by the arc. The low-pressure mercury arc lamps emit their maximum energy output at a wavelength of 253.7 nm.

Practical Applications. The major use of UV radiation is for the sterilization of potable water. Potable water is never bacteria free. There is always a low level of bacteria colonies in any potable water supply. This can be substantiated by the plate count (SPC). Therefore it is reasonable to expect that many processes using potable water desire or require close control or complete elimination of all bacteria in the system. Some of these applications include continuous UV exposure to recycled distilled water in the pharmaceutical and cosmetic industries, kidney dialysis water, and many others involving specialty products that require an almost sterile water supply. The beverage market is a large one for UV systems. They are used to sterilize make-up water for frozen juice concentrates, final rinse water for bottles and containers, and final treatment for bottled water. They are also used almost exclusively for cottage cheese wash water. Fish hatcheries have used UV to disinfect the water in order to avoid the toxic effect of chlorine.

Evidence of its widespread use is substantiated by the number of manufacturers engaged in this process: Aquafine Corp. (Sanitron Purifiers), California Ultraviolet Co., Capital Controls, Hanovia Ltd., Pure Water Systems Inc., Refco International, Steri-Tronics, Steroline, Ultra Dynamics Corp., and Ultraviolet Purification Systems, Inc.

Potable Water. The use of UV as an alternative to chlorine does not have a good record. The use of UV on cruise vessels has proved to be unreliable, and system inefficiency has led to serious gastrointestinal outbreaks on these vessels. The major complaint against UV for potable water systems is lack of proof of system reliability (absence of residual measurement) and lack of a persisting disinfecting residual.

In 1975 the EPA explored the use of UV for small public water supplies[60] by a grant to the Vermont State Dept. of Health. This was a demonstration grant designed to compare ozone and UV with chlorine as practical methods for disinfecting small water supplies. This report was definitely unfavorable to UV as a practical alternative to chlorine. The specific complaints were that both UV and ozone units are many times more costly than chlorination units, that they incur higher operation and maintenance costs, and that they are less reliable. Most important, though, was that neither ozone nor UV could provide a residual disinfectant to protect the water in the distribution system. For this project the capital cost for the UV system was $1995 versus $550 for a hypochlorinator and the annual operation and maintenance cost for a 20,000 GPD system and 2 mg/liter chlorine dosage was $141/yr for chlorine and $458/yr for UV. It is evident that public health agencies are not enthusiastic about the use of UV on public water supplies as an alternative to chlorine.

Wastewater. The situation concerning the use of UV for wastewater disinfection is surprisingly different from that of potable water. Here again the EPA, under the able direction of A. D. Venosa, planned and funded four investigations of UV as a wastewater disinfectant based upon the EPA discharge requirement of 200/100 ml MPN fecal coliforms. These projects included the plant scale investigations at St. Michaels Maryland,[61] Waldwick, New Jersey,[62] Port Richmond, New York,[63,64] and the biossay-dose measurement studies by Johnson et al.[65,66]

The results of these studies have generated a genuine interest in the acceptability of UV as an alternative to chlorine under certain circumstances described below. The EPA has been duly impressed with the performance of UV as a wastewater disinfectant. As of 1984 there have been about 100 systems funded, and of these about 50 are being installed and approximately 15 are in operation. This is indeed an encouraging start. It should be pointed out that effluent quality is all-important because suspended solids concentration is critical, as they shield the organism from the rays. It is the consensus that the suspended solids concentration should not exceed 20 mg/l for reliable operation. 10 mg/l is preferred.[68]

There are still many questions to be answered before standards for design, acceptability, and reliability can be formulated to the satisfaction of the EPA and the state health agencies.

Mode of Action:

General Discussion. The radiation energy of UV rays can be used to destroy microorganisms. In order to kill, the electromagnetic waves of ultraviolet irradiation must actually strike the organism. In this process some of the radiation energy is absorbed by the organism and other constituents in the medium surrounding the organism. There is sufficient evidence to conclude that if sufficient dosages of UV energy reach the organisms, water or wastewater can be disinfected to any degree required. The germicidal effect of ultraviolet energy is thought to be associated with its absorption by various organic molecular components essential to the functioning of cells. Energy dissipation by excitation, causing disruptions of unsaturated bonds, particularly of the purine and pyridimine components of nucleoproteins, appears to produce a progressive lethal biochemical change.[69] The UV treatment does not alter water chemically; nothing is being added except energy, which produces heat, resulting in a temperature rise in the treated water.

The UV rays penetrate the cell walls of microorganisms. In accordance with the laws of photochemical action, the energy which kills a bacterium is that absorbed by the organism. The only radiant energy effective in killing bacteria is that which reaches the bacteria; therefore, the water must be free of any particles that would act as a shield. This is one of the main disadvantages of the UV process; other disadvantages are the lack of a field test which readily establishes the efficiency of the process, and its inability to provide any residual disinfecting power.

The manufacturers of germicidal lamps caution in their literature that UV radiation will only kill invisible organisms. The implication is that it will not kill organisms visible to the naked eye.

Degree of Inactivation. The degree of inactivation by ultraviolet radiation is directly related to the UV dose applied to the water or wastewater. Dose is described as the product of the rate at which the energy is emitted (intensity) and the time the organism is exposed to the energy.

$$D = It \qquad (14\text{-}23)$$

where

D = dose, microwatt-seconds per cm²
I = irradiation, microwatts per cm²
t = time, seconds

Therefore the principal design factors for any UV system for disinfection are the intensity of radiant energy that is able to reach the organism and the time of exposure.

Germicidal Efficiency. It has been determined that the germicidal effect of ultraviolet rays is maximal at wavelengths between 250 and 265 nanometers. There is an abrupt decrease at 290–300 nm.[70,71] As a consequence, low-pressure mercury arc lamps have served as the ideal source of germicidal energy because their maximum energy output is at a wavelength of 253.7 nm.

Various investigations have shown a wide range of sensitivity of different microorganisms to ultraviolet energy. In 1959, Kawabata and Harada[72] reported the following contact times required to achieve a 99.9 percent kill (3-log reduction) at a fixed UV intensity for the following organisms:

E. coli	60 sec
Shigella	47 sec
S. Typhosa	49 sec
Streptococcus faecalis	165 sec
B. subtilis	240 sec
B. subtilis spores	369 sec

Ultraviolet radiation has also been shown to be effective in the inactivation of viruses. In 1965, Huff et al.[69] reported satisfactory results which included studies of several strains of polio virus, Echo 7, Coxsackie 9 viruses. The intensities varied from 7000 to 11,000 μW-sec/cm².

Hill et al. reported in 1971[73] that the infectivity loss of Polio 1 virus exceeded 3 logs after 15.7 seconds of UV exposure in continuous flowing seawater. Based upon these findings the use of UV should be highly effective in the disinfection of flowing seawater for use in artificial shellfish purification systems.

Owing to the long exposure times listed in the above (47–369 sec) investigations the data are misleading compared to what was known in 1984. The work done at St. Michaels, Maryland[61] used a 130-sec exposure and a correspondingly low

UV intensity to achieve the desired germicidal efficiency. Current designs are for high intensity and lower exposure times—6–10 seconds. There is no doubt that the germicidal efficiency of UV is predictable for a given species of organism on the basis of the UV intensity–exposure time product. In practice it will be necessary to prove and evaluate a given installation to confirm the design parameters.

There is no reference to the ability of ultraviolet radiation to destroy cysts. It is unlikely that UV would have any effect of this type of organism, as it is the consensus that UV radiation will not kill any organism which can be seen with the naked eye.

UV Dose Versus Bacterial Kill. This relationship is characterized by a mathematical model that assumes second-order kinetics when the coliform concentrations are in the range where disinfection usually takes place, as follows:

$$\frac{dN}{dt} = kN^2 I \qquad (14\text{-}24)$$

Integrated, this becomes

$$\frac{1}{N} - \frac{1}{N_0} = kIt \qquad (14\text{-}25)$$

where

N = coliform concentration, MPN/100 ml, at time t
N_0 = influent coliform concentration, MPN/100 ml
k = rate constant
I = the average ultraviolet intensity in the exposure chamber
t = exposure time

The influent coliform concentration is usually so much greater than the final concentration that the term $1/N_0$ becomes negligible and Eq. (14-25) can be simplified to:

$$\frac{1}{N} = kIt \qquad (14\text{-}26)$$

On the basis of 350 samplings conducted throughout a one year pilot program Scheible and Bassell[62] have been able to show quite favorable correlation between UV dose and coliform kill. Their data is illustrated by Fig. 14-8.

The mathematical model represented by Fig. 14-8 is:

$$\text{Effluent Fecal Coliform} = (1.26 \times 10^{13})\,(\text{UV dose})^{-2.27} \qquad (14\text{-}27)$$

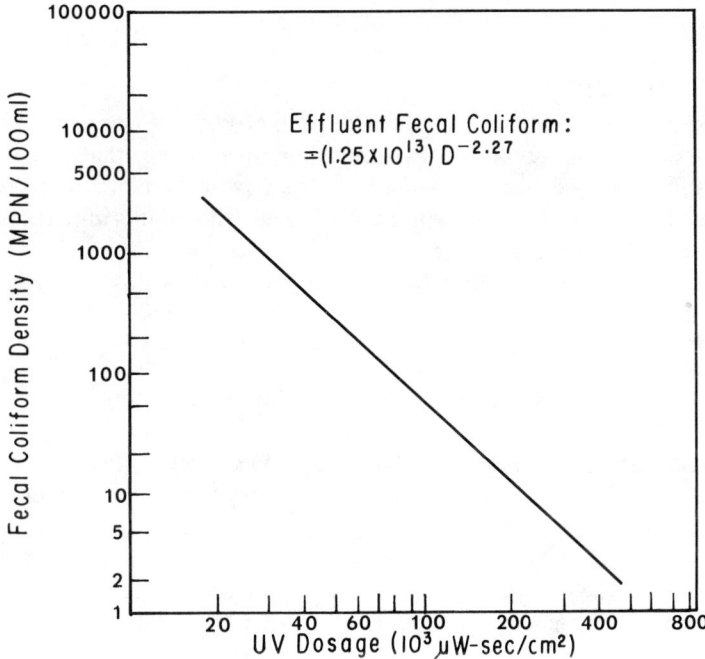

Fig. 14-8. UV dose versus fecal coliform destruction.[62]

To compute the U.V. dose required to achieve a fecal coliform concentration at time t, eq. 14-27 must be rearranged as follows:

$$(\text{UV dose})^{-2.27} = \frac{\text{Eff. FC}}{1.23 \times 10^{13}}$$

$$(\text{UV dose})^{+2.27} = \frac{1.23 \times 10^{10}}{\text{Eff. FC}}$$

$$[(\text{UV dose})^{+2.27}]^{0.4405} = \frac{(1.23 \times 10^{13})^{0.4405}}{(\text{Eff. FC})^{0.4405}}$$

$$\text{UV dose} = \frac{5.836 \times 10^5}{(\text{Eff. FC})^{0.4405}} \quad (14\text{-}28)$$

EXAMPLE. Compute UV dose to achieve 200/100 ml MPN fecal coliform using Eq. (14-28):

$$\text{UV dose} = \frac{5.836 \times 10^5}{(200)^{0.4405}}$$

$$= 56560 \ \mu \text{ W-sec./cm}^2$$

The dose used by Scheible and Bassell[62] was 60,000 μW-sec/cm². The above example is site specific because it is based upon an equation that is site specific. Eq. (14-28) was derived from a 10-month field study at the North Bergen County Wastewater Treatment Plant, Waldwick, NJ.[65] Nevertheless it is important because it demonstrates that a correlation between UV dose and organism destruction does exist. A more reliable and practical approach to the dose vs. bacterial kill is described under the heading "UV Facility Design" in this chapter.

Process Variables That Affect Disinfection Efficiency:

Ultraviolet Absorption. This may also be described as UV demand. There are various constitutents in wastewater which affect the rate of absorption of the ultraviolet energy as it passes through the wastewater in the disinfection chamber. UV energy is also absorbed within the reactor as it passes through quartz sleeves and teflon, and by the shadowing effect of the reactor walls. During the St. Michaels study[61] the absorption coefficient was found to vary from 0.052 to 0.722, a factor of 14, depending upon the effluent. The absorption coefficient was calculated for each observation from the equation

$$I/I_0 = e^{-K_a L} \tag{14-29}$$

where

I = radiation intensity at the bottom of disinfection chamber

I_0 = initial radiation intensity

K_a = absorption coefficient

L = distance in cm between points of measurement of I and I_0.

e = natural log base

At the North Bergen County Water Pollution Control Plant a different approach was used.[62] The investigators sought to find the average intensity of radiation based upon lamp design and configuration which also allowed for the effect of UV absorption as shown in the following equation:

$$I/I_0 = \frac{r}{r+L} e^{K_a L} \tag{14-30}$$

where

$$r = \text{radius of the lamp}$$
$$L = \text{distance from lamp surface}$$

The average absorbance coefficient (K_a sec^{-1}) was found to be 0.39 and the range over a 30-day period was from 0.22 to 0.55.

Each of these methods illustrates the process variable due to effluent quality and reactor design.

An entirely different approach was used by Scheible[63,64] at the Port Richmond Wastewater Treatment Plant, New York City investigation.

Effluent Quality. There are many factors that influence the efficiency of UV radiation energy all of which affect the absorption coefficient, i.e., which affect the transmittance efficiency. The St. Michaels study indicated the amount of COD might be a better indicator than turbidity. The turbidity ranged from 1.5 to 17 JTU and the COD from 15 to 65 mg/l. When the turbidity was 10 JTU and the COD 65 mg/l the absorption coefficient was at the highest level, 0.722; but at a turbidity of 17 JTU and a COD of 35 mg/l the absorption coefficient was only 0.356. This confuses the issue of turbidity as an effluent quality parameter.

Petrasek et al.[74] observed that both organic nitrogen and ammonia nitrogen interfered with the transmittance of UV radiation.

Johnson et al.[65,67] proved conclusively that filtration of a secondary effluent improved the disinfection efficiency significantly. Their study demonstrated that one of the chief factors limiting ideal UV radiation efficiency in practice was directly related to the protection of bacterial colonies inside sewage particles that represent suspended solids in the wastewater. This confirms the belief of Venosa[68] who summarized the experience gained from the comprehensive EPA research program that 20 mg/l suspended solids was the upper limit of practicality for UV as a disinfectant, and 10 mg/l SS was highly preferable. When compared to chlorine UV has much less penetrating power. Collins et al. found that suspended solids concentration as high as 100 mg/l did not interfere with the disinfection efficiency of chlorine.[77]

Photoreactivation. Until Kelner's report[75] in 1949 it was generally believed that when bacterial cells were exposed to UV light the DNA of these cells was permanently destroyed. Kelner reported partial reactivation of bacterial cells when exposed to visible light after absorbing UV light in lethal doses.

The damage incurred by UV absorption is the dimerization of thymine, one of the components of DNA. If this dimerization is not too extensive it can be reversed in some microorganisms by exposing the irradiated cells to visible light in the range of 330–500 nm.[59,63]

Scheible made a comprehensive analysis of photoreactivation effects at the North Bergen Co. plant scale investigation.[63] This was carried out for both total and fecal coliforms. In order to overcome the effects of photoreactivation based upon 200/100 ml MPN fecal coliform the UV dose would have to approximate 200×10^3 μW-sec/cm² as compared to 70×10^3 μW-sec/cm² for the nonphotoreactivated samples. This would represent a 460 percent increase in the system sizing and would have a direct effect upon the capital cost, which represents approximately 43 percent of the annual costs.

Toxicity of Radiation. Radiation technology has been used to destroy substances known to be responsible for foul tastes in potable water. Beta radiation is effective in the destruction of geosmin and phenols at a penetration distance of 10 cm; however, gamma radiation has been discovered to produce mutagenesis in the surviving viruses during potable-water and water-reuse treatment. Some of the side effects of UV on plants and animals include retardation of cell division, increase in the rate of mutation, erythema, chromosomal aberration, and changes in cellular viscosity.[59]

UV Facility Design:

Preface. The text under this heading must be separated into two principal categories: potable water and wastewater. This is the result of a long-term interest by the U.S. Public Health Service in the disinfection of water. The use of UV for wastewater disinfection to meet the EPA discharge standards is of recent origin (circa 1978) and therefore it is too soon to establish design standards. The new EPA Design Manual for UV Disinfection of Wastewater is expected to be published in 1985.

Potable Water. The following are excerpts from a policy statement on the use of the ultraviolet process by the U.S. Public Health Service, "Criteria for the Acceptability of an Ultra-Violet Disinfecting Unit."[76]

1. Ultraviolet radiation at a level of 253.7 nm must be applied at a minimum dosage of 16,000 microwatt-seconds per square centimeter at all points throughout the water disinfection chamber.
2. Maximum water depth in the chamber, measured from the tube surface to the chamber wall, shall not exceed three inches.
3. The ultraviolet tubes shall be:
    a. Jacketed so that a proper operating tube temperature of about 105°F is maintained, and
    b. the jacket shall be of quartz or high silica glass with similar optical characteristics.
4. A time delay mechanism shall be provided to permit a two-minute tube

warm-up period before water flows from the unit. One manufacturer recommends a 5-minute warm-up period.
5. The unit shall be designed to permit frequent mechanical cleaning of the water contact surface of the jacket without disassembly of the unit.
6. An automatic flow control valve, accurate within the expected pressure range, shall be installed to restrict flow to the maximum design flow of the treatment unit.
7. An accurately calibrated ultraviolet intensity meter, properly filtered to restrict its sensitivity to the disinfection spectrum, shall be installed in the wall of the disinfection chamber at the point of greatest water depth from the tube or tubes.
8. A flow diversion valve or automatic shut-off valve shall be installed which will permit flow into the potable water system only when at least the minimum ultra-violet dosage is applied. When power is not being supplied to the unit, the valve should be in a closed (fail-safe) position so as to prevent the flow of water into the potable system.
9. An automatic audible alarm shall be installed to warn of malfunction or impending shutdown if considered necessary by the Control or Regulatory agency.
10. The materials of construction shall not impart toxic materials into the water, either as a result of the presence of toxic constituents in the materials of construction or as a result of physical or chemical changes resulting from exposure to ultra-violet energy.
11. The unit shall be designed to protect the operator against electrical shock or excessive radiation.

As with any potable water treatment process, due consideration must be given to the reliability, economy, and competent operation of the disinfection process and related equipment, including:

1. Installation of the unit in a protected enclosure not subject to extremes of temperature which could cause malfunction
2. Provision of a spare ultraviolet tube and other necessary equipment to effect proper repair by qualified personnel properly instructed in the operation and maintenance of equipment
3. Frequent inspection of the unit and keeping a record of all operations, including maintenance problems

Of all the above items listed by the U.S. Public Health Service bulletin, probably the most important are: UV intensity meter and the fail-safe control.

One other item worthy of mention is personnel exposure to ultraviolet radiation. Workers in the area of ultraviolet radiation can suffer severe burns if not properly protected. Protection of the eyes is extremely important. Tolerance limits for exposure to ultraviolet have been set by the American Medical Association. Prolonged exposure can cause permanent damage to the eyesight.

The equipment manufacturers of ultraviolet purifiers recommend the following maintenance procedures:

1. Wipe the quartz jacket of the lamp at least once a month.
2. Let the ultraviolet lamps warm up for at least five minutes before allowing the use of treated water.
3. Lamps should be replaced when the intensity meter indicates less than 70 percent of the rated lamp intensity; in refrigerated or very cold water, replace at 50 percent of initial rate of intensity.
4. Sterilize the entire system, including the purifier and treated water system, prior to the use of the ultraviolet process.

Ultraviolet radiation has proved successful as a disinfectant under certain conditions. It can be used as such for small water supplies under the various requirements of the U.S. Public Health Service. It is far more expensive than chlorination, and so will find its application only under special circumstances that would not warrant the use of chlorine.

Wastewater. It is abundantly clear from the data collected by the EPA funded research projects ably reported by Scheible et al.[63,64] and Johnson and Qualls[66] that the manufacturers must take steps to provide the consumer with guaranteed "capacity" and "range" of operation for a prescribed accuracy. This would be akin to data sheets published by pump manufacturers which relate pump capacity versus total dynamic head, horsepower required, and calculated efficiencies.

Critical parameters for a UV reactor completely dependent upon manufacturer's design, type of lamp selection and configuration are: *average intensity,* the *dispersion coefficient,* and *hydraulic head loss.* The principal factors for an optimum design are described below.

1. *Intensity.* The rate of disinfection is directly related to the average intensity of the UV light. This is the rate at which the radiation energy is delivered to the wastewater by the UV source. This energy transmittance is dependent upon lamp selection (quartz-teflon, etc.) and lamp spacing. There is no practical way to measure the true intensity of a UV system in the field by the consumer. This must be done by the manufacturer of the UV reactor. Scheible et al.[65] have developed a mathematical model which calculates the intensity at any point within the UV reactor. This is used to estimate the average intensity emitted by any specific unit. The mathematical model described below is based upon the point source summation method (PSS) described by Jacob and Dranoff[78] and recently applied to UV disinfection systems by Johnson and Qualls.[66] The mathematical techniques rely upon the basic physical properties of the UV lamps, configuration of multilamp chambers, and assumed properties of the wastewater (absorbance coefficient).

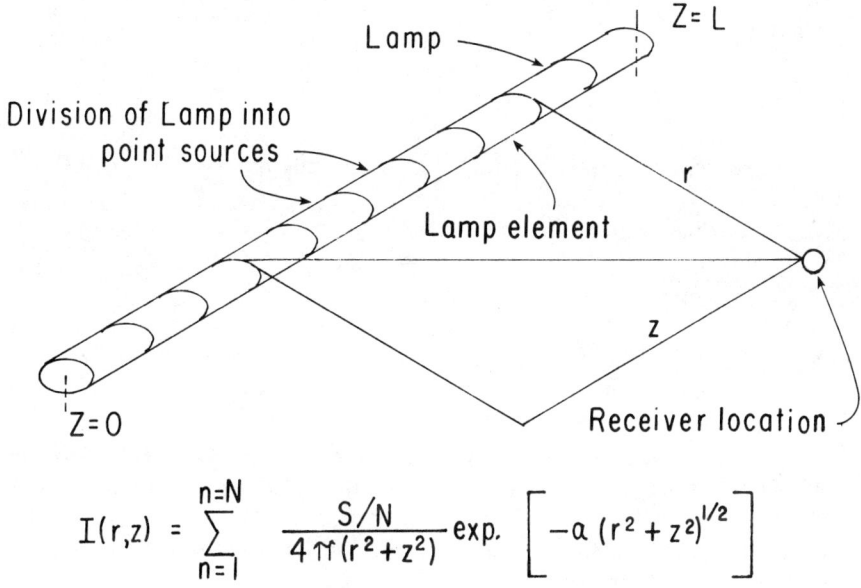

Fig. 14-9. Lamp geometry for point source intensity approximation.[65]

Scheible[64] describes the UV intensity at any given distance from a point source of energy as:

$$I = \left(\frac{S}{4\pi R^2}\right)e^{-aR} \qquad (14\text{-}31)$$

where:

I = intensity at a given point, Watts/cm²
S = UV energy output at source, Watts
a = absorbance coefficient, cm⁻¹ to the base e
R = distance from energy source to the point of receiver location

Figure 14-9 illustrates a tubular germicidal lamp which is treated as a series of point sources of energy. The UV intensity at a specific point is the sum of the intensities from the individual point sources. Scheible et al.[65] developed the following equation which represents UV intensity at a specific point (see Fig. 14-9) in the reactor:

$$I(r,z) = \sum_{n=1}^{n=N} \frac{S/N}{4\pi(r^2+z^2)} \exp[-a(r^2+z^2)^{1/2}] \qquad (14\text{-}32)$$

where

$$z = L\left(\frac{n-1}{N-1}\right)$$

N = the number of point sources into which the line source is divided.
L = lamp arc length
r,z = coordinates of the point receiver
a = absorbance coefficient of the wastewater, cm^{-1} to the base e
$R = \sqrt{r^2 + z^2}$

Equation (14-32) derives from an assumption that the point receiver located at (r,z) is spherical and infinitely small. Therefore energy radiating from any point source element of the lamp will strike the receiver normal to its surface.

To calculate the intensity at point (r,z), Eq. (14-32) must be solved N times and all the solutions summed. This mathematical exercise can be programmed into a computer or calculator. Then the average intensity of UV in a system can be estimated by averaging the computed intensities at a representative number of points in a cross-sectional plane of the UV reactor.

Equation (14-32) is capable of accounting for the absorption of energy as it passes through the wastewater, quartz sleeves, teflon, air, neighboring lamps, and reactor shadowing, for any lamp configuration where lamps are parallel to each other, any lamp rating output, any number of lamps per battery and any variation in energy output.

Figure 14-10 illustrates the calcualted values of average intensity for the two quartz systems installed at the Port Richmond WPCP, New York City.[64] The loss of efficiency displayed by the closely spaced unit is attributed to the shadowing effect of neighboring lamps. If lamp spacing allows UV energy to collide with a neighboring lamp it will be absorbed by that lamp.

Equation (14-32) demonstrates that the variables which directly affect the reactor intensity are: (1) UV source output, Watts; (2) a, absorption coefficient of wastewater (varies with quality); and (3) the distance R the energy will be required to travel through the wastewater. R will be a function of the lamp spacing which is governed by the absorbance characteristics of the wastewater to be treated.

2. *Hydraulic Characteristics of UV Reactor.* The efficiency of any UV reactor is intensely dependent upon the dispersion index (E_x), which is related directly to exposure time and lamp configuration (spacing). E_x can be determined in the same fashion as the dispersion index for a chlorine contact chamber—with a dye study (see Chapter 8). At ideal plug flow E_x is zero.

Assuming a one-dimensional forward (advective–dispersive) flow, Scheible[64] developed a mathematical expression that incorporates the dispersion index:

$$\frac{L}{L_0} = \exp\left[\left(\frac{ux}{2E_x}\right)\left\{1 - \left(1 + \frac{4kE_x}{u^2}\right)^{1/2}\right\}\right] \qquad (14\text{-}33)$$

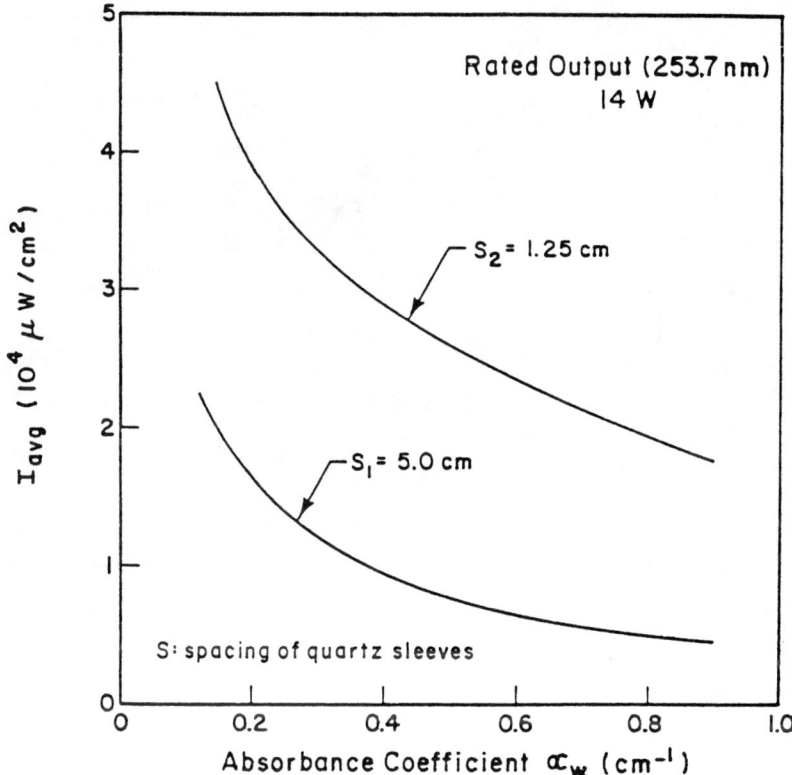

Fig. 14-10. Computed intensity for two different UV reactors at Point Richmond WPCP.[64]

where

E_x = dispersion coefficient, cm²/sec
L = concentration of organisms after UV exposure, MPN/100 ml
L_0 = concentration of organisms before UV exposure, MPN/100 ml
u = fluid velocity, cm/sec
x = forward distance traveled during exposure, cm
k = rate constant, sec⁻¹

Fig. 14-11 illustrates a series of solutions to Eq. (14-33), which demonstrate the sensitivity of a UV system disinfection efficiency to the dispersion coefficient.

Figure 14-12 illustrates the modern concept of advective–dispersive flow utilizing a lamp configuration to achieve the best possible dispersion coefficient at highest average intensity and minimum head loss through the reactor (6–8 inches). The longer the length x can be, the better the efficiency.

3. *Exposure Time* for a conventional modern system is the range of 10–30 sec.

1008 HANDBOOK OF CHLORINATION

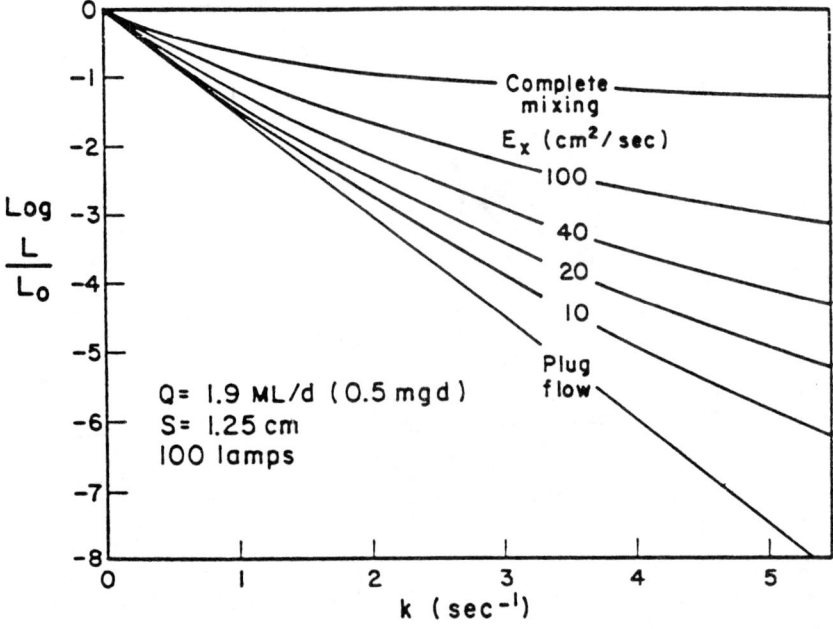

Fig. 14-11. Sensitivity of disinfection efficiency to the dispersion coefficient.[64]

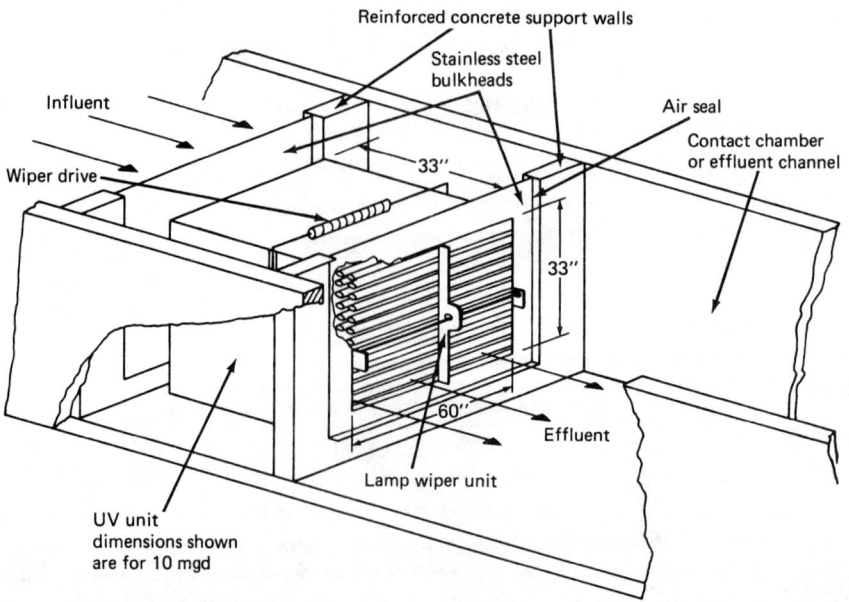

Fig. 14-12. Typical lamp configuration for a wastewater UV reactor where x = 33 in.

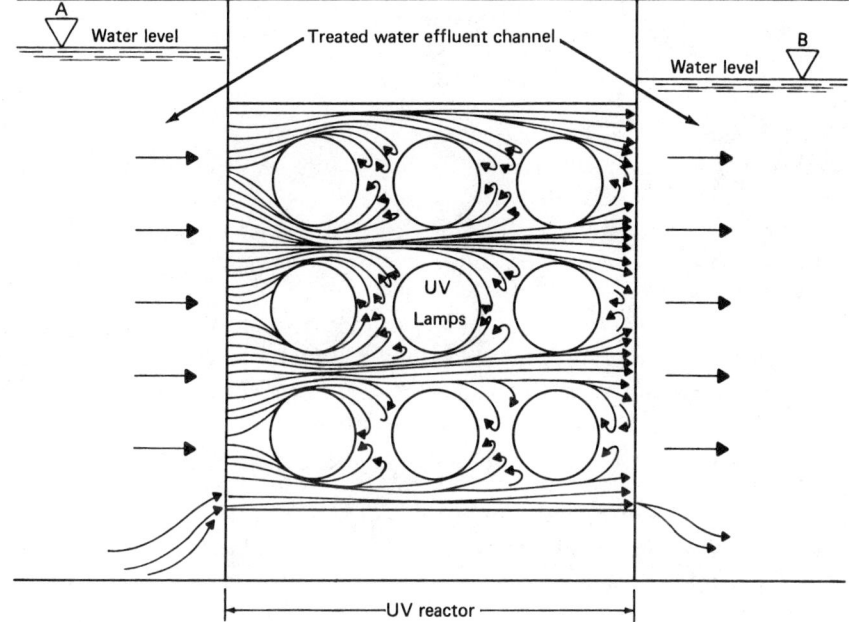

Note: Water level A-B = head loss through reactor

Fig. 14-13. Flow pattern showing turbulence through a typical UV reactor with perpendicular flow.

Exposure time is a function of lamp spacing. Closely spaced lamps have a high UV intensity. Teflon coated lamps require a longer exposure time because these lamps have a lower intensity.

Figure 14-13 illustrates the concept of turbulence through a typical UV reactor. The design of the reactor should strive for maximum turbulence at minimum head loss. Turbulence is stimulated by high velocities, closer lamp spacing, staggered lamp spacing, perpendicular flow to lamps and longer length.

Banks of lamps in series is preferred to banks in parallel.

4. *Wastewater Quality.* The single most damaging variable to UV efficiency is the concentration of suspended solids. Experience indicates that this parameter should be limited to 10–15 mg/l. Recently Scheible has modified Eq. (14-33) to account for this variable as follows:[79]

$$\frac{L}{L_0} = \exp\left[\left(\frac{ux}{2E_x}\right)\left\{1 - \sqrt{1 + \left(\frac{4kE_x}{u^2}\right)}\right\}\right] + C(SS)^2 \quad (14\text{-}34)$$

where SS is suspended solids, mg/l and

$$C = 0.25 \text{ for fecal coliforms}$$
$$C = 0.9 \text{ for total coliform}$$
$$k = 0.0000145 I^{1.3*}$$

This mathematical model was developed during the calibration of the Point Richmond UV reactors. It is now a predictive tool sensitive to all the process variables.

A total wastewater system should be designed to accommodate the combination of peak daily flow, maximum suspended solids concentration, and the peak daily absorbance coefficient.

5. *UV Lamps.* After having made a material selection and an arc length commitment, the next consideration is the provision for lamp aging. Fig. 14–14 illustrates the lamp output decay versus operating time. The serviceable life for germicidal lamps is usually accepted as 8000 hours. The lamp chosen in Fig. 14-14 has a 14.3 Watt nominal output when new. At 8000 hours the output deteriorates to about 5.8 Watts, so the deterioration is $5.8/14.3 = 0.41$. Therefore the original output of a given lamp deteriorates by a factor of 242 percent. This means that to satisfy the loss of energy over the life (8000 hours) of any lamp, the number of lamps required at time zero will have to be increased by a factor of 2.42. Ozone emission is another consideration. Use of lamps that produce excessive amounts of ozone should be avoided.

6. *Provisions for Cleaning.* Biofouling of the UV lamp surface in direct contact with the wastewater is a serious and never-ending maintenance problem with UV reactors. This is best studied by seeking operating experience from an existing installation. There are four categories of cleaning techniques: mechanical wiper, ultrasonics, high-pressure wash, and chemical cleaning with either acid, caustic, or detergent. This is a problem that must be dealt with regardless of cost.

7. *Control System.* The UV system must be equipped with controls capable of switching banks of lamps on or off to achieve the necessary dose proportional to flow. It is also desirable to have a slave lamp operating to check wastewater quality absorbance. Other features should include the ability to monitor lamp output with time, in-place intensity monitors, power meters, lamp operation indicators, elapsed time monitors, lamp temperature indicators, ballast panel temperature indicators, and appropriate alarms to assist the operators.

8. *Maintenance.* The reactor should be equipped with a system drain and have the ability to isolate modules. There should also be complete redundancy so that the system efficiency will not be impaired because of either routine or emergency maintenance. Lamps and ballasts must be accessible. An inventory of lamps, quartz sheaths, and ballasts should be maintained. Records should be kept of lamp use, lamp life, and replacement cycles for both lamps and ballasts.

9. *Safety Considerations.* Operators should be instructed about the dangers of UV. They should be provided with proper goggles and clothing to protect themselves

---

* Where I = computed intensity.

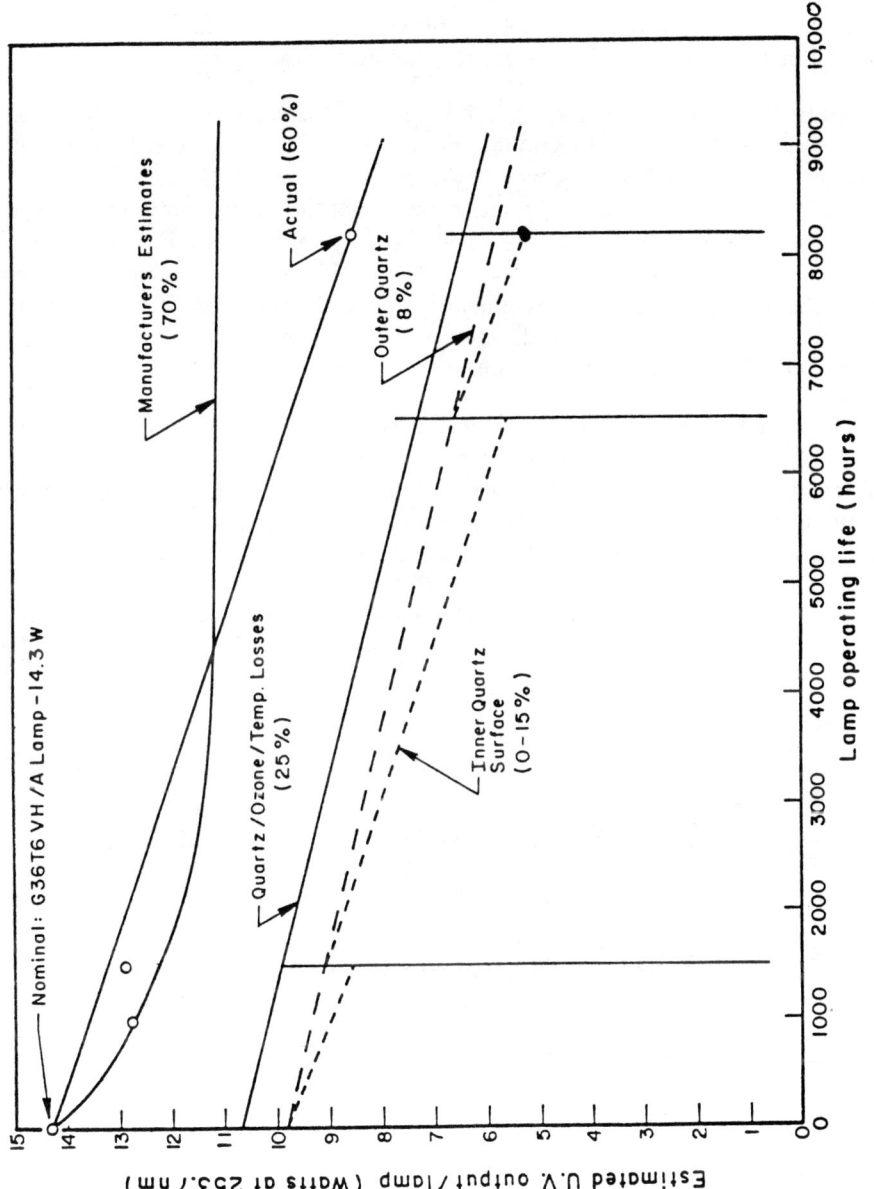

Fig. 14-14. Estimated UV output per lamp versus operating life.[64]

from UV radiation. The proximity of the water to electricity is so close that extra precautions *must* be taken. Safety equipment such as automatic power shut-off for hazardous circumstances should be provided. Proper storage and means of disposal of UV supplies should be provided in every system.

Reliability. The performance of a UV reactor is completely dependent upon the power supply and the functioning of the lamps. Therefore a redundant but separate power supply is necessary.

Since lamp failures require equipment shut down for repairs, provisions should be made to activate a redundant module while lamp replacement or other maintenance is in progress. There is at this time (1984) insufficient operating experience to conclude how much reactor redundancy should be provided. If a UV system is to be compared to a chlorine disinfection system complete redundancy should be provided, i.e., complete duplication of the UV reactor.

Cost. Scheible and Bassell[64] developed costs over a range of wastewater flows of 1–100 mgd. The following is an example for a 10 mgd plant which requires a 100-kW UV reactor system. The costs are based upon a peak to average power requirement ratio of 2.1; 50 percent replication at peak, 1.25 cm spacing; average absorption coefficient of 0.5 cm^{-1}; an NPDES requirement of 200/100 ml MPN fecal coliforms (max. 30 day mean); and EPA index = 330.

For 100 kW, total estimated capital costs are $720,000. This includes UV reactor, cleaning system, power supply and switch gear, instrumentation and controls, contact chamber structure, housing and ancillary equipment plus installation and start-up costs, and engineering contingencies:

Annual costs are:

Labor	10,400
Materials (Lamps, ballasts, etc.)	43,000
Power—416,000 kWh at $0.04545/kWh	19,700
Total O & M Costs	73,100

Assuming 20 years at 6 percent, capital recovery factor = 0.087, then amortized cost of capital investment = $60,800, which brings total annual cost to $134,000 or 3.6 cents/1000 gal.

SUMMARY

Bromine. *Bromine* is considered the most reactive oxidant of all the halogens, except fluorine which is not considered here. For this reason it has some desirable chemical characteristics. The hydrolysis of bromine in an aqueous solution is almost identical to chlorine. Bromine added to water reacts to form hypobromous acid and in the presence of ammonia nitrogen in the quantities usually found in wastewater bromamines are formed. The germicidal efficiency of these bromamines in a pH environment similar *to that of wastewater is practically equal to that of free chlorine.*

Bromine is highly reactive; therefore, its halogen demand is greatly distorted beyond its usefulness. So any system in wastewater treatment attempting to exploit the features of bromine should be arranged to use chlorine to minimize the "bromine demand."

Bromine as Br_2 is a difficult and hazardous chemical to handle because it is in liquid form at room temperature. This led to the development of bromine chloride which has the same vapor pressure characteristics as sulfur dioxide. This makes BrCl much more desirable than elemental bromine from the standpoint of chemical handling facilities.

Bromine Chloride. Bromine chloride has been investigated thoroughly as a practical substitute for elemental bromine and liquid-gas chlorine installations. Those investigations have shown conclusively that a new species of equipment will be required for the metering and control of BrCl. The factors of increased corrosivity of BrCl over chlorine and the delicate problem of BrCl dissociation have determined the necessity of fundamental design changes.

On-site generation of bromine from bromide salts using chlorine as the oxidizer has interesting possibilities. Injecting a bromide salt into a chlorinator solution discharge line (similar to the generation of chlorine dioxide) produces the formation of elemental bromine which hydrolyzes immediately to hypobromous acid. This concept can be used to great advantage at existing chlorination installations for the following reasons: the existing chlorination system can be arranged to produce a 2–5 min controlled chlorine residual of 0.3–0.5 mg/liter (instead of the usual 5–7 mg/liter). This sequesters the halogen demand. Following this is the formation of elemental bromine from the combination of the chlorine solution discharge with the bromide salt injection from the second chlorinator application. This becomes the second point of "chlorine" application downstream from the original point of chlorination. This system has many desirable features: it can utilize existing chlorination facilities; it eliminates the hazardous problems of handling bromine; and it can utilize the superior germicidal properties of bromine without sacrificing other design considerations. Moreover, the bromamine residual die-away phenomenon may be so rapid that debromination by SO_2 may not be required, and the contact time for these "hot" bromine residuals can be as low as 5 min and not longer than 15 min. This would bring contact chamber requirements to the lowest recommended for ozone which are the lowest for all disinfectants.

Iodine. Iodine has been used as a temporary expedient for the disinfection of small water supplies and as an alternative to chlorine in the treatment of swimming pools.

It has never been tried even on a pilot-plant experimental basis for the disinfection of wastewater. The cost and reliability of iodine supply makes it an impossible choice as a disinfectant for wastewaters or waters for reuse.

Ultraviolet Radiation. UV has significant disinfection qualities. This process has been used with success to *sterilize* potable water where the level of bacterial concentration is highly sensitive. This includes such requirements as occur in hospitals, pharmaceutical and cosmetic production and the beverage industry.

Use of UV as the primary disinfectant for potable water supplies has achieved little if any success. It has inherent liabilities. The most prominent are: lack of a persisting residual, and a low level of equipment reliability in the absence of continuous supervision. Other obstacles are the lack of equipment performance standards and the inability to translate water quality variables into reactor designs.

The use of UV as a primary disinfectant to meet EPA wastewater discharge requirements is enjoying considerable success. This is the result of EPA funded research under plant scale conditions. The use of UV disinfection for tertiary effluents appears to be most promising. The use of UV with ozone in these situations also has considerable merit. (See Chapter 13.)

REFERENCES

1. Mills, J. F., "Interhalogens and Halogen Mixtures as Disinfectants," Chapter 6 in J. D. Johnson (ed.), *Disinfection: Water and Wastewater,* Ann Arbor Science, Ann Arbor, MI, 1975.
2. Riley, J. P., and Skirrow, G., *Chemical Oceanography,* Vol. I, p. 36, Academic Press, New York, 1965.
3. "Bromine, Its Properties and Uses," Michigan Chem. Corp., Chicago, IL, 1958.
4. "Bromine Outlook Tied to Clean Air Rules," *Chem. and Engr. News,* p. 11, (Feb. 25, 1974).
5. Galal-Gorchev, Hend, and Morris, J. C., "Formation and Stability of Bromamide and Nitrogen Tribromide in Aqueous Solution," *Inorganic Chem.,* **4,** 899 (1965).
6. Johnson, D. J., and Overby, R., "Bromine and Bromamine Disinfection Chemistry," *ASCE J. San. Eng. Div.,* p. 617 (Oct. 1971).
7. La Pointe, T. F. Inman, G., and Johnson, J. D., "Kinetics of Tribromamine Decomposition," Chapter 15 in J. D. Johnson (ed.), *Disinfection: Water and Wastewater,* Ann Arbor Science, Ann Arbor, MI, 1975.
8. U.S. Patent 2,443,429, Procedure for Disinfecting Aqueous Liquid; Marks and Strandkov, and Wallace and Tiernan Co., Inc. Belleville, NJ, June 15, 1948.
9. Morrow, J. J., U.S. Patent No. 3,4.3,199 to Fischer and Porter Co., Warminster, PA, Nov. 26, 1968.
10. Mills, J. F., assignor to Dow Chemical Co. "Control of Microorganisms With Polyhalide Resins," U.S. Patent 3,462,363, Aug. 19, 1969.
11. Mills, J. F., Goodenough, R. D., and Nekervis, W. F., assignors to Dow Chemical Co., "Process for Treating Water With Bromine," U.S. Patent No. 3,316,173, April 25, 1967.
12. Goodenough, R. D., Mills, J. F., and Place, J., "Anion Exchange Resin (Polybromide Form) as a Source of Active Bromine for Water Disinfection," *Env. Sci. Tech.,* **3,** 354 (Sept. 1969).
13. Regunathan, P., and Brejcha, R. J., "A New Technology in Potable Water Disinfection for Offshore Rigs," paper presented Annual Mtg. Soc. of Petroleum Engrs. of AIME, Dallas, TX, Sept. 28– Oct. 1, 1975.
14. Brown, J. R., McLean, D. M., and Nixon, M. C. "Bromine Disinfection of a Large Swimming Pool," *Can. J. Public Health,* **55,** 251 (June 1964).

15. Derreumaux, A., "Process and Apparatus for Disinfecting or Sterilizing by the Combination of Chlorine and Other Halogens," French patent No. 2,171,890 (1973).
16. Sollo, F. W., Mueller, H. F., and Larson, T. E., "Prechlorination Enhances Disinfection by Bromine," paper presented at the 7th Int. Conf. on Water Poll. Res., Paris, France, Sept. 9–12, 1974.
17. Venosa, A. D., "Disinfection: State of the Art, Alternatives to Chlorination for Wastewater," paper presented at a Disinfection Seminar, Dept. of Ecology, Univ. of Washington, Seattle, WA, May 26–27, 1976.
18. Mills, J. F., and Oakes, B. D., "Bromine Chloride Less Corrosive than Bromine," *Chem. Eng.*, pp. 102–106 (Aug. 1973).
19. Mills, J. F., "Disinfection of Sewage by Chlorobromination," paper presented at Ann. Chem. Soc. Meeting, Dallas, TX, April 1973.
20. Cole, L. G., and Elverum, G. W., "Thermodynamic Properties of the Diatomic Interhalogens from Spectroscopic Data," *J. Chem. Phys.*, **20**, 1543 (1952).
21. Jolles, Z. E., "Bromine and Its Compounds," Ernest Benn Ltd., London, England, Chapter 1, 1966.
22. Ward, R. W., Giffin, R. D., De Graeve, G. M., and Stone, R. A., "Disinfection Efficiency and Residual Toxicity of Several Wastewater Disinfectants, Vol. I, Grandville, Mich., EPA Pre-Publication Report, Project No. S-802292 (1976).
23. Collins, H. F., Selleck, R. E., and White, G. C. "Problems in Obtaining Adequate Sewage Disinfection," *ASCE J. Env. Eng. Div.*, **97**, 549 (Oct. 1971).
24. Johnson, J. D., and Sun, W., "Bromine Disinfection of Wastewater," Chapter 9 in J. D. Johnson (ed.), *Disinfection: Water and Wastewater*, Ann Arbor Science, Ann Arbor, MI, 1975.
25. Bongers, L. H., O'Connor, T. P., and Burton, D. T., "Bromine chloride—An Alterative to Chlorine for Fouling Control in Condenser Cooling Systems," EPA Report No. 600/7-77-053, National Technical Information Service, Springfield, VA, May 1977.
26. Cole, S. A., "Experimental Equipment for Feeding Bromine Chloride," paper presented at the International Water Conference, Pittsburgh, PA, Oct. 31, 1974.
27. Task Force Report, "Disinfection of Wastewater, EPA-430/9-75-012," U.S. Env. Prot. Agency, Cincinnati, OH, March 1976.
28. Johnson, J. D. (Ed.), Disinfection: Water and Wastewater, Ann Arbor Science, Ann Arbor, MI, 1975.
29. Wyss, O., and Stockton, R. J., "The Germicidal Action of Bromine," *Arch. Biochem.*, **12**, 267 (1947).
30. Marks, H. C., and Strandkov, F. B., "Halogens and Their Mode of Action," *Ann. NY Acad. Sci.*, **53**, 163 (1950).
31. Johannesson, J. K., "Anomalous Bactericidal Action of Bromamine," *Nature*, **181**, 1799 (1958).
32. McKee, J. E., Brokaw, C. J., and McLaughlin, R. T., "Chemical and Colicidal Effects of Halogen in Sewage," *J. WPCF*, **32**, 795 (1960).
33. Koski, T. A., Stuart, L. S., and Ortenzio, L. F., "Comparison of Chlorine, Bromine, and Iodine as Disinfectants for Swimming Pool Water," *Appl. Micro.*, p. 276 (March, 1966).
34. Schaffer, R. B., and Mills, J. F., "Proceedings of the National Symposium on Quality Standards for Natural Waters," Univ. of Mich. Press, Ann Arbor, MI, p. 158 (1966).
35. Sollo, F. W., Mueller, H. F., Larson, T. E., and Johnson, J. D., "Bromine Disinfection of Wastewater Effluent," Chapter 8 in J. D. Johnson (ed.), *Disinfection: Water & Wastewater*, Ann Arbor Press, Ann Arbor, MI, 1975.
36. Kott, Y., "Effect of Halogens on Algae—III Field Experiment," *Water Research*, p. 265 (1969).
37. Derreumaux, A., CIFEC, Neuilly-sur-Seine, France, private communication, 1976.
38. Lange, A. L. and Kawczynski, Eliz., "Controlling Organics: The Contra Costa County Water District Experience," *J. AWWA*, **70**, 653 (Nov. 1978).
39. Mills, J. F., "A Spectrophotometric Method for Determining Microquantities of Various Halogen Species," Dow Chemical Co., Midland, MI, 1971.

40. Larson, T., and Sollo, F. W., "Determination of Free Bromine in Water," Annual Progress Report, U.S. Army Medical R & D Command, Contract No. DA-49-193-MD-2909 (1967).
41. Palin, A. T., "Analytical Control of Water Disinfection With Special Reference to Differential DPD Methods for Chlorine, Chlorine Dioxide, Bromine, Iodine and Ozone," *J. Inst. Water Eng.*, **28**, 139 (1974).
42. Anon., *Handbook of Chemistry and Physics*, 44th ed., pp. 1732–1744, Chemical Rubber Co., Cleveland, OH, 1962.
43. "Mineral Industry Surveys," U.S. Dept. of Interior Bureau of Mines, Washington, DC, 1975.
44. Lawrence, C. A., and Block, S. S., *Disinfection, Sterilization and Preservation*, Lea and Febiger, Philadelphia, PA, 1968.
45. Vergnoux, "Examinen rapide et sterilization des eaux pour les troupes en campagne," *L'Union Pharmaceutique* (France), pp. 194–201 (1915).
46. Chang, S. L., "Iodination of Water," U.S. Dept. H.E.W., Taft San. Engr. Center, Cincinnati, OH, 1966.
47. Pond, M. A., and Willard, W. R., "Emergency Iodine Sterilization for Small Samples of Drinking Water," *J. AWWA*, **29**, 1995 (1937).
48. Chang, S. L., and Morris, J. C., "Elemental Iodine as a Disinfectant for Drinking Water," *Ind. and Eng. Chem.*, **45**, 1009 (1953).
49. Morris, J. C., Chang, S. L., Fair, G. M., and Conant, G. H., Jr., "Disinfection of Drinking Water under Field Conditions," *Ind. and Eng. Chem.*, **45**, 1013 (1953).
50. Chang, S. L., "The Use of Iodine as a Water Disinfectant," *J. Am. Pharm. Assoc. Sc. Ed.*, **47**, 417 (1958).
51. Morgan, D. P., and Karpen, R. J., "Test of Chronic Toxicity of Iodine as Related to the Purification of Water," *U.S. Armed Forces Med. J.*, **4**, 725 (1953).
52. Black, A. P., Kinman, R. N., Thomas, W. C., Jr., Freund, G., and Bird, E. D., "Use of Iodine for Disinfection," *J. AWWA*, **57**, 1401 (1965).
53. Freund, G., Thomas, W. C., Jr., Bird, E. D., Kinman, R. N., and Black, A. P., "Effect of Iodinated Water Supplies on Thyroid Function," *J. Clin. Endocr.*, **26**, 619 (1966).
54. Wyss, O., and Strandkov, F. B., "The Germicidal Action of Iodine," *Arch. Biochem.*, **6**, 261 (1945).
55. Chambers, C. W., Kabler, P. W., Malaney, G., and Bryant, A., "Iodine as a Bactericide," *Soap and San. Chem.*, **28**, 149 (1952).
56. Clark, N. A., Berg, G., Kabler, P. W., and Chang, S. L., "Human Enteric Viruses in Water: Source, Survival and Removability," Int. Conf. Water Poll Res. London, Pergamon Press, 1962.
57. Gershenfeld, L., and Whitlin, B., "Free Halogens—A Comparative Study of Their Efficiencies as Bactericidal Agents," *Am. Jour. Pharm.*, **121**, 95 (1949).
58. Marks, H. C., and Strandkov, F. B., "Halogens and Their Mode of Action," *Ann. NY Acad Sci.*, **53**, 163 (1950).
59. Lawrence, C. A., and Block, S. S., *Disinfection, Sterilization and Preservation*, Lea and Febiger, Philadelphia, PA, 1968.
60. Witherell, L. E., Solomon, R. L., and Stone, K. M., "Ozone and Ultraviolet Radiation Disinfection for Small Community Water Systems," Municipal Env. Res. Lab, U.S. EPA report No. EPA 600/2–79–060, Cincinnati, OH, July 1979.
61. Roeber, J. A., and Hoot, F. M., "Ultraviolet Disinfection of Activated Sludge Effluent Discharging to Shellfish Waters," EPA 600/2–75–060, Municipal Environmental Research Lab., Cincinnati, OH, Dec. 1975.
62. Scheible, O. K., and Bassell, C. D., "Ultraviolet Disinfection of a Secondary Wastewater Treatment Plant," EPA Municipal Env. Res. Lab report EPA-600/S2–81–152, Cincinnati, OH, Sept. 1981.
63. Scheible, O. K., Forndran, A., and Leo, W. M., "Pilot Investigation of Ultraviolet Wastewater Disinfection at the New York City Port Richmond Plant," paper presented at Second National Symposium of Municipal Wastewater Disinfection, in Symposium Proceedings, EPA/600–9–83–009, Cincinnati, OH, July 1983.

64. Scheible, O. K., "Design and Operation of UV Systems," Presented at Ann. Conf. WPCF Atlanta, GA, Oct. 1983.
65. Johnson, J. D., Qualls, R. G., Aldrich, K. H., and Flynn, M. P., "UV Disinfection of Secondary Effluent: Dose Measurement and Filtration Effects," Second Ann. seminar, Wastewater Disinfection, sponsored by EPA, Orlando, FL, Jan. 26–28, 1982.
66. Qualls, R. G., and Johnson, J. D., "Bioassay and Dose Measurement in UV Disinfection," *Applied and Environmental Microbiol.*, p. 872 (March 1983).
67. Qualls, R. G., Flynn, M. P., and Johnson, J. D., *The Role of Suspended Particles in Ultraviolet Disinfection*, Univ. North Carolina, Chapel Hill, NC, 1983.
68. Venosa, A. D., private communication, Director, Wastewater Disinfection Research Program, May 1984.
69. Huff, C. B., Smith, B. S., Boring, W. D., and Clark, N. A., "Study of Ultraviolet Disinfection of Water and Factors in Treatment Efficiency," *Public Health Reports,* **80,** 695 (Aug. 1965).
70. Luckiesh, M., and Holladay, L. L., "Disinfection of Water by Means of Germicidal Lamps," *General Electric Review,* p. 45 (Apr. 1944).
71. Loofbourow, J. R., "The Effects of Ultraviolet Radiation on Cells," *Growth,* **12,** 77 (1948).
72. Kawabata, T., and Harada, T., "The Disinfection of Water by the Germicidal Lamp," *J. Illumination Soc.,* **36,** 89 (1959).
73. Hill, W. F. Jr., Akin, E. W., Benton, W. H., and Hambley, F. E., "Viral Disinfection of Estuarine Water by UV," *J. San. Engr. Div., ASCE,* p. 601 (Oct. 1971).
74. Petrasek, A. C., Jr., Andrews, D. C., and Wolf, H. W. "Ultraviolet Disinfection of Wastewater Effluents," paper presented at the Disinfection Seminar, AWWA Ann. Conference Anaheim, CA, May 8, 1977.
75. Kelner, A., "Effect of Visible Light on the Recovery of *Streptomyces griseus* conidia from Ultraviolet Irradiation Injury," *Proc. Nat. Acad. of Sci. U.S.,* **35,** 73 (1949).
76. Anon., "Policy Statement on the Use of Ultra-Violet Process for Disinfection of Water," Dept. of H.E.W., Div. of Env. Engr. and Food Prot., Washington, DC, Apr. 1, 1967.
77. Selleck, R. E., private communication, Univ. of California, Berkeley, CA, June 1984.
78. Jacob, S. M., and Dranoff, J. S., "Light Intensity Profiles in a Perfectly Mixed Photoreactor," *J. AIChE,* **16,** 359 (May 1970).
79. Scheible, O. K., "UV Disinfection," paper presented at the Disinfection of Water and Wastewater Seminar, University of Wisconsin Ext. Div., Milwaukee, WI, May 16–18, 1984.

Appendix

FRICTION LOSS FACTORS AND PIPE DATA

The nomograph in Fig. 15 allows a direct determination of the equivalent length of pipe for all appurtenances from the K factor. A list of K factors is shown in Table 1.

Table 1. Values of K for Determining "Equivalent Length of Pipe" from the Nomograph on Page 1051

Resistance of Valves and Fittings*

Type of Resistance	K	Type of Resistance	K
90° Ell std.	0.9	Outlet loss:	
90° Ell flanged	0.25	pipe to atmosphere	1.0
Ditto-long radius	0.2	main connection type	
		chlorine diffuser	1.0
Tee – std:	1.8	Valves in open position:	
	1.8	angle valve	5
	0.6	globe valve	10
45° Ell screwed	0.4	gate valve	0.2
Long radius flanged	0.2	plug valve	0.77
Sudden enlargement:		stem-type chlorine	
		gas valve	10
$d/D - 1/4$	0.92	diaphragm valve	2.3
$d/D - 1/2$	0.56	ball valve	0.8
$d/D - 3/4$	0.19	ordinary entrance	0.5
Sudden contraction:			
$d/D - 1/4$	0.42		
$d/D - 1/2$	0.33		
$d/D - 3/4$	0.19		

* Values given are based on consensus of Crane Company catalog #52; Bulletin 252, Univ. of Wisconsin; Giesecke & Badgett; "Hydraulics" by Daugherty; and Tentative Standards Hydraulic Institute.

The nomograph is by S. Labella* and was derived as follows:

$$Hf = K \frac{V^2}{2G} \quad \text{Darcy Equation} \tag{1}$$

---

* Reprinted from p. R-238 and R-239, 1967 WATER AND SEWAGE WORKS Reference Number.

Solving the Hazen Williams equation for Hf

$$Hf = \left[\frac{VL^{0.54}(4)^{0.64}}{131.8(d)^{0.63}}\right]^{1.85} \quad (2)$$

Substitute $V = Q/A$, where Q = discharge in gpm into equations (1) and (2) and equating (1) to (2), solve for L, which becomes:

$$L = \frac{K(Q^{0.15})d^{0.87}}{0.81}$$

Example: Find the equivalent length of 6" $C = 100$, in pipe for a valve with $K = 1.0$ and an average flow of 300 gpm.
Solution: Connect K and the diameter (6); find the point on the turning line. Connect the turning line point to the discharge (300). Read the answer, *14 ft*, on the length axis.

Table 2. Pipe Specifications

Nominal Diameter	Inside Diameter	A (in.²)	Volume (ft³/ft) Area (ft²)
Schedule 80 Steel for Gas and Liquid Chlorine Supply Lines			
¾	0.742	0.4324	0.00300
1	0.957	0.7193	0.00499
1¼	1.278	1.2813	0.00891
1½	1.500	1.767	0.01225
2	1.939	2.953	0.02050
Schedule 80 PVC for Chlorine Solution Piping and Injector Vacuum Lines			
½	0.546	0.2341	0.00163
¾	0.742	0.4324	0.00300
1	0.957	0.7193	0.00499
1¼	1.278	1.2813	0.00891
1½	1.500	1.767	0.01225
2	1.939	2.953	0.02050
2½	2.323	4.298	0.02942
3	2.900	6.605	0.04587
4	3.826	11.50	0.07986
6	5.761	26.07	0.1810
8	7.625	45.66	0.3171

Table 3. Friction Loss in PVC Pipe Schedule 80 (for Chlorine Solution Lines)

Velocity measured in ft/sec, Loss in feet of water head per 100 ft of pipe

Gal. min	1/2" Vel.	1/2" Loss	3/4" Vel.	3/4" Loss	1" Vel.	1" Loss	1 1/4" Vel.	1 1/4" Loss	1 1/2" Vel.	1 1/2" Loss	2" Vel.	2" Loss	2 1/2" Vel.	2 1/2" Loss	3" Vel.	3" Loss	3 1/2" Vel.	3 1/2" Loss	4" Vel.	4" Loss	Gal. min
2	2.74	6.72	1.48	1.51																	2
4	5.48	24.2	2.97	5.45	1.79	1.54															4
6	8.23	51.2	4.45	11.5	2.68	3.34	1.00	.39													6
8	11.0	86.9	5.94	19.6	3.57	5.69	1.50	.82	.73	.177											8
10	13.7	132.0	7.42	29.6	4.46	8.60	2.00	1.39	1.09	.375											10
12			8.91	41.5	5.36	12.0	2.50	2.10	1.45	.64	.65	.107									12
15			11.1	62.7	6.7	22.9	3.00	2.94	1.82	.96	.87	.183									15
18			13.4	87.9	8.03	25.5	3.76	4.45	2.18	1.35	1.09	.276	.61	.077							18
20	5" PIPE		14.8	107	8.92	30.9	4.50	6.25	2.72	2.04	1.30	.387	.76	.115							20
25					11.2	58.8	5.00	7.57	3.27	2.86	1.63	.585	.91	.161							25
30	.53	.025			13.4	65.3	6.25	11.4	3.63	3.47	1.96	.818	1.14	.243							30
35	.62	.034			15.6	86.9	7.50	16.0	4.55	5.25	2.17	.996	1.36	.340							35
40	.71	.043	6" PIPE		17.9	111	8.75	21.3	5.45	7.38	2.71	1.51	1.51	.414							40
45	.795	.054					10.0	27.3	6.38	9.78	3.26	2.11	1.9	.625							45
50	.88	.065	.62	.027			11.2	33.9	7.26	12.5	3.80	2.81	2.27	.874	.97	.039					50
55	.973	.078	.676	.032			12.5	41.3	8.26	15.6	4.35	3.59	2.65	1.16							55
60	1.06	.091	.74	.039			13.7	49.2	9.08	18.9	4.89	4.46	3.03	1.49	1.21	.055					60
65	1.15	.106	.80	.044			15.0	57.8	10.00	22.0	5.43	5.41	3.41	1.86	1.44	.083					65
70	1.23	.121	.86	.051			16.1	67.0	10.9	26.5	5.98	6.44	3.79	2.25	1.70	.116					70
75	1.33	.138	.923	.057					11.8	30.7	6.52	7.61	4.16	2.68	1.94	.140	.54	.035			75
80	1.41	.155	.98	.065			17.5	77.1	12.7	35.3	7.06	8.84	4.54	3.16	2.18	.212	.65	.056	.56	.037	80
85	1.50	.174	1.04	.072			18.8	87.4	13.6	40.1	7.61	10.1	4.92	3.66	2.42	.297	.72	.068	.695	.055	85
90	1.59	.193	1.11	.080			20.0	98.2	14.5	45.2	8.15	11.5	5.30	4.20	2.67	.396	.90	.103	.84	.077	90
95	1.67	.213	1.20	.089			21.2	110	15.4	50.3	8.69	12.9	5.68	4.79	2.92		1.08	.145	.973	.103	95
100	1.76	.234	1.23	.098			22.5	122	16.3	55.9	9.03	14.5	6.05	5.36	3.14	.507	1.26	.192	1.12	.132	100
110	1.95	.279	1.36	.117					17.2	62.0	9.78	16.1	6.43	6.02	1.25	.629	1.44	.246	1.25	.164	110
120	2.11	.329	1.48	.137	8" PIPE				18.2	68.2	10.3	17.8	6.81	6.53	1.94	.766	1.63	.306	1.40	.199	120
130	2.3	.381	1.60	.159					20.0	81.3	10.9	19.6	7.19	7.38	2.18	.912	1.80	.372	1.53	.237	130
140	2.47	.437	1.72	.182	.98	.047			21.8	95.4	12.0	23.4	7.57	8.13	2.42	1.07	1.99	.443	1.67	.279	140
150	2.65	.496	1.85	.207	1.05	.054			23.6	111	13.0	27.4	8.33	9.68	2.67	1.25	2.17	.522	1.81	.323	150
160	2.82	.559	1.97	.234	1.12	.059			25.4	127	14.1	31.8	9.08	11.4	2.92		2.35	.604	1.95	.371	160
170	3.0	.626	2.08	.261	1.19	.067					15.2	36.5	9.84	13.2	3.14		2.53	.691	2.08	.421	170
180	3.16	.696	2.22	.290	1.26	.074					16.3	41.5	10.6	15.1	3.39	1.43	2.70	.787	2.23	.475	180
190	3.36	.769	2.34	.321	1.33	.082					17.4	46.7	12.1	19.4	3.64	1.62	2.89	.888	2.34	.531	190
200	3.52	.846	2.46	.353	1.41	.090					18.5	52.2	12.9	21.7	3.88	1.83	3.05	.992	2.51	.592	200
220	3.88	1.01	2.71	.421	1.55	.108					19.6	58.3	13.6	24.1	4.10	2.04	3.25	1.10	2.64	.652	220
											20.6	64.4	14.4	26.6	4.33	2.27	3.42	1.21			
											21.7	70.5	15.1	29.3	4.57	2.51					
											23.9	84.1	16.7	34.9							
															4.85	2.76	3.67	1.34	2.79	.719	
															5.33	3.29	3.97	1.60	3.07	.855	
															5.80	3.87	4.33	1.88	3.35	1.00	
															6.30	4.48	4.69	2.18	3.63	1.16	
															6.80	5.12	5.05	2.50	3.91	1.33	
															7.27	5.87	5.41	2.84	4.19	1.52	
															7.75	6.58	5.78	3.20	4.47	1.71	
															8.20	7.37	6.14	3.58	4.75	1.91	
															8.60	8.18	6.50	3.97	5.02	2.12	
															9.20	9.05	6.85	4.39	5.30	2.35	
															9.70	9.96	7.22	4.84	5.58	2.58	
															10.6	11.9	7.94	5.78	6.14	3.08	

240	4.23	1.18	2.96	.484	1.69	.126	26.1	98.7	18.2	41.0	11.6	13.9	8.66	6.77	6.70	3.62	240
260	4.58	1.37	3.20	.573	1.83	.147	28.3	115	19.7	47.5	12.6	16.2	9.38	7.85	7.26	4.19	260
280	4.94	1.57	3.45	.658	1.97	.168			21.2	54.5	13.5	18.6	10.1	9.02	7.82	4.79	280
300	5.29	1.79	3.69	.747	2.11	.191			22.7	62.0	14.4	21.1	10.8	10.2	8.38	5.45	300
320	5.64	2.01	3.94	.841	2.24	.215			24.2	69.9	15.5	23.7	11.5	11.5	8.94	6.16	320
340	5.99	2.26	4.19	.940	2.39	.240			25.8	78.2	16.3	26.6	12.3	12.9	9.50	6.91	340
360	6.35	2.51	4.43	1.05	2.64	.261			27.2	86.9	17.4	29.5	13.0	14.3	10.0	7.66	360
380	6.70	2.77	4.68	1.16	2.68	.295			28.8	96.1	18.6	32.6	13.7	15.8	10.6	8.46	380
400	7.05	3.05	4.93	1.27	2.81	.325			30.3	106	19.4	35.9	14.4	17.4	11.2	9.31	400
450	7.95	3.79	5.54	1.58	3.16	.404					21.8	44.6	16.2	21.6	12.5	11.6	450
500	8.82	4.61	6.16	1.92	3.51	.493					23.2	54.1	18.1	26.3	14.0	14.1	500
550	9.70	5.50	6.77	2.29	3.86	.587					26.5	64.9	19.9	31.4	15.3	16.8	550

1022 HANDBOOK OF CHLORINATION

Table 4. Head-Loss Calculations for Saunders-Type Diaphragm Valves

Liquid flow formula:

$$C_v = Q\sqrt{\frac{G}{\Delta P}} \qquad (1)$$

Sizing formula from Fluid Controls Institute, Standard FCI 62-1:

$$Q = C_v\sqrt{\frac{\Delta P}{G}} \qquad (2)$$

$$\Delta P = G\left(\frac{Q}{C_v}\right)^2 \qquad (3)$$

Where: C_v = valve sizing constant
Q = flow in gallons per minute (U.S. gpm)
G = specific gravity of liquid (water = 1.0)
P = pressure drop $(P_1 - P_2)$ in psi

Valve Sizing Constant C_v

Valve Size	Rubber Lined	Plastic Lined	Solid Plastic
3/4"	9.0	7.0	10.0
1"	14	19.5	16.5
1¼"	24	19.5	19.2
1½"	39	40	30
2"	57	60	54
2½"	101	84	—
3"	180	140	110
4"	350	310	—
6"	890	580	—
8"	1670	1250	—

Example: 10 gpm through 1¼" rubber lined valve
Solution: From Table, C_v = 24
$P = (Q/C_v)^2 = (10/24)^2 = 0.174$ psi
0.174 psi × 2.31 = 0.4 ft H$_2$O head loss

CONTACT CHAMBER EVALUATION METHODS

Dispersion Index "d". An example of the computation of the Chemical Engineering Dispersion Index Number d where C = fluorometer units (dye concentration), and t = residence time (min.).*

t	C	$t \times C$	$t^2 \times C$
40	0	—	—
45	10	450	20,250
50	49	2450	122,500
55	78	4290	235,950
60	72	4320	259,200
65	61	3965	257,725
70	50	3500	245,000
75	37	2775	208,125
80	21	1680	134,400
85	15	1275	108,375
90	9	810	72,900
95	5	475	45,125
100	4	400	40,000
105	2	210	22,050
Sums	413	26,600	1,771,600

$$T = V/Q = 72.3 \text{ min.}$$

$$t_g = \frac{\Sigma tC}{\Sigma C} = \frac{26,600}{413} = 64.4 \text{ min.}$$

---

* *Courtesy* Endel Sepp California State Dept. of Health.

$$t_g^2 = 4148.2$$

$$\sigma_t^2 = \frac{\Sigma t^2 C}{\Sigma C} - t_g^2 = \frac{1,771,600}{413} - 4148.2$$

$$= 4289.6 - 4148.2 = 141.4$$

$$\sigma^2 = \sigma_t^2/t_g^2 = \frac{141.4}{4148.2} = 0.034$$

$$d = \sigma^2/2 = \frac{0.034}{2} = 0.017$$

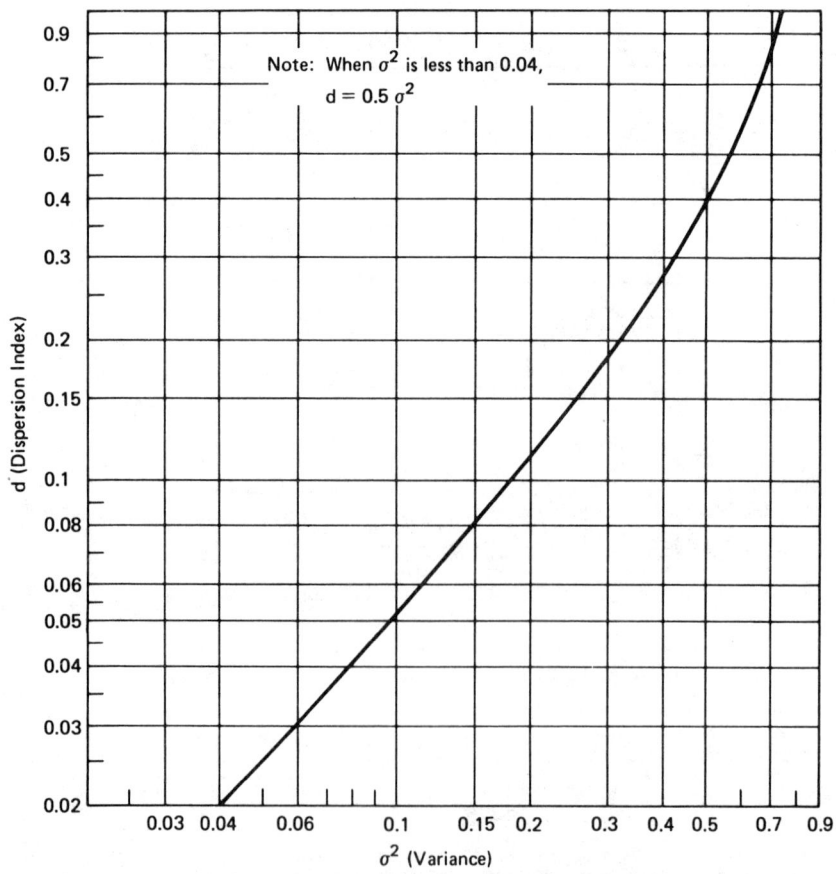

σ^2 (variance) versus "d" the chemical engineering index.

t_g = ave. contact time = time to centroid of curve.

Calculator Solution. The following is the keystroke sequence for a HP 21 calculator to determine the value of d for any given value of σ^2 based on the formula:

$$\sigma^2 = 2d - 2d^2(1 - e^{-1/d})$$

First assume $d = \tfrac{1}{2}\sigma^2$, then by trial and error find d by selecting values of d on either side of $\tfrac{1}{2}\sigma^2$, as follows:

$DSP,\ 3;\ d = $ approximately $0.5\sigma^2$

ENTER	ENTER	ENTER
$1/x$	CHS	e^x
1	$x \leftrightarrows y$	$-$
$x \rightleftarrows y$	ENTER	2
/////	y^x	2
$\times$	$\times$	$x \rightleftarrows y$
2	$\times$	$x \rightleftarrows y$
$-$	Read σ^2 value on the register	

Repeat this process until the value of d is within 0.005 for a given value of σ^2.

Morrill Index. An example of the Morrill Index computation, where

$t = $ residence time (min.)
$C = $ fluorometer units (dye concentration)

1026 HANDBOOK OF CHLORINATION

t	C	Cumulative C	$\dfrac{C}{\Sigma C}$ (percent)
0	0	—	—
20	10	10	6
25	15	25	14
30	25	50	28
35	30	80	44
40	60	140	78
45	30	170	94
50	10	180[a]	100

[a] $\Sigma C = 180$

The values of $C/\Sigma C$ are plotted against time on the probability scale.

The *Morrill Index* is the ratio of the time required for the passage of 90 percent of the dye to the time of passage of 10 percent of the dye.

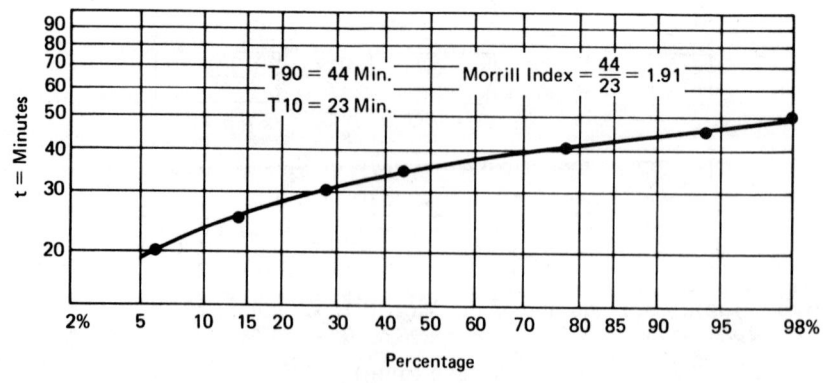

Morrill Index.

CONVERSION FACTORS

U.S. (English System)

1 GPM	= 1440 GPD
1 GPM	= 0.0022 cfs
1 gal.	= 0.832 Imperial gal.
1 gal.	= 0.1337 ft^3
1 ft^3	= 7.48 gal.
1 mgd	= 1.547 cfs
1 mgd	= 694.44 GPM
1 cfs	= 0.646 mgd
1 cfs	= 448.8 GPM
1 day	= 1440 min
1 day	= 86,400 sec
1 hp	= 550 ft-lb per sec
1 psi	= 2.04 in. Hg at 60°F
1 psi	= 2.31 ft H$_2$O

U.S. to Metric

1 in.	= 2.54 cm
1 ft	= 0.3048 meters
1 in.^{3}	= 16.387 cm^3
1 ft^3	= 0.0283 cu meters
1 oz	= 0.0295 liters
1 gal.	= 3.8754 liters
1 gal.	= 0.003785 cu meters
1 lb (mass)	= 0.45359 kilograms
1 oz	= 28.349 grams
1 watt hour	= 3600 joules
1 hp	= 746 W
1 lb (force)	= 4.448 Newtons
1 GPM	= 0.63 liters per sec
1 cfs	= 2446.6 cu meters per day
1 cfs	= 0.02832 cu meters per sec
1 mgd	= 3785.4 cu meters per day
1 psi	= 6.893 Kilo-Pascals
1 ppm (by wt.) × S.G.	= milligrams per liter

METRIC TO U.S. (ENGLISH SYSTEM)

To convert from metric to U.S. take the reciprocal of the metric values.

For example:
To convert from cubic meters to gal. multiply cubic meters by 1/0.003785.

$$\therefore 1 \text{ cu meter} \times \frac{1}{0.003785} = 264.2 \text{ gal.}$$

Temperature
Fahrenheit $= 1.8°C + 32$
Centigrade $= 5/9(°F - 32)$
Kelvin $= °C + 273.15$
Kelvin $= 5/9(°F + 459.67)$

Other Calculator Solutions. The following are the keystroke sequences for a HP 21 calculator for two often used equations in this text.

$$\Delta P \text{ (in. Hg)} = \frac{L \times 5.83 \times f \times W^2 \times 2.04}{10^9 \times \rho \times d^5} \qquad (3\text{-}1)^*$$

Solving for ΔP:

L | Enter ↑ | 5.83 | × | f | × | W | Enter ↑ | × | × | 2.04 | ×

EEX | 9 | ÷ | ρ | ÷ | d | Enter ↑ | 5 | ▨ | y^x | ÷ | ----→ ΔP

and

$$y/y_0 = [1 + 0.23\ ct]^{-3} \qquad (2\text{-}11)^\dagger$$

Solving for ct:

y_0 | Enter ↑ | y | ÷ | 3 | 1/x | ▨ | y^x | 1 | − | 0.23 | ÷ | ----→ ct

---

* Equation 3-1 is shown on page 665 this text.
† Equation 2-11 is in this text.

CONTROL SYSTEMS DEFINITIONS

Common terminology used in describing automatic control logic and functions are as follows:

Algorithm. This is a sequence of calculations performed to obtain a given result.
  Example. A proportional controller uses an algorithm. The controller output is $m = K_c E$, where K_c = proportional gain and E = error.

Analog. This is a representation of numerical values by means of physical variables such as voltage, current, resistance, rotation, etc.
  Example. The chlorine residual span of a continuous analyzer is 0–5 mg/l. This is represented as a 4–20 mA transmitter output signal where 4 mA = 0 mg/l and 20 mA = 5 mg/l.

Proportional Band. This describes the type of control action caused by a change in the process variable. The ability of a controller to produce a small or large output change (i.e., valve movement) for a small process variable change is described as a narrow or a wide proportional band adjustment. A wide band results from low gain and a narrow band from high gain.*

  The gain, or proportional band, is adjustable at the controller, but must be such that there will always be some finite error to produce a correction.

Reset Action (Integral). Proportional controllers used in chlorine residual control loops should be provided with reset action. Reset action combined with proportional band is known as PI—for proportional plus integral—because the reset action corresponds to a time integration or summation of the error.

  Reset action in a controller responds to both the duration and the magnitude of the error signal. Its output feeds into the proportional unit. This output increases in proportion to the error signal and the length of time the error signal exists. Whether the error signal is positive or negative, reset action will act upon the proportional controller until the error signal is zero. It supplements proportional action so as to wipe out the error signal. The error signal cannot be eliminated by proportional band alone, as described above.

Derivative Action. This mode of control action is often added to PI controllers. When this is done it becomes a PID controller. This mode of control has been called "rate action." It corresponds to the derivative of the error signal. Mathematician Klaf defines a derivative as: "the instantaneous rate of change of a function."

  Derivative action supplements the corrective action of proportional control by augmenting the signal by an amount proportional to the rate of change of the error signal. This is why it is sometimes referred to as "rate action."

---
* Gain is the numerical reciprocal of proportional band.

Summary. A PID control loop contains the following three important modes of control action. Each action can be adjusted to site-specific conditions or characteristics.

1. Proportional action with adjustable gain to obtain stability.
2. Reset action to compensate for load changes which wipes out the resulting error signal.
3. Derivative action (rate control) which speeds up control action when rapid load changes occur.

WATER VISCOSITY

Water viscosity is expressed in English units as absolute viscosity = lb-sec/ft² as follows:

$$2.735 \times 10^{-5} \text{ at } 50°F$$
$$2.05 \times 10^{-5} \text{ at } 70°F$$

The symbol for absolute viscosity is μ.

Water viscosity is also expressed as kinematic viscosity in English units as ft²/sec as follows:

$$1.41 \times 10^{-5} \text{ at } 50°F$$
$$1.06 \times 10^{-5} \text{ at } 70°F$$

The symbol for absolute viscosity is ν, where

$$\nu = \frac{\mu}{62.4}$$

Calculator (HP25) solution for velocity gradient G, where:

$$G = \sqrt{\frac{3 \times 550}{2.735 \times 10^{-5} \times 77.37}}$$

$\mu = 2.735 \times 10^{-5}$ lb-sec/ft²
hp = 3
$V = 77.37$ ft³

Solution:

3 | ENTER | 550 | X | 2.735 | ENTER | 5 | CHS | g · 10^x

X | ÷ | 77.37 | ÷ | $f \sqrt{x}$ = 883 sec^{-1}

THE EGAN JET DIFFUSER CONCEPT

The design of this type of jet mixing is based upon the notion that sufficient mixing of chlorine solution occurs when the velocity gradient G approximates 500.

The following are the calculations for the Egan diffuser retrofit described in Chapter 8 (see Figures 8-16 and 8-17).

Velocity Gradient G:

$$G = \left(\frac{550 \times P(\text{hp})}{\mu(\text{lb} - \text{sec}/\text{ft}^2) \times V(\text{ft}^3)} \right)^{1/2} \tag{8-11}$$

where:

$P =$ the power being dissipated in the mixing zone

$$= \frac{Q p h}{60 \times 550 \times \text{eff}}$$

with:

$Q =$ total injector discharge flow, gal/min
$p =$ water density, lb/gal
$h =$ head loss through diffuser, ft
$60 =$ sec/min
$550 =$ ft-lb/sec/hp
eff $=$ system efficiency

Continuing with Eq. (8-11):

$u =$ absolute fluid viscosity, lb-sec/ft^2
$V =$ process water volume in mixing zone, ft^3. (In this instance it is the distance between baffles B and $C = 3.34$ ft.)

Velocity Gradient of Jet System. Assume 3-¾" holes, where $Q = 67$ and flow per hole is approx. 22 gpm. This creates a diffuser head loss of approx. 4.03 ft. Therefore:

$$P = \frac{67 \times 4.03}{3960 \times 1.0} = 0.068 \text{ hp}$$

$$G = \left(\frac{550 \times 0.068}{2.5 \times 10^{-5} \times 41.8} \right)^{1/2} = 189 \text{ sec}^{-1}$$

Velocity Gradient of Baffles. Assume baffle A provides desired water surface level only—does not provide any mixing energy.

The mixing zone in this particular case is the volume contained between baffles B and C. Effluent flow is 10 mgd = 6944 gpm. This calculates to a velocity = 1.23 ft/sec, so $V^2/2\, gh = 0.023$ ft.

Assume efficiency of baffle energy generation to be 50 percent because the jet stream is centered in the upper half of the 4-ft RCP. Said another way, the cone of influence of baffle mixing extends to about half of the volume between baffles B and C. So

$$P = \frac{6944 \text{ gpm} \times 8.34 \text{ lb/ft}^3 \times 0.023 \text{ ft}}{60 \text{ sec} \times 550 \text{ ft-lb/min} \times 0.5}$$

$$= 0.08 \text{ hp}$$

and therefore

$$G = \left(\frac{550 \times 0.08}{2.5 \times 10^{-5} \times 41.8}\right)^{1/2}$$

$$= 205 \text{ sec}^{-1}$$

Velocity Gradient of Contraflow Effect. In order to perceive the mixing effect of contraflow, i.e., the diffuse jet opposing the process water flow, we must bear in mind that mixing results from turbulence, and turbulence is the manifestation of hydraulic energy loss. There are no proven mathematical relationships available to quantify this effect. However, Egan has offered the following approach by using what he calls "free-wheeling" assumptions.*

Referring to Fig. 8-17, the premise is that the diffuser jet stream and the opening created by baffle B act as orifices imparting jet stream energy in opposite directions. Some of the energy created will be cancelled out due to dissipation by the opposing velocity head energy. The remaining balance will be available for mixing energy. If the diffuser jet stream were directed downstream instead of upstream this differential energy would not be generated for mixing.

Egan defines these opposing energies as follows: (see Fig. 16): Assume the effect of the vertical component (V_B) is negligible. At baffle C, $V_B = 2.45$ ft/sec and $V = 16.1$ ft/sec.

Converting these values to velocity heads:

$$h_B = 0.068 \text{ ft}$$
$$h_J = 4.03$$

---

*See Ref. 96, Chapter 8.

Convert velocity heads to energy assuming time $t =$ one second for flash mixing duration.

Pounds of water imparted by diffusion jet:

$$67 \text{ gpm} \times \frac{1 \text{ min}}{60 \text{ sec}} \times 8.34 \text{ lb/gal} = 9.3 \text{ lb/sec}$$

Pounds of water imparted by baffle:

$$6944 \text{ gpm} \times \frac{1 \text{ min}}{60 \text{ sec}} \times 8.34 \text{ lb/gal} = 965.2 \text{ lb/sec}$$

Determine energy of both streams:

$$\text{energy} = \text{velocity head} \times \text{lb water/sec.}$$

Diffuser jet energy:

$$E_J = 4.03 \text{ ft} \times 9.31 \text{ lb/sec} = 37 \text{ ft-lb/sec}$$

Baffle energy:

$$E_B = 0.068 \text{ ft} \times 965.2 \text{ lb/sec} = 65.63 \text{ ft-lb/sec}$$

Net energy available for mixing:

$$\begin{aligned} E_M &- E_B + E_J \\ &= 65.63 \text{ ft-lb/sec} + (-37.52 \text{ ft-lb/sec}) \\ &= 28.11 \text{ ft-lb/sec} \end{aligned}$$

Convert to hp:

$$\frac{28.11}{550} = 0.051 \text{ hp}$$

$$G_C = \sqrt{\frac{550 \times 0.051}{2/5 \times 10^{-5} \times 20.9*}} = 232 \text{ sec}^{-1}$$

In the case of any diffuser acting as an obstruction to flow in a pipe or channel, will dissipate energy. This energy loss is used for mixing and can be calculated as additional velocity gradient.

---

* Use restricted flash mixing zone as volume

DEBYE—HUCKEL EQUATION CONSTANTS*

$$f_x = \text{antilog} \frac{AZ_x^2(I)^{1/2}}{1 + B\, a_x^0 (I)^{1/2}}$$

Temp. °C	Unit Volume of Solvent	
	A	B
0	0.4918	0.3248
5	0.4952	0.3256
10	0.4989	0.3264
15	0.5028	0.3273
20	0.5070	0.3282
25	0.5115	0.3291
30	0.5161	0.3301
35	0.5211	0.3312
40	0.5262	0.3323
45	0.5317	0.3334

In Chapter 4, Eq. (4-9a), the ionic radii for H^+ = 9.0 and for OCl^- — 4.0.

f_x = activity coefficient of the ionic species x
Z_x = ionic charge of the ionic species x
A = a constant which depends upon absolute temperature and dielectric constant of solvent
B = a constant which depends upon the absolute temperature and dielectric constant of the solvent
a_x^0 = approximate effective radius of ionic species x in aqueous solution (Å)
I = ionic strength of the solution (mol/l)
Å = angstroms

The values for unit weight of solvent (molality scale) can be obtained by multiplying the corresponding value for unit volume by the square root of the density of water at the approximate temperature.

* *Source:* Electric Power Research Inst., *Dechlorination Technology Manual,* EPRI, CS-3748, Palo Alto, CA, Nov. 1984.

GLOSSARY

Å	angstrom = 1×10^{-10} meters = 1×10^{-7} millimeters
AWT	advanced wastewater treatment
BOD	biochemical oxygen demand
B-P	breakpoint
Btu	British thermal units
CAC	combined available chlorine
CRC	combined residual chlorine
DPD	diethyl-*p*-phenylenediamine
DNA	deoxyribonucleic acid
DO	dissolved oxygen
FAC	free available chlorine
FACTS	free available chlorine test—syrigaldazine
FAS	ferrous ammonium sulfate
FRC	free residual chlorine
GAC	granular activated carbon
JTU	Jackson turbidity units
MPN	most probable number
μA	microamp = 0.001 milliamps
mA	milliamp = 0.001 amperes
micron	0.001 millimeters
μg	microgram = 0.001 milligram
mg	milligram = 0.001 grams
MLVSS	mixed liquor volatile suspended solids
mV	millivolts
nm	nanometer = 1×10^{-9} meter; nano = one billionth
NPDES	national pollutant discharge elimination system
ORP	oxidation-reduction potential (Redox)
OT	orthotolidine
OTA	orthotolidine arsenite
PAC	powdered activated carbon
PDWF	peak dry weather flow
PFU	plaque forming units
PSS	point summation sources
RAS	return activated sludge
RCP	reinforced concrete pipe
SNORT	stabilized neutral orthotolidine residual test
SPC	standard plate count
SS	suspended solids
TOC	total organic carbon
TRC	total residual chlorine
THMs	trihalomethanes
TTHM	total trihalomethane
WPCP	water pollution control plant
WWTP	wastewater treatment plant

Cubic Equation. The following cubic equation was used to generate the data shown in Tables 4-1 and 4-4 (Chapter 4). These data illustrate the molecular chlorine, hypochlorous acid, and hypochlorite ion equilibria under various conditions of concentration, temperature, and pH.

$$\left\{-\left(\frac{1}{K_D}\right)^2(H^+)^3 - K_A\left(\frac{1}{K_D}\right)^2(H^+)^2 + 4\left(\frac{1}{K_D}\right)K_3([H^+]+K_A)^2\right\}[HOCL]^3$$

$$\left\{-4\left(\frac{1}{K_D}\right)K_3T[H^+]([H^+]+K_A)\right\}[HOCL]^2$$

$$\left\{+\left(\frac{1}{K_D}\right)T[H^+]^2 + K_A + [H^+] + \left(\frac{1}{K_D}\right)K_3T^2[H^+]^2\right\}[HOCL] - T[H^+] = 0$$

Where

T = total halogen added
K_D = hydrolysis of Cl_2 at 25°C = 3.944×10^{-4}
K_A = acid dissociation of HOCL = 2.904×10^{-8}
K_3 = formation of trichloride ion = 0.191

REDOX POTENTIALS

Table 5. Oxidation-Reduction Potentials (E_0) at 25°C for Titrable Species of Halogens and Ozone

Reaction	Potential in Volts (E_0)
$O_3 + 2H^+ + 2e = O_2 + H_2O$	2.07
$HOCl + H^+ + 2e = Cl^- + H_2O$	1.49
$Cl_2 + 2e = 2Cl^-$	1.36
$HOBr + H^+ = 2e = Br^- + H_2O$	1.33
$O_3 + H_2O + 2e = O_2 + 2OH^-$	1.24
$ClO_2 + e = ClO_2^-$	1.15
$Br_2 + 2e = 2Br^-$	1.07
$HOI + H^+ + 2e = I^- + H_2O$	0.99
$ClO_2\ (aq) + e + ClO_2^-$	0.95
$OCl^- + H_2O + 2e = Cl^- + 2\ OH^-$	0.90
$OBr^- + H_2O + 2e = Br^- + OH^-$	0.70
$I_2 + 2e = 2I^-$	0.54
$I_3 + 2e = 3I^-$	0.53
$OI^- + H_2O + 2e = I^- + 2\ OH^-$	0.49

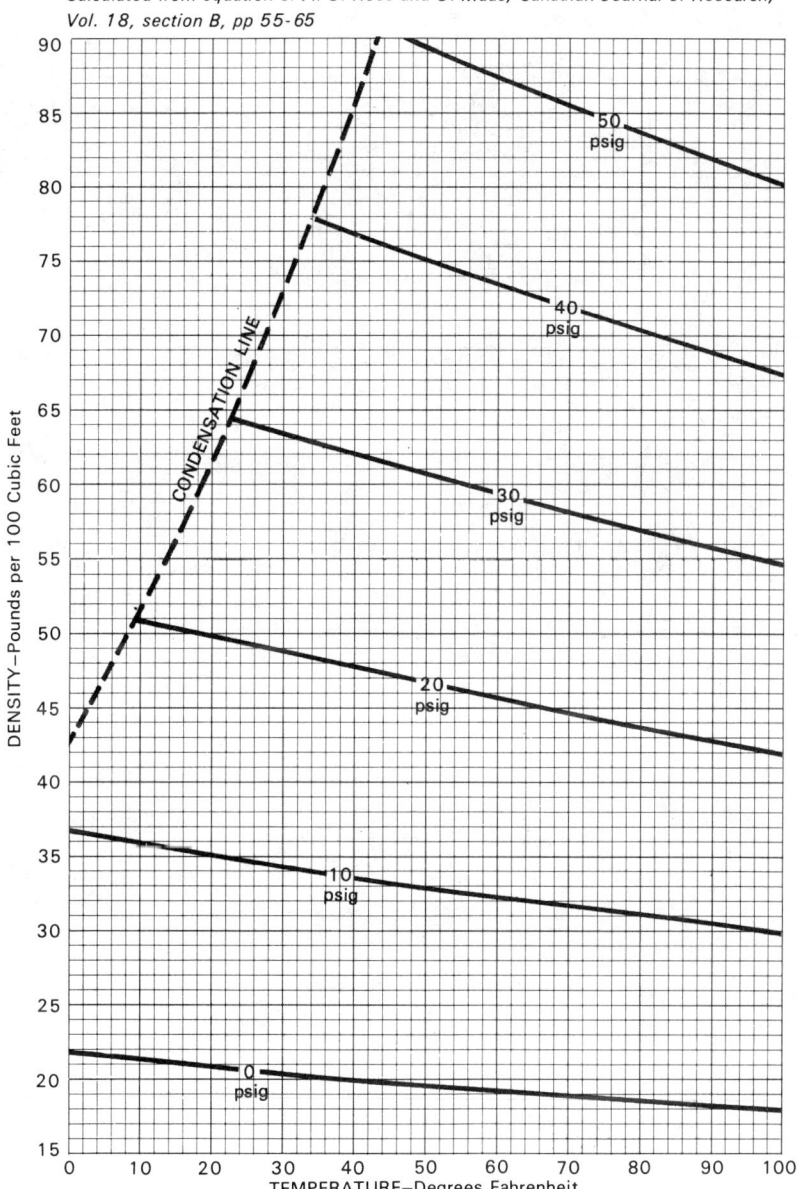

Fig. 1. Density of chlorine gas under pressure. Calculated from formula by A. S. Ross and O. Mass, "The Density of Gaseous Chlorine," Can. J. Res. 18B, 55–65 (1940).

Chlorine Gas Density

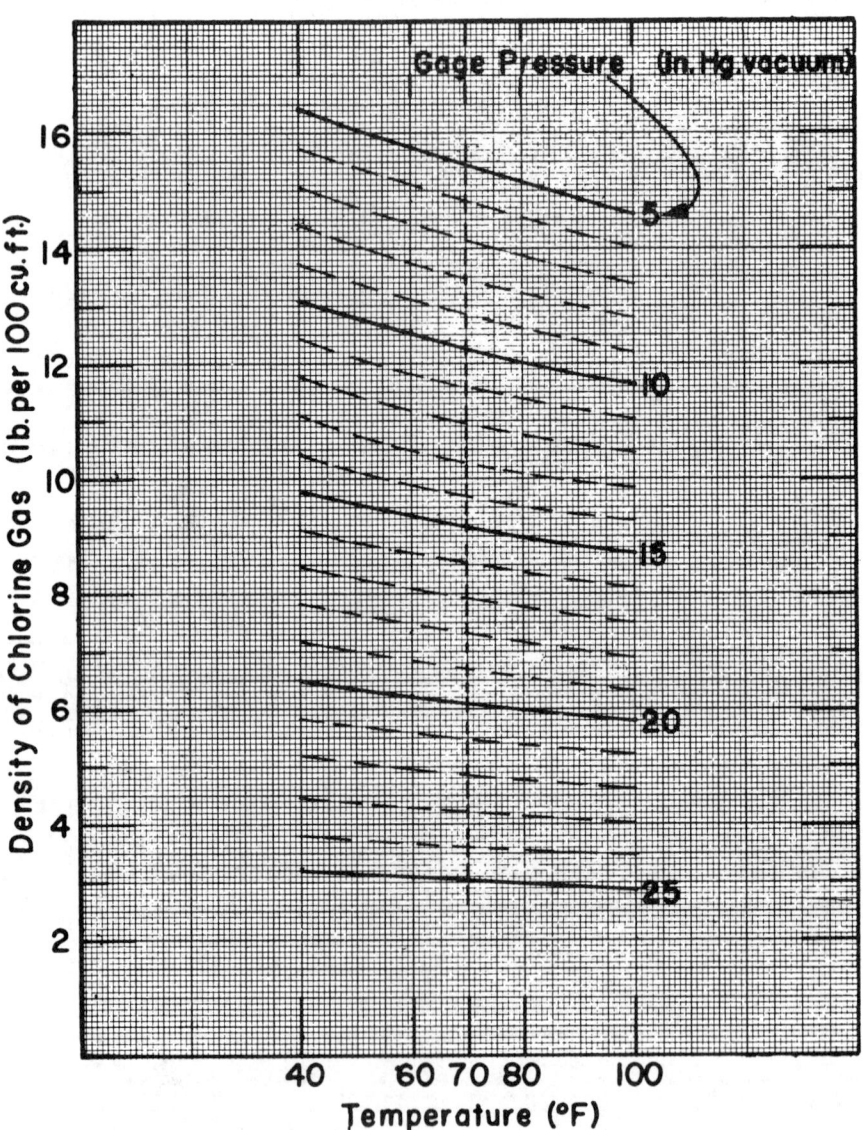

Fig. 2. Density of chlorine gas under vacuum (from Ross & Mass).

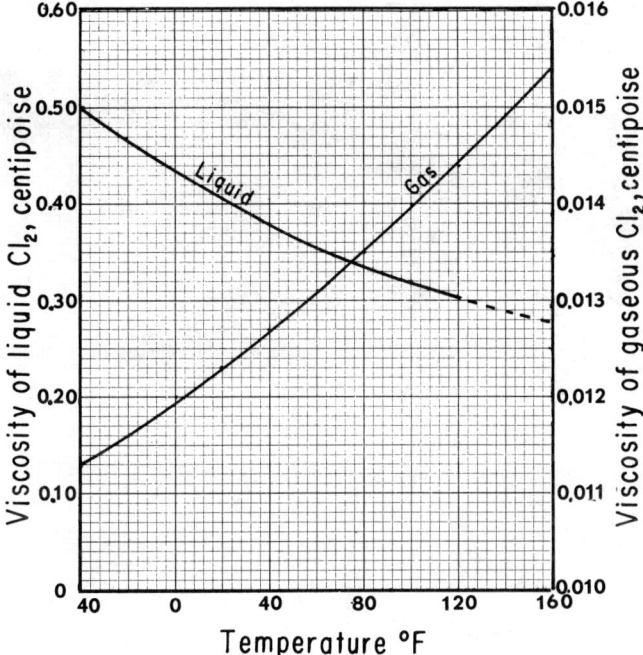

Fig. 3. Viscosity of chlorine liquid and gas (from Steacie and Johnson).

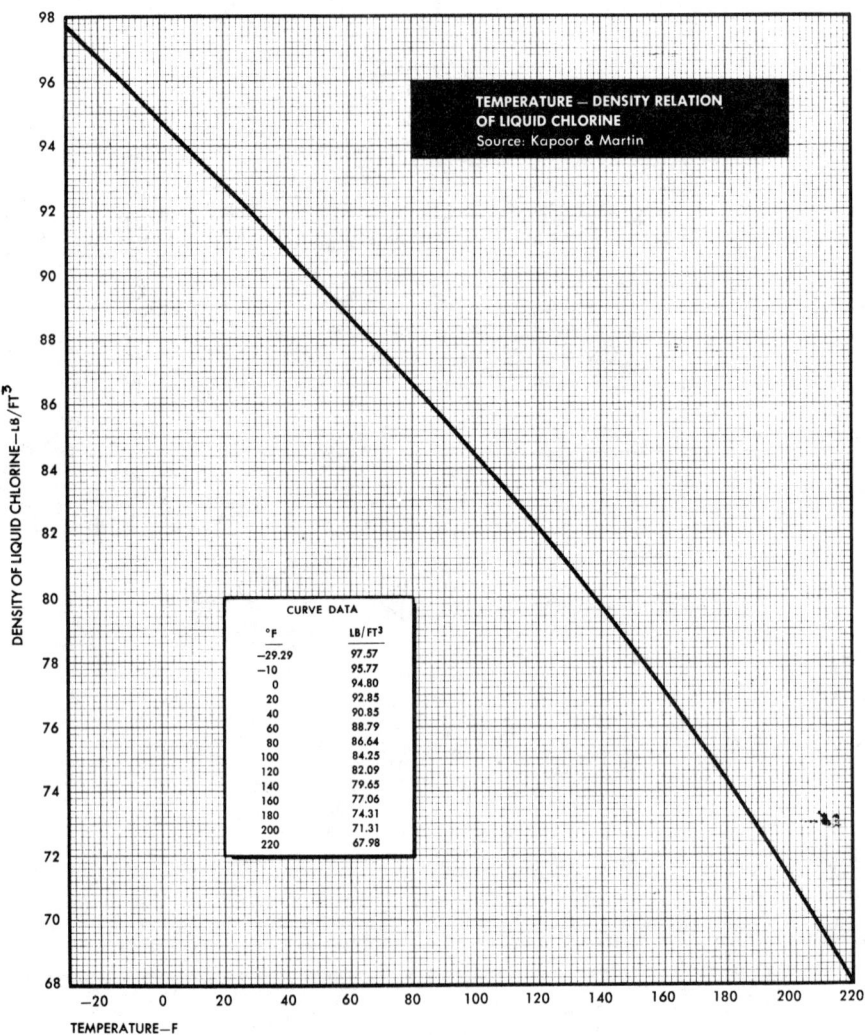

Fig. 4. Temperature-density relation of liquid chlorine (from Kapoor and Martin).

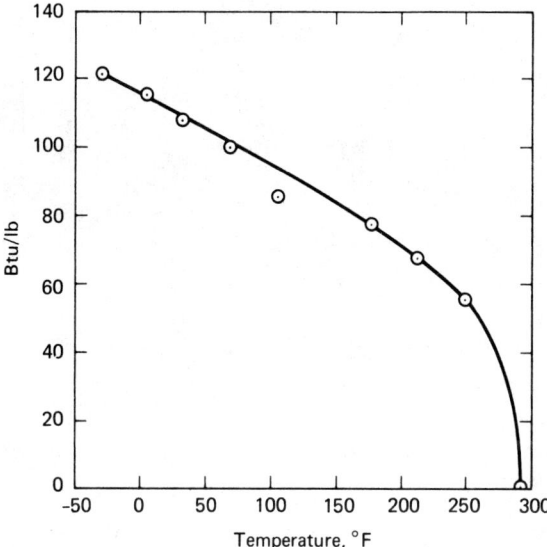

Fig. 5. Latent heat of vaporization of liquid chlorine (from Pellaton).

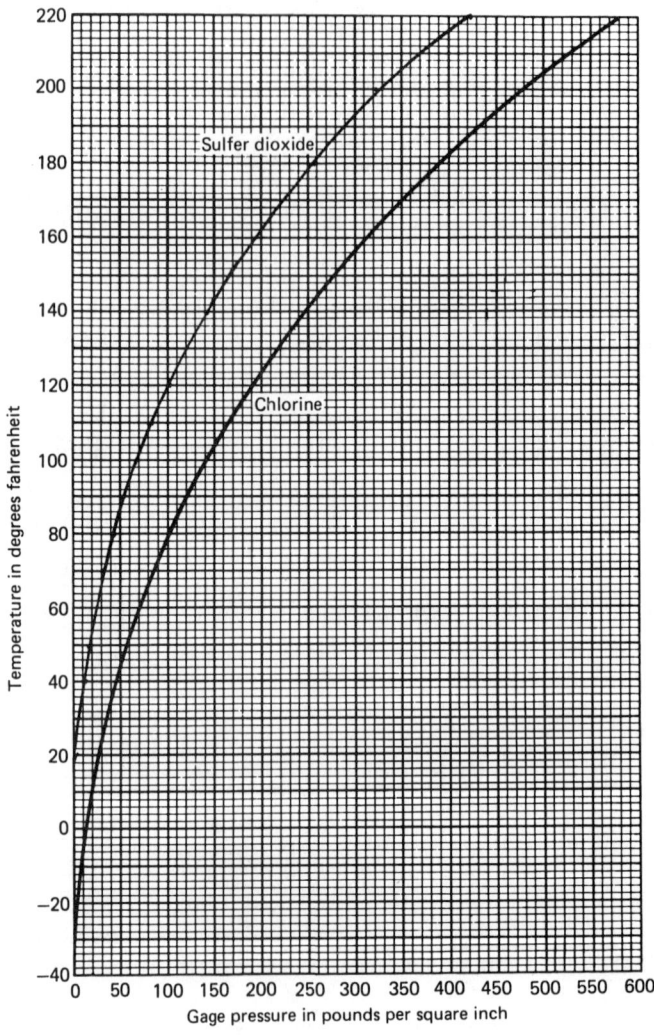

Fig. 6. Vapor pressure versus temperature of liquid chlorine and sulfur dioxide.

APPENDIX 1043

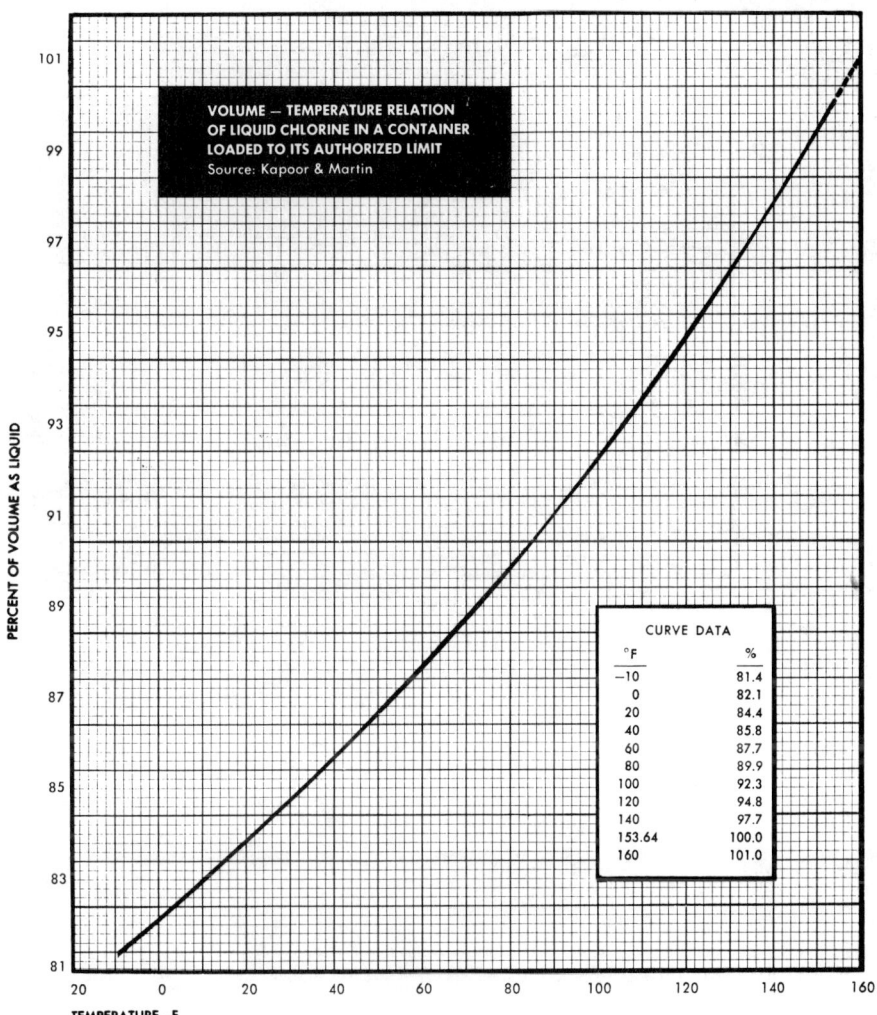

Fig. 7. Volume-temperature relation of liquid chlorine in container loaded to its authorized limit (from Kapoor and Martin.)

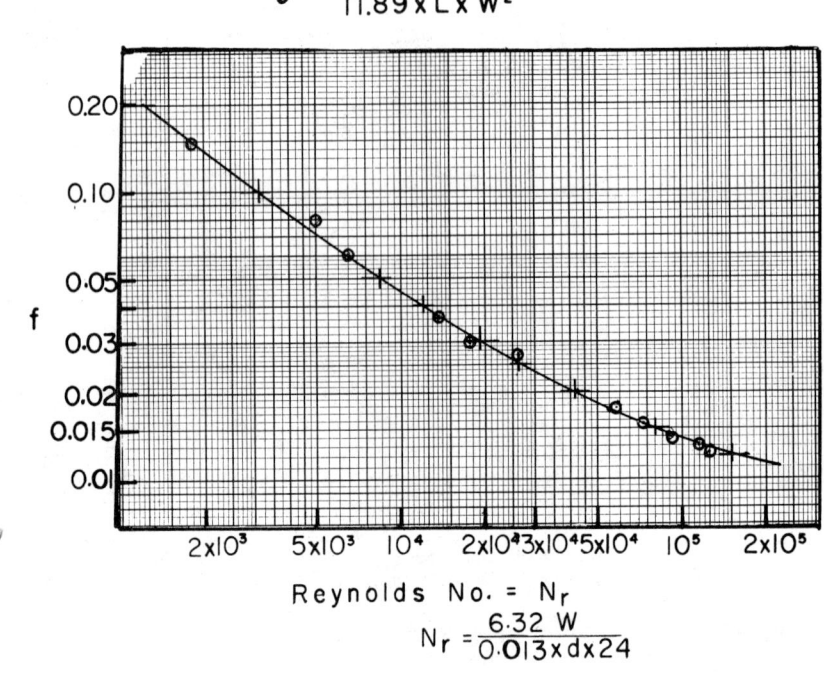

Fig. 8. Friction factor "f" as a function of Reynolds Number, N_r, for PVC injector vacuum lines—for either chlorine or sulfur dioxide gas flow.

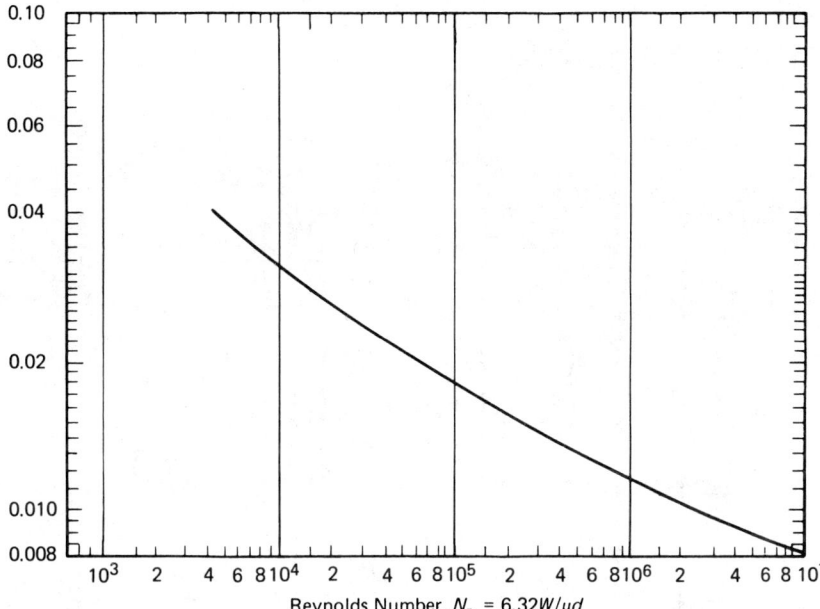

Fig. 9. Darcy-Weisbach friction factor "f" for fluid flow in PVC pipe (courtesy Alden Hydraulic Laboratory, Worcester Polytechnic Institute).

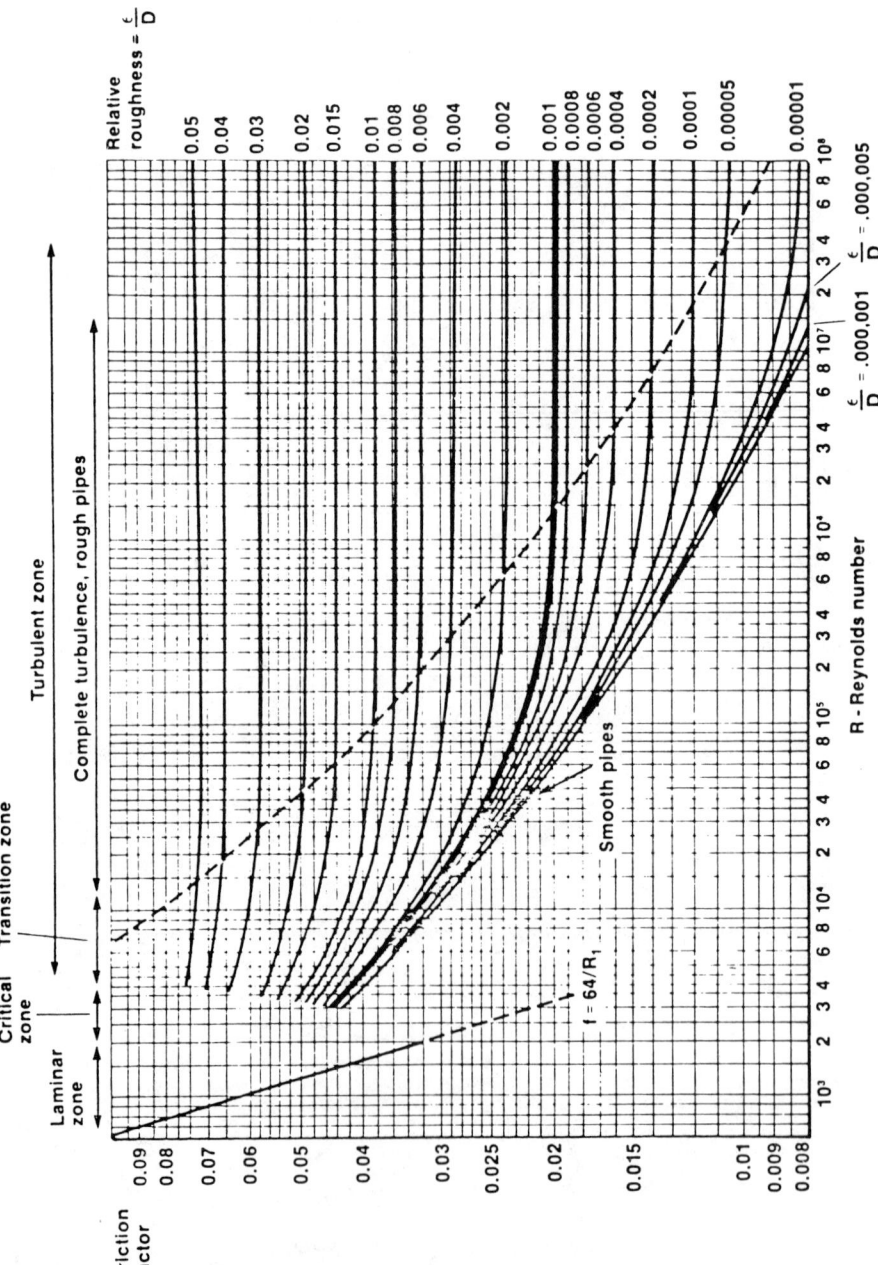

Fig. 10. Friction factor (f) as a function of Reynolds number and relative roughness (from L. F. Moody, "Friction Factors for Pipe Flow," *Trans. Am. Soc. Mech. Engrs.* 671 (1944).

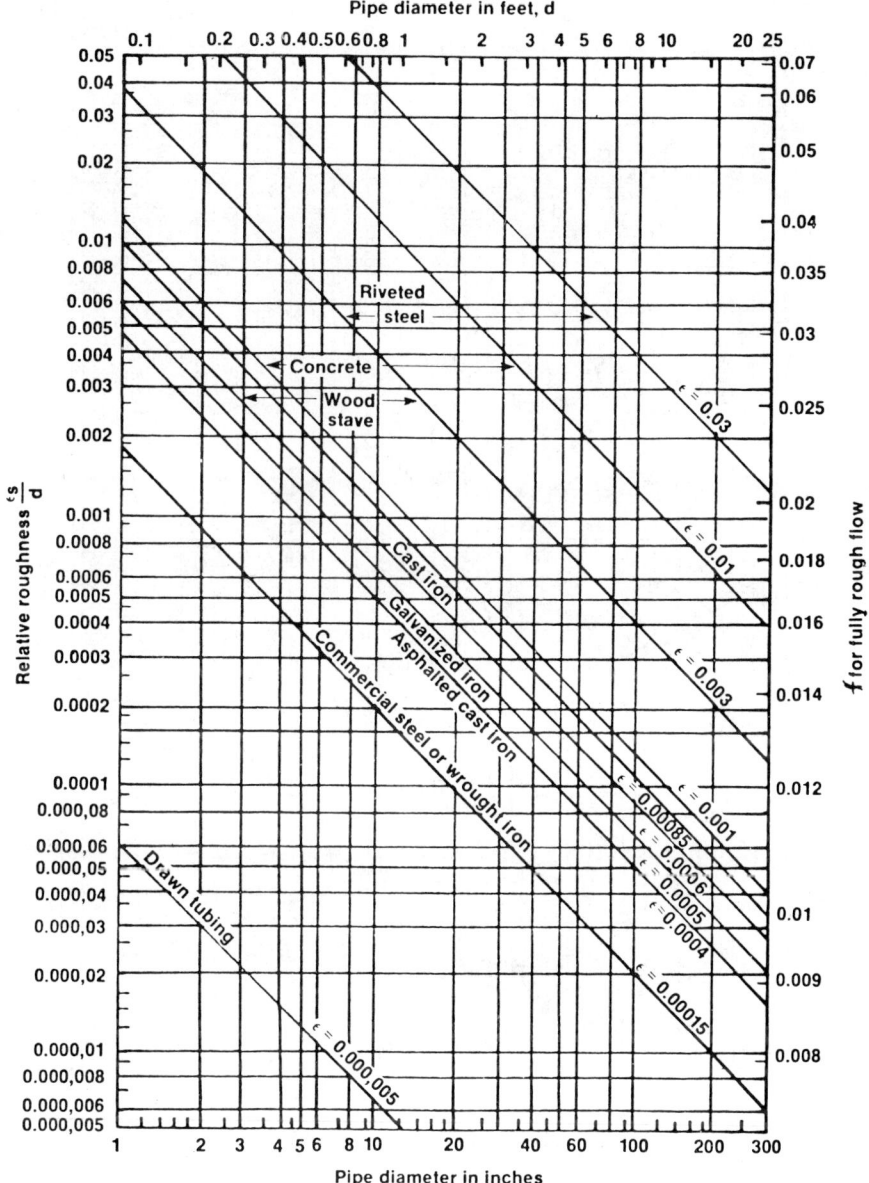

Fig. 11. Relative roughness E_s/d as a function of hydraulic diameter of pipe (after Moody—see Fig. 10).

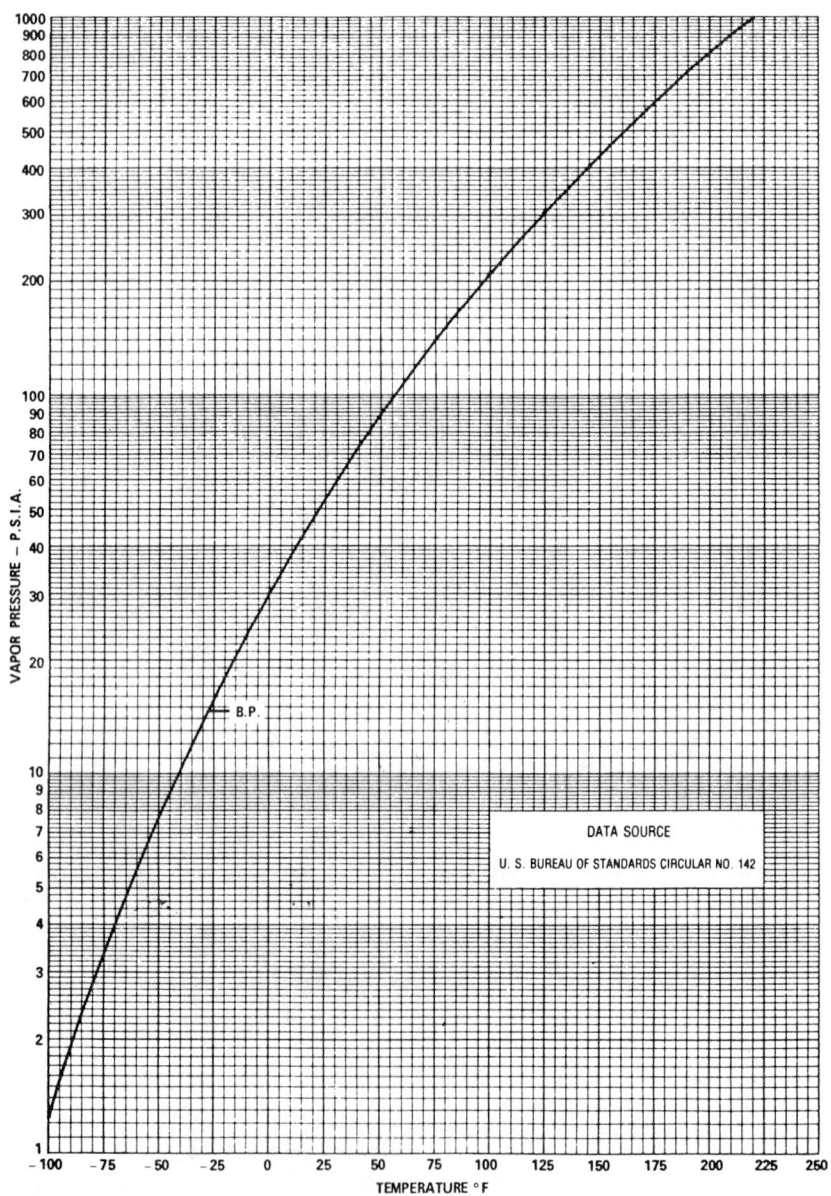

Fig. 12. Anhydrous ammonia vapor pressure versus temperature °F (U.S. Bureau of Standards Circular No. 142).

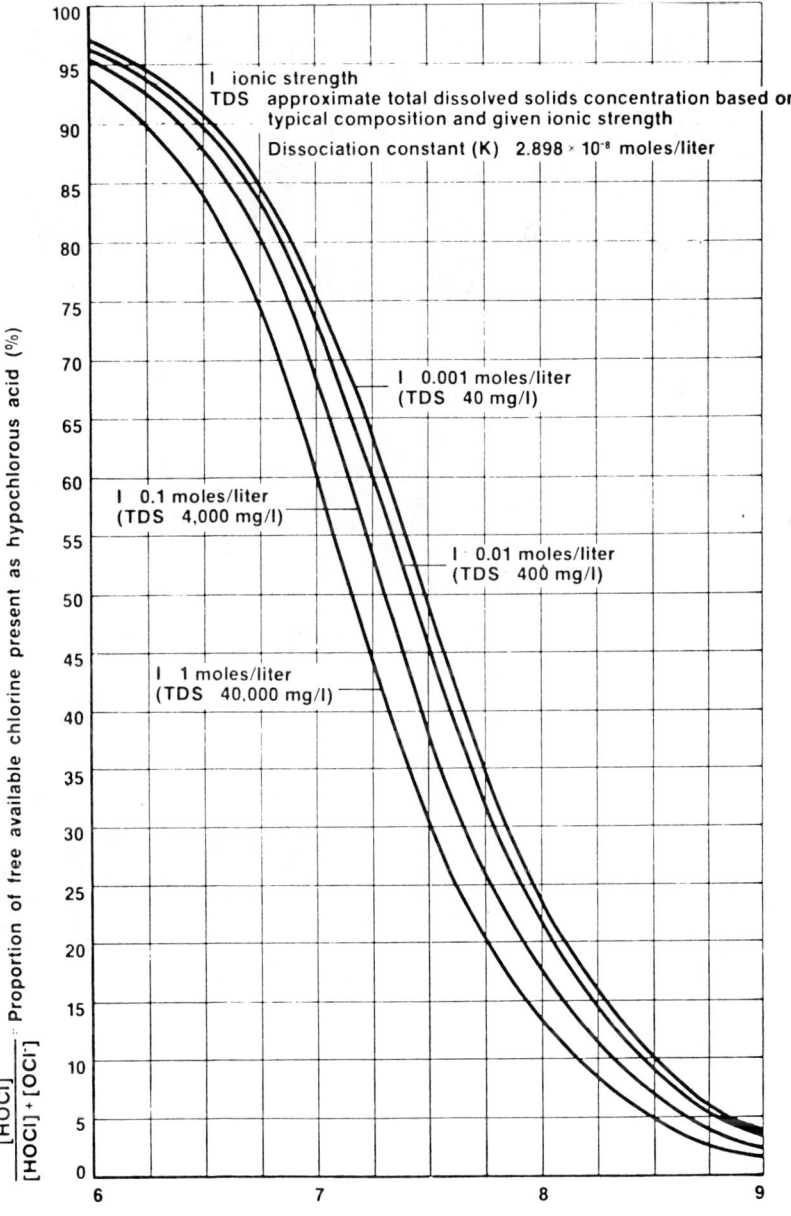

Fig. 13. Effect of pH and ionic strength on the dissociation of HOCl at 25°C.

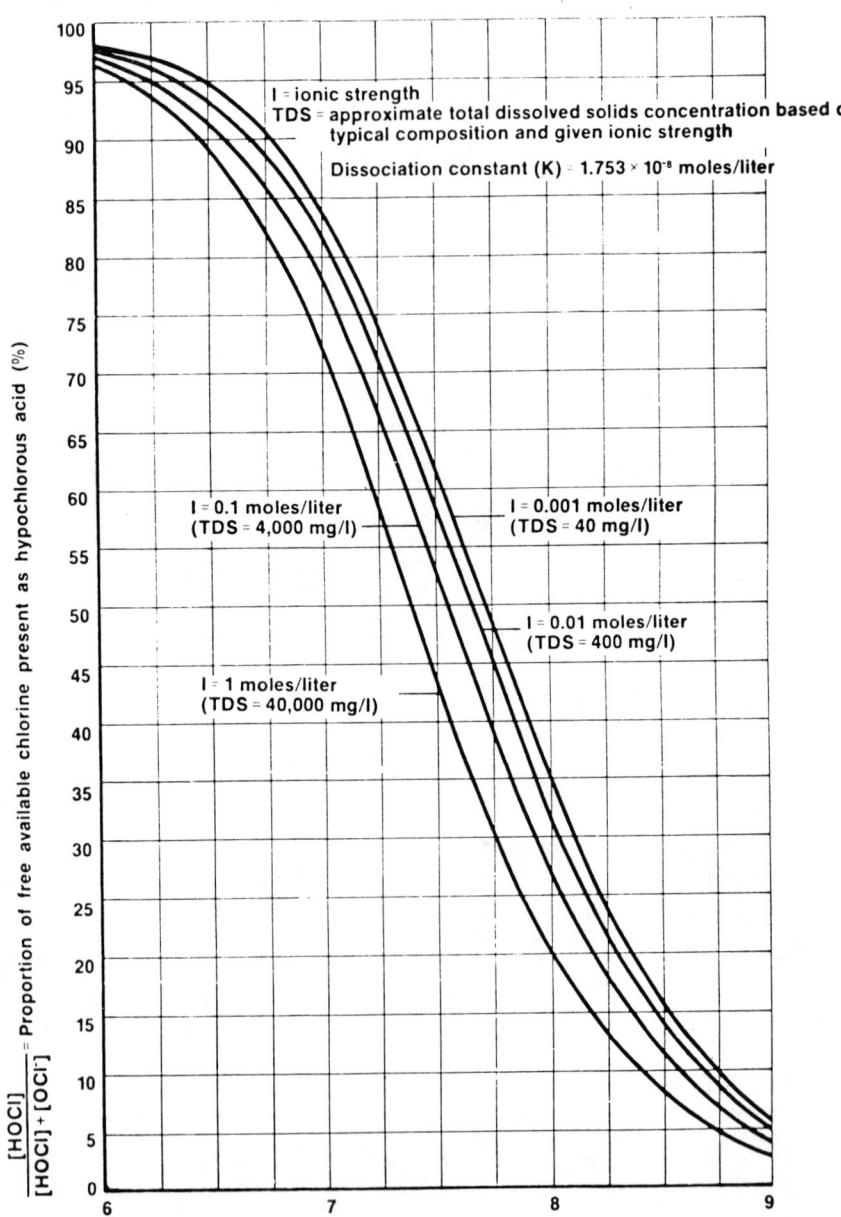

Fig. 14. Effect of pH and ionic strength on the dissociation of HOCl at 5°C.

APPENDIX 1051

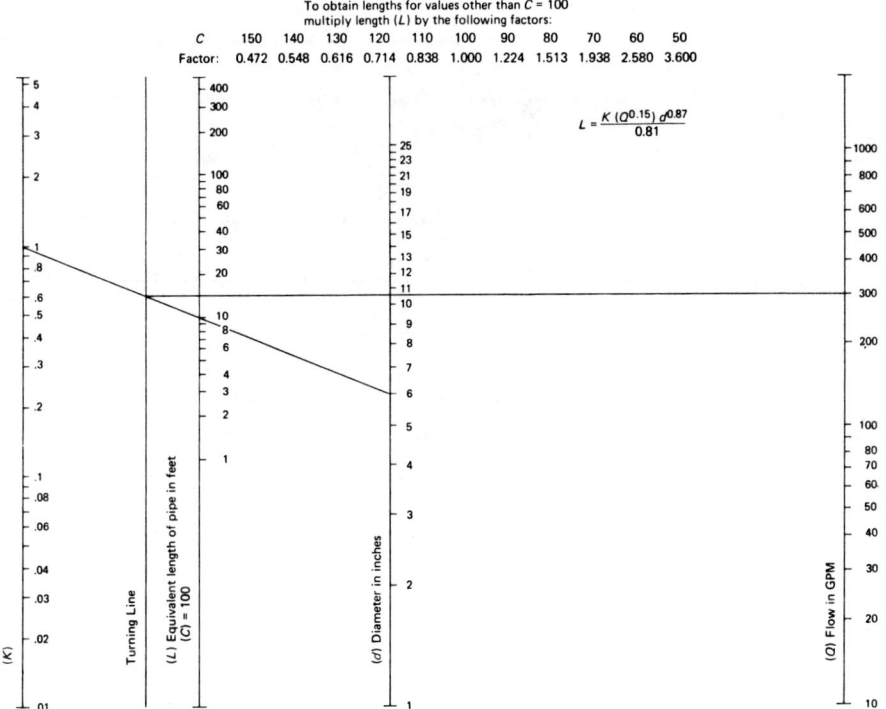

Fig. 15. Nomograph for calculating the equivalent length of pipe for fittings.

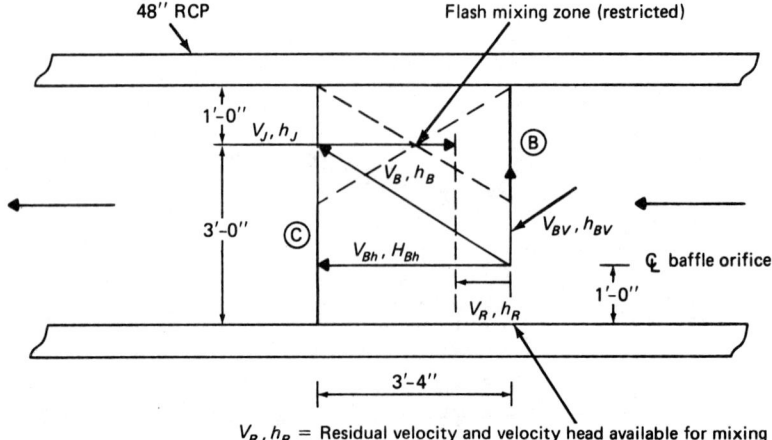

V_R, h_R = Residual velocity and velocity head available for mixing

Fig. 16. Free body diagram of contraflow effect in flash mixing zone. (Ref. 96 Ch. 8).

Index

Activated carbon, 333, 762–764
Aerochlorination, grease removal by, 428, 429
Algae, the principal taste producers in potable water, 333
Ammonia, anhydrous, 740–742
   characteristics which affect use in water treatment, 741–742
      corrosivity, 742
      heat of vaporization, 741
      reaction with water, 742
      solubility in water, 741
      water softening reaction, 742
      weight, 741
   physical and chemical characteristics, 740; Table 9-11, 741
   physiological effects, 742; Table 9-11, 743
   source, availability and uses, 740
   useful ammonia compounds, 740
   vapor pressure vs. temperature, Fig. 12, 104B
Ammonia-chlorine process, 292–305. *See also* Ammoniation facility
   American vs. European practice, 300–302
   blending chloraminated water with chlorinated water, 302, 303, Fig. 6-6a, 304; Fig. 6-6b, 404
   Disadvantages of, 297–300
      effect upon water quality in distribution system, 297
      effect upon kidney dialyses patients, 298
      effect upon the aquarium, 298
      general summary, 300
   factors affecting efficiency of the process, 296, 297
   general discussion, 293, 294
   germicidal efficiency, 295
   historical background, 292
   pretreatment problems, 297
   treatment plant research, 295
Ammoniation facility. *See also* Ammonia, anhydrous; Aqueous ammonia
   ammoniators, 746–749
      advantages of solution feed, 749
      control strategies, 746
      diffusers and solution lines, 749
      direct feed, 746
      injector system, 746–749
      solution feed, 746
   supply system, (anhydrous NH_3), 742–746
      cylinders, 742
      800 lb containers, 743
      evaporators, 744
      leak detectors, 745
      materials of construction, 745
      potential hazard of a major leak, 745
      storage tanks, 743, 744
Amperometric titrator, 231—239
   dual indicator electrode, 235, Fig. 5-4, 235; Fig. 5-5, 235, Fig. 5-6, 237
   historical development, 232
   operating characteristics, 238
   precision and sensitivity, 238
   preparation for titration, 240
   principal of titrator operation, 231
   single indicator electrode, 233, Fig. 5-2, 233; Fig. 5-3, 234; 235
Aqueous ammonia, 749, 751. *See also* Aqueous ammonia facility
   physical and chemical characteristics, 750
   potential hazards, 750
   source and availability, 749
Aqueous ammonia facility, 750–751
   metering and control system, 750, 751
      design considerations, 750
      diaphragm pumps, 751
      diffusers and solution lines, 751
      reliability provisions, 751
   supply system, 750
   storage tanks, 750

Beggiatoa, 340, 343, Fig. 6-18, 344
Bell-jar chlorinators, 282–283
Berthollet, 3
Biofouling, 338–358. *See also* Chlorination of distribution systems and transmission lines
Booster pumps. *See* Injector systems
Breakpoint curve, the, 165

1054 INDEX

Breakpoint phenomenon, 162
Breakpoint reaction, 167
Broad Street pump episode, 257
Bromides: on-site generation of bromine (Br_2), 968–972
   comparison with other methods, 971
      advantages, 971
      disadvantages, 972
   current practices in the U.S.A., 971
   system description, 968, Fig. 14-1, 969; 970
Bromine (Br_2), 962–972. *See also* Bromides; Seawater chlorination
   bromine (Br_2) facility design, 968
   bromine production, 963
      H. H. Dow process, 963
      Kubierschky process, 963
   chemistry of bromine in water and wastewater, 965
   effect of seawater bromides, 985
   occurrence, 962
   on-site generation of. *See* Bromides, on-site generation of bromine (Br_2)
   physical and chemical properties, 964
   preface, 962
   reactions with chlorine, 966
   summary, 1012
   use of bromine in water processes 966–968
      cooling water, 967
      potable water, 966
      swimming pools, 968
      wastewater, 967
Bromine chloride (BrCl), 972–983
   bromine chloride facility design, 975–980
      automatic controls, 978
      BrCl system evaluation, 976
      current practice, 975
      evaporators, 979
      injectors, 978
      materials of construction, 976
      metering and control equipment, 976, Fig. 14-3, 977
      safety equipment, 979
      solution lines and diffusers, 978
      storage and handling, 978
   bromo-organic compounds, 982, 983
   chemistry of bromine chloride, 974
   comparison with other methods, 979, 980
      advantages, 979
      costs, 980
      disadvantages, 980
   germicidal efficiency (Br_2 and BrCl), 980, 981
   measurement of bromine residuals, 983
      amperometric method, 984
      DPD method, 984
      general discussion, 983
   physical and chemical properties, 972, Fig. 14-2, 973
   preparation of BrCl, 973
   summary, 1013
   toxicity of bromine residuals, 981
Bromine residuals, measurement of, 983
Bubbly creek filter plant, 259

Calcium hypochlorite, 71–73
   granular, 71
      manufacturing techniques, 71
      storage of, 72
   liquid, 73, 74
      lime required, Table 2-4, 74
calculator solution, pipeline head-loss, 1028
Candida parapsilosis, 322
Castner's rocking cells, 4
Chemistry of chlorination, 150–213
   available chlorine, 158, Table 4-5, 159; 160
   breakpoint phenomenon, the, 162–177
      breakpoint curve, 165, Fig. 4-3, 166
      breakpoint reaction, 167, Fig. 4-4, 168; 169, 170, 171
      introduction, 162, 163
      organic nitrogen, significance of, 177–181, Fig. 4-9, 179
      pH and alkalinity consideration, 174
      Saunier's research, 174, Fig. 4-5, 175; Fig. 4-6, 175; Fig. 4-7, 176; Fig. 4-8, 177
      Selleck-Saunier breakpoint chemistry, 172, 173
      side-reactions of breakpoint chlorination, 173
   chlorine demand research by Feben and Taras, 181, Fig. 4-10, 182
      mathematical relationship, 182, 183
      practical significance, 184
   chlorine and nitrogenous compounds, effect of, 160–162, Fig. 4-1, 161; Fig. 4-2, 162
   chlorine and other inorganics, 185–190
      alkalinity, 185
      carbon, 186
      cyanide, 186
      hydrogen sulfide, 187, Table 4-6, 187; 188
      iron, 188
      manganese, 189
      methane, 189
      nitrites, 190

INDEX 1055

Chemistry of chlorination (cont.)
  electrochemical properties of chlorine residuals, 204–210, Fig. 4-14, 205; Fig. 4-15, 200; Fig. 4-16, 207
  ORP concept, practical application of, 207–210, Fig. 4-17, 208; Fig. 4-18, 209
  germicidal efficiency, investigators consensus of, 203, Fig. 4-13, 203
  germicidal significance of chlorine residuals, 194–204
    chloramines, discussion of, 197
    dichloramine, 199, 200
    hypochlorite ion, 196, 197
    hypochlorous acid, 195, 196
    introduction, 194, Fig. 4-12, 195
    monochloramine, 198, 199
    nitrogen trichloride, 200
    nuisance residuals, 202, 203
    organic chloramines, 201, 202
  hydrolysis of chlorine gas, 151–153, Table 4-1, 152; Table 4-2, 153
  hypochlorite solutions, 156
  hypochlorous acid, chemistry of, 152-156
    effect of pH, 153, Table 4-3, 154; Table 4-4, 155
    effect of ionic strength, 154, Fig. 13, 1049; Fig. 14, 1050
  Morris lethality coefficient, 204, Table 4-7, 205
  seawater chemistry, 190–194
    ammonia nitrogen, role of, 193
    bromides, effect of, 190, 191, Fig. 4-11, 192
    hypobromous acid, dissociation of, 192
    ionic strength, effect of, 193
    pH, effect of, 191, Fig. 13, 1049; Fig. 14, 1050
  valences of chlorine, 157
Chlorination. See also Chemistry of chlorination.
Chlorination, advances in, 303
Chlorination-dechlorination concept, 287
  photo decomposition by sunlight, 288
Chlorination of distribution systems and transmission lines, 338–358
  bacteriology of water systems, 340–349, Fig. 6-15 and 6-16, 342; Fig. 6-17, 343; Fig. 6-18, 344; Fig. 6-19, 346; Fig. 6-20, 347; Fig. 6-21, 348
    bacteria enumeration, 349
    coliform regrowth, 347
    iron bacteria, 341–343
    other types of biofouling, 345–347
    sulfur bacteria, 343–345
    tubercle formation, 345, Fig. 6-19, 346

  chlorine residuals to maintain bacterial quality, 351–356
    practical considerations, 354
    relay chlorination, 354–356
  control of biofouling in distribution systems, 349–351
    case histories, 350
    free chlorine vs. combined chlorine, 349
    implementing a cleanup program, 351
  general discussion, 338
  nature of the problem, 338, Fig. 6-14, 339; 340
  restoring pipeline capacity, 356–358
Chlorination of potable water, 256–393. See also Chlorination of potable water current practices and waterborne diseases
  historical background, 251–263
    early uses in water treatment, 257
    first uses in the USA, 258
    Hazen theorem, 263
    The Jersey City case, 259–262
    The Mills-Reincke phenomenon, 263
    typhoid fever and waterborne outbreaks, 262, 263
    use as a germicide, 256
  purpose, 256
  waterborne diseases, 263–278
Chlorination of potable water
  current practices, 317–323
    deficiencies in current practices, 320–323
      community water system survey, 320
      concern of viruses, 321
      proof of disinfection, 321–323
    review of modern practices, 317–320
      chlorine demand, 320
      chlorine dosages, 317–320
      chlorine residuals, 320
      contact time, 320
Chlorination of potable water, objectives of, 323–379
  chlorination of high pH waters, 371–373, Figs. 6-23 and 6-24, 372;
  coagulation, aid to, 358
  color removal, 370
  disinfection guidelines, 323–328, Fig. 6-11, 324; Fig. 6-12, 325; Table 6-3, 325; Fig. 6-13, 326; Table 6-4, 327
    chlorine-ammonia process, 328
    free residual process, 323, Fig. 6-11, 324, Table 6-3, 325, Fig. 6-12, 325, Fig. 6-13, 326, Table 6-4, 327
  disinfection of water mains and storage tanks, 377–379, Table 6-6, 378; Table 6-7, 379

1056 INDEX

Chlorination of potable water (*cont.*)
  disinfection of water mains and storage tanks
    (*cont.*)
    storage tanks, 378
    water mains, 377
  distribution systems and transmission lines,
    338–358. *See also* Chlorination of distribution systems and transmission lines
  filtration, aid to, 359, 360
  hydrogen sulfide control and removal, 360–365,
    Fig. 6-22, 362
  iron and manganese problem, 365–370
    nature and occurrence, 365, 366
    role of chlorine, 368–370
    significance, 366–368
  miscellaneous applications of chlorine, 373–377
    desalting plants, 373
    injection wells, 377
    reflecting pools, 373
    restoring wells, 374–377
  summary and recommendations, 379–382
    chlorine demand, 379, Fig. 6-25, 380
    chlorine residual analysis, 381
    controlling the chloramine process, 382
    monitoring residuals, 381, 382
    NCl_3 formation spells trouble, 381
    points of chlorine application, 381
Chlorination of potable water, taste and odor control, 328–338. *See also* Taste and odor control
Chlorination—dechlorination concept, 287
Chlorination equipment, development of, 279–285
  advances in equipment, 284
  historical background, 279; Fig. 6-1, 281; Fig.
    6-2, 283; Fig. 6-3, 284
Chlorination of wastewater, (chap. 7). *See* Wastewater chlorination
Chlorination practice, development of, 285–317
  advances in, 303
  ammonia-chlorine process, 292. *See also* Ammonia-chlorine process
  chlorination-dechlorination concept, 286, Fig.
    6-4, 289; Fig. 6-5, 290, Fig. 6-6, 291
  chlor-dechlor facility, 287
  complete dechlorination followed by rechlorination, 291
  controlling THM formation, 288
  operating installation, 288, Fig. 6-4, 289
  system description, 286, Fig. 6-4, 289
  free residual process, 305. *See also* Free residual process
  historical background, 285, 286

Chlorination stations
  housing, 735–738
    doors, 737
    drains, 737
    electrical, 738
    fire hazard, 735
    general, 735
    heating, 737
    inspection window, 737
    separation, 735
    space requirement, 735
    vents, 738
    ventilation, 736
    wind socks, 737
  individual deep well stations, 694, Fig. 9-54, 695
  multiple wells or pump stations, 697, Fig. 9-55, 696; Fig. 9-56, 698
  principal considerations, 694
  relay stations and transmission lines, 697, Fig.
    9-57, 699
  reliability provisions, 738–740
    chlorine residual analyzers, 739
    chlorine supply, 738
    power failure, 739
    standby equipment, 739
    water supply, 739
  reservoir outlets, 700–703
    flow reversal detection, 700
    flow reversal detection by residual analyzer,
      701, Fig. 9-58, 702
    general discussion, 700
    primary flowmeters, 700
  water treatment plants, 703–705
    chlorinator capacity, 704
    contact time, 704
    control system, 704
    general discussion, 703
    injector system, 705
    mixing, 704
    monitoring and control analyzers, 705
    points of application, 703
    safety equipment, 705
    solution lines, 705
Chlorination of wastewater. *See* Wastewater chlorination
Chlorinators, 655–681
  control strategies, 661–672
    compound-loop control, 668–672, Fig. 9-42,
      669; Table 9-8, 670; Table 9-9, 672
    direct residual control, 661–668, Fig. 9-40a,
      663; Fig. 9-41, 664
    Egan cluster-jet diffuser, Fig. 9-40a, 663

INDEX 1057

Chlorinators (cont.)
  control strategies (cont.)
    flow pacing, 661
    general discussion, 661
    loop time, 663, Table 9-7, 667
    sampling tap, 663
  development of spring loaded diaphragm concept, 656–660
    general discussion, 656
    remote vacuum, 659, Fig. 9-39, 659; Fig. 9-40, 660
    sonic flow concept, 658, Fig. 9-38, 658
    theory of operation, 657, Fig. 9-37, 657
  dual signal control systems, 672–680
    Capital Controls Co., 673, Fig. 9-43, 674; Fig. 9-44, 675; 676, 677
    Fischer and Porter Co., 677, Fig. 9-45, 678; 679
    introduction, 672
    Wallace & Tiernan Div., 679, Fig. 9-46, 680
  historical background, 655
    the bell-jar era, 656
    direct gas feed, 655
    suck-back phenomenon, 655
Chlorine, 1–61
  accidents, 42–51
    effect upon community, 50
    help in chlorine emergencies, 50
    notable consumer accidents, 47–50
    tanker trucks, 46
    transport accidents, 44–46
  discovery of, 1
  hazards of, 38–42
    characteristics of a major release, 41
    first aid, 39
    physiological response, 40
    toxic effects, 38
  impurities in, 29–34
    consequences of, 31
    NCl_3 occurrence, 32
    NCl_3 prevention, 33
    silica contamination, 33
    sources, 30
  ionization constant, 154
  leaks, 51–55
    frequency and magnitude, 51
    safety precautions, 53
  liquid, 2
  manufacture of, 2–29
    diaphragm cells, 14–21, Tables 1-2 and 1-3, 16; Fig. 1-5, 17; Fig. 1-6, 18; Fig. 1-7, 19
    dimensionally stable anodes, 6
    Dow Moving Bed process, 28
    electrolysis of hydrochloric acid solutions, 28
    electrolytic cell development, 5, Fig. 1-1, 6
    electrolytic process, history, 4
    HCl oxidation process, 27
    history, 2
    Hoechst-Uhde process, 28
    Kel-Chlor process, 28
    membrane cell, 7–14, Fig. 1-2, 9; Table 1-1 and 1-1a, 11; Fig. 1-3, 13; Fig. 1-4, 15
    mercury cells, 21–25, Fig. 1-8, 24; Table 1-4, 26
    salt process, 25
  physical and chemical properties, of, 34–38, Tables 1-5 and 1-6, 35, 1037–1043
    chemical reactions, 38
    density of vapor, 37, Fig. 1, 1037; Fig. 2, 1038
    latent heat of vaporization, 37, Fig. 5, 1041
    solubility of vapor in water, 37
    specific heat, 37
    temperature-density relation of liquid chlorine, Fig. 4, 1040
    vapor pressure vs. temperature of liquid chlorine, Fig. 6, 1042
    viscosity of liquid and vapor, 37, Fig. 3, 1039
    volume-temperature relation of liquid chlorine in container loaded to its authorized limit, 36, Fig. 7, 1043
  production, 55
    energy consumption, 55, Table 1-7, 56
  uses, 55–59
    chemical industry, 57
    power industry, 57
    water and wastewater treatment, 56
Chlorine demand, 302, 320
  definition of, 285
Chlorine dioxide, 833–891. See also: Chlorine dioxide, facility design; Chlorine dioxide generation by the chlorine-chlorite method; and Chlorine dioxide, other methods of generation.
  analytical measurements, 886, 887
    chlorine dioxide solutions, 886
    chlorine dioxide residuals, 886, 887
    continuous analyzers, 887
  chemistry of, 862–864
    ammonia, 863
    bromides, 864
    hyochlorous acid, 864
    nitrogen trichloride, 864
    ozone, 864
    sulfur dioxide, 863

1058 INDEX

Chlorine dioxide (cont.)
  dechlorination of, 879
  formation of halogenated organics, 864
  germicidal efficiency, 864–869, Fig. 12-9, 866; Fig. 12-10, 869; Fig. 12-11, 870
  health hazards from by-products, 869–871
  historical background, 833–836
    European practices, 834, 835
    Great Britain, practices, 836
    introduction, 833
    North American practice, 833, 834
  physical and chemical properties, 836–838
  summary, 887–888
    advantages of, 887
    disadvantages of, 888
  water reuse, 876–879
    food processing, 876, 877
    reclaimed wastewater, 877–879
  water treatment, uses, 871–879
    algae and decaying vegetation, 873, 874
    color control, 876
    control of phenolic tastes and odors, 871–873
    iron removal, 875
    manganese removal, 874
Chlorine dioxide, facility design, 880–886
  acid-chlorite system, 885
  chlorine-chlorite system, 880–885
    chlorine and chlorite mixed under pressure, 880–883
    chlorine and chlorite mixed under a vacuum, 883, 884
    chlorite storage tanks, 885
    general discussion, 880
Chlorine dioxide generation by the chlorine-chlorite method, 847–862
  analysis of solutions, 848, 849
    chemical reactions, 849
    general discussion, 848
    reaction steps, 849
  analytical procedure, 850–852, Table 12-3, 852
    chloride determination, 852
    titration summary, 851
  CIFEC system, 859, Fig. 12-7, 860
  Fischer and Porter system, 857
  general discussion, 847, 848
    effect of injector water alkalinity, 848
    effect of Na ClO_3 alkalinity, 848
    purity, 847
    yield, 848
  Olin system, 853–855, Fig. 12-4, 855
    Olin Water Services of Kansas City, 854
  Rio Linda Chemical Co. system, 859, Fig. 12-8, 861
  Wallace and Tiernan analytical procedure, 852, 853
  Wallace and Tiernan generating system, 855–857, Fig. 12-5, 856; Fig. 12-6, 858
Chlorine dioxide, other methods of generation, 838–847
  chloride reduction, 840
  commercial acid-chlorite generators, 843–847, Table 12-1, 844; Table 12-2, 845
    Fischer & Porter Co., 845, Fig. 12-3, 846; 847
    Rio Linda Chemical Co., 843, Table 12-1, 844; Table 12-2, 845
  introduction, 838
  JASZKA-CIP process, 841
  laboratory preparation, 841–843, Fig. 12-2, 842
  Mathieson (SO_2 reduction) process, 839, Fig. 12-1, 840
  Solvay (methanol) process, 839
Chlorine facilities design, 585–754 (chap. 9). See also: Chlorine leaks; Chlorine residual analyzers; Chlorine supply-system; Injector systems; Leak detectors
  how to deal with a major leak, 617–622, Fig. 9-23, 621
    containment with controlled release, 619
    definition of a major leak, 617
    leak scenario, 619
    scrubber system, 620, Fig. 9-23, 621
    spent caustic disposal, 622
Chlorine residuals, analytical methods, 214–255. See also Amperometric titrator
  amperometric titration method, 231–248, Fig. 5-2, 233; Fig. 5-3, 234; Fig. 5-4, 235, Fig. 5-5, 236, Fig. 5-6, 237
    Baker's alternate procedure, 248
    chemistry of amperometric method, 239
    chemistry of nitrite interference, 248
    determination of residual chlorine in wastewater, 245
    dichloramine, 242
    FAC residuals at short contact time, 244
    free and combined available chlorine, 243
    free available chlorine, 241
    monitoring normality of iodine solution, 246
    monochloramine, 242
    nitrogen trichloride, 243, 247
    sample calculations, 245, 246
    total available chlorine, 242
  current status of methods, 219–248

INDEX 1059

Chlorine residuals (cont.)
 development of methods, 214–219
 DPD-Colorimetric method I, 223
 DPD-colorimetric method II, 224, Fig. 5-1, 225
 DPD-Ferrous tetrimetric method, 229–231
 FACTS method, (syringaldazine), 222, 223
 iodometric method I, 225–227
  standardization of chlorine solutions, 227
 iodometric II, 227–229
  back titration for wastewaters, 228, 229
 leuco crystal violet, 221, 222
 nitrite interference, 228, 229, 247
  chemistry of, 247
 organic chloramine interference, 251
 Orthotolidine (O-T), 219–221
  drop dilution method, 220
 summary and recommendations, 249–252
  wastewater treatment, primary effluent, 250, secondary, 251.
  filtered tertiary, 251
  municipal supplies, 250
  organic chloramine interference, 251
  potable water treatment, small supplies, 249
Chlorine residual analyzers, 705–735
 Capital Controls Co., 716–718, Fig. 9-65, 717; Fig. 9-66, 718
 Delta Scientific (Xertex), 719–722
  historical background, 719–721, Fig. 9-67, 720
  model, 924, 721
  model, 925, 722
 designers evaluation, check list, 710, 711
  accuracy, 710
  precision, 710
  range, 711
  response time, 711
  sensitivity, 711
  species measurement, 710
  temperature compensation, 711
 EPCO: Clortect®, 727, Fig. 9-69; 728
 Fischer and Porter Co., 723–727
  Amperometric probe, 725, Fig. 9-68, 726
  Anachlor series 17B 4200-C, 723
  Anachlor series 17B 4200-2, 724
  Chlor-Trol (free res. Cl), 724
 general discussion and definitions, 708–710
  amperometric, 708
  polarography, 709, Fig. 9-60, 710
  potentiometry, 709
  voltammetry, 708
 Hach, 731–735, Fig. 9-70, 732; Fig. 9-71, 733; Fig. 9-72; 734

 historical background, 705–708, Fig. 9-59, 706
 IBM, EC/250, 729, Fig. 9-69; 728
 Orion Research Co., 729–731
 Unieoc (Rosemount) Co., 732
 Wallace and Tiernan Div., 711–716
  amperometric cell, 711–715, Fig. 9-61, 712; Fig. 9-62, 713
  general discussion, 711
  polarographic cell, 715, Fig. 9-63, 715; Fig. 9-64, 716
  potable water practice, 714
  wastewater practice, 715
Chlorine supply system, 585–634
 automatic switchover: ton containers, 606–611
  general discussion, 606
  liquid chlorine, 609, Fig. 9-17, 609; Fig. 9-18, 610
  pressure system, 606, Fig. 9-15, 607
  vacuum system, 607, Fig. 9-16, 608
 container withdrawal rates, 604–606, Table 9-1, 605
 container withdrawal factor, 605; Table 9-2, 605
 sample calculation, 605
 designers check list, 611
 flow of chlorine in pipes under pressure, 647–655, Fig. 9-34, 648; Fig. 9-35, 651; Fig. 9-36, 653
  gaseous chlorine, 650, Fig. 9-35, 651
  liquid chlorine, 647, Fig. 9-34, 648; Fig. 9-36, 653
 materials of construction, 632–634
  chemical reactions, 632
  construction, 634
  supply system, 633
  valves, 634
 100 and 150 lb cylinders, 585; Fig. 9-1, 586; Fig. 9-2, 588
  automatic switchover, 589, Fig. 9-3, 590
 reserve tank, 623, Fig. 9-24, 624, 625
 safety equipment, 640–647
  breathing apparatus, 640
  emergency container kits, 640, Fig. 9-31, 641
  expansion tanks, 646
 selection of container size, 611
 single unit tank cars, 612–627, Table 9-3, 613; Fig. 9-19, 613; Fig. 9-20, 614; Fig. 9-21, 614
  absorption tanks, 617
  air pad system, 615, Table 9-4, 9-5, 616; Fig. 9-22, 618

1060 INDEX

Chlorine supply system (*cont.*)
  single unit tank cars (*cont.*)
    automatic tank car valve shut-off system, 626, 627
    flexible connections, 614
    general description, 612
    header system, 615
    stationary storage tanks, 627–632, Fig. 9-25, 627; Fig. 9-26, 628; Fig. 9-27, 629; Fig. 9-28, 631; Fig. 9-29, 632
    air padding system, 630
    ejector system, 630
    general discussion, 627
    piping and header system, 630
    unloading system, 628
    weighing device, 630
  tank car supply monitoring system, 622, 623
    chlorine pressure switch, 623
    section methods for liquid exhaustion, 623
    general discussion, 622
  ton containers, 591–612, Fig. 9-14, 602–603
    alarms, 604
    expansion tanks, 600, Fig. 9-12, 599
    external pressure reducing valve, 600
    gages, 601
    gas filter, 600
    gas withdrawal, 594, 595, Fig. 9-10, 596
    general discussion, 591, Fig. 9-4, 591; Fig. 9-5, 592; Fig. 9-6; Fig. 9-7, 593
    handling equipment, 592, Fig. 9-7, 593; Fig. 9-8, 594; Fig. 9-9, 595
    liquid withdrawal, 596, Fig. 9-11, 597; 598; Fig. 9-12, 599
    platform scales, 599
    weighing devices, 599, Fig. 9-13, 600
Cloromat system (Ionics Inc), 125–129
  Cloromat generating system, Fig. 3-2, 126
  comparison with purchased hypochlorite, 128
  membrane cell, Fig. 3-1, 124
  operating costs, 127
  space requirements, 127
Collins mathematical model for wastewater disinfection, 511
Collins-Selleck model for wastewater disinfection, 513, Fig. 8-5, 514; Fig. 8-6, 515
Color removal by chlorine, 370
Concepts of wastewater disinfection, 476–477
  foreword, 476
  importance of a chlorination disinfection facility, 476
  importance of disinfection, 476, 485, Fig. 8-1, 479
  administration requirements, 482
  bacterial indicator concepts, 484, 485
  contemporary practices in the USA, 477
  current coliform requirements in California, 481
  disinfection efficiency: bacteria survival vs. percent kill, 483
  evolution of disinfection requirements, 478, Fig. 8-1, 479
  legal concepts, 477
  public health agency perspective, 476
  rationale for coliform concentration requirements, 480
  total coliforms versus fecal coliforms, 482
  the regrowth phenomenon, 497
  significance of regrowth, 497
  the clumping phenomenon, 497
Contact chambers, 556–576
  dispersion in long narrow structures, 569–572
    closed conduits and round pipes, 569
    open channels, 570
  distribution of residence time, 557–561
    analysis of residence time, 558, Table 8-3, 560, Fig. 8-28, 561
    effects of short circuiting, 558, Fig. 8-27, 559
    general theory, 557, Fig. 8-26
  effects of baffles, 564–566
    modification baffles, 564, Fig. 8-33, 567
    special structural design situations, 564, Fig. 8-31, 565; Fig. 8-32, 566
  effects of chamber configuration, 560–562
    circular chambers, 561
    depth to width ratio, 560
    length to width ratio, 560, Fig. 8-29, 562
    outfalls, 562, Fig. 8-30, 563
  evaluation of, 566–568
    dispersion index, 567
    general, 566
    Morrill index, 568
  evaluation methods, of 1023–1026
    calculator solution, 1025
    dispersion index, 1023–1025
    Morrill index, 1025, 1026
  function, 556
  optimum design considerations, 572–576
    Kennedy-Jenks engineers, 573
    optimum contact time, 574
    Sanitation Districts of Los Angeles County, 572
  other physical factors affecting residence time, 562–564

INDEX 1061

Contact chambers (cont.)
   other physical factors affecting residence time (cont.)
      turbulence, 564
      weirs, 563
      wind, 562
   provisions for cleaning, 566
   summary of requirements for initial mixing and contact chambers, 576
      contact chambers, 576
      initial mixing, 576
Control system definitions, 1029–1030
   algorithm, 1029
   analog, 1029
   derivative action, 1029
   reset action, 1029
   summary, 1030
Conversion factors, 1027, 1028
Cubic equation, 1036
Crenothrix, 340, 341, 342, Fig. 6-15
Cyanide destruction, 449–466. Also see Wastewater chlorination

Darnall's chlorinator, 279, 280, Fig. 6-1, 281
Debye-Huckel equation constants, 1034
Dechlorination, 755–788. See also Activated carbon; Dechlorination facility design
   activated carbon, 762–764
      chemistry of GAC dechlorination, 762
      filter bed design, 763
      mode of action, 762, 763
   current practices, 765–768
      potable water, 768
      reclaimed water, 768
      wastewater, 767, 768
   ferrous sulfate, 765
   hydrogen peroxide, 761
   introduction, 755, 756
   other methods, 765
   storage by, 764
   sulfur dioxide, 756–761
      operational characteristics, 758, 759
      properties, 756–759
      reactions with chlorine, 759
      reactions with D.O., 760
      sulfite compounds, 760, 761
      vapor pressure vs. temperature, Fig. 6, 1042
   wastewater research, 755, 756
Dechlorination facility design, 768–787
   bisulfite solutions, 786–787
      designers check list, 786
      design considerations, 786
      general discussion, 786
   contact time, 785
   control strategies, 773
      biased sample, Fig. 10-5, 777; 778
      feed forward, 773, Fig. 10-3, 774
      general discussion, 773
      ORP control, 782
      Renton system, 778–782, Fig. 10-6, 779; Fig. 10-7, 781
      two step system, 775, Fig. 10-4, 776
      zero residual system, 778
   evaporators, 772
   factors affecting system efficiency, 768
   flow of SO_2 in pipes, 772
   housing, 785
   injector system, 784
      diffusers, 748
      materials of construction, 748
      water requirements, 784
   leak detectors, 770–772
      Enterra, 772, Fig. 10-2, 773
      Interscan, Fig. 10-1, 770; 771, 772
   mixing, 784
   monitoring, 783, 784
   Orion sulfur dioxide electrode, 783
   reliability, 786
   sample lines, 785
   separation, 785
   sulfonators, 782
   sulfur dioxide supply system, 768–770
      accessory equipment, 769
      automatic switchover, 769
      flexible connections, 770
      general discussion, 768
      materials of construction, 769
      remote vacuum, 769
      safety equipment, 769
      storage tanks, 769
      ton container, gas withdrawal, 768
      ton container, liquid withdrawal, 769
De Nora Seaclor Systems, 145–147
   electrolyzer, 146
   prepared brine systems, 147
   seawater systems, 146, 147, Fig. 3-12, 146
Desulfouibrio desulfuricans, 343
Diamond Shamrock Corp., 6
Dichloramine, 337. See also Chemistry of chlorination
Disinfection guidelines for potable water, 323–328, Fig. 6-11, 324; Fig. 6-12, 325; Table 6-3, 325; Fig. 6-13, 326; Table 6-4, 327

1062 INDEX

Disinfection of wastewater, 476–584, (chap. 8). *See also* Wastewater disinfection
Disinfection of wastewater for reuse, 485–487
  California requirements, 486
    filtered effluent, 486
    importance of disinfection, 487
    other applications, 487
    primary effluent, 486
    secondary effluent, 46
Distribution systems and transmission lines, 338–358. *See also* Chlorination of distribution systems, etc.
DPD-colorimetric method, 215, 216, 218, 219
DPD-FAS tetrimetric method, 215, 216, 218, 219
Du Pont's research, 8

Eberthella typhosa, 264
Egan cluster-jet diffuser, Fig. 9-40a, 663
Egan jet diffuser concept, the, 1031–1033
  velocity gradient
    contraflow effect, 1032, 1033, Fig. 16, 1052
    baffles, of, 1032
    jet system, of, 1031
Electro Bleach Gas Co., 280, 281
Engelhard Industries, 129–133
  automatic dosage control, 133
  cost comparisons, 134
  cell configuration, 130, Fig. 3-3, 131
  chloropac generator, 130
  chloropac systems, 133
  raw material, 129
  seawater cell system, Fig. 3-4, 132
Entamoeba histolytica, 197, 266, 270
Enterobacter cloacae, 322
Enteroviruses, 276, 277
  coxsackie, 276
  ECHO, 277
  infectious hepatitis, 277
  poliovirus, 276
Evaporators, 634–640
  accessories, 636
  Beatty's mixture, 635
  cathodic protection and insulation, 636
  electric heater type, 635
  electrical requirements, 636
  emergency pressure relief system, 637–640, Fig. 9-30, 639
    absorption tank for relief valve vent system, 638
    current practice, 638
    historical background, 637
    monitoring relief system, 640
    relief valves, 638
    safety considerations, 638
  general discussion, 634
  hot water type, 635
  pressure reducing valve, 636
  steam type, 636

FACTS method, 218, 219
Faraday, Michael, 2
Faraday's law, 21
Ferrobacillus ferrooxidans, 343
Forecasting sulfide buildup, 408–412
  equations for partially filled pipes, 408–410
  force mains, 410–412
Foul air scrubbing. *See* Wastewater chlorination
Free residual process, 305–317, 323–328
  control and removal of THMs, 314–317
    chlorine demand, 316
    current control strategies, 314, 315
    max. THM potential, 316, Fig. 6-10, 317
    disinfection guidelines, 323, Fig. 6-11, 324; Table 6-3, 325; Fig. 6-12, 325; Fig. 6-13, 326; Table 6-4, 327
  effect of pH, 307, Fig. 6-8, 308
  formation of potentially hazardous compounds, 311–314, Table 6-2, 313
  photodecomposition by sunlight, 288
  reservoir effect of OCl^- ion, 308
  role of ammonia-N, 305–307, Fig. 6-7, 307
  role of organic-N, 309, Fig. 6-8, 309
Friction factor f as a function of Reynolds number
  pipe flow (variable roughness), Fig. 10, 1046
  PVC solution lines, for Fig. 9, 1045
  PVC vacuum lines, for Fig. 8, 1044
  relative roughness factors, Fig. 11, 1047
Friction loss factors. *See also* Head-loss
  Sch. 80 PVC pipe, Table 3, 1020, 1021
  valves and fittings, Table 1, 1018, Fig. 15, 1051
Fischer and Porter, 285

Gallionella, 340, 341, Fig. 6-17, 343
Gay-bowel syndrome, 274
Geosmin, 330, 331
Germicidal significance of chlorine residuals. *See* Chemistry of chlorination
Globaline tablets, 986
Glossary, 1035

Halazone tablets, 986
Head-loss calculations
  diaphragm valves, Table 4, 1022

INDEX 1063

Hooker Chemicals and Plastics, 7
Hydrogen sulfide in wastewater. See Odor problems in wastewater
Hypochlorination, 62–119. See also Large hypochlorite systems; On-site manufacture of hypochlorite; Calcium, Lithium and Sodium hypochlorite
  chemistry of hypochlorites, 63–75
Hypochlorination
  choice of chemical, 95
  choice of equipment, 75–96
Hypochlorinators, 75–96
  introduction, 75
  maintenance, 96
  shipboard installations, Fig. 2-13, 91
    historical background, 90
    recent waterborne outbreaks on cruise ships, 93
    U.S. Navy Bureau of Ships, 92
    U.S. Public Health Service, 92
  small water supplies, 75–90
    designers check list, 86
    gravity filter systems, 87
    gravity systems, 77–86
    gravity systems with power available, 86, 87, Fig. 2-11, 87; Fig. 2-12, 88
    meter control systems, 77–86, Fig. 2-6, Fig. 2-7, 80; Fig. 2-8, 82; Fig. 2-9, 83; Fig. 2-10, 85
    pumped supplies, 76, Fig. 2-3, 77; Fig. 2-4, 78; Fig. 2-5, 79
Hypochlorite manufacture in underdeveloped areas, 116, Fig. 2-18, 117
Hypochlorous acid, 337. See also Chemistry of chlorination

Induced breakpoint process, 294
Industrial wastes. See Cyanides
Initial mixing, 541–555
  diffusers as mixers, 547–549
    design example, 547
    grid type for open channel, 547, Fig. 8-18, 548
    general discussion, 547
  general discussion, 541
  historical background, 542, Fig. 8-15, 543
  hydraulic mixers, 550–556
    closed conduits, 550, Fig. 8-21, 551
    Fischer & Porter system, 555, Fig. 8-25, 556
    hydraulic jump, 551, Fig. 8-23, 552
    Pentech system, 553–555, Fig. 8-24, 553
    static mixer, Fig. 8-22, 552

  mechanical mixers, 549, 550, Fig. 8-19, 549; Fig. 8-20, 550
  velocity gradient, 544–546
    current practices, 545
    general discussion, 544
    recent developments, 545, Fig. 8-16, 545; Fig. 8-17, 546
  summary, requirements for, 576
Injector systems, 681–694
  booster pumps, 690–694
    centrifugal vs. turbine, 690, Fig. 9-52, 691
    example, 693
    high back-pressure conditions, 692, Fig. 9-53, 693
  designer's check list, 693
  general description, 681
  hydraulic gradient analysis, 683–686, Fig. 9-48, 684
    example I, Fig. 9-49, 685
    example II, 686, Fig. 9-50, 687
    general discussion, 683
  operating water supply, 681–683
    flow meters, 683
    pressure regulating valve, 683
    water pressure vs backpressure, 683
    water quantity, 681, Fig. 9-47, 682
  remote injectors, 686
  system description, 681
  vacuum line design, 687, Fig. 9-51, 688, 689, 690
Iodine, 985–994
  chemistry of iodination, 987–990, Fig. 14-4, 998; Tables 14-1, 14-2, 989
  comparison with chlorination, 992
  cost, 993
  determination of iodine residuals, 994
  germicidal efficiency, 990, Fig. 14-5, 991
  iodination facility, 992–994, Figs. 14-6, 14-7, 993
    wastewater, 992
  limitations of iodination, 992
  Montezuma's revenge, 987
  occurrence and production, 985
  physical and chemical characteristics, 985
  summary, 1013
  uses, 986
  water treatment, 986, 987
Iron and manganese problems, 365–370
  role of chlorine, 368–370
  iron removal, 368
  manganese removal, 369
Irreducible chlorine residual, 165, Fig. 4-3, 165

1064 INDEX

Javelle water, 3
Jersey City case, 259–262

Keinle, John, 280
Kellner cell, 22
Klebsiella, 322

Large hypochlorite systems, 96–105
  current practices, 96, 97
  hazards of hypochlorite, 105
  hypochlorite, quantities required, 98
  hypochlorite facility design, 98–105
    control system, 100, Fig. 2-14, 101; 102
    eductor system, 100, Fig. 2-14, 101
    gravity system, 99, 100
    monitoring the system, 103
    pumped systems, 98, 99
  materials of construction, 103–105
    diffusers, 104
    eductors, 105
    hypochlorite flow meters, 104
    piping, 104
    rate control valves, 105
    storage tanks, 103
  operating costs, 106, 107
  reliability, 107
Lavoisier's theory, 2
Leak detectors, 641–646
  Capital Controls, 6, 642, Fig. 9-32, 642
  Draeger Safety, 643, Fig. 9-33, 643
  Enterra, 644
  Fischer and Porter Co., 645
  leak locators, 646
  Wallace and Tiernan Div., 645, 646
LeBlanc soda process, 3
Leed's patent, 258
Leptothrix, 340, 341, 342
Lithium hypochlorite, 73
Loop time, 667

Maintenance; chlorination equipment, 814–830
  amperometric titrator, 829
  automatic switchover system, 822, 823
  booster pumps, 825
  chlorinators, 824, 825
  chlorine supply system, 814–818
    flexible connections, 817
    general discussion, 814
    locating leaks, 815, 816
    repairing leaks, 816
    safety equipment, 817
  cleaning and testing chemical piping, 818–820
    chemical detergent method, 819
    general discussion, 818
    pickling method, 819, 820
  control devices, 826
    electric, 826
    pneumatic, 826
  electrical switchgear, 829
  evaporators, 822–823
    cathodic protection system, 822
    cleaning the liquid chlorine vessel, 822–823
    immersion heaters, 822
    liquid chlorine vessel, 822
  filter, chlorine gas, 823
  leak detector, 818
  pressure reducing valve, 823
  remote vacuum system, 824
  residual analyzers, 826–829
    amperometric cell, 827
    electrodes, 827
    general discussion, 826
    inspections, 828–829
    recorder or indicator, 828
    sample lines, 827
  storage tanks, chlorine, 821, 822
  trouble shooting the supply system, 820, 821
Maintenance: dechlorination equipment, 830–832
  leak detection, 831
  sulfur dioxide control and metering system, 832
  sulfur dioxide supply system, 830, 831
    general discussion, 830
    moisture problem, 830
    preventive maintenance, 830
Malonic acid. *See* Chlorine dioxide measurement
Mathieson Chemical Co., 22
Methylene blue stability test, 416
Methyl orange method, 216, 217
MIB, 330, 331
Mixing. *See* Initial mixing and Diffusers as mixers
Monochloramine, 337. *See also* Chemistry of chlorination
MOTA method, 217
Mycobacterium phlei, 322
Mycobacterium fortuitum, 322

Nais, 272
Nascent oxygen theory, 194
Nematodes, 272, 273
Nevskia, 342
Nitrified effluents, 251, 515–521
Nitrite interference, 226, 228, 229, 247, 248
Nitrogen cycle, 160, Fig. 4-1, 161

INDEX 1065

Nitrogen trichloride, 32, 33, 200, 230, 243, 247, 306, 337, 434, Fig. 7-7, 434; 435
Nuisance residuals, 166, Fig. 4-9, 179; 180, 202

Odor problems in wastewater, 394–415. *See also* Forecasting sulfide buildup
  chlorine for odor control, 412–415
    chemistry, 413
    chlorine requirement, 413
    historical background, 412
    prechlorination facility, 414
    role of chlorine, 413
  hydrogen sulfide characteristics, 402, 403
  hydrogen sulfide chemistry, 404, 405
    category of sulfides important to H_2S generation, 405
    general discussion, Fig. 7-1, 404
  hydrogen sulfide generation, 405–412, Fig. 7-2, 407
    causes and occurrence, 405
    force mains, 410–412
    free flowing sewers, 406–410
  identifying the problem, 400, 401
    air sampling H_2S detectors, 401
    lead acetate impregnated tiles, 401, Table 7-1, 401
    significance, 399
    source of odors, 400
  source of hydrogen sulfide in wastewater, 403
    effect of sulfate concentration, 403
    other sulfide compounds in wastewater, 403
On-site generation of chlorine, 120–149. *See also:* Cloromat systems (Ionics Inc), DeNora Seaclor systems, Engelhard Industries, Sanilec Systems, and Wallace & Tiernan (OSEC).
  early experience in the USA, 120
  equipment development, 122
    cloromat system (Ionics Inc.), Fig. 3-1, 124; 125, Fig. 3-2, 126; 127, 128
    DeNora Seaclor systems, 145–147, Fig. 3-12, 146
    Engelhard Industries, 129–134, Fig. 3-3, 131; Fig. 3-4, 132; Fig. 3-5, 134
    Sanilec systems, 134–139, Fig. 3-6, 136; Fig. 3-7, 137
    Wallace and Tiernan (OSEC) systems, 134–145, Fig. 3-9, 141; Fig. 3-10, 142
  importance of raw material, 121
    brine systems, 122
    seawater composition, Table 3-1, 122
    seawater systems, 121

summary, 147–149
  choice of systems, 147
  health considerations, 148
  on-site capabilities, 147
  typical applications, 148
  wastewater treatment, 148
On-site manufacture of hypochlorite, 107–116
  potable water system, 108
  wastewater system 25 ton/day, 109–116
    chlorine facilities, 110–115, Fig. 2-16, 111; Fig. 2-17, 113
    cost comparison, 115, 116
    practical considerations, 116
    reliability, 115
    system description, 110
Operation: chlorination equipment, 789–810
  chlorine control and metering system, 803
    alarms, 809
    amperometric titrator, 808
    booster pumps, 804
    chlorinators, 805
    chlorine flow transmitter, 809
    chlorine residual analyzer, 806–808
    injector system, 803, 804
    safety equipment, 810
  chlorine leak detector, 802, 803
  chlorine supply system, 800–807
    100 and 150 lb containers, 800
    tank cars and storage tanks, 802
    ton containers, 801
  physical facilities inventory, 789
  start-up (gas system), 790–794
    checking chlorine supply system, 792
    checking vacuum relief system, 792
    four important steps, 790
    injector hydraulics, 793
    insufficient vacuum, 792
    procedure when a leak occurs, 791
    vacuum leaks, 794
  start-up (liquid system), 794–797
    evaporator, 794–796
    excessive chlorine pressure, 796, 797
  supply system preparation, 789
    introduction, 789
    small systems, 790
  to secure an evaporator, 798, Fig. 11-1, Fig. 11-2, 799
  to stop or secure an installation, 797–798
    four steps, 797
Operation, dechlorination equipment, 810, 814
  physical facilities, 810

1066 INDEX

Operation (*cont.*)
  sulfur dioxide control and metering system, 812
    analyzers, 814
    booster pumps, 812
    injector system, 812
    sulfonators, 812–814, Fig. 11-3, 813; Fig. 11-4, 813
  sulfur dioxide supply system, 810, 811
    general discussion, 810
    safety equipment, 810
    storage tanks, 811
    ton containers, 810
    supply system preparation, 810
    system start-up, 811
      superheat, 811
      supply pressure, 811
    to stop or secure a system, 812
Orustein patent, 280
OT method, 216, 217
OTA method, 215, 217
Oxidation reduction potentials (Redox), 1036. *See also* Electrochemical properties of chlorine
Ozone, 893–960, *see also* Ozone facility design; Ozone generators; Ozone system, operation and maintenance, Ozone, wastewater disinfection research
  chemistry of ozone, 894–902
    effect of ozone residuals upon chlorine and chlorine dioxide, 900
    inorganic reactions, 899
    introduction, 894
    organic reactions, 901
    ozonation of seawater, 900
    ozone as a bactericide, 894–896, Table 13-1, 896; Table 13-2, 896
    ozone as an oxidant, 897, 898
    ozone as a viricide, 896, 897
    physical and chemical properties, 898
    toxicity of, 901, 902
  comparative costs, 950–955
    cost studies, 951–955, Table 13-10, 952; Table 13-11, 952; Table 13-12, 953; Table 13-13, 953; Table 13-14, 954; Table 13-15, 955; Table 13-16, 955
    general considerations, 950, 951
  current practices: potable water, 902–908
    Austria, 906
    Canada, 904
    Federal Rep. of Germany, 905
    France, 904
    Great Britain, 907
    Netherlands, 906
    other installations, 908, Table 13-3, 908
    Switzerland, 905
    United States, 902, 903
  historical background, 893
  measurement of residuals, 945–950
    continuous residual recording, 947–950
    EPA investigation, 945
    introduction, 945
    modified amperometric method, 946
    standard methods, 947
  ozone-biological activated carbon process, 908–911
    historical background, 908
    process description, 909
    significant features, 909
  preozonation: potable water, 911
  preozonation: a tertiary process, 914—917
    fuial disinfection, 915, Fig. 13-1, 916
    general discussion, 914
    projected full-scale plant, 916, Table 13-6, 917
    virus destruction, 914, Table 13-4, 915; Table 13-5, 915; Fig. 13-2, 917
  summary, 955–957
  uses of in wastewater treatment, 911–916
    general discussion, 911, 912
    operating installations, 912–914
Ozone facility design, 934–944
  contacting system, 942–944, Figs. 13-12 13-13, 943; Fig. 13-14, 944
    contact time, 944
    general discussion, 942
    ozone destruction unit, 942
    safety considerations, 944
  disinfection guidelines, 938
  exhaust gas monitoring system, 941
  forecasting ozonator capacity, 938–941, Fig. 13-10, 939; Fig. 13-11, 940
    potable water, 938
    preozonation: tertiary effluents, 941
    wastewater: secondary effluents, 939
  general considerations, 934, 935
  general purpose facility, 935–937, Fig. 13-8, 936
  ozone demand and absorption, 938
  ozone dosage, metering and measurement, 941
  ozone systems, 935, Fig. 13-9, 937
    once through air, 935
    once through oxygen, 935
    oxygen recycle, 935
  photometer calibration, 942
  wastewater disinfection, 937
Ozone generators, 927–934
  introduction, 927

INDEX 1067

Ozone generators (cont.)
  operating characteristics, Table 13-9, 934
  ozonator manufacturers, 933
    comparison of systems, 934
    foreign, 934
    United States, 933
  theory of generation, 928–930
  three types of, 931, Fig. 13-5, 931
    cooled tube, 930, Fig. 13-6, 932; Fig. 13-7, 933
    OTTO plate, 930
    Lowther plate, 930, Fig. 13-5, 931
Ozone systems, operation and maintenance, 950
  air drying system, 95
  contactors, 950
  operating installation, study of, 950
  ozone destruct unit, 950
  ozone generators, 950
Ozone, wastewater disinfection research, 916–927
  application of ozone, 916–921
    absorbed dose, 920
    applied dose, 920
    solubility of ozone, 916–921, Table 13-7, 919
    transfer efficiency concept, 920
  control of disinfection efficiency, 921–922
    absorbed ozone dose, 921
    exhaust gas monitoring, 922
    factors affecting efficiency, 921
  forecasting coliform concentration in an ozonated effluent, 923
    effluent quality, 923
    filtered effluent, 923
    unfiltered secondary, 923, 924
  high level disinfection with ozone, 924
    Marlborough MA study, 924, Fig. 13-3, 925; Fig. 13-4, 926; Table 13-8, 927
  high level disinfection with ozone and UV, 926
  residual control, 922, 923
    continuous analyzers, 922, 923
    general discussion, 922

Pasteurella tularensis, 273
Pipe specifications, 1019, Table 2
Polarography, 709, Fig. 9-60, 710
Potentiometry, 709

Redox potentials, 1036
Residometer for DPD method, 225, Fig. 5-1

Safety in chlorine handling. See Chlorine leaks
Sanilec systems, 134–140
  brine cell, Fig. 3-8, 140
  brine system, 139
  seawater cell, Fig. 3-6, 136
  seawater system, 135, Fig. 3-7, 137
Scheele's gas, 1
Scott-Darcey process, 470
Seawater chlorination, 985
  effects of bromides, 985
Seawater, composition of, Table 3-1, 122
Siderophacus, 342
Slime forming organisms, 339, 340
SNORT method, 216, 217, 218
Snow, Dr. John, 257
Sodium hypochlorite
  characteristics, 66
  chlorine and caustic requirements, Table 2-1, 67
  dosage calculation, Table 2-2, 68
  freezing temperatures, Fig. 2-2, 71
  stability of solutions, 68–71, Fig. 2-1, 69; Table 2-3, 70
  suggested specifications, 71
Solvay Na^-NH_3 process, 3
Sphaerotilus, 340, 341, Fig. 6-16, 342
Sulfamic acid, 229
Sulfur bacteria, 343, 344
Sulfur dioxide. See also Dechlorination
  vapor pressure vs. temperature, Fig. 6, 1042

Taste and odor control, 328–338
  chlorination system requirements for taste and odor control, 336, 337
    residuals below which the various chlorine species are likely not to produce objectionable tastes, 337
  introduction, 328, 329
  man-made sources of, 331, Table 6-5, 332
  principal odor producing algae, 329–331
  role of chlorination, in, 332–335
    algae tastes and odors, 332
    industrial wastes, 333–335
    use of carbon with chlorine, 335
  seeking the solution to T/O problem, 335, 336
    activated carbon and potassium permaorganate, 336
    ammonia, 336
    chlorine dosage, 336
Tenant's bleach, 3
Thiobaccillus, 343
THM formation, control of, 288
Toxicity of chlorine residuals, 498–500
  general discussion, 498, 499
  need for dechlorination, 500
Typhoid Mary, 264

1068 INDEX

Ultra Violet radiation, 994–1012
  cost, 1012
  germicidal efficiency, 997
  introduction, 994–996
    germicidal lamps, 994
    physics of radiation, 994
    potable water, 995
    practical applications, 994
    wastewater, 995
  mode of action, 996, 997
    degree of inactivation, 996
    general discussion, 996
  photoreactivation, 1001
  process variables that affect disinfection efficiency, 1000, 1001
    effluent quality, 1001
    UV absorption, 1000
  reliability, 1012
  summary, 1014
  toxicity of radiation, 1002
  UV dose versus bacterial kill, 998–1000, Fig. 14-8, 999
  UV facility design, 1002–1012
    potable water, 1002–1004
    preface, 1002
    wastewater, 1004, Fig. 14-9, 1005, 1006; Fig. 14-10, 1007; Figs. 14-11, 14-12, 1009; Fig. 14-13, 1009; 1010, Fig. 14-14, 1011

Velocity gradient G. See also Egan jet diffuser concept; Initial mixing
  calculator solution of, 1030
Vibrio comma, 265
Viral gastroenteritis, 277
  AGI, 277
  Norwalk virus, 277, 278
  Rotavirus, 278
Viruses versus wastewater disinfection, 487–497
  disinfection processes, 491
    chlorine, 491, Fig. 8-2, 492; Fig. 8-3, 493
    comparison with other investigations, 494–496
    other conclusions, 492–494
    ozone, 491, Fig. 8-4, 494
    predisinfection effluent quality, 491
    results, 491
  future considerations for virus destruction, 496
  treatment system studies, 490
  virus hazard, the, 487
  virus inactivation, 490

  viruses in sewage contaminated potable water supplies, 489
Voltammetry, 708

Wallace & Tiernan, 280, 281
Wallace & Tiernan (OSEC), 139–145
  differences between seawater and brine models, 143
  electrolyzer, 139, Fig. 3-10, 142
  manufacturers recommendations, 143
  OSEC 2500 lb/day unit, Fig. 3-9, 141; Fig. 3-11, 144
Wastewater chlorination, 394–475. See also Odor problems in wastewater
  BOD reduction, 430–432
  chemistry of chlorine and wastewater, 396–399
    ammonia nitrogen, 396, 397
    chlorine demand, 399
    organic nitrogen, 397
    phenols, 398
    pickle liquor, 398
    sulfites, 398
    tannins, 398
  contemporary practices, 395
  cyanide wastes, 449–466
    chemistry of treatment, 451–458, Fig. 7-12, 455; Fig. 7-13, 456
    chlorination facility, 458, Fig. 7-14, 459; Fig. 7-15, 460, 461; Fig. 7-16, 463
    controlling the oxidation of cyanide by the chlorination process, 462–466, Fig. 7-16, 463
    occurrences, 449–451
    significance, 449
  foul air scrubbing, 436–442
    carbon columns, 442
    caustic towers, 442
    chlorine wet scrubbers, 438–443, Fig. 7-8, 437; Fig. 7-9, 438
    general discussion, 436, Fig. 7-8, 437
  grease removal, 428
  historical background, 394
  nitrogen removal, 432
    current practices, 433
    purpose, 432
  nitrogen removal by chlorine, 433–436, Fig. 7-7, 434
    introduction, 433
    Rancho Cordova project, 433, Fig. 7-7, 434
    Saunier's research, 435, 436
  other methods of H_2S generation control, 442–448

INDEX 1069

Wastewater chlorination (cont.)
  other methods of H_2S generation control (cont.)
    air-oxygen injection, 443, Fig. 7-10, 445; Fig. 7-11, 448
    caustic addition, 442
    hydrogen peroxide, 448
    methylene blue stability test, 416
    preventing septicity, 415
    wool scouring by hypochlorite, 469, 470
  Scott Darcey process, 470, 471
  sludge bulking, 417–426, Fig. 7-3, 418; Fig. 7-4, 420; Fig. 7-5, 421
    identifying the problem, 417
    microscopic observation, 419
    nature of the problem, 417
    sludge volume index, 418
  sludge bulking control by chlorination, 422
    chlorination system, 422–426; Fig. 7-6, 423
    historical background, 422
    operating plant example, 422–426
  sludge transport, 426–428
  textile wastes, 468, 469
  treatment and destruction of phenols, 467
    chlorination facility, 468
  trickling filters and high-rate recirculating biofilters, 429, 43
  uses of chlorine, 394
Wastewater chlorination, facility design, 524–579. See also Contact chambers; Diffusers as mixers and Initial mixing
  chlorinator sizing, 527
    chlorine demand, effect of, 528
    plant flow meter, influence of, 527
  chlorine control schemes, 529–531
    compound-loop, 530
    dual signal, 531
    flow pacing, 529
    general discussion, 524, 529
    multiplied signal, 530
    straight residual control, 531
  factors affecting system efficiency, 525
    elements of an optimum system, 525
    factors unique to a wastewater chlorination system, 524
  initial mixing, 541–555. See also Initial mixing
  injector systems, 531–540
    chlorine solution lines, 536
    diffusers, 536, Fig. 8-10, 537; Fig. 8-11, 538; Fig. 8-12, 539; Fig. 8-13, 539; Fig. 8-14, 540
    general description, 531

    hydraulic considerations, 532–535, Fig. 8-8, 534; Fig. 8-9, 535
    materials of construction, 540
    monitoring devices, 535
    water supply, 531
  optimized mobile plant study, 526
    description of project, 526
    results and conclusions of study, 526
  records and reports, 578
    physical facilities, 578
    records and reports, 578
    records of equipment inspection maintenance & repair, 579
    records of operation, 578
    reports, 579
  reliability provisions for, 576–579
    chlorine residual analyzers, 578
    chlorine supply, 577
    power failure, 577
    spare parts, 577
    standby equipment, 577
    water supply, 577
Wastewater disinfection, 476–584 (Chap. 8). See also: Concepts of wastewater disinfection, 476–597; Disinfection of wastewater for reuse, 485–487; Toxicity of chlorine residuals, 498–500; Viruses vs. wastewater disinfection, 487–497; Wastewater chlorination facility design, 524–584
  available methods, 500–504
    bromine, BrCl, and iodine, 502
    chlorine, 501
    chlorine deoxide, 502
    gamma radiation, 504
    introduction, 500
    ozone, 503
    UV radiation, 503
  comparative costs of the different methods of disinfection, 504, Table 8-1, 505
  summary, 505–507
Wastewater disinfection, chemistry and kinetics of chlorine, 508–524. See also: Collins mathematical model and Collins-Selleck mathematical model
  chlorine reactions with wastewater,
    ammonia-N, 508
    effluent quality as related to dosage, 511–515, Fig. 8-5, 514; Fig. 8-6, 515
    general discussion, 508
    nitrified effluents, 515–520, Fig. 8-7, 518
    nitrification summary, 520, 521
    organic-N, 509

Wastewater disinfection (*cont.*)
  formation of organic chloramine compounds, 521–524
    Montgomery MD simulation study, 524
    Occoquan watershed survey, 523
    Pomona virus study, 523
    Rancho Cordova investigation, 521–523
Wastewater reuse, 485–487
  California requirements, 486–487
    filtered effluents, 486
    importance of disinfection, 487
    other applications, 487
    primary effluents, 486
    secondary effluents, 486
Water viscosity, 1030
Waterborne diseases, 263–279. *See also* Chlorination of potable water
  acute gastrointestinal illnesses, 273
  amoebic dysentery, 266
  bacterial gastroenteritis, 266
    camphylobacter enteritis, 267
    pseudomonas, 268
    salmonellosis, 267
    shigellosis, 266
    toxigenic E. coli, 267
    yersina enteritis, 268
  cholera, 257, 265
  giardia, 269
  Legionnaire's disease, 270–272
  recently described conditions with conspicuous GI symptoms, 273
  schistosomiasis, 268, 269
  summary, 278
  tularemia, 273
  typhoid fever, 264
  viral diseases, 274–279. *See also* Enteroviruses and Viral gastroenteritis
    adenovirus, 276
    consensus organism indicator, 275
    coping with viruses in potable water, 275
    enteroviruses, 276
    general background, 274
    viral gastroenteritis, 277
  worms, 272
Watt, James, 3
Weldon process, 3
Woolf electrolytic process, 258